Instructor's Annotated Edition

Precalculus: A Concise Course

Ron Larson

The Pennsylvania State University
The Behrend College

Robert Hostetler

The Pennsylvania State University
The Behrend College

With the assistance of David C. Falvo

The Pennsylvania State University
The Behrend College

Houghton Mifflin Company Boston New York

Publisher: Richard Stratton
Sponsoring Editor: Cathy Cantin
Development Manager: Maureen Ross
Development Editor: Lisa Collette
Editorial Associate: Elizabeth Kassab
Supervising Editor: Karen Carter
Senior Project Editor: Patty Bergin
Editorial Assistant: Julia Keller
Art and Design Manager: Gary Crespo
Executive Marketing Manager: Brenda Bravener-Greville
Director of Manufacturing: Priscilla Manchester
Cover Design Manager: Tony Saizon

Cover Image: Ryuichi Okano/Photonica

Printed in the U.S.A.

Library of Congress Catalog Card Number: 2005933921

Instructor's exam copy:

ISBN 13: 978-0-618-62720-2
ISBN 10: 0-618-62720-0

For orders, use student text ISBNs:

ISBN 13: 978-0-618-62719-6
ISBN 10: 0-618-62719-7

123456789–DOW–10 09 08 07 06

Rational (Reciprocal) Function

$f(x) = \frac{1}{x}$

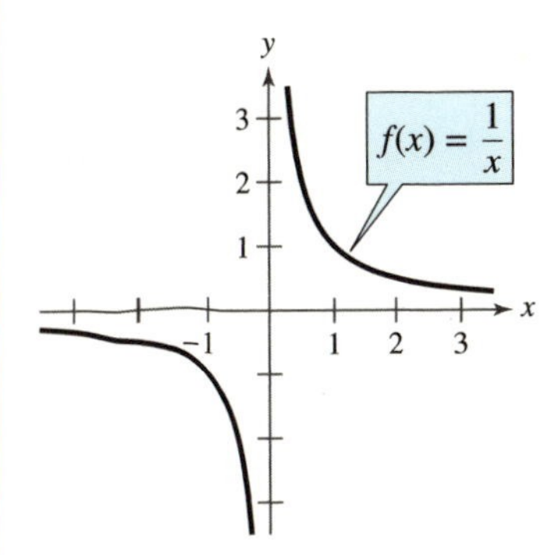

Domain: $(-\infty, 0) \cup (0, \infty)$
Range: $(-\infty, 0) \cup (0, \infty)$
No intercepts
Decreasing on $(-\infty, 0)$ and $(0, \infty)$
Odd function
Origin symmetry
Vertical asymptote: y-axis
Horizontal asymptote: x-axis

Exponential Function

$f(x) = a^x,\ a > 0,\ a \neq 1$

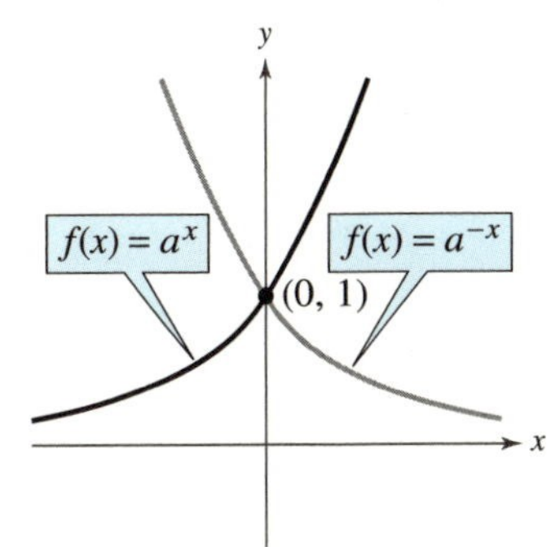

Domain: $(-\infty, \infty)$
Range: $(0, \infty)$
Intercept: $(0, 1)$
Increasing on $(-\infty, \infty)$
for $f(x) = a^x$
Decreasing on $(-\infty, \infty)$
for $f(x) = a^{-x}$
Horizontal asymptote: x-axis
Continuous

Logarithmic Function

$f(x) = \log_a x,\ a > 0,\ a \neq 1$

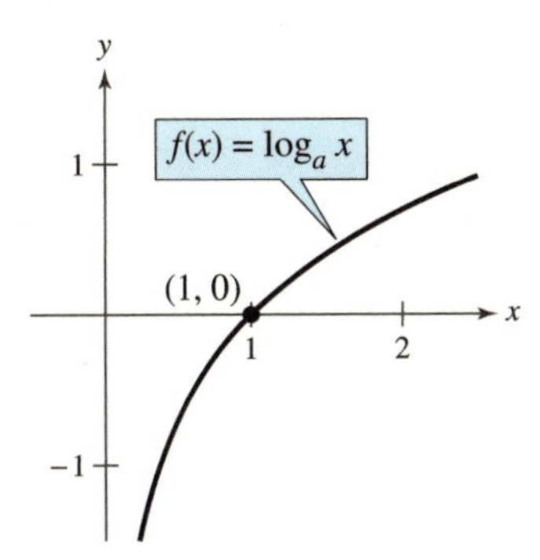

Domain: $(0, \infty)$
Range: $(-\infty, \infty)$
Intercept: $(1, 0)$
Increasing on $(0, \infty)$
Vertical asymptote: y-axis
Continuous
Reflection of graph of $f(x) = a^x$
in the line $y = x$

Sine Function

$f(x) = \sin x$

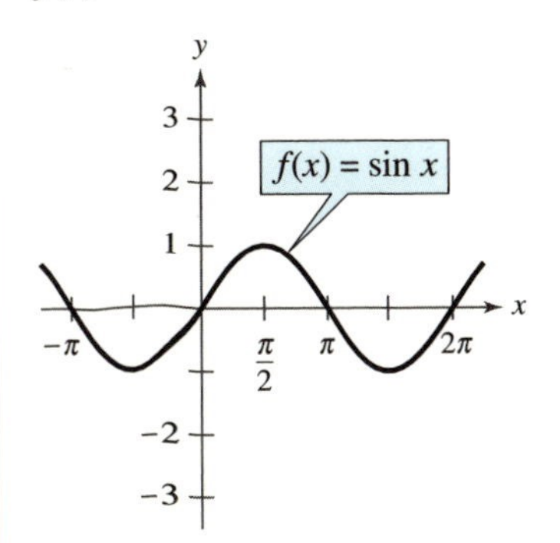

Domain: $(-\infty, \infty)$
Range: $[-1, 1]$
Period: 2π
x-intercepts: $(n\pi, 0)$
y-intercept: $(0, 0)$
Odd function
Origin symmetry

Cosine Function

$f(x) = \cos x$

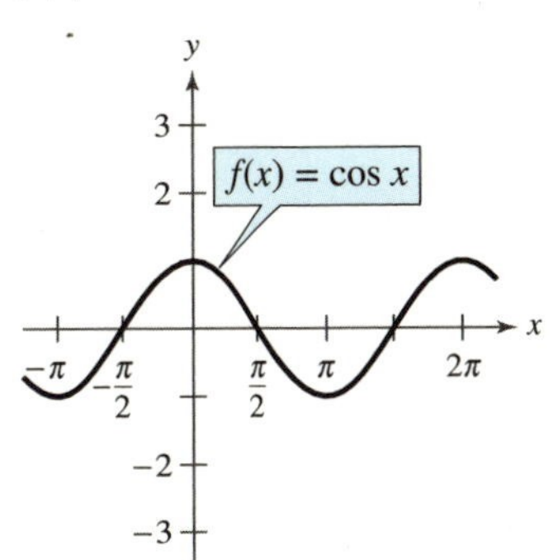

Domain: $(-\infty, \infty)$
Range: $[-1, 1]$
Period: 2π
x-intercepts: $\left(\frac{\pi}{2} + n\pi, 0\right)$
y-intercept: $(0, 1)$
Even function
y-axis symmetry

Tangent Function

$f(x) = \tan x$

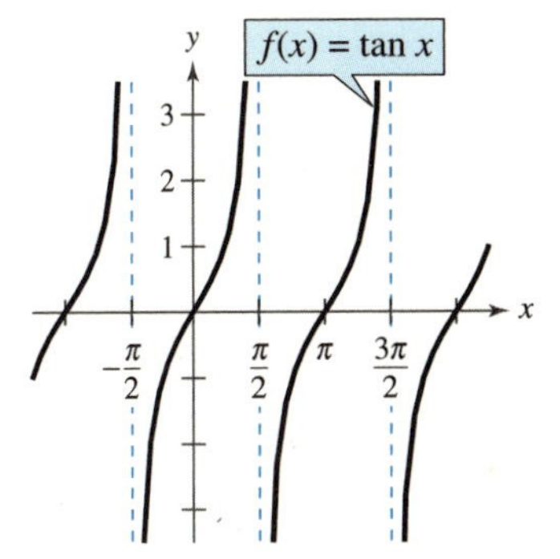

Domain: all $x \neq \frac{\pi}{2} + n\pi$
Range: $(-\infty, \infty)$
Period: π
x-intercepts: $(n\pi, 0)$
y-intercept: $(0, 0)$
Vertical asymptotes:

$$x = \frac{\pi}{2} + n\pi$$

Odd function
Origin symmetry

Cosecant Function

$f(x) = \csc x$

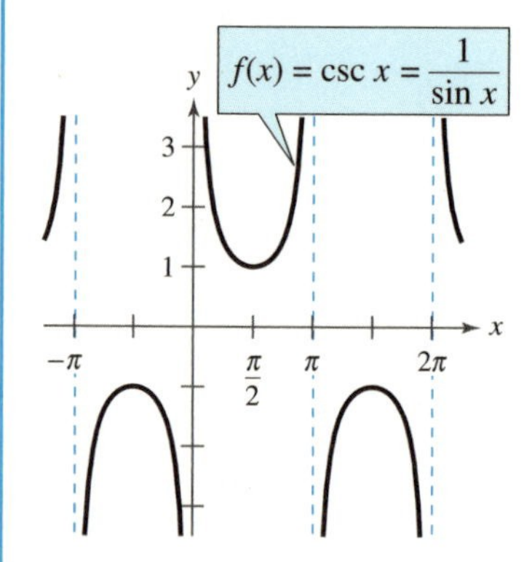

Domain: all $x \neq n\pi$
Range: $(-\infty, -1] \cup [1, \infty)$
Period: 2π
No intercepts
Vertical asymptotes: $x = n\pi$
Odd function
Origin symmetry

Secant Function

$f(x) = \sec x$

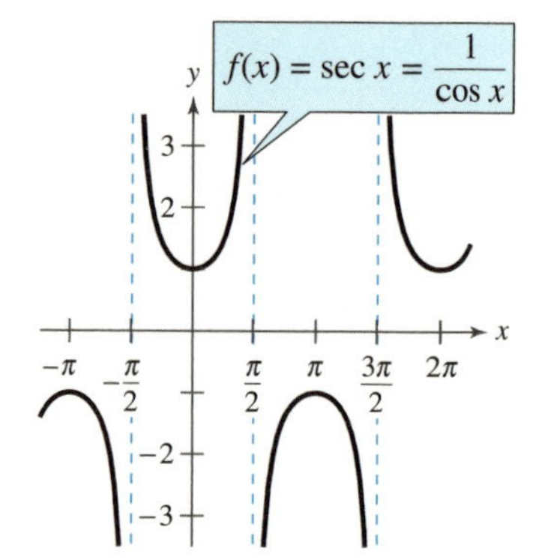

Domain: all $x \neq \dfrac{\pi}{2} + n\pi$
Range: $(-\infty, -1] \cup [1, \infty)$
Period: 2π
y-intercept: $(0, 1)$
Vertical asymptotes:

$$x = \frac{\pi}{2} + n\pi$$

Even function
y-axis symmetry

Cotangent Function

$f(x) = \cot x$

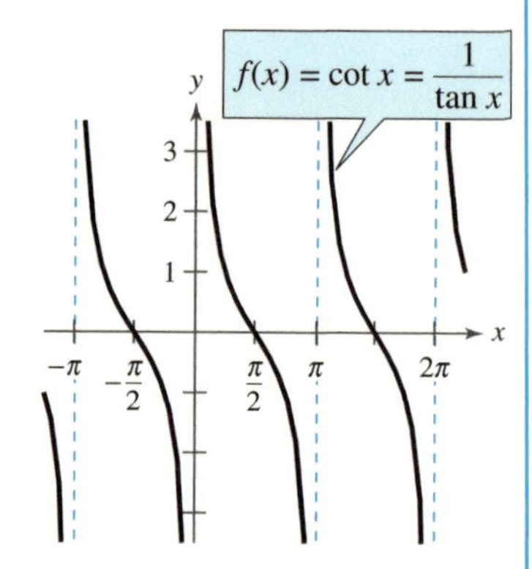

Domain: all $x \neq n\pi$
Range: $(-\infty, \infty)$
Period: π
x-intercepts: $\left(\dfrac{\pi}{2} + n\pi, 0\right)$
Vertical asymptotes: $x = n\pi$
Odd function
Origin symmetry

Inverse Sine Function

$f(x) = \arcsin x$

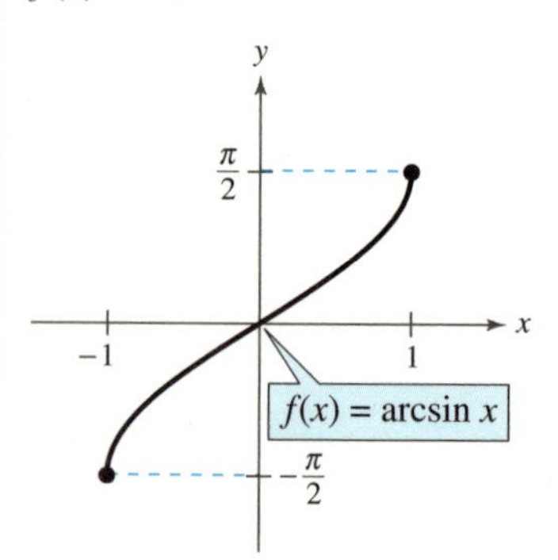

Domain: $[-1, 1]$
Range: $\left[-\dfrac{\pi}{2}, \dfrac{\pi}{2}\right]$
Intercept: $(0, 0)$
Odd function
Origin symmetry

Inverse Cosine Function

$f(x) = \arccos x$

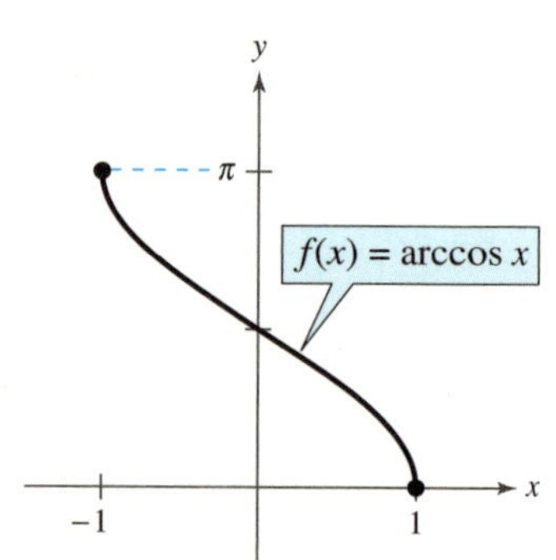

Domain: $[-1, 1]$
Range: $[0, \pi]$
y-intercept: $\left(0, \dfrac{\pi}{2}\right)$

Inverse Tangent Function

$f(x) = \arctan x$

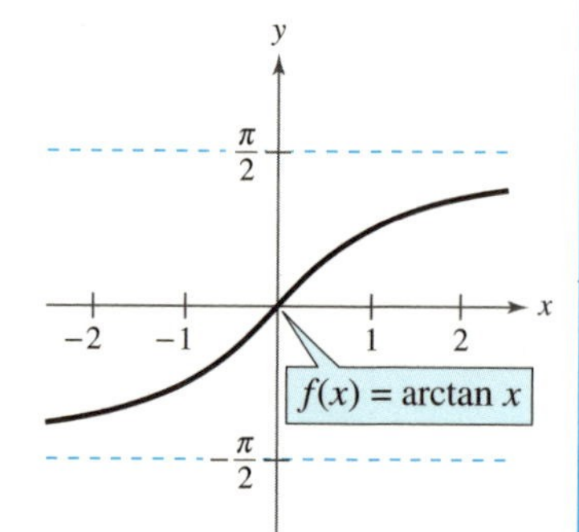

Domain: $(-\infty, \infty)$
Range: $\left(-\dfrac{\pi}{2}, \dfrac{\pi}{2}\right)$
Intercept: $(0, 0)$
Horizontal asymptotes:

$$y = \pm\frac{\pi}{2}$$

Odd function
Origin symmetry

Contents

CONTENTS

A Word from the Authors

Welcome to *Precalculus: A Concise Course*. We are pleased to present this new precalculus textbook in which we focus on making the mathematics accessible, supporting student success, and offering instructors flexible teaching options. The purpose of this text is to concisely cover the topics of precalculus. This text covers six chapters: *Functions and Their Graphs*, *Polynomial and Rational Functions*, *Exponential and Logarithmic Functions*, *Trigonometry*, *Analytic Trigonometry*, and *Topics in Analytic Geometry*. Should students need extra help or need to review material from intermediate algebra, the textbook website offers an appendix for review.

Accessible to Students

We have taken care to write this text with the student in mind. Paying careful attention to the presentation, we use precise mathematical language and a clear writing style to develop an effective learning tool. We believe that every student can learn mathematics, and we are committed to providing a text that makes the mathematics of the precalculus course accessible to all students.

Throughout the text, solutions to many examples are presented from multiple perspectives—algebraically, graphically, and numerically. The side-by-side format of this pedagogical feature helps students to see that a problem can be solved in more than one way and to see that different methods yield the same result. The side-by-side format also addresses many different learning styles.

We have found that many precalculus students grasp mathematical concepts more easily when they work with them in the context of real-life situations. Students have numerous opportunities to do this throughout the text. The *Make a Decision* feature further connects real-life data and applications and motivates students. They also offer students the opportunity to generate and analyze mathematical models from large data sets. To reinforce the concept of functions, each function is introduced at the first point of use in the text with a definition and description of basic characteristics. Also, all elementary functions are presented in a summary on the endpapers of the text for convenient reference.

We have carefully written and designed each page to make the book readable and accessible to students. For example, to avoid unnecessary page turning and disruptions to students' thought processes, each example and corresponding solution begins and ends on the same page.

Supports Student Success

During more than 30 years of teaching and writing, we have learned many things about the teaching and learning of mathematics. We have found that students are most successful when they know what they are expected to learn and why it is important to learn the concepts. With that in mind, we have incorporated a thematic study thread throughout *Precalculus: A Concise Course.*

Each chapter begins with a list of applications that are covered in the chapter and serve as a motivational tool by connecting section content to real-life situations. Using the same pedagogical theme, each section begins with a set of section learning objectives—*What You Should Learn*. These are followed by an engaging real-life application—*Why You Should Learn It*—that motivates students and illustrates an area where the mathematical concepts will be applied in an example or exercise in the section. The *Chapter Summary—What Did You Learn?*—at the end of each chapter is a section-by-section overview that ties the learning objectives from the chapter to sets of *Review Exercises* at the end of each chapter.

Throughout the text, other features further improve accessibility. *Study Tips* are provided throughout the text at point-of-use to reinforce concepts and to help students learn how to study mathematics. *Technology, Writing About Mathematics, Historical Notes,* and *Explorations* reinforce mathematical concepts. Each example with worked-out solution is now followed by a *Checkpoint*, which directs the student to work a similar exercise from the exercise set. The *Section Exercises* begin with a *Vocabulary Check*, which gives the students an opportunity to test their understanding of the important terms in the section. A *Prerequisite Skills Review* is offered at the beginning of each exercise set. *Synthesis Exercises* check students' conceptual understanding of the topics in each section. The *Make a Decision* exercises further connect real-life data and applications and motivate students. *Skills Review Exercises* provide additional practice with the concepts in the chapter or previous chapters. *Chapter Tests*, at the end of each chapter, and periodic *Cumulative Tests* offer students frequent opportunities for self-assessment and to develop strong study- and test-taking skills.

The use of technology also supports students with different learning styles. *Technology* notes are provided throughout the text at point-of-use. These notes call attention to the strengths and weaknesses of graphing technology, as well as offer alternative methods for solving or checking a problem using technology. These notes also direct students to the *Graphing Technology Guide*, on the textbook website, for keystroke support that is available for numerous calculator models. The use of technology is optional. This feature and related exercises can be omitted without the loss of continuity in coverage of topics.

Numerous additional text-specific resources are available to help students succeed in the precalculus course. These include "live" online tutoring, instructional DVDs, and a variety of other resources, such as tutorial support and self-assessment, which are available on the HM mathSpace® CD-ROM, the Web, and in Eduspace®. In addition, the *Online Notetaking Guide* is a notetaking guide that helps students organize their class notes and create an effective study and review tool.

Flexible Options for Instructors

From the time we first began writing textbooks in the early 1970s, we have always considered it a critical part of our role as authors to provide instructors with flexible programs. In addition to addressing a variety of learning styles, the optional features within the text allow instructors to design their courses to meet their instructional needs and the needs of their students. For example, the

Explorations throughout the text can be used as a quick introduction to concepts or as a way to reinforce student understanding.

Our goal when developing the exercise sets was to address a wide variety of learning styles and teaching preferences. The *Vocabulary Check* questions are provided at the beginning of every exercise set to help students learn proper mathematical terminology. In each exercise set we have included a variety of exercise types, including questions requiring writing and critical thinking, as well as real-data applications. The problems are carefully graded in difficulty from mastery of basic skills to more challenging exercises. Some of the more challenging exercises include the *Synthesis Exercises* that combine skills and are used to check for conceptual understanding and the *Make a Decision* exercises that further connect real-life data and applications and motivate students. *Skills Review Exercises*, placed at the end of each exercise set, reinforce previously learned skills. In addition, Houghton Mifflin's Eduspace® website offers instructors the option to assign homework and tests online—and also includes the ability to grade these assignments automatically.

Several other print and media resources are also available to support instructors. The *Online Instructor Success Organizer* includes suggested lesson plans and is an especially useful tool for larger departments that want all sections of a course to follow the same outline. The *Instructor's Edition* of the *Student Notetaking Guide* can be used as a lecture outline for every section of the text and includes additional examples for classroom discussion and important definitions. This is another valuable resource for schools trying to have consistent instruction and it can be used as a resource to support less experienced instructors. When used in conjunction with the *Student Notetaking Guide* these resources can save instructors preparation time and help students concentrate on important concepts.

Instructors who stress applications and problem solving, or exploration and technology, coupled with more traditional methods will be able to use this text successfully.

We hope you will enjoy using *Precalculus: A Concise Course* in your precalculus class.

Ron Larson
Robert Hostetler

Acknowledgments

We would like to thank the many people who have helped us prepare the text and the supplements package. Their encouragement, criticisms, and suggestions have been invaluable to us.

Reviewers

Arun Agarwal, *Grambling State University*; Jean Claude Antoine, *Bunker Hill Community College*; W. Edward Bolton, *East Georgia College*; Joanne Brunner, *Joliet Junior College*; Luajean Bryan, *Walker Valley High School*; Nancy Cholvin, *Antelope Valley College*; Amy Daniel, *University of New Orleans*; Nerissa Felder, *Polk Community College*; Kathi Fields, *Blue Ridge Community College*; Edward Green, *North Georgia College & State University*; Karen Guinn, *University of South Carolina Beaufort*; Duane Larson, *Bevill State Community College*; Babette Lowe, *Victoria College (TX)*; Rudy Maglio, *Northwestern University*; Antonio Mazza, *University of Toronto*; Robin McNally, *Reinhardt College*; Constance Meade, *College of Southern Idaho*; Matt Mitchell, *American River College*; Claude Moore, *Danville Community College*; Mark Naber, *Monroe Community College*; Paul Olsen, *Wesley College*; Yewande Olubummo, *Spelman College*; Claudia Pacioni, *Washington State University*; Gary Parker, *Blue Mountain Community College*; Kevin Ratliff, *Blue Ridge Community College*; Michael Simon, *Southern Connecticut State University*; Rick Simon, *University of La Verne*; Delores Smith, *Coppin State University*; Kostas Stroumbakis, *DeVry Institute of Technology*; Michael Tedder, *Jefferson Davis Community College*; Ellen Turnell, *North Harris College*; Pamela Weston, *Tennessee Wesleyan College*

We would like to thank the staff of Larson Texts, Inc. who assisted in preparing the manuscript, rendering the art package, and typesetting and proofreading the pages and supplements.

On a personal level, we are grateful to our wives, Deanna Gilbert Larson and Eloise Hostetler for their love, patience, and support. Also, a special thanks goes to R. Scott O'Neil.

If you have suggestions for improving this text, please feel free to write us. Over the past three decades we have received many useful comments from both instructors and students, and we value these very much.

Ron Larson
Robert Hostetler

Textbook Features and Highlights

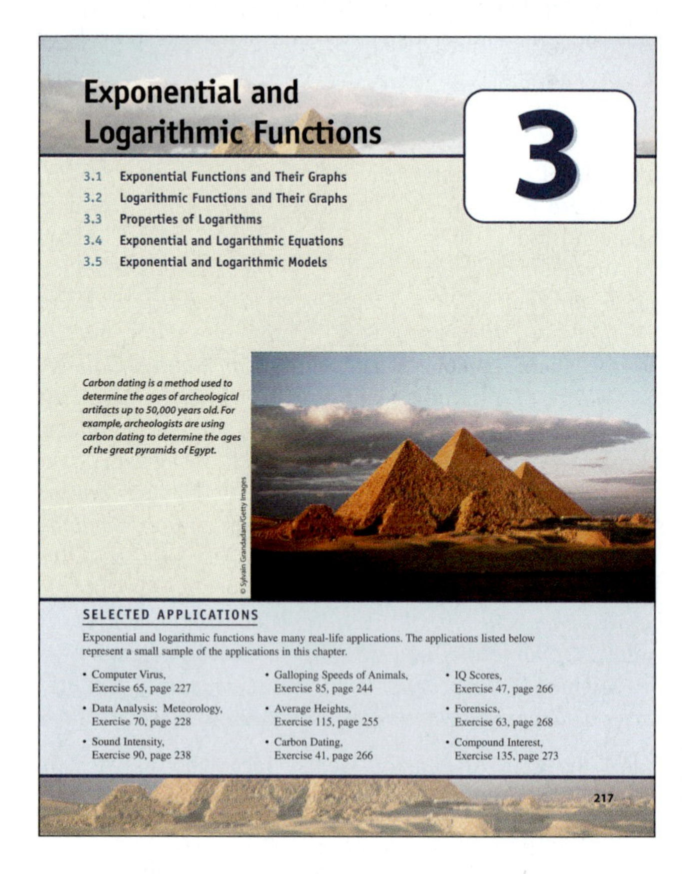

Exponential and Logarithmic Functions

3

3.1 Exponential Functions and Their Graphs
3.2 Logarithmic Functions and Their Graphs
3.3 Properties of Logarithms
3.4 Exponential and Logarithmic Equations
3.5 Exponential and Logarithmic Models

Carbon dating is a method used to determine the ages of archeological artifacts up to 50,000 years old. For example, archeologists are using carbon dating to determine the ages of the great pyramids of Egypt.

© Sylvain Grandadam/Getty Images

SELECTED APPLICATIONS

Exponential and logarithmic functions have many real-life applications. The applications listed below represent a small sample of the applications in this chapter.

- Computer Virus, Exercise 65, page 227
- Data Analysis: Meteorology, Exercise 70, page 228
- Sound Intensity, Exercise 90, page 238
- Galloping Speeds of Animals, Exercise 85, page 244
- Average Heights, Exercise 115, page 255
- Carbon Dating, Exercise 41, page 266
- IQ Scores, Exercise 47, page 266
- Forensics, Exercise 63, page 268
- Compound Interest, Exercise 135, page 273

217

• Chapter Opener

Each chapter begins with a comprehensive overview of the chapter concepts. The photograph and caption illustrate a real-life application of a key concept. Section references help students prepare for the chapter.

• Applications List

An abridged list of applications, covered in the chapter, serve as a motivational tool by connecting section content to real-life situations.

• "What You Should Learn" and "Why You Should Learn It"

Sections begin with *What You Should Learn*, an outline of the main concepts covered in the section, and *Why You Should Learn It*, a real-life application or mathematical reference that illustrates the relevance of the section content.

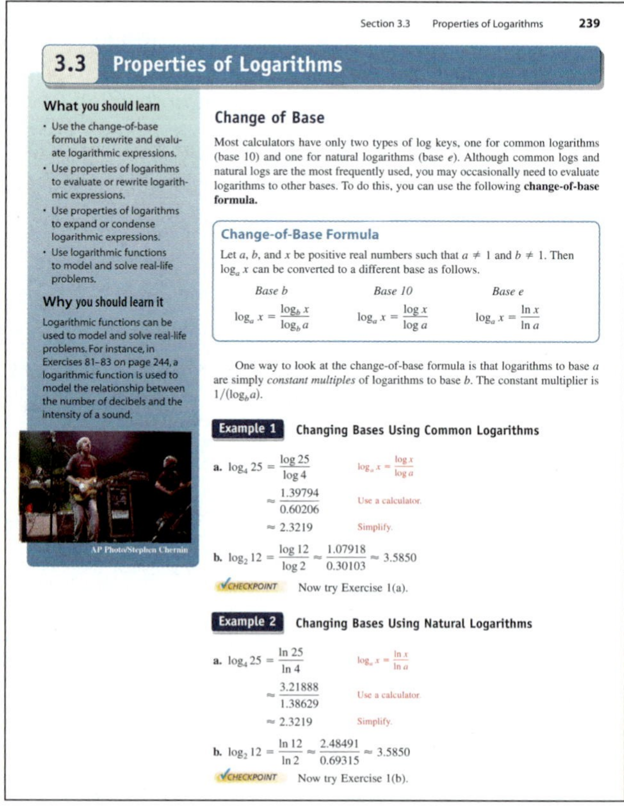

Section 3.3 Properties of Logarithms 239

3.3 Properties of Logarithms

What you should learn

- Use the change-of-base formula to rewrite and evaluate logarithmic expressions.
- Use properties of logarithms to evaluate or rewrite logarithmic expressions.
- Use properties of logarithms to expand or condense logarithmic expressions.
- Use logarithmic functions to model and solve real-life problems.

Why you should learn it

Logarithmic functions can be used to model and solve real-life problems. For instance, in Exercises 81–83 on page 244, a logarithmic function is used to model the relationship between the number of decibels and the intensity of a sound.

AP Photo/Stephen Chernin

Change of Base

Most calculators have only two types of log keys, one for common logarithms (base 10) and one for natural logarithms (base e). Although common logs and natural logs are the most frequently used, you may occasionally need to evaluate logarithms to other bases. To do this, you can use the following **change-of-base formula.**

Change-of-Base Formula

Let a, b, and x be positive real numbers such that $a \neq 1$ and $b \neq 1$. Then $\log_a x$ can be converted to a different base as follows.

Base b	*Base 10*	*Base e*
$\log_a x = \dfrac{\log_b x}{\log_b a}$	$\log_a x = \dfrac{\log x}{\log a}$	$\log_a x = \dfrac{\ln x}{\ln a}$

One way to look at the change-of-base formula is that logarithms to base a are simply *constant multiples* of logarithms to base b. The constant multiplier is $1/(\log_b a)$.

Example 1 Changing Bases Using Common Logarithms

a. $\log_4 25 = \dfrac{\log 25}{\log 4}$ $\log_a x = \dfrac{\log x}{\log a}$

$\approx \dfrac{1.39794}{0.60206}$ Use a calculator.

≈ 2.3219 Simplify.

b. $\log_2 12 = \dfrac{\log 12}{\log 2} \approx \dfrac{1.07918}{0.30103} \approx 3.5850$

CHECKPOINT Now try Exercise 1(a).

Example 2 Changing Bases Using Natural Logarithms

a. $\log_4 25 = \dfrac{\ln 25}{\ln 4}$ $\log_a x = \dfrac{\ln x}{\ln a}$

$\approx \dfrac{3.21888}{1.38629}$ Use a calculator.

≈ 2.3219 Simplify.

b. $\log_2 12 = \dfrac{\ln 12}{\ln 2} \approx \dfrac{2.48491}{0.69315} \approx 3.5850$

CHECKPOINT Now try Exercise 1(b).

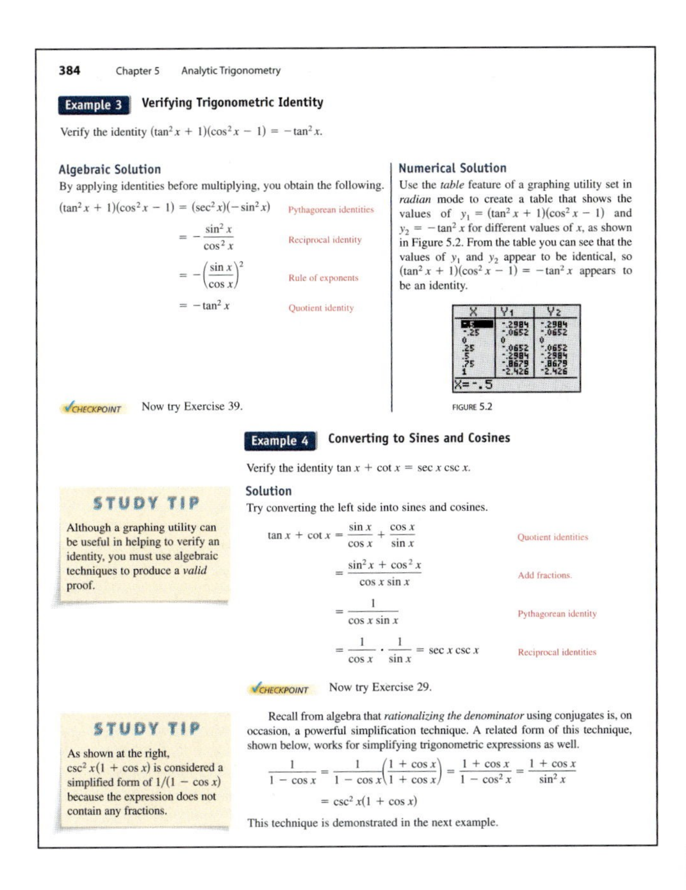

384 Chapter 5 Analytic Trigonometry

Example 3 **Verifying Trigonometric Identity**

Verify the identity $(\tan^2 x + 1)(\cos^2 x - 1) = -\tan^2 x$.

Algebraic Solution

By applying identities before multiplying, you obtain the following.

$$(\tan^2 x + 1)(\cos^2 x - 1) = (\sec^2 x)(-\sin^2 x) \quad \text{Pythagorean identities}$$
$$= -\frac{\sin^2 x}{\cos^2 x} \quad \text{Reciprocal identity}$$
$$= -\left(\frac{\sin x}{\cos x}\right)^2 \quad \text{Rule of exponents}$$
$$= -\tan^2 x \quad \text{Quotient identity}$$

Numerical Solution

Use the *table* feature of a graphing utility set in *radian* mode to create a table that shows the values of $y_1 = (\tan^2 x + 1)(\cos^2 x - 1)$ and $y_2 = -\tan^2 x$ for different values of x, as shown in Figure 5.2. From the table you can see that the values of y_1 and y_2 appear to be identical, so $(\tan^2 x + 1)(\cos^2 x - 1) = -\tan^2 x$ appears to be an identity.

X	Y1	Y2
-.5	-.2984	-.2984
-.25	-.0652	-.0652
0	0	0
.25	-.0652	-.0652
.5	-.2984	-.2984
.75	-.8679	-.8679
1	-2.426	-2.426

X=-.5

FIGURE 5.2

CHECKPOINT Now try Exercise 39.

Example 4 **Converting to Sines and Cosines**

Verify the identity $\tan x + \cot x = \sec x \csc x$.

Solution

Try converting the left side into sines and cosines.

$$\tan x + \cot x = \frac{\sin x}{\cos x} + \frac{\cos x}{\sin x} \quad \text{Quotient identities}$$
$$= \frac{\sin^2 x + \cos^2 x}{\cos x \sin x} \quad \text{Add fractions.}$$
$$= \frac{1}{\cos x \sin x} \quad \text{Pythagorean identity}$$
$$= \frac{1}{\cos x} \cdot \frac{1}{\sin x} = \sec x \csc x \quad \text{Reciprocal identities}$$

CHECKPOINT Now try Exercise 29.

STUDY TIP

Although a graphing utility can be useful in helping to verify an identity, you must use algebraic techniques to produce a *valid* proof.

Recall from algebra that *rationalizing the denominator* using conjugates is, on occasion, a powerful simplification technique. A related form of this technique, shown below, works for simplifying trigonometric expressions as well.

$$\frac{1}{1 - \cos x} = \frac{1}{1 - \cos x}\left(\frac{1 + \cos x}{1 + \cos x}\right) = \frac{1 + \cos x}{1 - \cos^2 x} = \frac{1 + \cos x}{\sin^2 x}$$
$$= \csc^2 x(1 + \cos x)$$

This technique is demonstrated in the next example.

STUDY TIP

As shown at the right, $\csc^2 x(1 + \cos x)$ is considered a simplified form of $1/(1 - \cos x)$ because the expression does not contain any fractions.

• Examples

Many examples present side-by-side solutions with multiple approaches—algebraic, graphical, and numerical. This format addresses a variety of learning styles and shows students that different solution methods yield the same result.

• Checkpoint

The *Checkpoint* directs students to work a similar problem in the exercise set for extra practice.

• Explorations

The *Exploration* engages students in active discovery of mathematical concepts, strengthens critical thinking skills, and helps them to develop an intuitive understanding of theoretical concepts.

• Study Tips

Study Tips reinforce concepts and help students learn how to study mathematics.

• Technology

The *Technology* feature gives instructions for graphing utilities at point of use.

• Additional Features

Additional carefully crafted learning tools, designed to connect concepts, are placed throughout the text. These learning tools include *Writing About Mathematics*, *Historical Notes,* and an extensive art program.

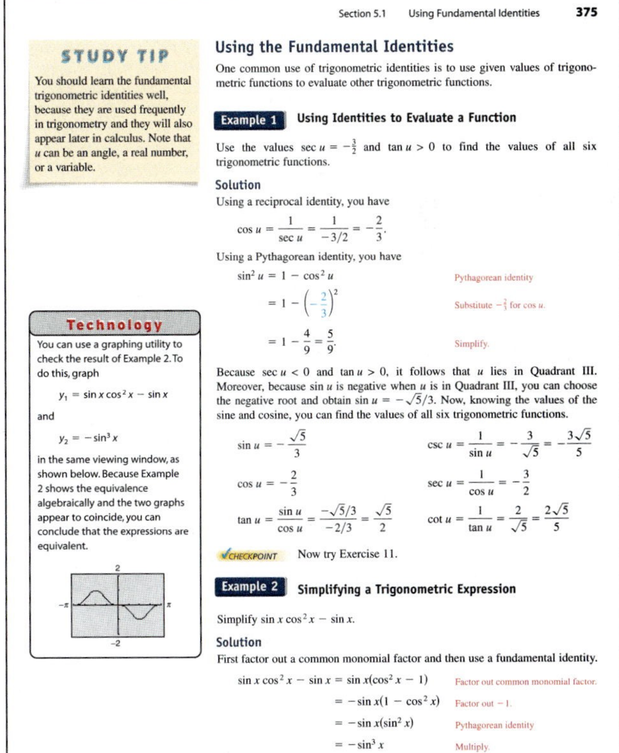

Section 5.1 Using Fundamental Identities 375

STUDY TIP

You should learn the fundamental trigonometric identities well, because they are used frequently in trigonometry and they will also appear later in calculus. Note that u can be an angle, a real number, or a variable.

Using the Fundamental Identities

One common use of trigonometric identities is to use given values of trigonometric functions to evaluate other trigonometric functions.

Example 1 **Using Identities to Evaluate a Function**

Use the values $\sec u = -\frac{3}{2}$ and $\tan u > 0$ to find the values of all six trigonometric functions.

Solution

Using a reciprocal identity, you have

$$\cos u = \frac{1}{\sec u} = \frac{1}{-3/2} = -\frac{2}{3}.$$

Using a Pythagorean identity, you have

$$\sin^2 u = 1 - \cos^2 u \quad \text{Pythagorean identity}$$
$$= 1 - \left(-\frac{2}{3}\right)^2 \quad \text{Substitute } -\tfrac{2}{3} \text{ for } \cos u.$$
$$= 1 - \frac{4}{9} = \frac{5}{9}. \quad \text{Simplify.}$$

Because $\sec u < 0$ and $\tan u > 0$, it follows that u lies in Quadrant III. Moreover, because $\sin u$ is negative when u is in Quadrant III, you can choose the negative root and obtain $\sin u = -\sqrt{5}/3$. Now, knowing the values of the sine and cosine, you can find the values of all six trigonometric functions.

$$\sin u = -\frac{\sqrt{5}}{3} \qquad \csc u = \frac{1}{\sin u} = -\frac{3}{\sqrt{5}} = -\frac{3\sqrt{5}}{5}$$
$$\cos u = -\frac{2}{3} \qquad \sec u = \frac{1}{\cos u} = -\frac{3}{2}$$
$$\tan u = \frac{\sin u}{\cos u} = \frac{-\sqrt{5}/3}{-2/3} = \frac{\sqrt{5}}{2} \qquad \cot u = \frac{1}{\tan u} = \frac{2}{\sqrt{5}} = \frac{2\sqrt{5}}{5}$$

CHECKPOINT Now try Exercise 11.

Technology

You can use a graphing utility to check the result of Example 2. To do this, graph

$$y_1 = \sin x \cos^2 x - \sin x$$

and

$$y_2 = -\sin^3 x$$

in the same viewing window, as shown below. Because Example 2 shows the equivalence algebraically and the two graphs appear to coincide, you can conclude that the expressions are equivalent.

Example 2 **Simplifying a Trigonometric Expression**

Simplify $\sin x \cos^2 x - \sin x$.

Solution

First factor out a common monomial factor and then use a fundamental identity.

$$\sin x \cos^2 x - \sin x = \sin x(\cos^2 x - 1) \quad \text{Factor out common monomial factor.}$$
$$= -\sin x(1 - \cos^2 x) \quad \text{Factor out } -1.$$
$$= -\sin x(\sin^2 x) \quad \text{Pythagorean identity}$$
$$= -\sin^3 x \quad \text{Multiply.}$$

CHECKPOINT Now try Exercise 45.

FEATURES

• Real-Life Applications

A wide variety of real-life applications, many using current real data, are integrated throughout the examples and exercises. The (globe icon) indicates an example that involves a real-life application.

• Algebra of Calculus

Throughout the text, special emphasis is given to the algebraic techniques used in calculus. Algebra of Calculus examples and exercises are integrated throughout the text and are identified by the symbol $\int$.

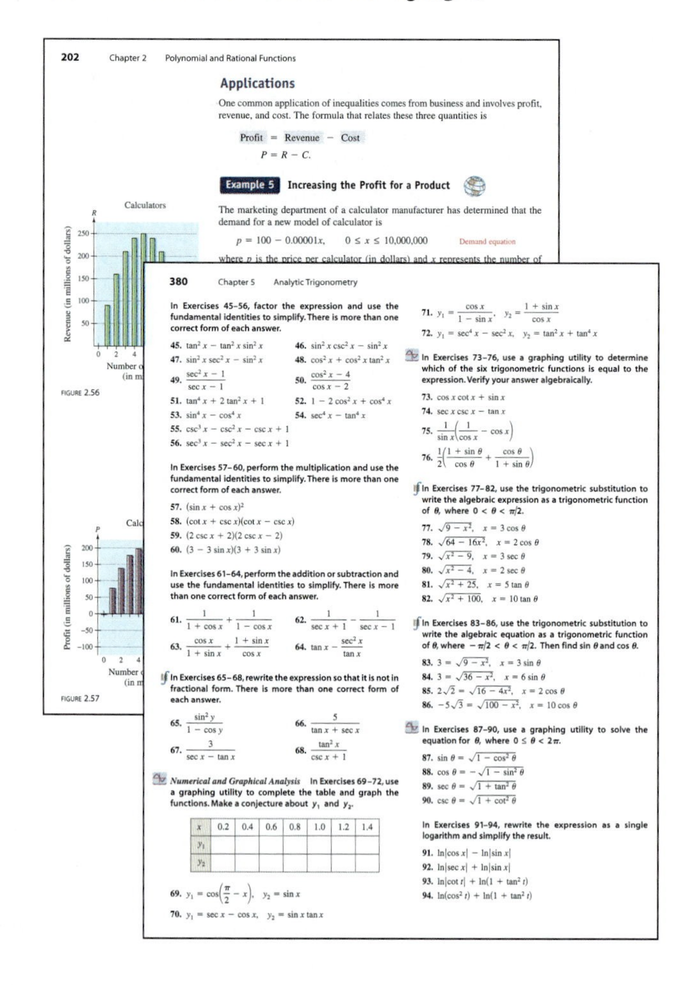

202 Chapter 2 Polynomial and Rational Functions

Applications

One common application of inequalities comes from business and involves profit, revenue, and cost. The formula that relates these three quantities is

Profit = Revenue − Cost

$P = R - C.$

Example 5 Increasing the Profit for a Product

The marketing department of a calculator manufacturer has determined that the demand for a new model of calculator is

$p = 100 - 0.00001x, \quad 0 \le x \le 10{,}000{,}000$ Demand equation

where p is the price per calculator (in dollars) and x represents the number of

FIGURE 2.56

FIGURE 2.57

380 Chapter 5 Analytic Trigonometry

In Exercises 45–56, factor the expression and use the fundamental identities to simplify. There is more than one correct form of each answer.

45. $\tan^2 x - \tan^2 x \sin^2 x$
46. $\sin^2 x \csc^2 x - \sin^2 x$
47. $\sin^2 x \sec^2 x - \sin^2 x$
48. $\cos^2 x + \cos^2 x \tan^2 x$
49. $\dfrac{\sec^2 x - 1}{\sec x - 1}$
50. $\dfrac{\cos^2 x - 4}{\cos x - 2}$
51. $\tan^4 x + 2\tan^2 x + 1$
52. $1 - 2\cos^2 x + \cos^4 x$
53. $\sin^4 x - \cos^4 x$
54. $\sec^4 x - \tan^4 x$
55. $\csc^3 x - \csc^2 x - \csc x + 1$
56. $\sec^3 x - \sec^2 x - \sec x + 1$

In Exercises 57–60, perform the multiplication and use the fundamental identities to simplify. There is more than one correct form of each answer.

57. $(\sin x + \cos x)^2$
58. $(\cot x + \csc x)(\cot x - \csc x)$
59. $(2\csc x + 2)(2\csc x - 2)$
60. $(3 - 3\sin x)(3 + 3\sin x)$

In Exercises 61–64, perform the addition or subtraction and use the fundamental identities to simplify. There is more than one correct form of each answer.

61. $\dfrac{1}{1+\cos x} + \dfrac{1}{1-\cos x}$
62. $\dfrac{1}{\sec x + 1} - \dfrac{1}{\sec x - 1}$
63. $\dfrac{\cos x}{1+\sin x} + \dfrac{1+\sin x}{\cos x}$
64. $\tan x - \dfrac{\sec^2 x}{\tan x}$

In Exercises 65–68, rewrite the expression so that it is not in fractional form. There is more than one correct form of each answer.

65. $\dfrac{\sin^2 y}{1 - \cos y}$
66. $\dfrac{5}{\tan x + \sec x}$
67. $\dfrac{3}{\sec x - \tan x}$
68. $\dfrac{\tan^2 x}{\csc x + 1}$

Numerical and Graphical Analysis In Exercises 69–72, use a graphing utility to complete the table and graph the functions. Make a conjecture about y_1 and y_2.

x	0.2	0.4	0.6	0.8	1.0	1.2	1.4
y_1							
y_2							

69. $y_1 = \cos\left(\dfrac{\pi}{2} - x\right), \quad y_2 = \sin x$
70. $y_1 = \sec x - \cos x, \quad y_2 = \sin x \tan x$
71. $y_1 = \dfrac{\cos x}{1 - \sin x}, \quad y_2 = \dfrac{1 + \sin x}{\cos x}$
72. $y_1 = \sec^4 x - \sec^2 x, \quad y_2 = \tan^2 x + \tan^4 x$

In Exercises 73–76, use a graphing utility to determine which of the six trigonometric functions is equal to the expression. Verify your answer algebraically.

73. $\cos x \cot x + \sin x$
74. $\sec x \csc x - \tan x$
75. $\dfrac{1}{\sin x}\left(\dfrac{1}{\cos x} - \cos x\right)$
76. $\dfrac{1}{2}\left(\dfrac{1+\sin\theta}{\cos\theta} + \dfrac{\cos\theta}{1+\sin\theta}\right)$

In Exercises 77–82, use the trigonometric substitution to write the algebraic expression as a trigonometric function of θ, where $0 < \theta < \pi/2$.

77. $\sqrt{9 - x^2}, \quad x = 3\cos\theta$
78. $\sqrt{64 - 16x^2}, \quad x = 2\cos\theta$
79. $\sqrt{x^2 - 9}, \quad x = 3\sec\theta$
80. $\sqrt{x^2 - 4}, \quad x = 2\sec\theta$
81. $\sqrt{x^2 + 25}, \quad x = 5\tan\theta$
82. $\sqrt{x^2 + 100}, \quad x = 10\tan\theta$

In Exercises 83–86, use the trigonometric substitution to write the algebraic equation as a trigonometric function of θ, where $-\pi/2 < \theta < \pi/2$. Then find $\sin\theta$ and $\cos\theta$.

83. $3 = \sqrt{9 - x^2}, \quad x = 3\sin\theta$
84. $3 = \sqrt{36 - x^2}, \quad x = 6\sin\theta$
85. $2\sqrt{2} = \sqrt{16 - 4x^2}, \quad x = 2\cos\theta$
86. $-5\sqrt{3} = \sqrt{100 - x^2}, \quad x = 10\cos\theta$

In Exercises 87–90, use a graphing utility to solve the equation for θ, where $0 \le \theta < 2\pi$.

87. $\sin\theta = \sqrt{1 - \cos^2\theta}$
88. $\cos\theta = -\sqrt{1 - \sin^2\theta}$
89. $\sec\theta = \sqrt{1 + \tan^2\theta}$
90. $\csc\theta = \sqrt{1 + \cot^2\theta}$

In Exercises 91–94, rewrite the expression as a single logarithm and simplify the result.

91. $\ln|\cos x| - \ln|\sin x|$
92. $\ln|\sec x| + \ln|\sin x|$
93. $\ln|\cot t| + \ln(1 + \tan^2 t)$
94. $\ln(\cos^2 t) + \ln(1 + \tan^2 t)$

• Section Exercises

The section exercise sets consist of a variety of computational, conceptual, and applied problems.

• Vocabulary Check

Section exercises begin with a *Vocabulary Check* that serves as a review of the important mathematical terms in each section.

• Prerequisite Skills Review

Extra practice and a review of algebra skills, needed to complete the section exercise sets, are offered to the students and available in Eduspace®.

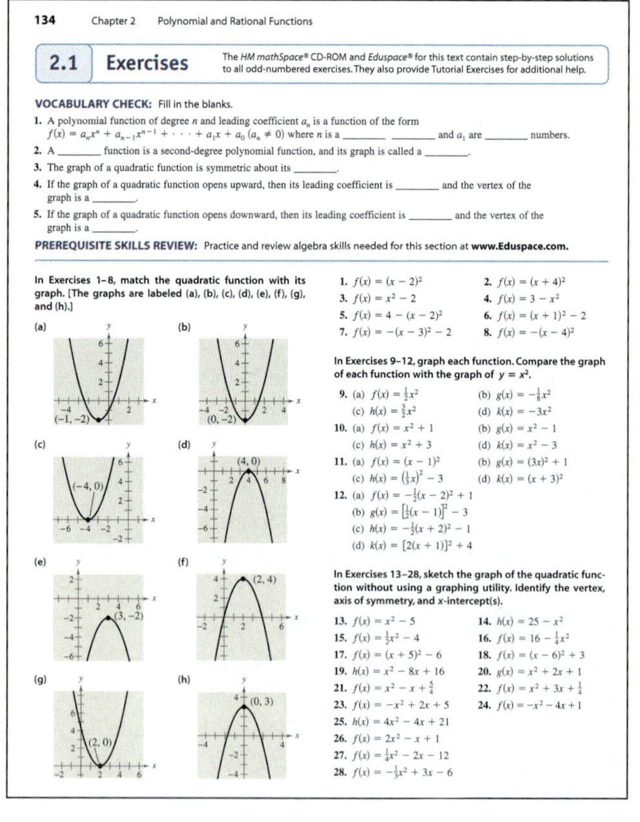

134 Chapter 2 Polynomial and Rational Functions

2.1 Exercises The *HM mathSpace®* CD-ROM and *Eduspace®* for this text contain step-by-step solutions to all odd-numbered exercises. They also provide Tutorial Exercises for additional help.

VOCABULARY CHECK: Fill in the blanks.

1. A polynomial function of degree n and leading coefficient a_n is a function of the form $f(x) = a_n x^n + a_{n-1}x^{n-1} + \cdots + a_1 x + a_0$ $(a_n \ne 0)$ where n is a ________ ________ and a_i are ________ numbers.
2. A ________ function is a second-degree polynomial function, and its graph is called a ________.
3. The graph of a quadratic function is symmetric about its ________.
4. If the graph of a quadratic function opens upward, then its leading coefficient is ________ and the vertex of the graph is a ________.
5. If the graph of a quadratic function opens downward, then its leading coefficient is ________ and the vertex of the graph is a ________.

PREREQUISITE SKILLS REVIEW: Practice and review algebra skills needed for this section at **www.Eduspace.com**.

In Exercises 1–8, match the quadratic function with its graph. [The graphs are labeled (a), (b), (c), (d), (e), (f), (g), and (h).]

1. $f(x) = (x - 2)^2$
2. $f(x) = (x + 4)^2$
3. $f(x) = x^2 - 2$
4. $f(x) = 3 - x^2$
5. $f(x) = 4 - (x - 2)^2$
6. $f(x) = (x + 1)^2 - 2$
7. $f(x) = -(x - 3)^2 - 2$
8. $f(x) = -(x - 4)^2$

In Exercises 9–12, graph each function. Compare the graph of each function with the graph of $y = x^2$.

9. (a) $f(x) = \frac{1}{2}x^2$ (b) $g(x) = -\frac{1}{8}x^2$ (c) $h(x) = \frac{3}{2}x^2$ (d) $k(x) = -3x^2$
10. (a) $f(x) = x^2 + 1$ (b) $g(x) = x^2 - 1$ (c) $h(x) = x^2 + 3$ (d) $k(x) = x^2 - 3$
11. (a) $f(x) = (x - 1)^2$ (b) $g(x) = (3x)^2 + 1$ (c) $h(x) = \left(\frac{1}{3}x\right)^2 - 3$ (d) $k(x) = (x + 3)^2$
12. (a) $f(x) = -\frac{1}{2}(x - 2)^2 + 1$ (b) $g(x) = \left[\frac{1}{2}(x - 1)\right]^2 - 3$ (c) $h(x) = -\frac{1}{2}(x + 2)^2 - 1$ (d) $k(x) = [2(x + 1)]^2 + 4$

In Exercises 13–28, sketch the graph of the quadratic function without using a graphing utility. Identify the vertex, axis of symmetry, and x-intercept(s).

13. $f(x) = x^2 - 5$
14. $h(x) = 25 - x^2$
15. $f(x) = \frac{1}{2}x^2 - 4$
16. $f(x) = 16 - \frac{1}{4}x^2$
17. $f(x) = (x + 5)^2 - 6$
18. $f(x) = (x - 6)^2 + 3$
19. $h(x) = x^2 - 8x + 16$
20. $g(x) = x^2 + 2x + 1$
21. $f(x) = x^2 - x + \frac{5}{4}$
22. $f(x) = x^2 + 3x + \frac{1}{4}$
23. $f(x) = -x^2 + 2x + 5$
24. $f(x) = -x^2 - 4x + 1$
25. $h(x) = 4x^2 - 4x + 21$
26. $f(x) = 2x^2 - x + 1$
27. $f(x) = \frac{1}{4}x^2 - 2x - 12$
28. $f(x) = -\frac{1}{3}x^2 + 3x - 6$

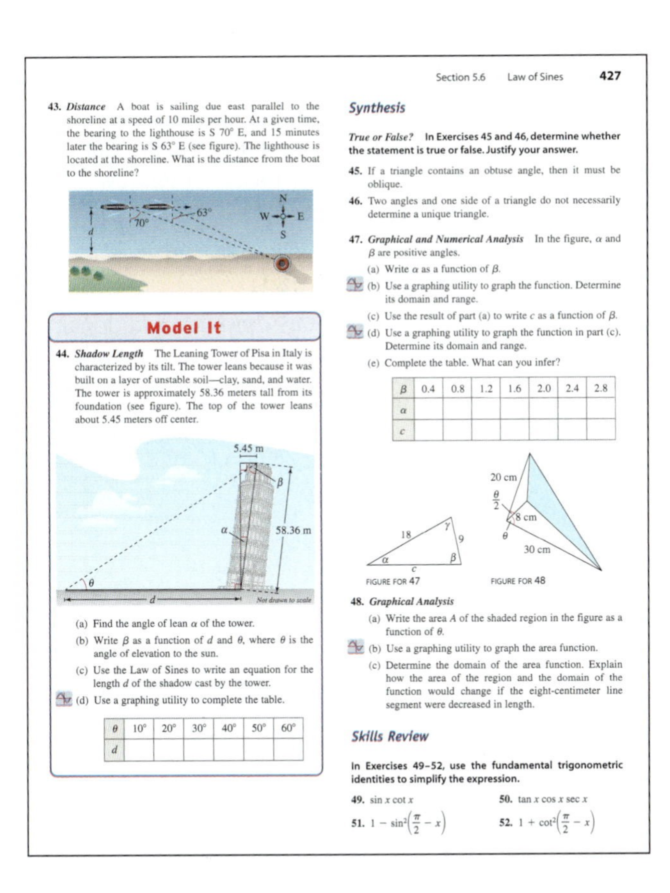

Section 5.6 Law of Sines 427

43. ***Distance*** A boat is sailing due east parallel to the shoreline at a speed of 10 miles per hour. At a given time, the bearing to the lighthouse is S 70° E, and 15 minutes later the bearing is S 63° E (see figure). The lighthouse is located at the shoreline. What is the distance from the boat to the shoreline?

Model It

44. ***Shadow Length*** The Leaning Tower of Pisa in Italy is characterized by its tilt. The tower leans because it was built on a layer of unstable soil—clay, sand, and water. The tower is approximately 58.36 meters tall from its foundation (see figure). The top of the tower leans about 5.45 meters off center.

(a) Find the angle of lean α of the tower.
(b) Write β as a function of d and θ, where θ is the angle of elevation to the sun.
(c) Use the Law of Sines to write an equation for the length d of the shadow cast by the tower.
(d) Use a graphing utility to complete the table.

θ	10°	20°	30°	40°	50°	60°
d						

Synthesis

True or False? **In Exercises 45 and 46, determine whether the statement is true or false. Justify your answer.**

45. If a triangle contains an obtuse angle, then it must be oblique.

46. Two angles and one side of a triangle do not necessarily determine a unique triangle.

47. ***Graphical and Numerical Analysis*** In the figure, α and β are positive angles.
(a) Write α as a function of β.
(b) Use a graphing utility to graph the function. Determine its domain and range.
(c) Use the result of part (a) to write c as a function of β.
(d) Use a graphing utility to graph the function in part (c). Determine its domain and range.
(e) Complete the table. What can you infer?

β	0.4	0.8	1.2	1.6	2.0	2.4	2.8
α							
c							

FIGURE FOR 47

FIGURE FOR 48

48. ***Graphical Analysis***
(a) Write the area A of the shaded region in the figure as a function of θ.
(b) Use a graphing utility to graph the area function.
(c) Determine the domain of the area function. Explain how the area of the region and the domain of the function would change if the eight-centimeter line segment were decreased in length.

Skills Review

In Exercises 49–52, use the fundamental trigonometric identities to simplify the expression.

49. $\sin x \cot x$
50. $\tan x \cos x \sec x$
51. $1 - \sin^2\left(\frac{\pi}{2} - x\right)$
52. $1 + \cot^2\left(\frac{\pi}{2} - x\right)$

• Model It

These multi-part applications that involve real data offer students the opportunity to generate and analyze mathematical models.

• Synthesis and Skills Review Exercises

Each exercise set concludes with the two types of exercises.

Synthesis exercises promote further exploration of mathematical concepts, critical thinking skills, and writing about mathematics. The exercises require students to show their understanding of the relationships between many concepts in the section.

Skills Review Exercises reinforce previously learned skills and concepts.

Make a Decision exercises, found in selected sections, further connect real-life data and applications and motivate students. They also offer students the opportunity to generate and analyze mathematical models from large data sets.

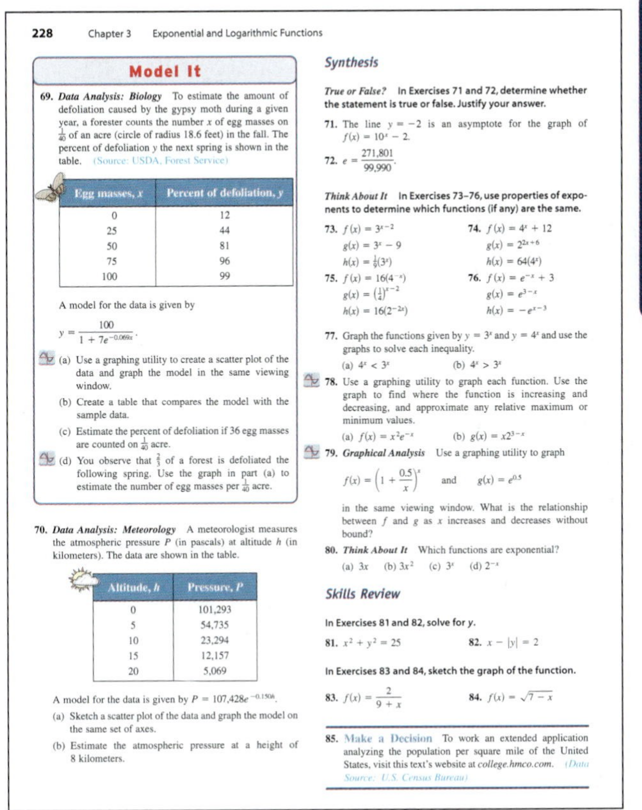

228 Chapter 3 Exponential and Logarithmic Functions

Model It

69. ***Data Analysis: Biology*** To estimate the amount of defoliation caused by the gypsy moth during a given year, a forester counts the number x of egg masses on $\frac{1}{40}$ of an acre (circle of radius 18.6 feet) in the fall. The percent of defoliation y the next spring is shown in the table. (Source: USDA, Forest Service)

Egg masses, x	Percent of defoliation, y
0	12
25	44
50	81
75	96
100	99

A model for the data is given by

$$y = \frac{100}{1 + 7e^{-0.069x}}.$$

(a) Use a graphing utility to create a scatter plot of the data and graph the model in the same viewing window.
(b) Create a table that compares the model with the sample data.
(c) Estimate the percent of defoliation if 36 egg masses are counted on $\frac{1}{40}$ acre.
(d) You observe that $\frac{2}{3}$ of a forest is defoliated the following spring. Use the graph in part (a) to estimate the number of egg masses per $\frac{1}{40}$ acre.

70. ***Data Analysis: Meteorology*** A meteorologist measures the atmospheric pressure P (in pascals) at altitude h (in kilometers). The data are shown in the table.

Altitude, h	Pressure, P
0	101,293
5	54,735
10	23,294
15	12,157
20	5,069

A model for the data is given by $P = 107{,}428e^{-0.150h}$.
(a) Sketch a scatter plot of the data and graph the model on the same set of axes.
(b) Estimate the atmospheric pressure at a height of 8 kilometers.

Synthesis

True or False? **In Exercises 71 and 72, determine whether the statement is true or false. Justify your answer.**

71. The line $y = -2$ is an asymptote for the graph of $f(x) = 10^x - 2$.

72. $e = \frac{271{,}801}{99{,}990}$.

Think About It **In Exercises 73–76, use properties of exponents to determine which functions (if any) are the same.**

73. $f(x) = 3^{x-2}$
$g(x) = 3^x - 9$
$h(x) = \frac{1}{9}(3^x)$

74. $f(x) = 4^x + 12$
$g(x) = 2^{2x+6}$
$h(x) = 64(4^x)$

75. $f(x) = 16(4^{-x})$
$g(x) = \left(\frac{1}{4}\right)^{x-2}$
$h(x) = 16(2^{-2x})$

76. $f(x) = e^{-x} + 3$
$g(x) = e^{3-x}$
$h(x) = -e^{x-3}$

77. Graph the functions given by $y = 3^x$ and $y = 4^x$ and use the graphs to solve each inequality.
(a) $4^x < 3^x$ (b) $4^x > 3^x$

78. Use a graphing utility to graph each function. Use the graph to find where the function is increasing and decreasing, and approximate any relative maximum or minimum values.
(a) $f(x) = x^2e^{-x}$ (b) $g(x) = x2^{3-x}$

79. ***Graphical Analysis*** Use a graphing utility to graph

$$f(x) = \left(1 + \frac{0.5}{x}\right)^x \quad \text{and} \quad g(x) = e^{0.5}$$

in the same viewing window. What is the relationship between f and g as x increases and decreases without bound?

80. ***Think About It*** Which functions are exponential?
(a) $3x$ (b) $3x^2$ (c) 3^x (d) 2^{-x}

Skills Review

In Exercises 81 and 82, solve for y.

81. $x^2 + y^2 = 25$
82. $x - |y| = 2$

In Exercises 83 and 84, sketch the graph of the function.

83. $f(x) = \frac{2}{9 + x}$
84. $f(x) = \sqrt{7 - x}$

85. Make a Decision To work an extended application analyzing the population per square mile of the United States, visit this text's website at *college.hmco.com*. (Data Source: U.S. Census Bureau)

FEATURES

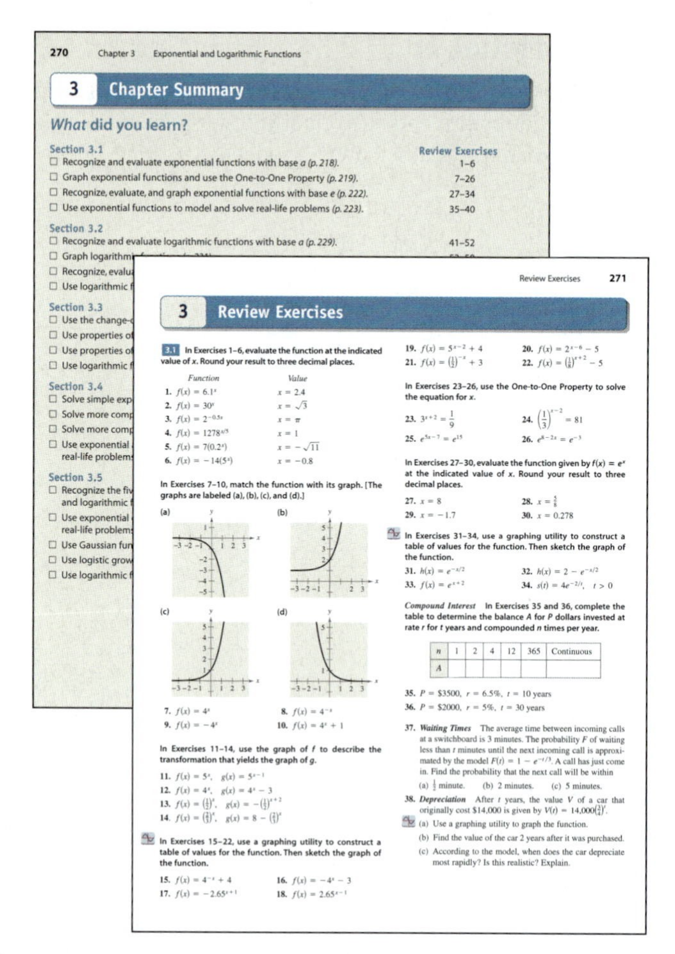

270 Chapter 3 Exponential and Logarithmic Functions

3 Chapter Summary

What did you learn?

Section 3.1	Review Exercises
☐ Recognize and evaluate exponential functions with base *a* (p. 218).	1–6
☐ Graph exponential functions and use the One-to-One Property (p. 219).	7–26
☐ Recognize, evaluate, and graph exponential functions with base *e* (p. 222).	27–34
☐ Use exponential functions to model and solve real-life problems (p. 223).	35–40

Section 3.2
☐ Recognize and evaluate logarithmic functions with base *a* (p. 229). 41–52
☐ Graph logarithm…
☐ Recognize, evalu…
☐ Use logarithmic f…

Section 3.3
☐ Use the change-o…
☐ Use properties of…
☐ Use properties of…
☐ Use logarithmic f…

Section 3.4
☐ Solve simple exp…
☐ Solve more comp…
☐ Solve more comp…
☐ Use exponential… real-life problem…

Section 3.5
☐ Recognize the fiv… and logarithmic f…
☐ Use exponential… real-life problem…
☐ Use Gaussian fun…
☐ Use logistic grow…
☐ Use logarithmic f…

Review Exercises 271

3 Review Exercises

3.1 In Exercises 1–6, evaluate the function at the indicated value of x. Round your result to three decimal places.

Function	Value
1. $f(x) = 6.1^x$	$x = 2.4$
2. $f(x) = 30^x$	$x = \sqrt{3}$
3. $f(x) = 2^{-0.5x}$	$x = \pi$
4. $f(x) = 1278^{x/5}$	$x = 1$
5. $f(x) = 7(0.2^x)$	$x = -\sqrt{11}$
6. $f(x) = -14(5^x)$	$x = -0.8$

In Exercises 7–10, match the function with its graph. [The graphs are labeled (a), (b), (c), and (d).]

(a) (b) (c) (d)

7. $f(x) = 4^x$ 8. $f(x) = 4^{-x}$
9. $f(x) = -4^x$ 10. $f(x) = 4^x + 1$

In Exercises 11–14, use the graph of f to describe the transformation that yields the graph of g.

11. $f(x) = 5^x$, $g(x) = 5^{x-1}$
12. $f(x) = 4^x$, $g(x) = 4^x - 3$
13. $f(x) = \left(\frac{1}{2}\right)^x$, $g(x) = -\left(\frac{1}{2}\right)^{x+2}$
14. $f(x) = \left(\frac{2}{3}\right)^x$, $g(x) = 8 - \left(\frac{2}{3}\right)^x$

In Exercises 15–22, use a graphing utility to construct a table of values for the function. Then sketch the graph of the function.

15. $f(x) = 4^{-x} + 4$ 16. $f(x) = -4^x - 3$
17. $f(x) = -2.65^{x+1}$ 18. $f(x) = 2.65^{x-1}$
19. $f(x) = 5^{x-2} + 4$ 20. $f(x) = 2^{x-6} - 5$
21. $f(x) = \left(\frac{1}{2}\right)^{-x} + 3$ 22. $f(x) = \left(\frac{1}{8}\right)^{x+2} - 5$

In Exercises 23–26, use the One-to-One Property to solve the equation for x.

23. $3^{x+2} = \frac{1}{9}$ 24. $\left(\frac{1}{3}\right)^{x-2} = 81$
25. $e^{5x-7} = e^{15}$ 26. $e^{8-2x} = e^{-3}$

In Exercises 27–30, evaluate the function given by $f(x) = e^x$ at the indicated value of x. Round your result to three decimal places.

27. $x = 8$ 28. $x = \frac{5}{8}$
29. $x = -1.7$ 30. $x = 0.278$

In Exercises 31–34, use a graphing utility to construct a table of values for the function. Then sketch the graph of the function.

31. $h(x) = e^{-x/2}$ 32. $h(x) = 2 - e^{-x/2}$
33. $f(x) = e^{x+2}$ 34. $s(t) = 4e^{-2/t}$, $t > 0$

Compound Interest In Exercises 35 and 36, complete the table to determine the balance A for P dollars invested at rate r for t years and compounded n times per year.

n	1	2	4	12	365	Continuous
A						

35. $P = \$3500$, $r = 6.5\%$, $t = 10$ years
36. $P = \$2000$, $r = 5\%$, $t = 30$ years

37. ***Waiting Times*** The average time between incoming calls at a switchboard is 3 minutes. The probability F of waiting less than t minutes until the next incoming call is approximated by the model $F(t) = 1 - e^{-t/3}$. A call has just come in. Find the probability that the next call will be within
(a) $\frac{1}{2}$ minute. (b) 2 minutes. (c) 5 minutes.

38. ***Depreciation*** After t years, the value V of a car that originally cost \$14,000 is given by $V(t) = 14{,}000\left(\frac{3}{4}\right)^t$.
(a) Use a graphing utility to graph the function.
(b) Find the value of the car 2 years after it was purchased.
(c) According to the model, when does the car depreciate most rapidly? Is this realistic? Explain.

• Chapter Summary

The *Chapter Summary* "*What Did You Learn?*" is a section-by-section overview that ties the learning objectives from the chapter to sets of Review Exercises for extra practice.

• Review Exercises

The chapter *Review Exercises* provide additional practice with the concepts covered in the chapter.

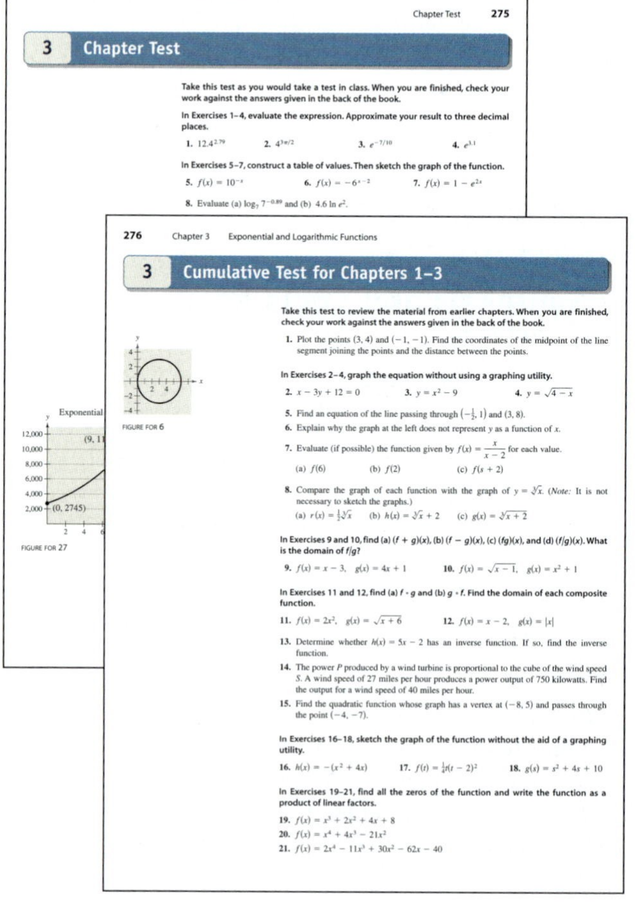

Chapter Test 275

3 Chapter Test

Take this test as you would take a test in class. When you are finished, check your work against the answers given in the back of the book.

In Exercises 1–4, evaluate the expression. Approximate your result to three decimal places.

1. $12.4^{2.79}$ 2. $4^{3\pi/2}$ 3. $e^{-7/10}$ 4. $e^{3.1}$

In Exercises 5–7, construct a table of values. Then sketch the graph of the function.

5. $f(x) = 10^{-x}$ 6. $f(x) = -6^{x-2}$ 7. $f(x) = 1 - e^{2x}$

8. Evaluate (a) $\log_7 7^{-0.89}$ and (b) $4.6 \ln e^2$.

FIGURE FOR 27

276 Chapter 3 Exponential and Logarithmic Functions

3 Cumulative Test for Chapters 1–3

Take this test to review the material from earlier chapters. When you are finished, check your work against the answers given in the back of the book.

1. Plot the points $(3, 4)$ and $(-1, -1)$. Find the coordinates of the midpoint of the line segment joining the points and the distance between the points.

In Exercises 2–4, graph the equation without using a graphing utility.

2. $x - 3y + 12 = 0$ 3. $y = x^2 - 9$ 4. $y = \sqrt{4 - x}$

5. Find an equation of the line passing through $\left(-\frac{1}{2}, 1\right)$ and $(3, 8)$.
6. Explain why the graph at the left does not represent y as a function of x.
7. Evaluate (if possible) the function given by $f(x) = \frac{x}{x-2}$ for each value.
(a) $f(6)$ (b) $f(2)$ (c) $f(s + 2)$
8. Compare the graph of each function with the graph of $y = \sqrt[3]{x}$. (*Note:* It is not necessary to sketch the graphs.)
(a) $r(x) = \frac{1}{2}\sqrt[3]{x}$ (b) $h(x) = \sqrt[3]{x} + 2$ (c) $g(x) = \sqrt[3]{x+2}$

FIGURE FOR 6

In Exercises 9 and 10, find (a) $(f + g)(x)$, (b) $(f - g)(x)$, (c) $(fg)(x)$, and (d) $(f/g)(x)$. What is the domain of f/g?

9. $f(x) = x - 3$, $g(x) = 4x + 1$ 10. $f(x) = \sqrt{x - 1}$, $g(x) = x^2 + 1$

In Exercises 11 and 12, find (a) $f \circ g$ and (b) $g \circ f$. Find the domain of each composite function.

11. $f(x) = 2x^2$, $g(x) = \sqrt{x + 6}$ 12. $f(x) = x - 2$, $g(x) = |x|$

13. Determine whether $h(x) = 5x - 2$ has an inverse function. If so, find the inverse function.
14. The power P produced by a wind turbine is proportional to the cube of the wind speed S. A wind speed of 27 miles per hour produces a power output of 750 kilowatts. Find the output for a wind speed of 40 miles per hour.
15. Find the quadratic function whose graph has a vertex at $(-8, 5)$ and passes through the point $(-4, -7)$.

In Exercises 16–18, sketch the graph of the function without the aid of a graphing utility.

16. $h(x) = -(x^2 + 4x)$ 17. $f(t) = \frac{1}{4}t(t - 2)^2$ 18. $g(s) = s^2 + 4s + 10$

In Exercises 19–21, find all the zeros of the function and write the function as a product of linear factors.

19. $f(x) = x^3 + 2x^2 + 4x + 8$
20. $f(x) = x^4 + 4x^3 - 21x^2$
21. $f(x) = 2x^4 - 11x^3 + 30x^2 - 62x - 40$

• Chapter Tests and Cumulative Tests

Chapter Tests, at the end of each chapter, and periodic *Cumulative Tests* offer students frequent opportunities for self-assessment and to develop strong study and test-taking skills.

Proofs in Mathematics

What does the word *proof* mean to you? In mathematics, the word *proof* is used to mean simply a valid argument. When you are proving a statement or theorem, you must use facts, definitions, and accepted properties in a logical order. You can also use previously proved theorems in your proof. For instance, the Distance Formula is used in the proof of the Midpoint Formula below. There are several different proof methods, which you will see in later chapters.

The Midpoint Formula *(p.)*
The midpoint of the line segment joining the points (x_1, y_1) and (x_2, y_2) is given by the Midpoint Formula

$$\text{Midpoint} = \left(\frac{x_1 + x_2}{2}, \frac{y_1 + y_2}{2}\right).$$

The Cartesian Plane
The Cartesian plane was named after the French mathematician René Descartes (1596–1650). While Descartes was lying in bed, he noticed a fly buzzing around on the square ceiling tiles. He discovered that the position of the fly could be described by which ceiling tile the fly landed on. This led to the development of the Cartesian plane. Descartes felt that a coordinate plane could be used to facilitate description of the positions of objects.

Proof
Using the figure, you must show that $d_1 = d_2$ and $d_1 + d_2 = d_3$.

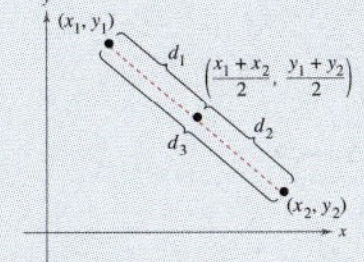

By the Distance Formula, you obtain

$$d_1 = \sqrt{\left(\frac{x_1 + x_2}{2} - x_1\right)^2 + \left(\frac{y_1 + y_2}{2} - y_1\right)^2}$$
$$= \frac{1}{2}\sqrt{(x_2 - x_1)^2 + (y_2 - y_1)^2}$$
$$d_2 = \sqrt{\left(x_2 - \frac{x_1 + x_2}{2}\right)^2 + \left(y_2 - \frac{y_1 + y_2}{2}\right)^2}$$
$$= \frac{1}{2}\sqrt{(x_2 - x_1)^2 + (y_2 - y_1)^2}$$
$$d_3 = \sqrt{(x_2 - x_1)^2 + (y_2 - y_1)^2}$$

So, it follows that $d_1 = d_2$ and $d_1 + d_2 = d_3$.

124

• Proofs in Mathematics

At the end of every chapter, proofs of important mathematical properties and theorems are presented as well as discussions of various proof techniques.

• P.S. Problem Solving

Each chapter concludes with a collection of thought-provoking and challenging exercises that further explore and expand upon the chapter concepts. These exercises have unusual characteristics that set them apart from traditional text exercises.

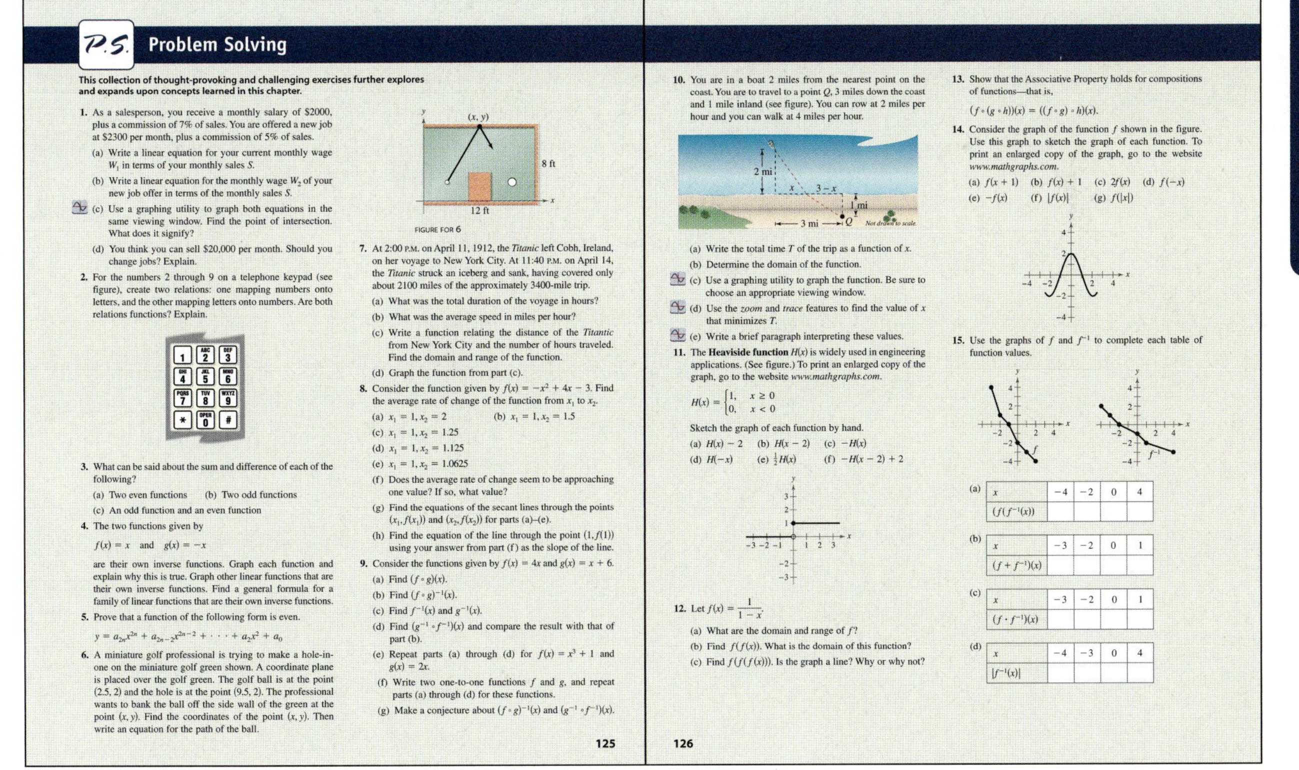

P.S. Problem Solving

This collection of thought-provoking and challenging exercises further explores and expands upon concepts learned in this chapter.

1. As a salesperson, you receive a monthly salary of \$2000, plus a commission of 7% of sales. You are offered a new job at \$2300 per month, plus a commission of 5% of sales.
   (a) Write a linear equation for your current monthly wage W_1 in terms of your monthly sales S.
   (b) Write a linear equation for the monthly wage W_2 of your new job offer in terms of the monthly sales S.
   (c) Use a graphing utility to graph both equations in the same viewing window. Find the point of intersection. What does it signify?
   (d) You think you can sell \$20,000 per month. Should you change jobs? Explain.
2. For the numbers 2 through 9 on a telephone keypad (see figure), create two relations: one mapping numbers onto letters, and the other mapping letters onto numbers. Are both relations functions? Explain.

3. What can be said about the sum and difference of each of the following?
   (a) Two even functions (b) Two odd functions
   (c) An odd function and an even function
4. The two functions given by
   $f(x) = x$ and $g(x) = -x$
   are their own inverse functions. Graph each function and explain why this is true. Graph other linear functions that are their own inverse functions. Find a general formula for a family of linear functions that are their own inverse functions.
5. Prove that a function of the following form is even.
   $y = a_{2n}x^{2n} + a_{2n-2}x^{2n-2} + \cdots + a_2x^2 + a_0$
6. A miniature golf professional is trying to make a hole-in-one on the miniature golf green shown. A coordinate plane is placed over the golf green. The golf ball is at the point (2.5, 2) and the hole is at the point (9.5, 2). The professional wants to bank the ball off the side wall of the green at the point (x, y). Find the coordinates of the point (x, y). Then write an equation for the path of the ball.

FIGURE FOR 6

7. At 2:00 P.M. on April 11, 1912, the *Titanic* left Cobh, Ireland, on her voyage to New York City. At 11:40 P.M. on April 14, the *Titanic* struck an iceberg and sank, having covered only about 2100 miles of the approximately 3400-mile trip.
   (a) What was the total duration of the voyage in hours?
   (b) What was the average speed in miles per hour?
   (c) Write a function relating the distance of the *Titantic* from New York City and the number of hours traveled. Find the domain and range of the function.
   (d) Graph the function from part (c).
8. Consider the function given by $f(x) = -x^2 + 4x - 3$. Find the average rate of change of the function from x_1 to x_2.
   (a) $x_1 = 1, x_2 = 2$ (b) $x_1 = 1, x_2 = 1.5$
   (c) $x_1 = 1, x_2 = 1.25$
   (d) $x_1 = 1, x_2 = 1.125$
   (e) $x_1 = 1, x_2 = 1.0625$
   (f) Does the average rate of change seem to be approaching one value? If so, what value?
   (g) Find the equations of the secant lines through the points $(x_1, f(x_1))$ and $(x_2, f(x_2))$ for parts (a)–(e).
   (h) Find the equation of the line through the point $(1, f(1))$ using your answer from part (f) as the slope of the line.
9. Consider the functions given by $f(x) = 4x$ and $g(x) = x + 6$.
   (a) Find $(f \circ g)(x)$.
   (b) Find $(f \circ g)^{-1}(x)$.
   (c) Find $f^{-1}(x)$ and $g^{-1}(x)$.
   (d) Find $(g^{-1} \circ f^{-1})(x)$ and compare the result with that of part (b).
   (e) Repeat parts (a) through (d) for $f(x) = x^3 + 1$ and $g(x) = 2x$.
   (f) Write two one-to-one functions f and g, and repeat parts (a) through (d) for these functions.
   (g) Make a conjecture about $(f \circ g)^{-1}(x)$ and $(g^{-1} \circ f^{-1})(x)$.

125

10. You are in a boat 2 miles from the nearest point on the coast. You are to travel to a point Q, 3 miles down the coast and 1 mile inland (see figure). You can row at 2 miles per hour and you can walk at 4 miles per hour.

   (a) Write the total time T of the trip as a function of x.
   (b) Determine the domain of the function.
   (c) Use a graphing utility to graph the function. Be sure to choose an appropriate viewing window.
   (d) Use the *zoom* and *trace* features to find the value of x that minimizes T.
   (e) Write a brief paragraph interpreting these values.
11. The **Heaviside function** $H(x)$ is widely used in engineering applications. (See figure.) To print an enlarged copy of the graph, go to the website *www.mathgraphs.com*.
   $$H(x) = \begin{cases} 1, & x \geq 0 \\ 0, & x < 0 \end{cases}$$
   Sketch the graph of each function by hand.
   (a) $H(x) - 2$ (b) $H(x - 2)$ (c) $-H(x)$
   (d) $H(-x)$ (e) $\frac{1}{2}H(x)$ (f) $-H(x - 2) + 2$
12. Let $f(x) = \dfrac{1}{1 - x}$.
   (a) What are the domain and range of f?
   (b) Find $f(f(x))$. What is the domain of this function?
   (c) Find $f(f(f(x)))$. Is the graph a line? Why or why not?
13. Show that the Associative Property holds for compositions of functions—that is,
   $(f \circ (g \circ h))(x) = ((f \circ g) \circ h)(x)$.
14. Consider the graph of the function f shown in the figure. Use this graph to sketch the graph of each function. To print an enlarged copy of the graph, go to the website *www.mathgraphs.com*.
   (a) $f(x + 1)$ (b) $f(x) + 1$ (c) $2f(x)$ (d) $f(-x)$
   (e) $-f(x)$ (f) $|f(x)|$ (g) $f(|x|)$
15. Use the graphs of f and f^{-1} to complete each table of function values.

(a)

x	-4	-2	0	4
$(f(f^{-1}(x))$				

(b)

x	-3	-2	0	1
$(f + f^{-1})(x)$				

(c)

x	-3	-2	0	1
$(f \cdot f^{-1})(x)$				

(d)

x	-4	-3	0	4
$\lvert f^{-1}(x) \rvert$				

126

FEATURES

Supplements

Supplements for the Instructor

Precalculus: A Concise Course has an extensive support package for the instructor that includes:

Instructor's Annotated Edition (IAE)

Online Complete Solutions Guide

Online Instructor Success Organizer

***Online Teaching Center*:** This free companion website contains an abundance of instructor resources.

***HM ClassPrep™ with HM Testing (powered by Diploma™)*:** This CD-ROM is a combination of two course management tools.

- *HM Testing* (powered by *Diploma*™) offers instructors a flexible and powerful tool for test generation and test management. Now supported by the Brownstone Research Group's market-leading *Diploma*™ software, this new version of *HM Testing* significantly improves on functionality and ease of use by offering all the tools needed to create, author, deliver, and customize multiple types of tests—including authoring and editing algorithmic questions. *Diploma*™ is currently in use at thousands of college and university campuses throughout the United States and Canada.
- HM ClassPrep™ also features supplements and text-specific resources for the instructor.

***Eduspace®*:** Eduspace®, powered by Blackboard®, is Houghton Mifflin's customizable and interactive online learning tool. Eduspace® provides instructors with online courses and content. By pairing the widely recognized tools of Blackboard® with quality, text-specific content from Houghton Mifflin Company, Eduspace® makes it easy for instructors to create all or part of a course online. This online learning tool also contains ready-to-use homework exercises, quizzes, tests, tutorials, and supplemental study materials.

Visit *www.eduspace.com* for more information.

***Eduspace® with eSolutions*:** Eduspace® with eSolutions combines all the features of Eduspace® with an electronic version of the textbook exercises and the complete solutions to the odd-numbered exercises, providing students with a convenient and comprehensive way to do homework and view course materials.

Supplements for the Student

Precalculus: A Concise Course has an extensive support package for the student that includes:

Study and Solutions Guide

Online Student Notetaking Guide

Instructional DVDs

***Online Study Center*:** This free companion website contains an abundance of student resources.

***HM mathSpace® CD-ROM*:** This tutorial CD-ROM provides opportunities for self-paced review and practice with algorithmically generated exercises and step-by-step solutions.

***Eduspace®*:** Eduspace®, powered by Blackboard®, is Houghton Mifflin's customizable and interactive online learning tool for instructors and students. Eduspace® is a text-specific, web-based learning environment that your instructor can use to offer students a combination of practice exercises, multimedia tutorials, video explanations, online algorithmic homework and more. Specific content is available 24 hours a day to help you succeed in your course.

***Eduspace® with eSolutions*:** Eduspace® with eSolutions combines all the features of Eduspace® with an electronic version of the textbook exercises and the complete solutions to the odd-numbered exercises. The result is a convenient and comprehensive way to do homework and view your course materials.

***Smarthinking®*:** Houghton Mifflin has partnered with Smarthinking® to provide an easy-to-use, effective, online tutorial service. Through state-of-the-art tools and whiteboard technology, students communicate in real-time with qualified e-instructors who can help the students understand difficult concepts and guide them through the problem-solving process while studying or completing homework.

Three levels of service are offered to the students.

Live Tutorial Help provides real-time, one-on-one instruction.

Question Submission allows students to submit questions to the tutor outside the scheduled hours and receive a reply usually within 24 hours.

Independent Study Resources connects students around-the-clock to additional educational resources, ranging from interactive websites to Frequently Asked Questions.

Visit *smarthinking.com* for more information.

**Limits apply; terms and hours of SMARTHINKING® service are subject to change.*

Functions and Their Graphs

1

Functions play a primary role in modeling real-life situations. The estimated growth in the number of digital music sales in the United States can be modeled by a cubic function.

SELECTED APPLICATIONS

Functions have many real-life applications. The applications listed below represent a small sample of the applications in this chapter.

- Data Analysis: Mail, Exercise 69, page 12
- Population Statistics, Exercise 75, page 24
- College Enrollment, Exercise 109, page 37
- Cost, Revenue, and Profit, Exercise 97, page 52
- Digital Music Sales, Exercise 89, page 64
- Fluid Flow, Exercise 70, page 68
- Fuel Use, Exercise 67, page 82
- Consumer Awareness, Exercise 68, page 92
- Diesel Mechanics, Exercise 83, page 102

1.1 Rectangular Coordinates

What you should learn

- Plot points in the Cartesian plane.
- Use the Distance Formula to find the distance between two points.
- Use the Midpoint Formula to find the midpoint of a line segment.
- Use a coordinate plane and geometric formulas to model and solve real-life problems.

Why you should learn it

The Cartesian plane can be used to represent relationships between two variables. For instance, in Exercise 60 on page 12, a graph represents the minimum wage in the United States from 1950 to 2004.

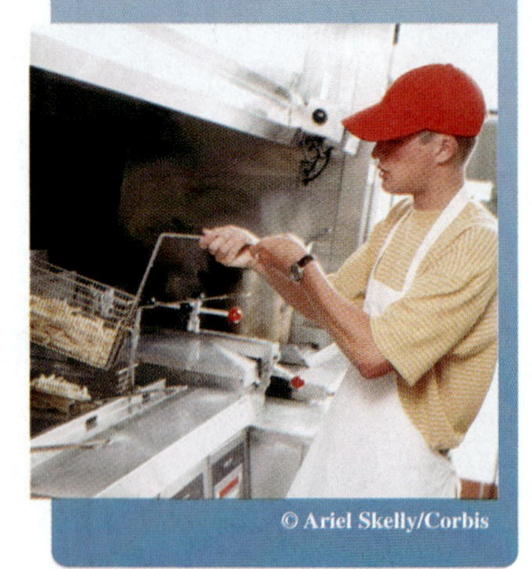

© Ariel Skelly/Corbis

The Cartesian Plane

Just as you can represent real numbers by points on a real number line, you can represent ordered pairs of real numbers by points in a plane called the **rectangular coordinate system,** or the **Cartesian plane,** named after the French mathematician René Descartes (1596–1650).

The Cartesian plane is formed by using two real number lines intersecting at right angles, as shown in Figure 1.1. The horizontal real number line is usually called the ***x*-axis,** and the vertical real number line is usually called the ***y*-axis.** The point of intersection of these two axes is the **origin,** and the two axes divide the plane into four parts called **quadrants.**

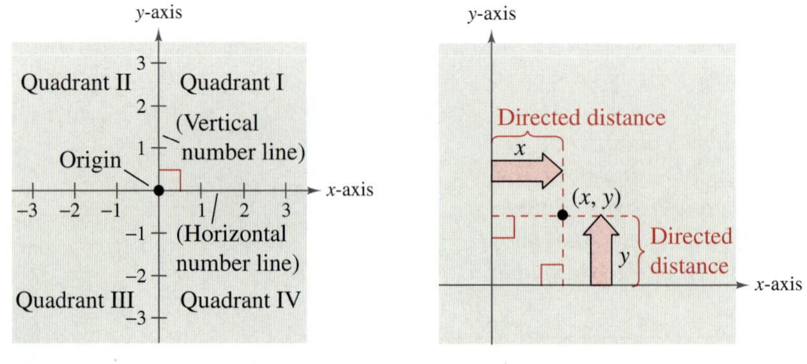

FIGURE 1.1 FIGURE 1.2

Each point in the plane corresponds to an **ordered pair** (x, y) of real numbers x and y, called **coordinates** of the point. The ***x*-coordinate** represents the directed distance from the y-axis to the point, and the ***y*-coordinate** represents the directed distance from the x-axis to the point, as shown in Figure 1.2.

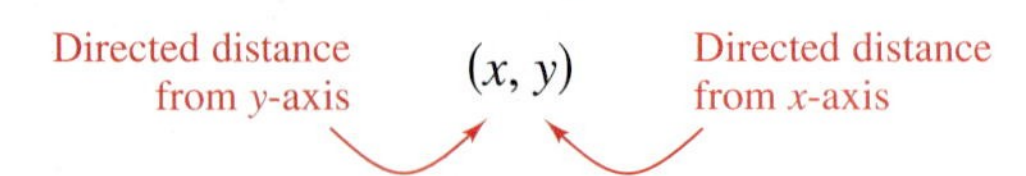

The notation (x, y) denotes both a point in the plane and an open interval on the real number line. The context will tell you which meaning is intended.

Example 1 Plotting Points in the Cartesian Plane

Plot the points $(-1, 2)$, $(3, 4)$, $(0, 0)$, $(3, 0)$, and $(-2, -3)$.

Solution

To plot the point $(-1, 2)$, imagine a vertical line through -1 on the x-axis and a horizontal line through 2 on the y-axis. The intersection of these two lines is the point $(-1, 2)$. The other four points can be plotted in a similar way, as shown in Figure 1.3.

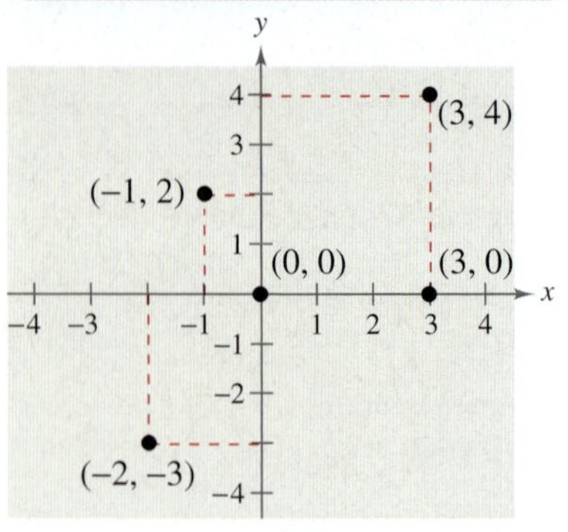

FIGURE 1.3

✓CHECKPOINT Now try Exercise 3.

The beauty of a rectangular coordinate system is that it allows you to *see* relationships between two variables. It would be difficult to overestimate the importance of Descartes's introduction of coordinates in the plane. Today, his ideas are in common use in virtually every scientific and business-related field.

Example 2 Sketching a Scatter Plot

From 1990 through 2003, the amounts A (in millions of dollars) spent on skiing equipment in the United States are shown in the table, where t represents the year. Sketch a scatter plot of the data. (Source: National Sporting Goods Association)

Year, t	Amount, A
1990	475
1991	577
1992	521
1993	569
1994	609
1995	562
1996	707
1997	723
1998	718
1999	648
2000	495
2001	476
2002	527
2003	464

Solution

To sketch a *scatter plot* of the data shown in the table, you simply represent each pair of values by an ordered pair (t, A) and plot the resulting points, as shown in Figure 1.4. For instance, the first pair of values is represented by the ordered pair (1990, 475). Note that the break in the t-axis indicates that the numbers between 0 and 1990 have been omitted.

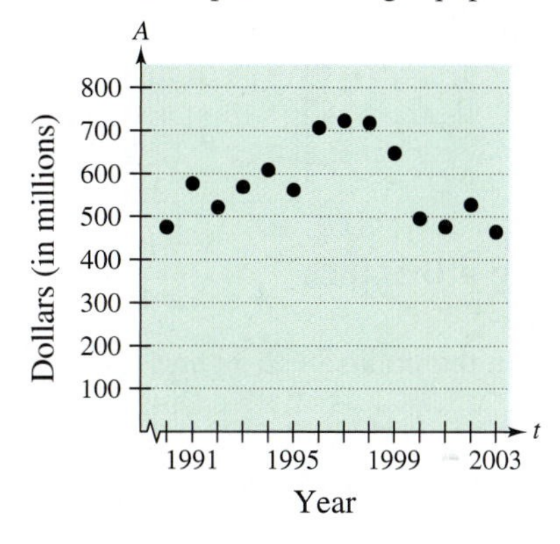

FIGURE 1.4

CHECKPOINT Now try Exercise 21.

In Example 2, you could have let $t = 1$ represent the year 1990. In that case, the horizontal axis would not have been broken, and the tick marks would have been labeled 1 through 14 (instead of 1990 through 2003).

Technology

The scatter plot in Example 2 is only one way to represent the data graphically. You could also represent the data using a bar graph and a line graph. If you have access to a graphing utility, try using it to represent graphically the data given in Example 2.

The *HM mathSpace®* CD-ROM and *Eduspace®* for this text contain additional resources related to the concepts discussed in this chapter.

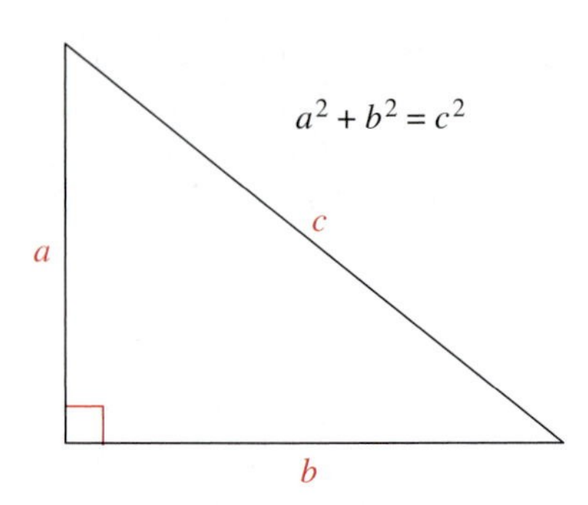

FIGURE 1.5

The Pythagorean Theorem and the Distance Formula

The following famous theorem is used extensively throughout this course.

Pythagorean Theorem

For a right triangle with hypotenuse of length c and sides of lengths a and b, you have $a^2 + b^2 = c^2$, as shown in Figure 1.5. (The converse is also true. That is, if $a^2 + b^2 = c^2$, then the triangle is a right triangle.)

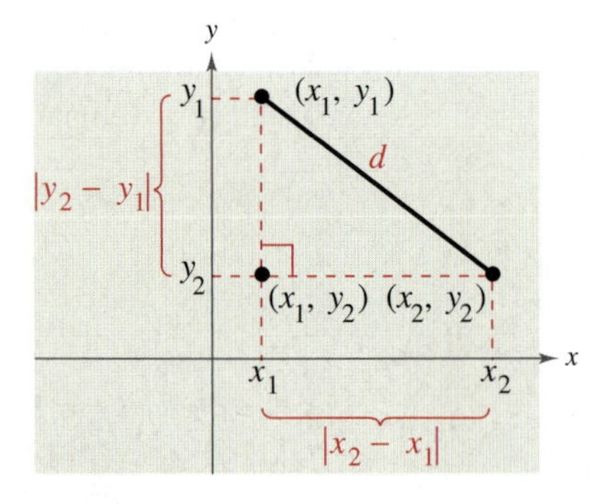

FIGURE 1.6

Suppose you want to determine the distance d between two points (x_1, y_1) and (x_2, y_2) in the plane. With these two points, a right triangle can be formed, as shown in Figure 1.6. The length of the vertical side of the triangle is $|y_2 - y_1|$, and the length of the horizontal side is $|x_2 - x_1|$. By the Pythagorean Theorem, you can write

$$d^2 = |x_2 - x_1|^2 + |y_2 - y_1|^2$$

$$d = \sqrt{|x_2 - x_1|^2 + |y_2 - y_1|^2} = \sqrt{(x_2 - x_1)^2 + (y_2 - y_1)^2}.$$

This result is the **Distance Formula.**

The Distance Formula

The distance d between the points (x_1, y_1) and (x_2, y_2) in the plane is

$$d = \sqrt{(x_2 - x_1)^2 + (y_2 - y_1)^2}.$$

Example 3 Finding a Distance

Find the distance between the points $(-2, 1)$ and $(3, 4)$.

Algebraic Solution

Let $(x_1, y_1) = (-2, 1)$ and $(x_2, y_2) = (3, 4)$. Then apply the Distance Formula.

$$d = \sqrt{(x_2 - x_1)^2 + (y_2 - y_1)^2}$$ Distance Formula

$$= \sqrt{[3 - (-2)]^2 + (4 - 1)^2}$$ Substitute for x_1, y_1, x_2, and y_2.

$$= \sqrt{(5)^2 + (3)^2}$$ Simplify.

$$= \sqrt{34}$$ Simplify.

$$\approx 5.83$$ Use a calculator.

So, the distance between the points is about 5.83 units. You can use the Pythagorean Theorem to check that the distance is correct.

$$d^2 \stackrel{?}{=} 3^2 + 5^2$$ Pythagorean Theorem

$$(\sqrt{34})^2 \stackrel{?}{=} 3^2 + 5^2$$ Substitute for d.

$$34 = 34$$ Distance checks. ✓

Graphical Solution

Use centimeter graph paper to plot the points $A(-2, 1)$ and $B(3, 4)$. Carefully sketch the line segment from A to B. Then use a centimeter ruler to measure the length of the segment.

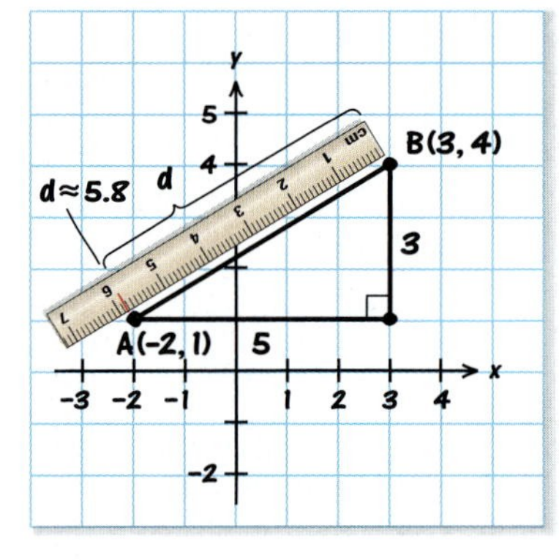

FIGURE 1.7

The line segment measures about 5.8 centimeters, as shown in Figure 1.7. So, the distance between the points is about 5.8 units.

CHECKPOINT Now try Exercises 31(a) and (b).

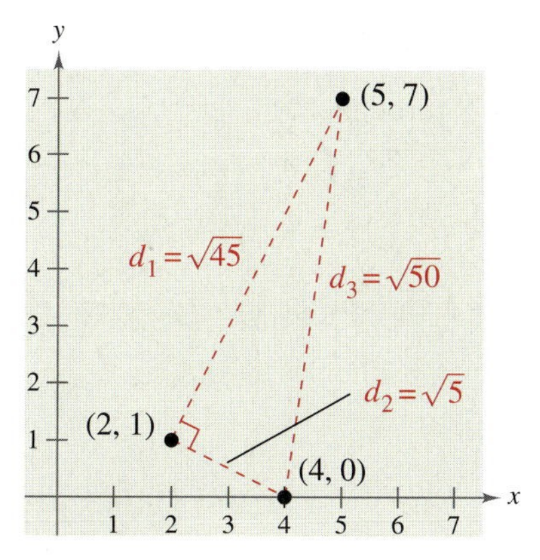

FIGURE 1.8

Example 4 Verifying a Right Triangle

Show that the points (2, 1), (4, 0), and (5, 7) are vertices of a right triangle.

Solution

The three points are plotted in Figure 1.8. Using the Distance Formula, you can find the lengths of the three sides as follows.

$$d_1 = \sqrt{(5-2)^2 + (7-1)^2} = \sqrt{9+36} = \sqrt{45}$$

$$d_2 = \sqrt{(4-2)^2 + (0-1)^2} = \sqrt{4+1} = \sqrt{5}$$

$$d_3 = \sqrt{(5-4)^2 + (7-0)^2} = \sqrt{1+49} = \sqrt{50}$$

Because

$$(d_1)^2 + (d_2)^2 = 45 + 5 = 50 = (d_3)^2$$

you can conclude by the Pythagorean Theorem that the triangle must be a right triangle.

CHECKPOINT Now try Exercise 41.

An overhead projector is useful for showing how to plot points and equations. Try projecting a grid onto the chalkboard and then plotting points on the chalkboard, or try using overhead markers and graph directly on the transparency. A viewscreen, a device used with an overhead projector to project a graphing calculator's screen image, is also useful.

The Midpoint Formula

To find the **midpoint** of the line segment that joins two points in a coordinate plane, you can simply find the average values of the respective coordinates of the two endpoints using the **Midpoint Formula.**

Exercises 43–46 on page 10 help develop a general understanding of the Midpoint Formula.

The Midpoint Formula

The midpoint of the line segment joining the points (x_1, y_1) and (x_2, y_2) is given by the Midpoint Formula

$$\text{Midpoint} = \left(\frac{x_1 + x_2}{2}, \frac{y_1 + y_2}{2}\right).$$

For a proof of the Midpoint Formula, see Proofs in Mathematics on page 124.

Example 5 Finding a Line Segment's Midpoint

Find the midpoint of the line segment joining the points $(-5, -3)$ and $(9, 3)$.

Solution

Let $(x_1, y_1) = (-5, -3)$ and $(x_2, y_2) = (9, 3)$.

$$\text{Midpoint} = \left(\frac{x_1 + x_2}{2}, \frac{y_1 + y_2}{2}\right) \quad \text{Midpoint Formula}$$

$$= \left(\frac{-5 + 9}{2}, \frac{-3 + 3}{2}\right) \quad \text{Substitute for } x_1, y_1, x_2, \text{ and } y_2.$$

$$= (2, 0) \quad \text{Simplify.}$$

The midpoint of the line segment is (2, 0), as shown in Figure 1.9.

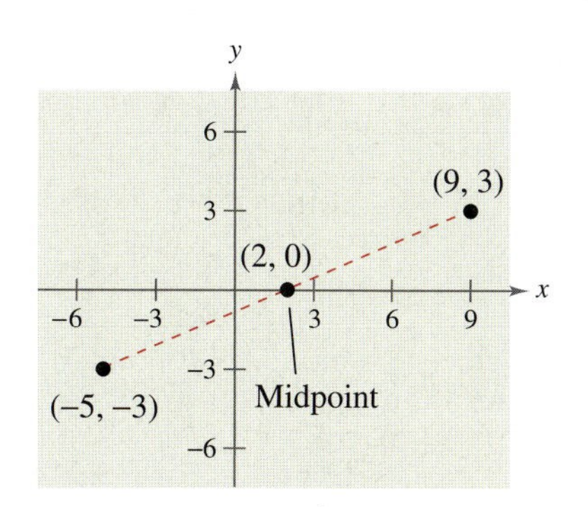

FIGURE 1.9

CHECKPOINT Now try Exercise 31(c).

Applications

Example 6 Finding the Length of a Pass 

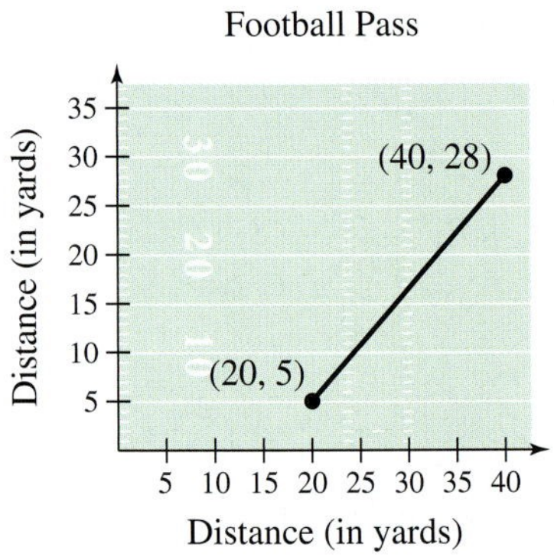

During the third quarter of the 2004 Sugar Bowl, the quarterback for Louisiana State University threw a pass from the 28-yard line, 40 yards from the sideline. The pass was caught by a wide receiver on the 5-yard line, 20 yards from the same sideline, as shown in Figure 1.10. How long was the pass?

FIGURE 1.10

Solution

You can find the length of the pass by finding the distance between the points (40, 28) and (20, 5).

$$\begin{aligned} d &= \sqrt{(x_2 - x_1)^2 + (y_2 - y_1)^2} && \text{Distance Formula} \\ &= \sqrt{(40 - 20)^2 + (28 - 5)^2} && \text{Substitute for } x_1, y_1, x_2, \text{ and } y_2. \\ &= \sqrt{400 + 529} && \text{Simplify.} \\ &= \sqrt{929} && \text{Simplify.} \\ &\approx 30 && \text{Use a calculator.} \end{aligned}$$

So, the pass was about 30 yards long.

CHECKPOINT Now try Exercise 47.

In Example 6, the scale along the goal line does not normally appear on a football field. However, when you use coordinate geometry to solve real-life problems, you are free to place the coordinate system in any way that is convenient for the solution of the problem.

Example 7 Estimating Annual Revenue 

FedEx Corporation had annual revenues of \$20.6 billion in 2002 and \$24.7 billion in 2004. Without knowing any additional information, what would you estimate the 2003 revenue to have been? (Source: FedEx Corp.)

Solution

One solution to the problem is to assume that revenue followed a linear pattern. With this assumption, you can estimate the 2003 revenue by finding the midpoint of the line segment connecting the points (2002, 20.6) and (2004, 24.7).

$$\begin{aligned} \text{Midpoint} &= \left(\frac{x_1 + x_2}{2}, \frac{y_1 + y_2}{2}\right) && \text{Midpoint Formula} \\ &= \left(\frac{2002 + 2004}{2}, \frac{20.6 + 24.7}{2}\right) && \text{Substitute for } x_1, y_1, x_2, \text{ and } y_2. \\ &= (2003, 22.65) && \text{Simplify.} \end{aligned}$$

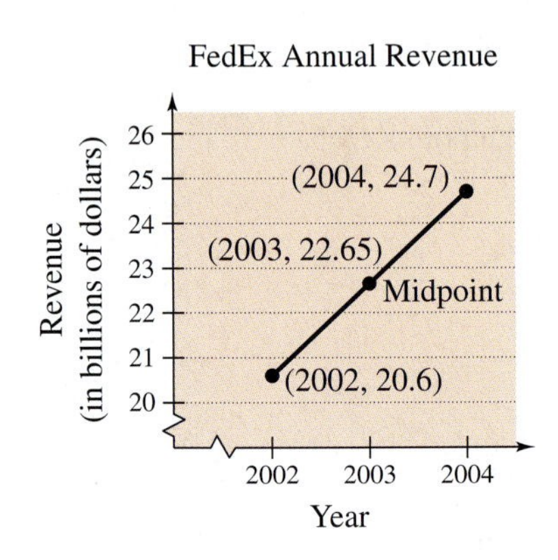

FIGURE 1.11

So, you would estimate the 2003 revenue to have been about \$22.65 billion, as shown in Figure 1.11. (The actual 2003 revenue was \$22.5 billion.)

CHECKPOINT Now try Exercise 49.

Paul Morrell

Much of computer graphics, including this computer-generated goldfish tessellation, consists of transformations of points in a coordinate plane. One type of transformation, a translation, is illustrated in Example 8. Other types include reflections, rotations, and stretches.

Activities

1. Set up a Cartesian plane and plot the points $(3, 0)$ and $(-4, 1)$.
2. Find the distance between $(3, 5)$ and $(-1, 2)$.

   Answer: 5
3. Find the midpoint of the line segment joining the points $(-1, -4)$ and $(3, -2)$.

   Answer: $(1, -3)$

Example 8 Translating Points in the Plane

The triangle in Figure 1.12 has vertices at the points $(-1, 2)$, $(1, -4)$, and $(2, 3)$. Shift the triangle three units to the right and two units upward and find the vertices of the shifted triangle, as shown in Figure 1.13.

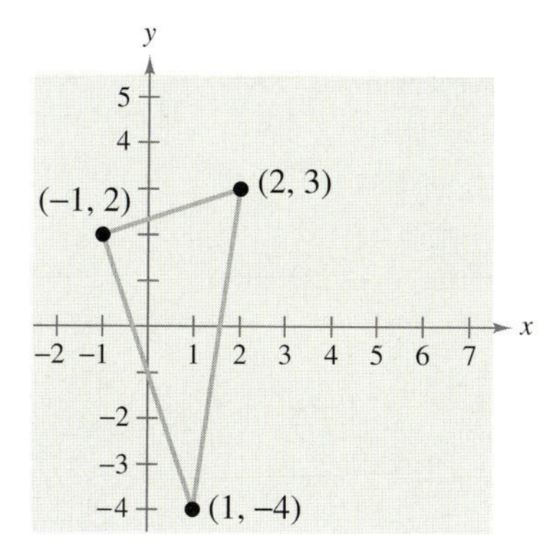

FIGURE 1.12

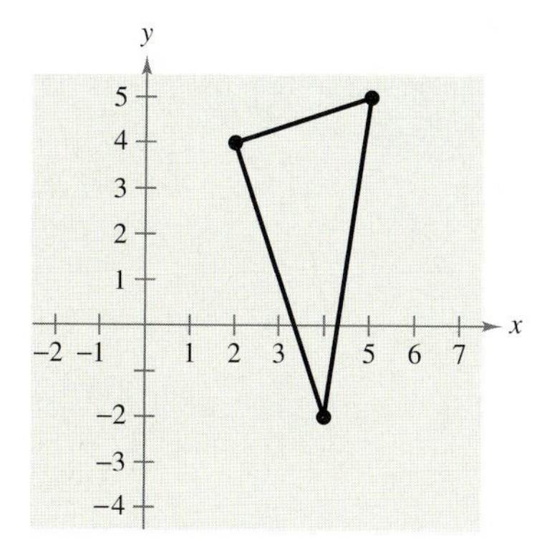

FIGURE 1.13

Solution

To shift the vertices three units to the right, add 3 to each of the x-coordinates. To shift the vertices two units upward, add 2 to each of the y-coordinates.

Original Point	*Translated Point*
$(-1, 2)$	$(-1 + 3, 2 + 2) = (2, 4)$
$(1, -4)$	$(1 + 3, -4 + 2) = (4, -2)$
$(2, 3)$	$(2 + 3, 3 + 2) = (5, 5)$

CHECKPOINT Now try Exercise 51.

The figures provided with Example 8 were not really essential to the solution. Nevertheless, it is strongly recommended that you develop the habit of including sketches with your solutions—even if they are not required.

The following geometric formulas are used at various times throughout this course. For your convenience, these formulas along with several others are also provided on the inside back cover of this text.

Common Formulas for Area A, Perimeter P, Circumference C, and Volume V

Rectangle	*Circle*	*Triangle*	*Rectangular Solid*	*Circular Cylinder*	*Sphere*
$A = lw$	$A = \pi r^2$	$A = \frac{1}{2}bh$	$V = lwh$	$V = \pi r^2 h$	$V = \frac{4}{3}\pi r^3$
$P = 2l + 2w$	$C = 2\pi r$	$P = a + b + c$			

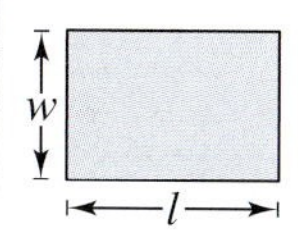

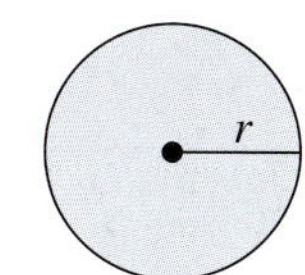

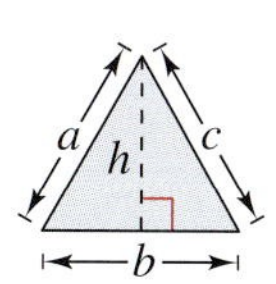

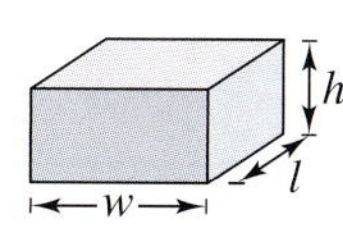

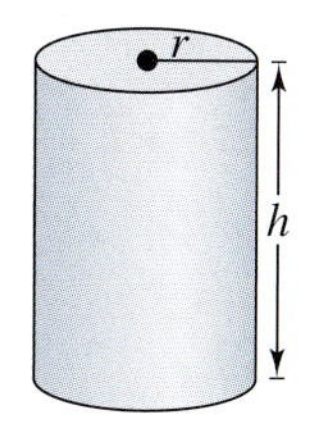

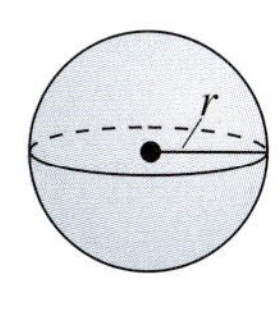

Example 9 Using a Geometric Formula

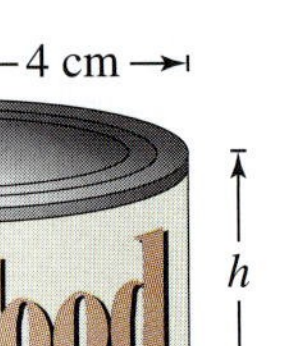

FIGURE 1.14

A cylindrical can has a volume of 200 cubic centimeters (cm^3) and a radius of 4 centimeters (cm), as shown in Figure 1.14. Find the height of the can.

Solution

The formula for the *volume of a cylinder* is $V = \pi r^2 h$. To find the height of the can, solve for h.

$$h = \frac{V}{\pi r^2}$$

Then, using $V = 200$ and $r = 4$, find the height.

$$h = \frac{200}{\pi(4)^2} \qquad \text{Substitute 200 for } V \text{ and 4 for } r.$$

$$= \frac{200}{16\pi} \qquad \text{Simplify denominator.}$$

$$\approx 3.98 \qquad \text{Use a calculator.}$$

Because the value of h was rounded in the solution, a check of the solution will not result in an equality. If the solution is valid, the expressions on each side of the equal sign will be approximately equal to each other.

$$V = \pi r^2 h \qquad \text{Write original equation.}$$

$$200 \stackrel{?}{\approx} \pi(4)^2(3.98) \qquad \text{Substitute 200 for } V\text{, 4 for } r\text{, and 3.98 for } h.$$

$$200 \approx 200.06 \qquad \text{Solution checks. ✓}$$

You can also use unit analysis to check that your answer is reasonable.

$$\frac{200 \text{ cm}^3}{16\pi \text{ cm}^2} \approx 3.98 \text{ cm}$$

✓CHECKPOINT Now try Exercise 63.

Alternative Writing About Mathematics

Use your school's library, the Internet, or some other reference source to locate a set of real data that can be considered as ordered pairs. Decide which variables should be the x- and y-variables, and explain why you made this decision. List each ordered pair. Sketch a scatter plot of the data, choosing appropriate axes and scale. Be sure to label each axis and give the units for each variable if necessary. Describe any apparent trends in the data. What can you learn from your graph?

WRITING ABOUT MATHEMATICS

Extending the Example Example 8 shows how to translate points in a coordinate plane. Write a short paragraph describing how each of the following transformed points is related to the original point.

Original Point	*Transformed Point*
(x, y)	$(-x, y)$
(x, y)	$(x, -y)$
(x, y)	$(-x, -y)$

1.1 Exercises

The *HM mathSpace®* CD-ROM and *Eduspace®* for this text contain step-by-step solutions to all odd-numbered exercises. They also provide Tutorial Exercises for additional help.

VOCABULARY CHECK

1. Match each term with its definition.

(a) x-axis	(i) point of intersection of vertical axis and horizontal axis
(b) y-axis	(ii) directed distance from the x-axis
(c) origin	(iii) directed distance from the y-axis
(d) quadrants	(iv) four regions of the coordinate plane
(e) x-coordinate	(v) horizontal real number line
(f) y-coordinate	(vi) vertical real number line

In Exercises 2–4, fill in the blanks.

2. An ordered pair of real numbers can be represented in a plane called the rectangular coordinate system or the ________ plane.

3. The ________ ________ is a result derived from the Pythagorean Theorem.

4. Finding the average values of the representative coordinates of the two endpoints of a line segment in a coordinate plane is also known as using the ________ ________.

PREREQUISITE SKILLS REVIEW: Practice and review algebra skills needed for this section at **www.Eduspace.com.**

In Exercises 1 and 2, approximate the coordinates of the points.

1.

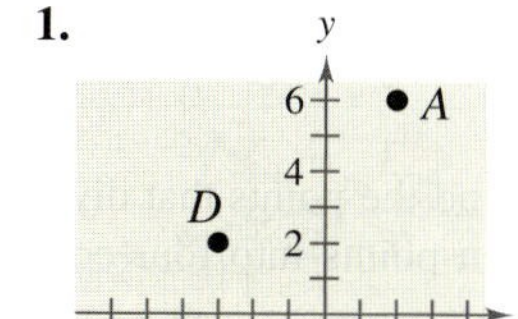

2.

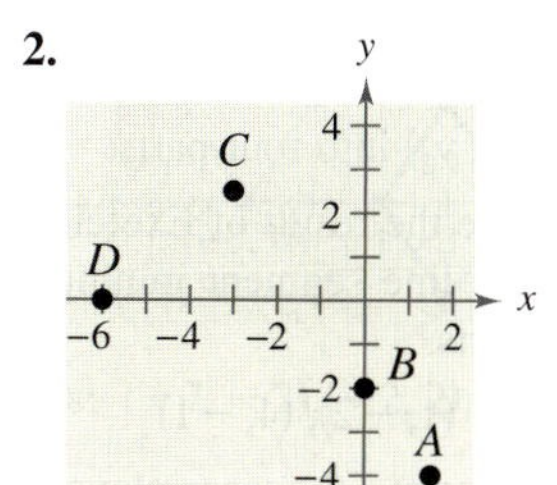

In Exercises 3–6, plot the points in the Cartesian plane.

3. $(-4, 2), (-3, -6), (0, 5), (1, -4)$

4. $(0, 0), (3, 1), (-2, 4), (1, -1)$

5. $(3, 8), (0.5, -1), (5, -6), (-2, 2.5)$

6. $\left(1, -\frac{1}{3}\right), \left(\frac{3}{4}, 3\right), (-3, 4), \left(-\frac{4}{3}, -\frac{3}{2}\right)$

In Exercises 7–10, find the coordinates of the point.

7. The point is located three units to the left of the y-axis and four units above the x-axis.

8. The point is located eight units below the x-axis and four units to the right of the y-axis.

9. The point is located five units below the x-axis and the coordinates of the point are equal.

10. The point is on the x-axis and 12 units to the left of the y-axis.

In Exercises 11–20, determine the quadrant(s) in which (x, y) is located so that the condition(s) is (are) satisfied.

11. $x > 0$ and $y < 0$

12. $x < 0$ and $y < 0$

13. $x = -4$ and $y > 0$

14. $x > 2$ and $y = 3$

15. $y < -5$

16. $x > 4$

17. $x < 0$ and $-y > 0$

18. $-x > 0$ and $y < 0$

19. $xy > 0$

20. $xy < 0$

In Exercises 21 and 22, sketch a scatter plot of the data shown in the table.

21. ***Number of Stores*** The table shows the number y of Wal-Mart stores for each year x from 1996 through 2003. (Source: Wal-Mart Stores, Inc.)

Year, x	Number of stores, y
1996	3054
1997	3406
1998	3599
1999	3985
2000	4189
2001	4414
2002	4688
2003	4906

1.2 Graphs of Equations

What you should learn

- Sketch graphs of equations.
- Find x- and y-intercepts of graphs of equations.
- Use symmetry to sketch graphs of equations.
- Find equations of and sketch graphs of circles.
- Use graphs of equations in solving real-life problems.

Why you should learn it

The graph of an equation can help you see relationships between real-life quantities. For example, in Exercise 75 on page 24, a graph can be used to estimate the life expectancies of children who are born in the years 2005 and 2010.

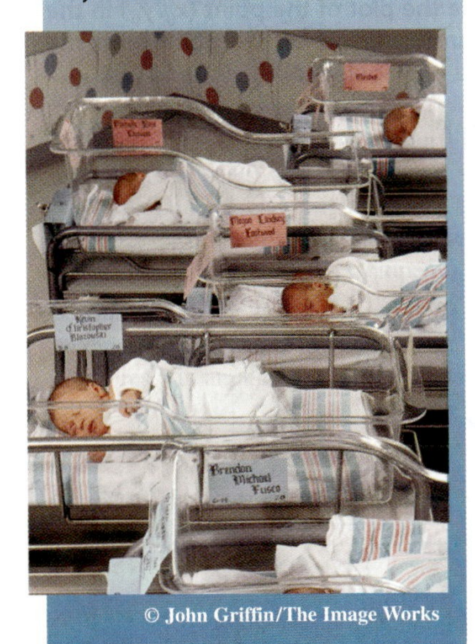

© John Griffin/The Image Works

The Graph of an Equation

In Section 1.1, you used a coordinate system to represent graphically the relationship between two quantities. There, the graphical picture consisted of a collection of points in a coordinate plane.

Frequently, a relationship between two quantities is expressed as an **equation in two variables.** For instance, $y = 7 - 3x$ is an equation in x and y. An ordered pair (a, b) is a **solution** or **solution point** of an equation in x and y if the equation is true when a is substituted for x and b is substituted for y. For instance, $(1, 4)$ is a solution of $y = 7 - 3x$ because $4 = 7 - 3(1)$ is a true statement.

In this section you will review some basic procedures for sketching the graph of an equation in two variables. The **graph of an equation** is the set of all points that are solutions of the equation.

Example 1 **Determining Solutions**

Determine whether (a) $(2, 13)$ and (b) $(-1, -3)$ are solutions of the equation $y = 10x - 7$.

Solution

a. $y = 10x - 7$ — Write original equation.

$13 \stackrel{?}{=} 10(2) - 7$ — Substitute 2 for x and 13 for y.

$13 = 13$ — $(2, 13)$ is a solution. ✓

Because the substitution does satisfy the original equation, you can conclude that the ordered pair $(2, 13)$ *is* a solution of the original equation.

b. $y = 10x - 7$ — Write original equation.

$-3 \stackrel{?}{=} 10(-1) - 7$ — Substitute -1 for x and -3 for y.

$-3 \neq -17$ — $(-1, -3)$ is not a solution.

Because the substitution does not satisfy the original equation, you can conclude that the ordered pair $(-1, -3)$ *is not* a solution of the original equation.

✓CHECKPOINT Now try Exercise 1.

The basic technique used for sketching the graph of an equation is the **point-plotting method.**

Sketching the Graph of an Equation by Point Plotting

1. If possible, rewrite the equation so that one of the variables is isolated on one side of the equation.
2. Make a table of values showing several solution points.
3. Plot these points on a rectangular coordinate system.
4. Connect the points with a smooth curve or line.

Example 2 Sketching the Graph of an Equation

Sketch the graph of

$$y = 7 - 3x.$$

Solution

Because the equation is already solved for y, construct a table of values that consists of several solution points of the equation. For instance, when $x = -1$,

$$y = 7 - 3(-1)$$
$$= 10$$

which implies that $(-1, 10)$ is a solution point of the graph.

x	$y = 7 - 3x$	(x, y)
-1	10	$(-1, 10)$
0	7	$(0, 7)$
1	4	$(1, 4)$
2	1	$(2, 1)$
3	-2	$(3, -2)$
4	-5	$(4, -5)$

From the table, it follows that

$$(-1, 10), (0, 7), (1, 4), (2, 1), (3, -2), \text{ and } (4, -5)$$

are solution points of the equation. After plotting these points, you can see that they appear to lie on a line, as shown in Figure 1.15. The graph of the equation is the line that passes through the six plotted points.

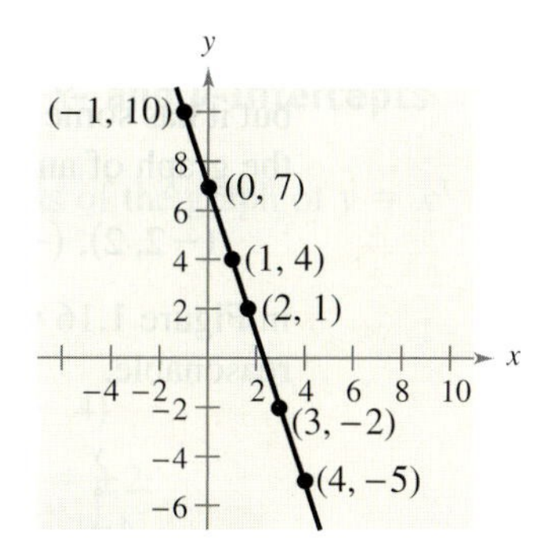

FIGURE 1.15

CHECKPOINT Now try Exercise 5.

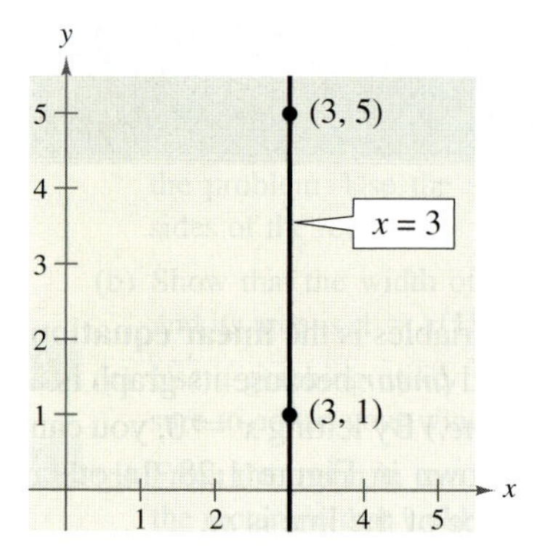

FIGURE 1.30 *Slope is undefined.*

Once you have determined the slope and the y-intercept of a line, it is a relatively simple matter to sketch its graph. In the next example, note that none of the lines is vertical. A vertical line has an equation of the form

$x = a$. Vertical line

The equation of a vertical line cannot be written in the form $y = mx + b$ because the slope of a vertical line is undefined, as indicated in Figure 1.30.

Common Error

Many students confuse the line $x = a$ with the point $x = a$ on the real number line, or the line $y = b$ with the point $y = b$. Point out to students that they need to be aware of the context in which $x = a$ or $y = b$ is presented to know whether it refers to the line in the plane or the point on the real number line.

Example 1 Graphing a Linear Equation

Sketch the graph of each linear equation.

a. $y = 2x + 1$

b. $y = 2$

c. $x + y = 2$

Solution

a. Because $b = 1$, the y-intercept is $(0, 1)$. Moreover, because the slope is $m = 2$, the line *rises* two units for each unit the line moves to the right, as shown in Figure 1.31.

b. By writing this equation in the form $y = (0)x + 2$, you can see that the y-intercept is $(0, 2)$ and the slope is zero. A zero slope implies that the line is horizontal—that is, it doesn't rise *or* fall, as shown in Figure 1.32.

c. By writing this equation in slope-intercept form

$x + y = 2$ Write original equation.

$y = -x + 2$ Subtract x from each side.

$y = (-1)x + 2$ Write in slope-intercept form.

you can see that the y-intercept is $(0, 2)$. Moreover, because the slope is $m = -1$, the line *falls* one unit for each unit the line moves to the right, as shown in Figure 1.33.

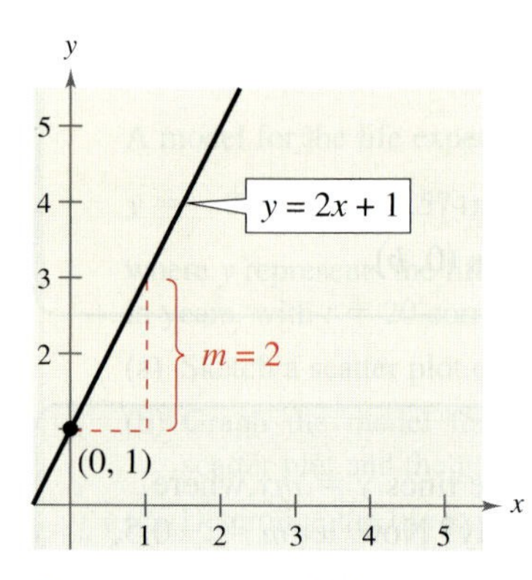

When m is positive, the line rises.
FIGURE 1.31

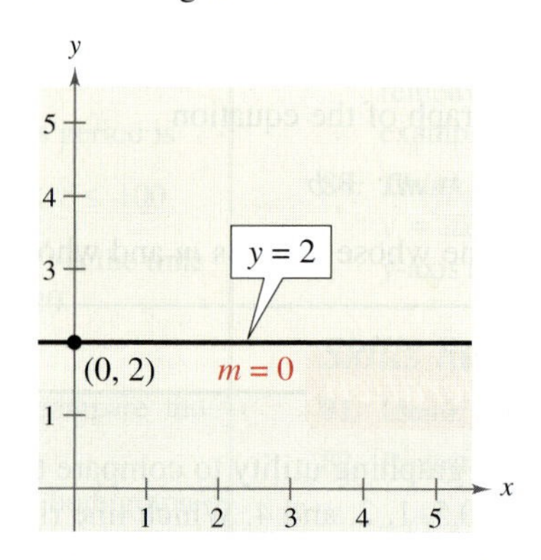

When m is 0, the line is horizontal.
FIGURE 1.32

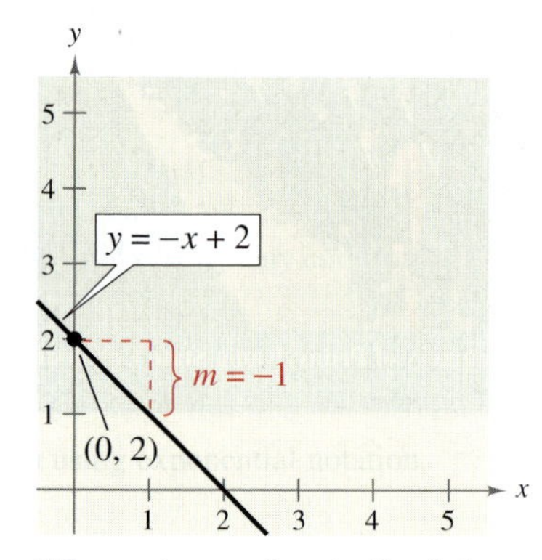

When m is negative, the line falls.
FIGURE 1.33

CHECKPOINT Now try Exercise 9.

Finding the Slope of a Line

Given an equation of a line, you can find its slope by writing the equation in slope-intercept form. If you are not given an equation, you can still find the slope of a line. For instance, suppose you want to find the slope of the line passing through the points (x_1, y_1) and (x_2, y_2), as shown in Figure 1.34. As you move from left to right along this line, a change of $(y_2 - y_1)$ units in the vertical direction corresponds to a change of $(x_2 - x_1)$ units in the horizontal direction.

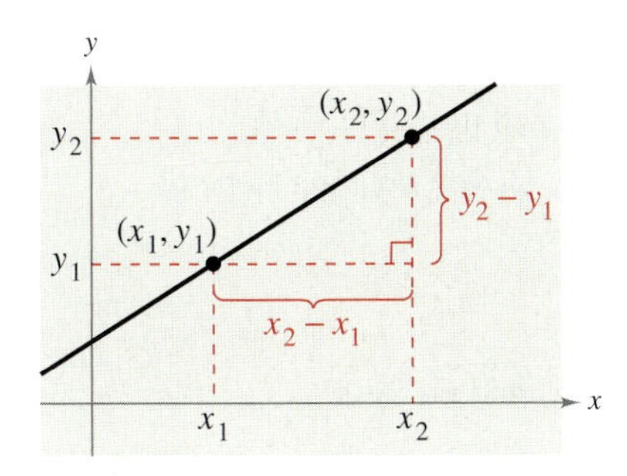

FIGURE 1.34

$$y_2 - y_1 = \text{the change in } y = \text{rise}$$

and

$$x_2 - x_1 = \text{the change in } x = \text{run}$$

The ratio of $(y_2 - y_1)$ to $(x_2 - x_1)$ represents the slope of the line that passes through the points (x_1, y_1) and (x_2, y_2).

$$\text{Slope} = \frac{\text{change in } y}{\text{change in } x}$$

$$= \frac{\text{rise}}{\text{run}}$$

$$= \frac{y_2 - y_1}{x_2 - x_1}$$

The Slope of a Line Passing Through Two Points

The **slope** m of the nonvertical line through (x_1, y_1) and (x_2, y_2) is

$$m = \frac{y_2 - y_1}{x_2 - x_1}$$

where $x_1 \neq x_2$.

When this formula is used for slope, the *order of subtraction* is important. Given two points on a line, you are free to label either one of them as (x_1, y_1) and the other as (x_2, y_2). However, once you have done this, you must form the numerator and denominator using the same order of subtraction.

$$\underbrace{m = \frac{y_2 - y_1}{x_2 - x_1}}_{\text{Correct}} \qquad \underbrace{m = \frac{y_1 - y_2}{x_1 - x_2}}_{\text{Correct}} \qquad \underbrace{\cancel{m = \frac{y_2 - y_1}{x_1 - x_2}}}_{\text{Incorrect}}$$

For instance, the slope of the line passing through the points (3, 4) and (5, 7) can be calculated as

$$m = \frac{7 - 4}{5 - 3} = \frac{3}{2}$$

or, reversing the subtraction order in both the numerator and denominator, as

$$m = \frac{4 - 7}{3 - 5} = \frac{-3}{-2} = \frac{3}{2}.$$

Example 2 Finding the Slope of a Line Through Two Points

Find the slope of the line passing through each pair of points.

a. $(-2, 0)$ and $(3, 1)$ **b.** $(-1, 2)$ and $(2, 2)$

c. $(0, 4)$ and $(1, -1)$ **d.** $(3, 4)$ and $(3, 1)$

Solution

a. Letting $(x_1, y_1) = (-2, 0)$ and $(x_2, y_2) = (3, 1)$, you obtain a slope of

$$m = \frac{y_2 - y_1}{x_2 - x_1} = \frac{1 - 0}{3 - (-2)} = \frac{1}{5}.$$

See Figure 1.35.

b. The slope of the line passing through $(-1, 2)$ and $(2, 2)$ is

$$m = \frac{2 - 2}{2 - (-1)} = \frac{0}{3} = 0.$$

See Figure 1.36.

c. The slope of the line passing through $(0, 4)$ and $(1, -1)$ is

$$m = \frac{-1 - 4}{1 - 0} = \frac{-5}{1} = -5.$$

See Figure 1.37.

d. The slope of the line passing through $(3, 4)$ and $(3, 1)$ is

$$m = \frac{1 - 4}{3 - 3} = \frac{-3}{0}.$$

See Figure 1.38.

Because division by 0 is undefined, the slope is undefined and the line is vertical.

Common Error

A common error when finding the slope of a line is combining x- and y-coordinates in either the numerator or denominator, or both, as in

$$m = \frac{y_2 - x_1}{x_2 - y_1}.$$

STUDY TIP

In Figures 1.35 to 1.38, note the relationships between slope and the orientation of the line.

a. Positive slope: line rises from left to right

b. Zero slope: line is horizontal

c. Negative slope: line falls from left to right

d. Undefined slope: line is vertical

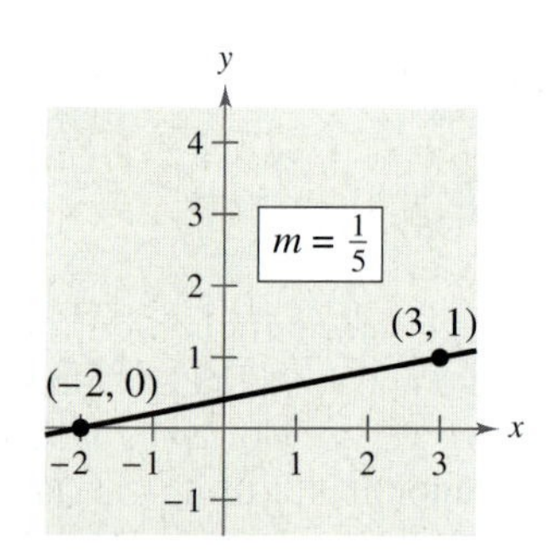

FIGURE 1.35

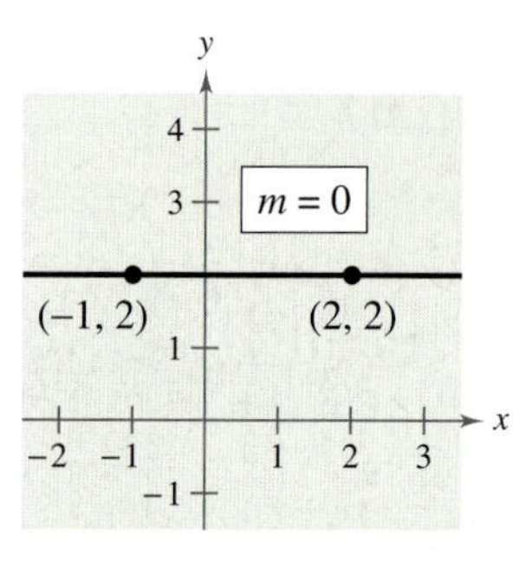

FIGURE 1.36

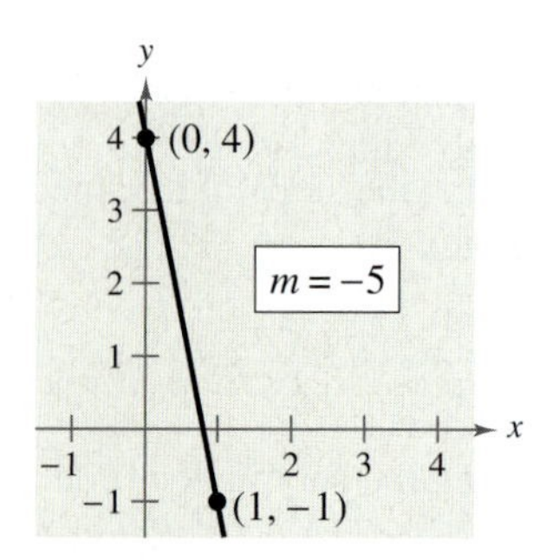

FIGURE 1.37

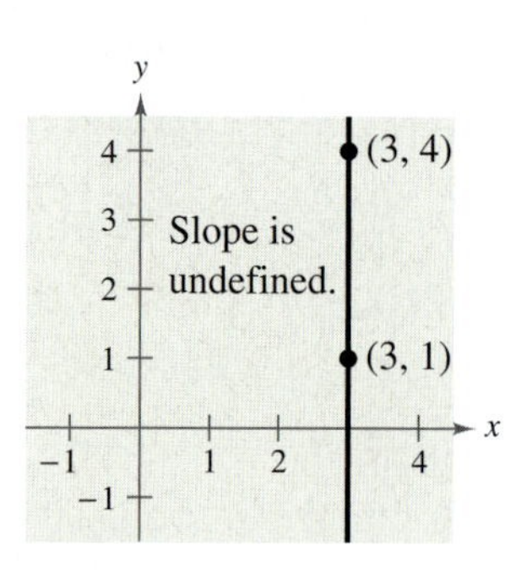

FIGURE 1.38

CHECKPOINT Now try Exercise 21.

Writing Linear Equations in Two Variables

If (x_1, y_1) is a point on a line of slope m and (x, y) is *any other* point on the line, then

$$\frac{y - y_1}{x - x_1} = m.$$

This equation, involving the variables x and y, can be rewritten in the form

$$y - y_1 = m(x - x_1)$$

which is the **point-slope form** of the equation of a line.

> **Point-Slope Form of the Equation of a Line**
>
> The equation of the line with slope m passing through the point (x_1, y_1) is
>
> $$y - y_1 = m(x - x_1).$$

The point-slope form is most useful for *finding* the equation of a line. You should remember this form.

Example 3 Using the Point-Slope Form

Find the slope-intercept form of the equation of the line that has a slope of 3 and passes through the point $(1, -2)$.

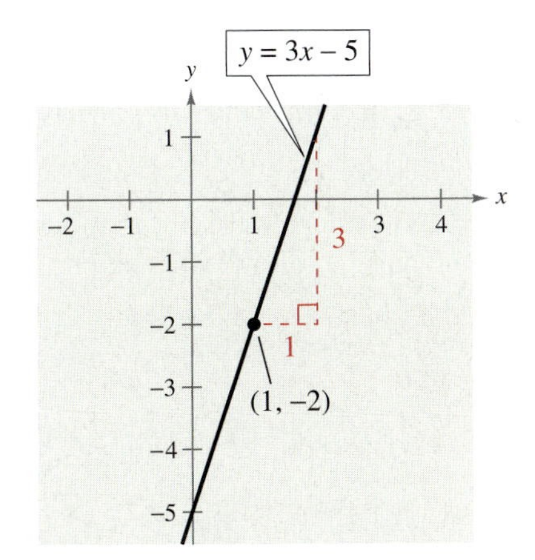

FIGURE 1.39

Solution

Use the point-slope form with $m = 3$ and $(x_1, y_1) = (1, -2)$.

$y - y_1 = m(x - x_1)$	Point-slope form
$y - (-2) = 3(x - 1)$	Substitute for m, x_1, and y_1.
$y + 2 = 3x - 3$	Simplify.
$y = 3x - 5$	Write in slope-intercept form.

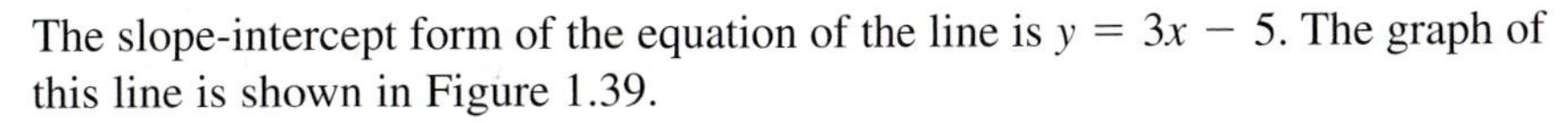

The slope-intercept form of the equation of the line is $y = 3x - 5$. The graph of this line is shown in Figure 1.39.

✓CHECKPOINT Now try Exercise 39.

> **STUDY TIP**
>
> When you find an equation of the line that passes through two given points, you only need to substitute the coordinates of one of the points into the point-slope form. It does not matter which point you choose because both points will yield the same result.

The point-slope form can be used to find an equation of the line passing through two points (x_1, y_1) and (x_2, y_2). To do this, first find the slope of the line

$$m = \frac{y_2 - y_1}{x_2 - x_1}, \qquad x_1 \neq x_2$$

and then use the point-slope form to obtain the equation

$$y - y_1 = \frac{y_2 - y_1}{x_2 - x_1}(x - x_1). \qquad \text{Two-point form}$$

This is sometimes called the **two-point form** of the equation of a line.

Exploration

Find d_1 and d_2 in terms of m_1 and m_2, respectively (see figure). Then use the Pythagorean Theorem to find a relationship between m_1 and m_2.

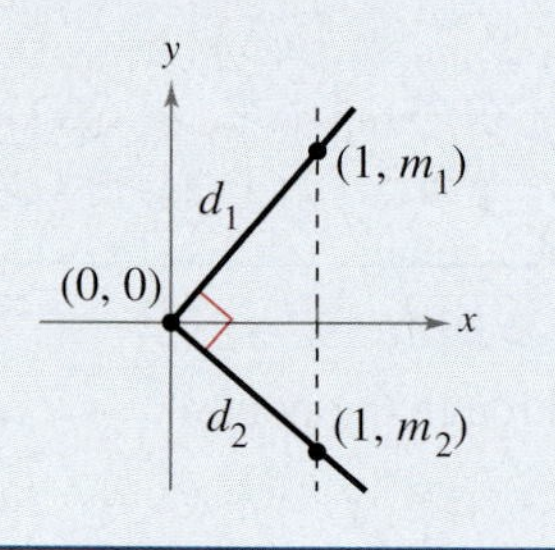

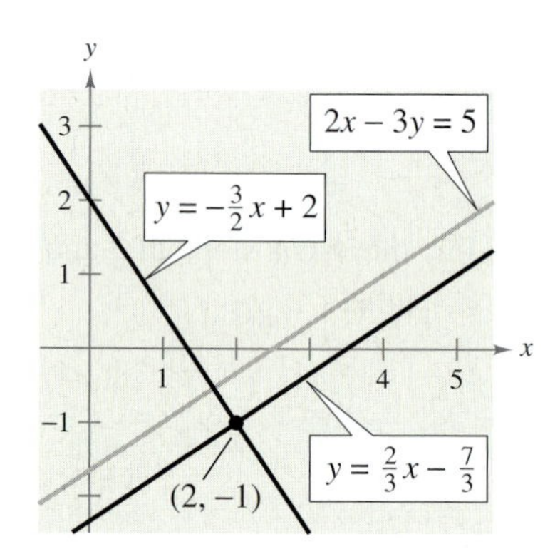

FIGURE 1.40

Technology

On a graphing utility, lines will not appear to have the correct slope unless you use a viewing window that has a square setting. For instance, try graphing the lines in Example 4 using the standard setting $-10 \le x \le 10$ and $-10 \le y \le 10$. Then reset the viewing window with the square setting $-9 \le x \le 9$ and $-6 \le y \le 6$. On which setting do the lines $y = \frac{2}{3}x - \frac{5}{3}$ and $y = -\frac{3}{2}x + 2$ appear to be perpendicular?

Parallel and Perpendicular Lines

Slope can be used to decide whether two nonvertical lines in a plane are parallel, perpendicular, or neither.

Parallel and Perpendicular Lines

1. Two distinct nonvertical lines are **parallel** if and only if their slopes are equal. That is, $m_1 = m_2$.
2. Two nonvertical lines are **perpendicular** if and only if their slopes are negative reciprocals of each other. That is, $m_1 = -1/m_2$.

Example 4 Finding Parallel and Perpendicular Lines

Find the slope-intercept forms of the equations of the lines that pass through the point $(2, -1)$ and are (a) parallel to and (b) perpendicular to the line $2x - 3y = 5$.

Solution

By writing the equation of the given line in slope-intercept form

$$2x - 3y = 5 \qquad \text{Write original equation.}$$

$$-3y = -2x + 5 \qquad \text{Subtract } 2x \text{ from each side.}$$

$$y = \frac{2}{3}x - \frac{5}{3} \qquad \text{Write in slope-intercept form.}$$

you can see that it has a slope of $m = \frac{2}{3}$, as shown in Figure 1.40.

a. Any line parallel to the given line must also have a slope of $\frac{2}{3}$. So, the line through $(2, -1)$ that is parallel to the given line has the following equation.

$$y - (-1) = \frac{2}{3}(x - 2) \qquad \text{Write in point-slope form.}$$

$$3(y + 1) = 2(x - 2) \qquad \text{Multiply each side by 3.}$$

$$3y + 3 = 2x - 4 \qquad \text{Distributive Property}$$

$$y = \frac{2}{3}x - \frac{7}{3} \qquad \text{Write in slope-intercept form.}$$

b. Any line perpendicular to the given line must have a slope of $-\frac{3}{2}$ (because $-\frac{3}{2}$ is the negative reciprocal of $\frac{2}{3}$). So, the line through $(2, -1)$ that is perpendicular to the given line has the following equation.

$$y - (-1) = -\frac{3}{2}(x - 2) \qquad \text{Write in point-slope form.}$$

$$2(y + 1) = -3(x - 2) \qquad \text{Multiply each side by 2.}$$

$$2y + 2 = -3x + 6 \qquad \text{Distributive Property}$$

$$y = -\frac{3}{2}x + 2 \qquad \text{Write in slope-intercept form.}$$

CHECKPOINT Now try Exercise 69.

Notice in Example 4 how the slope-intercept form is used to obtain information about the graph of a line, whereas the point-slope form is used to write the equation of a line.

Applications

In real-life problems, the slope of a line can be interpreted as either a *ratio* or a *rate*. If the x-axis and y-axis have the same unit of measure, then the slope has no units and is a **ratio.** If the x-axis and y-axis have different units of measure, then the slope is a **rate** or **rate of change.**

Example 5 Using Slope as a Ratio

The maximum recommended slope of a wheelchair ramp is $\frac{1}{12}$. A business is installing a wheelchair ramp that rises 22 inches over a horizontal length of 24 feet. Is the ramp steeper than recommended? (Source: *Americans with Disabilities Act Handbook*)

Solution

The horizontal length of the ramp is 24 feet or $12(24) = 288$ inches, as shown in Figure 1.41. So, the slope of the ramp is

$$\text{Slope} = \frac{\text{vertical change}}{\text{horizontal change}} = \frac{22 \text{ in.}}{288 \text{ in.}} \approx 0.076.$$

Because $\frac{1}{12} \approx 0.083$, the slope of the ramp is not steeper than recommended.

FIGURE 1.41

CHECKPOINT Now try Exercise 97.

Example 6 Using Slope as a Rate of Change

A kitchen appliance manufacturing company determines that the total cost in dollars of producing x units of a blender is

$$C = 25x + 3500. \qquad \text{Cost equation}$$

Describe the practical significance of the y-intercept and slope of this line.

Solution

The y-intercept $(0, 3500)$ tells you that the cost of producing zero units is \$3500. This is the *fixed cost* of production—it includes costs that must be paid regardless of the number of units produced. The slope of $m = 25$ tells you that the cost of producing each unit is \$25, as shown in Figure 1.42. Economists call the cost per unit the *marginal cost*. If the production increases by one unit, then the "margin," or extra amount of cost, is \$25. So, the cost increases at a rate of \$25 per unit.

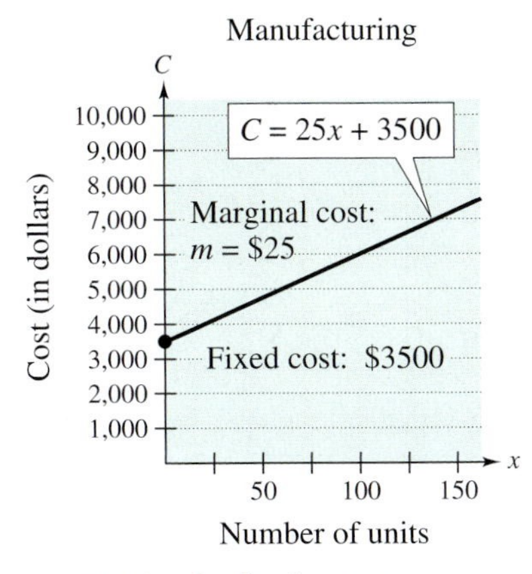

FIGURE 1.42 *Production cost*

CHECKPOINT Now try Exercise 101.

Activities

1. Write an equation of the line that passes through the points $(-2, 1)$ and $(3, 2)$.
   Answer: $x - 5y + 7 = 0$
2. Find the slope of the line that is perpendicular to the line $4x - 7y = 12$.
   Answer: $m = -\frac{7}{4}$
3. Write the equation of the vertical line that passes through the point $(3, 2)$.
   Answer: $x = 3$

Most business expenses can be deducted in the same year they occur. One exception is the cost of property that has a useful life of more than 1 year. Such costs must be *depreciated* (decreased in value) over the useful life of the property. If the *same amount* is depreciated each year, the procedure is called *linear* or *straight-line depreciation.* The *book value* is the difference between the original value and the total amount of depreciation accumulated to date.

Example 7 Straight-Line Depreciation

A college purchased exercise equipment worth \$12,000 for the new campus fitness center. The equipment has a useful life of 8 years. The salvage value at the end of 8 years is \$2000. Write a linear equation that describes the book value of the equipment each year.

Solution

Let V represent the value of the equipment at the end of year t. You can represent the initial value of the equipment by the data point $(0, 12{,}000)$ and the salvage value of the equipment by the data point $(8, 2000)$. The slope of the line is

$$m = \frac{2000 - 12{,}000}{8 - 0} = -\$1250$$

which represents the annual depreciation in *dollars per year.* Using the point-slope form, you can write the equation of the line as follows.

$$V - 12{,}000 = -1250(t - 0) \qquad \text{Write in point-slope form.}$$

$$V = -1250t + 12{,}000 \qquad \text{Write in slope-intercept form.}$$

The table shows the book value at the end of each year, and the graph of the equation is shown in Figure 1.43.

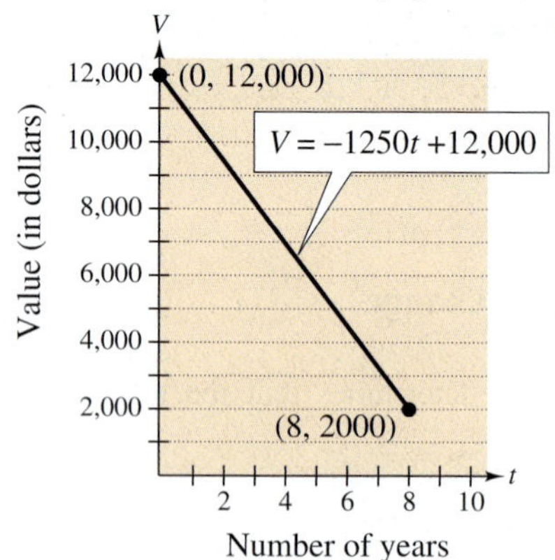

FIGURE 1.43 *Straight-line depreciation*

Year, t	Value, V
0	12,000
1	10,750
2	9,500
3	8,250
4	7,000
5	5,750
6	4,500
7	3,250
8	2,000

✓CHECKPOINT Now try Exercise 107.

In many real-life applications, the two data points that determine the line are often given in a disguised form. Note how the data points are described in Example 7.

Example 8 Predicting Sales per Share 

The sales per share for Starbucks Corporation were \$6.97 in 2001 and \$8.47 in 2002. Using only this information, write a linear equation that gives the sales per share in terms of the year. Then predict the sales per share for 2003. (Source: Starbucks Corporation)

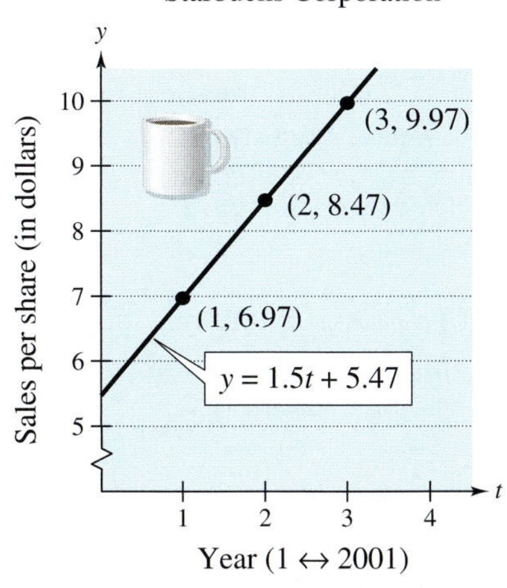

FIGURE 1.44

Solution

Let $t = 1$ represent 2001. Then the two given values are represented by the data points $(1, 6.97)$ and $(2, 8.47)$. The slope of the line through these points is

$$m = \frac{8.47 - 6.97}{2 - 1}$$

$$= 1.5.$$

Using the point-slope form, you can find the equation that relates the sales per share y and the year t to be

$$y - 6.97 = 1.5(t - 1) \qquad \text{Write in point-slope form.}$$

$$y = 1.5t + 5.47. \qquad \text{Write in slope-intercept form.}$$

According to this equation, the sales per share in 2003 was $y = 1.5(3) + 5.47 = \$9.97$, as shown in Figure 1.44. (In this case, the prediction is quite good—the actual sales per share in 2003 was \$10.35.)

Now try Exercise 109.

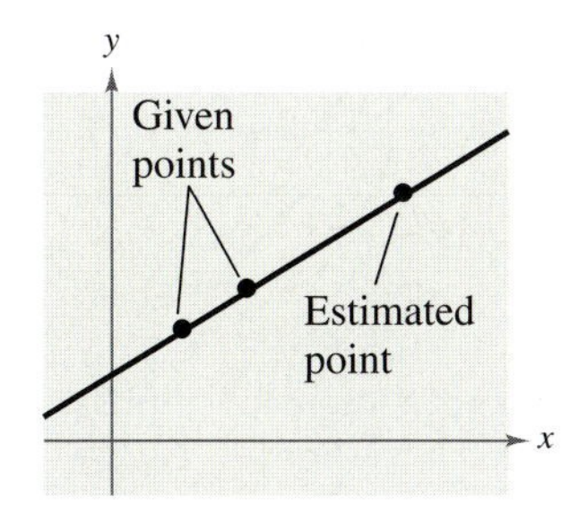

Linear extrapolation
FIGURE 1.45

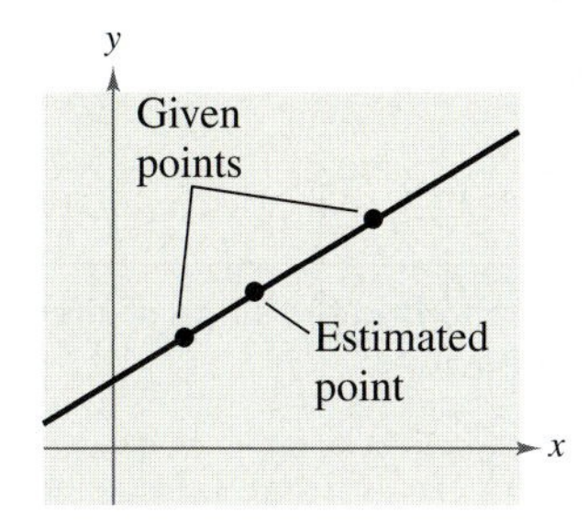

Linear interpolation
FIGURE 1.46

The prediction method illustrated in Example 8 is called **linear extrapolation.** Note in Figure 1.45 that an extrapolated point does not lie between the given points. When the estimated point lies between two given points, as shown in Figure 1.46, the procedure is called **linear interpolation.**

Because the slope of a vertical line is not defined, its equation cannot be written in slope-intercept form. However, every line has an equation that can be written in the **general form**

$$Ax + By + C = 0 \qquad \text{General form}$$

where A and B are not both zero. For instance, the vertical line given by $x = a$ can be represented by the general form $x - a = 0$.

Summary of Equations of Lines

1. General form: $Ax + By + C = 0$
2. Vertical line: $x = a$
3. Horizontal line: $y = b$
4. Slope-intercept form: $y = mx + b$
5. Point-slope form: $y - y_1 = m(x - x_1)$
6. Two-point form: $y - y_1 = \dfrac{y_2 - y_1}{x_2 - x_1}(x - x_1)$

1.3 Exercises

VOCABULARY CHECK:

In Exercises 1–6, fill in the blanks.

1. The simplest mathematical model for relating two variables is the ________ equation in two variables $y = mx + b$.
2. For a line, the ratio of the change in y to the change in x is called the ________ of the line.
3. Two lines are ________ if and only if their slopes are equal.
4. Two lines are ________ if and only if their slopes are negative reciprocals of each other.
5. When the x-axis and y-axis have different units of measure, the slope can be interpreted as a ________.
6. The prediction method ________ ________ is the method used to estimate a point on a line that does not lie between the given points.
7. Match each equation of a line with its form.

(a) $Ax + By + C = 0$	(i) Vertical line
(b) $x = a$	(ii) Slope-intercept form
(c) $y = b$	(iii) General form
(d) $y = mx + b$	(iv) Point-slope form
(e) $y - y_1 = m(x - x_1)$	(v) Horizontal line

PREREQUISITE SKILLS REVIEW: Practice and review algebra skills needed for this section at **www.Eduspace.com.**

In Exercises 1 and 2, identify the line that has each slope.

1. (a) $m = \frac{2}{3}$
   (b) m is undefined.
   (c) $m = -2$

2. (a) $m = 0$
   (b) $m = -\frac{3}{4}$
   (c) $m = 1$

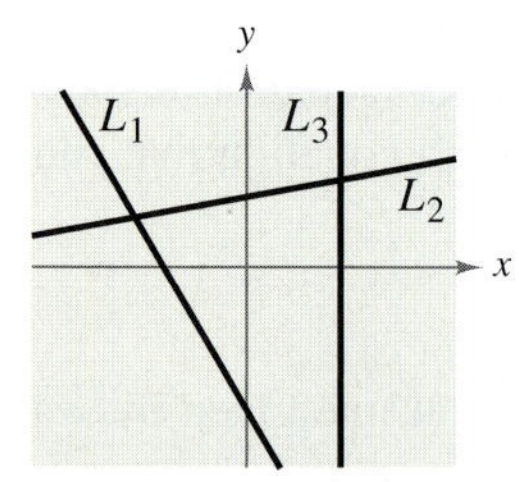

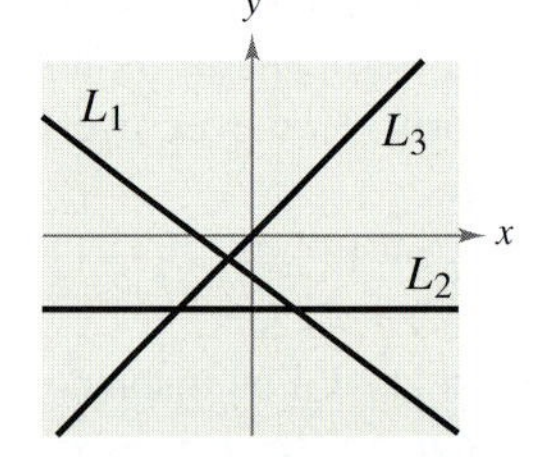

In Exercises 3 and 4, sketch the lines through the point with the indicated slopes on the same set of coordinate axes.

	Point	*Slopes*
3.	(2, 3)	(a) 0 (b) 1 (c) 2 (d) −3
4.	(−4, 1)	(a) 3 (b) −3 (c) $\frac{1}{2}$ (d) Undefined

In Exercises 5–8, estimate the slope of the line.

5.

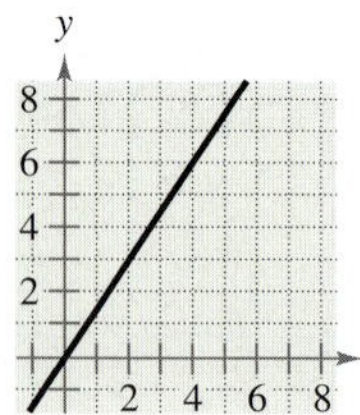

6.

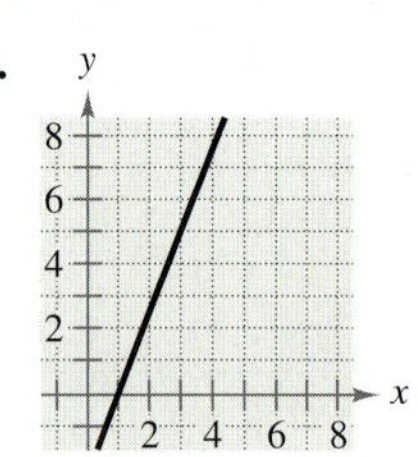

7.

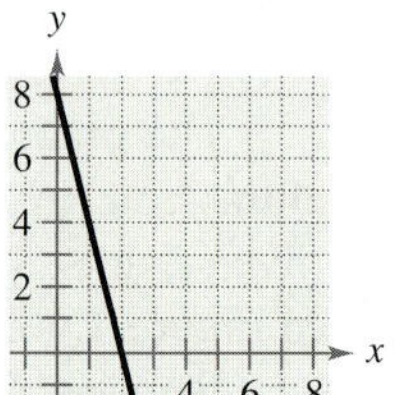

8.

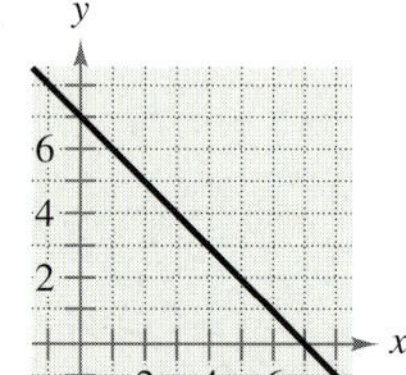

In Exercises 9–20, find the slope and y-intercept (if possible) of the equation of the line. Sketch the line.

9. $y = 5x + 3$
10. $y = x - 10$
11. $y = -\frac{1}{2}x + 4$
12. $y = -\frac{3}{2}x + 6$
13. $5x - 2 = 0$
14. $3y + 5 = 0$
15. $7x + 6y = 30$
16. $2x + 3y = 9$
17. $y - 3 = 0$
18. $y + 4 = 0$
19. $x + 5 = 0$
20. $x - 2 = 0$

In Exercises 21–28, plot the points and find the slope of the line passing through the pair of points.

21. $(-3, -2), (1, 6)$
22. $(2, 4), (4, -4)$
23. $(-6, -1), (-6, 4)$
24. $(0, -10), (-4, 0)$
25. $\left(\frac{11}{2}, -\frac{4}{3}\right), \left(-\frac{3}{2}, -\frac{1}{3}\right)$
26. $\left(\frac{7}{8}, \frac{3}{4}\right), \left(\frac{5}{4}, -\frac{1}{4}\right)$
27. $(4.8, 3.1), (-5.2, 1.6)$
28. $(-1.75, -8.3), (2.25, -2.6)$

In Exercises 29–38, use the point on the line and the slope of the line to find three additional points through which the line passes. (There are many correct answers.)

	Point	*Slope*
29.	$(2, 1)$	$m = 0$
30.	$(-4, 1)$	m is undefined.
31.	$(5, -6)$	$m = 1$
32.	$(10, -6)$	$m = -1$
33.	$(-8, 1)$	m is undefined.
34.	$(-3, -1)$	$m = 0$
35.	$(-5, 4)$	$m = 2$
36.	$(0, -9)$	$m = -2$
37.	$(7, -2)$	$m = \frac{1}{2}$
38.	$(-1, -6)$	$m = -\frac{1}{2}$

In Exercises 39–50, find the slope-intercept form of the equation of the line that passes through the given point and has the indicated slope. Sketch the line.

	Point	*Slope*
39.	$(0, -2)$	$m = 3$
40.	$(0, 10)$	$m = -1$
41.	$(-3, 6)$	$m = -2$
42.	$(0, 0)$	$m = 4$
43.	$(4, 0)$	$m = -\frac{1}{3}$
44.	$(-2, -5)$	$m = \frac{3}{4}$
45.	$(6, -1)$	m is undefined.
46.	$(-10, 4)$	m is undefined.
47.	$\left(4, \frac{5}{2}\right)$	$m = 0$
48.	$\left(-\frac{1}{2}, \frac{3}{2}\right)$	$m = 0$
49.	$(-5.1, 1.8)$	$m = 5$
50.	$(2.3, -8.5)$	$m = -\frac{5}{2}$

In Exercises 51–64, find the slope-intercept form of the equation of the line passing through the points. Sketch the line.

51. $(5, -1), (-5, 5)$

52. $(4, 3), (-4, -4)$

53. $(-8, 1), (-8, 7)$

54. $(-1, 4), (6, 4)$

55. $\left(2, \frac{1}{2}\right), \left(\frac{1}{2}, \frac{5}{4}\right)$

56. $(1, 1), \left(6, -\frac{2}{3}\right)$

57. $\left(-\frac{1}{10}, -\frac{3}{5}\right), \left(\frac{9}{10}, -\frac{9}{5}\right)$

58. $\left(\frac{3}{4}, \frac{3}{2}\right), \left(-\frac{4}{3}, \frac{7}{4}\right)$

59. $(1, 0.6), (-2, -0.6)$

60. $(-8, 0.6), (2, -2.4)$

61. $(2, -1), \left(\frac{1}{3}, -1\right)$

62. $\left(\frac{1}{5}, -2\right), (-6, -2)$

63. $\left(\frac{7}{3}, -8\right), \left(\frac{7}{3}, 1\right)$

64. $(1.5, -2), (1.5, 0.2)$

In Exercises 65–68, determine whether the lines L_1 and L_2 passing through the pairs of points are parallel, perpendicular, or neither.

65. L_1: $(0, -1), (5, 9)$
L_2: $(0, 3), (4, 1)$

66. L_1: $(-2, -1), (1, 5)$
L_2: $(1, 3), (5, -5)$

67. L_1: $(3, 6), (-6, 0)$
L_2: $(0, -1), \left(5, \frac{7}{3}\right)$

68. L_1: $(4, 8), (-4, 2)$
L_2: $(3, -5), \left(-1, \frac{1}{3}\right)$

In Exercises 69–78, write the slope-intercept forms of the equations of the lines through the given point (a) parallel to the given line and (b) perpendicular to the given line.

	Point	*Line*
69.	$(2, 1)$	$4x - 2y = 3$
70.	$(-3, 2)$	$x + y = 7$
71.	$\left(-\frac{2}{3}, \frac{7}{8}\right)$	$3x + 4y = 7$
72.	$\left(\frac{7}{8}, \frac{3}{4}\right)$	$5x + 3y = 0$
73.	$(-1, 0)$	$y = -3$
74.	$(4, -2)$	$y = 1$
75.	$(2, 5)$	$x = 4$
76.	$(-5, 1)$	$x = -2$
77.	$(2.5, 6.8)$	$x - y = 4$
78.	$(-3.9, -1.4)$	$6x + 2y = 9$

In Exercises 79–84, use the *intercept form* to find the equation of the line with the given intercepts. The intercept form of the equation of a line with intercepts $(a, 0)$ and $(0, b)$ is

$$\frac{x}{a} + \frac{y}{b} = 1, \quad a \neq 0, \; b \neq 0.$$

79. x-intercept: $(2, 0)$
y-intercept: $(0, 3)$

80. x-intercept: $(-3, 0)$
y-intercept: $(0, 4)$

81. x-intercept: $\left(-\frac{1}{6}, 0\right)$
y-intercept: $\left(0, -\frac{2}{3}\right)$

82. x-intercept: $\left(\frac{2}{3}, 0\right)$
y-intercept: $(0, -2)$

83. Point on line: $(1, 2)$
x-intercept: $(c, 0)$
y-intercept: $(0, c)$, $c \neq 0$

84. Point on line: $(-3, 4)$
x-intercept: $(d, 0)$
y-intercept: $(0, d)$, $d \neq 0$

Graphical Interpretation **In Exercises 85–88, identify any relationships that exist among the lines, and then use a graphing utility to graph the three equations in the same viewing window. Adjust the viewing window so that the slope appears visually correct—that is, so that parallel lines appear parallel and perpendicular lines appear to intersect at right angles.**

85. (a) $y = 2x$ (b) $y = -2x$ (c) $y = \frac{1}{2}x$

86. (a) $y = \frac{2}{3}x$ (b) $y = -\frac{3}{2}x$ (c) $y = \frac{2}{3}x + 2$

87. (a) $y = -\frac{1}{2}x$ (b) $y = -\frac{1}{2}x + 3$ (c) $y = 2x - 4$

88. (a) $y = x - 8$ (b) $y = x + 1$ (c) $y = -x + 3$

In Exercises 89–92, find a relationship between x and y such that (x, y) is equidistant (the same distance) from the two points.

89. $(4, -1), (-2, 3)$

90. $(6, 5), (1, -8)$

91. $\left(3, \frac{5}{2}\right), (-7, 1)$

92. $\left(-\frac{1}{2}, -4\right), \left(\frac{7}{2}, \frac{5}{4}\right)$

93. ***Sales*** The following are the slopes of lines representing annual sales y in terms of time x in years. Use the slopes to interpret any change in annual sales for a one-year increase in time.

(a) The line has a slope of $m = 135$.

(b) The line has a slope of $m = 0$.

(c) The line has a slope of $m = -40$.

94. ***Revenue*** The following are the slopes of lines representing daily revenues y in terms of time x in days. Use the slopes to interpret any change in daily revenues for a one-day increase in time.

(a) The line has a slope of $m = 400$.

(b) The line has a slope of $m = 100$.

(c) The line has a slope of $m = 0$.

95. ***Average Salary*** The graph shows the average salaries for senior high school principals from 1990 through 2002. (Source: Educational Research Service)

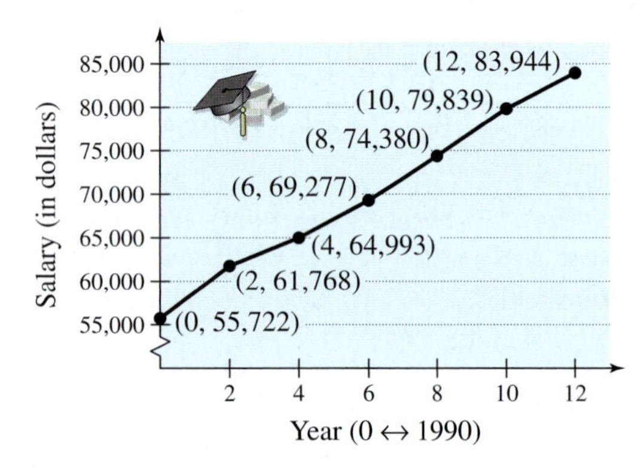

(a) Use the slopes to determine the time periods in which the average salary increased the greatest and the least.

(b) Find the slope of the line segment connecting the years 1990 and 2002.

(c) Interpret the meaning of the slope in part (b) in the context of the problem.

96. ***Net Profit*** The graph shows the net profits (in millions) for Applebee's International, Inc. for the years 1994 through 2003. (Source: Applebee's International, Inc.)

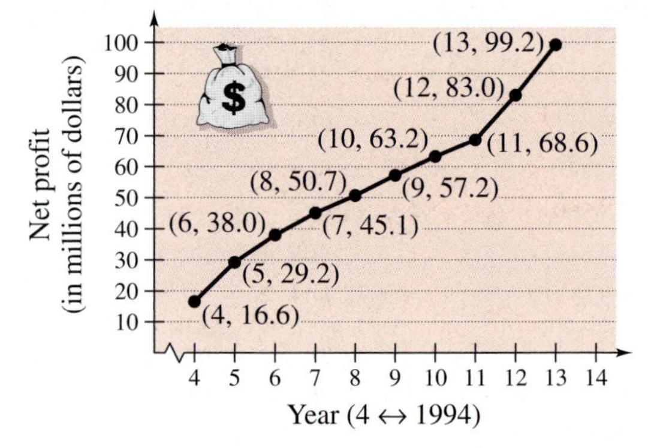

(a) Use the slopes to determine the years in which the net profit showed the greatest increase and the least increase.

(b) Find the slope of the line segment connecting the years 1994 and 2003.

(c) Interpret the meaning of the slope in part (b) in the context of the problem.

97. ***Road Grade*** You are driving on a road that has a 6% uphill grade (see figure). This means that the slope of the road is $\frac{6}{100}$. Approximate the amount of vertical change in your position if you drive 200 feet.

98. ***Road Grade*** From the top of a mountain road, a surveyor takes several horizontal measurements x and several vertical measurements y, as shown in the table (x and y are measured in feet).

x	300	600	900	1200	1500	1800	2100
y	-25	-50	-75	-100	-125	-150	-175

(a) Sketch a scatter plot of the data.

(b) Use a straightedge to sketch the line that you think best fits the data.

(c) Find an equation for the line you sketched in part (b).

(d) Interpret the meaning of the slope of the line in part (c) in the context of the problem.

(e) The surveyor needs to put up a road sign that indicates the steepness of the road. For instance, a surveyor would put up a sign that states "8% grade" on a road with a downhill grade that has a slope of $-\frac{8}{100}$. What should the sign state for the road in this problem?

Rate of Change **In Exercises 99 and 100, you are given the dollar value of a product in 2005 and the rate at which the value of the product is expected to change during the next 5 years. Use this information to write a linear equation that gives the dollar value V of the product in terms of the year t. (Let $t = 5$ represent 2005.)**

	2005 Value	*Rate*
99.	\$2540	\$125 decrease per year
100.	\$156	\$4.50 increase per year

Graphical Interpretation **In Exercises 101–104, match the description of the situation with its graph. Also determine the slope and y-intercept of each graph and interpret the slope and y-intercept in the context of the situation. [The graphs are labeled (a), (b), (c), and (d).]**

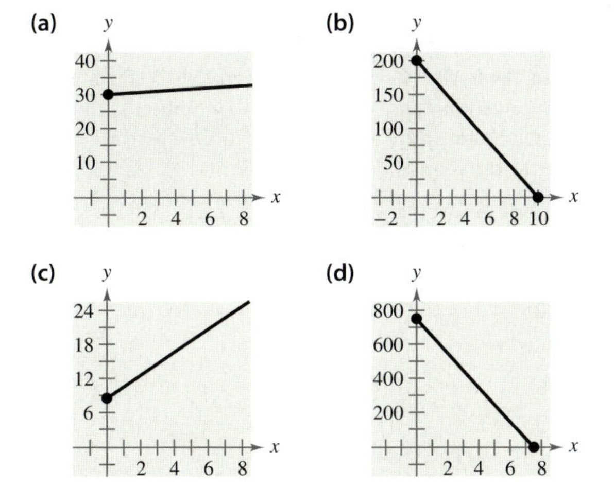

101. A person is paying \$20 per week to a friend to repay a \$200 loan.

102. An employee is paid \$8.50 per hour plus \$2 for each unit produced per hour.

103. A sales representative receives \$30 per day for food plus \$0.32 for each mile traveled.

104. A computer that was purchased for \$750 depreciates \$100 per year.

105. ***Cash Flow per Share*** The cash flow per share for the Timberland Co. was \$0.18 in 1995 and \$4.04 in 2003. Write a linear equation that gives the cash flow per share in terms of the year. Let $t = 5$ represent 1995. Then predict the cash flows for the years 2008 and 2010. (Source: The Timberland Co.)

106. ***Number of Stores*** In 1999 there were 4076 J.C. Penney stores and in 2003 there were 1078 stores. Write a linear equation that gives the number of stores in terms of the year. Let $t = 9$ represent 1999. Then predict the numbers of stores for the years 2008 and 2010. Are your answers reasonable? Explain. (Source: J.C. Penney Co.)

107. ***Depreciation*** A sub shop purchases a used pizza oven for \$875. After 5 years, the oven will have to be replaced. Write a linear equation giving the value V of the equipment during the 5 years it will be in use.

108. ***Depreciation*** A school district purchases a high-volume printer, copier, and scanner for \$25,000. After 10 years, the equipment will have to be replaced. Its value at that time is expected to be \$2000. Write a linear equation giving the value V of the equipment during the 10 years it will be in use.

109. ***College Enrollment*** The Pennsylvania State University had enrollments of 40,571 students in 2000 and 41,289 students in 2004 at its main campus in University Park, Pennsylvania. (Source: *Penn State Fact Book*)

(a) Assuming the enrollment growth is linear, find a linear model that gives the enrollment in terms of the year t, where $t = 0$ corresponds to 2000.

(b) Use your model from part (a) to predict the enrollments in 2008 and 2010.

(c) What is the slope of your model? Explain its meaning in the context of the situation.

110. ***College Enrollment*** The University of Florida had enrollments of 36,531 students in 1990 and 48,673 students in 2003. (Source: University of Florida)

(a) What was the average annual change in enrollment from 1990 to 2003?

(b) Use the average annual change in enrollment to estimate the enrollments in 1994, 1998, and 2002.

(c) Write the equation of a line that represents the given data. What is its slope? Interpret the slope in the context of the problem.

(d) Using the results of parts (a)–(c), write a short paragraph discussing the concepts of *slope* and *average rate of change*.

111. ***Sales*** A discount outlet is offering a 15% discount on all items. Write a linear equation giving the sale price S for an item with a list price L.

112. ***Hourly Wage*** A microchip manufacturer pays its assembly line workers \$11.50 per hour. In addition, workers receive a piecework rate of \$0.75 per unit produced. Write a linear equation for the hourly wage W in terms of the number of units x produced per hour.

113. ***Cost, Revenue, and Profit*** A roofing contractor purchases a shingle delivery truck with a shingle elevator for \$36,500. The vehicle requires an average expenditure of \$5.25 per hour for fuel and maintenance, and the operator is paid \$11.50 per hour.

(a) Write a linear equation giving the total cost C of operating this equipment for t hours. (Include the purchase cost of the equipment.)

(b) Assuming that customers are charged \$27 per hour of machine use, write an equation for the revenue R derived from t hours of use.

(c) Use the formula for profit $(P = R - C)$ to write an equation for the profit derived from t hours of use.

(d) Use the result of part (c) to find the break-even point—that is, the number of hours this equipment must be used to yield a profit of 0 dollars.

114. ***Rental Demand*** A real estate office handles an apartment complex with 50 units. When the rent per unit is \$580 per month, all 50 units are occupied. However, when the rent is \$625 per month, the average number of occupied units drops to 47. Assume that the relationship between the monthly rent p and the demand x is linear.

(a) Write the equation of the line giving the demand x in terms of the rent p.

(b) Use this equation to predict the number of units occupied when the rent is \$655.

(c) Predict the number of units occupied when the rent is \$595.

115. ***Geometry*** The length and width of a rectangular garden are 15 meters and 10 meters, respectively. A walkway of width x surrounds the garden.

(a) Draw a diagram that gives a visual representation of the problem.

(b) Write the equation for the perimeter y of the walkway in terms of x.

 (c) Use a graphing utility to graph the equation for the perimeter.

 (d) Determine the slope of the graph in part (c). For each additional one-meter increase in the width of the walkway, determine the increase in its perimeter.

116. ***Monthly Salary*** A pharmaceutical salesperson receives a monthly salary of \$2500 plus a commission of 7% of sales. Write a linear equation for the salesperson's monthly wage W in terms of monthly sales S.

117. ***Business Costs*** A sales representative of a company using a personal car receives \$120 per day for lodging and meals plus \$0.38 per mile driven. Write a linear equation giving the daily cost C to the company in terms of x, the number of miles driven.

118. ***Sports*** The median salaries (in thousands of dollars) for players on the Los Angeles Dodgers from 1996 to 2003 are shown in the scatter plot. Find the equation of the line that you think best fits these data. (Let y represent the median salary and let t represent the year, with $t = 6$ corresponding to 1996.) (Source: *USA TODAY*)

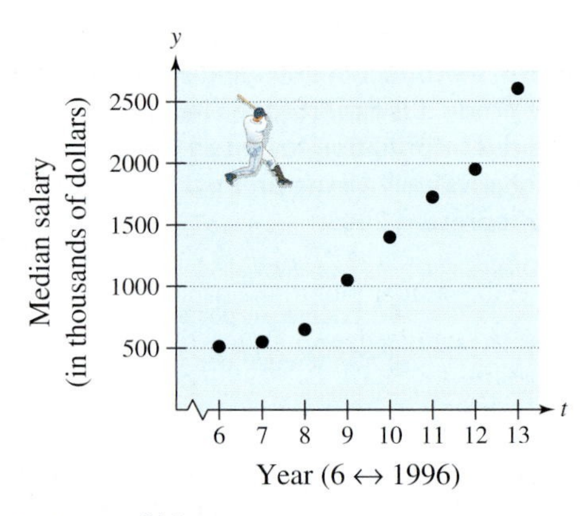

FIGURE FOR 118

Model It

119. ***Data Analysis: Cell Phone Suscribers*** The numbers of cellular phone suscribers y (in millions) in the United States from 1990 through 2002, where x is the year, are shown as data points (x, y). (Source: Cellular Telecommunications & Internet Association)

(1990, 5.3)
(1991, 7.6)
(1992, 11.0)
(1993, 16.0)
(1994, 24.1)
(1995, 33.8)
(1996, 44.0)
(1997, 55.3)
(1998, 69.2)
(1999, 86.0)
(2000, 109.5)
(2001, 128.4)
(2002, 140.8)

(a) Sketch a scatter plot of the data. Let $x = 0$ correspond to 1990.

(b) Use a straightedge to sketch the line that you think best fits the data.

(c) Find the equation of the line from part (b). Explain the procedure you used.

(d) Write a short paragraph explaining the meanings of the slope and y-intercept of the line in terms of the data.

(e) Compare the values obtained using your model with the actual values.

(f) Use your model to estimate the number of cellular phone suscribers in 2008.

120. ***Data Analysis: Average Scores*** An instructor gives regular 20-point quizzes and 100-point exams in an algebra course. Average scores for six students, given as data points (x, y) where x is the average quiz score and y is the average test score, are $(18, 87)$, $(10, 55)$, $(19, 96)$, $(16, 79)$, $(13, 76)$, and $(15, 82)$. [*Note:* There are many correct answers for parts (b)–(d).]

(a) Sketch a scatter plot of the data.

(b) Use a straightedge to sketch the line that you think best fits the data.

(c) Find an equation for the line you sketched in part (b).

(d) Use the equation in part (c) to estimate the average test score for a person with an average quiz score of 17.

(e) The instructor adds 4 points to the average test score of each student in the class. Describe the changes in the positions of the plotted points and the change in the equation of the line.

Synthesis

True or False? **In Exercises 121 and 122, determine whether the statement is true or false. Justify your answer.**

121. A line with a slope of $-\frac{5}{7}$ is steeper than a line with a slope of $-\frac{6}{7}$.

122. The line through $(-8, 2)$ and $(-1, 4)$ and the line through $(0, -4)$ and $(-7, 7)$ are parallel.

123. Explain how you could show that the points $A\ (2, 3)$, $B\ (2, 9)$, and $C\ (4, 3)$ are the vertices of a right triangle.

124. Explain why the slope of a vertical line is said to be undefined.

125. With the information shown in the graphs, is it possible to determine the slope of each line? Is it possible that the lines could have the same slope? Explain.

(a)

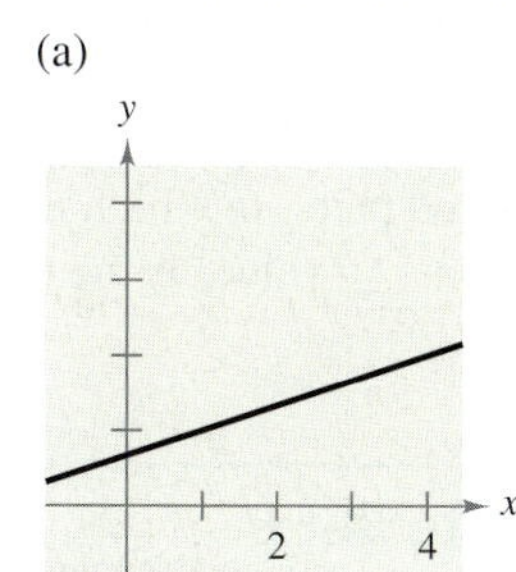

(b)

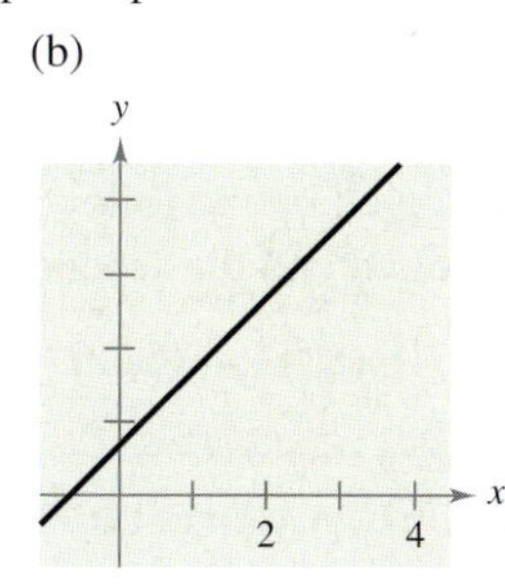

126. The slopes of two lines are -4 and $\frac{5}{2}$. Which is steeper? Explain.

127. The value V of a molding machine t years after it is purchased is

$$V = -4000t + 58{,}500, \quad 0 \le t \le 5.$$

Explain what the V-intercept and slope measure.

128. ***Think About It*** Is it possible for two lines with positive slopes to be perpendicular? Explain.

Skills Review

In Exercises 129–132, match the equation with its graph. [The graphs are labeled (a), (b), (c), and (d).]

(a)

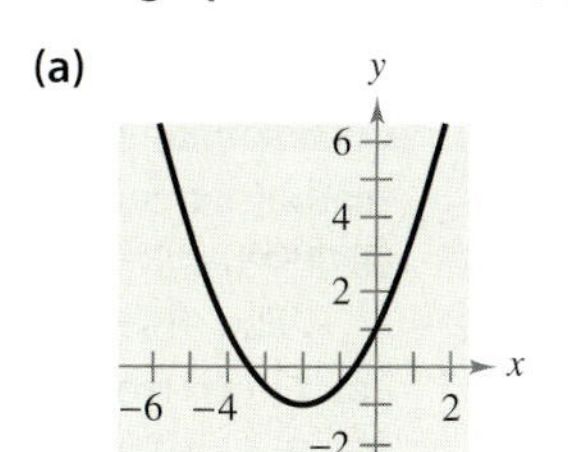

(b)

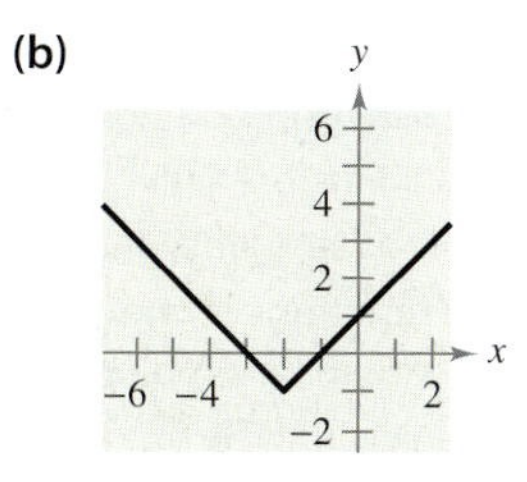

(c)

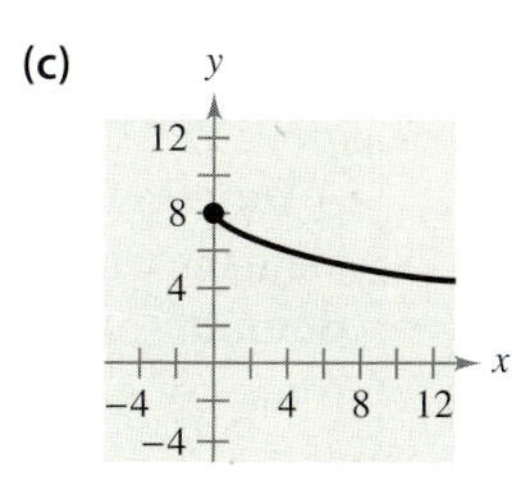

(d)

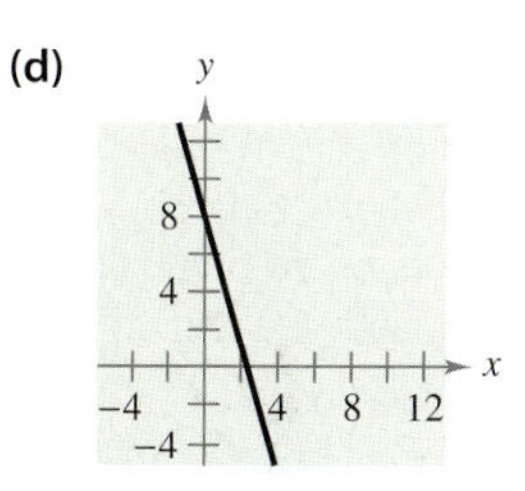

129. $y = 8 - 3x$

130. $y = 8 - \sqrt{x}$

131. $y = \frac{1}{2}x^2 + 2x + 1$

132. $y = |x + 2| - 1$

In Exercises 133–138, find all the solutions of the equation. Check your solution(s) in the original equation.

133. $-7(3 - x) = 14(x - 1)$

134. $\dfrac{8}{2x - 7} = \dfrac{4}{9 - 4x}$

135. $2x^2 - 21x + 49 = 0$

136. $x^2 - 8x + 3 = 0$

137. $\sqrt{x - 9} + 15 = 0$

138. $3x - 16\sqrt{x} + 5 = 0$

139. **Make a Decision** To work an extended application analyzing the numbers of bachelor's degrees earned by women in the United States from 1985 to 2002, visit this text's website at *college.hmco.com*. *(Data Source: U.S. Census Bureau)*

1.4 Functions

What you should learn

- Determine whether relations between two variables are functions.
- Use function notation and evaluate functions.
- Find the domains of functions.
- Use functions to model and solve real-life problems.
- Evaluate difference quotients.

Why you should learn it

Functions can be used to model and solve real-life problems. For instance, in Exercise 100 on page 52, you will use a function to model the force of water against the face of a dam.

Introduction to Functions

Many everyday phenomena involve two quantities that are related to each other by some rule of correspondence. The mathematical term for such a rule of correspondence is a **relation.** In mathematics, relations are often represented by mathematical equations and formulas. For instance, the simple interest I earned on \$1000 for 1 year is related to the annual interest rate r by the formula $I = 1000r$.

The formula $I = 1000r$ represents a special kind of relation that matches each item from one set with *exactly one* item from a different set. Such a relation is called a **function.**

Definition of Function

A **function** f from a set A to a set B is a relation that assigns to each element x in the set A exactly one element y in the set B. The set A is the **domain** (or set of inputs) of the function f, and the set B contains the **range** (or set of outputs).

To help understand this definition, look at the function that relates the time of day to the temperature in Figure 1.47.

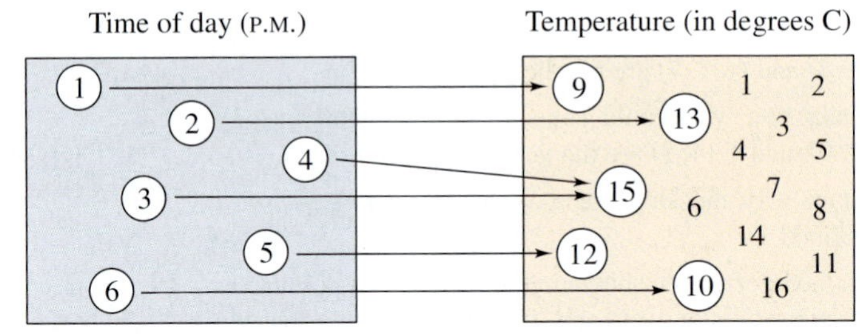

FIGURE 1.47

This function can be represented by the following ordered pairs, in which the first coordinate (x-value) is the input and the second coordinate (y-value) is the output.

$$\{(1, 9^\circ), (2, 13^\circ), (3, 15^\circ), (4, 15^\circ), (5, 12^\circ), (6, 10^\circ)\}$$

Characteristics of a Function from Set A to Set B

1. Each element in A must be matched with an element in B.
2. Some elements in B may not be matched with any element in A.
3. Two or more elements in A may be matched with the same element in B.
4. An element in A (the domain) cannot be matched with two different elements in B.

Have your students pay special attention to the concepts of function, domain, and range, because they will be used throughout this text and in calculus.

Functions are commonly represented in four ways.

Four Ways to Represent a Function

1. *Verbally* by a sentence that describes how the input variable is related to the output variable
2. *Numerically* by a table or a list of ordered pairs that matches input values with output values
3. *Graphically* by points on a graph in a coordinate plane in which the input values are represented by the horizontal axis and the output values are represented by the vertical axis
4. *Algebraically* by an equation in two variables

To determine whether or not a relation is a function, you must decide whether each input value is matched with exactly one output value. If any input value is matched with two or more output values, the relation is not a function.

Example 1 Testing for Functions

Determine whether the relation represents y as a function of x.

a. The input value x is the number of representatives from a state, and the output value y is the number of senators.

b.

Input, x	Output, y
2	11
2	10
3	8
4	5
5	1

c.

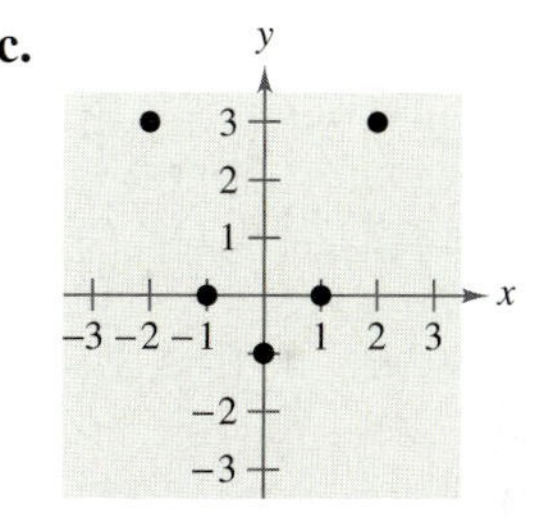

FIGURE 1.48

Solution

a. This verbal description *does* describe y as a function of x. Regardless of the value of x, the value of y is always 2. Such functions are called *constant functions*.

b. This table *does not* describe y as a function of x. The input value 2 is matched with two different y-values.

c. The graph in Figure 1.48 *does* describe y as a function of x. Each input value is matched with exactly one output value.

CHECKPOINT Now try Exercise 5.

Representing functions by sets of ordered pairs is common in *discrete mathematics*. In algebra, however, it is more common to represent functions by equations or formulas involving two variables. For instance, the equation

$$y = x^2 \qquad y \text{ is a function of } x.$$

represents the variable y as a function of the variable x. In this equation, x is

Historical Note
Leonhard Euler (1707–1783), a Swiss mathematician, is considered to have been the most prolific and productive mathematician in history. One of his greatest influences on mathematics was his use of symbols, or notation. The function notation $y = f(x)$ was introduced by Euler.

the **independent variable** and y is the **dependent variable.** The domain of the function is the set of all values taken on by the independent variable x, and the range of the function is the set of all values taken on by the dependent variable y.

Example 2 Testing for Functions Represented Algebraically

Which of the equations represent(s) y as a function of x?

a. $x^2 + y = 1$ **b.** $-x + y^2 = 1$

Solution

To determine whether y is a function of x, try to solve for y in terms of x.

a. Solving for y yields

$x^2 + y = 1$ Write original equation.

$y = 1 - x^2.$ Solve for y.

To each value of x there corresponds exactly one value of y. So, y is a function of x.

b. Solving for y yields

$-x + y^2 = 1$ Write original equation.

$y^2 = 1 + x$ Add x to each side.

$y = \pm\sqrt{1 + x}.$ Solve for y.

The $\pm$ indicates that to a given value of x there correspond two values of y. So, y is not a function of x.

✓CHECKPOINT Now try Exercise 15.

Function Notation

Understanding the concept of a function is essential. Be sure that students understand function notation. Frequently $f(x)$ is misinterpreted as "f times x" rather than "f of x."

When an equation is used to represent a function, it is convenient to name the function so that it can be referenced easily. For example, you know that the equation $y = 1 - x^2$ describes y as a function of x. Suppose you give this function the name "f." Then you can use the following **function notation.**

Input	*Output*	*Equation*
x	$f(x)$	$f(x) = 1 - x^2$

The symbol $f(x)$ is read as *the value of f at x* or simply *f of x*. The symbol $f(x)$ corresponds to the y-value for a given x. So, you can write $y = f(x)$. Keep in mind that f is the *name* of the function, whereas $f(x)$ is the *value* of the function at x. For instance, the function given by

$$f(x) = 3 - 2x$$

has *function values* denoted by $f(-1), f(0), f(2)$, and so on. To find these values, substitute the specified input values into the given equation.

For $x = -1$, $f(-1) = 3 - 2(-1) = 3 + 2 = 5.$

For $x = 0$, $f(0) = 3 - 2(0) = 3 - 0 = 3.$

For $x = 2$, $f(2) = 3 - 2(2) = 3 - 4 = -1.$

Although f is often used as a convenient function name and x is often used as the independent variable, you can use other letters. For instance,

$$f(x) = x^2 - 4x + 7, \quad f(t) = t^2 - 4t + 7, \quad \text{and} \quad g(s) = s^2 - 4s + 7$$

all define the same function. In fact, the role of the independent variable is that of a "placeholder." Consequently, the function could be described by

$$f(\quad) = (\quad)^2 - 4(\quad) + 7.$$

STUDY TIP

In Example 3, note that $g(x + 2)$ is not equal to $g(x) + g(2)$. In general, $g(u + v) \neq g(u) + g(v)$.

Students often have difficulty understanding how to evaluate piecewise-defined functions. You may want to use the following additional examples to demonstrate evaluation of piecewise-defined functions.

a. Evaluate at $x = 0, 1, 3$.

$$f(x) = \begin{cases} \dfrac{x}{2} + 1, & x \leq 1 \\ 3x + 2, & x > 1 \end{cases}$$

Solution

Because $x = 0$ is less than or equal to 1, use $f(x) = (x/2) + 1$ to obtain

$f(0) = \dfrac{0}{2} + 1 = 1.$

For $x = 1$, use $f(x) = (x/2) + 1$ to obtain

$f(1) = \dfrac{1}{2} + 1 = 1\dfrac{1}{2}.$

For $x = 3$, use $f(x) = 3x + 2$ to obtain

$f(3) = 3(3) + 2 = 11.$

b. Evaluate at $x = 0, 3, 5$.

$$f(x) = \begin{cases} x^2 + 3, & x < 2 \\ 7, & 2 \leq x \leq 4 \\ 2x - 1, & x > 4 \end{cases}$$

Solution

Because $x = 0$ is less than 2, use $f(x) = x^2 + 3$ to obtain

$f(0) = 0^2 + 3 = 3.$

For $x = 3$, use $f(x) = 7$ to obtain

$f(3) = 7.$

For $x = 5$, use $f(x) = 2x - 1$ to obtain

$f(5) = 2(5) - 1 = 9.$

Example 3 Evaluating a Function

Let $g(x) = -x^2 + 4x + 1$. Find each function value.

a. $g(2)$ **b.** $g(t)$ **c.** $g(x + 2)$

Solution

a. Replacing x with 2 in $g(x) = -x^2 + 4x + 1$ yields the following.

$$g(2) = -(2)^2 + 4(2) + 1 = -4 + 8 + 1 = 5$$

b. Replacing x with t yields the following.

$$g(t) = -(t)^2 + 4(t) + 1 = -t^2 + 4t + 1$$

c. Replacing x with $x + 2$ yields the following.

$$\begin{aligned} g(x + 2) &= -(x + 2)^2 + 4(x + 2) + 1 \\ &= -(x^2 + 4x + 4) + 4x + 8 + 1 \\ &= -x^2 - 4x - 4 + 4x + 8 + 1 \\ &= -x^2 + 5 \end{aligned}$$

CHECKPOINT Now try Exercise 29.

A function defined by two or more equations over a specified domain is called a **piecewise-defined function.**

Example 4 A Piecewise-Defined Function

Evaluate the function when $x = -1, 0$, and 1.

$$f(x) = \begin{cases} x^2 + 1, & x < 0 \\ x - 1, & x \geq 0 \end{cases}$$

Solution

Because $x = -1$ is less than 0, use $f(x) = x^2 + 1$ to obtain

$$f(-1) = (-1)^2 + 1 = 2.$$

For $x = 0$, use $f(x) = x - 1$ to obtain

$$f(0) = (0) - 1 = -1.$$

For $x = 1$, use $f(x) = x - 1$ to obtain

$$f(1) = (1) - 1 = 0.$$

CHECKPOINT Now try Exercise 35.

Technology

Use a graphing utility to graph the functions given by $y = \sqrt{4 - x^2}$ and $y = \sqrt{x^2 - 4}$. What is the domain of each function? Do the domains of these two functions overlap? If so, for what values do the domains overlap?

The Domain of a Function

The domain of a function can be described explicitly or it can be *implied* by the expression used to define the function. The **implied domain** is the set of all real numbers for which the expression is defined. For instance, the function given by

$$f(x) = \frac{1}{x^2 - 4}$$ Domain excludes x-values that result in division by zero.

has an implied domain that consists of all real x other than $x = \pm 2$. These two values are excluded from the domain because division by zero is undefined. Another common type of implied domain is that used to avoid even roots of negative numbers. For example, the function given by

$$f(x) = \sqrt{x}$$ Domain excludes x-values that result in even roots of negative numbers.

is defined only for $x \geq 0$. So, its implied domain is the interval $[0, \infty)$. In general, the domain of a function *excludes* values that would cause division by zero *or* that would result in the even root of a negative number.

Example 5 Finding the Domain of a Function

Find the domain of each function.

a. f: $\{(-3, 0), (-1, 4), (0, 2), (2, 2), (4, -1)\}$ **b.** $g(x) = \dfrac{1}{x + 5}$

c. Volume of a sphere: $V = \frac{4}{3}\pi r^3$ **d.** $h(x) = \sqrt{4 - x^2}$

Solution

a. The domain of f consists of all first coordinates in the set of ordered pairs.

Domain $= \{-3, -1, 0, 2, 4\}$

b. Excluding x-values that yield zero in the denominator, the domain of g is the set of all real numbers x except $x = -5$.

c. Because this function represents the volume of a sphere, the values of the radius r must be positive. So, the domain is the set of all real numbers r such that $r > 0$.

d. This function is defined only for x-values for which

$$4 - x^2 \geq 0.$$

By solving this inequality (see Section 2.7), you can conclude that $-2 \leq x \leq 2$. So, the domain is the interval $[-2, 2]$.

✓CHECKPOINT Now try Exercise 59.

In Example 5(c), note that the domain of a function may be implied by the physical context. For instance, from the equation

$$V = \frac{4}{3}\pi r^3$$

you would have no reason to restrict r to positive values, but the physical context implies that a sphere cannot have a negative or zero radius.

FIGURE 1.49

Applications

Example 6 The Dimensions of a Container 

You work in the marketing department of a soft-drink company and are experimenting with a new can for iced tea that is slightly narrower and taller than a standard can. For your experimental can, the ratio of the height to the radius is 4, as shown in Figure 1.49.

a. Write the volume of the can as a function of the radius r.

b. Write the volume of the can as a function of the height h.

Solution

a. $V(r) = \pi r^2 h = \pi r^2(4r) = 4\pi r^3$ — Write V as a function of r.

b. $V(h) = \pi\left(\frac{h}{4}\right)^2 h = \frac{\pi h^3}{16}$ — Write V as a function of h.

CHECKPOINT Now try Exercise 87.

Example 7 The Path of a Baseball 

A baseball is hit at a point 3 feet above ground at a velocity of 100 feet per second and an angle of 45°. The path of the baseball is given by the function

$$f(x) = -0.0032x^2 + x + 3$$

where y and x are measured in feet, as shown in Figure 1.50. Will the baseball clear a 10-foot fence located 300 feet from home plate?

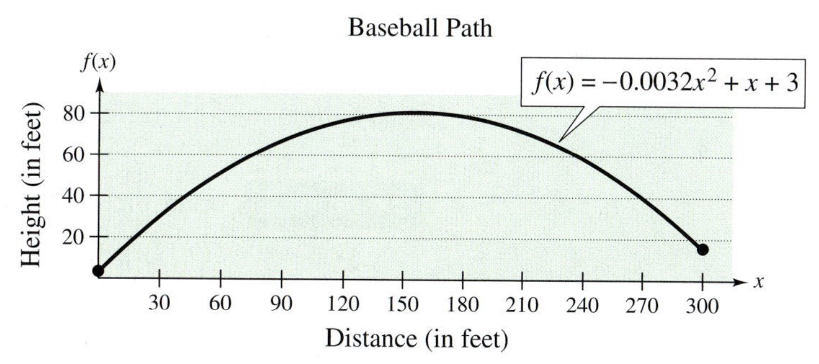

FIGURE 1.50

Solution

When $x = 300$, the height of the baseball is

$$f(300) = -0.0032(300)^2 + 300 + 3$$
$$= 15 \text{ feet.}$$

So, the baseball will clear the fence.

CHECKPOINT Now try Exercise 93.

In the equation in Example 7, the height of the baseball is a function of the distance from home plate.

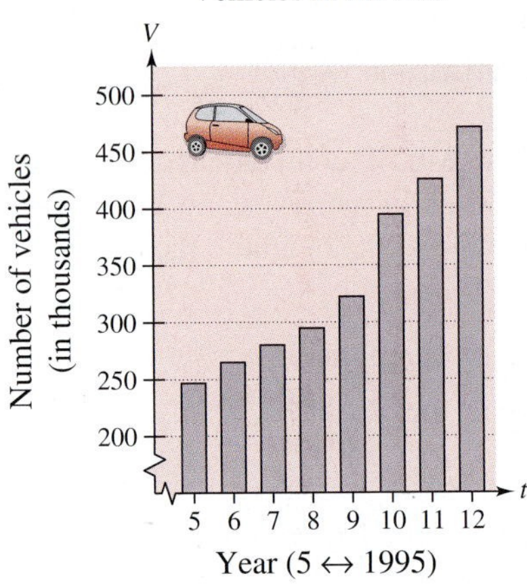

FIGURE 1.51

Example 8 Alternative-Fueled Vehicles

The number V (in thousands) of alternative-fueled vehicles in the United States increased in a linear pattern from 1995 to 1999, as shown in Figure 1.51. Then, in 2000, the number of vehicles took a jump and, until 2002, increased in a different linear pattern. These two patterns can be approximated by the function

$$V(t) = \begin{cases} 18.08t + 155.3 & 5 \le t \le 9 \\ 38.20t + 10.2, & 10 \le t \le 12 \end{cases}$$

where t represents the year, with $t = 5$ corresponding to 1995. Use this function to approximate the number of alternative-fueled vehicles for each year from 1995 to 2002. (Source: Science Applications International Corporation; Energy Information Administration)

Solution

From 1995 to 1999, use $V(t) = 18.08t + 155.3$.

$$\underbrace{245.7}_{1995} \quad \underbrace{263.8}_{1996} \quad \underbrace{281.9}_{1997} \quad \underbrace{299.9}_{1998} \quad \underbrace{318.0}_{1999}$$

From 2000 to 2002, use $V(t) = 38.20t + 10.2$.

$$\underbrace{392.2}_{2000} \quad \underbrace{430.4}_{2001} \quad \underbrace{468.6}_{2002}$$

CHECKPOINT Now try Exercise 95.

Difference Quotients

One of the basic definitions in calculus employs the ratio

$$\frac{f(x+h) - f(x)}{h}, \quad h \neq 0.$$

This ratio is called a **difference quotient,** as illustrated in Example 9.

Example 9 Evaluating a Difference Quotient

For $f(x) = x^2 - 4x + 7$, find $\dfrac{f(x+h) - f(x)}{h}$.

Solution

$$\begin{aligned} \frac{f(x+h) - f(x)}{h} &= \frac{[(x+h)^2 - 4(x+h) + 7] - (x^2 - 4x + 7)}{h} \\ &= \frac{x^2 + 2xh + h^2 - 4x - 4h + 7 - x^2 + 4x - 7}{h} \\ &= \frac{2xh + h^2 - 4h}{h} = \frac{h(2x + h - 4)}{h} = 2x + h - 4, \ h \neq 0 \end{aligned}$$

CHECKPOINT Now try Exercise 79.

The symbol $\int$ indicates an example or exercise that highlights algebraic techniques specifically used in calculus.

Activities

1. Evaluate $f(x) = 2 + 3x - x^2$ for
   a. $f(-3)$
   b. $f(x + 1)$
   c. $f(x + h) - f(x)$.

   Answers: a. -16
   b. $-x^2 + x + 4$
   c. $3h - 2hx - h^2$
2. Determine whether y is a function of x.

   $2x^3 + 3x^2y^2 + 1 = 0$

   Answer: No
3. Find the domain: $f(x) = \dfrac{3}{x+1}$.

   Answer: All real numbers $x \neq -1$

You may find it easier to calculate the difference quotient in Example 9 by first finding $f(x + h)$, and then substituting the resulting expression into the difference quotient, as follows.

$$f(x + h) = (x + h)^2 - 4(x + h) + 7 = x^2 + 2xh + h^2 - 4x - 4h + 7$$

$$\frac{f(x + h) - f(x)}{h} = \frac{(x^2 + 2xh + h^2 - 4x - 4h + 7) - (x^2 - 4x + 7)}{h}$$

$$= \frac{2xh + h^2 - 4h}{h} = \frac{h(2x + h - 4)}{h} = 2x + h - 4, h \neq 0$$

Summary of Function Terminology

Function: A **function** is a relationship between two variables such that to each value of the independent variable there corresponds exactly one value of the dependent variable.

Function Notation: $y = f(x)$

f is the *name* of the function.

y is the **dependent variable.**

x is the **independent variable.**

$f(x)$ is the *value of the function at* x.

Domain: The **domain** of a function is the set of all values (inputs) of the independent variable for which the function is defined. If x is in the domain of f, f is said to be *defined* at x. If x is not in the domain of f, f is said to be *undefined* at x.

Range: The **range** of a function is the set of all values (outputs) assumed by the dependent variable (that is, the set of all function values).

Implied Domain: If f is defined by an algebraic expression and the domain is not specified, the **implied domain** consists of all real numbers for which the expression is defined.

Writing About Mathematics

Everyday Functions In groups of two or three, identify common real-life functions. Consider everyday activities, events, and expenses, such as long distance telephone calls and car insurance. Here are two examples.

a. The statement, "Your happiness is a function of the grade you receive in this course" *is not* a correct mathematical use of the word "function." The word "happiness" is ambiguous.

b. The statement, "Your federal income tax is a function of your adjusted gross income" *is* a correct mathematical use of the word "function." Once you have determined your adjusted gross income, your income tax can be determined.

Describe your functions in words. Avoid using ambiguous words. Can you find an example of a piecewise-defined function?

1.4 Exercises

VOCABULARY CHECK: Fill in the blanks.

1. A relation that assigns to each element x from a set of inputs, or ________, exactly one element y in a set of outputs, or ________, is called a ________.
2. Functions are commonly represented in four different ways, ________, ________, ________, and ________.
3. For an equation that represents y as a function of x, the set of all values taken on by the ________ variable x is the domain, and the set of all values taken on by the ________ variable y is the range.
4. The function given by

$$f(x) = \begin{cases} 2x - 1, & x < 0 \\ x^2 + 4, & x \geq 0 \end{cases}$$

is an example of a ________ function.

5. If the domain of the function f is not given, then the set of values of the independent variable for which the expression is defined is called the ________ ________.
6. In calculus, one of the basic definitions is that of a ________ ________, given by $\dfrac{f(x+h) - f(x)}{h}$, $h \neq 0$.

PREREQUISITE SKILLS REVIEW: Practice and review algebra skills needed for this section at **www.Eduspace.com.**

In Exercises 1–4, is the relationship a function?

1. *Domain* *Range*

Domain: -2, -1, 0, 1, 2 Range: 5, 6, 7, 8

2. *Domain* *Range*

Domain: -2, -1, 0, 1, 2 Range: 3, 4, 5

3. *Domain* *Range*

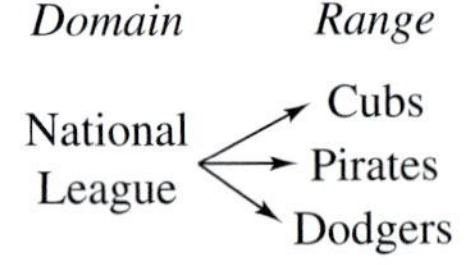

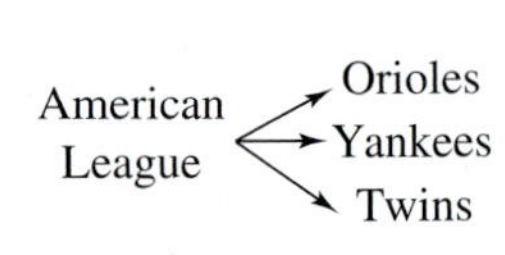

4. *Domain* (Year) *Range* (Number of North Atlantic tropical storms and hurricanes)

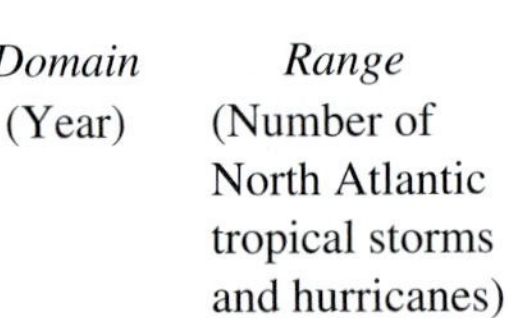

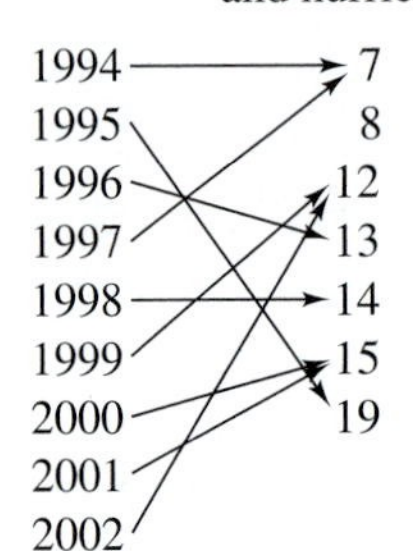

In Exercises 5–8, does the table describe a function? Explain your reasoning.

5.

Input value	-2	-1	0	1	2
Output value	-8	-1	0	1	8

6.

Input value	0	1	2	1	0
Output value	-4	-2	0	2	4

7.

Input value	10	7	4	7	10
Output value	3	6	9	12	15

8.

Input value	0	3	9	12	15
Output value	3	3	3	3	3

In Exercises 9 and 10, which sets of ordered pairs represent functions from A to B? Explain.

9. $A = \{0, 1, 2, 3\}$ and $B = \{-2, -1, 0, 1, 2\}$
   (a) $\{(0, 1), (1, -2), (2, 0), (3, 2)\}$
   (b) $\{(0, -1), (2, 2), (1, -2), (3, 0), (1, 1)\}$
   (c) $\{(0, 0), (1, 0), (2, 0), (3, 0)\}$
   (d) $\{(0, 2), (3, 0), (1, 1)\}$
10. $A = \{a, b, c\}$ and $B = \{0, 1, 2, 3\}$
   (a) $\{(a, 1), (c, 2), (c, 3), (b, 3)\}$
   (b) $\{(a, 1), (b, 2), (c, 3)\}$
   (c) $\{(1, a), (0, a), (2, c), (3, b)\}$
   (d) $\{(c, 0), (b, 0), (a, 3)\}$

Circulation of Newspapers **In Exercises 11 and 12, use the graph, which shows the circulation (in millions) of daily newspapers in the United States.** (Source: Editor & Publisher Company)

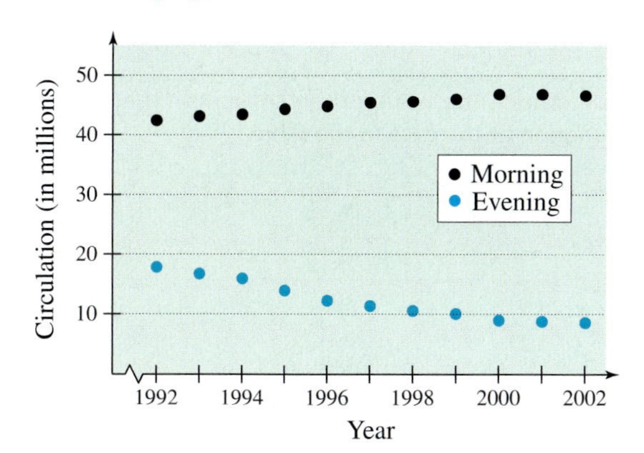

11. Is the circulation of morning newspapers a function of the year? Is the circulation of evening newspapers a function of the year? Explain.

12. Let $f(x)$ represent the circulation of evening newspapers in year x. Find $f(1998)$.

In Exercises 13–24, determine whether the equation represents *y* as a function of *x*.

13. $x^2 + y^2 = 4$ **14.** $x = y^2$

15. $x^2 + y = 4$ **16.** $x + y^2 = 4$

17. $2x + 3y = 4$ **18.** $(x - 2)^2 + y^2 = 4$

19. $y^2 = x^2 - 1$ **20.** $y = \sqrt{x + 5}$

21. $y = |4 - x|$ **22.** $|y| = 4 - x$

23. $x = 14$ **24.** $y = -75$

In Exercises 25–38, evaluate the function at each specified value of the independent variable and simplify.

25. $f(x) = 2x - 3$
(a) $f(1)$ (b) $f(-3)$ (c) $f(x - 1)$

26. $g(y) = 7 - 3y$
(a) $g(0)$ (b) $g\left(\frac{7}{3}\right)$ (c) $g(s + 2)$

27. $V(r) = \frac{4}{3}\pi r^3$
(a) $V(3)$ (b) $V\left(\frac{3}{2}\right)$ (c) $V(2r)$

28. $h(t) = t^2 - 2t$
(a) $h(2)$ (b) $h(1.5)$ (c) $h(x + 2)$

29. $f(y) = 3 - \sqrt{y}$
(a) $f(4)$ (b) $f(0.25)$ (c) $f(4x^2)$

30. $f(x) = \sqrt{x + 8} + 2$
(a) $f(-8)$ (b) $f(1)$ (c) $f(x - 8)$

31. $q(x) = \dfrac{1}{x^2 - 9}$
(a) $q(0)$ (b) $q(3)$ (c) $q(y + 3)$

32. $q(t) = \dfrac{2t^2 + 3}{t^2}$
(a) $q(2)$ (b) $q(0)$ (c) $q(-x)$

33. $f(x) = \dfrac{|x|}{x}$
(a) $f(2)$ (b) $f(-2)$ (c) $f(x - 1)$

34. $f(x) = |x| + 4$
(a) $f(2)$ (b) $f(-2)$ (c) $f(x^2)$

35. $f(x) = \begin{cases} 2x + 1, & x < 0 \\ 2x + 2, & x \ge 0 \end{cases}$
(a) $f(-1)$ (b) $f(0)$ (c) $f(2)$

36. $f(x) = \begin{cases} x^2 + 2, & x \le 1 \\ 2x^2 + 2, & x > 1 \end{cases}$
(a) $f(-2)$ (b) $f(1)$ (c) $f(2)$

37. $f(x) = \begin{cases} 3x - 1, & x < -1 \\ 4, & -1 \le x \le 1 \\ x^2, & x > 1 \end{cases}$
(a) $f(-2)$ (b) $f\left(-\frac{1}{2}\right)$ (c) $f(3)$

38. $f(x) = \begin{cases} 4 - 5x, & x \le -2 \\ 0, & -2 < x < 2 \\ x^2 + 1, & x > 2 \end{cases}$
(a) $f(-3)$ (b) $f(4)$ (c) $f(-1)$

In Exercises 39–44, complete the table.

39. $f(x) = x^2 - 3$

x	-2	-1	0	1	2
$f(x)$					

40. $g(x) = \sqrt{x - 3}$

x	3	4	5	6	7
$g(x)$					

41. $h(t) = \frac{1}{2}|t + 3|$

t	-5	-4	-3	-2	-1
$h(t)$					

42. $f(s) = \dfrac{|s - 2|}{s - 2}$

s	0	1	$\frac{3}{2}$	$\frac{5}{2}$	4
$f(s)$					

43. $f(x) = \begin{cases} -\frac{1}{2}x + 4, & x \le 0 \\ (x-2)^2, & x > 0 \end{cases}$

x	-2	-1	0	1	2
$f(x)$					

44. $f(x) = \begin{cases} 9 - x^2, & x < 3 \\ x - 3, & x \ge 3 \end{cases}$

x	1	2	3	4	5
$f(x)$					

In Exercises 45–52, find all real values of x such that $f(x) = 0$.

45. $f(x) = 15 - 3x$

46. $f(x) = 5x + 1$

47. $f(x) = \dfrac{3x - 4}{5}$

48. $f(x) = \dfrac{12 - x^2}{5}$

49. $f(x) = x^2 - 9$

50. $f(x) = x^2 - 8x + 15$

51. $f(x) = x^3 - x$

52. $f(x) = x^3 - x^2 - 4x + 4$

In Exercises 53–56, find the value(s) of x for which $f(x) = g(x)$.

53. $f(x) = x^2 + 2x + 1, \quad g(x) = 3x + 3$

54. $f(x) = x^4 - 2x^2, \quad g(x) = 2x^2$

55. $f(x) = \sqrt{3x} + 1, \quad g(x) = x + 1$

56. $f(x) = \sqrt{x} - 4, \quad g(x) = 2 - x$

In Exercises 57–70, find the domain of the function.

57. $f(x) = 5x^2 + 2x - 1$

58. $g(x) = 1 - 2x^2$

59. $h(t) = \dfrac{4}{t}$

60. $s(y) = \dfrac{3y}{y + 5}$

61. $g(y) = \sqrt{y - 10}$

62. $f(t) = \sqrt[3]{t + 4}$

63. $f(x) = \sqrt[4]{1 - x^2}$

64. $f(x) = \sqrt[4]{x^2 + 3x}$

65. $g(x) = \dfrac{1}{x} - \dfrac{3}{x + 2}$

66. $h(x) = \dfrac{10}{x^2 - 2x}$

67. $f(s) = \dfrac{\sqrt{s - 1}}{s - 4}$

68. $f(x) = \dfrac{\sqrt{x + 6}}{6 + x}$

69. $f(x) = \dfrac{x - 4}{\sqrt{x}}$

70. $f(x) = \dfrac{x - 5}{\sqrt{x^2 - 9}}$

In Exercises 71–74, assume that the domain of f is the set $A = \{-2, -1, 0, 1, 2\}$. Determine the set of ordered pairs that represents the function f.

71. $f(x) = x^2$

72. $f(x) = x^2 - 3$

73. $f(x) = |x| + 2$

74. $f(x) = |x + 1|$

Exploration **In Exercises 75–78, match the data with one of the following functions**

$$f(x) = cx, \; g(x) = cx^2, \; h(x) = c\sqrt{|x|}, \text{ and } r(x) = \frac{c}{x}$$

and determine the value of the constant c that will make the function fit the data in the table.

75.

x	-4	-1	0	1	4
y	-32	-2	0	-2	-32

76.

x	-4	-1	0	1	4
y	-1	$-\frac{1}{4}$	0	$\frac{1}{4}$	1

77.

x	-4	-1	0	1	4
y	-8	-32	Undef.	32	8

78.

x	-4	-1	0	1	4
y	6	3	0	3	6

In Exercises 79–86, find the difference quotient and simplify your answer.

79. $f(x) = x^2 - x + 1, \quad \dfrac{f(2 + h) - f(2)}{h}, \; h \ne 0$

80. $f(x) = 5x - x^2, \quad \dfrac{f(5 + h) - f(5)}{h}, \; h \ne 0$

81. $f(x) = x^3 + 3x, \quad \dfrac{f(x + h) - f(x)}{h}, \; h \ne 0$

82. $f(x) = 4x^2 - 2x, \quad \dfrac{f(x + h) - f(x)}{h}, \; h \ne 0$

83. $g(x) = \dfrac{1}{x^2}, \quad \dfrac{g(x) - g(3)}{x - 3}, \; x \ne 3$

84. $f(t) = \dfrac{1}{t - 2}, \quad \dfrac{f(t) - f(1)}{t - 1}, \; t \ne 1$

85. $f(x) = \sqrt{5x}, \quad \dfrac{f(x) - f(5)}{x - 5}, \; x \ne 5$

86. $f(x) = x^{2/3} + 1, \quad \dfrac{f(x) - f(8)}{x - 8}, \; x \ne 8$

87. ***Geometry*** Write the area A of a square as a function of its perimeter P.

88. ***Geometry*** Write the area A of a circle as a function of its circumference C.

The symbol $\int$ indicates an example or exercise that highlights algebraic techniques specifically used in calculus.

89. ***Maximum Volume*** An open box of maximum volume is to be made from a square piece of material 24 centimeters on a side by cutting equal squares from the corners and turning up the sides (see figure).

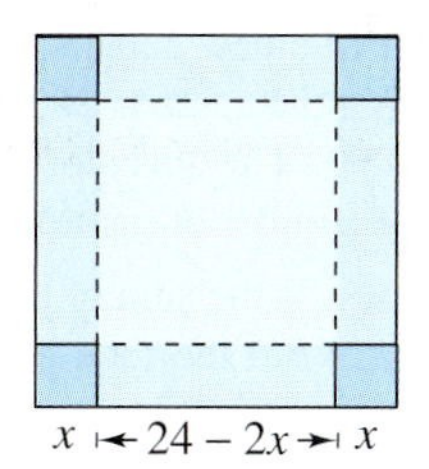

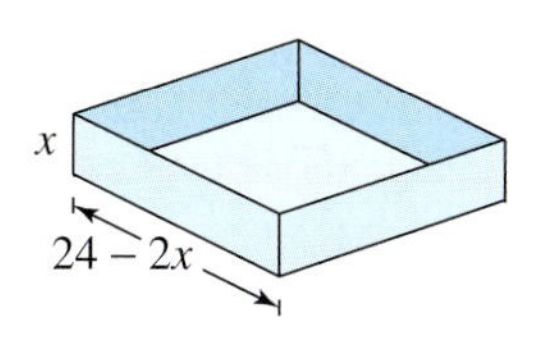

(a) The table shows the volume V (in cubic centimeters) of the box for various heights x (in centimeters). Use the table to estimate the maximum volume.

Height, x	1	2	3	4	5	6
Volume, V	484	800	972	1024	980	864

(b) Plot the points (x, V) from the table in part (a). Does the relation defined by the ordered pairs represent V as a function of x?

(c) If V is a function of x, write the function and determine its domain.

90. ***Maximum Profit*** The cost per unit in the production of a portable CD player is \$60. The manufacturer charges \$90 per unit for orders of 100 or less. To encourage large orders, the manufacturer reduces the charge by \$0.15 per CD player for each unit ordered in excess of 100 (for example, there would be a charge of \$87 per CD player for an order size of 120).

(a) The table shows the profit P (in dollars) for various numbers of units ordered, x. Use the table to estimate the maximum profit.

Units, x	110	120	130	140
Profit, P	3135	3240	3315	3360

Units, x	150	160	170
Profit, P	3375	3360	3315

(b) Plot the points (x, P) from the table in part (a). Does the relation defined by the ordered pairs represent P as a function of x?

(c) If P is a function of x, write the function and determine its domain.

91. ***Geometry*** A right triangle is formed in the first quadrant by the x- and y-axes and a line through the point $(2, 1)$ (see figure). Write the area A of the triangle as a function of x, and determine the domain of the function.

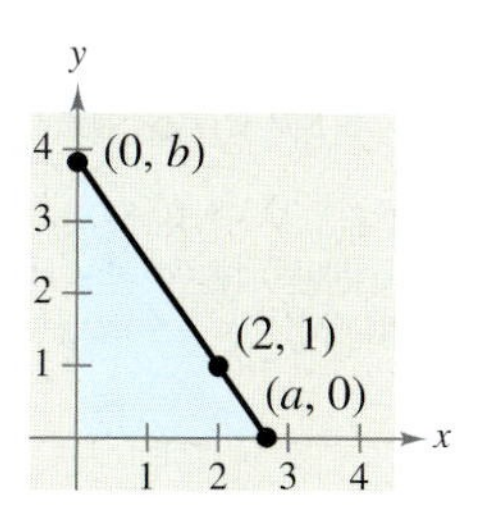

FIGURE FOR 91

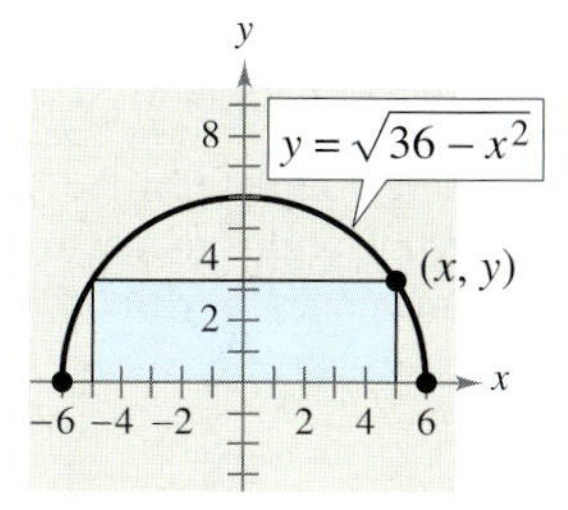

FIGURE FOR 92

92. ***Geometry*** A rectangle is bounded by the x-axis and the semicircle $y = \sqrt{36 - x^2}$ (see figure). Write the area A of the rectangle as a function of x, and determine the domain of the function.

93. ***Path of a Ball*** The height y (in feet) of a baseball thrown by a child is

$$y = -\frac{1}{10}x^2 + 3x + 6$$

where x is the horizontal distance (in feet) from where the ball was thrown. Will the ball fly over the head of another child 30 feet away trying to catch the ball? (Assume that the child who is trying to catch the ball holds a baseball glove at a height of 5 feet.)

94. ***Prescription Drugs*** The amounts d (in billions of dollars) spent on prescription drugs in the United States from 1991 to 2002 (see figure) can be approximated by the model

$$d(t) = \begin{cases} 5.0t + 37, & 1 \le t \le 7 \\ 18.7t - 64, & 8 \le t \le 12 \end{cases}$$

where t represents the year, with $t = 1$ corresponding to 1991. Use this model to find the amount spent on prescription drugs in each year from 1991 to 2002. (Source: U.S. Centers for Medicare & Medicaid Services)

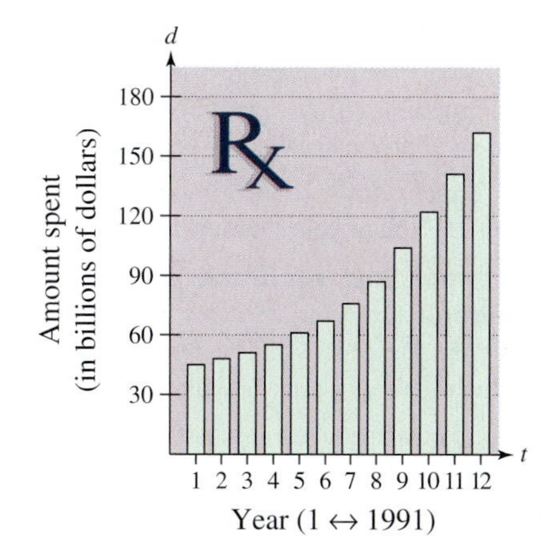

95. ***Average Price*** The average prices p (in thousands of dollars) of a new mobile home in the United States from 1990 to 2002 (see figure) can be approximated by the model

$$p(t) = \begin{cases} 0.182t^2 + 0.57t + 27.3, & 0 \le t \le 7 \\ 2.50t + 21.3, & 8 \le t \le 12 \end{cases}$$

where t represents the year, with $t = 0$ corresponding to 1990. Use this model to find the average price of a mobile home in each year from 1990 to 2002. (Source: U.S. Census Bureau)

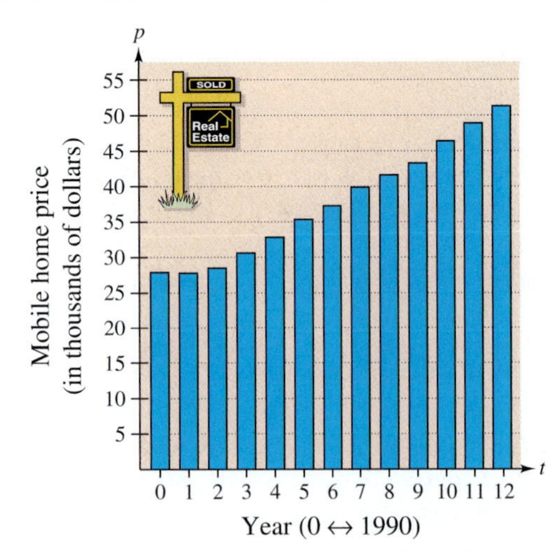

96. ***Postal Regulations*** A rectangular package to be sent by the U.S. Postal Service can have a maximum combined length and girth (perimeter of a cross section) of 108 inches (see figure).

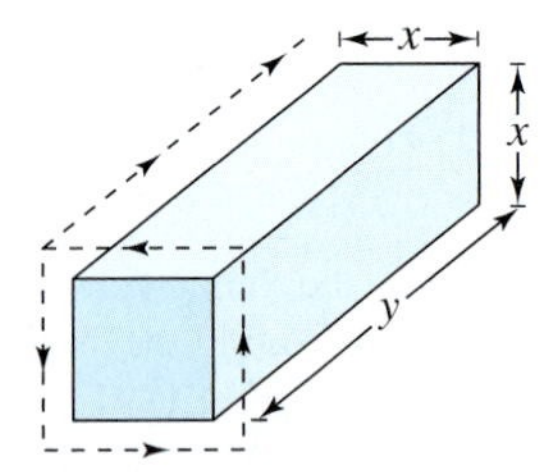

(a) Write the volume V of the package as a function of x. What is the domain of the function?

(b) Use a graphing utility to graph your function. Be sure to use an appropriate window setting.

(c) What dimensions will maximize the volume of the package? Explain your answer.

97. ***Cost, Revenue, and Profit*** A company produces a product for which the variable cost is \$12.30 per unit and the fixed costs are \$98,000. The product sells for \$17.98. Let x be the number of units produced and sold.

(a) The total cost for a business is the sum of the variable cost and the fixed costs. Write the total cost C as a function of the number of units produced.

(b) Write the revenue R as a function of the number of units sold.

(c) Write the profit P as a function of the number of units sold. (*Note:* $P = R - C$)

98. ***Average Cost*** The inventor of a new game believes that the variable cost for producing the game is \$0.95 per unit and the fixed costs are \$6000. The inventor sells each game for \$1.69. Let x be the number of games sold.

(a) The total cost for a business is the sum of the variable cost and the fixed costs. Write the total cost C as a function of the number of games sold.

(b) Write the average cost per unit $\overline{C} = C/x$ as a function of x.

99. ***Transportation*** For groups of 80 or more people, a charter bus company determines the rate per person according to the formula

$$\text{Rate} = 8 - 0.05(n - 80), \quad n \ge 80$$

where the rate is given in dollars and n is the number of people.

(a) Write the revenue R for the bus company as a function of n.

(b) Use the function in part (a) to complete the table. What can you conclude?

n	90	100	110	120	130	140	150
$R(n)$							

100. ***Physics*** The force F (in tons) of water against the face of a dam is estimated by the function $F(y) = 149.76\sqrt{10}y^{5/2}$, where y is the depth of the water (in feet).

(a) Complete the table. What can you conclude from the table?

y	5	10	20	30	40
$F(y)$					

(b) Use the table to approximate the depth at which the force against the dam is 1,000,000 tons.

(c) Find the depth at which the force against the dam is 1,000,000 tons algebraically.

101. ***Height of a Balloon*** A balloon carrying a transmitter ascends vertically from a point 3000 feet from the receiving station.

(a) Draw a diagram that gives a visual representation of the problem. Let h represent the height of the balloon and let d represent the distance between the balloon and the receiving station.

(b) Write the height of the balloon as a function of d. What is the domain of the function?

Model It

102. ***Wildlife*** The graph shows the numbers of threatened and endangered fish species in the world from 1996 through 2003. Let $f(t)$ represent the number of threatened and endangered fish species in the year t. (Source: U.S. Fish and Wildlife Service)

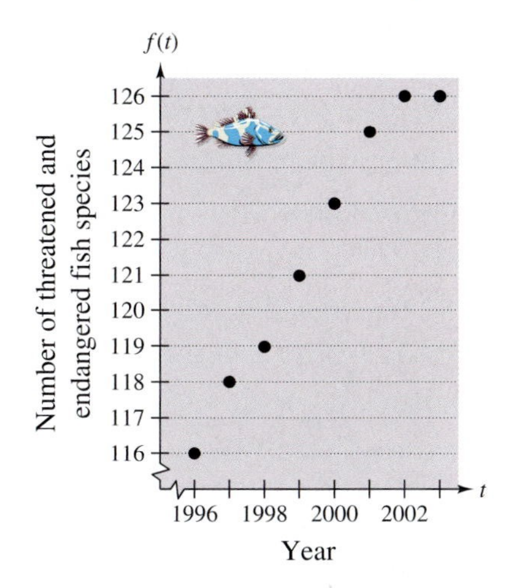

(a) Find $\dfrac{f(2003) - f(1996)}{2003 - 1996}$ and interpret the result in the context of the problem.

(b) Find a linear model for the data algebraically. Let N represent the number of threatened and endangered fish species and let $x = 6$ correspond to 1996.

(c) Use the model found in part (b) to complete the table.

x	6	7	8	9	10	11	12	13
N								

(d) Compare your results from part (c) with the actual data.

(e) Use a graphing utility to find a linear model for the data. Let $x = 6$ correspond to 1996. How does the model you found in part (b) compare with the model given by the graphing utility?

Synthesis

True or False? **In Exercises 103 and 104, determine whether the statement is true or false. Justify your answer.**

103. The domain of the function given by $f(x) = x^4 - 1$ is $(-\infty, \infty)$, and the range of $f(x)$ is $(0, \infty)$.

104. The set of ordered pairs $\{(-8, -2), (-6, 0), (-4, 0), (-2, 2), (0, 4), (2, -2)\}$ represents a function.

105. ***Writing*** In your own words, explain the meanings of *domain* and *range*.

106. ***Think About It*** Consider $f(x) = \sqrt{x - 2}$ and $g(x) = \sqrt[3]{x - 2}$. Why are the domains of f and g different?

In Exercises 107 and 108, determine whether the statements use the word *function* in ways that are mathematically correct. Explain your reasoning.

107. (a) The sales tax on a purchased item is a function of the selling price.

(b) Your score on the next algebra exam is a function of the number of hours you study the night before the exam.

108. (a) The amount in your savings account is a function of your salary.

(b) The speed at which a free-falling baseball strikes the ground is a function of the height from which it was dropped.

Skills Review

In Exercises 109–112, solve the equation.

109. $\dfrac{t}{3} + \dfrac{t}{5} = 1$

110. $\dfrac{3}{t} + \dfrac{5}{t} = 1$

111. $\dfrac{3}{x(x + 1)} - \dfrac{4}{x} = \dfrac{1}{x + 1}$

112. $\dfrac{12}{x} - 3 = \dfrac{4}{x} + 9$

In Exercises 113–116, find the equation of the line passing through the pair of points.

113. $(-2, -5), (4, -1)$

114. $(10, 0), (1, 9)$

115. $(-6, 5), (3, -5)$

116. $\left(-\frac{1}{2}, 3\right), \left(\frac{11}{2}, -\frac{1}{3}\right)$

1.5 Analyzing Graphs of Functions

What you should learn

- Use the Vertical Line Test for functions.
- Find the zeros of functions.
- Determine intervals on which functions are increasing or decreasing and determine relative maximum and relative minimum values of functions.
- Determine the average rate of change of a function.
- Identify even and odd functions.

Why you should learn it

Graphs of functions can help you visualize relationships between variables in real life. For instance, in Exercise 86 on page 64, you will use the graph of a function to represent visually the temperature for a city over a 24-hour period.

The Graph of a Function

In Section 1.4, you studied functions from an algebraic point of view. In this section, you will study functions from a graphical perspective.

The **graph of a function** f is the collection of ordered pairs $(x, f(x))$ such that x is in the domain of f. As you study this section, remember that

$x =$ the directed distance from the y-axis

$y = f(x) =$ the directed distance from the x-axis

as shown in Figure 1.52.

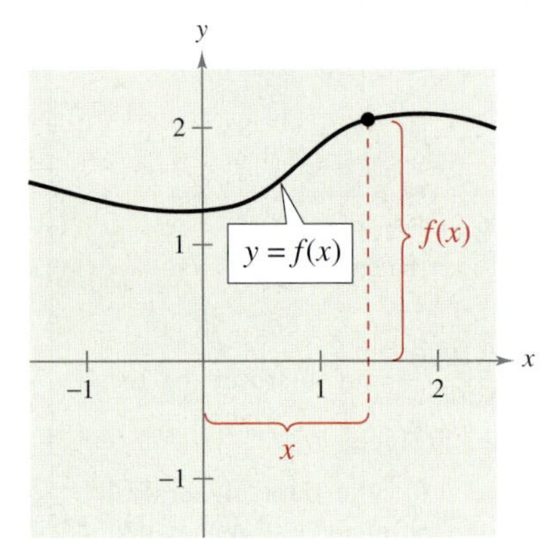

FIGURE 1.52

Example 1 **Finding the Domain and Range of a Function**

Use the graph of the function f, shown in Figure 1.53, to find (a) the domain of f, (b) the function values $f(-1)$ and $f(2)$, and (c) the range of f.

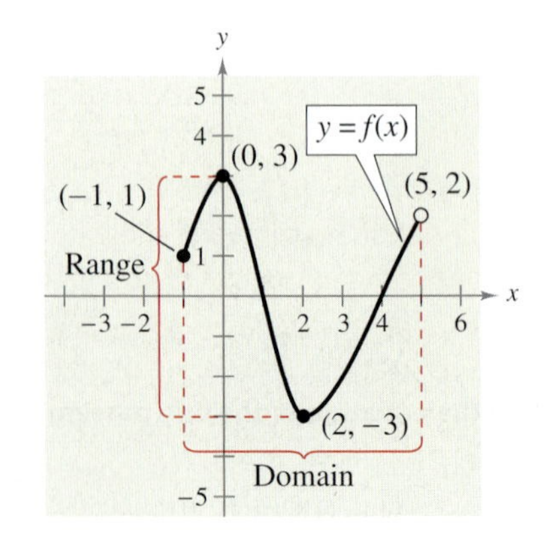

FIGURE 1.53

Solution

a. The closed dot at $(-1, 1)$ indicates that $x = -1$ is in the domain of f, whereas the open dot at $(5, 2)$ indicates that $x = 5$ is not in the domain. So, the domain of f is all x in the interval $[-1, 5)$.

b. Because $(-1, 1)$ is a point on the graph of f, it follows that $f(-1) = 1$. Similarly, because $(2, -3)$ is a point on the graph of f, it follows that $f(2) = -3$.

c. Because the graph does not extend below $f(2) = -3$ or above $f(0) = 3$, the range of f is the interval $[-3, 3]$.

CHECKPOINT Now try Exercise 1.

The use of dots (open or closed) at the extreme left and right points of a graph indicates that the graph does not extend beyond these points. If no such dots are shown, assume that the graph extends beyond these points.

By the definition of a function, at most one y-value corresponds to a given x-value. This means that the graph of a function cannot have two or more different points with the same x-coordinate, and no two points on the graph of a function can be vertically above or below each other. It follows, then, that a vertical line can intersect the graph of a function at most once. This observation provides a convenient visual test called the **Vertical Line Test** for functions.

Vertical Line Test for Functions

A set of points in a coordinate plane is the graph of y as a function of x if and only if no *vertical* line intersects the graph at more than one point.

Example 2 Vertical Line Test for Functions

Use the Vertical Line Test to decide whether the graphs in Figure 1.54 represent y as a function of x.

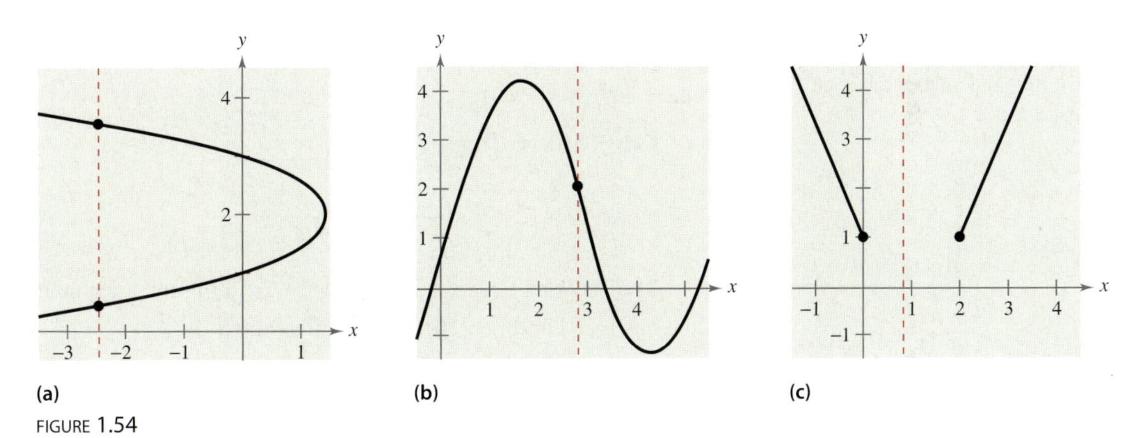

FIGURE 1.54

Solution

a. This *is not* a graph of y as a function of x, because you can find a vertical line that intersects the graph twice. That is, for a particular input x, there is more than one output y.

b. This *is* a graph of y as a function of x, because every vertical line intersects the graph at most once. That is, for a particular input x, there is at most one output y.

c. This *is* a graph of y as a function of x. (Note that if a vertical line does not intersect the graph, it simply means that the function is undefined for that particular value of x.) That is, for a particular input x, there is at most one output y.

✓CHECKPOINT Now try Exercise 9.

Zeros of a Function

If the graph of a function of x has an x-intercept at $(a, 0)$, then a is a **zero** of the function.

> **Zeros of a Function**
>
> The **zeros of a function** f of x are the x-values for which $f(x) = 0$.

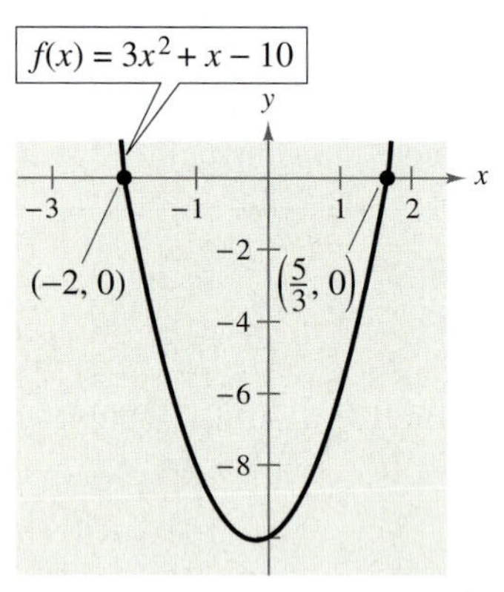

Zeros of f: $x = -2, x = \frac{5}{3}$

FIGURE 1.55

Example 3 Finding the Zeros of a Function

Find the zeros of each function.

a. $f(x) = 3x^2 + x - 10$ **b.** $g(x) = \sqrt{10 - x^2}$ **c.** $h(t) = \dfrac{2t - 3}{t + 5}$

Solution

To find the zeros of a function, set the function equal to zero and solve for the independent variable.

a.

$$3x^2 + x - 10 = 0 \qquad \text{Set } f(x) \text{ equal to 0.}$$

$$(3x - 5)(x + 2) = 0 \qquad \text{Factor.}$$

$$3x - 5 = 0 \;\Rightarrow\; x = \frac{5}{3} \qquad \text{Set 1st factor equal to 0.}$$

$$x + 2 = 0 \;\Rightarrow\; x = -2 \qquad \text{Set 2nd factor equal to 0.}$$

The zeros of f are $x = \frac{5}{3}$ and $x = -2$. In Figure 1.55, note that the graph of f has $\left(\frac{5}{3}, 0\right)$ and $(-2, 0)$ as its x-intercepts.

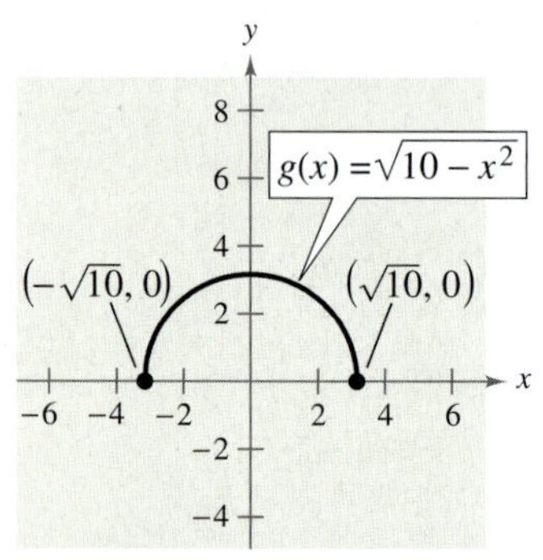

Zeros of g: $x = \pm\sqrt{10}$

FIGURE 1.56

b.

$$\sqrt{10 - x^2} = 0 \qquad \text{Set } g(x) \text{ equal to 0.}$$

$$10 - x^2 = 0 \qquad \text{Square each side.}$$

$$10 = x^2 \qquad \text{Add } x^2 \text{ to each side.}$$

$$\pm\sqrt{10} = x \qquad \text{Extract square roots.}$$

The zeros of g are $x = -\sqrt{10}$ and $x = \sqrt{10}$. In Figure 1.56, note that the graph of g has $\left(-\sqrt{10}, 0\right)$ and $\left(\sqrt{10}, 0\right)$ as its x-intercepts.

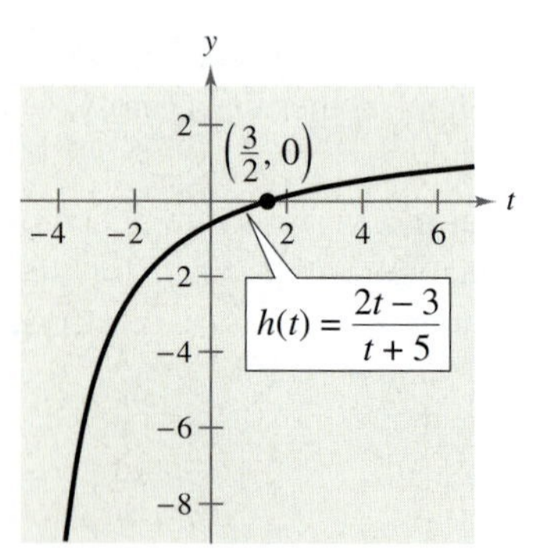

Zero of h: $t = \frac{3}{2}$

FIGURE 1.57

c.

$$\frac{2t - 3}{t + 5} = 0 \qquad \text{Set } h(t) \text{ equal to 0.}$$

$$2t - 3 = 0 \qquad \text{Multiply each side by } t + 5.$$

$$2t = 3 \qquad \text{Add 3 to each side.}$$

$$t = \frac{3}{2} \qquad \text{Divide each side by 2.}$$

The zero of h is $t = \frac{3}{2}$. In Figure 1.57, note that the graph of h has $\left(\frac{3}{2}, 0\right)$ as its t-intercept.

CHECKPOINT Now try Exercise 15.

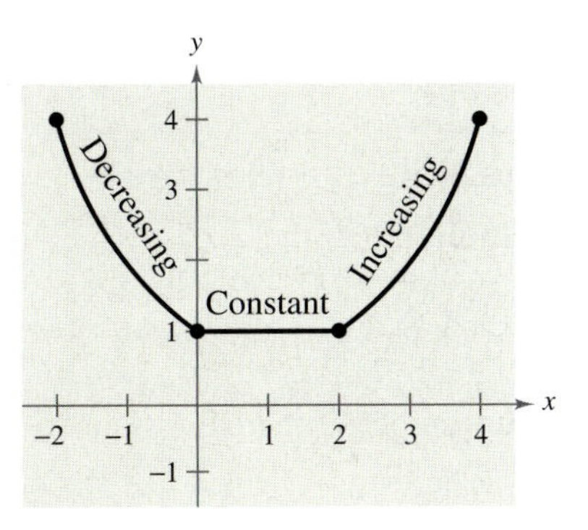

FIGURE 1.58

Increasing and Decreasing Functions

The more you know about the graph of a function, the more you know about the function itself. Consider the graph shown in Figure 1.58. As you move from *left to right*, this graph falls from $x = -2$ to $x = 0$, is constant from $x = 0$ to $x = 2$, and rises from $x = 2$ to $x = 4$.

Increasing, Decreasing, and Constant Functions

A function f is **increasing** on an interval if, for any x_1 and x_2 in the interval, $x_1 < x_2$ implies $f(x_1) < f(x_2)$.

A function f is **decreasing** on an interval if, for any x_1 and x_2 in the interval, $x_1 < x_2$ implies $f(x_1) > f(x_2)$.

A function f is **constant** on an interval if, for any x_1 and x_2 in the interval, $f(x_1) = f(x_2)$.

Example 4 Increasing and Decreasing Functions

Use the graphs in Figure 1.59 to describe the increasing or decreasing behavior of each function.

Solution

a. This function is increasing over the entire real line.

b. This function is increasing on the interval $(-\infty, -1)$, decreasing on the interval $(-1, 1)$, and increasing on the interval $(1, \infty)$.

c. This function is increasing on the interval $(-\infty, 0)$, constant on the interval $(0, 2)$, and decreasing on the interval $(2, \infty)$.

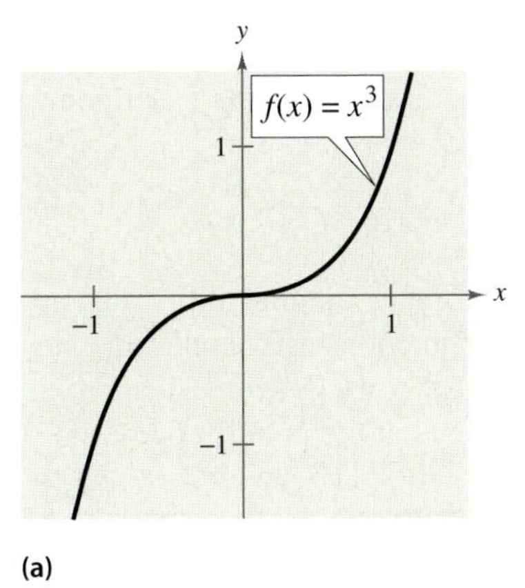

(a)

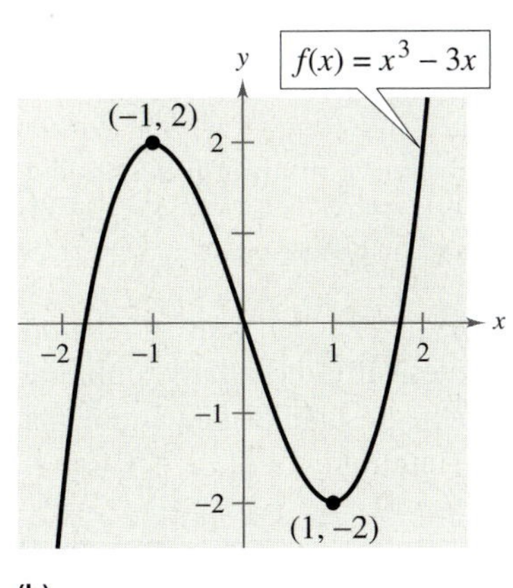

(b)

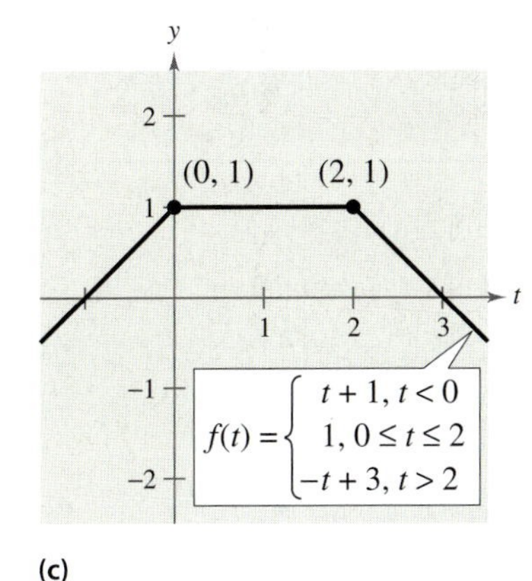

(c)

FIGURE 1.59

Now try Exercise 33.

To help you decide whether a function is increasing, decreasing, or constant on an interval, you can evaluate the function for several values of x. However, calculus is needed to determine, for certain, all intervals on which a function is increasing, decreasing, or constant.

STUDY TIP

A relative minimum or relative maximum is also referred to as a *local* minimum or *local* maximum.

The points at which a function changes its increasing, decreasing, or constant behavior are helpful in determining the **relative minimum** or **relative maximum** values of the function.

Definitions of Relative Minimum and Relative Maximum

A function value $f(a)$ is called a **relative minimum** of f if there exists an interval (x_1, x_2) that contains a such that

$$x_1 < x < x_2 \quad \text{implies} \quad f(a) \le f(x).$$

A function value $f(a)$ is called a **relative maximum** of f if there exists an interval (x_1, x_2) that contains a such that

$$x_1 < x < x_2 \quad \text{implies} \quad f(a) \ge f(x).$$

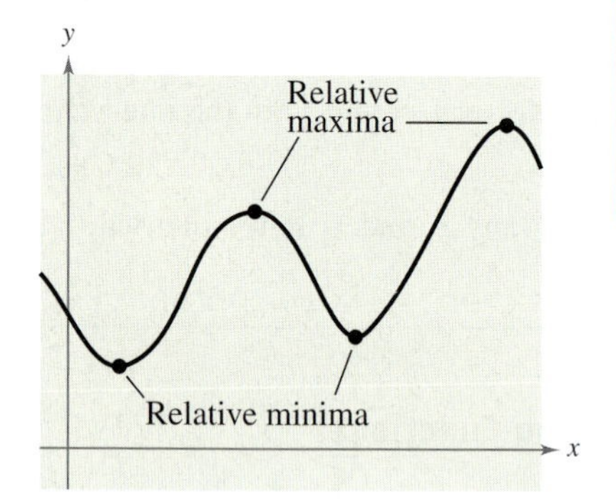

FIGURE 1.60

Figure 1.60 shows several different examples of relative minima and relative maxima. In Section 2.1, you will study a technique for finding the *exact point* at which a second-degree polynomial function has a relative minimum or relative maximum. For the time being, however, you can use a graphing utility to find reasonable approximations of these points.

Example 5 Approximating a Relative Minimum

Use a graphing utility to approximate the relative minimum of the function given by $f(x) = 3x^2 - 4x - 2$.

Solution

The graph of f is shown in Figure 1.61. By using the *zoom* and *trace* features or the *minimum* feature of a graphing utility, you can estimate that the function has a relative minimum at the point

$(0.67, -3.33)$. Relative minimum

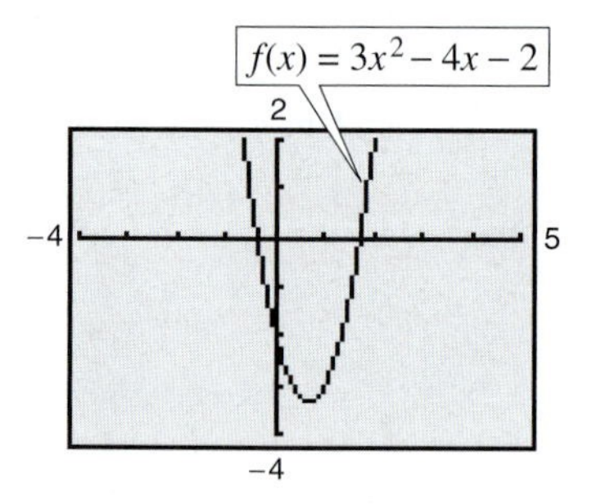

FIGURE 1.61

Later, in Section 2.1, you will be able to determine that the exact point at which the relative minimum occurs is $\left(\frac{2}{3}, -\frac{10}{3}\right)$.

CHECKPOINT Now try Exercise 49.

You can also use the *table* feature of a graphing utility to approximate numerically the relative minimum of the function in Example 5. Using a table that begins at 0.6 and increments the value of x by 0.01, you can approximate that the minimum of $f(x) = 3x^2 - 4x - 2$ occurs at the point $(0.67, -3.33)$.

Technology

If you use a graphing utility to estimate the x- and y-values of a relative minimum or relative maximum, the *zoom* feature will often produce graphs that are nearly flat. To overcome this problem, you can manually change the vertical setting of the viewing window. The graph will stretch vertically if the values of Ymin and Ymax are closer together.

Average Rate of Change

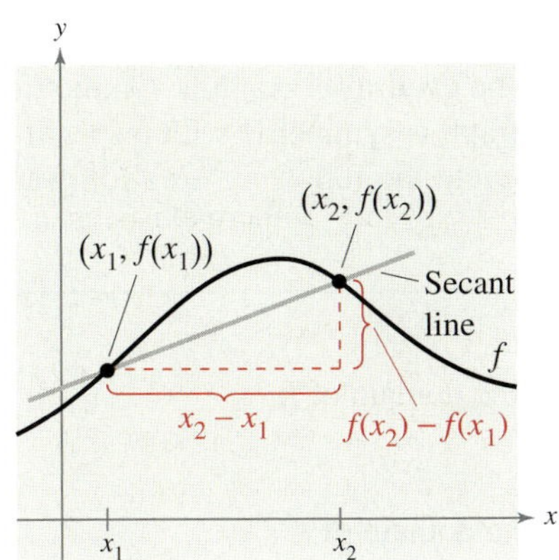

FIGURE 1.62

In Section 1.3, you learned that the slope of a line can be interpreted as a *rate of change*. For a nonlinear graph whose slope changes at each point, the **average rate of change** between any two points $(x_1, f(x_1))$ and $(x_2, f(x_2))$ is the slope of the line through the two points (see Figure 1.62). The line through the two points is called the **secant line,** and the slope of this line is denoted as m_{sec}.

$$\text{Average rate of change of } f \text{ from } x_1 \text{ to } x_2 = \frac{f(x_2) - f(x_1)}{x_2 - x_1}$$
$$= \frac{\text{change in } y}{\text{change in } x}$$
$$= m_{sec}$$

Example 6 Average Rate of Change of a Function

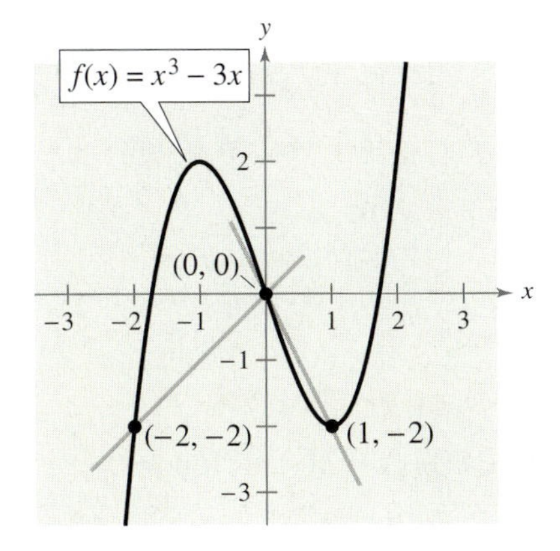

FIGURE 1.63

Find the average rates of change of $f(x) = x^3 - 3x$ (a) from $x_1 = -2$ to $x_2 = 0$ and (b) from $x_1 = 0$ to $x_2 = 1$ (see Figure 1.63).

Solution

a. The average rate of change of f from $x_1 = -2$ to $x_2 = 0$ is

$$\frac{f(x_2) - f(x_1)}{x_2 - x_1} = \frac{f(0) - f(-2)}{0 - (-2)} = \frac{0 - (-2)}{2} = 1.$$

Secant line has positive slope.

b. The average rate of change of f from $x_1 = 0$ to $x_2 = 1$ is

$$\frac{f(x_2) - f(x_1)}{x_2 - x_1} = \frac{f(1) - f(0)}{1 - 0} = \frac{-2 - 0}{1} = -2.$$

Secant line has negative slope.

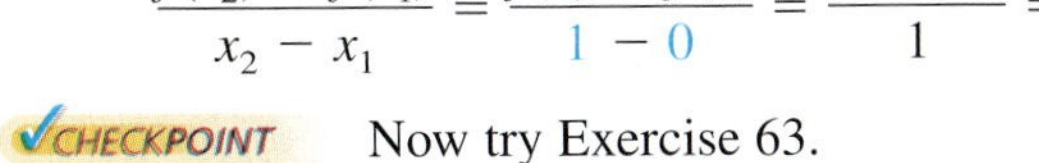

CHECKPOINT Now try Exercise 63.

Example 7 Finding Average Speed

The distance s (in feet) a moving car is from a stoplight is given by the function $s(t) = 20t^{3/2}$, where t is the time (in seconds). Find the average speed of the car (a) from $t_1 = 0$ to $t_2 = 4$ seconds and (b) from $t_1 = 4$ to $t_2 = 9$ seconds.

Solution

a. The average speed of the car from $t_1 = 0$ to $t_2 = 4$ seconds is

$$\frac{s(t_2) - s(t_1)}{t_2 - t_1} = \frac{s(4) - s(0)}{4 - (0)} = \frac{160 - 0}{4} = 40 \text{ feet per second.}$$

b. The average speed of the car from $t_1 = 4$ to $t_2 = 9$ seconds is

$$\frac{s(t_2) - s(t_1)}{t_2 - t_1} = \frac{s(9) - s(4)}{9 - 4} = \frac{540 - 160}{5} = 76 \text{ feet per second.}$$

CHECKPOINT Now try Exercise 89.

Exploration

Use the information in Example 7 to find the average speed of the car from $t_1 = 0$ to $t_2 = 9$ seconds. Explain why the result is less than the value obtained in part (b).

Even and Odd Functions

In Section 1.2, you studied different types of symmetry of a graph. In the terminology of functions, a function is said to be **even** if its graph is symmetric with respect to the y-axis and to be **odd** if its graph is symmetric with respect to the origin. The symmetry tests in Section 1.2 yield the following tests for even and odd functions.

Tests for Even and Odd Functions

A function $y = f(x)$ is **even** if, for each x in the domain of f,

$$f(-x) = f(x).$$

A function $y = f(x)$ is **odd** if, for each x in the domain of f,

$$f(-x) = -f(x).$$

Exploration

Graph each of the functions with a graphing utility. Determine whether the function is *even*, *odd*, or *neither*.

$f(x) = x^2 - x^4$

$g(x) = 2x^3 + 1$

$h(x) = x^5 - 2x^3 + x$

$j(x) = 2 - x^6 - x^8$

$k(x) = x^5 - 2x^4 + x - 2$

$p(x) = x^9 + 3x^5 - x^3 + x$

What do you notice about the equations of functions that are odd? What do you notice about the equations of functions that are even? Can you describe a way to identify a function as odd or even by inspecting the equation? Can you describe a way to identify a function as neither odd nor even by inspecting the equation?

Example 8 Even and Odd Functions

a. The function $g(x) = x^3 - x$ is odd because $g(-x) = -g(x)$, as follows.

$g(-x) = (-x)^3 - (-x)$ Substitute $-x$ for x.

$= -x^3 + x$ Simplify.

$= -(x^3 - x)$ Distributive Property

$= -g(x)$ Test for odd function

b. The function $h(x) = x^2 + 1$ is even because $h(-x) = h(x)$, as follows.

$h(-x) = (-x)^2 + 1$ Substitute $-x$ for x.

$= x^2 + 1$ Simplify.

$= h(x)$ Test for even function

The graphs and symmetry of these two functions are shown in Figure 1.64.

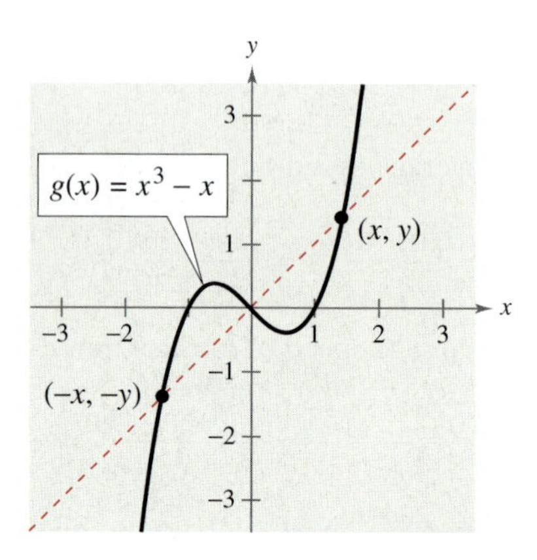

(a) Symmetric to origin: Odd Function

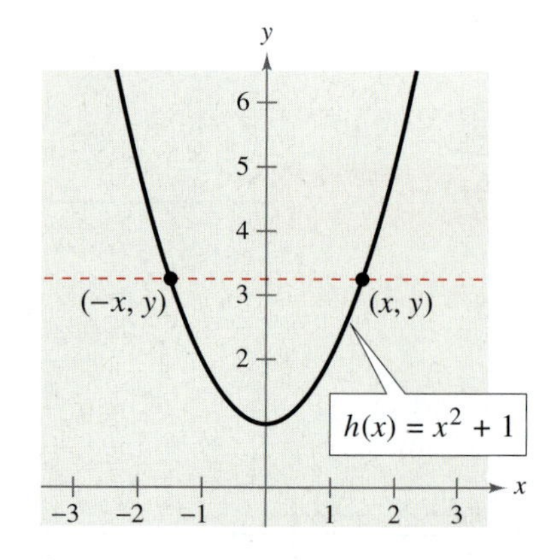

(b) Symmetric to y-axis: Even Function

FIGURE 1.64

CHECKPOINT Now try Exercise 71.

Additional Example

Is the function given by $f(x) = x^3 - 1$ even, odd, or neither?

Solution

Substituting $-x$ for x,

$f(-x) = (-x)^3 - 1 = -x^3 - 1.$

Because $f(x) = x^3 - 1$ and $f(-x) = -x^3 - 1$, $f(-x) \neq f(x)$ and $f(-x) \neq -f(x)$. The function is neither even nor odd.

1.5 Exercises

VOCABULARY CHECK: Fill in the blanks.

1. The graph of a function f is the collection of ________ ________ or $(x, f(x))$ such that x is in the domain of f.
2. The ________ ________ ________ is used to determine whether the graph of an equation is a function of y in terms of x.
3. The ________ of a function f are the values of x for which $f(x) = 0$.
4. A function f is ________ on an interval if, for any x_1 and x_2 in the interval, $x_1 < x_2$ implies $f(x_1) > f(x_2)$.
5. A function value $f(a)$ is a relative ________ of f if there exists an interval (x_1, x_2) containing a such that $x_1 < x < x_2$ implies $f(a) \geq f(x)$.
6. The ________ ________ ________ ________ between any two points $(x_1, f(x_1))$ and $(x_2, f(x_2))$ is the slope of the line through the two points, and this line is called the ________ line.
7. A function f is ________ if for the each x in the domain of f, $f(-x) = -f(x)$.
8. A function f is ________ if its graph is symmetric with respect to the y-axis.

PREREQUISITE SKILLS REVIEW: Practice and review algebra skills needed for this section at **www.Eduspace.com.**

In Exercises 1–4, use the graph of the function to find the domain and range of *f*.

1.

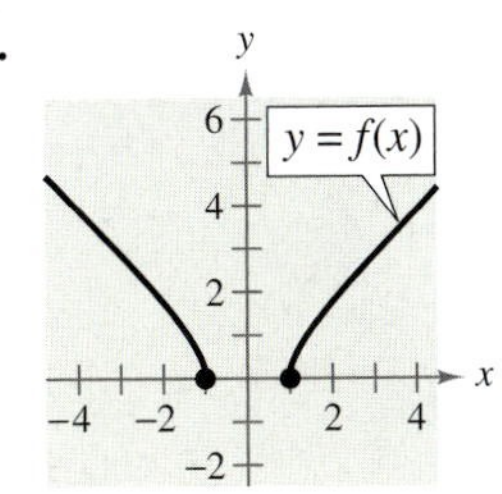

2.

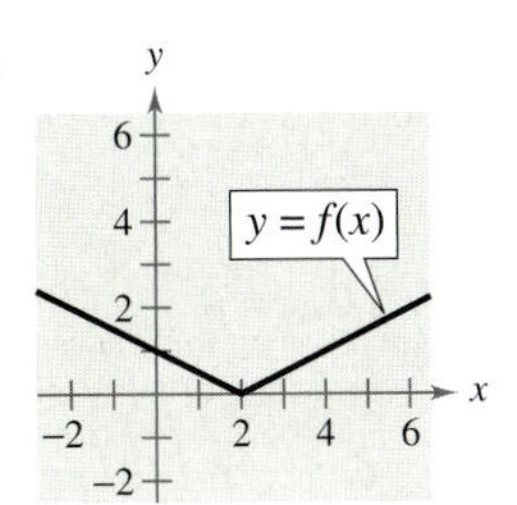

3.

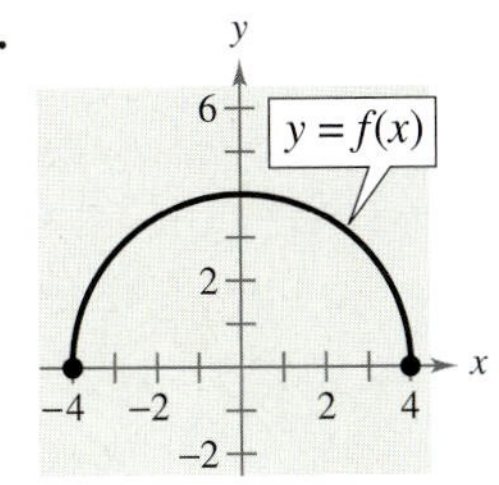

4.

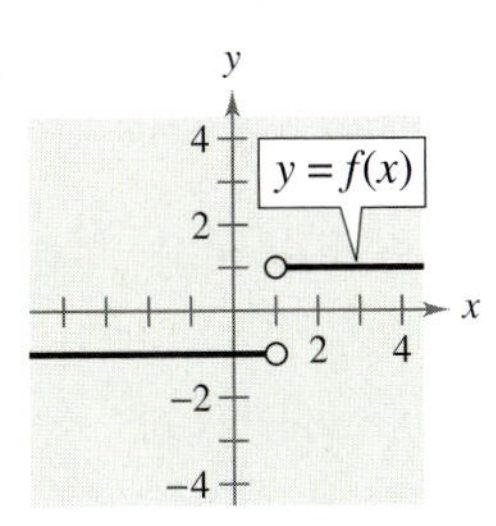

In Exercises 5–8, use the graph of the function to find the indicated function values.

5. (a) $f(-2)$ (b) $f(-1)$ (c) $f\left(\frac{1}{2}\right)$ (d) $f(1)$

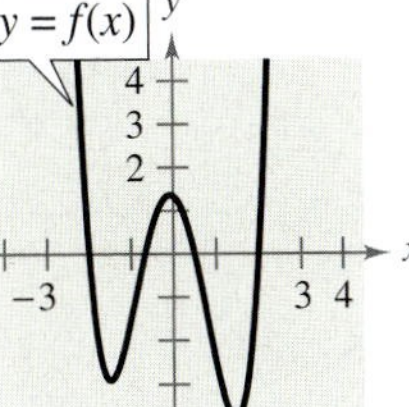
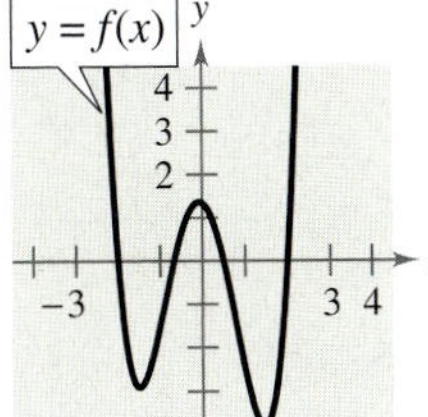

6. (a) $f(-1)$ (b) $f(2)$ (c) $f(0)$ (d) $f(1)$

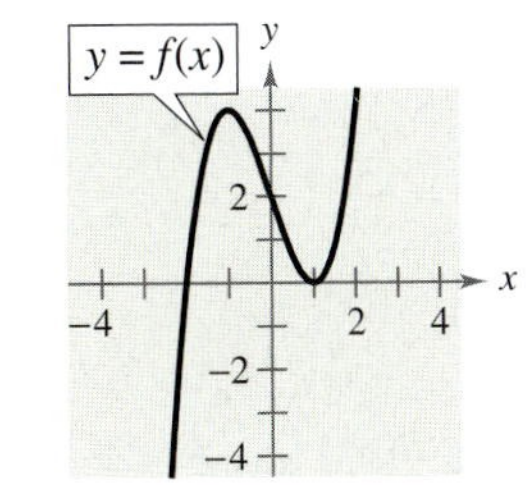

7. (a) $f(-2)$ (b) $f(1)$ (c) $f(0)$ (d) $f(2)$

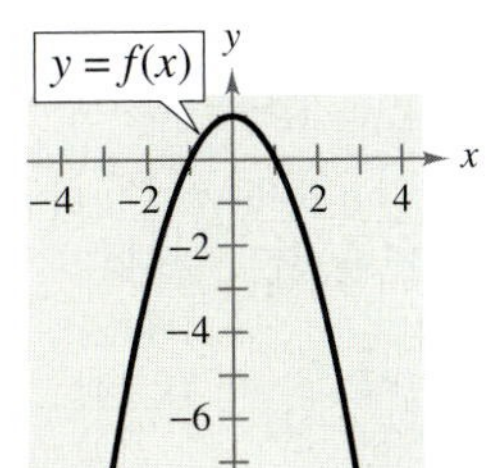

8. (a) $f(2)$ (b) $f(1)$ (c) $f(3)$ (d) $f(-1)$

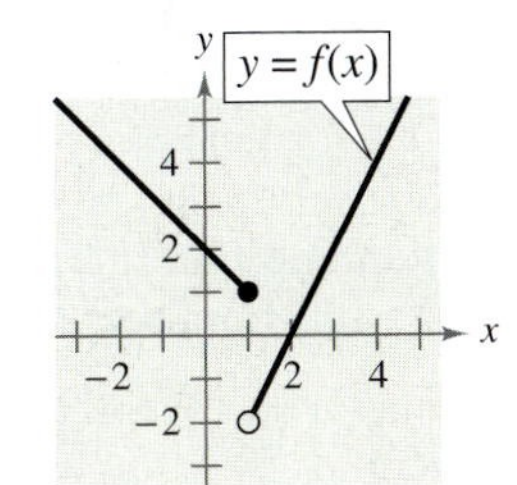

In Exercises 9–14, use the Vertical Line Test to determine whether *y* is a function of *x*. To print an enlarged copy of the graph, go to the website *www.mathgraphs.com*.

9. $y = \frac{1}{2}x^2$

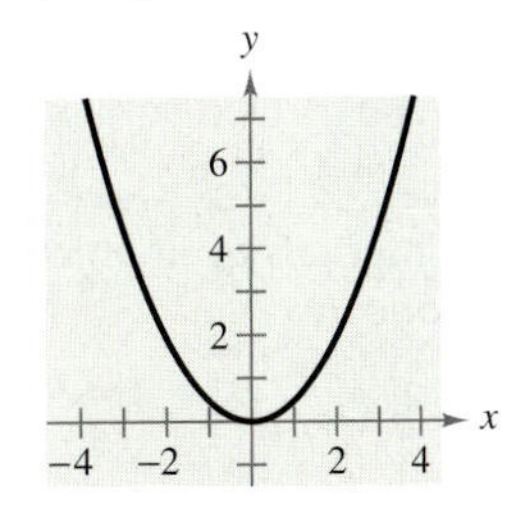

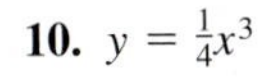
10. $y = \frac{1}{4}x^3$

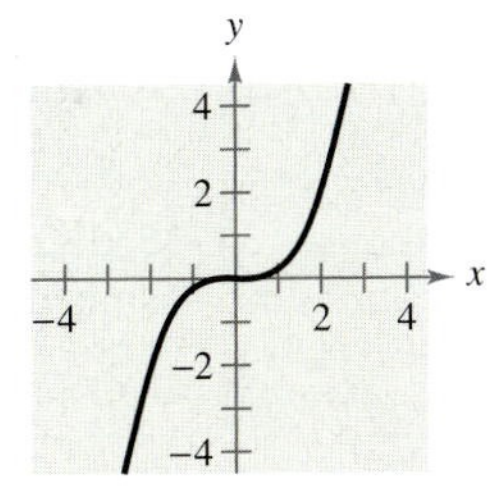

11. $x - y^2 = 1$

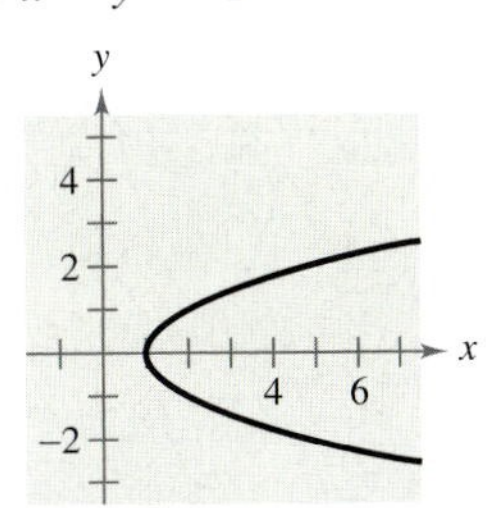

12. $x^2 + y^2 = 25$

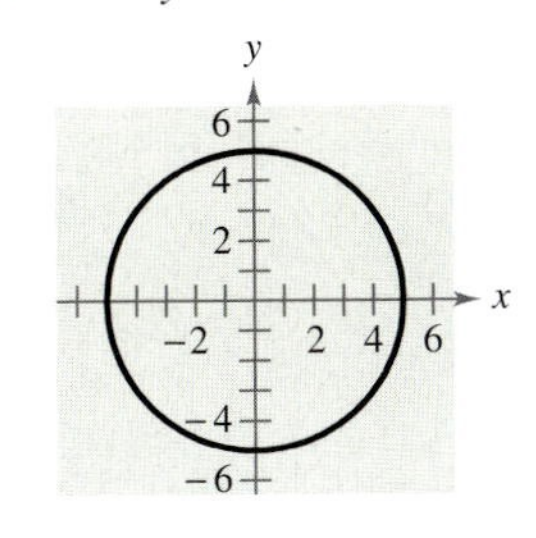

13. $x^2 = 2xy - 1$

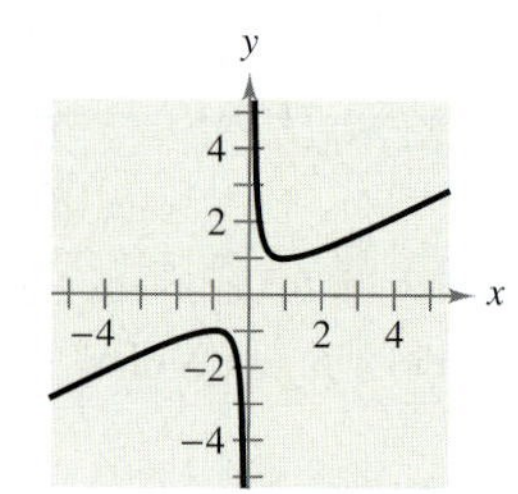

14. $x = |y + 2|$

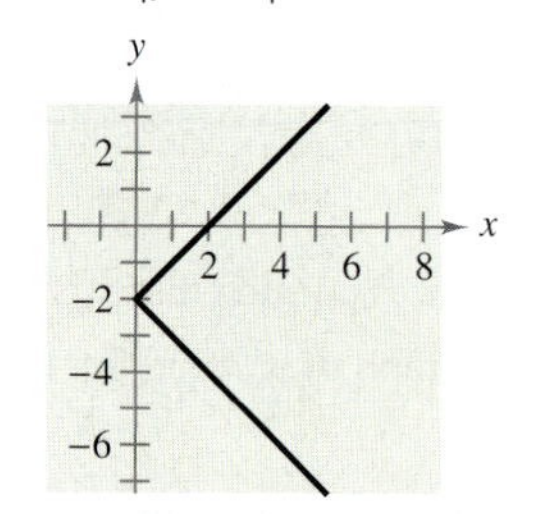

In Exercises 15–24, find the zeros of the function algebraically.

15. $f(x) = 2x^2 - 7x - 30$

16. $f(x) = 3x^2 + 22x - 16$

17. $f(x) = \dfrac{x}{9x^2 - 4}$

18. $f(x) = \dfrac{x^2 - 9x + 14}{4x}$

19. $f(x) = \frac{1}{2}x^3 - x$

20. $f(x) = x^3 - 4x^2 - 9x + 36$

21. $f(x) = 4x^3 - 24x^2 - x + 6$

22. $f(x) = 9x^4 - 25x^2$

23. $f(x) = \sqrt{2x} - 1$

24. $f(x) = \sqrt{3x + 2}$

In Exercises 25–30, (a) use a graphing utility to graph the function and find the zeros of the function and (b) verify your results from part (a) algebraically.

25. $f(x) = 3 + \dfrac{5}{x}$

26. $f(x) = x(x - 7)$

27. $f(x) = \sqrt{2x + 11}$

28. $f(x) = \sqrt{3x - 14} - 8$

29. $f(x) = \dfrac{3x - 1}{x - 6}$

30. $f(x) = \dfrac{2x^2 - 9}{3 - x}$

In Exercises 31–38, determine the intervals over which the function is increasing, decreasing, or constant.

31. $f(x) = \frac{3}{2}x$

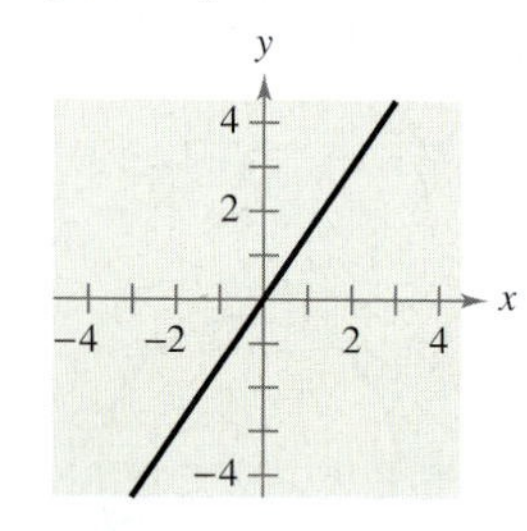

32. $f(x) = x^2 - 4x$

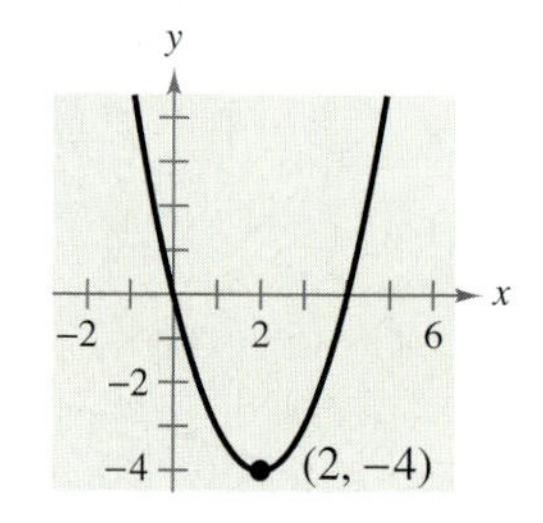

33. $f(x) = x^3 - 3x^2 + 2$

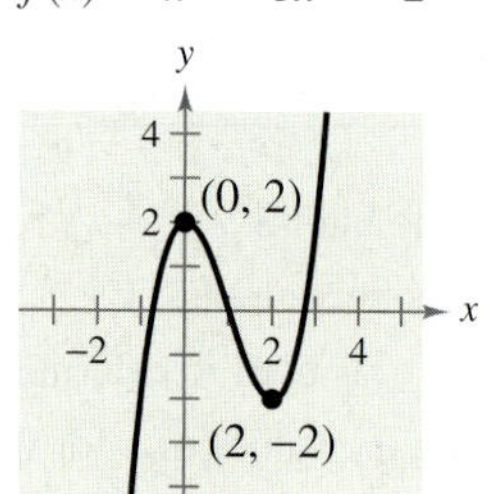

34. $f(x) = \sqrt{x^2 - 1}$

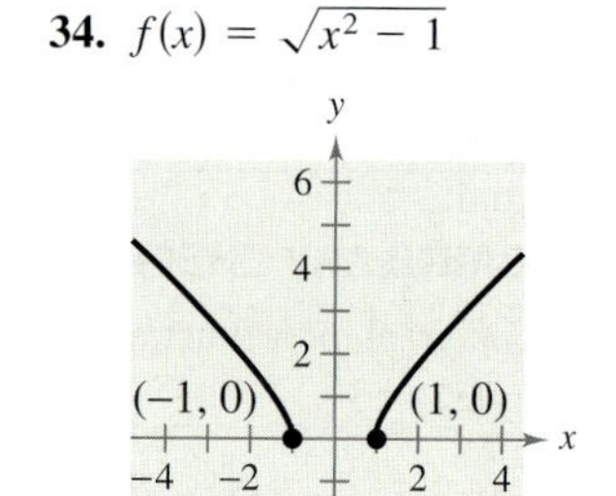

35. $f(x) = \begin{cases} x + 3, & x \le 0 \\ 3, & 0 < x \le 2 \\ 2x + 1, & x > 2 \end{cases}$

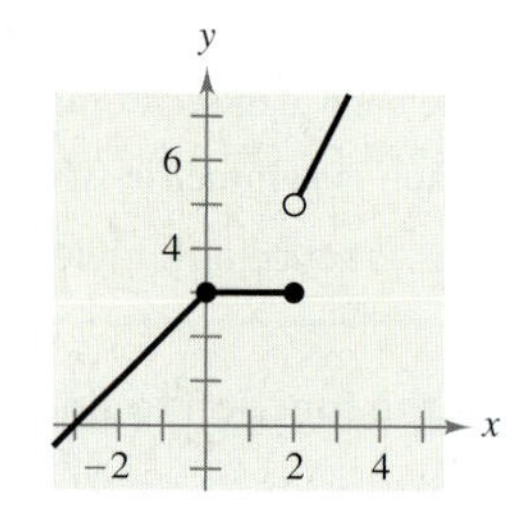

36. $f(x) = \begin{cases} 2x + 1, & x \le -1 \\ x^2 - 2, & x > -1 \end{cases}$

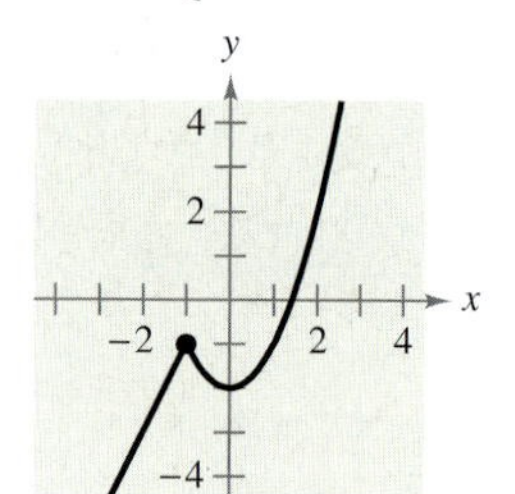

37. $f(x) = |x + 1| + |x - 1|$

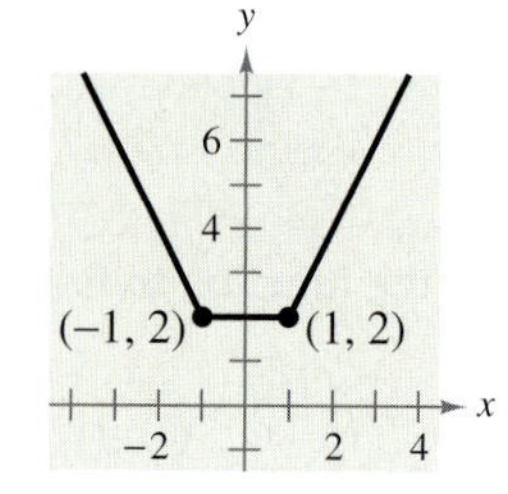

38. $f(x) = \dfrac{x^2 + x + 1}{x + 1}$

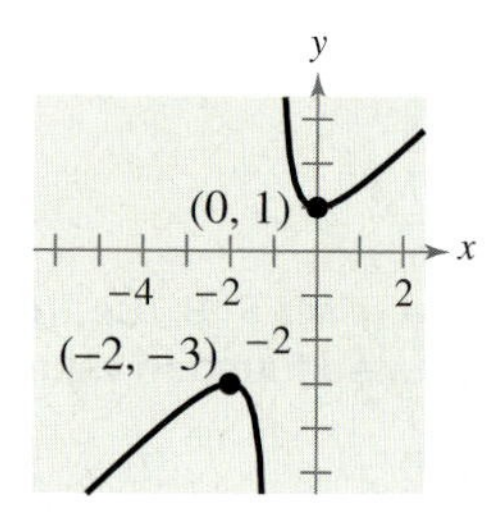

In Exercises 39–48, (a) use a graphing utility to graph the function and visually determine the intervals over which the function is increasing, decreasing, or constant, and (b) make a table of values to verify whether the function is increasing, decreasing, or constant over the intervals you identified in part (a).

39. $f(x) = 3$ **40.** $g(x) = x$

41. $g(s) = \dfrac{s^2}{4}$ **42.** $h(x) = x^2 - 4$

43. $f(t) = -t^4$ **44.** $f(x) = 3x^4 - 6x^2$

45. $f(x) = \sqrt{1-x}$ **46.** $f(x) = x\sqrt{x+3}$

47. $f(x) = x^{3/2}$ **48.** $f(x) = x^{2/3}$

In Exercises 49–54, use a graphing utility to graph the function and approximate (to two decimal places) any relative minimum or relative maximum values.

49. $f(x) = (x-4)(x+2)$ **50.** $f(x) = 3x^2 - 2x - 5$

51. $f(x) = -x^2 + 3x - 2$ **52.** $f(x) = -2x^2 + 9x$

53. $f(x) = x(x-2)(x+3)$

54. $f(x) = x^3 - 3x^2 - x + 1$

In Exercises 55–62, graph the function and determine the interval(s) for which $f(x) \geq 0$.

55. $f(x) = 4 - x$ **56.** $f(x) = 4x + 2$

57. $f(x) = x^2 + x$ **58.** $f(x) = x^2 - 4x$

59. $f(x) = \sqrt{x-1}$ **60.** $f(x) = \sqrt{x+2}$

61. $f(x) = -(1 + |x|)$ **62.** $f(x) = \frac{1}{2}(2 + |x|)$

In Exercises 63–70, find the average rate of change of the function from x_1 to x_2.

	Function	*x-Values*
63.	$f(x) = -2x + 15$	$x_1 = 0, x_2 = 3$
64.	$f(x) = 3x + 8$	$x_1 = 0, x_2 = 3$
65.	$f(x) = x^2 + 12x - 4$	$x_1 = 1, x_2 = 5$
66.	$f(x) = x^2 - 2x + 8$	$x_1 = 1, x_2 = 5$
67.	$f(x) = x^3 - 3x^2 - x$	$x_1 = 1, x_2 = 3$
68.	$f(x) = -x^3 + 6x^2 + x$	$x_1 = 1, x_2 = 6$
69.	$f(x) = -\sqrt{x-2} + 5$	$x_1 = 3, x_2 = 11$
70.	$f(x) = -\sqrt{x+1} + 3$	$x_1 = 3, x_2 = 8$

In Exercises 71–76, determine whether the function is even, odd, or neither. Then describe the symmetry.

71. $f(x) = x^6 - 2x^2 + 3$ **72.** $h(x) = x^3 - 5$

73. $g(x) = x^3 - 5x$ **74.** $f(x) = x\sqrt{1-x^2}$

75. $f(t) = t^2 + 2t - 3$ **76.** $g(s) = 4s^{2/3}$

In Exercises 77–80, write the height h of the rectangle as a function of x.

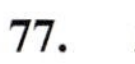

77.

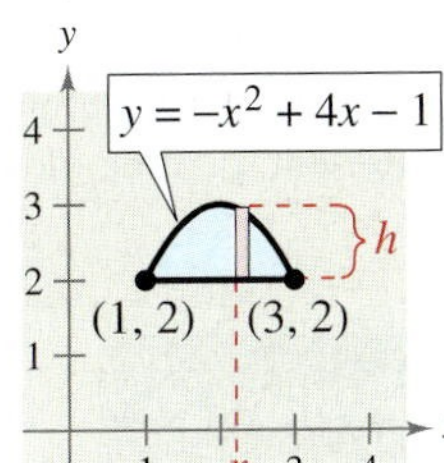

78.

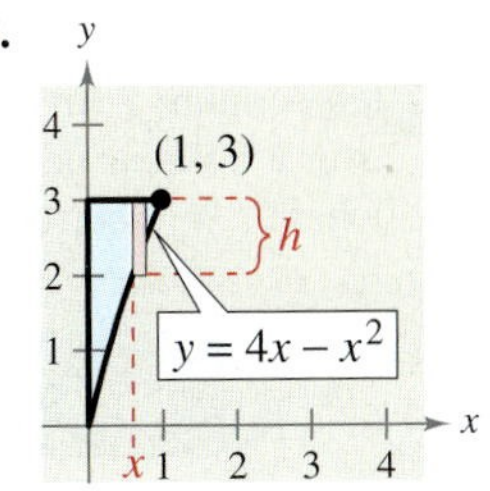

79.

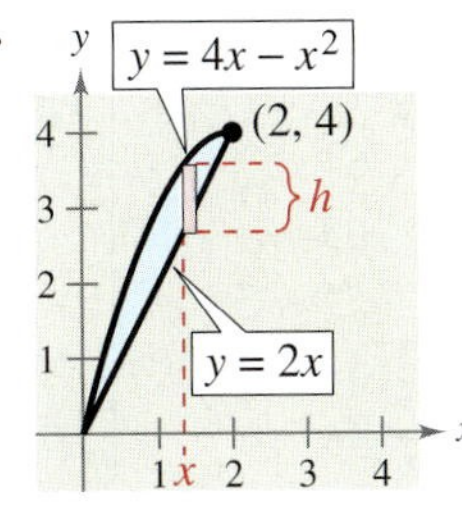

80.

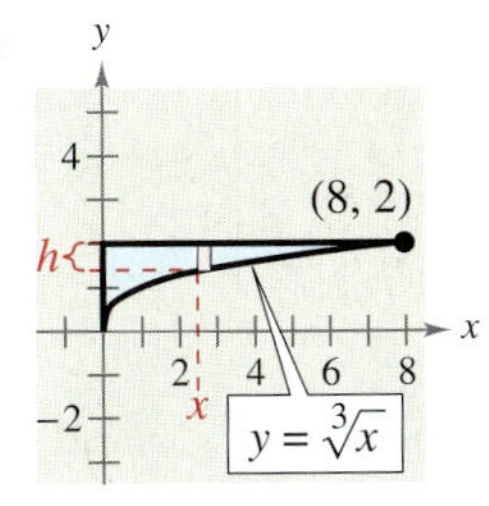

In Exercises 81–84, write the length L of the rectangle as a function of y.

81.

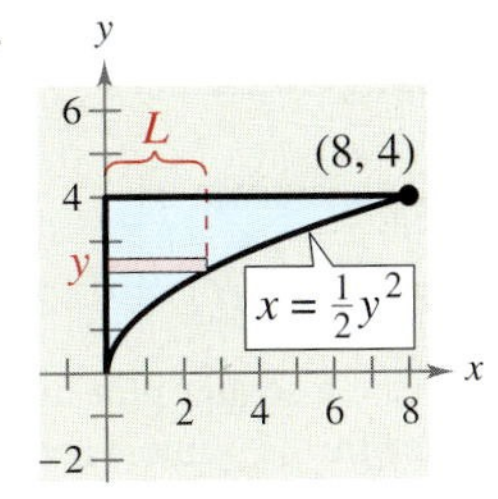

82.

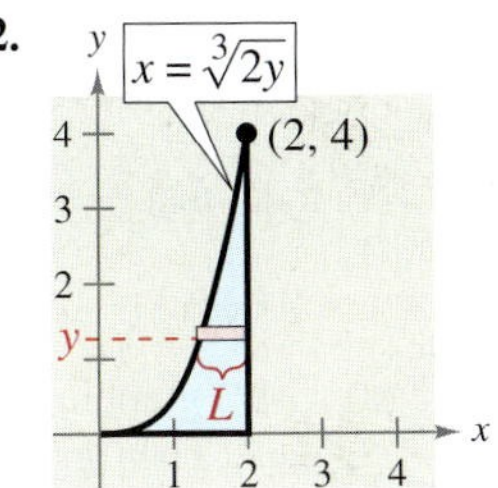

83.

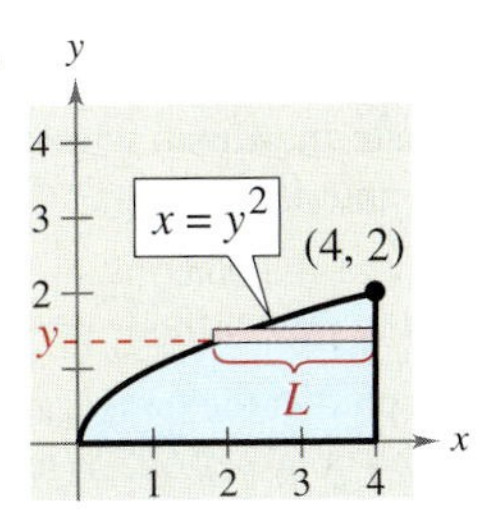

84.

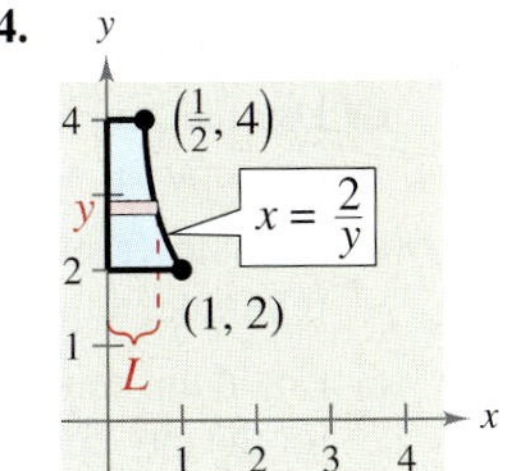

85. ***Electronics*** The number of lumens (time rate of flow of light) L from a fluorescent lamp can be approximated by the model

$$L = -0.294x^2 + 97.744x - 664.875, \quad 20 \leq x \leq 90$$

where x is the wattage of the lamp.

(a) Use a graphing utility to graph the function.

(b) Use the graph from part (a) to estimate the wattage necessary to obtain 2000 lumens.

Model It

86. ***Data Analysis: Temperature*** The table shows the temperature y (in degrees Fahrenheit) of a certain city over a 24-hour period. Let x represent the time of day, where $x = 0$ corresponds to 6 A.M.

Time, x	Temperature, y
0	34
2	50
4	60
6	64
8	63
10	59
12	53
14	46
16	40
18	36
20	34
22	37
24	45

A model that represents these data is given by

$$y = 0.026x^3 - 1.03x^2 + 10.2x + 34, \quad 0 \le x \le 24.$$

(a) Use a graphing utility to create a scatter plot of the data. Then graph the model in the same viewing window.

(b) How well does the model fit the data?

(c) Use the graph to approximate the times when the temperature was increasing and decreasing.

(d) Use the graph to approximate the maximum and minimum temperatures during this 24-hour period.

(e) Could this model be used to predict the temperature for the city during the next 24-hour period? Why or why not?

87. ***Coordinate Axis Scale*** Each function models the specified data for the years 1995 through 2005, with $t = 5$ corresponding to 1995. Estimate a reasonable scale for the vertical axis (e.g., hundreds, thousands, millions, etc.) of the graph and justify your answer. (There are many correct answers.)

(a) $f(t)$ represents the average salary of college professors.

(b) $f(t)$ represents the U.S. population.

(c) $f(t)$ represents the percent of the civilian work force that is unemployed.

88. ***Geometry*** Corners of equal size are cut from a square with sides of length 8 meters (see figure).

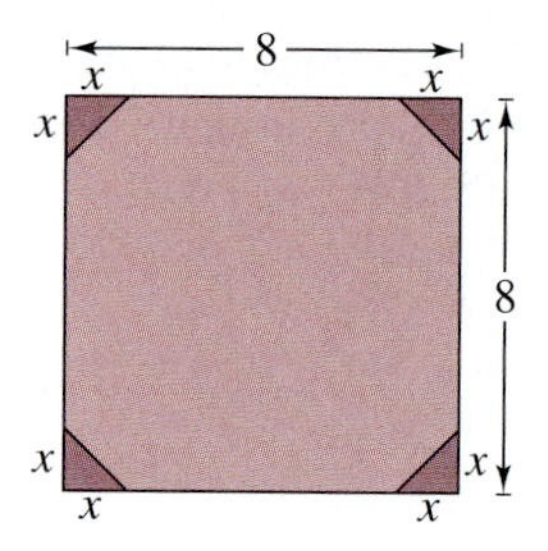

(a) Write the area A of the resulting figure as a function of x. Determine the domain of the function.

(b) Use a graphing utility to graph the area function over its domain. Use the graph to find the range of the function.

(c) Identify the figure that would result if x were chosen to be the maximum value in the domain of the function. What would be the length of each side of the figure?

89. ***Digital Music Sales*** The estimated revenues r (in billions of dollars) from sales of digital music from 2002 to 2007 can be approximated by the model

$$r = 15.639t^3 - 104.75t^2 + 303.5t - 301, \quad 2 \le t \le 7$$

where t represents the year, with $t = 2$ corresponding to 2002. (Source: *Fortune*)

(a) Use a graphing utility to graph the model.

(b) Find the average rate of change of the model from 2002 to 2007. Interpret your answer in the context of the problem.

90. ***Foreign College Students*** The numbers of foreign students F (in thousands) enrolled in colleges in the United States from 1992 to 2002 can be approximated by the model

$$F = 0.004t^4 + 0.46t^2 + 431.6, \quad 2 \le t \le 12$$

where t represents the year, with $t = 2$ corresponding to 1992. (Source: Institute of International Education)

(a) Use a graphing utility to graph the model.

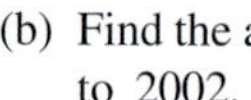

(b) Find the average rate of change of the model from 1992 to 2002. Interpret your answer in the context of the problem.

(c) Find the five-year time periods when the rate of change was the greatest and the least.

Physics **In Exercises 91–96, (a) use the position equation $s = -16t^2 + v_0t + s_0$ to write a function that represents the situation, (b) use a graphing utility to graph the function, (c) find the average rate of change of the function from t_1 to t_2, (d) interpret your answer to part (c) in the context of the problem, (e) find the equation of the secant line through t_1 and t_2, and (f) graph the secant line in the same viewing window as your position function.**

91. An object is thrown upward from a height of 6 feet at a velocity of 64 feet per second.

$t_1 = 0, t_2 = 3$

92. An object is thrown upward from a height of 6.5 feet at a velocity of 72 feet per second.

$t_1 = 0, t_2 = 4$

93. An object is thrown upward from ground level at a velocity of 120 feet per second.

$t_1 = 3, t_2 = 5$

94. An object is thrown upward from ground level at a velocity of 96 feet per second.

$t_1 = 2, t_2 = 5$

95. An object is dropped from a height of 120 feet.

$t_1 = 0, t_2 = 2$

96. An object is dropped from a height of 80 feet.

$t_1 = 1, t_2 = 2$

Synthesis

True or False? **In Exercises 97 and 98, determine whether the statement is true or false. Justify your answer.**

97. A function with a square root cannot have a domain that is the set of real numbers.

98. It is possible for an odd function to have the interval $[0, \infty)$ as its domain.

99. If f is an even function, determine whether g is even, odd, or neither. Explain.

(a) $g(x) = -f(x)$

(b) $g(x) = f(-x)$

(c) $g(x) = f(x) - 2$

(d) $g(x) = f(x - 2)$

100. ***Think About It*** Does the graph in Exercise 11 represent x as a function of y? Explain.

Think About It **In Exercises 101–104, find the coordinates of a second point on the graph of a function f if the given point is on the graph and the function is (a) even and (b) odd.**

101. $\left(-\frac{3}{2}, 4\right)$

102. $\left(-\frac{5}{3}, -7\right)$

103. $(4, 9)$

104. $(5, -1)$

105. ***Writing*** Use a graphing utility to graph each function. Write a paragraph describing any similarities and differences you observe among the graphs.

(a) $y = x$ (b) $y = x^2$

(c) $y = x^3$ (d) $y = x^4$

(e) $y = x^5$ (f) $y = x^6$

106. ***Conjecture*** Use the results of Exercise 105 to make a conjecture about the graphs of $y = x^7$ and $y = x^8$. Use a graphing utility to graph the functions and compare the results with your conjecture.

Skills Review

In Exercises 107–110, solve the equation.

107. $x^2 - 10x = 0$

108. $100 - (x - 5)^2 = 0$

109. $x^3 - x = 0$

110. $16x^2 - 40x + 25 = 0$

In Exercises 111–114, evaluate the function at each specified value of the independent variable and simplify.

111. $f(x) = 5x - 8$

(a) $f(9)$ (b) $f(-4)$ (c) $f(x - 7)$

112. $f(x) = x^2 - 10x$

(a) $f(4)$ (b) $f(-8)$ (c) $f(x - 4)$

113. $f(x) = \sqrt{x - 12} - 9$

(a) $f(12)$ (b) $f(40)$ (c) $f(-\sqrt{36})$

114. $f(x) = x^4 - x - 5$

(a) $f(-1)$ (b) $f\left(\frac{1}{2}\right)$ (c) $f(2\sqrt{3})$

In Exercises 115 and 116, find the difference quotient and simplify your answer.

115. $f(x) = x^2 - 2x + 9, \quad \dfrac{f(3 + h) - f(3)}{h}, \quad h \neq 0$

116. $f(x) = 5 + 6x - x^2, \quad \dfrac{f(6 + h) - f(6)}{h}, \quad h \neq 0$

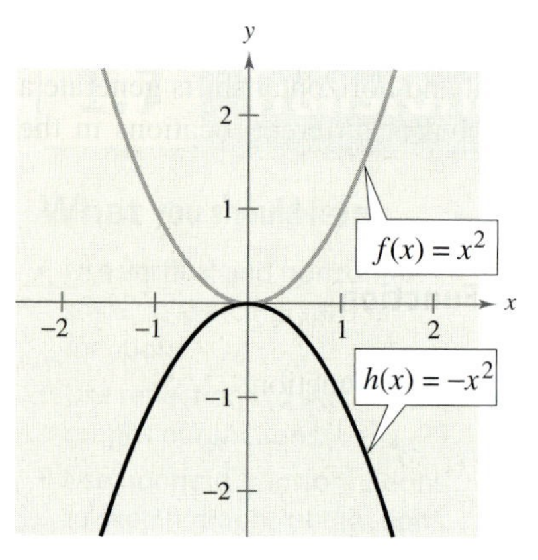

FIGURE 1.80

Reflecting Graphs

The second common type of transformation is a **reflection.** For instance, if you consider the x-axis to be a mirror, the graph of

$$h(x) = -x^2$$

is the mirror image (or reflection) of the graph of

$$f(x) = x^2,$$

as shown in Figure 1.80.

Reflections in the Coordinate Axes

Reflections in the coordinate axes of the graph of $y = f(x)$ are represented as follows.

1. Reflection in the x-axis: $h(x) = -f(x)$

2. Reflection in the y-axis: $h(x) = f(-x)$

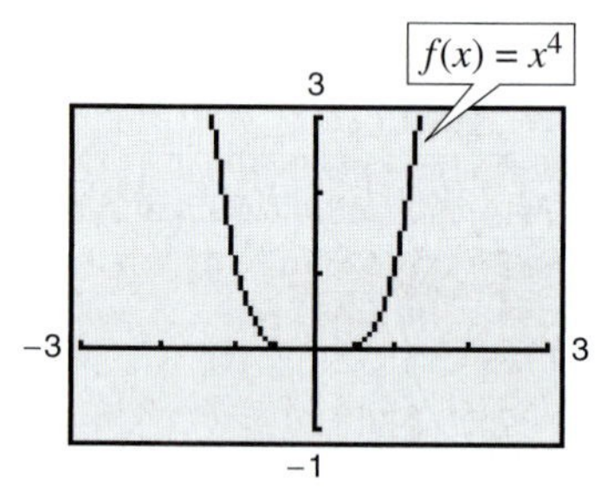

FIGURE 1.81

Example 2 Finding Equations from Graphs

The graph of the function given by

$$f(x) = x^4$$

is shown in Figure 1.81. Each of the graphs in Figure 1.82 is a transformation of the graph of f. Find an equation for each of these functions.

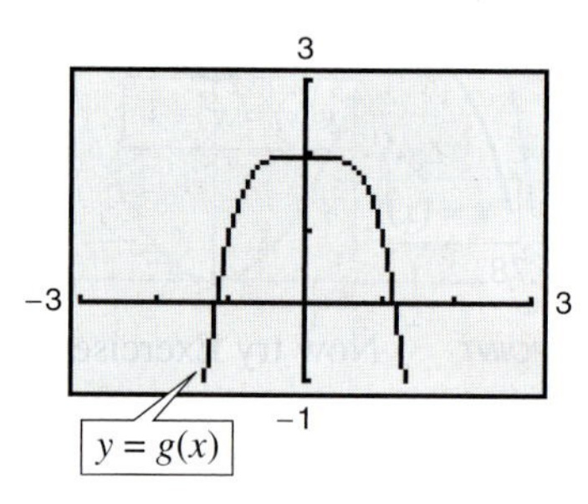

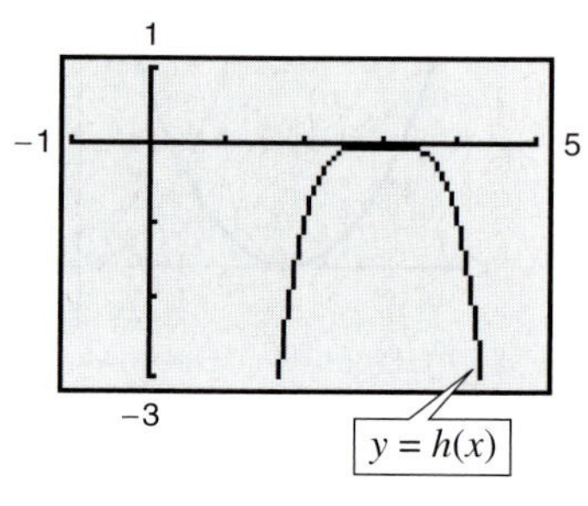

(a) (b)

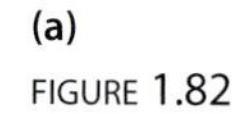

FIGURE 1.82

Solution

a. The graph of g is a reflection in the x-axis *followed by* an upward shift of two units of the graph of $f(x) = x^4$. So, the equation for g is

$$g(x) = -x^4 + 2.$$

b. The graph of h is a horizontal shift of three units to the right *followed by* a reflection in the x-axis of the graph of $f(x) = x^4$. So, the equation for h is

$$h(x) = -(x - 3)^4.$$

CHECKPOINT Now try Exercise 9.

Exploration

Reverse the order of transformations in Example 2(a). Do you obtain the same graph? Do the same for Example 2(b). Do you obtain the same graph? Explain.

Example 3 Reflections and Shifts

Compare the graph of each function with the graph of $f(x) = \sqrt{x}$.

a. $g(x) = -\sqrt{x}$ **b.** $h(x) = \sqrt{-x}$ **c.** $k(x) = -\sqrt{x+2}$

Algebraic Solution

a. The graph of g is a reflection of the graph of f in the x-axis because

$$g(x) = -\sqrt{x}$$
$$= -f(x).$$

b. The graph of h is a reflection of the graph of f in the y-axis because

$$h(x) = \sqrt{-x}$$
$$= f(-x).$$

c. The graph of k is a left shift of two units followed by a reflection in the x-axis because

$$k(x) = -\sqrt{x+2}$$
$$= -f(x+2).$$

✓CHECKPOINT Now try Exercise 19.

Graphical Solution

a. Graph f and g on the same set of coordinate axes. From the graph in Figure 1.83, you can see that the graph of g is a reflection of the graph of f in the x-axis.

b. Graph f and h on the same set of coordinate axes. From the graph in Figure 1.84, you can see that the graph of h is a reflection of the graph of f in the y-axis.

c. Graph f and k on the same set of coordinate axes. From the graph in Figure 1.85, you can see that the graph of k is a left shift of two units of the graph of f, followed by a reflection in the x-axis.

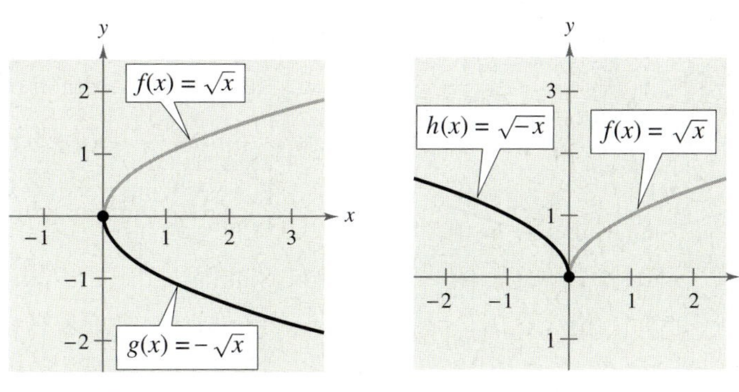

FIGURE 1.83

FIGURE 1.84

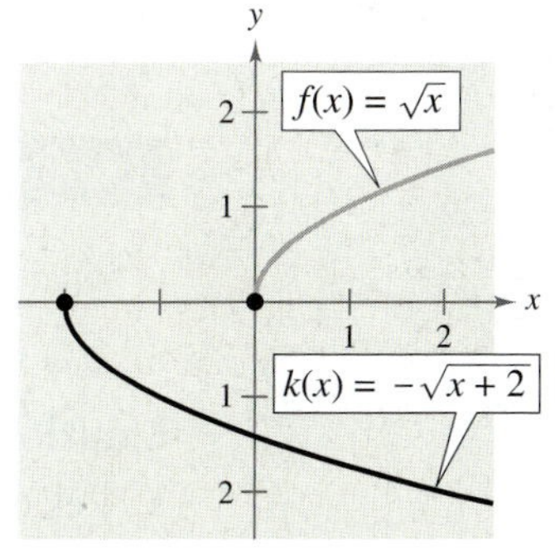

FIGURE 1.85

Activities

1. How are the graphs of $f(x)$ and $g(x) = -f(x)$ related?
   Answer: They are reflections of each other in the x-axis.
2. Compare the graph of $f(x) = |x|$ with the graph of $g(x) = |x - 9|$.
   Answer: $g(x)$ is $f(x)$ shifted to the right nine units.

When sketching the graphs of functions involving square roots, remember that the domain must be restricted to exclude negative numbers inside the radical. For instance, here are the domains of the functions in Example 3.

Domain of $g(x) = -\sqrt{x}$: $\quad x \geq 0$

Domain of $h(x) = \sqrt{-x}$: $\quad x \leq 0$

Domain of $k(x) = -\sqrt{x+2}$: $\quad x \geq -2$

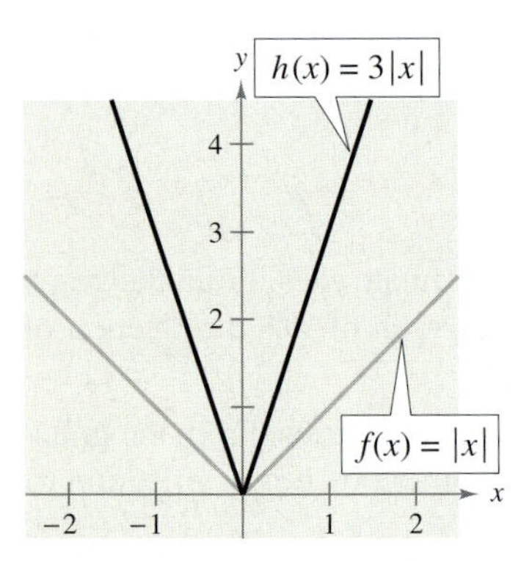

FIGURE 1.86

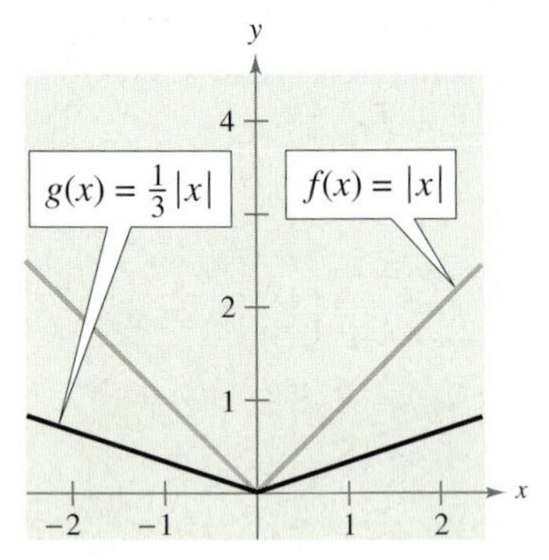

FIGURE 1.87

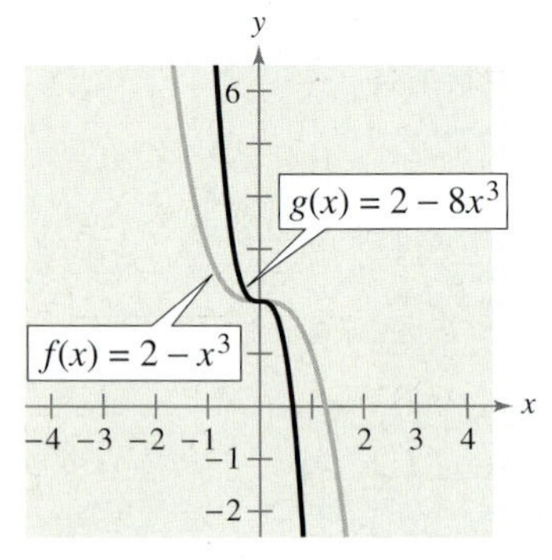

FIGURE 1.88

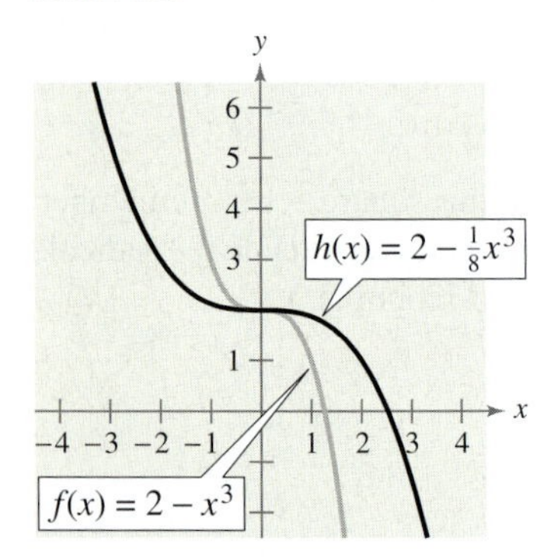

FIGURE 1.89

Nonrigid Transformations

Horizontal shifts, vertical shifts, and reflections are **rigid transformations** because the basic shape of the graph is unchanged. These transformations change only the *position* of the graph in the coordinate plane. **Nonrigid transformations** are those that cause a *distortion*—a change in the shape of the original graph. For instance, a nonrigid transformation of the graph of $y = f(x)$ is represented by $g(x) = cf(x)$, where the transformation is a **vertical stretch** if $c > 1$ and a **vertical shrink** if $0 < c < 1$. Another nonrigid transformation of the graph of $y = f(x)$ is represented by $h(x) = f(cx)$, where the transformation is a **horizontal shrink** if $c > 1$ and a **horizontal stretch** if $0 < c < 1$.

Example 4 Nonrigid Transformations

Compare the graph of each function with the graph of $f(x) = |x|$.

a. $h(x) = 3|x|$ **b.** $g(x) = \frac{1}{3}|x|$

Solution

a. Relative to the graph of $f(x) = |x|$, the graph of

$$h(x) = 3|x| = 3f(x)$$

is a vertical stretch (each y-value is multiplied by 3) of the graph of f. (See Figure 1.86.)

b. Similarly, the graph of

$$g(x) = \frac{1}{3}|x| = \frac{1}{3}f(x)$$

is a vertical shrink (each y-value is multiplied by $\frac{1}{3}$) of the graph of f. (See Figure 1.87.)

CHECKPOINT Now try Exercise 23.

Example 5 Nonrigid Transformations

Compare the graph of each function with the graph of $f(x) = 2 - x^3$.

a. $g(x) = f(2x)$ **b.** $h(x) = f\left(\frac{1}{2}x\right)$

Solution

a. Relative to the graph of $f(x) = 2 - x^3$, the graph of

$$g(x) = f(2x) = 2 - (2x)^3 = 2 - 8x^3$$

is a horizontal shrink $(c > 1)$ of the graph of f. (See Figure 1.88.)

b. Similarly, the graph of

$$h(x) = f\left(\frac{1}{2}x\right) = 2 - \left(\frac{1}{2}x\right)^3 = 2 - \frac{1}{8}x^3$$

is a horizontal stretch $(0 < c < 1)$ of the graph of f. (See Figure 1.89.)

CHECKPOINT Now try Exercise 27.

1.7 Exercises

VOCABULARY CHECK:

In Exercises 1–5, fill in the blanks.

1. Horizontal shifts, vertical shifts, and reflections are called ________ transformations.
2. A reflection in the x-axis of $y = f(x)$ is represented by $h(x) =$ ________, while a reflection in the y-axis of $y = f(x)$ is represented by $h(x) =$ ________.
3. Transformations that cause a distortion in the shape of the graph of $y = f(x)$ are called ________ transformations.
4. A nonrigid transformation of $y = f(x)$ represented by $h(x) = f(cx)$ is a ________ ________ if $c > 1$ and a ________ ________ if $0 < c < 1$.
5. A nonrigid transformation of $y = f(x)$ represented by $g(x) = cf(x)$ is a ________ ________ if $c > 1$ and a ________ ________ if $0 < c < 1$.
6. Match the rigid transformation of $y = f(x)$ with the correct representation of the graph of h, where $c > 0$.

(a) $h(x) = f(x) + c$ (i) A horizontal shift of f, c units to the right

(b) $h(x) = f(x) - c$ (ii) A vertical shift of f, c units downward

(c) $h(x) = f(x + c)$ (iii) A horizontal shift of f, c units to the left

(d) $h(x) = f(x - c)$ (iv) A vertical shift of f, c units upward

PREREQUISITE SKILLS REVIEW: Practice and review algebra skills needed for this section at **www.Eduspace.com.**

1. For each function, sketch (on the same set of coordinate axes) a graph of each function for $c = -1, 1,$ and 3.
   (a) $f(x) = |x| + c$
   (b) $f(x) = |x - c|$
   (c) $f(x) = |x + 4| + c$
2. For each function, sketch (on the same set of coordinate axes) a graph of each function for $c = -3, -1, 1,$ and 3.
   (a) $f(x) = \sqrt{x} + c$
   (b) $f(x) = \sqrt{x - c}$
   (c) $f(x) = \sqrt{x - 3} + c$
3. For each function, sketch (on the same set of coordinate axes) a graph of each function for $c = -2, 0,$ and 2.
   (a) $f(x) = [\![x]\!] + c$
   (b) $f(x) = [\![x + c]\!]$
   (c) $f(x) = [\![x - 1]\!] + c$
4. For each function, sketch (on the same set of coordinate axes) a graph of each function for $c = -3, -1, 1,$ and 3.
   (a) $f(x) = \begin{cases} x^2 + c, & x < 0 \\ -x^2 + c, & x \geq 0 \end{cases}$
   (b) $f(x) = \begin{cases} (x + c)^2, & x < 0 \\ -(x + c)^2, & x \geq 0 \end{cases}$

In Exercises 5–8, use the graph of *f* to sketch each graph. To print an enlarged copy of the graph go to the website *www.mathgraphs.com*.

5. (a) $y = f(x) + 2$
   (b) $y = f(x - 2)$
   (c) $y = 2f(x)$
   (d) $y = -f(x)$
   (e) $y = f(x + 3)$
   (f) $y = f(-x)$
   (g) $y = f\left(\frac{1}{2}x\right)$
6. (a) $y = f(-x)$
   (b) $y = f(x) + 4$
   (c) $y = 2f(x)$
   (d) $y = -f(x - 4)$
   (e) $y = f(x) - 3$
   (f) $y = -f(x) - 1$
   (g) $y = f(2x)$

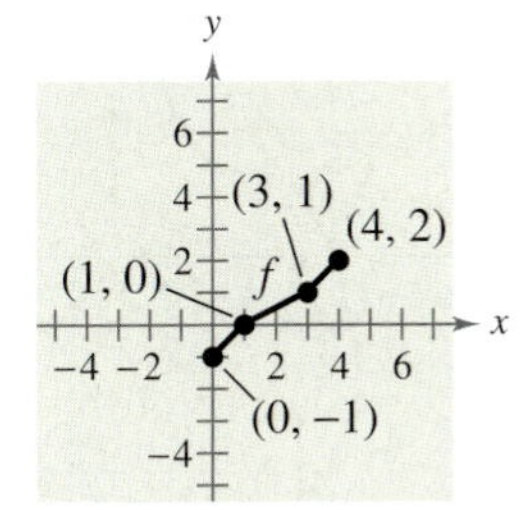

FIGURE FOR 5

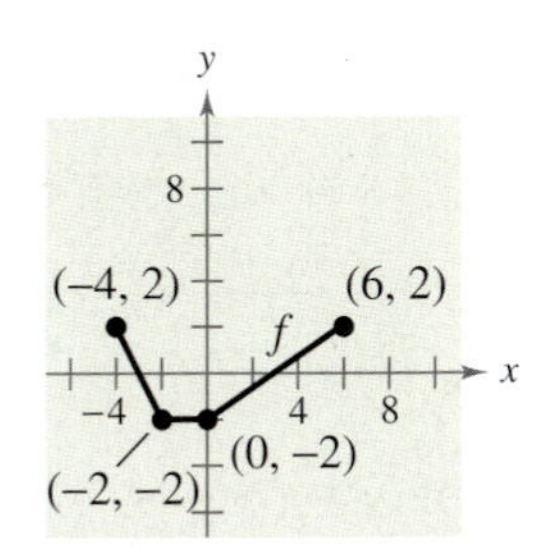

FIGURE FOR 6

7. (a) $y = f(x) - 1$
   (b) $y = f(x - 1)$
   (c) $y = f(-x)$
   (d) $y = f(x + 1)$
   (e) $y = -f(x - 2)$
   (f) $y = \frac{1}{2}f(x)$
   (g) $y = f(2x)$
8. (a) $y = f(x - 5)$
   (b) $y = -f(x) + 3$
   (c) $y = \frac{1}{3}f(x)$
   (d) $y = -f(x + 1)$
   (e) $y = f(-x)$
   (f) $y = f(x) - 10$
   (g) $y = f\left(\frac{1}{3}x\right)$

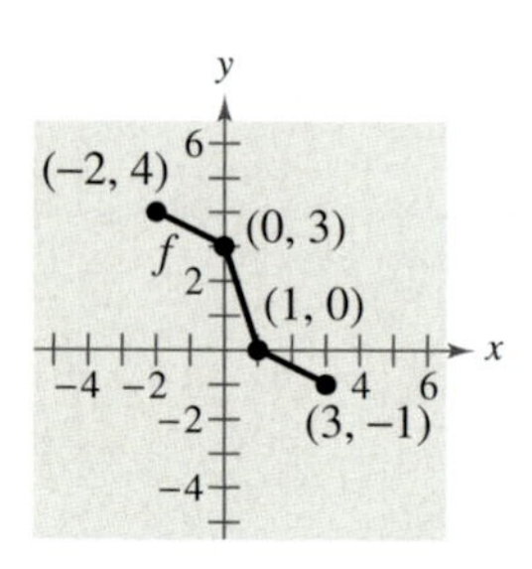

FIGURE FOR 7

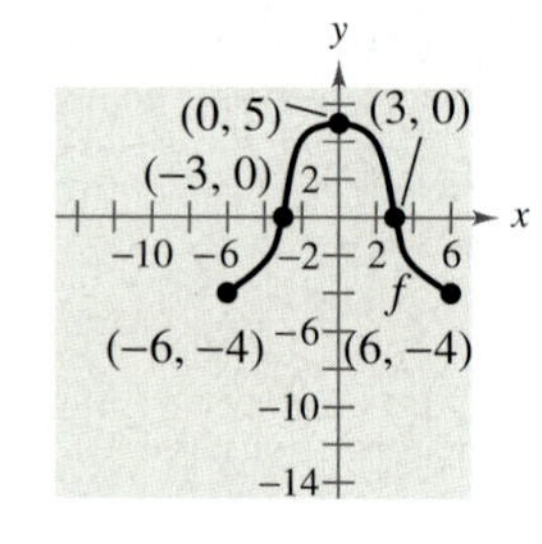

FIGURE FOR 8

9. Use the graph of $f(x) = x^2$ to write an equation for each function whose graph is shown.

(a)

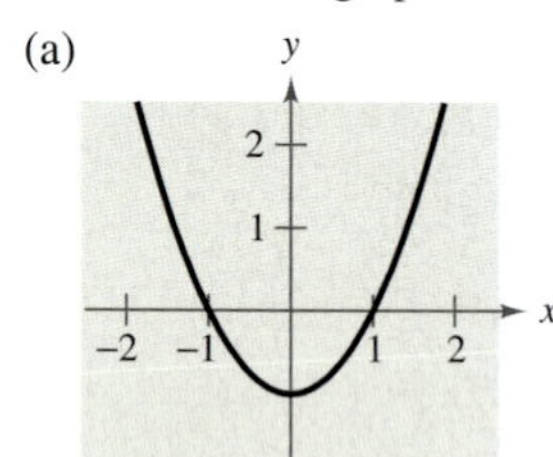

(b)

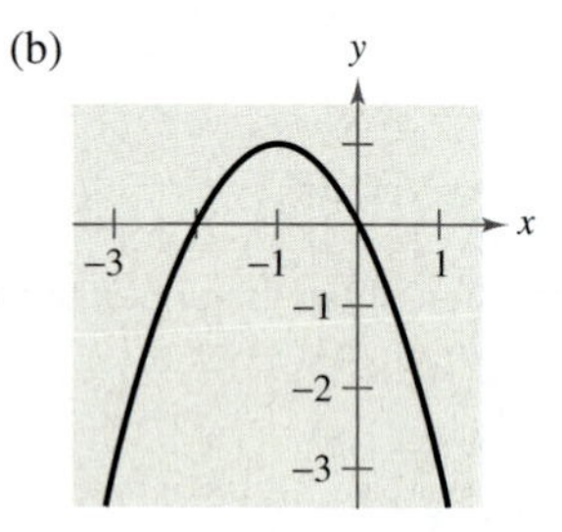

(c)

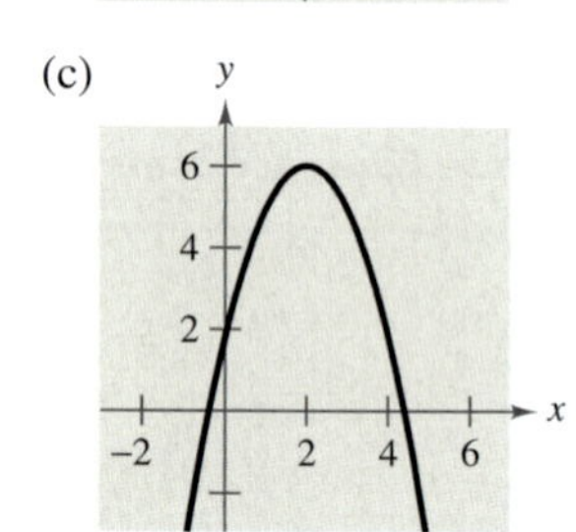

(d)

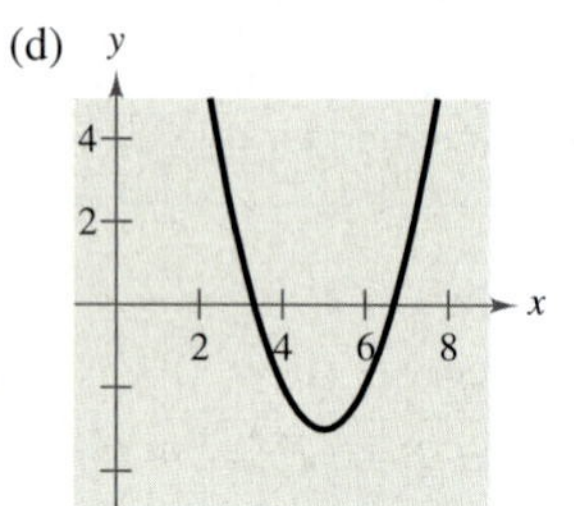

10. Use the graph of $f(x) = x^3$ to write an equation for each function whose graph is shown.

(a)

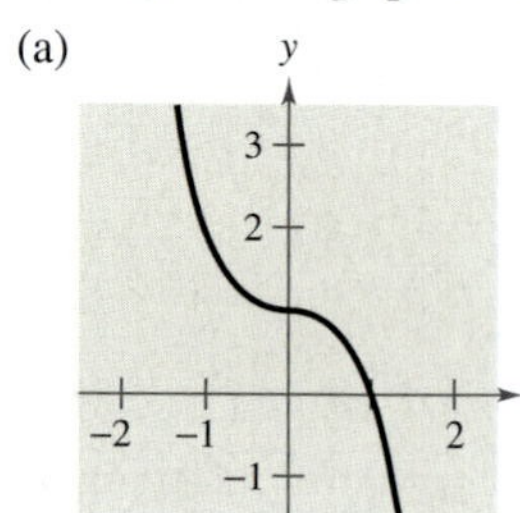

(b)

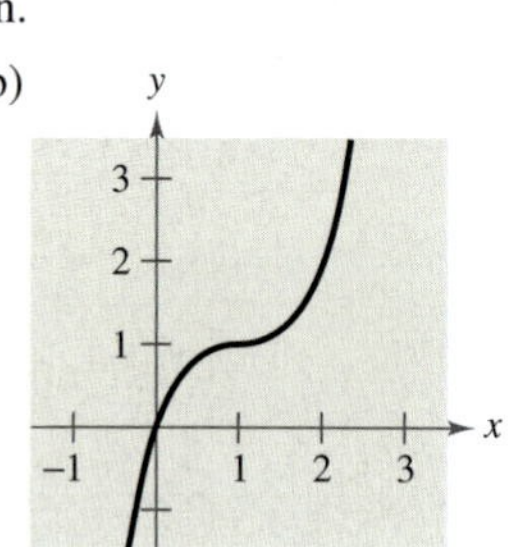

(c)

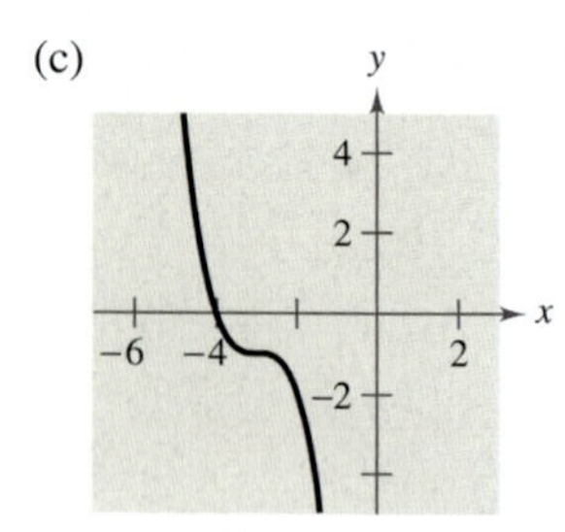

(d)

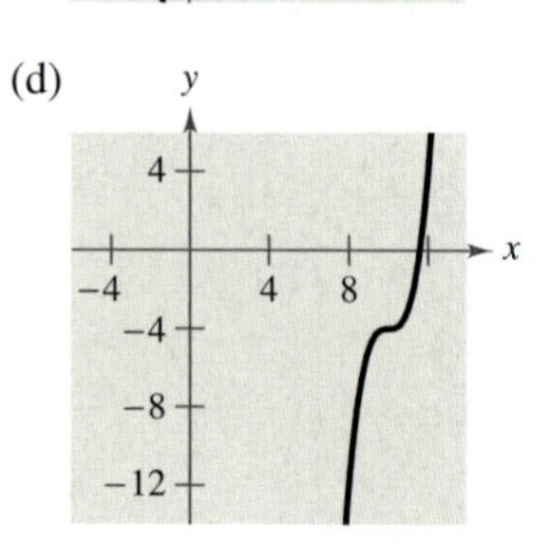

11. Use the graph of $f(x) = |x|$ to write an equation for each function whose graph is shown.

(a)

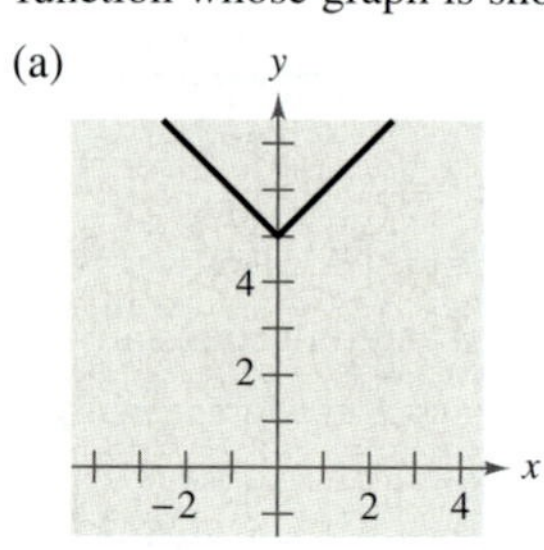

(b)

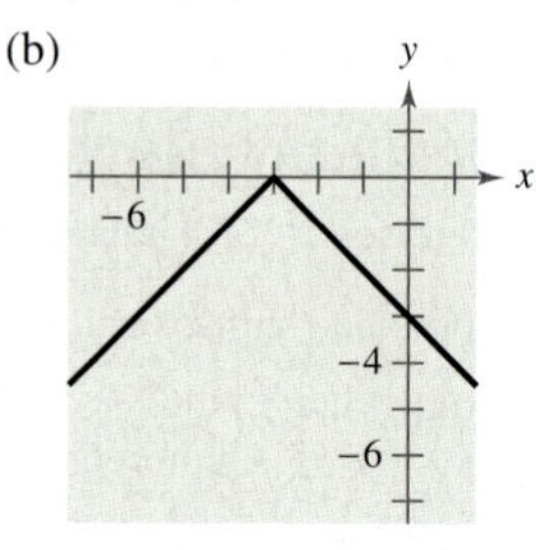

(c)

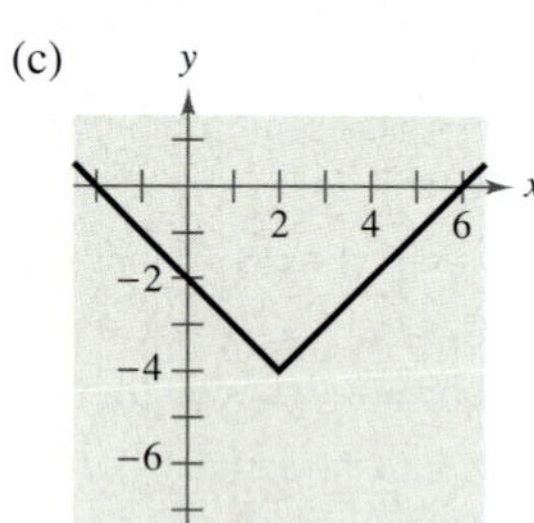

(d)

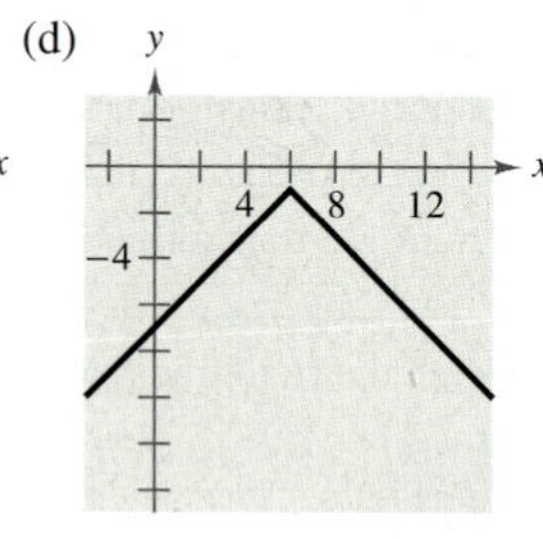

12. Use the graph of $f(x) = \sqrt{x}$ to write an equation for each function whose graph is shown.

(a)

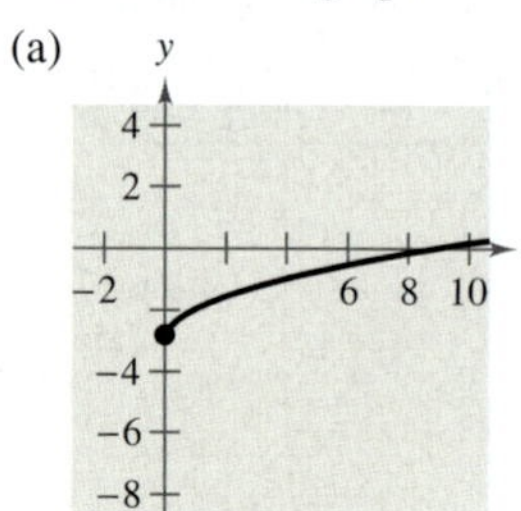

(b)

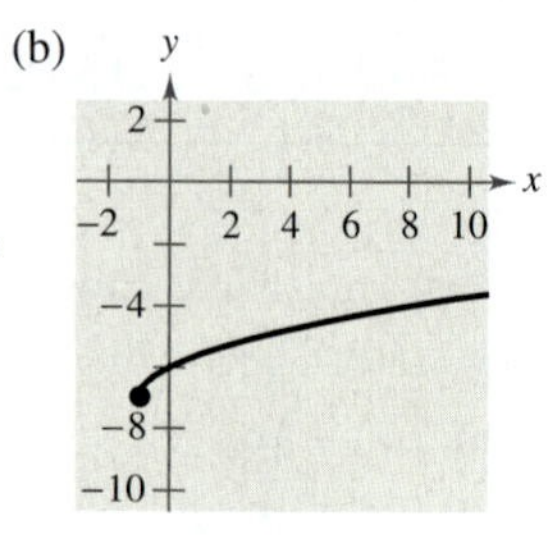

(c)

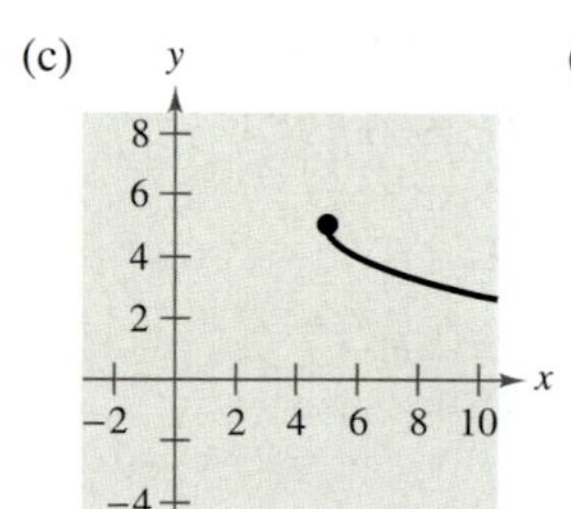

(d)

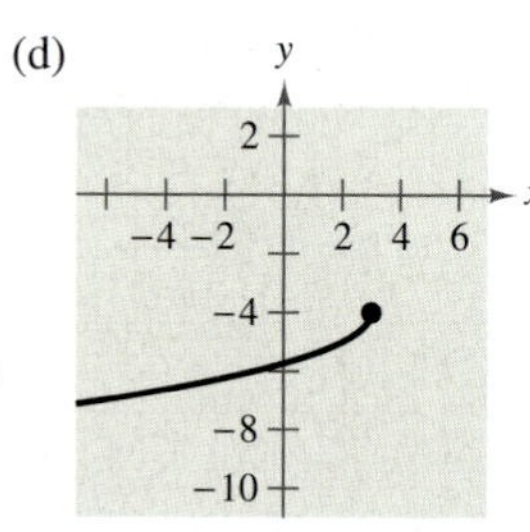

In Exercises 13–18, identify the parent function and the transformation shown in the graph. Write an equation for the function shown in the graph.

13.

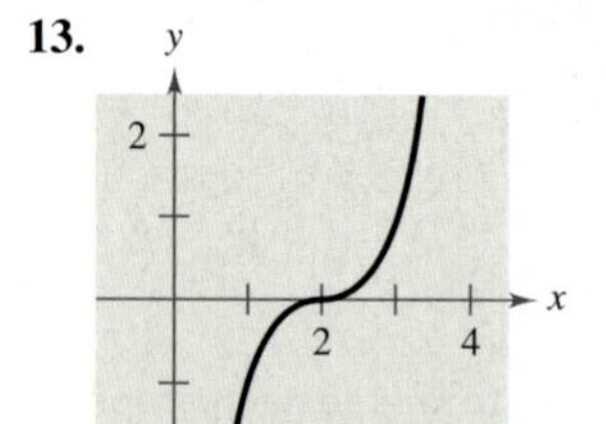

14.

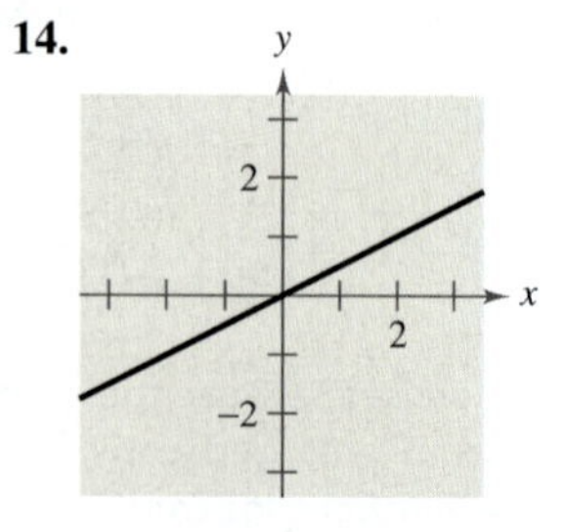

15.

16.

17.

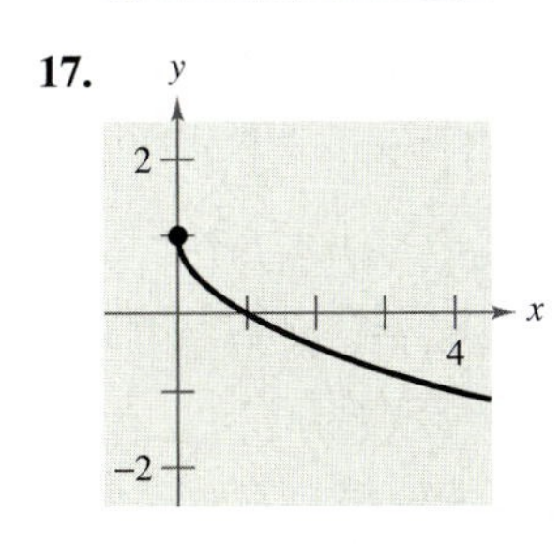

18.

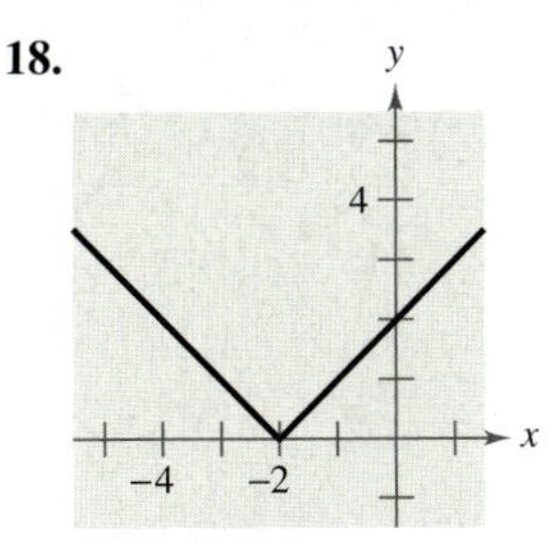

In Exercises 19–42, g is related to one of the parent functions described in this chapter. (a) Identify the parent function f. (b) Describe the sequence of tranformations from f to g. (c) Sketch the graph of g. (d) Use function notation to write g in terms of f.

19. $g(x) = 12 - x^2$
20. $g(x) = (x - 8)^2$
21. $g(x) = x^3 + 7$
22. $g(x) = -x^3 - 1$
23. $g(x) = \frac{2}{3}x^2 + 4$
24. $g(x) = 2(x - 7)^2$
25. $g(x) = 2 - (x + 5)^2$
26. $g(x) = -(x + 10)^2 + 5$
27. $g(x) = \sqrt{3x}$
28. $g(x) = \sqrt{\frac{1}{4}x}$
29. $g(x) = (x - 1)^3 + 2$
30. $g(x) = (x + 3)^3 - 10$
31. $g(x) = -|x| - 2$
32. $g(x) = 6 - |x + 5|$
33. $g(x) = -|x + 4| + 8$
34. $g(x) = |-x + 3| + 9$
35. $g(x) = 3 - [\![x]\!]$
36. $g(x) = 2[\![x + 5]\!]$
37. $g(x) = \sqrt{x - 9}$
38. $g(x) = \sqrt{x + 4} + 8$
39. $g(x) = \sqrt{7 - x} - 2$
40. $g(x) = -\sqrt{x + 1} - 6$
41. $g(x) = \sqrt{\frac{1}{2}x} - 4$
42. $g(x) = \sqrt{3x} + 1$

In Exercises 43–50, write an equation for the function that is described by the given characteristics.

43. The shape of $f(x) = x^2$, but moved two units to the right and eight units downward

44. The shape of $f(x) = x^2$, but moved three units to the left, seven units upward, and reflected in the x-axis

45. The shape of $f(x) = x^3$, but moved 13 units to the right

46. The shape of $f(x) = x^3$, but moved six units to the left, six units downward, and reflected in the y-axis

47. The shape of $f(x) = |x|$, but moved 10 units upward and reflected in the x-axis

48. The shape of $f(x) = |x|$, but moved one unit to the left and seven units downward

49. The shape of $f(x) = \sqrt{x}$, but moved six units to the left and reflected in both the x-axis and the y-axis

50. The shape of $f(x) = \sqrt{x}$, but moved nine units downward and reflected in both the x-axis and the y-axis

51. Use the graph of $f(x) = x^2$ to write an equation for each function whose graph is shown.

(a)

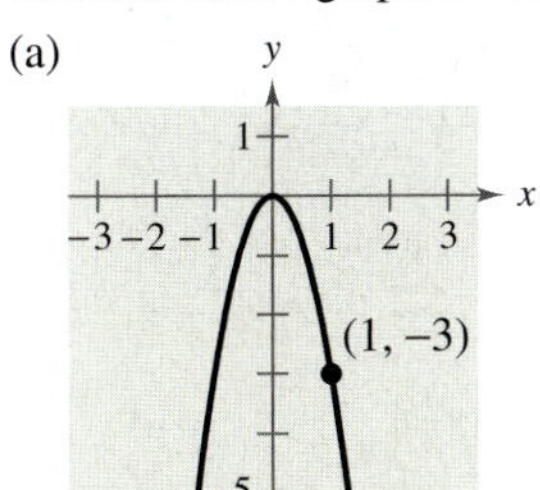

(b)

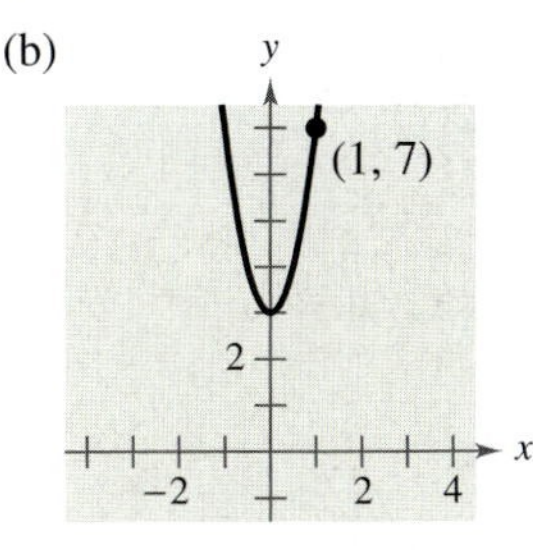

52. Use the graph of $f(x) = x^3$ to write an equation for each function whose graph is shown.

(a)

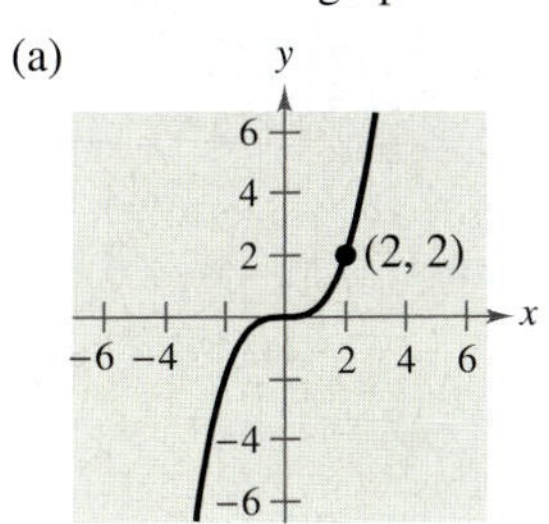

(b)

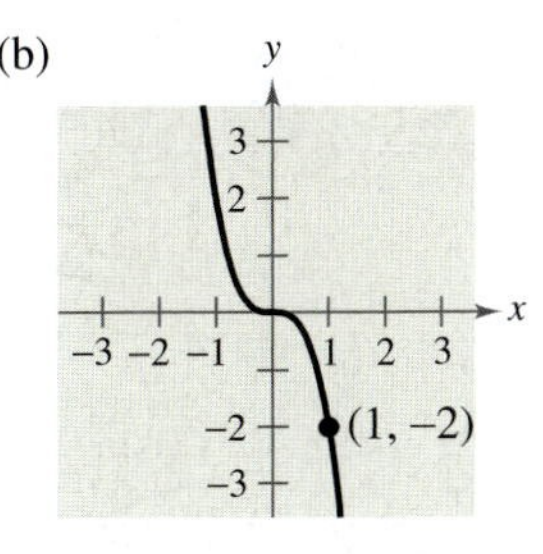

53. Use the graph of $f(x) = |x|$ to write an equation for each function whose graph is shown.

(a)

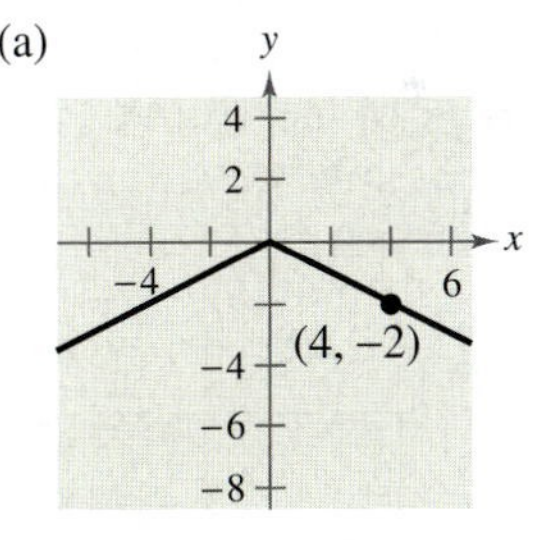

(b)

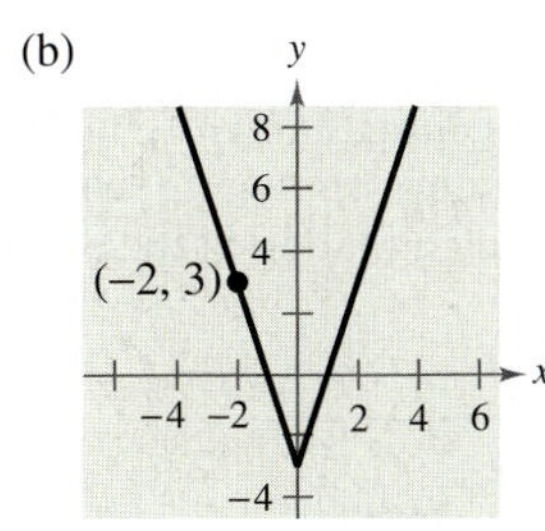

54. Use the graph of $f(x) = \sqrt{x}$ to write an equation for each function whose graph is shown.

(a)

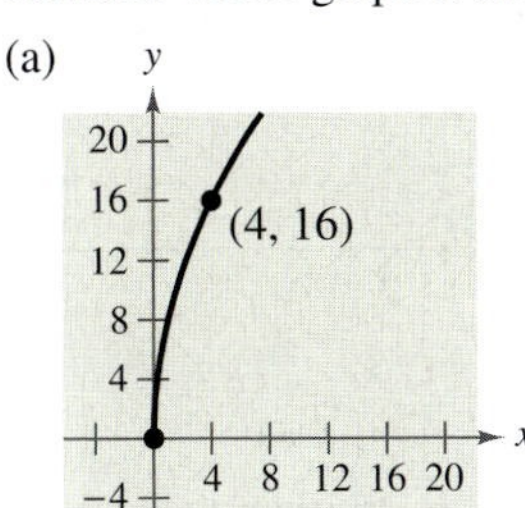

(b)

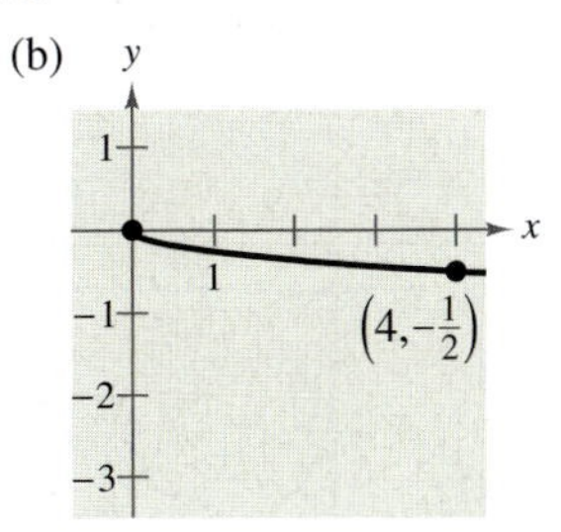

In Exercises 55–60, identify the parent function and the transformation shown in the graph. Write an equation for the function shown in the graph. Then use a graphing utility to verify your answer.

55.

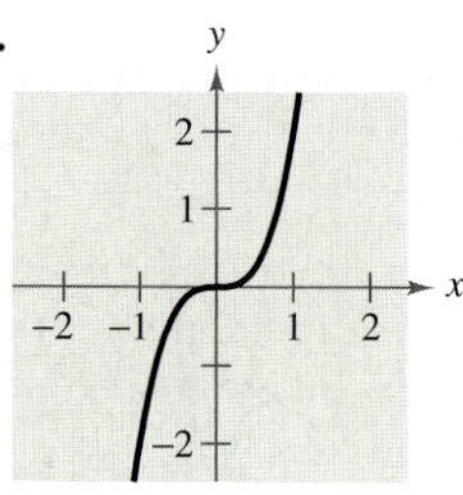

56.

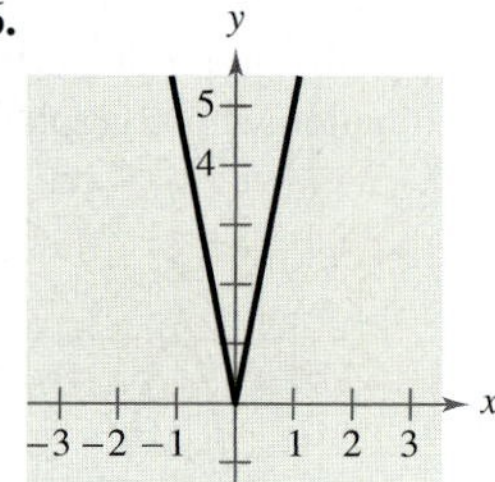

57.

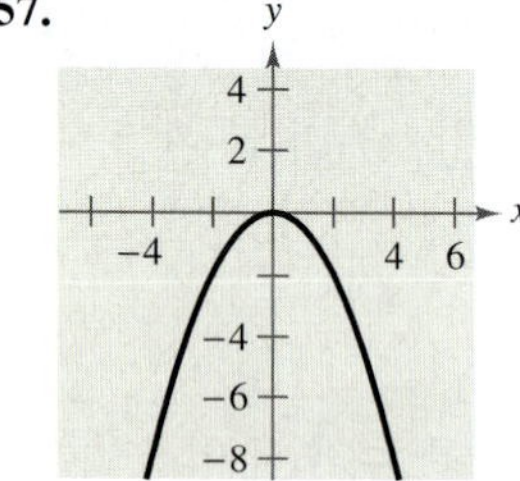

58.

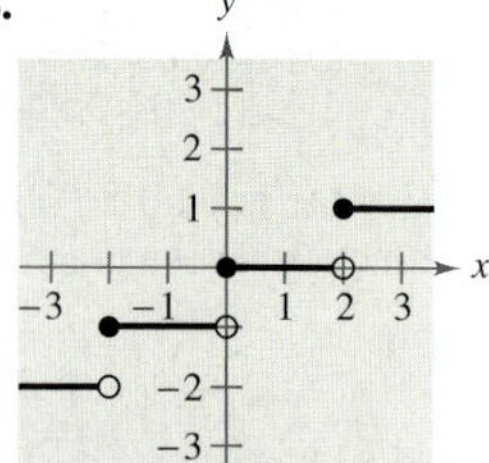

59.

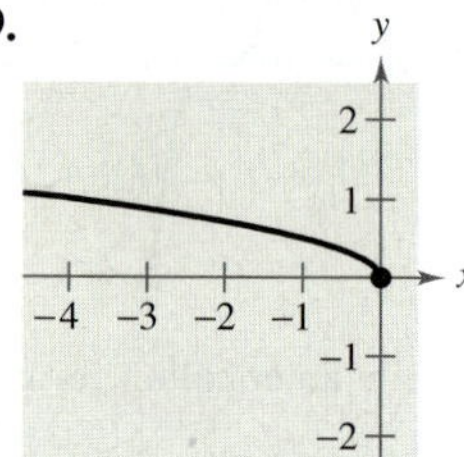

60.

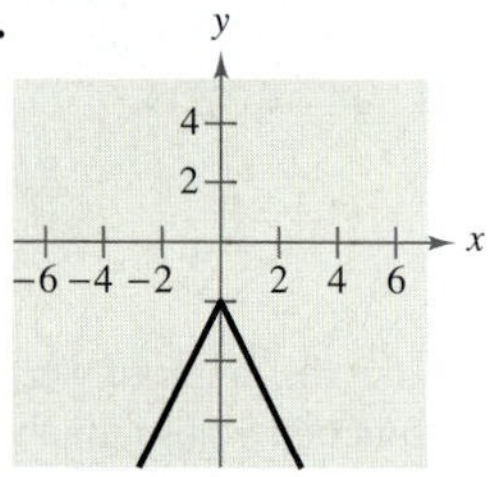

Graphical Analysis **In Exercises 61–64, use the viewing window shown to write a possible equation for the transformation of the parent function.**

61.

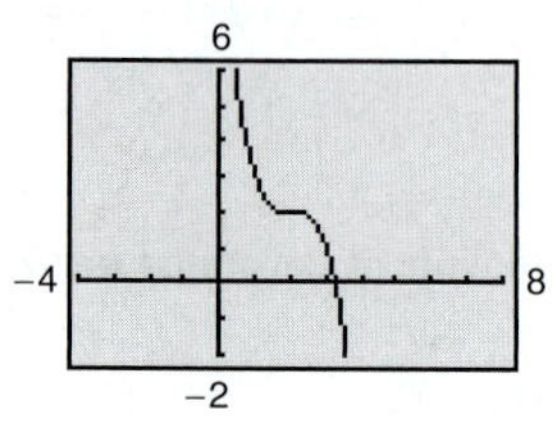

62.

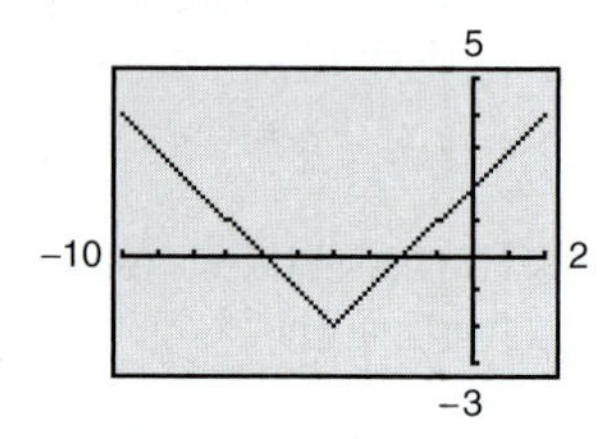

63.

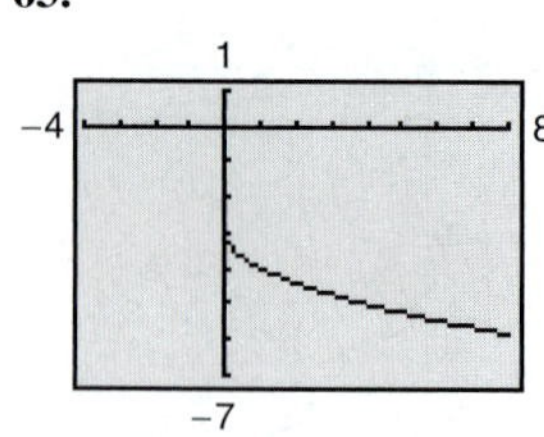

64.

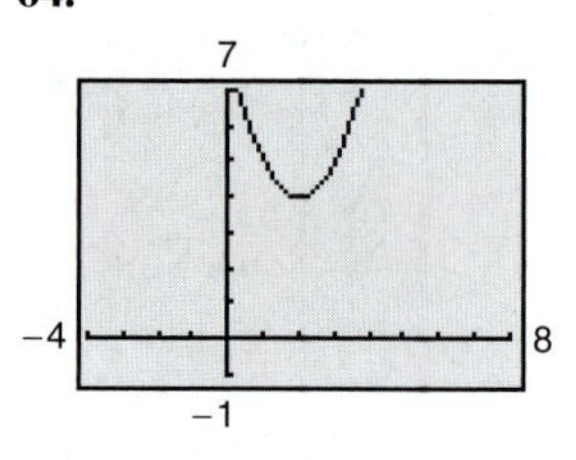

Graphical Reasoning **In Exercises 65 and 66, use the graph of *f* to sketch the graph of *g*. To print an enlarged copy of the graph, go to the website *www.mathgraphs.com*.**

65.

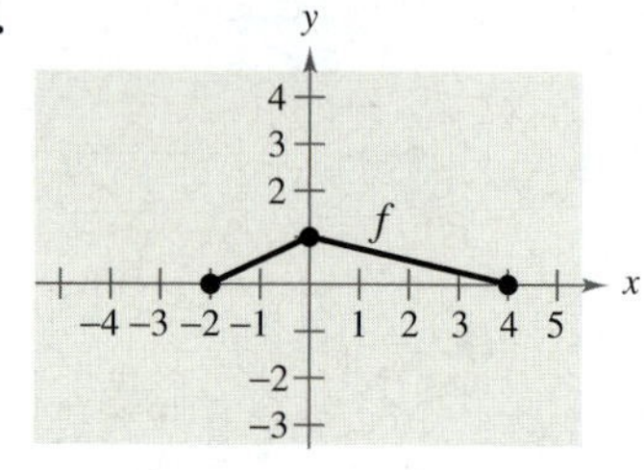

(a) $g(x) = f(x) + 2$
(b) $g(x) = f(x) - 1$
(c) $g(x) = f(-x)$
(d) $g(x) = -2f(x)$
(e) $g(x) = f(4x)$
(f) $g(x) = f\left(\frac{1}{2}x\right)$

66.

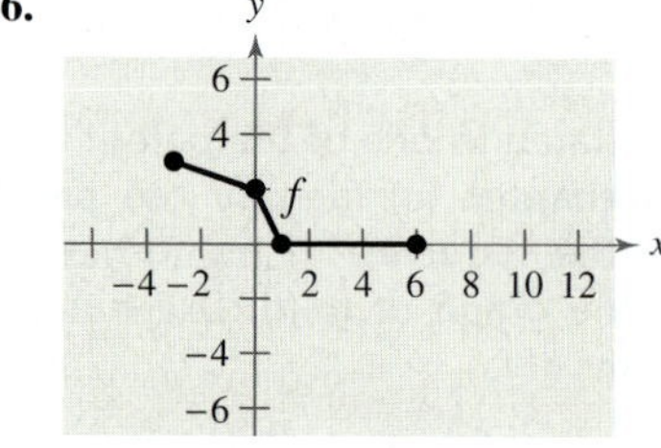

(a) $g(x) = f(x) - 5$
(b) $g(x) = f(x) + \frac{1}{2}$
(c) $g(x) = f(-x)$
(d) $g(x) = -4f(x)$
(e) $g(x) = f(2x) + 1$
(f) $g(x) = f\left(\frac{1}{4}x\right) - 2$

Model It

67. *Fuel Use* The amounts of fuel F (in billions of gallons) used by trucks from 1980 through 2002 can be approximated by the function

$$F = f(t) = 20.6 + 0.035t^2, \quad 0 \le t \le 22$$

where t represents the year, with $t = 0$ corresponding to 1980. (Source: U.S. Federal Highway Administration)

(a) Describe the transformation of the parent function $f(x) = x^2$. Then sketch the graph over the specified domain.

(b) Find the average rate of change of the function from 1980 to 2002. Interpret your answer in the context of the problem.

(c) Rewrite the function so that $t = 0$ represents 1990. Explain how you got your answer.

(d) Use the model from part (c) to predict the amount of fuel used by trucks in 2010. Does your answer seem reasonable? Explain.

68. ***Finance*** The amounts M (in trillions of dollars) of mortgage debt outstanding in the United States from 1990 through 2002 can be approximated by the function

$$M = f(t) = 0.0054(t + 20.396)^2, \quad 0 \le t \le 12$$

where t represents the year, with $t = 0$ corresponding to 1990. (Source: Board of Governors of the Federal Reserve System)

(a) Describe the transformation of the parent function $f(x) = x^2$. Then sketch the graph over the specified domain.

(b) Rewrite the function so that $t = 0$ represents 2000. Explain how you got your answer.

Synthesis

True or False? **In Exercises 69 and 70, determine whether the statement is true or false. Justify your answer.**

69. The graphs of

$f(x) = |x| + 6$ and $f(x) = |-x| + 6$

are identical.

70. If the graph of the parent function $f(x) = x^2$ is moved six units to the right, three units upward, and reflected in the x-axis, then the point $(-2, 19)$ will lie on the graph of the transformation.

71. ***Describing Profits*** Management originally predicted that the profits from the sales of a new product would be approximated by the graph of the function f shown. The actual profits are shown by the function g along with a verbal description. Use the concepts of transformations of graphs to write g in terms of f.

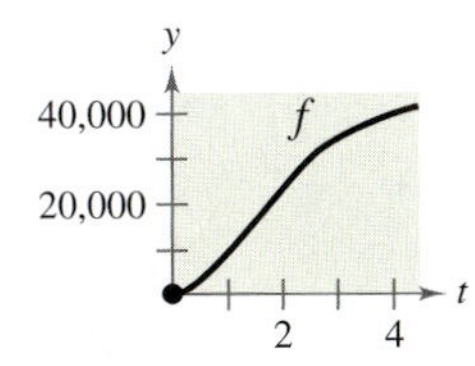

(a) The profits were only three-fourths as large as expected.

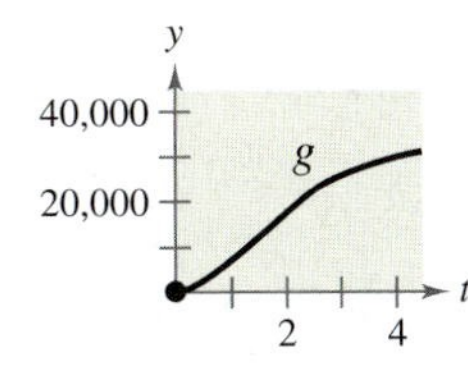

(b) The profits were consistently \$10,000 greater than predicted.

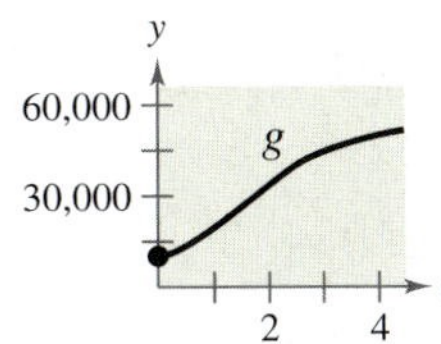

(c) There was a two-year delay in the introduction of the product. After sales began, profits grew as expected.

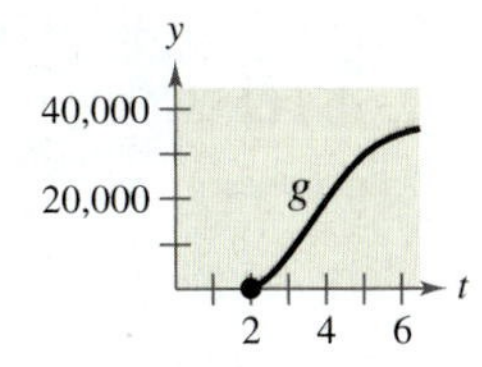

72. Explain why the graph of $y = -f(x)$ is a reflection of the graph of $y = f(x)$ about the x-axis.

73. The graph of $y = f(x)$ passes through the points $(0, 1)$, $(1, 2)$, and $(2, 3)$. Find the corresponding points on the graph of $y = f(x + 2) - 1$.

74. ***Think About It*** You can use either of two methods to graph a function: plotting points or translating a parent function as shown in this section. Which method of graphing do you prefer to use for each function? Explain.

(a) $f(x) = 3x^2 - 4x + 1$ (b) $f(x) = 2(x - 1)^2 - 6$

Skills Review

In Exercises 75–82, perform the operation and simplify.

75. $\dfrac{4}{x} + \dfrac{4}{1 - x}$

76. $\dfrac{2}{x + 5} - \dfrac{2}{x - 5}$

77. $\dfrac{3}{x - 1} - \dfrac{2}{x(x - 1)}$

78. $\dfrac{x}{x - 5} + \dfrac{1}{2}$

79. $(x - 4)\left(\dfrac{1}{\sqrt{x^2 - 4}}\right)$

80. $\left(\dfrac{x}{x^2 - 4}\right)\left(\dfrac{x^2 - x - 2}{x^2}\right)$

81. $(x^2 - 9) \div \left(\dfrac{x + 3}{5}\right)$

82. $\left(\dfrac{x}{x^2 - 3x - 28}\right) \div \left(\dfrac{x^2 + 3x}{x^2 + 5x + 4}\right)$

In Exercises 83 and 84, evaluate the function at the specified values of the independent variable and simplify.

83. $f(x) = x^2 - 6x + 11$

(a) $f(-3)$ (b) $f\left(-\frac{1}{2}\right)$ (c) $f(x - 3)$

84. $f(x) = \sqrt{x + 10} - 3$

(a) $f(-10)$ (b) $f(26)$ (c) $f(x - 10)$

In Exercises 85–88, find the domain of the function.

85. $f(x) = \dfrac{2}{11 - x}$

86. $f(x) = \dfrac{\sqrt{x - 3}}{x - 8}$

87. $f(x) = \sqrt{81 - x^2}$

88. $f(x) = \sqrt[3]{4 - x^2}$

1.8 Combinations of Functions: Composite Functions

What you should learn

- Add, subtract, multiply, and divide functions.
- Find the composition of one function with another function.
- Use combinations and compositions of functions to model and solve real-life problems.

Why you should learn it

Compositions of functions can be used to model and solve real-life problems. For instance, in Exercise 68 on page 92, compositions of functions are used to determine the price of a new hybrid car.

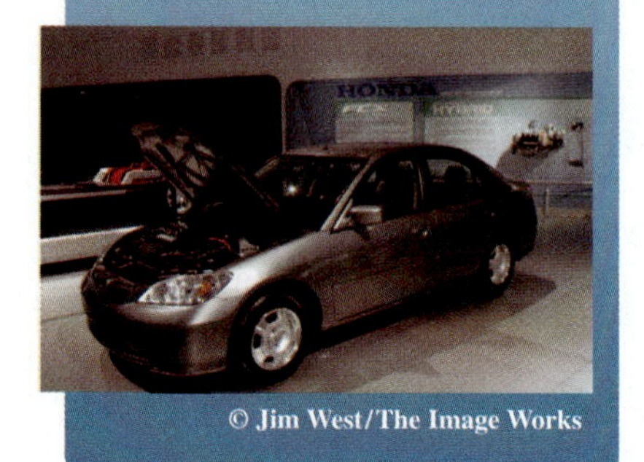

© Jim West/The Image Works

Arithmetic Combinations of Functions

Just as two real numbers can be combined by the operations of addition, subtraction, multiplication, and division to form other real numbers, two *functions* can be combined to create new functions. For example, the functions given by $f(x) = 2x - 3$ and $g(x) = x^2 - 1$ can be combined to form the sum, difference, product, and quotient of f and g.

$$f(x) + g(x) = (2x - 3) + (x^2 - 1)$$
$$= x^2 + 2x - 4 \qquad \text{Sum}$$
$$f(x) - g(x) = (2x - 3) - (x^2 - 1)$$
$$= -x^2 + 2x - 2 \qquad \text{Difference}$$
$$f(x)g(x) = (2x - 3)(x^2 - 1)$$
$$= 2x^3 - 3x^2 - 2x + 3 \qquad \text{Product}$$
$$\frac{f(x)}{g(x)} = \frac{2x - 3}{x^2 - 1}, \quad x \neq \pm 1 \qquad \text{Quotient}$$

The domain of an **arithmetic combination** of functions f and g consists of all real numbers that are common to the domains of f and g. In the case of the quotient $f(x)/g(x)$, there is the further restriction that $g(x) \neq 0$.

Sum, Difference, Product, and Quotient of Functions

Let f and g be two functions with overlapping domains. Then, for all x common to both domains, the *sum*, *difference*, *product*, and *quotient* of f and g are defined as follows.

1. *Sum:* $(f + g)(x) = f(x) + g(x)$
2. *Difference:* $(f - g)(x) = f(x) - g(x)$
3. *Product:* $(fg)(x) = f(x) \cdot g(x)$
4. *Quotient:* $\left(\frac{f}{g}\right)(x) = \frac{f(x)}{g(x)}, \quad g(x) \neq 0$

Example 1 Finding the Sum of Two Functions

Given $f(x) = 2x + 1$ and $g(x) = x^2 + 2x - 1$, find $(f + g)(x)$.

Solution

$$(f + g)(x) = f(x) + g(x) = (2x + 1) + (x^2 + 2x - 1) = x^2 + 4x$$

CHECKPOINT Now try Exercise 5(a).

Additional Examples

a. Given $f(x) = x + 5$ and $g(x) = 3x$, find $(fg)(x)$.

Solution

$$\begin{aligned}(fg)(x) &= f(x) \cdot g(x) \\ &= (x + 5)(3x) \\ &= 3x^2 + 15x\end{aligned}$$

b. Given $f(x) = \frac{1}{x}$ and $g(x) = \frac{x}{x + 1}$, find $(gf)(x)$.

Solution

$$\begin{aligned}(gf)(x) &= g(x) \cdot f(x) \\ &= \left(\frac{x}{x + 1}\right)\left(\frac{1}{x}\right) \\ &= \frac{1}{x + 1}, \quad x \neq 0\end{aligned}$$

Example 2 Finding the Difference of Two Functions

Given $f(x) = 2x + 1$ and $g(x) = x^2 + 2x - 1$, find $(f - g)(x)$. Then evaluate the difference when $x = 2$.

Solution

The difference of f and g is

$$\begin{aligned}(f - g)(x) &= f(x) - g(x) \\ &= (2x + 1) - (x^2 + 2x - 1) \\ &= -x^2 + 2.\end{aligned}$$

When $x = 2$, the value of this difference is

$$\begin{aligned}(f - g)(2) &= -(2)^2 + 2 \\ &= -2.\end{aligned}$$

CHECKPOINT Now try Exercise 5(b).

In Examples 1 and 2, both f and g have domains that consist of all real numbers. So, the domains of $(f + g)$ and $(f - g)$ are also the set of all real numbers. Remember that any restrictions on the domains of f and g must be considered when forming the sum, difference, product, or quotient of f and g.

Example 3 Finding the Domains of Quotients of Functions

Find $\left(\frac{f}{g}\right)(x)$ and $\left(\frac{g}{f}\right)(x)$ for the functions given by

$$f(x) = \sqrt{x} \qquad \text{and} \qquad g(x) = \sqrt{4 - x^2}.$$

Then find the domains of f/g and g/f.

Solution

The quotient of f and g is

$$\left(\frac{f}{g}\right)(x) = \frac{f(x)}{g(x)} = \frac{\sqrt{x}}{\sqrt{4 - x^2}}$$

and the quotient of g and f is

$$\left(\frac{g}{f}\right)(x) = \frac{g(x)}{f(x)} = \frac{\sqrt{4 - x^2}}{\sqrt{x}}.$$

The domain of f is $[0, \infty)$ and the domain of g is $[-2, 2]$. The intersection of these domains is $[0, 2]$. So, the domains of $\left(\frac{f}{g}\right)$ and $\left(\frac{g}{f}\right)$ are as follows.

$$\text{Domain of } \left(\frac{f}{g}\right): [0, 2) \qquad \text{Domain of } \left(\frac{g}{f}\right): (0, 2]$$

Note that the domain of (f/g) includes $x = 0$, but not $x = 2$, because $x = 2$ yields a zero in the denominator, whereas the domain of (g/f) includes $x = 2$, but not $x = 0$, because $x = 0$ yields a zero in the denominator.

CHECKPOINT Now try Exercise 5(d).

Composition of Functions

Another way of combining two functions is to form the **composition** of one with the other. For instance, if $f(x) = x^2$ and $g(x) = x + 1$, the composition of f with g is

$$f(g(x)) = f(x + 1)$$
$$= (x + 1)^2.$$

This composition is denoted as $(f \circ g)$ and reads as "f composed with g."

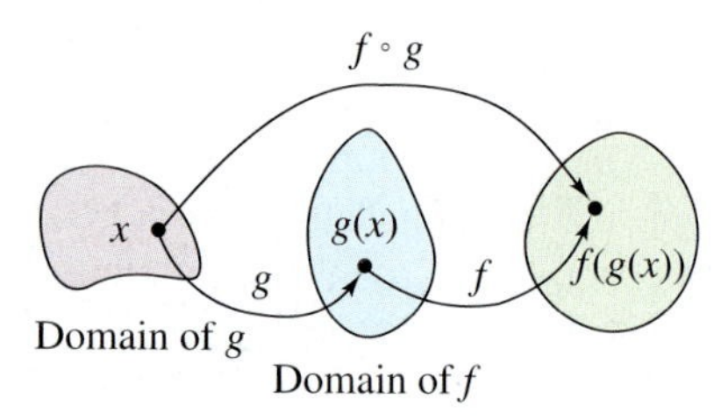

FIGURE 1.90

Definition of Composition of Two Functions

The **composition** of the function f with the function g is

$$(f \circ g)(x) = f(g(x)).$$

The domain of $(f \circ g)$ is the set of all x in the domain of g such that $g(x)$ is in the domain of f. (See Figure 1.90.)

Example 4 Composition of Functions

Given $f(x) = x + 2$ and $g(x) = 4 - x^2$, find the following.

a. $(f \circ g)(x)$ **b.** $(g \circ f)(x)$ **c.** $(g \circ f)(-2)$

Solution

a. The composition of f with g is as follows.

$$(f \circ g)(x) = f(g(x)) \quad \text{Definition of } f \circ g$$
$$= f(4 - x^2) \quad \text{Definition of } g(x)$$
$$= (4 - x^2) + 2 \quad \text{Definition of } f(x)$$
$$= -x^2 + 6 \quad \text{Simplify.}$$

b. The composition of g with f is as follows.

$$(g \circ f)(x) = g(f(x)) \quad \text{Definition of } g \circ f$$
$$= g(x + 2) \quad \text{Definition of } f(x)$$
$$= 4 - (x + 2)^2 \quad \text{Definition of } g(x)$$
$$= 4 - (x^2 + 4x + 4) \quad \text{Expand.}$$
$$= -x^2 - 4x \quad \text{Simplify.}$$

Note that, in this case, $(f \circ g)(x) \neq (g \circ f)(x)$.

c. Using the result of part (b), you can write the following.

$$(g \circ f)(-2) = -(-2)^2 - 4(-2) \quad \text{Substitute.}$$
$$= -4 + 8 \quad \text{Simplify.}$$
$$= 4 \quad \text{Simplify.}$$

CHECKPOINT Now try Exercise 31.

STUDY TIP

The following tables of values help illustrate the composition $(f \circ g)(x)$ given in Example 4.

x	0	1	2	3
$g(x)$	4	3	0	-5

$g(x)$	4	3	0	-5
$f(g(x))$	6	5	2	-3

x	0	1	2	3
$f(g(x))$	6	5	2	-3

Note that the first two tables can be combined (or "composed") to produce the values given in the third table.

Technology

You can use a graphing utility to determine the domain of a composition of functions. For the composition in Example 5, enter the function composition as

$$y = \left(\sqrt{9 - x^2}\right)^2 - 9.$$

You should obtain the graph shown below. Use the *trace* feature to determine that the x-coordinates of points on the graph extend from -3 to 3. So, the domain of $(f \circ g)(x)$ is $-3 \le x \le 3$.

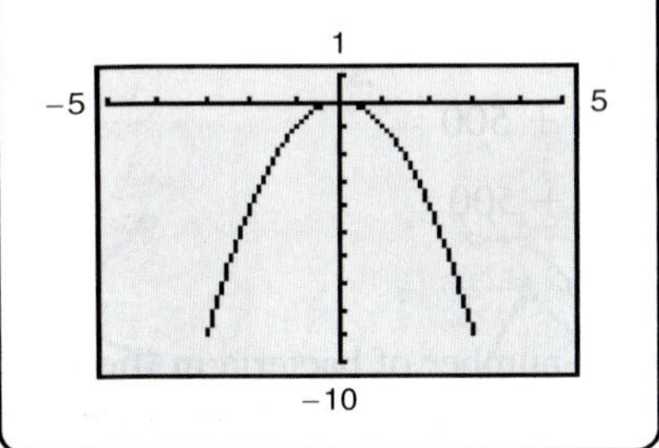

Activities

1. Given $f(x) = 3x^2 + 2$ and $g(x) = 2x$, find $f \circ g$.
   Answer: $(f \circ g)(x) = 12x^2 + 2$
2. Given the functions
   $$f(x) = \frac{1}{x - 2} \text{ and } g(x) = \sqrt{x},$$
   find the composition of f with g. Then find the domain of the composition.
   Answer: $(f \circ g)(x) = \dfrac{1}{\sqrt{x} - 2}$. The domain of $(f \circ g)$ is the set of all nonnegative real numbers except $x = 4$.
3. Find two functions f and g such that $(f \circ g)(x) = h(x)$. (There are many correct answers.)
   a. $h(x) = \dfrac{1}{\sqrt{3x + 1}}$
   Answer: $f(x) = \dfrac{1}{\sqrt{x}}$ and $g(x) = 3x + 1$
   b. $h(x) = (2x + 3)^4$
   Answer: $f(x) = x^4$ and $g(x) = 2x + 3$

Example 5 Finding the Domain of a Composite Function

Given $f(x) = x^2 - 9$ and $g(x) = \sqrt{9 - x^2}$, find the composition $(f \circ g)(x)$. Then find the domain of $(f \circ g)$.

Solution

$$\begin{aligned}(f \circ g)(x) &= f(g(x)) \\ &= f\left(\sqrt{9 - x^2}\right) \\ &= \left(\sqrt{9 - x^2}\right)^2 - 9 \\ &= 9 - x^2 - 9 \\ &= -x^2\end{aligned}$$

From this, it might appear that the domain of the composition is the set of all real numbers. This, however is not true. Because the domain of f is the set of all real numbers and the domain of g is $-3 \le x \le 3$, the domain of $(f \circ g)$ is $-3 \le x \le 3$.

✓CHECKPOINT Now try Exercise 35.

In Examples 4 and 5, you formed the composition of two given functions. In calculus, it is also important to be able to identify two functions that make up a given composite function. For instance, the function h given by

$$h(x) = (3x - 5)^3$$

is the composition of f with g, where $f(x) = x^3$ and $g(x) = 3x - 5$. That is,

$$h(x) = (3x - 5)^3 = [g(x)]^3 = f(g(x)).$$

Basically, to "decompose" a composite function, look for an "inner" function and an "outer" function. In the function h above, $g(x) = 3x - 5$ is the inner function and $f(x) = x^3$ is the outer function.

Example 6 Decomposing a Composite Function 

Write the function given by $h(x) = \dfrac{1}{(x - 2)^2}$ as a composition of two functions.

Solution

One way to write h as a composition of two functions is to take the inner function to be $g(x) = x - 2$ and the outer function to be

$$f(x) = \frac{1}{x^2} = x^{-2}.$$

Then you can write

$$h(x) = \frac{1}{(x - 2)^2} = (x - 2)^{-2} = f(x - 2) = f(g(x)).$$

✓CHECKPOINT Now try Exercise 47.

Section 1.6

☐ Identify and graph linear, squaring *(p. 66)*, cubic, square root, reciprocal *(p. 68)*, step, and other piecewise-defined functions *(p. 69)*.	103–114
☐ Recognize graphs of parent functions *(p. 70)*.	115, 116

Section 1.7

☐ Use vertical and horizontal shifts to sketch graphs of functions *(p. 74)*.	117–120
☐ Use reflections to sketch graphs of functions *(p. 76)*.	121–126
☐ Use nonrigid transformations to sketch graphs of functions *(p. 78)*.	127–130

Section 1.8

☐ Add, subtract, multiply, and divide functions *(p. 84)*.	131, 132
☐ Find the composition of one function with another function *(p. 86)*.	133–136
☐ Use combinations and compositions of functions to model and solve real-life problems *(p. 88)*.	137, 138

Section 1.9

☐ Find inverse functions informally and verify that two functions are inverse functions of each other *(p. 93)*.	139, 140
☐ Use graphs of functions to determine whether functions have inverse functions *(p. 95)*.	141, 142
☐ Use the Horizontal Line Test to determine if functions are one-to-one *(p. 96)*.	143–146
☐ Find inverse functions algebraically *(p. 97)*.	147–152

Section 1.10

☐ Use mathematical models to approximate sets of data points *(p. 103)*.	153
☐ Use the *regression* feature of a graphing utility to find the equation of a least squares regression line *(p. 104)*.	154
☐ Write mathematical models for direct variation *(p. 105)*.	155
☐ Write mathematical models for direct variation as an *n*th power *(p. 106)*.	156, 157
☐ Write mathematical models for inverse variation *(p. 107)*.	158, 159
☐ Write mathematical models for joint variation *(p. 108)*.	160

1 Review Exercises

1.1 In Exercises 1 and 2, plot the points in the Cartesian plane.

1. $(2, 2), (0, -4), (-3, 6), (-1, -7)$
2. $(5, 0), (8, 1), (4, -2), (-3, -3)$

In Exercises 3 and 4, determine the quadrant(s) in which (x, y) is located so that the condition(s) is (are) satisfied.

3. $x > 0$ and $y = -2$
4. $y > 0$

In Exercises 5–8, (a) plot the points, (b) find the distance between the points, and (c) find the midpoint of the line segment joining the points.

5. $(-3, 8), (1, 5)$
6. $(-2, 6), (4, -3)$
7. $(5.6, 0), (0, 8.2)$
8. $(0, -1.2), (-3.6, 0)$

In Exercises 9 and 10, the polygon is shifted to a new position in the plane. Find the coordinates of the vertices of the polygon in its new position.

9. Original coordinates of vertices:

   $(4, 8), (6, 8), (4, 3), (6, 3)$

   Shift: three units downward, two units to the left

10. Original coordinates of vertices:

    $(0, 1), (3, 3), (0, 5), (-3, 3)$

    Shift: five units upward, four units to the left

11. ***Sales*** The Cheesecake Factory had annual sales of \$539.1 million in 2001 and \$773.8 million in 2003. Use the Midpoint Formula to estimate the sales in 2002. (Source: The Cheesecake Factory, Inc.)

12. ***Meteorology*** The apparent temperature is a measure of relative discomfort to a person from heat and high humidity. The table shows the actual temperatures x (in degrees Fahrenheit) versus the apparent temperatures y (in degrees Fahrenheit) for a relative humidity of 75%.

x	70	75	80	85	90	95	100
y	70	77	85	95	109	130	150

    (a) Sketch a scatter plot of the data shown in the table.

    (b) Find the change in the apparent temperature when the actual temperature changes from 70°F to 100°F.

13. ***Geometry*** The volume of a globe is about 47,712.94 cubic centimeters. Find the radius of the globe.

14. ***Geometry*** The volume of a rectangular package is 2304 cubic inches. The length of the package is 3 times its width, and the height is 1.5 times its width.

    (a) Draw a diagram that represents the problem. Label the height, width, and length accordingly.

    (b) Find the dimensions of the package.

1.2 In Exercises 15–18, complete a table of values. Use the solution points to sketch the graph of the equation.

15. $y = 3x - 5$
16. $y = -\frac{1}{2}x + 2$
17. $y = x^2 - 3x$
18. $y = 2x^2 - x - 9$

In Exercises 19–24, sketch the graph *by hand*.

19. $y - 2x - 3 = 0$
20. $3x + 2y + 6 = 0$
21. $y = \sqrt{5 - x}$
22. $y = \sqrt{x + 2}$
23. $y + 2x^2 = 0$
24. $y = x^2 - 4x$

In Exercises 25–28, find the x- and y-intercepts of the graph of the equation.

25. $y = 2x + 7$
26. $y = |x + 1| - 3$
27. $y = (x - 3)^2 - 4$
28. $y = x\sqrt{4 - x^2}$

In Exercises 29–36, use the algebraic tests to check for symmetry with respect to both axes and the origin. Then sketch the graph of the equation.

29. $y = -4x + 1$
30. $y = 5x - 6$
31. $y = 5 - x^2$
32. $y = x^2 - 10$
33. $y = x^3 + 3$
34. $y = -6 - x^3$
35. $y = \sqrt{x + 5}$
36. $y = |x| + 9$

In Exercises 71–74, find two positive real numbers whose product is a maximum.

71. The sum is 110.

72. The sum is S.

73. The sum of the first and twice the second is 24.

74. The sum of the first and three times the second is 42.

75. ***Numerical, Graphical, and Analytical Analysis*** A rancher has 200 feet of fencing to enclose two adjacent rectangular corrals (see figure).

(a) Write the area A of the corral as a function of x.

(b) Create a table showing possible values of x and the corresponding areas of the corral. Use the table to estimate the dimensions that will produce the maximum enclosed area.

(c) Use a graphing utility to graph the area function. Use the graph to approximate the dimensions that will produce the maximum enclosed area.

(d) Write the area function in standard form to find analytically the dimensions that will produce the maximum area.

(e) Compare your results from parts (b), (c), and (d).

76. ***Geometry*** An indoor physical fitness room consists of a rectangular region with a semicircle on each end (see figure). The perimeter of the room is to be a 200-meter single-lane running track.

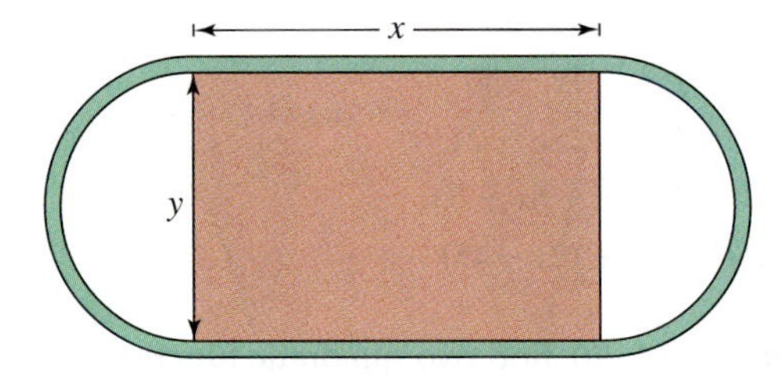

(a) Determine the radius of the semicircular ends of the room. Determine the distance, in terms of y, around the inside edge of the two semicircular parts of the track.

(b) Use the result of part (a) to write an equation, in terms of x and y, for the distance traveled in one lap around the track. Solve for y.

(c) Use the result of part (b) to write the area A of the rectangular region as a function of x. What dimensions will produce a maximum area of the rectangle?

77. ***Path of a Diver*** The path of a diver is given by

$$y = -\frac{4}{9}x^2 + \frac{24}{9}x + 12$$

where y is the height (in feet) and x is the horizontal distance from the end of the diving board (in feet). What is the maximum height of the diver?

78. ***Height of a Ball*** The height y (in feet) of a punted football is given by

$$y = -\frac{16}{2025}x^2 + \frac{9}{5}x + 1.5$$

where x is the horizontal distance (in feet) from the point at which the ball is punted (see figure).

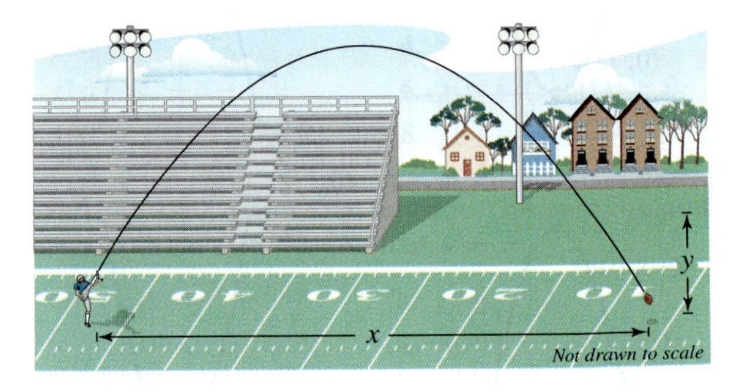

(a) How high is the ball when it is punted?

(b) What is the maximum height of the punt?

(c) How long is the punt?

79. ***Minimum Cost*** A manufacturer of lighting fixtures has daily production costs of

$$C = 800 - 10x + 0.25x^2$$

where C is the total cost (in dollars) and x is the number of units produced. How many fixtures should be produced each day to yield a minimum cost?

80. ***Minimum Cost*** A textile manufacturer has daily production costs of

$$C = 100{,}000 - 110x + 0.045x^2$$

where C is the total cost (in dollars) and x is the number of units produced. How many units should be produced each day to yield a minimum cost?

81. ***Maximum Profit*** The profit P (in dollars) for a company that produces antivirus and system utilities software is

$$P = -0.0002x^2 + 140x - 250{,}000$$

where x is the number of units sold. What sales level will yield a maximum profit?

82. ***Maximum Profit*** The profit P (in hundreds of dollars) that a company makes depends on the amount x (in hundreds of dollars) the company spends on advertising according to the model

$$P = 230 + 20x - 0.5x^2.$$

What expenditure for advertising will yield a maximum profit?

83. ***Maximum Revenue*** The total revenue R earned (in thousands of dollars) from manufacturing handheld video games is given by

$$R(p) = -25p^2 + 1200p$$

where p is the price per unit (in dollars).

(a) Find the revenue earned for each price per unit given below.

$20

$25

$30

(b) Find the unit price that will yield a maximum revenue. What is the maximum revenue? Explain your results.

84. ***Maximum Revenue*** The total revenue R earned per day (in dollars) from a pet-sitting service is given by

$$R(p) = -12p^2 + 150p$$

where p is the price charged per pet (in dollars).

(a) Find the revenue earned for each price per pet given below.

$4

$6

$8

(b) Find the price that will yield a maximum revenue. What is the maximum revenue? Explain your results.

85. ***Graphical Analysis*** From 1960 to 2003, the per capita consumption C of cigarettes by Americans (age 18 and older) can be modeled by

$$C = 4299 - 1.8t - 1.36t^2, \quad 0 \le t \le 43$$

where t is the year, with $t = 0$ corresponding to 1960. (Source: *Tobacco Outlook Report*)

(a) Use a graphing utility to graph the model.

(b) Use the graph of the model to approximate the maximum average annual consumption. Beginning in 1966, all cigarette packages were required by law to carry a health warning. Do you think the warning had any effect? Explain.

(c) In 2000, the U.S. population (age 18 and over) was 209,128,094. Of those, about 48,308,590 were smokers. What was the average annual cigarette consumption *per smoker* in 2000? What was the average daily cigarette consumption *per smoker*?

Model It

86. ***Data Analysis*** The numbers y (in thousands) of hairdressers and cosmetologists in the United States for the years 1994 through 2002 are shown in the table. (Source: U.S. Bureau of Labor Statistics)

Year	Number of hairdressers and cosmetologists, y
1994	753
1995	750
1996	737
1997	748
1998	763
1999	784
2000	820
2001	854
2002	908

(a) Use a graphing utility to create a scatter plot of the data. Let x represent the year, with $x = 4$ corresponding to 1994.

(b) Use the *regression* feature of a graphing utility to find a quadratic model for the data.

(c) Use a graphing utility to graph the model in the same viewing window as the scatter plot. How well does the model fit the data?

(d) Use the *trace* feature of the graphing utility to approximate the year in which the number of hairdressers and cosmetologists was the least.

(e) Verify your answer to part (d) algebraically.

(f) Use the model to predict the number of hairdressers and cosmetologists in 2008.

87. ***Wind Drag*** The number of horsepower y required to overcome wind drag on an automobile is approximated by

$$y = 0.002s^2 + 0.005s - 0.029, \quad 0 \le s \le 100$$

where s is the speed of the car (in miles per hour).

(a) Use a graphing utility to graph the function.

(b) Graphically estimate the maximum speed of the car if the power required to overcome wind drag is not to exceed 10 horsepower. Verify your estimate algebraically.

STUDY TIP

For power functions given by $f(x) = x^n$, if n is even, then the graph of the function is symmetric with respect to the y-axis, and if n is odd, then the graph of the function is symmetric with respect to the origin.

The polynomial functions that have the simplest graphs are monomials of the form $f(x) = x^n$, where n is an integer greater than zero. From Figure 2.13, you can see that when n is *even*, the graph is similar to the graph of $f(x) = x^2$, and when n is *odd*, the graph is similar to the graph of $f(x) = x^3$. Moreover, the greater the value of n, the flatter the graph near the origin. Polynomial functions of the form $f(x) = x^n$ are often referred to as **power functions.**

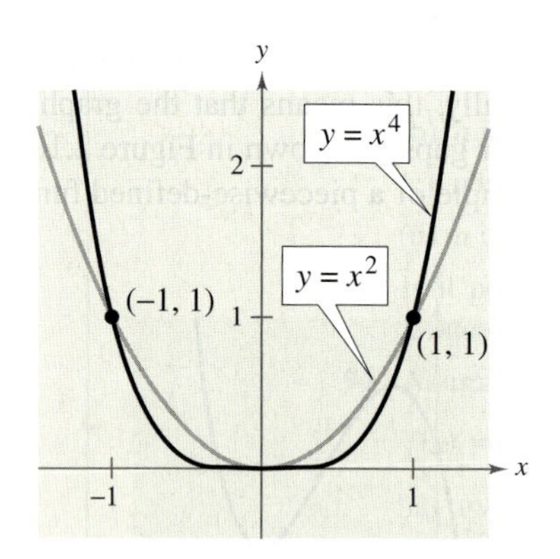

(a) If n is even, the graph of $y = x^n$ touches the axis at the x-intercept.

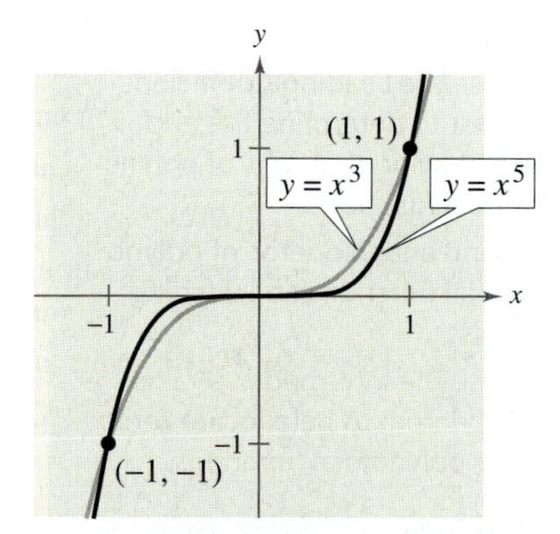

(b) If n is odd, the graph of $y = x^n$ crosses the axis at the x-intercept.

FIGURE 2.13

Example 1 Sketching Transformations of Monomial Functions

Sketch the graph of each function.

a. $f(x) = -x^5$ **b.** $h(x) = (x + 1)^4$

Solution

a. Because the degree of $f(x) = -x^5$ is odd, its graph is similar to the graph of $y = x^3$. In Figure 2.14, note that the negative coefficient has the effect of reflecting the graph in the x-axis.

b. The graph of $h(x) = (x + 1)^4$, as shown in Figure 2.15, is a left shift by one unit of the graph of $y = x^4$.

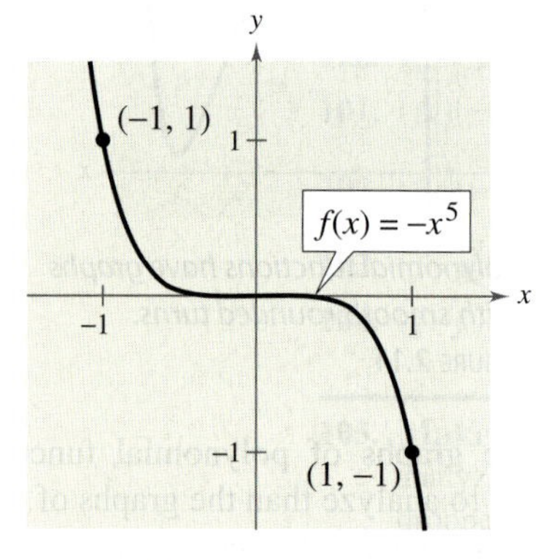

FIGURE 2.14

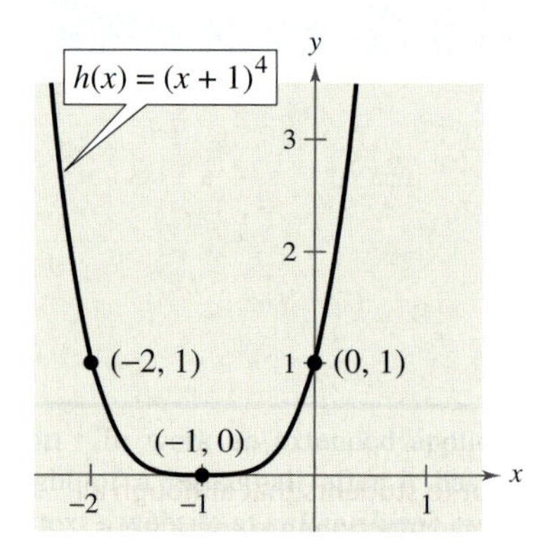

FIGURE 2.15

CHECKPOINT Now try Exercise 9.

Exploration

For each function, identify the degree of the function and whether the degree of the function is even or odd. Identify the leading coefficient and whether the leading coefficient is positive or negative. Use a graphing utility to graph each function. Describe the relationship between the degree and the sign of the leading coefficient of the function and the right-hand and left-hand behavior of the graph of the function.

a. $f(x) = x^3 - 2x^2 - x + 1$

b. $f(x) = 2x^5 + 2x^2 - 5x + 1$

c. $f(x) = -2x^5 - x^2 + 5x + 3$

d. $f(x) = -x^3 + 5x - 2$

e. $f(x) = 2x^2 + 3x - 4$

f. $f(x) = x^4 - 3x^2 + 2x - 1$

g. $f(x) = x^2 + 3x + 2$

STUDY TIP

The notation "$f(x) \to -\infty$ as $x \to -\infty$" indicates that the graph falls to the left. The notation "$f(x) \to \infty$ as $x \to \infty$" indicates that the graph rises to the right.

A review of the shapes of the graphs of polynomial functions of degrees 0, 1, and 2 may be used to illustrate the Leading Coefficient Test.

The Leading Coefficient Test

In Example 1, note that both graphs eventually rise or fall without bound as x moves to the right. Whether the graph of a polynomial function eventually rises or falls can be determined by the function's degree (even or odd) and by its leading coefficient, as indicated in the **Leading Coefficient Test.**

Leading Coefficient Test

As x moves without bound to the left or to the right, the graph of the polynomial function $f(x) = a_n x^n + \cdot \cdot \cdot + a_1 x + a_0$ eventually rises or falls in the following manner.

1. When n is *odd:*

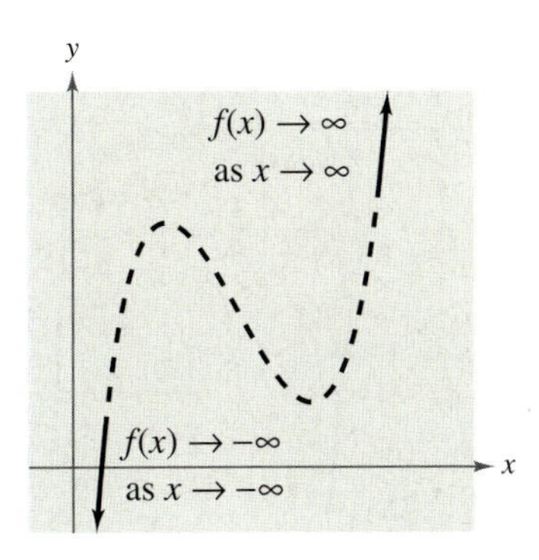

If the leading coefficient is positive ($a_n > 0$), the graph falls to the left and rises to the right.

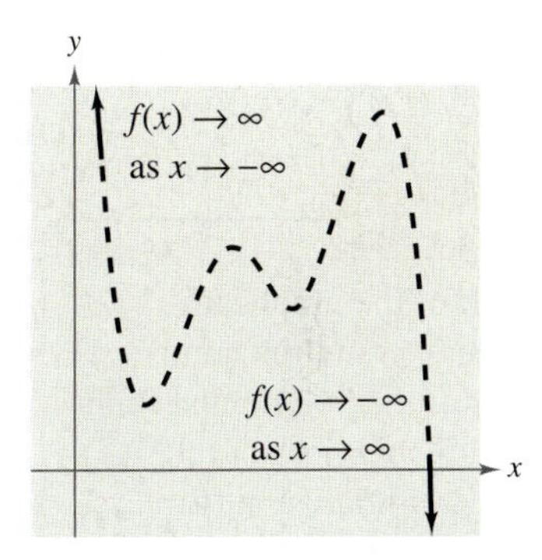

If the leading coefficient is negative ($a_n < 0$), the graph rises to the left and falls to the right.

2. When n is *even:*

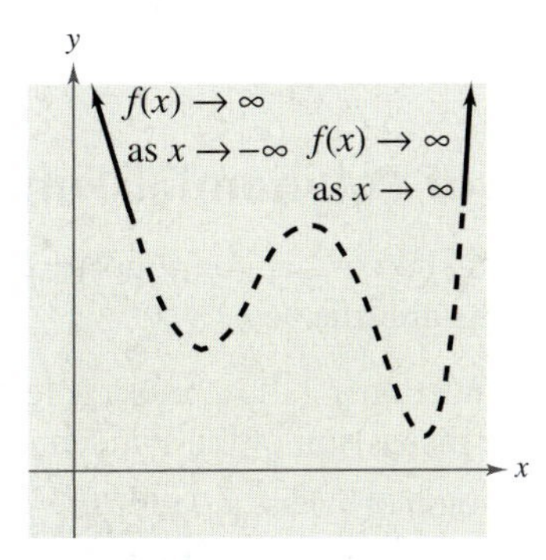

If the leading coefficient is positive ($a_n > 0$), the graph rises to the left and right.

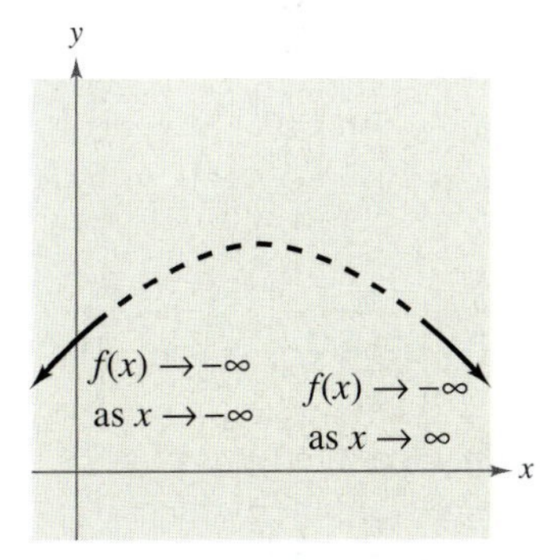

If the leading coefficient is negative ($a_n < 0$), the graph falls to the left and right.

The dashed portions of the graphs indicate that the test determines *only* the right-hand and left-hand behavior of the graph.

STUDY TIP

A polynomial function is written in **standard form** if its terms are written in descending order of exponents from left to right. Before applying the Leading Coefficient Test to a polynomial function, it is a good idea to check that the polynomial function is written in standard form.

Example 2 Applying the Leading Coefficient Test

Describe the right-hand and left-hand behavior of the graph of each function.

a. $f(x) = -x^3 + 4x$ **b.** $f(x) = x^4 - 5x^2 + 4$ **c.** $f(x) = x^5 - x$

Solution

a. Because the degree is odd and the leading coefficient is negative, the graph rises to the left and falls to the right, as shown in Figure 2.16.

b. Because the degree is even and the leading coefficient is positive, the graph rises to the left and right, as shown in Figure 2.17.

c. Because the degree is odd and the leading coefficient is positive, the graph falls to the left and rises to the right, as shown in Figure 2.18.

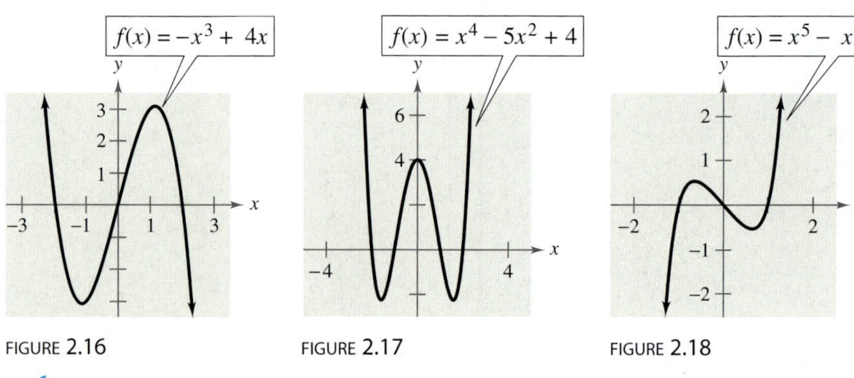

FIGURE 2.16 FIGURE 2.17 FIGURE 2.18

CHECKPOINT Now try Exercise 15.

Exploration

For each of the graphs in Example 2, count the number of zeros of the polynomial function and the number of relative minima and relative maxima. Compare these numbers with the degree of the polynomial. What do you observe?

In Example 2, note that the Leading Coefficient Test tells you only whether the graph *eventually* rises or falls to the right or left. Other characteristics of the graph, such as intercepts and minimum and maximum points, must be determined by other tests.

Zeros of Polynomial Functions

It can be shown that for a polynomial function f of degree n, the following statements are true.

1. The function f has, at most, n real zeros. (You will study this result in detail in the discussion of the Fundamental Theorem of Algebra in Section 2.5.)
2. The graph of f has, at most, $n - 1$ turning points. (Turning points, also called relative minima or relative maxima, are points at which the graph changes from increasing to decreasing or vice versa.)

STUDY TIP

Remember that the *zeros* of a function of x are the x-values for which the function is zero.

Finding the zeros of polynomial functions is one of the most important problems in algebra. There is a strong interplay between graphical and algebraic approaches to this problem. Sometimes you can use information about the graph of a function to help find its zeros, and in other cases you can use information about the zeros of a function to help sketch its graph. Finding zeros of polynomial functions is closely related to factoring and finding x-intercepts.

Real Zeros of Polynomial Functions

If f is a polynomial function and a is a real number, the following statements are equivalent.

1. $x = a$ is a *zero* of the function f.
2. $x = a$ is a *solution* of the polynomial equation $f(x) = 0$.
3. $(x - a)$ is a *factor* of the polynomial $f(x)$.
4. $(a, 0)$ is an *x-intercept* of the graph of f.

Example 3 Finding the Zeros of a Polynomial Function

Find all real zeros of

$$f(x) = -2x^4 + 2x^2.$$

Then determine the number of turning points of the graph of the function.

Algebraic Solution

To find the real zeros of the function, set $f(x)$ equal to zero and solve for x.

$$-2x^4 + 2x^2 = 0$$ Set $f(x)$ equal to 0.

$$-2x^2(x^2 - 1) = 0$$ Remove common monomial factor.

$$-2x^2(x - 1)(x + 1) = 0$$ Factor completely.

So, the real zeros are $x = 0$, $x = 1$, and $x = -1$. Because the function is a fourth-degree polynomial, the graph of f can have at most $4 - 1 = 3$ turning points.

Graphical Solution

Use a graphing utility to graph $y = -2x^4 + 2x^2$. In Figure 2.19, the graph appears to have zeros at $(0, 0)$, $(1, 0)$, and $(-1, 0)$. Use the *zero* or *root* feature, or the *zoom* and *trace* features, of the graphing utility to verify these zeros. So, the real zeros are $x = 0$, $x = 1$, and $x = -1$. From the figure, you can see that the graph has three turning points. This is consistent with the fact that a fourth-degree polynomial can have at most three turning points.

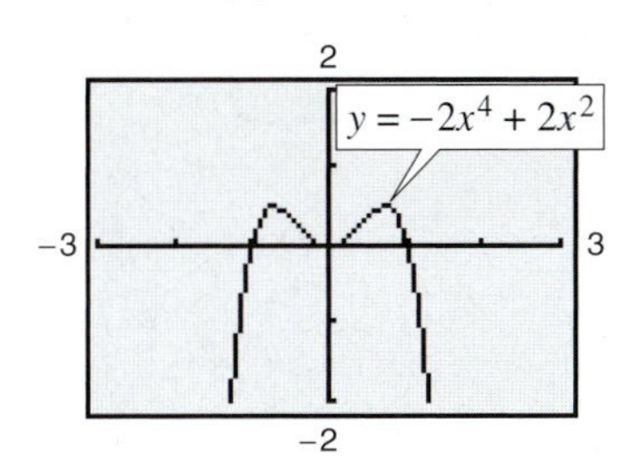

FIGURE 2.19

CHECKPOINT Now try Exercise 27.

In Example 3, note that because k is even, the factor $-2x^2$ yields the *repeated* zero $x = 0$. The graph touches the x-axis at $x = 0$, as shown in Figure 2.19.

Repeated Zeros

A factor $(x - a)^k$, $k > 1$, yields a **repeated zero** $x = a$ of **multiplicity** k.

1. If k is odd, the graph *crosses* the x-axis at $x = a$.
2. If k is even, the graph *touches* the x-axis (but does not cross the x-axis) at $x = a$.

Technology

Example 4 uses an *algebraic approach* to describe the graph of the function. A graphing utility is a complement to this approach. Remember that an important aspect of using a graphing utility is to find a viewing window that shows all significant features of the graph. For instance, the viewing window in part (a) illustrates all of the significant features of the function in Example 4.

a.

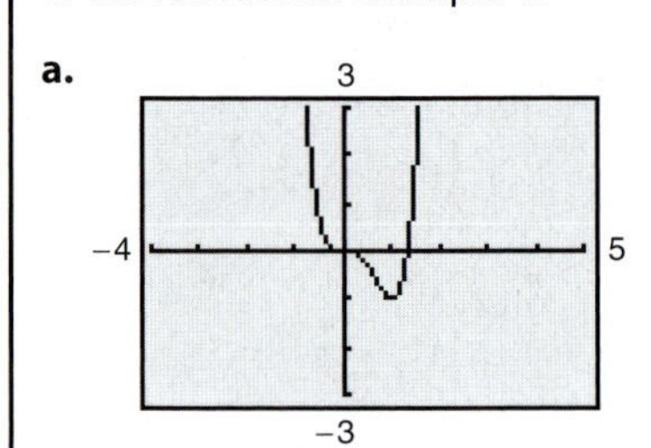

b.

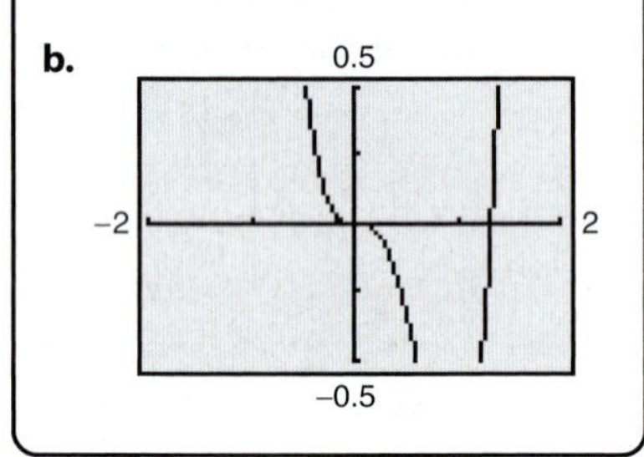

To graph polynomial functions, you can use the fact that a polynomial function can change signs only at its zeros. Between two consecutive zeros, a polynomial must be entirely positive or entirely negative. This means that when the real zeros of a polynomial function are put in order, they divide the real number line into intervals in which the function has no sign changes. These resulting intervals are **test intervals** in which a representative x-value in the interval is chosen to determine if the value of the polynomial function is positive (the graph lies above the x-axis) or negative (the graph lies below the x-axis).

Example 4 Sketching the Graph of a Polynomial Function

Sketch the graph of $f(x) = 3x^4 - 4x^3$.

Solution

1. *Apply the Leading Coefficient Test.* Because the leading coefficient is positive and the degree is even, you know that the graph eventually rises to the left and to the right (see Figure 2.20).
2. *Find the Zeros of the Polynomial.* By factoring $f(x) = 3x^4 - 4x^3$ as $f(x) = x^3(3x - 4)$, you can see that the zeros of f are $x = 0$ and $x = \frac{4}{3}$ (both of odd multiplicity). So, the x-intercepts occur at $(0, 0)$ and $\left(\frac{4}{3}, 0\right)$. Add these points to your graph, as shown in Figure 2.20.
3. *Plot a Few Additional Points.* Use the zeros of the polynomial to find the test intervals. In each test interval, choose a representative x-value and evaluate the polynomial function, as shown in the table.

Test interval	Representative x-value	Value of f	Sign	Point on graph
$(-\infty, 0)$	-1	$f(-1) = 7$	Positive	$(-1, 7)$
$\left(0, \frac{4}{3}\right)$	1	$f(1) = -1$	Negative	$(1, -1)$
$\left(\frac{4}{3}, \infty\right)$	1.5	$f(1.5) = 1.6875$	Positive	$(1.5, 1.6875)$

4. *Draw the Graph.* Draw a continuous curve through the points, as shown in Figure 2.21. Because both zeros are of odd multiplicity, you know that the graph should cross the x-axis at $x = 0$ and $x = \frac{4}{3}$.

STUDY TIP

If you are unsure of the shape of a portion of the graph of a polynomial function, plot some additional points, such as the point $(0.5, -0.3125)$ as shown in Figure 2.21.

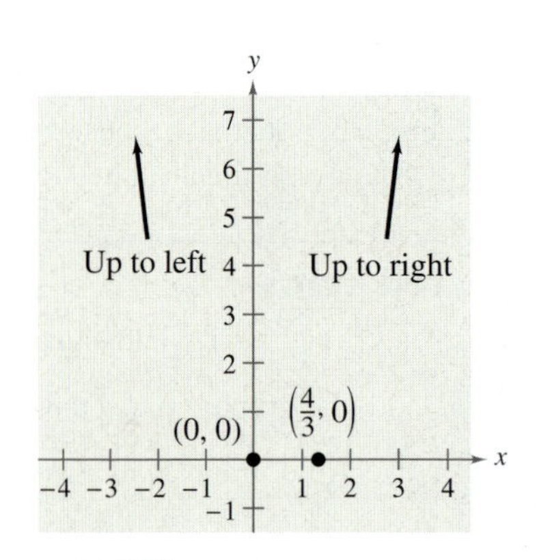

FIGURE 2.20

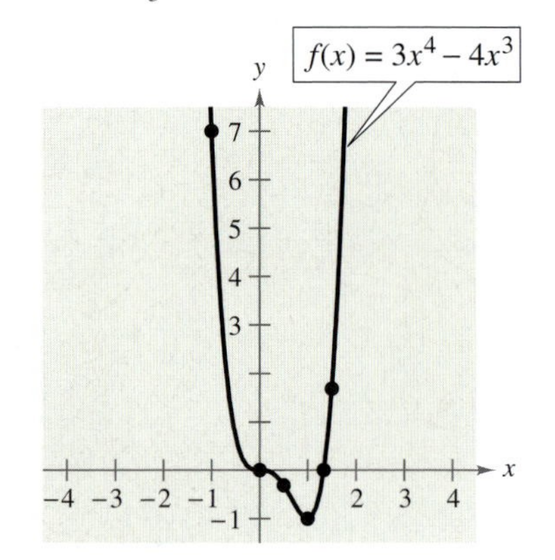

FIGURE 2.21

CHECKPOINT Now try Exercise 67.

Example 5 Sketching the Graph of a Polynomial Function

Sketch the graph of $f(x) = -2x^3 + 6x^2 - \frac{9}{2}x$.

Solution

1. *Apply the Leading Coefficient Test.* Because the leading coefficient is negative and the degree is odd, you know that the graph eventually rises to the left and falls to the right (see Figure 2.22).
2. *Find the Zeros of the Polynomial.* By factoring

$$\begin{aligned} f(x) &= -2x^3 + 6x^2 - \tfrac{9}{2}x \\ &= -\tfrac{1}{2}x(4x^2 - 12x + 9) \\ &= -\tfrac{1}{2}x(2x - 3)^2 \end{aligned}$$

   you can see that the zeros of f are $x = 0$ (odd multiplicity) and $x = \frac{3}{2}$ (even multiplicity). So, the x-intercepts occur at $(0, 0)$ and $\left(\frac{3}{2}, 0\right)$. Add these points to your graph, as shown in Figure 2.22.
3. *Plot a Few Additional Points.* Use the zeros of the polynomial to find the test intervals. In each test interval, choose a representative x-value and evaluate the polynomial function, as shown in the table.

Test interval	Representative x-value	Value of f	Sign	Point on graph
$(-\infty, 0)$	-0.5	$f(-0.5) = 4$	Positive	$(-0.5, 4)$
$\left(0, \frac{3}{2}\right)$	0.5	$f(0.5) = -1$	Negative	$(0.5, -1)$
$\left(\frac{3}{2}, \infty\right)$	2	$f(2) = -1$	Negative	$(2, -1)$

4. *Draw the Graph.* Draw a continuous curve through the points, as shown in Figure 2.23. As indicated by the multiplicities of the zeros, the graph crosses the x-axis at $(0, 0)$ but does not cross the x-axis at $\left(\frac{3}{2}, 0\right)$.

STUDY TIP

Observe in Example 5 that the sign of $f(x)$ is positive to the left of and negative to the right of the zero $x = 0$. Similarly, the sign of $f(x)$ is negative to the left and to the right of the zero $x = \frac{3}{2}$. This suggests that if the zero of a polynomial function is of *odd* multiplicity, then the sign of $f(x)$ changes from one side of the zero to the other side. If the zero is of *even* multiplicity, then the sign of $f(x)$ does not change from one side of the zero to the other side.

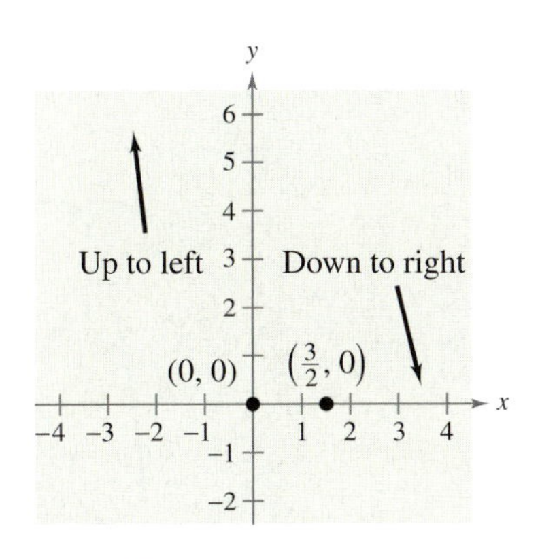

FIGURE 2.22

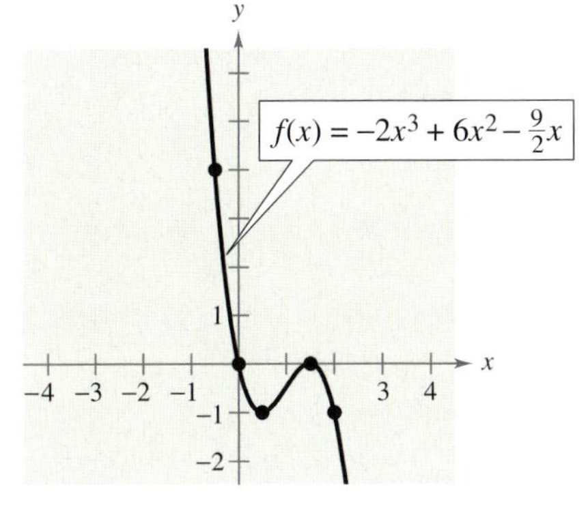

FIGURE 2.23

CHECKPOINT Now try Exercise 69.

Activities

1. Find all of the real zeros of $f(x) = 6x^4 - 33x^3 - 18x^2$.
   Answer: $-\frac{1}{2}, 0, 6$
2. Determine the right-hand and left-hand behavior of $f(x) = 6x^4 - 33x^3 - 18x^2$.
   Answer: The graph rises to the left and to the right.
3. Find a polynomial function of degree 3 that has zeros of 0, 2, and $-\frac{1}{3}$.
   Answer: $f(x) = 3x^3 - 5x^2 - 2x$

The Intermediate Value Theorem 

The next theorem, called the **Intermediate Value Theorem,** illustrates the existence of real zeros of polynomial functions. This theorem implies that if $(a, f(a))$ and $(b, f(b))$ are two points on the graph of a polynomial function such that $f(a) \neq f(b)$, then for any number d between $f(a)$ and $f(b)$ there must be a number c between a and b such that $f(c) = d$. (See Figure 2.24.)

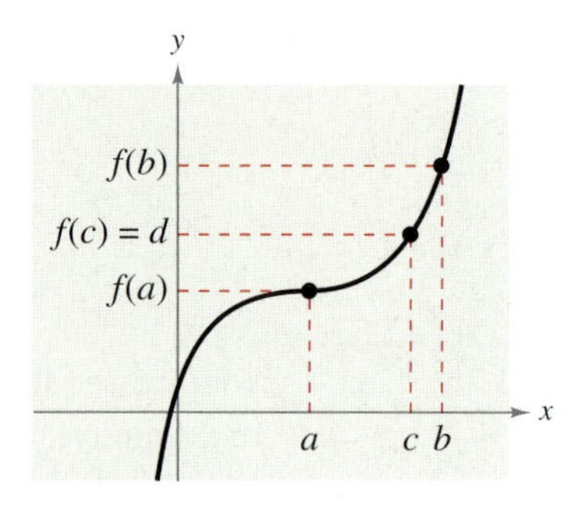

FIGURE 2.24

Intermediate Value Theorem

Let a and b be real numbers such that $a < b$. If f is a polynomial function such that $f(a) \neq f(b)$, then, in the interval $[a, b]$, f takes on every value between $f(a)$ and $f(b)$.

The Intermediate Value Theorem helps you locate the real zeros of a polynomial function in the following way. If you can find a value $x = a$ at which a polynomial function is positive, and another value $x = b$ at which it is negative, you can conclude that the function has at least one real zero between these two values. For example, the function given by $f(x) = x^3 + x^2 + 1$ is negative when $x = -2$ and positive when $x = -1$. Therefore, it follows from the Intermediate Value Theorem that f must have a real zero somewhere between -2 and -1, as shown in Figure 2.25.

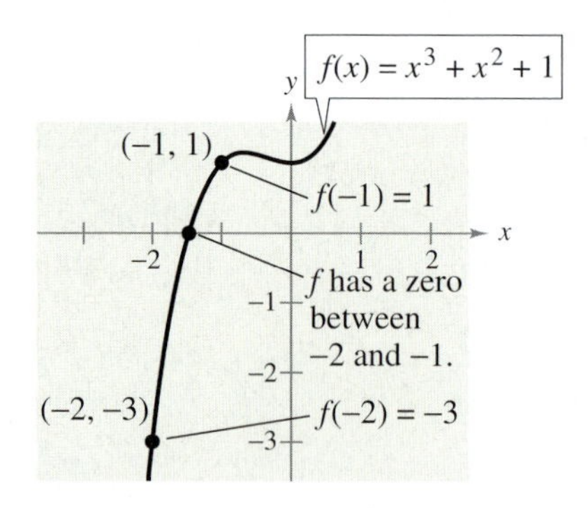

FIGURE 2.25

By continuing this line of reasoning, you can approximate any real zeros of a polynomial function to any desired accuracy. This concept is further demonstrated in Example 6.

The approximation process in Example 6 adapts very well to a graphing utility. Have students repeatedly use the *zoom* and *trace* features to find that the real zero of $f(x) = x^3 - x^2 + 1$ occurs between -0.755 and -0.754.

Example 6 Approximating a Zero of a Polynomial Function

Use the Intermediate Value Theorem to approximate the real zero of

$$f(x) = x^3 - x^2 + 1.$$

Solution

Begin by computing a few function values, as follows.

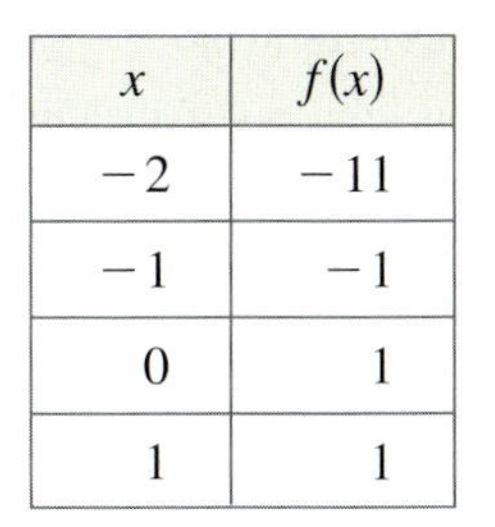

x	$f(x)$
-2	-11
-1	-1
0	1
1	1

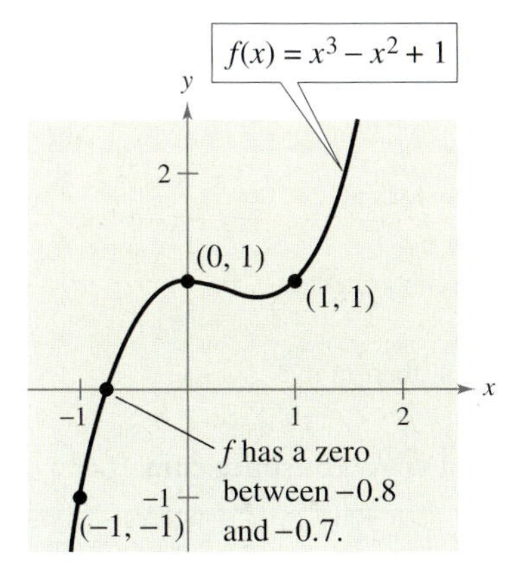

FIGURE 2.26

Because $f(-1)$ is negative and $f(0)$ is positive, you can apply the Intermediate Value Theorem to conclude that the function has a zero between -1 and 0. To pinpoint this zero more closely, divide the interval $[-1, 0]$ into tenths and evaluate the function at each point. When you do this, you will find that

$$f(-0.8) = -0.152 \qquad \text{and} \qquad f(-0.7) = 0.167.$$

So, f must have a zero between -0.8 and -0.7, as shown in Figure 2.26. For a more accurate approximation, compute function values between $f(-0.8)$ and $f(-0.7)$ and apply the Intermediate Value Theorem again. By continuing this process, you can approximate this zero to any desired accuracy.

CHECKPOINT Now try Exercise 85.

Technology

You can use the *table* feature of a graphing utility to approximate the zeros of a polynomial function. For instance, for the function given by

$$f(x) = -2x^3 - 3x^2 + 3$$

create a table that shows the function values for $-20 \leq x \leq 20$, as shown in the first table at the right. Scroll through the table looking for consecutive function values that differ in sign. From the table, you can see that $f(0)$ and $f(1)$ differ in sign. So, you can conclude from the Intermediate Value Theorem that the function has a zero between 0 and 1. You can adjust your table to show function values for $0 \leq x \leq 1$ using increments of 0.1, as shown in the second table at the right. By scrolling through the table you can see that $f(0.8)$ and $f(0.9)$ differ in sign. So, the function has a zero between 0.8 and 0.9. If you repeat this process several times, you should obtain $x \approx 0.806$ as the zero of the function. Use the *zero* or *root* feature of a graphing utility to confirm this result.

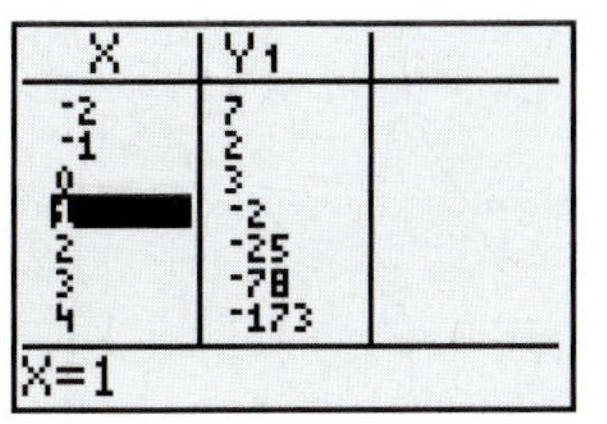

X	Y1
-2	7
-1	2
0	3
1	-2
2	-25
3	-78
4	-173

X=1

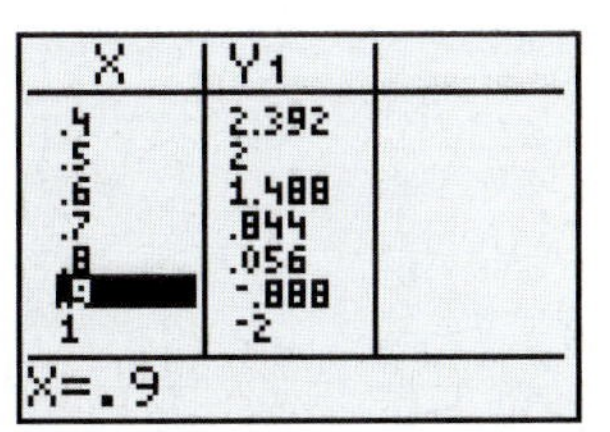

X	Y1
.4	2.392
.5	2
.6	1.488
.7	.844
.8	.056
.9	-.888
1	-2

X=.9

2.2 Exercises

VOCABULARY CHECK: Fill in the blanks.

1. The graphs of all polynomial functions are ________, which means that the graphs have no breaks, holes, or gaps.
2. The ________ ________ ________ is used to determine the left-hand and right-hand behavior of the graph of a polynomial function.
3. A polynomial function of degree n has at most ________ real zeros and at most ________ turning points.
4. If $x = a$ is a zero of a polynomial function f, then the following three statements are true.
   (a) $x = a$ is a ________ of the polynomial equation $f(x) = 0$.
   (b) ________ is a factor of the polynomial $f(x)$.
   (c) $(a, 0)$ is an ________ of the graph f.
5. If a real zero of a polynomial function is of even multiplicity, then the graph of f ________ the x-axis at $x = a$, and if it is of odd multiplicity then the graph of f ________ the x-axis at $x = a$.
6. A polynomial function is written in ________ form if its terms are written in descending order of exponents from left to right.
7. The ________ ________ Theorem states that if f is a polynomial function such that $f(a) \neq f(b)$, then in the interval $[a, b]$, f takes on every value between $f(a)$ and $f(b)$.

PREREQUISITE SKILLS REVIEW: Practice and review algebra skills needed for this section at **www.Eduspace.com.**

In Exercises 1–8, match the polynomial function with its graph. [The graphs are labeled (a), (b), (c), (d), (e), (f), (g), and (h).]

(a)

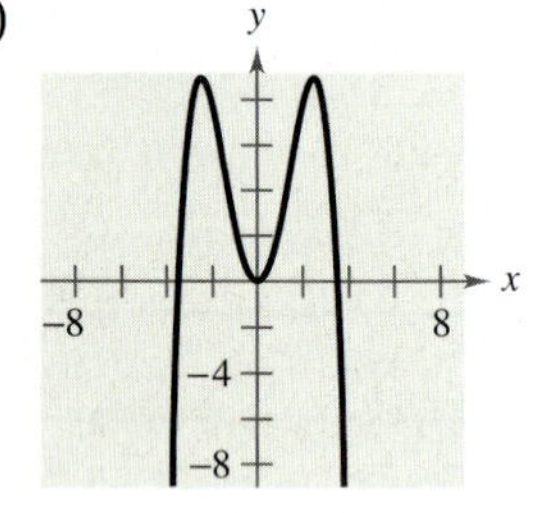

(b)

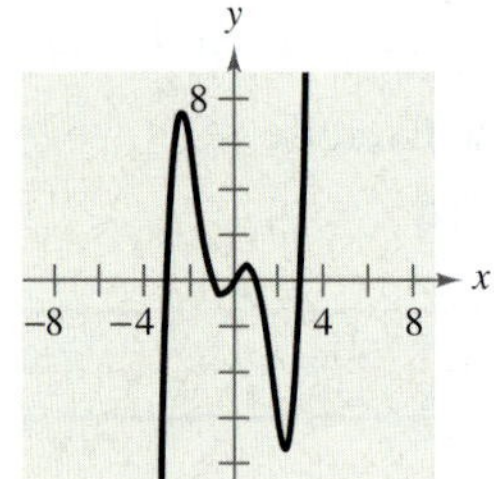

(c)

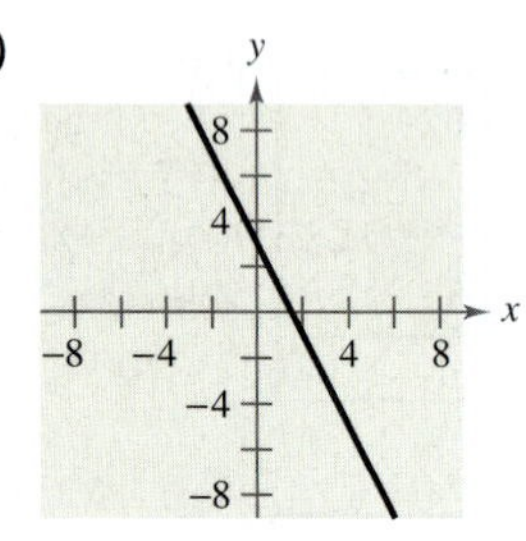

(d)

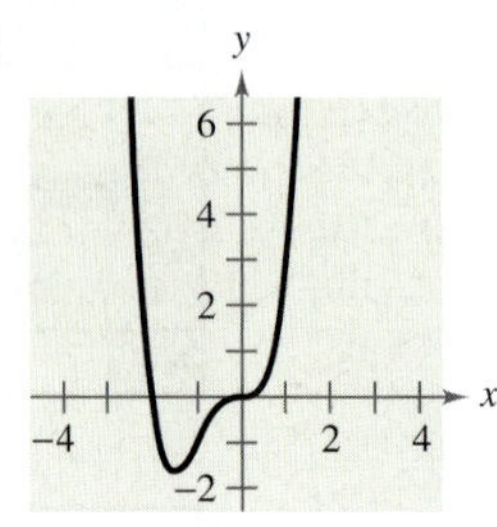

(e)

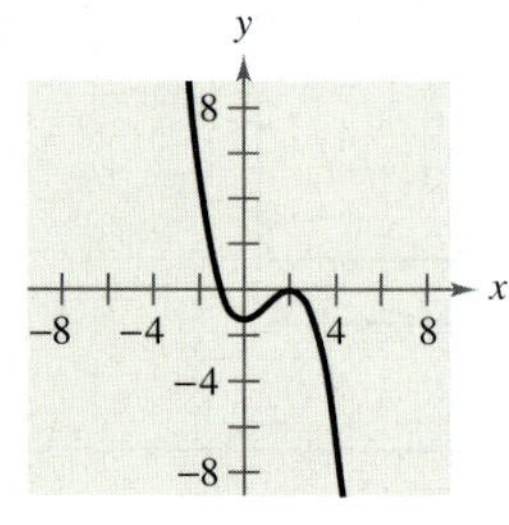

(f)

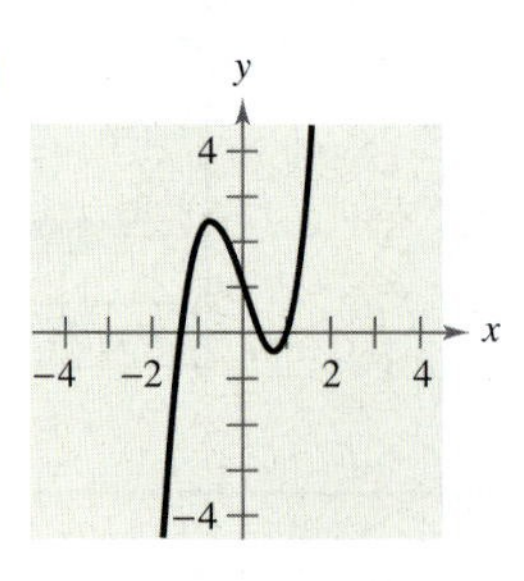

(g)

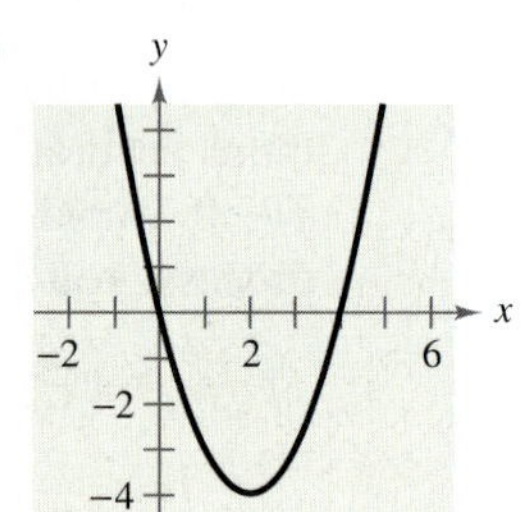

(h)

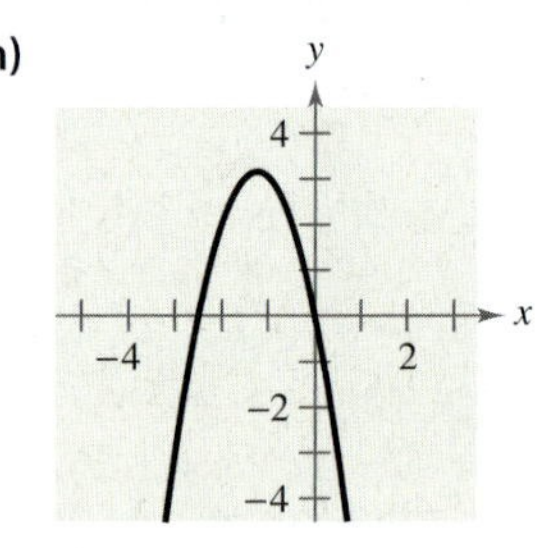

1. $f(x) = -2x + 3$
2. $f(x) = x^2 - 4x$
3. $f(x) = -2x^2 - 5x$
4. $f(x) = 2x^3 - 3x + 1$
5. $f(x) = -\frac{1}{4}x^4 + 3x^2$
6. $f(x) = -\frac{1}{3}x^3 + x^2 - \frac{4}{3}$
7. $f(x) = x^4 + 2x^3$
8. $f(x) = \frac{1}{5}x^5 - 2x^3 + \frac{9}{5}x$

In Exercises 9–12, sketch the graph of $y = x^n$ and each transformation.

9. $y = x^3$
   (a) $f(x) = (x - 2)^3$
   (b) $f(x) = x^3 - 2$
   (c) $f(x) = -\frac{1}{2}x^3$
   (d) $f(x) = (x - 2)^3 - 2$
10. $y = x^5$
    (a) $f(x) = (x + 1)^5$
    (b) $f(x) = x^5 + 1$
    (c) $f(x) = 1 - \frac{1}{2}x^5$
    (d) $f(x) = -\frac{1}{2}(x + 1)^5$
11. $y = x^4$
    (a) $f(x) = (x + 3)^4$
    (b) $f(x) = x^4 - 3$
    (c) $f(x) = 4 - x^4$
    (d) $f(x) = \frac{1}{2}(x - 1)^4$
    (e) $f(x) = (2x)^4 + 1$
    (f) $f(x) = \left(\frac{1}{2}x\right)^4 - 2$

12. $y = x^6$

(a) $f(x) = -\frac{1}{8}x^6$
(b) $f(x) = (x + 2)^6 - 4$
(c) $f(x) = x^6 - 4$
(d) $f(x) = -\frac{1}{4}x^6 + 1$
(e) $f(x) = \left(\frac{1}{4}x\right)^6 - 2$
(f) $f(x) = (2x)^6 - 1$

In Exercises 13–22, describe the right-hand and left-hand behavior of the graph of the polynomial function.

13. $f(x) = \frac{1}{3}x^3 + 5x$
14. $f(x) = 2x^2 - 3x + 1$
15. $g(x) = 5 - \frac{7}{2}x - 3x^2$
16. $h(x) = 1 - x^6$
17. $f(x) = -2.1x^5 + 4x^3 - 2$
18. $f(x) = 2x^5 - 5x + 7.5$
19. $f(x) = 6 - 2x + 4x^2 - 5x^3$
20. $f(x) = \dfrac{3x^4 - 2x + 5}{4}$
21. $h(t) = -\frac{2}{3}(t^2 - 5t + 3)$
22. $f(s) = -\frac{7}{8}(s^3 + 5s^2 - 7s + 1)$

Graphical Analysis **In Exercises 23–26, use a graphing utility to graph the functions f and g in the same viewing window. Zoom out sufficiently far to show that the right-hand and left-hand behaviors of f and g appear identical.**

23. $f(x) = 3x^3 - 9x + 1, \quad g(x) = 3x^3$
24. $f(x) = -\frac{1}{3}(x^3 - 3x + 2), \quad g(x) = -\frac{1}{3}x^3$
25. $f(x) = -(x^4 - 4x^3 + 16x), \quad g(x) = -x^4$
26. $f(x) = 3x^4 - 6x^2, \quad g(x) = 3x^4$

In Exercises 27–42, (a) find all the real zeros of the polynomial function, (b) determine the multiplicity of each zero and the number of turning points of the graph of the function, and (c) use a graphing utility to graph the function and verify your answers.

27. $f(x) = x^2 - 25$
28. $f(x) = 49 - x^2$
29. $h(t) = t^2 - 6t + 9$
30. $f(x) = x^2 + 10x + 25$
31. $f(x) = \frac{1}{3}x^2 + \frac{1}{3}x - \frac{2}{3}$
32. $f(x) = \frac{1}{2}x^2 + \frac{5}{2}x - \frac{3}{2}$
33. $f(x) = 3x^3 - 12x^2 + 3x$
34. $g(x) = 5x(x^2 - 2x - 1)$
35. $f(t) = t^3 - 4t^2 + 4t$
36. $f(x) = x^4 - x^3 - 20x^2$
37. $g(t) = t^5 - 6t^3 + 9t$
38. $f(x) = x^5 + x^3 - 6x$
39. $f(x) = 5x^4 + 15x^2 + 10$
40. $f(x) = 2x^4 - 2x^2 - 40$
41. $g(x) = x^3 + 3x^2 - 4x - 12$
42. $f(x) = x^3 - 4x^2 - 25x + 100$

Graphical Analysis **In Exercises 43–46, (a) use a graphing utility to graph the function, (b) use the graph to approximate any x-intercepts of the graph, (c) set $y = 0$ and solve the resulting equation, and (d) compare the results of part (c) with any x-intercepts of the graph.**

43. $y = 4x^3 - 20x^2 + 25x$
44. $y = 4x^3 + 4x^2 - 8x + 8$
45. $y = x^5 - 5x^3 + 4x$
46. $y = \frac{1}{4}x^3(x^2 - 9)$

In Exercises 47–56, find a polynomial function that has the given zeros. (There are many correct answers.)

47. 0, 10
48. 0, -3
49. 2, -6
50. -4, 5
51. 0, -2, -3
52. 0, 2, 5
53. 4, -3, 3, 0
54. -2, -1, 0, 1, 2
55. $1 + \sqrt{3}$, $1 - \sqrt{3}$
56. 2, $4 + \sqrt{5}$, $4 - \sqrt{5}$

In Exercises 57–66, find a polynomial of degree n that has the given zero(s). (There are many correct answers.)

	Zero(s)	*Degree*
57.	$x = -2$	$n = 2$
58.	$x = -8, -4$	$n = 2$
59.	$x = -3, 0, 1$	$n = 3$
60.	$x = -2, 4, 7$	$n = 3$
61.	$x = 0, \sqrt{3}, -\sqrt{3}$	$n = 3$
62.	$x = 9$	$n = 3$
63.	$x = -5, 1, 2$	$n = 4$
64.	$x = -4, -1, 3, 6$	$n = 4$
65.	$x = 0, -4$	$n = 5$
66.	$x = -3, 1, 5, 6$	$n = 5$

In Exercises 67–80, sketch the graph of the function by (a) applying the Leading Coefficient Test, (b) finding the zeros of the polynomial, (c) plotting sufficient solution points, and (d) drawing a continuous curve through the points.

67. $f(x) = x^3 - 9x$
68. $g(x) = x^4 - 4x^2$
69. $f(t) = \frac{1}{4}(t^2 - 2t + 15)$
70. $g(x) = -x^2 + 10x - 16$
71. $f(x) = x^3 - 3x^2$
72. $f(x) = 1 - x^3$
73. $f(x) = 3x^3 - 15x^2 + 18x$
74. $f(x) = -4x^3 + 4x^2 + 15x$
75. $f(x) = -5x^2 - x^3$
76. $f(x) = -48x^2 + 3x^4$
77. $f(x) = x^2(x - 4)$
78. $h(x) = \frac{1}{3}x^3(x - 4)^2$
79. $g(t) = -\frac{1}{4}(t - 2)^2(t + 2)^2$
80. $g(x) = \frac{1}{10}(x + 1)^2(x - 3)^3$

In Exercises 81–84, use a graphing utility to graph the function. Use the *zero* or *root* feature to approximate the real zeros of the function. Then determine the multiplicity of each zero.

81. $f(x) = x^3 - 4x$

82. $f(x) = \frac{1}{4}x^4 - 2x^2$

83. $g(x) = \frac{1}{5}(x + 1)^2(x - 3)(2x - 9)$

84. $h(x) = \frac{1}{5}(x + 2)^2(3x - 5)^2$

In Exercises 85–88, use the Intermediate Value Theorem and the *table* feature of a graphing utility to find intervals one unit in length in which the polynomial function is guaranteed to have a zero. Adjust the table to approximate the zeros of the function. Use the *zero* or *root* feature of a graphing utility to verify your results.

85. $f(x) = x^3 - 3x^2 + 3$

86. $f(x) = 0.11x^3 - 2.07x^2 + 9.81x - 6.88$

87. $g(x) = 3x^4 + 4x^3 - 3$

88. $h(x) = x^4 - 10x^2 + 3$

89. ***Numerical and Graphical Analysis*** An open box is to be made from a square piece of material, 36 inches on a side, by cutting equal squares with sides of length x from the corners and turning up the sides (see figure).

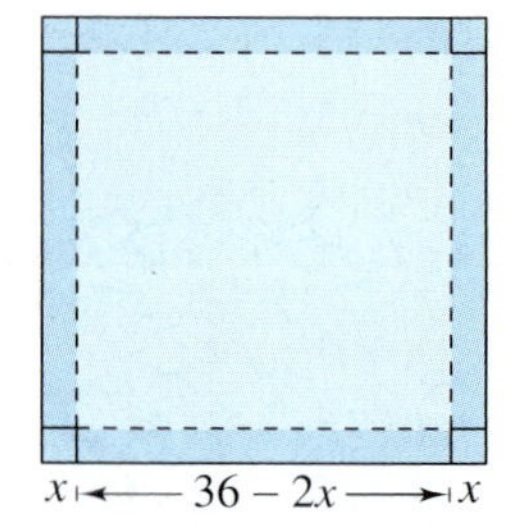

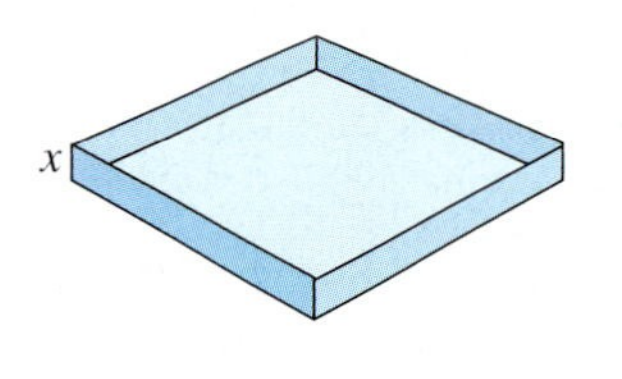

(a) Verify that the volume of the box is given by the function

$$V(x) = x(36 - 2x)^2.$$

(b) Determine the domain of the function.

(c) Use a graphing utility to create a table that shows the box height x and the corresponding volumes V. Use the table to estimate the dimensions that will produce a maximum volume.

(d) Use a graphing utility to graph V and use the graph to estimate the value of x for which $V(x)$ is maximum. Compare your result with that of part (c).

90. ***Maximum Volume*** An open box with locking tabs is to be made from a square piece of material 24 inches on a side. This is to be done by cutting equal squares from the corners and folding along the dashed lines shown in the figure.

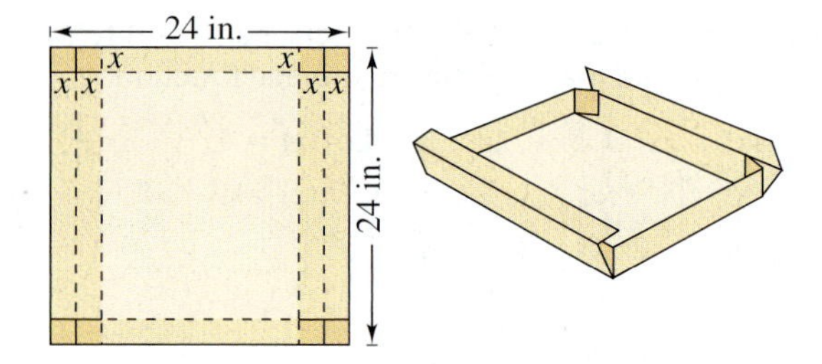

(a) Verify that the volume of the box is given by the function

$$V(x) = 8x(6 - x)(12 - x).$$

(b) Determine the domain of the function V.

(c) Sketch a graph of the function and estimate the value of x for which $V(x)$ is maximum.

91. ***Construction*** A roofing contractor is fabricating gutters from 12-inch aluminum sheeting. The contractor plans to use an aluminum siding folding press to create the gutter by creasing equal lengths for the sidewalls (see figure).

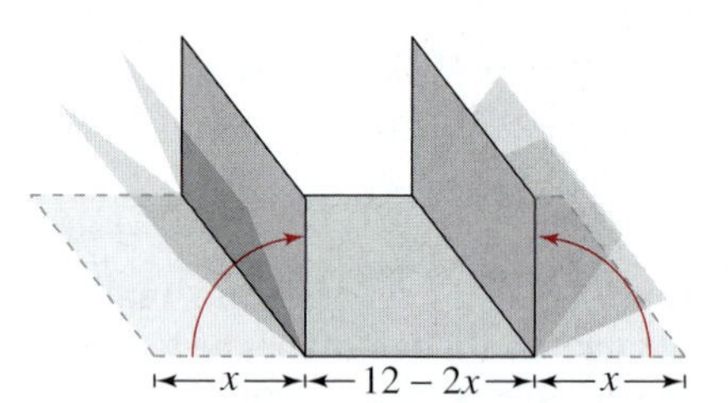

(a) Let x represent the height of the sidewall of the gutter. Write a function A that represents the cross-sectional area of the gutter.

(b) The length of the aluminum sheeting is 16 feet. Write a function V that represents the volume of one run of gutter in terms of x.

(c) Determine the domain of the function in part (b).

(d) Use a graphing utility to create a table that shows the sidewall height x and the corresponding volumes V. Use the table to estimate the dimensions that will produce a maximum volume.

(e) Use a graphing utility to graph V. Use the graph to estimate the value of x for which $V(x)$ is a maximum. Compare your result with that of part (d).

(f) Would the value of x change if the aluminum sheeting were of different lengths? Explain.

92. ***Construction*** An industrial propane tank is formed by adjoining two hemispheres to the ends of a right circular cylinder. The length of the cylindrical portion of the tank is four times the radius of the hemispherical components (see figure).

(a) Write a function that represents the total volume V of the tank in terms of r.

(b) Find the domain of the function.

 (c) Use a graphing utility to graph the function.

 (d) The total volume of the tank is to be 120 cubic feet. Use the graph from part (c) to estimate the radius and length of the cylindrical portion of the tank.

 ***Data Analysis: Home Prices*** **In Exercise 93–96, use the table, which shows the median prices (in thousands of dollars) of new privately owned U.S. homes in the Midwest y_1 and in the South y_2 for the years 1997 through 2003. The data can be approximated by the following models.**

$$y_1 = 0.139t^3 - 4.42t^2 + 51.1t - 39$$

$$y_2 = 0.056t^3 - 1.73t^2 + 23.8t + 29$$

In the models, t represents the year, with $t = 7$ corresponding to 1997. (Source: U.S. Census Bureau; U.S. Department of Housing and Urban Development)

Year, t	y_1	y_2
7	150	130
8	158	136
9	164	146
10	170	148
11	173	155
12	178	163
13	184	168

93. Use a graphing utility to plot the data and graph the model for y_1 in the same viewing window. How closely does the model represent the data?

94. Use a graphing utility to plot the data and graph the model for y_2 in the same viewing window. How closely does the model represent the data?

95. Use the models to predict the median prices of a new privately owned home in both regions in 2008. Do your answers seem reasonable? Explain.

96. Use the graphs of the models in Exercises 93 and 94 to write a short paragraph about the relationship between the median prices of homes in the two regions.

Model It

97. ***Tree Growth*** The growth of a red oak tree is approximated by the function

$$G = -0.003t^3 + 0.137t^2 + 0.458t - 0.839$$

where G is the height of the tree (in feet) and t $(2 \le t \le 34)$ is its age (in years).

(a) Use a graphing utility to graph the function. (*Hint:* Use a viewing window in which $-10 \le x \le 45$ and $-5 \le y \le 60$.)

(b) Estimate the age of the tree when it is growing most rapidly. This point is called the *point of diminishing returns* because the increase in size will be less with each additional year.

(c) Using calculus, the point of diminishing returns can also be found by finding the vertex of the parabola given by

$$y = -0.009t^2 + 0.274t + 0.458.$$

Find the vertex of this parabola.

(d) Compare your results from parts (b) and (c).

98. ***Revenue*** The total revenue R (in millions of dollars) for a company is related to its advertising expense by the function

$$R = \frac{1}{100{,}000}(-x^3 + 600x^2), \qquad 0 \le x \le 400$$

where x is the amount spent on advertising (in tens of thousands of dollars). Use the graph of this function, shown in the figure, to estimate the point on the graph at which the function is increasing most rapidly. This point is called the *point of diminishing returns* because any expense above this amount will yield less return per dollar invested in advertising.

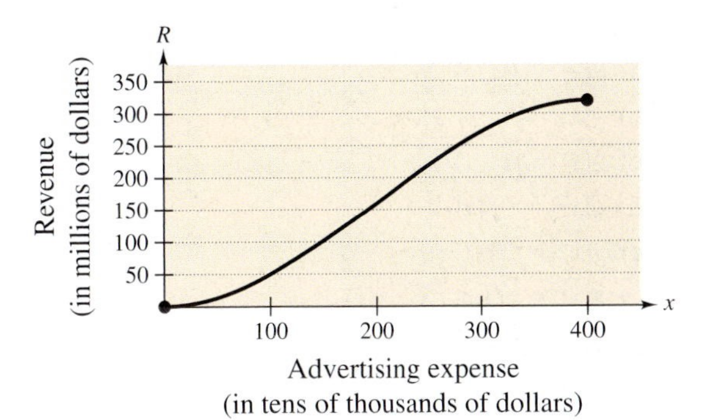

Synthesis

True or False? **In Exercises 99–101, determine whether the statement is true or false. Justify your answer.**

99. A fifth-degree polynomial can have five turning points in its graph.

100. It is possible for a sixth-degree polynomial to have only one solution.

101. The graph of the function given by

$$f(x) = 2 + x - x^2 + x^3 - x^4 + x^5 + x^6 - x^7$$

rises to the left and falls to the right.

102. ***Graphical Analysis*** For each graph, describe a polynomial function that could represent the graph. (Indicate the degree of the function and the sign of its leading coefficient.)

(a)

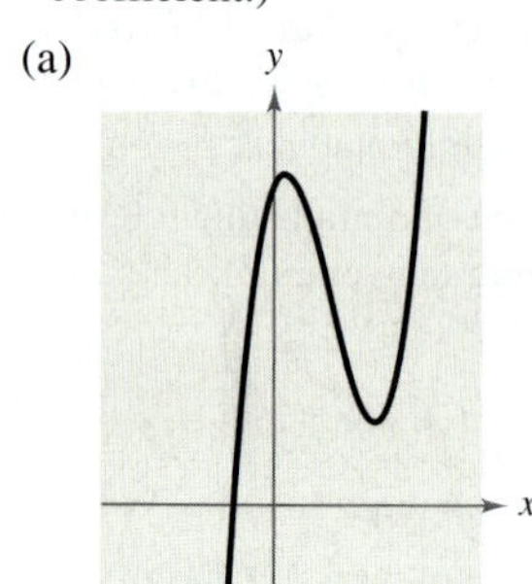

(b)

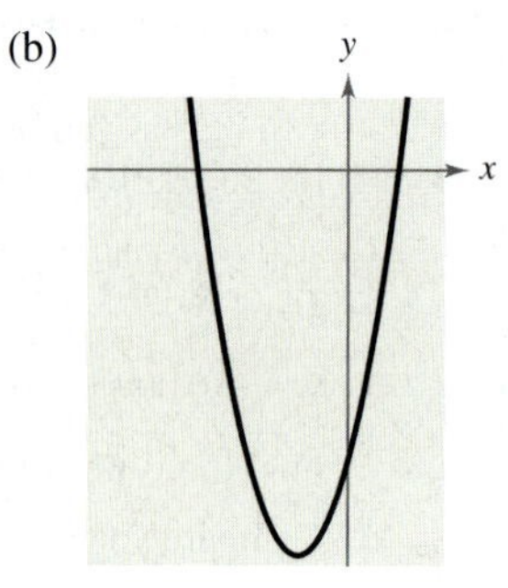

(c)

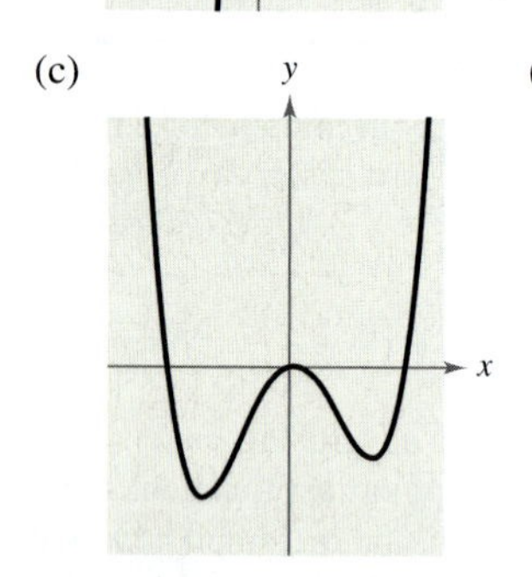

(d)

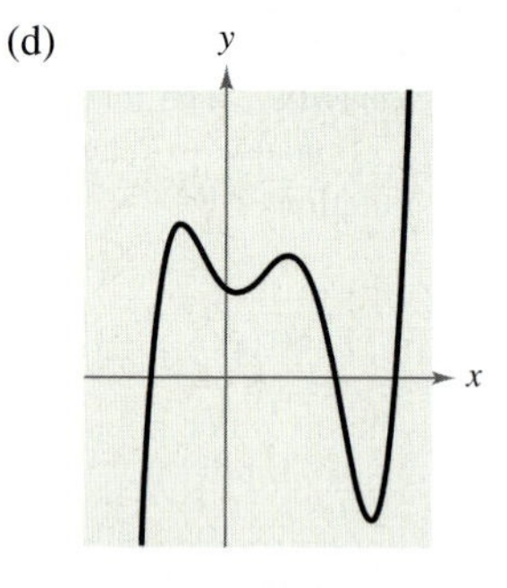

103. ***Graphical Reasoning*** Sketch a graph of the function given by $f(x) = x^4$. Explain how the graph of each function g differs (if it does) from the graph of each function f. Determine whether g is odd, even, or neither.

(a) $g(x) = f(x) + 2$

(b) $g(x) = f(x + 2)$

(c) $g(x) = f(-x)$

(d) $g(x) = -f(x)$

(e) $g(x) = f\left(\frac{1}{2}x\right)$

(f) $g(x) = \frac{1}{2}f(x)$

(g) $g(x) = f\left(x^{3/4}\right)$

(h) $g(x) = (f \circ f)(x)$

104. ***Exploration*** Explore the transformations of the form $g(x) = a(x - h)^5 + k$.

(a) Use a graphing utility to graph the functions given by

$$y_1 = -\frac{1}{3}(x - 2)^5 + 1$$

and

$$y_2 = \frac{3}{5}(x + 2)^5 - 3.$$

Determine whether the graphs are increasing or decreasing. Explain.

(b) Will the graph of g always be increasing or decreasing? If so, is this behavior determined by a, h, or k? Explain.

(c) Use a graphing utility to graph the function given by

$$H(x) = x^5 - 3x^3 + 2x + 1.$$

Use the graph and the result of part (b) to determine whether H can be written in the form $H(x) = a(x - h)^5 + k$. Explain.

Skills Review

In Exercises 105–108, factor the expression completely.

105. $5x^2 + 7x - 24$

106. $6x^3 - 61x^2 + 10x$

107. $4x^4 - 7x^3 - 15x^2$

108. $y^3 + 216$

In Exercises 109–112, solve the equation by factoring.

109. $2x^2 - x - 28 = 0$

110. $3x^2 - 22x - 16 = 0$

111. $12x^2 + 11x - 5 = 0$

112. $x^2 + 24x + 144 = 0$

In Exercises 113–116, solve the equation by completing the square.

113. $x^2 - 2x - 21 = 0$

114. $x^2 - 8x + 2 = 0$

115. $2x^2 + 5x - 20 = 0$

116. $3x^2 + 4x - 9 = 0$

In Exercises 117–122, describe the transformation from a common function that occurs in $f(x)$. Then sketch its graph.

117. $f(x) = (x + 4)^2$

118. $f(x) = 3 - x^2$

119. $f(x) = \sqrt{x + 1} - 5$

120. $f(x) = 7 - \sqrt{x - 6}$

121. $f(x) = 2[\![x]\!] + 9$

122. $f(x) = 10 - \frac{1}{3}[\![x + 3]\!]$

2.3 Polynomial and Synthetic Division

What you should learn

- Use long division to divide polynomials by other polynomials.
- Use synthetic division to divide polynomials by binomials of the form $(x - k)$.
- Use the Remainder Theorem and the Factor Theorem.

Why you should learn it

Synthetic division can help you evaluate polynomial functions. For instance, in Exercise 73 on page 160, you will use synthetic division to determine the number of U.S. military personnel in 2008.

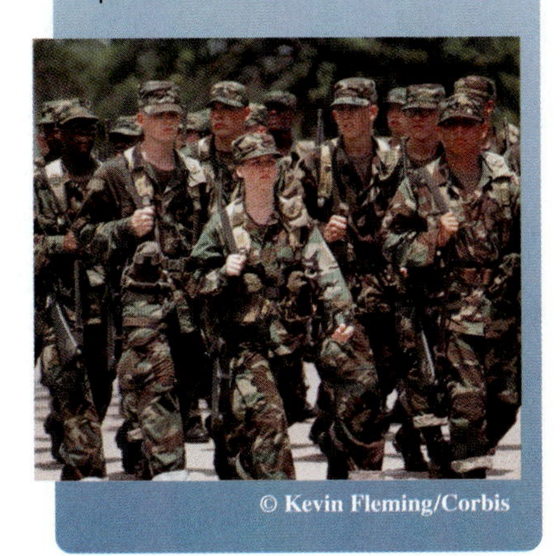

© Kevin Fleming/Corbis

Long Division of Polynomials

In this section, you will study two procedures for *dividing* polynomials. These procedures are especially valuable in factoring and finding the zeros of polynomial functions. To begin, suppose you are given the graph of

$$f(x) = 6x^3 - 19x^2 + 16x - 4.$$

Notice that a zero of f occurs at $x = 2$, as shown in Figure 2.27. Because $x = 2$ is a zero of f, you know that $(x - 2)$ is a factor of $f(x)$. This means that there exists a second-degree polynomial $q(x)$ such that

$$f(x) = (x - 2) \cdot q(x).$$

To find $q(x)$, you can use **long division,** as illustrated in Example 1.

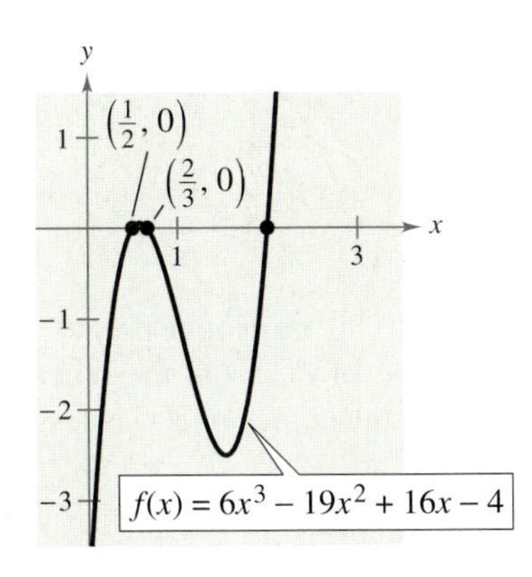

FIGURE 2.27

Example 1 **Long Division of Polynomials**

Divide $6x^3 - 19x^2 + 16x - 4$ by $x - 2$, and use the result to factor the polynomial completely.

Solution

Think $\frac{6x^3}{x} = 6x^2$.

Think $\frac{-7x^2}{x} = -7x$.

Think $\frac{2x}{x} = 2$.

$$
\begin{array}{r l l}
 & 6x^2 - 7x + 2 & \\
x - 2 \,) & 6x^3 - 19x^2 + 16x - 4 & \\
 & \underline{6x^3 - 12x^2} & \text{Multiply: } 6x^2(x - 2). \\
 & -7x^2 + 16x & \text{Subtract.} \\
 & \underline{-7x^2 + 14x} & \text{Multiply: } -7x(x - 2). \\
 & 2x - 4 & \text{Subtract.} \\
 & \underline{2x - 4} & \text{Multiply: } 2(x - 2). \\
 & 0 & \text{Subtract.}
\end{array}
$$

From this division, you can conclude that

$$6x^3 - 19x^2 + 16x - 4 = (x - 2)(6x^2 - 7x + 2)$$

and by factoring the quadratic $6x^2 - 7x + 2$, you have

$$6x^3 - 19x^2 + 16x - 4 = (x - 2)(2x - 1)(3x - 2).$$

Note that this factorization agrees with the graph shown in Figure 2.27 in that the three x-intercepts occur at $x = 2$, $x = \frac{1}{2}$, and $x = \frac{2}{3}$.

✓CHECKPOINT Now try Exercise 5.

Note that one of the many uses of polynomial division is to write a function as a sum of terms to find slant asymptotes (see Section 2.6). This is a skill that is also used frequently in calculus.

In Example 1, $x - 2$ is a factor of the polynomial $6x^3 - 19x^2 + 16x - 4$, and the long division process produces a remainder of zero. Often, long division will produce a nonzero remainder. For instance, if you divide $x^2 + 3x + 5$ by $x + 1$, you obtain the following.

$$
\begin{array}{rl}
 & x + 2 \quad \leftarrow \text{Quotient} \\
\text{Divisor} \rightarrow \; x + 1 \; \big) & x^2 + 3x + 5 \quad \leftarrow \text{Dividend} \\
 & \underline{x^2 + x} \\
 & 2x + 5 \\
 & \underline{2x + 2} \\
 & 3 \quad \leftarrow \text{Remainder}
\end{array}
$$

In fractional form, you can write this result as follows.

$$\underbrace{\frac{\overbrace{x^2 + 3x + 5}^{\text{Dividend}}}{x + 1}}_{\text{Divisor}} = \overbrace{x + 2}^{\text{Quotient}} + \underbrace{\frac{\overset{\text{Remainder}}{3}}{x + 1}}_{\text{Divisor}}$$

Have students identify the dividend, divisor, quotient, and remainder when dividing polynomials.

This implies that

$$x^2 + 3x + 5 = (x + 1)(x + 2) + 3 \qquad \text{Multiply each side by } (x + 1).$$

which illustrates the following theorem, called the **Division Algorithm.**

The Division Algorithm

If $f(x)$ and $d(x)$ are polynomials such that $d(x) \neq 0$, and the degree of $d(x)$ is less than or equal to the degree of $f(x)$, there exist unique polynomials $q(x)$ and $r(x)$ such that

$$f(x) = d(x)q(x) + r(x)$$

Dividend: $f(x)$; Divisor: $d(x)$; Quotient: $q(x)$; Remainder: $r(x)$

where $r(x) = 0$ *or* the degree of $r(x)$ is less than the degree of $d(x)$. If the remainder $r(x)$ is zero, $d(x)$ *divides evenly* into $f(x)$.

The Division Algorithm can also be written as

$$\frac{f(x)}{d(x)} = q(x) + \frac{r(x)}{d(x)}.$$

In the Division Algorithm, the rational expression $f(x)/d(x)$ is **improper** because the degree of $f(x)$ is greater than or equal to the degree of $d(x)$. On the other hand, the rational expression $r(x)/d(x)$ is **proper** because the degree of $r(x)$ is less than the degree of $d(x)$.

Before you apply the Division Algorithm, follow these steps.

1. Write the dividend and divisor in descending powers of the variable.
2. Insert placeholders with zero coefficients for missing powers of the variable.

Example 2 Long Division of Polynomials

Divide $x^3 - 1$ by $x - 1$.

Solution

Because there is no x^2-term or x-term in the dividend, you need to line up the subtraction by using zero coefficients (or leaving spaces) for the missing terms.

$$\begin{array}{rl}
 & x^2 + x + 1 \\
x - 1 \,\big) & \overline{x^3 + 0x^2 + 0x - 1} \\
 & \underline{x^3 - x^2} \\
 & \quad\;\; x^2 + 0x \\
 & \quad\;\; \underline{x^2 - x} \\
 & \qquad\quad\;\; x - 1 \\
 & \qquad\quad\;\; \underline{x - 1} \\
 & \qquad\qquad\quad 0
\end{array}$$

So, $x - 1$ divides evenly into $x^3 - 1$, and you can write

$$\frac{x^3 - 1}{x - 1} = x^2 + x + 1, \quad x \neq 1.$$

✓CHECKPOINT Now try Exercise 13.

You can check the result of Example 2 by multiplying.

$$(x - 1)(x^2 + x + 1) = x^3 + x^2 + x - x^2 - x - 1 = x^3 - 1$$

Example 3 Long Division of Polynomials

Divide $2x^4 + 4x^3 - 5x^2 + 3x - 2$ by $x^2 + 2x - 3$.

Solution

$$\begin{array}{rl}
 & 2x^2 \qquad\qquad + 1 \\
x^2 + 2x - 3 \,\big) & \overline{2x^4 + 4x^3 - 5x^2 + 3x - 2} \\
 & \underline{2x^4 + 4x^3 - 6x^2} \\
 & \qquad\qquad\quad x^2 + 3x - 2 \\
 & \qquad\qquad\quad \underline{x^2 + 2x - 3} \\
 & \qquad\qquad\qquad\quad\;\; x + 1
\end{array}$$

Note that the first subtraction eliminated two terms from the dividend. When this happens, the quotient skips a term. You can write the result as

$$\frac{2x^4 + 4x^3 - 5x^2 + 3x - 2}{x^2 + 2x - 3} = 2x^2 + 1 + \frac{x + 1}{x^2 + 2x - 3}.$$

✓CHECKPOINT Now try Exercise 15.

Remind students that when division yields a remainder, it is important that they write the remainder term correctly.

Point out to students that a graphing utility can be used to check the answer to a polynomial division problem. When students graph both the original polynomial division problem and the answer in the same viewing window, the graphs should coincide.

Synthetic Division

There is a nice shortcut for long division of polynomials when dividing by divisors of the form $x - k$. This shortcut is called **synthetic division.** The pattern for synthetic division of a cubic polynomial is summarized as follows. (The pattern for higher-degree polynomials is similar.)

Synthetic Division (for a Cubic Polynomial)

To divide $ax^3 + bx^2 + cx + d$ by $x - k$, use the following pattern.

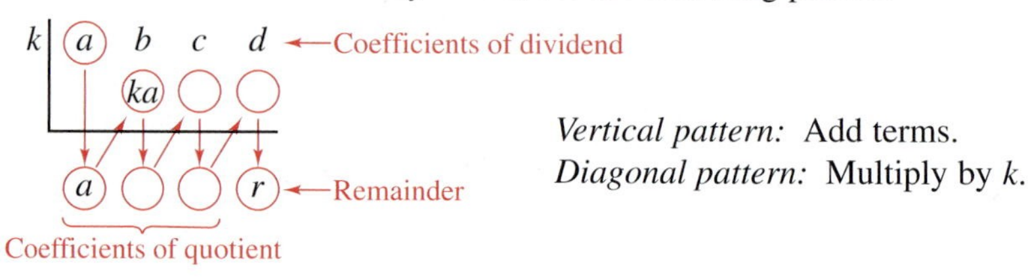

Synthetic division works only for divisors of the form $x - k$. [Remember that $x + k = x - (-k)$.] You cannot use synthetic division to divide a polynomial by a quadratic such as $x^2 - 3$.

Example 4 Using Synthetic Division

Use synthetic division to divide $x^4 - 10x^2 - 2x + 4$ by $x + 3$.

Solution

You should set up the array as follows. Note that a zero is included for the missing x^3-term in the dividend.

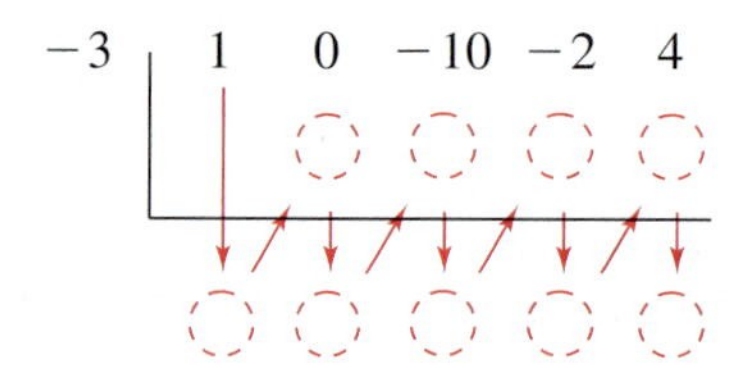

Then, use the synthetic division pattern by adding terms in columns and multiplying the results by -3.

Divisor: $x + 3$ Dividend: $x^4 - 10x^2 - 2x + 4$

-3	1	0	-10	-2	4
		-3	9	3	-3
	1	-3	-1	1	1 ← Remainder: 1

Quotient: $x^3 - 3x^2 - x + 1$

So, you have

$$\frac{x^4 - 10x^2 - 2x + 4}{x + 3} = x^3 - 3x^2 - x + 1 + \frac{1}{x + 3}.$$

CHECKPOINT Now try Exercise 19.

The Remainder and Factor Theorems

The remainder obtained in the synthetic division process has an important interpretation, as described in the **Remainder Theorem.**

> **The Remainder Theorem**
>
> If a polynomial $f(x)$ is divided by $x - k$, the remainder is
>
> $r = f(k)$.

For a proof of the Remainder Theorem, see Proofs in Mathematics on page 213.

The Remainder Theorem tells you that synthetic division can be used to evaluate a polynomial function. That is, to evaluate a polynomial function $f(x)$ when $x = k$, divide $f(x)$ by $x - k$. The remainder will be $f(k)$, as illustrated in Example 5.

Example 5 Using the Remainder Theorem

Use the Remainder Theorem to evaluate the following function at $x = -2$.

$$f(x) = 3x^3 + 8x^2 + 5x - 7$$

Solution

Using synthetic division, you obtain the following.

-2	3	8	5	-7
		-6	-4	-2
	3	2	1	-9

Because the remainder is $r = -9$, you can conclude that

$$f(-2) = -9. \qquad r = f(k)$$

This means that $(-2, -9)$ is a point on the graph of f. You can check this by substituting $x = -2$ in the original function.

Check

$$f(-2) = 3(-2)^3 + 8(-2)^2 + 5(-2) - 7$$
$$= 3(-8) + 8(4) - 10 - 7 = -9$$

CHECKPOINT Now try Exercise 45.

Additional Example

Use the Remainder Theorem to evaluate $f(x) = 4x^2 - 10x - 21$ when $x = 5$.

Solution

Using synthetic division, you obtain the following.

5	4	-10	-21
		20	50
	4	10	29

Because the remainder is 29, you can conclude that $f(5) = 29$.

Another important theorem is the **Factor Theorem,** stated below. This theorem states that you can test to see whether a polynomial has $(x - k)$ as a factor by evaluating the polynomial at $x = k$. If the result is 0, $(x - k)$ is a factor.

> **The Factor Theorem**
>
> A polynomial $f(x)$ has a factor $(x - k)$ if and only if $f(k) = 0$.

For a proof of the Factor Theorem, see Proofs in Mathematics on page 213.

Example 6 **Factoring a Polynomial: Repeated Division**

Show that $(x - 2)$ and $(x + 3)$ are factors of

$$f(x) = 2x^4 + 7x^3 - 4x^2 - 27x - 18.$$

Then find the remaining factors of $f(x)$.

Solution

Using synthetic division with the factor $(x - 2)$, you obtain the following.

$$\begin{array}{r|rrrrr} 2 & 2 & 7 & -4 & -27 & -18 \\ & & 4 & 22 & 36 & 18 \\ \hline & 2 & 11 & 18 & 9 & 0 \end{array}$$

0 remainder, so $f(2) = 0$ and $(x - 2)$ is a factor.

Take the result of this division and perform synthetic division again using the factor $(x + 3)$.

$$\begin{array}{r|rrrr} -3 & 2 & 11 & 18 & 9 \\ & & -6 & -15 & -9 \\ \hline & 2 & 5 & 3 & 0 \end{array}$$

0 remainder, so $f(-3) = 0$ and $(x + 3)$ is a factor.

Because the resulting quadratic expression factors as

$$2x^2 + 5x + 3 = (2x + 3)(x + 1)$$

the complete factorization of $f(x)$ is

$$f(x) = (x - 2)(x + 3)(2x + 3)(x + 1).$$

Note that this factorization implies that f has four real zeros:

$$x = 2, x = -3, x = -\tfrac{3}{2}, \text{ and } x = -1.$$

This is confirmed by the graph of f, which is shown in Figure 2.28.

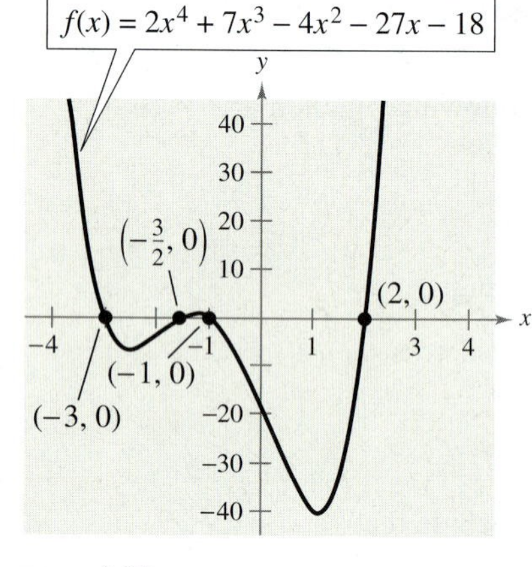

FIGURE 2.28

CHECKPOINT Now try Exercise 57.

Uses of the Remainder in Synthetic Division

The remainder r, obtained in the synthetic division of $f(x)$ by $x - k$, provides the following information.

1. The remainder r gives the value of f at $x = k$. That is, $r = f(k)$.
2. If $r = 0$, $(x - k)$ is a factor of $f(x)$.
3. If $r = 0$, $(k, 0)$ is an x-intercept of the graph of f.

Throughout this text, the importance of developing several problem-solving strategies is emphasized. In the exercises for this section, try using more than one strategy to solve several of the exercises. For instance, if you find that $x - k$ divides evenly into $f(x)$ (with no remainder), try sketching the graph of f. You should find that $(k, 0)$ is an x-intercept of the graph.

Activities

1. Divide using long division.

   $$\frac{4x^5 - x^3 + 2x^2 - x}{2x + 1}$$

   Answer: $2x^4 - x^3 + x - 1 + \dfrac{1}{2x + 1}$

2. Use synthetic division to determine if $(x + 3)$ is a factor of $f(x) = 3x^3 + 4x^2 - 18x - 3$.

   Answer: No, it is not.

3. Use the Remainder Theorem to evaluate $f(-3)$ for $f(x) = 2x^3 - 4x^2 + 1$.

   Answer: -89

2.3 Exercises

VOCABULARY CHECK:

1. Two forms of the Division Algorithm are shown below. Identify and label each term or function.

$$f(x) = d(x)q(x) + r(x) \qquad \frac{f(x)}{d(x)} = q(x) + \frac{r(x)}{d(x)}$$

In Exercises 2–5, fill in the blanks.

2. The rational expression $p(x)/q(x)$ is called ________ if the degree of the numerator is greater than or equal to that of the denominator, and is called ________ if the degree of the numerator is less than that of the denominator.
3. An alternative method to long division of polynomials is called ________ ________, in which the divisor must be of the form $x - k$.
4. The ________ Theorem states that a polynomial $f(x)$ has a factor $(x - k)$ if and only if $f(k) = 0$.
5. The ________ Theorem states that if a polynomial $f(x)$ is divided by $x - k$, the remainder is $r = f(k)$.

PREREQUISITE SKILLS REVIEW: Practice and review algebra skills needed for this section at **www.Eduspace.com.**

Analytical Analysis **In Exercises 1 and 2, use long division to verify that $y_1 = y_2$.**

1. $y_1 = \dfrac{x^2}{x + 2}, \quad y_2 = x - 2 + \dfrac{4}{x + 2}$
2. $y_1 = \dfrac{x^4 - 3x^2 - 1}{x^2 + 5}, \quad y_2 = x^2 - 8 + \dfrac{39}{x^2 + 5}$

Graphical Analysis **In Exercises 3 and 4, (a) use a graphing utility to graph the two equations in the same viewing window, (b) use the graphs to verify that the expressions are equivalent, and (c) use long division to verify the results algebraically.**

3. $y_1 = \dfrac{x^5 - 3x^3}{x^2 + 1}, \quad y_2 = x^3 - 4x + \dfrac{4x}{x^2 + 1}$
4. $y_1 = \dfrac{x^3 - 2x^2 + 5}{x^2 + x + 1}, \quad y_2 = x - 3 + \dfrac{2(x + 4)}{x^2 + x + 1}$

In Exercises 5–18, use long division to divide.

5. $(2x^2 + 10x + 12) \div (x + 3)$
6. $(5x^2 - 17x - 12) \div (x - 4)$
7. $(4x^3 - 7x^2 - 11x + 5) \div (4x + 5)$
8. $(6x^3 - 16x^2 + 17x - 6) \div (3x - 2)$
9. $(x^4 + 5x^3 + 6x^2 - x - 2) \div (x + 2)$
10. $(x^3 + 4x^2 - 3x - 12) \div (x - 3)$
11. $(7x + 3) \div (x + 2)$
12. $(8x - 5) \div (2x + 1)$
13. $(6x^3 + 10x^2 + x + 8) \div (2x^2 + 1)$
14. $(x^3 - 9) \div (x^2 + 1)$
15. $(x^4 + 3x^2 + 1) \div (x^2 - 2x + 3)$
16. $(x^5 + 7) \div (x^3 - 1)$
17. $\dfrac{x^4}{(x - 1)^3}$
18. $\dfrac{2x^3 - 4x^2 - 15x + 5}{(x - 1)^2}$

In Exercises 19–36, use synthetic division to divide.

19. $(3x^3 - 17x^2 + 15x - 25) \div (x - 5)$
20. $(5x^3 + 18x^2 + 7x - 6) \div (x + 3)$
21. $(4x^3 - 9x + 8x^2 - 18) \div (x + 2)$
22. $(9x^3 - 16x - 18x^2 + 32) \div (x - 2)$
23. $(-x^3 + 75x - 250) \div (x + 10)$
24. $(3x^3 - 16x^2 - 72) \div (x - 6)$
25. $(5x^3 - 6x^2 + 8) \div (x - 4)$
26. $(5x^3 + 6x + 8) \div (x + 2)$
27. $\dfrac{10x^4 - 50x^3 - 800}{x - 6}$
28. $\dfrac{x^5 - 13x^4 - 120x + 80}{x + 3}$
29. $\dfrac{x^3 + 512}{x + 8}$
30. $\dfrac{x^3 - 729}{x - 9}$
31. $\dfrac{-3x^4}{x - 2}$
32. $\dfrac{-3x^4}{x + 2}$
33. $\dfrac{180x - x^4}{x - 6}$
34. $\dfrac{5 - 3x + 2x^2 - x^3}{x + 1}$
35. $\dfrac{4x^3 + 16x^2 - 23x - 15}{x + \frac{1}{2}}$
36. $\dfrac{3x^3 - 4x^2 + 5}{x - \frac{3}{2}}$

In Exercises 37–44, write the function in the form $f(x) = (x - k)q(x) + r$ for the given value of k, and demonstrate that $f(k) = r$.

	Function	*Value of k*
37.	$f(x) = x^3 - x^2 - 14x + 11$	$k = 4$
38.	$f(x) = x^3 - 5x^2 - 11x + 8$	$k = -2$

	Function	*Value of k*
39.	$f(x) = 15x^4 + 10x^3 - 6x^2 + 14$	$k = -\frac{2}{3}$
40.	$f(x) = 10x^3 - 22x^2 - 3x + 4$	$k = \frac{1}{5}$
41.	$f(x) = x^3 + 3x^2 - 2x - 14$	$k = \sqrt{2}$
42.	$f(x) = x^3 + 2x^2 - 5x - 4$	$k = -\sqrt{5}$
43.	$f(x) = -4x^3 + 6x^2 + 12x + 4$	$k = 1 - \sqrt{3}$
44.	$f(x) = -3x^3 + 8x^2 + 10x - 8$	$k = 2 + \sqrt{2}$

In Exercises 45–48, use synthetic division to find each function value. Verify your answers using another method.

45. $f(x) = 4x^3 - 13x + 10$
(a) $f(1)$ (b) $f(-2)$ (c) $f\left(\frac{1}{2}\right)$ (d) $f(8)$

46. $g(x) = x^6 - 4x^4 + 3x^2 + 2$
(a) $g(2)$ (b) $g(-4)$ (c) $g(3)$ (d) $g(-1)$

47. $h(x) = 3x^3 + 5x^2 - 10x + 1$
(a) $h(3)$ (b) $h\left(\frac{1}{3}\right)$ (c) $h(-2)$ (d) $h(-5)$

48. $f(x) = 0.4x^4 - 1.6x^3 + 0.7x^2 - 2$
(a) $f(1)$ (b) $f(-2)$ (c) $f(5)$ (d) $f(-10)$

In Exercises 49–56, use synthetic division to show that *x* is a solution of the third-degree polynomial equation, and use the result to factor the polynomial completely. List all real solutions of the equation.

	Polynomial Equation	*Value of x*
49.	$x^3 - 7x + 6 = 0$	$x = 2$
50.	$x^3 - 28x - 48 = 0$	$x = -4$
51.	$2x^3 - 15x^2 + 27x - 10 = 0$	$x = \frac{1}{2}$
52.	$48x^3 - 80x^2 + 41x - 6 = 0$	$x = \frac{2}{3}$
53.	$x^3 + 2x^2 - 3x - 6 = 0$	$x = \sqrt{3}$
54.	$x^3 + 2x^2 - 2x - 4 = 0$	$x = \sqrt{2}$
55.	$x^3 - 3x^2 + 2 = 0$	$x = 1 + \sqrt{3}$
56.	$x^3 - x^2 - 13x - 3 = 0$	$x = 2 - \sqrt{5}$

In Exercises 57–64, (a) verify the given factors of the function *f*, (b) find the remaining factors of *f*, (c) use your results to write the complete factorization of *f*, (d) list all real zeros of *f*, and (e) confirm your results by using a graphing utility to graph the function.

	Function	*Factors*
57.	$f(x) = 2x^3 + x^2 - 5x + 2$	$(x + 2), (x - 1)$
58.	$f(x) = 3x^3 + 2x^2 - 19x + 6$	$(x + 3), (x - 2)$
59.	$f(x) = x^4 - 4x^3 - 15x^2 + 58x - 40$	$(x - 5), (x + 4)$
60.	$f(x) = 8x^4 - 14x^3 - 71x^2 - 10x + 24$	$(x + 2), (x - 4)$
61.	$f(x) = 6x^3 + 41x^2 - 9x - 14$	$(2x + 1), (3x - 2)$
62.	$f(x) = 10x^3 - 11x^2 - 72x + 45$	$(2x + 5), (5x - 3)$
63.	$f(x) = 2x^3 - x^2 - 10x + 5$	$(2x - 1), (x + \sqrt{5})$
64.	$f(x) = x^3 + 3x^2 - 48x - 144$	$(x + 4\sqrt{3}), (x + 3)$

Graphical Analysis **In Exercises 65–68, (a) use the *zero* or *root* feature of a graphing utility to approximate the zeros of the function accurate to three decimal places, (b) determine one of the exact zeros, and (c) use synthetic division to verify your result from part (b), and then factor the polynomial completely.**

65. $f(x) = x^3 - 2x^2 - 5x + 10$

66. $g(x) = x^3 - 4x^2 - 2x + 8$

67. $h(t) = t^3 - 2t^2 - 7t + 2$

68. $f(s) = s^3 - 12s^2 + 40s - 24$

In Exercises 69–72, simplify the rational expression by using long division or synthetic division.

69. $\dfrac{4x^3 - 8x^2 + x + 3}{2x - 3}$

70. $\dfrac{x^3 + x^2 - 64x - 64}{x + 8}$

71. $\dfrac{x^4 + 6x^3 + 11x^2 + 6x}{x^2 + 3x + 2}$

72. $\dfrac{x^4 + 9x^3 - 5x^2 - 36x + 4}{x^2 - 4}$

Model It

73. ***Data Analysis: Military Personnel*** The numbers M (in thousands) of United States military personnel on active duty for the years 1993 through 2003 are shown in the table, where t represents the year, with $t = 3$ corresponding to 1993. (Source: U.S. Department of Defense)

Year, t	Military personnel, M
3	1705
4	1611
5	1518
6	1472
7	1439
8	1407
9	1386
10	1384
11	1385
12	1412
13	1434

Model It (continued)

(a) Use a graphing utility to create a scatter plot of the data.

(b) Use the *regression* feature of the graphing utility to find a cubic model for the data. Graph the model in the same viewing window as the scatter plot.

(c) Use the model to create a table of estimated values of M. Compare the model with the original data.

(d) Use synthetic division to evaluate the model for the year 2008. Even though the model is relatively accurate for estimating the given data, would you use this model to predict the number of military personnel in the future? Explain.

74. ***Data Analysis: Cable Television*** The average monthly basic rates R (in dollars) for cable television in the United States for the years 1992 through 2002 are shown in the table, where t represents the year, with $t = 2$ corresponding to 1992. (Source: Kagan Research LLC)

Year, t	Basic rate, R
2	19.08
3	19.39
4	21.62
5	23.07
6	24.41
7	26.48
8	27.81
9	28.92
10	30.37
11	32.87
12	34.71

(a) Use a graphing utility to create a scatter plot of the data.

(b) Use the *regression* feature of the graphing utility to find a cubic model for the data. Then graph the model in the same viewing window as the scatter plot. Compare the model with the data.

(c) Use synthetic division to evaluate the model for the year 2008.

Synthesis

True or False? **In Exercises 75–77, determine whether the statement is true or false. Justify your answer.**

75. If $(7x + 4)$ is a factor of some polynomial function f, then $\frac{4}{7}$ is a zero of f.

76. $(2x - 1)$ is a factor of the polynomial

$$6x^6 + x^5 - 92x^4 + 45x^3 + 184x^2 + 4x - 48.$$

77. The rational expression

$$\frac{x^3 + 2x^2 - 13x + 10}{x^2 - 4x - 12}$$

is improper.

78. ***Exploration*** Use the form $f(x) = (x - k)q(x) + r$ to create a cubic function that (a) passes through the point $(2, 5)$ and rises to the right, and (b) passes through the point $(-3, 1)$ and falls to the right. (There are many correct answers.)

Think About It **In Exercises 79 and 80, perform the division by assuming that n is a positive integer.**

79. $\dfrac{x^{3n} + 9x^{2n} + 27x^n + 27}{x^n + 3}$

80. $\dfrac{x^{3n} - 3x^{2n} + 5x^n - 6}{x^n - 2}$

81. ***Writing*** Briefly explain what it means for a divisor to divide evenly into a dividend.

82. ***Writing*** Briefly explain how to check polynomial division, and justify your reasoning. Give an example.

Exploration **In Exercises 83 and 84, find the constant c such that the denominator will divide evenly into the numerator.**

83. $\dfrac{x^3 + 4x^2 - 3x + c}{x - 5}$

84. $\dfrac{x^5 - 2x^2 + x + c}{x + 2}$

Think About It **In Exercises 85 and 86, answer the questions about the division $f(x) \div (x - k)$, where $f(x) = (x + 3)^2(x - 3)(x + 1)^3$.**

85. What is the remainder when $k = -3$? Explain.

86. If it is necessary to find $f(2)$, is it easier to evaluate the function directly or to use synthetic division? Explain.

Skills Review

In Exercises 87–92, use any method to solve the quadratic equation.

87. $9x^2 - 25 = 0$

88. $16x^2 - 21 = 0$

89. $5x^2 - 3x - 14 = 0$

90. $8x^2 - 22x + 15 = 0$

91. $2x^2 + 6x + 3 = 0$

92. $x^2 + 3x - 3 = 0$

In Exercises 93–96, find a polynomial function that has the given zeros. (There are many correct answers.)

93. $0, 3, 4$

94. $-6, 1$

95. $-3, 1 + \sqrt{2}, 1 - \sqrt{2}$

96. $1, -2, 2 + \sqrt{3}, 2 - \sqrt{3}$

2.4 Complex Numbers

What you should learn

- Use the imaginary unit i to write complex numbers.
- Add, subtract, and multiply complex numbers.
- Use complex conjugates to write the quotient of two complex numbers in standard form.
- Find complex solutions of quadratic equations.

Why you should learn it

You can use complex numbers to model and solve real-life problems in electronics. For instance, in Exercise 83 on page 168, you will learn how to use complex numbers to find the impedance of an electrical circuit.

The Imaginary Unit i

You have learned that some quadratic equations have no real solutions. For instance, the quadratic equation $x^2 + 1 = 0$ has no real solution because there is no real number x that can be squared to produce -1. To overcome this deficiency, mathematicians created an expanded system of numbers using the **imaginary unit i,** defined as

$$i = \sqrt{-1} \qquad \text{Imaginary unit}$$

where $i^2 = -1$. By adding real numbers to real multiples of this imaginary unit, the set of **complex numbers** is obtained. Each complex number can be written in the **standard form $a + bi$.** For instance, the standard form of the complex number $-5 + \sqrt{-9}$ is $-5 + 3i$ because

$$-5 + \sqrt{-9} = -5 + \sqrt{3^2(-1)} = -5 + 3\sqrt{-1} = -5 + 3i.$$

In the standard form $a + bi$, the real number a is called the **real part** of the **complex number $a + bi$,** and the number bi (where b is a real number) is called the **imaginary part** of the complex number.

Definition of a Complex Number

If a and b are real numbers, the number $a + bi$ is a **complex number,** and it is said to be written in **standard form.** If $b = 0$, the number $a + bi = a$ is a real number. If $b \neq 0$, the number $a + bi$ is called an **imaginary number.** A number of the form bi, where $b \neq 0$, is called a **pure imaginary number.**

The set of real numbers is a subset of the set of complex numbers, as shown in Figure 2.29. This is true because every real number a can be written as a complex number using $b = 0$. That is, for every real number a, you can write $a = a + 0i$.

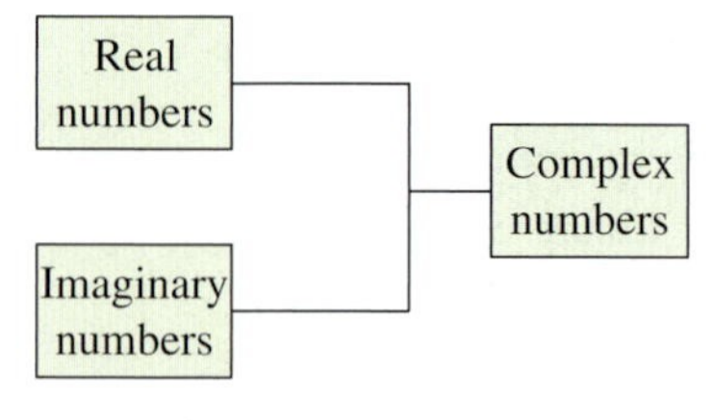

FIGURE 2.29

Equality of Complex Numbers

Two complex numbers $a + bi$ and $c + di$, written in standard form, are equal to each other

$$a + bi = c + di \qquad \text{Equality of two complex numbers}$$

if and only if $a = c$ and $b = d$.

Operations with Complex Numbers

To add (or subtract) two complex numbers, you add (or subtract) the real and imaginary parts of the numbers separately.

For each operation on complex numbers, you can show the parallel operation on polynomials.

Addition and Subtraction of Complex Numbers

If $a + bi$ and $c + di$ are two complex numbers written in standard form, their sum and difference are defined as follows.

Sum: $(a + bi) + (c + di) = (a + c) + (b + d)i$

Difference: $(a + bi) - (c + di) = (a - c) + (b - d)i$

The **additive identity** in the complex number system is zero (the same as in the real number system). Furthermore, the **additive inverse** of the complex number $a + bi$ is

$$-(a + bi) = -a - bi.$$ Additive inverse

So, you have

$$(a + bi) + (-a - bi) = 0 + 0i = 0.$$

Example 1 Adding and Subtracting Complex Numbers

a. $(4 + 7i) + (1 - 6i) = 4 + 7i + 1 - 6i$ Remove parentheses.

$= (4 + 1) + (7i - 6i)$ Group like terms.

$= 5 + i$ Write in standard form.

b. $(1 + 2i) - (4 + 2i) = 1 + 2i - 4 - 2i$ Remove parentheses.

$= (1 - 4) + (2i - 2i)$ Group like terms.

$= -3 + 0$ Simplify.

$= -3$ Write in standard form.

c. $3i - (-2 + 3i) - (2 + 5i) = 3i + 2 - 3i - 2 - 5i$

$= (2 - 2) + (3i - 3i - 5i)$

$= 0 - 5i$

$= -5i$

d. $(3 + 2i) + (4 - i) - (7 + i) = 3 + 2i + 4 - i - 7 - i$

$= (3 + 4 - 7) + (2i - i - i)$

$= 0 + 0i$

$= 0$

CHECKPOINT Now try Exercise 17.

Note in Examples 1(b) and 1(d) that the sum of two complex numbers can be a real number.

Exploration

Complete the following.

$i^1 = i$	$i^7 = \square$
$i^2 = -1$	$i^8 = \square$
$i^3 = -i$	$i^9 = \square$
$i^4 = 1$	$i^{10} = \square$
$i^5 = \square$	$i^{11} = \square$
$i^6 = \square$	$i^{12} = \square$

What pattern do you see? Write a brief description of how you would find i raised to any positive integer power.

Many of the properties of real numbers are valid for complex numbers as well. Here are some examples.

Associative Properties of Addition and Multiplication

Commutative Properties of Addition and Multiplication

Distributive Property of Multiplication Over Addition

Notice below how these properties are used when two complex numbers are multiplied.

$$(a + bi)(c + di) = a(c + di) + bi(c + di) \quad \text{Distributive Property}$$
$$= ac + (ad)i + (bc)i + (bd)i^2 \quad \text{Distributive Property}$$
$$= ac + (ad)i + (bc)i + (bd)(-1) \quad i^2 = -1$$
$$= ac - bd + (ad)i + (bc)i \quad \text{Commutative Property}$$
$$= (ac - bd) + (ad + bc)i \quad \text{Associative Property}$$

Rather than trying to memorize this multiplication rule, you should simply remember how the Distributive Property is used to multiply two complex numbers.

STUDY TIP

The procedure described above is similar to multiplying two polynomials and combining like terms, as in the FOIL Method shown in Appendix A.3. For instance, you can use the FOIL Method to multiply the two complex numbers from Example 2(b).

$$(2 - i)(4 + 3i) = \overset{F}{8} + \overset{O}{6i} - \overset{I}{4i} - \overset{L}{3i^2}$$

Example 2 Multiplying Complex Numbers

a. $4(-2 + 3i) = 4(-2) + 4(3i)$ Distributive Property

$= -8 + 12i$ Simplify.

b. $(2 - i)(4 + 3i) = 2(4 + 3i) - i(4 + 3i)$ Distributive Property

$= 8 + 6i - 4i - 3i^2$ Distributive Property

$= 8 + 6i - 4i - 3(-1)$ $i^2 = -1$

$= (8 + 3) + (6i - 4i)$ Group like terms.

$= 11 + 2i$ Write in standard form.

c. $(3 + 2i)(3 - 2i) = 3(3 - 2i) + 2i(3 - 2i)$ Distributive Property

$= 9 - 6i + 6i - 4i^2$ Distributive Property

$= 9 - 6i + 6i - 4(-1)$ $i^2 = -1$

$= 9 + 4$ Simplify.

$= 13$ Write in standard form.

d. $(3 + 2i)^2 = (3 + 2i)(3 + 2i)$ Square of a binomial

$= 3(3 + 2i) + 2i(3 + 2i)$ Distributive Property

$= 9 + 6i + 6i + 4i^2$ Distributive Property

$= 9 + 6i + 6i + 4(-1)$ $i^2 = -1$

$= 9 + 12i - 4$ Simplify.

$= 5 + 12i$ Write in standard form.

CHECKPOINT Now try Exercise 27.

Complex Conjugates

Notice in Example 2(c) that the product of two complex numbers can be a real number. This occurs with pairs of complex numbers of the form $a + bi$ and $a - bi$, called **complex conjugates.**

A comparison with the method of rationalizing denominators (Appendix A.2) may be helpful.

$$\begin{aligned}(a + bi)(a - bi) &= a^2 - abi + abi - b^2i^2 \\ &= a^2 - b^2(-1) \\ &= a^2 + b^2\end{aligned}$$

Example 3 Multiplying Conjugates

Multiply each complex number by its complex conjugate.

a. $1 + i$ **b.** $4 - 3i$

Solution

a. The complex conjugate of $1 + i$ is $1 - i$.

$$(1 + i)(1 - i) = 1^2 - i^2 = 1 - (-1) = 2$$

b. The complex conjugate of $4 - 3i$ is $4 + 3i$.

$$(4 - 3i)(4 + 3i) = 4^2 - (3i)^2 = 16 - 9i^2 = 16 - 9(-1) = 25$$

CHECKPOINT Now try Exercise 37.

To write the quotient of $a + bi$ and $c + di$ in standard form, where c and d are not both zero, multiply the numerator and denominator by the complex conjugate of the *denominator* to obtain

$$\begin{aligned}\frac{a + bi}{c + di} &= \frac{a + bi}{c + di}\left(\frac{c - di}{c - di}\right) \\ &= \frac{(ac + bd) + (bc - ad)i}{c^2 + d^2}. \quad \text{Standard form}\end{aligned}$$

STUDY TIP

Note that when you multiply the numerator and denominator of a quotient of complex numbers by

$$\frac{c - di}{c - di}$$

you are actually multiplying the quotient by a form of 1. You are not changing the original expression, you are only creating an expression that is equivalent to the original expression.

Example 4 Writing a Quotient of Complex Numbers in Standard Form

$$\begin{aligned}\frac{2 + 3i}{4 - 2i} &= \frac{2 + 3i}{4 - 2i}\left(\frac{4 + 2i}{4 + 2i}\right) && \text{Multiply numerator and denominator by complex conjugate of denominator.} \\ &= \frac{8 + 4i + 12i + 6i^2}{16 - 4i^2} && \text{Expand.} \\ &= \frac{8 - 6 + 16i}{16 + 4} && i^2 = -1 \\ &= \frac{2 + 16i}{20} && \text{Simplify.} \\ &= \frac{1}{10} + \frac{4}{5}i && \text{Write in standard form.}\end{aligned}$$

CHECKPOINT Now try Exercise 49.

Complex Solutions of Quadratic Equations

When using the Quadratic Formula to solve a quadratic equation, you often obtain a result such as $\sqrt{-3}$, which you know is not a real number. By factoring out $i = \sqrt{-1}$, you can write this number in standard form.

$$\sqrt{-3} = \sqrt{3(-1)} = \sqrt{3}\sqrt{-1} = \sqrt{3}i$$

The number $\sqrt{3}i$ is called the *principal square root* of -3.

STUDY TIP

The definition of principal square root uses the rule

$$\sqrt{ab} = \sqrt{a}\sqrt{b}$$

for $a > 0$ and $b < 0$. This rule is not valid if *both* a and b are negative. For example,

$$\sqrt{-5}\sqrt{-5} = \sqrt{5(-1)}\sqrt{5(-1)}$$
$$= \sqrt{5}i\sqrt{5}i$$
$$= \sqrt{25}i^2$$
$$= 5i^2 = -5$$

whereas

$$\sqrt{(-5)(-5)} = \sqrt{25} = 5.$$

To avoid problems with square roots of negative numbers, be sure to convert complex numbers to standard form *before* multiplying.

Principal Square Root of a Negative Number

If a is a positive number, the **principal square root** of the negative number $-a$ is defined as

$$\sqrt{-a} = \sqrt{a}i.$$

Example 5 Writing Complex Numbers in Standard Form

a. $\sqrt{-3}\sqrt{-12} = \sqrt{3}i\sqrt{12}i = \sqrt{36}i^2 = 6(-1) = -6$

b. $\sqrt{-48} - \sqrt{-27} = \sqrt{48}i - \sqrt{27}i = 4\sqrt{3}i - 3\sqrt{3}i = \sqrt{3}i$

c. $(-1 + \sqrt{-3})^2 = (-1 + \sqrt{3}i)^2$

$$= (-1)^2 - 2\sqrt{3}i + (\sqrt{3})^2(i^2)$$
$$= 1 - 2\sqrt{3}i + 3(-1)$$
$$= -2 - 2\sqrt{3}i$$

CHECKPOINT Now try Exercise 59.

Example 6 Complex Solutions of a Quadratic Equation

Solve (a) $x^2 + 4 = 0$ and (b) $3x^2 - 2x + 5 = 0$.

Solution

a. $x^2 + 4 = 0$ — Write original equation.

$x^2 = -4$ — Subtract 4 from each side.

$x = \pm 2i$ — Extract square roots.

b. $3x^2 - 2x + 5 = 0$ — Write original equation.

$x = \dfrac{-(-2) \pm \sqrt{(-2)^2 - 4(3)(5)}}{2(3)}$ — Quadratic Formula

$= \dfrac{2 \pm \sqrt{-56}}{6}$ — Simplify.

$= \dfrac{2 \pm 2\sqrt{14}i}{6}$ — Write $\sqrt{-56}$ in standard form.

$= \dfrac{1}{3} \pm \dfrac{\sqrt{14}}{3}i$ — Write in standard form.

CHECKPOINT Now try Exercise 65.

Activities

1. Find the product and write the result in standard form.
   $(4 - \sqrt{-9})(2 + \sqrt{-9})$
   Answer: $17 + 6i$
2. Write $\dfrac{3 + i}{i}$ in standard form.
   Answer: $1 - 3i$
3. Use the Quadratic Formula to solve $2x^2 - 5x + 7 = 0$.
   Answer: $x = \dfrac{5}{4} \pm \dfrac{\sqrt{31}}{4}i$

2.4 Exercises

VOCABULARY CHECK:

1. Match the type of complex number with its definition.

(a) Real Number (i) $a + bi, \quad a \neq 0, \quad b \neq 0$

(b) Imaginary number (ii) $a + bi, \quad a = 0, \quad b \neq 0$

(c) Pure imaginary number (iii) $a + bi, \quad b = 0$

In Exercises 2–5, fill in the blanks.

2. The imaginary unit i is defined as $i =$ ________, where $i^2 =$ ________.
3. If a is a positive number, the ________ ________ root of the negative number $-a$ is defined as $\sqrt{-a} = \sqrt{a}\, i$.
4. The numbers $a + bi$ and $a - bi$ are called ________ ________, and their product is a real number $a^2 + b^2$.

PREREQUISITE SKILLS REVIEW: Practice and review algebra skills needed for this section at **www.Eduspace.com.**

In Exercises 1–4, find real numbers *a* and *b* such that the equation is true.

1. $a + bi = -10 + 6i$
2. $a + bi = 13 + 4i$
3. $(a - 1) + (b + 3)i = 5 + 8i$
4. $(a + 6) + 2bi = 6 - 5i$

In Exercises 5–16, write the complex number in standard form.

5. $4 + \sqrt{-9}$
6. $3 + \sqrt{-16}$
7. $2 - \sqrt{-27}$
8. $1 + \sqrt{-8}$
9. $\sqrt{-75}$
10. $\sqrt{-4}$
11. 8
12. 45
13. $-6i + i^2$
14. $-4i^2 + 2i$
15. $\sqrt{-0.09}$
16. $\sqrt{-0.0004}$

In Exercises 17–26, perform the addition or subtraction and write the result in standard form.

17. $(5 + i) + (6 - 2i)$
18. $(13 - 2i) + (-5 + 6i)$
19. $(8 - i) - (4 - i)$
20. $(3 + 2i) - (6 + 13i)$
21. $\left(-2 + \sqrt{-8}\right) + \left(5 - \sqrt{-50}\right)$
22. $\left(8 + \sqrt{-18}\right) - \left(4 + 3\sqrt{2}i\right)$
23. $13i - (14 - 7i)$
24. $22 + (-5 + 8i) + 10i$
25. $-\left(\frac{3}{2} + \frac{5}{2}i\right) + \left(\frac{5}{3} + \frac{11}{3}i\right)$
26. $(1.6 + 3.2i) + (-5.8 + 4.3i)$

In Exercises 27–36, perform the operation and write the result in standard form.

27. $(1 + i)(3 - 2i)$
28. $(6 - 2i)(2 - 3i)$
29. $6i(5 - 2i)$
30. $-8i(9 + 4i)$
31. $\left(\sqrt{14} + \sqrt{10}i\right)\left(\sqrt{14} - \sqrt{10}i\right)$
32. $\left(\sqrt{3} + \sqrt{15}i\right)\left(\sqrt{3} - \sqrt{15}i\right)$
33. $(4 + 5i)^2$
34. $(2 - 3i)^2$
35. $(2 + 3i)^2 + (2 - 3i)^2$
36. $(1 - 2i)^2 - (1 + 2i)^2$

In Exercises 37–44, write the complex conjugate of the complex number. Then multiply the number by its complex conjugate.

37. $6 + 3i$
38. $7 - 12i$
39. $-1 - \sqrt{5}i$
40. $-3 + \sqrt{2}i$
41. $\sqrt{-20}$
42. $\sqrt{-15}$
43. $\sqrt{8}$
44. $1 + \sqrt{8}$

In Exercises 45–54, write the quotient in standard form.

45. $\dfrac{5}{i}$
46. $-\dfrac{14}{2i}$
47. $\dfrac{2}{4 - 5i}$
48. $\dfrac{5}{1 - i}$
49. $\dfrac{3 + i}{3 - i}$
50. $\dfrac{6 - 7i}{1 - 2i}$
51. $\dfrac{6 - 5i}{i}$
52. $\dfrac{8 + 16i}{2i}$
53. $\dfrac{3i}{(4 - 5i)^2}$
54. $\dfrac{5i}{(2 + 3i)^2}$

In Exercises 55–58, perform the operation and write the result in standard form.

55. $\dfrac{2}{1 + i} - \dfrac{3}{1 - i}$
56. $\dfrac{2i}{2 + i} + \dfrac{5}{2 - i}$
57. $\dfrac{i}{3 - 2i} + \dfrac{2i}{3 + 8i}$
58. $\dfrac{1 + i}{i} - \dfrac{3}{4 - i}$

In Exercises 59–64, write the complex number in standard form.

59. $\sqrt{-6} \cdot \sqrt{-2}$

60. $\sqrt{-5} \cdot \sqrt{-10}$

61. $\left(\sqrt{-10}\right)^2$

62. $\left(\sqrt{-75}\right)^2$

63. $\left(3 + \sqrt{-5}\right)\left(7 - \sqrt{-10}\right)$

64. $\left(2 - \sqrt{-6}\right)^2$

In Exercises 65–74, use the Quadratic Formula to solve the quadratic equation.

65. $x^2 - 2x + 2 = 0$

66. $x^2 + 6x + 10 = 0$

67. $4x^2 + 16x + 17 = 0$

68. $9x^2 - 6x + 37 = 0$

69. $4x^2 + 16x + 15 = 0$

70. $16t^2 - 4t + 3 = 0$

71. $\frac{3}{2}x^2 - 6x + 9 = 0$

72. $\frac{7}{8}x^2 - \frac{3}{4}x + \frac{5}{16} = 0$

73. $1.4x^2 - 2x - 10 = 0$

74. $4.5x^2 - 3x + 12 = 0$

In Exercises 75–82, simplify the complex number and write it in standard form.

75. $-6i^3 + i^2$

76. $4i^2 - 2i^3$

77. $-5i^5$

78. $(-i)^3$

79. $\left(\sqrt{-75}\right)^3$

80. $\left(\sqrt{-2}\right)^6$

81. $\dfrac{1}{i^3}$

82. $\dfrac{1}{(2i)^3}$

Model It

83. ***Impedance*** The opposition to current in an electrical circuit is called its impedance. The impedance z in a parallel circuit with two pathways satisfies the equation

$$\frac{1}{z} = \frac{1}{z_1} + \frac{1}{z_2}$$

where z_1 is the impedance (in ohms) of pathway 1 and z_2 is the impedance of pathway 2.

(a) The impedance of each pathway in a parallel circuit is found by adding the impedances of all components in the pathway. Use the table to find z_1 and z_2.

(b) Find the impedance z.

	Resistor	Inductor	Capacitor
Symbol	$a\Omega$	$b\Omega$	$c\Omega$
Impedance	a	bi	$-ci$

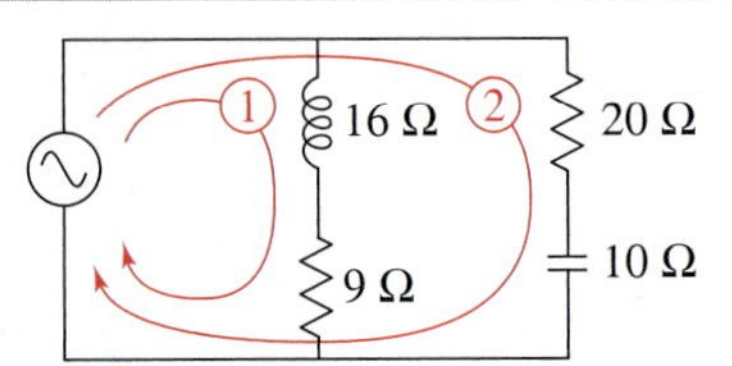

84. Cube each complex number.

(a) 2 (b) $-1 + \sqrt{3}i$ (c) $-1 - \sqrt{3}i$

85. Raise each complex number to the fourth power.

(a) 2 (b) -2 (c) $2i$ (d) $-2i$

86. Write each of the powers of i as i, $-i$, 1, or -1.

(a) i^{40} (b) i^{25} (c) i^{50} (d) i^{67}

Synthesis

True or False? **In Exercises 87–89, determine whether the statement is true or false. Justify your answer.**

87. There is no complex number that is equal to its complex conjugate.

88. $-i\sqrt{6}$ is a solution of $x^4 - x^2 + 14 = 56$.

89. $i^{44} + i^{150} - i^{74} - i^{109} + i^{61} = -1$

90. ***Error Analysis*** Describe the error.

$$\sqrt{-6}\sqrt{-6} = \sqrt{(-6)(-6)} = \sqrt{36} = 6$$

91. ***Proof*** Prove that the complex conjugate of the product of two complex numbers $a_1 + b_1 i$ and $a_2 + b_2 i$ is the product of their complex conjugates.

92. ***Proof*** Prove that the complex conjugate of the sum of two complex numbers $a_1 + b_1 i$ and $a_2 + b_2 i$ is the sum of their complex conjugates.

Skills Review

In Exercises 93–96, perform the operation and write the result in standard form.

93. $(4 + 3x) + (8 - 6x - x^2)$

94. $(x^3 - 3x^2) - (6 - 2x - 4x^2)$

95. $\left(3x - \frac{1}{2}\right)(x + 4)$

96. $(2x - 5)^2$

In Exercises 97–100, solve the equation and check your solution.

97. $-x - 12 = 19$

98. $8 - 3x = -34$

99. $4(5x - 6) - 3(6x + 1) = 0$

100. $5[x - (3x + 11)] = 20x - 15$

101. ***Volume of an Oblate Spheroid***

Solve for a: $V = \frac{4}{3}\pi a^2 b$

102. ***Newton's Law of Universal Gravitation***

Solve for r: $F = \alpha \dfrac{m_1 m_2}{r^2}$

103. ***Mixture Problem*** A five-liter container contains a mixture with a concentration of 50%. How much of this mixture must be withdrawn and replaced by 100% concentrate to bring the mixture up to 60% concentration?

2.5 Zeros of Polynomial Functions

What you should learn

- Use the Fundamental Theorem of Algebra to determine the number of zeros of polynomial functions.
- Find rational zeros of polynomial functions.
- Find conjugate pairs of complex zeros.
- Find zeros of polynomials by factoring.
- Use Descartes's Rule of Signs and the Upper and Lower Bound Rules to find zeros of polynomials.

Why you should learn it

Finding zeros of polynomial functions is an important part of solving real-life problems. For instance, in Exercise 112 on page 182, the zeros of a polynomial function can help you analyze the attendance at women's college basketball games.

The Fundamental Theorem of Algebra

You know that an nth-degree polynomial can have at most n real zeros. In the complex number system, this statement can be improved. That is, in the complex number system, every nth-degree polynomial function has *precisely* n zeros. This important result is derived from the **Fundamental Theorem of Algebra,** first proved by the German mathematician Carl Friedrich Gauss (1777–1855).

The Fundamental Theorem of Algebra

If $f(x)$ is a polynomial of degree n, where $n > 0$, then f has at least one zero in the complex number system.

Using the Fundamental Theorem of Algebra and the equivalence of zeros and factors, you obtain the **Linear Factorization Theorem.**

Linear Factorization Theorem

If $f(x)$ is a polynomial of degree n, where $n > 0$, then f has precisely n linear factors

$$f(x) = a_n(x - c_1)(x - c_2) \cdot \cdot \cdot (x - c_n)$$

where $c_1, c_2, . . . , c_n$ are complex numbers.

For a proof of the Linear Factorization Theorem, see Proofs in Mathematics on page 214.

Note that the Fundamental Theorem of Algebra and the Linear Factorization Theorem tell you only that the zeros or factors of a polynomial exist, not how to find them. Such theorems are called *existence theorems.*

Example 1 Zeros of Polynomial Functions

a. The first-degree polynomial $f(x) = x - 2$ has exactly *one* zero: $x = 2$.

b. Counting multiplicity, the second-degree polynomial function

$$f(x) = x^2 - 6x + 9 = (x - 3)(x - 3)$$

has exactly *two* zeros: $x = 3$ and $x = 3$. (This is called a *repeated zero.*)

c. The third-degree polynomial function

$$f(x) = x^3 + 4x = x(x^2 + 4) = x(x - 2i)(x + 2i)$$

has exactly *three* zeros: $x = 0$, $x = 2i$, and $x = -2i$.

d. The fourth-degree polynomial function

$$f(x) = x^4 - 1 = (x - 1)(x + 1)(x - i)(x + i)$$

has exactly *four* zeros: $x = 1$, $x = -1$, $x = i$, and $x = -i$.

✓CHECKPOINT Now try Exercise 1.

STUDY TIP

Recall that in order to find the zeros of a function $f(x)$, set $f(x)$ equal to 0 and solve the resulting equation for x. For instance, the function in Example 1(a) has a zero at $x = 2$ because

$$x - 2 = 0$$
$$x = 2.$$

Finding zeros of polynomial functions is a very important concept in algebra. This is a good place to discuss the fact that polynomials do not necessarily have rational zeros but may have zeros that are irrational or complex.

Historical Note
Although they were not contemporaries, Jean Le Rond d'Alembert (1717–1783) worked independently of Carl Gauss in trying to prove the Fundamental Theorem of Algebra. His efforts were such that, in France, the Fundamental Theorem of Algebra is frequently known as the Theorem of d'Alembert.

The Rational Zero Test

The **Rational Zero Test** relates the possible rational zeros of a polynomial (having integer coefficients) to the leading coefficient and to the constant term of the polynomial.

The Rational Zero Test

If the polynomial $f(x) = a_nx^n + a_{n-1}x^{n-1} + \cdot \cdot \cdot + a_2x^2 + a_1x + a_0$ has *integer* coefficients, every rational zero of f has the form

$$\text{Rational zero} = \frac{p}{q}$$

where p and q have no common factors other than 1, and

$p =$ a factor of the constant term a_0

$q =$ a factor of the leading coefficient a_n.

To use the Rational Zero Test, you should first list all rational numbers whose numerators are factors of the constant term and whose denominators are factors of the leading coefficient.

$$\text{Possible rational zeros} = \frac{\text{factors of constant term}}{\text{factors of leading coefficient}}$$

Having formed this list of *possible rational zeros*, use a trial-and-error method to determine which, if any, are actual zeros of the polynomial. Note that when the leading coefficient is 1, the possible rational zeros are simply the factors of the constant term.

Example 2 Rational Zero Test with Leading Coefficient of 1

Find the rational zeros of

$$f(x) = x^3 + x + 1.$$

Solution

Because the leading coefficient is 1, the possible rational zeros are ± 1, the factors of the constant term. By testing these possible zeros, you can see that neither works.

$$f(1) = (1)^3 + 1 + 1$$
$$= 3$$
$$f(-1) = (-1)^3 + (-1) + 1$$
$$= -1$$

So, you can conclude that the given polynomial has *no* rational zeros. Note from the graph of f in Figure 2.30 that f does have one real zero between -1 and 0. However, by the Rational Zero Test, you know that this real zero is *not* a rational number.

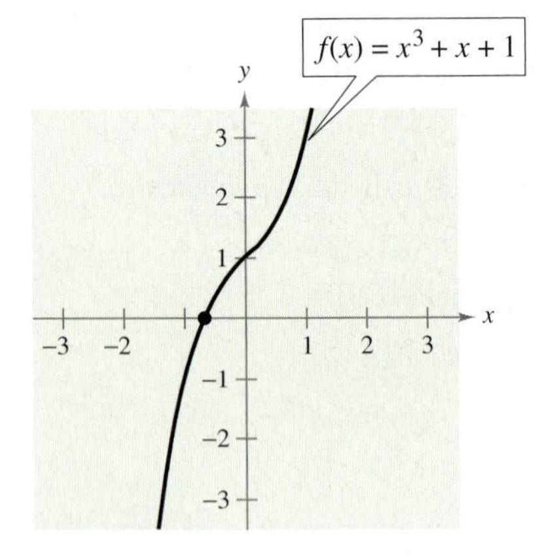

FIGURE 2.30

CHECKPOINT Now try Exercise 7.

STUDY TIP

When the list of possible rational zeros is small, as in Example 2, it may be quicker to test the zeros by evaluating the function. When the list of possible rational zeros is large, as in Example 3, it may be quicker to use a different approach to test the zeros, such as using synthetic division or sketching a graph.

Example 3 Rational Zero Test with Leading Coefficient of 1

Find the rational zeros of $f(x) = x^4 - x^3 + x^2 - 3x - 6$.

Solution

Because the leading coefficient is 1, the possible rational zeros are the factors of the constant term.

Possible rational zeros: $\pm 1, \pm 2, \pm 3, \pm 6$

By applying synthetic division successively, you can determine that $x = -1$ and $x = 2$ are the only two rational zeros.

$$\begin{array}{r|rrrrr} -1 & 1 & -1 & 1 & -3 & -6 \\ & & -1 & 2 & -3 & 6 \\ \hline & 1 & -2 & 3 & -6 & 0 \end{array}$$

0 remainder, so $x = -1$ is a zero.

$$\begin{array}{r|rrrr} 2 & 1 & -2 & 3 & -6 \\ & & 2 & 0 & 6 \\ \hline & 1 & 0 & 3 & 0 \end{array}$$

0 remainder, so $x = 2$ is a zero.

So, $f(x)$ factors as

$$f(x) = (x + 1)(x - 2)(x^2 + 3).$$

Because the factor $(x^2 + 3)$ produces no real zeros, you can conclude that $x = -1$ and $x = 2$ are the only *real* zeros of f, which is verified in Figure 2.31.

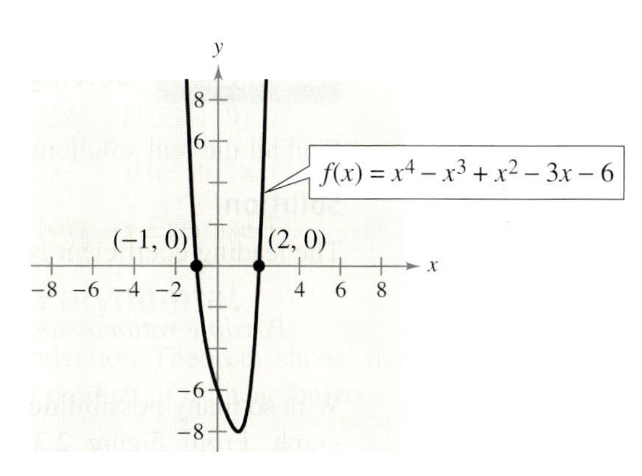

FIGURE 2.31

CHECKPOINT Now try Exercise 11.

If the leading coefficient of a polynomial is not 1, the list of possible rational zeros can increase dramatically. In such cases, the search can be shortened in several ways: (1) a programmable calculator can be used to speed up the calculations; (2) a graph, drawn either by hand or with a graphing utility, can give a good estimate of the locations of the zeros; (3) the Intermediate Value Theorem along with a table generated by a graphing utility can give approximations of zeros; and (4) synthetic division can be used to test the possible rational zeros.

Finding the first zero is often the most difficult part. After that, the search is simplified by working with the lower-degree polynomial obtained in synthetic division, as shown in Example 3.

Additional Example

List the possible rational zeros of $f(x) = x^3 + 8x^2 + 40x - 525$.

Solution

The leading coefficient is 1, so the possible rational zeros are $\pm 1, \pm 3, \pm 5, \pm 7, \pm 15, \pm 21, \pm 25, \pm 35, \pm 75, \pm 105, \pm 175$, and ± 525. To decide which possible rational zeros should be tested using synthetic division, graph the function. From the graph, you can see that the zero is positive and less than 10, so the only values of x that should be tested are $x = 1$, $x = 3$, $x = 5$, and $x = 7$.

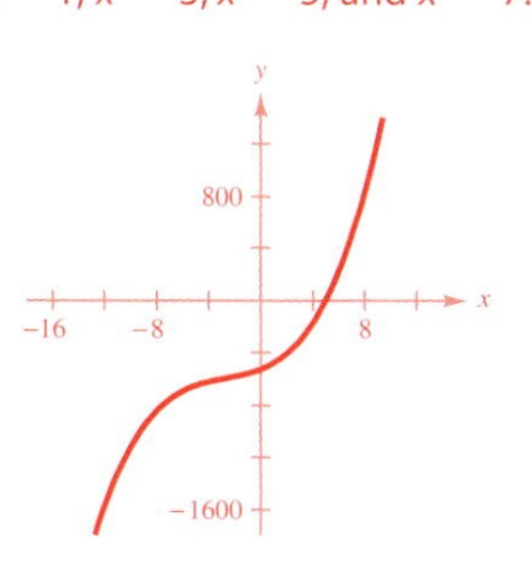

You may want to remind students that a graphing calculator is helpful in determining real zeros, which in turn are useful in finding the complex zeros.

A quadratic factor with no real zeros is said to be *prime* or **irreducible over the reals.** Be sure you see that this is not the same as being *irreducible over the rationals*. For example, the quadratic $x^2 + 1 = (x - i)(x + i)$ is irreducible over the reals (and therefore over the rationals). On the other hand, the quadratic $x^2 - 2 = (x - \sqrt{2})(x + \sqrt{2})$ is irreducible over the rationals but *reducible* over the reals.

Example 7 Finding the Zeros of a Polynomial Function

Find all the zeros of $f(x) = x^4 - 3x^3 + 6x^2 + 2x - 60$ given that $1 + 3i$ is a zero of f.

Algebraic Solution

Because complex zeros occur in conjugate pairs, you know that $1 - 3i$ is also a zero of f. This means that both

$$[x - (1 + 3i)] \quad \text{and} \quad [x - (1 - 3i)]$$

are factors of f. Multiplying these two factors produces

$$\begin{aligned}[x - (1 + 3i)][x - (1 - 3i)] &= [(x - 1) - 3i][(x - 1) + 3i] \\ &= (x - 1)^2 - 9i^2 \\ &= x^2 - 2x + 10.\end{aligned}$$

Using long division, you can divide $x^2 - 2x + 10$ into f to obtain the following.

$$\begin{array}{r} x^2 - x - 6 \\ x^2 - 2x + 10 \overline{\smash{)}\, x^4 - 3x^3 + 6x^2 + 2x - 60} \\ \underline{x^4 - 2x^3 + 10x^2} \\ -x^3 - 4x^2 + 2x \\ \underline{-x^3 + 2x^2 - 10x} \\ -6x^2 + 12x - 60 \\ \underline{-6x^2 + 12x - 60} \\ 0 \end{array}$$

So, you have

$$\begin{aligned}f(x) &= (x^2 - 2x + 10)(x^2 - x - 6) \\ &= (x^2 - 2x + 10)(x - 3)(x + 2)\end{aligned}$$

and you can conclude that the zeros of f are $x = 1 + 3i$, $x = 1 - 3i$, $x = 3$, and $x = -2$.

CHECKPOINT Now try Exercise 47.

Graphical Solution

Because complex zeros always occur in conjugate pairs, you know that $1 - 3i$ is also a zero of f. Because the polynomial is a fourth-degree polynomial, you know that there are at most two other zeros of the function. Use a graphing utility to graph

$$y = x^4 - 3x^3 + 6x^2 + 2x - 60$$

as shown in Figure 2.33.

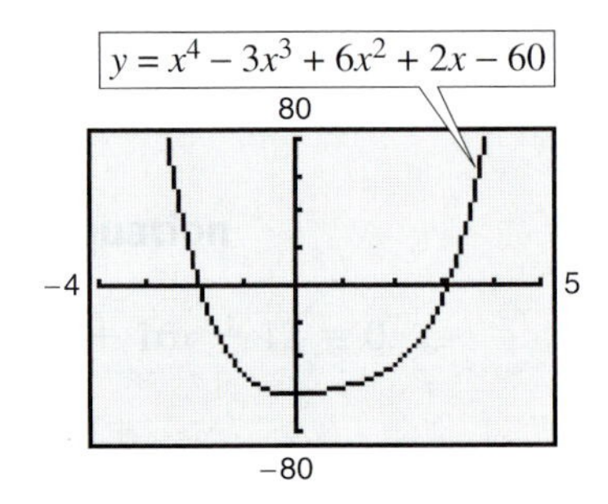

FIGURE 2.33

You can see that -2 and 3 appear to be zeros of the graph of the function. Use the *zero* or *root* feature or the *zoom* and *trace* features of the graphing utility to confirm that $x = -2$ and $x = 3$ are zeros of the graph. So, you can conclude that the zeros of f are $x = 1 + 3i$, $x = 1 - 3i$, $x = 3$, and $x = -2$.

In Example 7, if you were not told that $1 + 3i$ is a zero of f, you could still find all zeros of the function by using synthetic division to find the real zeros -2 and 3. Then you could factor the polynomial as $(x + 2)(x - 3)(x^2 - 2x + 10)$. Finally, by using the Quadratic Formula, you could determine that the zeros are $x = -2$, $x = 3$, $x = 1 + 3i$, and $x = 1 - 3i$.

STUDY TIP

In Example 8, the fifth-degree polynomial function has three real zeros. In such cases, you can use the *zoom* and *trace* features or the *zero* or *root* feature of a graphing utility to approximate the real zeros. You can then use these real zeros to determine the complex zeros algebraically.

Example 8 shows how to find all the zeros of a polynomial function, including complex zeros.

Example 8 Finding the Zeros of a Polynomial Function

Write $f(x) = x^5 + x^3 + 2x^2 - 12x + 8$ as the product of linear factors, and list all of its zeros.

Solution

The possible rational zeros are $\pm 1, \pm 2, \pm 4$, and ± 8. Synthetic division produces the following.

$$\begin{array}{r|rrrrrr} 1 & 1 & 0 & 1 & 2 & -12 & 8 \\ & & 1 & 1 & 2 & 4 & -8 \\ \hline & 1 & 1 & 2 & 4 & -8 & 0 \end{array} \quad \longrightarrow \text{1 is a zero.}$$

$$\begin{array}{r|rrrrr} -2 & 1 & 1 & 2 & 4 & -8 \\ & & -2 & 2 & -8 & 8 \\ \hline & 1 & -1 & 4 & -4 & 0 \end{array} \quad \longrightarrow -2 \text{ is a zero.}$$

So, you have

$$\begin{aligned} f(x) &= x^5 + x^3 + 2x^2 - 12x + 8 \\ &= (x - 1)(x + 2)(x^3 - x^2 + 4x - 4). \end{aligned}$$

You can factor $x^3 - x^2 + 4x - 4$ as $(x - 1)(x^2 + 4)$, and by factoring $x^2 + 4$ as

$$\begin{aligned} x^2 - (-4) &= (x - \sqrt{-4})(x + \sqrt{-4}) \\ &= (x - 2i)(x + 2i) \end{aligned}$$

you obtain

$$f(x) = (x - 1)(x - 1)(x + 2)(x - 2i)(x + 2i)$$

which gives the following five zeros of f.

$$x = 1, x = 1, x = -2, x = 2i, \quad \text{and} \quad x = -2i$$

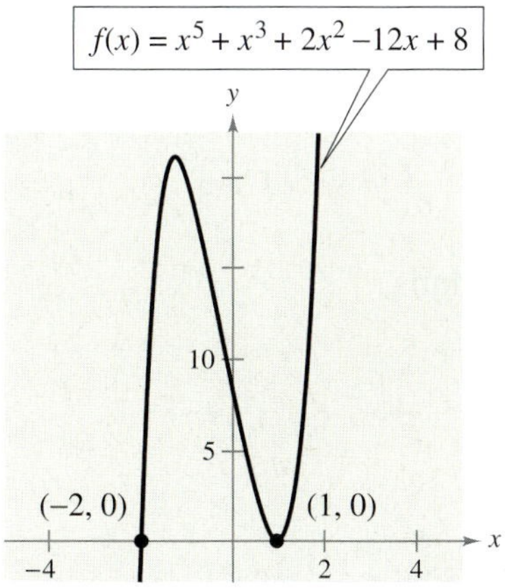

From the graph of f shown in Figure 2.34, you can see that the *real* zeros are the only ones that appear as x-intercepts. Note that $x = 1$ is a repeated zero.

$f(x) = x^5 + x^3 + 2x^2 - 12x + 8$

FIGURE 2.34

CHECKPOINT Now try Exercise 63.

Technology

You can use the *table* feature of a graphing utility to help you determine which of the possible rational zeros are zeros of the polynomial in Example 8. The table should be set to *ask* mode. Then enter each of the possible rational zeros in the table. When you do this, you will see that there are two rational zeros, -2 and 1, as shown at the right.

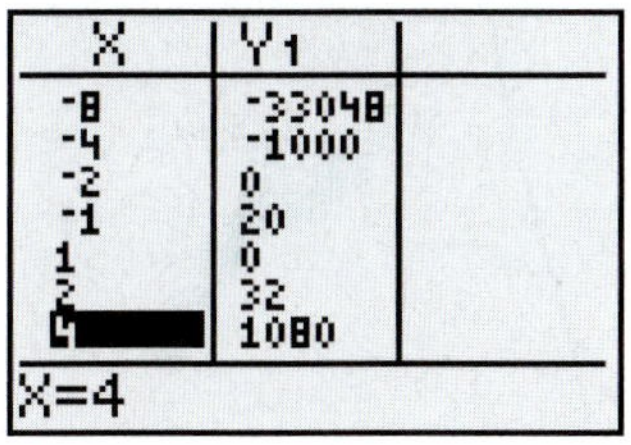

X	Y1
-8	-33048
-4	-1000
-2	0
-1	20
1	0
2	32
4	1080

X=4

Other Tests for Zeros of Polynomials

You know that an nth-degree polynomial function can have *at most* n real zeros. Of course, many nth-degree polynomials do not have that many real zeros. For instance, $f(x) = x^2 + 1$ has no real zeros, and $f(x) = x^3 + 1$ has only one real zero. The following theorem, called **Descartes's Rule of Signs,** sheds more light on the number of real zeros of a polynomial.

Descartes's Rule of Signs

Let $f(x) = a_n x^n + a_{n-1}x^{n-1} + \cdots + a_2x^2 + a_1x + a_0$ be a polynomial with real coefficients and $a_0 \neq 0$.

1. The number of *positive real zeros* of f is either equal to the number of variations in sign of $f(x)$ or less than that number by an even integer.

2. The number of *negative real zeros* of f is either equal to the number of variations in sign of $f(-x)$ or less than that number by an even integer.

A **variation in sign** means that two consecutive coefficients have opposite signs.

When using Descartes's Rule of Signs, a zero of multiplicity k should be counted as k zeros. For instance, the polynomial $x^3 - 3x + 2$ has two variations in sign, and so has either two positive or no positive real zeros. Because

$$x^3 - 3x + 2 = (x - 1)(x - 1)(x + 2)$$

you can see that the two positive real zeros are $x = 1$ of multiplicity 2.

Example 9 Using Descartes's Rule of Signs

Describe the possible real zeros of

$$f(x) = 3x^3 - 5x^2 + 6x - 4.$$

Solution

The original polynomial has *three* variations in sign.

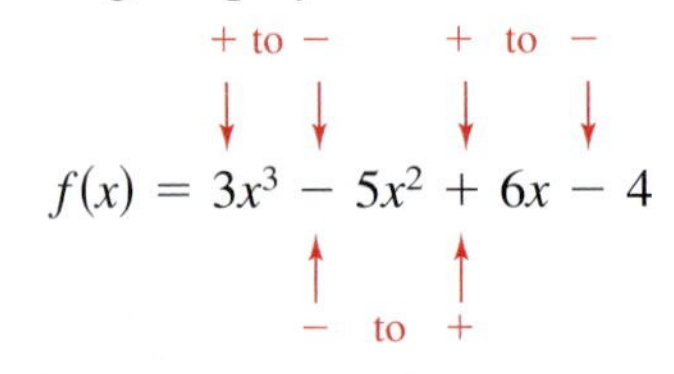

The polynomial

$$f(-x) = 3(-x)^3 - 5(-x)^2 + 6(-x) - 4$$
$$= -3x^3 - 5x^2 - 6x - 4$$

has no variations in sign. So, from Descartes's Rule of Signs, the polynomial $f(x) = 3x^3 - 5x^2 + 6x - 4$ has either three positive real zeros or one positive real zero, and has no negative real zeros. From the graph in Figure 2.35, you can see that the function has only one real zero (it is a positive number, near $x = 1$).

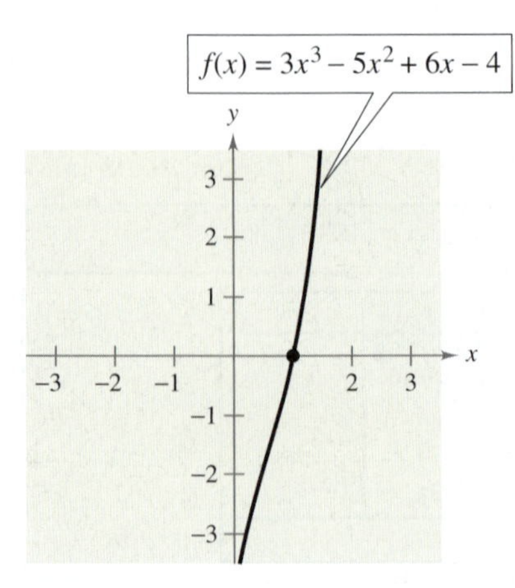

FIGURE 2.35

CHECKPOINT Now try Exercise 79.

Another test for zeros of a polynomial function is related to the sign pattern in the last row of the synthetic division array. This test can give you an upper or lower bound of the real zeros of f. A real number b is an **upper bound** for the real zeros of f if no zeros are greater than b. Similarly, b is a **lower bound** if no real zeros of f are less than b.

Upper and Lower Bound Rules

Let $f(x)$ be a polynomial with real coefficients and a positive leading coefficient. Suppose $f(x)$ is divided by $x - c$, using synthetic division.

1. If $c > 0$ and each number in the last row is either positive or zero, c is an **upper bound** for the real zeros of f.
2. If $c < 0$ and the numbers in the last row are alternately positive and negative (zero entries count as positive or negative), c is a **lower bound** for the real zeros of f.

Example 10 Finding the Zeros of a Polynomial Function

Find the real zeros of $f(x) = 6x^3 - 4x^2 + 3x - 2$.

Solution

The possible real zeros are as follows.

$$\frac{\text{Factors of } 2}{\text{Factors of } 6} = \frac{\pm 1, \pm 2}{\pm 1, \pm 2, \pm 3, \pm 6} = \pm 1, \pm\frac{1}{2}, \pm\frac{1}{3}, \pm\frac{1}{6}, \pm\frac{2}{3}, \pm 2$$

The original polynomial $f(x)$ has three variations in sign. The polynomial

$$f(-x) = 6(-x)^3 - 4(-x)^2 + 3(-x) - 2$$
$$= -6x^3 - 4x^2 - 3x - 2$$

has no variations in sign. As a result of these two findings, you can apply Descartes's Rule of Signs to conclude that there are three positive real zeros or one positive real zero, and no negative zeros. Trying $x = 1$ produces the following.

In Example 10, notice how the Rational Zero Test, Descartes's Rule of Signs, and the Upper and Lower Bound Rules may be used together in a search for all real zeros of a polynomial function.

$$\begin{array}{r|rrrr} 1 & 6 & -4 & 3 & -2 \\ & & 6 & 2 & 5 \\ \hline & 6 & 2 & 5 & 3 \end{array}$$

So, $x = 1$ is not a zero, but because the last row has all positive entries, you know that $x = 1$ is an upper bound for the real zeros. So, you can restrict the search to zeros between 0 and 1. By trial and error, you can determine that $x = \frac{2}{3}$ is a zero. So,

$$f(x) = \left(x - \frac{2}{3}\right)(6x^2 + 3).$$

Because $6x^2 + 3$ has no real zeros, it follows that $x = \frac{2}{3}$ is the only real zero.

CHECKPOINT Now try Exercise 87.

Before concluding this section, here are two additional hints that can help you find the real zeros of a polynomial.

1. If the terms of $f(x)$ have a common monomial factor, it should be factored out before applying the tests in this section. For instance, by writing

$$f(x) = x^4 - 5x^3 + 3x^2 + x$$
$$= x(x^3 - 5x^2 + 3x + 1)$$

you can see that $x = 0$ is a zero of f and that the remaining zeros can be obtained by analyzing the cubic factor.

2. If you are able to find all but two zeros of $f(x)$, you can always use the Quadratic Formula on the remaining quadratic factor. For instance, if you succeeded in writing

$$f(x) = x^4 - 5x^3 + 3x^2 + x$$
$$= x(x - 1)(x^2 - 4x - 1)$$

you can apply the Quadratic Formula to $x^2 - 4x - 1$ to conclude that the two remaining zeros are $x = 2 + \sqrt{5}$ and $x = 2 - \sqrt{5}$.

Example 11 Using a Polynomial Model

You are designing candle-making kits. Each kit contains 25 cubic inches of candle wax and a mold for making a pyramid-shaped candle. You want the height of the candle to be 2 inches less than the length of each side of the candle's square base. What should the dimensions of your candle mold be?

Solution

The volume of a pyramid is $V = \frac{1}{3}Bh$, where B is the area of the base and h is the height. The area of the base is x^2 and the height is $(x - 2)$. So, the volume of the pyramid is $V = \frac{1}{3}x^2(x - 2)$. Substituting 25 for the volume yields the following.

$$25 = \frac{1}{3}x^2(x - 2)$$ Substitute 25 for V.

$$75 = x^3 - 2x^2$$ Multiply each side by 3.

$$0 = x^3 - 2x^2 - 75$$ Write in general form.

The possible rational solutions are $x = \pm 1, \pm 3, \pm 5, \pm 15, \pm 25, \pm 75$. Use synthetic division to test some of the possible solutions. Note that in this case, it makes sense to test only positive x-values. Using synthetic division, you can determine that $x = 5$ is a solution.

$$\begin{array}{r|rrrr} 5 & 1 & -2 & 0 & -75 \\ & & 5 & 15 & 75 \\ \hline & 1 & 3 & 15 & 0 \end{array}$$

The other two solutions, which satisfy $x^2 + 3x + 15 = 0$, are imaginary and can be discarded. You can conclude that the base of the candle mold should be 5 inches by 5 inches and the height of the mold should be $5 - 2 = 3$ inches.

CHECKPOINT Now try Exercise 107.

Activities

1. Write as a product of linear factors: $f(x) = x^4 - 16$.
   Answer: $(x - 2)(x + 2)(x - 2i)(x + 2i)$
2. Find a third-degree polynomial with integer coefficients that has $2, 3 + i$, and $3 - i$ as zeros.
   Answer: $x^3 - 8x^2 + 22x - 20$
3. Use the zero $x = 2i$ to find all the zeros of $f(x) = x^4 - x^3 - 2x^2 - 4x - 24$.
   Answer: $-2, 3, 2i, -2i$

2.5 Exercises

VOCABULARY CHECK: Fill in the blanks.

1. The ________ ________ of ________ states that if $f(x)$ is a polynomial of degree n $(n > 0)$, then f has at least one zero in the complex number system.
2. The ________ ________ ________ states that if $f(x)$ is a polynomial of degree n $(n > 0)$, then f has precisely n linear factors $f(x) = a_n(x - c_1)(x - c_2) \cdot \cdot \cdot (x - c_n)$ where $c_1, c_2, . . . , c_n$ are complex numbers.
3. The test that gives a list of the possible rational zeros of a polynomial function is called the ________ ________ Test.
4. If $a + bi$ is a complex zero of a polynomial with real coefficients, then so is its ________, $a - bi$.
5. A quadratic factor that cannot be factored further as a product of linear factors containing real numbers is said to be ________ over the ________.
6. The theorem that can be used to determine the possible numbers of positive real zeros and negative real zeros of a function is called ________ ________ of ________.
7. A real number b is a(n) ________ bound for the real zeros of f if no real zeros are less than b, and is a(n) ________ bound if no real zeros are greater than b.

PREREQUISITE SKILLS REVIEW: Practice and review algebra skills needed for this section at **www.Eduspace.com.**

In Exercises 1–6, find all the zeros of the function.

1. $f(x) = x(x - 6)^2$
2. $f(x) = x^2(x + 3)(x^2 - 1)$
3. $g(x) = (x - 2)(x + 4)^3$
4. $f(x) = (x + 5)(x - 8)^2$
5. $f(x) = (x + 6)(x + i)(x - i)$
6. $h(t) = (t - 3)(t - 2)(t - 3i)(t + 3i)$

In Exercises 7–10, use the Rational Zero Test to list all possible rational zeros of *f*. Verify that the zeros of *f* shown on the graph are contained in the list.

7. $f(x) = x^3 + 3x^2 - x - 3$

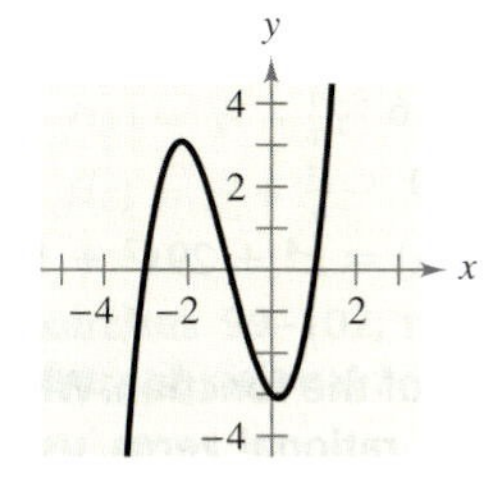

8. $f(x) = x^3 - 4x^2 - 4x + 16$

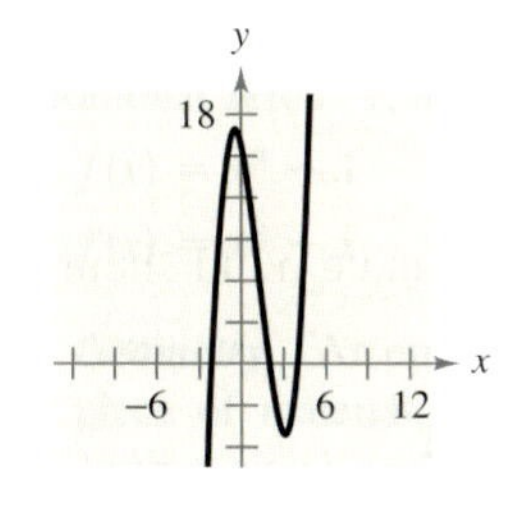

9. $f(x) = 2x^4 - 17x^3 + 35x^2 + 9x - 45$

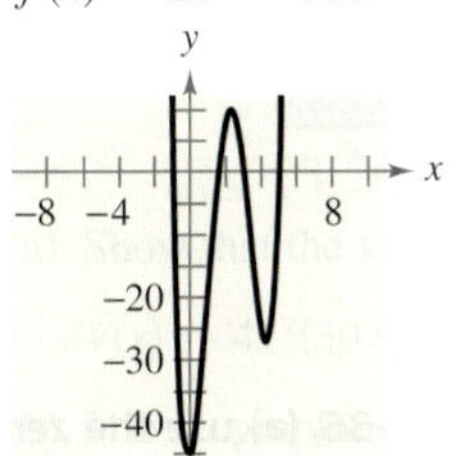

10. $f(x) = 4x^5 - 8x^4 - 5x^3 + 10x^2 + x - 2$

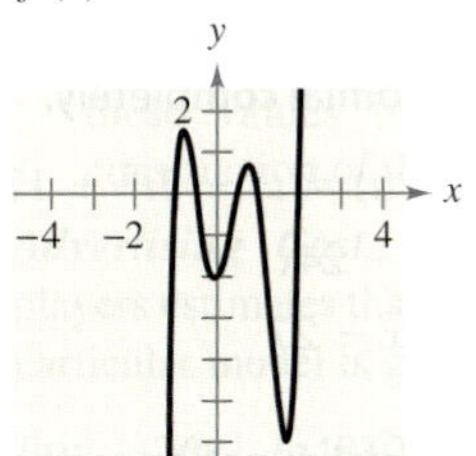

In Exercises 11–20, find all the rational zeros of the function.

11. $f(x) = x^3 - 6x^2 + 11x - 6$
12. $f(x) = x^3 - 7x - 6$
13. $g(x) = x^3 - 4x^2 - x + 4$
14. $h(x) = x^3 - 9x^2 + 20x - 12$
15. $h(t) = t^3 + 12t^2 + 21t + 10$
16. $p(x) = x^3 - 9x^2 + 27x - 27$
17. $C(x) = 2x^3 + 3x^2 - 1$
18. $f(x) = 3x^3 - 19x^2 + 33x - 9$
19. $f(x) = 9x^4 - 9x^3 - 58x^2 + 4x + 24$
20. $f(x) = 2x^4 - 15x^3 + 23x^2 + 15x - 25$

2.6 Rational Functions

What you should learn

- Find the domains of rational functions.
- Find the horizontal and vertical asymptotes of graphs of rational functions.
- Analyze and sketch graphs of rational functions.
- Sketch graphs of rational functions that have slant asymptotes.
- Use rational functions to model and solve real-life problems.

Why you should learn it

Rational functions can be used to model and solve real-life problems relating to business. For instance, in Exercise 79 on page 196, a rational function is used to model average speed over a distance.

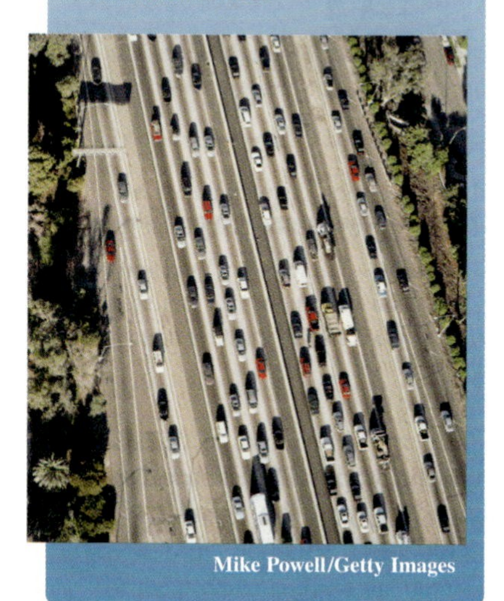

Mike Powell/Getty Images

Introduction

A **rational function** can be written in the form

$$f(x) = \frac{N(x)}{D(x)}$$

where $N(x)$ and $D(x)$ are polynomials and $D(x)$ is not the zero polynomial.

In general, the *domain* of a rational function of x includes all real numbers except x-values that make the denominator zero. Much of the discussion of rational functions will focus on their graphical behavior near the x-values excluded from the domain.

Example 1 Finding the Domain of a Rational Function

Find the domain of $f(x) = \frac{1}{x}$ and discuss the behavior of f near any excluded x-values.

Solution

Because the denominator is zero when $x = 0$, the domain of f is all real numbers except $x = 0$. To determine the behavior of f near this excluded value, evaluate $f(x)$ to the left and right of $x = 0$, as indicated in the following tables.

x	-1	-0.5	-0.1	-0.01	-0.001	$\longrightarrow 0$
$f(x)$	-1	-2	-10	-100	-1000	$\longrightarrow -\infty$

x	$0 \longleftarrow$	0.001	0.01	0.1	0.5	1
$f(x)$	$\infty \longleftarrow$	1000	100	10	2	1

Note that as x approaches 0 *from the left*, $f(x)$ decreases without bound. In contrast, as x approaches 0 *from the right*, $f(x)$ increases without bound. The graph of f is shown in Figure 2.36.

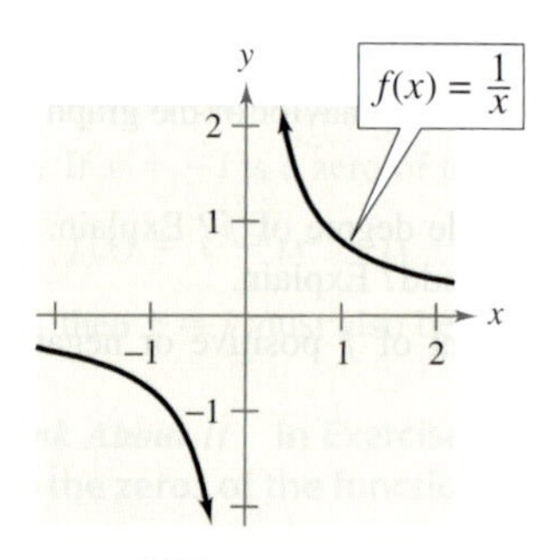

FIGURE 2.36

✓CHECKPOINT Now try Exercise 1.

STUDY TIP

Note that the rational function given by $f(x) = 1/x$ is also referred to as the reciprocal function discussed in Section 1.6.

FIGURE 2.37

Horizontal and Vertical Asymptotes

In Example 1, the behavior of f near $x = 0$ is denoted as follows.

$$\underbrace{f(x) \longrightarrow -\infty \text{ as } x \longrightarrow 0^-}_{\substack{f(x) \text{ decreases without bound} \\ \text{as } x \text{ approaches 0 from the left.}}} \qquad \underbrace{f(x) \longrightarrow \infty \text{ as } x \longrightarrow 0^+}_{\substack{f(x) \text{ increases without bound} \\ \text{as } x \text{ approaches 0 from the right.}}}$$

The line $x = 0$ is a **vertical asymptote** of the graph of f, as shown in Figure 2.37. From this figure, you can see that the graph of f also has a **horizontal asymptote**—the line $y = 0$. This means that the values of $f(x) = 1/x$ approach zero as x increases or decreases without bound.

$$\underbrace{f(x) \longrightarrow 0 \text{ as } x \longrightarrow -\infty}_{\substack{f(x) \text{ approaches 0 as } x \\ \text{decreases without bound.}}} \qquad \underbrace{f(x) \longrightarrow 0 \text{ as } x \longrightarrow \infty}_{\substack{f(x) \text{ approaches 0 as } x \\ \text{increases without bound.}}}$$

Definitions of Vertical and Horizontal Asymptotes

1. The line $x = a$ is a **vertical asymptote** of the graph of f if

$$f(x) \longrightarrow \infty \quad \text{or} \quad f(x) \longrightarrow -\infty$$

as $x \longrightarrow a$, either from the right or from the left.

2. The line $y = b$ is a **horizontal asymptote** of the graph of f if

$$f(x) \longrightarrow b$$

as $x \longrightarrow \infty$ or $x \longrightarrow -\infty$.

Additional Examples

State the domain of each function.

a. $f(x) = \dfrac{3x}{x + 10}$

b. $f(x) = \dfrac{x + 1}{(x + 2)(x - 6)}$

Solution

a. The domain is all real numbers except $x = -10$.

b. The domain is all real numbers except $x = -2$ and $x = 6$.

Eventually (as $x \longrightarrow \infty$ or $x \longrightarrow -\infty$), the distance between the horizontal asymptote and the points on the graph must approach zero. Figure 2.38 shows the horizontal and vertical asymptotes of the graphs of three rational functions.

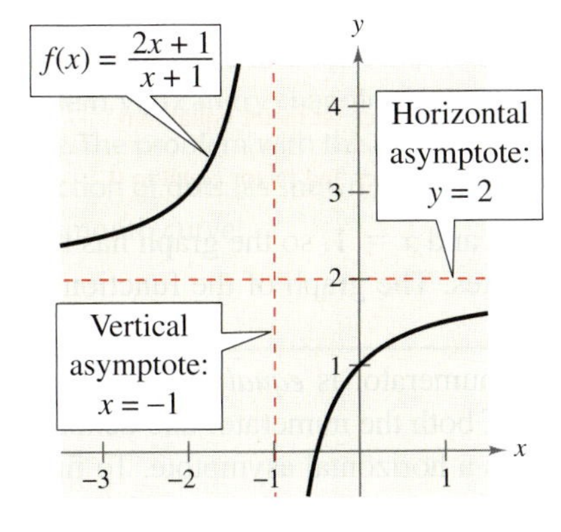

(a)

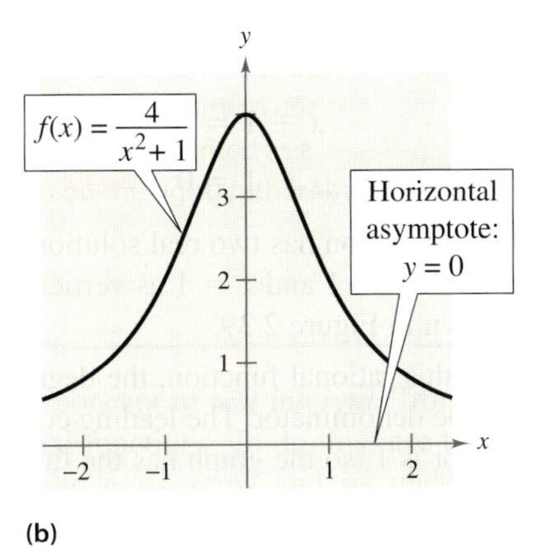

(b)

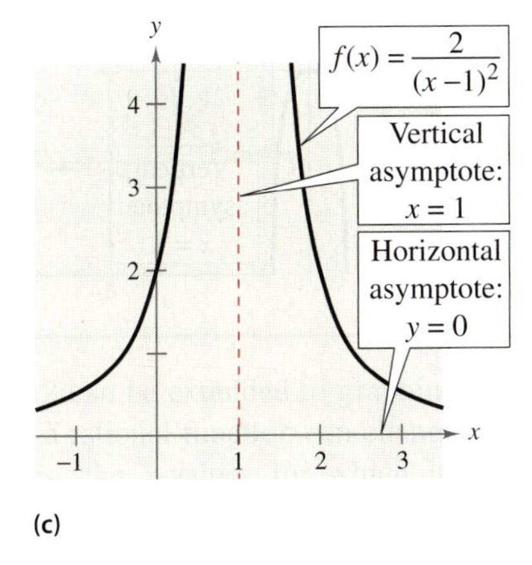

(c)

FIGURE 2.38

The graphs of $f(x) = 1/x$ in Figure 2.37 and $f(x) = (2x + 1)/(x + 1)$ in Figure 2.38(a) are **hyperbolas.** You will study hyperbolas in Section 10.4.

STUDY TIP

You can use transformations to help you sketch graphs of rational functions. For instance, the graph of g in Example 3 is a vertical stretch and a right shift of the graph of $f(x) = 1/x$ because

$$g(x) = \frac{3}{x-2} = 3\left(\frac{1}{x-2}\right) = 3f(x-2).$$

Example 3 Sketching the Graph of a Rational Function

Sketch the graph of $g(x) = \dfrac{3}{x-2}$ and state its domain.

Solution

y-intercept: $\left(0, -\frac{3}{2}\right)$, because $g(0) = -\frac{3}{2}$

x-intercept: None, because $3 \neq 0$

Vertical asymptote: $x = 2$, zero of denominator

Horizontal asymptote: $y = 0$, because degree of $N(x) <$ degree of $D(x)$

Additional points:

Test interval	Representative x-value	Value of g	Sign	Point on graph
$(-\infty, 2)$	-4	$g(-4) = -0.5$	Negative	$(-4, -0.5)$
$(2, \infty)$	3	$g(3) = 3$	Positive	$(3, 3)$

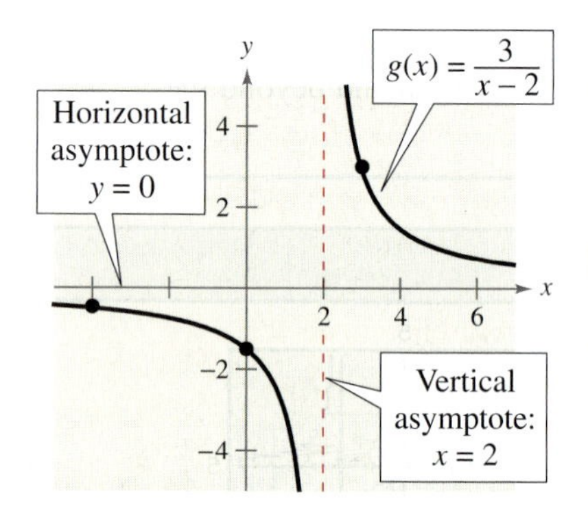

FIGURE 2.40

By plotting the intercepts, asymptotes, and a few additional points, you can obtain the graph shown in Figure 2.40. The domain of g is all real numbers x except $x = 2$.

CHECKPOINT Now try Exercise 27.

Example 4 Sketching the Graph of a Rational Function

Sketch the graph of

$$f(x) = \frac{2x-1}{x}$$

and state its domain.

Solution

y-intercept: None, because $x = 0$ is not in the domain

x-intercept: $\left(\frac{1}{2}, 0\right)$, because $2x - 1 = 0$

Vertical asymptote: $x = 0$, zero of denominator

Horizontal asymptote: $y = 2$, because degree of $N(x) =$ degree of $D(x)$

Additional points:

Test interval	Representative x-value	Value of f	Sign	Point on graph
$(-\infty, 0)$	-1	$f(-1) = 3$	Positive	$(-1, 3)$
$\left(0, \frac{1}{2}\right)$	$\frac{1}{4}$	$f\left(\frac{1}{4}\right) = -2$	Negative	$\left(\frac{1}{4}, -2\right)$
$\left(\frac{1}{2}, \infty\right)$	4	$f(4) = 1.75$	Positive	$(4, 1.75)$

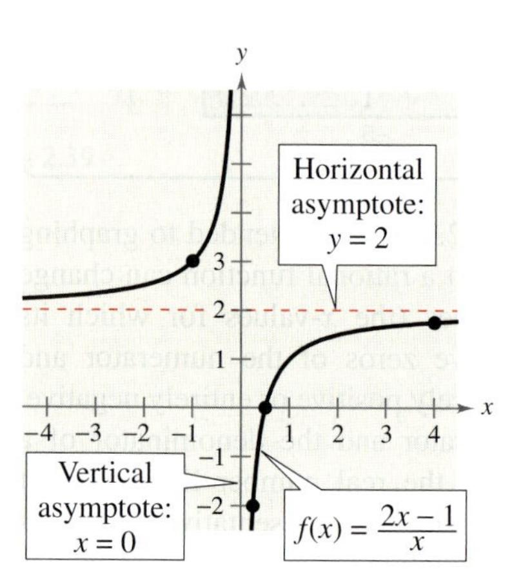

FIGURE 2.41

By plotting the intercepts, asymptotes, and a few additional points, you can obtain the graph shown in Figure 2.41. The domain of f is all real numbers x except $x = 0$.

CHECKPOINT Now try Exercise 31.

Example 5 Sketching the Graph of a Rational Function

Sketch the graph of $f(x) = x/(x^2 - x - 2)$.

Solution

Factoring the denominator, you have $f(x) = \dfrac{x}{(x + 1)(x - 2)}$.

y-intercept: $(0, 0)$, because $f(0) = 0$

x-intercept: $(0, 0)$

Vertical asymptotes: $x = -1$, $x = 2$, zeros of denominator

Horizontal asymptote: $y = 0$, because degree of $N(x) <$ degree of $D(x)$

Additional points:

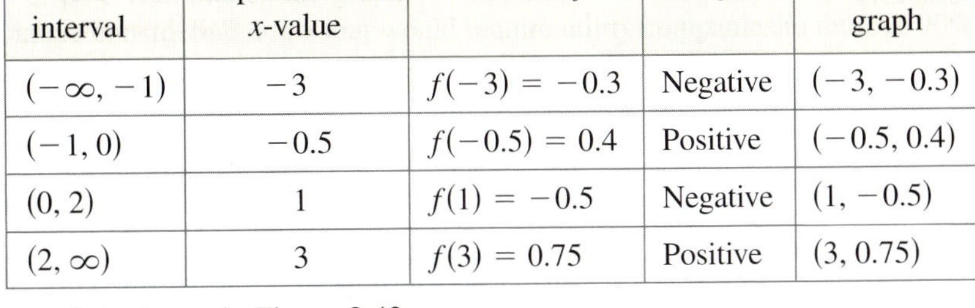

Test interval	Representative x-value	Value of f	Sign	Point on graph
$(-\infty, -1)$	-3	$f(-3) = -0.3$	Negative	$(-3, -0.3)$
$(-1, 0)$	-0.5	$f(-0.5) = 0.4$	Positive	$(-0.5, 0.4)$
$(0, 2)$	1	$f(1) = -0.5$	Negative	$(1, -0.5)$
$(2, \infty)$	3	$f(3) = 0.75$	Positive	$(3, 0.75)$

The graph is shown in Figure 2.42.

FIGURE 2.42

CHECKPOINT Now try Exercise 35.

STUDY TIP

If you are unsure of the shape of a portion of the graph of a rational function, plot some additional points. Also note that when the numerator and the denominator of a rational function have a common factor, the graph of the function has a *hole* at the zero of the common factor (see Example 6).

Example 6 A Rational Function with Common Factors

Sketch the graph of $f(x) = (x^2 - 9)/(x^2 - 2x - 3)$.

Solution

By factoring the numerator and denominator, you have

$$f(x) = \frac{x^2 - 9}{x^2 - 2x - 3} = \frac{(x - 3)(x + 3)}{(x - 3)(x + 1)} = \frac{x + 3}{x + 1}, \quad x \neq 3.$$

y-intercept: $(0, 3)$, because $f(0) = 3$

x-intercept: $(-3, 0)$, because $f(-3) = 0$

Vertical asymptote: $x = -1$, zero of (simplified) denominator

Horizontal asymptote: $y = 1$, because degree of $N(x) =$ degree of $D(x)$

Additional points:

Test interval	Representative x-value	Value of f	Sign	Point on graph
$(-\infty, -3)$	-4	$f(-4) = 0.33$	Positive	$(-4, -0.33)$
$(-3, -1)$	-2	$f(-2) = -1$	Negative	$(-2, -1)$
$(-1, \infty)$	2	$f(2) = 1.67$	Positive	$(2, 1.67)$

The graph is shown in Figure 2.43. Notice that there is a hole in the graph at $x = 3$ because the function is not defined when $x = 3$.

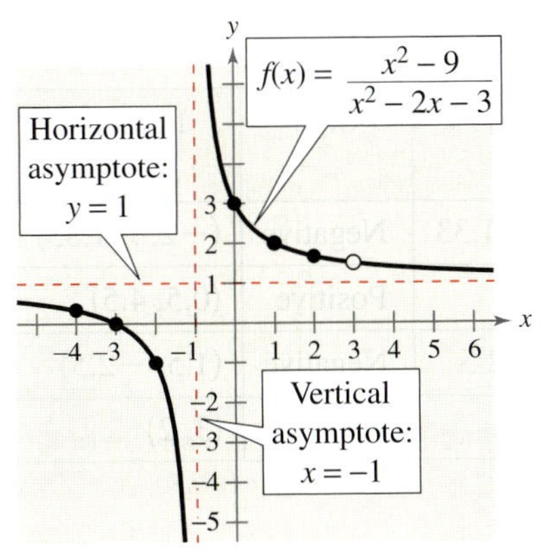

FIGURE 2.43 HOLE AT $x = 3$

CHECKPOINT Now try Exercise 41.

In Exercises 21–26, find the domain of the function and identify any horizontal and vertical asymptotes.

21. $f(x) = \dfrac{x-4}{x^2-16}$

22. $f(x) = \dfrac{x+3}{x^2-9}$

23. $f(x) = \dfrac{x^2-1}{x^2-2x-3}$

24. $f(x) = \dfrac{x^2-4}{x^2-3x+2}$

25. $f(x) = \dfrac{x^2-3x-4}{2x^2+x-1}$

26. $f(x) = \dfrac{6x^2-11x+3}{6x^2-7x-3}$

In Exercises 27–46, (a) state the domain of the function, (b) identify all intercepts, (c) find any vertical and horizontal asymptotes, and (d) plot additional solution points as needed to sketch the graph of the rational function.

27. $f(x) = \dfrac{1}{x+2}$

28. $f(x) = \dfrac{1}{x-3}$

29. $h(x) = \dfrac{-1}{x+2}$

30. $g(x) = \dfrac{1}{3-x}$

31. $C(x) = \dfrac{5+2x}{1+x}$

32. $P(x) = \dfrac{1-3x}{1-x}$

33. $f(x) = \dfrac{x^2}{x^2+9}$

34. $f(t) = \dfrac{1-2t}{t}$

35. $g(s) = \dfrac{s}{s^2+1}$

36. $f(x) = -\dfrac{1}{(x-2)^2}$

37. $h(x) = \dfrac{x^2-5x+4}{x^2-4}$

38. $g(x) = \dfrac{x^2-2x-8}{x^2-9}$

39. $f(x) = \dfrac{2x^2-5x-3}{x^3-2x^2-x+2}$

40. $f(x) = \dfrac{x^2-x-2}{x^3-2x^2-5x+6}$

41. $f(x) = \dfrac{x^2+3x}{x^2+x-6}$

42. $f(x) = \dfrac{5(x+4)}{x^2+x-12}$

43. $f(x) = \dfrac{2x^2-5x+2}{2x^2-x-6}$

44. $f(x) = \dfrac{3x^2-8x+4}{2x^2-3x-2}$

45. $f(t) = \dfrac{t^2-1}{t+1}$

46. $f(x) = \dfrac{x^2-16}{x-4}$

Analytical, Numerical, and Graphical Analysis **In Exercises 47–50, do the following.**

(a) Determine the domains of f and g.

(b) Simplify f and find any vertical asymptotes of the graph of f.

(c) Compare the functions by completing the table.

(d) Use a graphing utility to graph f and g in the same viewing window.

(e) Explain why the graphing utility may not show the difference in the domains of f and g.

47. $f(x) = \dfrac{x^2-1}{x+1}, \quad g(x) = x-1$

x	-3	-2	-1.5	-1	-0.5	0	1
$f(x)$							
$g(x)$							

48. $f(x) = \dfrac{x^2(x-2)}{x^2-2x}, \quad g(x) = x$

x	-1	0	1	1.5	2	2.5	3
$f(x)$							
$g(x)$							

49. $f(x) = \dfrac{x-2}{x^2-2x}, \quad g(x) = \dfrac{1}{x}$

x	-0.5	0	0.5	1	1.5	2	3
$f(x)$							
$g(x)$							

50. $f(x) = \dfrac{2x-6}{x^2-7x+12}, \quad g(x) = \dfrac{2}{x-4}$

x	0	1	2	3	4	5	6
$f(x)$							
$g(x)$							

In Exercises 51–64, (a) state the domain of the function, (b) identify all intercepts, (c) identify any vertical and slant asymptotes, and (d) plot additional solution points as needed to sketch the graph of the rational function.

51. $h(x) = \dfrac{x^2-4}{x}$

52. $g(x) = \dfrac{x^2+5}{x}$

53. $f(x) = \dfrac{2x^2+1}{x}$

54. $f(x) = \dfrac{1-x^2}{x}$

55. $g(x) = \dfrac{x^2+1}{x}$

56. $h(x) = \dfrac{x^2}{x-1}$

57. $f(t) = -\dfrac{t^2+1}{t+5}$

58. $f(x) = \dfrac{x^2}{3x+1}$

59. $f(x) = \dfrac{x^3}{x^2-1}$

60. $g(x) = \dfrac{x^3}{2x^2-8}$

61. $f(x) = \dfrac{x^2-x+1}{x-1}$

62. $f(x) = \dfrac{2x^2-5x+5}{x-2}$

63. $f(x) = \dfrac{2x^3 - x^2 - 2x + 1}{x^2 + 3x + 2}$

64. $f(x) = \dfrac{2x^3 + x^2 - 8x - 4}{x^2 - 3x + 2}$

In Exercises 65–68, use a graphing utility to graph the rational function. Give the domain of the function and identify any asymptotes. Then zoom out sufficiently far so that the graph appears as a line. Identify the line.

65. $f(x) = \dfrac{x^2 + 5x + 8}{x + 3}$

66. $f(x) = \dfrac{2x^2 + x}{x + 1}$

67. $g(x) = \dfrac{1 + 3x^2 - x^3}{x^2}$

68. $h(x) = \dfrac{12 - 2x - x^2}{2(4 + x)}$

Graphical Reasoning **In Exercises 69–72, (a) use the graph to determine any x-intercepts of the graph of the rational function and (b) set $y = 0$ and solve the resulting equation to confirm your result in part (a).**

69. $y = \dfrac{x + 1}{x - 3}$

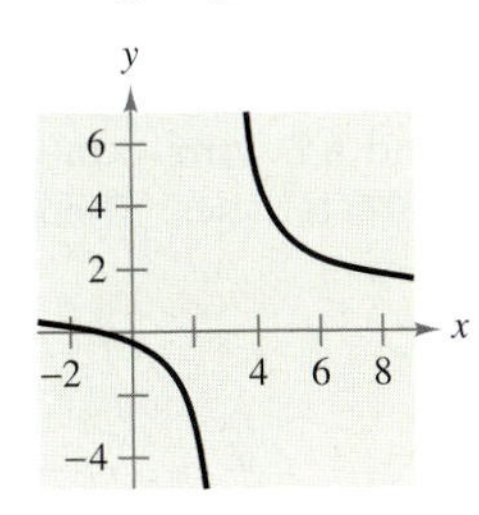

70. $y = \dfrac{2x}{x - 3}$

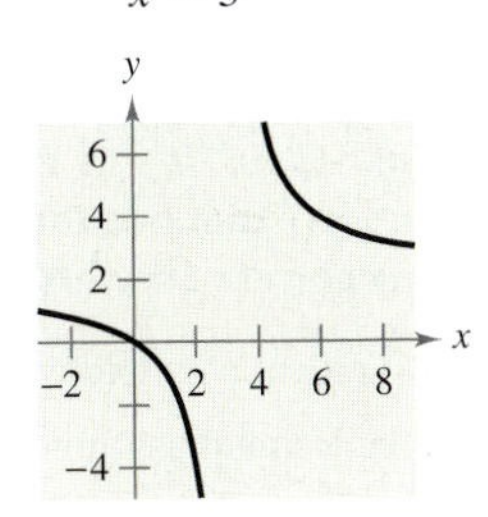

71. $y = \dfrac{1}{x} - x$

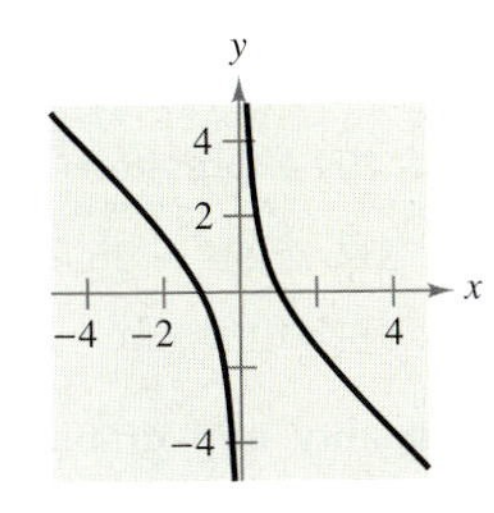

72. $y = x - 3 + \dfrac{2}{x}$

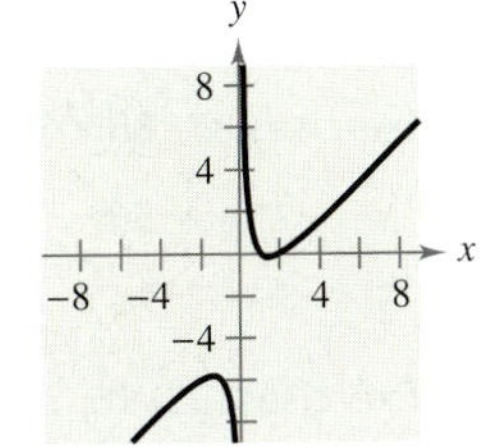

73. ***Pollution*** The cost C (in millions of dollars) of removing $p\%$ of the industrial and municipal pollutants discharged into a river is given by

$$C = \frac{255p}{100 - p}, \quad 0 \le p < 100.$$

(a) Use a graphing utility to graph the cost function.

(b) Find the costs of removing 10%, 40%, and 75% of the pollutants.

(c) According to this model, would it be possible to remove 100% of the pollutants? Explain.

74. ***Recycling*** In a pilot project, a rural township is given recycling bins for separating and storing recyclable products. The cost C (in dollars) for supplying bins to $p\%$ of the population is given by

$$C = \frac{25{,}000p}{100 - p}, \quad 0 \le p < 100.$$

(a) Use a graphing utility to graph the cost function.

(b) Find the costs of supplying bins to 15%, 50%, and 90% of the population.

(c) According to this model, would it be possible to supply bins to 100% of the residents? Explain.

75. ***Population Growth*** The game commission introduces 100 deer into newly acquired state game lands. The population N of the herd is modeled by

$$N = \frac{20(5 + 3t)}{1 + 0.04t}, \quad t \ge 0$$

where t is the time in years (see figure).

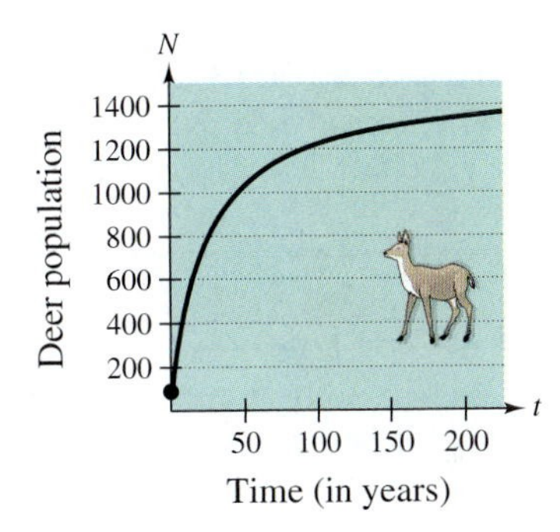

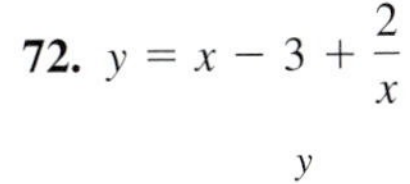

(a) Find the populations when $t = 5$, $t = 10$, and $t = 25$.

(b) What is the limiting size of the herd as time increases?

76. ***Concentration of a Mixture*** A 1000-liter tank contains 50 liters of a 25% brine solution. You add x liters of a 75% brine solution to the tank.

(a) Show that the concentration C, the proportion of brine to total solution, in the final mixture is

$$C = \frac{3x + 50}{4(x + 50)}.$$

(b) Determine the domain of the function based on the physical constraints of the problem.

(c) Sketch a graph of the concentration function.

(d) As the tank is filled, what happens to the rate at which the concentration of brine is increasing? What percent does the concentration of brine appear to approach?

77. ***Page Design*** A page that is x inches wide and y inches high contains 30 square inches of print. The top and bottom margins are 1 inch deep and the margins on each side are 2 inches wide (see figure).

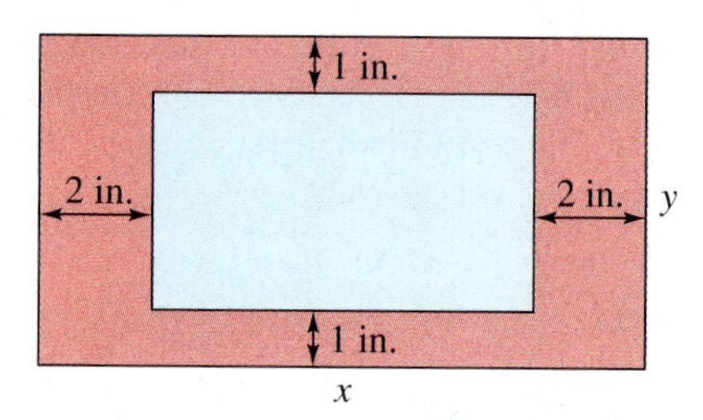

(a) Show that the total area A on the page is

$$A = \frac{2x(x + 11)}{x - 4}.$$

(b) Determine the domain of the function based on the physical constraints of the problem.

(c) Use a graphing utility to graph the area function and approximate the page size for which the least amount of paper will be used. Verify your answer numerically using the *table* feature of the graphing utility.

78. ***Page Design*** A rectangular page is designed to contain 64 square inches of print. The margins at the top and bottom of the page are each 1 inch deep. The margins on each side are $1\frac{1}{2}$ inches wide. What should the dimensions of the page be so that the least amount of paper is used?

Model It

79. ***Average Speed*** A driver averaged 50 miles per hour on the round trip between Akron, Ohio, and Columbus, Ohio, 100 miles away. The average speeds for going and returning were x and y miles per hour, respectively.

(a) Show that $y = \dfrac{25x}{x - 25}$.

(b) Determine the vertical and horizontal asymptotes of the graph of the function.

(c) Use a graphing utility to graph the function.

(d) Complete the table.

x	30	35	40	45	50	55	60
y							

(e) Are the results in the table what you expected? Explain.

(f) Is it possible to average 20 miles per hour in one direction and still average 50 miles per hour on the round trip? Explain.

80. ***Sales*** The sales S (in millions of dollars) for the Yankee Candle Company in the years 1998 through 2003 are shown in the table. (Source: The Yankee Candle Company)

1998	184.5	1999	256.6	2000	338.8
2001	379.8	2002	444.8	2003	508.6

A model for these data is given by

$$S = \frac{5.816t^2 - 130.68}{0.004t^2 + 1.00}, \quad 8 \le t \le 13$$

where t represents the year, with $t = 8$ corresponding to 1998.

(a) Use a graphing utility to plot the data and graph the model in the same viewing window. How well does the model fit the data?

(b) Use the model to estimate the sales for the Yankee Candle Company in 2008.

(c) Would this model be useful for estimating sales after 2008? Explain.

Synthesis

True or False? **In Exercises 81 and 82, determine whether the statement is true or false. Justify your answer.**

81. A polynomial can have infinitely many vertical asymptotes.

82. The graph of a rational function can never cross one of its asymptotes.

Think About It **In Exercises 83 and 84, write a rational function f that has the specified characteristics. (There are many correct answers.)**

83. Vertical asymptote: None

Horizontal asymptote: $y = 2$

84. Vertical asymptote: $x = -2, x = 1$

Horizontal asymptote: None

Skills Review

In Exercises 85–88, completely factor the expression.

85. $x^2 - 15x + 56$

86. $3x^2 + 23x - 36$

87. $x^3 - 5x^2 + 4x - 20$

88. $x^3 + 6x^2 - 2x - 12$

In Exercises 93–96, solve the inequality and graph the solution on the real number line.

89. $10 - 3x \le 0$

90. $5 - 2x > 5(x + 1)$

91. $|4(x - 2)| < 20$

92. $\frac{1}{2}|2x + 3| \ge 5$

93. **Make a Decision** To work an extended application analyzing the total manpower of the Department of Defense, visit this text's website at *college.hmco.com.* *(Data Source: U.S. Census Bureau)*

2.7 Nonlinear Inequalities

What you should learn

- Solve polynomial inequalities.
- Solve rational inequalities.
- Use inequalities to model and solve real-life problems.

Why you should learn it

Inequalities can be used to model and solve real-life problems. For instance, in Exercise 73 on page 205, a polynomial inequality is used to model the percent of households that own a television and have cable in the United States.

Polynomial Inequalities

To solve a polynomial inequality such as $x^2 - 2x - 3 < 0$, you can use the fact that a polynomial can change signs only at its zeros (the x-values that make the polynomial equal to zero). Between two consecutive zeros, a polynomial must be entirely positive or entirely negative. This means that when the real zeros of a polynomial are put in order, they divide the real number line into intervals in which the polynomial has no sign changes. These zeros are the **critical numbers** of the inequality, and the resulting intervals are the **test intervals** for the inequality. For instance, the polynomial above factors as

$$x^2 - 2x - 3 = (x + 1)(x - 3)$$

and has two zeros, $x = -1$ and $x = 3$. These zeros divide the real number line into three test intervals:

$$(-\infty, -1), \quad (-1, 3), \quad \text{and} \quad (3, \infty). \qquad \text{(See Figure 2.51.)}$$

So, to solve the inequality $x^2 - 2x - 3 < 0$, you need only test one value from each of these test intervals to determine whether the value satisfies the original inequality. If so, you can conclude that the interval is a solution of the inequality.

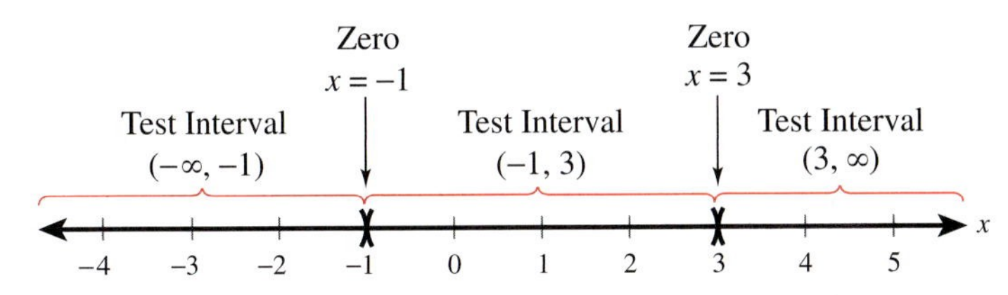

FIGURE 2.51 *Three test intervals for* $x^2 - 2x - 3$

You can use the same basic approach to determine the test intervals for any polynomial.

Finding Test Intervals for a Polynomial

To determine the intervals on which the values of a polynomial are entirely negative or entirely positive, use the following steps.

1. Find all real zeros of the polynomial, and arrange the zeros in increasing order (from smallest to largest). These zeros are the critical numbers of the polynomial.
2. Use the critical numbers of the polynomial to determine its test intervals.
3. Choose one representative x-value in each test interval and evaluate the polynomial at that value. If the value of the polynomial is negative, the polynomial will have negative values for every x-value in the interval. If the value of the polynomial is positive, the polynomial will have positive values for every x-value in the interval.

In Exercises 75 and 76, write the quotient in standard form.

75. $\dfrac{6+i}{4-i}$

76. $\dfrac{3+2i}{5+i}$

In Exercises 77 and 78, perform the operation and write the result in standard form.

77. $\dfrac{4}{2-3i}+\dfrac{2}{1+i}$

78. $\dfrac{1}{2+i}-\dfrac{5}{1+4i}$

In Exercises 79–82, find all solutions of the equation.

79. $3x^2+1=0$

80. $2+8x^2=0$

81. $x^2-2x+10=0$

82. $6x^2+3x+27=0$

2.5 In Exercises 83–88, find all the zeros of the function.

83. $f(x)=3x(x-2)^2$

84. $f(x)=(x-4)(x+9)^2$

85. $f(x)=x^2-9x+8$

86. $f(x)=x^3+6x$

87. $f(x)=(x+4)(x-6)(x-2i)(x+2i)$

88. $f(x)=(x-8)(x-5)^2(x-3+i)(x-3-i)$

In Exercises 89 and 90, use the Rational Zero Test to list all possible rational zeros of *f*.

89. $f(x)=-4x^3+8x^2-3x+15$

90. $f(x)=3x^4+4x^3-5x^2-8$

In Exercises 91–96, find all the rational zeros of the function.

91. $f(x)=x^3-2x^2-21x-18$

92. $f(x)=3x^3-20x^2+7x+30$

93. $f(x)=x^3-10x^2+17x-8$

94. $f(x)=x^3+9x^2+24x+20$

95. $f(x)=x^4+x^3-11x^2+x-12$

96. $f(x)=25x^4+25x^3-154x^2-4x+24$

In Exercises 97 and 98, find a polynomial function with real coefficients that has the given zeros. (There are many correct answers.)

97. $\frac{2}{3}, 4, \sqrt{3}i$

98. $2, -3, 1-2i$

In Exercises 99–102, use the given zero to find all the zeros of the function.

	Function	*Zero*
99.	$f(x)=x^3-4x^2+x-4$	i
100.	$h(x)=-x^3+2x^2-16x+32$	$-4i$
101.	$g(x)=2x^4-3x^3-13x^2+37x-15$	$2+i$
102.	$f(x)=4x^4-11x^3+14x^2-6x$	$1-i$

In Exercises 103–106, find all the zeros of the function and write the polynomial as a product of linear factors.

103. $f(x)=x^3+4x^2-5x$

104. $g(x)=x^3-7x^2+36$

105. $g(x)=x^4+4x^3-3x^2+40x+208$

106. $f(x)=x^4+8x^3+8x^2-72x-153$

In Exercises 107 and 108, use Descartes's Rule of Signs to determine the possible numbers of positive and negative zeros of the function.

107. $g(x)=5x^3+3x^2-6x+9$

108. $h(x)=-2x^5+4x^3-2x^2+5$

In Exercises 109 and 110, use synthetic division to verify the upper and lower bounds of the real zeros of *f*.

109. $f(x)=4x^3-3x^2+4x-3$

(a) Upper: $x=1$

(b) Lower: $x=-\frac{1}{4}$

110. $f(x)=2x^3-5x^2-14x+8$

(a) Upper: $x=8$

(b) Lower: $x=-4$

2.6 In Exercises 111–114, find the domain of the rational function.

111. $f(x)=\dfrac{5x}{x+12}$

112. $f(x)=\dfrac{3x^2}{1+3x}$

113. $f(x)=\dfrac{8}{x^2-10x+24}$

114. $f(x)=\dfrac{x^2+x-2}{x^2+4}$

In Exercises 115–118, identify any horizontal or vertical asymptotes.

115. $f(x)=\dfrac{4}{x+3}$

116. $f(x)=\dfrac{2x^2+5x-3}{x^2+2}$

117. $h(x)=\dfrac{2x-10}{x^2-2x-15}$

118. $h(x)=\dfrac{x^3-4x^2}{x^2+3x+2}$

In Exercises 119–130, (a) state the domain of the function, (b) identify all intercepts, (c) find any vertical and horizontal asymptotes, and (d) plot additional solution points as needed to sketch the graph of the rational function.

119. $f(x)=\dfrac{-5}{x^2}$

120. $f(x)=\dfrac{4}{x}$

121. $g(x)=\dfrac{2+x}{1-x}$

122. $h(x)=\dfrac{x-3}{x-2}$

123. $p(x)=\dfrac{x^2}{x^2+1}$

124. $f(x)=\dfrac{2x}{x^2+4}$

125. $f(x)=\dfrac{x}{x^2+1}$

126. $h(x)=\dfrac{4}{(x-1)^2}$

127. $f(x) = \dfrac{-6x^2}{x^2 + 1}$

128. $y = \dfrac{2x^2}{x^2 - 4}$

129. $f(x) = \dfrac{6x^2 - 11x + 3}{3x^2 - x}$

130. $f(x) = \dfrac{6x^2 - 7x + 2}{4x^2 - 1}$

In Exercises 131–134, (a) state the domain of the function, (b) identify all intercepts, (c) identify any vertical and slant asymptotes, and (d) plot additional solution points as needed to sketch the graph of the rational function.

131. $f(x) = \dfrac{2x^3}{x^2 + 1}$

132. $f(x) = \dfrac{x^2 + 1}{x + 1}$

133. $f(x) = \dfrac{3x^3 - 2x^2 - 3x + 2}{3x^2 - x - 4}$

134. $f(x) = \dfrac{3x^3 - 4x^2 - 12x + 16}{3x^2 + 5x - 2}$

135. ***Average Cost*** A business has a production cost of $C = 0.5x + 500$ for producing x units of a product. The average cost per unit, $\overline{C}$, is given by

$$\overline{C} = \frac{C}{x} = \frac{0.5x + 500}{x}, \quad x > 0.$$

Determine the average cost per unit as x increases without bound. (Find the horizontal asymptote.)

136. ***Seizure of Illegal Drugs*** The cost C (in millions of dollars) for the federal government to seize $p\%$ of an illegal drug as it enters the country is given by

$$C = \frac{528p}{100 - p}, \quad 0 \le p < 100.$$

(a) Use a graphing utility to graph the cost function.

(b) Find the costs of seizing 25%, 50%, and 75% of the drug.

(c) According to this model, would it be possible to seize 100% of the drug?

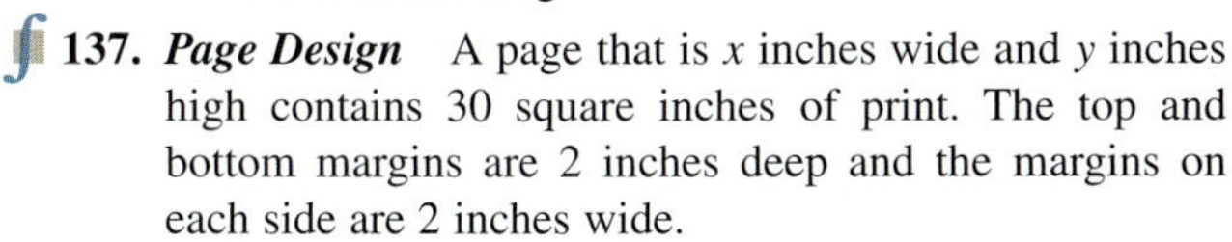

137. ***Page Design*** A page that is x inches wide and y inches high contains 30 square inches of print. The top and bottom margins are 2 inches deep and the margins on each side are 2 inches wide.

(a) Draw a diagram that gives a visual representation of the problem.

(b) Show that the total area A on the page is

$$A = \frac{2x(2x + 7)}{x - 4}.$$

(c) Determine the domain of the function based on the physical constraints of the problem.

(d) Use a graphing utility to graph the area function and approximate the page size for which the least amount of paper will be used. Verify your answer numerically using the *table* feature of the graphing utility.

138. ***Photosynthesis*** The amount y of CO_2 uptake (in milligrams per square decimeter per hour) at optimal temperatures and with the natural supply of CO_2 is approximated by the model

$$y = \frac{18.47x - 2.96}{0.23x + 1}, \quad x > 0$$

where x is the light intensity (in watts per square meter). Use a graphing utility to graph the function and determine the limiting amount of CO_2 uptake.

2.7 In Exercises 139–146, solve the inequality.

139. $6x^2 + 5x < 4$

140. $2x^2 + x \ge 15$

141. $x^3 - 16x \ge 0$

142. $12x^3 - 20x^2 < 0$

143. $\dfrac{2}{x + 1} \le \dfrac{3}{x - 1}$

144. $\dfrac{x - 5}{3 - x} < 0$

145. $\dfrac{x^2 + 7x + 12}{x} \ge 0$

146. $\dfrac{1}{x - 2} > \dfrac{1}{x}$

147. ***Investment*** P dollars invested at interest rate r compounded annually increases to an amount

$$A = P(1 + r)^2$$

in 2 years. An investment of \$5000 is to increase to an amount greater than \$5500 in 2 years. The interest rate must be greater than what percent?

148. ***Population of a Species*** A biologist introduces 200 ladybugs into a crop field. The population P of the ladybugs is approximated by the model

$$P = \frac{1000(1 + 3t)}{5 + t}$$

where t is the time in days. Find the time required for the population to increase to at least 2000 ladybugs.

Synthesis

True or False? **In Exercises 149 and 150, determine whether the statement is true or false. Justify your answer.**

149. A fourth-degree polynomial with real coefficients can have -5, $-8i$, $4i$, and 5 as its zeros.

150. The domain of a rational function can never be the set of all real numbers.

151. ***Writing*** Explain how to determine the maximum or minimum value of a quadratic function.

152. ***Writing*** Explain the connections among factors of a polynomial, zeros of a polynomial function, and solutions of a polynomial equation.

153. ***Writing*** Describe what is meant by an asymptote of a graph.

2 Chapter Test

Take this test as you would take a test in class. When you are finished, check your work against the answers given in the back of the book.

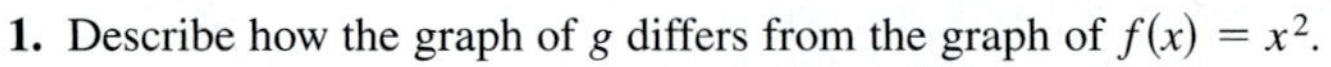

1. Describe how the graph of g differs from the graph of $f(x) = x^2$.
   (a) $g(x) = 2 - x^2$ (b) $g(x) = \left(x - \frac{3}{2}\right)^2$
2. Find an equation of the parabola shown in the figure at the left.

FIGURE FOR 2

3. The path of a ball is given by $y = -\frac{1}{20}x^2 + 3x + 5$, where y is the height (in feet) of the ball and x is the horizontal distance (in feet) from where the ball was thrown.
   (a) Find the maximum height of the ball.
   (b) Which number determines the height at which the ball was thrown? Does changing this value change the coordinates of the maximum height of the ball? Explain.
4. Determine the right-hand and left-hand behavior of the graph of the function $h(t) = -\frac{3}{4}t^5 + 2t^2$. Then sketch its graph.
5. Divide using long division.
   $$\frac{3x^3 + 4x - 1}{x^2 + 1}$$
6. Divide using synthetic division.
   $$\frac{2x^4 - 5x^2 - 3}{x - 2}$$
7. Use synthetic division to show that $x = \sqrt{3}$ is a zero of the function given by
   $f(x) = 4x^3 - x^2 - 12x + 3.$
   Use the result to factor the polynomial function completely and list all the real zeros of the function.
8. Perform each operation and write the result in standard form.
   (a) $10i - \left(3 + \sqrt{-25}\right)$ (b) $\left(2 + \sqrt{3}i\right)\left(2 - \sqrt{3}i\right)$
9. Write the quotient in standard form: $\dfrac{5}{2 + i}$.

In Exercises 10 and 11, find a polynomial function with real coefficients that has the given zeros. (There are many correct answers.)

10. $0, 3, 3 + i, 3 - i$
11. $1 + \sqrt{3}i, 1 - \sqrt{3}i, 2, 2$

In Exercises 12 and 13, find all the zeros of the function.

12. $f(x) = x^3 + 2x^2 + 5x + 10$
13. $f(x) = x^4 - 9x^2 - 22x - 24$

In Exercises 14–16, identify any intercepts and asymptotes of the graph the function. Then sketch a graph of the function.

14. $h(x) = \dfrac{4}{x^2} - 1$
15. $f(x) = \dfrac{2x^2 - 5x - 12}{x^2 - 16}$
16. $g(x) = \dfrac{x^2 + 2}{x - 1}$

In Exercises 17 and 18, solve the inequality. Sketch the solution set on the real number line.

17. $2x^2 + 5x > 12$
18. $\dfrac{2}{x} > \dfrac{5}{x + 6}$

Proofs in Mathematics

These two pages contain proofs of four important theorems about polynomial functions. The first two theorems are from Section 2.3, and the second two theorems are from Section 2.5.

The Remainder Theorem *(p. 157)*

If a polynomial $f(x)$ is divided by $x - k$, the remainder is

$$r = f(k).$$

Proof

From the Division Algorithm, you have

$$f(x) = (x - k)q(x) + r(x)$$

and because either $r(x) = 0$ or the degree of $r(x)$ is less than the degree of $x - k$, you know that $r(x)$ must be a constant. That is, $r(x) = r$. Now, by evaluating $f(x)$ at $x = k$, you have

$$\begin{aligned} f(k) &= (k - k)q(k) + r \\ &= (0)q(k) + r = r. \end{aligned}$$

To be successful in algebra, it is important that you understand the connection among *factors* of a polynomial, *zeros* of a polynomial function, and *solutions* or *roots* of a polynomial equation. The Factor Theorem is the basis for this connection.

The Factor Theorem *(p. 157)*

A polynomial $f(x)$ has a factor $(x - k)$ if and only if $f(k) = 0$.

Proof

Using the Division Algorithm with the factor $(x - k)$, you have

$$f(x) = (x - k)q(x) + r(x).$$

By the Remainder Theorem, $r(x) = r = f(k)$, and you have

$$f(x) = (x - k)q(x) + f(k)$$

where $q(x)$ is a polynomial of lesser degree than $f(x)$. If $f(k) = 0$, then

$$f(x) = (x - k)q(x)$$

and you see that $(x - k)$ is a factor of $f(x)$. Conversely, if $(x - k)$ is a factor of $f(x)$, division of $f(x)$ by $(x - k)$ yields a remainder of 0. So, by the Remainder Theorem, you have $f(k) = 0$.

Proofs in Mathematics

Linear Factorization Theorem (p. 169)

If $f(x)$ is a polynomial of degree n, where $n > 0$, then f has precisely n linear factors

$$f(x) = a_n(x - c_1)(x - c_2) \cdot \cdot \cdot (x - c_n)$$

where $c_1, c_2, \ldots, c_n$ are complex numbers.

Proof

Using the Fundamental Theorem of Algebra, you know that f must have at least one zero, c_1. Consequently, $(x - c_1)$ is a factor of $f(x)$, and you have

$$f(x) = (x - c_1)f_1(x).$$

If the degree of $f_1(x)$ is greater than zero, you again apply the Fundamental Theorem to conclude that f_1 must have a zero c_2, which implies that

$$f(x) = (x - c_1)(x - c_2)f_2(x).$$

It is clear that the degree of $f_1(x)$ is $n - 1$, that the degree of $f_2(x)$ is $n - 2$, and that you can repeatedly apply the Fundamental Theorem n times until you obtain

$$f(x) = a_n(x - c_1)(x - c_2) \cdot \cdot \cdot (x - c_n)$$

where a_n is the leading coefficient of the polynomial $f(x)$.

The Fundamental Theorem of Algebra

The Linear Factorization Theorem is closely related to the Fundamental Theorem of Algebra. The Fundamental Theorem of Algebra has a long and interesting history. In the early work with polynomial equations, The Fundamental Theorem of Algebra was thought to have been not true, because imaginary solutions were not considered. In fact, in the very early work by mathematicians such as Abu al-Khwarizmi (c. 800 A.D.), negative solutions were also not considered.

Once imaginary numbers were accepted, several mathematicians attempted to give a general proof of the Fundamental Theorem of Algebra. These included Gottfried von Leibniz (1702), Jean d'Alembert (1746), Leonhard Euler (1749), Joseph-Louis Lagrange (1772), and Pierre Simon Laplace (1795). The mathematician usually credited with the first correct proof of the Fundamental Theorem of Algebra is Carl Friedrich Gauss, who published the proof in his doctoral thesis in 1799.

Factors of a Polynomial (p. 173)

Every polynomial of degree $n > 0$ with real coefficients can be written as the product of linear and quadratic factors with real coefficients, where the quadratic factors have no real zeros.

Proof

To begin, you use the Linear Factorization Theorem to conclude that $f(x)$ can be *completely* factored in the form

$$f(x) = d(x - c_1)(x - c_2)(x - c_3) \cdot \cdot \cdot (x - c_n).$$

If each c_i is real, there is nothing more to prove. If any c_i is complex ($c_i = a + bi$, $b \neq 0$), then, because the coefficients of $f(x)$ are real, you know that the conjugate $c_j = a - bi$ is also a zero. By multiplying the corresponding factors, you obtain

$$\begin{aligned}(x - c_i)(x - c_j) &= [x - (a + bi)][x - (a - bi)] \\ &= x^2 - 2ax + (a^2 + b^2)\end{aligned}$$

where each coefficient is real.

P.S. Problem Solving

This collection of thought-provoking and challenging exercises further explores and expands upon concepts learned in this chapter.

1. Show that if $f(x) = ax^3 + bx^2 + cx + d$ then $f(k) = r$, where $r = ak^3 + bk^2 + ck + d$ using long division. In other words, verify the Remainder Theorem for a third-degree polynomial function.

2. In 2000 B.C., the Babylonians solved polynomial equations by referring to tables of values. One such table gave the values of $y^3 + y^2$. To be able to use this table, the Babylonians sometimes had to manipulate the equation as shown below.

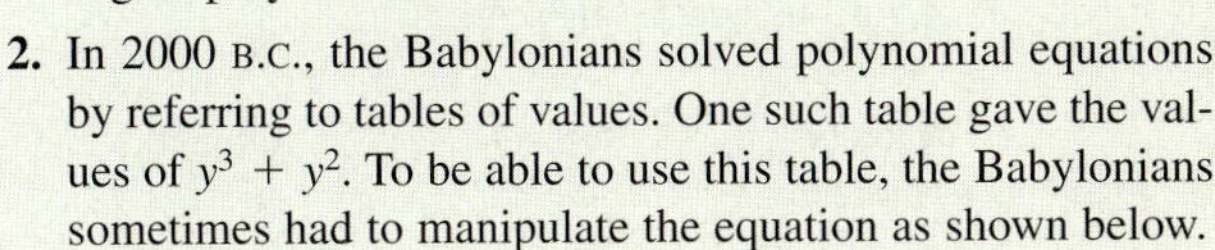

$ax^3 + bx^2 = c$ Original equation

$$\frac{a^3x^3}{b^3} + \frac{a^2x^2}{b^2} = \frac{a^2c}{b^3}$$ Multiply each side by $\frac{a^2}{b^3}$.

$$\left(\frac{ax}{b}\right)^3 + \left(\frac{ax}{b}\right)^2 = \frac{a^2c}{b^3}$$ Rewrite.

Then they would find $(a^2c)/b^3$ in the $y^3 + y^2$ column of the table. Because they knew that the corresponding y-value was equal to $(ax)/b$, they could conclude that $x = (by)/a$.

(a) Calculate $y^3 + y^2$ for $y = 1, 2, 3, \ldots, 10$. Record the values in a table.

Use the table from part (a) and the method above to solve each equation.

(b) $x^3 + x^2 = 252$

(c) $x^3 + 2x^2 = 288$

(d) $3x^3 + x^2 = 90$

(e) $2x^3 + 5x^2 = 2500$

(f) $7x^3 + 6x^2 = 1728$

(g) $10x^3 + 3x^2 = 297$

Using the methods from this chapter, verify your solution to each equation.

3. At a glassware factory, molten cobalt glass is poured into molds to make paperweights. Each mold is a rectangular prism whose height is 3 inches greater than the length of each side of the square base. A machine pours 20 cubic inches of liquid glass into each mold. What are the dimensions of the mold?

4. Determine whether the statement is true or false. If false, provide one or more reasons why the statement is false and correct the statement. Let $f(x) = ax^3 + bx^2 + cx + d$, $a \neq 0$, and let $f(2) = -1$. Then

$$\frac{f(x)}{x+1} = q(x) + \frac{2}{x+1}$$

where $q(x)$ is a second-degree polynomial.

5. The parabola shown in the figure has an equation of the form $y = ax^2 + bx + c$. Find the equation of this parabola by the following methods. (a) Find the equation analytically. (b) Use the *regression* feature of a graphing utility to find the equation.

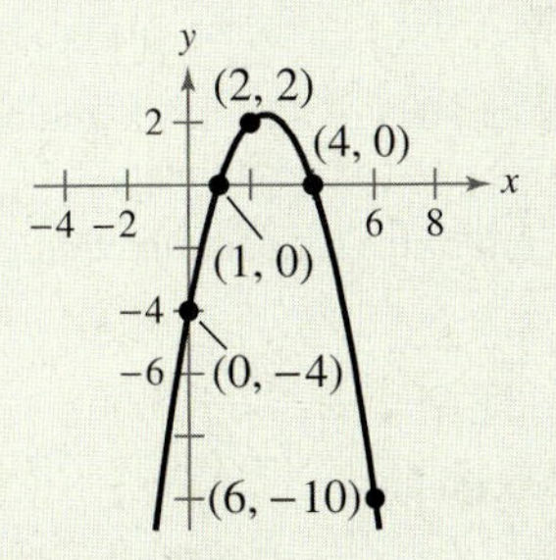

6. One of the fundamental themes of calculus is to find the slope of the tangent line to a curve at a point. To see how this can be done, consider the point (2, 4) on the graph of the quadratic function $f(x) = x^2$.

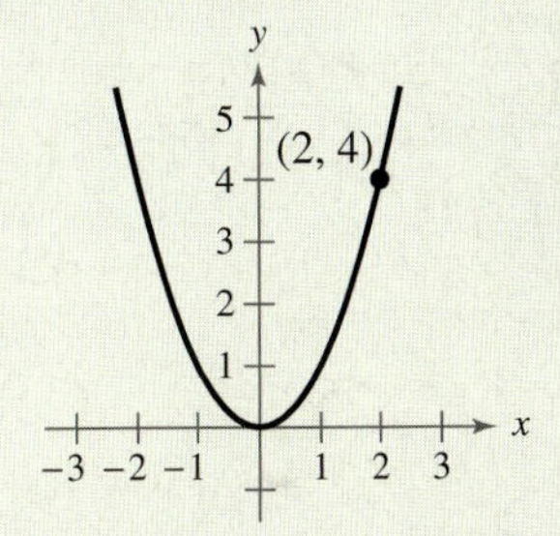

(a) Find the slope of the line joining (2, 4) and (3, 9). Is the slope of the tangent line at (2, 4) greater than or less than the slope of the line through (2, 4) and (3, 9)?

(b) Find the slope of the line joining (2, 4) and (1, 1). Is the slope of the tangent line at (2, 4) greater than or less than the slope of the line through (2, 4) and (1, 1)?

(c) Find the slope of the line joining (2, 4) and (2.1, 4.41). Is the slope of the tangent line at (2, 4) greater than or less than the slope of the line through (2, 4) and (2.1, 4.41)?

(d) Find the slope of the line joining (2, 4) and $(2 + h, f(2 + h))$ in terms of the nonzero number h.

(e) Evaluate the slope formula from part (d) for $h = -1$, 1, and 0.1. Compare these values with those in parts (a)–(c).

(f) What can you conclude the slope of the tangent line at (2, 4) to be? Explain your answer.

7. Use the form $f(x) = (x - k)q(x) + r$ to create a cubic function that (a) passes through the point $(2, 5)$ and rises to the right and (b) passes through the point $(-3, 1)$ and falls to the right. (There are many correct answers.)

8. The multiplicative inverse of z is a complex number z_m such that $z \cdot z_m = 1$. Find the multiplicative inverse of each complex number.

   (a) $z = 1 + i$ (b) $z = 3 - i$ (c) $z = -2 + 8i$

9. Prove that the product of a complex number $a + bi$ and its complex conjugate is a real number.

10. Match the graph of the rational function given by

$$f(x) = \frac{ax + b}{cx + d}$$

with the given conditions.

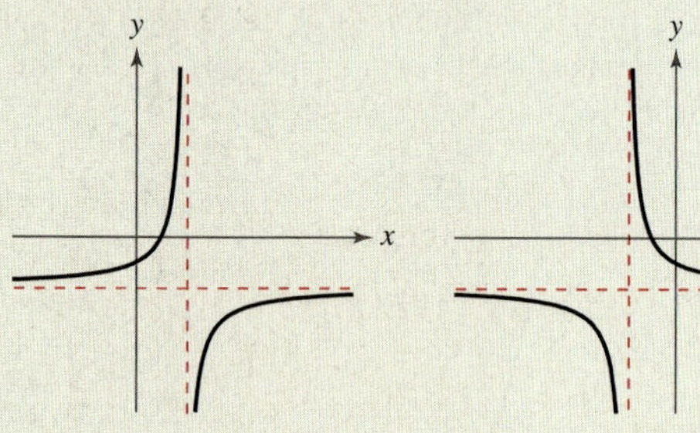

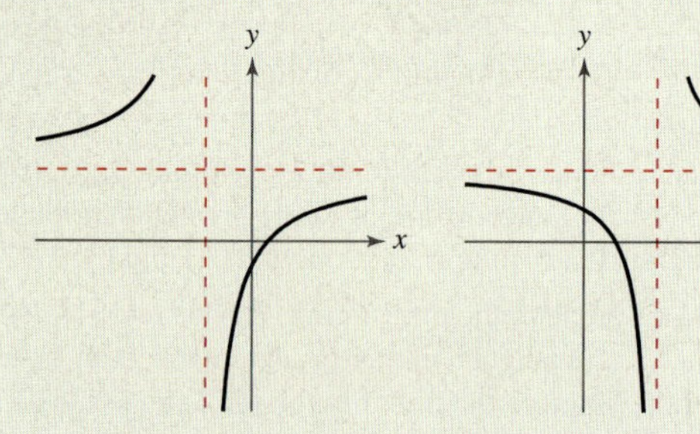

(i)	(ii)	(iii)	(iv)
$a > 0$	$a > 0$	$a < 0$	$a > 0$
$b < 0$	$b > 0$	$b > 0$	$b < 0$
$c > 0$	$c < 0$	$c > 0$	$c > 0$
$d < 0$	$d < 0$	$d < 0$	$d > 0$

11. Consider the function given by

$$f(x) = \frac{ax}{(x - b)^2}.$$

   (a) Determine the effect on the graph of f if $b \neq 0$ and a is varied. Consider cases in which a is positive and a is negative.

   (b) Determine the effect on the graph of f if $a \neq 0$ and b is varied.

12. The endpoints of the interval over which distinct vision is possible is called the *near point* and *far point* of the eye (see figure). With increasing age, these points normally change. The table shows the approximate near points y (in inches) for various ages x (in years).

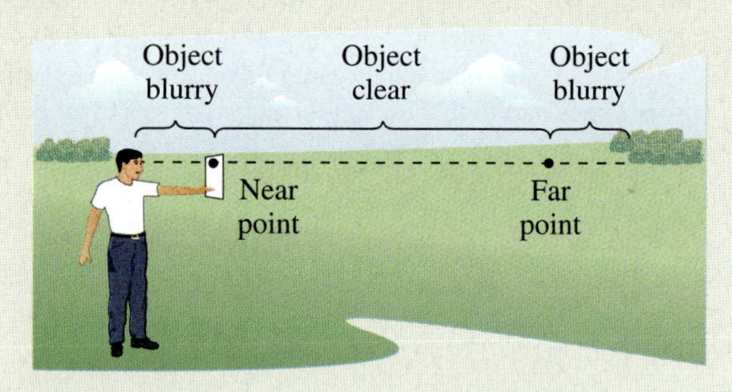

FIGURE FOR 12

Age, x	Near point, y
16	3.0
32	4.7
44	9.8
50	19.7
60	39.4

   (a) Use the *regression* feature of a graphing utility to find a quadratic model for the data. Use a graphing utility to plot the data and graph the model in the same viewing window.

   (b) Find a rational model for the data. Take the reciprocals of the near points to generate the points $(x, 1/y)$. Use the *regression* feature of a graphing utility to find a linear model for the data. The resulting line has the form

$$\frac{1}{y} = ax + b.$$

   Solve for y. Use a graphing utility to plot the data and graph the model in the same viewing window.

   (c) Use the *table* feature of a graphing utility to create a table showing the predicted near point based on each model for each of the ages in the original table. How well do the models fit the original data?

   (d) Use both models to estimate the near point for a person who is 25 years old. Which model is a better fit?

   (e) Do you think either model can be used to predict the near point for a person who is 70 years old? Explain.

Exponential and Logarithmic Functions

3

Carbon dating is a method used to determine the ages of archeological artifacts up to 50,000 years old. For example, archeologists are using carbon dating to determine the ages of the great pyramids of Egypt.

SELECTED APPLICATIONS

Exponential and logarithmic functions have many real-life applications. The applications listed below represent a small sample of the applications in this chapter.

- Computer Virus, Exercise 65, page 227
- Data Analysis: Meteorology, Exercise 70, page 228
- Sound Intensity, Exercise 90, page 238
- Galloping Speeds of Animals, Exercise 85, page 244
- Average Heights, Exercise 115, page 255
- Carbon Dating, Exercise 41, page 266
- IQ Scores, Exercise 47, page 266
- Forensics, Exercise 63, page 268
- Compound Interest, Exercise 135, page 273

3.1 Exponential Functions and Their Graphs

What you should learn

- Recognize and evaluate exponential functions with base a.
- Graph exponential functions and use the One-to-One Property.
- Recognize, evaluate, and graph exponential functions with base e.
- Use exponential functions to model and solve real-life problems.

Why you should learn it

Exponential functions can be used to model and solve real-life problems. For instance, in Exercise 70 on page 228, an exponential function is used to model the atmospheric pressure at different altitudes.

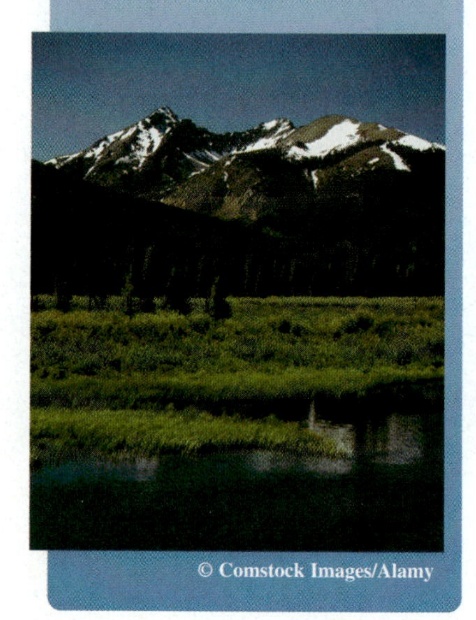

© Comstock Images/Alamy

Exponential Functions

So far, this text has dealt mainly with **algebraic functions,** which include polynomial functions and rational functions. In this chapter, you will study two types of nonalgebraic functions—*exponential functions* and *logarithmic functions*. These functions are examples of **transcendental functions.**

> **Definition of Exponential Function**
>
> The **exponential function f with base a** is denoted by
>
> $$f(x) = a^x$$
>
> where $a > 0$, $a \neq 1$, and x is any real number.

The base $a = 1$ is excluded because it yields $f(x) = 1^x = 1$. This is a constant function, not an exponential function.

You have evaluated a^x for integer and rational values of x. For example, you know that $4^3 = 64$ and $4^{1/2} = 2$. However, to evaluate 4^x for any real number x, you need to interpret forms with *irrational* exponents. For the purposes of this text, it is sufficient to think of

$$a^{\sqrt{2}} \quad (\text{where } \sqrt{2} \approx 1.41421356)$$

as the number that has the successively closer approximations

$$a^{1.4}, a^{1.41}, a^{1.414}, a^{1.4142}, a^{1.41421}, \ldots .$$

Example 1 Evaluating Exponential Functions

Use a calculator to evaluate each function at the indicated value of x.

	Function	*Value*
a.	$f(x) = 2^x$	$x = -3.1$
b.	$f(x) = 2^{-x}$	$x = \pi$
c.	$f(x) = 0.6^x$	$x = \frac{3}{2}$

Solution

	Function Value	*Graphing Calculator Keystrokes*	*Display*
a.	$f(-3.1) = 2^{-3.1}$	2 [^] [(-)] 3.1 [ENTER]	0.1166291
b.	$f(\pi) = 2^{-\pi}$	2 [^] [(-)] π [ENTER]	0.1133147
c.	$f(\frac{3}{2}) = (0.6)^{3/2}$	.6 [^] [(] 3 [÷] 2 [)] [ENTER]	0.4647580

CHECKPOINT Now try Exercise 1.

When evaluating exponential functions with a calculator, remember to enclose fractional exponents in parentheses. Because the calculator follows the order of operations, parentheses are crucial in order to obtain the correct result.

The *HM mathSpace®* CD-ROM and *Eduspace®* for this text contain additional resources related to the concepts discussed in this chapter.

Exploration

Note that an exponential function $f(x) = a^x$ is a constant raised to a variable power, whereas a power function $g(x) = x^n$ is a variable raised to a constant power. Use a graphing utility to graph each pair of functions in the same viewing window. Describe any similarities and differences in the graphs.

a. $y_1 = 2^x, y_2 = x^2$

b. $y_1 = 3^x, y_2 = x^3$

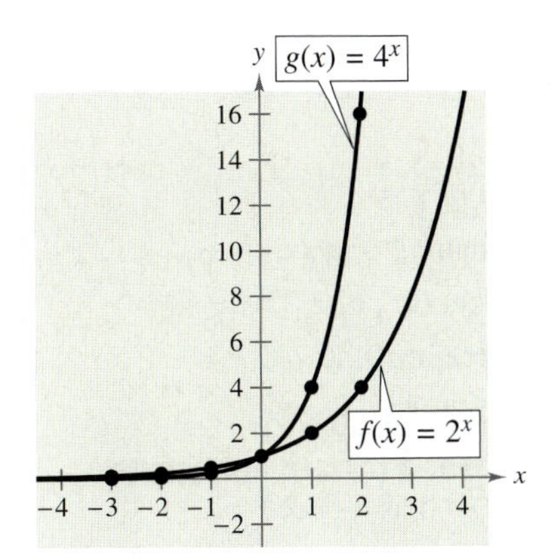

FIGURE 3.1

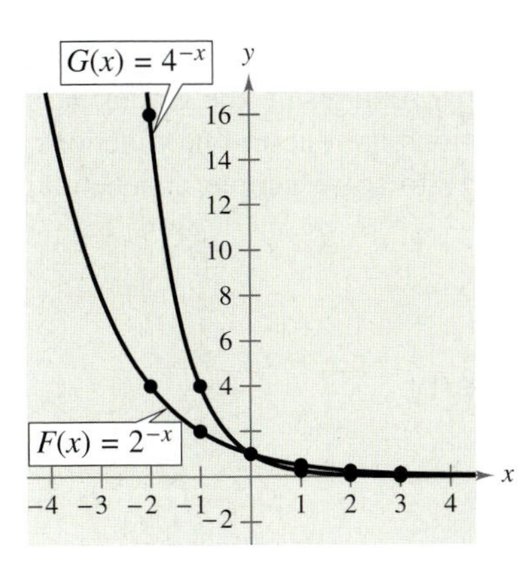

FIGURE 3.2

Graphs of Exponential Functions

The graphs of all exponential functions have similar characteristics, as shown in Examples 2, 3, and 5.

Example 2 Graphs of $y = a^x$

In the same coordinate plane, sketch the graph of each function.

a. $f(x) = 2^x$ **b.** $g(x) = 4^x$

Solution

The table below lists some values for each function, and Figure 3.1 shows the graphs of the two functions. Note that both graphs are increasing. Moreover, the graph of $g(x) = 4^x$ is increasing more rapidly than the graph of $f(x) = 2^x$.

x	-3	-2	-1	0	1	2
2^x	$\frac{1}{8}$	$\frac{1}{4}$	$\frac{1}{2}$	1	2	4
4^x	$\frac{1}{64}$	$\frac{1}{16}$	$\frac{1}{4}$	1	4	16

CHECKPOINT Now try Exercise 11.

The table in Example 2 was evaluated by hand. You could, of course, use a graphing utility to construct tables with even more values.

Example 3 Graphs of $y = a^{-x}$

In the same coordinate plane, sketch the graph of each function.

a. $F(x) = 2^{-x}$ **b.** $G(x) = 4^{-x}$

Solution

The table below lists some values for each function, and Figure 3.2 shows the graphs of the two functions. Note that both graphs are decreasing. Moreover, the graph of $G(x) = 4^{-x}$ is decreasing more rapidly than the graph of $F(x) = 2^{-x}$.

x	-2	-1	0	1	2	3
2^{-x}	4	2	1	$\frac{1}{2}$	$\frac{1}{4}$	$\frac{1}{8}$
4^{-x}	16	4	1	$\frac{1}{4}$	$\frac{1}{16}$	$\frac{1}{64}$

CHECKPOINT Now try Exercise 13.

In Example 3, note that by using one of the properties of exponents, the functions $F(x) = 2^{-x}$ and $G(x) = 4^{-x}$ can be rewritten with positive exponents.

$$F(x) = 2^{-x} = \frac{1}{2^x} = \left(\frac{1}{2}\right)^x \quad \text{and} \quad G(x) = 4^{-x} = \frac{1}{4^x} = \left(\frac{1}{4}\right)^x$$

Comparing the functions in Examples 2 and 3, observe that

$$F(x) = 2^{-x} = f(-x) \qquad \text{and} \qquad G(x) = 4^{-x} = g(-x).$$

Consequently, the graph of F is a reflection (in the y-axis) of the graph of f. The graphs of G and g have the same relationship. The graphs in Figures 3.1 and 3.2 are typical of the exponential functions $y = a^x$ and $y = a^{-x}$. They have one y-intercept and one horizontal asymptote (the x-axis), and they are continuous. The basic characteristics of these exponential functions are summarized in Figures 3.3 and 3.4.

STUDY TIP

Notice that the range of an exponential function is $(0, \infty)$, which means that $a^x > 0$ for all values of x.

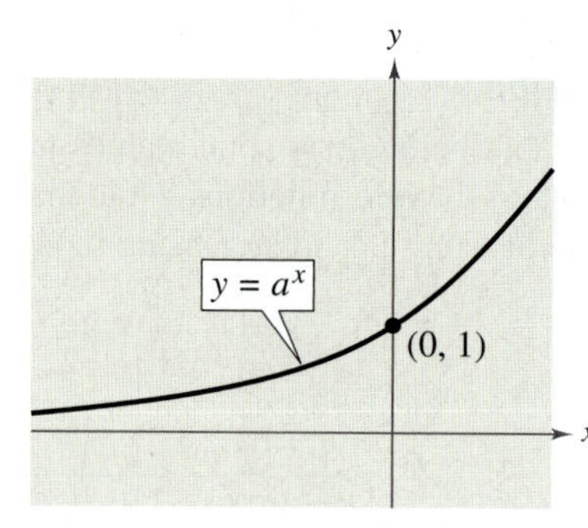

FIGURE 3.3

Graph of $y = a^x$, $a > 1$

- Domain: $(-\infty, \infty)$
- Range: $(0, \infty)$
- Intercept: $(0, 1)$
- Increasing
- x-axis is a horizontal asymptote ($a^x \rightarrow 0$ as $x \rightarrow -\infty$)
- Continuous

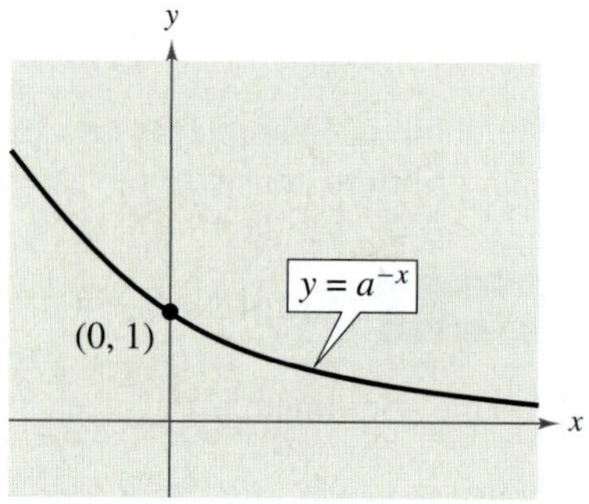

FIGURE 3.4

Graph of $y = a^{-x}$, $a > 1$

- Domain: $(-\infty, \infty)$
- Range: $(0, \infty)$
- Intercept: $(0, 1)$
- Decreasing
- x-axis is a horizontal asymptote ($a^{-x} \rightarrow 0$ as $x \rightarrow \infty$)
- Continuous

From Figures 3.3 and 3.4, you can see that the graph of an exponential function is always increasing or always decreasing. As a result, the graphs pass the Horizontal Line Test, and therefore the functions are one-to-one functions. You can use the following **One-to-One Property** to solve simple exponential equations.

For $a > 0$ and $a \neq 1$, $a^x = a^y$ if and only if $x = y$. — One-to-One Property

Example 4 Using the One-to-One Property

a. $9 = 3^{x+1}$ — Original equation

$3^2 = 3^{x+1}$ — $9 = 3^2$

$2 = x + 1$ — One-to-One Property

$1 = x$ — Solve for x.

b. $\left(\frac{1}{2}\right)^x = 8 \Rightarrow 2^{-x} = 2^3 \Rightarrow x = -3$

CHECKPOINT Now try Exercise 45.

In the following example, notice how the graph of $y = a^x$ can be used to sketch the graphs of functions of the form $f(x) = b \pm a^{x+c}$.

Example 5 Transformations of Graphs of Exponential Functions

Each of the following graphs is a transformation of the graph of $f(x) = 3^x$.

a. Because $g(x) = 3^{x+1} = f(x + 1)$, the graph of g can be obtained by shifting the graph of f one unit to the *left*, as shown in Figure 3.5.

b. Because $h(x) = 3^x - 2 = f(x) - 2$, the graph of h can be obtained by shifting the graph of f *downward* two units, as shown in Figure 3.6.

c. Because $k(x) = -3^x = -f(x)$, the graph of k can be obtained by *reflecting* the graph of f in the x-axis, as shown in Figure 3.7.

d. Because $j(x) = 3^{-x} = f(-x)$, the graph of j can be obtained by *reflecting* the graph of f in the y-axis, as shown in Figure 3.8.

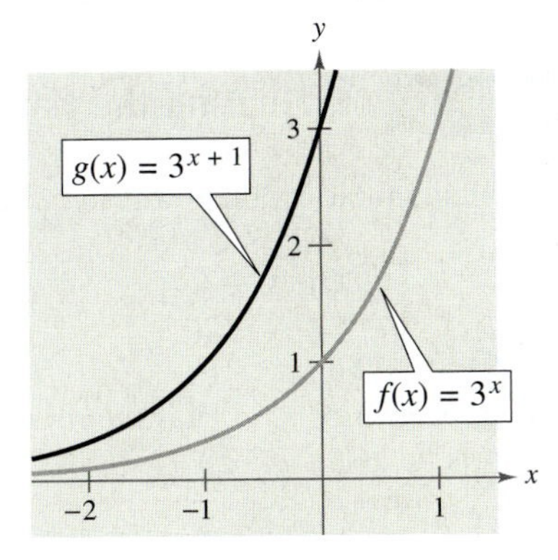

FIGURE 3.5 *Horizontal shift*

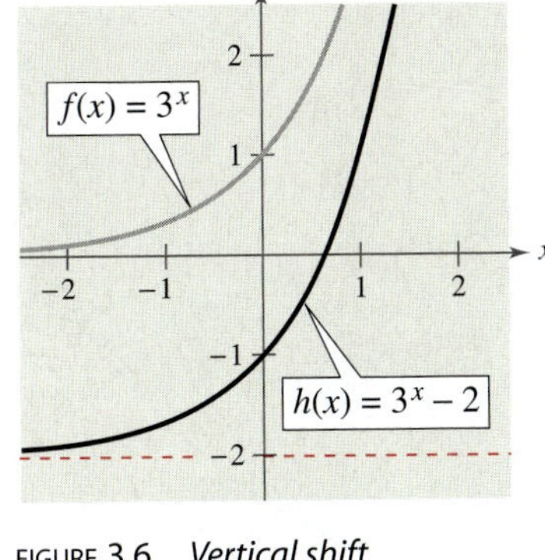

FIGURE 3.6 *Vertical shift*

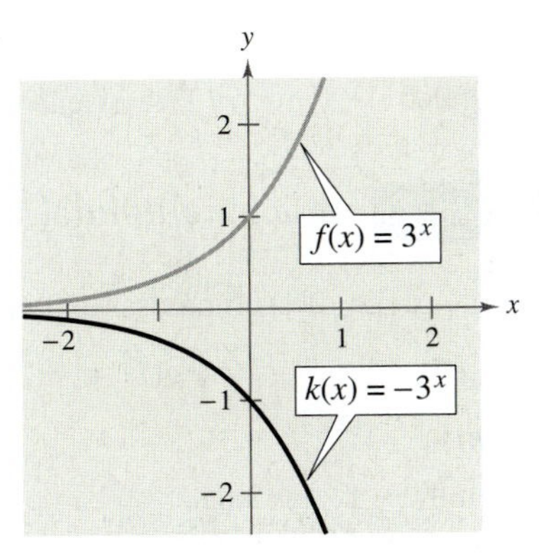

FIGURE 3.7 *Reflection in x-axis*

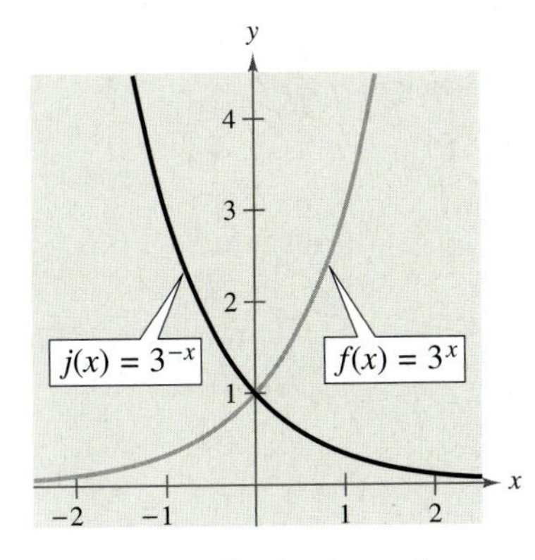

FIGURE 3.8 *Reflection in y-axis*

CHECKPOINT Now try Exercise 17.

Notice that the transformations in Figures 3.5, 3.7, and 3.8 keep the x-axis as a horizontal asymptote, but the transformation in Figure 3.6 yields a new horizontal asymptote of $y = -2$. Also, be sure to note how the y-intercept is affected by each transformation.

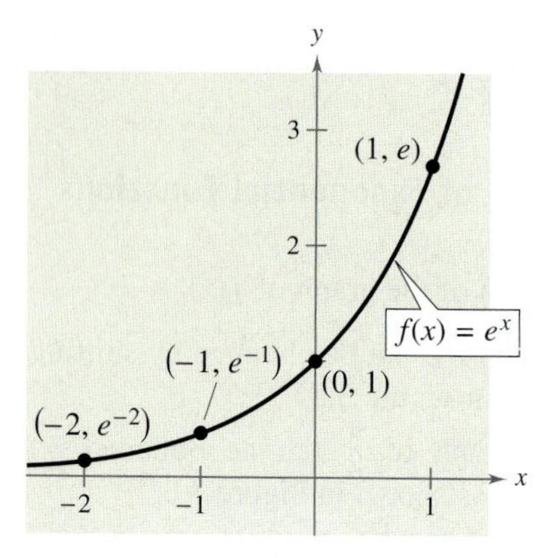

FIGURE 3.9

The Natural Base e

In many applications, the most convenient choice for a base is the irrational number

$$e \approx 2.718281828 \ldots.$$

This number is called the **natural base.** The function given by $f(x) = e^x$ is called the **natural exponential function.** Its graph is shown in Figure 3.9. Be sure you see that for the exponential function $f(x) = e^x$, e is the constant $2.718281828 \ldots$, whereas x is the variable.

Exploration

Use a graphing utility to graph $y_1 = (1 + 1/x)^x$ and $y_2 = e$ in the same viewing window. Using the *trace* feature, explain what happens to the graph of y_1 as x increases.

Example 6 Evaluating the Natural Exponential Function

Use a calculator to evaluate the function given by $f(x) = e^x$ at each indicated value of x.

a. $x = -2$ **b.** $x = -1$ **c.** $x = 0.25$ **d.** $x = -0.3$

Solution

Function Value	Graphing Calculator Keystrokes	Display
a. $f(-2) = e^{-2}$	[e^x] [(−)] 2 [ENTER]	0.1353353
b. $f(-1) = e^{-1}$	[e^x] [(−)] 1 [ENTER]	0.3678794
c. $f(0.25) = e^{0.25}$	[e^x] 0.25 [ENTER]	1.2840254
d. $f(-0.3) = e^{-0.3}$	[e^x] [(−)] 0.3 [ENTER]	0.7408182

CHECKPOINT Now try Exercise 27.

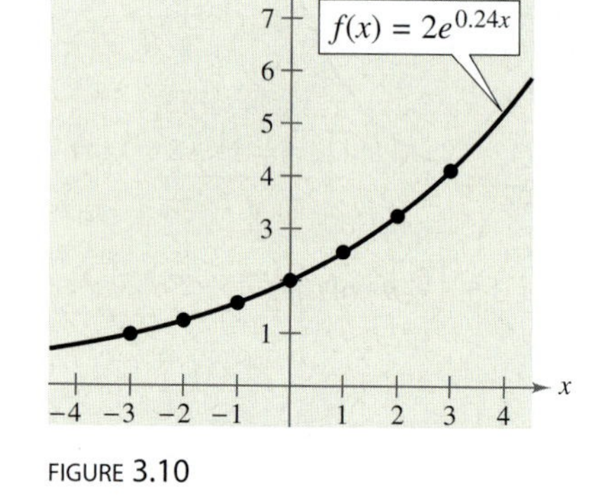

FIGURE 3.10

Example 7 Graphing Natural Exponential Functions

Sketch the graph of each natural exponential function.

a. $f(x) = 2e^{0.24x}$ **b.** $g(x) = \frac{1}{2}e^{-0.58x}$

Solution

To sketch these two graphs, you can use a graphing utility to construct a table of values, as shown below. After constructing the table, plot the points and connect them with smooth curves, as shown in Figures 3.10 and 3.11. Note that the graph in Figure 3.10 is increasing, whereas the graph in Figure 3.11 is decreasing.

x	-3	-2	-1	0	1	2	3
$f(x)$	0.974	1.238	1.573	2.000	2.542	3.232	4.109
$g(x)$	2.849	1.595	0.893	0.500	0.280	0.157	0.088

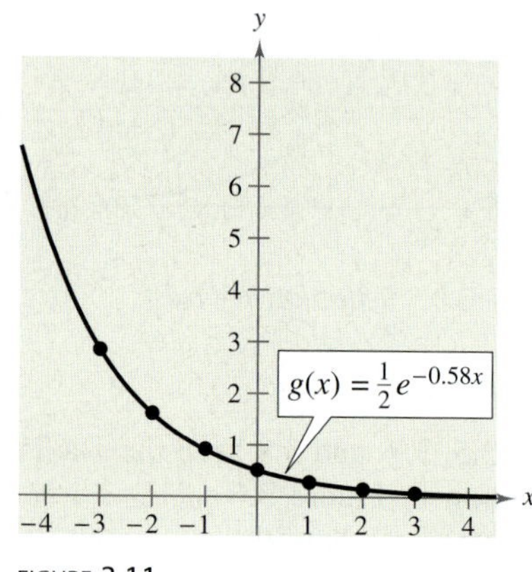

FIGURE 3.11

CHECKPOINT Now try Exercise 35.

Exploration

Use the formula

$$A = P\left(1 + \frac{r}{n}\right)^{nt}$$

to calculate the amount in an account when $P = \$3000$, $r = 6\%$, $t = 10$ years, and compounding is done (a) by the day, (b) by the hour, (c) by the minute, and (d) by the second. Does increasing the number of compoundings per year result in unlimited growth of the amount in the account? Explain.

Applications

One of the most familiar examples of exponential growth is that of an investment earning *continuously compounded interest*. Using exponential functions, you can *develop* a formula for interest compounded n times per year and show how it leads to continuous compounding.

Suppose a principal P is invested at an annual interest rate r, compounded once a year. If the interest is added to the principal at the end of the year, the new balance P_1 is

$$\begin{aligned} P_1 &= P + Pr \\ &= P(1 + r). \end{aligned}$$

This pattern of multiplying the previous principal by $1 + r$ is then repeated each successive year, as shown below.

Year	*Balance After Each Compounding*
0	$P = P$
1	$P_1 = P(1 + r)$
2	$P_2 = P_1(1 + r) = P(1 + r)(1 + r) = P(1 + r)^2$
3	$P_3 = P_2(1 + r) = P(1 + r)^2(1 + r) = P(1 + r)^3$
$\vdots$	$\vdots$
t	$P_t = P(1 + r)^t$

To accommodate more frequent (quarterly, monthly, or daily) compounding of interest, let n be the number of compoundings per year and let t be the number of years. Then the rate per compounding is r/n and the account balance after t years is

$$A = P\left(1 + \frac{r}{n}\right)^{nt}. \qquad \text{Amount (balance) with } n \text{ compoundings per year}$$

If you let the number of compoundings n increase without bound, the process approaches what is called **continuous compounding.** In the formula for n compoundings per year, let $m = n/r$. This produces

$$\begin{aligned} A &= P\left(1 + \frac{r}{n}\right)^{nt} && \text{Amount with } n \text{ compoundings per year} \\ &= P\left(1 + \frac{r}{mr}\right)^{mrt} && \text{Substitute } mr \text{ for } n. \\ &= P\left(1 + \frac{1}{m}\right)^{mrt} && \text{Simplify.} \\ &= P\left[\left(1 + \frac{1}{m}\right)^{m}\right]^{rt}. && \text{Property of exponents} \end{aligned}$$

m	$\left(1 + \frac{1}{m}\right)^m$
1	2
10	2.59374246
100	2.704813829
1,000	2.716923932
10,000	2.718145927
100,000	2.718268237
1,000,000	2.718280469
10,000,000	2.718281693
↓	↓
∞	e

As m increases without bound, the table at the left shows that $[1 + (1/m)]^m \to e$ as $m \to \infty$. From this, you can conclude that the formula for continuous compounding is

$$A = Pe^{rt}. \qquad \text{Substitute } e \text{ for } (1 + 1/m)^m.$$

STUDY TIP

Be sure you see that the annual interest rate must be written in decimal form. For instance, 6% should be written as 0.06.

Activities

1. Sketch the graphs of the functions $f(x) = e^x$ and $g(x) = 1 + e^x$ on the same coordinate system.

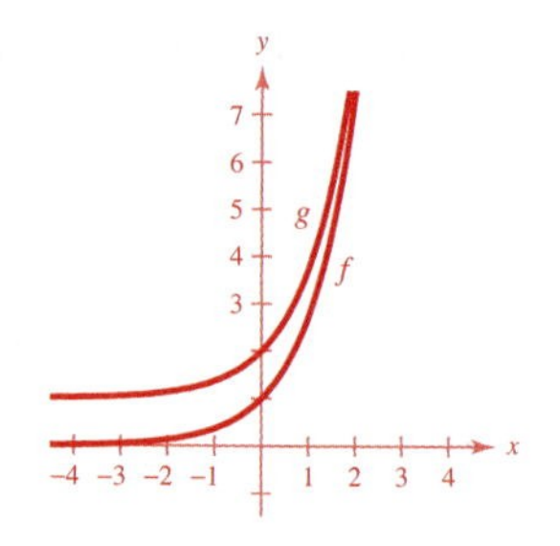

2. Determine the balance A at the end of 20 years if \$1500 is invested at 6.5% interest and the interest is compounded (a) quarterly and (b) continuously.
   Answer: (a) \$5446.73 (b) \$5503.95
3. The number of fruit flies in an experimental population after t hours is given by $Q(t) = 20e^{0.03t}, t \geq 0$.
   a. Find the initial number of fruit flies in the population.
   b. How large is the population of fruit flies after 72 hours?
   Answer: (a) 20 flies (b) 173 flies

Group Activity

The sequence $3, 6, 9, 12, 15, \ldots$ is given by $f(n) = 3n$ and is an example of linear growth. The sequence $3, 9, 27, 81, 243, \ldots$ is given by $f(n) = 3^n$ and is an example of exponential growth. Explain the difference between these two types of growth. For each of the following sequences, indicate whether the sequence represents linear growth or exponential growth, and find a linear or exponential function that represents the sequence. Give several other examples of linear and exponential growth.

a. $\frac{1}{2}, \frac{1}{4}, \frac{1}{8}, \frac{1}{16}, \frac{1}{32}, \ldots$
b. $4, 8, 12, 16, 20, \ldots$
c. $\frac{2}{3}, \frac{4}{3}, 2, \frac{8}{3}, \frac{10}{3}, 4, \ldots$
d. $5, 25, 125, 625, \ldots$

Formulas for Compound Interest

After t years, the balance A in an account with principal P and annual interest rate r (in decimal form) is given by the following formulas.

1. For n compoundings per year: $A = P\left(1 + \frac{r}{n}\right)^{nt}$
2. For continuous compounding: $A = Pe^{rt}$

Example 8 Compound Interest

A total of \$12,000 is invested at an annual interest rate of 9%. Find the balance after 5 years if it is compounded

a. quarterly.
b. monthly.
c. continuously.

Solution

a. For quarterly compounding, you have $n = 4$. So, in 5 years at 9%, the balance is

$$A = P\left(1 + \frac{r}{n}\right)^{nt} \quad \text{Formula for compound interest}$$

$$= 12{,}000\left(1 + \frac{0.09}{4}\right)^{4(5)} \quad \text{Substitute for } P, r, n, \text{ and } t.$$

$$\approx \$18{,}726.11. \quad \text{Use a calculator.}$$

b. For monthly compounding, you have $n = 12$. So, in 5 years at 9%, the balance is

$$A = P\left(1 + \frac{r}{n}\right)^{nt} \quad \text{Formula for compound interest}$$

$$= 12{,}000\left(1 + \frac{0.09}{12}\right)^{12(5)} \quad \text{Substitute for } P, r, n, \text{ and } t.$$

$$\approx \$18{,}788.17. \quad \text{Use a calculator.}$$

c. For continuous compounding, the balance is

$$A = Pe^{rt} \quad \text{Formula for continuous compounding}$$

$$= 12{,}000e^{0.09(5)} \quad \text{Substitute for } P, r, \text{ and } t.$$

$$\approx \$18{,}819.75. \quad \text{Use a calculator.}$$

CHECKPOINT Now try Exercise 53.

In Example 8, note that continuous compounding yields more than quarterly or monthly compounding. This is typical of the two types of compounding. That is, for a given principal, interest rate, and time, continuous compounding will always yield a larger balance than compounding n times a year.

Example 9 Radioactive Decay 

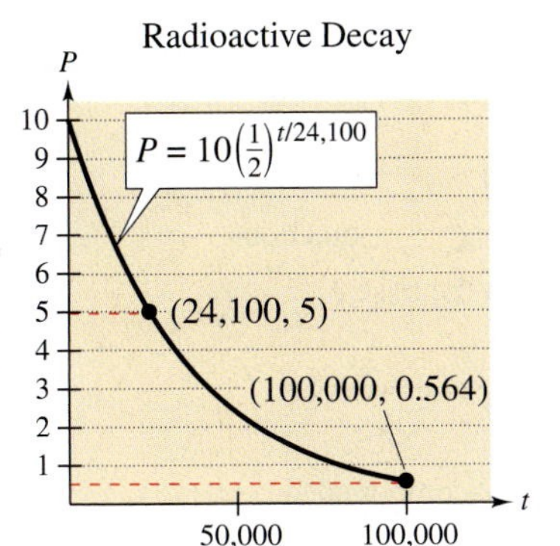

FIGURE 3.12

In 1986, a nuclear reactor accident occurred in Chernobyl in what was then the Soviet Union. The explosion spread highly toxic radioactive chemicals, such as plutonium, over hundreds of square miles, and the government evacuated the city and the surrounding area. To see why the city is now uninhabited, consider the model

$$P = 10\left(\frac{1}{2}\right)^{t/24,100}$$

which represents the amount of plutonium P that remains (from an initial amount of 10 pounds) after t years. Sketch the graph of this function over the interval from $t = 0$ to $t = 100,000$, where $t = 0$ represents 1986. How much of the 10 pounds will remain in the year 2010? How much of the 10 pounds will remain after 100,000 years?

Solution

The graph of this function is shown in Figure 3.12. Note from this graph that plutonium has a *half-life* of about 24,100 years. That is, after 24,100 years, *half* of the original amount will remain. After another 24,100 years, one-quarter of the original amount will remain, and so on. In the year 2010 ($t = 24$), there will still be

$$P = 10\left(\frac{1}{2}\right)^{24/24,100} \approx 10\left(\frac{1}{2}\right)^{0.0009959} \approx 9.993 \text{ pounds}$$

of plutonium remaining. After 100,000 years, there will still be

$$P = 10\left(\frac{1}{2}\right)^{100,000/24,100} \approx 10\left(\frac{1}{2}\right)^{4.1494} \approx 0.564 \text{ pound}$$

of plutonium remaining.

CHECKPOINT Now try Exercise 67.

Writing About Mathematics Suggestion:

One way your students might approach this problem is to create a table, covering x-values from -2 through 3, for each of the functions and compare this table with the given tables. If this method is used, you might consider dividing your class into groups of three or six and having the groups assign one or two functions to each member. They should then pool their results and work cooperatively to determine that each function has a y-intercept of $(0, 8)$.

Another approach is a graphical one: the groups can create scatter plots of the data shown in the table and compare them with sketches of the graphs of the given functions. Consider assigning students to groups of four and giving the responsibility for sketching three graphs to each group member.

WRITING ABOUT MATHEMATICS

Identifying Exponential Functions Which of the following functions generated the two tables below? Discuss how you were able to decide. What do these functions have in common? Are any of them the same? If so, explain why.

a. $f_1(x) = 2^{(x+3)}$ **b.** $f_2(x) = 8\left(\frac{1}{2}\right)^x$ **c.** $f_3(x) = \left(\frac{1}{2}\right)^{(x-3)}$

d. $f_4(x) = \left(\frac{1}{2}\right)^x + 7$ **e.** $f_5(x) = 7 + 2^x$ **f.** $f_6(x) = (8)2^x$

x	-1	0	1	2	3
$g(x)$	7.5	8	9	11	15

x	-2	-1	0	1	2
$h(x)$	32	16	8	4	2

Create two different exponential functions of the forms $y = a(b)^x$ and $y = c^x + d$ with y-intercepts of $(0, -3)$.

3.1 Exercises

The *HM mathSpace®* CD-ROM and *Eduspace®* for this text contain step-by-step solutions to all odd-numbered exercises. They also provide Tutorial Exercises for additional help.

VOCABULARY CHECK: Fill in the blanks.

1. Polynomials and rational functions are examples of ________ functions.
2. Exponential and logarithmic functions are examples of nonalgebraic functions, also called ________ functions.
3. The exponential function given by $f(x) = e^x$ is called the ________ ________ function, and the base e is called the ________ base.
4. To find the amount A in an account after t years with principal P and an annual interest rate r compounded n times per year, you can use the formula ________.
5. To find the amount A in an account after t years with principal P and an annual interest rate r compounded continuously, you can use the formula ________.

PREREQUISITE SKILLS REVIEW: Practice and review algebra skills needed for this section at **www.Eduspace.com.**

In Exercises 1–6, evaluate the function at the indicated value of *x*. Round your result to three decimal places.

	Function	*Value*
1.	$f(x) = 3.4^x$	$x = 5.6$
2.	$f(x) = 2.3^x$	$x = \frac{3}{2}$
3.	$f(x) = 5^x$	$x = -\pi$
4.	$f(x) = \left(\frac{2}{3}\right)^{5x}$	$x = \frac{3}{10}$
5.	$g(x) = 5000(2^x)$	$x = -1.5$
6.	$f(x) = 200(1.2)^{12x}$	$x = 24$

In Exercises 7–10, match the exponential function with its graph. [The graphs are labeled (a), (b), (c), and (d).]

(a)

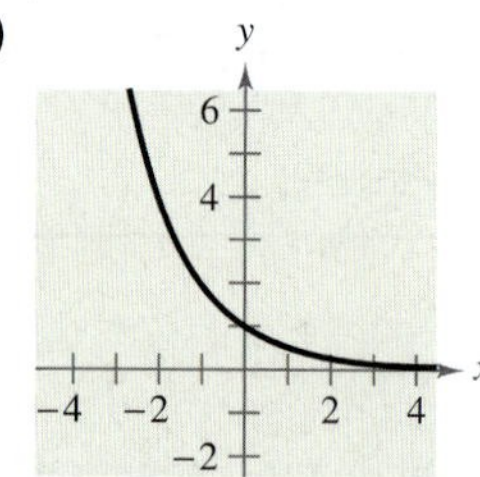

(b)

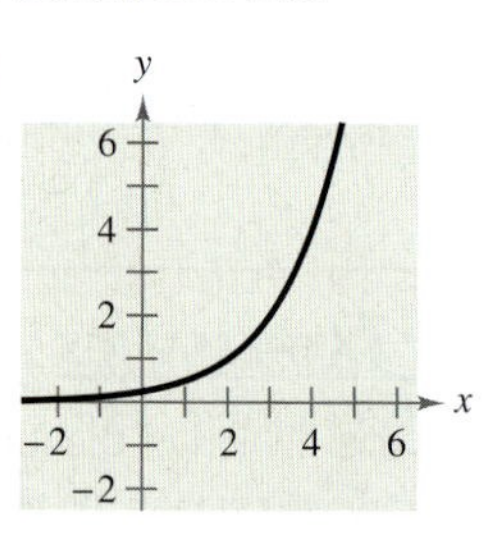

(c)

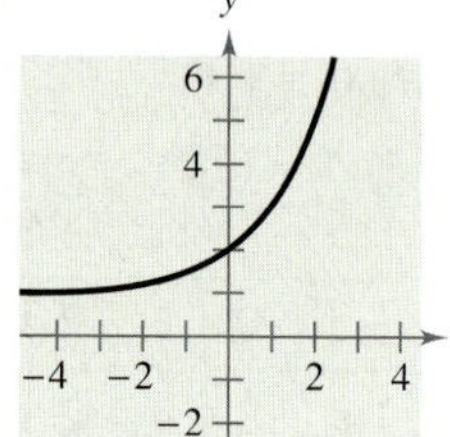

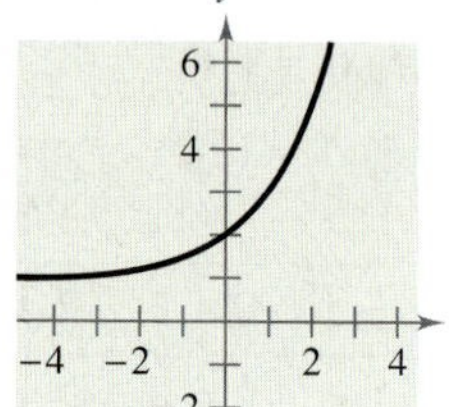

(d)

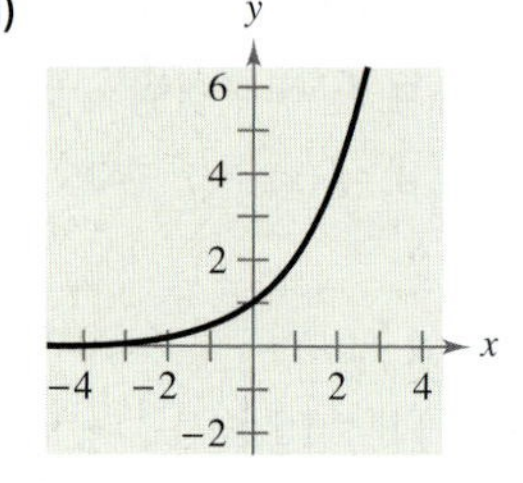

7. $f(x) = 2^x$
8. $f(x) = 2^x + 1$
9. $f(x) = 2^{-x}$
10. $f(x) = 2^{x-2}$

In Exercises 11–16, use a graphing utility to construct a table of values for the function. Then sketch the graph of the function.

11. $f(x) = \left(\frac{1}{2}\right)^x$
12. $f(x) = \left(\frac{1}{2}\right)^{-x}$
13. $f(x) = 6^{-x}$
14. $f(x) = 6^x$
15. $f(x) = 2^{x-1}$
16. $f(x) = 4^{x-3} + 3$

In Exercises 17–22, use the graph of *f* to describe the transformation that yields the graph of *g*.

17. $f(x) = 3^x, \quad g(x) = 3^{x-4}$
18. $f(x) = 4^x, \quad g(x) = 4^x + 1$
19. $f(x) = -2^x, \quad g(x) = 5 - 2^x$
20. $f(x) = 10^x, \quad g(x) = 10^{-x+3}$
21. $f(x) = \left(\frac{7}{2}\right)^x, \quad g(x) = -\left(\frac{7}{2}\right)^{-x+6}$
22. $f(x) = 0.3^x, \quad g(x) = -0.3^x + 5$

In Exercises 23–26, use a graphing utility to graph the exponential function.

23. $y = 2^{-x^2}$
24. $y = 3^{-|x|}$
25. $y = 3^{x-2} + 1$
26. $y = 4^{x+1} - 2$

In Exercises 27–32, evaluate the function at the indicated value of *x*. Round your result to three decimal places.

	Function	*Value*
27.	$h(x) = e^{-x}$	$x = \frac{3}{4}$
28.	$f(x) = e^x$	$x = 3.2$
29.	$f(x) = 2e^{-5x}$	$x = 10$
30.	$f(x) = 1.5e^{x/2}$	$x = 240$
31.	$f(x) = 5000e^{0.06x}$	$x = 6$
32.	$f(x) = 250e^{0.05x}$	$x = 20$

In Exercises 33–38, use a graphing utility to construct a table of values for the function. Then sketch the graph of the function.

33. $f(x) = e^x$

34. $f(x) = e^{-x}$

35. $f(x) = 3e^{x+4}$

36. $f(x) = 2e^{-0.5x}$

37. $f(x) = 2e^{x-2} + 4$

38. $f(x) = 2 + e^{x-5}$

In Exercises 39–44, use a graphing utility to graph the exponential function.

39. $y = 1.08^{-5x}$

40. $y = 1.08^{5x}$

41. $s(t) = 2e^{0.12t}$

42. $s(t) = 3e^{-0.2t}$

43. $g(x) = 1 + e^{-x}$

44. $h(x) = e^{x-2}$

In Exercise 45–52, use the One-to-One Property to solve the equation for *x*.

45. $3^{x+1} = 27$

46. $2^{x-3} = 16$

47. $2^{x-2} = \frac{1}{32}$

48. $\left(\frac{1}{5}\right)^{x+1} = 125$

49. $e^{3x+2} = e^3$

50. $e^{2x-1} = e^4$

51. $e^{x^2-3} = e^{2x}$

52. $e^{x^2+6} = e^{5x}$

Compound Interest **In Exercises 53–56, complete the table to determine the balance *A* for *P* dollars invested at rate *r* for *t* years and compounded *n* times per year.**

n	1	2	4	12	365	Continuous
A						

53. $P = \$2500, r = 2.5\%, t = 10$ years

54. $P = \$1000, r = 4\%, t = 10$ years

55. $P = \$2500, r = 3\%, t = 20$ years

56. $P = \$1000, r = 6\%, t = 40$ years

Compound Interest **In Exercises 57–60, complete the table to determine the balance *A* for $12,000 invested at rate *r* for *t* years, compounded continuously.**

t	10	20	30	40	50
A					

57. $r = 4\%$

58. $r = 6\%$

59. $r = 6.5\%$

60. $r = 3.5\%$

61. ***Trust Fund*** On the day of a child's birth, a deposit of $25,000 is made in a trust fund that pays 8.75% interest, compounded continuously. Determine the balance in this account on the child's 25th birthday.

62. ***Trust Fund*** A deposit of $5000 is made in a trust fund that pays 7.5% interest, compounded continuously. It is specified that the balance will be given to the college from which the donor graduated after the money has earned interest for 50 years. How much will the college receive?

63. ***Inflation*** If the annual rate of inflation averages 4% over the next 10 years, the approximate costs C of goods or services during any year in that decade will be modeled by $C(t) = P(1.04)^t$, where t is the time in years and P is the present cost. The price of an oil change for your car is presently $23.95. Estimate the price 10 years from now.

64. ***Demand*** The demand equation for a product is given by

$$p = 5000\left(1 - \frac{4}{4 + e^{-0.002x}}\right)$$

where p is the price and x is the number of units.

(a) Use a graphing utility to graph the demand function for $x > 0$ and $p > 0$.

(b) Find the price p for a demand of $x = 500$ units.

(c) Use the graph in part (a) to approximate the greatest price that will still yield a demand of at least 600 units.

65. ***Computer Virus*** The number V of computers infected by a computer virus increases according to the model $V(t) = 100e^{4.6052t}$, where t is the time in hours. Find (a) $V(1)$, (b) $V(1.5)$, and (c) $V(2)$.

66. ***Population*** The population P (in millions) of Russia from 1996 to 2004 can be approximated by the model $P = 152.26e^{-0.0039t}$, where t represents the year, with $t = 6$ corresponding to 1996. (Source: Census Bureau, International Data Base)

(a) According to the model, is the population of Russia increasing or decreasing? Explain.

(b) Find the population of Russia in 1998 and 2000.

(c) Use the model to predict the population of Russia in 2010.

67. ***Radioactive Decay*** Let Q represent a mass of radioactive radium (^{226}Ra) (in grams), whose half-life is 1599 years. The quantity of radium present after t years is $Q = 25\left(\frac{1}{2}\right)^{t/1599}$.

(a) Determine the initial quantity (when $t = 0$).

(b) Determine the quantity present after 1000 years.

(c) Use a graphing utility to graph the function over the interval $t = 0$ to $t = 5000$.

68. ***Radioactive Decay*** Let Q represent a mass of carbon 14 (^{14}C) (in grams), whose half-life is 5715 years. The quantity of carbon 14 present after t years is $Q = 10\left(\frac{1}{2}\right)^{t/5715}$.

(a) Determine the initial quantity (when $t = 0$).

(b) Determine the quantity present after 2000 years.

(c) Sketch the graph of this function over the interval $t = 0$ to $t = 10{,}000$.

Model It

69. ***Data Analysis: Biology*** To estimate the amount of defoliation caused by the gypsy moth during a given year, a forester counts the number x of egg masses on $\frac{1}{40}$ of an acre (circle of radius 18.6 feet) in the fall. The percent of defoliation y the next spring is shown in the table. (Source: USDA, Forest Service)

Egg masses, x	Percent of defoliation, y
0	12
25	44
50	81
75	96
100	99

A model for the data is given by

$$y = \frac{100}{1 + 7e^{-0.069x}}.$$

(a) Use a graphing utility to create a scatter plot of the data and graph the model in the same viewing window.

(b) Create a table that compares the model with the sample data.

(c) Estimate the percent of defoliation if 36 egg masses are counted on $\frac{1}{40}$ acre.

(d) You observe that $\frac{2}{3}$ of a forest is defoliated the following spring. Use the graph in part (a) to estimate the number of egg masses per $\frac{1}{40}$ acre.

70. ***Data Analysis: Meteorology*** A meteorologist measures the atmospheric pressure P (in pascals) at altitude h (in kilometers). The data are shown in the table.

Altitude, h	Pressure, P
0	101,293
5	54,735
10	23,294
15	12,157
20	5,069

A model for the data is given by $P = 107{,}428e^{-0.150h}$.

(a) Sketch a scatter plot of the data and graph the model on the same set of axes.

(b) Estimate the atmospheric pressure at a height of 8 kilometers.

Synthesis

True or False? **In Exercises 71 and 72, determine whether the statement is true or false. Justify your answer.**

71. The line $y = -2$ is an asymptote for the graph of $f(x) = 10^x - 2$.

72. $e = \dfrac{271{,}801}{99{,}990}$.

Think About It **In Exercises 73–76, use properties of exponents to determine which functions (if any) are the same.**

73. $f(x) = 3^{x-2}$
$g(x) = 3^x - 9$
$h(x) = \frac{1}{9}(3^x)$

74. $f(x) = 4^x + 12$
$g(x) = 2^{2x+6}$
$h(x) = 64(4^x)$

75. $f(x) = 16(4^{-x})$
$g(x) = \left(\frac{1}{4}\right)^{x-2}$
$h(x) = 16(2^{-2x})$

76. $f(x) = e^{-x} + 3$
$g(x) = e^{3-x}$
$h(x) = -e^{x-3}$

77. Graph the functions given by $y = 3^x$ and $y = 4^x$ and use the graphs to solve each inequality.

(a) $4^x < 3^x$ (b) $4^x > 3^x$

78. Use a graphing utility to graph each function. Use the graph to find where the function is increasing and decreasing, and approximate any relative maximum or minimum values.

(a) $f(x) = x^2e^{-x}$ (b) $g(x) = x2^{3-x}$

79. ***Graphical Analysis*** Use a graphing utility to graph

$$f(x) = \left(1 + \frac{0.5}{x}\right)^x \quad \text{and} \quad g(x) = e^{0.5}$$

in the same viewing window. What is the relationship between f and g as x increases and decreases without bound?

80. ***Think About It*** Which functions are exponential?

(a) $3x$ (b) $3x^2$ (c) 3^x (d) 2^{-x}

Skills Review

In Exercises 81 and 82, solve for y.

81. $x^2 + y^2 = 25$

82. $x - |y| = 2$

In Exercises 83 and 84, sketch the graph of the function.

83. $f(x) = \dfrac{2}{9 + x}$

84. $f(x) = \sqrt{7 - x}$

85. **Make a Decision** To work an extended application analyzing the population per square mile of the United States, visit this text's website at *college.hmco.com*. *(Data Source: U.S. Census Bureau)*

3.2 Logarithmic Functions and Their Graphs

What you should learn

- Recognize and evaluate logarithmic functions with base *a*.
- Graph logarithmic functions.
- Recognize, evaluate, and graph natural logarithmic functions.
- Use logarithmic functions to model and solve real-life problems.

Why you should learn it

Logarithmic functions are often used to model scientific observations. For instance, in Exercise 89 on page 238, a logarithmic function is used to model human memory.

© Ariel Skelley/Corbis

Logarithmic Functions

In Section 1.9, you studied the concept of an inverse function. There, you learned that if a function is one-to-one—that is, if the function has the property that no horizontal line intersects the graph of the function more than once—the function must have an inverse function. By looking back at the graphs of the exponential functions introduced in Section 3.1, you will see that every function of the form $f(x) = a^x$ passes the Horizontal Line Test and therefore must have an inverse function. This inverse function is called the **logarithmic function with base *a*.**

Definition of Logarithmic Function with Base *a*

For $x > 0$, $a > 0$, and $a \neq 1$,

$$y = \log_a x \text{ if and only if } x = a^y.$$

The function given by

$$f(x) = \log_a x \qquad \text{Read as "log base } a \text{ of } x."$$

is called the **logarithmic function with base *a*.**

The equations

$$y = \log_a x \qquad \text{and} \qquad x = a^y$$

are equivalent. The first equation is in logarithmic form and the second is in exponential form. For example, the logarithmic equation $2 = \log_3 9$ can be rewritten in exponential form as $9 = 3^2$. The exponential equation $5^3 = 125$ can be rewritten in logarithmic form as $\log_5 125 = 3$.

When evaluating logarithms, remember that *a logarithm is an exponent.* This means that $\log_a x$ is the exponent to which a must be raised to obtain x. For instance, $\log_2 8 = 3$ because 2 must be raised to the third power to get 8.

Example 1 Evaluating Logarithms

Use the definition of logarithmic function to evaluate each logarithm at the indicated value of x.

a. $f(x) = \log_2 x, \quad x = 32$ **b.** $f(x) = \log_3 x, \quad x = 1$

c. $f(x) = \log_4 x, \quad x = 2$ **d.** $f(x) = \log_{10} x, \quad x = \frac{1}{100}$

Solution

a. $f(32) = \log_2 32 = 5$ because $2^5 = 32$.

b. $f(1) = \log_3 1 = 0$ because $3^0 = 1$.

c. $f(2) = \log_4 2 = \frac{1}{2}$ because $4^{1/2} = \sqrt{4} = 2$.

d. $f\left(\frac{1}{100}\right) = \log_{10} \frac{1}{100} = -2$ because $10^{-2} = \frac{1}{10^2} = \frac{1}{100}$.

✓CHECKPOINT Now try Exercise 17.

STUDY TIP

Remember that a logarithm is an exponent. So, to evaluate the logarithmic expression $\log_a x$, you need to ask the question, "To what power must a be raised to obtain x?"

Exploration

Complete the table for $f(x) = 10^x$.

x	-2	-1	0	1	2
$f(x)$					

Complete the table for $f(x) = \log x$.

x	$\frac{1}{100}$	$\frac{1}{10}$	1	10	100
$f(x)$					

Compare the two tables. What is the relationship between $f(x) = 10^x$ and $f(x) = \log x$?

The logarithmic function with base 10 is called the **common logarithmic function.** It is denoted by $\log_{10}$ or simply by log. On most calculators, this function is denoted by [LOG]. Example 2 shows how to use a calculator to evaluate common logarithmic functions. You will learn how to use a calculator to calculate logarithms to any base in the next section.

Example 2 Evaluating Common Logarithms on a Calculator

Use a calculator to evaluate the function given by $f(x) = \log x$ at each value of x.

a. $x = 10$ **b.** $x = \frac{1}{3}$ **c.** $x = 2.5$ **d.** $x = -2$

Solution

	Function Value	*Graphing Calculator Keystrokes*	*Display*
a.	$f(10) = \log 10$	[LOG] 10 [ENTER]	1
b.	$f(\frac{1}{3}) = \log \frac{1}{3}$	[LOG] [(] 1 [÷] 3 [)] [ENTER]	-0.4771213
c.	$f(2.5) = \log 2.5$	[LOG] 2.5 [ENTER]	0.3979400
d.	$f(-2) = \log(-2)$	[LOG] [(−)] 2 [ENTER]	ERROR

Note that the calculator displays an error message (or a complex number) when you try to evaluate $\log(-2)$. The reason for this is that there is no real number power to which 10 can be raised to obtain -2.

✓CHECKPOINT Now try Exercise 23.

The following properties follow directly from the definition of the logarithmic function with base a.

The logarithmic function can be one of the most difficult concepts for students to understand. Remind students that a logarithm is an exponent. Converting back and forth from logarithmic form to exponential form supports this concept.

Properties of Logarithms

1. $\log_a 1 = 0$ because $a^0 = 1$.
2. $\log_a a = 1$ because $a^1 = a$.
3. $\log_a a^x = x$ and $a^{\log_a x} = x$ — Inverse Properties
4. If $\log_a x = \log_a y$, then $x = y$. — One-to-One Property

Example 3 Using Properties of Logarithms

a. Simplify: $\log_4 1$ **b.** Simplify: $\log_{\sqrt{7}} \sqrt{7}$ **c.** Simplify: $6^{\log_6 20}$

Solution

a. Using Property 1, it follows that $\log_4 1 = 0$.

b. Using Property 2, you can conclude that $\log_{\sqrt{7}} \sqrt{7} = 1$.

c. Using the Inverse Property (Property 3), it follows that $6^{\log_6 20} = 20$.

✓CHECKPOINT Now try Exercise 27.

You can use the One-to-One Property (Property 4) to solve simple logarithmic equations, as shown in Example 4.

Example 4 **Using the One-to-One Property**

a. $\log_3 x = \log_3 12$ — Original equation

$x = 12$ — One-to-One Property

b. $\log(2x + 1) = \log x \implies 2x + 1 = x \implies x = -1$

c. $\log_4(x^2 - 6) = \log_4 10 \implies x^2 - 6 = 10 \implies x^2 = 16 \implies x = \pm 4$

CHECKPOINT Now try Exercise 79.

Graphs of Logarithmic Functions

To sketch the graph of $y = \log_a x$, you can use the fact that the graphs of inverse functions are reflections of each other in the line $y = x$.

Example 5 **Graphs of Exponential and Logarithmic Functions**

In the same coordinate plane, sketch the graph of each function.

a. $f(x) = 2^x$ **b.** $g(x) = \log_2 x$

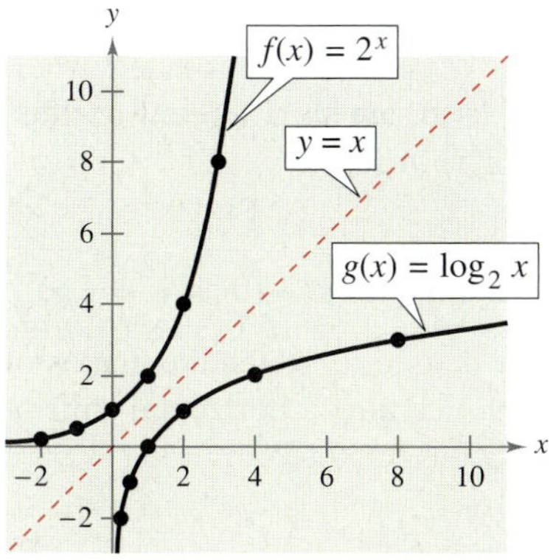

Solution

a. For $f(x) = 2^x$, construct a table of values. By plotting these points and connecting them with a smooth curve, you obtain the graph shown in Figure 3.13.

x	-2	-1	0	1	2	3
$f(x) = 2^x$	$\frac{1}{4}$	$\frac{1}{2}$	1	2	4	8

FIGURE 3.13

b. Because $g(x) = \log_2 x$ is the inverse function of $f(x) = 2^x$, the graph of g is obtained by plotting the points $(f(x), x)$ and connecting them with a smooth curve. The graph of g is a reflection of the graph of f in the line $y = x$, as shown in Figure 3.13.

CHECKPOINT Now try Exercise 31.

Example 6 **Sketching the Graph of a Logarithmic Function**

Sketch the graph of the common logarithmic function $f(x) = \log x$. Identify the vertical asymptote.

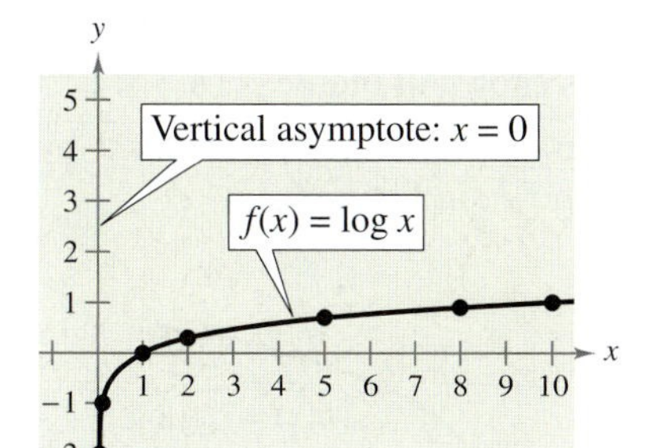

FIGURE 3.14

Solution

Begin by constructing a table of values. Note that some of the values can be obtained without a calculator by using the Inverse Property of Logarithms. Others require a calculator. Next, plot the points and connect them with a smooth curve, as shown in Figure 3.14. The vertical asymptote is $x = 0$ (y-axis).

	Without calculator				With calculator		
x	$\frac{1}{100}$	$\frac{1}{10}$	1	10	2	5	8
$f(x) = \log x$	-2	-1	0	1	0.301	0.699	0.903

CHECKPOINT Now try Exercise 37.

The nature of the graph in Figure 3.14 is typical of functions of the form $f(x) = \log_a x$, $a > 1$. They have one x-intercept and one vertical asymptote. Notice how slowly the graph rises for $x > 1$. The basic characteristics of logarithmic graphs are summarized in Figure 3.15.

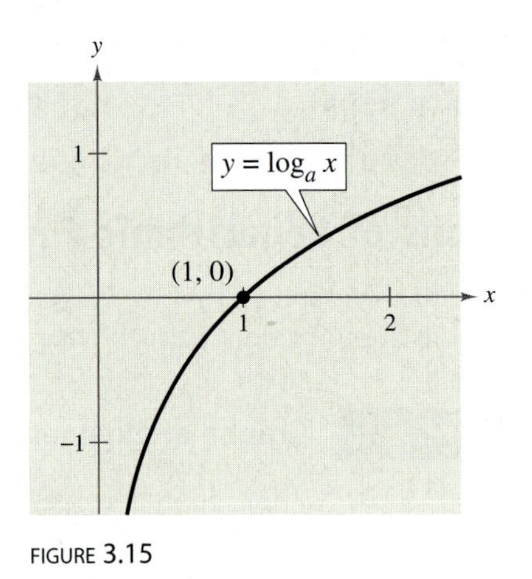

FIGURE 3.15

Graph of $y = \log_a x$, $a > 1$

- Domain: $(0, \infty)$
- Range: $(-\infty, \infty)$
- x-intercept: $(1, 0)$
- Increasing
- One-to-one, therefore has an inverse function
- y-axis is a vertical asymptote ($\log_a x \rightarrow -\infty$ as $x \rightarrow 0^+$).
- Continuous
- Reflection of graph of $y = a^x$ about the line $y = x$

The basic characteristics of the graph of $f(x) = a^x$ are shown below to illustrate the inverse relation between $f(x) = a^x$ and $g(x) = \log_a x$.

- Domain: $(-\infty, \infty)$
- Range: $(0, \infty)$
- y-intercept: $(0,1)$
- x-axis is a horizontal asymptote ($a^x \rightarrow 0$ as $x \rightarrow -\infty$).

In the next example, the graph of $y = \log_a x$ is used to sketch the graphs of functions of the form $f(x) = b \pm \log_a(x + c)$. Notice how a horizontal shift of the graph results in a horizontal shift of the vertical asymptote.

STUDY TIP

You can use your understanding of transformations to identify vertical asymptotes of logarithmic functions. For instance, in Example 7(a) the graph of $g(x) = f(x - 1)$ shifts the graph of $f(x)$ one unit to the right. So, the vertical asymptote of $g(x)$ is $x = 1$, one unit to the right of the vertical asymptote of the graph of $f(x)$.

Example 7 Shifting Graphs of Logarithmic Functions

The graph of each of the functions is similar to the graph of $f(x) = \log x$.

a. Because $g(x) = \log(x - 1) = f(x - 1)$, the graph of g can be obtained by shifting the graph of f one unit to the right, as shown in Figure 3.16.

b. Because $h(x) = 2 + \log x = 2 + f(x)$, the graph of h can be obtained by shifting the graph of f two units upward, as shown in Figure 3.17.

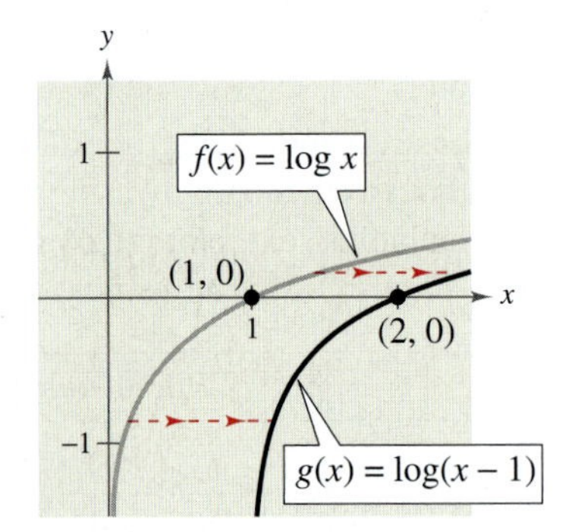

FIGURE 3.16

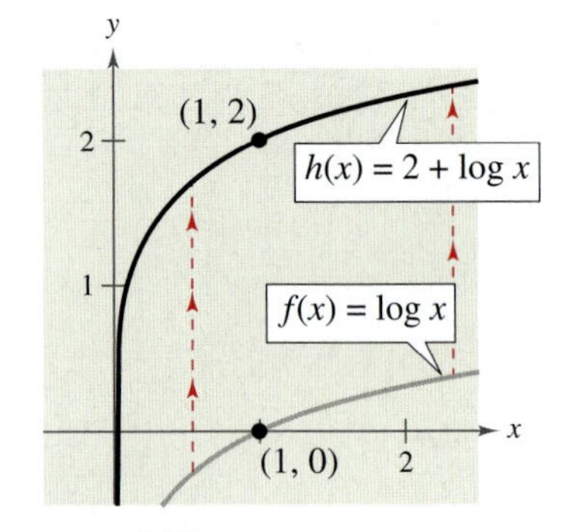

FIGURE 3.17

✓CHECKPOINT Now try Exercise 39.

The Natural Logarithmic Function

By looking back at the graph of the natural exponential function introduced in Section 3.1 on page 388, you will see that $f(x) = e^x$ is one-to-one and so has an inverse function. This inverse function is called the **natural logarithmic function** and is denoted by the special symbol $\ln x$, read as "the natural log of x" or "el en of x." Note that the natural logarithm is written without a base. The base is understood to be e.

The Natural Logarithmic Function

The function defined by

$$f(x) = \log_e x = \ln x, \quad x > 0$$

is called the **natural logarithmic function.**

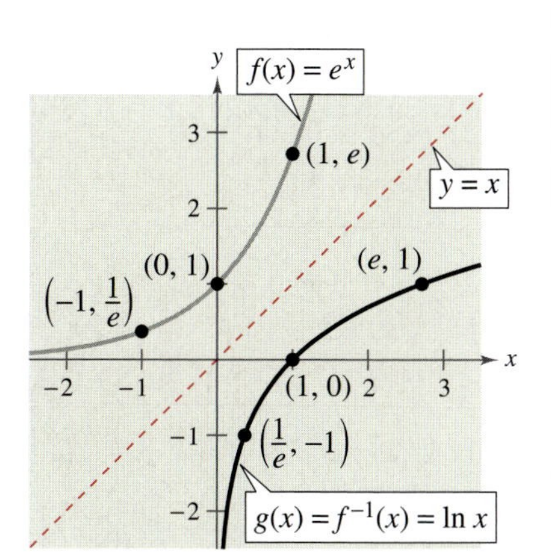

Reflection of graph of $f(x) = e^x$ about the line $y = x$
FIGURE 3.18

The definition above implies that the natural logarithmic function and the natural exponential function are inverse functions of each other. So, every logarithmic equation can be written in an equivalent exponential form and every exponential equation can be written in logarithmic form. That is, $y = \ln x$ and $x = e^y$ are equivalent equations.

Because the functions given by $f(x) = e^x$ and $g(x) = \ln x$ are inverse functions of each other, their graphs are reflections of each other in the line $y = x$. This reflective property is illustrated in Figure 3.18.

On most calculators, the natural logarithm is denoted by [LN], as illustrated in Example 8.

STUDY TIP

Notice that as with every other logarithmic function, the domain of the natural logarithmic function is the set of *positive real numbers*—be sure you see that $\ln x$ is not defined for zero or for negative numbers.

Example 8 Evaluating the Natural Logarithmic Function

Use a calculator to evaluate the function given by $f(x) = \ln x$ for each value of x.

a. $x = 2$ **b.** $x = 0.3$ **c.** $x = -1$ **d.** $x = 1 + \sqrt{2}$

Solution

Function Value	Graphing Calculator Keystrokes	Display
a. $f(2) = \ln 2$	[LN] 2 [ENTER]	0.6931472
b. $f(0.3) = \ln 0.3$	[LN] .3 [ENTER]	–1.2039728
c. $f(-1) = \ln(-1)$	[LN] [(–)] 1 [ENTER]	ERROR
d. $f(1 + \sqrt{2}) = \ln(1 + \sqrt{2})$	[LN] [(] 1 [+] [√] 2 [)] [ENTER]	0.8813736

CHECKPOINT Now try Exercise 61.

In Example 8, be sure you see that $\ln(-1)$ gives an error message on most calculators. (Some calculators may display a complex number.) This occurs because the domain of $\ln x$ is the set of positive real numbers (see Figure 3.18). So, $\ln(-1)$ is undefined.

The four properties of logarithms listed on page 230 are also valid for natural logarithms.

Properties of Natural Logarithms

1. $\ln 1 = 0$ because $e^0 = 1$.
2. $\ln e = 1$ because $e^1 = e$.
3. $\ln e^x = x$ and $e^{\ln x} = x$ — Inverse Properties
4. If $\ln x = \ln y$, then $x = y$. — One-to-One Property

Example 9 Using Properties of Natural Logarithms

Use the properties of natural logarithms to simplify each expression.

a. $\ln \dfrac{1}{e}$ **b.** $e^{\ln 5}$ **c.** $\dfrac{\ln 1}{3}$ **d.** $2 \ln e$

Solution

a. $\ln \dfrac{1}{e} = \ln e^{-1} = -1$ Inverse Property

b. $e^{\ln 5} = 5$ Inverse Property

c. $\dfrac{\ln 1}{3} = \dfrac{0}{3} = 0$ Property 1

d. $2 \ln e = 2(1) = 2$ Property 2

CHECKPOINT Now try Exercise 65.

Example 10 Finding the Domains of Logarithmic Functions

Find the domain of each function.

a. $f(x) = \ln(x - 2)$ **b.** $g(x) = \ln(2 - x)$ **c.** $h(x) = \ln x^2$

Solution

a. Because $\ln(x - 2)$ is defined only if $x - 2 > 0$, it follows that the domain of f is $(2, \infty)$. The graph of f is shown in Figure 3.19.

b. Because $\ln(2 - x)$ is defined only if $2 - x > 0$, it follows that the domain of g is $(-\infty, 2)$. The graph of g is shown in Figure 3.20.

c. Because $\ln x^2$ is defined only if $x^2 > 0$, it follows that the domain of h is all real numbers except $x = 0$. The graph of h is shown in Figure 3.21.

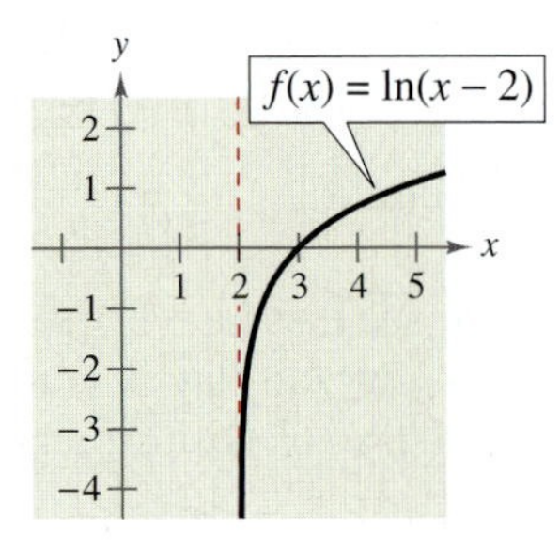

FIGURE 3.19

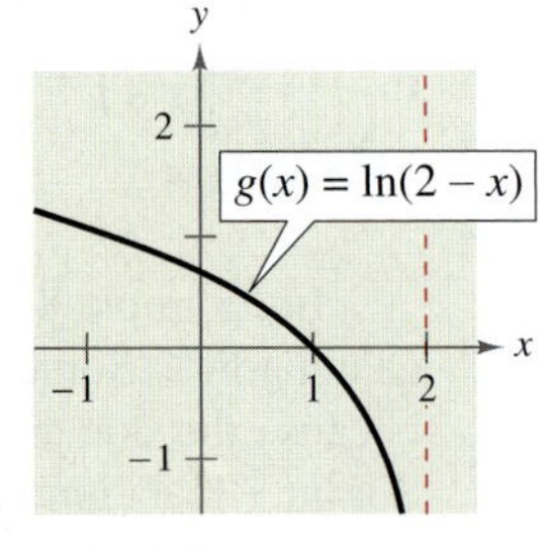

FIGURE 3.20

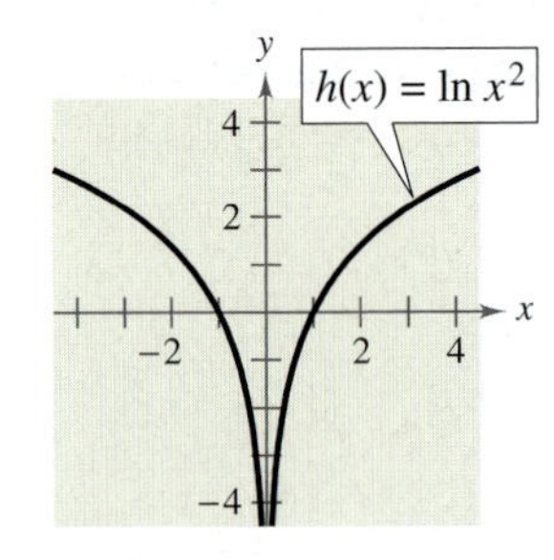

FIGURE 3.21

CHECKPOINT Now try Exercise 69.

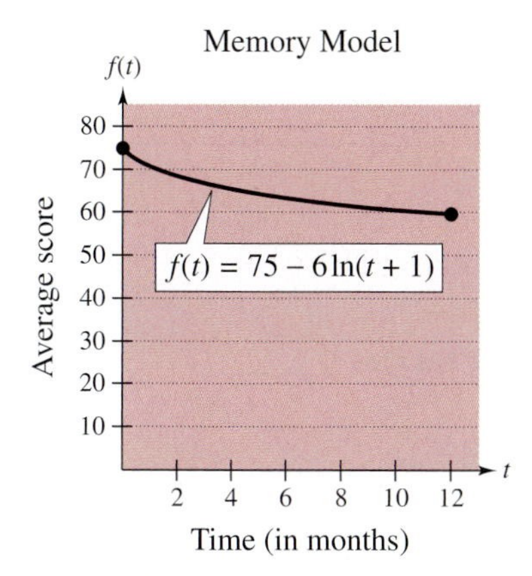

FIGURE 3.22

Application

Example 11 Human Memory Model 

Students participating in a psychology experiment attended several lectures on a subject and were given an exam. Every month for a year after the exam, the students were retested to see how much of the material they remembered. The average scores for the group are given by the *human memory model*

$$f(t) = 75 - 6\ln(t + 1), \quad 0 \le t \le 12$$

where t is the time in months. The graph of f is shown in Figure 3.22.

a. What was the average score on the original $(t = 0)$ exam?

b. What was the average score at the end of $t = 2$ months?

c. What was the average score at the end of $t = 6$ months?

Solution

a. The original average score was

$f(0) = 75 - 6\ln(0 + 1)$	Substitute 0 for t.
$= 75 - 6\ln 1$	Simplify.
$= 75 - 6(0)$	Property of natural logarithms
$= 75.$	Solution

b. After 2 months, the average score was

$f(2) = 75 - 6\ln(2 + 1)$	Substitute 2 for t.
$= 75 - 6\ln 3$	Simplify.
$\approx 75 - 6(1.0986)$	Use a calculator.
$\approx 68.4.$	Solution

c. After 6 months, the average score was

$f(6) = 75 - 6\ln(6 + 1)$	Substitute 6 for t.
$= 75 - 6\ln 7$	Simplify.
$\approx 75 - 6(1.9459)$	Use a calculator.
$\approx 63.3.$	Solution

CHECKPOINT Now try Exercise 89.

Alternative Writing About Mathematics

Use a graphing utility to graph $f(x) = \ln x$. How will the graphs of $h(x) = \ln x + 5$, $j(x) = \ln(x - 3)$, and $l(x) = \ln x - 4$ differ from the graph of f?

How will the basic graph of f be affected when a constant c is introduced: $g(x) = c \ln x$? Use a graphing utility to graph g with several different positive values of c, and summarize the effect of c.

Writing About Mathematics

Analyzing a Human Memory Model Use a graphing utility to determine the time in months when the average score in Example 11 was 60. Explain your method of solving the problem. Describe another way that you can use a graphing utility to determine the answer.

3.2 Exercises

VOCABULARY CHECK: Fill in the blanks.

1. The inverse function of the exponential function given by $f(x) = a^x$ is called the ________ function with base a.
2. The common logarithmic function has base ________ .
3. The logarithmic function given by $f(x) = \ln x$ is called the ________ logarithmic function and has base ________.
4. The Inverse Property of logarithms and exponentials states that $\log_a a^x = x$ and ________.
5. The One-to-One Property of natural logarithms states that if $\ln x = \ln y$, then ________.

PREREQUISITE SKILLS REVIEW: Practice and review algebra skills needed for this section at **www.Eduspace.com.**

In Exercises 1–8, write the logarithmic equation in exponential form. For example, the exponential form of $\log_5 25 = 2$ is $5^2 = 25$.

1. $\log_4 64 = 3$
2. $\log_3 81 = 4$
3. $\log_7 \frac{1}{49} = -2$
4. $\log \frac{1}{1000} = -3$
5. $\log_{32} 4 = \frac{2}{5}$
6. $\log_{16} 8 = \frac{3}{4}$
7. $\log_{36} 6 = \frac{1}{2}$
8. $\log_8 4 = \frac{2}{3}$

In Exercises 9–16, write the exponential equation in logarithmic form. For example, the logarithmic form of $2^3 = 8$ is $\log_2 8 = 3$.

9. $5^3 = 125$
10. $8^2 = 64$
11. $81^{1/4} = 3$
12. $9^{3/2} = 27$
13. $6^{-2} = \frac{1}{36}$
14. $4^{-3} = \frac{1}{64}$
15. $7^0 = 1$
16. $10^{-3} = 0.001$

In Exercises 17–22, evaluate the function at the indicated value of x without using a calculator.

	Function	*Value*
17.	$f(x) = \log_2 x$	$x = 16$
18.	$f(x) = \log_{16} x$	$x = 4$
19.	$f(x) = \log_7 x$	$x = 1$
20.	$f(x) = \log x$	$x = 10$
21.	$g(x) = \log_a x$	$x = a^2$
22.	$g(x) = \log_b x$	$x = b^{-3}$

In Exercises 23–26, use a calculator to evaluate $f(x) = \log x$ at the indicated value of x. Round your result to three decimal places.

23. $x = \frac{4}{5}$
24. $x = \frac{1}{500}$
25. $x = 12.5$
26. $x = 75.25$

In Exercises 27–30, use the properties of logarithms to simplify the expression.

27. $\log_3 3^4$
28. $\log_{1.5} 1$
29. $\log_\pi \pi$
30. $9^{\log_9 15}$

In Exercises 31–38, find the domain, x-intercept, and vertical asymptote of the logarithmic function and sketch its graph.

31. $f(x) = \log_4 x$
32. $g(x) = \log_6 x$
33. $y = -\log_3 x + 2$
34. $h(x) = \log_4(x - 3)$
35. $f(x) = -\log_6(x + 2)$
36. $y = \log_5(x - 1) + 4$
37. $y = \log\left(\frac{x}{5}\right)$
38. $y = \log(-x)$

In Exercises 39–44, use the graph of $g(x) = \log_3 x$ to match the given function with its graph. Then describe the relationship between the graphs of f and g. [The graphs are labeled (a), (b), (c), (d), (e), and (f).]

(a)

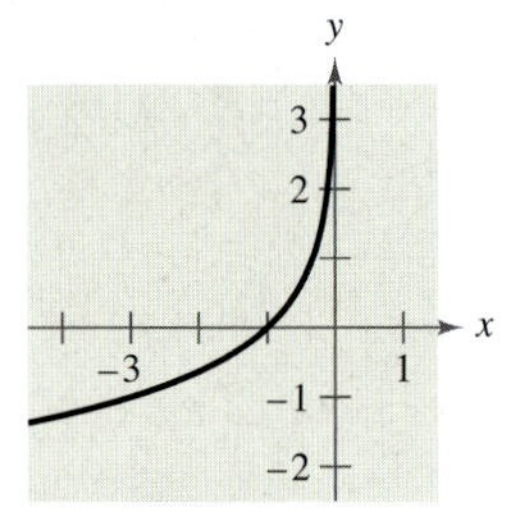

(b)

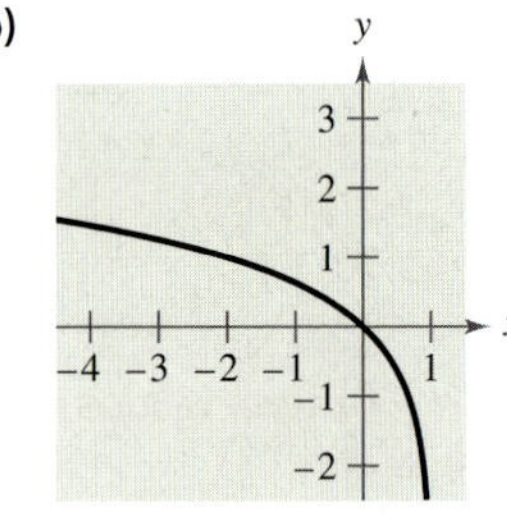

(c)

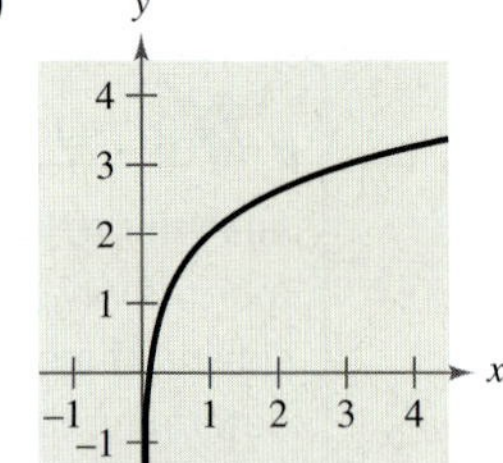

(d)

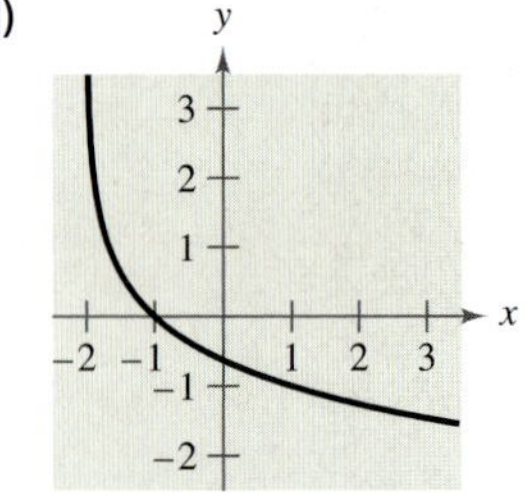

(e)

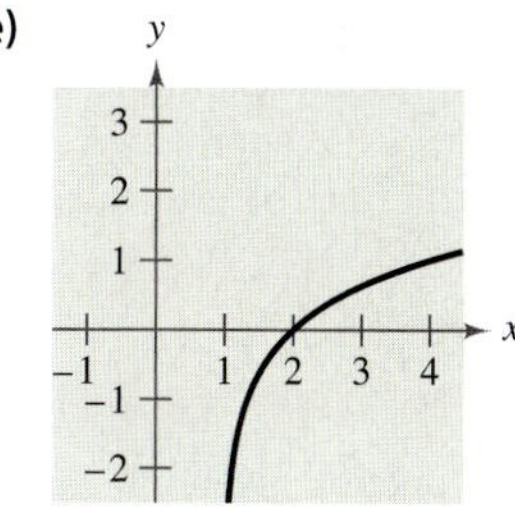

(f)

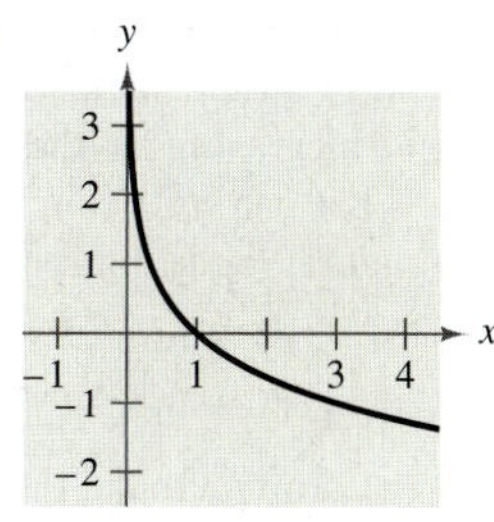

39. $f(x) = \log_3 x + 2$
40. $f(x) = -\log_3 x$
41. $f(x) = -\log_3(x + 2)$
42. $f(x) = \log_3(x - 1)$
43. $f(x) = \log_3(1 - x)$
44. $f(x) = -\log_3(-x)$

In Exercises 45–52, write the logarithmic equation in exponential form.

45. $\ln \frac{1}{2} = -0.693 \ldots$
46. $\ln \frac{2}{5} = -0.916 \ldots$
47. $\ln 4 = 1.386 \ldots$
48. $\ln 10 = 2.302 \ldots$
49. $\ln 250 = 5.521 \ldots$
50. $\ln 679 = 6.520 \ldots$
51. $\ln 1 = 0$
52. $\ln e = 1$

In Exercises 53–60, write the exponential equation in logarithmic form.

53. $e^3 = 20.0855 \ldots$
54. $e^2 = 7.3890 \ldots$
55. $e^{1/2} = 1.6487 \ldots$
56. $e^{1/3} = 1.3956 \ldots$
57. $e^{-0.5} = 0.6065 \ldots$
58. $e^{-4.1} = 0.0165 \ldots$
59. $e^x = 4$
60. $e^{2x} = 3$

In Exercises 61–64, use a calculator to evaluate the function at the indicated value of x. Round your result to three decimal places.

	Function	*Value*
61.	$f(x) = \ln x$	$x = 18.42$
62.	$f(x) = 3 \ln x$	$x = 0.32$
63.	$g(x) = 2 \ln x$	$x = 0.75$
64.	$g(x) = -\ln x$	$x = \frac{1}{2}$

In Exercises 65–68, evaluate $g(x) = \ln x$ at the indicated value of x without using a calculator.

65. $x = e^3$
66. $x = e^{-2}$
67. $x = e^{-2/3}$
68. $x = e^{-5/2}$

In Exercises 69–72, find the domain, x-intercept, and vertical asymptote of the logarithmic function and sketch its graph.

69. $f(x) = \ln(x - 1)$
70. $h(x) = \ln(x + 1)$
71. $g(x) = \ln(-x)$
72. $f(x) = \ln(3 - x)$

In Exercises 73–78, use a graphing utility to graph the function. Be sure to use an appropriate viewing window.

73. $f(x) = \log(x + 1)$
74. $f(x) = \log(x - 1)$
75. $f(x) = \ln(x - 1)$
76. $f(x) = \ln(x + 2)$
77. $f(x) = \ln x + 2$
78. $f(x) = 3 \ln x - 1$

In Exercises 79–86, use the One-to-One Property to solve the equation for x.

79. $\log_2(x + 1) = \log_2 4$
80. $\log_2(x - 3) = \log_2 9$
81. $\log(2x + 1) = \log 15$
82. $\log(5x + 3) = \log 12$
83. $\ln(x + 2) = \ln 6$
84. $\ln(x - 4) = \ln 2$
85. $\ln(x^2 - 2) = \ln 23$
86. $\ln(x^2 - x) = \ln 6$

Model It

87. ***Monthly Payment*** The model

$$t = 12.542 \ln\left(\frac{x}{x - 1000}\right), \quad x > 1000$$

approximates the length of a home mortgage of \$150,000 at 8% in terms of the monthly payment. In the model, t is the length of the mortgage in years and x is the monthly payment in dollars (see figure).

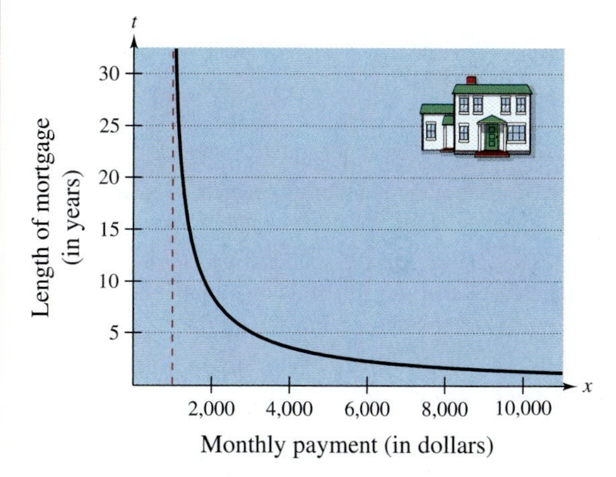

(a) Use the model to approximate the lengths of a \$150,000 mortgage at 8% when the monthly payment is \$1100.65 and when the monthly payment is \$1254.68.

(b) Approximate the total amounts paid over the term of the mortgage with a monthly payment of \$1100.65 and with a monthly payment of \$1254.68.

(c) Approximate the total interest charges for a monthly payment of \$1100.65 and for a monthly payment of \$1254.68.

(d) What is the vertical asymptote for the model? Interpret its meaning in the context of the problem.

88. ***Compound Interest*** A principal P, invested at $9\frac{1}{2}\%$ and compounded continuously, increases to an amount K times the original principal after t years, where t is given by $t = (\ln K)/0.095$.

(a) Complete the table and interpret your results.

K	1	2	4	6	8	10	12
t							

(b) Sketch a graph of the function.

89. ***Human Memory Model*** Students in a mathematics class were given an exam and then retested monthly with an equivalent exam. The average scores for the class are given by the human memory model $f(t) = 80 - 17\log(t + 1)$, $0 \le t \le 12$ where t is the time in months.

(a) Use a graphing utility to graph the model over the specified domain.

(b) What was the average score on the original exam ($t = 0$)?

(c) What was the average score after 4 months?

(d) What was the average score after 10 months?

90. ***Sound Intensity*** The relationship between the number of decibels β and the intensity of a sound I in watts per square meter is

$$\beta = 10\log\left(\frac{I}{10^{-12}}\right).$$

(a) Determine the number of decibels of a sound with an intensity of 1 watt per square meter.

(b) Determine the number of decibels of a sound with an intensity of 10^{-2} watt per square meter.

(c) The intensity of the sound in part (a) is 100 times as great as that in part (b). Is the number of decibels 100 times as great? Explain.

Synthesis

True or False? **In Exercises 91 and 92, determine whether the statement is true or false. Justify your answer.**

91. You can determine the graph of $f(x) = \log_6 x$ by graphing $g(x) = 6^x$ and reflecting it about the x-axis.

92. The graph of $f(x) = \log_3 x$ contains the point (27, 3).

In Exercises 93–96, sketch the graph of *f* and *g* and describe the relationship between the graphs of *f* and *g*. What is the relationship between the functions *f* and *g*?

93. $f(x) = 3^x$, $g(x) = \log_3 x$

94. $f(x) = 5^x$, $g(x) = \log_5 x$

95. $f(x) = e^x$, $g(x) = \ln x$

96. $f(x) = 10^x$, $g(x) = \log x$

97. ***Graphical Analysis*** Use a graphing utility to graph f and g in the same viewing window and determine which is increasing at the greater rate as x approaches $+\infty$. What can you conclude about the rate of growth of the natural logarithmic function?

(a) $f(x) = \ln x$, $g(x) = \sqrt{x}$

(b) $f(x) = \ln x$, $g(x) = \sqrt[4]{x}$

98. (a) Complete the table for the function given by

$$f(x) = \frac{\ln x}{x}.$$

x	1	5	10	10^2	10^4	10^6
$f(x)$						

(b) Use the table in part (a) to determine what value $f(x)$ approaches as x increases without bound.

(c) Use a graphing utility to confirm the result of part (b).

99. ***Think About It*** The table of values was obtained by evaluating a function. Determine which of the statements may be true and which must be false.

x	1	2	8
y	0	1	3

(a) y is an exponential function of x.

(b) y is a logarithmic function of x.

(c) x is an exponential function of y.

(d) y is a linear function of x.

100. ***Writing*** Explain why $\log_a x$ is defined only for $0 < a < 1$ and $a > 1$.

In Exercises 101 and 102, (a) use a graphing utility to graph the function, (b) use the graph to determine the intervals in which the function is increasing and decreasing, and (c) approximate any relative maximum or minimum values of the function.

101. $f(x) = |\ln x|$

102. $h(x) = \ln(x^2 + 1)$

Skills Review

In Exercises 103–108, evaluate the function for $f(x) = 3x + 2$ and $g(x) = x^3 - 1$.

103. $(f + g)(2)$

104. $(f - g)(-1)$

105. $(fg)(6)$

106. $\left(\frac{f}{g}\right)(0)$

107. $(f \circ g)(7)$

108. $(g \circ f)(-3)$

3.3 Properties of Logarithms

What you should learn

- Use the change-of-base formula to rewrite and evaluate logarithmic expressions.
- Use properties of logarithms to evaluate or rewrite logarithmic expressions.
- Use properties of logarithms to expand or condense logarithmic expressions.
- Use logarithmic functions to model and solve real-life problems.

Why you should learn it

Logarithmic functions can be used to model and solve real-life problems. For instance, in Exercises 81–83 on page 244, a logarithmic function is used to model the relationship between the number of decibels and the intensity of a sound.

AP Photo/Stephen Chernin

Change of Base

Most calculators have only two types of log keys, one for common logarithms (base 10) and one for natural logarithms (base e). Although common logs and natural logs are the most frequently used, you may occasionally need to evaluate logarithms to other bases. To do this, you can use the following **change-of-base formula.**

Change-of-Base Formula

Let a, b, and x be positive real numbers such that $a \neq 1$ and $b \neq 1$. Then $\log_a x$ can be converted to a different base as follows.

Base b	*Base 10*	*Base e*
$\log_a x = \dfrac{\log_b x}{\log_b a}$	$\log_a x = \dfrac{\log x}{\log a}$	$\log_a x = \dfrac{\ln x}{\ln a}$

One way to look at the change-of-base formula is that logarithms to base a are simply *constant multiples* of logarithms to base b. The constant multiplier is $1/(\log_b a)$.

Example 1 Changing Bases Using Common Logarithms

a. $\log_4 25 = \dfrac{\log 25}{\log 4}$ $\log_a x = \dfrac{\log x}{\log a}$

$\approx \dfrac{1.39794}{0.60206}$ Use a calculator.

≈ 2.3219 Simplify.

b. $\log_2 12 = \dfrac{\log 12}{\log 2} \approx \dfrac{1.07918}{0.30103} \approx 3.5850$

CHECKPOINT Now try Exercise 1(a).

Example 2 Changing Bases Using Natural Logarithms

a. $\log_4 25 = \dfrac{\ln 25}{\ln 4}$ $\log_a x = \dfrac{\ln x}{\ln a}$

$\approx \dfrac{3.21888}{1.38629}$ Use a calculator.

≈ 2.3219 Simplify.

b. $\log_2 12 = \dfrac{\ln 12}{\ln 2} \approx \dfrac{2.48491}{0.69315} \approx 3.5850$

CHECKPOINT Now try Exercise 1(b).

Encourage your students to know these properties well. They will be used for solving logarithmic and exponential equations, as well as in calculus.

Properties of Logarithms

You know from the preceding section that the logarithmic function with base a is the *inverse function* of the exponential function with base a. So, it makes sense that the properties of exponents should have corresponding properties involving logarithms. For instance, the exponential property $a^0 = 1$ has the corresponding logarithmic property $\log_a 1 = 0$.

STUDY TIP

There is no general property that can be used to rewrite $\log_a(u \pm v)$. Specifically, $\log_a(u + v)$ is *not* equal to $\log_a u + \log_a v$.

Properties of Logarithms

Let a be a positive number such that $a \neq 1$, and let n be a real number. If u and v are positive real numbers, the following properties are true.

	Logarithm with Base a	*Natural Logarithm*
1. Product Property:	$\log_a(uv) = \log_a u + \log_a v$	$\ln(uv) = \ln u + \ln v$
2. Quotient Property:	$\log_a \frac{u}{v} = \log_a u - \log_a v$	$\ln \frac{u}{v} = \ln u - \ln v$
3. Power Property:	$\log_a u^n = n \log_a u$	$\ln u^n = n \ln u$

Remind your students to note the domain when applying properties of logarithms to a logarithmic function. For example, the domain of $f(x) = \ln x^2$ is all real numbers $x \neq 0$, whereas the domain of $g(x) = 2 \ln x$ is all real numbers $x > 0$.

For proofs of the properties listed above, see Proofs in Mathematics on page 278.

Example 3 Using Properties of Logarithms

Write each logarithm in terms of ln 2 and ln 3.

a. $\ln 6$ **b.** $\ln \frac{2}{27}$

Solution

a. $\ln 6 = \ln(2 \cdot 3)$ Rewrite 6 as $2 \cdot 3$.

$= \ln 2 + \ln 3$ Product Property

b. $\ln \frac{2}{27} = \ln 2 - \ln 27$ Quotient Property

$= \ln 2 - \ln 3^3$ Rewrite 27 as 3^3.

$= \ln 2 - 3 \ln 3$ Power Property

CHECKPOINT Now try Exercise 17.

Historical Note

John Napier, a Scottish mathematician, developed logarithms as a way to simplify some of the tedious calculations of his day. Beginning in 1594, Napier worked about 20 years on the invention of logarithms. Napier was only partially successful in his quest to simplify tedious calculations. Nonetheless, the development of logarithms was a step forward and received immediate recognition.

Example 4 Using Properties of Logarithms

Find the exact value of each expression without using a calculator.

a. $\log_5 \sqrt[3]{5}$ **b.** $\ln e^6 - \ln e^2$

Solution

a. $\log_5 \sqrt[3]{5} = \log_5 5^{1/3} = \frac{1}{3} \log_5 5 = \frac{1}{3}(1) = \frac{1}{3}$

b. $\ln e^6 - \ln e^2 = \ln \frac{e^6}{e^2} = \ln e^4 = 4 \ln e = 4(1) = 4$

CHECKPOINT Now try Exercise 23.

Rewriting Logarithmic Expressions

The properties of logarithms are useful for rewriting logarithmic expressions in forms that simplify the operations of algebra. This is true because these properties convert complicated products, quotients, and exponential forms into simpler sums, differences, and products, respectively.

A common error made in expanding logarithmic expressions is to rewrite $\log ax^n$ as $n \log ax$ instead of as $\log a + n \log x$.

Example 5 Expanding Logarithmic Expressions

Expand each logarithmic expression.

a. $\log_4 5x^3y$ **b.** $\ln \frac{\sqrt{3x-5}}{7}$

Solution

a. $\log_4 5x^3y = \log_4 5 + \log_4 x^3 + \log_4 y$ — Product Property

$= \log_4 5 + 3\log_4 x + \log_4 y$ — Power Property

b. $\ln \frac{\sqrt{3x-5}}{7} = \ln \frac{(3x-5)^{1/2}}{7}$ — Rewrite using rational exponent.

$= \ln(3x-5)^{1/2} - \ln 7$ — Quotient Property

$= \frac{1}{2}\ln(3x-5) - \ln 7$ — Power Property

CHECKPOINT Now try Exercise 47.

Exploration

Use a graphing utility to graph the functions given by

$$y_1 = \ln x - \ln(x-3)$$

and

$$y_2 = \ln \frac{x}{x-3}$$

in the same viewing window. Does the graphing utility show the functions with the same domain? If so, should it? Explain your reasoning.

In Example 5, the properties of logarithms were used to *expand* logarithmic expressions. In Example 6, this procedure is reversed and the properties of logarithms are used to *condense* logarithmic expressions.

A common error made in condensing logarithmic expressions is to rewrite $\log x - \log y$ as $\frac{\log x}{\log y}$ instead of as $\log \frac{x}{y}$.

Example 6 Condensing Logarithmic Expressions

Condense each logarithmic expression.

a. $\frac{1}{2}\log x + 3\log(x+1)$ **b.** $2\ln(x+2) - \ln x$

c. $\frac{1}{3}[\log_2 x + \log_2(x+1)]$

Solution

a. $\frac{1}{2}\log x + 3\log(x+1) = \log x^{1/2} + \log(x+1)^3$ — Power Property

$= \log\left[\sqrt{x}(x+1)^3\right]$ — Product Property

b. $2\ln(x+2) - \ln x = \ln(x+2)^2 - \ln x$ — Power Property

$= \ln \frac{(x+2)^2}{x}$ — Quotient Property

c. $\frac{1}{3}[\log_2 x + \log_2(x+1)] = \frac{1}{3}\{\log_2[x(x+1)]\}$ — Product Property

$= \log_2[x(x+1)]^{1/3}$ — Power Property

$= \log_2 \sqrt[3]{x(x+1)}$ — Rewrite with a radical.

CHECKPOINT Now try Exercise 69.

Application

One method of determining how the x- and y-values for a set of nonlinear data are related is to take the natural logarithm of each of the x- and y-values. If the points are graphed and fall on a line, then you can determine that the x- and y-values are related by the equation

$$\ln y = m \ln x$$

where m is the slope of the line.

Example 7 Finding a Mathematical Model

The table shows the mean distance x and the period (the time it takes a planet to orbit the sun) y for each of the six planets that are closest to the sun. In the table, the mean distance is given in terms of astronomical units (where Earth's mean distance is defined as 1.0), and the period is given in years. Find an equation that relates y and x.

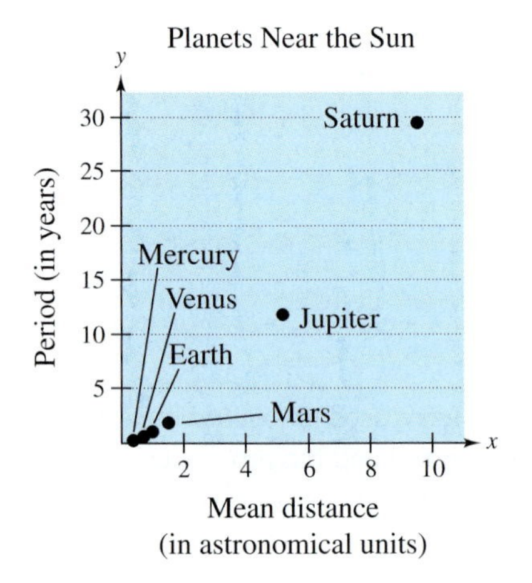

FIGURE 3.23

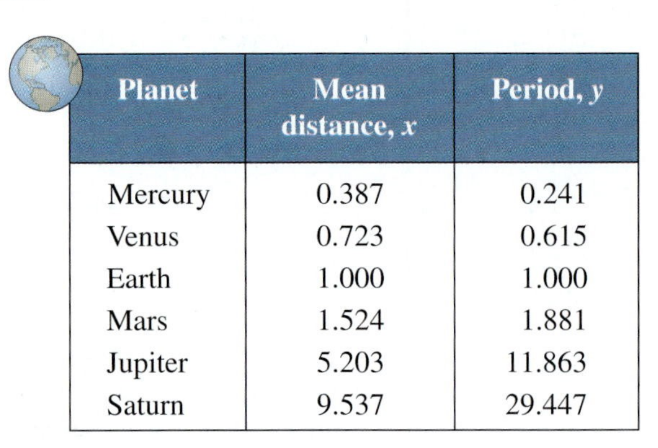

Planet	Mean distance, x	Period, y
Mercury	0.387	0.241
Venus	0.723	0.615
Earth	1.000	1.000
Mars	1.524	1.881
Jupiter	5.203	11.863
Saturn	9.537	29.447

Solution

The points in the table above are plotted in Figure 3.23. From this figure it is not clear how to find an equation that relates y and x. To solve this problem, take the natural logarithm of each of the x- and y-values in the table. This produces the following results.

Planet	Mercury	Venus	Earth	Mars	Jupiter	Saturn
$\ln x$	-0.949	-0.324	0.000	0.421	1.649	2.255
$\ln y$	-1.423	-0.486	0.000	0.632	2.473	3.383

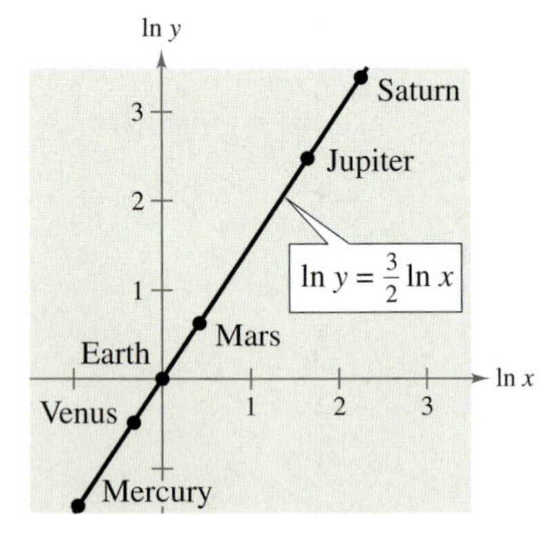

FIGURE 3.24

Now, by plotting the points in the second table, you can see that all six of the points appear to lie in a line (see Figure 3.24). Choose any two points to determine the slope of the line. Using the two points (0.421, 0.632) and (0, 0), you can determine that the slope of the line is

$$m = \frac{0.632 - 0}{0.421 - 0} \approx 1.5 = \frac{3}{2}.$$

By the point-slope form, the equation of the line is $Y = \frac{3}{2}X$, where $Y = \ln y$ and $X = \ln x$. You can therefore conclude that $\ln y = \frac{3}{2} \ln x$.

CHECKPOINT Now try Exercise 85.

3.3 Exercises

VOCABULARY CHECK:

In Exercises 1 and 2, fill in the blanks.

1. To evaluate a logarithm to any base, you can use the ________ formula.
2. The change-of-base formula for base e is given by $\log_a x =$ ________.

In Exercises 3–5, match the property of logarithms with its name.

3. $\log_a(uv) = \log_a u + \log_a v$ (a) Power Property
4. $\ln u^n = n \ln u$ (b) Quotient Property
5. $\log_a \frac{u}{v} = \log_a u - \log_a v$ (c) Product Property

PREREQUISITE SKILLS REVIEW: Practice and review algebra skills needed for this section at **www.Eduspace.com.**

In Exercises 1–8, rewrite the logarithm as a ratio of (a) common logarithms and (b) natural logarithms.

1. $\log_5 x$
2. $\log_3 x$
3. $\log_{1/5} x$
4. $\log_{1/3} x$
5. $\log_x \frac{3}{10}$
6. $\log_x \frac{3}{4}$
7. $\log_{2.6} x$
8. $\log_{7.1} x$

In Exercises 9–16, evaluate the logarithm using the change-of-base formula. Round your result to three decimal places.

9. $\log_3 7$
10. $\log_7 4$
11. $\log_{1/2} 4$
12. $\log_{1/4} 5$
13. $\log_9 0.4$
14. $\log_{20} 0.125$
15. $\log_{15} 1250$
16. $\log_3 0.015$

In Exercises 17–22, use the properties of logarithms to rewrite and simplify the logarithmic expression.

17. $\log_4 8$
18. $\log_2(4^2 \cdot 3^4)$
19. $\log_5 \frac{1}{250}$
20. $\log \frac{9}{300}$
21. $\ln(5e^6)$
22. $\ln \frac{6}{e^2}$

In Exercises 23–38, find the exact value of the logarithmic expression without using a calculator. (If this is not possible, state the reason.)

23. $\log_3 9$
24. $\log_5 \frac{1}{125}$
25. $\log_2 \sqrt[4]{8}$
26. $\log_6 \sqrt[3]{6}$
27. $\log_4 16^{1.2}$
28. $\log_3 81^{-0.2}$
29. $\log_3(-9)$
30. $\log_2(-16)$
31. $\ln e^{4.5}$
32. $3 \ln e^4$
33. $\ln \frac{1}{\sqrt{e}}$
34. $\ln \sqrt[4]{e^3}$
35. $\ln e^2 + \ln e^5$
36. $2 \ln e^6 - \ln e^5$
37. $\log_5 75 - \log_5 3$
38. $\log_4 2 + \log_4 32$

In Exercises 39–60, use the properties of logarithms to expand the expression as a sum, difference, and/or constant multiple of logarithms. (Assume all variables are positive.)

39. $\log_4 5x$
40. $\log_3 10z$
41. $\log_8 x^4$
42. $\log_{10} \frac{y}{2}$
43. $\log_5 \frac{5}{x}$
44. $\log_6 \frac{1}{z^3}$
45. $\ln \sqrt{z}$
46. $\ln \sqrt[3]{t}$
47. $\ln xyz^2$
48. $\log 4x^2 y$
49. $\ln z(z - 1)^2, \; z > 1$
50. $\ln\left(\frac{x^2 - 1}{x^3}\right), \; x > 1$
51. $\log_2 \frac{\sqrt{a - 1}}{9}, \; a > 1$
52. $\ln \frac{6}{\sqrt{x^2 + 1}}$
53. $\ln \sqrt[3]{\frac{x}{y}}$
54. $\ln \sqrt{\frac{x^2}{y^3}}$
55. $\ln \frac{x^4 \sqrt{y}}{z^5}$
56. $\log_2 \frac{\sqrt{x}\, y^4}{z^4}$
57. $\log_5 \frac{x^2}{y^2 z^3}$
58. $\log_{10} \frac{xy^4}{z^5}$
59. $\ln \sqrt[4]{x^3(x^2 + 3)}$
60. $\ln \sqrt{x^2(x + 2)}$

Additional Example

Solve $2^x = 10$.

Solution

$$2^x = 10$$
$$\ln 2^x = \ln 10$$
$$x \ln 2 = \ln 10$$
$$x = \frac{\ln 10}{\ln 2} \approx 3.322$$

Note: Using the change-of-base formula or the definition of a logarithmic function, you could write this solution as $x = \log_2 10$.

STUDY TIP

Remember that to evaluate a logarithm such as $\log_3 7.5$, you need to use the change-of-base formula.

$$\log_3 7.5 = \frac{\ln 7.5}{\ln 3} \approx 1.834$$

To ensure that students first solve for the unknown variable algebraically and then use their calculators, you can require both exact algebraic solutions and approximate numerical answers.

Example 4 Solving an Exponential Equation

Solve $2(3^{2t-5}) - 4 = 11$ and approximate the result to three decimal places.

Solution

$2(3^{2t-5}) - 4 = 11$ Write original equation.

$2(3^{2t-5}) = 15$ Add 4 to each side.

$3^{2t-5} = \frac{15}{2}$ Divide each side by 2.

$\log_3 3^{2t-5} = \log_3 \frac{15}{2}$ Take log (base 3) of each side.

$2t - 5 = \log_3 \frac{15}{2}$ Inverse Property

$2t = 5 + \log_3 7.5$ Add 5 to each side.

$t = \frac{5}{2} + \frac{1}{2}\log_3 7.5$ Divide each side by 2.

$t \approx 3.417$ Use a calculator.

The solution is $t = \frac{5}{2} + \frac{1}{2}\log_3 7.5 \approx 3.417$. Check this in the original equation.

✓CHECKPOINT Now try Exercise 53.

When an equation involves two or more exponential expressions, you can still use a procedure similar to that demonstrated in Examples 2, 3, and 4. However, the algebra is a bit more complicated.

Example 5 Solving an Exponential Equation of Quadratic Type

Solve $e^{2x} - 3e^x + 2 = 0$.

Algebraic Solution

$e^{2x} - 3e^x + 2 = 0$ Write original equation.

$(e^x)^2 - 3e^x + 2 = 0$ Write in quadratic form.

$(e^x - 2)(e^x - 1) = 0$ Factor.

$e^x - 2 = 0$ Set 1st factor equal to 0.

$x = \ln 2$ Solution

$e^x - 1 = 0$ Set 2nd factor equal to 0.

$x = 0$ Solution

The solutions are $x = \ln 2 \approx 0.693$ and $x = 0$. Check these in the original equation.

✓CHECKPOINT Now try Exercise 67.

Graphical Solution

Use a graphing utility to graph $y = e^{2x} - 3e^x + 2$. Use the *zero* or *root* feature or the *zoom* and *trace* features of the graphing utility to approximate the values of x for which $y = 0$. In Figure 3.25, you can see that the zeros occur at $x = 0$ and at $x \approx 0.693$. So, the solutions are $x = 0$ and $x \approx 0.693$.

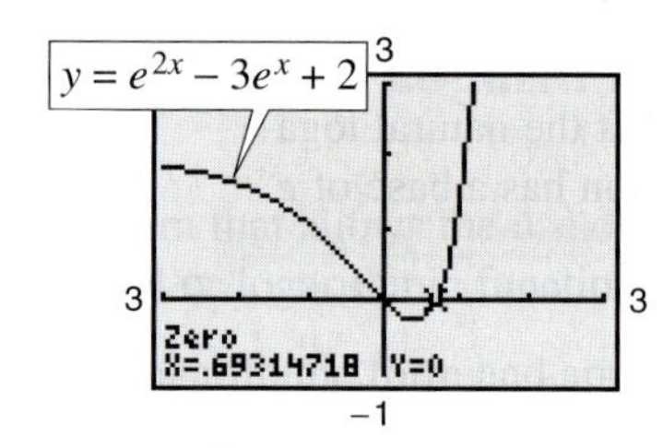

FIGURE 3.25

Solving Logarithmic Equations

To solve a logarithmic equation, you can write it in exponential form.

$\ln x = 3$ Logarithmic form

$e^{\ln x} = e^3$ Exponentiate each side.

$x = e^3$ Exponential form

This procedure is called *exponentiating* each side of an equation.

Example 6 Solving Logarithmic Equations

a. $\ln x = 2$ Original equation

$e^{\ln x} = e^2$ Exponentiate each side.

$x = e^2$ Inverse Property

b. $\log_3(5x - 1) = \log_3(x + 7)$ Original equation

$5x - 1 = x + 7$ One-to-One Property

$4x = 8$ Add $-x$ and 1 to each side.

$x = 2$ Divide each side by 4.

c. $\log_6(3x + 14) - \log_6 5 = \log_6 2x$ Original equation

$\log_6\left(\frac{3x + 14}{5}\right) = \log_6 2x$ Quotient Property of Logarithms

$\frac{3x + 14}{5} = 2x$ One-to-One Property

$3x + 14 = 10x$ Cross multiply.

$-7x = -14$ Isolate x.

$x = 2$ Divide each side by -7.

✓CHECKPOINT Now try Exercise 77.

STUDY TIP

Remember to check your solutions in the original equation when solving equations to verify that the answer is correct and to make sure that the answer lies in the domain of the original equation.

Example 7 Solving a Logarithmic Equation

Solve $5 + 2 \ln x = 4$ and approximate the result to three decimal places.

Solution

$5 + 2 \ln x = 4$ Write original equation.

$2 \ln x = -1$ Subtract 5 from each side.

$\ln x = -\frac{1}{2}$ Divide each side by 2.

$e^{\ln x} = e^{-1/2}$ Exponentiate each side.

$x = e^{-1/2}$ Inverse Property

$x \approx 0.607$ Use a calculator.

✓CHECKPOINT Now try Exercise 85.

Activities

1. Solve for x: $7^x = 3$.

   Answer: $x = \frac{\ln 3}{\ln 7} \approx 0.5646$

2. Solve for x: $\log(x + 4) + \log(x + 1) = 1$.

   Answer: $x = 1$ ($x = -6$ is not in the domain.)

118. ***Data Analysis*** An object at a temperature of 160°C was removed from a furnace and placed in a room at 20°C. The temperature T of the object was measured each hour h and recorded in the table. A model for the data is given by $T = 20[1 + 7(2^{-h})]$. The graph of this model is shown in the figure.

Hour, h	Temperature, T
0	160°
1	90°
2	56°
3	38°
4	29°
5	24°

(a) Use the graph to identify the horizontal asymptote of the model and interpret the asymptote in the context of the problem.

(b) Use the model to approximate the time when the temperature of the object was 100°C.

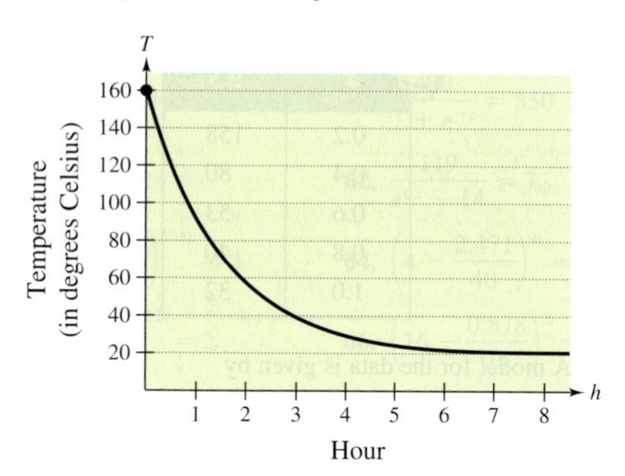

Synthesis

True or False? **In Exercises 119–122, rewrite each verbal statement as an equation. Then decide whether the statement is true or false. Justify your answer.**

119. The logarithm of the product of two numbers is equal to the sum of the logarithms of the numbers.

120. The logarithm of the sum of two numbers is equal to the product of the logarithms of the numbers.

121. The logarithm of the difference of two numbers is equal to the difference of the logarithms of the numbers.

122. The logarithm of the quotient of two numbers is equal to the difference of the logarithms of the numbers.

123. ***Think About It*** Is it possible for a logarithmic equation to have more than one extraneous solution? Explain.

124. ***Finance*** You are investing P dollars at an annual interest rate of r, compounded continuously, for t years. Which of the following would result in the highest value of the investment? Explain your reasoning.

(a) Double the amount you invest.

(b) Double your interest rate.

(c) Double the number of years.

125. ***Think About It*** Are the times required for the investments in Exercises 107 and 108 to quadruple twice as long as the times for them to double? Give a reason for your answer and verify your answer algebraically.

126. ***Writing*** Write two or three sentences stating the general guidelines that you follow when solving (a) exponential equations and (b) logarithmic equations.

Skills Review

In Exercises 127–130, simplify the expression.

127. $\sqrt{48x^2y^5}$

128. $\sqrt{32} - 2\sqrt{25}$

129. $\sqrt[3]{25} \cdot \sqrt[3]{15}$

130. $\dfrac{3}{\sqrt{10} - 2}$

In Exercises 131–134, sketch a graph of the function.

131. $f(x) = |x| + 9$

132. $f(x) = |x + 2| - 8$

133. $g(x) = \begin{cases} 2x, & x < 0 \\ -x^2 + 4, & x \geq 0 \end{cases}$

134. $g(x) = \begin{cases} x - 3, & x \leq -1 \\ x^2 + 1, & x > -1 \end{cases}$

In Exercises 135–138, evaluate the logarithm using the change-of-base formula. Approximate your result to three decimal places.

135. $\log_6 9$

136. $\log_3 4$

137. $\log_{3/4} 5$

138. $\log_8 22$

3.5 Exponential and Logarithmic Models

What you should learn

- Recognize the five most common types of models involving exponential and logarithmic functions.
- Use exponential growth and decay functions to model and solve real-life problems.
- Use Gaussian functions to model and solve real-life problems.
- Use logistic growth functions to model and solve real-life problems.
- Use logarithmic functions to model and solve real-life problems.

Why you should learn it

Exponential growth and decay models are often used to model the population of a country. For instance, in Exercise 36 on page 265, you will use exponential growth and decay models to compare the populations of several countries.

Alan Becker/Getty Images

Introduction

The five most common types of mathematical models involving exponential functions and logarithmic functions are as follows.

1. **Exponential growth model:** $y = ae^{bx}, \quad b > 0$
2. **Exponential decay model:** $y = ae^{-bx}, \quad b > 0$
3. **Gaussian model:** $y = ae^{-(x-b)^2/c}$
4. **Logistic growth model:** $y = \dfrac{a}{1 + be^{-rx}}$
5. **Logarithmic models:** $y = a + b \ln x, \quad y = a + b \log x$

The basic shapes of the graphs of these functions are shown in Figure 3.29.

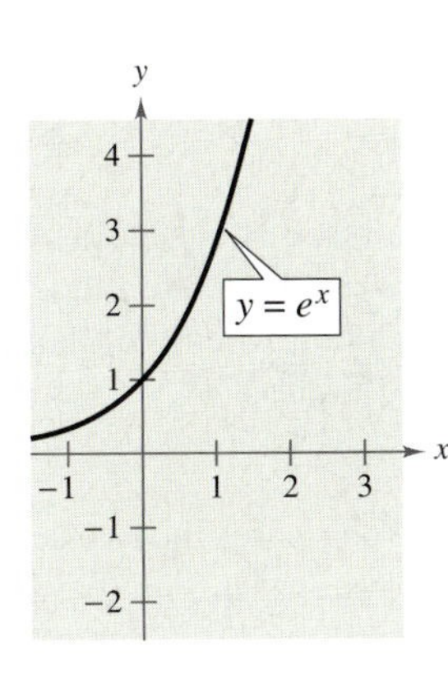

EXPONENTIAL GROWTH MODEL

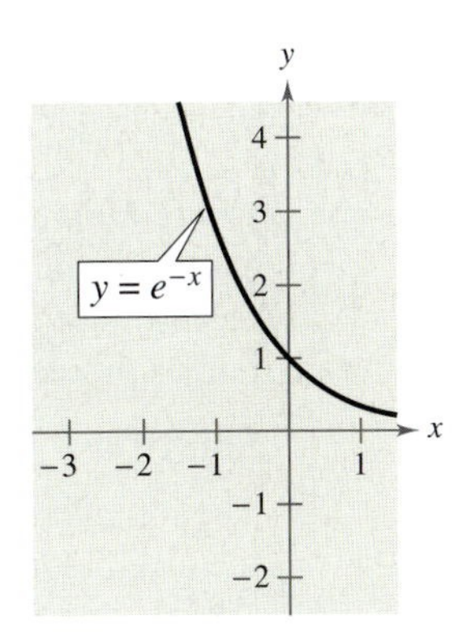

EXPONENTIAL DECAY MODEL

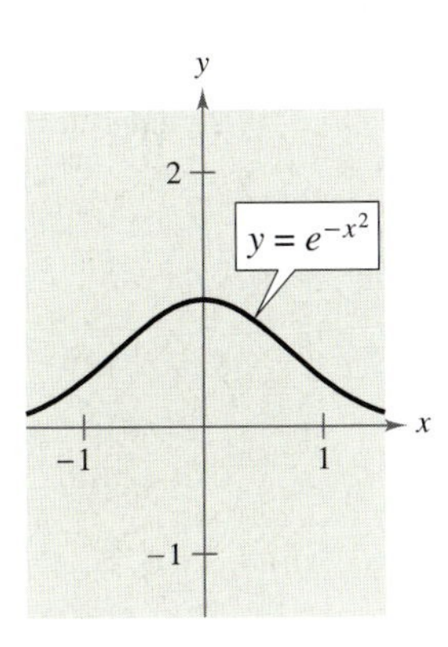

GAUSSIAN MODEL

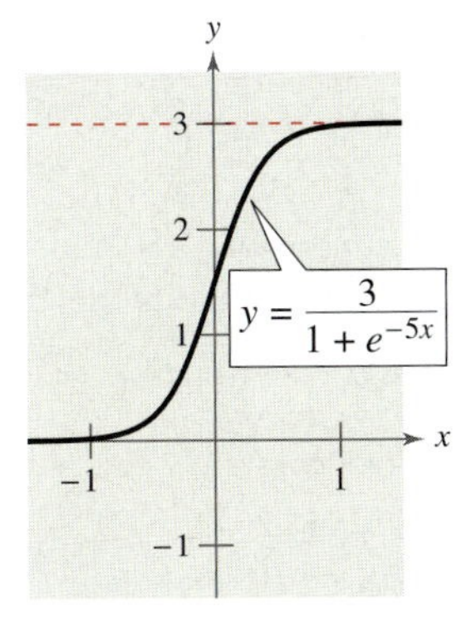

LOGISTIC GROWTH MODEL

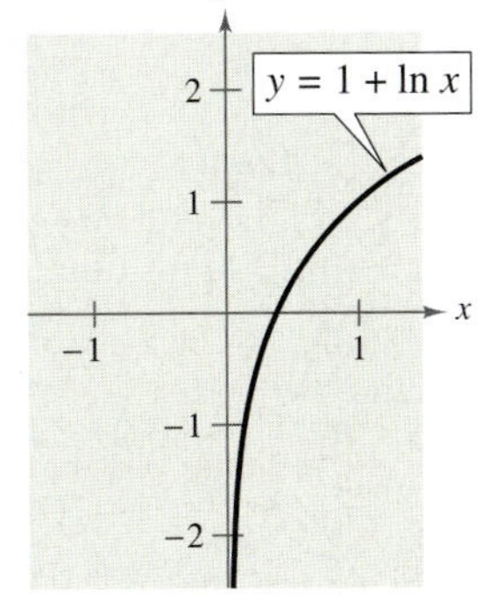

NATURAL LOGARITHMIC MODEL

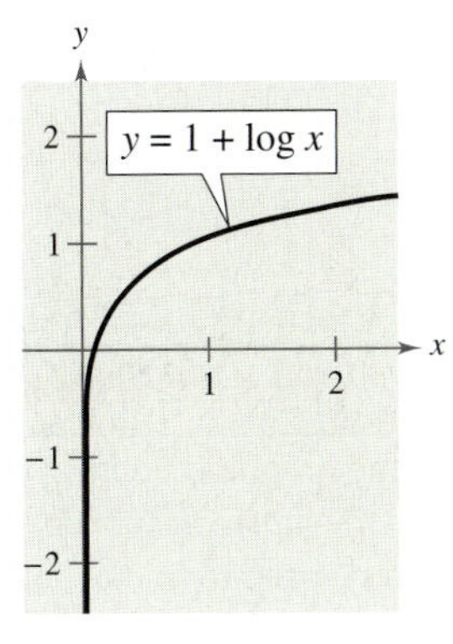

COMMON LOGARITHMIC MODEL

FIGURE 3.29

You can often gain quite a bit of insight into a situation modeled by an exponential or logarithmic function by identifying and interpreting the function's asymptotes. Use the graphs in Figure 3.29 to identify the asymptotes of the graph of each function.

This section shows students real-world applications of logarithmic and exponential functions.

Exponential Growth and Decay

Example 1 Digital Television

Estimates of the numbers (in millions) of U.S. households with digital television from 2003 through 2007 are shown in the table. The scatter plot of the data is shown in Figure 3.30. (Source: eMarketer)

Year	Households
2003	44.2
2004	49.0
2005	55.5
2006	62.5
2007	70.3

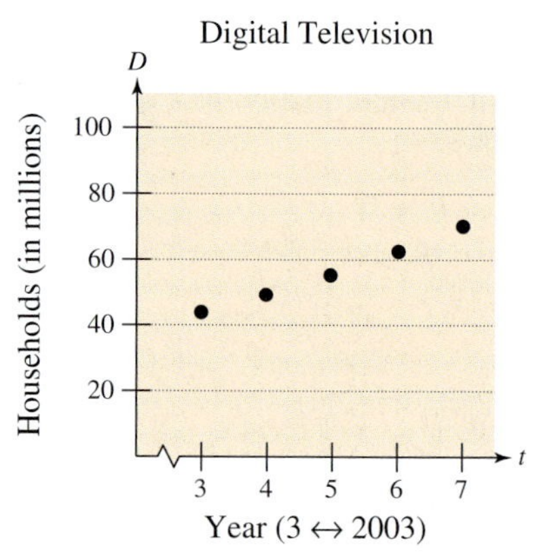

FIGURE 3.30

An exponential growth model that approximates these data is given by

$$D = 30.92e^{0.1171t}, \quad 3 \le t \le 7$$

where D is the number of households (in millions) and $t = 3$ represents 2003. Compare the values given by the model with the estimates shown in the table. According to this model, when will the number of U.S. households with digital television reach 100 million?

Solution

The following table compares the two sets of figures. The graph of the model and the original data are shown in Figure 3.31.

Year	2003	2004	2005	2006	2007
Households	44.2	49.0	55.5	62.5	70.3
Model	43.9	49.4	55.5	62.4	70.2

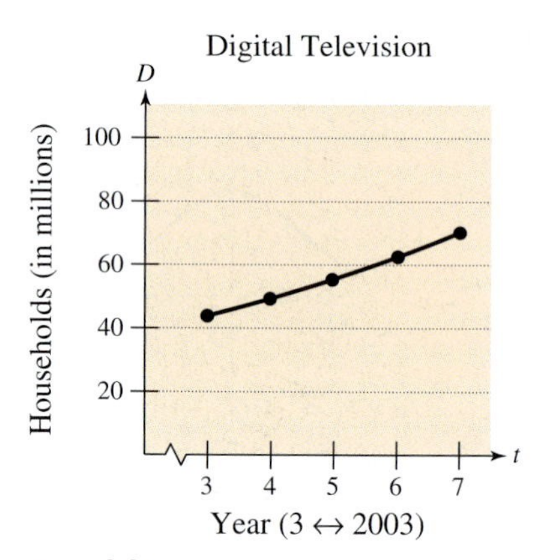

FIGURE 3.31

To find when the number of U.S. households with digital television will reach 100 million, let $D = 100$ in the model and solve for t.

$$30.92e^{0.1171t} = D \quad \text{Write original model.}$$

$$30.92e^{0.1171t} = 100 \quad \text{Let } D = 100.$$

$$e^{0.1171t} \approx 3.2342 \quad \text{Divide each side by 30.92.}$$

$$\ln e^{0.1171t} \approx \ln 3.2342 \quad \text{Take natural log of each side.}$$

$$0.1171t \approx 1.1738 \quad \text{Inverse Property}$$

$$t \approx 10.0 \quad \text{Divide each side by 0.1171.}$$

According to the model, the number of U.S. households with digital television will reach 100 million in 2010.

CHECKPOINT Now try Exercise 35.

Technology

Some graphing utilities have an *exponential regression* feature that can be used to find exponential models that represent data. If you have such a graphing utility, try using it to find an exponential model for the data given in Example 1. How does your model compare with the model given in Example 1?

Additional Example

Radioactive iodine is a by-product of some types of nuclear reactors. Its half-life is about 60 days. That is, after 60 days, a given amount of radioactive iodine will have decayed to half the original amount. Suppose a contained nuclear accident occurs and gives off an initial amount C of radioactive iodine.

a. Write an equation for the amount of radioactive iodine present at any time t following the accident.

b. How long will it take for the radioactive iodine to decay to a level of 20% of the original amount?

Solution

a. Knowing that half the original amount remains after 60 days, you can use the exponential decay model $y = ae^{-bt}$ to obtain

$$\frac{1}{2}C = Ce^{-b(60)}$$

$$\frac{1}{2} = e^{-60b}$$

$$-\ln 2 = -60b$$

$$b = \frac{\ln 2}{60} \approx 0.0116.$$

So, $y = Ce^{-0.0116t}$.

b. The time required for the radioactive iodine to decay to 20% of the original amount is

$$Ce^{-0.0116t} = (0.2)C$$

$$e^{-0.0116t} = 0.2$$

$$-0.0116t = \ln 0.2$$

$$t = \frac{\ln 0.2}{-0.0116} \approx 139 \text{ days.}$$

In Example 1, you were given the exponential growth model. But suppose this model were not given; how could you find such a model? One technique for doing this is demonstrated in Example 2.

Example 2 Modeling Population Growth 

In a research experiment, a population of fruit flies is increasing according to the law of exponential growth. After 2 days there are 100 flies, and after 4 days there are 300 flies. How many flies will there be after 5 days?

Solution

Let y be the number of flies at time t. From the given information, you know that $y = 100$ when $t = 2$ and $y = 300$ when $t = 4$. Substituting this information into the model $y = ae^{bt}$ produces

$$100 = ae^{2b} \quad \text{and} \quad 300 = ae^{4b}.$$

To solve for b, solve for a in the first equation.

$$100 = ae^{2b} \quad \Rightarrow \quad a = \frac{100}{e^{2b}} \qquad \text{Solve for } a \text{ in the first equation.}$$

Then substitute the result into the second equation.

$$300 = ae^{4b} \qquad \text{Write second equation.}$$

$$300 = \left(\frac{100}{e^{2b}}\right)e^{4b} \qquad \text{Substitute } 100/e^{2b} \text{ for } a.$$

$$\frac{300}{100} = e^{2b} \qquad \text{Divide each side by 100.}$$

$$\ln 3 = 2b \qquad \text{Take natural log of each side.}$$

$$\frac{1}{2}\ln 3 = b \qquad \text{Solve for } b.$$

Using $b = \frac{1}{2}\ln 3$ and the equation you found for a, you can determine that

$$a = \frac{100}{e^{2[(1/2)\ln 3]}} \qquad \text{Substitute } \tfrac{1}{2}\ln 3 \text{ for } b.$$

$$= \frac{100}{e^{\ln 3}} \qquad \text{Simplify.}$$

$$= \frac{100}{3} \qquad \text{Inverse Property}$$

$$\approx 33.33. \qquad \text{Simplify.}$$

So, with $a \approx 33.33$ and $b = \frac{1}{2}\ln 3 \approx 0.5493$, the exponential growth model is

$$y = 33.33e^{0.5493t}$$

as shown in Figure 3.32. This implies that, after 5 days, the population will be

$$y = 33.33e^{0.5493(5)} \approx 520 \text{ flies.}$$

CHECKPOINT Now try Exercise 37.

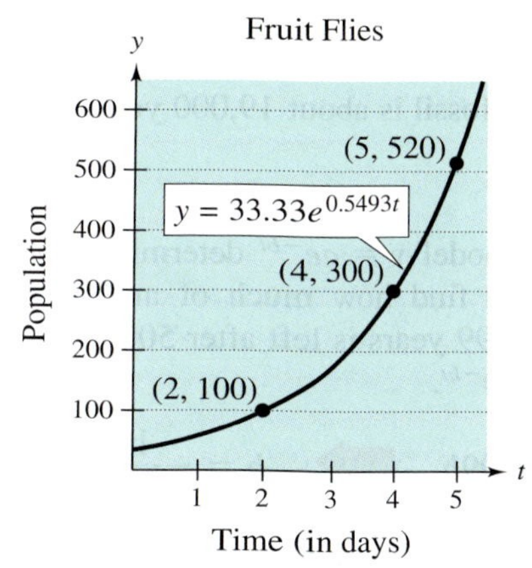

FIGURE 3.32

4.1 Radian and Degree Measure

What you should learn

- Describe angles.
- Use radian measure.
- Use degree measure.
- Use angles to model and solve real-life problems.

Why you should learn it

You can use angles to model and solve real-life problems. For instance, in Exercise 108 on page 293, you are asked to use angles to find the speed of a bicycle.

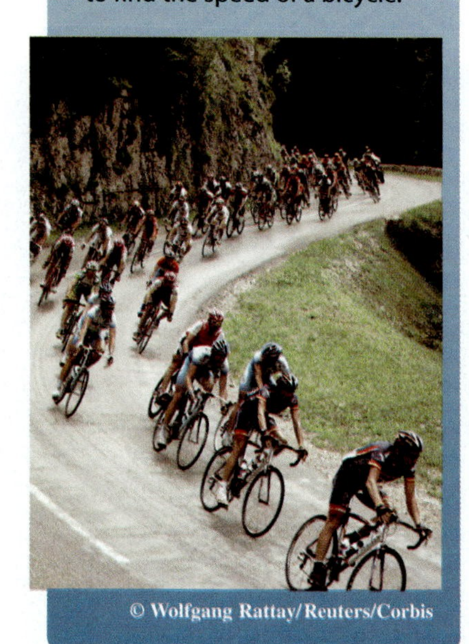

© Wolfgang Rattay/Reuters/Corbis

Angles

As derived from the Greek language, the word **trigonometry** means "measurement of triangles." Initially, trigonometry dealt with relationships among the sides and angles of triangles and was used in the development of astronomy, navigation, and surveying. With the development of calculus and the physical sciences in the 17th century, a different perspective arose—one that viewed the classic trigonometric relationships as *functions* with the set of real numbers as their domains. Consequently, the applications of trigonometry expanded to include a vast number of physical phenomena involving rotations and vibrations. These phenomena include sound waves, light rays, planetary orbits, vibrating strings, pendulums, and orbits of atomic particles.

The approach in this text incorporates *both* perspectives, starting with angles and their measure.

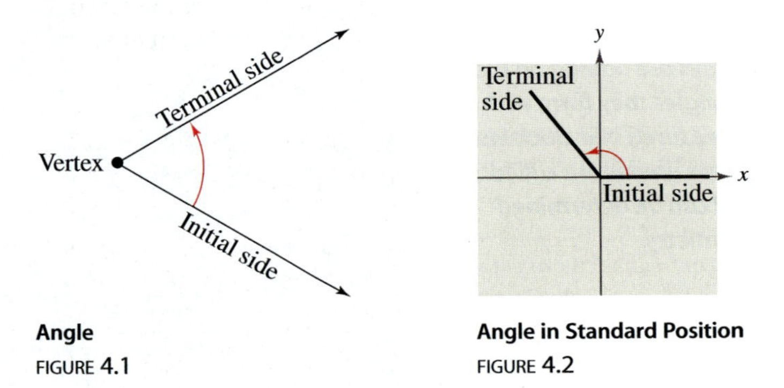

Angle
FIGURE 4.1

Angle in Standard Position
FIGURE 4.2

An **angle** is determined by rotating a ray (half-line) about its endpoint. The starting position of the ray is the **initial side** of the angle, and the position after rotation is the **terminal side,** as shown in Figure 4.1. The endpoint of the ray is the **vertex** of the angle. This perception of an angle fits a coordinate system in which the origin is the vertex and the initial side coincides with the positive x-axis. Such an angle is in **standard position,** as shown in Figure 4.2. **Positive angles** are generated by counterclockwise rotation, and **negative angles** by clockwise rotation, as shown in Figure 4.3. Angles are labeled with Greek letters α (alpha), β (beta), and θ (theta), as well as uppercase letters A, B, and C. In Figure 4.4, note that angles α and β have the same initial and terminal sides. Such angles are **coterminal.**

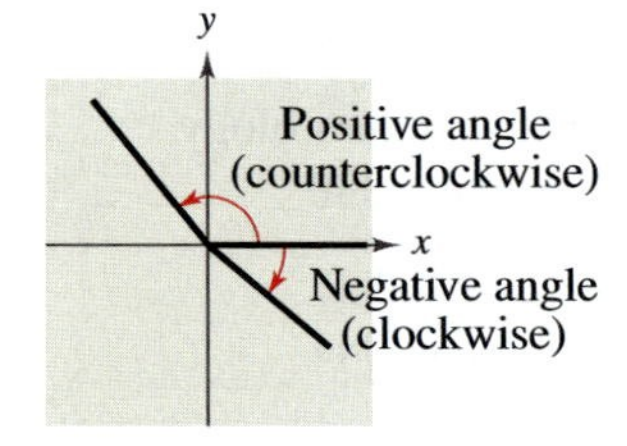

FIGURE 4.3

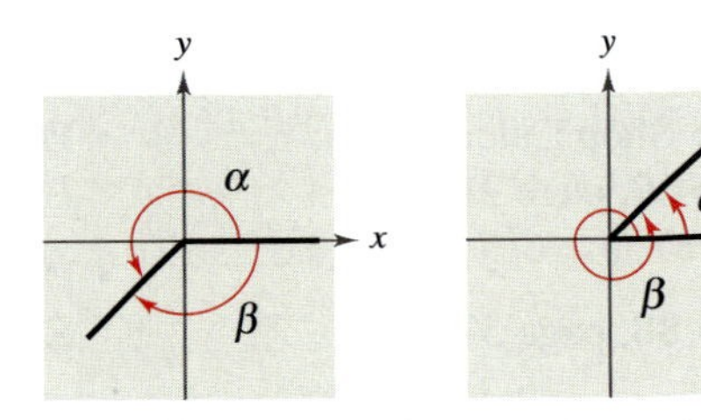

FIGURE 4.4 *Coterminal Angles*

The *HM mathSpace®* CD-ROM and *Eduspace®* for this text contain additional resources related to the concepts discussed in this chapter.

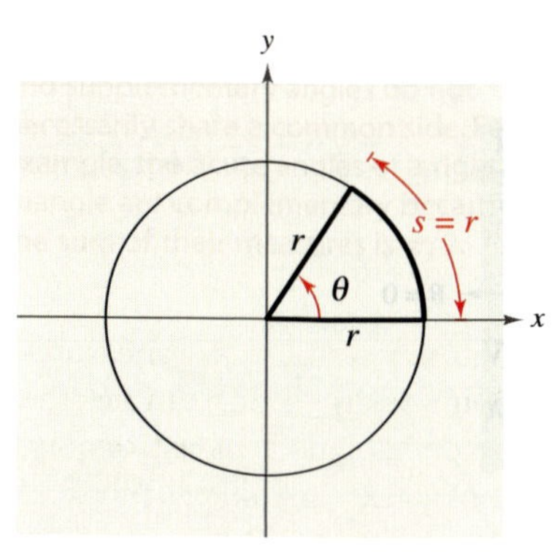

Arc length = radius when θ = 1 radian
FIGURE 4.5

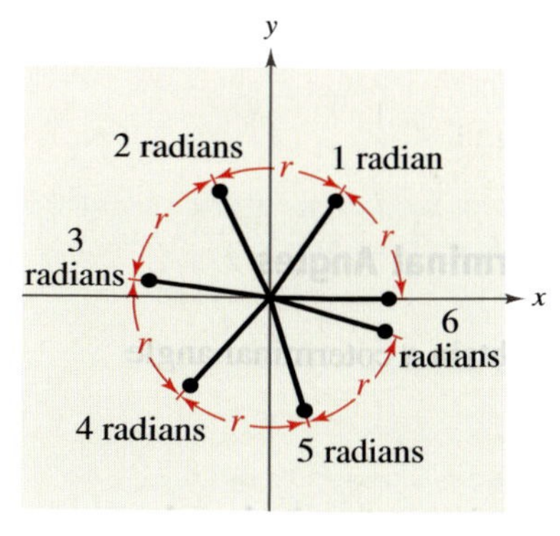

FIGURE 4.6

STUDY TIP

One revolution around a circle of radius r corresponds to an angle of 2π radians because

$$\theta = \frac{s}{r} = \frac{2\pi r}{r} = 2\pi \text{ radians.}$$

Radian Measure

The **measure of an angle** is determined by the amount of rotation from the initial side to the terminal side. One way to measure angles is in *radians*. This type of measure is especially useful in calculus. To define a radian, you can use a **central angle** of a circle, one whose vertex is the center of the circle, as shown in Figure 4.5.

Definition of Radian

One **radian** is the measure of a central angle θ that intercepts an arc s equal in length to the radius r of the circle. See Figure 4.5. Algebraically, this means that

$$\theta = \frac{s}{r}$$

where θ is measured in radians.

Because the circumference of a circle is $2\pi r$ units, it follows that a central angle of one full revolution (counterclockwise) corresponds to an arc length of

$$s = 2\pi r.$$

Moreover, because $2\pi \approx 6.28$, there are just over six radius lengths in a full circle, as shown in Figure 4.6. Because the units of measure for s and r are the same, the ratio s/r has no units—it is simply a real number.

Because the radian measure of an angle of one full revolution is 2π, you can obtain the following.

$$\frac{1}{2} \text{ revolution} = \frac{2\pi}{2} = \pi \text{ radians}$$

$$\frac{1}{4} \text{ revolution} = \frac{2\pi}{4} = \frac{\pi}{2} \text{ radians}$$

$$\frac{1}{6} \text{ revolution} = \frac{2\pi}{6} = \frac{\pi}{3} \text{ radians}$$

These and other common angles are shown in Figure 4.7.

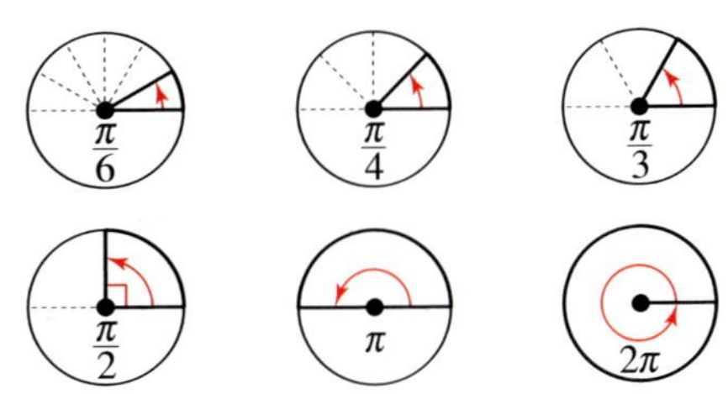

FIGURE 4.7

Recall that the four quadrants in a coordinate system are numbered I, II, III, and IV. Figure 4.8 on page 284 shows which angles between 0 and 2π lie in each of the four quadrants. Note that angles between 0 and $\pi/2$ are **acute** angles and angles between $\pi/2$ and π are **obtuse** angles.

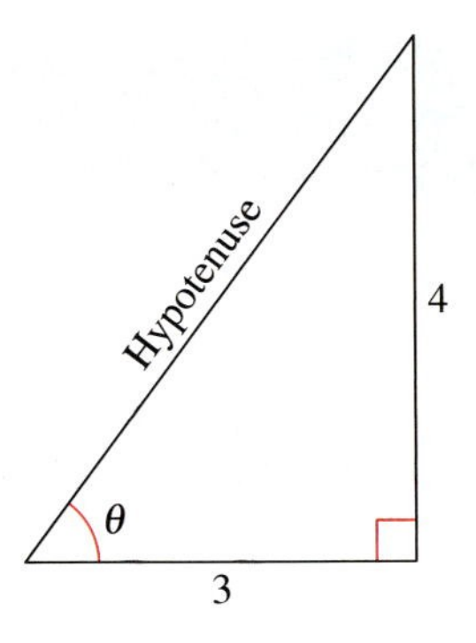

FIGURE 4.27

You may wish to review the Pythagorean Theorem before presenting the examples in this section.

Example 1 **Evaluating Trigonometric Functions**

Use the triangle in Figure 4.27 to find the values of the six trigonometric functions of θ.

Solution

By the Pythagorean Theorem, $(\text{hyp})^2 = (\text{opp})^2 + (\text{adj})^2$, it follows that

$$\begin{aligned}\text{hyp} &= \sqrt{4^2 + 3^2}\\ &= \sqrt{25}\\ &= 5.\end{aligned}$$

So, the six trigonometric functions of θ are

$$\sin\theta = \frac{\text{opp}}{\text{hyp}} = \frac{4}{5} \qquad \csc\theta = \frac{\text{hyp}}{\text{opp}} = \frac{5}{4}$$

$$\cos\theta = \frac{\text{adj}}{\text{hyp}} = \frac{3}{5} \qquad \sec\theta = \frac{\text{hyp}}{\text{adj}} = \frac{5}{3}$$

$$\tan\theta = \frac{\text{opp}}{\text{adj}} = \frac{4}{3} \qquad \cot\theta = \frac{\text{adj}}{\text{opp}} = \frac{3}{4}.$$

✓CHECKPOINT Now try Exercise 3.

Historical Note
Georg Joachim Rhaeticus (1514–1576) was the leading Teutonic mathematical astronomer of the 16th century. He was the first to define the trigonometric functions as ratios of the sides of a right triangle.

In Example 1, you were given the lengths of two sides of the right triangle, but not the angle θ. Often, you will be asked to find the trigonometric functions of a *given* acute angle θ. To do this, construct a right triangle having θ as one of its angles.

Example 2 **Evaluating Trigonometric Functions of 45°**

Find the values of sin 45°, cos 45°, and tan 45°.

Solution

Construct a right triangle having 45° as one of its acute angles, as shown in Figure 4.28. Choose the length of the adjacent side to be 1. From geometry, you know that the other acute angle is also 45°. So, the triangle is isosceles and the length of the opposite side is also 1. Using the Pythagorean Theorem, you find the length of the hypotenuse to be $\sqrt{2}$.

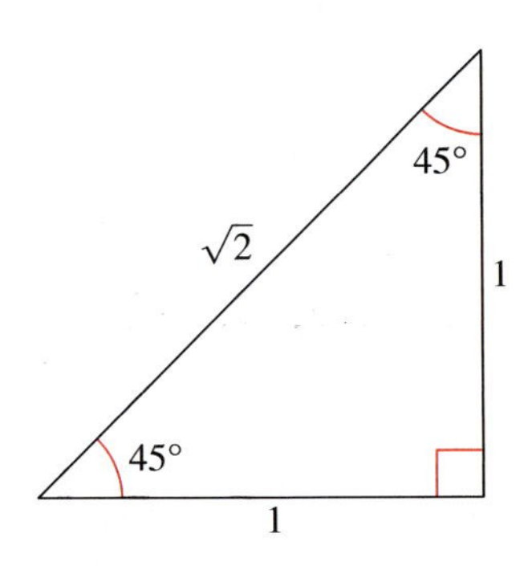

FIGURE 4.28

$$\sin 45° = \frac{\text{opp}}{\text{hyp}} = \frac{1}{\sqrt{2}} = \frac{\sqrt{2}}{2}$$

$$\cos 45° = \frac{\text{adj}}{\text{hyp}} = \frac{1}{\sqrt{2}} = \frac{\sqrt{2}}{2}$$

$$\tan 45° = \frac{\text{opp}}{\text{adj}} = \frac{1}{1} = 1$$

✓CHECKPOINT Now try Exercise 17.

STUDY TIP

Because the angles 30°, 45°, and 60° ($\pi/6$, $\pi/4$, and $\pi/3$) occur frequently in trigonometry, you should learn to construct the triangles shown in Figures 4.28 and 4.29.

Consider having your students construct the triangle in Figure 4.29 with angles in the corresponding radian measures, then find the six trigonometric functions for each of the acute angles.

Technology

You can use a calculator to convert the answers in Example 3 to decimals. However, the radical form is the exact value and in most cases, the exact value is preferred.

The triangles in Figures 4.27, 4.28, and 4.29 are useful problem-solving aids. Encourage your students to draw diagrams when they solve problems similar to those in Examples 1, 2, and 3.

Example 3 Evaluating Trigonometric Functions of 30° and 60°

Use the equilateral triangle shown in Figure 4.29 to find the values of sin 60°, cos 60°, sin 30°, and cos 30°.

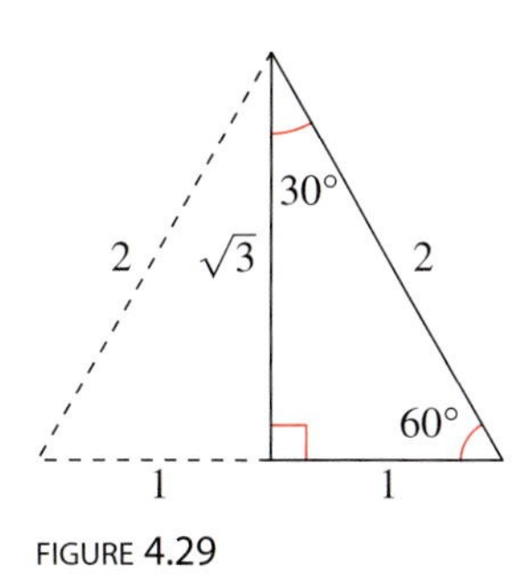

FIGURE 4.29

Solution

Use the Pythagorean Theorem and the equilateral triangle in Figure 4.29 to verify the lengths of the sides shown in the figure. For $\theta = 60°$, you have adj $= 1$, opp $= \sqrt{3}$, and hyp $= 2$. So,

$$\sin 60° = \frac{\text{opp}}{\text{hyp}} = \frac{\sqrt{3}}{2} \quad \text{and} \quad \cos 60° = \frac{\text{adj}}{\text{hyp}} = \frac{1}{2}.$$

For $\theta = 30°$, adj $= \sqrt{3}$, opp $= 1$, and hyp $= 2$. So,

$$\sin 30° = \frac{\text{opp}}{\text{hyp}} = \frac{1}{2} \quad \text{and} \quad \cos 30° = \frac{\text{adj}}{\text{hyp}} = \frac{\sqrt{3}}{2}.$$

CHECKPOINT Now try Exercise 19.

Sines, Cosines, and Tangents of Special Angles

$$\sin 30° = \sin\frac{\pi}{6} = \frac{1}{2} \qquad \cos 30° = \cos\frac{\pi}{6} = \frac{\sqrt{3}}{2} \qquad \tan 30° = \tan\frac{\pi}{6} = \frac{\sqrt{3}}{3}$$

$$\sin 45° = \sin\frac{\pi}{4} = \frac{\sqrt{2}}{2} \qquad \cos 45° = \cos\frac{\pi}{4} = \frac{\sqrt{2}}{2} \qquad \tan 45° = \tan\frac{\pi}{4} = 1$$

$$\sin 60° = \sin\frac{\pi}{3} = \frac{\sqrt{3}}{2} \qquad \cos 60° = \cos\frac{\pi}{3} = \frac{1}{2} \qquad \tan 60° = \tan\frac{\pi}{3} = \sqrt{3}$$

In the box, note that $\sin 30° = \frac{1}{2} = \cos 60°$. This occurs because 30° and 60° are complementary angles. In general, it can be shown from the right triangle definitions that *cofunctions of complementary angles are equal*. That is, if θ is an acute angle, the following relationships are true.

$$\sin(90° - \theta) = \cos\theta \qquad \cos(90° - \theta) = \sin\theta$$

$$\tan(90° - \theta) = \cot\theta \qquad \cot(90° - \theta) = \tan\theta$$

$$\sec(90° - \theta) = \csc\theta \qquad \csc(90° - \theta) = \sec\theta$$

4.3 Exercises

VOCABULARY CHECK:

1. Match the trigonometric function with its right triangle definition.

(a) Sine (b) Cosine (c) Tangent (d) Cosecant (e) Secant (f) Cotangent

(i) $\frac{\text{hypotenuse}}{\text{adjacent}}$ (ii) $\frac{\text{adjacent}}{\text{opposite}}$ (iii) $\frac{\text{hypotenuse}}{\text{opposite}}$ (iv) $\frac{\text{adjacent}}{\text{hypotenuse}}$ (v) $\frac{\text{opposite}}{\text{hypotenuse}}$ (vi) $\frac{\text{opposite}}{\text{adjacent}}$

In Exercises 2 and 3, fill in the blanks.

2. Relative to the angle θ, the three sides of a right triangle are the ________ side, the ________ side, and the ________.
3. An angle that measures from the horizontal upward to an object is called the angle of ________, whereas an angle that measures from the horizontal downward to an object is called the angle of ________.

PREREQUISITE SKILLS REVIEW: Practice and review algebra skills needed for this section at **www.Eduspace.com.**

In Exercises 1–4, find the exact values of the six trigonometric functions of the angle θ shown in the figure. (Use the Pythagorean Theorem to find the third side of the triangle.)

1.

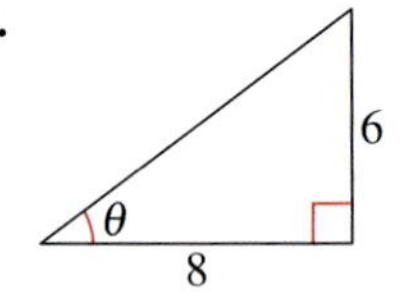

2.

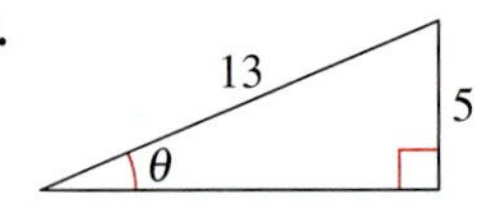

3.

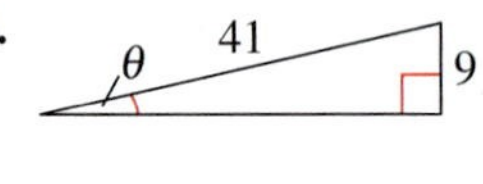

4.

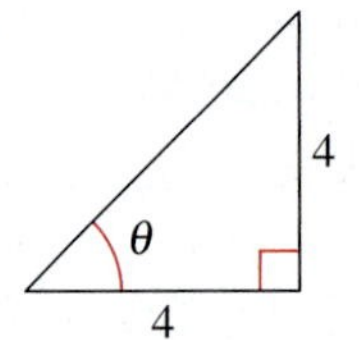

In Exercises 5–8, find the exact values of the six trigonometric functions of the angle θ for each of the two triangles. Explain why the function values are the same.

5.

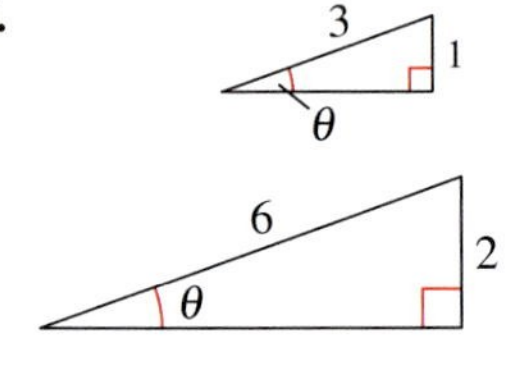

6.

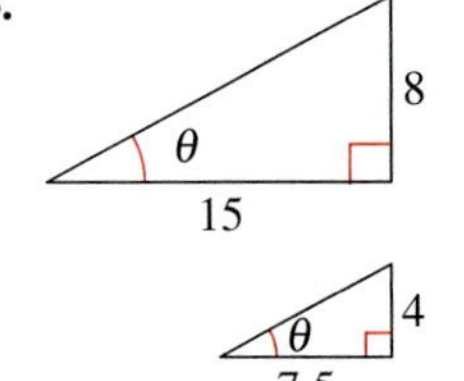

7.

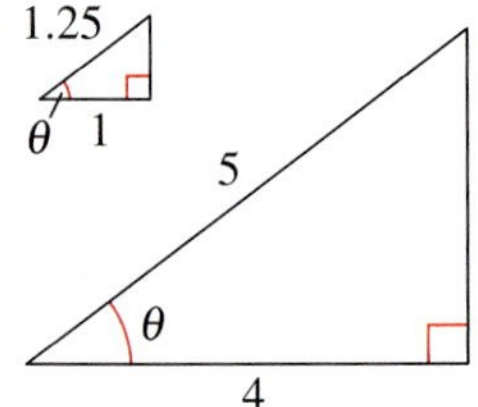

8.

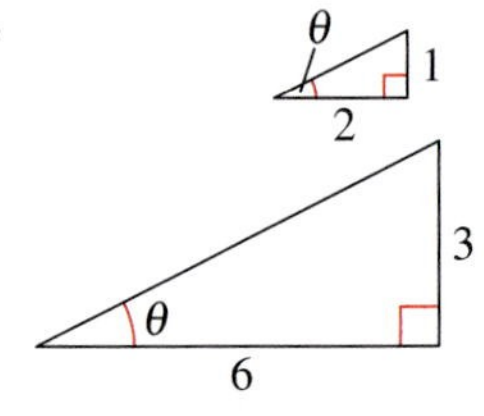

In Exercises 9–16, sketch a right triangle corresponding to the trigonometric function of the acute angle θ. Use the Pythagorean Theorem to determine the third side and then find the other five trigonometric functions of θ.

9. $\sin\theta = \frac{3}{4}$
10. $\cos\theta = \frac{5}{7}$
11. $\sec\theta = 2$
12. $\cot\theta = 5$
13. $\tan\theta = 3$
14. $\sec\theta = 6$
15. $\cot\theta = \frac{3}{2}$
16. $\csc\theta = \frac{17}{4}$

In Exercises 17–26, construct an appropriate triangle to complete the table. ($0 \le \theta \le 90°$, $0 \le \theta \le \pi/2$)

	Function	*θ (deg)*	*θ (rad)*	*Function Value*
17.	sin	30°		
18.	cos	45°		
19.	tan		$\frac{\pi}{3}$	
20.	sec		$\frac{\pi}{4}$	
21.	cot			$\frac{\sqrt{3}}{3}$
22.	csc			$\sqrt{2}$
23.	cos		$\frac{\pi}{6}$	
24.	sin		$\frac{\pi}{4}$	
25.	cot			1
26.	tan			$\frac{\sqrt{3}}{3}$

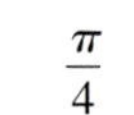

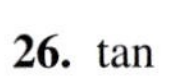

In Exercises 27–32, use the given function value(s), and trigonometric identities (including the cofunction identities), to find the indicated trigonometric functions.

27. $\sin 60^\circ = \dfrac{\sqrt{3}}{2}$, $\cos 60^\circ = \dfrac{1}{2}$

(a) $\tan 60^\circ$ (b) $\sin 30^\circ$

(c) $\cos 30^\circ$ (d) $\cot 60^\circ$

28. $\sin 30^\circ = \dfrac{1}{2}$, $\tan 30^\circ = \dfrac{\sqrt{3}}{3}$

(a) $\csc 30^\circ$ (b) $\cot 60^\circ$

(c) $\cos 30^\circ$ (d) $\cot 30^\circ$

29. $\csc \theta = \dfrac{\sqrt{13}}{2}$, $\sec \theta = \dfrac{\sqrt{13}}{3}$

(a) $\sin \theta$ (b) $\cos \theta$

(c) $\tan \theta$ (d) $\sec(90^\circ - \theta)$

30. $\sec \theta = 5$, $\tan \theta = 2\sqrt{6}$

(a) $\cos \theta$ (b) $\cot \theta$

(c) $\cot(90^\circ - \theta)$ (d) $\sin \theta$

31. $\cos \alpha = \frac{1}{3}$

(a) $\sec \alpha$ (b) $\sin \alpha$

(c) $\cot \alpha$ (d) $\sin(90^\circ - \alpha)$

32. $\tan \beta = 5$

(a) $\cot \beta$ (b) $\cos \beta$

(c) $\tan(90^\circ - \beta)$ (d) $\csc \beta$

In Exercises 33–42, use trigonometric identities to transform the left side of the equation into the right side ($0 < \theta < \pi/2$).

33. $\tan \theta \cot \theta = 1$

34. $\cos \theta \sec \theta = 1$

35. $\tan \alpha \cos \alpha = \sin \alpha$

36. $\cot \alpha \sin \alpha = \cos \alpha$

37. $(1 + \cos \theta)(1 - \cos \theta) = \sin^2 \theta$

38. $(1 + \sin \theta)(1 - \sin \theta) = \cos^2 \theta$

39. $(\sec \theta + \tan \theta)(\sec \theta - \tan \theta) = 1$

40. $\sin^2 \theta - \cos^2 \theta = 2 \sin^2 \theta - 1$

41. $\dfrac{\sin \theta}{\cos \theta} + \dfrac{\cos \theta}{\sin \theta} = \csc \theta \sec \theta$

42. $\dfrac{\tan \beta + \cot \beta}{\tan \beta} = \csc^2 \beta$

In Exercises 43–52, use a calculator to evaluate each function. Round your answers to four decimal places. (Be sure the calculator is in the correct angle mode.)

43. (a) $\sin 10^\circ$ (b) $\cos 80^\circ$

44. (a) $\tan 23.5^\circ$ (b) $\cot 66.5^\circ$

45. (a) $\sin 16.35^\circ$ (b) $\csc 16.35^\circ$

46. (a) $\cos 16^\circ\, 18'$ (b) $\sin 73^\circ\, 56'$

47. (a) $\sec 42^\circ\, 12'$ (b) $\csc 48^\circ\, 7'$

48. (a) $\cos 4^\circ\, 50'\, 15''$ (b) $\sec 4^\circ\, 50'\, 15''$

49. (a) $\cot 11^\circ\, 15'$ (b) $\tan 11^\circ\, 15'$

50. (a) $\sec 56^\circ\, 8\, 10''$ (b) $\cos 56^\circ\, 8\, 10''$

51. (a) $\csc 32^\circ\, 40'\, 3''$ (b) $\tan 44^\circ\, 28\, 16''$

52. (a) $\sec\left(\frac{9}{5} \cdot 20 + 32\right)^\circ$ (b) $\cot\left(\frac{9}{5} \cdot 30 + 32\right)^\circ$

In Exercises 53–58, find the values of θ in degrees ($0^\circ < \theta < 90^\circ$) and radians ($0 < \theta < \pi/2$) without the aid of a calculator.

53. (a) $\sin \theta = \dfrac{1}{2}$ (b) $\csc \theta = 2$

54. (a) $\cos \theta = \dfrac{\sqrt{2}}{2}$ (b) $\tan \theta = 1$

55. (a) $\sec \theta = 2$ (b) $\cot \theta = 1$

56. (a) $\tan \theta = \sqrt{3}$ (b) $\cos \theta = \dfrac{1}{2}$

57. (a) $\csc \theta = \dfrac{2\sqrt{3}}{3}$ (b) $\sin \theta = \dfrac{\sqrt{2}}{2}$

58. (a) $\cot \theta = \dfrac{\sqrt{3}}{3}$ (b) $\sec \theta = \sqrt{2}$

In Exercises 59–62, solve for *x*, *y*, or *r* as indicated.

59. Solve for x.

30

30°

x

60. Solve for y.

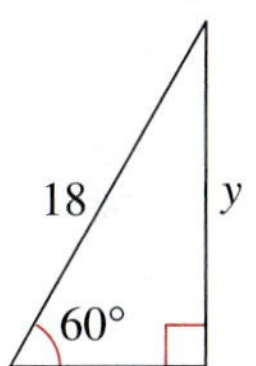

61. Solve for x.

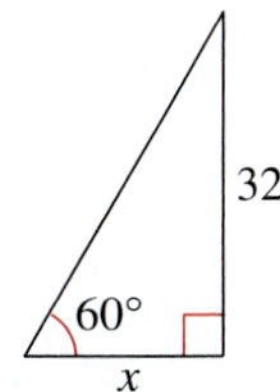

62. Solve for r.

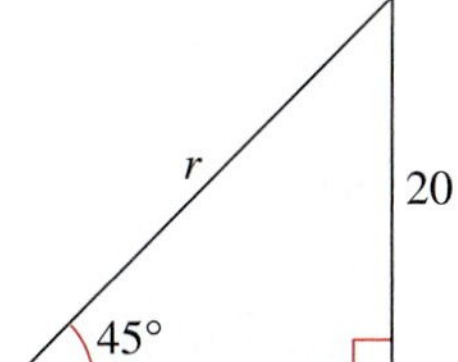

63. ***Empire State Building*** You are standing 45 meters from the base of the Empire State Building. You estimate that the angle of elevation to the top of the 86th floor (the observatory) is 82°. If the total height of the building is another 123 meters above the 86th floor, what is the approximate height of the building? One of your friends is on the 86th floor. What is the distance between you and your friend?

64. ***Height*** A six-foot person walks from the base of a broadcasting tower directly toward the tip of the shadow cast by the tower. When the person is 132 feet from the tower and 3 feet from the tip of the shadow, the person's shadow starts to appear beyond the tower's shadow.

(a) Draw a right triangle that gives a visual representation of the problem. Show the known quantities of the triangle and use a variable to indicate the height of the tower.

(b) Use a trigonometric function to write an equation involving the unknown quantity.

(c) What is the height of the tower?

65. ***Angle of Elevation*** You are skiing down a mountain with a vertical height of 1500 feet. The distance from the top of the mountain to the base is 3000 feet. What is the angle of elevation from the base to the top of the mountain?

66. ***Width of a River*** A biologist wants to know the width w of a river so in order to properly set instruments for studying the pollutants in the water. From point A, the biologist walks downstream 100 feet and sights to point C (see figure). From this sighting, it is determined that $\theta = 54°$. How wide is the river?

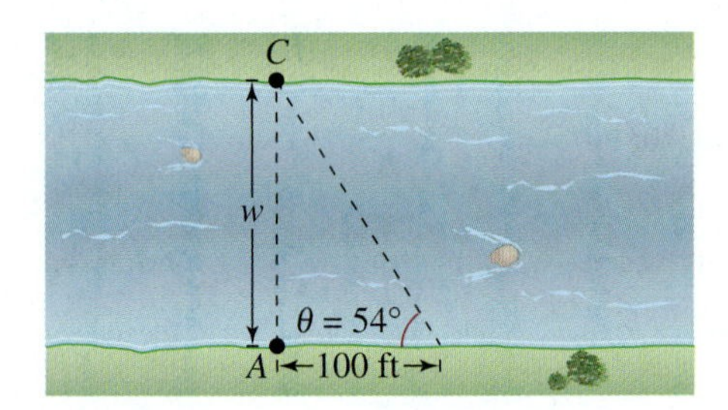

67. ***Length*** A steel cable zip-line is being constructed for a competition on a reality television show. One end of the zip-line is attached to a platform on top of a 150-foot pole. The other end of the zip-line is attached to the top of a 5-foot stake. The angle of elevation to the platform is 23° (see figure).

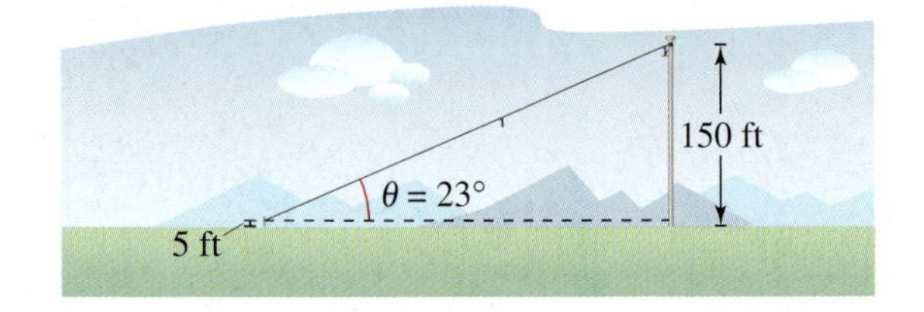

(a) How long is the zip-line?

(b) How far is the stake from the pole?

(c) Contestants take an average of 6 seconds to reach the ground from the top of the zip-line. At what rate are contestants moving down the line? At what rate are they dropping vertically?

68. ***Height of a Mountain*** In traveling across flat land, you notice a mountain directly in front of you. Its angle of elevation (to the peak) is 3.5°. After you drive 13 miles closer to the mountain, the angle of elevation is 9°. Approximate the height of the mountain.

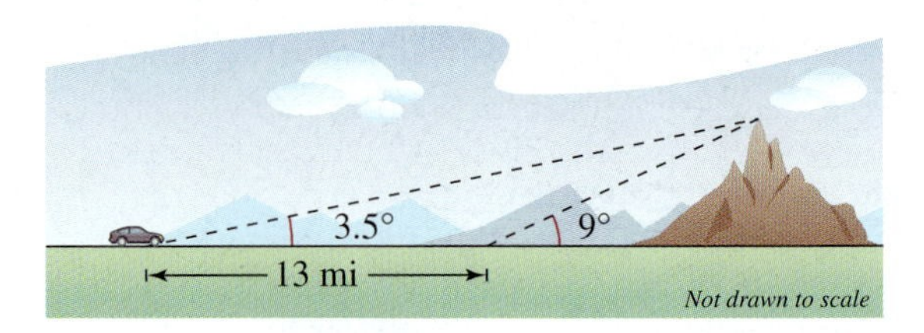

69. ***Machine Shop Calculations*** A steel plate has the form of one-fourth of a circle with a radius of 60 centimeters. Two two-centimeter holes are to be drilled in the plate positioned as shown in the figure. Find the coordinates of the center of each hole.

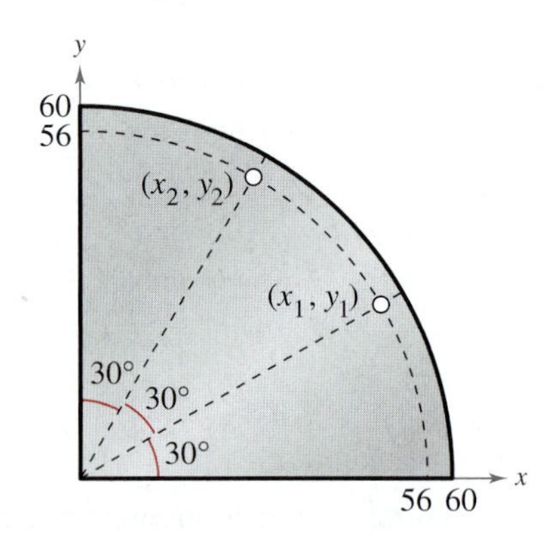

70. ***Machine Shop Calculations*** A tapered shaft has a diameter of 5 centimeters at the small end and is 15 centimeters long (see figure). The taper is 3°. Find the diameter d of the large end of the shaft.

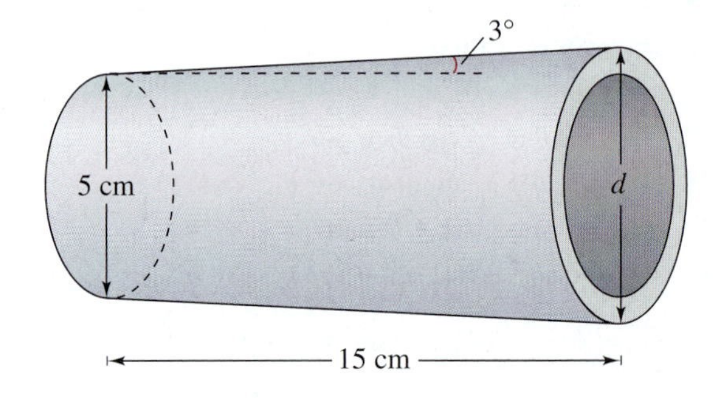

Model It

71. ***Height*** A 20-meter line is used to tether a helium-filled balloon. Because of a breeze, the line makes an angle of approximately 85° with the ground.

(a) Draw a right triangle that gives a visual representation of the problem. Show the known quantities of the triangle and use a variable to indicate the height of the balloon.

(b) Use a trigonometric function to write an equation involving the unknown quantity.

(c) What is the height of the balloon?

(d) The breeze becomes stronger and the angle the balloon makes with the ground decreases. How does this affect the triangle you drew in part (a)?

(e) Complete the table, which shows the heights (in meters) of the balloon for decreasing angle measures θ.

Angle, θ	80°	70°	60°	50°
Height				

Angle, θ	40°	30°	20°	10°
Height				

(f) As the angle the balloon makes with the ground approaches 0°, how does this affect the height of the balloon? Draw a right triangle to explain your reasoning.

72. ***Geometry*** Use a compass to sketch a quarter of a circle of radius 10 centimeters. Using a protractor, construct an angle of 20° in standard position (see figure). Drop a perpendicular line from the point of intersection of the terminal side of the angle and the arc of the circle. By actual measurement, calculate the coordinates (x, y) of the point of intersection and use these measurements to approximate the six trigonometric functions of a 20° angle.

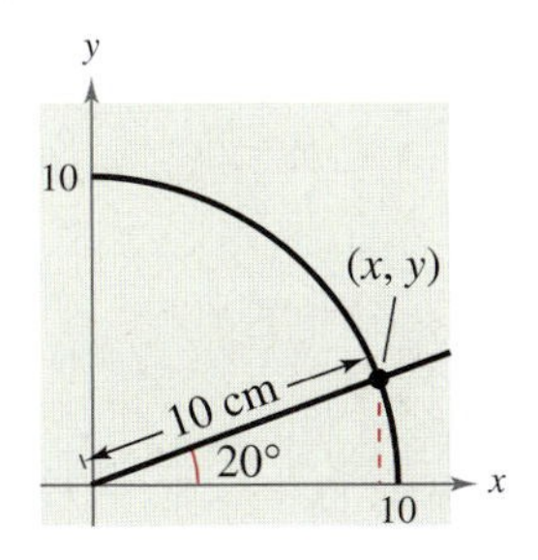

Synthesis

True or False? **In Exercises 73–78, determine whether the statement is true or false. Justify your answer.**

73. $\sin 60° \csc 60° = 1$

74. $\sec 30° = \csc 60°$

75. $\sin 45° + \cos 45° = 1$

76. $\cot^2 10° - \csc^2 10° = -1$

77. $\dfrac{\sin 60°}{\sin 30°} = \sin 2°$

78. $\tan[(5°)^2] = \tan^2(5°)$

79. ***Writing*** In right triangle trigonometry, explain why $\sin 30° = \frac{1}{2}$ regardless of the size of the triangle.

80. ***Think About It*** You are given only the value $\tan \theta$. Is it possible to find the value of $\sec \theta$ without finding the measure of θ? Explain.

81. ***Exploration***

(a) Complete the table.

θ	0.1	0.2	0.3	0.4	0.5
$\sin \theta$					

(b) Is θ or $\sin \theta$ greater for θ in the interval $(0, 0.5]$?

(c) As θ approaches 0, how do θ and $\sin \theta$ compare? Explain.

82. ***Exploration***

(a) Complete the table.

θ	0°	18°	36°	54°	72°	90°
$\sin \theta$						
$\cos \theta$						

(b) Discuss the behavior of the sine function for θ in the range from 0° to 90°.

(c) Discuss the behavior of the cosine function for θ in the range from 0° to 90°.

(d) Use the definitions of the sine and cosine functions to explain the results of parts (b) and (c).

Skills Review

In Exercises 83–86, perform the operations and simplify.

83. $\dfrac{x^2 - 6x}{x^2 + 4x - 12} \cdot \dfrac{x^2 + 12x + 36}{x^2 - 36}$

84. $\dfrac{2t^2 + 5t - 12}{9 - 4t^2} \div \dfrac{t^2 - 16}{4t^2 + 12t + 9}$

85. $\dfrac{3}{x + 2} - \dfrac{2}{x - 2} + \dfrac{x}{x^2 + 4x + 4}$

86. $\dfrac{\left(\dfrac{3}{x} - \dfrac{1}{4}\right)}{\left(\dfrac{12}{x} - 1\right)}$

4.4 Trigonometric Functions of Any Angle

What you should learn

- Evaluate trigonometric functions of any angle.
- Use reference angles to evaluate trigonometric functions.
- Evaluate trigonometric functions of real numbers.

Why you should learn it

You can use trigonometric functions to model and solve real-life problems. For instance, in Exercise 87 on page 319, you can use trigonometric functions to model the monthly normal temperatures in New York City and Fairbanks, Alaska.

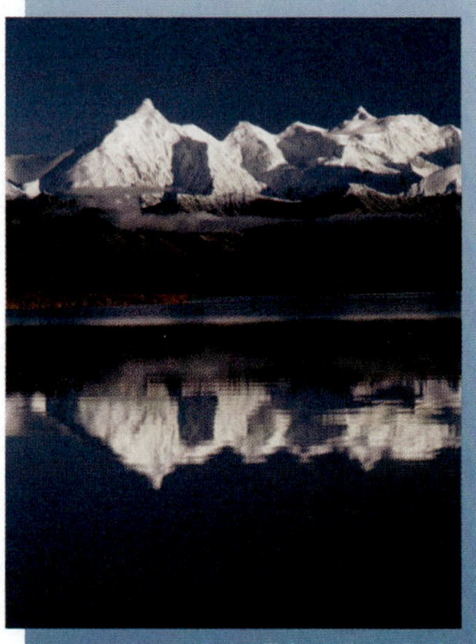

James Urbach/SuperStock

Introduction

In Section 4.3, the definitions of trigonometric functions were restricted to acute angles. In this section, the definitions are extended to cover *any* angle. If θ is an *acute* angle, these definitions coincide with those given in the preceding section.

Definitions of Trigonometric Functions of Any Angle

Let θ be an angle in standard position with (x, y) a point on the terminal side of θ and $r = \sqrt{x^2 + y^2} \neq 0$.

$$\sin\theta = \frac{y}{r} \qquad \cos\theta = \frac{x}{r}$$

$$\tan\theta = \frac{y}{x},\quad x \neq 0 \qquad \cot\theta = \frac{x}{y},\quad y \neq 0$$

$$\sec\theta = \frac{r}{x},\quad x \neq 0 \qquad \csc\theta = \frac{r}{y},\quad y \neq 0$$

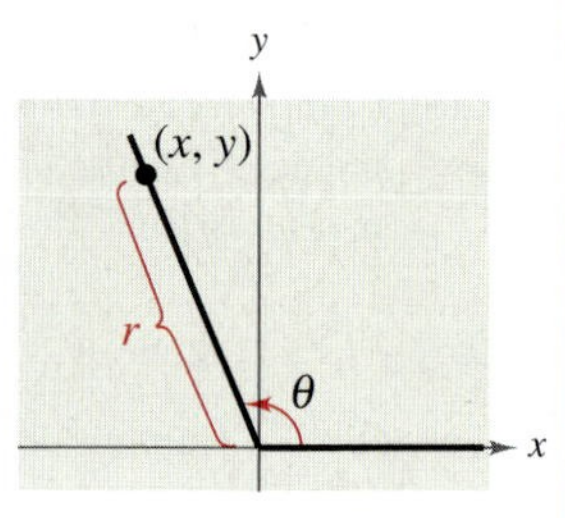

Because $r = \sqrt{x^2 + y^2}$ *cannot* be zero, it follows that the sine and cosine functions are defined for any real value of θ. However, if $x = 0$, the tangent and secant of θ are undefined. For example, the tangent of 90° is undefined. Similarly, if $y = 0$, the cotangent and cosecant of θ are undefined.

Example 1 Evaluating Trigonometric Functions

Let $(-3, 4)$ be a point on the terminal side of θ. Find the sine, cosine, and tangent of θ.

Solution

Referring to Figure 4.36, you can see that $x = -3$, $y = 4$, and

$$r = \sqrt{x^2 + y^2} = \sqrt{(-3)^2 + 4^2} = \sqrt{25} = 5.$$

So, you have the following.

$$\sin\theta = \frac{y}{r} = \frac{4}{5}$$

$$\cos\theta = \frac{x}{r} = -\frac{3}{5}$$

$$\tan\theta = \frac{y}{x} = -\frac{4}{3}$$

CHECKPOINT Now try Exercise 1.

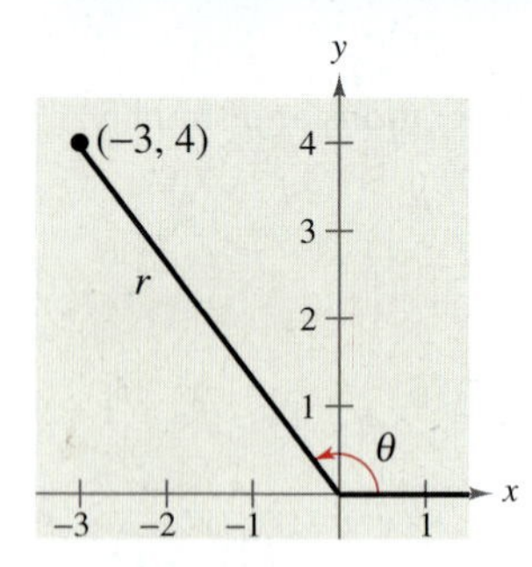

FIGURE 4.36

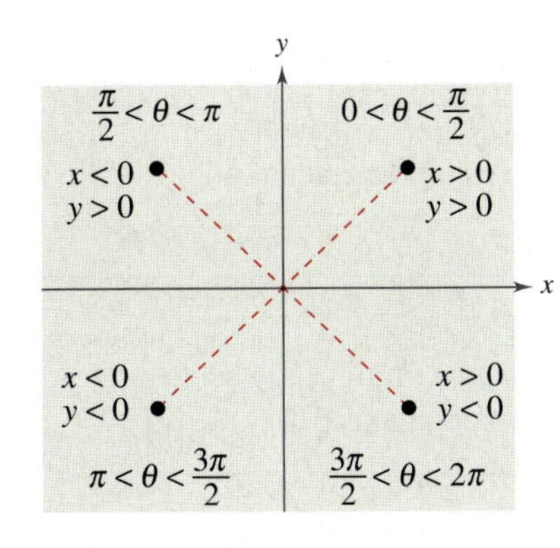

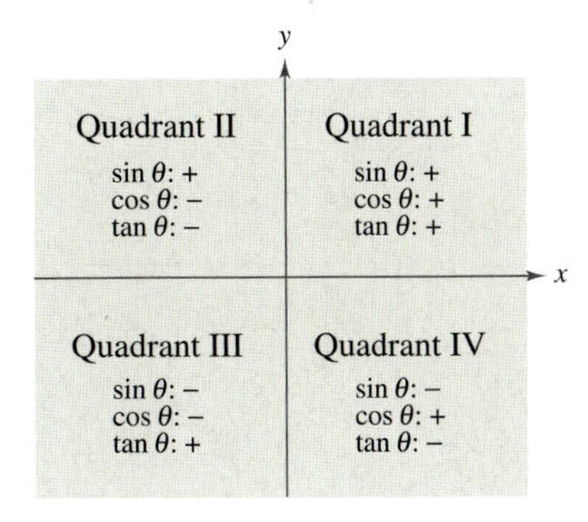

FIGURE 4.37

The *signs* of the trigonometric functions in the four quadrants can be determined easily from the definitions of the functions. For instance, because $\cos \theta = x/r$, it follows that $\cos \theta$ is positive wherever $x > 0$, which is in Quadrants I and IV. (Remember, r is always positive.) In a similar manner, you can verify the results shown in Figure 4.37.

Example 2 Evaluating Trigonometric Functions

Given $\tan \theta = -\frac{5}{4}$ and $\cos \theta > 0$, find $\sin \theta$ and $\sec \theta$.

Solution

Note that θ lies in Quadrant IV because that is the only quadrant in which the tangent is negative and the cosine is positive. Moreover, using

$$\tan \theta = \frac{y}{x} = -\frac{5}{4}$$

and the fact that y is negative in Quadrant IV, you can let $y = -5$ and $x = 4$. So, $r = \sqrt{16 + 25} = \sqrt{41}$ and you have

$$\sin \theta = \frac{y}{r} = \frac{-5}{\sqrt{41}}$$

$$\approx -0.7809$$

$$\sec \theta = \frac{r}{x} = \frac{\sqrt{41}}{4}$$

$$\approx 1.6008.$$

CHECKPOINT Now try Exercise 17.

Example 3 Trigonometric Functions of Quadrant Angles

Evaluate the cosine and tangent functions at the four quadrant angles $0, \frac{\pi}{2}, \pi$, and $\frac{3\pi}{2}$.

Solution

To begin, choose a point on the terminal side of each angle, as shown in Figure 4.38. For each of the four points, $r = 1$, and you have the following.

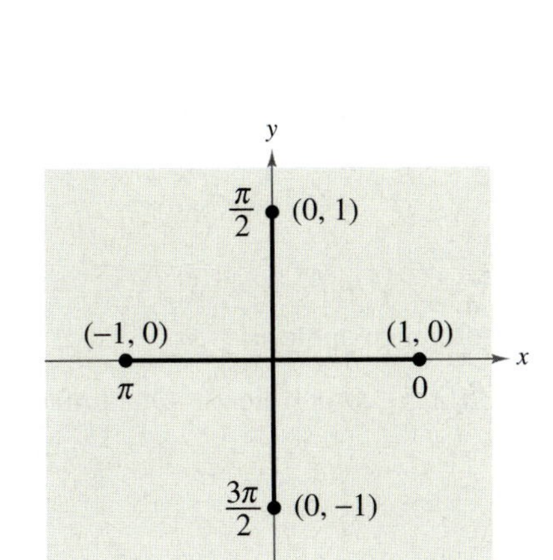

FIGURE 4.38

$$\cos 0 = \frac{x}{r} = \frac{1}{1} = 1 \qquad \tan 0 = \frac{y}{x} = \frac{0}{1} = 0 \qquad (x, y) = (1, 0)$$

$$\cos \frac{\pi}{2} = \frac{x}{r} = \frac{0}{1} = 0 \qquad \tan \frac{\pi}{2} = \frac{y}{x} = \frac{1}{0} \Rightarrow \text{undefined} \qquad (x, y) = (0, 1)$$

$$\cos \pi = \frac{x}{r} = \frac{-1}{1} = -1 \qquad \tan \pi = \frac{y}{x} = \frac{0}{-1} = 0 \qquad (x, y) = (-1, 0)$$

$$\cos \frac{3\pi}{2} = \frac{x}{r} = \frac{0}{1} = 0 \qquad \tan \frac{3\pi}{2} = \frac{y}{x} = \frac{-1}{0} \Rightarrow \text{undefined} \qquad (x, y) = (0, -1)$$

CHECKPOINT Now try Exercise 29.

Reference Angles

Sketching several angles with their reference angles may help reinforce the fact that the reference angle is the acute angle formed with the horizontal.

The values of the trigonometric functions of angles greater than 90° (or less than 0°) can be determined from their values at corresponding acute angles called **reference angles.**

Definition of Reference Angle

Let θ be an angle in standard position. Its **reference angle** is the acute angle θ' formed by the terminal side of θ and the horizontal axis.

Figure 4.39 shows the reference angles for θ in Quadrants II, III, and IV.

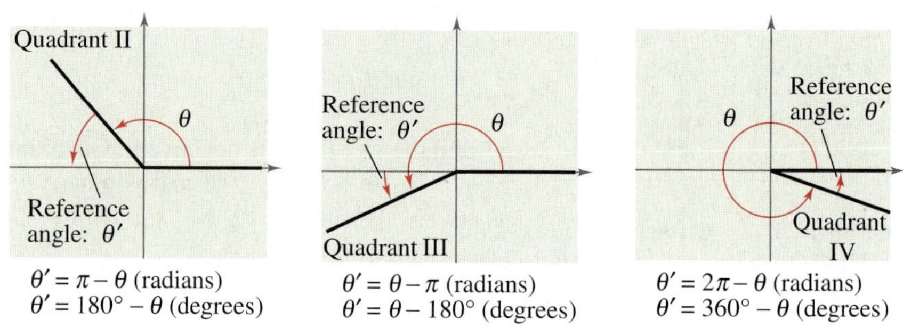

FIGURE 4.39

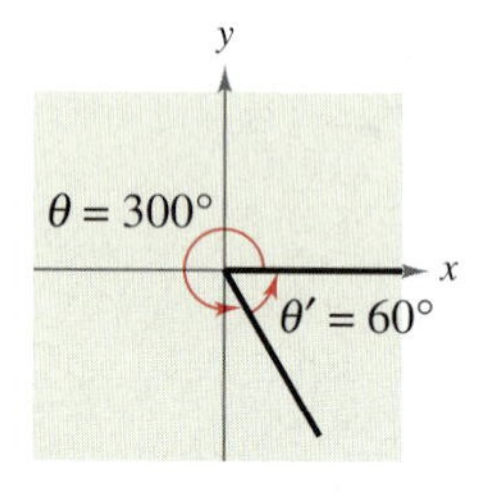

FIGURE 4.40

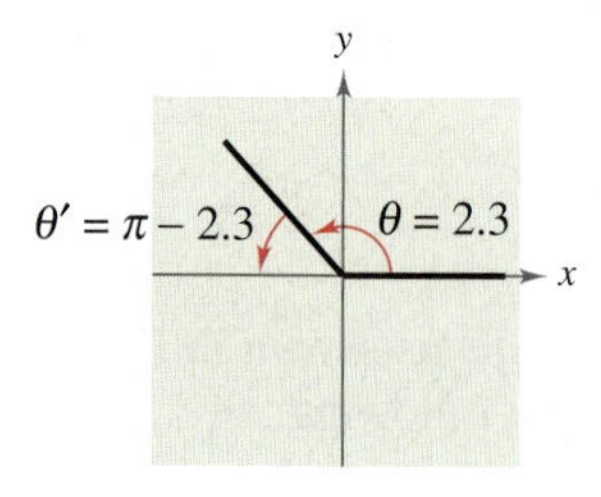

FIGURE 4.41

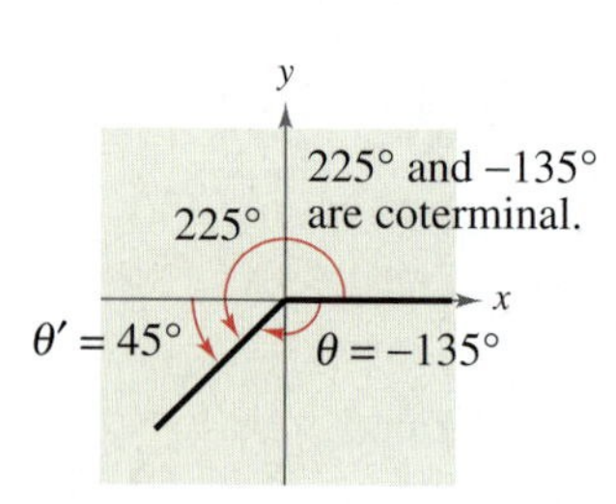

FIGURE 4.42

Example 4 Finding Reference Angles

Find the reference angle θ'.

a. $\theta = 300°$ **b.** $\theta = 2.3$ **c.** $\theta = -135°$

Solution

a. Because 300° lies in Quadrant IV, the angle it makes with the x-axis is

$$\theta' = 360° - 300°$$
$$= 60°. \quad \text{Degrees}$$

Figure 4.40 shows the angle $\theta = 300°$ and its reference angle $\theta' = 60°$.

b. Because 2.3 lies between $\pi/2 \approx 1.5708$ and $\pi \approx 3.1416$, it follows that it is in Quadrant II and its reference angle is

$$\theta' = \pi - 2.3$$
$$\approx 0.8416. \quad \text{Radians}$$

Figure 4.41 shows the angle $\theta = 2.3$ and its reference angle $\theta' = \pi - 2.3$.

c. First, determine that $-135°$ is coterminal with 225°, which lies in Quadrant III. So, the reference angle is

$$\theta' = 225° - 180°$$
$$= 45°. \quad \text{Degrees}$$

Figure 4.42 shows the angle $\theta = -135°$ and its reference angle $\theta' = 45°$.

✓CHECKPOINT Now try Exercise 37.

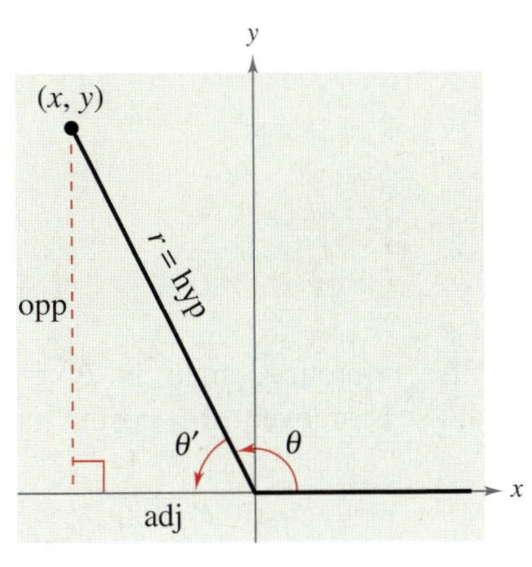

opp = $|y|$, adj = $|x|$
FIGURE 4.43

Trigonometric Functions of Real Numbers

To see how a reference angle is used to evaluate a trigonometric function, consider the point (x, y) on the terminal side of θ, as shown in Figure 4.43. By definition, you know that

$$\sin \theta = \frac{y}{r} \quad \text{and} \quad \tan \theta = \frac{y}{x}.$$

For the right triangle with acute angle θ' and sides of lengths $|x|$ and $|y|$, you have

$$\sin \theta' = \frac{\text{opp}}{\text{hyp}} = \frac{|y|}{r}$$

and

$$\tan \theta' = \frac{\text{opp}}{\text{adj}} = \frac{|y|}{|x|}.$$

So, it follows that $\sin \theta$ and $\sin \theta'$ are equal, *except possibly in sign*. The same is true for $\tan \theta$ and $\tan \theta'$ *and* for the other four trigonometric functions. In all cases, the sign of the function value can be determined by the quadrant in which θ lies.

Evaluating Trigonometric Functions of Any Angle

To find the value of a trigonometric function of any angle θ:

1. Determine the function value for the associated reference angle θ'.
2. Depending on the quadrant in which θ lies, affix the appropriate sign to the function value.

STUDY TIP

Learning the table of values at the right is worth the effort because doing so will increase both your efficiency and your confidence. Here is a pattern for the sine function that may help you remember the values.

θ	0°	30°	45°	60°	90°
$\sin \theta$	$\frac{\sqrt{0}}{2}$	$\frac{\sqrt{1}}{2}$	$\frac{\sqrt{2}}{2}$	$\frac{\sqrt{3}}{2}$	$\frac{\sqrt{4}}{2}$

Reverse the order to get cosine values of the same angles.

By using reference angles and the special angles discussed in the preceding section, you can greatly extend the scope of *exact* trigonometric values. For instance, knowing the function values of 30° means that you know the function values of all angles for which 30° is a reference angle. For convenience, the table below shows the exact values of the trigonometric functions of special angles and quadrant angles.

Trigonometric Values of Common Angles

θ (degrees)	0°	30°	45°	60°	90°	180°	270°
θ (radians)	0	$\frac{\pi}{6}$	$\frac{\pi}{4}$	$\frac{\pi}{3}$	$\frac{\pi}{2}$	π	$\frac{3\pi}{2}$
$\sin \theta$	0	$\frac{1}{2}$	$\frac{\sqrt{2}}{2}$	$\frac{\sqrt{3}}{2}$	1	0	-1
$\cos \theta$	1	$\frac{\sqrt{3}}{2}$	$\frac{\sqrt{2}}{2}$	$\frac{1}{2}$	0	-1	0
$\tan \theta$	0	$\frac{\sqrt{3}}{3}$	1	$\sqrt{3}$	Undef.	0	Undef.

Example 5 Using Reference Angles

Evaluate each trigonometric function.

a. $\cos \dfrac{4\pi}{3}$ **b.** $\tan(-210°)$ **c.** $\csc \dfrac{11\pi}{4}$

Solution

a. Because $\theta = 4\pi/3$ lies in Quadrant III, the reference angle is $\theta' = (4\pi/3) - \pi = \pi/3$, as shown in Figure 4.44. Moreover, the cosine is negative in Quadrant III, so

$$\cos \frac{4\pi}{3} = (-) \cos \frac{\pi}{3}$$
$$= -\frac{1}{2}.$$

Emphasize the importance of reference angles in evaluating trigonometric functions of angles greater than 90°.

b. Because $-210° + 360° = 150°$, it follows that $-210°$ is coterminal with the second-quadrant angle 150°. So, the reference angle is $\theta' = 180° - 150° = 30°$, as shown in Figure 4.45. Finally, because the tangent is negative in Quadrant II, you have

$$\tan(-210°) = (-) \tan 30°$$
$$= -\frac{\sqrt{3}}{3}.$$

c. Because $(11\pi/4) - 2\pi = 3\pi/4$, it follows that $11\pi/4$ is coterminal with the second-quadrant angle $3\pi/4$. So, the reference angle is $\theta' = \pi - (3\pi/4) = \pi/4$, as shown in Figure 4.46. Because the cosecant is positive in Quadrant II, you have

$$\csc \frac{11\pi}{4} = (+) \csc \frac{\pi}{4}$$
$$= \frac{1}{\sin(\pi/4)}$$
$$= \sqrt{2}.$$

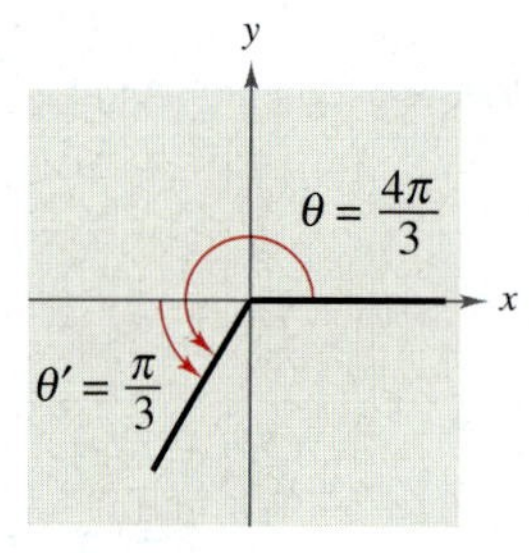

FIGURE 4.44

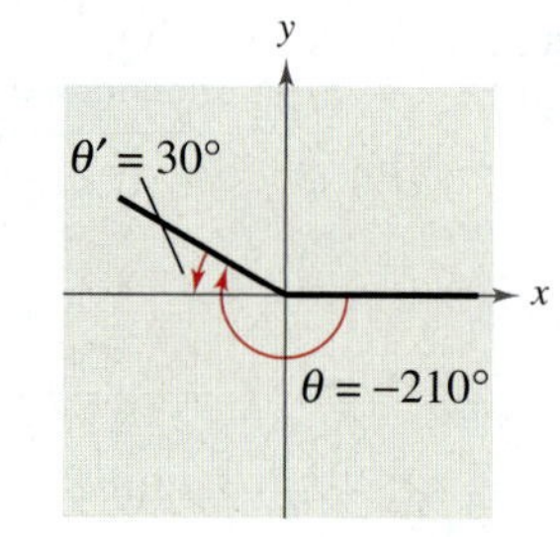

FIGURE 4.45

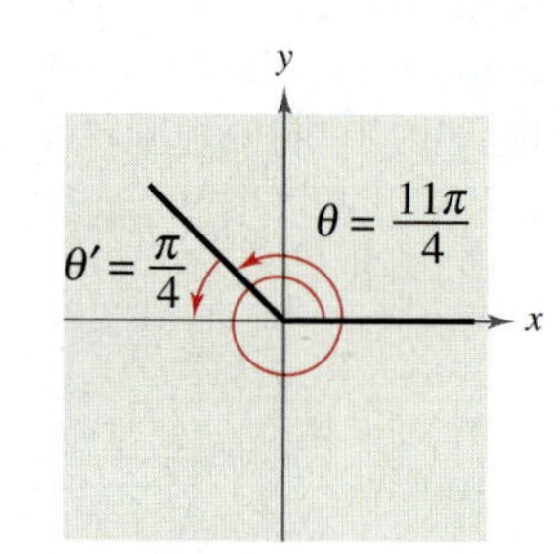

FIGURE 4.46

CHECKPOINT Now try Exercise 51.

Example 6 Using Trigonometric Identities

Let θ be an angle in Quadrant II such that $\sin\theta = \frac{1}{3}$. Find (a) $\cos\theta$ and (b) $\tan\theta$ by using trigonometric identities.

Solution

a. Using the Pythagorean identity $\sin^2\theta + \cos^2\theta = 1$, you obtain

$$\left(\frac{1}{3}\right)^2 + \cos^2\theta = 1 \qquad \text{Substitute } \tfrac{1}{3} \text{ for } \sin\theta.$$

$$\cos^2\theta = 1 - \frac{1}{9} = \frac{8}{9}.$$

Because $\cos\theta < 0$ in Quadrant II, you can use the negative root to obtain

$$\cos\theta = -\frac{\sqrt{8}}{\sqrt{9}} = -\frac{2\sqrt{2}}{3}.$$

b. Using the trigonometric identity $\tan\theta = \dfrac{\sin\theta}{\cos\theta}$, you obtain

$$\tan\theta = \frac{1/3}{-2\sqrt{2}/3} \qquad \text{Substitute for } \sin\theta \text{ and } \cos\theta.$$

$$= -\frac{1}{2\sqrt{2}} = -\frac{\sqrt{2}}{4}.$$

CHECKPOINT Now try Exercise 59.

Students often have difficulty determining angles, especially when the functions given are csc, sec, and/or cot. Have your students rewrite the expression in terms of sin, cos, or tan, whichever is applicable, before evaluating.

Activities

1. Determine the exact values of the six trigonometric functions of the angle in standard position whose terminal side contains the point $(-3, -7)$.
   Answer:
   $\sin\theta = -\frac{7}{\sqrt{58}}$ $\quad \csc\theta = -\frac{\sqrt{58}}{7}$
   $\cos\theta = -\frac{3}{\sqrt{58}}$ $\quad \sec\theta = -\frac{\sqrt{58}}{3}$
   $\tan\theta = \frac{7}{3}$ $\quad \cot\theta = \frac{3}{7}$
2. For the angle $\theta = -135°$, find the reference angle θ', and sketch θ and θ' in standard position.
   Answer: $\theta' = 45°$

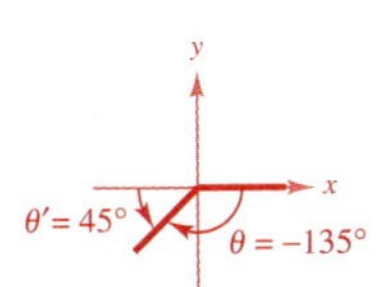

3. Find two values of θ, $0 \le \theta < 2\pi$, that satisfy the equation $\tan\theta = -1$. Do not use your calculator.
   Answer: $\theta = \frac{3\pi}{4}, \frac{7\pi}{4}$

You can use a calculator to evaluate trigonometric functions, as shown in the next example.

Example 7 Using a Calculator

Use a calculator to evaluate each trigonometric function.

a. $\cot 410°$ **b.** $\sin(-7)$ **c.** $\sec\dfrac{\pi}{9}$

Solution

	Function	*Mode*	*Calculator Keystrokes*	*Display*
a.	$\cot 410°$	Degree	(TAN (410)) x^{-1} ENTER	0.8390996
b.	$\sin(-7)$	Radian	SIN ((−) 7) ENTER	−0.6569866
c.	$\sec\frac{\pi}{9}$	Radian	(COS (π ÷ 9)) x^{-1} ENTER	1.0641778

CHECKPOINT Now try Exercise 69.

4.4 Exercises

VOCABULARY CHECK:

In Exercises 1–6, let θ be an angle in standard position, with (x, y) a point on the terminal side of θ and $r\sqrt{x^2 + y^2} \neq 0$.

1. $\sin \theta =$ ________ **2.** $\dfrac{r}{y} =$ ________

3. $\tan \theta =$ ________ **4.** $\sec \theta =$ ________

5. $\dfrac{x}{r} =$ ________ **6.** $\dfrac{x}{y} =$ ________

7. The acute positive angle that is formed by the terminal side of the angle θ and the horizontal axis is called the ________ angle of θ and is denoted by θ'.

PREREQUISITE SKILLS REVIEW: Practice and review algebra skills needed for this section at **www.Eduspace.com.**

In Exercises 1–4, determine the exact values of the six trigonometric functions of the angle θ.

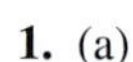

1. (a)

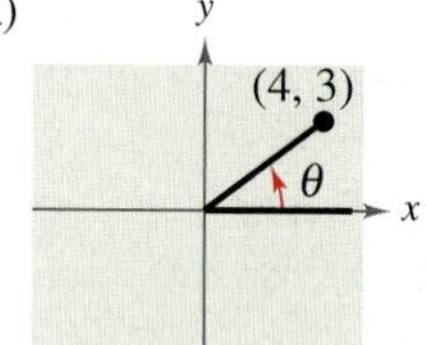

(b)

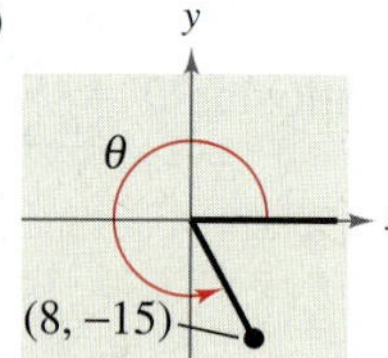

2. (a)

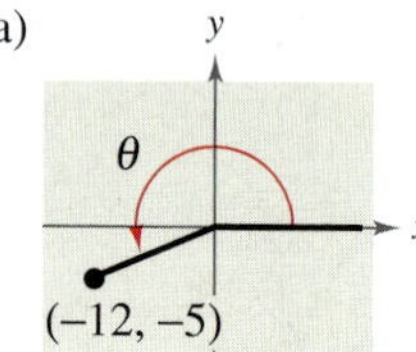

(b)

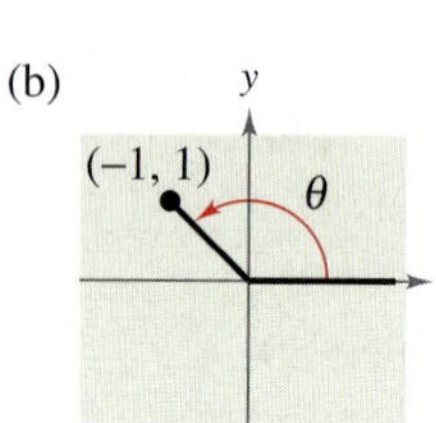

3. (a)

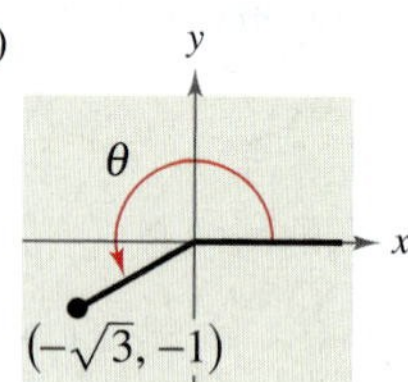

(b)

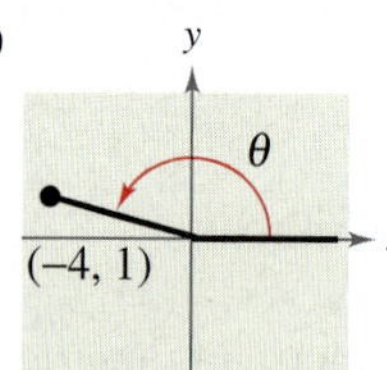

4. (a)

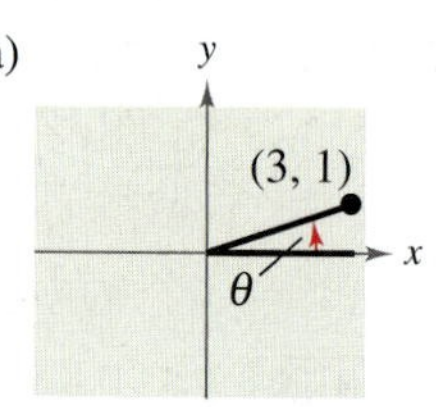

(b)

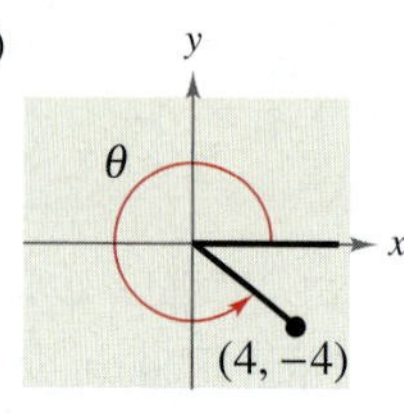

In Exercises 5–10, the point is on the terminal side of an angle in standard position. Determine the exact values of the six trigonometric functions of the angle.

5. $(7, 24)$ **6.** $(8, 15)$

7. $(-4, 10)$ **8.** $(-5, -2)$

9. $(-3.5, 6.8)$ **10.** $\left(3\frac{1}{2}, -7\frac{3}{4}\right)$

In Exercises 11–14, state the quadrant in which θ lies.

11. $\sin \theta < 0$ and $\cos \theta < 0$

12. $\sin \theta > 0$ and $\cos \theta > 0$

13. $\sin \theta > 0$ and $\tan \theta < 0$

14. $\sec \theta > 0$ and $\cot \theta < 0$

In Exercises 15–24, find the values of the six trigonometric functions of θ with the given constraint.

	Function Value	*Constraint*
15.	$\sin \theta = \frac{3}{5}$	θ lies in Quadrant II.
16.	$\cos \theta = -\frac{4}{5}$	θ lies in Quadrant III.
17.	$\tan \theta = -\frac{15}{8}$	$\sin \theta < 0$
18.	$\cos \theta = \frac{8}{17}$	$\tan \theta < 0$
19.	$\cot \theta = -3$	$\cos \theta > 0$
20.	$\csc \theta = 4$	$\cot \theta < 0$
21.	$\sec \theta = -2$	$\sin \theta > 0$
22.	$\sin \theta = 0$	$\sec \theta = -1$
23.	$\cot \theta$ is undefined.	$\pi/2 \leq \theta \leq 3\pi/2$
24.	$\tan \theta$ is undefined.	$\pi \leq \theta \leq 2\pi$

In Exercises 25–28, the terminal side of θ lies on the given line in the specified quadrant. Find the values of the six trigonometric functions of θ by finding a point on the line.

	Line	*Quadrant*
25.	$y = -x$	II
26.	$y = \frac{1}{3}x$	III
27.	$2x - y = 0$	III
28.	$4x + 3y = 0$	IV

In Exercises 29–36, evaluate the trigonometric function of the quadrant angle.

29. $\sin \pi$

30. $\csc \frac{3\pi}{2}$

31. $\sec \frac{3\pi}{2}$

32. $\sec \pi$

33. $\sin \frac{\pi}{2}$

34. $\cot \pi$

35. $\csc \pi$

36. $\cot \frac{\pi}{2}$

In Exercises 37–44, find the reference angle θ', and sketch θ and θ' in standard position.

37. $\theta = 203°$

38. $\theta = 309°$

39. $\theta = -245°$

40. $\theta = -145°$

41. $\theta = \frac{2\pi}{3}$

42. $\theta = \frac{7\pi}{4}$

43. $\theta = 3.5$

44. $\theta = \frac{11\pi}{3}$

In Exercises 45–58, evaluate the sine, cosine, and tangent of the angle without using a calculator.

45. $225°$

46. $300°$

47. $750°$

48. $-405°$

49. $-150°$

50. $-840°$

51. $\frac{4\pi}{3}$

52. $\frac{\pi}{4}$

53. $-\frac{\pi}{6}$

54. $-\frac{\pi}{2}$

55. $\frac{11\pi}{4}$

56. $\frac{10\pi}{3}$

57. $-\frac{3\pi}{2}$

58. $-\frac{25\pi}{4}$

In Exercises 59–64, find the indicated trigonometric value in the specified quadrant.

Function	*Quadrant*	*Trigonometric Value*
59. $\sin \theta = -\frac{3}{5}$	IV	$\cos \theta$
60. $\cot \theta = -3$	II	$\sin \theta$
61. $\tan \theta = \frac{3}{2}$	III	$\sec \theta$
62. $\csc \theta = -2$	IV	$\cot \theta$
63. $\cos \theta = \frac{5}{8}$	I	$\sec \theta$
64. $\sec \theta = -\frac{9}{4}$	III	$\tan \theta$

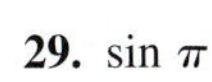

In Exercises 65–80, use a calculator to evaluate the trigonometric function. Round your answer to four decimal places. (Be sure the calculator is set in the correct angle mode.)

65. $\sin 10°$

66. $\sec 225°$

67. $\cos(-110°)$

68. $\csc(-330°)$

69. $\tan 304°$

70. $\cot 178°$

71. $\sec 72°$

72. $\tan(-188°)$

73. $\tan 4.5$

74. $\cot 1.35$

75. $\tan \frac{\pi}{9}$

76. $\tan\left(-\frac{\pi}{9}\right)$

77. $\sin(-0.65)$

78. $\sec 0.29$

79. $\cot\left(-\frac{11\pi}{8}\right)$

80. $\csc\left(-\frac{15\pi}{14}\right)$

In Exercises 81–86, find two solutions of the equation. Give your answers in degrees ($0° \le \theta < 360°$) and in radians ($0 \le \theta < 2\pi$). Do not use a calculator.

81. (a) $\sin \theta = \frac{1}{2}$ (b) $\sin \theta = -\frac{1}{2}$

82. (a) $\cos \theta = \frac{\sqrt{2}}{2}$ (b) $\cos \theta = -\frac{\sqrt{2}}{2}$

83. (a) $\csc \theta = \frac{2\sqrt{3}}{3}$ (b) $\cot \theta = -1$

84. (a) $\sec \theta = 2$ (b) $\sec \theta = -2$

85. (a) $\tan \theta = 1$ (b) $\cot \theta = -\sqrt{3}$

86. (a) $\sin \theta = \frac{\sqrt{3}}{2}$ (b) $\sin \theta = -\frac{\sqrt{3}}{2}$

Model It

87. ***Data Analysis: Meteorology*** The table shows the monthly normal temperatures (in degrees Fahrenheit) for selected months for New York City (N) and Fairbanks, Alaska (F). (Source: National Climatic Data Center)

Month	New York City, N	Fairbanks, F
January	33	−10
April	52	32
July	77	62
October	58	24
December	38	−6

(a) Use the *regression* feature of a graphing utility to find a model of the form

$$y = a \sin(bt + c) + d$$

for each city. Let t represent the month, with $t = 1$ corresponding to January.

Model It (continued)

(b) Use the models from part (a) to find the monthly normal temperatures for the two cities in February, March, May, June, August, September, and November.

(c) Compare the models for the two cities.

88. ***Sales*** A company that produces snowboards, which are seasonal products, forecasts monthly sales over the next 2 years to be

$$S = 23.1 + 0.442t + 4.3\cos\frac{\pi t}{6}$$

where S is measured in thousands of units and t is the time in months, with $t = 1$ representing January 2006. Predict sales for each of the following months.

(a) February 2006

(b) February 2007

(c) June 2006

(d) June 2007

89. ***Harmonic Motion*** The displacement from equilibrium of an oscillating weight suspended by a spring is given by

$$y(t) = 2\cos 6t$$

where y is the displacement (in centimeters) and t is the time (in seconds). Find the displacement when (a) $t = 0$, (b) $t = \frac{1}{4}$, and (c) $t = \frac{1}{2}$.

90. ***Harmonic Motion*** The displacement from equilibrium of an oscillating weight suspended by a spring and subject to the damping effect of friction is given by

$$y(t) = 2e^{-t}\cos 6t$$

where y is the displacement (in centimeters) and t is the time (in seconds). Find the displacement when (a) $t = 0$, (b) $t = \frac{1}{4}$, and (c) $t = \frac{1}{2}$.

91. ***Electric Circuits*** The current I (in amperes) when 100 volts is applied to a circuit is given by

$$I = 5e^{-2t}\sin t$$

where t is the time (in seconds) after the voltage is applied. Approximate the current at $t = 0.7$ second after the voltage is applied.

92. ***Distance*** An airplane, flying at an altitude of 6 miles, is on a flight path that passes directly over an observer (see figure). If θ is the angle of elevation from the observer to the plane, find the distance d from the observer to the plane when (a) $\theta = 30°$, (b) $\theta = 90°$, and (c) $\theta = 120°$.

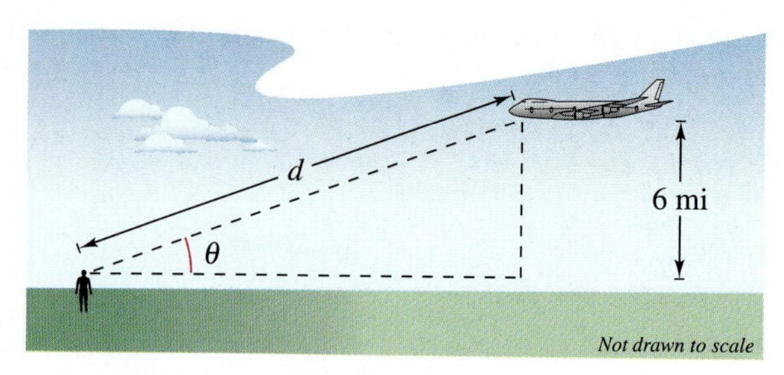

FIGURE FOR 92

Synthesis

True or False? **In Exercises 93 and 94, determine whether the statement is true or false. Justify your answer.**

93. In each of the four quadrants, the signs of the secant function and sine function will be the same.

94. To find the reference angle for an angle θ (given in degrees), find the integer n such that $0 \le 360°n - \theta \le 360°$. The difference $360°n - \theta$ is the reference angle.

95. ***Writing*** Consider an angle in standard position with $r = 12$ centimeters, as shown in the figure. Write a short paragraph describing the changes in the values of x, y, $\sin\theta$, $\cos\theta$, and $\tan\theta$ as θ increases continuously from 0° to 90°.

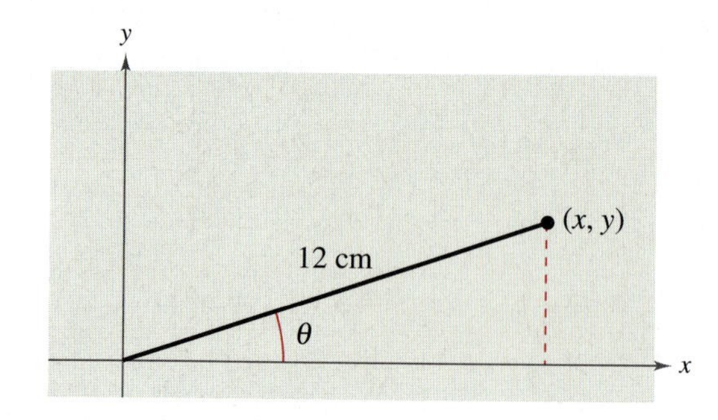

96. ***Writing*** Explain how reference angles are used to find the trigonometric functions of obtuse angles.

Skills Review

In Exercises 97–106, graph the function. Identify the domain and any intercepts and asymptotes of the function.

97. $y = x^2 + 3x - 4$

98. $y = 2x^2 - 5x$

99. $f(x) = x^3 + 8$

100. $g(x) = x^4 + 2x^2 - 3$

101. $f(x) = \dfrac{x - 7}{x^2 + 4x + 4}$

102. $h(x) = \dfrac{x^2 - 1}{x + 5}$

103. $y = 2^{x-1}$

104. $y = 3^{x+1} + 2$

105. $y = \ln x^4$

106. $y = \log_{10}(x + 2)$

4.5 Graphs of Sine and Cosine Functions

What you should learn

- Sketch the graphs of basic sine and cosine functions.
- Use amplitude and period to help sketch the graphs of sine and cosine functions.
- Sketch translations of the graphs of sine and cosine functions.
- Use sine and cosine functions to model real-life data.

Why you should learn it

Sine and cosine functions are often used in scientific calculations. For instance, in Exercise 73 on page 330, you can use a trigonometric function to model the airflow of your respiratory cycle.

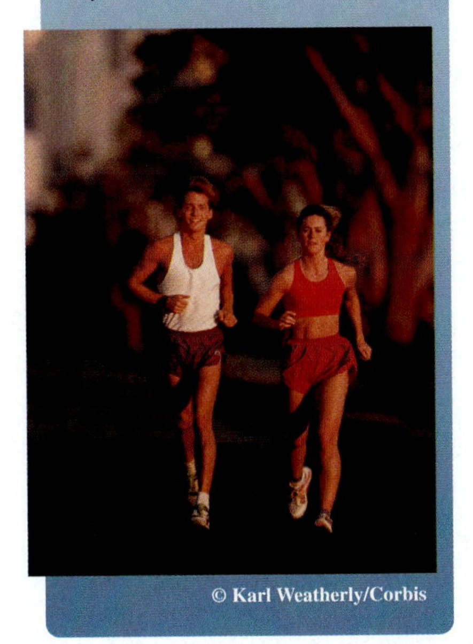

© Karl Weatherly/Corbis

Basic Sine and Cosine Curves

In this section, you will study techniques for sketching the graphs of the sine and cosine functions. The graph of the sine function is a **sine curve.** In Figure 4.47, the black portion of the graph represents one period of the function and is called **one cycle** of the sine curve. The gray portion of the graph indicates that the basic sine curve repeats indefinitely in the positive and negative directions. The graph of the cosine function is shown in Figure 4.48.

Recall from Section 4.2 that the domain of the sine and cosine functions is the set of all real numbers. Moreover, the range of each function is the interval $[-1, 1]$, and each function has a period of 2π. Do you see how this information is consistent with the basic graphs shown in Figures 4.47 and 4.48?

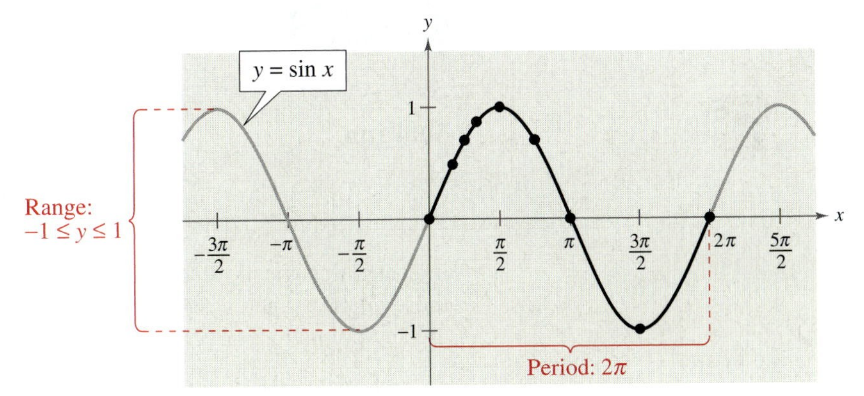

FIGURE 4.47

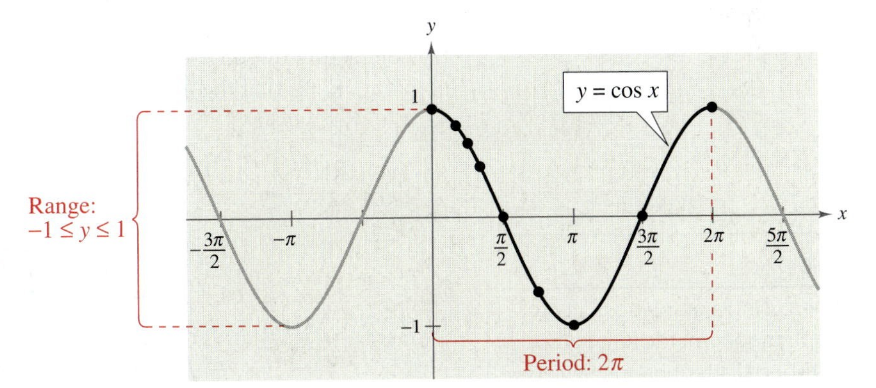

FIGURE 4.48

Note in Figures 4.47 and 4.48 that the sine curve is symmetric with respect to the *origin*, whereas the cosine curve is symmetric with respect to the *y-axis*. These properties of symmetry follow from the fact that the sine function is odd and the cosine function is even.

To sketch the graphs of the basic sine and cosine functions by hand, it helps to note five **key points** in one period of each graph: the *intercepts*, *maximum points*, and *minimum points* (see Figure 4.49).

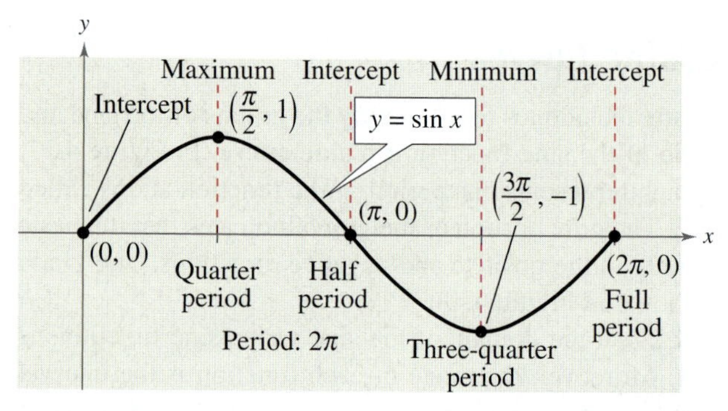

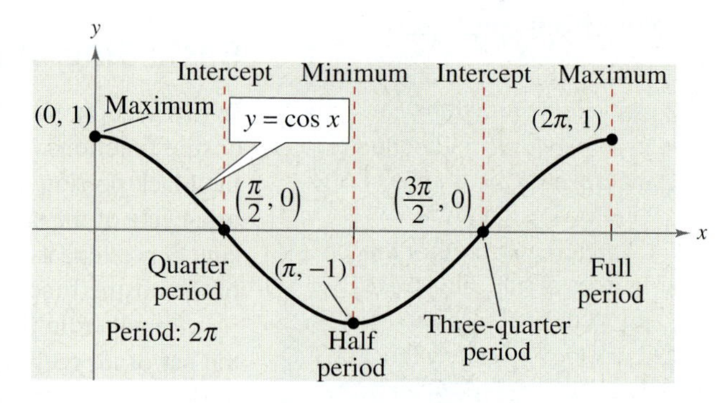

FIGURE 4.49

Example 1 Using Key Points to Sketch a Sine Curve

Sketch the graph of $y = 2 \sin x$ on the interval $[-\pi, 4\pi]$.

Solution

Note that

$$y = 2 \sin x = 2(\sin x)$$

indicates that the y-values for the key points will have twice the magnitude of those on the graph of $y = \sin x$. Divide the period 2π into four equal parts to get the key points for $y = 2 \sin x$.

Intercept	*Maximum*	*Intercept*	*Minimum*		*Intercept*
$(0, 0)$,	$\left(\frac{\pi}{2}, 2\right)$,	$(\pi, 0)$,	$\left(\frac{3\pi}{2}, -2\right)$,	and	$(2\pi, 0)$

By connecting these key points with a smooth curve and extending the curve in both directions over the interval $[-\pi, 4\pi]$, you obtain the graph shown in Figure 4.50.

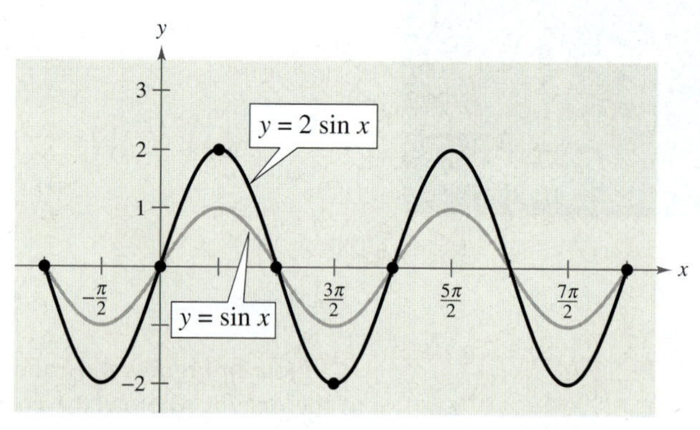

FIGURE 4.50

Technology

When using a graphing utility to graph trigonometric functions, pay special attention to the viewing window you use. For instance, try graphing $y = [\sin(10x)]/10$ in the standard viewing window in *radian* mode. What do you observe? Use the *zoom* feature to find a viewing window that displays a good view of the graph.

✓CHECKPOINT Now try Exercise 35.

To graph the examples in this section, your students must know the basic graphs of $y = \sin x$ and $y = \cos x$. For example, to sketch the graph of $y = 3 \sin x$, your students must be able to identify that because $a = 3$, the amplitude is 3 times the amplitude of $y = \sin x$.

To help students learn how to determine and locate key points (intercepts, minimums, maximums), have them mark each of the points on their graphs and then check their graphs using a graphing utility.

Amplitude and Period

In the remainder of this section you will study the graphic effect of each of the constants a, b, c, and d in equations of the forms

$$y = d + a \sin(bx - c)$$

and

$$y = d + a \cos(bx - c).$$

A quick review of the transformations you studied in Section 1.7 should help in this investigation.

The constant factor a in $y = a \sin x$ acts as a *scaling factor*—a *vertical stretch* or *vertical shrink* of the basic sine curve. If $|a| > 1$, the basic sine curve is stretched, and if $|a| < 1$, the basic sine curve is shrunk. The result is that the graph of $y = a \sin x$ ranges between $-a$ and a instead of between -1 and 1. The absolute value of a is the **amplitude** of the function $y = a \sin x$. The range of the function $y = a \sin x$ for $a > 0$ is $-a \le y \le a$.

Definition of Amplitude of Sine and Cosine Curves

The **amplitude** of $y = a \sin x$ and $y = a \cos x$ represents half the distance between the maximum and minimum values of the function and is given by

$$\text{Amplitude} = |a|.$$

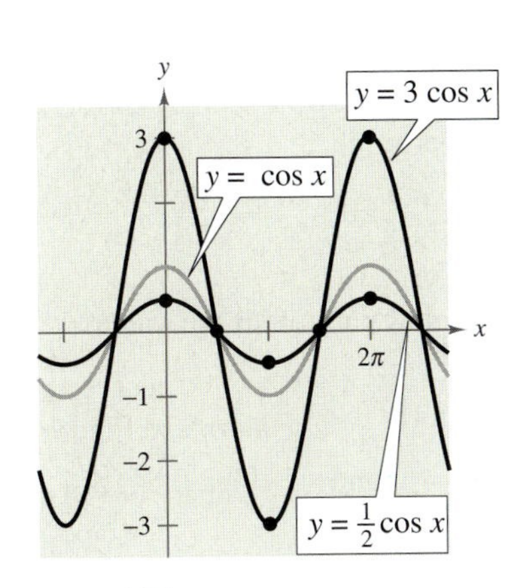

FIGURE 4.51

Example 2 Scaling: Vertical Shrinking and Stretching

On the same coordinate axes, sketch the graph of each function.

a. $y = \dfrac{1}{2} \cos x$ **b.** $y = 3 \cos x$

Solution

a. Because the amplitude of $y = \frac{1}{2} \cos x$ is $\frac{1}{2}$, the maximum value is $\frac{1}{2}$ and the minimum value is $-\frac{1}{2}$. Divide one cycle, $0 \le x \le 2\pi$, into four equal parts to get the key points

Maximum	*Intercept*	*Minimum*	*Intercept*		*Maximum*
$\left(0, \frac{1}{2}\right)$,	$\left(\frac{\pi}{2}, 0\right)$,	$\left(\pi, -\frac{1}{2}\right)$,	$\left(\frac{3\pi}{2}, 0\right)$,	and	$\left(2\pi, \frac{1}{2}\right)$.

b. A similar analysis shows that the amplitude of $y = 3 \cos x$ is 3, and the key points are

Maximum	*Intercept*	*Minimum*	*Intercept*		*Maximum*
$(0, 3)$,	$\left(\frac{\pi}{2}, 0\right)$,	$(\pi, -3)$,	$\left(\frac{3\pi}{2}, 0\right)$,	and	$(2\pi, 3)$.

The graphs of these two functions are shown in Figure 4.51. Notice that the graph of $y = \frac{1}{2} \cos x$ is a vertical *shrink* of the graph of $y = \cos x$ and the graph of $y = 3 \cos x$ is a vertical *stretch* of the graph of $y = \cos x$.

CHECKPOINT Now try Exercise 37.

Exploration

Sketch the graph of $y = \cos bx$ for $b = \frac{1}{2}$, 2, and 3. How does the value of b affect the graph? How many complete cycles occur between 0 and 2π for each value of b?

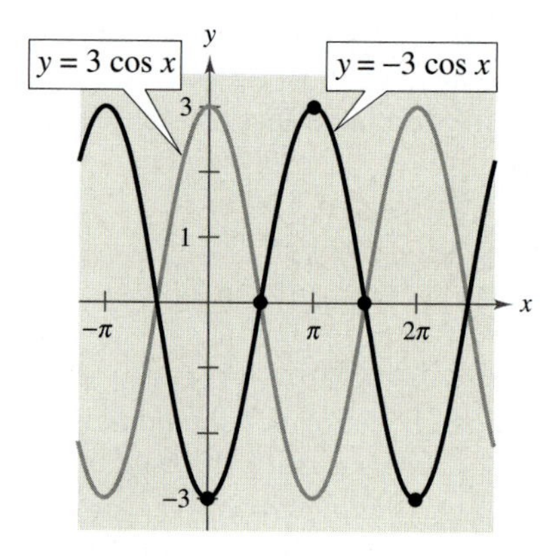

FIGURE 4.52

You know from Section 1.7 that the graph of $y = -f(x)$ is a **reflection** in the x-axis of the graph of $y = f(x)$. For instance, the graph of $y = -3 \cos x$ is a reflection of the graph of $y = 3 \cos x$, as shown in Figure 4.52.

Because $y = a \sin x$ completes one cycle from $x = 0$ to $x = 2\pi$, it follows that $y = a \sin bx$ completes one cycle from $x = 0$ to $x = 2\pi/b$.

Period of Sine and Cosine Functions

Let b be a positive real number. The **period** of $y = a \sin bx$ and $y = a \cos bx$ is given by

$$\text{Period} = \frac{2\pi}{b}.$$

Note that if $0 < b < 1$, the period of $y = a \sin bx$ is greater than 2π and represents a *horizontal stretching* of the graph of $y = a \sin x$. Similarly, if $b > 1$, the period of $y = a \sin bx$ is less than 2π and represents a *horizontal shrinking* of the graph of $y = a \sin x$. If b is negative, the identities $\sin(-x) = -\sin x$ and $\cos(-x) = \cos x$ are used to rewrite the function.

Exploration

Sketch the graph of

$$y = \sin(x - c)$$

where $c = -\pi/4, 0,$ and $\pi/4$. How does the value of c affect the graph?

Example 3 Scaling: Horizontal Stretching

Sketch the graph of $y = \sin \dfrac{x}{2}$.

Solution

The amplitude is 1. Moreover, because $b = \frac{1}{2}$, the period is

$$\frac{2\pi}{b} = \frac{2\pi}{\frac{1}{2}} = 4\pi. \qquad \text{Substitute for } b.$$

Now, divide the period-interval $[0, 4\pi]$ into four equal parts with the values π, 2π, and 3π to obtain the key points on the graph.

Intercept	*Maximum*	*Intercept*	*Minimum*		*Intercept*
$(0, 0)$,	$(\pi, 1)$,	$(2\pi, 0)$,	$(3\pi, -1)$,	and	$(4\pi, 0)$

The graph is shown in Figure 4.53.

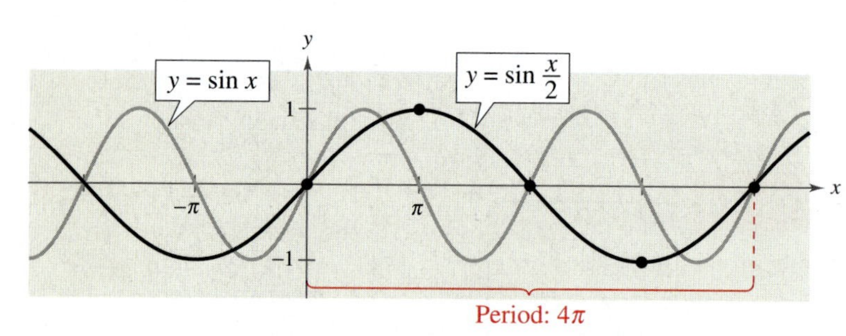

FIGURE 4.53

CHECKPOINT Now try Exercise 39.

STUDY TIP

In general, to divide a period-interval into four equal parts, successively add "period/4," starting with the left endpoint of the interval. For instance, for the period-interval $[-\pi/6, \pi/2]$ of length $2\pi/3$, you would successively add

$$\frac{2\pi/3}{4} = \frac{\pi}{6}$$

to get $-\pi/6, 0, \pi/6, \pi/3,$ and $\pi/2$ as the x-values for the key points on the graph.

Translations of Sine and Cosine Curves

The constant c in the general equations

$$y = a\sin(bx - c) \qquad \text{and} \qquad y = a\cos(bx - c)$$

creates a *horizontal translation* (shift) of the basic sine and cosine curves. Comparing $y = a\sin bx$ with $y = a\sin(bx - c)$, you find that the graph of $y = a\sin(bx - c)$ completes one cycle from $bx - c = 0$ to $bx - c = 2\pi$. By solving for x, you can find the interval for one cycle to be

$$\underbrace{\frac{c}{b}}_{\text{Left endpoint}} \le x \le \underbrace{\frac{c}{b} + \underbrace{\frac{2\pi}{b}}_{\text{Period}}}_{\text{Right endpoint}}.$$

This implies that the period of $y = a\sin(bx - c)$ is $2\pi/b$, and the graph of $y = a\sin bx$ is shifted by an amount c/b. The number c/b is the **phase shift.**

> **Graphs of Sine and Cosine Functions**
>
> The graphs of $y = a\sin(bx - c)$ and $y = a\cos(bx - c)$ have the following characteristics. (Assume $b > 0$.)
>
> $$\text{Amplitude} = |a| \qquad \text{Period} = \frac{2\pi}{b}$$
>
> The left and right endpoints of a one-cycle interval can be determined by solving the equations $bx - c = 0$ and $bx - c = 2\pi$.

Example 4 **Horizontal Translation**

Sketch the graph of $y = \frac{1}{2}\sin\left(x - \frac{\pi}{3}\right)$.

Solution

The amplitude is $\frac{1}{2}$ and the period is 2π. By solving the equations

$$x - \frac{\pi}{3} = 0 \quad \Rightarrow \quad x = \frac{\pi}{3}$$

and

$$x - \frac{\pi}{3} = 2\pi \quad \Rightarrow \quad x = \frac{7\pi}{3}$$

you see that the interval $[\pi/3, 7\pi/3]$ corresponds to one cycle of the graph. Dividing this interval into four equal parts produces the key points

Intercept	Maximum	Intercept	Minimum		Intercept
$\left(\frac{\pi}{3}, 0\right)$,	$\left(\frac{5\pi}{6}, \frac{1}{2}\right)$,	$\left(\frac{4\pi}{3}, 0\right)$,	$\left(\frac{11\pi}{6}, -\frac{1}{2}\right)$,	and	$\left(\frac{7\pi}{3}, 0\right)$.

The graph is shown in Figure 4.54.

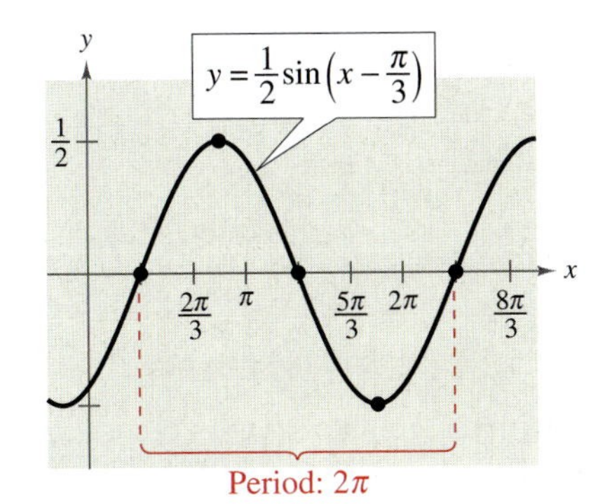

FIGURE 4.54

CHECKPOINT Now try Exercise 45.

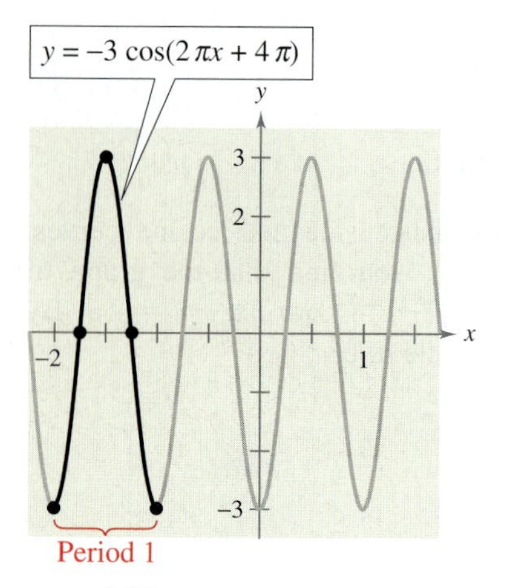

FIGURE 4.55

Example 5 Horizontal Translation

Sketch the graph of

$$y = -3\cos(2\pi x + 4\pi).$$

Solution

The amplitude is 3 and the period is $2\pi/2\pi = 1$. By solving the equations

$$2\pi x + 4\pi = 0$$
$$2\pi x = -4\pi$$
$$x = -2$$

and

$$2\pi x + 4\pi = 2\pi$$
$$2\pi x = -2\pi$$
$$x = -1$$

you see that the interval $[-2, -1]$ corresponds to one cycle of the graph. Dividing this interval into four equal parts produces the key points

Minimum	*Intercept*	*Maximum*	*Intercept*		*Minimum*
$(-2, -3)$,	$\left(-\frac{7}{4}, 0\right)$,	$\left(-\frac{3}{2}, 3\right)$,	$\left(-\frac{5}{4}, 0\right)$,	and	$(-1, -3)$.

The graph is shown in Figure 4.55.

✓CHECKPOINT Now try Exercise 47.

Activities

1. Describe the relationship between the graphs of $f(x) = \sin x$ and $g(x) = 3\sin(2x + 1)$.
   Answer: The amplitude of the basic sine curve is 1, whereas the amplitude of g is 3. The period of the basic sine curve is 2π, whereas the period of g is π. Lastly, the graph of g has a phase shift $\frac{1}{2}$ unit to the left of the graph of $f(x) = \sin x$.
2. Determine the amplitude, period, and phase shift of
   $y = \frac{1}{2}\cos(\pi x - 1)$.
   Answer: Amplitude: $\frac{1}{2}$;
   period: 2;
   phase shift $\frac{1}{\pi}$

The final type of transformation is the *vertical translation* caused by the constant d in the equations

$$y = d + a\sin(bx - c)$$

and

$$y = d + a\cos(bx - c).$$

The shift is d units upward for $d > 0$ and d units downward for $d < 0$. In other words, the graph oscillates about the horizontal line $y = d$ instead of about the x-axis.

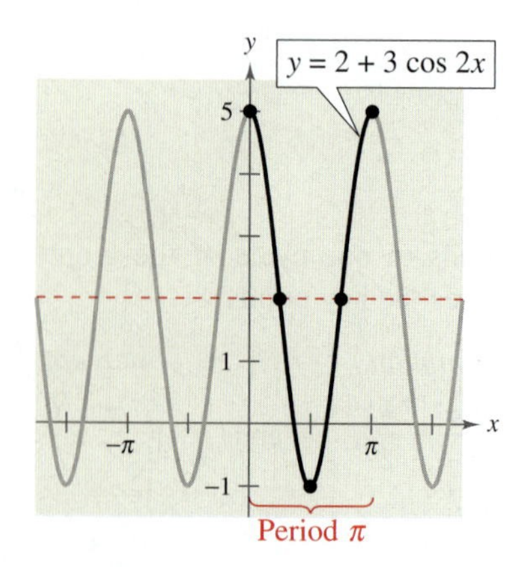

FIGURE 4.56

Example 6 Vertical Translation

Sketch the graph of

$$y = 2 + 3\cos 2x.$$

Solution

The amplitude is 3 and the period is π. The key points over the interval $[0, \pi]$ are

$$(0, 5), \quad \left(\frac{\pi}{4}, 2\right), \quad \left(\frac{\pi}{2}, -1\right), \quad \left(\frac{3\pi}{4}, 2\right), \quad \text{and} \quad (\pi, 5).$$

The graph is shown in Figure 4.56. Compared with the graph of $f(x) = 3\cos 2x$, the graph of $y - 2 + 3\cos 2x$ is shifted upward two units.

✓CHECKPOINT Now try Exercise 53.

Mathematical Modeling

Sine and cosine functions can be used to model many real-life situations, including electric currents, musical tones, radio waves, tides, and weather patterns.

Time, t	Depth, y
Midnight	3.4
2 A.M.	8.7
4 A.M.	11.3
6 A.M.	9.1
8 A.M.	3.8
10 A.M.	0.1
Noon	1.2

Example 7 Finding a Trigonometric Model

Throughout the day, the depth of water at the end of a dock in Bar Harbor, Maine varies with the tides. The table shows the depths (in feet) at various times during the morning. (Source: Nautical Software, Inc.)

a. Use a trigonometric function to model the data.

b. Find the depths at 9 A.M. and 3 P.M.

c. A boat needs at least 10 feet of water to moor at the dock. During what times in the afternoon can it safely dock?

Solution

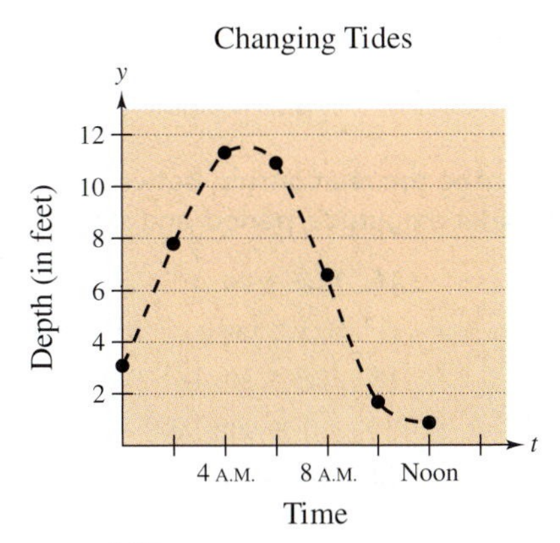

FIGURE 4.57

a. Begin by graphing the data, as shown in Figure 4.57. You can use either a sine or cosine model. Suppose you use a cosine model of the form

$$y = a\cos(bt - c) + d.$$

The difference between the maximum height and the minimum height of the graph is twice the amplitude of the function. So, the amplitude is

$$a = \frac{1}{2}[(\text{maximum depth}) - (\text{minimum depth})] = \frac{1}{2}(11.3 - 0.1) = 5.6.$$

The cosine function completes one half of a cycle between the times at which the maximum and minimum depths occur. So, the period is

$$p = 2[(\text{time of min. depth}) - (\text{time of max. depth})] = 2(10 - 4) = 12$$

which implies that $b = 2\pi/p \approx 0.524$. Because high tide occurs 4 hours after midnight, consider the left endpoint to be $c/b = 4$, so $c \approx 2.094$. Moreover, because the average depth is $\frac{1}{2}(11.3 + 0.1) = 5.7$, it follows that $d = 5.7$. So, you can model the depth with the function given by

$$y = 5.6\cos(0.524t - 2.094) + 5.7.$$

b. The depths at 9 A.M. and 3 P.M. are as follows.

$$y = 5.6\cos(0.524 \cdot 9 - 2.094) + 5.7$$

$$\approx 0.84 \text{ foot} \qquad \text{9 A.M.}$$

$$y = 5.6\cos(0.524 \cdot 15 - 2.094) + 5.7$$

$$\approx 10.57 \text{ feet} \qquad \text{3 P.M.}$$

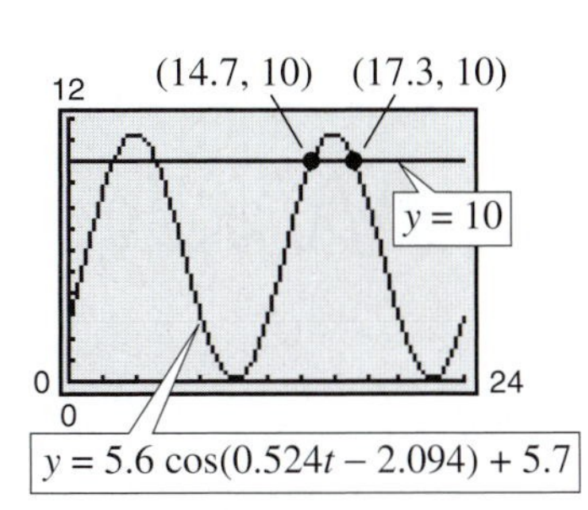

FIGURE 4.58

c. To find out when the depth y is at least 10 feet, you can graph the model with the line $y = 10$ using a graphing utility, as shown in Figure 4.58. Using the *intersect* feature, you can determine that the depth is at least 10 feet between 2:42 P.M. ($t \approx 14.7$) and 5:18 P.M. ($t \approx 17.3$).

CHECKPOINT Now try Exercise 77.

4.5 Exercises

VOCABULARY CHECK: Fill in the blanks.

1. One period of a sine or cosine function function is called one ________ of the sine curve or cosine curve.
2. The ________ of a sine or cosine curve represents half the distance between the maximum and minimum values of the function.
3. The period of a sine or cosine function is given by ________.
4. For the function given by $y = a \sin(bx - c)$, $\frac{c}{b}$ represents the ________ ________ of the graph of the function.
5. For the function given by $y = d + a \cos(bx - c)$, d represents a ________ ________ of the graph of the function.

PREREQUISITE SKILLS REVIEW: Practice and review algebra skills needed for this section at **www.Eduspace.com.**

In Exercises 1–14, find the period and amplitude.

1. $y = 3 \sin 2x$

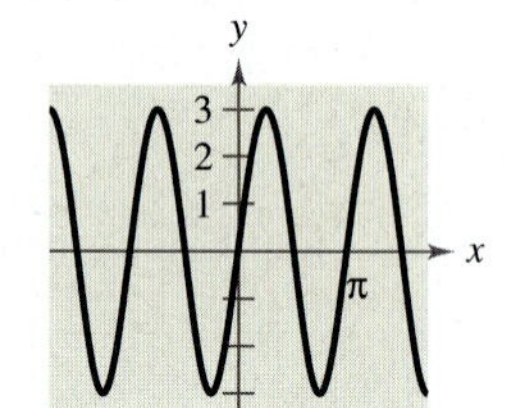

2. $y = 2 \cos 3x$

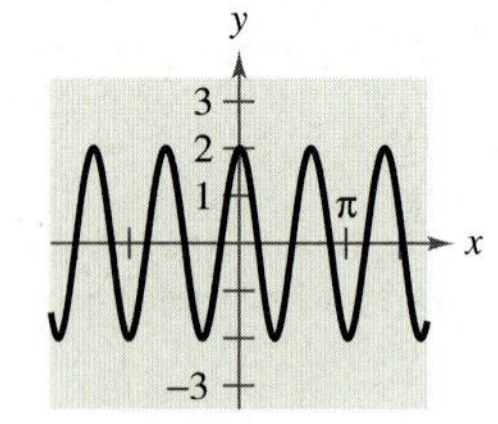

3. $y = \frac{5}{2} \cos \frac{x}{2}$

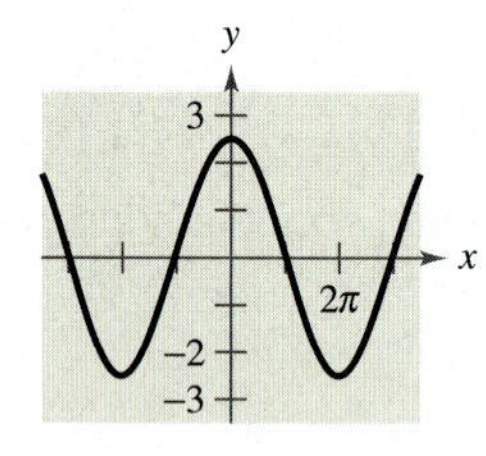

4. $y = -3 \sin \frac{x}{3}$

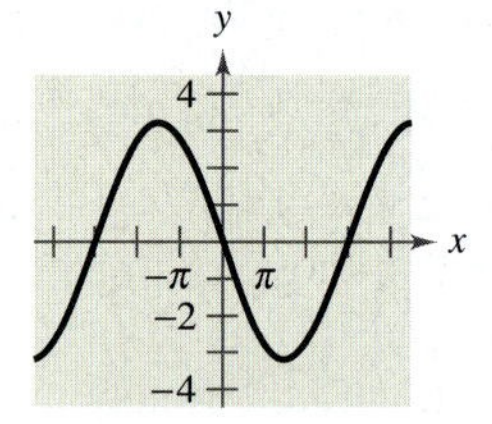

5. $y = \frac{1}{2} \sin \frac{\pi x}{3}$

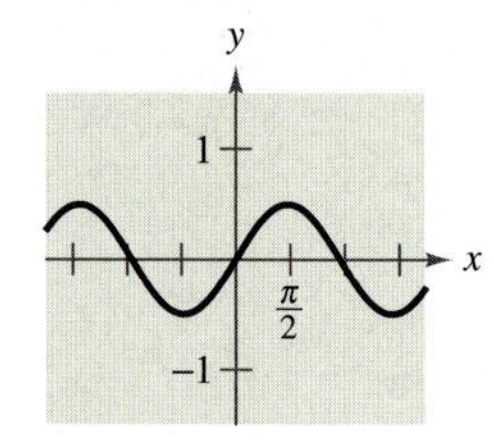

6. $y = \frac{3}{2} \cos \frac{\pi x}{2}$

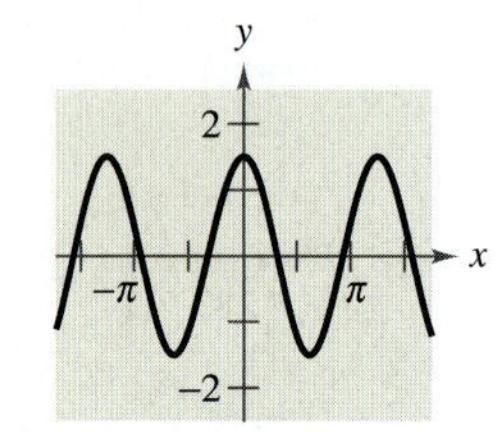

7. $y = -2 \sin x$
8. $y = -\cos \frac{2x}{3}$
9. $y = 3 \sin 10x$
10. $y = \frac{1}{3} \sin 8x$
11. $y = \frac{1}{2} \cos \frac{2x}{3}$
12. $y = \frac{5}{2} \cos \frac{x}{4}$
13. $y = \frac{1}{4} \sin 2\pi x$
14. $y = \frac{2}{3} \cos \frac{\pi x}{10}$

In Exercises 15–22, describe the relationship between the graphs of *f* and *g*. Consider amplitude, period, and shifts.

15. $f(x) = \sin x$
    $g(x) = \sin(x - \pi)$
16. $f(x) = \cos x$
    $g(x) = \cos(x + \pi)$
17. $f(x) = \cos 2x$
    $g(x) = -\cos 2x$
18. $f(x) = \sin 3x$
    $g(x) = \sin(-3x)$
19. $f(x) = \cos x$
    $g(x) = \cos 2x$
20. $f(x) = \sin x$
    $g(x) = \sin 3x$
21. $f(x) = \sin 2x$
    $g(x) = 3 + \sin 2x$
22. $f(x) = \cos 4x$
    $g(x) = -2 + \cos 4x$

In Exercises 23–26, describe the relationship between the graphs of *f* and *g*. Consider amplitude, period, and shifts.

23.

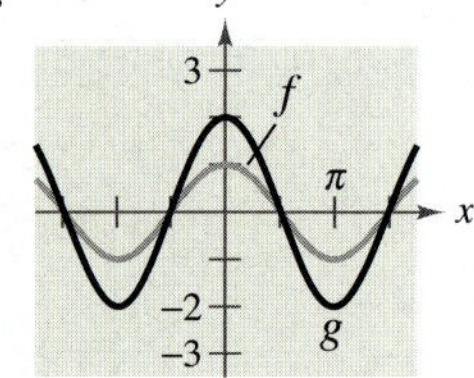

24.

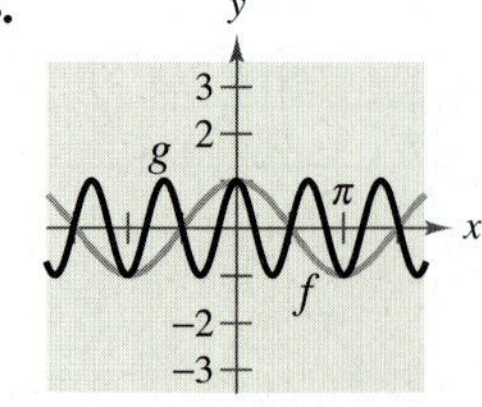

25.

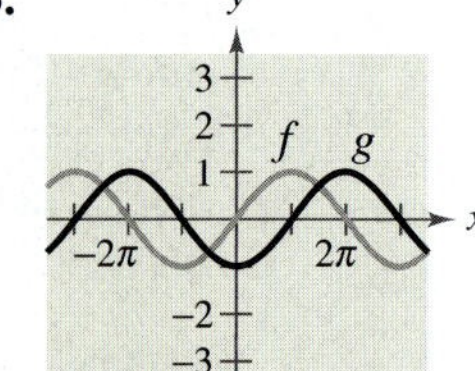

26.

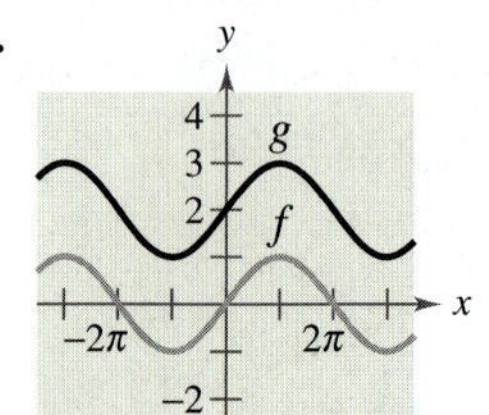

In Exercises 27–34, graph *f* and *g* on the same set of coordinate axes. (Include two full periods.)

27. $f(x) = -2 \sin x$
$g(x) = 4 \sin x$

28. $f(x) = \sin x$
$g(x) = \sin \dfrac{x}{3}$

29. $f(x) = \cos x$
$g(x) = 1 + \cos x$

30. $f(x) = 2 \cos 2x$
$g(x) = -\cos 4x$

31. $f(x) = -\dfrac{1}{2} \sin \dfrac{x}{2}$
$g(x) = 3 - \dfrac{1}{2} \sin \dfrac{x}{2}$

32. $f(x) = 4 \sin \pi x$
$g(x) = 4 \sin \pi x - 3$

33. $f(x) = 2 \cos x$
$g(x) = 2 \cos(x + \pi)$

34. $f(x) = -\cos x$
$g(x) = -\cos(x - \pi)$

In Exercises 35–56, sketch the graph of the function. (Include two full periods.)

35. $y = 3 \sin x$

36. $y = \frac{1}{4} \sin x$

37. $y = \frac{1}{3} \cos x$

38. $y = 4 \cos x$

39. $y = \cos \dfrac{x}{2}$

40. $y = \sin 4x$

41. $y = \cos 2\pi x$

42. $y = \sin \dfrac{\pi x}{4}$

43. $y = -\sin \dfrac{2\pi x}{3}$

44. $y = -10 \cos \dfrac{\pi x}{6}$

45. $y = \sin\left(x - \dfrac{\pi}{4}\right)$

46. $y = \sin(x - \pi)$

47. $y = 3 \cos(x + \pi)$

48. $y = 4 \cos\left(x + \dfrac{\pi}{4}\right)$

49. $y = 2 - \sin \dfrac{2\pi x}{3}$

50. $y = -3 + 5 \cos \dfrac{\pi t}{12}$

51. $y = 2 + \frac{1}{10} \cos 60\pi x$

52. $y = 2 \cos x - 3$

53. $y = 3 \cos(x + \pi) - 3$

54. $y = 4 \cos\left(x + \dfrac{\pi}{4}\right) + 4$

55. $y = \dfrac{2}{3} \cos\left(\dfrac{x}{2} - \dfrac{\pi}{4}\right)$

56. $y = -3 \cos(6x + \pi)$

In Exercises 57–62, use a graphing utility to graph the function. Include two full periods. Be sure to choose an appropriate viewing window.

57. $y = -2 \sin(4x + \pi)$

58. $y = -4 \sin\left(\dfrac{2}{3}x - \dfrac{\pi}{3}\right)$

59. $y = \cos\left(2\pi x - \dfrac{\pi}{2}\right) + 1$

60. $y = 3 \cos\left(\dfrac{\pi x}{2} + \dfrac{\pi}{2}\right) - 2$

61. $y = -0.1 \sin\left(\dfrac{\pi x}{10} + \pi\right)$

62. $y = \frac{1}{100} \sin 120\pi t$

Graphical Reasoning **In Exercises 63–66, find *a* and *d* for the function $f(x) = a \cos x + d$ such that the graph of *f* matches the figure.**

63.

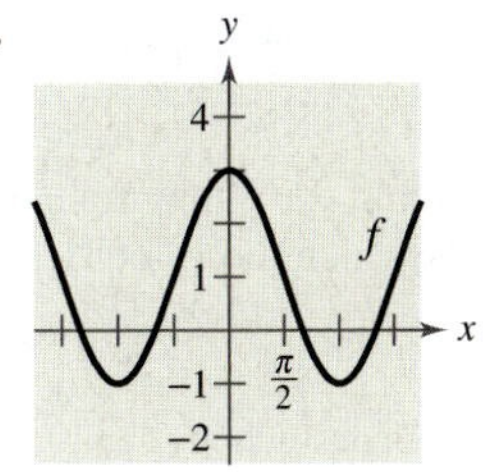

64.

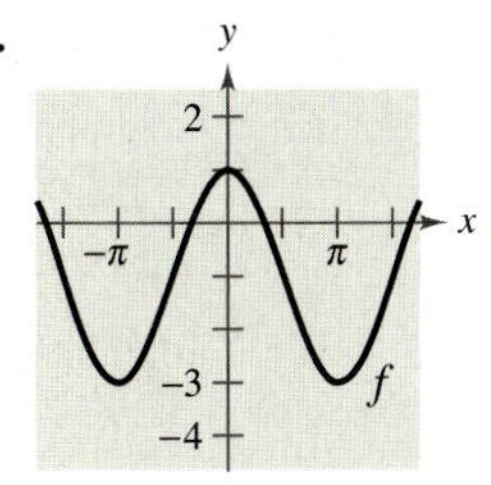

65.

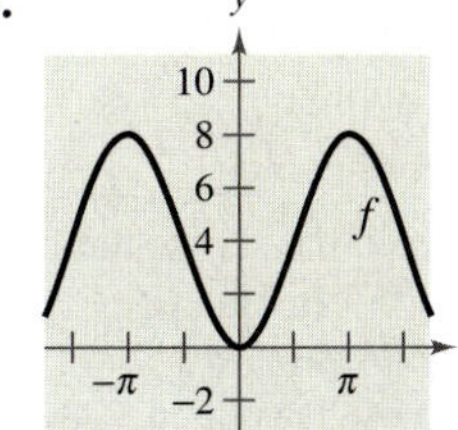

66.

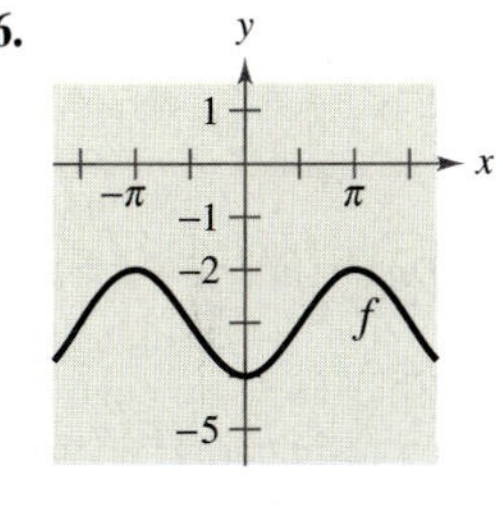

Graphical Reasoning **In Exercises 67–70, find *a*, *b*, and *c* for the function $f(x) = a \sin(bx - c)$ such that the graph of *f* matches the figure.**

67.

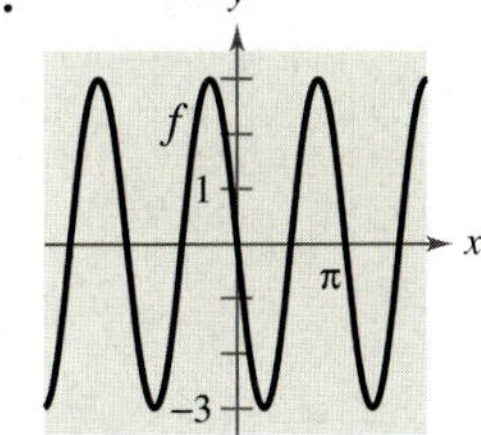

68.

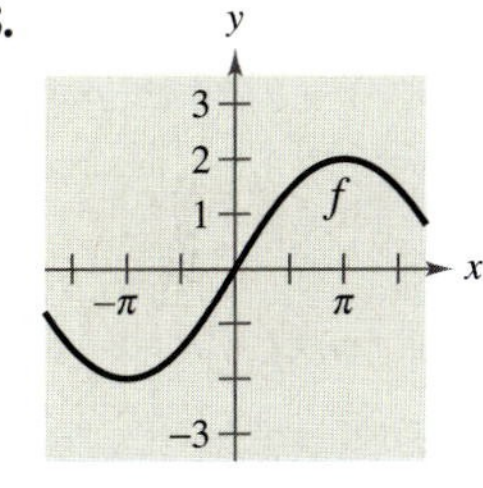

69.

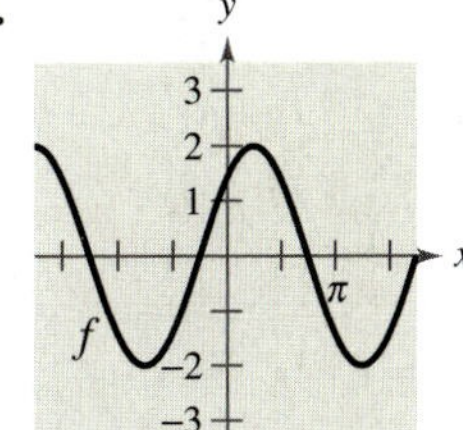

70.

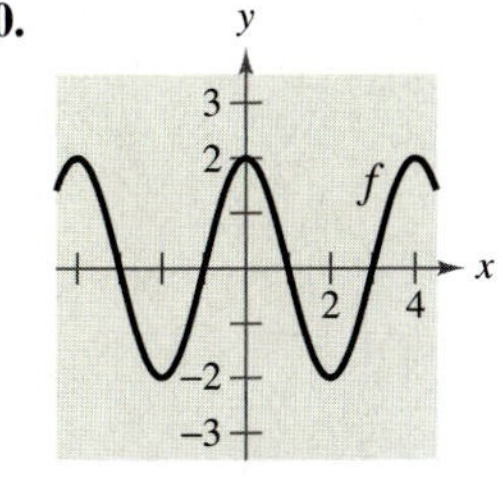

In Exercises 71 and 72, use a graphing utility to graph y_1 and y_2 in the interval $[-2\pi, 2\pi]$. Use the graphs to find real numbers x such that $y_1 = y_2$.

71. $y_1 = \sin x$
$y_2 = -\frac{1}{2}$

72. $y_1 = \cos x$
$y_2 = -1$

73. *Respiratory Cycle* For a person at rest, the velocity v (in liters per second) of air flow during a respiratory cycle (the time from the beginning of one breath to the beginning of the next) is given by $v = 0.85 \sin \frac{\pi t}{3}$, where t is the time (in seconds). (Inhalation occurs when $v > 0$, and exhalation occurs when $v < 0$.)

(a) Find the time for one full respiratory cycle.

(b) Find the number of cycles per minute.

(c) Sketch the graph of the velocity function.

74. *Respiratory Cycle* After exercising for a few minutes, a person has a respiratory cycle for which the velocity of air flow is approximated by $v = 1.75 \sin \frac{\pi t}{2}$, where t is the time (in seconds). (Inhalation occurs when $v > 0$, and exhalation occurs when $v < 0$.)

(a) Find the time for one full respiratory cycle.

(b) Find the number of cycles per minute.

(c) Sketch the graph of the velocity function.

75. *Data Analysis: Meteorology* The table shows the maximum daily high temperatures for Tallahassee T and Chicago C (in degrees Fahrenheit) for month t, with $t = 1$ corresponding to January. (Source: National Climatic Data Center)

Month, t	Tallahassee, T	Chicago, C
1	63.8	29.6
2	67.4	34.7
3	74.0	46.1
4	80.0	58.0
5	86.5	69.9
6	90.9	79.2
7	92.0	83.5
8	91.5	81.2
9	88.5	73.9
10	81.2	62.1
11	72.9	47.1
12	65.8	34.4

(a) A model for the temperature in Tallahassee is given by

$$T(t) = 77.90 + 14.10 \cos\left(\frac{\pi t}{6} - 3.67\right).$$

Find a trigonometric model for Chicago.

(b) Use a graphing utility to graph the data points and the model for the temperatures in Tallahassee. How well does the model fit the data?

(c) Use a graphing utility to graph the data points and the model for the temperatures in Chicago. How well does the model fit the data?

(d) Use the models to estimate the average maximum temperature in each city. Which term of the models did you use? Explain.

(e) What is the period of each model? Are the periods what you expected? Explain.

(f) Which city has the greater variability in temperature throughout the year? Which factor of the models determines this variability? Explain.

76. *Health* The function given by $P = 100 - 20 \cos \frac{5\pi t}{3}$ approximates the blood pressure P (in millimeters) of mercury at time t (in seconds) for a person at rest.

(a) Find the period of the function.

(b) Find the number of heartbeats per minute.

77. *Piano Tuning* When tuning a piano, a technician strikes a tuning fork for the A above middle C and sets up a wave motion that can be approximated by $y = 0.001 \sin 880\pi t$, where t is the time (in seconds).

(a) What is the period of the function?

(b) The frequency f is given by $f = 1/p$. What is the frequency of the note?

Model It

78. *Data Analysis: Astronomy* The percent y of the moon's face that is illuminated on day x of the year 2007, where $x = 1$ represents January 1, is shown in the table. (Source: U.S. Naval Observatory)

x	y
3	1.0
11	0.5
19	0.0
26	0.5
32	1.0
40	0.5

(a) Create a scatter plot of the data.

(b) Find a trigonometric model that fits the data.

(c) Add the graph of your model in part (b) to the scatter plot. How well does the model fit the data?

(d) What is the period of the model?

(e) Estimate the moon's percent illumination for March 12, 2007.

79. ***Fuel Consumption*** The daily consumption C (in gallons) of diesel fuel on a farm is modeled by

$$C = 30.3 + 21.6 \sin\left(\frac{2\pi t}{365} + 10.9\right)$$

where t is the time (in days), with $t = 1$ corresponding to January 1.

(a) What is the period of the model? Is it what you expected? Explain.

(b) What is the average daily fuel consumption? Which term of the model did you use? Explain.

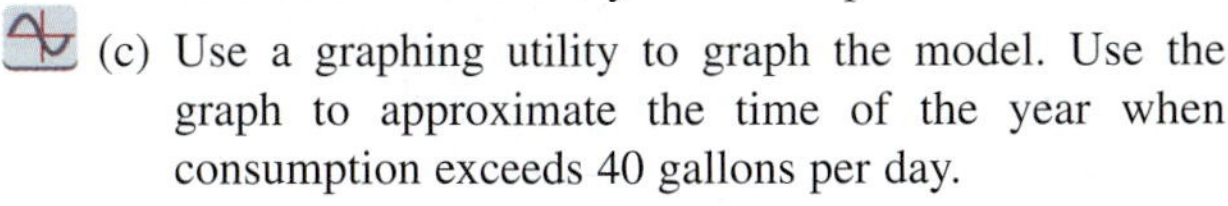

(c) Use a graphing utility to graph the model. Use the graph to approximate the time of the year when consumption exceeds 40 gallons per day.

80. ***Ferris Wheel*** A Ferris wheel is built such that the height h (in feet) above ground of a seat on the wheel at time t (in seconds) can be modeled by

$$h(t) = 53 + 50 \sin\left(\frac{\pi}{10}t - \frac{\pi}{2}\right).$$

(a) Find the period of the model. What does the period tell you about the ride?

(b) Find the amplitude of the model. What does the amplitude tell you about the ride?

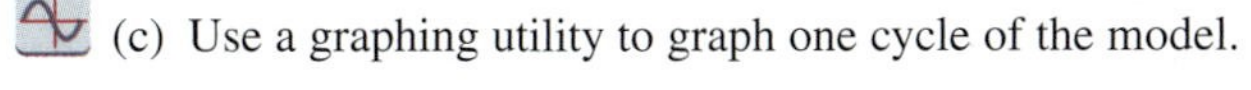

(c) Use a graphing utility to graph one cycle of the model.

Synthesis

True or False? **In Exercises 81–83, determine whether the statement is true or false. Justify your answer.**

81. The graph of the function given by $f(x) = \sin(x + 2\pi)$ translates the graph of $f(x) = \sin x$ exactly one period to the right so that the two graphs look identical.

82. The function given by $y = \frac{1}{2}\cos 2x$ has an amplitude that is twice that of the function given by $y = \cos x$.

83. The graph of $y = -\cos x$ is a reflection of the graph of $y = \sin(x + \pi/2)$ in the x-axis.

84. ***Writing*** Use a graphing utility to graph the function given by $y = d + a\sin(bx - c)$, for several different values of a, b, c, and d. Write a paragraph describing the changes in the graph corresponding to changes in each constant.

Conjecture **In Exercises 85 and 86, graph f and g on the same set of coordinate axes. Include two full periods. Make a conjecture about the functions.**

85. $f(x) = \sin x, \quad g(x) = \cos\left(x - \frac{\pi}{2}\right)$

86. $f(x) = \sin x, \quad g(x) = -\cos\left(x + \frac{\pi}{2}\right)$

87. ***Exploration*** Using calculus, it can be shown that the sine and cosine functions can be approximated by the polynomials

$$\sin x \approx x - \frac{x^3}{3!} + \frac{x^5}{5!} \quad \text{and} \quad \cos x \approx 1 - \frac{x^2}{2!} + \frac{x^4}{4!}$$

where x is in radians.

(a) Use a graphing utility to graph the sine function and its polynomial approximation in the same viewing window. How do the graphs compare?

(b) Use a graphing utility to graph the cosine function and its polynomial approximation in the same viewing window. How do the graphs compare?

(c) Study the patterns in the polynomial approximations of the sine and cosine functions and predict the next term in each. Then repeat parts (a) and (b). How did the accuracy of the approximations change when an additional term was added?

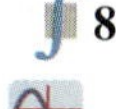

88. ***Exploration*** Use the polynomial approximations for the sine and cosine functions in Exercise 87 to approximate the following function values. Compare the results with those given by a calculator. Is the error in the approximation the same in each case? Explain.

(a) $\sin\frac{1}{2}$ (b) $\sin 1$ (c) $\sin\frac{\pi}{6}$

(d) $\cos(-0.5)$ (e) $\cos 1$ (f) $\cos\frac{\pi}{4}$

Skills Review

In Exercises 89–92, use the properties of logarithms to write the expression as a sum, difference, and/or constant multiple of a logarithm.

89. $\log_{10}\sqrt{x - 2}$

90. $\log_2[x^2(x - 3)]$

91. $\ln\dfrac{t^3}{t - 1}$

92. $\ln\sqrt{\dfrac{z}{z^2 + 1}}$

In Exercises 93–96, write the expression as the logarithm of a single quantity.

93. $\frac{1}{2}(\log_{10} x + \log_{10} y)$

94. $2\log_2 x + \log_2(xy)$

95. $\ln 3x - 4\ln y$

96. $\frac{1}{2}(\ln 2x - 2\ln x) + 3\ln x$

---

97. Make a Decision To work an extended application analyzing the normal daily maximum temperature and normal precipitation in Honolulu, Hawaii, visit this text's website at *college.hmco.com*. *(Data Source: NOAA)*

4.6 Graphs of Other Trigonometric Functions

What you should learn

- Sketch the graphs of tangent functions.
- Sketch the graphs of cotangent functions.
- Sketch the graphs of secant and cosecant functions.
- Sketch the graphs of damped trigonometric functions.

Why you should learn it

Trigonometric functions can be used to model real-life situations such as the distance from a television camera to a unit in a parade as in Exercise 76 on page 341.

Photodisc/Getty Images

Graph of the Tangent Function

Recall that the tangent function is odd. That is, $\tan(-x) = -\tan x$. Consequently, the graph of $y = \tan x$ is symmetric with respect to the origin. You also know from the identity $\tan x = \sin x/\cos x$ that the tangent is undefined for values at which $\cos x = 0$. Two such values are $x = \pm\pi/2 \approx \pm 1.5708$.

x	$-\frac{\pi}{2}$	-1.57	-1.5	$-\frac{\pi}{4}$	0	$\frac{\pi}{4}$	1.5	1.57	$\frac{\pi}{2}$
$\tan x$	Undef.	-1255.8	-14.1	-1	0	1	14.1	1255.8	Undef.

As indicated in the table, $\tan x$ increases without bound as x approaches $\pi/2$ from the left, and decreases without bound as x approaches $-\pi/2$ from the right. So, the graph of $y = \tan x$ has *vertical asymptotes* at $x = \pi/2$ and $x = -\pi/2$, as shown in Figure 4.59. Moreover, because the period of the tangent function is π, vertical asymptotes also occur when $x = \pi/2 + n\pi$, where n is an integer. The domain of the tangent function is the set of all real numbers other than $x = \pi/2 + n\pi$, and the range is the set of all real numbers.

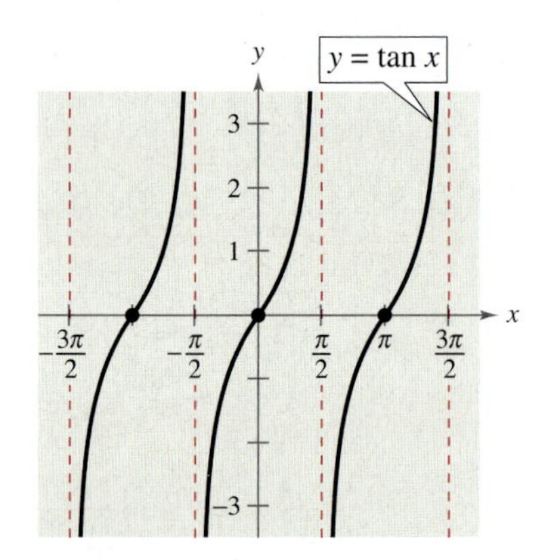

FIGURE 4.59

PERIOD: π
DOMAIN: ALL $x \neq \frac{\pi}{2} + n\pi$
RANGE: $(-\infty, \infty)$
VERTICAL ASYMPTOTES: $x = \frac{\pi}{2} + n\pi$

Sketching the graph of $y = a\tan(bx - c)$ is similar to sketching the graph of $y = a\sin(bx - c)$ in that you locate key points that identify the intercepts and asymptotes. Two consecutive vertical asymptotes can be found by solving the equations

$$bx - c = -\frac{\pi}{2} \quad \text{and} \quad bx - c = \frac{\pi}{2}.$$

The midpoint between two consecutive vertical asymptotes is an x-intercept of the graph. The period of the function $y = a\tan(bx - c)$ is the distance between two consecutive vertical asymptotes. The amplitude of a tangent function is not defined. After plotting the asymptotes and the x-intercept, plot a few additional points between the two asymptotes and sketch one cycle. Finally, sketch one or two additional cycles to the left and right.

Consider reviewing period, range, and domain for all six trigonometric functions, especially emphasizing the difference between the periods of the tangent and cotangent functions.

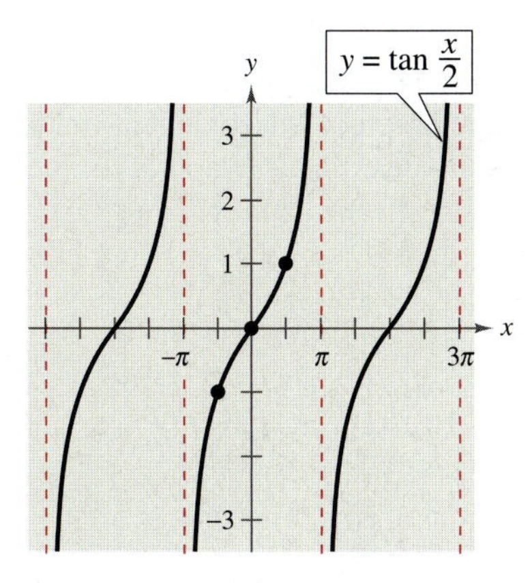

FIGURE 4.60

Example 1 Sketching the Graph of a Tangent Function

Sketch the graph of $y = \tan \frac{x}{2}$.

Solution

By solving the equations

$$\frac{x}{2} = -\frac{\pi}{2} \quad \text{and} \quad \frac{x}{2} = \frac{\pi}{2}$$

$$x = -\pi \qquad\qquad x = \pi$$

you can see that two consecutive vertical asymptotes occur at $x = -\pi$ and $x = \pi$. Between these two asymptotes, plot a few points, including the x-intercept, as shown in the table. Three cycles of the graph are shown in Figure 4.60.

x	$-\pi$	$-\frac{\pi}{2}$	0	$\frac{\pi}{2}$	π
$\tan \frac{x}{2}$	Undef.	-1	0	1	Undef.

Now try Exercise 7.

Example 2 Sketching the Graph of a Tangent Function

Sketch the graph of $y = -3 \tan 2x$.

Solution

By solving the equations

$$2x = -\frac{\pi}{2} \quad \text{and} \quad 2x = \frac{\pi}{2}$$

$$x = -\frac{\pi}{4} \qquad\qquad x = \frac{\pi}{4}$$

you can see that two consecutive vertical asymptotes occur at $x = -\pi/4$ and $x = \pi/4$. Between these two asymptotes, plot a few points, including the x-intercept, as shown in the table. Three cycles of the graph are shown in Figure 4.61.

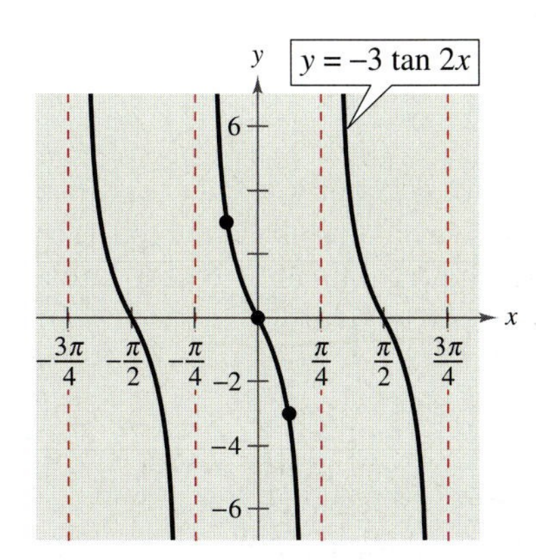

FIGURE 4.61

x	$-\frac{\pi}{4}$	$-\frac{\pi}{8}$	0	$\frac{\pi}{8}$	$\frac{\pi}{4}$
$-3 \tan 2x$	Undef.	3	0	-3	Undef.

CHECKPOINT Now try Exercise 9.

By comparing the graphs in Examples 1 and 2, you can see that the graph of $y = a \tan(bx - c)$ increases between consecutive vertical asymptotes when $a > 0$, and decreases between consecutive vertical asymptotes when $a < 0$. In other words, the graph for $a < 0$ is a reflection in the x-axis of the graph for $a > 0$.

Graph of the Cotangent Function

The graph of the cotangent function is similar to the graph of the tangent function. It also has a period of π. However, from the identity

$$y = \cot x = \frac{\cos x}{\sin x}$$

you can see that the cotangent function has vertical asymptotes when $\sin x$ is zero, which occurs at $x = n\pi$, where n is an integer. The graph of the cotangent function is shown in Figure 4.62. Note that two consecutive vertical asymptotes of the graph of $y = a\cot(bx - c)$ can be found by solving the equations $bx - c = 0$ and $bx - c = \pi$.

Technology

Some graphing utilities have difficulty graphing trigonometric functions that have vertical asymptotes. Your graphing utility may connect parts of the graphs of tangent, cotangent, secant, and cosecant functions that are not supposed to be connected. To eliminate this problem, change the mode of the graphing utility to *dot* mode.

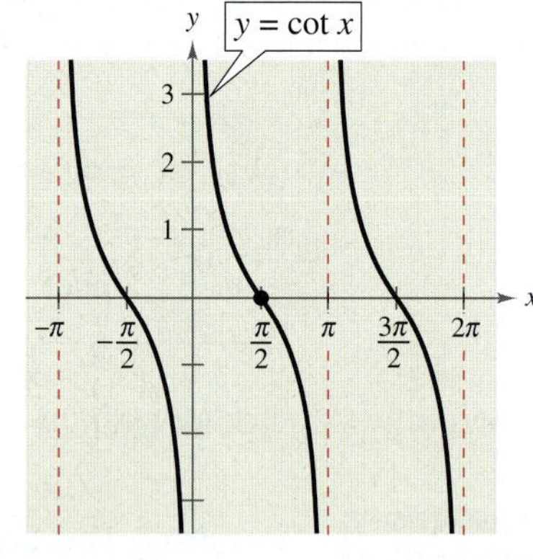

FIGURE 4.62

PERIOD: π

DOMAIN: ALL $x \neq n\pi$

RANGE: $(-\infty, \infty)$

VERTICAL ASYMPTOTES: $x = n\pi$

Example 3 Sketching the Graph of a Cotangent Function

Sketch the graph of $y = 2\cot\frac{x}{3}$.

Solution

By solving the equations

$$\frac{x}{3} = 0 \quad \text{and} \quad \frac{x}{3} = \pi$$

$$x = 0 \qquad\qquad x = 3\pi$$

you can see that two consecutive vertical asymptotes occur at $x = 0$ and $x = 3\pi$. Between these two asymptotes, plot a few points, including the x-intercept, as shown in the table. Three cycles of the graph are shown in Figure 4.63. Note that the period is 3π, the distance between consecutive asymptotes.

x	0	$\frac{3\pi}{4}$	$\frac{3\pi}{2}$	$\frac{9\pi}{4}$	3π
$2\cot\frac{x}{3}$	Undef.	2	0	-2	Undef.

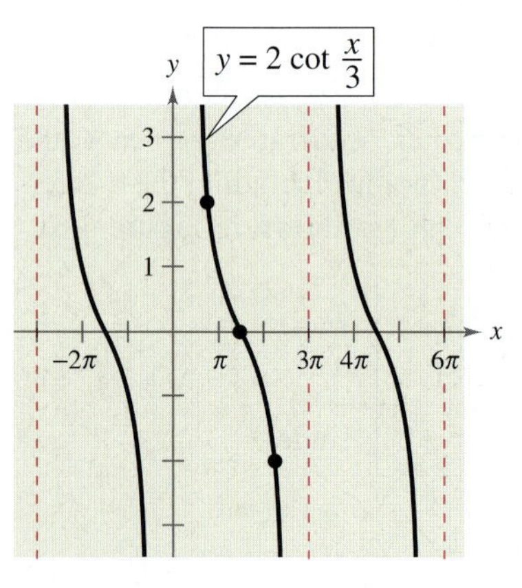

FIGURE 4.63

CHECKPOINT Now try Exercise 19.

Graphs of the Reciprocal Functions

Using a graphing utility to graph cosecant and secant functions can help reinforce the need to use the reciprocal functions.

The graphs of the two remaining trigonometric functions can be obtained from the graphs of the sine and cosine functions using the reciprocal identities

$$\csc x = \frac{1}{\sin x} \quad \text{and} \quad \sec x = \frac{1}{\cos x}.$$

For instance, at a given value of x, the y-coordinate of $\sec x$ is the reciprocal of the y-coordinate of $\cos x$. Of course, when $\cos x = 0$, the reciprocal does not exist. Near such values of x, the behavior of the secant function is similar to that of the tangent function. In other words, the graphs of

$$\tan x = \frac{\sin x}{\cos x} \quad \text{and} \quad \sec x = \frac{1}{\cos x}$$

have vertical asymptotes at $x = \pi/2 + n\pi$, where n is an integer, and the cosine is zero at these x-values. Similarly,

$$\cot x = \frac{\cos x}{\sin x} \quad \text{and} \quad \csc x = \frac{1}{\sin x}$$

have vertical asymptotes where $\sin x = 0$—that is, at $x = n\pi$.

To sketch the graph of a secant or cosecant function, you should first make a sketch of its reciprocal function. For instance, to sketch the graph of $y = \csc x$, first sketch the graph of $y = \sin x$. Then take reciprocals of the y-coordinates to obtain points on the graph of $y = \csc x$. This procedure is used to obtain the graphs shown in Figure 4.64.

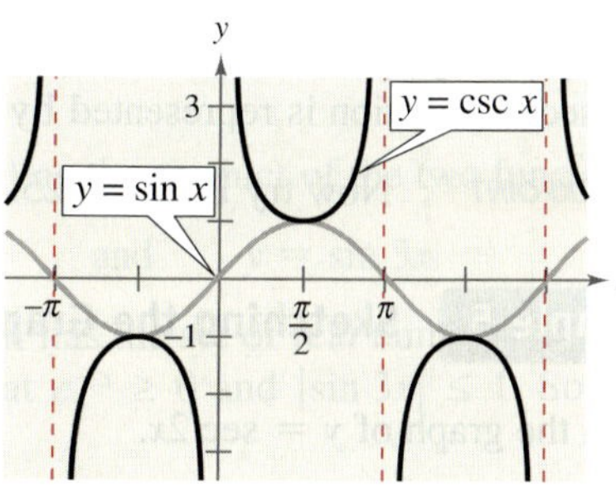

PERIOD: 2π
DOMAIN: ALL $x \neq n\pi$
RANGE: $(-\infty, -1] \cup [1, \infty)$
VERTICAL ASYMPTOTES: $x = n\pi$
SYMMETRY: ORIGIN

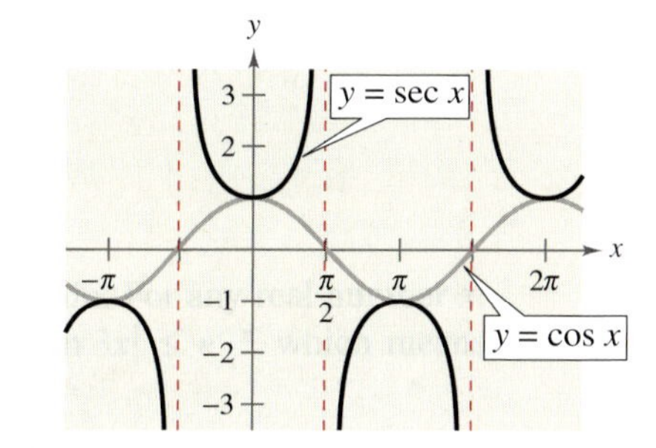

PERIOD: 2π
DOMAIN: ALL $x \neq \frac{\pi}{2} + n\pi$
RANGE: $(-\infty, -1] \cup [1, \infty)$
VERTICAL ASYMPTOTES: $x = \frac{\pi}{2} + n\pi$
SYMMETRY: y-AXIS

FIGURE 4.64

In comparing the graphs of the cosecant and secant functions with those of the sine and cosine functions, note that the "hills" and "valleys" are interchanged. For example, a hill (or maximum point) on the sine curve corresponds to a valley (a relative minimum) on the cosecant curve, and a valley (or minimum point) on the sine curve corresponds to a hill (a relative maximum) on the cosecant curve, as shown in Figure 4.65. Additionally, x-intercepts of the sine and cosine functions become vertical asymptotes of the cosecant and secant functions, respectively (see Figure 4.65).

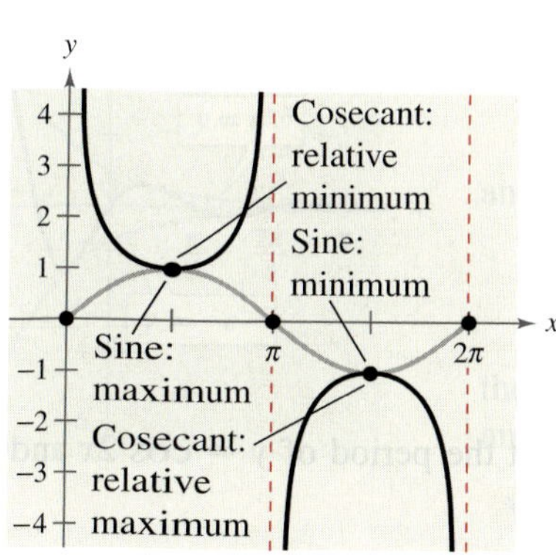

FIGURE 4.65

23. *Angle of Elevation* The height of an outdoor basketball backboard is $12\frac{1}{2}$ feet, and the backboard casts a shadow $17\frac{1}{3}$ feet long.

(a) Draw a right triangle that gives a visual representation of the problem. Label the known and unknown quantities.

(b) Use a trigonometric function to write an equation involving the unknown quantity.

(c) Find the angle of elevation of the sun.

24. *Angle of Depression* A Global Positioning System satellite orbits 12,500 miles above Earth's surface (see figure). Find the angle of depression from the satellite to the horizon. Assume the radius of Earth is 4000 miles.

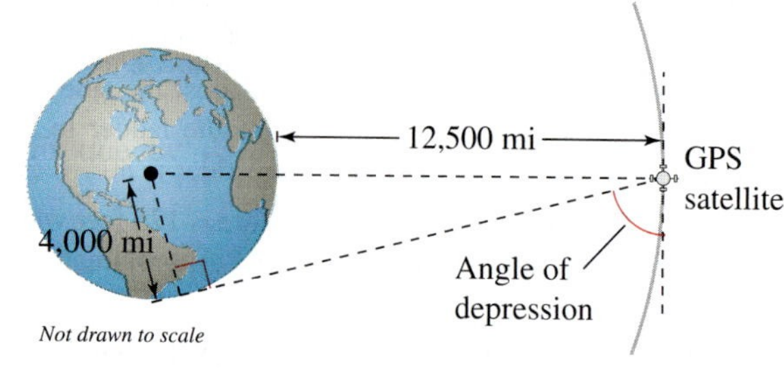

25. *Angle of Depression* A cellular telephone tower that is 150 feet tall is placed on top of a mountain that is 1200 feet above sea level. What is the angle of depression from the top of the tower to a cell phone user who is 5 horizontal miles away and 400 feet above sea level?

26. *Airplane Ascent* During takeoff, an airplane's angle of ascent is 18° and its speed is 275 feet per second.

(a) Find the plane's altitude after 1 minute.

(b) How long will it take the plane to climb to an altitude of 10,000 feet?

27. *Mountain Descent* A sign on a roadway at the top of a mountain indicates that for the next 4 miles the grade is 10.5° (see figure). Find the change in elevation over that distance for a car descending the mountain.

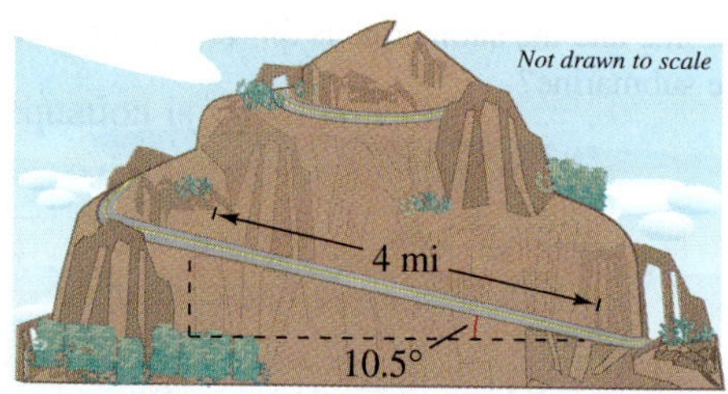

28. *Mountain Descent* A roadway sign at the top of a mountain indicates that for the next 4 miles the grade is 12%. Find the angle of the grade and the change in elevation over the 4 miles for a car descending the mountain.

29. *Navigation* An airplane flying at 600 miles per hour has a bearing of 52°. After flying for 1.5 hours, how far north and how far east will the plane have traveled from its point of departure?

30. *Navigation* A jet leaves Reno, Nevada and is headed toward Miami, Florida at a bearing of 100°. The distance between the two cities is approximately 2472 miles.

(a) How far north and how far west is Reno relative to Miami?

(b) If the jet is to return directly to Reno from Miami, at what bearing should it travel?

31. *Navigation* A ship leaves port at noon and has a bearing of S 29° W. The ship sails at 20 knots.

(a) How many nautical miles south and how many nautical miles west will the ship have traveled by 6:00 P.M.?

(b) At 6:00 P.M., the ship changes course to due west. Find the ship's bearing and distance from the port of departure at 7:00 P.M.

32. *Navigation* A privately owned yacht leaves a dock in Myrtle Beach, South Carolina and heads toward Freeport in the Bahamas at a bearing of S 1.4° E. The yacht averages a speed of 20 knots over the 428 nautical-mile trip.

(a) How long will it take the yacht to make the trip?

(b) How far east and south is the yacht after 12 hours?

(c) If a plane leaves Myrtle Beach to fly to Freeport, what bearing should be taken?

33. *Surveying* A surveyor wants to find the distance across a swamp (see figure). The bearing from A to B is N 32° W. The surveyor walks 50 meters from A, and at the point C the bearing to B is N 68° W. Find (a) the bearing from A to C and (b) the distance from A to B.

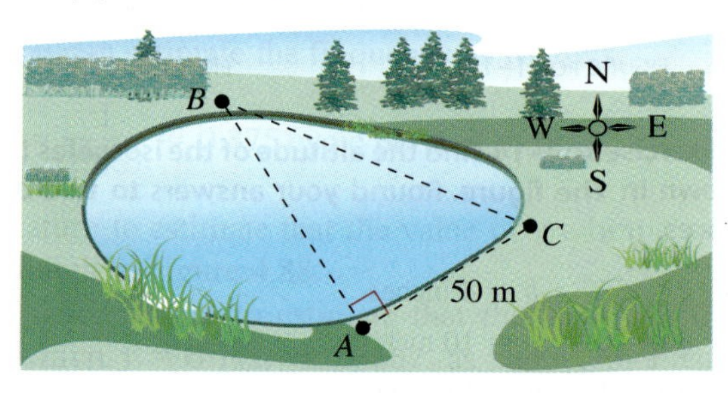

34. *Location of a Fire* Two fire towers are 30 kilometers apart, where tower A is due west of tower B. A fire is spotted from the towers, and the bearings from A and B are E 14° N and W 34° N, respectively (see figure). Find the distance d of the fire from the line segment AB.

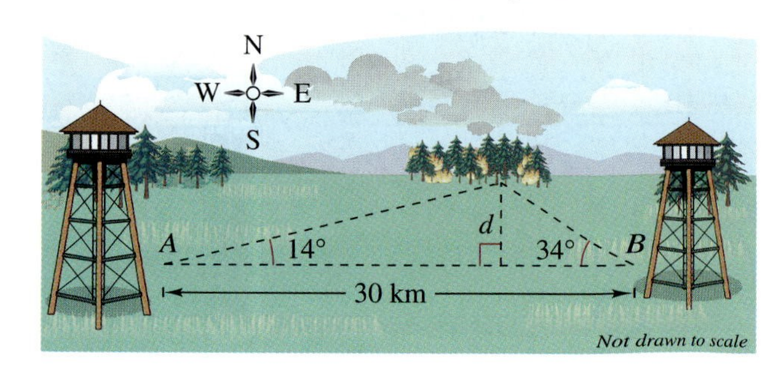

35. ***Navigation*** A ship is 45 miles east and 30 miles south of port. The captain wants to sail directly to port. What bearing should be taken?

36. ***Navigation*** An airplane is 160 miles north and 85 miles east of an airport. The pilot wants to fly directly to the airport. What bearing should be taken?

37. ***Distance*** An observer in a lighthouse 350 feet above sea level observes two ships directly offshore. The angles of depression to the ships are 4° and 6.5° (see figure). How far apart are the ships?

38. ***Distance*** A passenger in an airplane at an altitude of 10 kilometers sees two towns directly to the east of the plane. The angles of depression to the towns are 28° and 55° (see figure). How far apart are the towns?

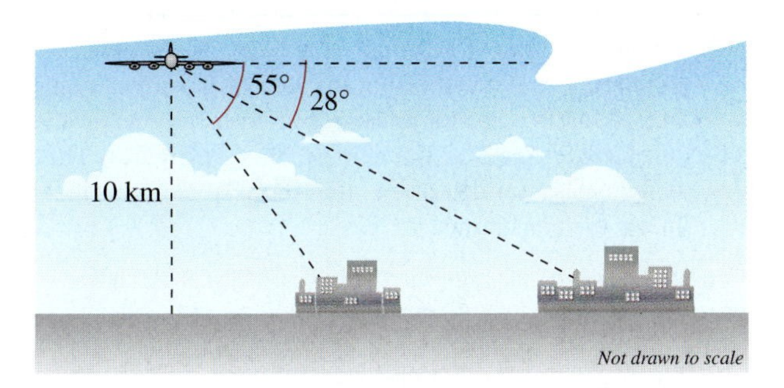

39. ***Altitude*** A plane is observed approaching your home and you assume that its speed is 550 miles per hour. The angle of elevation of the plane is 16° at one time and 57° one minute later. Approximate the altitude of the plane.

40. ***Height*** While traveling across flat land, you notice a mountain directly in front of you. The angle of elevation to the peak is 2.5°. After you drive 17 miles closer to the mountain, the angle of elevation is 9°. Approximate the height of the mountain.

Geometry **In Exercises 41 and 42, find the angle α between two nonvertical lines L_1 and L_2. The angle α satisfies the equation**

$$\tan \alpha = \left| \frac{m_2 - m_1}{1 + m_2 m_1} \right|$$

where m_1 and m_2 are the slopes of L_1 and L_2, respectively. (Assume that $m_1 m_2 \neq -1$.)

41. L_1: $3x - 2y = 5$
L_2: $x + y = 1$

42. L_1: $2x - y = 8$
L_2: $x - 5y = -4$

43. ***Geometry*** Determine the angle between the diagonal of a cube and the diagonal of its base, as shown in the figure.

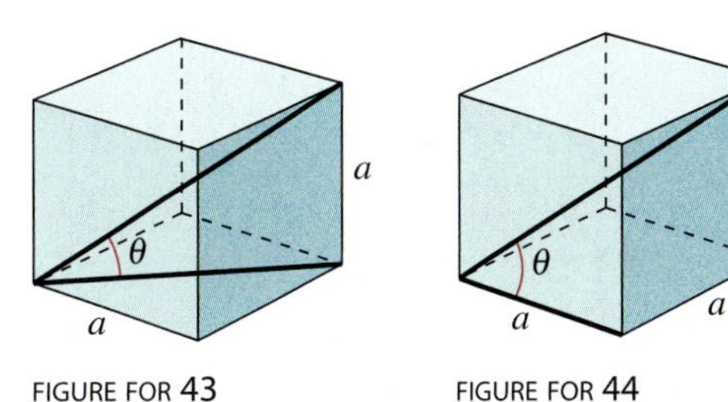

FIGURE FOR 43 FIGURE FOR 44

44. ***Geometry*** Determine the angle between the diagonal of a cube and its edge, as shown in the figure.

45. ***Geometry*** Find the length of the sides of a regular pentagon inscribed in a circle of radius 25 inches.

46. ***Geometry*** Find the length of the sides of a regular hexagon inscribed in a circle of radius 25 inches.

47. ***Hardware*** Write the distance y across the flat sides of a hexagonal nut as a function of r, as shown in the figure.

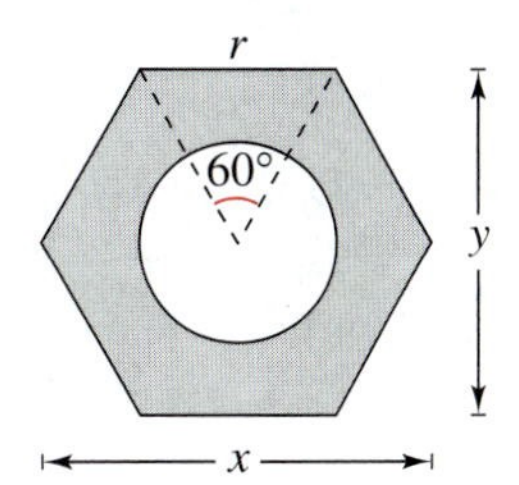

48. ***Bolt Holes*** The figure shows a circular piece of sheet metal that has a diameter of 40 centimeters and contains 12 equally spaced bolt holes. Determine the straight-line distance between the centers of consecutive bolt holes.

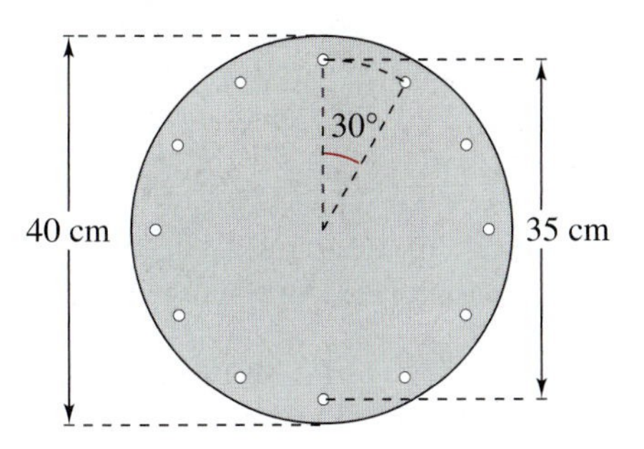

Trusses **In Exercises 49 and 50, find the lengths of all the unknown members of the truss.**

49.

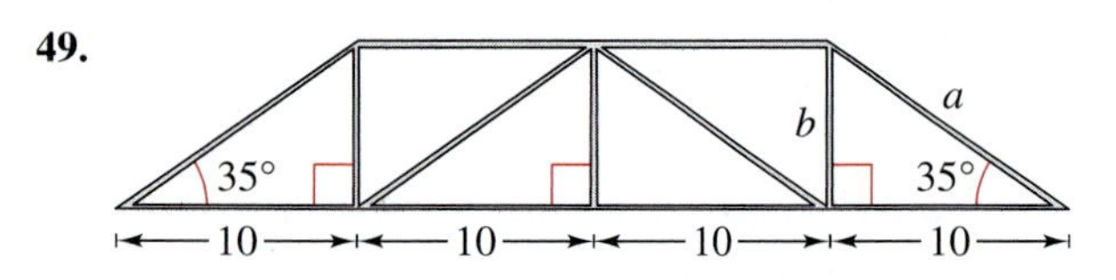

50.

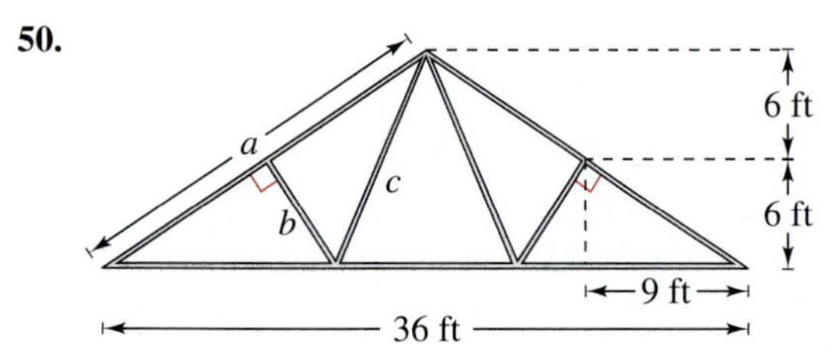

Harmonic Motion **In Exercises 51–54, find a model for simple harmonic motion satisfying the specified conditions.**

	Displacement $(t = 0)$	*Amplitude*	*Period*
51.	0	4 centimeters	2 seconds
52.	0	3 meters	6 seconds
53.	3 inches	3 inches	1.5 seconds
54.	2 feet	2 feet	10 seconds

Harmonic Motion **In Exercises 55–58, for the simple harmonic motion described by the trigonometric function, find (a) the maximum displacement, (b) the frequency, (c) the value of d when $t = 5$, and (d) the least positive value of t for which $d = 0$. Use a graphing utility to verify your results.**

55. $d = 4 \cos 8\pi t$

56. $d = \frac{1}{2} \cos 20\pi t$

57. $d = \frac{1}{16} \sin 120\pi t$

58. $d = \frac{1}{64} \sin 792\pi t$

59. ***Tuning Fork*** A point on the end of a tuning fork moves in simple harmonic motion described by $d = a \sin \omega t$. Find ω given that the tuning fork for middle C has a frequency of 264 vibrations per second.

60. ***Wave Motion*** A buoy oscillates in simple harmonic motion as waves go past. It is noted that the buoy moves a total of 3.5 feet from its low point to its high point (see figure), and that it returns to its high point every 10 seconds. Write an equation that describes the motion of the buoy if its high point is at $t = 0$.

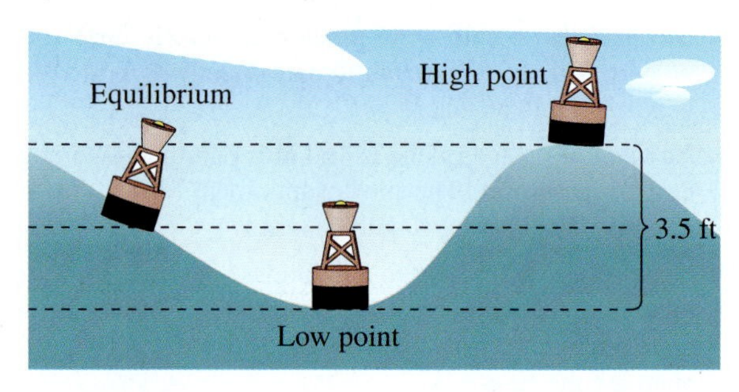

FIGURE FOR 60

61. ***Oscillation of a Spring*** A ball that is bobbing up and down on the end of a spring has a maximum displacement of 3 inches. Its motion (in ideal conditions) is modeled by $y = \frac{1}{4} \cos 16t$ $(t > 0)$, where y is measured in feet and t is the time in seconds.

(a) Graph the function.

(b) What is the period of the oscillations?

(c) Determine the first time the weight passes the point of equilibrium $(y = 0)$.

Model It

62. ***Numerical and Graphical Analysis*** A two-meter-high fence is 3 meters from the side of a grain storage bin. A grain elevator must reach from ground level outside the fence to the storage bin (see figure). The objective is to determine the shortest elevator that meets the constraints.

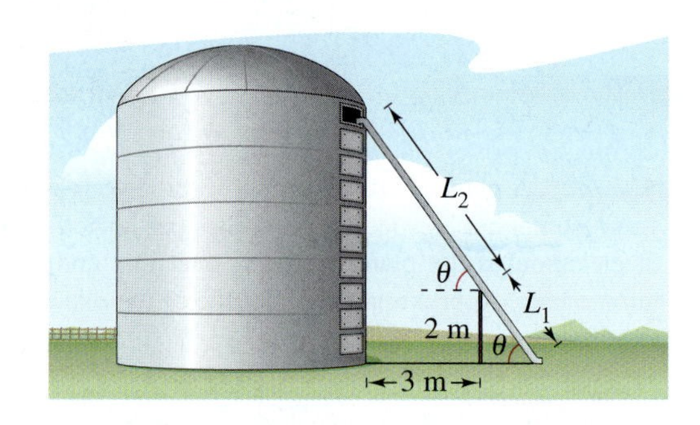

(a) Complete four rows of the table.

θ	L_1	L_2	$L_1 + L_2$
0.1	$\dfrac{2}{\sin 0.1}$	$\dfrac{3}{\cos 0.1}$	23.0
0.2	$\dfrac{2}{\sin 0.2}$	$\dfrac{3}{\cos 0.2}$	13.1

Model It (continued)

(b) Use a graphing utility to generate additional rows of the table. Use the table to estimate the minimum length of the elevator.

(c) Write the length $L_1 + L_2$ as a function of θ.

(d) Use a graphing utility to graph the function. Use the graph to estimate the minimum length. How does your estimate compare with that of part (b)?

63. ***Numerical and Graphical Analysis*** The cross section of an irrigation canal is an isosceles trapezoid of which three of the sides are 8 feet long (see figure). The objective is to find the angle θ that maximizes the area of the cross section. [*Hint:* The area of a trapezoid is $(h/2)(b_1 + b_2)$.]

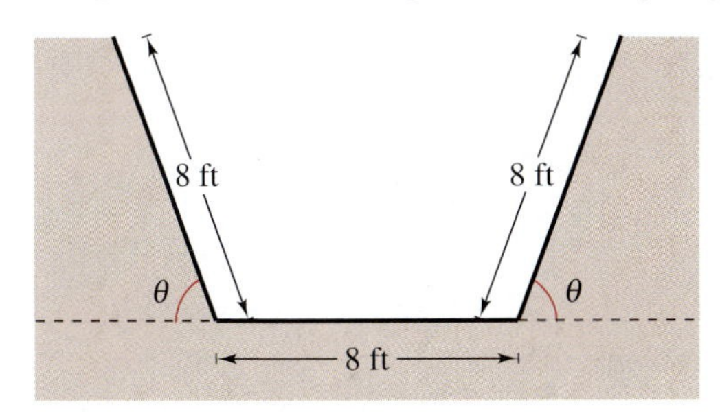

(a) Complete seven additional rows of the table.

Base 1	Base 2	Altitude	Area
8	$8 + 16 \cos 10°$	$8 \sin 10°$	22.1
8	$8 + 16 \cos 20°$	$8 \sin 20°$	42.5

(b) Use a graphing utility to generate additional rows of the table. Use the table to estimate the maximum cross-sectional area.

(c) Write the area A as a function of θ.

(d) Use a graphing utility to graph the function. Use the graph to estimate the maximum cross-sectional area. How does your estimate compare with that of part (b)?

64. ***Data Analysis*** The table shows the average sales S (in millions of dollars) of an outerwear manufacturer for each month t, where $t = 1$ represents January.

Time, t	1	2	3	4	5	6
Sales, s	13.46	11.15	8.00	4.85	2.54	1.70

Time, t	7	8	9	10	11	12
Sales, s	2.54	4.85	8.00	11.15	13.46	14.3

(a) Create a scatter plot of the data.

(b) Find a trigonometric model that fits the data. Graph the model with your scatter plot. How well does the model fit the data?

(c) What is the period of the model? Do you think it is reasonable given the context? Explain your reasoning.

(d) Interpret the meaning of the model's amplitude in the context of the problem.

Synthesis

True or False? **In Exercises 65 and 66, determine whether the statement is true or false. Justify your answer.**

65. The Leaning Tower of Pisa is not vertical, but if you know the exact angle of elevation θ to the 191-foot tower when you stand near it, then you can determine the exact distance to the tower d by using the formula

$$\tan \theta = \frac{191}{d}.$$

66. For the harmonic motion of a ball bobbing up and down on the end of a spring, one period can be described as the length of one coil of the spring.

67. ***Writing*** Is it true that N 24° E means 24 degrees north of east? Explain.

68. ***Writing*** Explain the difference between bearings used in nautical navigation and bearings used in air navigation.

Skills Review

In Exercises 69–72, write the slope-intercept form of the equation of the line with the specified characteristics. Then sketch the line.

69. $m = 4$, passes through $(-1, 2)$

70. $m = -\frac{1}{2}$, passes through $\left(\frac{1}{3}, 0\right)$

71. Passes through $(-2, 6)$ and $(3, 2)$

72. Passes through $\left(\frac{1}{4}, -\frac{2}{3}\right)$ and $\left(-\frac{1}{2}, \frac{1}{3}\right)$

4 Chapter Summary

What did you learn?

Section 4.1	Review Exercises
☐ Describe angles *(p. 282)*.	1, 2
☐ Use radian measure *(p. 283)*.	3–6, 11–18
☐ Use degree measure *(p. 285)*.	7–18
☐ Use angles to model and solve real-life problems *(p. 287)*.	19–24

Section 4.2	
☐ Identify a unit circle and describe its relationship to real numbers *(p. 294)*.	25–28
☐ Evaluate trigonometric functions using the unit circle *(p. 295)*.	29–32
☐ Use domain and period to evaluate sine and cosine functions *(p. 297)*.	33–36
☐ Use a calculator to evaluate trigonometric functions *(p. 298)*.	37–40

Section 4.3	
☐ Evaluate trigonometric functions of acute angles *(p. 301)*.	41–44
☐ Use the fundamental trigonometric identities *(p. 304)*.	45–48
☐ Use a calculator to evaluate trigonometric functions *(p. 305)*.	49–54
☐ Use trigonometric functions to model and solve real-life problems *(p. 306)*.	55, 56

Section 4.4	
☐ Evaluate trigonometric functions of any angle *(p. 312)*.	57–70
☐ Use reference angles to evaluate trigonometric functions *(p. 314)*.	71–82
☐ Evaluate trigonometric functions of real numbers *(p. 315)*.	83–88

Section 4.5	
☐ Use amplitude and period to help sketch the graphs of sine and cosine functions *(p. 323)*.	89–92
☐ Sketch translations of the graphs of sine and cosine functions *(p. 325)*.	93–96
☐ Use sine and cosine functions to model real-life data *(p. 327)*.	97, 98

Section 4.6	
☐ Sketch the graphs of tangent *(p. 332)* and cotangent *(p. 334)* functions.	99–102
☐ Sketch the graphs of secant and cosecant functions *(p. 335)*.	103–106
☐ Sketch the graphs of damped trigonometric functions *(p. 337)*.	107, 108

Section 4.7	
☐ Evaluate and graph the inverse sine function *(p. 343)*.	109–114, 123, 126
☐ Evaluate and graph the other inverse trigonometric functions *(p. 345)*.	115–122, 124, 125
☐ Evaluate compositions of trigonometric functions *(p. 347)*.	127–132

Section 4.8	
☐ Solve real-life problems involving right triangles *(p. 353)*.	133, 134
☐ Solve real-life problems involving directional bearings *(p. 355)*.	135
☐ Solve real-life problems involving harmonic motion *(p. 356)*.	136

4 Review Exercises

4.1 In Exercises 1 and 2, estimate the angle to the nearest one-half radian.

1. 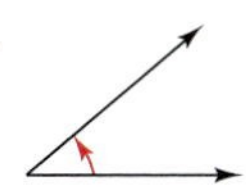

2. 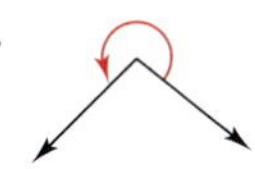

In Exercises 3–10, (a) sketch the angle in standard position, (b) determine the quadrant in which the angle lies, and (c) determine one positive and one negative coterminal angle.

3. $\frac{11\pi}{4}$ **4.** $\frac{2\pi}{9}$

5. $-\frac{4\pi}{3}$ **6.** $-\frac{23\pi}{3}$

7. $70°$ **8.** $280°$

9. $-110°$ **10.** $-405°$

In Exercises 11–14, convert the angle measure from degrees to radians. Round your answer to three decimal places.

11. $480°$ **12.** $-127.5°$

13. $-33° 45'$ **14.** $196° 77'$

In Exercises 15–18, convert the angle measure from radians to degrees. Round your answer to three decimal places.

15. $\frac{5\pi}{7}$ **16.** $-\frac{11\pi}{6}$

17. -3.5 **18.** 5.7

19. ***Arc Length*** Find the length of the arc on a circle with a radius of 20 inches intercepted by a central angle of $138°$.

20. ***Arc Length*** Find the length of the arc on a circle with a radius of 11 meters intercepted by a central angle of $60°$.

21. ***Phonograph*** Compact discs have all but replaced phonograph records. Phonograph records are vinyl discs that rotate on a turntable. A typical record album is 12 inches in diameter and plays at $33\frac{1}{3}$ revolutions per minute.

(a) What is the angular speed of a record album?

(b) What is the linear speed of the outer edge of a record album?

22. ***Bicycle*** At what speed is a bicyclist traveling when his 27-inch-diameter tires are rotating at an angular speed of 5π radians per second?

23. ***Circular Sector*** Find the area of the sector of a circle with a radius of 18 inches and central angle $\theta = 120°$.

24. ***Circular Sector*** Find the area of the sector of a circle with a radius of 6.5 millimeters and central angle $\theta = 5\pi/6$.

4.2 In Exercises 25–28, find the point (x, y) on the unit circle that corresponds to the real number t.

25. $t = \frac{2\pi}{3}$ **26.** $t = \frac{3\pi}{4}$

27. $t = \frac{5\pi}{6}$ **28.** $t = -\frac{4\pi}{3}$

In Exercises 29–32, evaluate (if possible) the six trigonometric functions of the real number.

29. $t = \frac{7\pi}{6}$ **30.** $t = \frac{\pi}{4}$

31. $t = -\frac{2\pi}{3}$ **32.** $t = 2\pi$

In Exercises 33–36, evaluate the trigonometric function using its period as an aid.

33. $\sin \frac{11\pi}{4}$ **34.** $\cos 4\pi$

35. $\sin\left(-\frac{17\pi}{6}\right)$ **36.** $\cos\left(-\frac{13\pi}{3}\right)$

In Exercises 37–40, use a calculator to evaluate the trigonometric function. Round your answer to four decimal places.

37. $\tan 33$ **38.** $\csc 10.5$

39. $\sec \frac{12\pi}{5}$ **40.** $\sin\left(-\frac{\pi}{9}\right)$

4.3 In Exercises 41–44, find the exact values of the six trigonometric functions of the angle θ shown in the figure.

41. 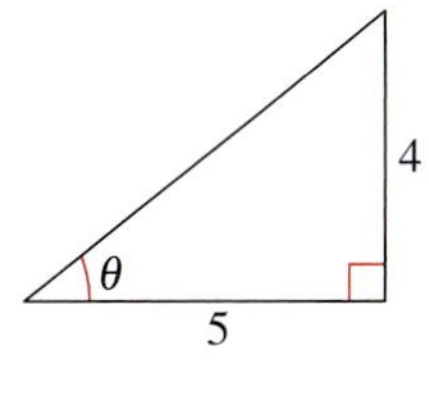

42. 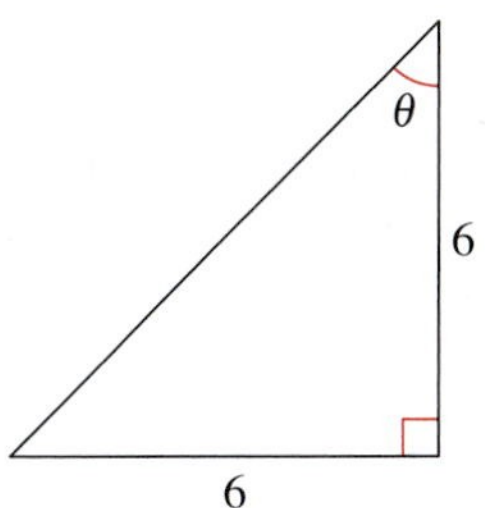

43. 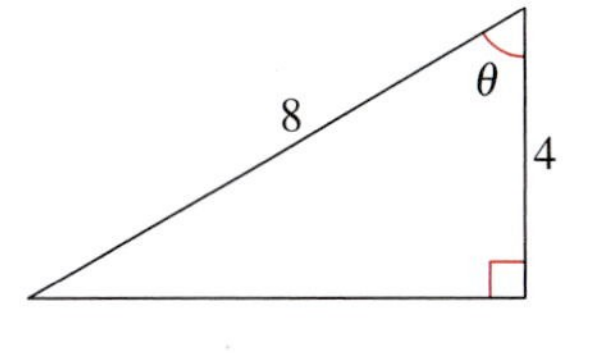

44. 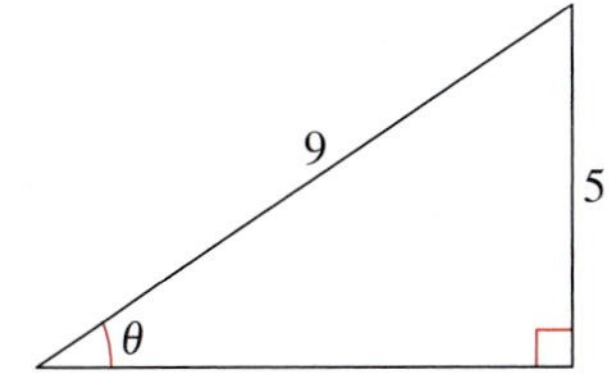

In Exercises 45–48, use the given function value and trigonometric identities (including the cofunction identities) to find the indicated trigonometric functions.

45. $\sin\theta = \frac{1}{3}$ (a) $\csc\theta$ (b) $\cos\theta$ (c) $\sec\theta$ (d) $\tan\theta$

46. $\tan\theta = 4$ (a) $\cot\theta$ (b) $\sec\theta$ (c) $\cos\theta$ (d) $\csc\theta$

47. $\csc\theta = 4$ (a) $\sin\theta$ (b) $\cos\theta$ (c) $\sec\theta$ (d) $\tan\theta$

48. $\csc\theta = 5$ (a) $\sin\theta$ (b) $\cot\theta$ (c) $\tan\theta$ (d) $\sec(90° - \theta)$

In Exercises 49–54, use a calculator to evaluate the trigonometric function. Round your answer to four decimal places.

49. $\tan 33°$

50. $\csc 11°$

51. $\sin 34.2°$

52. $\sec 79.3°$

53. $\cot 15°14'$

54. $\cos 78°11'58''$

55. ***Railroad Grade*** A train travels 3.5 kilometers on a straight track with a grade of $1°\,10'$ (see figure). What is the vertical rise of the train in that distance?

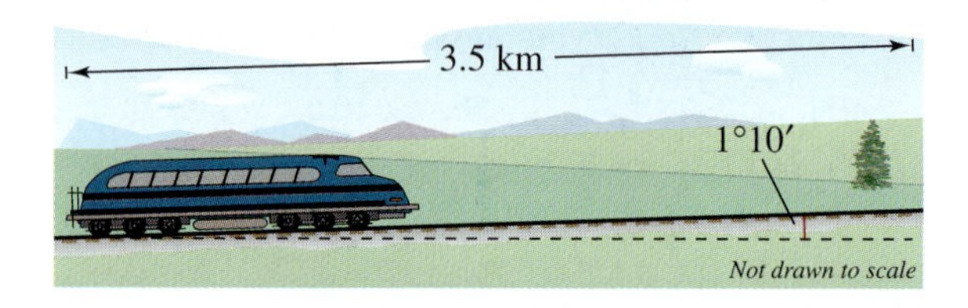

56. ***Guy Wire*** A guy wire runs from the ground to the top of a 25-foot telephone pole. The angle formed between the wire and the ground is 52°. How far from the base of the pole is the wire attached to the ground?

4.4 **In Exercises 57–64, the point is on the terminal side of an angle θ in standard position. Determine the exact values of the six trigonometric functions of the angle θ.**

57. $(12, 16)$

58. $(3, -4)$

59. $\left(\frac{2}{3}, \frac{5}{2}\right)$

60. $\left(-\frac{10}{3}, -\frac{2}{3}\right)$

61. $(-0.5, 4.5)$

62. $(0.3, 0.4)$

63. $(x, 4x),\ x > 0$

64. $(-2x, -3x),\ x > 0$

In Exercises 65–70, find the values of the six trigonometric functions of θ.

Function Value	*Constraint*
65. $\sec\theta = \frac{6}{5}$	$\tan\theta < 0$
66. $\csc\theta = \frac{3}{2}$	$\cos\theta < 0$
67. $\sin\theta = \frac{3}{8}$	$\cos\theta < 0$
68. $\tan\theta = \frac{5}{4}$	$\cos\theta < 0$
69. $\cos\theta = -\frac{2}{5}$	$\sin\theta > 0$
70. $\sin\theta = -\frac{2}{4}$	$\cos\theta > 0$

In Exercises 71–74, find the reference angle θ', and sketch θ and θ' in standard position.

71. $\theta = 264°$

72. $\theta = 635°$

73. $\theta = -\dfrac{6\pi}{5}$

74. $\theta = \dfrac{17\pi}{3}$

In Exercises 75–82, evaluate the sine, cosine, and tangent of the angle without using a calculator.

75. $\dfrac{\pi}{3}$

76. $\dfrac{\pi}{4}$

77. $-\dfrac{7\pi}{3}$

78. $-\dfrac{5\pi}{4}$

79. $495°$

80. $-150°$

81. $-240°$

82. $315°$

In Exercises 83–88, use a calculator to evaluate the trigonometric function. Round your answer to four decimal places.

83. $\sin 4$

84. $\tan 3$

85. $\sin(-3.2)$

86. $\cot(-4.8)$

85. $\sin\dfrac{12\pi}{5}$

88. $\tan\left(-\dfrac{25\pi}{7}\right)$

4.5 **In Exercises 89–96, sketch the graph of the function. Include two full periods.**

89. $y = \sin x$

90. $y = \cos x$

91. $f(x) = 5\sin\dfrac{2x}{5}$

92. $f(x) = 8\cos\left(-\dfrac{x}{4}\right)$

93. $y = 2 + \sin x$

94. $y = -4 - \cos\pi x$

95. $g(t) = \frac{5}{2}\sin(t - \pi)$

96. $g(t) = 3\cos(t + \pi)$

97. ***Sound Waves*** Sound waves can be modeled by sine functions of the form $y = a\sin bx$, where x is measured in seconds.

(a) Write an equation of a sound wave whose amplitude is 2 and whose period is $\frac{1}{264}$ second.

(b) What is the frequency of the sound wave described in part (a)?

98. ***Data Analysis: Meteorology*** The times S of sunset (Greenwich Mean Time) at 40° north latitude on the 15th of each month are: 1(16:59), 2(17:35), 3(18:06), 4(18:38), 5(19:08), 6(19:30), 7(19:28), 8(18:57), 9(18:09), 10(17:21), 11(16:44), 12(16:36). The month is represented by t, with $t = 1$ corresponding to January. A model (in which minutes have been converted to the decimal parts of an hour) for the data is

$$S(t) = 18.09 + 1.41 \sin\left(\frac{\pi t}{6} + 4.60\right).$$

(a) Use a graphing utility to graph the data points and the model in the same viewing window.

(b) What is the period of the model? Is it what you expected? Explain.

(c) What is the amplitude of the model? What does it represent in the model? Explain.

4.6 **In Exercises 99–106, sketch a graph of the function. Include two full periods.**

99. $f(x) = \tan x$

100. $f(t) = \tan\left(t - \frac{\pi}{4}\right)$

101. $f(x) = \cot x$

102. $g(t) = 2 \cot 2t$

103. $f(x) = \sec x$

104. $h(t) = \sec\left(t - \frac{\pi}{4}\right)$

105. $f(x) = \csc x$

106. $f(t) = 3 \csc\left(2t + \frac{\pi}{4}\right)$

In Exercises 107 and 108, use a graphing utility to graph the function and the damping factor of the function in the same viewing window. Describe the behavior of the function as x increases without bound.

107. $f(x) = x \cos x$

108. $g(x) = x^4 \cos x$

4.7 **In Exercises 109–114, evaluate the expression. If necessary, round your answer to two decimal places.**

109. $\arcsin\left(-\frac{1}{2}\right)$

110. $\arcsin(-1)$

111. $\arcsin 0.4$

112. $\arcsin 0.213$

113. $\sin^{-1}(-0.44)$

114. $\sin^{-1} 0.89$

In Exercises 115–118, evaluate the expression without the aid of a calculator.

115. $\arccos \frac{\sqrt{3}}{2}$

116. $\arccos \frac{\sqrt{2}}{2}$

117. $\cos^{-1}(-1)$

118. $\cos^{-1} \frac{\sqrt{3}}{2}$

In Exercises 119–122, use a calculator to evaluate the expression. Round your answer to two decimal places.

119. $\arccos 0.324$

120. $\arccos(-0.888)$

121. $\tan^{-1}(-1.5)$

122. $\tan^{-1} 8.2$

In Exercises 123–126, use a graphing utility to graph the function.

123. $f(x) = 2 \arcsin x$

124. $f(x) = 3 \arccos x$

125. $f(x) = \arctan \frac{x}{2}$

126. $f(x) = -\arcsin 2x$

In Exercises 127–130, find the exact value of the expression.

127. $\cos\left(\arctan \frac{3}{4}\right)$

128. $\tan\left(\arccos \frac{3}{5}\right)$

129. $\sec\left(\arctan \frac{12}{5}\right)$

130. $\cot\left[\arcsin\left(-\frac{12}{13}\right)\right]$

In Exercises 131 and 132, write an algebraic expression that is equivalent to the expression.

131. $\tan\left(\arccos \frac{x}{2}\right)$

132. $\sec[\arcsin(x - 1)]$

4.8 **133.** ***Angle of Elevation*** The height of a radio transmission tower is 70 meters, and it casts a shadow of length 30 meters (see figure). Find the angle of elevation of the sun.

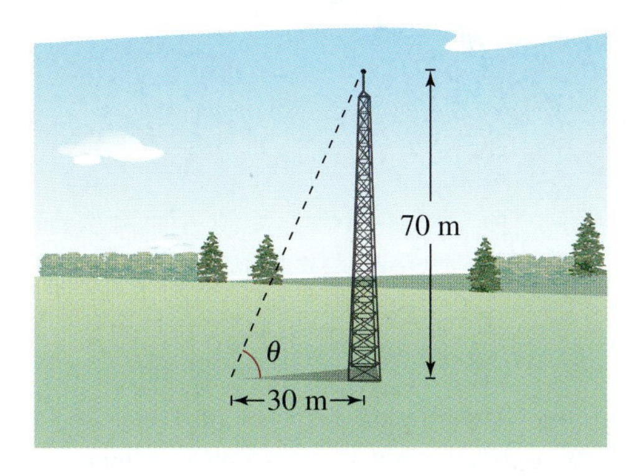

134. ***Height*** Your football has landed at the edge of the roof of your school building. When you are 25 feet from the base of the building, the angle of elevation to your football is 21°. How high off the ground is your football?

135. ***Distance*** From city A to city B, a plane flies 650 miles at a bearing of 48°. From city B to city C, the plane flies 810 miles at a bearing of 115°. Find the distance from city A to city C and the bearing from city A to city C.

136. ***Wave Motion*** Your fishing bobber oscillates in simple harmonic motion from the waves in the lake where you fish. Your bobber moves a total of 1.5 inches from its high point to its low point and returns to its high point every 3 seconds. Write an equation modeling the motion of your bobber if it is at its high point at time $t = 0$.

Synthesis

True or False? **In Exercises 137–140, determine whether the statement is true or false. Justify your answer.**

137. The tangent function is often useful for modeling simple harmonic motion.

138. The inverse sine function $y = \arcsin x$ cannot be defined as a function over any interval that is greater than the interval defined as $-\pi/2 \le y \le \pi/2$.

139. $y = \sin \theta$ is not a function because $\sin 30° = \sin 150°$.

140. Because $\tan 3\pi/4 = -1$, $\arctan(-1) = 3\pi/4$.

In Exercises 141–144, match the function $y = a \sin bx$ with its graph. Base your selection solely on your interpretation of the constants a and b. Explain your reasoning. [The graphs are labeled (a), (b), (c), and (d).]

(a)

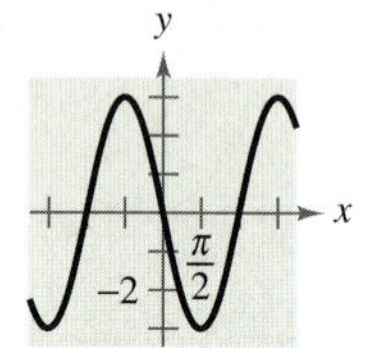

(b)

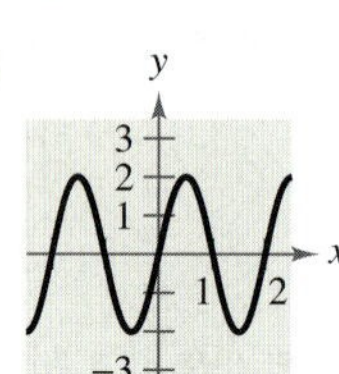

(c)

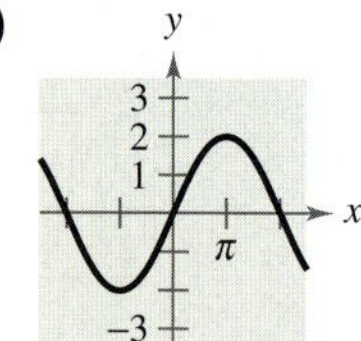

(d)

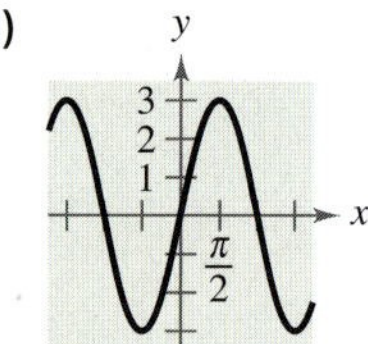

141. $y = 3 \sin x$

142. $y = -3 \sin x$

143. $y = 2 \sin \pi x$

144. $y = 2 \sin \dfrac{x}{2}$

145. ***Writing*** Describe the behavior of $f(\theta) = \sec \theta$ at the zeros of $g(\theta) = \cos \theta$. Explain your reasoning.

146. ***Conjecture***

(a) Use a graphing utility to complete the table.

θ	0.1	0.4	0.7	1.0	1.3
$\tan\left(\theta - \frac{\pi}{2}\right)$					
$-\cot \theta$					

(b) Make a conjecture about the relationship between $\tan\left(\theta - \dfrac{\pi}{2}\right)$ and $-\cot \theta$.

147. ***Writing*** When graphing the sine and cosine functions, determining the amplitude is part of the analysis. Explain why this is not true for the other four trigonometric functions.

148. ***Oscillation of a Spring*** A weight is suspended from a ceiling by a steel spring. The weight is lifted (positive direction) from the equilibrium position and released. The resulting motion of the weight is modeled by

$$y = Ae^{-kt} \cos bt = \tfrac{1}{5}e^{-t/10} \cos 6t$$

where y is the distance in feet from equilibrium and t is the time in seconds. The graph of the function is shown in the figure. For each of the following, describe the change in the system without graphing the resulting function.

(a) A is changed from $\frac{1}{5}$ to $\frac{1}{3}$.

(b) k is changed from $\frac{1}{10}$ to $\frac{1}{3}$.

(c) b is changed from 6 to 9.

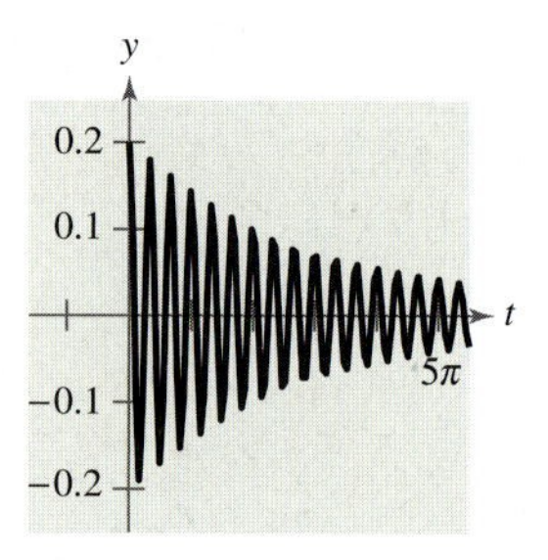

149. ***Graphical Reasoning*** The formulas for the area of a circular sector and arc length are $A = \frac{1}{2}r^2\theta$ and $s = r\theta$, respectively. (r is the radius and θ is the angle measured in radians.)

(a) For $\theta = 0.8$, write the area and arc length as functions of r. What is the domain of each function? Use a graphing utility to graph the functions. Use the graphs to determine which function changes more rapidly as r increases. Explain.

(b) For $r = 10$ centimeters, write the area and arc length as functions of θ. What is the domain of each function? Use a graphing utility to graph and identify the functions.

150. ***Writing*** Describe a real-life application that can be represented by a simple harmonic motion model and is different from any that you've seen in this chapter. Explain which function you would use to model your application and why. Explain how you would determine the amplitude, period, and frequency of the model for your application.

4 Chapter Test

Take this test as you would take a test in class. When you are finished, check your work against the answers given in the back of the book.

FIGURE FOR 4

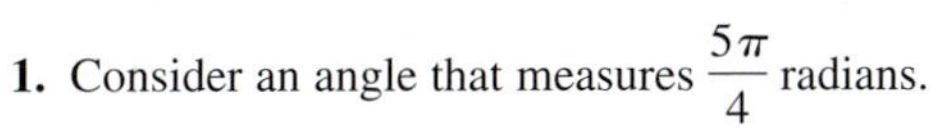

1. Consider an angle that measures $\frac{5\pi}{4}$ radians.
   (a) Sketch the angle in standard position.
   (b) Determine two coterminal angles (one positive and one negative).
   (c) Convert the angle to degree measure.
2. A truck is moving at a rate of 90 kilometers per hour, and the diameter of its wheels is 1 meter. Find the angular speed of the wheels in radians per minute.
3. A water sprinkler sprays water on a lawn over a distance of 25 feet and rotates through an angle of 130°. Find the area of the lawn watered by the sprinkler.
4. Find the exact values of the six trigonometric functions of the angle θ shown in the figure.
5. Given that $\tan \theta = \frac{3}{2}$, find the other five trigonometric functions of θ.
6. Determine the reference angle θ' of the angle $\theta = 290^\circ$ and sketch θ and θ' in standard position.
7. Determine the quadrant in which θ lies if $\sec \theta < 0$ and $\tan \theta > 0$.
8. Find two exact values of θ in degrees $(0 \le \theta < 360^\circ)$ if $\cos \theta = -\sqrt{3}/2$. (Do not use a calculator.)
9. Use a calculator to approximate two values of θ in radians $(0 \le \theta < 2\pi)$ if $\csc \theta = 1.030$. Round the results to two decimal places.

In Exercises 10 and 11, find the remaining five trigonometric functions of θ satisfying the conditions.

10. $\cos \theta = \frac{3}{5}, \quad \tan \theta < 0$
11. $\sec \theta = -\frac{17}{8}, \quad \sin \theta > 0$

In Exercises 12 and 13, sketch the graph of the function. (Include two full periods.)

12. $g(x) = -2 \sin\left(x - \frac{\pi}{4}\right)$
13. $f(\alpha) = \frac{1}{2} \tan 2\alpha$

In Exercises 14 and 15, use a graphing utility to graph the function. If the function is periodic, find its period.

14. $y = \sin 2\pi x + 2 \cos \pi x$
15. $y = 6e^{-0.12t} \cos(0.25t), \quad 0 \le t \le 32$

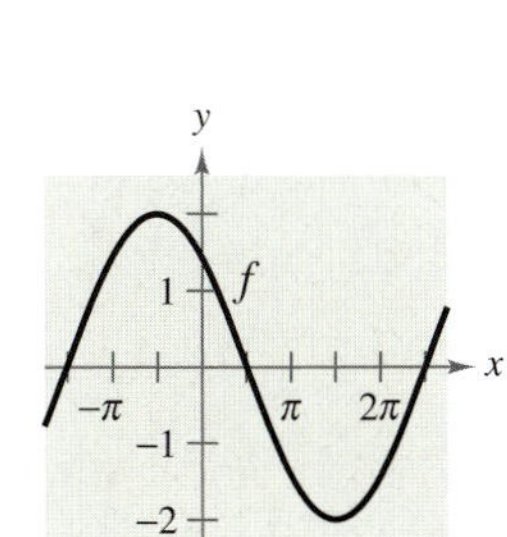

FIGURE FOR 16

16. Find a, b, and c for the function $f(x) = a \sin(bx + c)$ such that the graph of f matches the figure.
17. Find the exact value of $\tan\left(\arccos \frac{2}{3}\right)$ without the aid of a calculator.
18. Graph the function $f(x) = 2 \arcsin\left(\frac{1}{2}x\right)$.
19. A plane is 80 miles south and 95 miles east of Cleveland Hopkins International Airport. What bearing should be taken to fly directly to the airport?
20. Write the equation for the simple harmonic motion of a ball on a spring that starts at its lowest point of 6 inches below equilibrium, bounces to its maximum height of 6 inches above equilibrium, and returns to its lowest point in a total of 2 seconds.

Proofs in Mathematics

The Pythagorean Theorem

The Pythagorean Theorem is one of the most famous theorems in mathematics. More than 100 different proofs now exist. James A. Garfield, the twentieth president of the United States, developed a proof of the Pythagorean Theorem in 1876. His proof, shown below, involved the fact that a trapezoid can be formed from two congruent right triangles and an isosceles right triangle.

The Pythagorean Theorem

In a right triangle, the sum of the squares of the lengths of the legs is equal to the square of the length of the hypotenuse, where a and b are the legs and c is the hypotenuse.

$$a^2 + b^2 = c^2$$

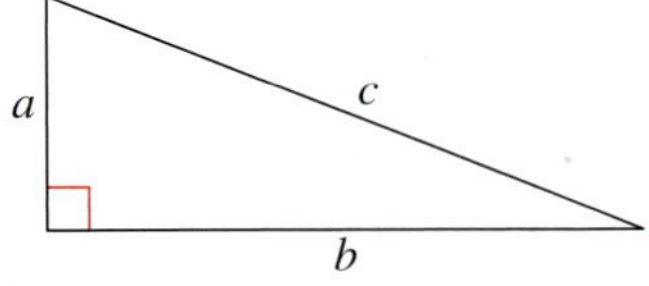

Proof

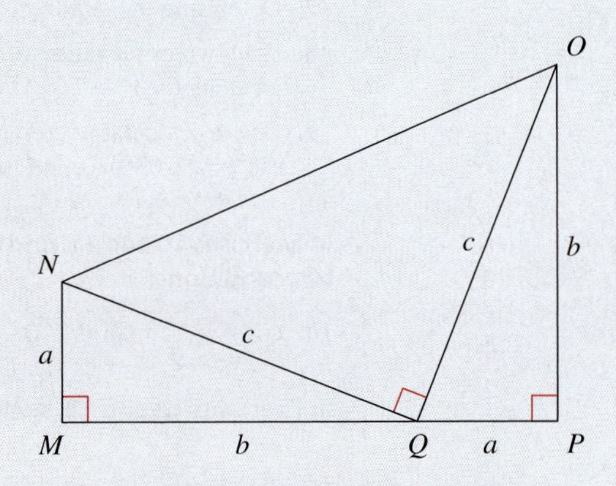

$$\text{Area of trapezoid } MNOP = \text{Area of } \triangle MNQ + \text{Area of } \triangle PQO + \text{Area of } \triangle NOQ$$

$$\frac{1}{2}(a + b)(a + b) = \frac{1}{2}ab + \frac{1}{2}ab + \frac{1}{2}c^2$$

$$\frac{1}{2}(a + b)(a + b) = ab + \frac{1}{2}c^2$$

$$(a + b)(a + b) = 2ab + c^2$$

$$a^2 + 2ab + b^2 = 2ab + c^2$$

$$a^2 + b^2 = c^2$$

P.S. Problem Solving

This collection of thought-provoking and challenging exercises further explores and expands upon concepts learned in this chapter.

1. The restaurant at the top of the Space Needle in Seattle, Washington is circular and has a radius of 47.25 feet. The dining part of the restaurant revolves, making about one complete revolution every 48 minutes. A dinner party was seated at the edge of the revolving restaurant at 6:45 P.M. and was finished at 8:57 P.M.

(a) Find the angle through which the dinner party rotated.

(b) Find the distance the party traveled during dinner.

2. A bicycle's gear ratio is the number of times the freewheel turns for every one turn of the chainwheel (see figure). The table shows the numbers of teeth in the freewheel and chainwheel for the first five gears of an 18-speed touring bicycle. The chainwheel completes one rotation for each gear. Find the angle through which the freewheel turns for each gear. Give your answers in both degrees and radians.

Gear number	Number of teeth in freewheel	Number of teeth in chainwheel
1	32	24
2	26	24
3	22	24
4	32	40
5	19	24

3. A surveyor in a helicopter is trying to determine the width of an island, as shown in the figure.

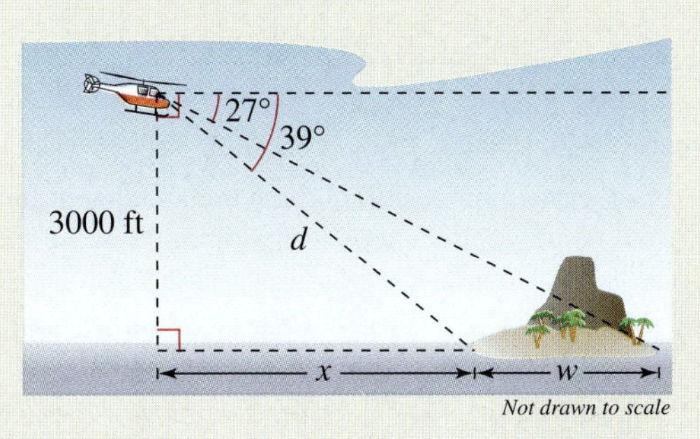

(a) What is the shortest distance d the helicopter would have to travel to land on the island?

(b) What is the horizontal distance x that the helicopter would have to travel before it would be directly over the nearer end of the island?

(c) Find the width w of the island. Explain how you obtained your answer.

4. Use the figure below.

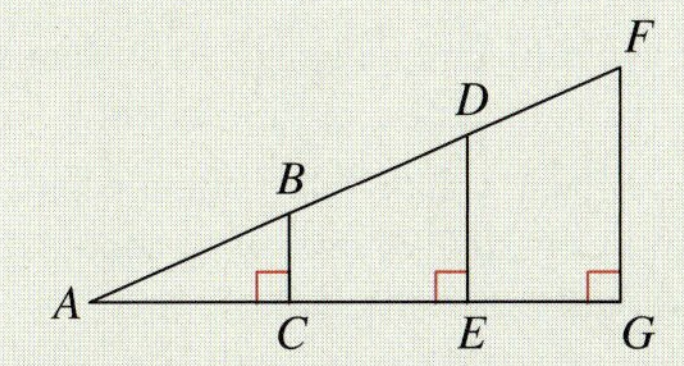

(a) Explain why $\triangle ABC$, $\triangle ADE$, and $\triangle AFG$ are similar triangles.

(b) What does similarity imply about the ratios

$$\frac{BC}{AB}, \frac{DE}{AD}, \text{ and } \frac{FG}{AF}?$$

(c) Does the value of sin A depend on which triangle from part (a) is used to calculate it? Would the value of sin A change if it were found using a different right triangle that was similar to the three given triangles?

(d) Do your conclusions from part (c) apply to the other five trigonometric functions? Explain.

5. Use a graphing utility to graph h, and use the graph to decide whether h is even, odd, or neither.

(a) $h(x) = \cos^2 x$

(b) $h(x) = \sin^2 x$

6. If f is an even function and g is an odd function, use the results of Exercise 5 to make a conjecture about h, where

(a) $h(x) = [f(x)]^2$

(b) $h(x) = [g(x)]^2$.

7. The model for the height h (in feet) of a Ferris wheel car is

$$h = 50 + 50 \sin 8\pi t$$

where t is the time (in minutes). (The Ferris wheel has a radius of 50 feet.) This model yields a height of 50 feet when $t = 0$. Alter the model so that the height of the car is 1 foot when $t = 0$.

8. The pressure P (in millimeters of mercury) against the walls of the blood vessels of a patient is modeled by

$$P = 100 - 20\cos\left(\frac{8\pi}{3}t\right)$$

where t is time (in seconds).

(a) Use a graphing utility to graph the model.

(b) What is the period of the model? What does the period tell you about this situation?

(c) What is the amplitude of the model? What does it tell you about this situation?

(d) If one cycle of this model is equivalent to one heartbeat, what is the pulse of this patient?

(e) If a physician wants this patient's pulse rate to be 64 beats per minute or less, what should the period be? What should the coefficient of t be?

9. A popular theory that attempts to explain the ups and downs of everyday life states that each of us has three cycles, called biorhythms, which begin at birth. These three cycles can be modeled by sine waves.

Physical (23 days): $P = \sin\frac{2\pi t}{23}, \quad t \geq 0$

Emotional (28 days): $E = \sin\frac{2\pi t}{28}, \quad t \geq 0$

Intellectual (33 days): $I = \sin\frac{2\pi t}{33}, \quad t \geq 0$

where t is the number of days since birth. Consider a person who was born on July 20, 1986.

(a) Use a graphing utility to graph the three models in the same viewing window for $7300 \leq t \leq 7380$.

(b) Describe the person's biorhythms during the month of September 2006.

(c) Calculate the person's three energy levels on September 22, 2006.

10. (a) Use a graphing utility to graph the functions given by

$$f(x) = 2\cos 2x + 3\sin 3x$$

and

$$g(x) = 2\cos 2x + 3\sin 4x.$$

(b) Use the graphs from part (a) to find the period of each function.

(c) If α and β are positive integers, is the function given by

$$h(x) = A\cos\alpha x + B\sin\beta x$$

periodic? Explain your reasoning.

11. Two trigonometric functions f and g have periods of 2, and their graphs intersect at $x = 5.35$.

(a) Give one smaller and one larger positive value of x at which the functions have the same value.

(b) Determine one negative value of x at which the graphs intersect.

(c) Is it true that $f(13.35) = g(-4.65)$? Explain your reasoning.

12. The function f is periodic, with period c. So, $f(t + c) = f(t)$. Are the following equal? Explain.

(a) $f(t - 2c) = f(t)$ (b) $f\left(t + \frac{1}{2}c\right) = f\left(\frac{1}{2}t\right)$

(c) $f\left(\frac{1}{2}(t + c)\right) = f\left(\frac{1}{2}t\right)$

13. If you stand in shallow water and look at an object below the surface of the water, the object will look farther away from you than it really is. This is because when light rays pass between air and water, the water refracts, or bends, the light rays. The index of refraction for water is 1.333. This is the ratio of the sine of θ_1 and the sine of θ_2 (see figure).

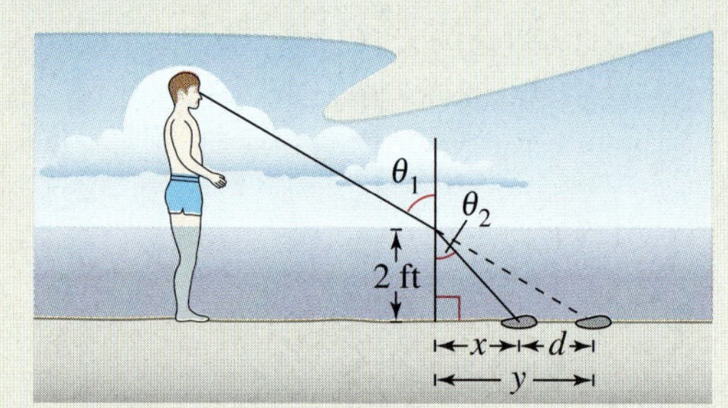

(a) You are standing in water that is 2 feet deep and are looking at a rock at angle $\theta_1 = 60°$ (measured from a line perpendicular to the surface of the water). Find θ_2.

(b) Find the distances x and y.

(c) Find the distance d between where the rock is and where it appears to be.

(d) What happens to d as you move closer to the rock? Explain your reasoning.

14. In calculus, it can be shown that the arctangent function can be approximated by the polynomial

$$\arctan x \approx x - \frac{x^3}{3} + \frac{x^5}{5} - \frac{x^7}{7}$$

where x is in radians.

(a) Use a graphing utility to graph the arctangent function and its polynomial approximation in the same viewing window. How do the graphs compare?

(b) Study the pattern in the polynomial approximation of the arctangent function and guess the next term. Then repeat part (a). How does the accuracy of the approximation change when additional terms are added?

Analytic Trigonometry

5

Concepts of trigonometry can be used to model the height above ground of a seat on a Ferris wheel.

SELECTED APPLICATIONS

Trigonometric equations and identities have many real-life applications. The applications listed below represent a small sample of the applications in this chapter.

- Friction, Exercise 99, page 381
- Shadow Length, Exercise 56, page 388
- Ferris Wheel, Exercise 75, page 398
- Data Analysis: Unemployment Rate, Exercise 76, page 398
- Harmonic Motion, Exercise 75, page 405
- Mach Number, Exercise 121, page 417
- Bridge Design, Exercise 39, page 426
- Surveying, Exercise 31, page 433
- Projectile Motion, Exercise 101, page 438

5.1 Using Fundamental Identities

What you should learn

- Recognize and write the fundamental trigonometric identities.
- Use the fundamental trigonometric identities to evaluate trigonometric functions, simplify trigonometric expressions, and rewrite trigonometric expressions.

Why you should learn it

Fundamental trigonometric identities can be used to simplify trigonometric expressions. For instance, in Exercise 99 on page 381, you can use trigonometric identities to simplify an expression for the coefficient of friction.

Introduction

In Chapter 4, you studied the basic definitions, properties, graphs, and applications of the individual trigonometric functions. In this chapter, you will learn how to use the fundamental identities to do the following.

1. Evaluate trigonometric functions.
2. Simplify trigonometric expressions.
3. Develop additional trigonometric identities.
4. Solve trigonometric equations.

Fundamental Trigonometric Identities

Reciprocal Identities

$$\sin u = \frac{1}{\csc u} \qquad \cos u = \frac{1}{\sec u} \qquad \tan u = \frac{1}{\cot u}$$

$$\csc u = \frac{1}{\sin u} \qquad \sec u = \frac{1}{\cos u} \qquad \cot u = \frac{1}{\tan u}$$

Quotient Identities

$$\tan u = \frac{\sin u}{\cos u} \qquad \cot u = \frac{\cos u}{\sin u}$$

Pythagorean Identities

$$\sin^2 u + \cos^2 u = 1 \qquad 1 + \tan^2 u = \sec^2 u \qquad 1 + \cot^2 u = \csc^2 u$$

Cofunction Identities

$$\sin\left(\frac{\pi}{2} - u\right) = \cos u \qquad \cos\left(\frac{\pi}{2} - u\right) = \sin u$$

$$\tan\left(\frac{\pi}{2} - u\right) = \cot u \qquad \cot\left(\frac{\pi}{2} - u\right) = \tan u$$

$$\sec\left(\frac{\pi}{2} - u\right) = \csc u \qquad \csc\left(\frac{\pi}{2} - u\right) = \sec u$$

Even/Odd Identities

$$\sin(-u) = -\sin u \qquad \cos(-u) = \cos u \qquad \tan(-u) = -\tan u$$

$$\csc(-u) = -\csc u \qquad \sec(-u) = \sec u \qquad \cot(-u) = -\cot u$$

Pythagorean identities are sometimes used in radical form such as

$$\sin u = \pm\sqrt{1 - \cos^2 u}$$

or

$$\tan u = \pm\sqrt{\sec^2 u - 1}$$

where the sign depends on the choice of u.

The *HM mathSpace®* CD-ROM and *Eduspace®* for this text contain additional resources related to the concepts discussed in this chapter.

Using the Fundamental Identities

One common use of trigonometric identities is to use given values of trigonometric functions to evaluate other trigonometric functions.

Example 1 Using Identities to Evaluate a Function

Use the values $\sec u = -\frac{3}{2}$ and $\tan u > 0$ to find the values of all six trigonometric functions.

Solution

Using a reciprocal identity, you have

$$\cos u = \frac{1}{\sec u} = \frac{1}{-3/2} = -\frac{2}{3}.$$

Using a Pythagorean identity, you have

$$\sin^2 u = 1 - \cos^2 u \quad \text{Pythagorean identity}$$

$$= 1 - \left(-\frac{2}{3}\right)^2 \quad \text{Substitute } -\tfrac{2}{3} \text{ for } \cos u.$$

$$= 1 - \frac{4}{9} = \frac{5}{9}. \quad \text{Simplify.}$$

Because $\sec u < 0$ and $\tan u > 0$, it follows that u lies in Quadrant III. Moreover, because $\sin u$ is negative when u is in Quadrant III, you can choose the negative root and obtain $\sin u = -\sqrt{5}/3$. Now, knowing the values of the sine and cosine, you can find the values of all six trigonometric functions.

$$\sin u = -\frac{\sqrt{5}}{3} \qquad \csc u = \frac{1}{\sin u} = -\frac{3}{\sqrt{5}} = -\frac{3\sqrt{5}}{5}$$

$$\cos u = -\frac{2}{3} \qquad \sec u = \frac{1}{\cos u} = -\frac{3}{2}$$

$$\tan u = \frac{\sin u}{\cos u} = \frac{-\sqrt{5}/3}{-2/3} = \frac{\sqrt{5}}{2} \qquad \cot u = \frac{1}{\tan u} = \frac{2}{\sqrt{5}} = \frac{2\sqrt{5}}{5}$$

CHECKPOINT Now try Exercise 11.

Example 2 Simplifying a Trigonometric Expression

Simplify $\sin x \cos^2 x - \sin x$.

Solution

First factor out a common monomial factor and then use a fundamental identity.

$$\sin x \cos^2 x - \sin x = \sin x(\cos^2 x - 1) \quad \text{Factor out common monomial factor.}$$

$$= -\sin x(1 - \cos^2 x) \quad \text{Factor out } -1.$$

$$= -\sin x(\sin^2 x) \quad \text{Pythagorean identity}$$

$$= -\sin^3 x \quad \text{Multiply.}$$

CHECKPOINT Now try Exercise 45.

STUDY TIP

You should learn the fundamental trigonometric identities well, because they are used frequently in trigonometry and they will also appear later in calculus. Note that u can be an angle, a real number, or a variable.

Technology

You can use a graphing utility to check the result of Example 2. To do this, graph

$$y_1 = \sin x \cos^2 x - \sin x$$

and

$$y_2 = -\sin^3 x$$

in the same viewing window, as shown below. Because Example 2 shows the equivalence algebraically and the two graphs appear to coincide, you can conclude that the expressions are equivalent.

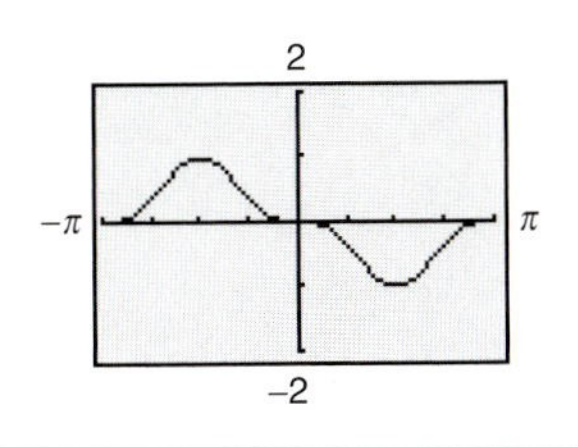

Remind students that they must use an algebraic approach to prove that two expressions are equivalent. A graphical approach can only confirm that the simplification found using algebraic techniques is correct.

5.2 Verifying Trigonometric Identities

What you should learn

- Verify trigonometric identities.

Why you should learn it

You can use trigonometric identities to rewrite trigonometric equations that model real-life situations. For instance, in Exercise 56 on page 388, you can use trigonometric identities to simplify the equation that models the length of a shadow cast by a gnomon (a device used to tell time).

Robert Ginn/PhotoEdit

Introduction

In this section, you will study techniques for verifying trigonometric identities. In the next section, you will study techniques for solving trigonometric equations. The key to verifying identities *and* solving equations is the ability to use the fundamental identities and the rules of algebra to rewrite trigonometric expressions.

Remember that a *conditional equation* is an equation that is true for only some of the values in its domain. For example, the conditional equation

$$\sin x = 0 \qquad \text{Conditional equation}$$

is true only for $x = n\pi$, where n is an integer. When you find these values, you are *solving* the equation.

On the other hand, an equation that is true for all real values in the domain of the variable is an *identity*. For example, the familiar equation

$$\sin^2 x = 1 - \cos^2 x \qquad \text{Identity}$$

is true for all real numbers x. So, it is an identity.

Verifying Trigonometric Identities

Although there are similarities, verifying that a trigonometric equation is an identity is quite different from solving an equation. There is no well-defined set of rules to follow in verifying trigonometric identities, and the process is best learned by practice.

Guidelines for Verifying Trigonometric Identities

1. Work with one side of the equation at a time. It is often better to work with the more complicated side first.
2. Look for opportunities to factor an expression, add fractions, square a binomial, or create a monomial denominator.
3. Look for opportunities to use the fundamental identities. Note which functions are in the final expression you want. Sines and cosines pair up well, as do secants and tangents, and cosecants and cotangents.
4. If the preceding guidelines do not help, try converting all terms to sines and cosines.
5. Always try *something*. Even paths that lead to dead ends provide insights.

You may want to review the distinctions among expressions, equations, and identities. Have your students look at some algebraic identities and conditional equations before starting this section. It is important for them to understand what it means to verify an identity and not try to solve it as an equation.

Verifying trigonometric identities is a useful process if you need to convert a trigonometric expression into a form that is more useful algebraically. When you verify an identity, you cannot *assume* that the two sides of the equation are equal because you are trying to verify that they *are* equal. As a result, when verifying identities, you cannot use operations such as adding the same quantity to each side of the equation or cross multiplication.

Example 1 Verifying a Trigonometric Identity

Verify the identity $\dfrac{\sec^2\theta - 1}{\sec^2\theta} = \sin^2\theta$.

STUDY TIP

Remember that an identity is only true for all real values in the domain of the variable. For instance, in Example 1 the identity is not true when $\theta = \pi/2$ because $\sec^2\theta$ is not defined when $\theta = \pi/2$.

Solution

Because the left side is more complicated, start with it.

$$\frac{\sec^2\theta - 1}{\sec^2\theta} = \frac{(\tan^2\theta + 1) - 1}{\sec^2\theta} \qquad \text{Pythagorean identity}$$

$$= \frac{\tan^2\theta}{\sec^2\theta} \qquad \text{Simplify.}$$

$$= \tan^2\theta(\cos^2\theta) \qquad \text{Reciprocal identity}$$

$$= \frac{\sin^2\theta}{\cancel{(\cos^2\theta)}}\cancel{(\cos^2\theta)} \qquad \text{Quotient identity}$$

$$= \sin^2\theta \qquad \text{Simplify.}$$

Notice how the identity is verified. You start with the left side of the equation (the more complicated side) and use the fundamental trigonometric identities to simplify it until you obtain the right side.

✓CHECKPOINT Now try Exercise 5.

Encourage your students to identify the reasoning behind each solution step in the examples of this section while covering the comment lines. This will help students to recognize and remember the fundamental trigonometric identities.

There is more than one way to verify an identity. Here is another way to verify the identity in Example 1.

$$\frac{\sec^2\theta - 1}{\sec^2\theta} = \frac{\sec^2\theta}{\sec^2\theta} - \frac{1}{\sec^2\theta} \qquad \text{Rewrite as the difference of fractions.}$$

$$= 1 - \cos^2\theta \qquad \text{Reciprocal identity}$$

$$= \sin^2\theta \qquad \text{Pythagorean identity}$$

Example 2 Combining Fractions Before Using Identities

Verify the identity $\dfrac{1}{1 - \sin\alpha} + \dfrac{1}{1 + \sin\alpha} = 2\sec^2\alpha$.

Solution

$$\frac{1}{1 - \sin\alpha} + \frac{1}{1 + \sin\alpha} = \frac{1 + \sin\alpha + 1 - \sin\alpha}{(1 - \sin\alpha)(1 + \sin\alpha)} \qquad \text{Add fractions.}$$

$$= \frac{2}{1 - \sin^2\alpha} \qquad \text{Simplify.}$$

$$= \frac{2}{\cos^2\alpha} \qquad \text{Pythagorean identity}$$

$$= 2\sec^2\alpha \qquad \text{Reciprocal identity}$$

✓CHECKPOINT Now try Exercise 19.

In Exercises 39–46, (a) use a graphing utility to graph each side of the equation to determine whether the equation is an identity, (b) use the *table* feature of a graphing utility to determine whether the equation is an identity, and (c) confirm the results of parts (a) and (b) algebraically.

39. $2\sec^2 x - 2\sec^2 x \sin^2 x - \sin^2 x - \cos^2 x = 1$

40. $\csc x(\csc x - \sin x) + \dfrac{\sin x - \cos x}{\sin x} + \cot x = \csc^2 x$

41. $2 + \cos^2 x - 3\cos^4 x = \sin^2 x(3 + 2\cos^2 x)$

42. $\tan^4 x + \tan^2 x - 3 = \sec^2 x(4\tan^2 x - 3)$

43. $\csc^4 x - 2\csc^2 x + 1 = \cot^4 x$

44. $(\sin^4 \beta - 2\sin^2 \beta + 1)\cos \beta = \cos^5 \beta$

45. $\dfrac{\cos x}{1 - \sin x} = \dfrac{1 - \sin x}{\cos x}$ **46.** $\dfrac{\cot \alpha}{\csc \alpha + 1} = \dfrac{\csc \alpha + 1}{\cot \alpha}$

In Exercises 47–50, verify the identity.

47. $\tan^5 x = \tan^3 x \sec^2 x - \tan^3 x$

48. $\sec^4 x \tan^2 x = (\tan^2 x + \tan^4 x)\sec^2 x$

49. $\cos^3 x \sin^2 x = (\sin^2 x - \sin^4 x)\cos x$

50. $\sin^4 x + \cos^4 x = 1 - 2\cos^2 x + 2\cos^4 x$

In Exercises 51–54, use the cofunction identities to evaluate the expression without the aid of a calculator.

51. $\sin^2 25° + \sin^2 65°$ **52.** $\cos^2 55° + \cos^2 35°$

53. $\cos^2 20° + \cos^2 52° + \cos^2 38° + \cos^2 70°$

54. $\sin^2 12° + \sin^2 40° + \sin^2 50° + \sin^2 78°$

55. ***Rate of Change*** The rate of change of the function $f(x) = \sin x + \csc x$ with respect to change in the variable x is given by the expression $\cos x - \csc x \cot x$. Show that the expression for the rate of change can also be $-\cos x \cot^2 x$.

Model It

56. ***Shadow Length*** The length s of a shadow cast by a vertical gnomon (a device used to tell time) of height h when the angle of the sun above the horizon is θ (see figure) can be modeled by the equation

$$s = \frac{h\sin(90° - \theta)}{\sin \theta}.$$

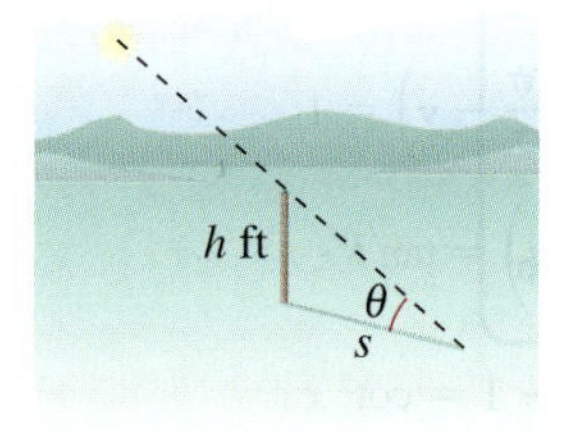

Model It (continued)

(a) Verify that the equation for s is equal to $h \cot \theta$.

(b) Use a graphing utility to complete the table. Let $h = 5$ feet.

θ	10°	20°	30°	40°	50°
s					

θ	60°	70°	80°	90°
s				

(c) Use your table from part (b) to determine the angles of the sun for which the length of the shadow is the greatest and the least.

(d) Based on your results from part (c), what time of day do you think it is when the angle of the sun above the horizon is 90°?

Synthesis

True or False? **In Exercises 57 and 58, determine whether the statement is true or false. Justify your answer.**

57. The equation $\sin^2 \theta + \cos^2 \theta = 1 + \tan^2 \theta$ is an identity, because $\sin^2(0) + \cos^2(0) = 1$ and $1 + \tan^2(0) = 1$.

58. The equation $1 + \tan^2 \theta = 1 + \cot^2 \theta$ is *not* an identity, because it is true that $1 + \tan^2(\pi/6) = 1\frac{1}{3}$, and $1 + \cot^2(\pi/6) = 4$.

Think About It **In Exercises 59 and 60, explain why the equation is not an identity and find one value of the variable for which the equation is not true.**

59. $\sin \theta = \sqrt{1 - \cos^2 \theta}$

60. $\tan \theta = \sqrt{\sec^2 \theta - 1}$

Skills Review

In Exercises 61–64, perform the operation and simplify.

61. $(2 + 3i) - \sqrt{-26}$ **62.** $(2 - 5i)^2$

63. $\sqrt{-16}\left(1 + \sqrt{-4}\right)$ **64.** $(3 + 2i)^3$

In Exercises 65–68, use the Quadratic Formula to solve the quadratic equation.

65. $x^2 + 6x - 12 = 0$ **66.** $x^2 + 5x - 7 = 0$

67. $3x^2 - 6x - 12 = 0$ **68.** $8x^2 - 4x - 3 = 0$

5.3 Solving Trigonometric Equations

What you should learn

- Use standard algebraic techniques to solve trigonometric equations.
- Solve trigonometric equations of quadratic type.
- Solve trigonometric equations involving multiple angles.
- Use inverse trigonometric functions to solve trigonometric equations.

Why you should learn it

You can use trigonometric equations to solve a variety of real-life problems. For instance, in Exercise 72 on page 398, you can solve a trigonometric equation to help answer questions about monthly sales of skiing equipment.

Tom Stillo/Index Stock Imagery

Introduction

To solve a trigonometric equation, use standard algebraic techniques such as collecting like terms and factoring. Your preliminary goal in solving a trigonometric equation is to *isolate* the trigonometric function involved in the equation. For example, to solve the equation $2 \sin x = 1$, divide each side by 2 to obtain

$$\sin x = \frac{1}{2}.$$

To solve for x, note in Figure 5.3 that the equation $\sin x = \frac{1}{2}$ has solutions $x = \pi/6$ and $x = 5\pi/6$ in the interval $[0, 2\pi)$. Moreover, because $\sin x$ has a period of 2π, there are infinitely many other solutions, which can be written as

$$x = \frac{\pi}{6} + 2n\pi \quad \text{and} \quad x = \frac{5\pi}{6} + 2n\pi \qquad \text{General solution}$$

where n is an integer, as shown in Figure 5.3.

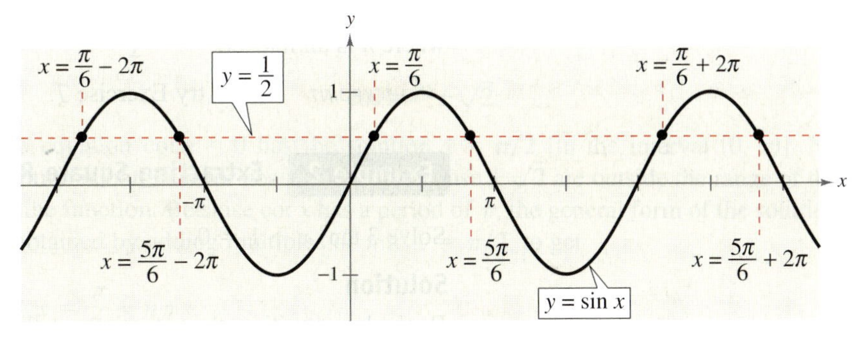

FIGURE 5.3

Another way to show that the equation $\sin x = \frac{1}{2}$ has infinitely many solutions is indicated in Figure 5.4. Any angles that are coterminal with $\pi/6$ or $5\pi/6$ will also be solutions of the equation.

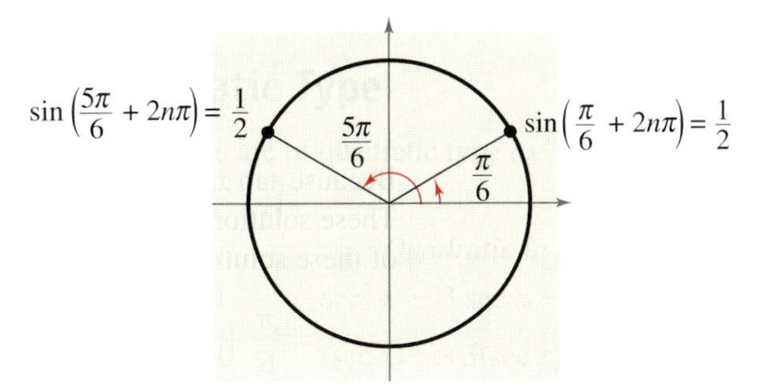

FIGURE 5.4

When solving trigonometric equations, you should write your answer(s) using exact values rather than decimal approximations.

69. ***Harmonic Motion*** A weight is oscillating on the end of a spring (see figure). The position of the weight relative to the point of equilibrium is given by $y = \frac{1}{12}(\cos 8t - 3 \sin 8t)$, where y is the displacement (in meters) and t is the time (in seconds). Find the times when the weight is at the point of equilibrium $(y = 0)$ for $0 \le t \le 1$.

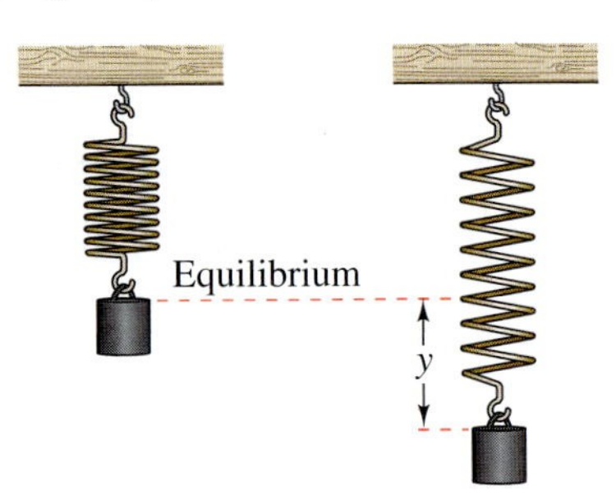

70. ***Damped Harmonic Motion*** The displacement from equilibrium of a weight oscillating on the end of a spring is given by $y = 1.56e^{-0.22t}\cos 4.9t$, where y is the displacement (in feet) and t is the time (in seconds). Use a graphing utility to graph the displacement function for $0 \le t \le 10$. Find the time beyond which the displacement does not exceed 1 foot from equilibrium.

71. ***Sales*** The monthly sales S (in thousands of units) of a seasonal product are approximated by

$$S = 74.50 + 43.75 \sin \frac{\pi t}{6}$$

where t is the time (in months), with $t = 1$ corresponding to January. Determine the months when sales exceed 100,000 units.

72. ***Sales*** The monthly sales S (in hundreds of units) of skiing equipment at a sports store are approximated by

$$S = 58.3 + 32.5 \cos \frac{\pi t}{6}$$

where t is the time (in months), with $t = 1$ corresponding to January. Determine the months when sales exceed 7500 units.

73. ***Projectile Motion*** A batted baseball leaves the bat at an angle of θ with the horizontal and an initial velocity of $v_0 = 100$ feet per second. The ball is caught by an outfielder 300 feet from home plate (see figure). Find θ if the range r of a projectile is given by $r = \frac{1}{32}v_0^2 \sin 2\theta$.

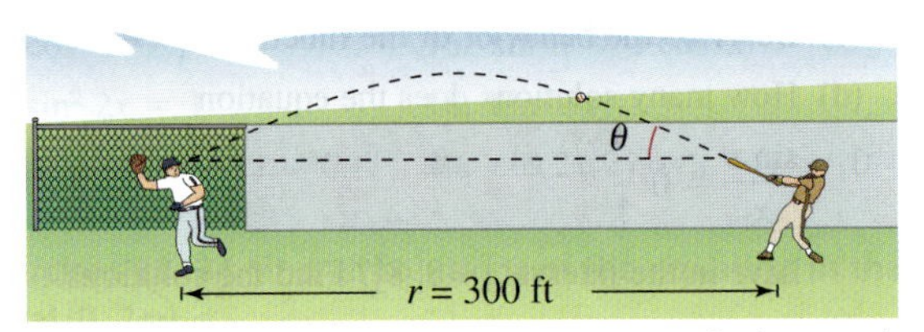

Not drawn to scale

74. ***Projectile Motion*** A sharpshooter intends to hit a target at a distance of 1000 yards with a gun that has a muzzle velocity of 1200 feet per second (see figure). Neglecting air resistance, determine the gun's minimum angle of elevation θ if the range r is given by

$$r = \frac{1}{32}v_0^2 \sin 2\theta.$$

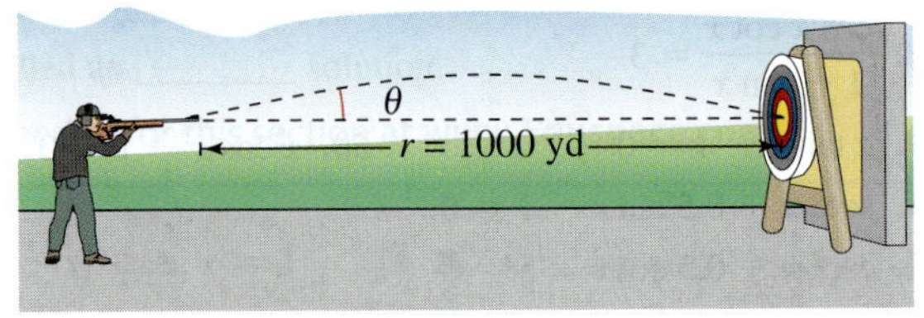

Not drawn to scale

75. ***Ferris Wheel*** A Ferris wheel is built such that the height h (in feet) above ground of a seat on the wheel at time t (in minutes) can be modeled by

$$h(t) = 53 + 50 \sin\left(\frac{\pi}{16}t - \frac{\pi}{2}\right).$$

The wheel makes one revolution every 32 seconds. The ride begins when $t = 0$.

(a) During the first 32 seconds of the ride, when will a person on the Ferris wheel be 53 feet above ground?

(b) When will a person be at the top of the Ferris wheel for the first time during the ride? If the ride lasts 160 seconds, how many times will a person be at the top of the ride, and at what times?

Model It

76. ***Data Analysis: Unemployment Rate*** The table shows the unemployment rates r in the United States for selected years from 1990 through 2004. The time t is measured in years, with $t = 0$ corresponding to 1990. (Source: U.S. Bureau of Labor Statistics)

Time, t	Rate, r
0	5.6
2	7.5
4	6.1
6	5.4

Time, t	Rate, r
8	4.5
10	4.0
12	5.8
14	5.5

(a) Create a scatter plot of the data.

Model It (continued)

(b) Which of the following models best represents the data? Explain your reasoning.

(1) $r = 1.24 \sin(0.47t + 0.40) + 5.45$

(2) $r = 1.24 \sin(0.47t - 0.01) + 5.45$

(3) $r = \sin(0.10t + 5.61) + 4.80$

(4) $r = 896 \sin(0.57t - 2.05) + 6.48$

(c) What term in the model gives the average unemployment rate? What is the rate?

(d) Economists study the lengths of business cycles such as unemployment rates. Based on this short span of time, use the model to find the length of this cycle.

(e) Use the model to estimate the next time the unemployment rate will be 5% or less.

77. ***Geometry*** The area of a rectangle (see figure) inscribed in one arc of the graph of $y = \cos x$ is given by

$$A = 2x \cos x, \quad 0 < x < \frac{\pi}{2}.$$

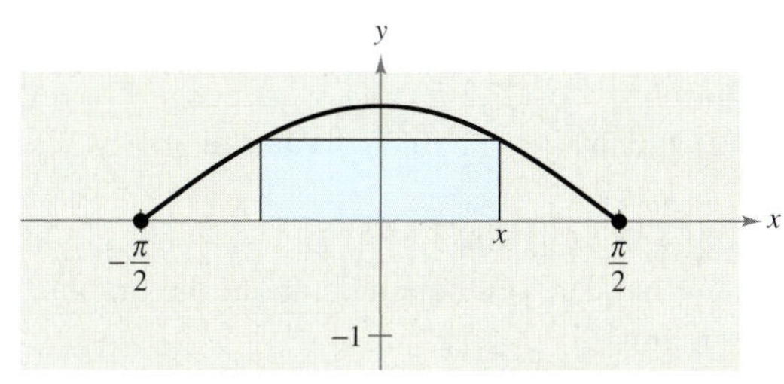

(a) Use a graphing utility to graph the area function, and approximate the area of the largest inscribed rectangle.

(b) Determine the values of x for which $A \geq 1$.

78. ***Quadratic Approximation*** Consider the function given by $f(x) = 3 \sin(0.6x - 2)$.

(a) Approximate the zero of the function in the interval $[0, 6]$.

(b) A quadratic approximation agreeing with f at $x = 5$ is $g(x) = -0.45x^2 + 5.52x - 13.70$. Use a graphing utility to graph f and g in the same viewing window. Describe the result.

(c) Use the Quadratic Formula to find the zeros of g. Compare the zero in the interval $[0, 6]$ with the result of part (a).

Synthesis

True or False? **In Exercises 79 and 80, determine whether the statement is true or false. Justify your answer.**

79. The equation $2 \sin 4t - 1 = 0$ has four times the number of solutions in the interval $[0, 2\pi)$ as the equation $2 \sin t - 1 = 0$.

80. If you correctly solve a trigonometric equation to the statement $\sin x = 3.4$, then you can finish solving the equation by using an inverse function.

In Exercises 81 and 82, use the graph to approximate the number of points of intersection of the graphs of y_1 and y_2.

81. $y_1 = 2 \sin x$
$y_2 = 3x + 1$

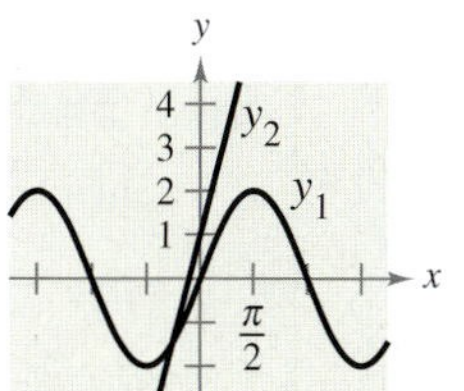

82. $y_1 = 2 \sin x$
$y_2 = \frac{1}{2}x + 1$

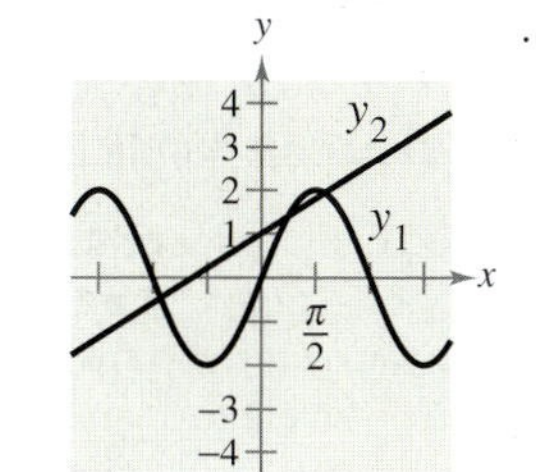

Skills Review

In Exercises 83 and 84, solve triangle *ABC* by finding all missing angle measures and side lengths.

83.

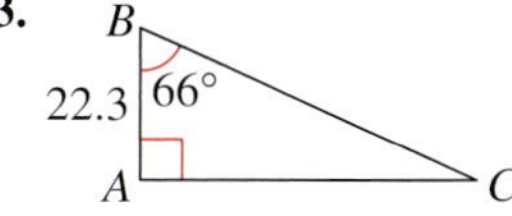

84.

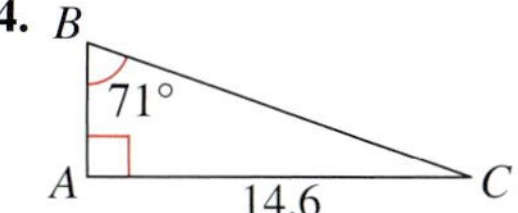

In Exercises 85–88, use reference angles to find the exact values of the sine, cosine, and tangent of the angle with the given measure.

85. $390°$ **86.** $600°$

87. $-1845°$ **88.** $-1410°$

89. ***Angle of Depression*** Find the angle of depression from the top of a lighthouse 250 feet above water level to the water line of a ship 2 miles offshore.

90. ***Height*** From a point 100 feet in front of a public library, the angles of elevation to the base of the flagpole and the top of the pole are $28°$ and $39° 45'$, respectively. The flagpole is mounted on the front of the library's roof. Find the height of the flagpole.

91. **Make a Decision** To work an extended application analyzing the normal daily high temperatures in Phoenix and in Seattle, visit this text's website at *college.hmco.com.* *(Data Source: NOAA)*

5.4 Sum and Difference Formulas

What you should learn

- Use sum and difference formulas to evaluate trigonometric functions, verify identities, and solve trigonometric equations.

Why you should learn it

You can use identities to rewrite trigonometric expressions. For instance, in Exercise 75 on page 405, you can use an identity to rewrite a trigonometric expression in a form that helps you analyze a harmonic motion equation.

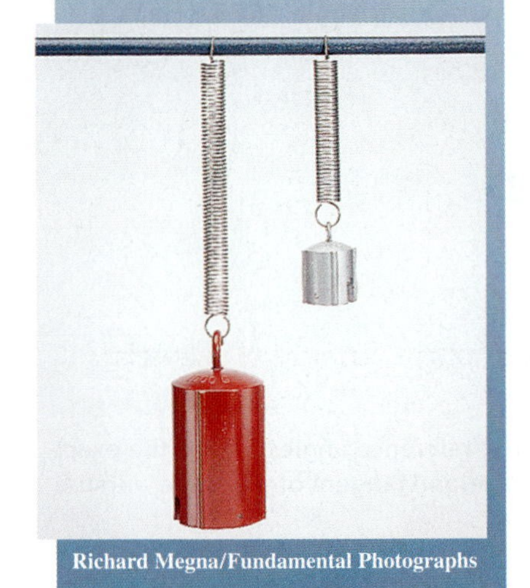

Richard Megna/Fundamental Photographs

Using Sum and Difference Formulas

In this and the following section, you will study the uses of several trigonometric identities and formulas.

Sum and Difference Formulas

$$\sin(u + v) = \sin u \cos v + \cos u \sin v$$

$$\sin(u - v) = \sin u \cos v - \cos u \sin v$$

$$\cos(u + v) = \cos u \cos v - \sin u \sin v$$

$$\cos(u - v) = \cos u \cos v + \sin u \sin v$$

$$\tan(u + v) = \frac{\tan u + \tan v}{1 - \tan u \tan v}$$

$$\tan(u - v) = \frac{\tan u - \tan v}{1 + \tan u \tan v}$$

For a proof of the sum and difference formulas, see Proofs in Mathematics on page 442.

Exploration

Use a graphing utility to graph $y_1 = \cos(x + 2)$ and $y_2 = \cos x + \cos 2$ in the same viewing window. What can you conclude about the graphs? Is it true that $\cos(x + 2) = \cos x + \cos 2$?

Use a graphing utility to graph $y_1 = \sin(x + 4)$ and $y_2 = \sin x + \sin 4$ in the same viewing window. What can you conclude about the graphs? Is it true that $\sin(x + 4) = \sin x + \sin 4$?

Examples 1 and 2 show how **sum and difference formulas** can be used to find exact values of trigonometric functions involving sums or differences of special angles.

Example 1 Evaluating a Trigonometric Function

Find the exact value of $\cos 75°$.

Solution

To find the *exact* value of $\cos 75°$, use the fact that $75° = 30° + 45°$. Consequently, the formula for $\cos(u + v)$ yields

$$\cos 75° = \cos(30° + 45°)$$

$$= \cos 30° \cos 45° - \sin 30° \sin 45°$$

$$= \frac{\sqrt{3}}{2}\left(\frac{\sqrt{2}}{2}\right) - \frac{1}{2}\left(\frac{\sqrt{2}}{2}\right) = \frac{\sqrt{6} - \sqrt{2}}{4}.$$

Try checking this result on your calculator. You will find that $\cos 75° \approx 0.259$.

✓CHECKPOINT Now try Exercise 1.

The Granger Collection, New York

Historical Note
Hipparchus, considered the most eminent of Greek astronomers, was born about 160 B.C. in Nicaea. He was credited with the invention of trigonometry. He also derived the sum and difference formulas for $\sin(A \pm B)$ and $\cos(A \pm B)$.

Example 2 Evaluating a Trigonometric Expression

Find the exact value of $\sin \frac{\pi}{12}$.

Solution

Using the fact that

$$\frac{\pi}{12} = \frac{\pi}{3} - \frac{\pi}{4}$$

together with the formula for $\sin(u - v)$, you obtain

$$\begin{aligned}
\sin \frac{\pi}{12} &= \sin\left(\frac{\pi}{3} - \frac{\pi}{4}\right) \\
&= \sin \frac{\pi}{3} \cos \frac{\pi}{4} - \cos \frac{\pi}{3} \sin \frac{\pi}{4} \\
&= \frac{\sqrt{3}}{2}\left(\frac{\sqrt{2}}{2}\right) - \frac{1}{2}\left(\frac{\sqrt{2}}{2}\right) \\
&= \frac{\sqrt{6} - \sqrt{2}}{4}.
\end{aligned}$$

CHECKPOINT Now try Exercise 3.

Example 3 Evaluating a Trigonometric Expression

Find the exact value of $\sin 42^\circ \cos 12^\circ - \cos 42^\circ \sin 12^\circ$.

Solution

Recognizing that this expression fits the formula for $\sin(u - v)$, you can write

$$\begin{aligned}
\sin 42^\circ \cos 12^\circ - \cos 42^\circ \sin 12^\circ &= \sin(42^\circ - 12^\circ) \\
&= \sin 30^\circ \\
&= \tfrac{1}{2}.
\end{aligned}$$

CHECKPOINT Now try Exercise 31.

Example 4 An Application of a Sum Formula

Write $\cos(\arctan 1 + \arccos x)$ as an algebraic expression.

Solution

This expression fits the formula for $\cos(u + v)$. Angles $u = \arctan 1$ and $v = \arccos x$ are shown in Figure 5.7. So

$$\begin{aligned}
\cos(u + v) &= \cos(\arctan 1) \cos(\arccos x) - \sin(\arctan 1) \sin(\arccos x) \\
&= \frac{1}{\sqrt{2}} \cdot x - \frac{1}{\sqrt{2}} \cdot \sqrt{1 - x^2} \\
&= \frac{x - \sqrt{1 - x^2}}{\sqrt{2}}.
\end{aligned}$$

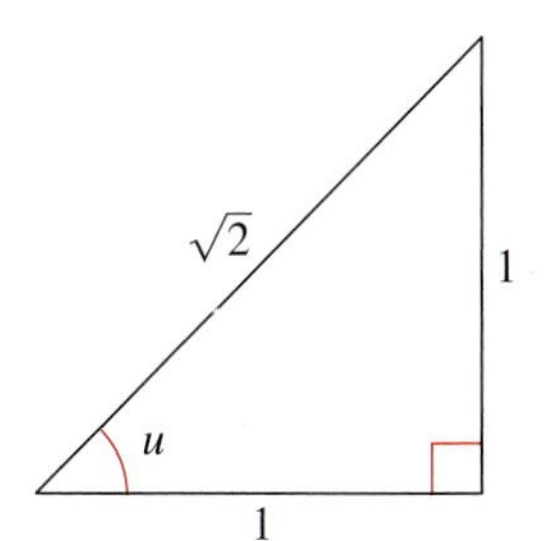

FIGURE 5.7

CHECKPOINT Now try Exercise 51.

Example 5 shows how to use a difference formula to prove the cofunction identity

$$\cos\left(\frac{\pi}{2} - x\right) = \sin x.$$

Example 5 Proving a Cofunction Identity

Prove the cofunction identity $\cos\left(\frac{\pi}{2} - x\right) = \sin x$.

Solution

Using the formula for $\cos(u - v)$, you have

$$\cos\left(\frac{\pi}{2} - x\right) = \cos\frac{\pi}{2}\cos x + \sin\frac{\pi}{2}\sin x$$

$$= (0)(\cos x) + (1)(\sin x) = \sin x.$$

CHECKPOINT Now try Exercise 55.

Sum and difference formulas can be used to rewrite expressions such as

$$\sin\left(\theta + \frac{n\pi}{2}\right) \quad \text{and} \quad \cos\left(\theta + \frac{n\pi}{2}\right), \quad \text{where } n \text{ is an integer}$$

as expressions involving only $\sin\theta$ or $\cos\theta$. The resulting formulas are called **reduction formulas.**

Example 6 Deriving Reduction Formulas

Simplify each expression.

a. $\cos\left(\theta - \frac{3\pi}{2}\right)$ **b.** $\tan(\theta + 3\pi)$

Solution

a. Using the formula for $\cos(u - v)$, you have

$$\cos\left(\theta - \frac{3\pi}{2}\right) = \cos\theta\cos\frac{3\pi}{2} + \sin\theta\sin\frac{3\pi}{2}$$

$$= (\cos\theta)(0) + (\sin\theta)(-1)$$

$$= -\sin\theta.$$

b. Using the formula for $\tan(u + v)$, you have

$$\tan(\theta + 3\pi) = \frac{\tan\theta + \tan 3\pi}{1 - \tan\theta\tan 3\pi}$$

$$= \frac{\tan\theta + 0}{1 - (\tan\theta)(0)}$$

$$= \tan\theta.$$

CHECKPOINT Now try Exercise 65.

Activities

1. Use the sum and difference formulas to find the exact value of cos 15°.
   Answer: $\frac{\sqrt{6} + \sqrt{2}}{4}$
2. Rewrite the expression using the sum and difference formulas.
   $\frac{\tan 40^\circ + \tan 10^\circ}{1 - \tan 40^\circ \tan 10^\circ}$
   Answer: $\tan(40^\circ + 10^\circ) = \tan 50^\circ$
3. Verify the identity
   $\sin\left(\frac{\pi}{2} - \theta\right) = \cos\theta.$
   Answer:
   $$\sin\left(\frac{\pi}{2} - \theta\right) = \sin\frac{\pi}{2}\cos\theta - \cos\frac{\pi}{2}\sin\theta$$
   $$= (1)\cos\theta - (0)\sin\theta$$
   $$= \cos\theta$$

Example 7 Solving a Trigonometric Equation

Find all solutions of $\sin\left(x + \frac{\pi}{4}\right) + \sin\left(x - \frac{\pi}{4}\right) = -1$ in the interval $[0, 2\pi)$.

Solution

Using sum and difference formulas, rewrite the equation as

$$\sin x \cos \frac{\pi}{4} + \cos x \sin \frac{\pi}{4} + \sin x \cos \frac{\pi}{4} - \cos x \sin \frac{\pi}{4} = -1$$

$$2 \sin x \cos \frac{\pi}{4} = -1$$

$$2(\sin x)\left(\frac{\sqrt{2}}{2}\right) = -1$$

$$\sin x = -\frac{1}{\sqrt{2}}$$

$$\sin x = -\frac{\sqrt{2}}{2}.$$

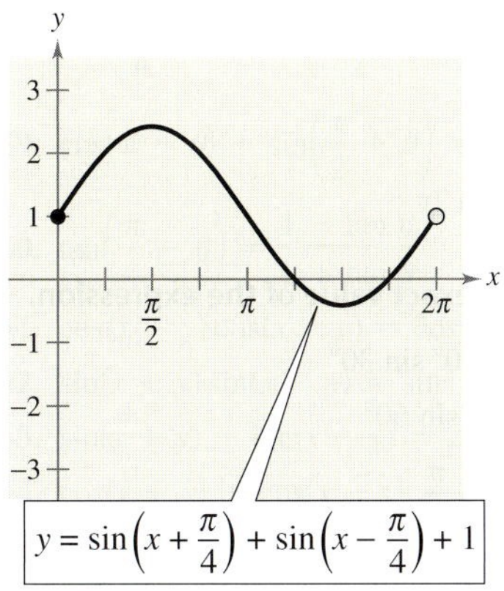

So, the only solutions in the interval $[0, 2\pi)$ are

$$x = \frac{5\pi}{4} \quad \text{and} \quad x = \frac{7\pi}{4}.$$

You can confirm this graphically by sketching the graph of

$$y = \sin\left(x + \frac{\pi}{4}\right) + \sin\left(x - \frac{\pi}{4}\right) + 1 \text{ for } 0 \le x < 2\pi,$$

as shown in Figure 5.8. From the graph you can see that the x-intercepts are $5\pi/4$ and $7\pi/4$.

$y = \sin\left(x + \frac{\pi}{4}\right) + \sin\left(x - \frac{\pi}{4}\right) + 1$

FIGURE 5.8

CHECKPOINT Now try Exercise 69.

The next example was taken from calculus. It is used to derive the derivative of the sine function.

Example 8 An Application from Calculus

Verify that

$$\frac{\sin(x + h) - \sin x}{h} = (\cos x)\left(\frac{\sin h}{h}\right) - (\sin x)\left(\frac{1 - \cos h}{h}\right)$$

where $h \neq 0$.

Solution

Using the formula for $\sin(u + v)$, you have

$$\frac{\sin(x + h) - \sin x}{h} = \frac{\sin x \cos h + \cos x \sin h - \sin x}{h}$$

$$= \frac{\cos x \sin h - \sin x(1 - \cos h)}{h}$$

$$= (\cos x)\left(\frac{\sin h}{h}\right) - (\sin x)\left(\frac{1 - \cos h}{h}\right).$$

CHECKPOINT Now try Exercise 91.

Half-Angle Formulas

You can derive some useful alternative forms of the power-reducing formulas by replacing u with $u/2$. The results are called **half-angle formulas.**

Half-Angle Formulas

$$\sin \frac{u}{2} = \pm \sqrt{\frac{1 - \cos u}{2}}$$

$$\cos \frac{u}{2} = \pm \sqrt{\frac{1 + \cos u}{2}}$$

$$\tan \frac{u}{2} = \frac{1 - \cos u}{\sin u} = \frac{\sin u}{1 + \cos u}$$

The signs of $\sin \frac{u}{2}$ and $\cos \frac{u}{2}$ depend on the quadrant in which $\frac{u}{2}$ lies.

Example 6 **Using a Half-Angle Formula**

Find the exact value of $\sin 105°$.

Solution

Begin by noting that 105° is half of 210°. Then, using the half-angle formula for $\sin(u/2)$ and the fact that 105° lies in Quadrant II, you have

$$\sin 105° = \sqrt{\frac{1 - \cos 210°}{2}}$$

$$= \sqrt{\frac{1 - (-\cos 30°)}{2}}$$

$$= \sqrt{\frac{1 + (\sqrt{3}/2)}{2}}$$

$$= \frac{\sqrt{2 + \sqrt{3}}}{2}.$$

The positive square root is chosen because $\sin \theta$ is positive in Quadrant II.

CHECKPOINT Now try Exercise 41.

Use your calculator to verify the result obtained in Example 6. That is, evaluate $\sin 105°$ and $\left(\sqrt{2 + \sqrt{3}}\right)/2$.

$$\sin 105° \approx 0.9659258$$

$$\frac{\sqrt{2 + \sqrt{3}}}{2} \approx 0.9659258$$

You can see that both values are approximately 0.9659258.

STUDY TIP

To find the exact value of a trigonometric function with an angle measure in D°M′S″ form using a half-angle formula, first convert the angle measure to decimal degree form. Then multiply the resulting angle measure by 2.

Example 7 Solving a Trigonometric Equation

Find all solutions of $2 - \sin^2 x = 2\cos^2 \frac{x}{2}$ in the interval $[0, 2\pi)$.

Algebraic Solution

$$2 - \sin^2 x = 2\cos^2 \frac{x}{2}$$ Write original equation.

$$2 - \sin^2 x = 2\left(\pm\sqrt{\frac{1 + \cos x}{2}}\right)^2$$ Half-angle formula

$$2 - \sin^2 x = 2\left(\frac{1 + \cos x}{2}\right)$$ Simplify.

$$2 - \sin^2 x = 1 + \cos x$$ Simplify.

$$2 - (1 - \cos^2 x) = 1 + \cos x$$ Pythagorean identity

$$\cos^2 x - \cos x = 0$$ Simplify.

$$\cos x(\cos x - 1) = 0$$ Factor.

By setting the factors $\cos x$ and $\cos x - 1$ equal to zero, you find that the solutions in the interval $[0, 2\pi)$ are

$$x = \frac{\pi}{2}, \quad x = \frac{3\pi}{2}, \quad \text{and} \quad x = 0.$$

CHECKPOINT Now try Exercise 59.

Graphical Solution

Use a graphing utility set in *radian* mode to graph $y = 2 - \sin^2 x - 2\cos^2(x/2)$, as shown in Figure 5.11. Use the *zero* or *root* feature or the *zoom* and *trace* features to approximate the x-intercepts in the interval $[0, 2\pi)$ to be

$$x = 0, x \approx 1.571 \approx \frac{\pi}{2}, \text{ and } x \approx 4.712 \approx \frac{3\pi}{2}.$$

These values are the approximate solutions of $2 - \sin^2 x - 2\cos^2(x/2) = 0$ in the interval $[0, 2\pi)$.

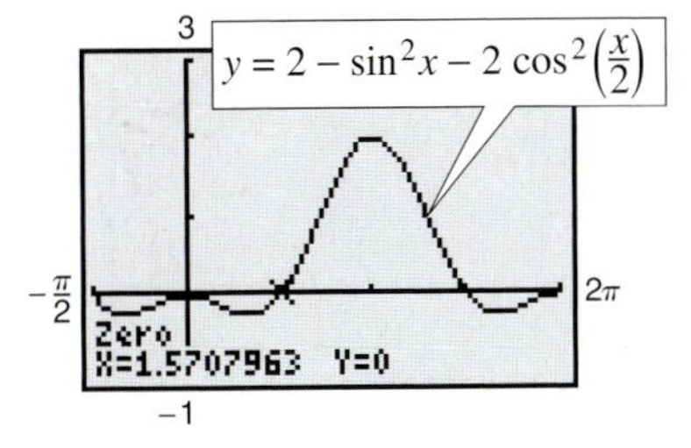

FIGURE 5.11

Product-to-Sum Formulas

A common error is to write

$$2\cos^2 \frac{x}{2}$$

as $\cos^2 x$, rather than to use the correct identity

$$2\cos^2 \frac{x}{2} = 2\left(\frac{1 + \cos x}{2}\right).$$

Each of the following **product-to-sum formulas** is easily verified using the sum and difference formulas discussed in the preceding section.

Product-to-Sum Formulas

$$\sin u \sin v = \frac{1}{2}[\cos(u - v) - \cos(u + v)]$$

$$\cos u \cos v = \frac{1}{2}[\cos(u - v) + \cos(u + v)]$$

$$\sin u \cos v = \frac{1}{2}[\sin(u + v) + \sin(u - v)]$$

$$\cos u \sin v = \frac{1}{2}[\sin(u + v) - \sin(u - v)]$$

Product-to-sum formulas are used in calculus to evaluate integrals involving the products of sines and cosines of two different angles.

Example 8 Writing Products as Sums

Rewrite the product $\cos 5x \sin 4x$ as a sum or difference.

Solution

Using the appropriate product-to-sum formula, you obtain

$$\cos 5x \sin 4x = \tfrac{1}{2}[\sin(5x + 4x) - \sin(5x - 4x)]$$
$$= \tfrac{1}{2}\sin 9x - \tfrac{1}{2}\sin x.$$

CHECKPOINT Now try Exercise 67.

Occasionally, it is useful to reverse the procedure and write a sum of trigonometric functions as a product. This can be accomplished with the following **sum-to-product formulas.**

Activities

1. Find the exact values of $\sin 2u$, $\cos 2u$, and $\tan 2u$ given $\cot u = -5$, $\frac{3\pi}{2} < u < 2\pi$.

   Answer: $\sin 2u = -\frac{5}{13}$

   $\cos 2u = \frac{12}{13}$

   $\tan 2u = -\frac{5}{12}$

2. Find all solutions of the equation in the interval $[0, 2\pi)$.

   $4 \sin x \cos x = 1$

   Answer: $x = \frac{\pi}{12}, \frac{5\pi}{12}, \frac{13\pi}{12}, \frac{17\pi}{12}$

3. Rewrite as a sum: $2 \sin\theta \sin\phi$.

   Answer: $\cos(\theta - \phi) - \cos(\theta + \phi)$

Sum-to-Product Formulas

$$\sin u + \sin v = 2\sin\left(\frac{u+v}{2}\right)\cos\left(\frac{u-v}{2}\right)$$

$$\sin u - \sin v = 2\cos\left(\frac{u+v}{2}\right)\sin\left(\frac{u-v}{2}\right)$$

$$\cos u + \cos v = 2\cos\left(\frac{u+v}{2}\right)\cos\left(\frac{u-v}{2}\right)$$

$$\cos u - \cos v = -2\sin\left(\frac{u+v}{2}\right)\sin\left(\frac{u-v}{2}\right)$$

For a proof of the sum-to-product formulas, see Proofs in Mathematics on page 444.

Example 9 Using a Sum-to-Product Formula

Find the exact value of $\cos 195° + \cos 105°$.

Solution

Using the appropriate sum-to-product formula, you obtain

$$\cos 195° + \cos 105° = 2\cos\left(\frac{195° + 105°}{2}\right)\cos\left(\frac{195° - 105°}{2}\right)$$
$$= 2\cos 150° \cos 45°$$
$$= 2\left(-\frac{\sqrt{3}}{2}\right)\left(\frac{\sqrt{2}}{2}\right)$$
$$= -\frac{\sqrt{6}}{2}.$$

CHECKPOINT Now try Exercise 83.

Example 10 Solving a Trigonometric Equation

Solve $\sin 5x + \sin 3x = 0$.

Solution

$$\sin 5x + \sin 3x = 0 \qquad \text{Write original equation.}$$

$$2 \sin\left(\frac{5x + 3x}{2}\right) \cos\left(\frac{5x - 3x}{2}\right) = 0 \qquad \text{Sum-to-product formula}$$

$$2 \sin 4x \cos x = 0 \qquad \text{Simplify.}$$

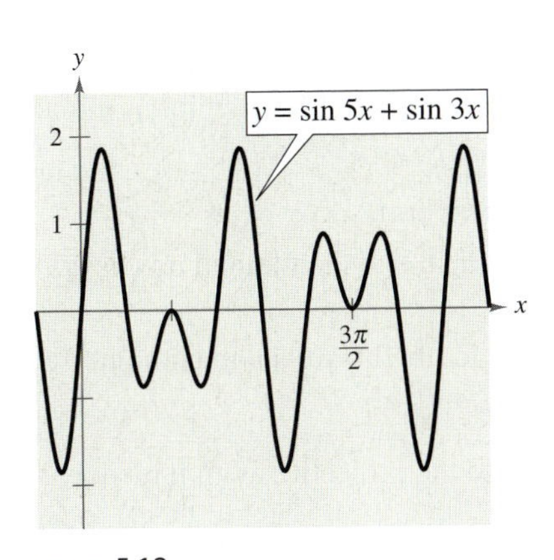

FIGURE 5.12

By setting the factor $2 \sin 4x$ equal to zero, you can find that the solutions in the interval $[0, 2\pi)$ are

$$x = 0, \frac{\pi}{4}, \frac{\pi}{2}, \frac{3\pi}{4}, \pi, \frac{5\pi}{4}, \frac{3\pi}{2}, \frac{7\pi}{4}.$$

The equation $\cos x = 0$ yields no additional solutions, and you can conclude that the solutions are of the form

$$x = \frac{n\pi}{4}$$

where n is an integer. You can confirm this graphically by sketching the graph of $y = \sin 5x + \sin 3x$, as shown in Figure 5.12. From the graph you can see that the x-intercepts occur at multiples of $\pi/4$.

CHECKPOINT Now try Exercise 87.

Example 11 Verifying a Trigonometric Identity

Verify the identity

$$\frac{\sin t + \sin 3t}{\cos t + \cos 3t} = \tan 2t.$$

Solution

Using appropriate sum-to-product formulas, you have

$$\frac{\sin t + \sin 3t}{\cos t + \cos 3t} = \frac{2 \sin\left(\frac{t + 3t}{2}\right)\cos\left(\frac{t - 3t}{2}\right)}{2 \cos\left(\frac{t + 3t}{2}\right)\cos\left(\frac{t - 3t}{2}\right)}$$

$$= \frac{2 \sin(2t)\cos(-t)}{2 \cos(2t)\cos(-t)}$$

$$= \frac{\sin 2t}{\cos 2t}$$

$$= \tan 2t.$$

CHECKPOINT Now try Exercise 105.

Application

Example 12 Projectile Motion 

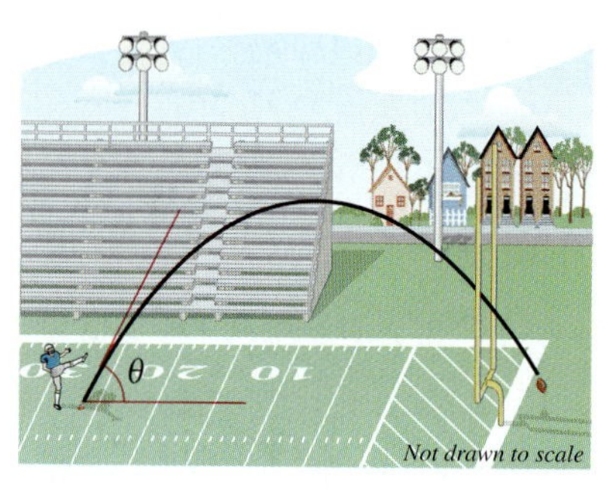

FIGURE 5.13

Ignoring air resistance, the range of a projectile fired at an angle θ with the horizontal and with an initial velocity of v_0 feet per second is given by

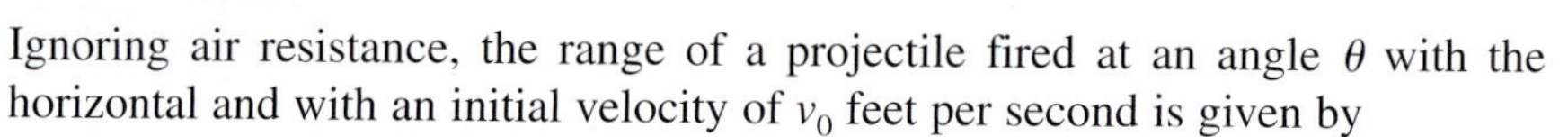

$$r = \frac{1}{16}v_0^2 \sin\theta\cos\theta$$

where r is the horizontal distance (in feet) that the projectile will travel. A place kicker for a football team can kick a football from ground level with an initial velocity of 80 feet per second (see Figure 5.13).

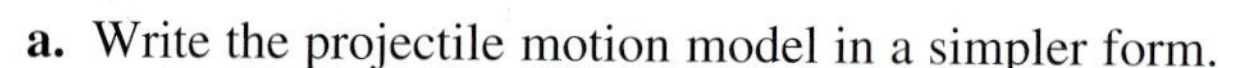

a. Write the projectile motion model in a simpler form.

b. At what angle must the player kick the football so that the football travels 200 feet?

c. For what angle is the horizontal distance the football travels a maximum?

Solution

a. You can use a double-angle formula to rewrite the projectile motion model as

$$r = \frac{1}{32}v_0^2(2\sin\theta\cos\theta)$$ Rewrite original projectile motion model.

$$= \frac{1}{32}v_0^2 \sin 2\theta.$$ Rewrite model using a double-angle formula.

b. $$r = \frac{1}{32}v_0^2 \sin 2\theta$$ Write projectile motion model.

$$200 = \frac{1}{32}(80)^2 \sin 2\theta$$ Substitute 200 for r and 80 for v_0.

$$200 = 200 \sin 2\theta$$ Simplify.

$$1 = \sin 2\theta$$ Divide each side by 200.

You know that $2\theta = \pi/2$, so dividing this result by 2 produces $\theta = \pi/4$. Because $\pi/4 = 45°$, you can conclude that the player must kick the football at an angle of 45° so that the football will travel 200 feet.

c. From the model $r = 200 \sin 2\theta$ you can see that the amplitude is 200. So the maximum range is $r = 200$ feet. From part (b), you know that this corresponds to an angle of 45°. Therefore, kicking the football at an angle of 45° will produce a maximum horizontal distance of 200 feet.

✓CHECKPOINT Now try Exercise 119.

WRITING ABOUT MATHEMATICS

Deriving an Area Formula Describe how you can use a double-angle formula or a half-angle formula to derive a formula for the area of an isosceles triangle. Use a labeled sketch to illustrate your derivation. Then write two examples that show how your formula can be used.

5.5 Exercises

VOCABULARY CHECK: Fill in the blank to complete the trigonometric formula.

1. $\sin 2u =$ ________
2. $\dfrac{1 + \cos 2u}{2} =$ ________
3. $\cos 2u =$ ________
4. $\dfrac{1 - \cos 2u}{1 + \cos 2u} =$ ________
5. $\sin \dfrac{u}{2} =$ ________
6. $\tan \dfrac{u}{2} =$ ________
7. $\cos u \cos v =$ ________
8. $\sin u \cos v =$ ________
9. $\sin u + \sin v =$ ________
10. $\cos u - \cos v =$ ________

PREREQUISITE SKILLS REVIEW: Practice and review algebra skills needed for this section at **www.Eduspace.com.**

In Exercises 1–8, use the figure to find the exact value of the trigonometric function.

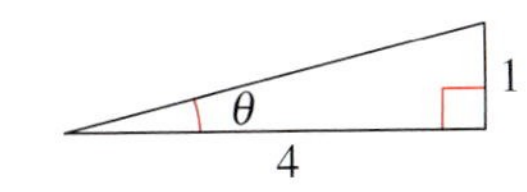

1. $\sin \theta$
2. $\tan \theta$
3. $\cos 2\theta$
4. $\sin 2\theta$
5. $\tan 2\theta$
6. $\sec 2\theta$
7. $\csc 2\theta$
8. $\cot 2\theta$

In Exercises 9–18, find the exact solutions of the equation in the interval $[0, 2\pi)$.

9. $\sin 2x - \sin x = 0$
10. $\sin 2x + \cos x = 0$
11. $4 \sin x \cos x = 1$
12. $\sin 2x \sin x = \cos x$
13. $\cos 2x - \cos x = 0$
14. $\cos 2x + \sin x = 0$
15. $\tan 2x - \cot x = 0$
16. $\tan 2x - 2 \cos x = 0$
17. $\sin 4x = -2 \sin 2x$
18. $(\sin 2x + \cos 2x)^2 = 1$

In Exercises 19–22, use a double-angle formula to rewrite the expression.

19. $6 \sin x \cos x$
20. $6 \cos^2 x - 3$
21. $4 - 8 \sin^2 x$
22. $(\cos x + \sin x)(\cos x - \sin x)$

In Exercises 23–28, find the exact values of $\sin 2u$, $\cos 2u$, and $\tan 2u$ using the double-angle formulas.

23. $\sin u = -\dfrac{4}{5}, \quad \pi < u < \dfrac{3\pi}{2}$
24. $\cos u = -\dfrac{2}{3}, \quad \dfrac{\pi}{2} < u < \pi$
25. $\tan u = \dfrac{3}{4}, \quad 0 < u < \dfrac{\pi}{2}$
26. $\cot u = -4, \quad \dfrac{3\pi}{2} < u < 2\pi$
27. $\sec u = -\dfrac{5}{2}, \quad \dfrac{\pi}{2} < u < \pi$
28. $\csc u = 3, \quad \dfrac{\pi}{2} < u < \pi$

In Exercises 29–34, use the power-reducing formulas to rewrite the expression in terms of the first power of the cosine.

29. $\cos^4 x$
30. $\sin^8 x$
31. $\sin^2 x \cos^2 x$
32. $\sin^4 x \cos^4 x$
33. $\sin^2 x \cos^4 x$
34. $\sin^4 x \cos^2 x$

In Exercises 35–40, use the figure to find the exact value of the trigonometric function.

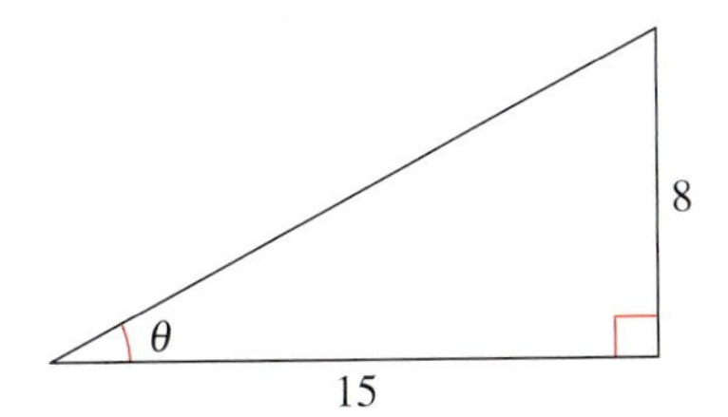

35. $\cos \dfrac{\theta}{2}$
36. $\sin \dfrac{\theta}{2}$
37. $\tan \dfrac{\theta}{2}$
38. $\sec \dfrac{\theta}{2}$
39. $\csc \dfrac{\theta}{2}$
40. $\cot \dfrac{\theta}{2}$

122. ***Railroad Track*** When two railroad tracks merge, the overlapping portions of the tracks are in the shapes of circular arcs (see figure). The radius of each arc r (in feet) and the angle θ are related by

$$\frac{x}{2} = 2r \sin^2 \frac{\theta}{2}.$$

Write a formula for x in terms of $\cos \theta$.

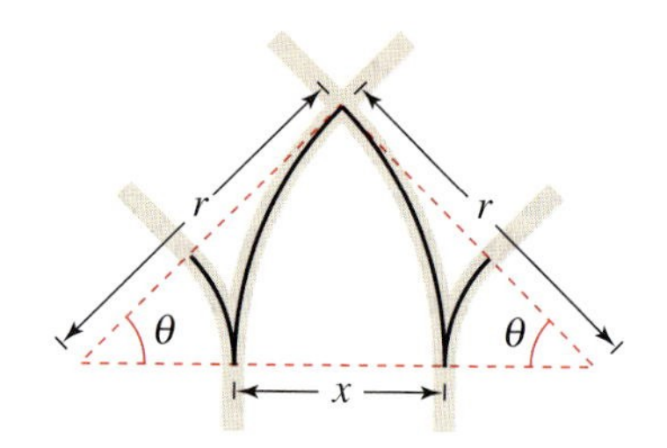

Synthesis

True or False? **In Exercises 123 and 124, determine whether the statement is true or false. Justify your answer.**

123. Because the sine function is an odd function, for a negative number u, $\sin 2u = -2 \sin u \cos u$.

124. $\sin \frac{u}{2} = -\sqrt{\frac{1 - \cos u}{2}}$ when u is in the second quadrant.

In Exercises 125 and 126, (a) use a graphing utility to graph the function and approximate the maximum and minimum points on the graph in the interval $[0, 2\pi)$ and (b) solve the trigonometric equation and verify that its solutions are the x-coordinates of the maximum and minimum points of f. (Calculus is required to find the trigonometric equation.)

	Function	*Trigonometric Equation*
125.	$f(x) = 4 \sin \frac{x}{2} + \cos x$	$2 \cos \frac{x}{2} - \sin x = 0$
126.	$f(x) = \cos 2x - 2 \sin x$	$-2 \cos x(2 \sin x + 1) = 0$

127. ***Exploration*** Consider the function given by

$$f(x) = \sin^4 x + \cos^4 x.$$

(a) Use the power-reducing formulas to write the function in terms of cosine to the first power.

(b) Determine another way of rewriting the function. Use a graphing utility to rule out incorrectly rewritten functions.

(c) Add a trigonometric term to the function so that it becomes a perfect square trinomial. Rewrite the function as a perfect square trinomial minus the term that you added. Use a graphing utility to rule out incorrectly rewritten functions.

(d) Rewrite the result of part (c) in terms of the sine of a double angle. Use a graphing utility to rule out incorrectly rewritten functions.

(e) When you rewrite a trigonometric expression, the result may not be the same as a friend's. Does this mean that one of you is wrong? Explain.

128. ***Conjecture*** Consider the function given by

$$f(x) = 2 \sin x\left(2 \cos^2 \frac{x}{2} - 1\right).$$

(a) Use a graphing utility to graph the function.

(b) Make a conjecture about the function that is an identity with f.

(c) Verify your conjecture analytically.

Skills Review

In Exercises 129–132, (a) plot the points, (b) find the distance between the points, and (c) find the midpoint of the line segment connecting the points.

129. $(5, 2), (-1, 4)$

130. $(-4, -3), (6, 10)$

131. $\left(0, \frac{1}{2}\right), \left(\frac{4}{3}, \frac{5}{2}\right)$

132. $\left(\frac{1}{3}, \frac{2}{3}\right), \left(-1, -\frac{3}{2}\right)$

In Exercises 133–136, find (if possible) the complement and supplement of each angle.

133. (a) $55°$ (b) $162°$

134. (a) $109°$ (b) $78°$

135. (a) $\frac{\pi}{18}$ (b) $\frac{9\pi}{20}$

136. (a) 0.95 (b) 2.76

137. ***Profit*** The total profit for a car manufacturer in October was 16% higher than it was in September. The total profit for the 2 months was \$507,600. Find the profit for each month.

138. ***Mixture Problem*** A 55-gallon barrel contains a mixture with a concentration of 30%. How much of this mixture must be withdrawn and replaced by 100% concentrate to bring the mixture up to 50% concentration?

139. ***Distance*** A baseball diamond has the shape of a square in which the distance between each of the consecutive bases is 90 feet. Approximate the straight-line distance from home plate to second base.

5.6 Law of Sines

What you should learn

- Use the Law of Sines to solve oblique triangles (AAS or ASA).
- Use the Law of Sines to solve oblique triangles (SSA).
- Find the areas of oblique triangles.
- Use the Law of Sines to model and solve real-life problems.

Why you should learn it

You can use the Law of Sines to solve real-life problems involving oblique triangles. For instance, in Exercise 44 on page 427, you can use the Law of Sines to determine the length of the shadow of the Leaning Tower of Pisa.

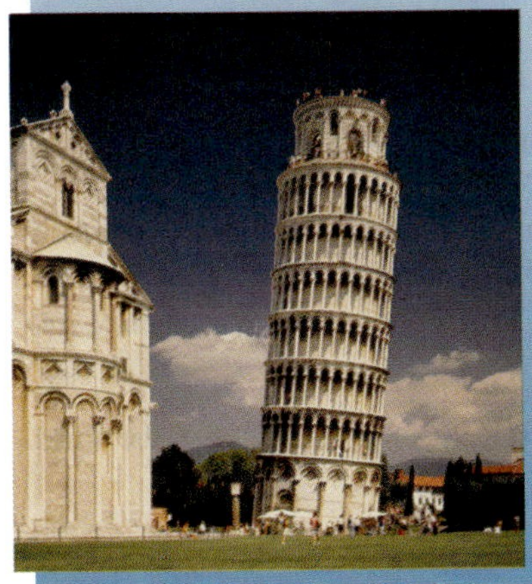

Hideo Kurihara/Getty Images

Introduction

In Chapter 4, you studied techniques for solving right triangles. In this section and the next, you will solve **oblique triangles**—triangles that have no right angles. As standard notation, the angles of a triangle are labeled A, B, and C, and their opposite sides are labeled a, b, and c, as shown in Figure 5.14.

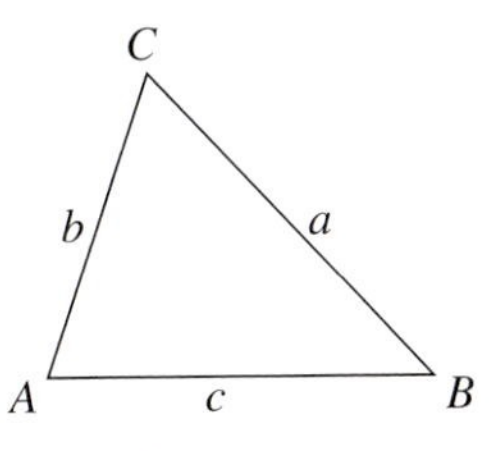

FIGURE 5.14

To solve an oblique triangle, you need to know the measure of at least one side and any two other parts of the triangle—either two sides, two angles, or one angle and one side. This breaks down into the following four cases.

1. Two angles and any side (AAS or ASA)
2. Two sides and an angle opposite one of them (SSA)
3. Three sides (SSS)
4. Two sides and their included angle (SAS)

The first two cases can be solved using the **Law of Sines,** whereas the last two cases require the Law of Cosines (see Section 5.7).

Law of Sines

If ABC is a triangle with sides a, b, and c, then

$$\frac{a}{\sin A} = \frac{b}{\sin B} = \frac{c}{\sin C}.$$

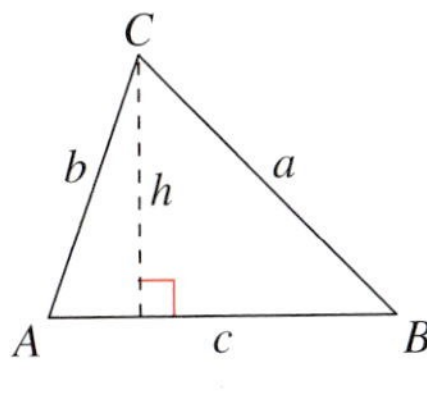

A is acute.

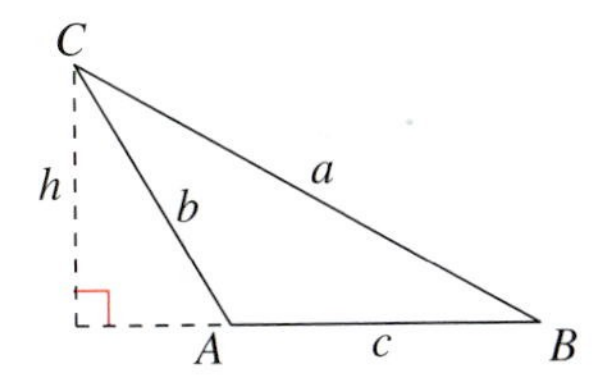

A is obtuse.

The Law of Sines can also be written in the reciprocal form

$$\frac{\sin A}{a} = \frac{\sin B}{b} = \frac{\sin C}{c}.$$

For a proof of the Law of Sines, see Proofs in Mathematics on page 444.

The *HM mathSpace*® CD-ROM and *Eduspace*® for this text contain additional resources related to the concepts discussed in this chapter.

5.7 Law of Cosines

What you should learn

- Use the Law of Cosines to solve oblique triangles (SSS or SAS).
- Use the Law of Cosines to model and solve real-life problems.
- Use Heron's Area Formula to find the area of a triangle.

Why you should learn it

You can use the Law of Cosines to solve real-life problems involving oblique triangles. For instance, in Exercise 31 on page 433, you can use the Law of Cosines to approximate the length of a marsh.

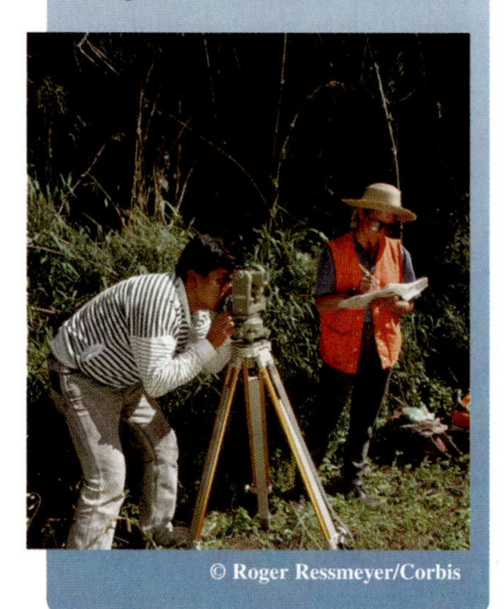

Introduction

Two cases remain in the list of conditions needed to solve an oblique triangle—SSS and SAS. If you are given three sides (SSS), or two sides and their included angle (SAS), none of the ratios in the Law of Sines would be complete. In such cases, you can use the **Law of Cosines.**

Law of Cosines

Standard Form	*Alternative Form*
$a^2 = b^2 + c^2 - 2bc \cos A$	$\cos A = \dfrac{b^2 + c^2 - a^2}{2bc}$
$b^2 = a^2 + c^2 - 2ac \cos B$	$\cos B = \dfrac{a^2 + c^2 - b^2}{2ac}$
$c^2 = a^2 + b^2 - 2ab \cos C$	$\cos C = \dfrac{a^2 + b^2 - c^2}{2ab}$

For a proof of the Law of Cosines, see Proofs in Mathematics on page 445.

Example 1 Three Sides of a Triangle—SSS

Find the three angles of the triangle in Figure 5.24.

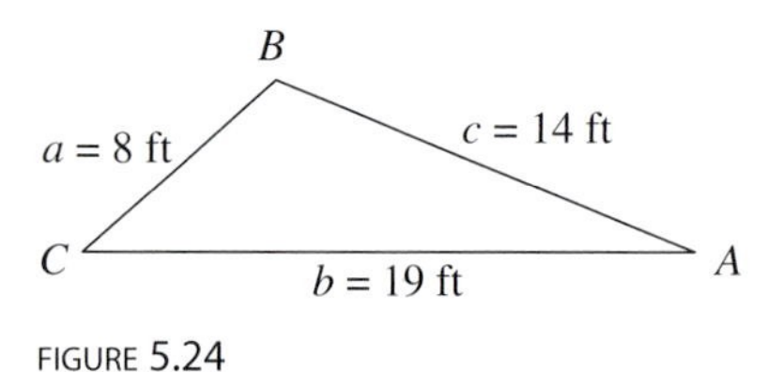

FIGURE 5.24

Solution

It is a good idea first to find the angle opposite the longest side—side b in this case. Using the alternative form of the Law of Cosines, you find that

$$\cos B = \frac{a^2 + c^2 - b^2}{2ac} = \frac{8^2 + 14^2 - 19^2}{2(8)(14)} \approx -0.45089.$$

Because $\cos B$ is negative, you know that B is an *obtuse* angle given by $B \approx 116.80°$. At this point, it is simpler to use the Law of Sines to determine A.

$$\sin A = a\left(\frac{\sin B}{b}\right) \approx 8\left(\frac{\sin 116.80°}{19}\right) \approx 0.37583$$

Because B is obtuse, A must be acute, because a triangle can have, at most, one obtuse angle. So, $A \approx 22.08°$ and $C \approx 180° - 22.08° - 116.80° = 41.12°$.

✓CHECKPOINT Now try Exercise 1.

In cases where the Law of Cosines must be used, encourage your students to solve for the largest angle first, then finish the problem using either the Law of Sines or the Law of Cosines.

> **Exploration**
>
> What familiar formula do you obtain when you use the third form of the Law of Cosines
>
> $$c^2 = a^2 + b^2 - 2ab \cos C$$
>
> and you let $C = 90°$? What is the relationship between the Law of Cosines and this formula?

Do you see why it was wise to find the largest angle *first* in Example 1? Knowing the cosine of an angle, you can determine whether the angle is acute or obtuse. That is,

$$\cos \theta > 0 \quad \text{for} \quad 0° < \theta < 90° \qquad \text{Acute}$$

$$\cos \theta < 0 \quad \text{for} \quad 90° < \theta < 180°. \qquad \text{Obtuse}$$

So, in Example 1, once you found that angle B was obtuse, you knew that angles A and C were both acute. If the largest angle is acute, the remaining two angles are acute also.

Example 2 Two Sides and the Included Angle—SAS

Find the remaining angles and side of the triangle in Figure 5.25.

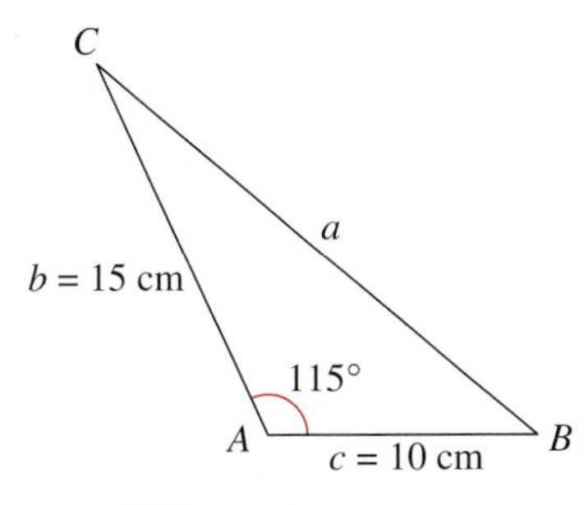

FIGURE 5.25

Solution

Use the Law of Cosines to find the unknown side a in the figure.

$$a^2 = b^2 + c^2 - 2bc \cos A$$

$$a^2 = 15^2 + 10^2 - 2(15)(10) \cos 115°$$

$$a^2 \approx 451.79$$

$$a \approx 21.26$$

Because $a \approx 21.26$ centimeters, you now know the ratio $\sin A/a$ and you can use the reciprocal form of the Law of Sines to solve for B.

$$\frac{\sin B}{b} = \frac{\sin A}{a}$$

$$\sin B = b\left(\frac{\sin A}{a}\right)$$

$$= 15\left(\frac{\sin 115°}{21.26}\right)$$

$$\approx 0.63945$$

So, $B = \arcsin 0.63945 \approx 39.75°$ and $C \approx 180° - 115° - 39.75° = 25.25°$.

✓CHECKPOINT Now try Exercise 3.

> **STUDY TIP**
>
> When solving an oblique triangle given three sides, you use the alternative form of the Law of Cosines to solve for an angle. When solving an oblique triangle given two sides and their included angle, you use the standard form of the Law of Cosines to solve for an unknown.

Applications

Example 3 An Application of the Law of Cosines

The pitcher's mound on a women's softball field is 43 feet from home plate and the distance between the bases is 60 feet, as shown in Figure 5.26. (The pitcher's mound is not halfway between home plate and second base.) How far is the pitcher's mound from first base?

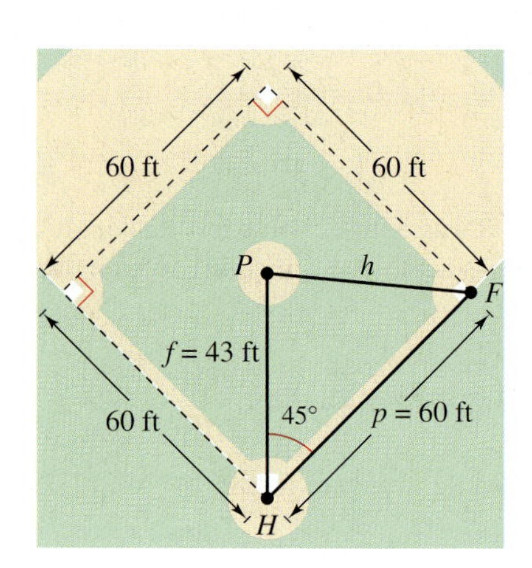

FIGURE 5.26

Solution

In triangle HPF, $H = 45°$ (line HP bisects the right angle at H), $f = 43$, and $p = 60$. Using the Law of Cosines for this SAS case, you have

$$h^2 = f^2 + p^2 - 2fp \cos H$$

$$= 43^2 + 60^2 - 2(43)(60) \cos 45° \approx 1800.3$$

So, the approximate distance from the pitcher's mound to first base is

$$h \approx \sqrt{1800.3} \approx 42.43 \text{ feet.}$$

CHECKPOINT Now try Exercise 31.

Example 4 An Application of the Law of Cosines

A ship travels 60 miles due east, then adjusts its course northward, as shown in Figure 5.27. After traveling 80 miles in that direction, the ship is 139 miles from its point of departure. Describe the bearing from point B to point C.

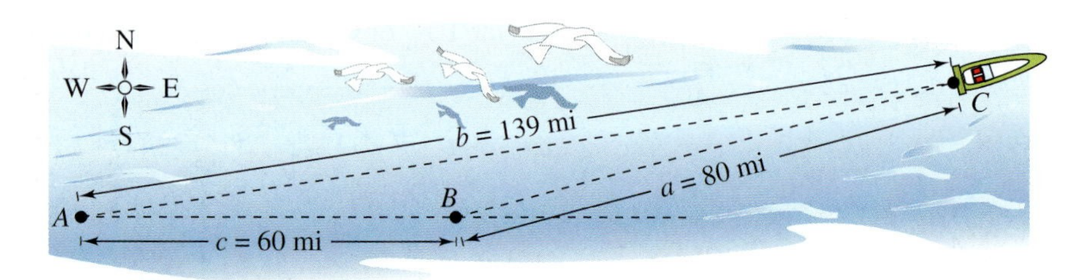

FIGURE 5.27

Solution

You have $a = 80$, $b = 139$, and $c = 60$; so, using the alternative form of the Law of Cosines, you have

$$\cos B = \frac{a^2 + c^2 - b^2}{2ac}$$

$$= \frac{80^2 + 60^2 - 139^2}{2(80)(60)}$$

$$\approx -0.97094.$$

So, $B \approx \arccos(-0.97094) \approx 166.15°$, and thus the bearing measured from due north from point B to point C is $166.15° - 90° = 76.15°$, or N 76.15° E.

CHECKPOINT Now try Exercise 37.

Historical Note

Heron of Alexandria (c. 100 B.C.) was a Greek geometer and inventor. His works describe how to find the areas of triangles, quadrilaterals, regular polygons having 3 to 12 sides, and circles as well as the surface areas and volumes of three-dimensional objects.

Heron's Area Formula

The Law of Cosines can be used to establish the following formula for the area of a triangle. This formula is called **Heron's Area Formula** after the Greek mathematician Heron (c. 100 B.C.).

Heron's Area Formula

Given any triangle with sides of lengths a, b, and c, the area of the triangle is

$$\text{Area} = \sqrt{s(s-a)(s-b)(s-c)}$$

where $s = (a + b + c)/2$.

For a proof of Heron's Area Formula, see Proofs in Mathematics on page 446.

Activities

1. Determine whether the Law of Sines or the Law of Cosines is needed to solve each of the triangles.
   a. $A = 15°, B = 58°, c = 94$
   b. $a = 96, b = 43, A = 105°$
   c. $a = 24, b = 16, c = 29$
   d. $a = 15, c = 42, B = 49°$
   Answer: a. Law of Sines, b. Law of Sines, c. Law of Cosines, d. Law of Cosines
2. Solve the triangle: $a = 31, b = 52, c = 28$.
   Answer: $A = 29.8°, B = 123.5°, C = 26.7°$
3. Use Heron's Area Formula to find the area of a triangle with sides of lengths $a = 31, b = 52$, and $c = 28$.
   Answer: Area $\approx$ 361.8 square units

Example 5 Using Heron's Area Formula

Find the area of a triangle having sides of lengths $a = 43$ meters, $b = 53$ meters, and $c = 72$ meters.

Solution

Because $s = (a + b + c)/2 = 168/2 = 84$, Heron's Area Formula yields

$$\begin{aligned}\text{Area} &= \sqrt{s(s-a)(s-b)(s-c)}\\ &= \sqrt{84(41)(31)(12)} \approx 1131.89 \text{ square meters.}\end{aligned}$$

✓CHECKPOINT Now try Exercise 47.

You have now studied three different formulas for the area of a triangle.

Standard Formula $\quad \text{Area} = \frac{1}{2}bh$

Oblique Triangle $\quad \text{Area} = \frac{1}{2}bc \sin A = \frac{1}{2}ab \sin C = \frac{1}{2}ac \sin B$

Heron's Area Formula $\quad \text{Area} = \sqrt{s(s-a)(s-b)(s-c)}$

Writing About Mathematics

The Area of a Triangle Use the most appropriate formula to find the area of each triangle below. Show your work and give your reasons for choosing each formula.

a.
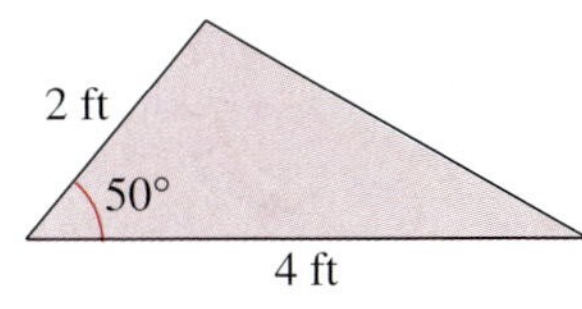

b.
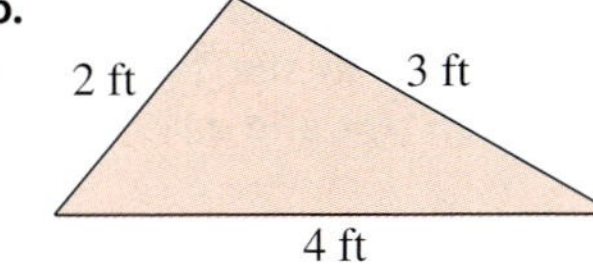

c.
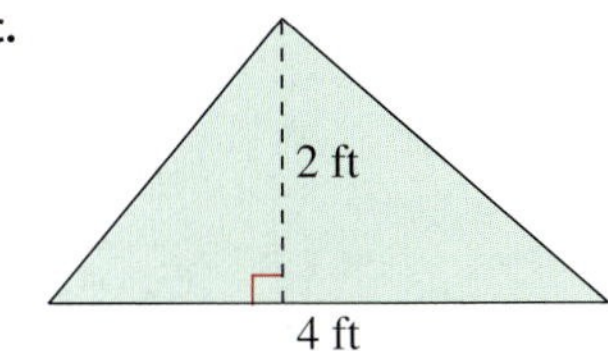

d.
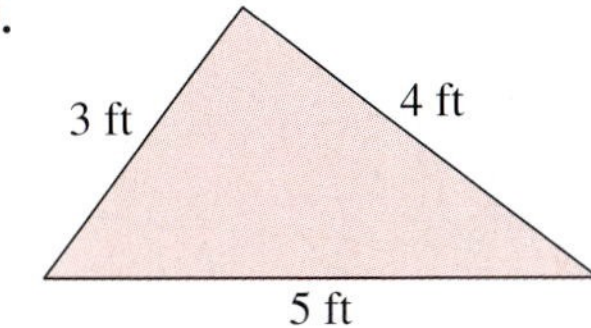

Writing About Mathematics Suggestion:

You may want to ask students to draw the diagram of a triangle (labeling whatever sides and/or angles are necessary) that is well-suited to each area formula. Then ask students to exchange triangles with a partner to find the area of each.

5.7 Exercises

VOCABULARY CHECK: Fill in the blanks.

1. If you are given three sides of a triangle, you would use the Law of ________ to find the three angles of the triangle.

2. The standard form of the Law of Cosines for $\cos B = \dfrac{a^2 + c^2 - b^2}{2ac}$ is ________ .

3. The Law of Cosines can be used to establish a formula for finding the area of a triangle called ________ ________ Formula.

PREREQUISITE SKILLS REVIEW: Practice and review algebra skills needed for this section at **www.Eduspace.com.**

In Exercises 1–16, use the Law of Cosines to solve the triangle. Round your answers to two decimal places.

1. 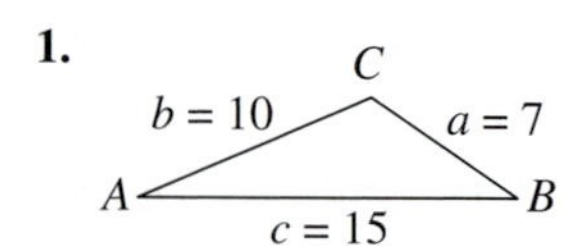

2. 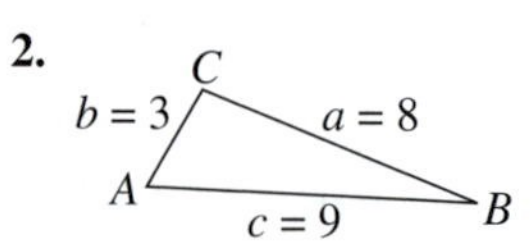

3. 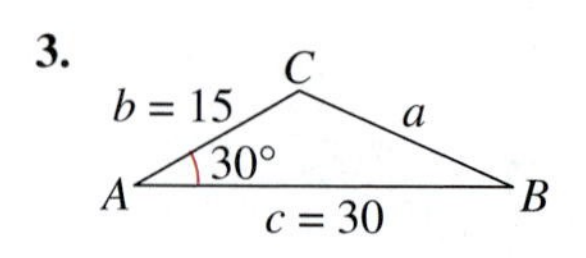

4. 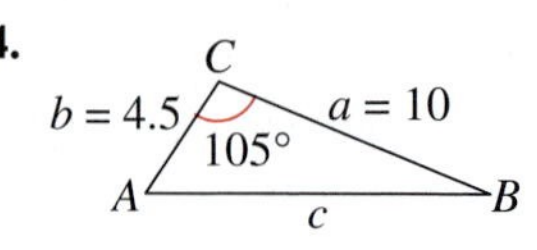

5. $a = 11, \quad b = 14, \quad c = 20$
6. $a = 55, \quad b = 25, \quad c = 72$
7. $a = 75.4, \quad b = 52, \quad c = 52$
8. $a = 1.42, \quad b = 0.75, \quad c = 1.25$
9. $A = 135°, \quad b = 4, \quad c = 9$
10. $A = 55°, \quad b = 3, \quad c = 10$
11. $B = 10° \, 35', \quad a = 40, \quad c = 30$
12. $B = 75° \, 20', \quad a = 6.2, \quad c = 9.5$
13. $B = 125° \, 40', \quad a = 32, \quad c = 32$
14. $C = 15° \, 15', \quad a = 6.25, \quad b = 2.15$
15. $C = 43°, \quad a = \frac{4}{9}, \quad b = \frac{7}{9}$
16. $C = 103°, \quad a = \frac{3}{8}, \quad b = \frac{3}{4}$

In Exercises 17–22, complete the table by solving the parallelogram shown in the figure. (The lengths of the diagonals are given by c and d.)

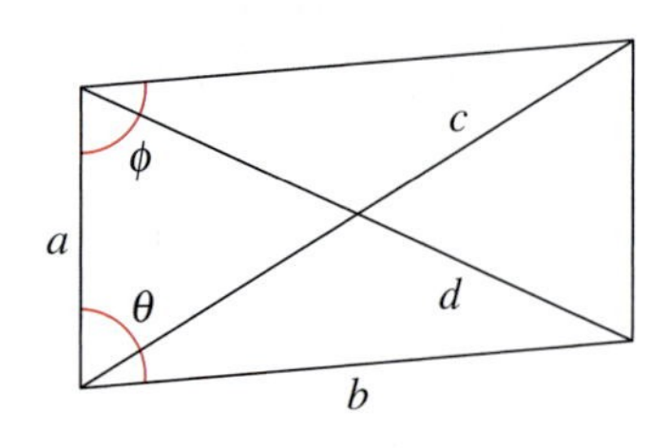

	a	b	c	d	θ	ϕ
17.	5	8			45°	
18.	25	35				120°
19.	10	14	20			
20.	40	60		80		
21.	15		25	20		
22.		25	50	35		

In Exercises 23–28, use Heron's Area Formula to find the area of the triangle.

23. $a = 5, \quad b = 7, \quad c = 10$
24. $a = 12, \quad b = 15, \quad c = 9$
25. $a = 2.5, \quad b = 10.2, \quad c = 9$
26. $a = 75.4, \quad b = 52, \quad c = 52$
27. $a = 12.32, \quad b = 8.46, \quad c = 15.05$
28. $a = 3.05, \quad b = 0.75, \quad c = 2.45$

29. ***Navigation*** A boat race runs along a triangular course marked by buoys A, B, and C. The race starts with the boats headed west for 3700 meters. The other two sides of the course lie to the north of the first side, and their lengths are 1700 meters and 3000 meters. Draw a figure that gives a visual representation of the problem, and find the bearings for the last two legs of the race.

30. ***Navigation*** A plane flies 810 miles from Franklin to Centerville with a bearing of 75°. Then it flies 648 miles from Centerville to Rosemount with a bearing of 32°. Draw a figure that visually represents the problem, and find the straight-line distance and bearing from Franklin to Rosemount.

31. *Surveying* To approximate the length of a marsh, a surveyor walks 250 meters from point A to point B, then turns 75° and walks 220 meters to point C (see figure). Approximate the length AC of the marsh.

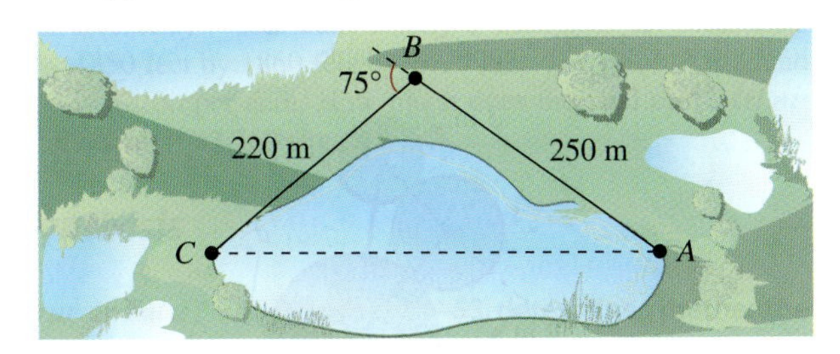

32. *Surveying* A triangular parcel of land has 115 meters of frontage, and the other boundaries have lengths of 76 meters and 92 meters. What angles does the frontage make with the two other boundaries?

33. *Surveying* A triangular parcel of ground has sides of lengths 725 feet, 650 feet, and 575 feet. Find the measure of the largest angle.

34. *Streetlight Design* Determine the angle θ in the design of the streetlight shown in the figure.

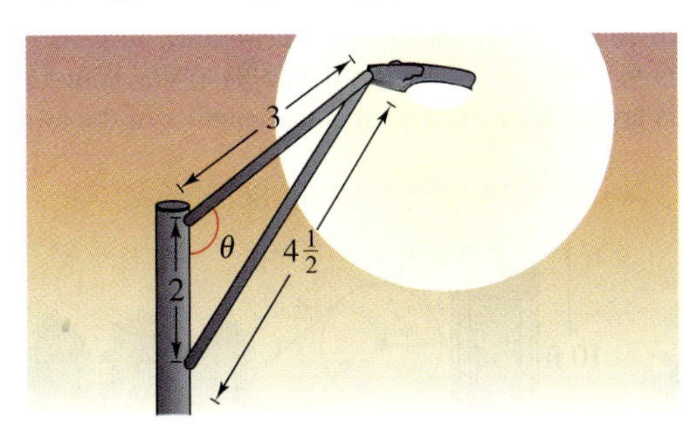

35. *Distance* Two ships leave a port at 9 A.M. One travels at a bearing of N 53° W at 12 miles per hour, and the other travels at a bearing of S 67° W at 16 miles per hour. Approximate how far apart they are at noon that day.

36. *Length* A 100-foot vertical tower is to be erected on the side of a hill that makes a 6° angle with the horizontal (see figure). Find the length of each of the two guy wires that will be anchored 75 feet uphill and downhill from the base of the tower.

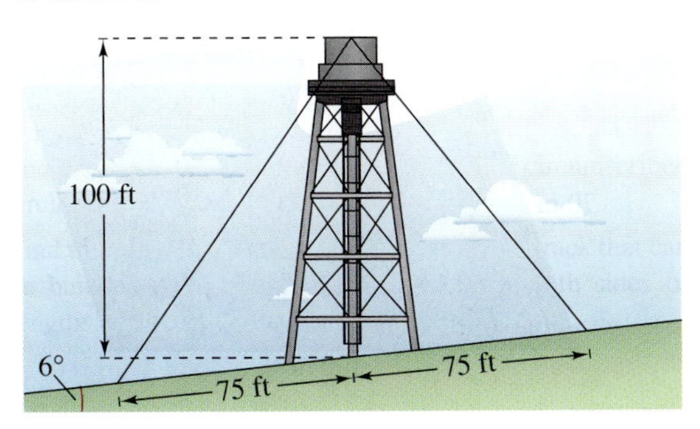

37. *Navigation* On a map, Orlando is 178 millimeters due south of Niagara Falls, Denver is 273 millimeters from Orlando, and Denver is 235 millimeters from Niagara Falls (see figure).

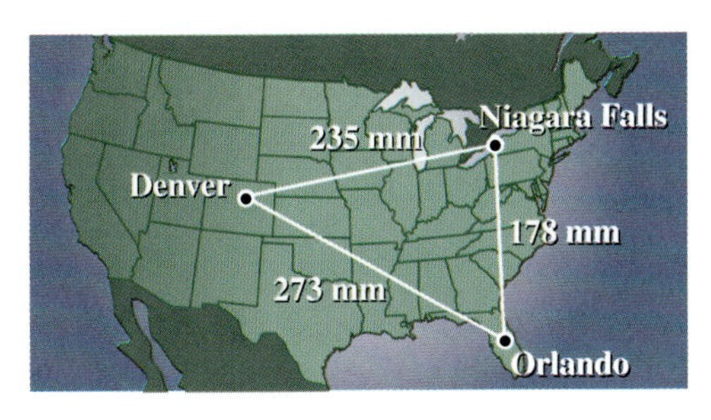

(a) Find the bearing of Denver from Orlando.

(b) Find the bearing of Denver from Niagara Falls.

38. *Navigation* On a map, Minneapolis is 165 millimeters due west of Albany, Phoenix is 216 millimeters from Minneapolis, and Phoenix is 368 millimeters from Albany (see figure).

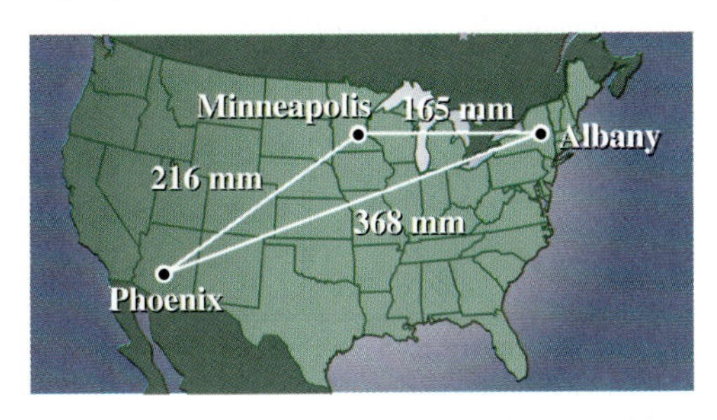

(a) Find the bearing of Minneapolis from Phoenix.

(b) Find the bearing of Albany from Phoenix.

39. *Baseball* On a baseball diamond with 90-foot sides, the pitcher's mound is 60.5 feet from home plate. How far is it from the pitcher's mound to third base?

40. *Baseball* The baseball player in center field is playing approximately 330 feet from the television camera that is behind home plate. A batter hits a fly ball that goes to the wall 420 feet from the camera (see figure). The camera turns 8° to follow the play. Approximately how far does the center fielder have to run to make the catch?

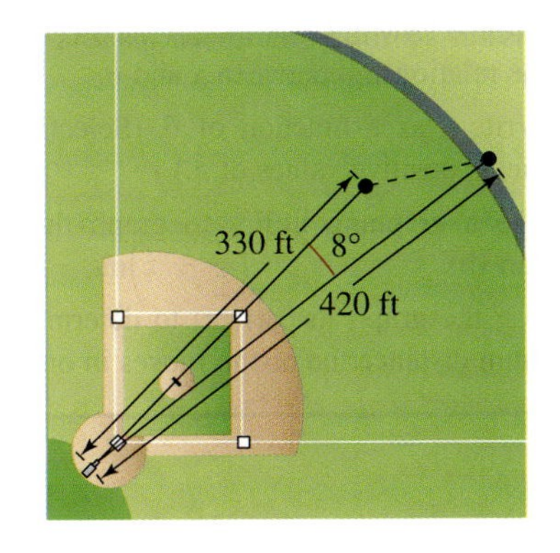

In Exercises 55–58, write the expression as the sine, cosine, or tangent of an angle.

55. $\sin 60° \cos 45° - \cos 60° \sin 45°$

56. $\cos 45° \cos 120° - \sin 45° \sin 120°$

57. $\dfrac{\tan 25° + \tan 10°}{1 - \tan 25° \tan 10°}$ **58.** $\dfrac{\tan 68° - \tan 115°}{1 + \tan 68° \tan 115°}$

In Exercises 59–64, find the exact value of the trigonometric function given that $\sin u = \frac{3}{4}$ and $\cos v = -\frac{5}{13}$. (Both u and v are in Quadrant II.)

59. $\sin(u + v)$ **60.** $\tan(u + v)$

61. $\cos(u - v)$ **62.** $\sin(u - v)$

63. $\cos(u + v)$ **64.** $\tan(u - v)$

In Exercises 65–70, verify the identity.

65. $\cos\left(x + \dfrac{\pi}{2}\right) = -\sin x$ **66.** $\sin\left(x - \dfrac{3\pi}{2}\right) = \cos x$

67. $\cot\left(\dfrac{\pi}{2} - x\right) = \tan x$ **68.** $\sin(\pi - x) = \sin x$

69. $\cos 3x = 4 \cos^3 x - 3 \cos x$

70. $\dfrac{\sin(\alpha + \beta)}{\cos \alpha \cos \beta} = \tan \alpha + \tan\beta$

In Exercises 71–74, find all solutions of the equation in the interval $[0, 2\pi)$.

71. $\sin\left(x + \dfrac{\pi}{4}\right) - \sin\left(x - \dfrac{\pi}{4}\right) = 1$

72. $\cos\left(x + \dfrac{\pi}{6}\right) - \cos\left(x - \dfrac{\pi}{6}\right) = 1$

73. $\sin\left(x + \dfrac{\pi}{2}\right) - \sin\left(x - \dfrac{\pi}{2}\right) = \sqrt{3}$

74. $\cos\left(x + \dfrac{3\pi}{4}\right) - \cos\left(x - \dfrac{3\pi}{4}\right) = 0$

5.5 In Exercises 75 and 76, find the exact values of $\sin 2u$, $\cos 2u$, and $\tan 2u$ using the double-angle formulas.

75. $\sin u = -\dfrac{4}{5}, \quad \pi < u < \dfrac{3\pi}{2}$

76. $\cos u = -\dfrac{2}{\sqrt{5}}, \quad \dfrac{\pi}{2} < u < \pi$

In Exercises 77 and 78, use double-angle formulas to verify the identity algebraically and use a graphing utility to confirm your result graphically.

77. $\sin 4x = 8 \cos^3 x \sin x - 4 \cos x \sin x$

78. $\tan^2 x = \dfrac{1 - \cos 2x}{1 + \cos 2x}$

In Exercises 79–82, use the power-reducing formulas to rewrite the expression in terms of the first power of the cosine.

79. $\tan^2 2x$ **80.** $\cos^2 3x$

81. $\sin^2 x \tan^2 x$ **82.** $\cos^2 x \tan^2 x$

In Exercises 83–86, use the half-angle formulas to determine the exact values of the sine, cosine, and tangent of the angle.

83. $-75°$ **84.** $15°$

85. $\dfrac{19\pi}{12}$ **86.** $-\dfrac{17\pi}{12}$

In Exercises 87–90, find the exact values of $\sin(u/2)$, $\cos(u/2)$, and $\tan(u/2)$ using the half-angle formulas.

87. $\sin u = \frac{3}{5}, \quad 0 < u < \pi/2$

88. $\tan u = \frac{5}{8}, \quad \pi < u < 3\pi/2$

89. $\cos u = -\frac{2}{7}, \quad \pi/2 < u < \pi$

90. $\sec u = -6, \quad \pi/2 < u < \pi$

In Exercises 91 and 92, use the half-angle formulas to simplify the expression.

91. $-\sqrt{\dfrac{1 + \cos 10x}{2}}$ **92.** $\dfrac{\sin 6x}{1 + \cos 6x}$

In Exercises 93–96, use the product-to-sum formulas to write the product as a sum or difference.

93. $\cos \dfrac{\pi}{6} \sin \dfrac{\pi}{6}$ **94.** $6 \sin 15° \sin 45°$

95. $\cos 5\theta \cos 3\theta$ **96.** $4 \sin 3\alpha \cos 2\alpha$

In Exercises 97–100, use the sum-to-product formulas to write the sum or difference as a product.

97. $\sin 4\theta - \sin 2\theta$

98. $\cos 3\theta + \cos 2\theta$

99. $\cos\left(x + \dfrac{\pi}{6}\right) - \cos\left(x - \dfrac{\pi}{6}\right)$

100. $\sin\left(x + \dfrac{\pi}{4}\right) - \sin\left(x - \dfrac{\pi}{4}\right)$

101. ***Projectile Motion*** A baseball leaves the hand of the person at first base at an angle of θ with the horizontal and at an initial velocity of $v_0 = 80$ feet per second. The ball is caught by the person at second base 100 feet away. Find θ if the range r of a projectile is

$$r = \frac{1}{32} v_0^2 \sin 2\theta.$$

102. *Geometry* A trough for feeding cattle is 4 meters long and its cross sections are isosceles triangles with the two equal sides being $\frac{1}{2}$ meter (see figure). The angle between the two sides is θ.

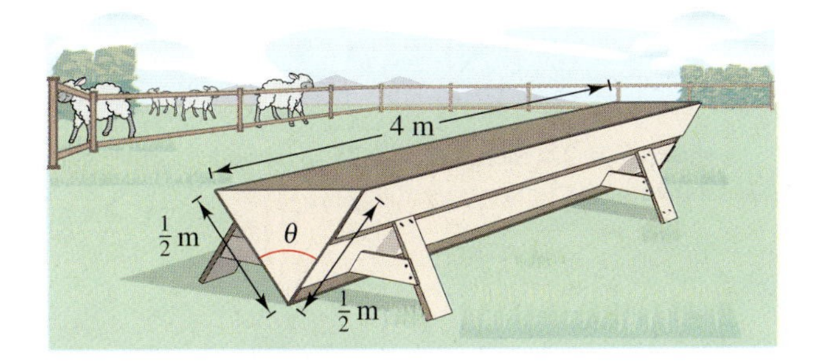

(a) Write the trough's volume as a function of $\frac{\theta}{2}$.

(b) Write the volume of the trough as a function of θ and determine the value of θ such that the volume is maximum.

Harmonic Motion **In Exercises 103–106, use the following information. A weight is attached to a spring suspended vertically from a ceiling. When a driving force is applied to the system, the weight moves vertically from its equilibrium position, and this motion is described by the model**

$$y = 1.5 \sin 8t - 0.5 \cos 8t$$

where y is the distance from equilibrium (in feet) and t is the time (in seconds).

103. Use a graphing utility to graph the model.

104. Write the model in the form

$$y = \sqrt{a^2 + b^2} \sin(Bt + C).$$

105. Find the amplitude of the oscillations of the weight.

106. Find the frequency of the oscillations of the weight.

5.6 In Exercises 107–116, use the Law of Sines to solve (if possible) the triangle. If two solutions exist, find both. Round your answers to two decimal places.

107. $B = 72°,\ C = 82°,\ b = 54$

108. $B = 10°,\ C = 20°,\ c = 33$

109. $A = 16°,\ B = 98°,\ c = 8.4$

110. $A = 95°,\ B = 45°,\ c = 104.8$

111. $A = 24°,\ C = 48°,\ b = 27.5$

112. $B = 64°,\ C = 36°,\ a = 367$

113. $B = 150°,\ b = 30,\ c = 10$

114. $B = 150°,\ a = 10,\ b = 3$

115. $A = 75°,\ a = 51.2,\ b = 33.7$

116. $B = 25°,\ a = 6.2,\ b = 4$

In Exercises 117–120, find the area of the triangle having the indicated angle and sides.

117. $A = 27°,\ b = 5,\ c = 7$

118. $B = 80°,\ a = 4,\ c = 8$

119. $C = 123°,\ a = 16,\ b = 5$

120. $A = 11°,\ b = 22,\ c = 21$

121. *Height* From a certain distance, the angle of elevation to the top of a building is 17°. At a point 50 meters closer to the building, the angle of elevation is 31°. Approximate the height of the building.

122. *Geometry* Find the length of the side w of the parallelogram.

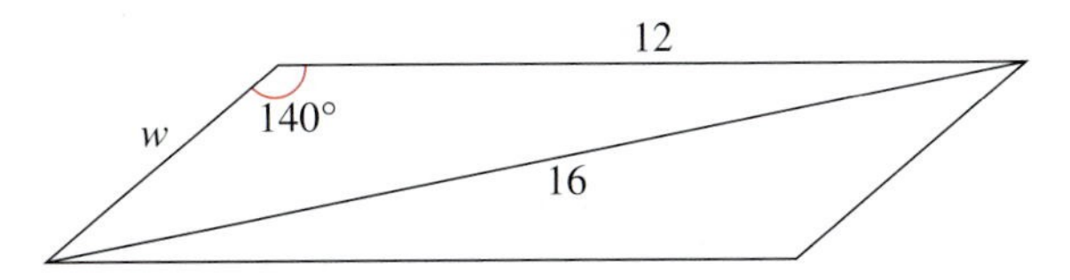

123. *Height* A tree stands on a hillside of slope 28° from the horizontal. From a point 75 feet down the hill, the angle of elevation to the top of the tree is 45° (see figure). Find the height of the tree.

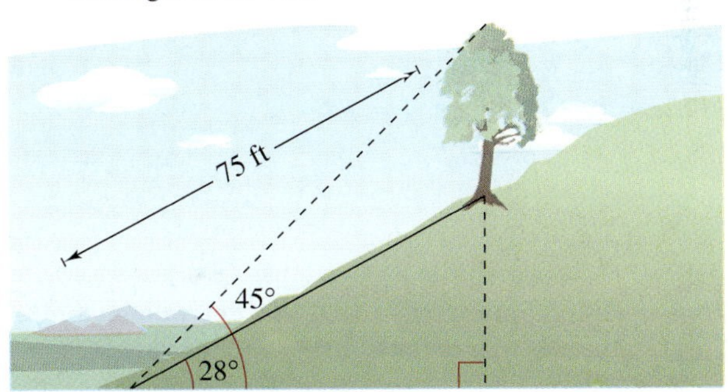

124. *River Width* A surveyor finds that a tree on the opposite bank of a river, flowing due east, has a bearing of N 22° 30′E from a certain point and a bearing of N 15° W from a point 400 feet downstream. Find the width of the river.

5.7 In Exercises 125–132, use the Law of Cosines to solve the triangle. Round your answers to two decimal places.

125. $a = 5,\ b = 8,\ c = 10$

126. $a = 80,\ b = 60,\ c = 100$

127. $a = 2.5,\ b = 5.0,\ c = 4.5$

128. $a = 16.4,\ b = 8.8,\ c = 12.2$

129. $B = 110°,\ a = 4,\ c = 4$

130. $B = 150°,\ a = 10,\ c = 20$

131. $C = 43°,\ a = 22.5,\ b = 31.4$

132. $A = 62°,\ b = 11.34,\ c = 19.52$

133. ***Surveying*** To approximate the length of a marsh, a surveyor walks 425 meters from point A to point B. Then the surveyor turns 65° and walks 300 meters to point C (see figure). Approximate the length AC of the marsh.

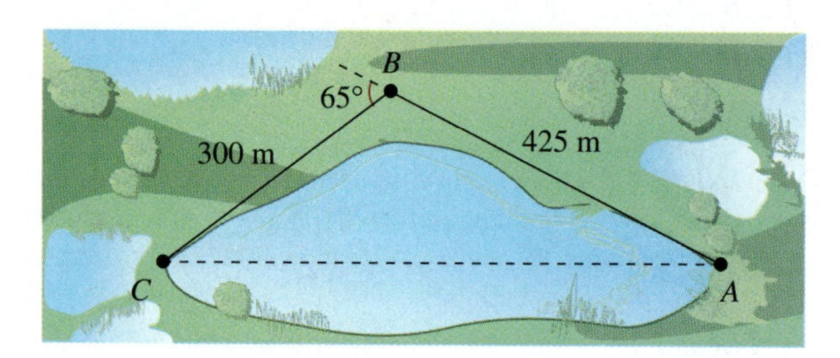

134. ***Navigation*** Two planes leave Raleigh-Durham Airport at approximately the same time. One is flying 425 miles per hour at a bearing of 355°, and the other is flying 530 miles per hour at a bearing of 67°. Draw a figure that gives a visual representation of the problem and determine the distance between the planes after they have flown for 2 hours.

In Exercises 135–138, use Heron's Area Formula to find the area of the triangle.

135. $a = 4,\ b = 5,\ c = 7$

136. $a = 15,\ b = 8,\ c = 10$

137. $a = 12.3,\ b = 15.8,\ c = 3.7$

138. $a = 38.1,\ b = 26.7,\ c = 19.4$

Synthesis

True or False? **In Exercises 139–144, determine whether the statement is true or false. Justify your answer.**

139. If $\frac{\pi}{2} < \theta < \pi$, then $\cos \frac{\theta}{2} < 0$.

140. $\sin(x + y) = \sin x + \sin y$

141. $4 \sin(-x) \cos(-x) = -2 \sin 2x$

142. $4 \sin 45° \cos 15° = 1 + \sqrt{3}$

143. The Law of Sines is true if one of the angles in the triangle is a right angle.

144. When the Law of Sines is used, the solution is always unique.

145. List the reciprocal identities, quotient identities, and Pythagorean identities from memory.

146. ***Think About It*** If a trigonometric equation has an infinite number of solutions, is it true that the equation is an identity? Explain.

147. ***Think About It*** Explain why you know from observation that the equation $a \sin x - b = 0$ has no solution if $|a| < |b|$.

148. ***Surface Area*** The surface area of a honeycomb is given by the equation

$$S = 6hs + \frac{3}{2}s^2\left(\frac{\sqrt{3} - \cos \theta}{\sin \theta}\right),\quad 0 < \theta \le 90°$$

where $h = 2.4$ inches, $s = 0.75$ inch, and θ is the angle shown in the figure.

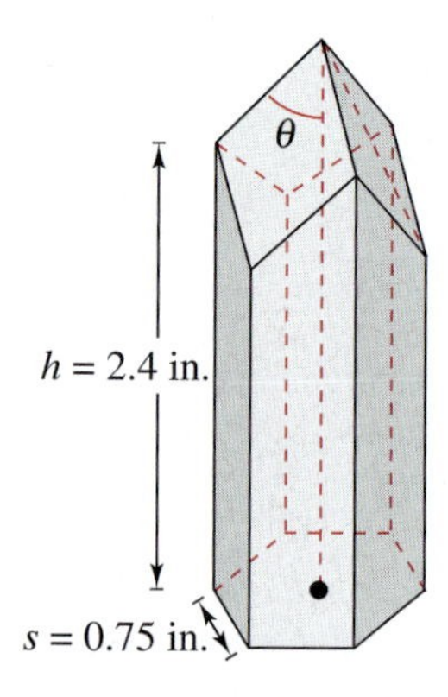

(a) For what value(s) of θ is the surface area 12 square inches?

(b) What value of θ gives the minimum surface area?

In Exercises 149 and 150, use the graphs of y_1 and y_2 to determine how to change one function to form the identity $y_1 = y_2$.

149. $y_1 = \sec^2\left(\frac{\pi}{2} - x\right)$

$y_2 = \cot^2 x$

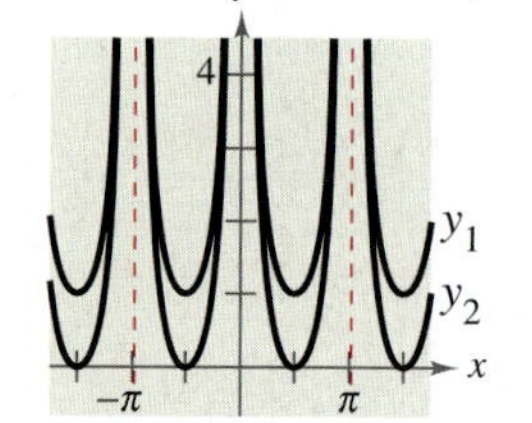

150. $y_1 = \frac{\cos 3x}{\cos x}$

$y_2 = (2 \sin x)^2$

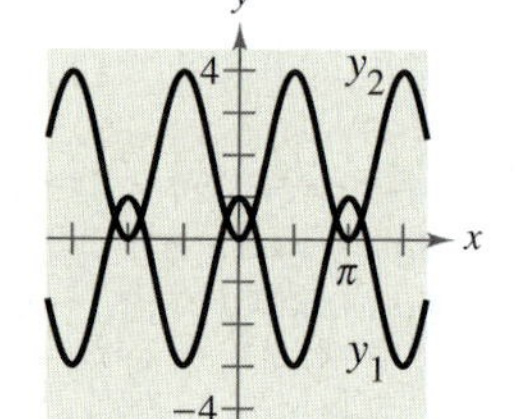

In Exercises 151 and 152, use the *zero* or *root* feature of a graphing utility to approximate the solutions of the equation.

151. $y = \sqrt{x + 3} + 4 \cos x$

152. $y = 2 - \frac{1}{2}x^2 + 3 \sin \frac{\pi x}{2}$

5 Chapter Test

Take this test as you would take a test in class. When you are finished, check your work against the answers given in the back of the book.

1. If $\tan \theta = \frac{3}{2}$ and $\cos \theta < 0$, use the fundamental identities to evaluate the other five trigonometric functions of θ.
2. Use the fundamental identities to simplify $\csc^2 \beta(1 - \cos^2 \beta)$.
3. Factor and simplify $\dfrac{\sec^4 x - \tan^4 x}{\sec^2 x + \tan^2 x}$.
4. Add and simplify $\dfrac{\cos \theta}{\sin \theta} + \dfrac{\sin \theta}{\cos \theta}$.

In Exercises 5–10, verify the identity.

5. $\sin \theta \sec \theta = \tan \theta$
6. $\sec^2 x \tan^2 x + \sec^2 x = \sec^4 x$
7. $\dfrac{\csc \alpha + \sec \alpha}{\sin \alpha + \cos \alpha} = \cot \alpha + \tan \alpha$
8. $\cos\left(x + \dfrac{\pi}{2}\right) = -\sin x$
9. $\sin(n\pi + \theta) = (-1)^n \sin \theta$, n is an integer.
10. $(\sin x + \cos x)^2 = 1 + \sin 2x$

11. Rewrite $\sin^4 x \tan^2 x$ in terms of the first power of the cosine.
12. Use a half-angle formula to simplify the expression $\dfrac{\sin 4\theta}{1 + \cos 4\theta}$.
13. Write $4 \cos 2\theta \sin 4\theta$ as a sum or difference.
14. Write $\sin 3\theta - \sin 4\theta$ as a product.

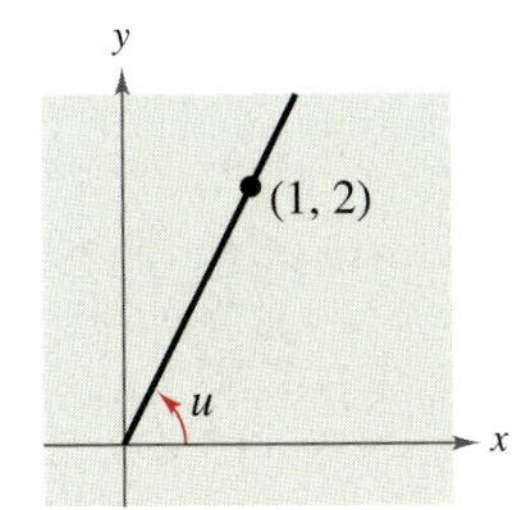

FIGURE FOR 20

In Exercises 15–18, find all solutions of the equation in the interval $[0, 2\pi)$.

15. $\tan^2 x + \tan x = 0$
16. $\sin 2\alpha - \cos \alpha = 0$
17. $4 \cos^2 x - 3 = 0$
18. $\csc^2 x - \csc x - 2 = 0$

19. Find the exact value of $\cos 105°$ using the fact that $105° = 135° - 30°$.
20. Use the figure to find the exact values of $\sin 2u$, $\cos 2u$, and $\tan 2u$.

In Exercises 21–26, use the information to solve the triangle. If two solutions exist, find both solutions. Round your answers to two decimal places.

21. $A = 24°,\ B = 68°,\ a = 12.2$
22. $B = 104°,\ C = 33°,\ a = 18.1$
23. $A = 24°,\ a = 11.2,\ b = 13.4$
24. $a = 4.0,\ b = 7.3,\ c = 12.4$
25. $B = 100°,\ a = 15,\ b = 23$
26. $C = 123°,\ a = 41,\ b = 57$
27. Cheyenne, Wyoming has a latitude of 41°N. At this latitude, the position of the sun at sunrise can be modeled by

$$D = 31 \sin\left(\frac{2\pi}{365}t - 1.4\right)$$

where t is the time (in days) and $t = 1$ represents January 1. In this model, D represents the number of degrees north or south of due east that the sun rises. Use a graphing utility to determine the days on which the sun is more than 20° north of due east at sunrise.

28. A triangular parcel of land has borders of lengths 60 meters, 70 meters, and 82 meters. Find the area of the parcel of land.
29. An airplane flies 370 miles from point A to point B with a bearing of 24°. It then flies 240 miles from point B to point C with a bearing of 37° (see figure). Find the distance and bearing from point A to point C.

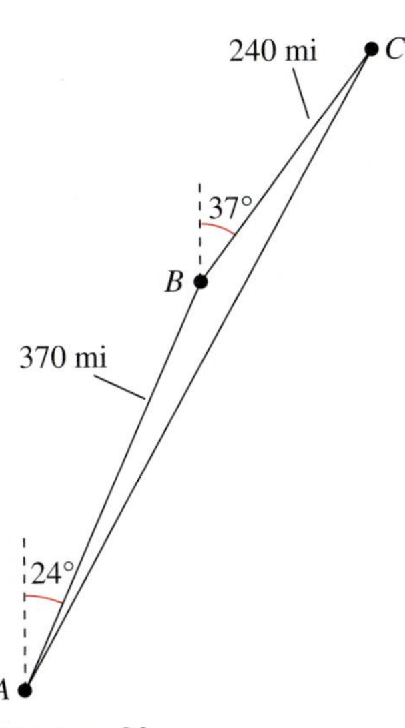

FIGURE FOR 29

Proofs in Mathematics

Sum and Difference Formulas (p. 400)

$$\sin(u + v) = \sin u \cos v + \cos u \sin v$$

$$\sin(u - v) = \sin u \cos v - \cos u \sin v$$

$$\cos(u + v) = \cos u \cos v - \sin u \sin v$$

$$\cos(u - v) = \cos u \cos v + \sin u \sin v$$

$$\tan(u + v) = \frac{\tan u + \tan v}{1 - \tan u \tan v}$$

$$\tan(u - v) = \frac{\tan u - \tan v}{1 + \tan u \tan v}$$

Proof

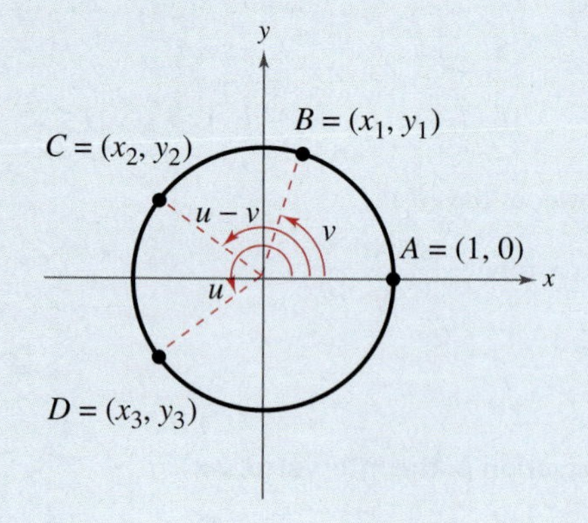

You can use the figures at the left for the proofs of the formulas for $\cos(u \pm v)$. In the top figure, let A be the point $(1, 0)$ and then use u and v to locate the points $B = (x_1, y_1)$, $C = (x_2, y_2)$, and $D = (x_3, y_3)$ on the unit circle. So, $x_i^2 + y_i^2 = 1$ for $i = 1, 2$, and 3. For convenience, assume that $0 < v < u < 2\pi$. In the bottom figure, note that arcs AC and BD have the same length. So, line segments AC and BD are also equal in length, which implies that

$$\sqrt{(x_2 - 1)^2 + (y_2 - 0)^2} = \sqrt{(x_3 - x_1)^2 + (y_3 - y_1)^2}$$

$$x_2^2 - 2x_2 + 1 + y_2^2 = x_3^2 - 2x_1x_3 + x_1^2 + y_3^2 - 2y_1y_3 + y_1^2$$

$$(x_2^2 + y_2^2) + 1 - 2x_2 = (x_3^2 + y_3^2) + (x_1^2 + y_1^2) - 2x_1x_3 - 2y_1y_3$$

$$1 + 1 - 2x_2 = 1 + 1 - 2x_1x_3 - 2y_1y_3$$

$$x_2 = x_3x_1 + y_3y_1.$$

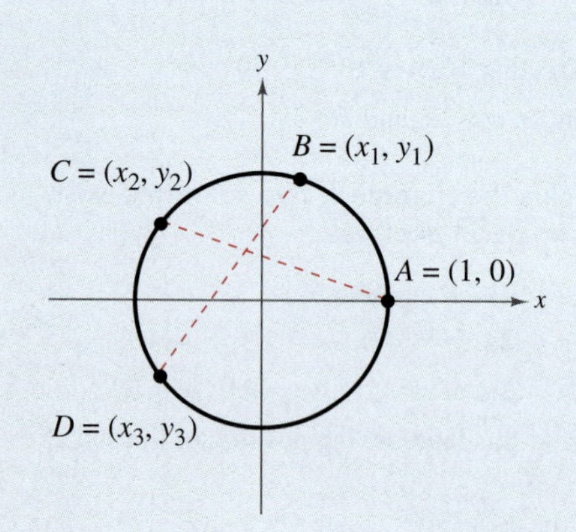

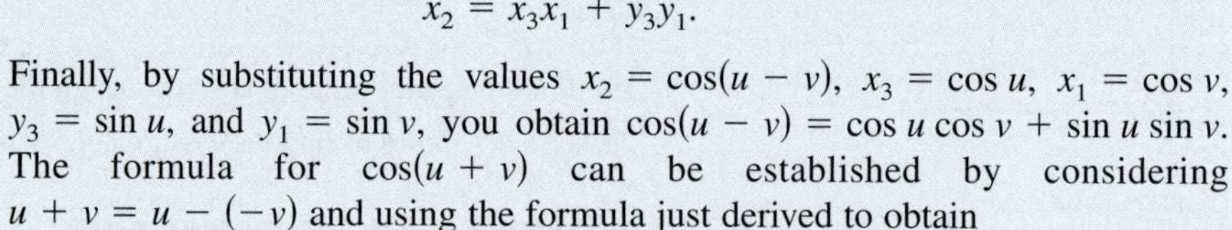

Finally, by substituting the values $x_2 = \cos(u - v)$, $x_3 = \cos u$, $x_1 = \cos v$, $y_3 = \sin u$, and $y_1 = \sin v$, you obtain $\cos(u - v) = \cos u \cos v + \sin u \sin v$. The formula for $\cos(u + v)$ can be established by considering $u + v = u - (-v)$ and using the formula just derived to obtain

$$\cos(u + v) = \cos[u - (-v)] = \cos u \cos(-v) + \sin u \sin(-v)$$

$$= \cos u \cos v - \sin u \sin v.$$

You can use the sum and difference formulas for sine and cosine to prove the formulas for $\tan(u \pm v)$.

$$\tan(u \pm v) = \frac{\sin(u \pm v)}{\cos(u \pm v)} \qquad \text{Quotient identity}$$

$$= \frac{\sin u \cos v \pm \cos u \sin v}{\cos u \cos v \mp \sin u \sin v} \qquad \text{Sum and difference formulas}$$

$$= \frac{\dfrac{\sin u \cos v \pm \cos u \sin v}{\cos u \cos v}}{\dfrac{\cos u \cos v \mp \sin u \sin v}{\cos u \cos v}} \qquad \text{Divide numerator and denominator by } \cos u \cos v.$$

$$= \frac{\dfrac{\sin u \cos v}{\cos u \cos v} \pm \dfrac{\cos u \sin v}{\cos u \cos v}}{\dfrac{\cos u \cos v}{\cos u \cos v} \mp \dfrac{\sin u \sin v}{\cos u \cos v}}$$ Write as separate fractions.

$$= \frac{\dfrac{\sin u}{\cos u} \pm \dfrac{\sin v}{\cos v}}{1 \mp \dfrac{\sin u}{\cos u} \cdot \dfrac{\sin v}{\cos v}}$$ Product of fractions

$$= \frac{\tan u \pm \tan v}{1 \mp \tan u \tan v}$$ Quotient identity

---

Trigonometry and Astronomy

Trigonometry was used by early astronomers to calculate measurements in the universe. Trigonometry was used to calculate the circumference of Earth and the distance from Earth to the moon. Another major accomplishment in astronomy using trigonometry was computing distances to stars.

Double-Angle Formulas *(p. 407)*

$$\sin 2u = 2 \sin u \cos u \qquad \cos 2u = \cos^2 u - \sin^2 u$$

$$\tan 2u = \frac{2 \tan u}{1 - \tan^2 u} \qquad = 2\cos^2 u - 1 = 1 - 2\sin^2 u$$

Proof

To prove all three formulas, let $v = u$ in the corresponding sum formulas.

$$\sin 2u = \sin(u + u) = \sin u \cos u + \cos u \sin u = 2 \sin u \cos u$$

$$\cos 2u = \cos(u + u) = \cos u \cos u - \sin u \sin u = \cos^2 u - \sin^2 u$$

$$\tan 2u = \tan(u + u) = \frac{\tan u + \tan u}{1 - \tan u \tan u} = \frac{2 \tan u}{1 - \tan^2 u}$$

Power-Reducing Formulas *(p. 409)*

$$\sin^2 u = \frac{1 - \cos 2u}{2} \qquad \cos^2 u = \frac{1 + \cos 2u}{2} \qquad \tan^2 u = \frac{1 - \cos 2u}{1 + \cos 2u}$$

Proof

To prove the first formula, solve for $\sin^2 u$ in the double-angle formula $\cos 2u = 1 - 2\sin^2 u$, as follows.

$$\cos 2u = 1 - 2\sin^2 u$$ Write double-angle formula.

$$2\sin^2 u = 1 - \cos 2u$$ Subtract $\cos 2u$ from, and add $2 \sin^2 u$ to, each side.

$$\sin^2 u = \frac{1 - \cos 2u}{2}$$ Divide each side by 2.

In a similar way you can prove the second formula, by solving for $\cos^2 u$ in the double-angle formula $\cos 2u = 2\cos^2 u - 1$. To prove the third formula, use a quotient identity, as follows.

$$\tan^2 u = \frac{\sin^2 u}{\cos^2 u} = \frac{\dfrac{1 - \cos 2u}{2}}{\dfrac{1 + \cos 2u}{2}} = \frac{1 - \cos 2u}{1 + \cos 2u}$$

Sum-to-Product Formulas *(p. 412)*

$$\sin u + \sin v = 2\sin\left(\frac{u + v}{2}\right)\cos\left(\frac{u - v}{2}\right)$$

$$\sin u - \sin v = 2\cos\left(\frac{u + v}{2}\right)\sin\left(\frac{u - v}{2}\right)$$

$$\cos u + \cos v = 2\cos\left(\frac{u + v}{2}\right)\cos\left(\frac{u - v}{2}\right)$$

$$\cos u - \cos v = -2\sin\left(\frac{u + v}{2}\right)\sin\left(\frac{u - v}{2}\right)$$

Proof

To prove the first formula, let $x = u + v$ and $y = u - v$. Then substitute $u = (x + y)/2$ and $v = (x - y)/2$ in the product-to-sum formula.

$$\sin u \cos v = \frac{1}{2}[\sin(u + v) + \sin(u - v)]$$

$$\sin\left(\frac{x + y}{2}\right)\cos\left(\frac{x - y}{2}\right) = \frac{1}{2}(\sin x + \sin y)$$

$$2\sin\left(\frac{x + y}{2}\right)\cos\left(\frac{x - y}{2}\right) = \sin x + \sin y$$

The other sum-to-product formulas can be proved in a similar manner.

Law of Sines *(p. 419)*

If ABC is a triangle with sides a, b, and c, then $\dfrac{a}{\sin A} = \dfrac{b}{\sin B} = \dfrac{c}{\sin C}$.

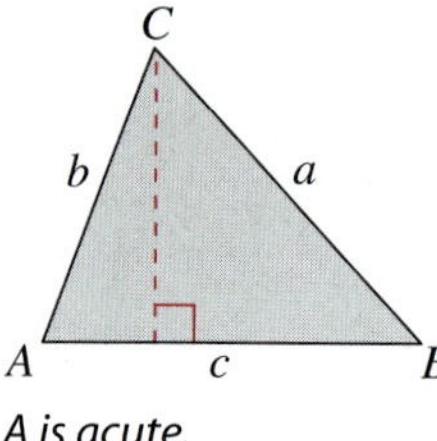

A is acute.

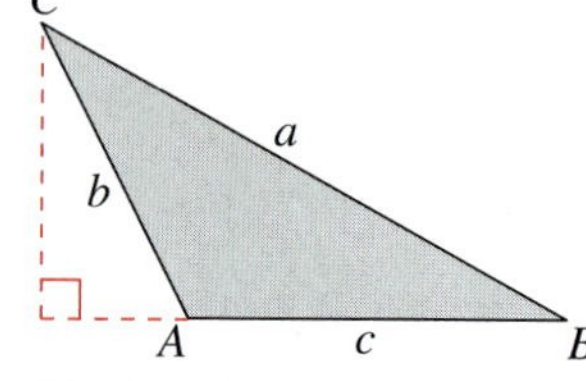

A is obtuse.

Proof

Let h be the altitude of either triangle found in the figure on the previous page. Then you have

$$\sin A = \frac{h}{b} \quad \text{or} \quad h = b \sin A \qquad \text{and} \qquad \sin B = \frac{h}{a} \quad \text{or} \quad h = a \sin B.$$

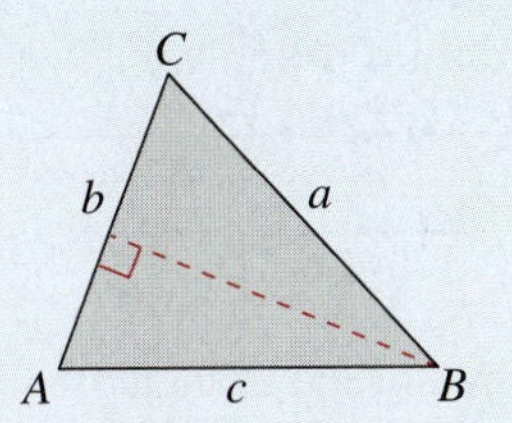

A is acute.

Equating these two values of h, you have

$$a \sin B = b \sin A \qquad \text{or} \qquad \frac{a}{\sin A} = \frac{b}{\sin B}.$$

Note that $\sin A \neq 0$ and $\sin B \neq 0$ because no angle of a triangle can have a measure of $0°$ or $180°$. In a similar manner, construct an altitude from vertex B to side AC (extended in the obtuse triangle), as shown at the left. Then you have

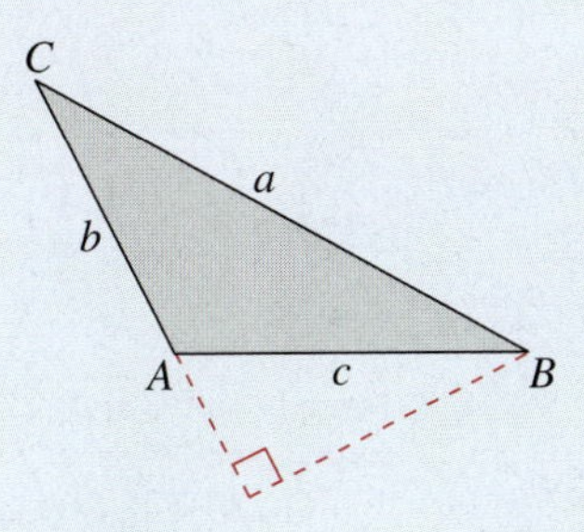

A is obtuse.

$$\sin A = \frac{h}{c} \quad \text{or} \quad h = c \sin A \qquad \text{and} \qquad \sin C = \frac{h}{a} \quad \text{or} \quad h = a \sin C.$$

Equating these two values of h, you have

$$a \sin C = c \sin A \qquad \text{or} \qquad \frac{a}{\sin A} = \frac{c}{\sin C}.$$

By the Transitive Property of Equality you know that

$$\frac{a}{\sin A} = \frac{b}{\sin B} = \frac{c}{\sin C}.$$

So, the Law of Sines is established.

Law of Cosines *(p. 428)*

Standard Form	*Alternative Form*
$a^2 = b^2 + c^2 - 2bc \cos A$	$\cos A = \dfrac{b^2 + c^2 - a^2}{2bc}$
$b^2 = a^2 + c^2 - 2ac \cos B$	$\cos B = \dfrac{a^2 + c^2 - b^2}{2ac}$
$c^2 = a^2 + b^2 - 2ab \cos C$	$\cos C = \dfrac{a^2 + b^2 - c^2}{2ab}$

Proof

To prove the first formula, consider the triangle at the left, which has three acute angles. Note that vertex B has coordinates $(c, 0)$. Furthermore, C has coordinates (x, y), where $x = b \cos A$ and $y = b \sin A$. Because a is the distance from vertex C to vertex B, it follows that

$$a = \sqrt{(x - c)^2 + (y - 0)^2} \qquad \text{Distance Formula}$$

$$a^2 = (x - c)^2 + (y - 0)^2 \qquad \text{Square each side.}$$

$$a^2 = (b \cos A - c)^2 + (b \sin A)^2$$ Substitute for x and y.

$$a^2 = b^2 \cos^2 A - 2bc \cos A + c^2 + b^2 \sin^2 A$$ Expand.

$$a^2 = b^2(\sin^2 A + \cos^2 A) + c^2 - 2bc \cos A$$ Factor out b^2.

$$a^2 = b^2 + c^2 - 2bc \cos A.$$ $\sin^2 A + \cos^2 A = 1$

Similar arguments can be used to establish the second and third formulas.

Heron's Area Formula *(p. 431)*

Given any triangle with sides of lengths a, b, and c, the area of the triangle is

$$\text{Area} = \sqrt{s(s-a)(s-b)(s-c)}$$

where $s = \dfrac{(a+b+c)}{2}$.

Proof

From Section 5.6, you know that

$$\text{Area} = \frac{1}{2}bc \sin A$$ Formula for the area of an oblique triangle

$$(\text{Area})^2 = \frac{1}{4}b^2c^2 \sin^2 A$$ Square each side.

$$\text{Area} = \sqrt{\frac{1}{4}b^2c^2 \sin^2 A}$$ Take the square root of each side.

$$= \sqrt{\frac{1}{4}b^2c^2(1 - \cos^2 A)}$$ Pythagorean Identity

$$= \sqrt{\left[\frac{1}{2}bc(1 + \cos A)\right]\left[\frac{1}{2}bc(1 - \cos A)\right]}.$$ Factor.

Using the Law of Cosines, you can show that

$$\frac{1}{2}bc(1 + \cos A) = \frac{a+b+c}{2} \cdot \frac{-a+b+c}{2}$$

and

$$\frac{1}{2}bc(1 - \cos A) = \frac{a-b+c}{2} \cdot \frac{a+b-c}{2}.$$

Letting $s = (a+b+c)/2$, these two equations can be rewritten as

$$\frac{1}{2}bc(1 + \cos A) = s(s-a) \quad \text{and} \quad \frac{1}{2}bc(1 - \cos A) = (s-b)(s-c).$$

By substituting into the last formula for area, you can conclude that

$$\text{Area} = \sqrt{s(s-a)(s-b)(s-c)}.$$

P.S. Problem Solving

This collection of thought-provoking and challenging exercises further explores and expands upon concepts learned in this chapter.

1. (a) Write each of the other trigonometric functions of θ in terms of $\sin \theta$.

(b) Write each of the other trigonometric functions of θ in terms of $\cos \theta$.

2. Verify that for all integers n, $\cos\left[\frac{(2n+1)\pi}{2}\right] = 0$.

3. Verify that for all integers n, $\sin\left[\frac{(12n+1)\pi}{6}\right] = \frac{1}{2}$.

4. A particular sound wave is modeled by

$$p(t) = \frac{1}{4\pi}(p_1(t) + 30p_2(t) + p_3(t) + p_5(t) + 30p_6(t))$$

where $p_n(t) = \frac{1}{n}\sin(524n\pi t)$, and t is the time (in seconds).

(a) Find the sine components $p_n(t)$ and use a graphing utility to graph each component. Then verify the graph of p that is shown.

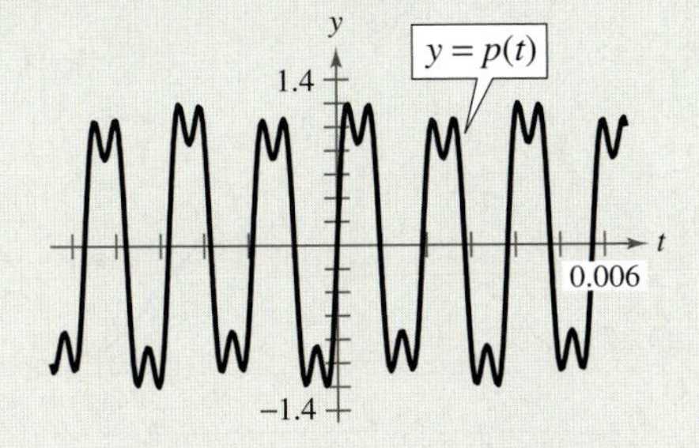

(b) Find the period of each sine component of p. Is p periodic? If so, what is its period?

(c) Use the *zero* or *root* feature or the *zoom* and *trace* features of a graphing utility to find the t-intercepts of the graph of p over one cycle.

(d) Use the *maximum* and *minimum* features of a graphing utility to approximate the absolute maximum and absolute minimum values of p over one cycle.

5. Three squares of side s are placed side by side (see figure). Make a conjecture about the relationship between the sum $u + v$ and w. Prove your conjecture by using the identity for the tangent of the sum of two angles.

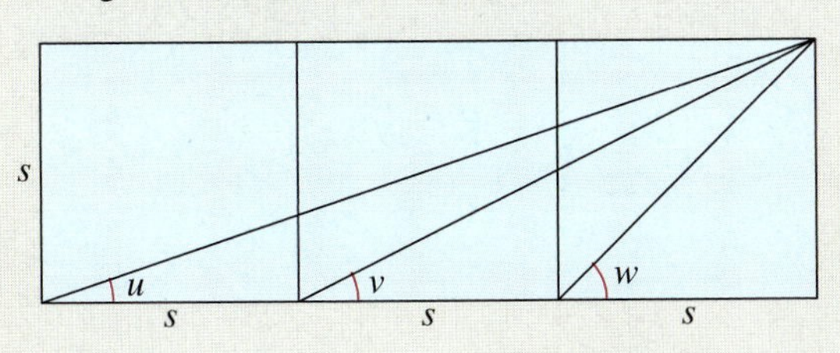

6. The path traveled by an object (neglecting air resistance) that is projected at an initial height of h_0 feet, an initial velocity of v_0 feet per second, and an initial angle θ is given by

$$y = -\frac{16}{v_0^2 \cos^2 \theta}x^2 + (\tan \theta)x + h_0$$

where x and y are measured in feet. Find a formula for the maximum height of an object projected from ground level at velocity v_0 and angle θ. To do this, find half of the horizontal distance

$$\frac{1}{32}v_0^2 \sin 2\theta$$

and then substitute it for x in the general model for the path of a projectile (where $h_0 = 0$).

7. Use the figure to derive the formulas for

$$\sin\frac{\theta}{2}, \cos\frac{\theta}{2}, \text{ and } \tan\frac{\theta}{2}$$

where θ is an acute angle.

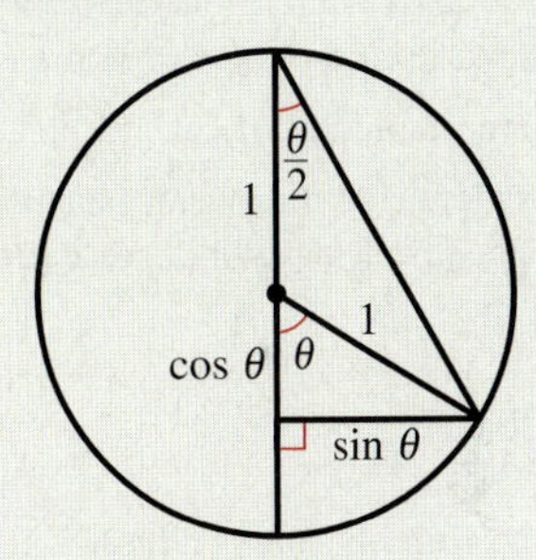

8. The force F (in pounds) on a person's back when he or she bends over at an angle θ is modeled by

$$F = \frac{0.6W \sin(\theta + 90°)}{\sin 12°}$$

where W is the person's weight (in pounds).

(a) Simplify the model.

(b) Use a graphing utility to graph the model, where $W = 185$ and $0° < \theta < 90°$.

(c) At what angle is the force a maximum? At what angle is the force a minimum?

9. The number of hours of daylight that occur at any location on Earth depends on the time of year and the latitude of the location. The following equations model the numbers of hours of daylight in Seward, Alaska (60° latitude) and New Orleans, Louisiana (30° latitude).

$$D = 12.2 - 6.4 \cos\left[\frac{\pi(t + 0.2)}{182.6}\right] \qquad \text{Seward}$$

$$D = 12.2 - 1.9 \cos\left[\frac{\pi(t + 0.2)}{182.6}\right] \qquad \text{New Orleans}$$

In these models, D represents the number of hours of daylight and t represents the day, with $t = 0$ corresponding to January 1.

(a) Use a graphing utility to graph both models in the same viewing window. Use a viewing window of $0 \le t \le 365$.

(b) Find the days of the year on which both cities receive the same amount of daylight. What are these days called?

(c) Which city has the greater variation in the number of daylight hours? Which constant in each model would you use to determine the difference between the greatest and least numbers of hours of daylight?

(d) Determine the period of each model.

10. The tide, or depth of the ocean near the shore, changes throughout the day. The water depth d (in feet) of a bay can be modeled by

$$d = 35 - 28 \cos \frac{\pi}{6.2} t$$

where t is the time in hours, with $t = 0$ corresponding to 12:00 A.M.

(a) Algebraically find the times at which the high and low tides occur.

(b) Algebraically find the time(s) at which the water depth is 3.5 feet.

(c) Use a graphing utility to verify your results from parts (a) and (b).

11. Find the solution of each inequality in the interval $[0, 2\pi]$.

(a) $\sin x \ge 0.5$ (b) $\cos x \le -0.5$

(c) $\tan x < \sin x$ (d) $\cos x \ge \sin x$

12. (a) Write a sum formula for $\sin(u + v + w)$.

(b) Write a sum formula for $\tan(u + v + w)$.

13. (a) Derive a formula for $\cos 3\theta$.

(b) Derive a formula for $\cos 4\theta$.

14. The heights h (in inches) of pistons 1 and 2 in an automobile engine can be modeled by

$$h_1 = 3.75 \sin 733t + 7.5$$

and

$$h_2 = 3.75 \sin 733\left(t + \frac{4\pi}{3}\right) + 7.5$$

where t is measured in seconds.

(a) Use a graphing utility to graph the heights of these two pistons in the same viewing window for $0 \le t \le 1$.

(b) How often are the pistons at the same height?

15. In the figure, a beam of light is directed at the blue mirror, reflected to the red mirror, and then reflected back to the blue mirror. Find the distance PT that the light travels from the red mirror back to the blue mirror.

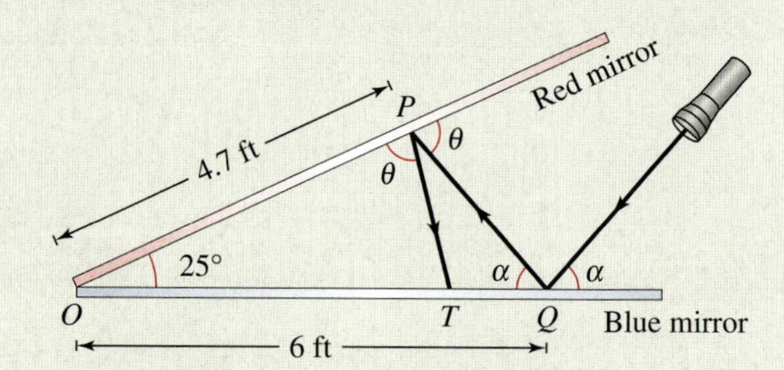

16. A triathlete sets a course to swim S 25° E from a point on shore to a buoy $\frac{3}{4}$ mile away. After swimming 300 yards through a strong current, the triathlete is off course at a bearing of S 35° E. Find the bearing and distance the triathlete needs to swim to correct her course.

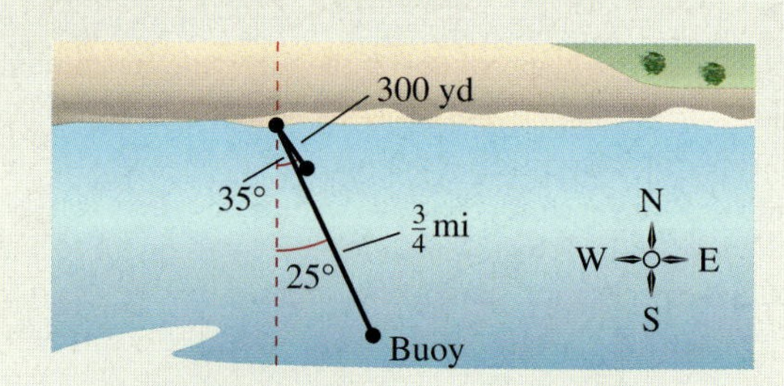

Topics in Analytic Geometry

6

The nine planets move about the sun in elliptical orbits. You can use the techniques presented in this chapter to determine the distances between the planets and the center of the sun.

Kauko Helavuo/Getty Images

SELECTED APPLICATIONS

Analytic geometry concepts have many real-life applications. The applications listed below represent a small sample of the applications in this chapter.

- Inclined Plane, Exercise 56, page 456
- Revenue, Exercise 59, page 463
- Architecture, Exercise 57, page 473
- Satellite Orbit, Exercise 60, page 474
- LORAN, Exercise 42, page 483
- Running Path, Exercise 44, page 484
- Projectile Motion, Exercises 57 and 58, page 491
- Planetary Motion, Exercises 51–56, page 512
- Locating an Explosion, Exercise 40, page 516

6.1 Lines

What you should learn

- Find the inclination of a line.
- Find the angle between two lines.
- Find the distance between a point and a line.

Why you should learn it

The inclination of a line can be used to measure heights indirectly. For instance, in Exercise 56 on page 456, the inclination of a line can be used to determine the change in elevation from the base to the top of the Johnstown Inclined Plane.

AP/Wide World Photos

Inclination of a Line

In Section 1.3, you learned that the graph of the linear equation

$$y = mx + b$$

is a nonvertical line with slope m and y-intercept $(0, b)$. There, the slope of a line was described as the rate of change in y with respect to x. In this section, you will look at the slope of a line in terms of the angle of inclination of the line.

Every nonhorizontal line must intersect the x-axis. The angle formed by such an intersection determines the **inclination** of the line, as specified in the following definition.

Definition of Inclination

The **inclination** of a nonhorizontal line is the positive angle θ (less than π) measured counterclockwise from the x-axis to the line. (See Figure 6.1.)

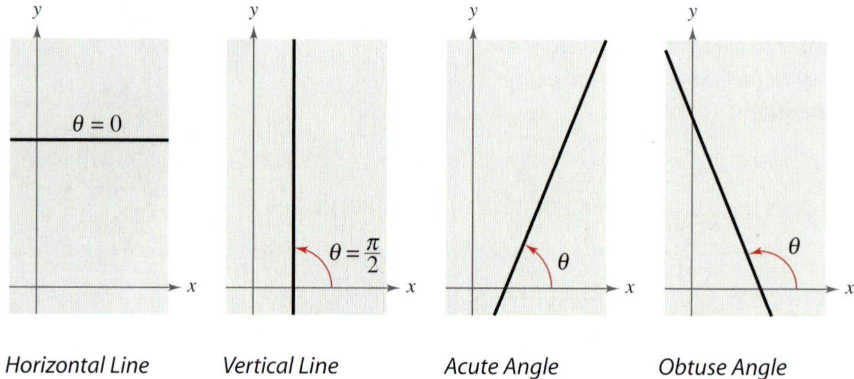

Horizontal Line *Vertical Line* *Acute Angle* *Obtuse Angle*

FIGURE 6.1

The inclination of a line is related to its slope in the following manner.

Inclination and Slope

If a nonvertical line has inclination θ and slope m, then

$$m = \tan \theta.$$

For a proof of the relation between inclination and slope, see Proofs in Mathematics on page 522.

The *HM mathSpace®* CD-ROM and *Eduspace®* for this text contain additional resources related to the concepts discussed in this chapter.

Example 1 Finding the Inclination of a Line

Find the inclination of the line $2x + 3y = 6$.

Solution

The slope of this line is $m = -\frac{2}{3}$. So, its inclination is determined from the equation

$$\tan \theta = -\frac{2}{3}.$$

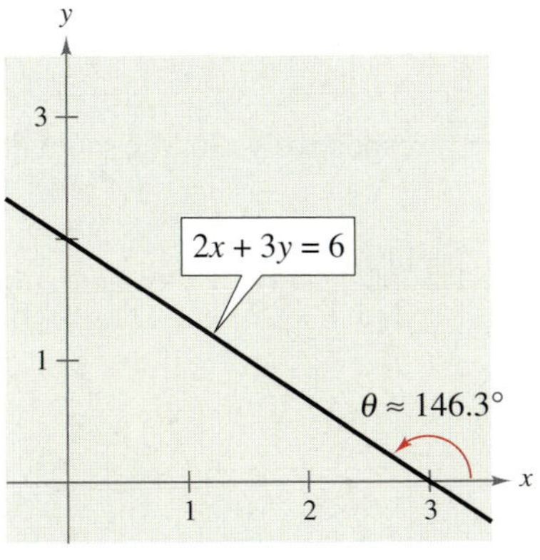

FIGURE 6.2

From Figure 6.2, it follows that $\frac{\pi}{2} < \theta < \pi$. This means that

$$\begin{aligned}\theta &= \pi + \arctan\left(-\frac{2}{3}\right)\\ &\approx \pi + (-0.588)\\ &= \pi - 0.588\\ &\approx 2.554.\end{aligned}$$

The angle of inclination is about 2.554 radians or about 146.3°.

Now try Exercise 19.

The Angle Between Two Lines

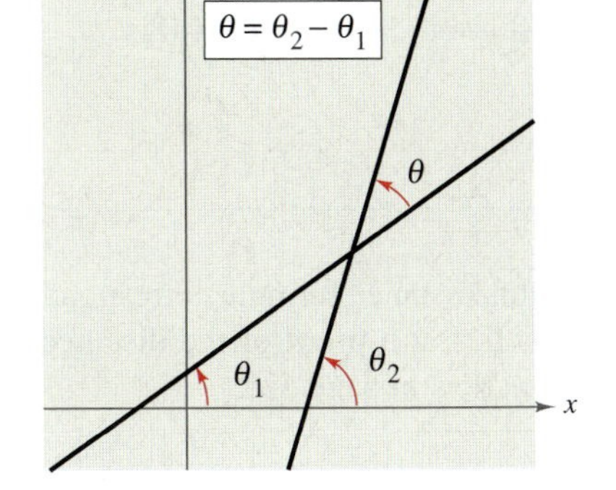

FIGURE 6.3

Two distinct lines in a plane are either parallel or intersecting. If they intersect and are nonperpendicular, their intersection forms two pairs of opposite angles. One pair is acute and the other pair is obtuse. The smaller of these angles is called the **angle between the two lines.** As shown in Figure 6.3, you can use the inclinations of the two lines to find the angle between the two lines. If two lines have inclinations θ_1 and θ_2, where $\theta_1 < \theta_2$ and $\theta_2 - \theta_1 < \pi/2$, the angle between the two lines is

$$\theta = \theta_2 - \theta_1.$$

You can use the formula for the tangent of the difference of two angles

$$\begin{aligned}\tan \theta &= \tan(\theta_2 - \theta_1)\\ &= \frac{\tan \theta_2 - \tan \theta_1}{1 + \tan \theta_1 \tan \theta_2}\end{aligned}$$

to obtain the formula for the angle between two lines.

Angle Between Two Lines

If two nonperpendicular lines have slopes m_1 and m_2, the angle between the two lines is

$$\tan \theta = \left|\frac{m_2 - m_1}{1 + m_1 m_2}\right|.$$

Example 2 Finding the Angle Between Two Lines

Find the angle between the two lines.

Line 1: $2x - y - 4 = 0$ *Line 2:* $3x + 4y - 12 = 0$

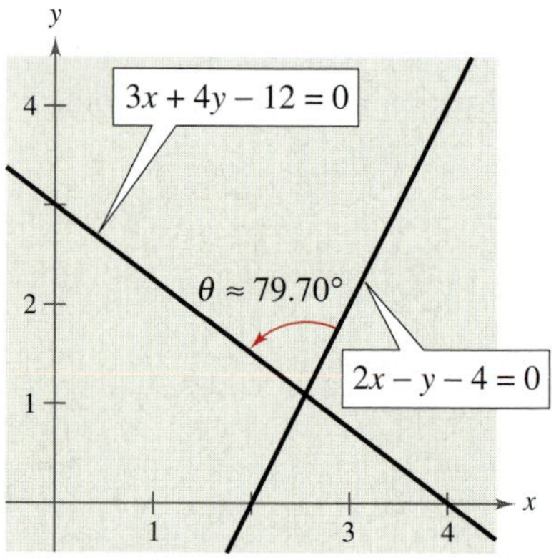

Solution

The two lines have slopes of $m_1 = 2$ and $m_2 = -\frac{3}{4}$, respectively. So, the tangent of the angle between the two lines is

$$\tan\theta = \left|\frac{m_2 - m_1}{1 + m_1 m_2}\right| = \left|\frac{(-3/4) - 2}{1 + (2)(-3/4)}\right| = \left|\frac{-11/4}{-2/4}\right| = \frac{11}{2}.$$

Finally, you can conclude that the angle is

$$\theta = \arctan\frac{11}{2} \approx 1.391 \text{ radians} \approx 79.70°$$

as shown in Figure 6.4.

FIGURE 6.4

✓CHECKPOINT Now try Exercise 27.

The Distance Between a Point and a Line

Finding the distance between a line and a point not on the line is an application of perpendicular lines. This distance is defined as the length of the perpendicular line segment joining the point and the line, as shown in Figure 6.5.

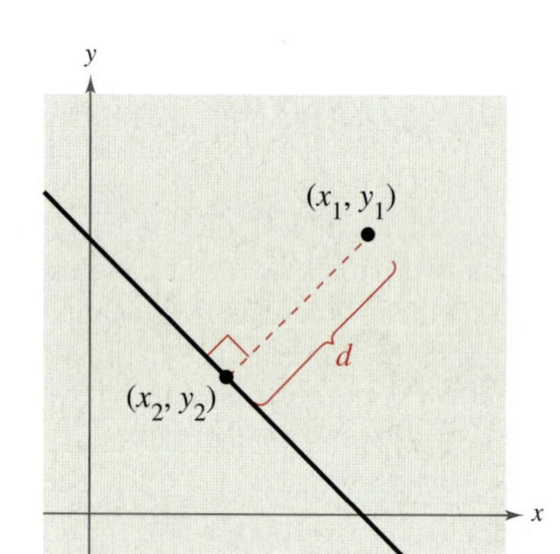

FIGURE 6.5

Distance Between a Point and a Line

The distance between the point (x_1, y_1) and the line $Ax + By + C = 0$ is

$$d = \frac{|Ax_1 + By_1 + C|}{\sqrt{A^2 + B^2}}.$$

Remember that the values of A, B, and C in this distance formula correspond to the general equation of a line, $Ax + By + C = 0$. For a proof of the distance between a point and a line, see Proofs in Mathematics on page 522.

Example 3 Finding the Distance Between a Point and a Line

Find the distance between the point $(4, 1)$ and the line $y = 2x + 1$.

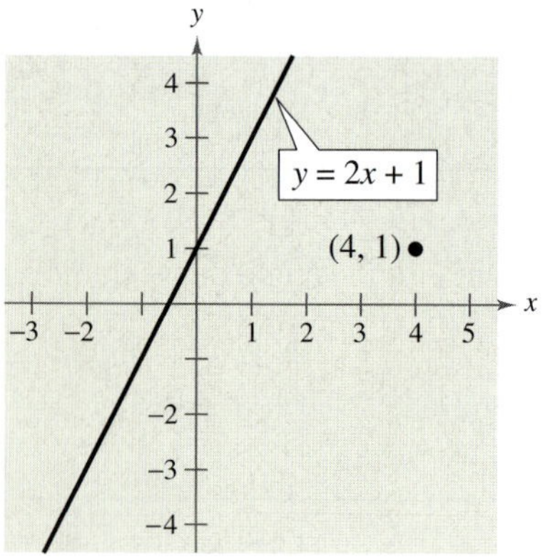

Solution

The general form of the equation is

$$-2x + y - 1 = 0.$$

So, the distance between the point and the line is

$$d = \frac{|-2(4) + 1(1) + (-1)|}{\sqrt{(-2)^2 + 1^2}} = \frac{8}{\sqrt{5}} \approx 3.58 \text{ units.}$$

The line and the point are shown in Figure 6.6.

FIGURE 6.6

✓CHECKPOINT Now try Exercise 39.

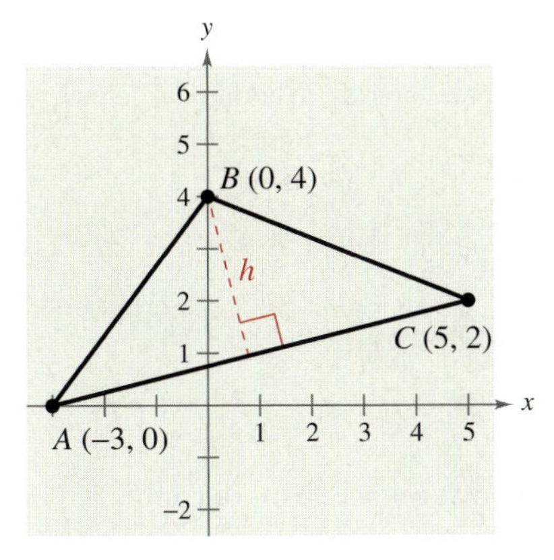

FIGURE 6.7

Activities

1. Find the inclination of the line $5x - 4y = 20$.
   Answer: $\theta = 0.896$ radian or $51.34°$
2. Find the angle θ between the lines $x + 2y = 5$ and $3x - y = 6$.
   Answer: $\theta = 1.429$ radians or $81.87°$
3. Find the distance between the point $(3, -1)$ and the line $4x - 3y - 12 = 0$.
   Answer: $d = \frac{3}{5}$ unit

Group Activity: Graphing Utility

Put your students in groups of two. Ask each group to write a graphing calculator program that finds the angle between two lines based on user input. Ask each group to demonstrate the program on an example or exercise in this section.

Example 4 An Application of Two Distance Formulas

Figure 6.7 shows a triangle with vertices $A(-3, 0)$, $B(0, 4)$, and $C(5, 2)$.

a. Find the altitude h from vertex B to side AC.

b. Find the area of the triangle.

Solution

a. To find the altitude, use the formula for the distance between line AC and the point $(0, 4)$. The equation of line AC is obtained as follows.

Slope: $m = \dfrac{2 - 0}{5 - (-3)} = \dfrac{2}{8} = \dfrac{1}{4}$

Equation: $y - 0 = \dfrac{1}{4}(x + 3)$ — Point-slope form

$4y = x + 3$ — Multiply each side by 4.

$x - 4y + 3 = 0$ — General form

So, the distance between this line and the point $(0, 4)$ is

$$\text{Altitude} = h = \frac{|1(0) + (-4)(4) + 3|}{\sqrt{1^2 + (-4)^2}} = \frac{13}{\sqrt{17}} \text{ units.}$$

b. Using the formula for the distance between two points, you can find the length of the base AC to be

$b = \sqrt{[5 - (-3)]^2 + (2 - 0)^2}$ — Distance Formula

$= \sqrt{8^2 + 2^2}$ — Simplify.

$= \sqrt{68}$ — Simplify.

$= 2\sqrt{17}$ units. — Simplify.

Finally, the area of the triangle in Figure 10.7 is

$A = \dfrac{1}{2}bh$ — Formula for the area of a triangle

$= \dfrac{1}{2}\left(2\sqrt{17}\right)\left(\dfrac{13}{\sqrt{17}}\right)$ — Substitute for b and h.

$= 13$ square units. — Simplify.

CHECKPOINT Now try Exercise 45.

WRITING ABOUT MATHEMATICS

Inclination and the Angle Between Two Lines Discuss why the inclination of a line can be an angle that is larger than $\pi/2$, but the angle between two lines cannot be larger than $\pi/2$. Decide whether the following statement is true or false: "The inclination of a line is the angle between the line and the x-axis." Explain.

6.1 Exercises

The *HM mathSpace®* CD-ROM and *Eduspace®* for this text contain step-by-step solutions to all odd-numbered exercises. They also provide Tutorial Exercises for additional help.

VOCABULARY CHECK: Fill in the blanks.

1. The ________ of a nonhorizontal line is the positive angle θ (less than π) measured counterclockwise from the x-axis to the line.
2. If a nonvertical line has inclination θ and slope m, then $m =$ ________ .
3. If two nonperpendicular lines have slopes m_1 and m_2, the angle between the two lines is $\tan\theta =$ ________ .
4. The distance between the point (x_1, y_1) and the line $Ax + By + C = 0$ is given by $d =$ ________ .

PREREQUISITE SKILLS REVIEW: Practice and review algebra skills needed for this section at **www.Eduspace.com.**

In Exercises 1–8, find the slope of the line with inclination θ.

1.

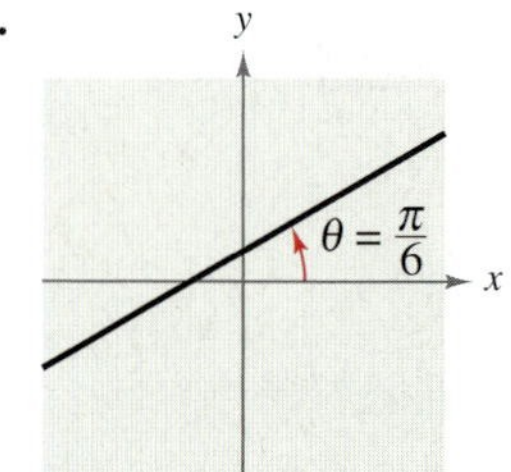

2.

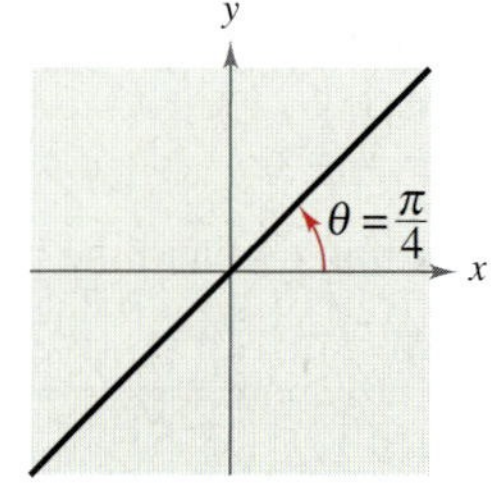

3.

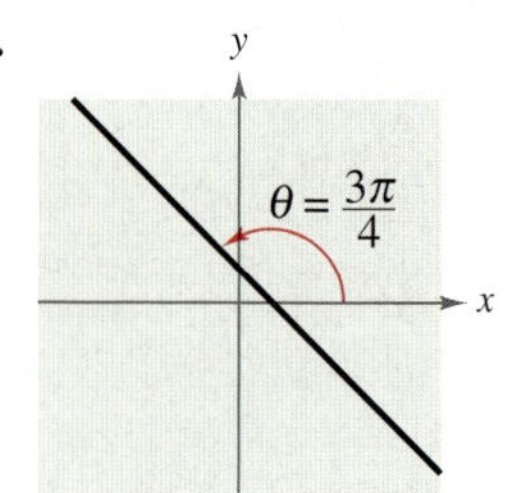

4.

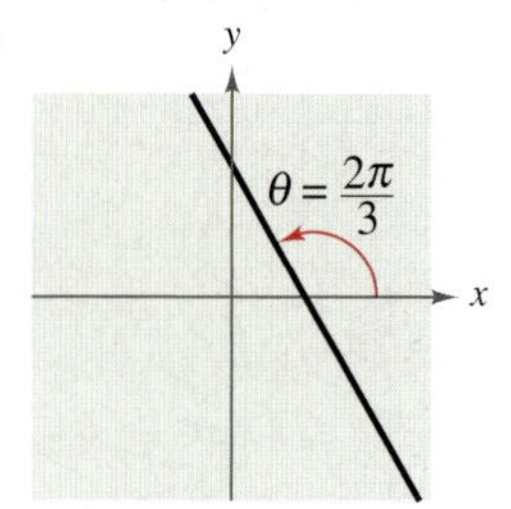

5. $\theta = \dfrac{\pi}{3}$ radians

6. $\theta = \dfrac{5\pi}{6}$ radians

7. $\theta = 1.27$ radians

8. $\theta = 2.88$ radians

In Exercises 9–14, find the inclination θ (in radians and degrees) of the line with a slope of m.

9. $m = -1$

10. $m = -2$

11. $m = 1$

12. $m = 2$

13. $m = \frac{3}{4}$

14. $m = -\frac{5}{2}$

In Exercises 15–18, find the inclination θ (in radians and degrees) of the line passing through the points.

15. $(6, 1), (10, 8)$

16. $(12, 8), (-4, -3)$

17. $(-2, 20), (10, 0)$

18. $(0, 100), (50, 0)$

In Exercises 19–22, find the inclination θ (in radians and degrees) of the line.

19. $6x - 2y + 8 = 0$

20. $4x + 5y - 9 = 0$

21. $5x + 3y = 0$

22. $x - y - 10 = 0$

In Exercises 23–32, find the angle θ (in radians and degrees) between the lines.

23. $3x + y = 3$
$x - y = 2$

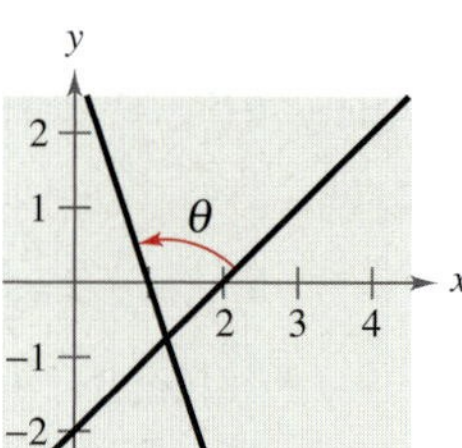

24. $x + 3y = 2$
$x - 2y = -3$

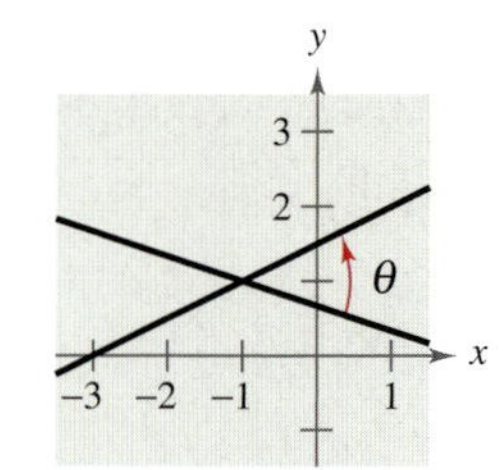

25. $x - y = 0$
$3x - 2y = -1$

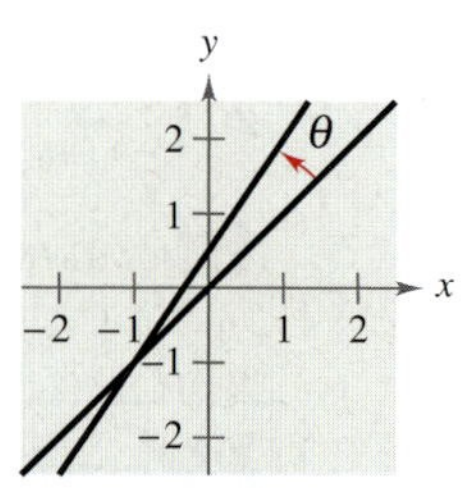

26. $2x - y = 2$
$4x + 3y = 24$

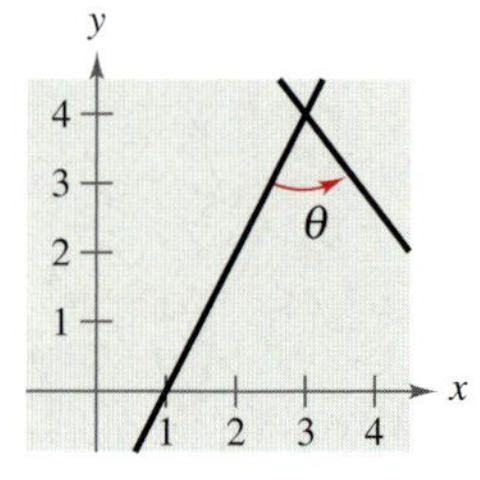

27. $x - 2y = 7$
$6x + 2y = 5$

28. $5x + 2y = 16$
$3x - 5y = -1$

29. $x + 2y = 8$
$x - 2y = 2$

30. $3x - 5y = 3$
$3x + 5y = 12$

31. $0.05x - 0.03y = 0.21$
$0.07x + 0.02y = 0.16$

32. $0.02x - 0.05y = -0.19$
$0.03x + 0.04y = 0.52$

Angle Measurement **In Exercises 33–36, find the slope of each side of the triangle and use the slopes to find the measures of the interior angles.**

33.

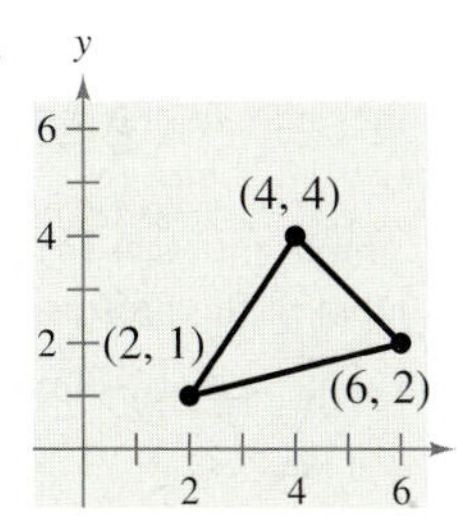

34.

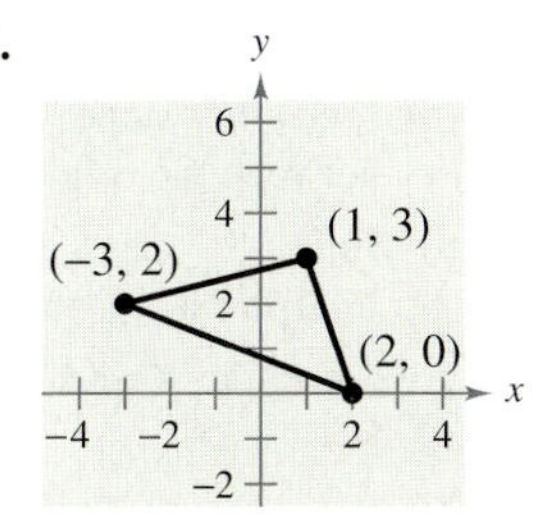

35.

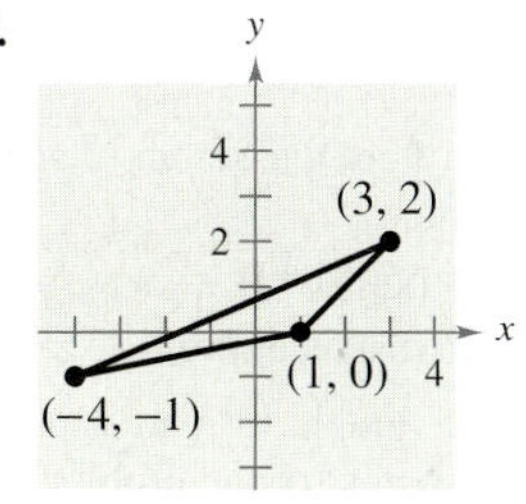

36.

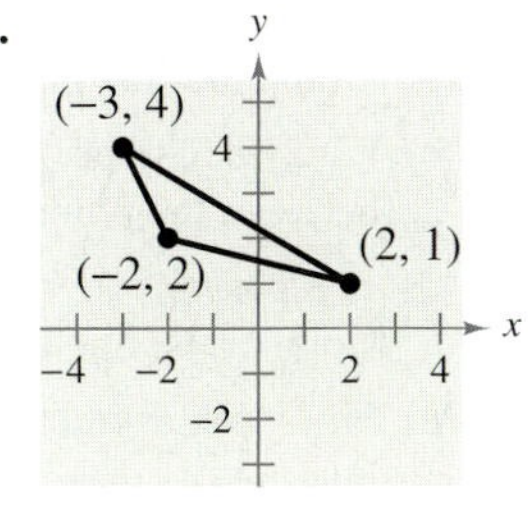

In Exercises 37–44, find the distance between the point and the line.

	Point	*Line*
37.	(0, 0)	$4x + 3y = 0$
38.	(0, 0)	$2x - y = 4$
39.	(2, 3)	$4x + 3y = 10$
40.	(−2, 1)	$x - y = 2$
41.	(6, 2)	$x + 1 = 0$
42.	(10, 8)	$y - 4 = 0$
43.	(0, 8)	$6x - y = 0$
44.	(4, 2)	$x - y = 20$

In Exercises 45–48, the points represent the vertices of a triangle. (a) Draw triangle *ABC* in the coordinate plane, (b) find the altitude from vertex *B* of the triangle to side *AC*, and (c) find the area of the triangle.

45. $A = (0, 0), B = (1, 4), C = (4, 0)$

46. $A = (0, 0), B = (4, 5), C = (5, -2)$

47. $A = \left(-\frac{1}{2}, \frac{1}{2}\right), B = (2, 3), C = \left(\frac{5}{2}, 0\right)$

48. $A = (-4, -5), B = (3, 10), C = (6, 12)$

In Exercises 49 and 50, find the distance between the parallel lines.

49. $x + y = 1$
$x + y = 5$

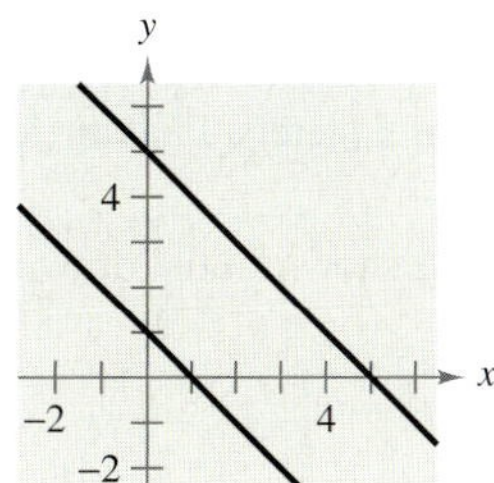

50. $3x - 4y = 1$
$3x - 4y = 10$

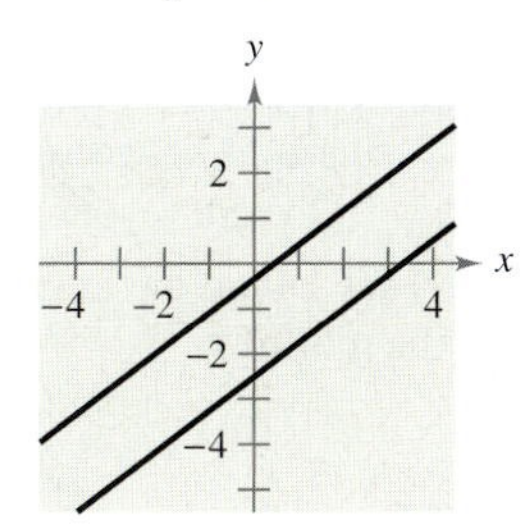

51. ***Road Grade*** A straight road rises with an inclination of 0.10 radian from the horizontal (see figure). Find the slope of the road and the change in elevation over a two-mile stretch of the road.

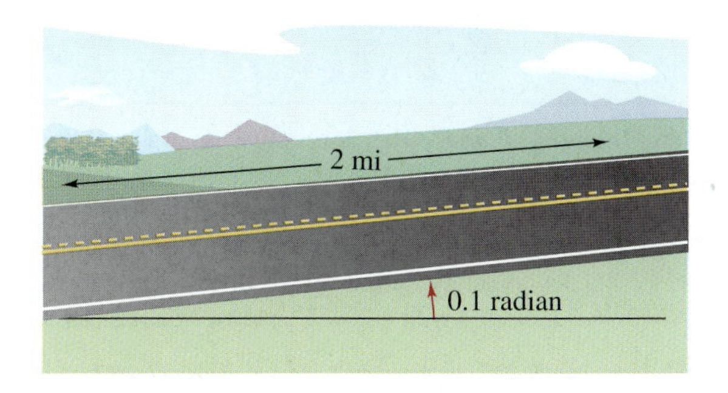

52. ***Road Grade*** A straight road rises with an inclination of 0.20 radian from the horizontal. Find the slope of the road and the change in elevation over a one-mile stretch of the road.

53. ***Pitch of a Roof*** A roof has a rise of 3 feet for every horizontal change of 5 feet (see figure). Find the inclination of the roof.

54. *Conveyor Design* A moving conveyor is built so that it rises 1 meter for each 3 meters of horizontal travel.

(a) Draw a diagram that gives a visual representation of the problem.

(b) Find the inclination of the conveyor.

(c) The conveyor runs between two floors in a factory. The distance between the floors is 5 meters. Find the length of the conveyor.

55. *Truss* Find the angles α and β shown in the drawing of the roof truss.

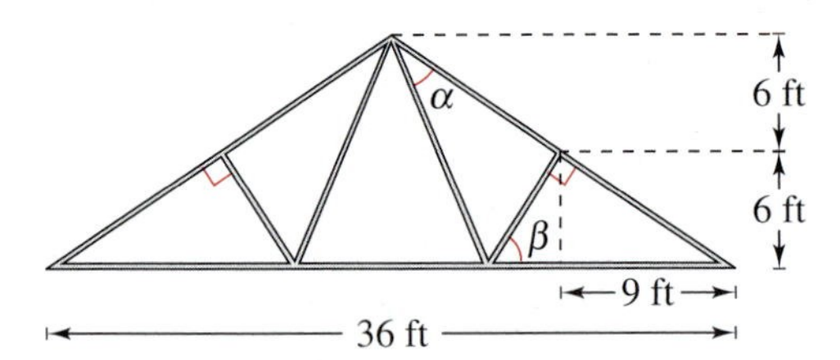

Model It

56. *Inclined Plane* The Johnstown Inclined Plane in Johnstown, Pennsylvania is an inclined railway that was designed to carry people to the hilltop community of Westmont. It also proved useful in carrying people and vehicles to safety during severe floods. The railway is 896.5 feet long with a 70.9% uphill grade (see figure).

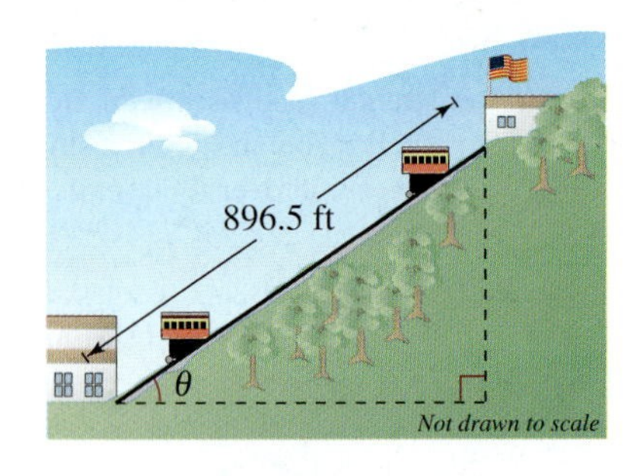

(a) Find the inclination θ of the railway.

(b) Find the change in elevation from the base to the top of the railway.

(c) Using the origin of a rectangular coordinate system as the base of the inclined plane, find the equation of the line that models the railway track.

(d) Sketch a graph of the equation you found in part (c).

Synthesis

True or False? **In Exercises 57 and 58, determine whether the statement is true or false. Justify your answer.**

57. A line that has an inclination greater than $\pi/2$ radians has a negative slope.

58. To find the angle between two lines whose angles of inclination θ_1 and θ_2 are known, substitute θ_1 and θ_2 for m_1 and m_2, respectively, in the formula for the angle between two lines.

59. *Exploration* Consider a line with slope m and y-intercept $(0, 4)$.

(a) Write the distance d between the origin and the line as a function of m.

(b) Graph the function in part (a).

(c) Find the slope that yields the maximum distance between the origin and the line.

(d) Find the asymptote of the graph in part (b) and interpret its meaning in the context of the problem.

60. *Exploration* Consider a line with slope m and y-intercept $(0, 4)$.

(a) Write the distance d between the point $(3, 1)$ and the line as a function of m.

(b) Graph the function in part (a).

(c) Find the slope that yields the maximum distance between the point and the line.

(d) Is it possible for the distance to be 0? If so, what is the slope of the line that yields a distance of 0?

(e) Find the asymptote of the graph in part (b) and interpret its meaning in the context of the problem.

Skills Review

In Exercises 61–66, find all x-intercepts and y-intercepts of the graph of the quadratic function.

61. $f(x) = (x - 7)^2$

62. $f(x) = (x + 9)^2$

63. $f(x) = (x - 5)^2 - 5$

64. $f(x) = (x + 11)^2 + 12$

65. $f(x) = x^2 - 7x - 1$

66. $f(x) = x^2 + 9x - 22$

In Exercises 67–72, write the quadratic function in standard form by completing the square. Identify the vertex of the function.

67. $f(x) = 3x^2 + 2x - 16$

68. $f(x) = 2x^2 - x - 21$

69. $f(x) = 5x^2 + 34x - 7$

70. $f(x) = -x^2 - 8x - 15$

71. $f(x) = 6x^2 - x - 12$

72. $f(x) = -8x^2 - 34x - 21$

In Exercises 73–76, graph the quadratic function.

73. $f(x) = (x - 4)^2 + 3$

74. $f(x) = 6 - (x + 1)^2$

75. $g(x) = 2x^2 - 3x + 1$

76. $g(x) = -x^2 + 6x - 8$

6.2 Introduction to Conics: Parabolas

What you should learn

- Recognize a conic as the intersection of a plane and a double-napped cone.
- Write equations of parabolas in standard form and graph parabolas.
- Use the reflective property of parabolas to solve real-life problems.

Why you should learn it

Parabolas can be used to model and solve many types of real-life problems. For instance, in Exercise 62 on page 464, a parabola is used to model the cables of the Golden Gate Bridge.

Cosmo Condina/Getty Images

Conics

Conic sections were discovered during the classical Greek period, 600 to 300 B.C. The early Greeks were concerned largely with the geometric properties of conics. It was not until the 17th century that the broad applicability of conics became apparent and played a prominent role in the early development of calculus.

A **conic section** (or simply **conic**) is the intersection of a plane and a double-napped cone. Notice in Figure 6.8 that in the formation of the four basic conics, the intersecting plane does not pass through the vertex of the cone. When the plane does pass through the vertex, the resulting figure is a **degenerate conic,** as shown in Figure 6.9.

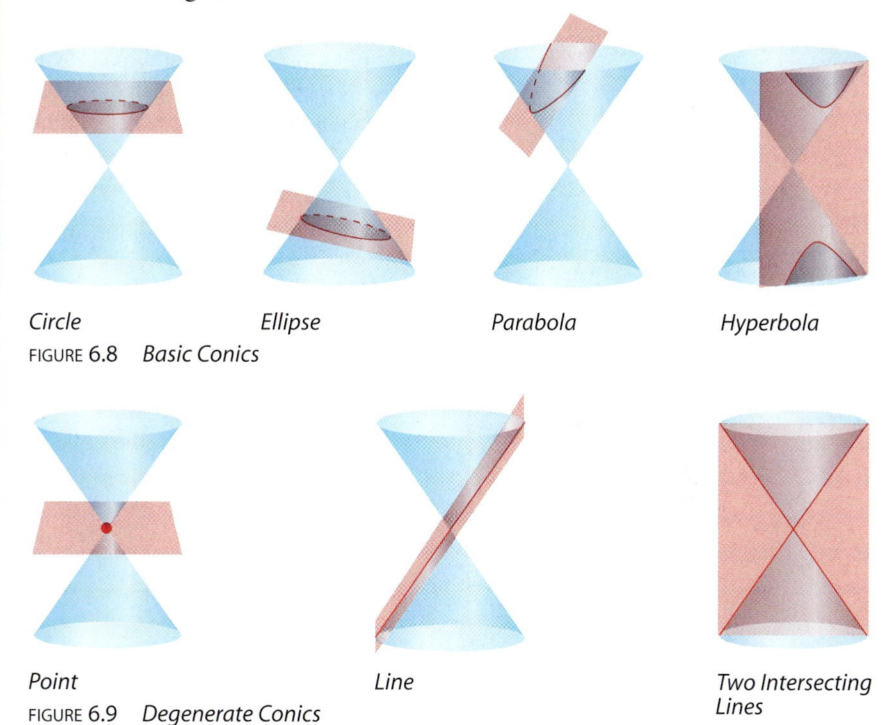

Circle *Ellipse* *Parabola* *Hyperbola*

FIGURE 6.8 *Basic Conics*

Point *Line* *Two Intersecting Lines*

FIGURE 6.9 *Degenerate Conics*

There are several ways to approach the study of conics. You could begin by defining conics in terms of the intersections of planes and cones, as the Greeks did, or you could define them algebraically, in terms of the general second-degree equation

$$Ax^2 + Bxy + Cy^2 + Dx + Ey + F = 0.$$

However, you will study a third approach, in which each of the conics is defined as a **locus** (collection) of points satisfying a geometric property. For example, in Section 1.2, you learned that a circle is defined as the collection of all points (x, y) that are equidistant from a fixed point (h, k). This leads to the standard form of the equation of a circle

$$(x - h)^2 + (y - k)^2 = r^2. \qquad \text{Equation of circle}$$

This study of conics is from a locus-of-points approach, which leads to the development of the standard equation for each conic. Your students should know the standard equations of all conics well. Make sure they understand the relationship of h and k to the horizontal and vertical shifts.

Parabolas

In Section 2.1, you learned that the graph of the quadratic function

$$f(x) = ax^2 + bx + c$$

is a parabola that opens upward or downward. The following definition of a parabola is more general in the sense that it is independent of the orientation of the parabola.

Definition of Parabola

A **parabola** is the set of all points (x, y) in a plane that are equidistant from a fixed line **(directrix)** and a fixed point **(focus)** not on the line.

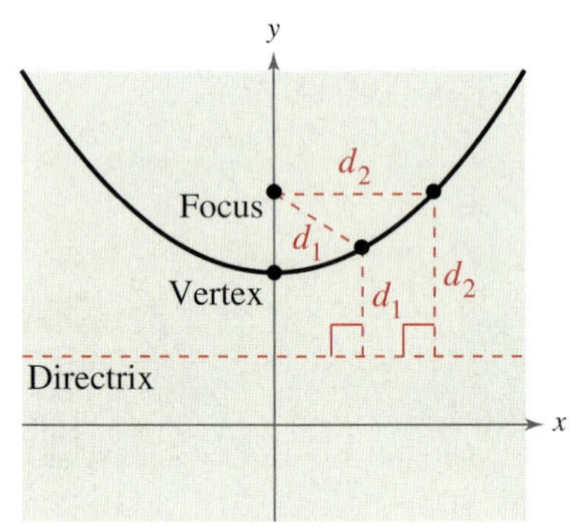

FIGURE 6.10 *Parabola*

The midpoint between the focus and the directrix is called the **vertex,** and the line passing through the focus and the vertex is called the **axis** of the parabola. Note in Figure 6.10 that a parabola is symmetric with respect to its axis. Using the definition of a parabola, you can derive the following **standard form** of the equation of a parabola whose directrix is parallel to the x-axis or to the y-axis.

Standard Equation of a Parabola

The **standard form of the equation of a parabola** with vertex at (h, k) is as follows.

$$(x - h)^2 = 4p(y - k), \quad p \neq 0$$ Vertical axis, directrix: $y = k - p$

$$(y - k)^2 = 4p(x - h), \quad p \neq 0$$ Horizontal axis, directrix: $x = h - p$

The focus lies on the axis p units (*directed distance*) from the vertex. If the vertex is at the origin $(0, 0)$, the equation takes one of the following forms.

$$x^2 = 4py$$ Vertical axis

$$y^2 = 4px$$ Horizontal axis

See Figure 6.11.

For a proof of the standard form of the equation of a parabola, see Proofs in Mathematics on page 523.

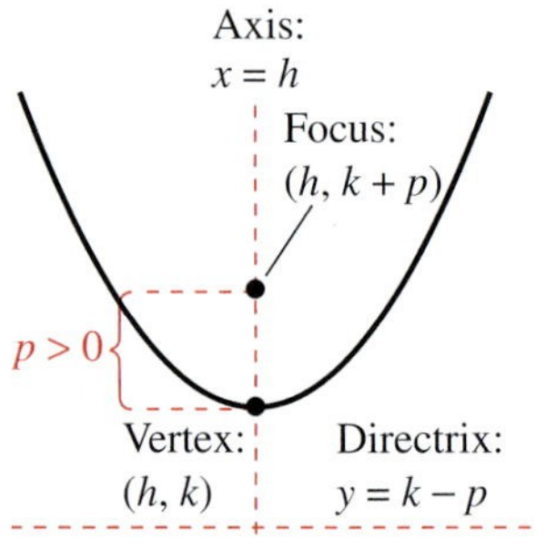

(a) $(x - h)^2 = 4p(y - k)$
Vertical axis: $p > 0$

Axis: $x = h$
Directrix: $y = k - p$
Vertex: (h, k)
$p < 0$
Focus:
$(h, k + p)$

(b) $(x - h)^2 = 4p(y - k)$
Vertical axis: $p < 0$

Directrix:
$x = h - p$
$p > 0$
Axis:
$y = k$
Focus:
$(h + p, k)$
Vertex: (h, k)

(c) $(y - k)^2 = 4p(x - h)$
Horizontal axis: $p > 0$

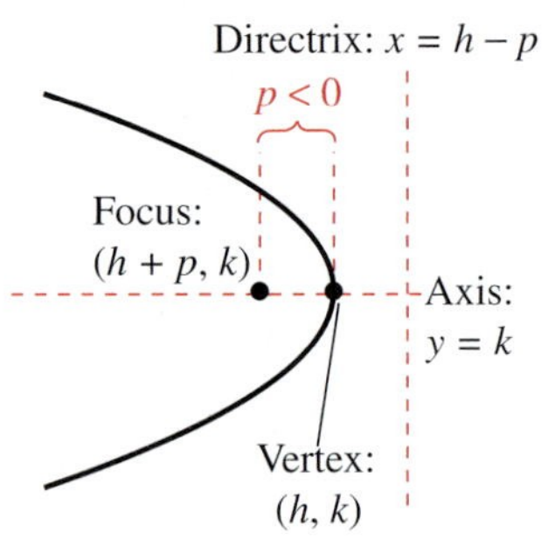

(d) $(y - k)^2 = 4p(x - h)$
Horizontal axis: $p < 0$

FIGURE 6.11

Technology

Use a graphing utility to confirm the equation found in Example 1. In order to graph the equation, you may have to use two separate equations:

$y_1 = \sqrt{8x}$ Upper part

and

$y_2 = -\sqrt{8x}.$ Lower part

Example 1 Vertex at the Origin

Find the standard equation of the parabola with vertex at the origin and focus $(2, 0)$.

Solution

The axis of the parabola is horizontal, passing through $(0, 0)$ and $(2, 0)$, as shown in Figure 6.12.

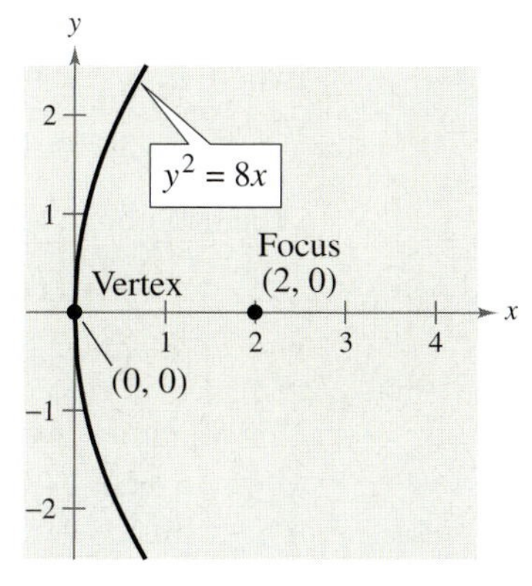

FIGURE 6.12

So, the standard form is $y^2 = 4px$, where $h = 0$, $k = 0$, and $p = 2$. So, the equation is $y^2 = 8x$.

CHECKPOINT Now try Exercise 33.

STUDY TIP

You may want to review the technique of completing the square found in Appendix A.5 (on our website *college.hmco.com*), which will be used to rewrite each of the conics in standard form.

Example 2 Finding the Focus of a Parabola

Find the focus of the parabola given by $y = -\frac{1}{2}x^2 - x + \frac{1}{2}$.

Solution

To find the focus, convert to standard form by completing the square.

$$y = -\tfrac{1}{2}x^2 - x + \tfrac{1}{2} \qquad \text{Write original equation.}$$

$$-2y = x^2 + 2x - 1 \qquad \text{Multiply each side by } -2.$$

$$1 - 2y = x^2 + 2x \qquad \text{Add 1 to each side.}$$

$$1 + 1 - 2y = x^2 + 2x + 1 \qquad \text{Complete the square.}$$

$$2 - 2y = x^2 + 2x + 1 \qquad \text{Combine like terms.}$$

$$-2(y - 1) = (x + 1)^2 \qquad \text{Standard form}$$

Comparing this equation with

$$(x - h)^2 = 4p(y - k)$$

you can conclude that $h = -1$, $k = 1$, and $p = -\frac{1}{2}$. Because p is negative, the parabola opens downward, as shown in Figure 6.13. So, the focus of the parabola is $(h, k + p) = \left(-1, \frac{1}{2}\right)$.

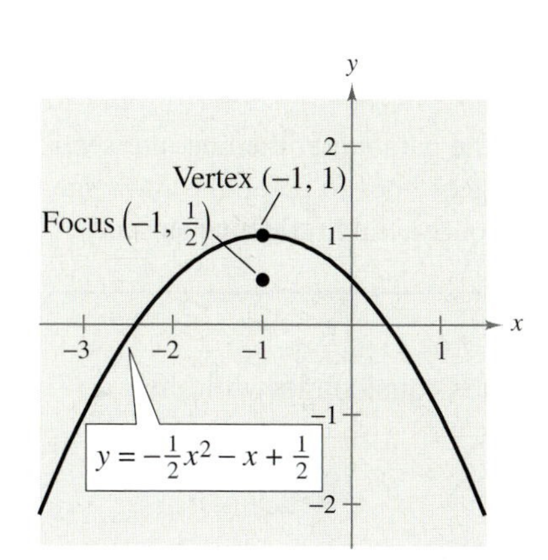

FIGURE 6.13

CHECKPOINT Now try Exercise 21.

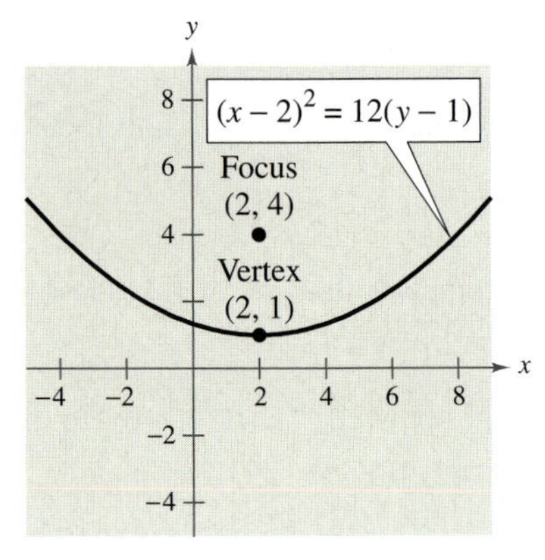

FIGURE 6.14

Example 3 Finding the Standard Equation of a Parabola

Find the standard form of the equation of the parabola with vertex $(2, 1)$ and focus $(2, 4)$.

Solution

Because the axis of the parabola is vertical, passing through $(2, 1)$ and $(2, 4)$, consider the equation

$$(x-h)^2 = 4p(y-k)$$

where $h = 2$, $k = 1$, and $p = 4 - 1 = 3$. So, the standard form is

$$(x-2)^2 = 12(y-1).$$

You can obtain the more common quadratic form as follows.

$$(x-2)^2 = 12(y-1) \quad \text{Write original equation.}$$

$$x^2 - 4x + 4 = 12y - 12 \quad \text{Multiply.}$$

$$x^2 - 4x + 16 = 12y \quad \text{Add 12 to each side.}$$

$$\frac{1}{12}(x^2 - 4x + 16) = y \quad \text{Divide each side by 12.}$$

The graph of this parabola is shown in Figure 6.14.

CHECKPOINT Now try Exercise 45.

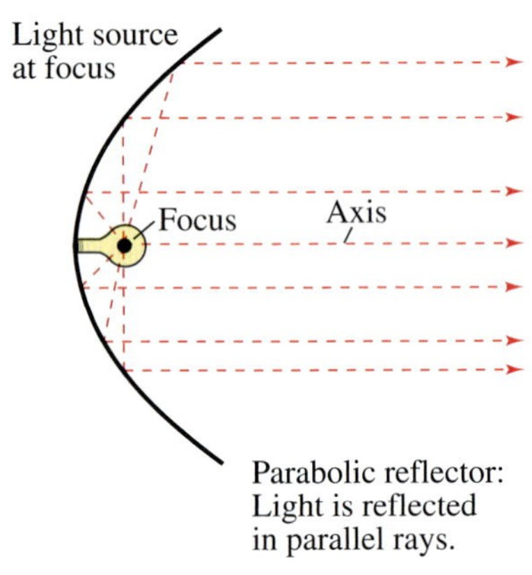

FIGURE 6.15

Application

A line segment that passes through the focus of a parabola and has endpoints on the parabola is called a **focal chord.** The specific focal chord perpendicular to the axis of the parabola is called the **latus rectum.**

Parabolas occur in a wide variety of applications. For instance, a parabolic reflector can be formed by revolving a parabola around its axis. The resulting surface has the property that all incoming rays parallel to the axis are reflected through the focus of the parabola. This is the principle behind the construction of the parabolic mirrors used in reflecting telescopes. Conversely, the light rays emanating from the focus of a parabolic reflector used in a flashlight are all parallel to one another, as shown in Figure 6.15.

A line is **tangent** to a parabola at a point on the parabola if the line intersects, but does not cross, the parabola at the point. Tangent lines to parabolas have special properties related to the use of parabolas in constructing reflective surfaces.

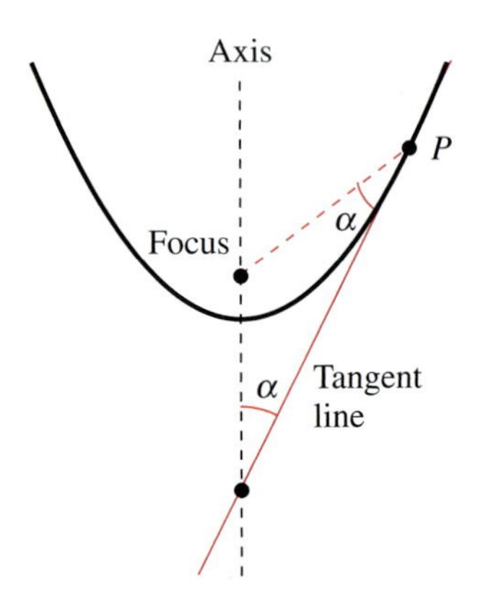

FIGURE 6.16

Reflective Property of a Parabola

The tangent line to a parabola at a point P makes equal angles with the following two lines (see Figure 6.16).

1. The line passing through P and the focus
2. The axis of the parabola

Example 4 Finding the Tangent Line at a Point on a Parabola

Find the equation of the tangent line to the parabola given by $y = x^2$ at the point $(1, 1)$.

Solution

For this parabola, $p = \frac{1}{4}$ and the focus is $\left(0, \frac{1}{4}\right)$, as shown in Figure 6.17. You can find the y-intercept $(0, b)$ of the tangent line by equating the lengths of the two sides of the isosceles triangle shown in Figure 6.17:

$$d_1 = \frac{1}{4} - b$$

and

$$d_2 = \sqrt{(1 - 0)^2 + \left[1 - \left(\frac{1}{4}\right)\right]^2} = \frac{5}{4}.$$

Note that $d_1 = \frac{1}{4} - b$ rather than $b - \frac{1}{4}$. The order of subtraction for the distance is important because the distance must be positive. Setting $d_1 = d_2$ produces

$$\frac{1}{4} - b = \frac{5}{4}$$

$$b = -1.$$

So, the slope of the tangent line is

$$m = \frac{1 - (-1)}{1 - 0} = 2$$

and the equation of the tangent line in slope-intercept form is

$$y = 2x - 1.$$

✓CHECKPOINT Now try Exercise 55.

FIGURE 6.17

Technology

Use a graphing utility to confirm the result of Example 4. By graphing

$y_1 = x^2$ and $y_2 = 2x - 1$

in the same viewing window, you should be able to see that the line touches the parabola at the point $(1, 1)$.

WRITING ABOUT MATHEMATICS

Television Antenna Dishes Cross sections of television antenna dishes are parabolic in shape. Use the figure shown to write a paragraph explaining why these dishes are parabolic.

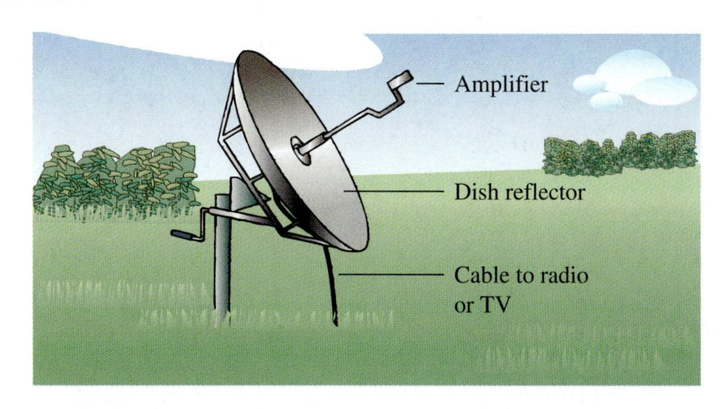

Activities

1. Find the vertex, focus, and directrix of the parabola $x^2 - 6x - 4y + 5 = 0$.
   Answer: Vertex $(3, -1)$; Focus $(3, 0)$; Directrix $y = -2$
2. Find the standard form of the equation of the parabola with vertex $(4, 0)$ and directrix $x = 5$.
   Answer: $y^2 = -4(x - 4)$
3. Find an equation of the tangent line to the parabola $y = 2x^2$ at the point $(1, 2)$.
   Answer: $y = 4x - 2$

6.2 Exercises

VOCABULARY CHECK: Fill in the blanks.

1. A ________ is the intersection of a plane and a double-napped cone.
2. A collection of points satisfying a geometric property can also be referred to as a ________ of points.
3. A ________ is defined as the set of all points (x, y) in a plane that are equidistant from a fixed line, called the ________, and a fixed point, called the ________, not on the line.
4. The line that passes through the focus and vertex of a parabola is called the ________ of the parabola.
5. The ________ of a parabola is the midpoint between the focus and the directrix.
6. A line segment that passes through the focus of a parabola and has endpoints on the parabola is called a ________ ________ .
7. A line is ________ to a parabola at a point on the parabola if the line intersects, but does not cross, the parabola at the point.

PREREQUISITE SKILLS REVIEW: Practice and review algebra skills needed for this section at **www.Eduspace.com.**

In Exercises 1–4, describe in words how a plane could intersect with the double-napped cone shown to form the conic section.

1. Circle
2. Ellipse
3. Parabola
4. Hyperbola

In Exercises 5–10, match the equation with its graph. [The graphs are labeled (a), (b), (c), (d), (e), and (f).]

(a)

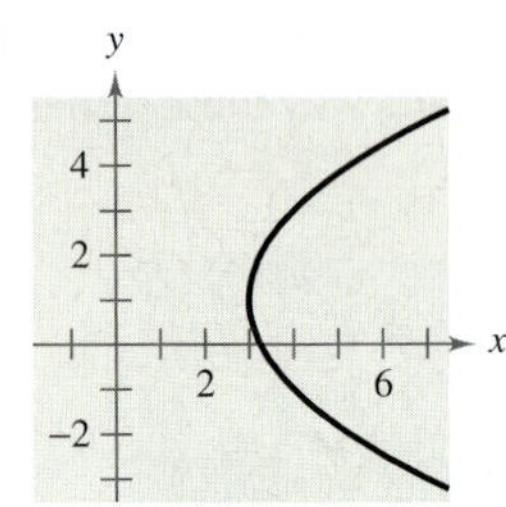

(b)

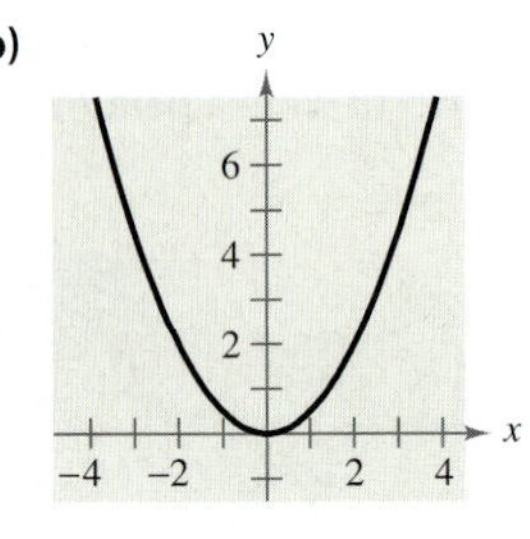

(c)

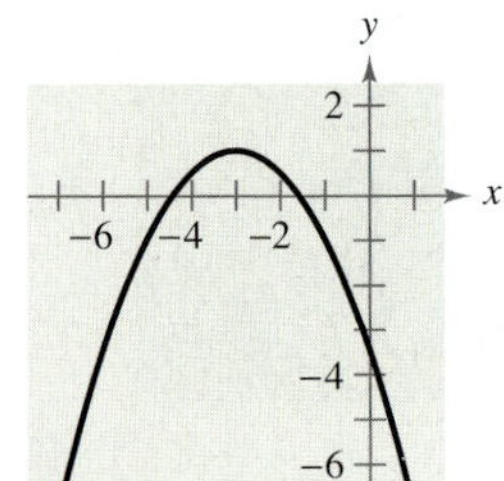

(d)

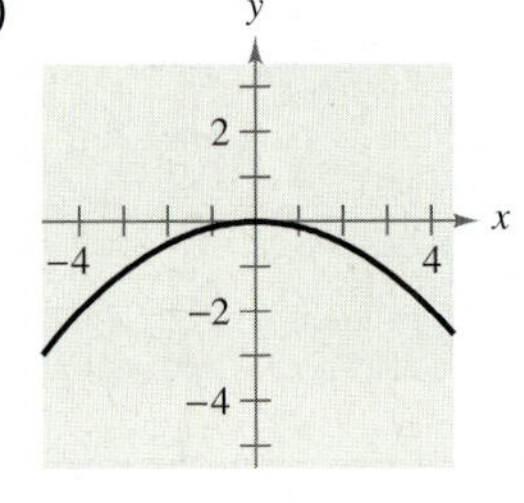

(e)

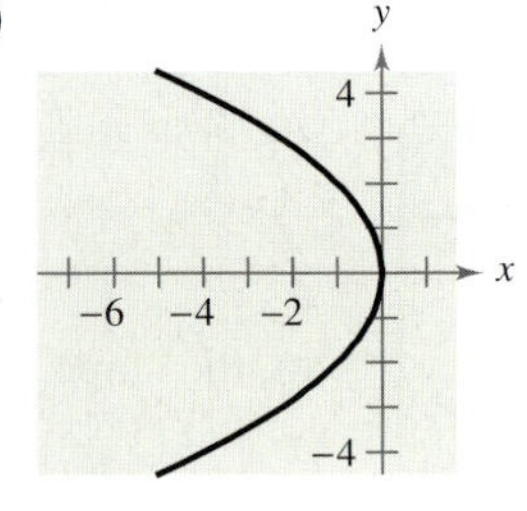

(f)

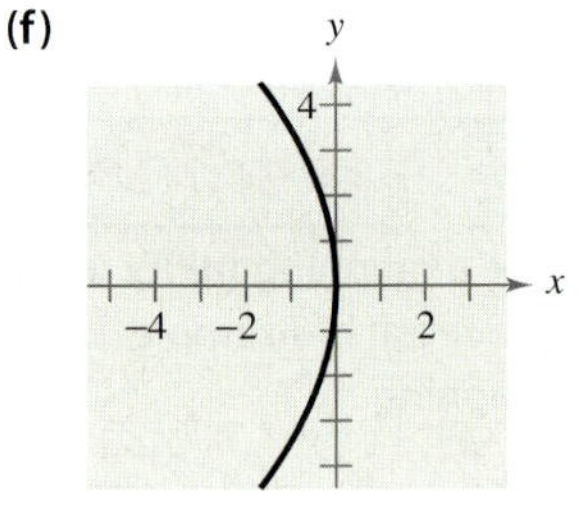

5. $y^2 = -4x$
6. $x^2 = 2y$
7. $x^2 = -8y$
8. $y^2 = -12x$
9. $(y - 1)^2 = 4(x - 3)$
10. $(x + 3)^2 = -2(y - 1)$

In Exercises 11–24, find the vertex, focus, and directrix of the parabola and sketch its graph.

11. $y = \frac{1}{2}x^2$
12. $y = -2x^2$
13. $y^2 = -6x$
14. $y^2 = 3x$
15. $x^2 + 6y = 0$
16. $x + y^2 = 0$
17. $(x - 1)^2 + 8(y + 2) = 0$
18. $(x + 5) + (y - 1)^2 = 0$
19. $\left(x + \frac{3}{2}\right)^2 = 4(y - 2)$
20. $\left(x + \frac{1}{2}\right)^2 = 4(y - 1)$
21. $y = \frac{1}{4}(x^2 - 2x + 5)$
22. $x = \frac{1}{4}(y^2 + 2y + 33)$
23. $y^2 + 6y + 8x + 25 = 0$
24. $y^2 - 4y - 4x = 0$

In Exercises 25–28, find the vertex, focus, and directrix of the parabola. Use a graphing utility to graph the parabola.

25. $x^2 + 4x + 6y - 2 = 0$
26. $x^2 - 2x + 8y + 9 = 0$
27. $y^2 + x + y = 0$
28. $y^2 - 4x - 4 = 0$

In Exercises 29–40, find the standard form of the equation of the parabola with the given characteristic(s) and vertex at the origin.

29.

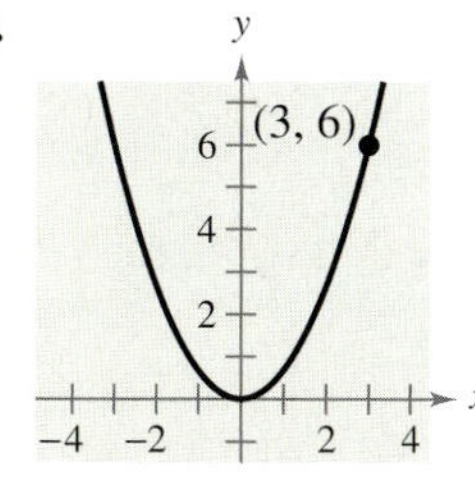

30.

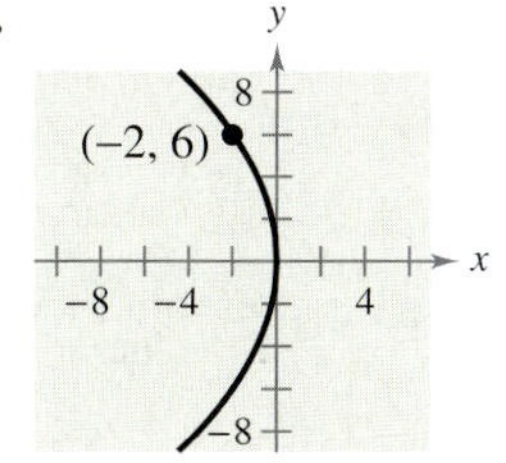

31. Focus: $\left(0, -\frac{3}{2}\right)$

32. Focus: $\left(\frac{5}{2}, 0\right)$

33. Focus: $(-2, 0)$

34. Focus: $(0, -2)$

35. Directrix: $y = -1$

36. Directrix: $y = 3$

37. Directrix: $x = 2$

38. Directrix: $x = -3$

39. Horizontal axis and passes through the point $(4, 6)$

40. Vertical axis and passes through the point $(-3, -3)$

In Exercises 41–50, find the standard form of the equation of the parabola with the given characteristics.

41.

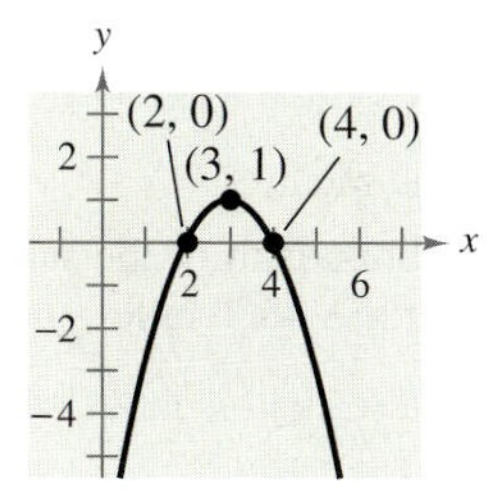

42.

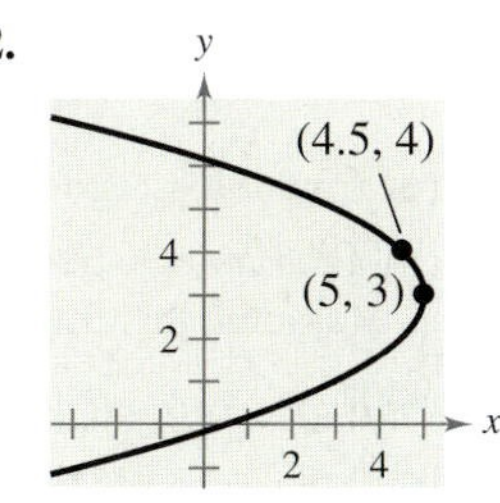

43.

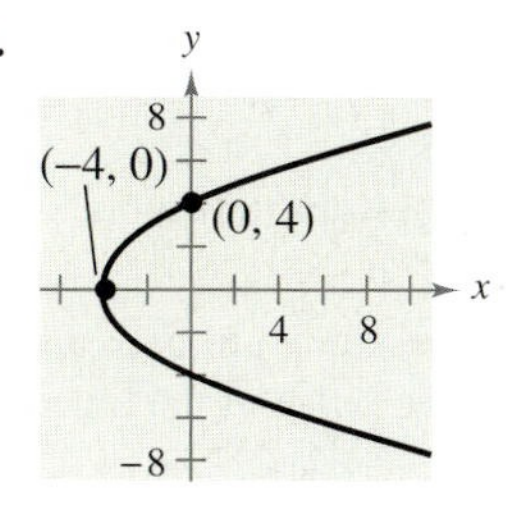

44.

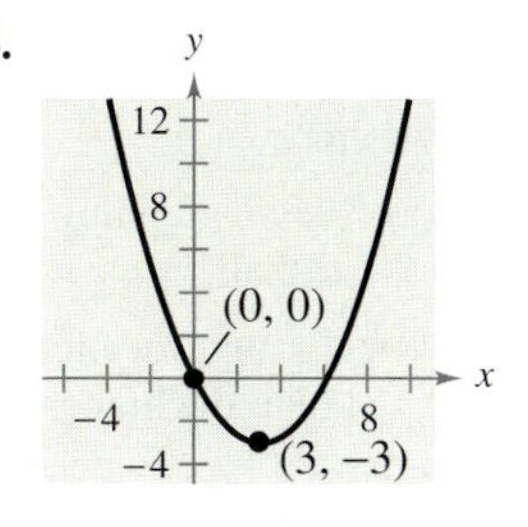

45. Vertex: $(5, 2)$; focus: $(3, 2)$

46. Vertex: $(-1, 2)$; focus: $(-1, 0)$

47. Vertex: $(0, 4)$; directrix: $y = 2$

48. Vertex: $(-2, 1)$; directrix: $x = 1$

49. Focus: $(2, 2)$; directrix: $x = -2$

50. Focus: $(0, 0)$; directrix: $y = 8$

In Exercises 51 and 52, change the equation of the parabola so that its graph matches the description.

51. $(y - 3)^2 = 6(x + 1)$; upper half of parabola

52. $(y + 1)^2 = 2(x - 4)$; lower half of parabola

In Exercises 53 and 54, the equations of a parabola and a tangent line to the parabola are given. Use a graphing utility to graph both equations in the same viewing window. Determine the coordinates of the point of tangency.

	Parabola	*Tangent Line*
53.	$y^2 - 8x = 0$	$x - y + 2 = 0$
54.	$x^2 + 12y = 0$	$x + y - 3 = 0$

In Exercises 55–58, find an equation of the tangent line to the parabola at the given point, and find the *x*-intercept of the line.

55. $x^2 = 2y,\ (4, 8)$

56. $x^2 = 2y,\ \left(-3, \frac{9}{2}\right)$

57. $y = -2x^2,\ (-1, -2)$

58. $y = -2x^2,\ (2, -8)$

59. ***Revenue*** The revenue R (in dollars) generated by the sale of x units of a patio furniture set is given by

$$(x - 106)^2 = -\frac{4}{5}(R - 14{,}045).$$

Use a graphing utility to graph the function and approximate the number of sales that will maximize revenue.

60. ***Revenue*** The revenue R (in dollars) generated by the sale of x units of a digital camera is given by

$$(x - 135)^2 = -\frac{5}{7}(R - 25{,}515).$$

Use a graphing utility to graph the function and approximate the number of sales that will maximize revenue.

61. ***Satellite Antenna*** The receiver in a parabolic television dish antenna is 4.5 feet from the vertex and is located at the focus (see figure). Write an equation for a cross section of the reflector. (Assume that the dish is directed upward and the vertex is at the origin.)

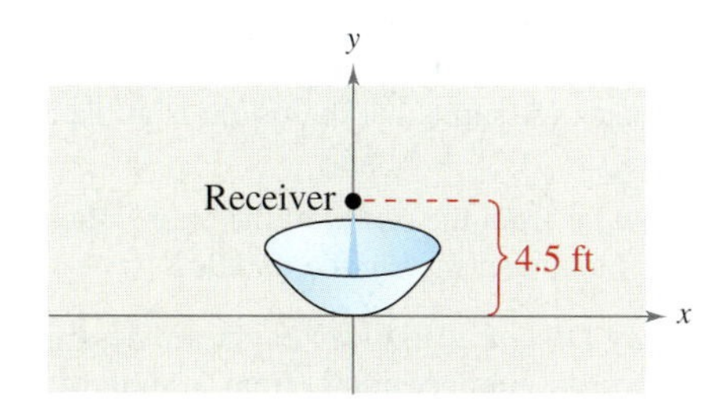

Model It

62. ***Suspension Bridge*** Each cable of the Golden Gate Bridge is suspended (in the shape of a parabola) between two towers that are 1280 meters apart. The top of each tower is 152 meters above the roadway. The cables touch the roadway midway between the towers.

(a) Draw a sketch of the bridge. Locate the origin of a rectangular coordinate system at the center of the roadway. Label the coordinates of the known points.

(b) Write an equation that models the cables.

(c) Complete the table by finding the height y of the suspension cables over the roadway at a distance of x meters from the center of the bridge.

Distance, x	Height, y
0	
250	
400	
500	
1000	

63. ***Road Design*** Roads are often designed with parabolic surfaces to allow rain to drain off. A particular road that is 32 feet wide is 0.4 foot higher in the center than it is on the sides (see figure).

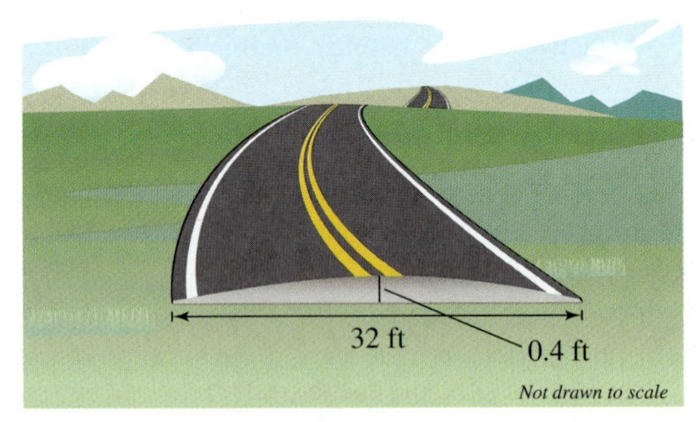

Cross section of road surface

(a) Find an equation of the parabola that models the road surface. (Assume that the origin is at the center of the road.)

(b) How far from the center of the road is the road surface 0.1 foot lower than in the middle?

64. ***Highway Design*** Highway engineers design a parabolic curve for an entrance ramp from a straight street to an interstate highway (see figure). Find an equation of the parabola.

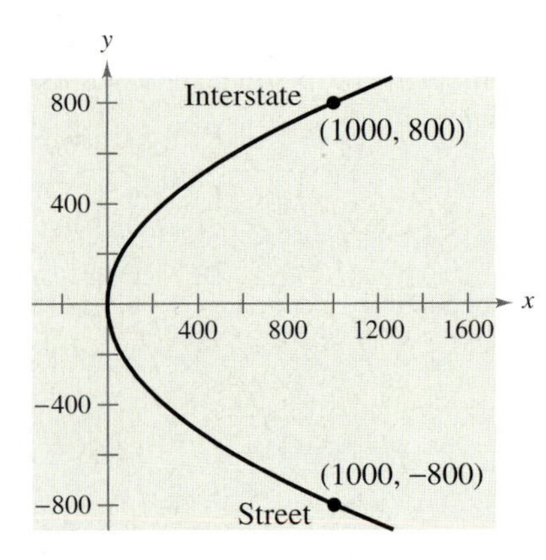

FIGURE FOR 64

65. ***Satellite Orbit*** A satellite in a 100-mile-high circular orbit around Earth has a velocity of approximately 17,500 miles per hour. If this velocity is multiplied by $\sqrt{2}$, the satellite will have the minimum velocity necessary to escape Earth's gravity and it will follow a parabolic path with the center of Earth as the focus (see figure).

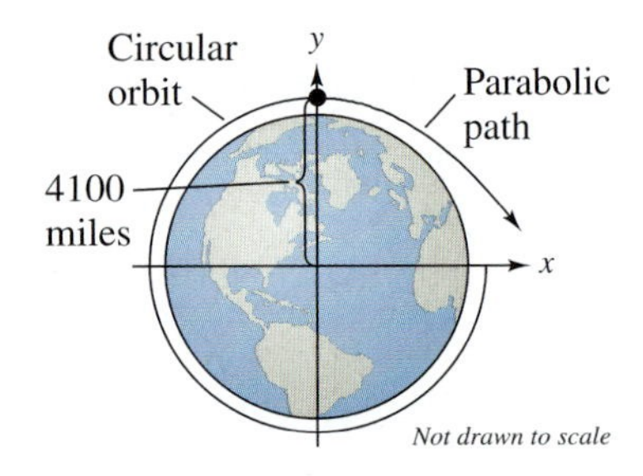

(a) Find the escape velocity of the satellite.

(b) Find an equation of the parabolic path of the satellite (assume that the radius of Earth is 4000 miles).

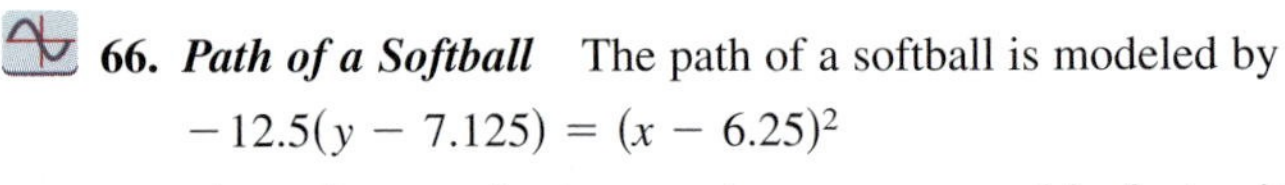

66. ***Path of a Softball*** The path of a softball is modeled by

$$-12.5(y - 7.125) = (x - 6.25)^2$$

where the coordinates x and y are measured in feet, with $x = 0$ corresponding to the position from which the ball was thrown.

(a) Use a graphing utility to graph the trajectory of the softball.

(b) Use the *trace* feature of the graphing utility to approximate the highest point and the range of the trajectory.

Projectile Motion **In Exercises 67 and 68, consider the path of a projectile projected horizontally with a velocity of v feet per second at a height of s feet, where the model for the path is**

$$x^2 = -\frac{v^2}{16}(y - s).$$

In this model (in which air resistance is disregarded), y is the height (in feet) of the projectile and x is the horizontal distance (in feet) the projectile travels.

67. A ball is thrown from the top of a 75-foot tower with a velocity of 32 feet per second.

(a) Find the equation of the parabolic path.

(b) How far does the ball travel horizontally before striking the ground?

68. A cargo plane is flying at an altitude of 30,000 feet and a speed of 540 miles per hour. A supply crate is dropped from the plane. How many *feet* will the crate travel horizontally before it hits the ground?

Synthesis

True or False? **In Exercises 69 and 70, determine whether the statement is true or false. Justify your answer.**

69. It is possible for a parabola to intersect its directrix.

70. If the vertex and focus of a parabola are on a horizontal line, then the directrix of the parabola is vertical.

71. ***Exploration*** Consider the parabola $x^2 = 4py$.

(a) Use a graphing utility to graph the parabola for $p = 1$, $p = 2$, $p = 3$, and $p = 4$. Describe the effect on the graph when p increases.

(b) Locate the focus for each parabola in part (a).

(c) For each parabola in part (a), find the length of the chord passing through the focus and parallel to the directrix (see figure). How can the length of this chord be determined directly from the standard form of the equation of the parabola?

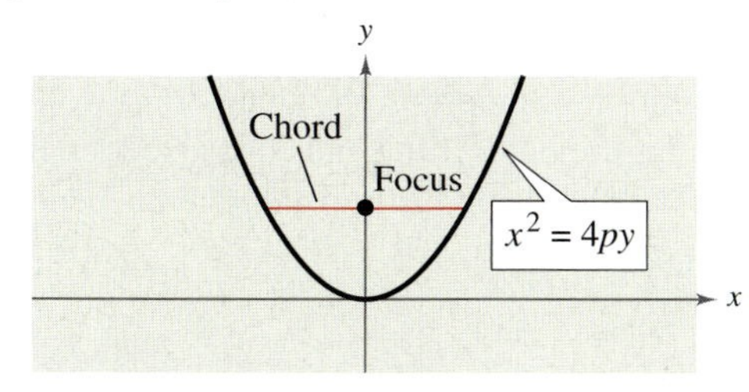

(d) Explain how the result of part (c) can be used as a sketching aid when graphing parabolas.

72. ***Geometry*** The area of the shaded region in the figure is

$$A = \frac{8}{3}p^{1/2}b^{3/2}.$$

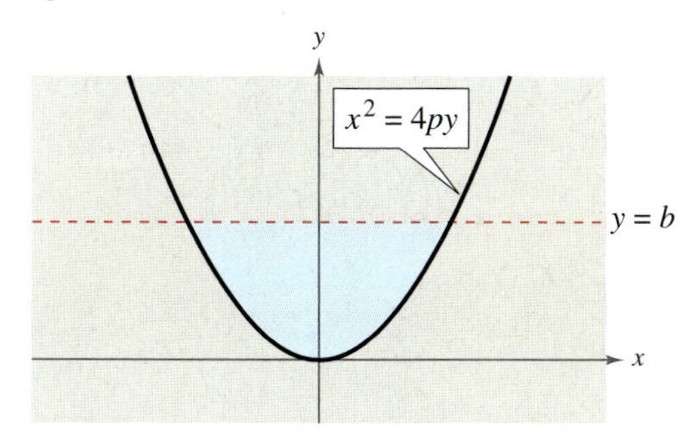

(a) Find the area when $p = 2$ and $b = 4$.

(b) Give a geometric explanation of why the area approaches 0 as p approaches 0.

73. ***Exploration*** Let (x_1, y_1) be the coordinates of a point on the parabola $x^2 = 4py$. The equation of the line tangent to the parabola at the point is

$$y - y_1 = \frac{x_1}{2p}(x - x_1).$$

What is the slope of the tangent line?

74. ***Writing*** In your own words, state the reflective property of a parabola.

Skills Review

In Exercises 75–78, list the possible rational zeros of f given by the Rational Zero Test.

75. $f(x) = x^3 - 2x^2 + 2x - 4$

76. $f(x) = 2x^3 + 4x^2 - 3x + 10$

77. $f(x) = 2x^5 + x^2 + 16$

78. $f(x) = 3x^3 - 12x + 22$

79. Find a polynomial with real coefficients that has the zeros $3, 2 + i$, and $2 - i$.

80. Find all the zeros of

$$f(x) = 2x^3 - 3x^2 + 50x - 75$$

if one of the zeros is $x = \frac{3}{2}$.

81. Find all the zeros of the function

$$g(x) = 6x^4 + 7x^3 - 29x^2 - 28x + 20$$

if two of the zeros are $x = \pm 2$.

82. Use a graphing utility to graph the function given by

$$h(x) = 2x^4 + x^3 - 19x^2 - 9x + 9.$$

Use the graph to approximate the zeros of h.

In Exercises 83–90, use the information to solve the triangle. Round your answers to two decimal places.

83. $A = 35°, a = 10, b = 7$

84. $B = 54°, b = 18, c = 11$

85. $A = 40°, B = 51°, c = 3$

86. $B = 26°, C = 104°, a = 19$

87. $a = 7, b = 10, c = 16$

88. $a = 58, b = 28, c = 75$

89. $A = 65°, b = 5, c = 12$

90. $B = 71°, a = 21, c = 29$

6.3 Ellipses

What you should learn

- Write equations of ellipses in standard form and graph ellipses.
- Use properties of ellipses to model and solve real-life problems.
- Find eccentricities of ellipses.

Why you should learn it

Ellipses can be used to model and solve many types of real-life problems. For instance, in Exercise 59 on page 473, an ellipse is used to model the orbit of Halley's comet.

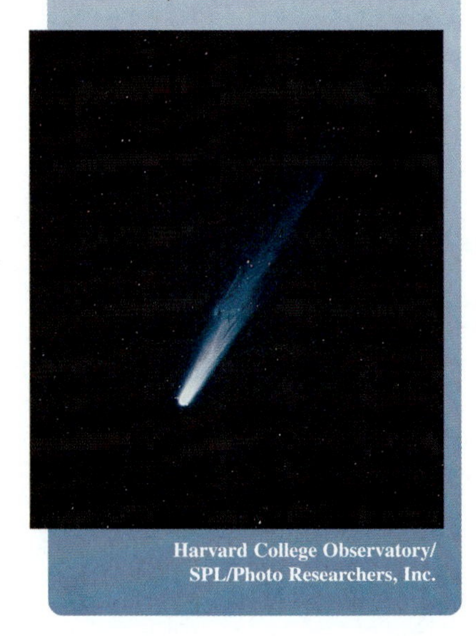

Harvard College Observatory/ SPL/Photo Researchers, Inc.

Introduction

The second type of conic is called an **ellipse,** and is defined as follows.

Definition of Ellipse

An **ellipse** is the set of all points (x, y) in a plane, the sum of whose distances from two distinct fixed points **(foci)** is constant. See Figure 6.18.

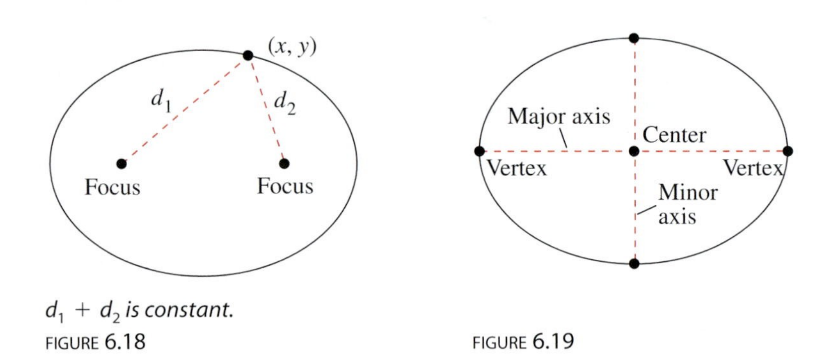

$d_1 + d_2$ *is constant.*

FIGURE 6.18

FIGURE 6.19

The line through the foci intersects the ellipse at two points called **vertices.** The chord joining the vertices is the **major axis,** and its midpoint is the **center** of the ellipse. The chord perpendicular to the major axis at the center is the **minor axis** of the ellipse. See Figure 6.19.

You can visualize the definition of an ellipse by imagining two thumbtacks placed at the foci, as shown in Figure 6.20. If the ends of a fixed length of string are fastened to the thumbtacks and the string is *drawn taut* with a pencil, the path traced by the pencil will be an ellipse.

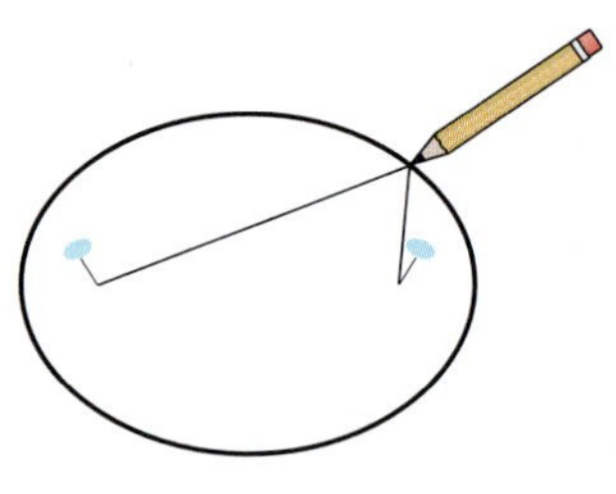

FIGURE 6.20

To derive the standard form of the equation of an ellipse, consider the ellipse in Figure 6.21 with the following points: center, (h, k); vertices, $(h \pm a, k)$; foci, $(h \pm c, k)$. Note that the center is the midpoint of the segment joining the foci.

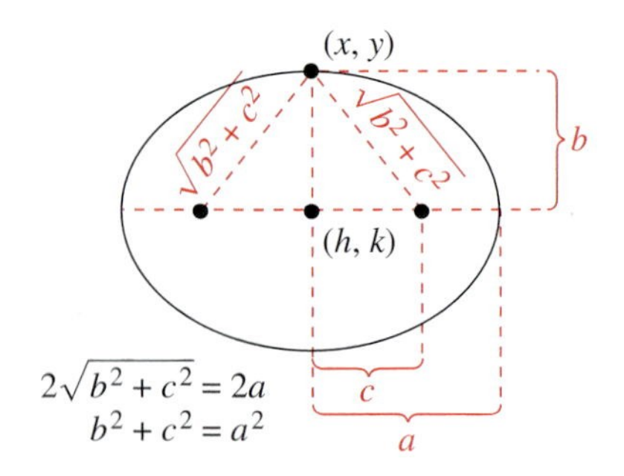

$$2\sqrt{b^2 + c^2} = 2a$$
$$b^2 + c^2 = a^2$$

FIGURE 6.21

When discussing ellipses, you might also choose to discuss the latera recta as background for Exercises 62–66.

The sum of the distances from any point on the ellipse to the two foci is constant. Using a vertex point, this constant sum is

$$(a + c) + (a - c) = 2a \qquad \text{Length of major axis}$$

or simply the length of the major axis. Now, if you let (x, y) be *any* point on the ellipse, the sum of the distances between (x, y) and the two foci must also be $2a$. That is,

$$\sqrt{[x - (h - c)]^2 + (y - k)^2} + \sqrt{[x - (h + c)]^2 + (y - k)^2} = 2a.$$

Finally, in Figure 6.21, you can see that $b^2 = a^2 - c^2$, which implies that the equation of the ellipse is

$$b^2(x - h)^2 + a^2(y - k)^2 = a^2b^2$$

$$\frac{(x - h)^2}{a^2} + \frac{(y - k)^2}{b^2} = 1.$$

You would obtain a similar equation in the derivation by starting with a vertical major axis. Both results are summarized as follows.

STUDY TIP

Consider the equation of the ellipse

$$\frac{(x - h)^2}{a^2} + \frac{(y - k)^2}{b^2} = 1.$$

If you let $a = b$, then the equation can be rewritten as

$$(x - h)^2 + (y - k)^2 = a^2$$

which is the standard form of the equation of a circle with radius $r = a$ (see Section 1.2). Geometrically, when $a = b$ for an ellipse, the major and minor axes are of equal length, and so the graph is a circle.

Standard Equation of an Ellipse

The **standard form of the equation of an ellipse,** with center (h, k) and major and minor axes of lengths $2a$ and $2b$, respectively, where $0 < b < a$, is

$$\frac{(x - h)^2}{a^2} + \frac{(y - k)^2}{b^2} = 1 \qquad \text{Major axis is horizontal.}$$

$$\frac{(x - h)^2}{b^2} + \frac{(y - k)^2}{a^2} = 1. \qquad \text{Major axis is vertical.}$$

The foci lie on the major axis, c units from the center, with $c^2 = a^2 - b^2$. If the center is at the origin $(0, 0)$, the equation takes one of the following forms.

$$\frac{x^2}{a^2} + \frac{y^2}{b^2} = 1 \quad \text{Major axis is horizontal.} \qquad \frac{x^2}{b^2} + \frac{y^2}{a^2} = 1 \quad \text{Major axis is vertical.}$$

Figure 6.22 shows both the horizontal and vertical orientations for an ellipse.

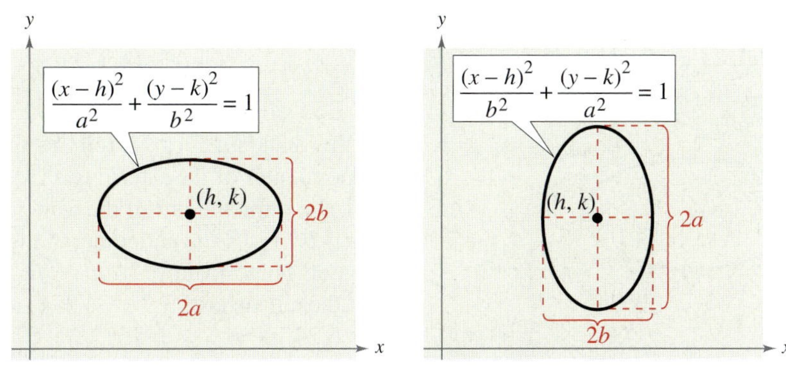

Major axis is horizontal. *Major axis is vertical.*

FIGURE 6.22

Example 1 Finding the Standard Equation of an Ellipse

Find the standard form of the equation of the ellipse having foci at $(0, 1)$ and $(4, 1)$ and a major axis of length 6, as shown in Figure 6.23.

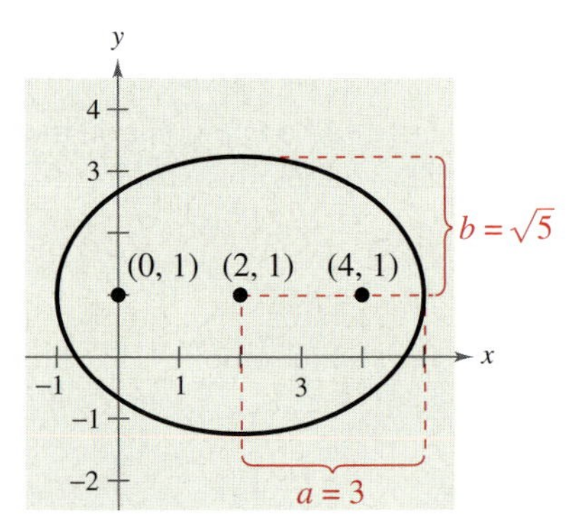

FIGURE 6.23

Solution

Because the foci occur at $(0, 1)$ and $(4, 1)$, the center of the ellipse is $(2, 1)$ and the distance from the center to one of the foci is $c = 2$. Because $2a = 6$, you know that $a = 3$. Now, from $c^2 = a^2 - b^2$, you have

$$b = \sqrt{a^2 - c^2} = \sqrt{3^2 - 2^2} = \sqrt{5}.$$

Because the major axis is horizontal, the standard equation is

$$\frac{(x-2)^2}{3^2} + \frac{(y-1)^2}{(\sqrt{5})^2} = 1.$$

This equation simplifies to

$$\frac{(x-2)^2}{9} + \frac{(y-1)^2}{5} = 1.$$

CHECKPOINT Now try Exercise 49.

Remind your students that completing the square must be performed twice to write the equation of the ellipse in standard form in Example 2.

Example 2 Sketching an Ellipse

Sketch the ellipse given by $x^2 + 4y^2 + 6x - 8y + 9 = 0$.

Solution

Begin by writing the original equation in standard form. In the fourth step, note that 9 and 4 are added to *both* sides of the equation when completing the squares.

$$x^2 + 4y^2 + 6x - 8y + 9 = 0 \quad \text{Write original equation.}$$

$$(x^2 + 6x + \quad) + (4y^2 - 8y + \quad) = -9 \quad \text{Group terms.}$$

$$(x^2 + 6x + \quad) + 4(y^2 - 2y + \quad) = -9 \quad \text{Factor 4 out of } y\text{-terms.}$$

$$(x^2 + 6x + 9) + 4(y^2 - 2y + 1) = -9 + 9 + 4(1)$$

$$(x+3)^2 + 4(y-1)^2 = 4 \quad \text{Write in completed square form.}$$

$$\frac{(x+3)^2}{4} + \frac{(y-1)^2}{1} = 1 \quad \text{Divide each side by 4.}$$

$$\frac{(x+3)^2}{2^2} + \frac{(y-1)^2}{1^2} = 1 \quad \text{Write in standard form.}$$

From this standard form, it follows that the center is $(h, k) = (-3, 1)$. Because the denominator of the x-term is $a^2 = 2^2$, the endpoints of the major axis lie two units to the right and left of the center. Similarly, because the denominator of the y-term is $b^2 = 1^2$, the endpoints of the minor axis lie one unit up and down from the center. Now, from $c^2 = a^2 - b^2$, you have $c = \sqrt{2^2 - 1^2} = \sqrt{3}$. So, the foci of the ellipse are $(-3 - \sqrt{3}, 1)$ and $(-3 + \sqrt{3}, 1)$. The ellipse is shown in Figure 6.24.

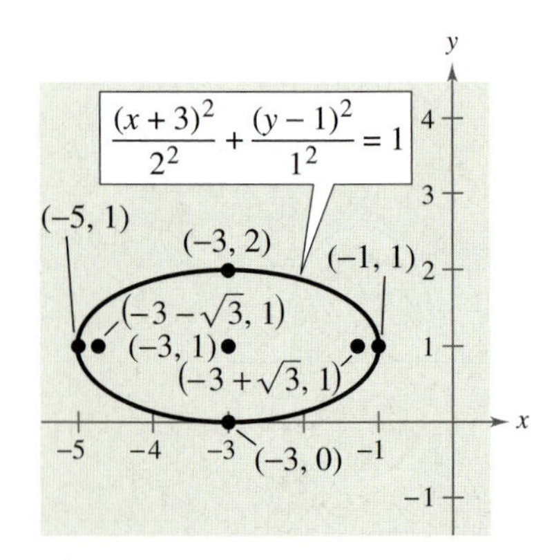

FIGURE 6.24

CHECKPOINT Now try Exercise 25.

Example 3 Analyzing an Ellipse

Find the center, vertices, and foci of the ellipse $4x^2 + y^2 - 8x + 4y - 8 = 0$.

Solution

By completing the square, you can write the original equation in standard form.

$$4x^2 + y^2 - 8x + 4y - 8 = 0 \quad \text{Write original equation.}$$

$$(4x^2 - 8x + \quad) + (y^2 + 4y + \quad) = 8 \quad \text{Group terms.}$$

$$4(x^2 - 2x + \quad) + (y^2 + 4y + \quad) = 8 \quad \text{Factor 4 out of } x\text{-terms.}$$

$$4(x^2 - 2x + 1) + (y^2 + 4y + 4) = 8 + 4(1) + 4$$

$$4(x-1)^2 + (y+2)^2 = 16 \quad \text{Write in completed square form.}$$

$$\frac{(x-1)^2}{4} + \frac{(y+2)^2}{16} = 1 \quad \text{Divide each side by 16.}$$

$$\frac{(x-1)^2}{2^2} + \frac{(y+2)^2}{4^2} = 1 \quad \text{Write in standard form.}$$

The major axis is vertical, where $h = 1$, $k = -2$, $a = 4$, $b = 2$, and

$$c = \sqrt{a^2 - b^2} = \sqrt{16 - 4} = \sqrt{12} = 2\sqrt{3}.$$

So, you have the following.

Center: $(1, -2)$ Vertices: $(1, -6)$, $(1, 2)$ Foci: $(1, -2 - 2\sqrt{3})$, $(1, -2 + 2\sqrt{3})$

The graph of the ellipse is shown in Figure 6.25.

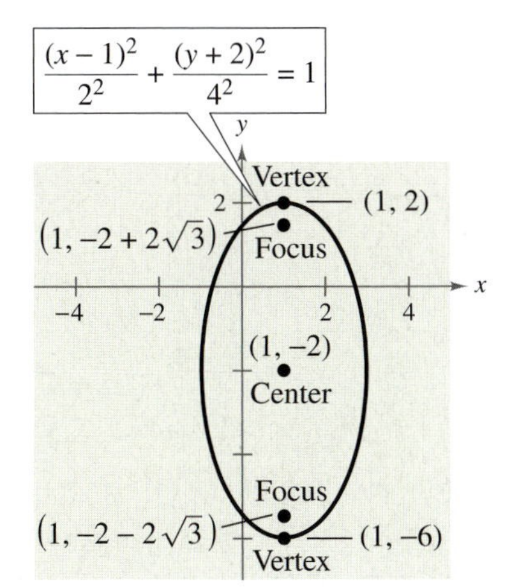

FIGURE 6.25

CHECKPOINT Now try Exercise 29.

Technology

You can use a graphing utility to graph an ellipse by graphing the upper and lower portions in the same viewing window. For instance, to graph the ellipse in Example 3, first solve for y to get

$$y_1 = -2 + 4\sqrt{1 - \frac{(x-1)^2}{4}} \quad \text{and} \quad y_2 = -2 - 4\sqrt{1 - \frac{(x-1)^2}{4}}.$$

Use a viewing window in which $-6 \le x \le 9$ and $-7 \le y \le 3$. You should obtain the graph shown below.

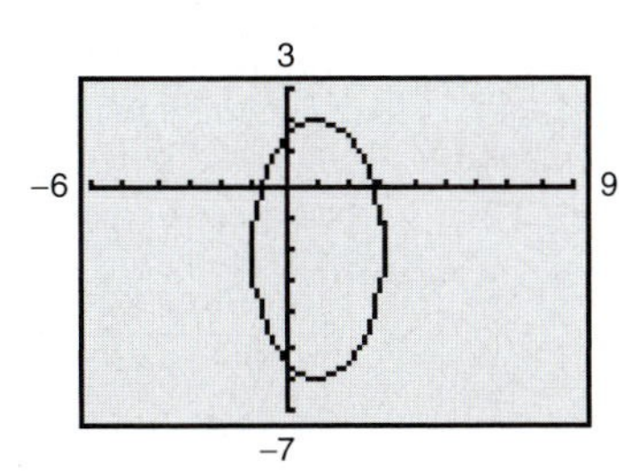

Application

Ellipses have many practical and aesthetic uses. For instance, machine gears, supporting arches, and acoustic designs often involve elliptical shapes. The orbits of satellites and planets are also ellipses. Example 4 investigates the elliptical orbit of the moon about Earth.

Example 4 An Application Involving an Elliptical Orbit

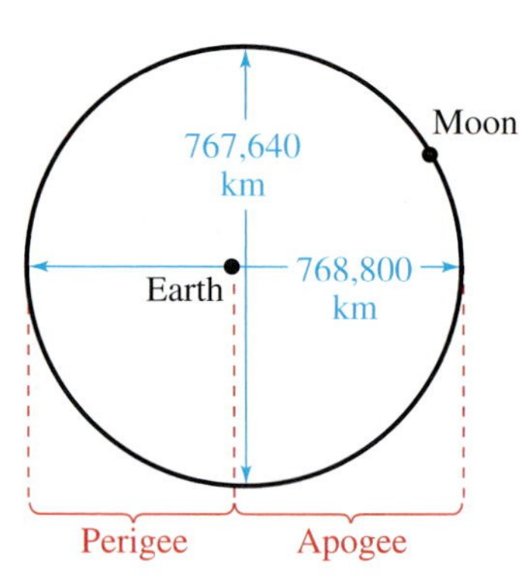

FIGURE 6.26

The moon travels about Earth in an elliptical orbit with Earth at one focus, as shown in Figure 6.26. The major and minor axes of the orbit have lengths of 768,800 kilometers and 767,640 kilometers, respectively. Find the greatest and smallest distances (the *apogee* and *perigee*), respectively from Earth's center to the moon's center.

Solution

Because $2a = 768{,}800$ and $2b = 767{,}640$, you have

$$a = 384{,}400 \text{ and } b = 383{,}820$$

which implies that

$$\begin{aligned} c &= \sqrt{a^2 - b^2} \\ &= \sqrt{384{,}400^2 - 383{,}820^2} \\ &\approx 21{,}108. \end{aligned}$$

So, the greatest distance between the center of Earth and the center of the moon is

$$a + c \approx 384{,}400 + 21{,}108 = 405{,}508 \text{ kilometers}$$

and the smallest distance is

$$a - c \approx 384{,}400 - 21{,}108 = 363{,}292 \text{ kilometers.}$$

Now try Exercise 59.

STUDY TIP

Note in Example 4 and Figure 6.26 that Earth *is not* the center of the moon's orbit.

Eccentricity

One of the reasons it was difficult for early astronomers to detect that the orbits of the planets are ellipses is that the foci of the planetary orbits are relatively close to their centers, and so the orbits are nearly circular. To measure the ovalness of an ellipse, you can use the concept of **eccentricity.**

Ask students to make a conjecture about the eccentricity of a circle before going further.

Definition of Eccentricity

The **eccentricity** e of an ellipse is given by the ratio

$$e = \frac{c}{a}.$$

Note that $0 < e < 1$ for *every* ellipse.

Activities

1. Find the center, foci, vertices, and eccentricity of the ellipse
$\frac{(x-2)^2}{25} + \frac{(y+1)^2}{9} = 1.$
Answer: Center $(2, -1)$; Foci $(-2, -1), (6, -1)$; Vertices $(-3, -1), (7, -1)$; $e = \frac{4}{5}$
2. Rewrite the equation of the ellipse in standard form:
$9x^2 + 5y^2 + 36x - 30y + 36 = 0.$
Answer: $\frac{(x+2)^2}{(\sqrt{5})^2} + \frac{(y-3)^2}{3^2} = 1$

To see how this ratio is used to describe the shape of an ellipse, note that because the foci of an ellipse are located along the major axis between the vertices and the center, it follows that

$$0 < c < a.$$

For an ellipse that is nearly circular, the foci are close to the center and the ratio c/a is small, as shown in Figure 6.27. On the other hand, for an elongated ellipse, the foci are close to the vertices, and the ratio c/a is close to 1, as shown in Figure 6.28.

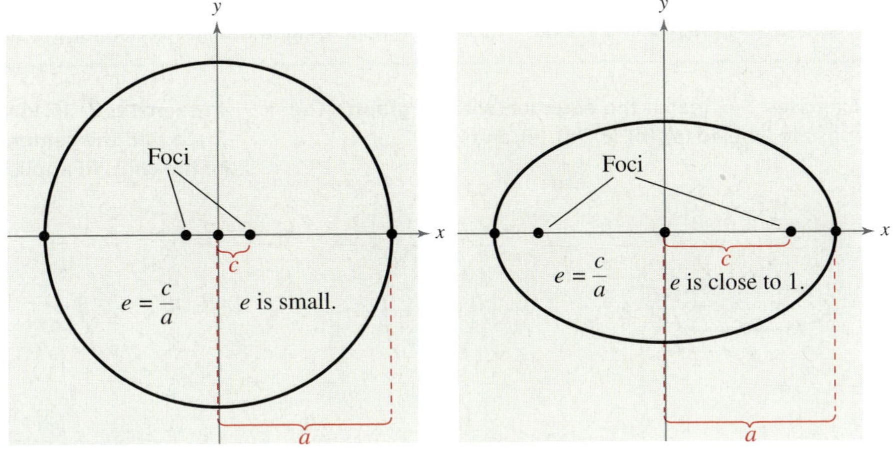

FIGURE 6.27

FIGURE 6.28

The time it takes Saturn to orbit the sun is equal to 29.4 Earth years.

The orbit of the moon has an eccentricity of $e \approx 0.0549$, and the eccentricities of the nine planetary orbits are as follows.

Mercury:	$e \approx 0.2056$	Saturn:	$e \approx 0.0542$
Venus:	$e \approx 0.0068$	Uranus:	$e \approx 0.0472$
Earth:	$e \approx 0.0167$	Neptune:	$e \approx 0.0086$
Mars:	$e \approx 0.0934$	Pluto:	$e \approx 0.2488$
Jupiter:	$e \approx 0.0484$		

Writing about Mathematics

Ellipses and Circles

a. Show that the equation of an ellipse can be written as

$$\frac{(x-h)^2}{a^2} + \frac{(y-k)^2}{a^2(1-e^2)} = 1.$$

b. For the equation in part (a), let $a = 4$, $h = 1$, and $k = 2$, and use a graphing utility to graph the ellipse for $e = 0.95$, $e = 0.75$, $e = 0.5$, $e = 0.25$, and $e = 0.1$. Discuss the changes in the shape of the ellipse as e approaches 0.

c. Make a conjecture about the shape of the graph in part (b) when $e = 0$. What is the equation of this ellipse? What is another name for an ellipse with an eccentricity of 0?

6.3 Exercises

VOCABULARY CHECK: Fill in the blanks.

1. An ________ is the set of all points (x, y) in a plane, the sum of whose distances from two distinct fixed points, called ________, is constant.
2. The chord joining the vertices of an ellipse is called the ________ ________, and its midpoint is the ________ of the ellipse.
3. The chord perpendicular to the major axis at the center of the ellipse is called the ________ ________ of the ellipse.
4. The concept of ________ is used to measure the ovalness of an ellipse.

PREREQUISITE SKILLS REVIEW: Practice and review algebra skills needed for this section at **www.Eduspace.com.**

In Exercises 1–6, match the equation with its graph. [The graphs are labeled (a), (b), (c), (d), (e), and (f).]

(a)

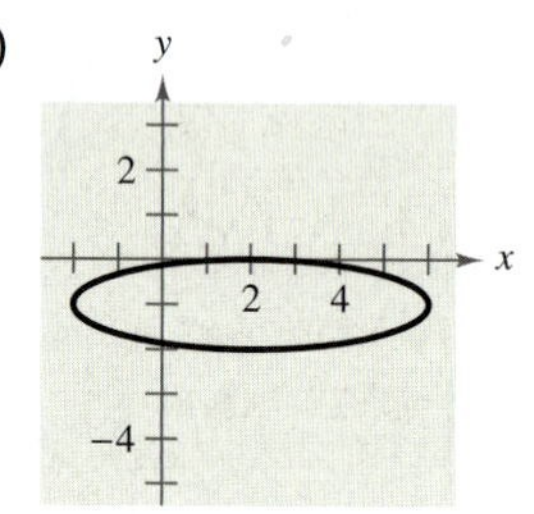

(b)

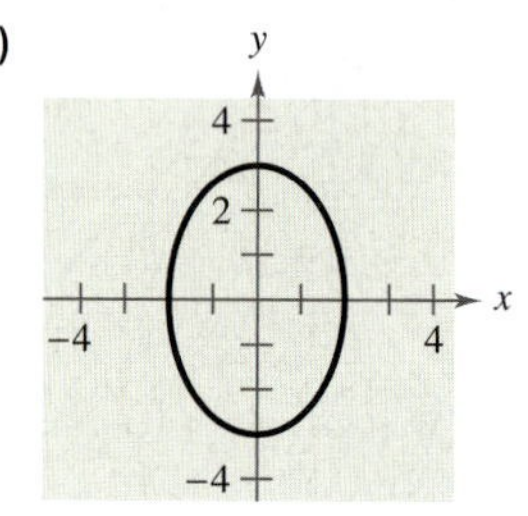

(c)

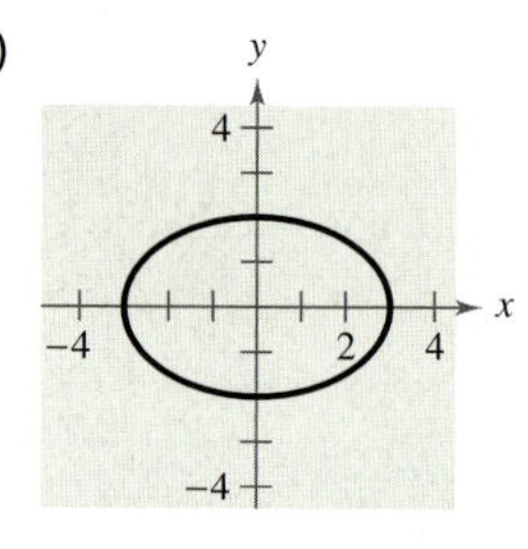

(d)

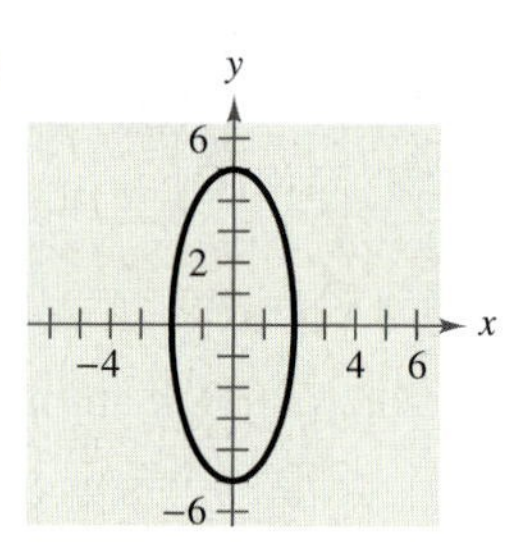

(e)

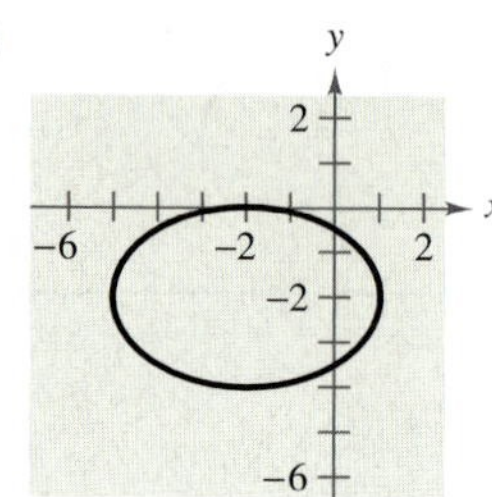

(f)

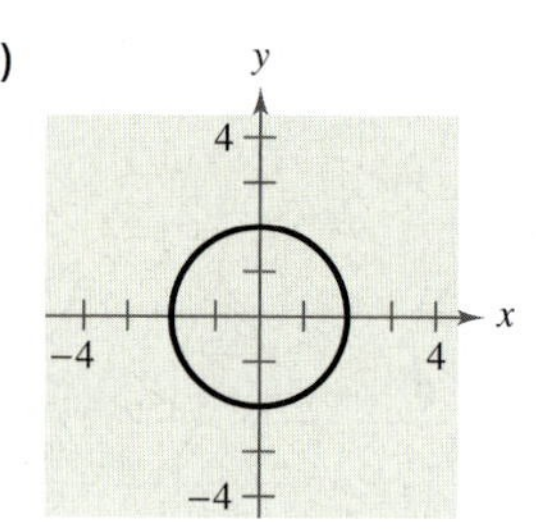

1. $\dfrac{x^2}{4} + \dfrac{y^2}{9} = 1$
2. $\dfrac{x^2}{9} + \dfrac{y^2}{4} = 1$
3. $\dfrac{x^2}{4} + \dfrac{y^2}{25} = 1$
4. $\dfrac{x^2}{4} + \dfrac{y^2}{4} = 1$
5. $\dfrac{(x-2)^2}{16} + (y+1)^2 = 1$
6. $\dfrac{(x+2)^2}{9} + \dfrac{(y+2)^2}{4} = 1$

In Exercises 7–30, identify the conic as a circle or an ellipse. Then find the center, radius, vertices, foci, and eccentricity of the conic (if applicable), and sketch its graph.

7. $\dfrac{x^2}{25} + \dfrac{y^2}{16} = 1$
8. $\dfrac{x^2}{81} + \dfrac{y^2}{144} = 1$
9. $\dfrac{x^2}{25} + \dfrac{y^2}{25} = 1$
10. $\dfrac{x^2}{9} + \dfrac{y^2}{9} = 1$
11. $\dfrac{x^2}{5} + \dfrac{y^2}{9} = 1$
12. $\dfrac{x^2}{64} + \dfrac{y^2}{28} = 1$
13. $\dfrac{(x+3)^2}{16} + \dfrac{(y-5)^2}{25} = 1$
14. $\dfrac{(x-4)^2}{12} + \dfrac{(y+3)^2}{16} = 1$
15. $\dfrac{x^2}{4/9} + \dfrac{(y+1)^2}{4/9} = 1$
16. $\dfrac{(x+5)^2}{9/4} + (y-1)^2 = 1$
17. $(x+2)^2 + \dfrac{(y+4)^2}{1/4} = 1$
18. $\dfrac{(x-3)^2}{25/4} + \dfrac{(y-1)^2}{25/4} = 1$
19. $9x^2 + 4y^2 + 36x - 24y + 36 = 0$
20. $9x^2 + 4y^2 - 54x + 40y + 37 = 0$
21. $x^2 + y^2 - 2x + 4y - 31 = 0$
22. $x^2 + 5y^2 - 8x - 30y - 39 = 0$
23. $3x^2 + y^2 + 18x - 2y - 8 = 0$
24. $6x^2 + 2y^2 + 18x - 10y + 2 = 0$
25. $x^2 + 4y^2 - 6x + 20y - 2 = 0$
26. $x^2 + y^2 - 4x + 6y - 3 = 0$
27. $9x^2 + 9y^2 + 18x - 18y + 14 = 0$
28. $16x^2 + 25y^2 - 32x + 50y + 16 = 0$
29. $9x^2 + 25y^2 - 36x - 50y + 60 = 0$
30. $16x^2 + 16y^2 - 64x + 32y + 55 = 0$

In Exercises 31–34, use a graphing utility to graph the ellipse. Find the center, foci, and vertices. (Recall that it may be necessary to solve the equation for *y* and obtain two equations.)

31. $5x^2 + 3y^2 = 15$
32. $3x^2 + 4y^2 = 12$
33. $12x^2 + 20y^2 - 12x + 40y - 37 = 0$
34. $36x^2 + 9y^2 + 48x - 36y - 72 = 0$

In Exercises 35–42, find the standard form of the equation of the ellipse with the given characteristics and center at the origin.

35.

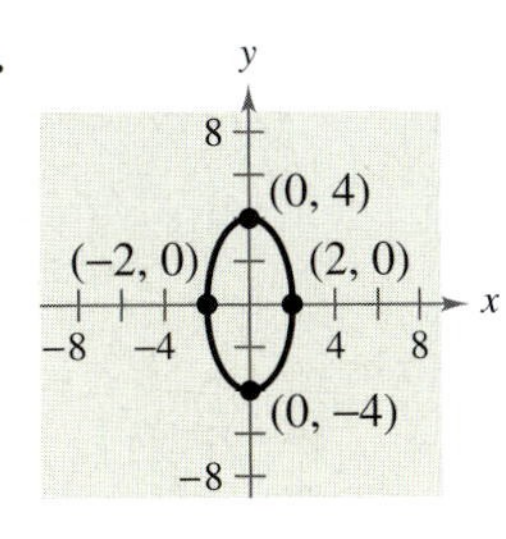

36.

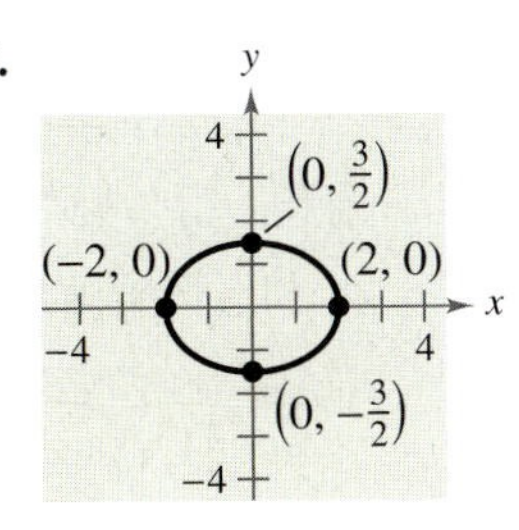

37. Vertices: $(\pm 6, 0)$; foci: $(\pm 2, 0)$

38. Vertices: $(0, \pm 8)$; foci: $(0, \pm 4)$

39. Foci: $(\pm 5, 0)$; major axis of length 12

40. Foci: $(\pm 2, 0)$; major axis of length 8

41. Vertices: $(0, \pm 5)$; passes through the point $(4, 2)$

42. Major axis vertical; passes through the points $(0, 4)$ and $(2, 0)$

In Exercises 43–54, find the standard form of the equation of the ellipse with the given characteristics.

43.

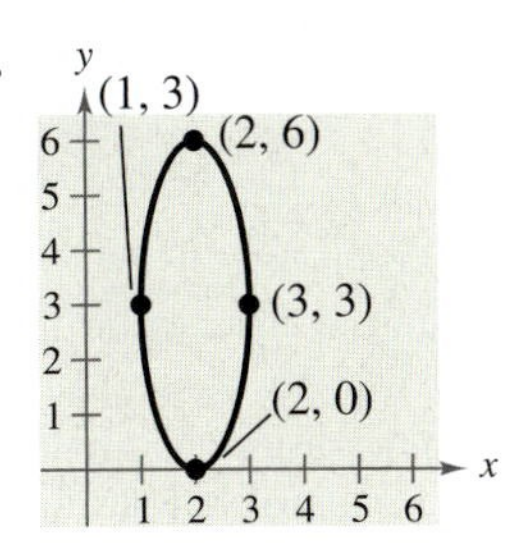

44.

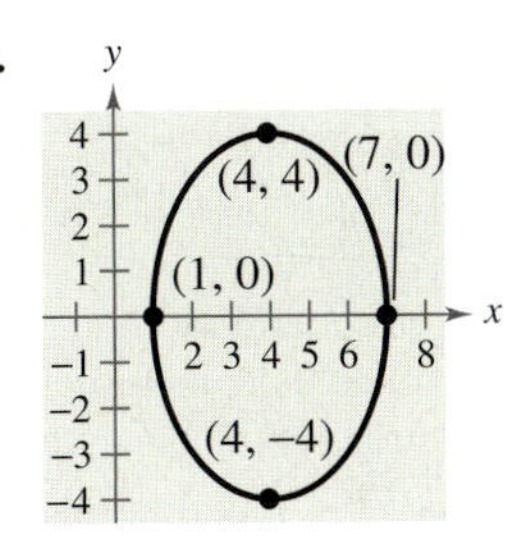

45.

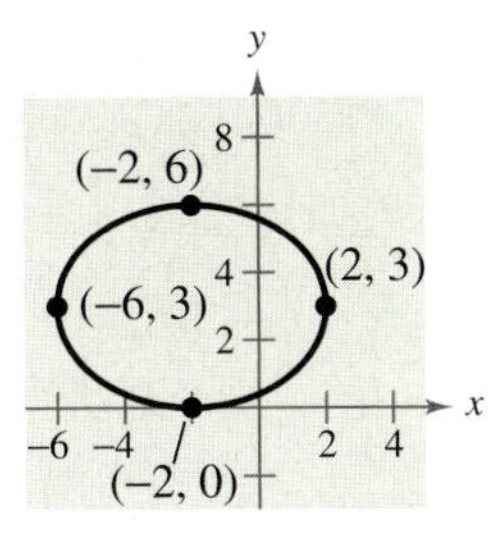

46.

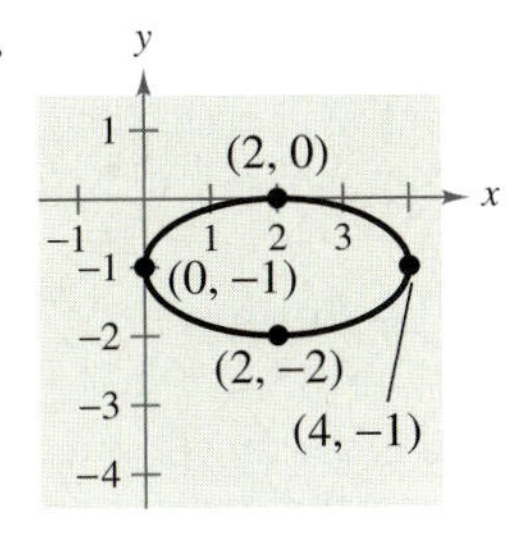

47. Vertices: $(0, 4)$, $(4, 4)$; minor axis of length 2

48. Foci: $(0, 0)$, $(4, 0)$; major axis of length 8

49. Foci: $(0, 0)$, $(0, 8)$; major axis of length 16

50. Center: $(2, -1)$; vertex: $\left(2, \frac{1}{2}\right)$; minor axis of length 2

51. Center: $(0, 4)$; $a = 2c$; vertices: $(-4, 4)$, $(4, 4)$

52. Center: $(3, 2)$; $a = 3c$; foci: $(1, 2)$, $(5, 2)$

53. Vertices: $(0, 2)$, $(4, 2)$; endpoints of the minor axis: $(2, 3)$, $(2, 1)$

54. Vertices: $(5, 0)$, $(5, 12)$; endpoints of the minor axis: $(1, 6)$, $(9, 6)$

55. Find an equation of the ellipse with vertices $(\pm 5, 0)$ and eccentricity $e = \frac{3}{5}$.

56. Find an equation of the ellipse with vertices $(0, \pm 8)$ and eccentricity $e = \frac{1}{2}$.

57. ***Architecture*** A semielliptical arch over a tunnel for a one-way road through a mountain has a major axis of 50 feet and a height at the center of 10 feet.

(a) Draw a rectangular coordinate system on a sketch of the tunnel with the center of the road entering the tunnel at the origin. Identify the coordinates of the known points.

(b) Find an equation of the semielliptical arch over the tunnel.

(c) You are driving a moving truck that has a width of 8 feet and a height of 9 feet. Will the moving truck clear the opening of the arch?

58. ***Architecture*** A fireplace arch is to be constructed in the shape of a semiellipse. The opening is to have a height of 2 feet at the center and a width of 6 feet along the base (see figure). The contractor draws the outline of the ellipse using tacks as described at the beginning of this section. Give the required positions of the tacks and the length of the string.

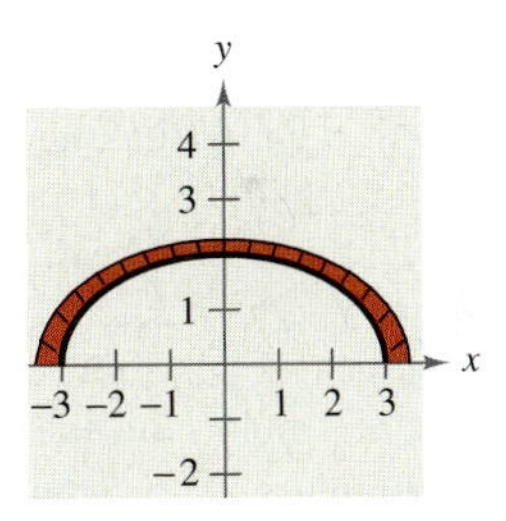

Model It

59. ***Comet Orbit*** Halley's comet has an elliptical orbit, with the sun at one focus. The eccentricity of the orbit is approximately 0.967. The length of the major axis of the orbit is approximately 35.88 astronomical units. (An astronomical unit is about 93 million miles.)

(a) Find an equation of the orbit. Place the center of the orbit at the origin, and place the major axis on the x-axis.

 (b) Use a graphing utility to graph the equation of the orbit.

(c) Find the greatest (aphelion) and smallest (perihelion) distances from the sun's center to the comet's center.

60. ***Satellite Orbit*** The first artificial satellite to orbit Earth was Sputnik I (launched by the former Soviet Union in 1957). Its highest point above Earth's surface was 947 kilometers, and its lowest point was 228 kilometers (see figure). The center of Earth was the focus of the elliptical orbit, and the radius of Earth is 6378 kilometers. Find the eccentricity of the orbit.

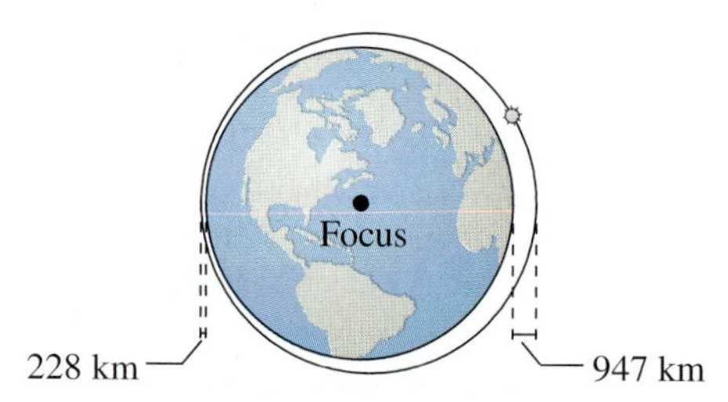

61. ***Motion of a Pendulum*** The relation between the velocity y (in radians per second) of a pendulum and its angular displacement θ from the vertical can be modeled by a semiellipse. A 12-centimeter pendulum crests $(y = 0)$ when the angular displacement is -0.2 radian and 0.2 radian. When the pendulum is at equilibrium $(\theta = 0)$, the velocity is -1.6 radians per second.

(a) Find an equation that models the motion of the pendulum. Place the center at the origin.

(b) Graph the equation from part (a).

(c) Which half of the ellipse models the motion of the pendulum?

62. ***Geometry*** A line segment through a focus of an ellipse with endpoints on the ellipse and perpendicular to the major axis is called a **latus rectum** of the ellipse. Therefore, an ellipse has two latera recta. Knowing the length of the latera recta is helpful in sketching an ellipse because it yields other points on the curve (see figure). Show that the length of each latus rectum is $2b^2/a$.

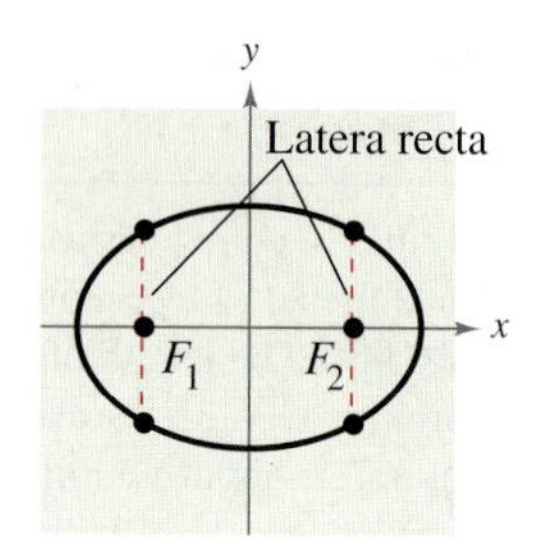

In Exercises 63–66, sketch the graph of the ellipse, using latera recta (see Exercise 62).

63. $\dfrac{x^2}{9} + \dfrac{y^2}{16} = 1$

64. $\dfrac{x^2}{4} + \dfrac{y^2}{1} = 1$

65. $5x^2 + 3y^2 = 15$

66. $9x^2 + 4y^2 = 36$

Synthesis

True or False? **In Exercises 67 and 68, determine whether the statement is true or false. Justify your answer.**

67. The graph of $x^2 + 4y^4 - 4 = 0$ is an ellipse.

68. It is easier to distinguish the graph of an ellipse from the graph of a circle if the eccentricity of the ellipse is large (close to 1).

69. ***Exploration*** Consider the ellipse

$$\frac{x^2}{a^2} + \frac{y^2}{b^2} = 1, \quad a + b = 20.$$

(a) The area of the ellipse is given by $A = \pi ab$. Write the area of the ellipse as a function of a.

(b) Find the equation of an ellipse with an area of 264 square centimeters.

(c) Complete the table using your equation from part (a), and make a conjecture about the shape of the ellipse with maximum area.

a	8	9	10	11	12	13
A						

(d) Use a graphing utility to graph the area function and use the graph to support your conjecture in part (c).

70. ***Think About It*** At the beginning of this section it was noted that an ellipse can be drawn using two thumbtacks, a string of fixed length (greater than the distance between the two tacks), and a pencil. If the ends of the string are fastened at the tacks and the string is drawn taut with a pencil, the path traced by the pencil is an ellipse.

(a) What is the length of the string in terms of a?

(b) Explain why the path is an ellipse.

Skills Review

In Exercises 71–74, use the power-reducing formulas to rewrite the expression in terms of the first power of the cosine.

71. $\sin^6 x$

72. $\cos^6 x$

73. $\cos^4 2x$

74. $\sin^4 2x$

In Exercises 75–78, find the area of the triangle with the given characteristics.

75. $A = 35°, b = 24, c = 52$

76. $a = 7.1, b = 4.8, c = 10.2$

77. $a = 14.4, b = 9.8, c = 9.8$

78. $C = 108°, a = 8, b = 12$

6.4 Hyperbolas

What you should learn

- Write equations of hyperbolas in standard form.
- Find asymptotes of and graph hyperbolas.
- Use properties of hyperbolas to solve real-life problems.
- Classify conics from their general equations.

Why you should learn it

Hyperbolas can be used to model and solve many types of real-life problems. For instance, in Exercise 42 on page 483, hyperbolas are used in long distance radio navigation for aircraft and ships.

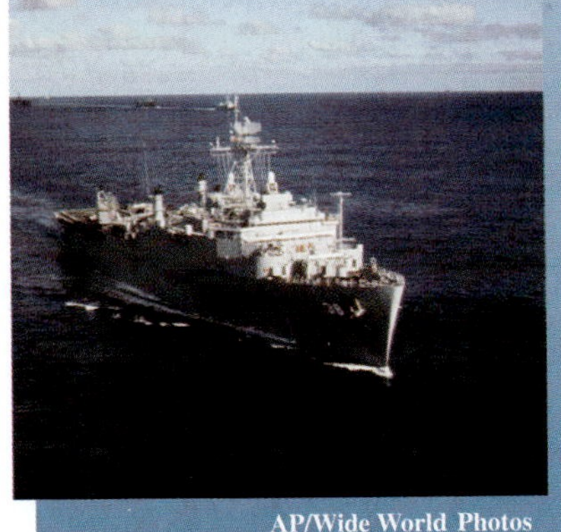

AP/Wide World Photos

Introduction

The third type of conic is called a **hyperbola.** The definition of a hyperbola is similar to that of an ellipse. The difference is that for an ellipse the *sum* of the distances between the foci and a point on the ellipse is fixed, whereas for a hyperbola the *difference* of the distances between the foci and a point on the hyperbola is fixed.

Definition of Hyperbola

A **hyperbola** is the set of all points (x, y) in a plane, the difference of whose distances from two distinct fixed points **(foci)** is a positive constant. See Figure 6.29.

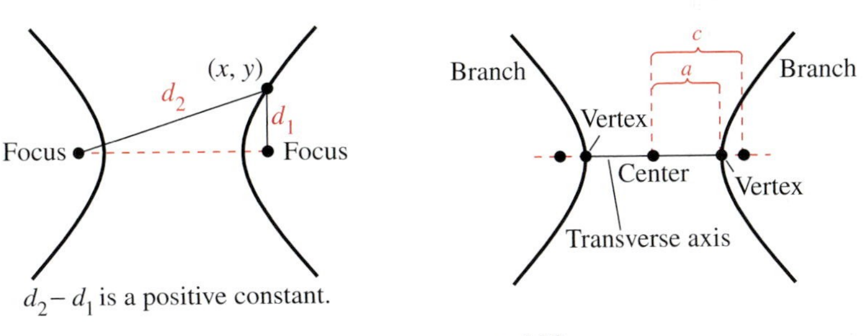

FIGURE 6.29

FIGURE 6.30

The graph of a hyperbola has two disconnected **branches.** The line through the two foci intersects the hyperbola at its two **vertices.** The line segment connecting the vertices is the **transverse axis,** and the midpoint of the transverse axis is the **center** of the hyperbola. See Figure 6.30. The development of the standard form of the equation of a hyperbola is similar to that of an ellipse. Note in the definition below that a, b, and c are related differently for hyperbolas than for ellipses.

Standard Equation of a Hyperbola

The **standard form of the equation of a hyperbola** with center (h, k) is

$$\frac{(x-h)^2}{a^2} - \frac{(y-k)^2}{b^2} = 1 \qquad \text{Transverse axis is horizontal.}$$

$$\frac{(y-k)^2}{a^2} - \frac{(x-h)^2}{b^2} = 1. \qquad \text{Transverse axis is vertical.}$$

The vertices are a units from the center, and the foci are c units from the center. Moreover, $c^2 = a^2 + b^2$. If the center of the hyperbola is at the origin $(0, 0)$, the equation takes one of the following forms.

$$\frac{x^2}{a^2} - \frac{y^2}{b^2} = 1 \qquad \text{Transverse axis is horizontal.} \qquad \frac{y^2}{a^2} - \frac{x^2}{b^2} = 1 \qquad \text{Transverse axis is vertical.}$$

Figure 6.31 shows both the horizontal and vertical orientations for a hyperbola.

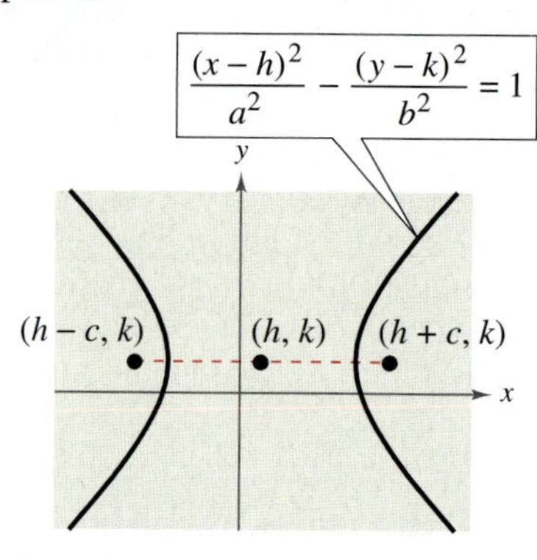

Transverse axis is horizontal.

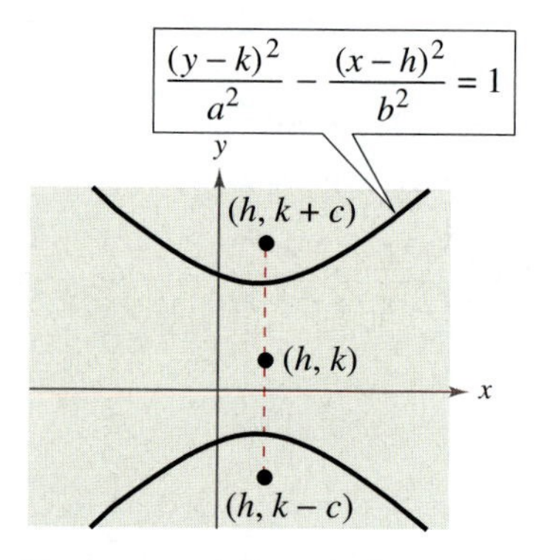

Transverse axis is vertical.

FIGURE 6.31

Example 1 Finding the Standard Equation of a Hyperbola

Find the standard form of the equation of the hyperbola with foci $(-1, 2)$ and $(5, 2)$ and vertices $(0, 2)$ and $(4, 2)$.

Solution

By the Midpoint Formula, the center of the hyperbola occurs at the point $(2, 2)$. Furthermore, $c = 5 - 2 = 3$ and $a = 4 - 2 = 2$, and it follows that

$$b = \sqrt{c^2 - a^2} = \sqrt{3^2 - 2^2} = \sqrt{9 - 4} = \sqrt{5}.$$

So, the hyperbola has a horizontal transverse axis and the standard form of the equation is

$$\frac{(x-2)^2}{2^2} - \frac{(y-2)^2}{(\sqrt{5})^2} = 1.$$

See Figure 6.32.

This equation simplifies to

$$\frac{(x-2)^2}{4} - \frac{(y-2)^2}{5} = 1.$$

> **STUDY TIP**
>
> When finding the standard form of the equation of any conic, it is helpful to sketch a graph of the conic with the given characteristics.

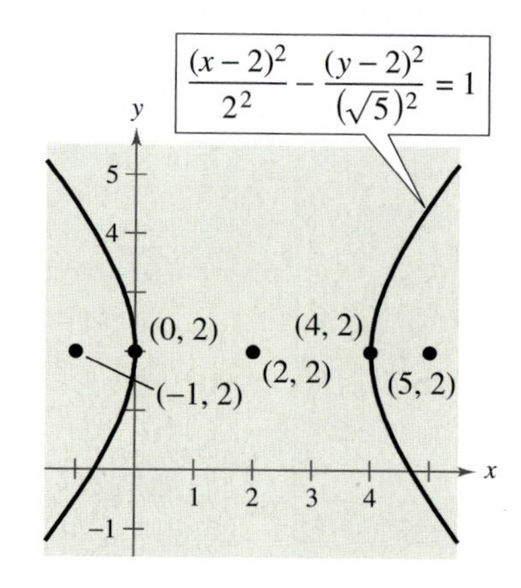

FIGURE 6.32

✓CHECKPOINT Now try Exercise 27.

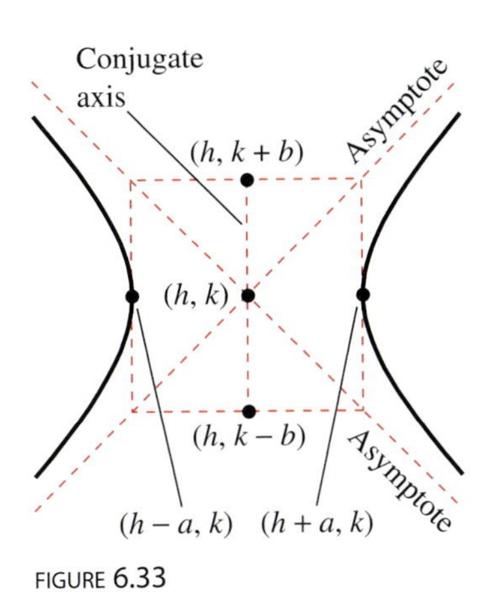

FIGURE 6.33

Asymptotes of a Hyperbola

Each hyperbola has two **asymptotes** that intersect at the center of the hyperbola, as shown in Figure 6.33. The asymptotes pass through the vertices of a rectangle of dimensions $2a$ by $2b$, with its center at (h, k). The line segment of length $2b$ joining $(h, k + b)$ and $(h, k - b)$ [or $(h + b, k)$ and $(h - b, k)$] is the **conjugate axis** of the hyperbola.

Asymptotes of a Hyperbola

The equations of the asymptotes of a hyperbola are

$$y = k \pm \frac{b}{a}(x - h) \qquad \text{Transverse axis is horizontal.}$$

$$y = k \pm \frac{a}{b}(x - h). \qquad \text{Transverse axis is vertical.}$$

Example 2 Using Asymptotes to Sketch a Hyperbola

Sketch the hyperbola whose equation is $4x^2 - y^2 = 16$.

Solution

Divide each side of the original equation by 16, and rewrite the equation in standard form.

$$\frac{x^2}{2^2} - \frac{y^2}{4^2} = 1 \qquad \text{Write in standard form.}$$

From this, you can conclude that $a = 2$, $b = 4$, and the transverse axis is horizontal. So, the vertices occur at $(-2, 0)$ and $(2, 0)$, and the endpoints of the conjugate axis occur at $(0, -4)$ and $(0, 4)$. Using these four points, you are able to sketch the rectangle shown in Figure 6.34. Now, from $c^2 = a^2 + b^2$, you have $c = \sqrt{2^2 + 4^2} = \sqrt{20} = 2\sqrt{5}$. So, the foci of the hyperbola are $\left(-2\sqrt{5}, 0\right)$ and $\left(2\sqrt{5}, 0\right)$. Finally, by drawing the asymptotes through the corners of this rectangle, you can complete the sketch shown in Figure 6.35. Note that the asymptotes are $y = 2x$ and $y = -2x$.

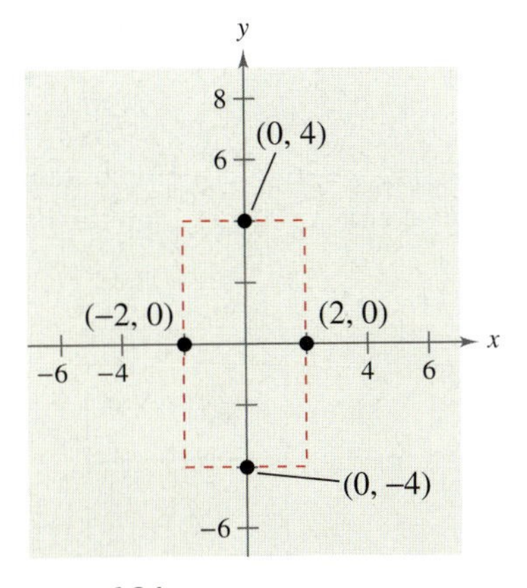

FIGURE 6.34

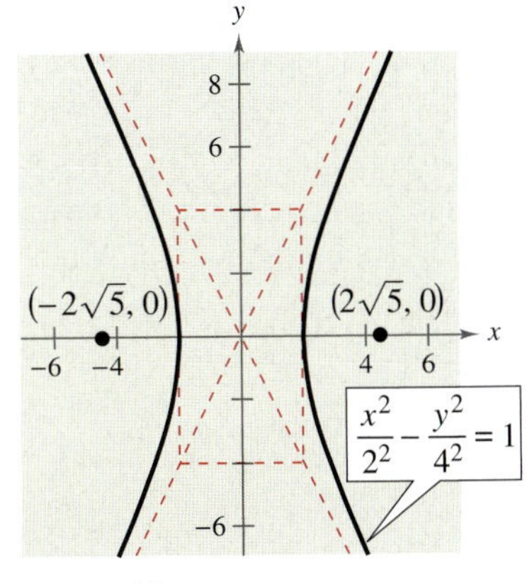

FIGURE 6.35

CHECKPOINT Now try Exercise 7.

Example 3 Finding the Asymptotes of a Hyperbola

Sketch the hyperbola given by $4x^2 - 3y^2 + 8x + 16 = 0$ and find the equations of its asymptotes and the foci.

Solution

$$4x^2 - 3y^2 + 8x + 16 = 0 \qquad \text{Write original equation.}$$

$$(4x^2 + 8x) - 3y^2 = -16 \qquad \text{Group terms.}$$

$$4(x^2 + 2x) - 3y^2 = -16 \qquad \text{Factor 4 from } x\text{-terms.}$$

$$4(x^2 + 2x + 1) - 3y^2 = -16 + 4 \qquad \text{Add 4 to each side.}$$

$$4(x + 1)^2 - 3y^2 = -12 \qquad \text{Write in completed square form.}$$

$$-\frac{(x + 1)^2}{3} + \frac{y^2}{4} = 1 \qquad \text{Divide each side by } -12.$$

$$\frac{y^2}{2^2} - \frac{(x + 1)^2}{(\sqrt{3})^2} = 1 \qquad \text{Write in standard form.}$$

From this equation you can conclude that the hyperbola has a vertical transverse axis, centered at $(-1, 0)$, has vertices $(-1, 2)$ and $(-1, -2)$, and has a conjugate axis with endpoints $(-1 - \sqrt{3}, 0)$ and $(-1 + \sqrt{3}, 0)$. To sketch the hyperbola, draw a rectangle through these four points. The asymptotes are the lines passing through the corners of the rectangle. Using $a = 2$ and $b = \sqrt{3}$, you can conclude that the equations of the asymptotes are

$$y = \frac{2}{\sqrt{3}}(x + 1) \qquad \text{and} \qquad y = -\frac{2}{\sqrt{3}}(x + 1).$$

Finally, you can determine the foci by using the equation $c^2 = a^2 + b^2$. So, youhave $c = \sqrt{2^2 + (\sqrt{3})^2} = \sqrt{7}$, and the foci are $(-1, -2 - \sqrt{7})$ and $(-1, -2 + \sqrt{7})$. The hyperbola is shown in Figure 6.36.

FIGURE 6.36

CHECKPOINT Now try Exercise 13.

Technology

You can use a graphing utility to graph a hyperbola by graphing the upper and lower portions in the same viewing window. For instance, to graph the hyperbola in Example 3, first solve for y to get

$$y_1 = 2\sqrt{1 + \frac{(x + 1)^2}{3}} \qquad \text{and} \qquad y_2 = -2\sqrt{1 + \frac{(x + 1)^2}{3}}.$$

Use a viewing window in which $-9 \le x \le 9$ and $-6 \le y \le 6$. You should obtain the graph shown below. Notice that the graphing utility does not draw the asymptotes. However, if you trace along the branches, you will see that the values of the hyperbola approach the asymptotes.

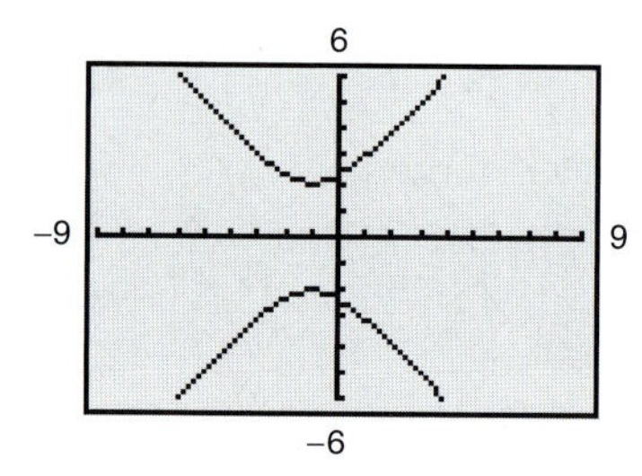

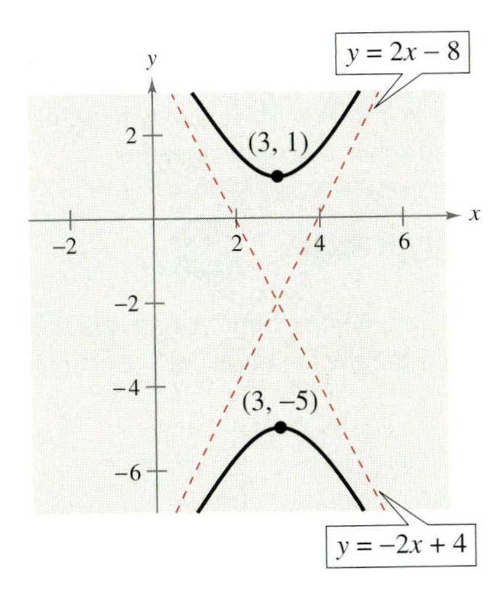

FIGURE 6.37

Example 4 Using Asymptotes to Find the Standard Equation

Find the standard form of the equation of the hyperbola having vertices $(3, -5)$ and $(3, 1)$ and having asymptotes

$$y = 2x - 8 \quad \text{and} \quad y = -2x + 4$$

as shown in Figure 6.37.

Solution

By the Midpoint Formula, the center of the hyperbola is $(3, -2)$. Furthermore, the hyperbola has a vertical transverse axis with $a = 3$. From the original equations, you can determine the slopes of the asymptotes to be

$$m_1 = 2 = \frac{a}{b} \quad \text{and} \quad m_2 = -2 = -\frac{a}{b}$$

and, because $a = 3$ you can conclude

$$2 = \frac{a}{b} \quad \Rightarrow \quad 2 = \frac{3}{b} \quad \Rightarrow \quad b = \frac{3}{2}.$$

So, the standard form of the equation is

$$\frac{(y+2)^2}{3^2} - \frac{(x-3)^2}{\left(\frac{3}{2}\right)^2} = 1.$$

CHECKPOINT Now try Exercise 35.

As with ellipses, the *eccentricity* of a hyperbola is

$$e = \frac{c}{a} \qquad \text{Eccentricity}$$

and because $c > a$, it follows that $e > 1$. If the eccentricity is large, the branches of the hyperbola are nearly flat, as shown in Figure 6.38. If the eccentricity is close to 1, the branches of the hyperbola are more narrow, as shown in Figure 6.39.

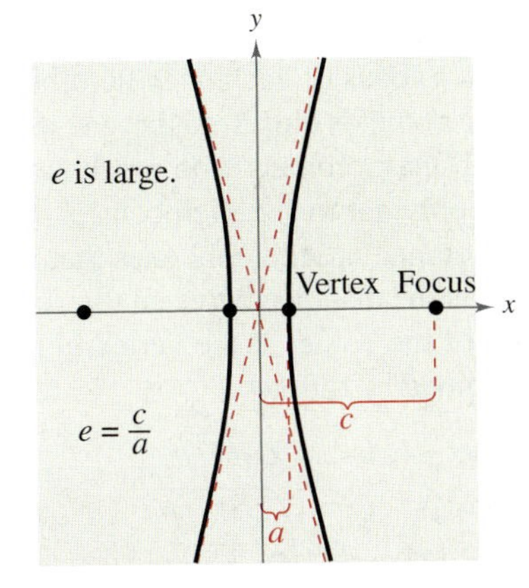

FIGURE 6.38

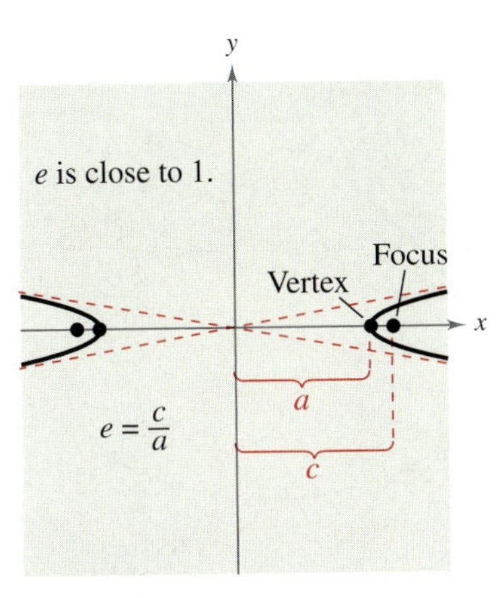

FIGURE 6.39

Additional Example

Find the standard form of the equation of the hyperbola having vertices $(-4, 2)$ and $(1, 2)$ and having foci $(-7, 2)$ and $(4, 2)$.

Answer: $\dfrac{\left(x+\frac{3}{2}\right)^2}{\left(\frac{5}{2}\right)^2} - \dfrac{(y-2)^2}{(\sqrt{24})^2} = 1$

Applications

The following application was developed during World War II. It shows how the properties of hyperbolas can be used in radar and other detection systems.

Example 5 An Application Involving Hyperbolas

Two microphones, 1 mile apart, record an explosion. Microphone A receives the sound 2 seconds before microphone B. Where did the explosion occur? (Assume sound travels at 1100 feet per second.)

Solution

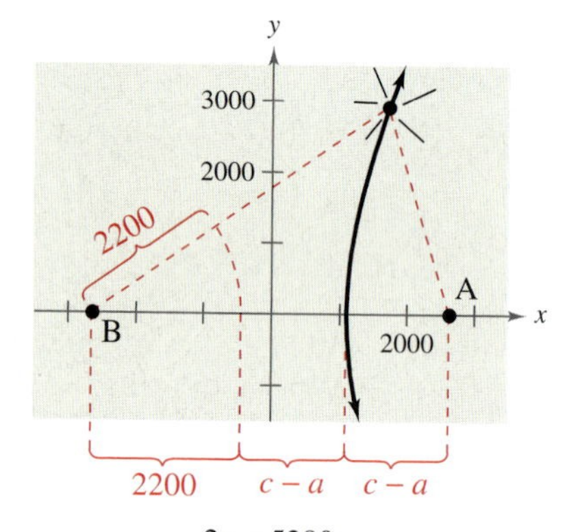

FIGURE 6.40

Assuming sound travels at 1100 feet per second, you know that the explosion took place 2200 feet farther from B than from A, as shown in Figure 6.40. The locus of all points that are 2200 feet closer to A than to B is one branch of the hyperbola

$$\frac{x^2}{a^2} - \frac{y^2}{b^2} = 1$$

where

$$c = \frac{5280}{2} = 2640$$

and

$$a = \frac{2200}{2} = 1100.$$

So, $b^2 = c^2 - a^2 = 2640^2 - 1100^2 = 5{,}759{,}600$, and you can conclude that the explosion occurred somewhere on the right branch of the hyperbola

$$\frac{x^2}{1{,}210{,}000} - \frac{y^2}{5{,}759{,}600} = 1.$$

CHECKPOINT Now try Exercise 41.

Another interesting application of conic sections involves the orbits of comets in our solar system. Of the 610 comets identified prior to 1970, 245 have elliptical orbits, 295 have parabolic orbits, and 70 have hyperbolic orbits. The center of the sun is a focus of each of these orbits, and each orbit has a vertex at the point where the comet is closest to the sun, as shown in Figure 6.41. Undoubtedly, there have been many comets with parabolic or hyperbolic orbits that were not identified. We only get to see such comets *once*. Comets with elliptical orbits, such as Halley's comet, are the only ones that remain in our solar system.

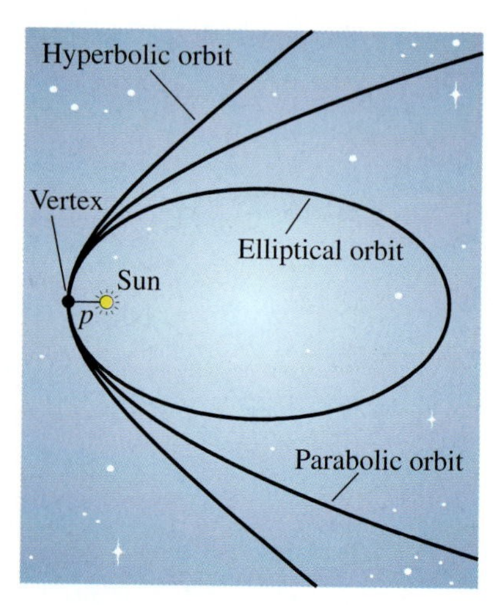

FIGURE 6.41

If p is the distance between the vertex and the focus (in meters), and v is the velocity of the comet at the vertex in (meters per second), then the type of orbit is determined as follows.

1. Ellipse: $v < \sqrt{2GM/p}$
2. Parabola: $v = \sqrt{2GM/p}$
3. Hyperbola: $v > \sqrt{2GM/p}$

In each of these relations, $M = 1.989 \times 10^{30}$ kilograms (the mass of the sun) and $G \approx 6.67 \times 10^{-11}$ cubic meter per kilogram-second squared (the universal gravitational constant).

Activities

1. Find the standard form of the equation of the hyperbola with asymptotes $y = \pm 2x$ and vertices $(0, \pm 2)$.

   Answer: $\frac{y^2}{2^2} - \frac{x^2}{1^2} = 1$

2. Classify the graph of each equation.
   a. $3x^2 - 2y^2 + 4y - 3 = 0$
   b. $2y^2 - 3x + 2 = 0$
   c. $x^2 + 4y^2 - 2x - 3 = 0$
   d. $x^2 - 2x + 4y - 1 = 0$

   Answer: (a) Hyperbola, (b) Parabola, (c) Ellipse, (d) Parabola

General Equations of Conics

Classifying a Conic from Its General Equation

The graph of $Ax^2 + Cy^2 + Dx + Ey + F = 0$ is one of the following.

1. *Circle:* $A = C$
2. *Parabola:* $AC = 0$ — $A = 0$ or $C = 0$, but not both.
3. *Ellipse:* $AC > 0$ — A and C have like signs.
4. *Hyperbola:* $AC < 0$ — A and C have unlike signs.

The test above is valid *if* the graph is a conic. The test does not apply to equations such as $x^2 + y^2 = -1$, whose graph is not a conic.

Example 6 Classifying Conics from General Equations

Classify the graph of each equation.

a. $4x^2 - 9x + y - 5 = 0$
b. $4x^2 - y^2 + 8x - 6y + 4 = 0$
c. $2x^2 + 4y^2 - 4x + 12y = 0$
d. $2x^2 + 2y^2 - 8x + 12y + 2 = 0$

Solution

a. For the equation $4x^2 - 9x + y - 5 = 0$, you have

$$AC = 4(0) = 0. \qquad \text{Parabola}$$

So, the graph is a parabola.

b. For the equation $4x^2 - y^2 + 8x - 6y + 4 = 0$, you have

$$AC = 4(-1) < 0. \qquad \text{Hyperbola}$$

So, the graph is a hyperbola.

c. For the equation $2x^2 + 4y^2 - 4x + 12y = 0$, you have

$$AC = 2(4) > 0. \qquad \text{Ellipse}$$

So, the graph is an ellipse.

d. For the equation $2x^2 + 2y^2 - 8x + 12y + 2 = 0$, you have

$$A = C = 2. \qquad \text{Circle}$$

So, the graph is a circle.

✓CHECKPOINT Now try Exercise 49.

The Granger Collection

Historical Note

Caroline Herschel (1750–1848) was the first woman to be credited with detecting a new comet. During her long life, this English astronomer discovered a total of eight new comets.

WRITING ABOUT MATHEMATICS

Sketching Conics Sketch each of the conics described in Example 6. Write a paragraph describing the procedures that allow you to sketch the conics efficiently.

6.4 Exercises

VOCABULARY CHECK: Fill in the blanks.

1. A ________ is the set of all points (x, y) in a plane, the difference of whose distances from two distinct fixed points, called ________, is a positive constant.
2. The graph of a hyperbola has two disconnected parts called ________.
3. The line segment connecting the vertices of a hyperbola is called the ________ ________, and the midpoint of the line segment is the ________ of the hyperbola.
4. Each hyperbola has two ________ that intersect at the center of the hyperbola.
5. The general form of the equation of a conic is given by ________.

PREREQUISITE SKILLS REVIEW: Practice and review algebra skills needed for this section at **www.Eduspace.com.**

In Exercises 1–4, match the equation with its graph. [The graphs are labeled (a), (b), (c), and (d).]

(a)

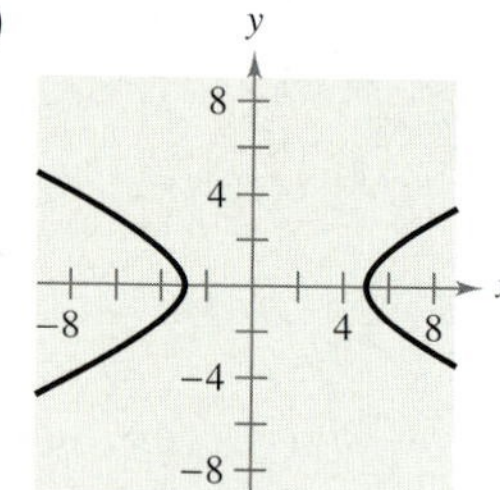

(b)

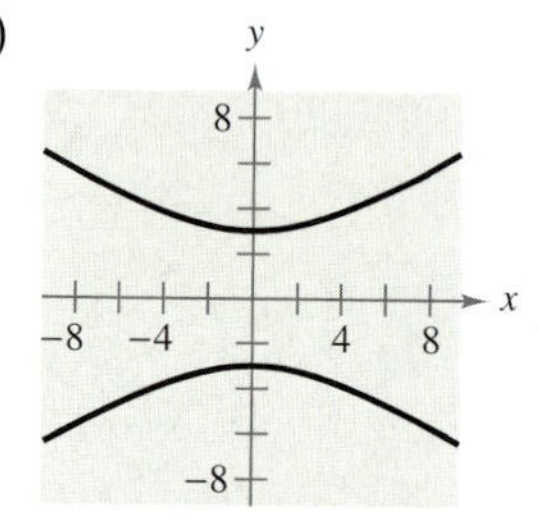

(c)

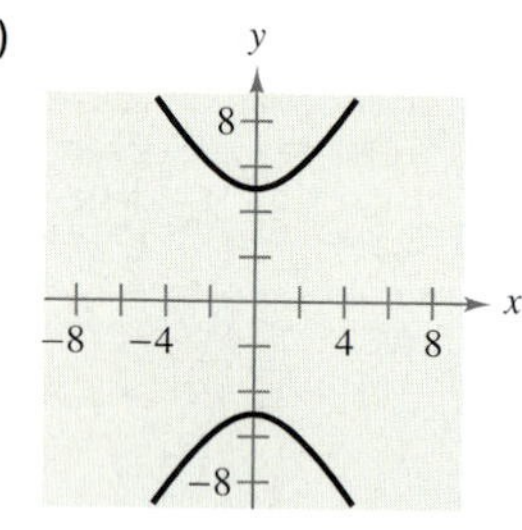

(d)

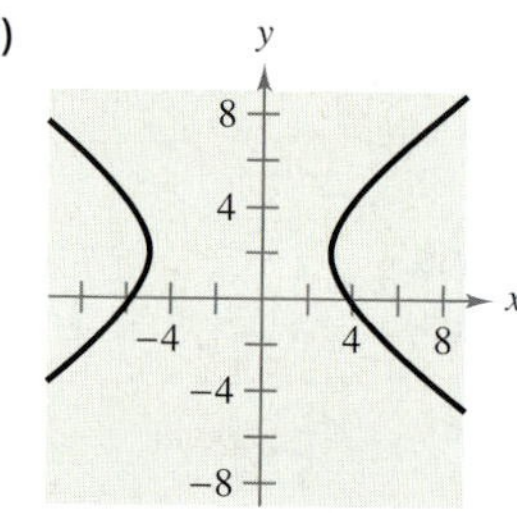

1. $\dfrac{y^2}{9} - \dfrac{x^2}{25} = 1$
2. $\dfrac{y^2}{25} - \dfrac{x^2}{9} = 1$
3. $\dfrac{(x-1)^2}{16} - \dfrac{y^2}{4} = 1$
4. $\dfrac{(x+1)^2}{16} - \dfrac{(y-2)^2}{9} = 1$

In Exercises 5–16, find the center, vertices, foci, and the equations of the asymptotes of the hyperbola, and sketch its graph using the asymptotes as an aid.

5. $x^2 - y^2 = 1$
6. $\dfrac{x^2}{9} - \dfrac{y^2}{25} = 1$
7. $\dfrac{y^2}{25} - \dfrac{x^2}{81} = 1$
8. $\dfrac{x^2}{36} - \dfrac{y^2}{4} = 1$
9. $\dfrac{(x-1)^2}{4} - \dfrac{(y+2)^2}{1} = 1$
10. $\dfrac{(x+3)^2}{144} - \dfrac{(y-2)^2}{25} = 1$
11. $\dfrac{(y+6)^2}{1/9} - \dfrac{(x-2)^2}{1/4} = 1$
12. $\dfrac{(y-1)^2}{1/4} - \dfrac{(x+3)^2}{1/16} = 1$
13. $9x^2 - y^2 - 36x - 6y + 18 = 0$
14. $x^2 - 9y^2 + 36y - 72 = 0$
15. $x^2 - 9y^2 + 2x - 54y - 80 = 0$
16. $16y^2 - x^2 + 2x + 64y + 63 = 0$

In Exercises 17–20, find the center, vertices, foci, and the equations of the asymptotes of the hyperbola. Use a graphing utility to graph the hyperbola and its asymptotes.

17. $2x^2 - 3y^2 = 6$
18. $6y^2 - 3x^2 = 18$
19. $9y^2 - x^2 + 2x + 54y + 62 = 0$
20. $9x^2 - y^2 + 54x + 10y + 55 = 0$

In Exercises 21–26, find the standard form of the equation of the hyperbola with the given characteristics and center at the origin.

21. Vertices: $(0, \pm 2)$; foci: $(0, \pm 4)$
22. Vertices: $(\pm 4, 0)$; foci: $(\pm 6, 0)$
23. Vertices: $(\pm 1, 0)$; asymptotes: $y = \pm 5x$
24. Vertices: $(0, \pm 3)$; asymptotes: $y = \pm 3x$
25. Foci: $(0, \pm 8)$; asymptotes: $y = \pm 4x$
26. Foci: $(\pm 10, 0)$; asymptotes: $y = \pm \frac{3}{4}x$

In Exercises 27–38, find the standard form of the equation of the hyperbola with the given characteristics.

27. Vertices: $(2, 0), (6, 0)$; foci: $(0, 0), (8, 0)$
28. Vertices: $(2, 3), (2, -3)$; foci: $(2, 6), (2, -6)$

29. Vertices: $(4, 1), (4, 9)$; foci: $(4, 0), (4, 10)$

30. Vertices: $(-2, 1), (2, 1)$; foci: $(-3, 1), (3, 1)$

31. Vertices: $(2, 3), (2, -3)$;
passes through the point $(0, 5)$

32. Vertices: $(-2, 1), (2, 1)$;
passes through the point $(5, 4)$

33. Vertices: $(0, 4), (0, 0)$;
passes through the point $(\sqrt{5}, -1)$

34. Vertices: $(1, 2), (1, -2)$;
passes through the point $(0, \sqrt{5})$

35. Vertices: $(1, 2), (3, 2)$;
asymptotes: $y = x,\ y = 4 - x$

36. Vertices: $(3, 0), (3, 6)$;
asymptotes: $y = 6 - x,\ y = x$

37. Vertices: $(0, 2), (6, 2)$;
asymptotes: $y = \frac{2}{3}x,\ y = 4 - \frac{2}{3}x$

38. Vertices: $(3, 0), (3, 4)$;
asymptotes: $y = \frac{2}{3}x,\ y = 4 - \frac{2}{3}x$

39. ***Art*** A sculpture has a hyperbolic cross section (see figure).

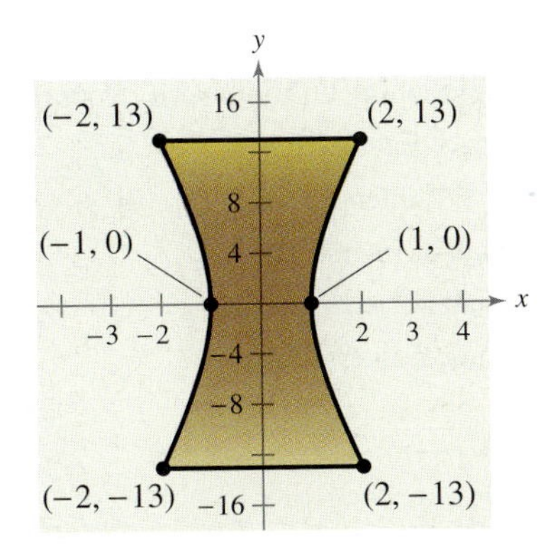

(a) Write an equation that models the curved sides of the sculpture.

(b) Each unit in the coordinate plane represents 1 foot. Find the width of the sculpture at a height of 5 feet.

40. ***Sound Location*** You and a friend live 4 miles apart (on the same "east-west" street) and are talking on the phone. You hear a clap of thunder from lightning in a storm, and 18 seconds later your friend hears the thunder. Find an equation that gives the possible places where the lightning could have occurred. (Assume that the coordinate system is measured in feet and that sound travels at 1100 feet per second.)

41. ***Sound Location*** Three listening stations located at $(3300, 0)$, $(3300, 1100)$, and $(-3300, 0)$ monitor an explosion. The last two stations detect the explosion 1 second and 4 seconds after the first, respectively. Determine the coordinates of the explosion. (Assume that the coordinate system is measured in feet and that sound travels at 100 feet per second.)

Model It

42. ***LORAN*** Long distance radio navigation for aircraft and ships uses synchronized pulses transmitted by widely separated transmitting stations. These pulses travel at the speed of light (186,000 miles per second). The difference in the times of arrival of these pulses at an aircraft or ship is constant on a hyperbola having the transmitting stations as foci. Assume that two stations, 300 miles apart, are positioned on the rectangular coordinate system at points with coordinates $(-150, 0)$ and $(150, 0)$, and that a ship is traveling on a hyperbolic path with coordinates $(x, 75)$ (see figure).

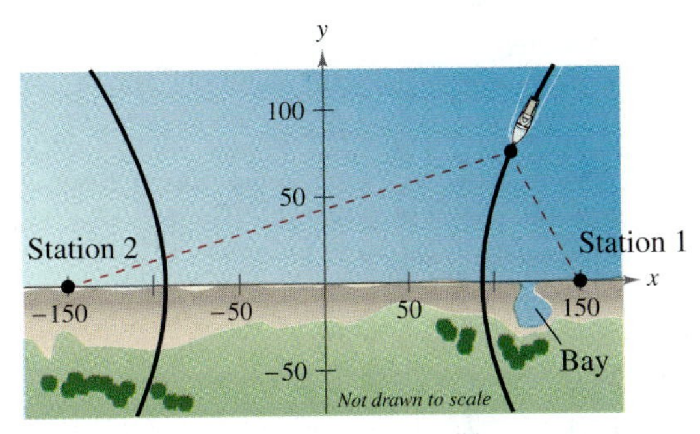

(a) Find the x-coordinate of the position of the ship if the time difference between the pulses from the transmitting stations is 1000 microseconds (0.001 second).

(b) Determine the distance between the ship and station 1 when the ship reaches the shore.

(c) The ship wants to enter a bay located between the two stations. The bay is 30 miles from station 1. What should the time difference be between the pulses?

(d) The ship is 60 miles offshore when the time difference in part (c) is obtained. What is the position of the ship?

43. ***Hyperbolic Mirror*** A hyperbolic mirror (used in some telescopes) has the property that a light ray directed at a focus will be reflected to the other focus. The focus of a hyperbolic mirror (see figure) has coordinates (24, 0). Find the vertex of the mirror if the mount at the top edge of the mirror has coordinates (24, 24).

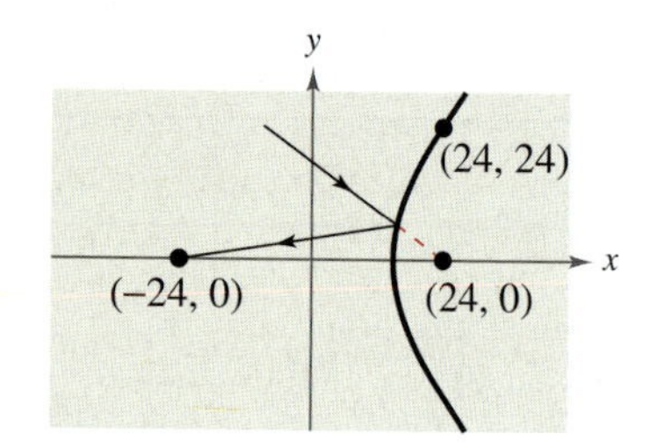

44. ***Running Path*** Let (0, 0) represent a water fountain located in a city park. Each day you run through the park along a path given by

$$x^2 + y^2 - 200x - 52{,}500 = 0$$

where x and y are measured in meters.

(a) What type of conic is your path? Explain your reasoning.

(b) Write the equation of the path in standard form. Sketch a graph of the equation.

(c) After you run, you walk to the water fountain. If you stop running at $(-100, 150)$, how far must you walk for a drink of water?

In Exercises 45–60, classify the graph of the equation as a circle, a parabola, an ellipse, or a hyperbola.

45. $x^2 + y^2 - 6x + 4y + 9 = 0$

46. $x^2 + 4y^2 - 6x + 16y + 21 = 0$

47. $4x^2 - y^2 - 4x - 3 = 0$

48. $y^2 - 6y - 4x + 21 = 0$

49. $y^2 - 4x^2 + 4x - 2y - 4 = 0$

50. $x^2 + y^2 - 4x + 6y - 3 = 0$

51. $x^2 - 4x - 8y + 2 = 0$

52. $4x^2 + y^2 - 8x + 3 = 0$

53. $4x^2 + 3y^2 + 8x - 24y + 51 = 0$

54. $4y^2 - 2x^2 - 4y - 8x - 15 = 0$

55. $25x^2 - 10x - 200y - 119 = 0$

56. $4y^2 + 4x^2 - 24x + 35 = 0$

57. $4x^2 + 16y^2 - 4x - 32y + 1 = 0$

58. $2y^2 + 2x + 2y + 1 = 0$

59. $100x^2 + 100y^2 - 100x + 400y + 409 = 0$

60. $4x^2 - y^2 + 4x + 2y - 1 = 0$

Synthesis

True or False? **In Exercises 61 and 62, determine whether the statement is true or false. Justify your answer.**

61. In the standard form of the equation of a hyperbola, the larger the ratio of b to a, the larger the eccentricity of the hyperbola.

62. In the standard form of the equation of a hyperbola, the trivial solution of two intersecting lines occurs when $b = 0$.

63. Consider a hyperbola centered at the origin with a horizontal transverse axis. Use the definition of a hyperbola to derive its standard form.

64. ***Writing*** Explain how the central rectangle of a hyperbola can be used to sketch its asymptotes.

65. ***Think About It*** Change the equation of the hyperbola so that its graph is the bottom half of the hyperbola.

$$9x^2 - 54x - 4y^2 + 8y + 41 = 0$$

66. ***Exploration*** A circle and a parabola can have 0, 1, 2, 3, or 4 points of intersection. Sketch the circle given by $x^2 + y^2 = 4$. Discuss how this circle could intersect a parabola with an equation of the form $y = x^2 + C$. Then find the values of C for each of the five cases described below. Use a graphing utility to verify your results.

(a) No points of intersection

(b) One point of intersection

(c) Two points of intersection

(d) Three points of intersection

(e) Four points of intersection

Skills Review

In Exercises 67–72, factor the polynomial completely.

67. $x^3 - 16x$

68. $x^2 + 14x + 49$

69. $2x^3 - 24x^2 + 72x$

70. $6x^3 - 11x^2 - 10x$

71. $16x^3 + 54$

72. $4 - x + 4x^2 - x^3$

In Exercises 73–76, sketch a graph of the function. Include two full periods.

73. $y = 2\cos x + 1$

74. $y = \sin \pi x$

75. $y = \tan 2x$

76. $y = -\frac{1}{2}\sec x$

6.5 Parametric Equations

What you should learn

- Evaluate sets of parametric equations for given values of the parameter.
- Sketch curves that are represented by sets of parametric equations.
- Rewrite sets of parametric equations as single rectangular equations by eliminating the parameter.
- Find sets of parametric equations for graphs.

Why you should learn it

Parametric equations are useful for modeling the path of an object. For instance, in Exercise 59 on page 491, you will use a set of parametric equations to model the path of a baseball.

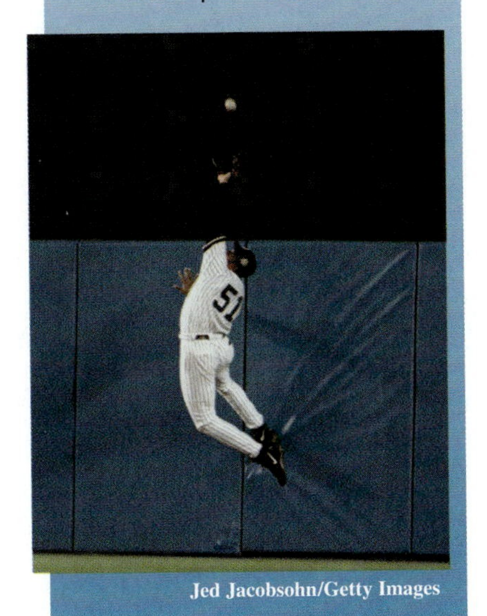

Jed Jacobsohn/Getty Images

Plane Curves

Up to this point you have been representing a graph by a single equation involving the *two* variables x and y. In this section, you will study situations in which it is useful to introduce a *third* variable to represent a curve in the plane.

To see the usefulness of this procedure, consider the path followed by an object that is propelled into the air at an angle of 45°. If the initial velocity of the object is 48 feet per second, it can be shown that the object follows the parabolic path

$$y = -\frac{x^2}{72} + x \qquad \text{Rectangular equation}$$

as shown in Figure 6.42. However, this equation does not tell the whole story. Although it does tell you *where* the object has been, it doesn't tell you *when* the object was at a given point (x, y) on the path. To determine this time, you can introduce a third variable t, called a **parameter.** It is possible to write both x and y as functions of t to obtain the **parametric equations**

$$x = 24\sqrt{2}t \qquad \text{Parametric equation for } x$$

$$y = -16t^2 + 24\sqrt{2}t. \qquad \text{Parametric equation for } y$$

From this set of equations you can determine that at time $t = 0$, the object is at the point $(0, 0)$. Similarly, at time $t = 1$, the object is at the point $\left(24\sqrt{2}, 24\sqrt{2} - 16\right)$, and so on, as shown in Figure 6.42.

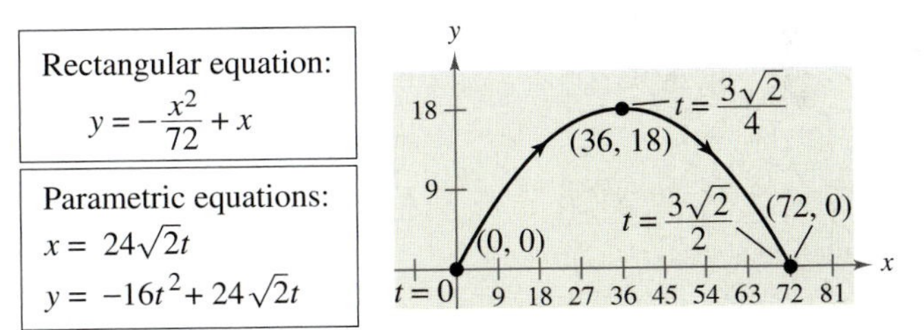

Curvilinear Motion: Two Variables for Position, One Variable for Time

FIGURE 6.42

For this particular motion problem, x and y are continuous functions of t, and the resulting path is a **plane curve.** (Recall that a *continuous function* is one whose graph can be traced without lifting the pencil from the paper.)

Definition of Plane Curve

If f and g are continuous functions of t on an interval I, the set of ordered pairs $(f(t), g(t))$ is a **plane curve** C. The equations

$$x = f(t) \qquad \text{and} \qquad y = g(t)$$

are **parametric equations** for C, and t is the **parameter.**

Point out to your students the importance of knowing the orientation of a curve, and thus the usefulness of parametric equations.

Sketching a Plane Curve

When sketching a curve represented by a pair of parametric equations, you still plot points in the xy-plane. Each set of coordinates (x, y) is determined from a value chosen for the parameter t. Plotting the resulting points in the order of *increasing* values of t traces the curve in a specific direction. This is called the **orientation** of the curve.

Example 1 Sketching a Curve

Sketch the curve given by the parametric equations

$$x = t^2 - 4 \quad \text{and} \quad y = \frac{t}{2}, \quad -2 \le t \le 3.$$

Solution

Using values of t in the interval, the parametric equations yield the points (x, y) shown in the table.

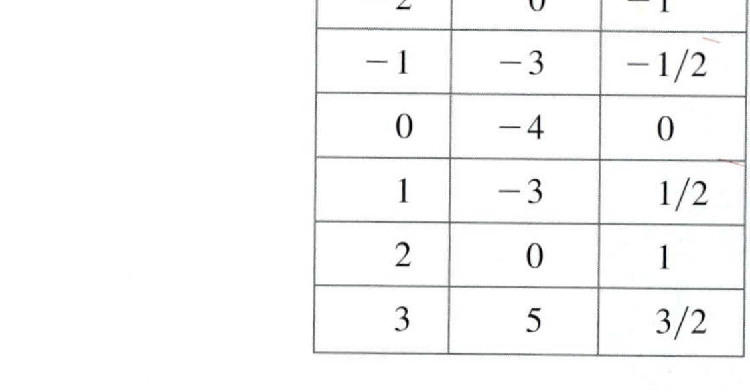

t	x	y
-2	0	-1
-1	-3	$-1/2$
0	-4	0
1	-3	$1/2$
2	0	1
3	5	$3/2$

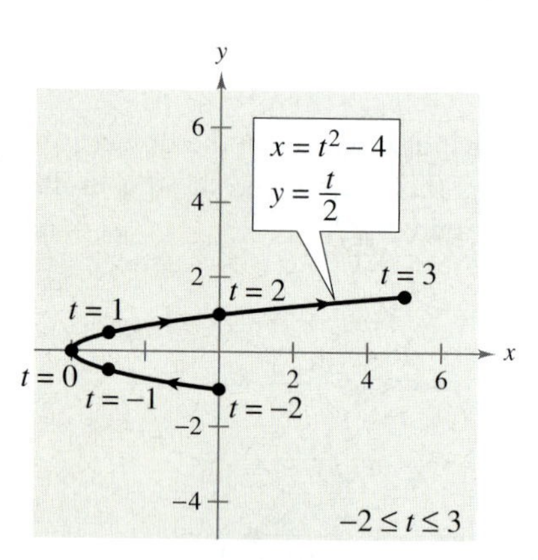

FIGURE 6.43

By plotting these points in the order of increasing t, you obtain the curve C shown in Figure 6.43. Note that the arrows on the curve indicate its orientation as t increases from -2 to 3. So, if a particle were moving on this curve, it would start at $(0, -1)$ and then move along the curve to the point $\left(5, \frac{3}{2}\right)$.

CHECKPOINT Now try Exercises 1(a) and (b).

Note that the graph shown in Figure 6.43 does not define y as a function of x. This points out one benefit of parametric equations—they can be used to represent graphs that are more general than graphs of functions.

It often happens that two different sets of parametric equations have the same graph. For example, the set of parametric equations

$$x = 4t^2 - 4 \quad \text{and} \quad y = t, \quad -1 \le t \le \frac{3}{2}$$

has the same graph as the set given in Example 1. However, by comparing the values of t in Figures 6.43 and 6.44, you see that this second graph is traced out more *rapidly* (considering t as time) than the first graph. So, in applications, different parametric representations can be used to represent various *speeds* at which objects travel along a given path.

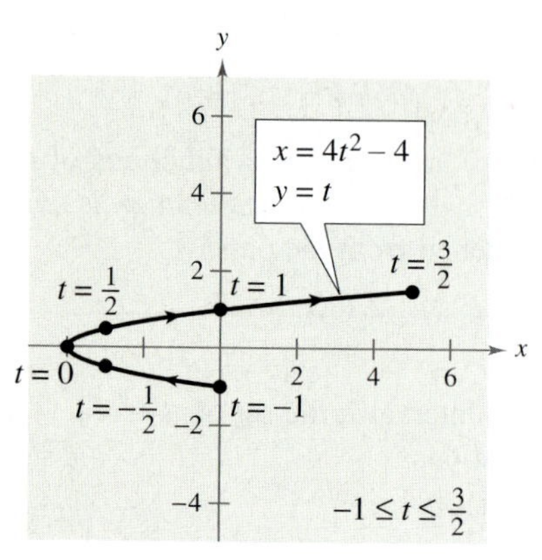

FIGURE 6.44

Eliminating the Parameter

Example 1 uses simple point plotting to sketch the curve. This tedious process can sometimes be simplified by finding a rectangular equation (in x and y) that has the same graph. This process is called **eliminating the parameter.**

Parametric equations	→	Solve for t in one equation.	→	Substitute in other equation.	→	Rectangular equation
$x = t^2 - 4$ $y = t/2$		$t = 2y$		$x = (2y)^2 - 4$		$x = 4y^2 - 4$

Emphasize that converting equations from parametric to rectangular form is primarily an aid in graphing.

Now you can recognize that the equation $x = 4y^2 - 4$ represents a parabola with a horizontal axis and vertex $(-4, 0)$.

When converting equations from parametric to rectangular form, you may need to alter the domain of the rectangular equation so that its graph matches the graph of the parametric equations. Such a situation is demonstrated in Example 2.

Exploration

Most graphing utilities have a *parametric* mode. If yours does, enter the parametric equations from Example 2. Over what values should you let t vary to obtain the graph shown in Figure 6.45?

Example 2 Eliminating the Parameter

Sketch the curve represented by the equations

$$x = \frac{1}{\sqrt{t+1}} \quad \text{and} \quad y = \frac{t}{t+1}$$

by eliminating the parameter and adjusting the domain of the resulting rectangular equation.

Solution

Solving for t in the equation for x produces

$$x = \frac{1}{\sqrt{t+1}} \quad \Longrightarrow \quad x^2 = \frac{1}{t+1}$$

which implies that

$$t = \frac{1 - x^2}{x^2}.$$

Now, substituting in the equation for y, you obtain the rectangular equation

$$y = \frac{t}{t+1} = \frac{\frac{(1-x^2)}{x^2}}{\left[\frac{(1-x^2)}{x^2}\right] + 1} = \frac{\frac{1-x^2}{x^2}}{\frac{1-x^2}{x^2} + 1} \cdot \frac{x^2}{x^2} = 1 - x^2.$$

From this rectangular equation, you can recognize that the curve is a parabola that opens downward and has its vertex at $(0, 1)$. Also, this rectangular equation is defined for all values of x, but from the parametric equation for x you can see that the curve is defined only when $t > -1$. This implies that you should restrict the domain of x to positive values, as shown in Figure 6.45.

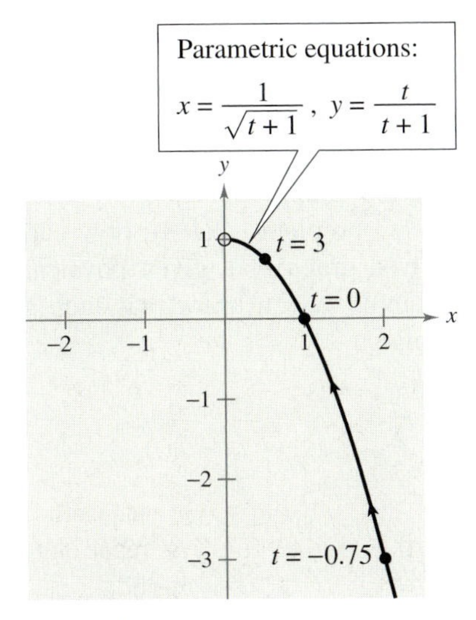

FIGURE 6.45

CHECKPOINT Now try Exercise 1(c).

STUDY TIP

To eliminate the parameter in equations involving trigonometric functions, try using the identities

$$\sin^2 \theta + \cos^2 \theta = 1$$

$$\sec^2 \theta - \tan^2 \theta = 1$$

as shown in Example 3.

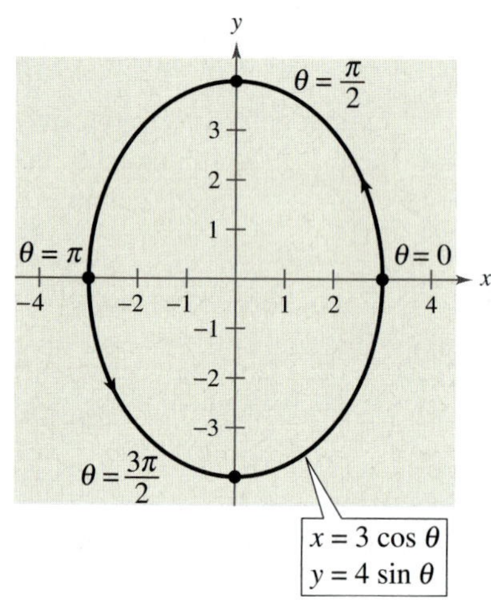

FIGURE 6.46

It is not necessary for the parameter in a set of parametric equations to represent time. The next example uses an *angle* as the parameter.

Example 3 **Eliminating an Angle Parameter**

Sketch the curve represented by

$$x = 3 \cos \theta \quad \text{and} \quad y = 4 \sin \theta, \quad 0 \le \theta \le 2\pi$$

by eliminating the parameter.

Solution

Begin by solving for $\cos \theta$ and $\sin \theta$ in the equations.

$$\cos \theta = \frac{x}{3} \quad \text{and} \quad \sin \theta = \frac{y}{4}$$ Solve for $\cos \theta$ and $\sin \theta$.

Use the identity $\sin^2 \theta + \cos^2 \theta = 1$ to form an equation involving only x and y.

$$\cos^2 \theta + \sin^2 \theta = 1$$ Pythagorean identity

$$\left(\frac{x}{3}\right)^2 + \left(\frac{y}{4}\right)^2 = 1$$ Substitute $\frac{x}{3}$ for $\cos \theta$ and $\frac{y}{4}$ for $\sin \theta$.

$$\frac{x^2}{9} + \frac{y^2}{16} = 1$$ Rectangular equation

From this rectangular equation, you can see that the graph is an ellipse centered at $(0, 0)$, with vertices $(0, 4)$ and $(0, -4)$ and minor axis of length $2b = 6$, as shown in Figure 6.46. Note that the elliptic curve is traced out *counterclockwise* as θ varies from 0 to 2π.

✓CHECKPOINT Now try Exercise 13.

In Examples 2 and 3, it is important to realize that eliminating the parameter is primarily an *aid to curve sketching*. If the parametric equations represent the path of a moving object, the graph alone is not sufficient to describe the object's motion. You still need the parametric equations to tell you the *position*, *direction*, and *speed* at a given time.

Finding Parametric Equations for a Graph

You have been studying techniques for sketching the graph represented by a set of parametric equations. Now consider the *reverse* problem—that is, how can you find a set of parametric equations for a given graph or a given physical description? From the discussion following Example 1, you know that such a representation is not unique. That is, the equations

$$x = 4t^2 - 4 \quad \text{and} \quad y = t, \ -1 \le t \le \frac{3}{2}$$

produced the same graph as the equations

$$x = t^2 - 4 \quad \text{and} \quad y = \frac{t}{2}, \ -2 \le t \le 3.$$

This is further demonstrated in Example 4.

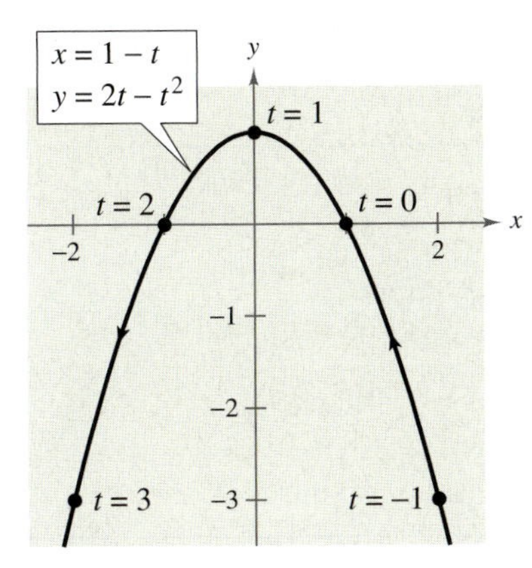

FIGURE 6.47

Example 4 Finding Parametric Equations for a Graph

Find a set of parametric equations to represent the graph of $y = 1 - x^2$, using the following parameters.

a. $t = x$ **b.** $t = 1 - x$

Solution

a. Letting $t = x$, you obtain the parametric equations

$$x = t \quad \text{and} \quad y = 1 - x^2 = 1 - t^2.$$

b. Letting $t = 1 - x$, you obtain the parametric equations

$$x = 1 - t \quad \text{and} \quad y = 1 - x^2 = 1 - (1 - t)^2 = 2t - t^2.$$

In Figure 6.47, note how the resulting curve is oriented by the increasing values of t. For part (a), the curve would have the opposite orientation.

CHECKPOINT Now try Exercise 37.

Point out that a single rectangular equation can have many different parametric representations. To reinforce this, demonstrate along with parts (a) and (b) of Example 4 the parametric equation representations of the graph of $y = 1 - x^2$ using the parameters $t = 2x$ and $t = 2 - 3x$. A graphing utility can be a helpful tool in demonstrating that each of these representations yields the same graph.

Example 5 Parametric Equations for a Cycloid

Describe the **cycloid** traced out by a point P on the circumference of a circle of radius a as the circle rolls along a straight line in a plane.

Solution

As the parameter, let θ be the measure of the circle's rotation, and let the point $P = (x, y)$ begin at the origin. When $\theta = 0$, P is at the origin; when $\theta = \pi$, P is at a maximum point $(\pi a, 2a)$; and when $\theta = 2\pi$, P is back on the x-axis at $(2\pi a, 0)$. From Figure 6.48, you can see that $\angle APC = 180° - \theta$. So, you have

$$\sin \theta = \sin(180° - \theta) = \sin(\angle APC) = \frac{AC}{a} = \frac{BD}{a}$$

$$\cos \theta = -\cos(180° - \theta) = -\cos(\angle APC) = \frac{AP}{-a}$$

which implies that $AP = -a \cos \theta$ and $BD = a \sin \theta$. Because the circle rolls along the x-axis, you know that $OD = \overparen{PD} = a\theta$. Furthermore, because $BA = DC = a$, you have

$$x = OD - BD = a\theta - a \sin \theta \quad \text{and} \quad y = BA + AP = a - a \cos \theta.$$

So, the parametric equations are $x = a(\theta - \sin \theta)$ and $y = a(1 - \cos \theta)$.

STUDY TIP

In Example 5, $\overparen{PD}$ represents the arc of the circle between points P and D.

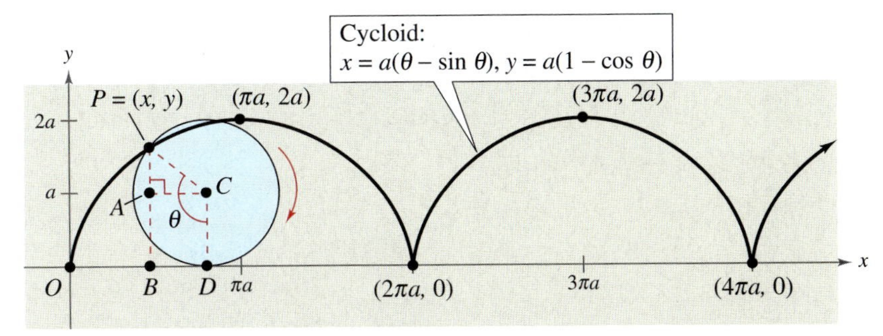

FIGURE 6.48

CHECKPOINT Now try Exercise 63.

Technology

Use a graphing utility in *parametric* mode to obtain a graph similar to Figure 6.48 by graphing the following equations.

$$X_{1T} = T - \sin T$$

$$Y_{1T} = 1 - \cos T$$

6.5 Exercises

VOCABULARY CHECK: Fill in the blanks.

1. If f and g are continuous functions of t on an interval I, the set of ordered pairs $(f(t), g(t))$ is a ________ ________ C. The equations $x = f(t)$ and $y = g(t)$ are ________ equations for C, and t is the ________.
2. The ________ of a curve is the direction in which the curve is traced out for increasing values of the parameter.
3. The process of converting a set of parametric equations to a corresponding rectangular equation is called ________ the ________.

PREREQUISITE SKILLS REVIEW: Practice and review algebra skills needed for this section at **www.Eduspace.com.**

1. Consider the parametric equations $x = \sqrt{t}$ and $y = 3 - t$.
   (a) Create a table of x- and y-values using $t = 0, 1, 2, 3,$ and 4.
   (b) Plot the points (x, y) generated in part (a), and sketch a graph of the parametric equations.
   (c) Find the rectangular equation by eliminating the parameter. Sketch its graph. How do the graphs differ?
2. Consider the parametric equations $x = 4\cos^2\theta$ and $y = 2\sin\theta$.
   (a) Create a table of x- and y-values using $\theta = -\pi/2, -\pi/4, 0, \pi/4,$ and $\pi/2$.
   (b) Plot the points (x, y) generated in part (a), and sketch a graph of the parametric equations.
   (c) Find the rectangular equation by eliminating the parameter. Sketch its graph. How do the graphs differ?

In Exercises 3–22, (a) sketch the curve represented by the parametric equations (indicate the orientation of the curve) and (b) eliminate the parameter and write the corresponding rectangular equation whose graph represents the curve. Adjust the domain of the resulting rectangular equation if necessary.

3. $x = 3t - 3$, $y = 2t + 1$
4. $x = 3 - 2t$, $y = 2 + 3t$
5. $x = \frac{1}{4}t$, $y = t^2$
6. $x = t$, $y = t^3$
7. $x = t + 2$, $y = t^2$
8. $x = \sqrt{t}$, $y = 1 - t$
9. $x = t + 1$, $y = \dfrac{t}{t + 1}$
10. $x = t - 1$, $y = \dfrac{t}{t - 1}$
11. $x = 2(t + 1)$, $y = |t - 2|$
12. $x = |t - 1|$, $y = t + 2$
13. $x = 3\cos\theta$, $y = 3\sin\theta$
14. $x = 2\cos\theta$, $y = 3\sin\theta$
15. $x = 4\sin 2\theta$, $y = 2\cos 2\theta$
16. $x = \cos\theta$, $y = 2\sin 2\theta$
17. $x = 4 + 2\cos\theta$, $y = -1 + \sin\theta$
18. $x = 4 + 2\cos\theta$, $y = 2 + 3\sin\theta$
19. $x = e^{-t}$, $y = e^{3t}$
20. $x = e^{2t}$, $y = e^{t}$
21. $x = t^3$, $y = 3\ln t$
22. $x = \ln 2t$, $y = 2t^2$

In Exercises 23 and 24, determine how the plane curves differ from each other.

23. (a) $x = t$, $y = 2t + 1$
    (b) $x = \cos\theta$, $y = 2\cos\theta + 1$
    (c) $x = e^{-t}$, $y = 2e^{-t} + 1$
    (d) $x = e^{t}$, $y = 2e^{t} + 1$
24. (a) $x = t$, $y = t^2 - 1$
    (b) $x = t^2$, $y = t^4 - 1$
    (c) $x = \sin t$, $y = \sin^2 t - 1$
    (d) $x = e^{t}$, $y = e^{2t} - 1$

In Exercises 25–28, eliminate the parameter and obtain the standard form of the rectangular equation.

25. Line through (x_1, y_1) and (x_2, y_2):
    $x = x_1 + t(x_2 - x_1),\ y = y_1 + t(y_2 - y_1)$
26. Circle: $x = h + r\cos\theta,\ y = k + r\sin\theta$
27. Ellipse: $x = h + a\cos\theta,\ y = k + b\sin\theta$
28. Hyperbola: $x = h + a\sec\theta,\ y = k + b\tan\theta$

In Exercises 29–36, use the results of Exercises 25–28 to find a set of parametric equations for the line or conic.

29. Line: passes through $(0, 0)$ and $(6, -3)$
30. Line: passes through $(2, 3)$ and $(6, -3)$
31. Circle: center: $(3, 2)$; radius: 4

32. Circle: center: $(-3, 2)$; radius: 5

33. Ellipse: vertices: $(\pm 4, 0)$; foci: $(\pm 3, 0)$

34. Ellipse: vertices: $(4, 7), (4, -3)$; foci: $(4, 5), (4, -1)$

35. Hyperbola: vertices: $(\pm 4, 0)$; foci: $(\pm 5, 0)$

36. Hyperbola: vertices: $(\pm 2, 0)$; foci: $(\pm 4, 0)$

In Exercises 37–44, find a set of parametric equations for the rectangular equation using (a) $t = x$ and (b) $t = 2 - x$.

37. $y = 3x - 2$

38. $x = 3y - 2$

39. $y = x^2$

40. $y = x^3$

41. $y = x^2 + 1$

42. $y = 2 - x$

43. $y = \dfrac{1}{x}$

44. $y = \dfrac{1}{2x}$

In Exercises 45–52, use a graphing utility to graph the curve represented by the parametric equations.

45. Cycloid: $x = 4(\theta - \sin\theta),\ y = 4(1 - \cos\theta)$

46. Cycloid: $x = \theta + \sin\theta,\ y = 1 - \cos\theta$

47. Prolate cycloid: $x = \theta - \frac{3}{2}\sin\theta,\ y = 1 - \frac{3}{2}\cos\theta$

48. Prolate cycloid: $x = 2\theta - 4\sin\theta,\ y = 2 - 4\cos\theta$

49. Hypocycloid: $x = 3\cos^3\theta,\ y = 3\sin^3\theta$

50. Curtate cycloid: $x = 8\theta - 4\sin\theta,\ y = 8 - 4\cos\theta$

51. Witch of Agnesi: $x = 2\cot\theta,\ y = 2\sin^2\theta$

52. Folium of Descartes: $x = \dfrac{3t}{1 + t^3},\ y = \dfrac{3t^2}{1 + t^3}$

In Exercises 53–56, match the parametric equations with the correct graph and describe the domain and range. [The graphs are labeled (a), (b), (c), and (d).]

(a)

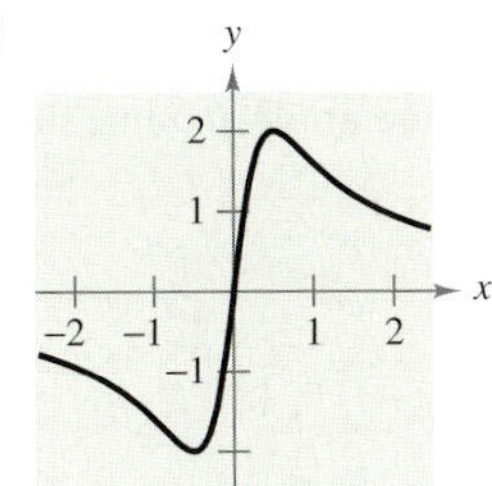

(b)

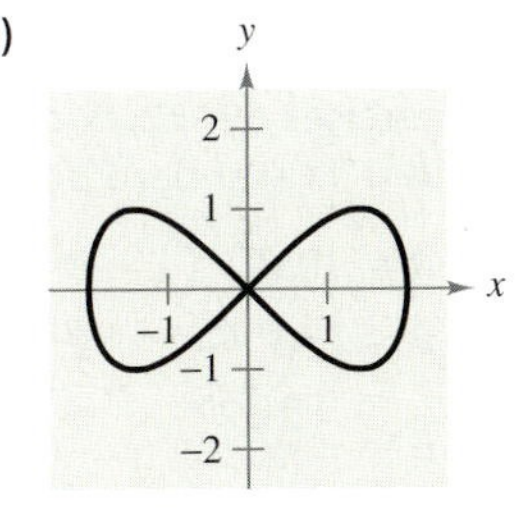

(c)

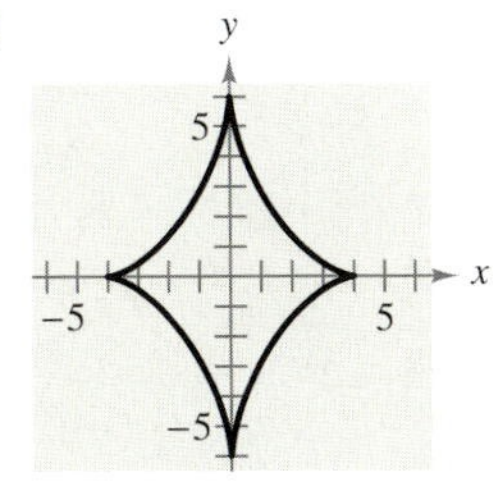

(d)

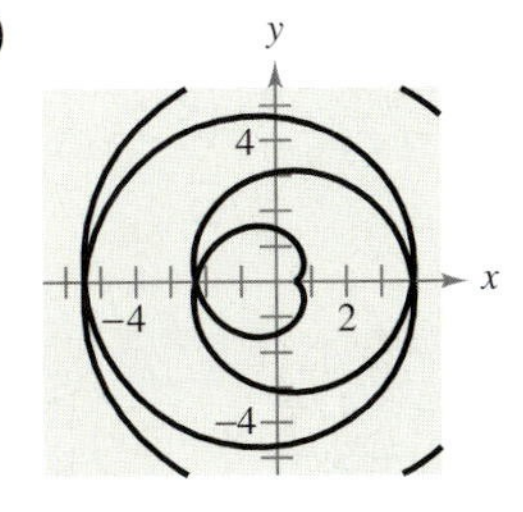

53. Lissajous curve: $x = 2\cos\theta,\ y = \sin 2\theta$

54. Evolute of ellipse: $x = 4\cos^3\theta,\ y = 6\sin^3\theta$

55. Involute of circle: $x = \frac{1}{2}(\cos\theta + \theta\sin\theta)$
$y = \frac{1}{2}(\sin\theta - \theta\cos\theta)$

56. Serpentine curve: $x = \frac{1}{2}\cot\theta,\ y = 4\sin\theta\cos\theta$

Projectile Motion **A projectile is launched at a height of h feet above the ground at an angle of θ with the horizontal. The initial velocity is v_0 feet per second and the path of the projectile is modeled by the parametric equations**

$$x = (v_0 \cos\theta)t \quad \text{and} \quad y = h + (v_0 \sin\theta)t - 16t^2.$$

In Exercises 57 and 58, use a graphing utility to graph the paths of a projectile launched from ground level at each value of θ and v_0. For each case, use the graph to approximate the maximum height and the range of the projectile.

57. (a) $\theta = 60°,\ v_0 = 88$ feet per second
(b) $\theta = 60°,\ v_0 = 132$ feet per second
(c) $\theta = 45°,\ v_0 = 88$ feet per second
(d) $\theta = 45°,\ v_0 = 132$ feet per second

58. (a) $\theta = 15°,\ v_0 = 60$ feet per second
(b) $\theta = 15°,\ v_0 = 100$ feet per second
(c) $\theta = 30°,\ v_0 = 60$ feet per second
(d) $\theta = 30°,\ v_0 = 100$ feet per second

Model It

59. ***Sports*** The center field fence in Yankee Stadium is 7 feet high and 408 feet from home plate. A baseball is hit at a point 3 feet above the ground. It leaves the bat at an angle of θ degrees with the horizontal at a speed of 100 miles per hour (see figure).

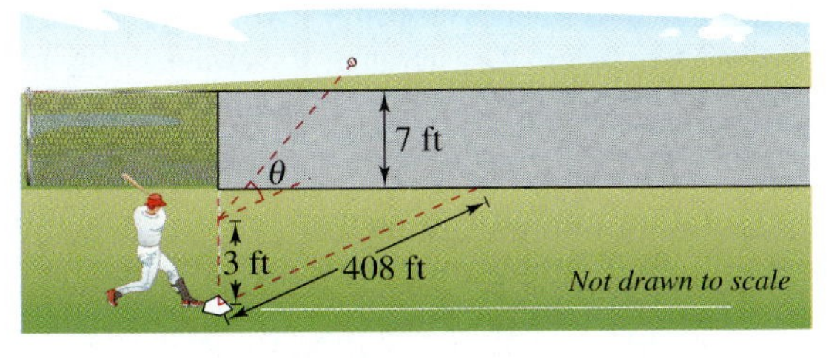

(a) Write a set of parametric equations that model the path of the baseball.

(b) Use a graphing utility to graph the path of the baseball when $\theta = 15°$. Is the hit a home run?

(c) Use a graphing utility to graph the path of the baseball when $\theta = 23°$. Is the hit a home run?

(d) Find the minimum angle required for the hit to be a home run.

60. ***Sports*** An archer releases an arrow from a bow at a point 5 feet above the ground. The arrow leaves the bow at an angle of 10° with the horizontal and at an initial speed of 240 feet per second.

(a) Write a set of parametric equations that model the path of the arrow.

(b) Assuming the ground is level, find the distance the arrow travels before it hits the ground. (Ignore air resistance.)

(c) Use a graphing utility to graph the path of the arrow and approximate its maximum height.

(d) Find the total time the arrow is in the air.

61. ***Projectile Motion*** Eliminate the parameter t from the parametric equations

$$x = (v_0 \cos \theta)t \quad \text{and} \quad y = h + (v_0 \sin \theta)t - 16t^2$$

for the motion of a projectile to show that the rectangular equation is

$$y = -\frac{16 \sec^2 \theta}{v_0^2}x^2 + (\tan \theta)x + h.$$

62. ***Path of a Projectile*** The path of a projectile is given by the rectangular equation

$$y = 7 + x - 0.02x^2.$$

(a) Use the result of Exercise 61 to find h, v_0, and θ. Find the parametric equations of the path.

(b) Use a graphing utility to graph the rectangular equation for the path of the projectile. Confirm your answer in part (a) by sketching the curve represented by the parametric equations.

(c) Use a graphing utility to approximate the maximum height of the projectile and its range.

63. ***Curtate Cycloid*** A wheel of radius a units rolls along a straight line without slipping. The curve traced by a point P that is b units from the center $(b < a)$ is called a **curtate cycloid** (see figure). Use the angle θ shown in the figure to find a set of parametric equations for the curve.

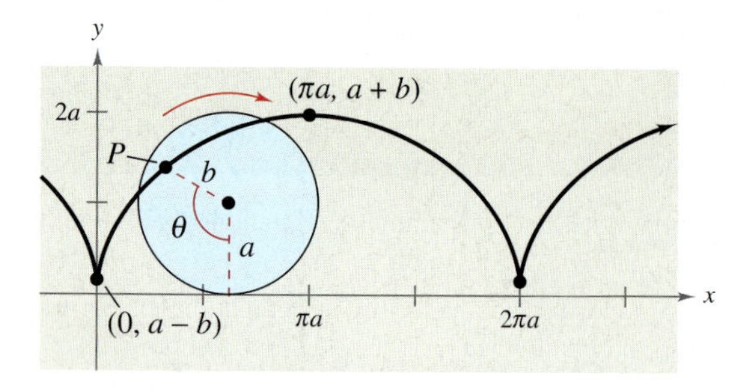

64. ***Epicycloid*** A circle of radius one unit rolls around the outside of a circle of radius two units without slipping. The curve traced by a point on the circumference of the smaller circle is called an **epicycloid** (see figure). Use the angle θ shown in the figure to find a set of parametric equations for the curve.

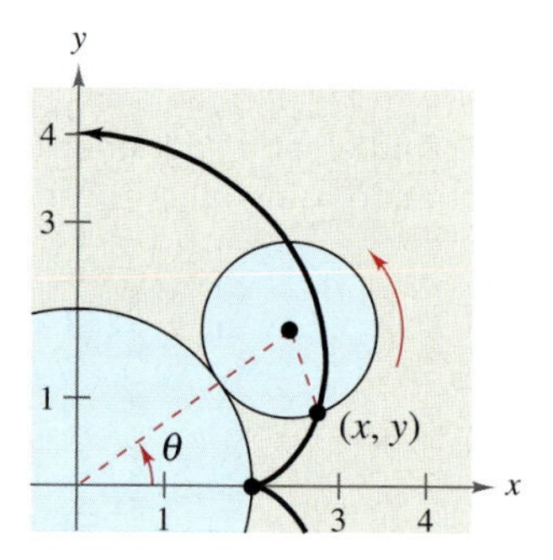

Synthesis

True or False? **In Exercises 65 and 66, determine whether the statement is true or false. Justify your answer.**

65. The two sets of parametric equations $x = t$, $y = t^2 + 1$ and $x = 3t$, $y = 9t^2 + 1$ have the same rectangular equation.

66. The graph of the parametric equations $x = t^2$ and $y = t^2$ is the line $y = x$.

67. ***Writing*** Write a short paragraph explaining why parametric equations are useful.

68. ***Writing*** Explain the process of sketching a plane curve given by parametric equations. What is meant by the orientation of the curve?

Skills Review

In Exercises 69–72, find the reference angle θ', and sketch θ and θ' in standard position.

69. $\theta = 105°$

70. $\theta = 230°$

71. $\theta = -\dfrac{2\pi}{3}$

72. $\theta = \dfrac{5\pi}{6}$

6.6 Polar Coordinates

What you should learn

- Plot points on the polar coordinate system.
- Convert points from rectangular to polar form and vice versa.
- Convert equations from rectangular to polar form and vice versa.

Why you should learn it

Polar coordinates offer a different mathematical perspective on graphing. For instance, in Exercises 1–8 on page 497, you are asked to find multiple representations of polar coordinates.

Introduction

So far, you have been representing graphs of equations as collections of points (x, y) on the rectangular coordinate system, where x and y represent the directed distances from the coordinate axes to the point (x, y). In this section, you will study a different system called the **polar coordinate system.**

To form the polar coordinate system in the plane, fix a point O, called the **pole** (or **origin**), and construct from O an initial ray called the **polar axis,** as shown in Figure 6.49. Then each point P in the plane can be assigned **polar coordinates** (r, θ) as follows.

1. $r =$ *directed distance* from O to P
2. $\theta =$ *directed angle*, counterclockwise from polar axis to segment $\overline{OP}$

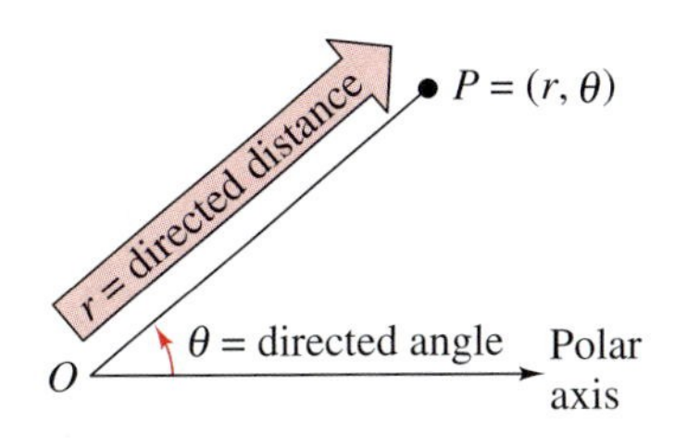

FIGURE 6.49

Example 1 **Plotting Points on the Polar Coordinate System**

a. The point $(r, \theta) = (2, \pi/3)$ lies two units from the pole on the terminal side of the angle $\theta = \pi/3$, as shown in Figure 6.50.

b. The point $(r, \theta) = (3, -\pi/6)$ lies three units from the pole on the terminal side of the angle $\theta = -\pi/6$, as shown in Figure 6.51.

c. The point $(r, \theta) = (3, 11\pi/6)$ coincides with the point $(3, -\pi/6)$, as shown in Figure 6.52.

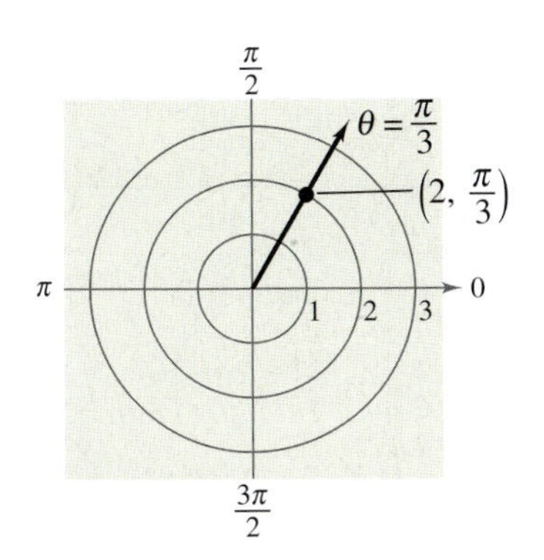

FIGURE 6.50

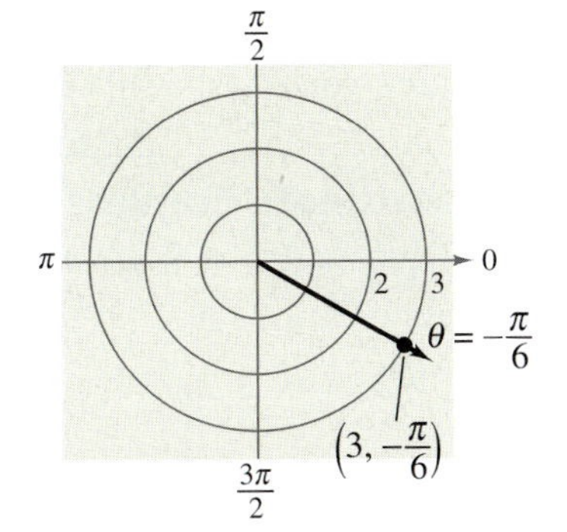

FIGURE 6.51

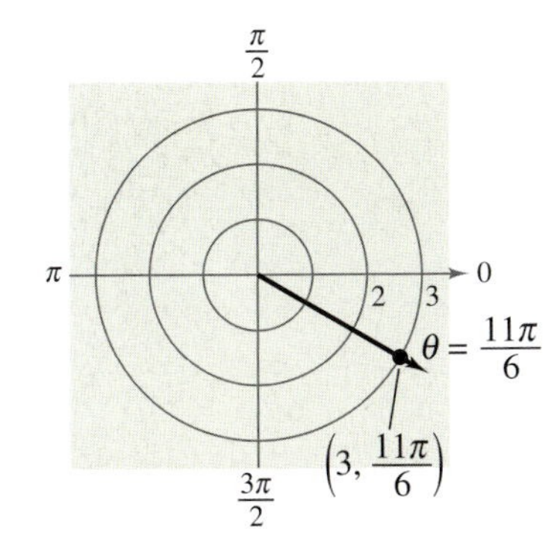

FIGURE 6.52

✓CHECKPOINT Now try Exercise 1.

Exploration

Most graphing calculators have a *polar* graphing mode. If yours does, graph the equation $r = 3$. (Use a setting in which $-6 \le x \le 6$ and $-4 \le y \le 4$.) You should obtain a circle of radius 3.

a. Use the *trace* feature to cursor around the circle. Can you locate the point $(3, 5\pi/4)$?

b. Can you find other polar representations of the point $(3, 5\pi/4)$? If so, explain how you did it.

In rectangular coordinates, each point (x, y) has a unique representation. This is not true for polar coordinates. For instance, the coordinates (r, θ) and $(r, \theta + 2\pi)$ represent the same point, as illustrated in Example 1. Another way to obtain multiple representations of a point is to use negative values for r. Because r is a *directed distance*, the coordinates (r, θ) and $(-r, \theta + \pi)$ represent the same point. In general, the point (r, θ) can be represented as

$$(r, \theta) = (r, \theta \pm 2n\pi) \qquad \text{or} \qquad (r, \theta) = (-r, \theta \pm (2n + 1)\pi)$$

where n is any integer. Moreover, the pole is represented by $(0, \theta)$, where θ is any angle.

Example 2 Multiple Representations of Points

Plot the point $(3, -3\pi/4)$ and find three additional polar representations of this point, using $-2\pi < \theta < 2\pi$.

Solution

The point is shown in Figure 6.53. Three other representations are as follows.

$$\left(3, -\frac{3\pi}{4} + 2\pi\right) = \left(3, \frac{5\pi}{4}\right)$$ Add 2π to θ.

$$\left(-3, -\frac{3\pi}{4} - \pi\right) = \left(-3, -\frac{7\pi}{4}\right)$$ Replace r by $-r$; subtract π from θ.

$$\left(-3, -\frac{3\pi}{4} + \pi\right) = \left(-3, \frac{\pi}{4}\right)$$ Replace r by $-r$; add π to θ.

CHECKPOINT Now try Exercise 3.

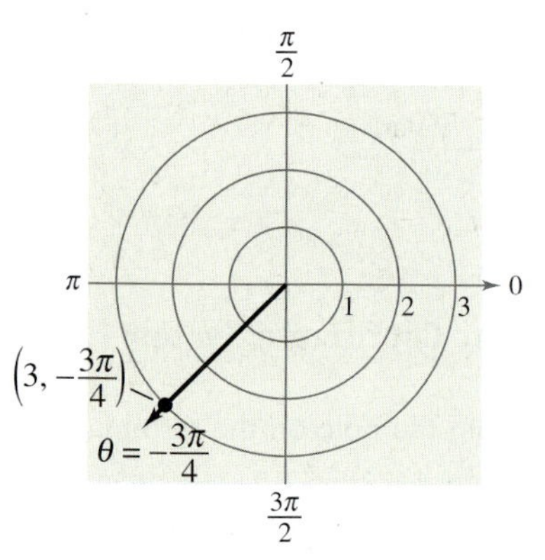

FIGURE 6.53

Coordinate Conversion

To establish the relationship between polar and rectangular coordinates, let the polar axis coincide with the positive x-axis and the pole with the origin, as shown in Figure 6.54. Because (x, y) lies on a circle of radius r, it follows that $r^2 = x^2 + y^2$. Moreover, for $r > 0$, the definitions of the trigonometric functions imply that

$$\tan\theta = \frac{y}{x}, \qquad \cos\theta = \frac{x}{r}, \qquad \text{and} \qquad \sin\theta = \frac{y}{r}.$$

If $r < 0$, you can show that the same relationships hold.

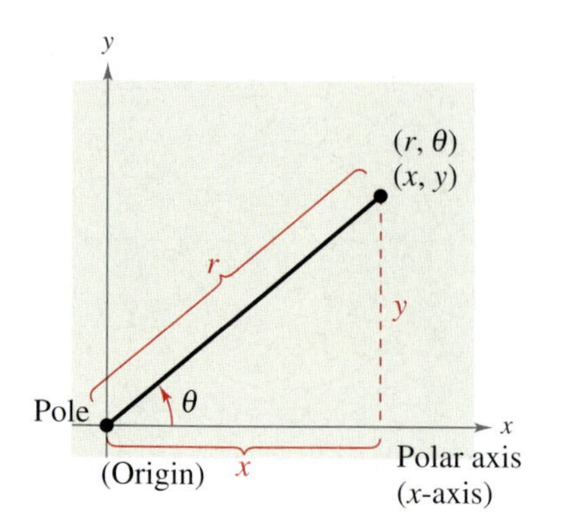

FIGURE 6.54

Coordinate Conversion

The polar coordinates (r, θ) are related to the rectangular coordinates (x, y) as follows.

Polar-to-Rectangular	*Rectangular-to-Polar*
$x = r\cos\theta$	$\tan\theta = \dfrac{y}{x}$
$y = r\sin\theta$	$r^2 = x^2 + y^2$

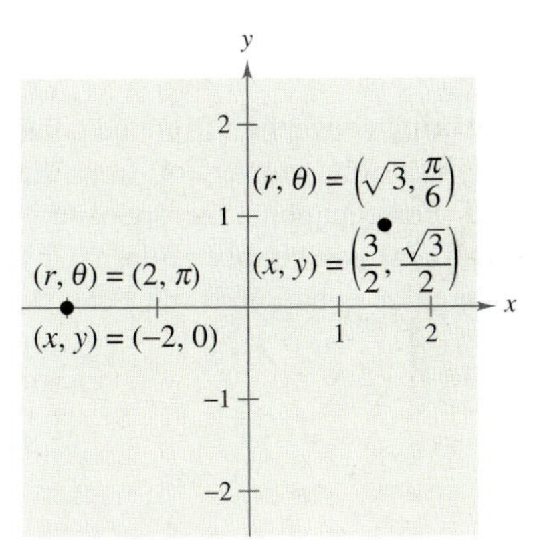

FIGURE 6.55

Activities

1. Find three additional polar representations of the point $\left(-2, \frac{\pi}{3}\right)$.

   Answer: $\left(2, \frac{4\pi}{3}\right)$, $\left(2, -\frac{2\pi}{3}\right)$, and $\left(-2, -\frac{5\pi}{3}\right)$

2. Convert the point $(-4, 2)$ from rectangular to polar form.

   Answer: $(-2\sqrt{5}, -0.4636)$ or $(2\sqrt{5}, 2.6779)$

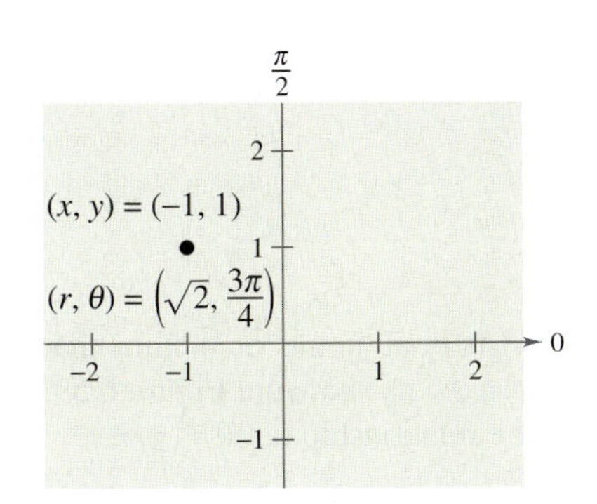

FIGURE 6.56

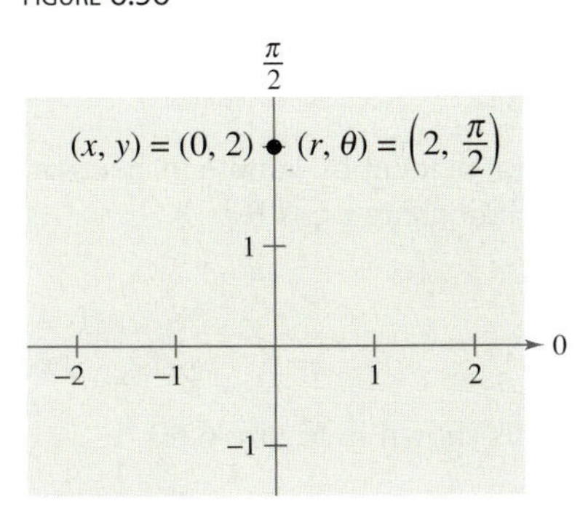

FIGURE 6.57

Example 3 Polar-to-Rectangular Conversion

Convert each point to rectangular coordinates.

a. $(2, \pi)$ **b.** $\left(\sqrt{3}, \frac{\pi}{6}\right)$

Solution

a. For the point $(r, \theta) = (2, \pi)$, you have the following.

$$x = r \cos \theta = 2 \cos \pi = -2$$

$$y = r \sin \theta = 2 \sin \pi = 0$$

The rectangular coordinates are $(x, y) = (-2, 0)$. (See Figure 6.55.)

b. For the point $(r, \theta) = \left(\sqrt{3}, \frac{\pi}{6}\right)$, you have the following.

$$x = \sqrt{3} \cos \frac{\pi}{6} = \sqrt{3}\left(\frac{\sqrt{3}}{2}\right) = \frac{3}{2}$$

$$y = \sqrt{3} \sin \frac{\pi}{6} = \sqrt{3}\left(\frac{1}{2}\right) = \frac{\sqrt{3}}{2}$$

The rectangular coordinates are $(x, y) = \left(\frac{3}{2}, \frac{\sqrt{3}}{2}\right)$. (See Figure 6.55.)

CHECKPOINT Now try Exercise 13.

Example 4 Rectangular-to-Polar Conversion

Convert each point to polar coordinates.

a. $(-1, 1)$ **b.** $(0, 2)$

Solution

a. For the second-quadrant point $(x, y) = (-1, 1)$, you have

$$\tan \theta = \frac{y}{x} = -1$$

$$\theta = \frac{3\pi}{4}.$$

Because θ lies in the same quadrant as (x, y), use positive r.

$$r = \sqrt{x^2 + y^2} = \sqrt{(-1)^2 + (1)^2} = \sqrt{2}$$

So, *one* set of polar coordinates is $(r, \theta) = (\sqrt{2}, 3\pi/4)$, as shown in Figure 6.56.

b. Because the point $(x, y) = (0, 2)$ lies on the positive y-axis, choose

$$\theta = \frac{\pi}{2} \quad \text{and} \quad r = 2.$$

This implies that *one* set of polar coordinates is $(r, \theta) = (2, \pi/2)$, as shown in Figure 6.57.

CHECKPOINT Now try Exercise 19.

Multiple representations of points and equations in the polar system can cause confusion. You may want to discuss several examples.

Activities

1. Convert the polar equation $r = 3\cos\theta$ to rectangular form.
   Answer: $x^2 + y^2 - 3x = 0$
2. Convert the rectangular equation $x = 4$ to polar form.
   Answer: $r = 4\sec\theta$

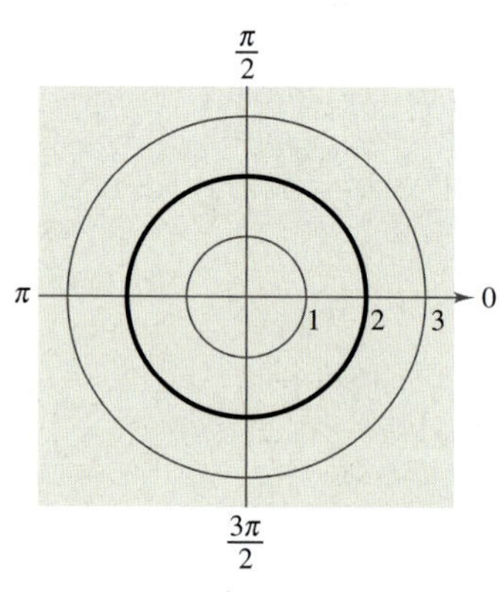

FIGURE 6.58

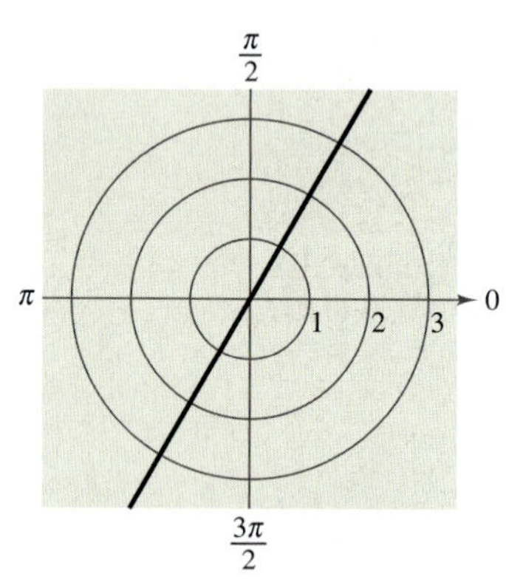

FIGURE 6.59

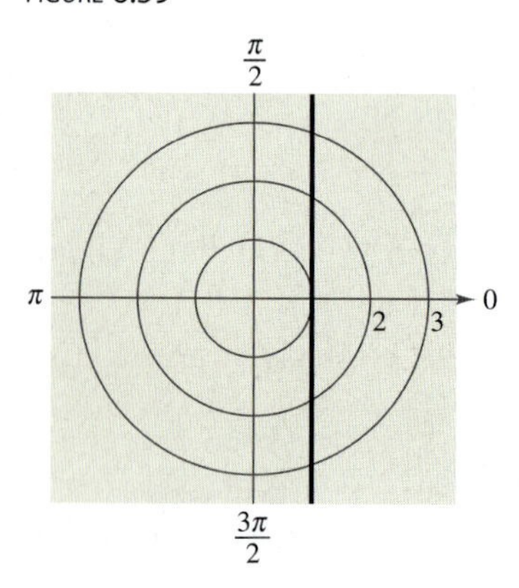

FIGURE 6.60

Equation Conversion

By comparing Examples 3 and 4, you can see that point conversion from the polar to the rectangular system is straightforward, whereas point conversion from the rectangular to the polar system is more involved. For equations, the opposite is true. To convert a rectangular equation to polar form, you simply replace x by $r\cos\theta$ and y by $r\sin\theta$. For instance, the rectangular equation $y = x^2$ can be written in polar form as follows.

$$y = x^2 \qquad \text{Rectangular equation}$$

$$r\sin\theta = (r\cos\theta)^2 \qquad \text{Polar equation}$$

$$r = \sec\theta\tan\theta \qquad \text{Simplest form}$$

On the other hand, converting a polar equation to rectangular form requires considerable ingenuity.

Example 5 demonstrates several polar-to-rectangular conversions that enable you to sketch the graphs of some polar equations.

Example 5 Converting Polar Equations to Rectangular Form

Describe the graph of each polar equation and find the corresponding rectangular equation.

a. $r = 2$ **b.** $\theta = \dfrac{\pi}{3}$ **c.** $r = \sec\theta$

Solution

a. The graph of the polar equation $r = 2$ consists of all points that are two units from the pole. In other words, this graph is a circle centered at the origin with a radius of 2, as shown in Figure 6.58. You can confirm this by converting to rectangular form, using the relationship $r^2 = x^2 + y^2$.

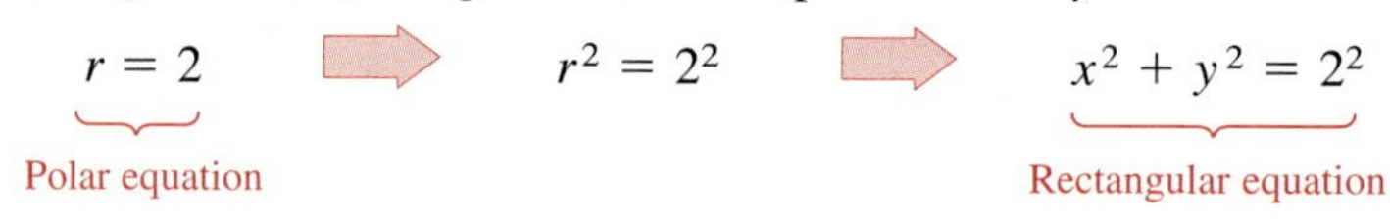

b. The graph of the polar equation $\theta = \pi/3$ consists of all points on the line that makes an angle of $\pi/3$ with the positive polar axis, as shown in Figure 6.59. To convert to rectangular form, make use of the relationship $\tan\theta = y/x$.

$$\underbrace{\theta = \frac{\pi}{3}}_{\text{Polar equation}} \Rightarrow \tan\theta = \sqrt{3} \Rightarrow \underbrace{y = \sqrt{3}x}_{\text{Rectangular equation}}$$

c. The graph of the polar equation $r = \sec\theta$ is not evident by simple inspection, so convert to rectangular form by using the relationship $r\cos\theta = x$.

$$\underbrace{r = \sec\theta}_{\text{Polar equation}} \Rightarrow r\cos\theta = 1 \Rightarrow \underbrace{x = 1}_{\text{Rectangular equation}}$$

Now you see that the graph is a vertical line, as shown in Figure 6.60.

✓CHECKPOINT Now try Exercise 65.

6.6 Exercises

VOCABULARY CHECK: Fill in the blanks.

1. The origin of the polar coordinate system is called the ________.
2. For the point (r, θ), r is the ________ ________ from O to P and θ is the ________ ________ counterclockwise from the polar axis to the line segment $\overline{OP}$.
3. To plot the point (r, θ), use the ________ coordinate system.
4. The polar coordinates (r, θ) are related to the rectangular coordinates (x, y) as follows:

$x =$ ________ $\tan \theta =$ ________

$y =$ ________ $r^2 =$ ________

PREREQUISITE SKILLS REVIEW: Practice and review algebra skills needed for this section at **www.Eduspace.com.**

In Exercises 1–8, plot the point given in polar coordinates and find two additional polar representations of the point, using $-2\pi < \theta < 2\pi$.

1. $\left(4, -\frac{\pi}{3}\right)$ **2.** $\left(-1, -\frac{3\pi}{4}\right)$

3. $\left(0, -\frac{7\pi}{6}\right)$ **4.** $\left(16, \frac{5\pi}{2}\right)$

5. $\left(\sqrt{2}, 2.36\right)$ **6.** $(-3, -1.57)$

7. $\left(2\sqrt{2}, 4.71\right)$ **8.** $(-5, -2.36)$

In Exercises 9–16, a point in polar coordinates is given. Convert the point to rectangular coordinates.

9. $\left(3, \frac{\pi}{2}\right)$

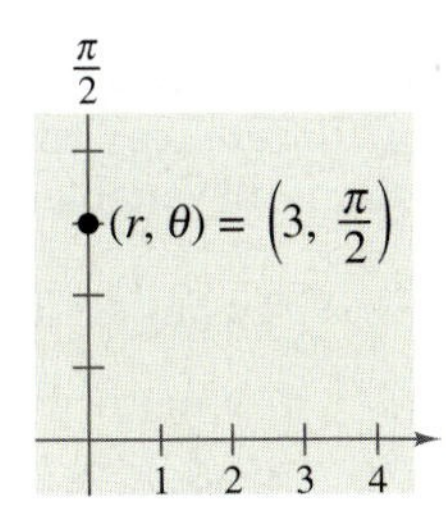

10. $\left(3, \frac{3\pi}{2}\right)$

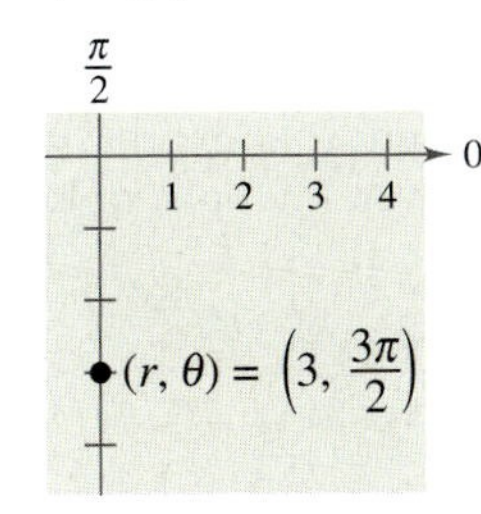

11. $\left(-1, \frac{5\pi}{4}\right)$

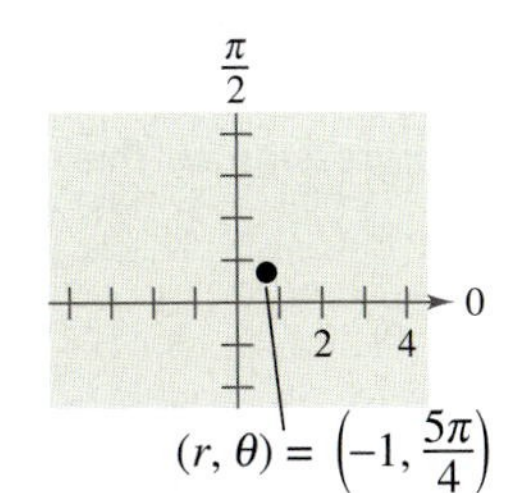

12. $(0, -\pi)$

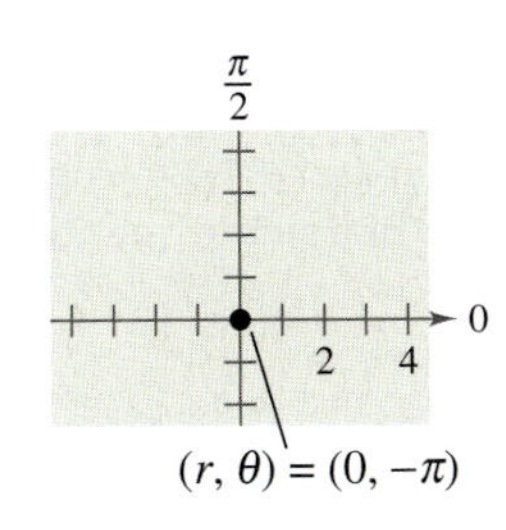

13. $\left(2, \frac{3\pi}{4}\right)$ **14.** $\left(-2, \frac{7\pi}{6}\right)$

15. $(-2.5, 1.1)$ **16.** $(8.25, 3.5)$

In Exercises 17–26, a point in rectangular coordinates is given. Convert the point to polar coordinates.

17. $(1, 1)$ **18.** $(-3, -3)$

19. $(-6, 0)$ **20.** $(0, -5)$

21. $(-3, 4)$ **22.** $(3, -1)$

23. $\left(-\sqrt{3}, -\sqrt{3}\right)$ **24.** $\left(\sqrt{3}, -1\right)$

25. $(6, 9)$ **26.** $(5, 12)$

In Exercises 27–32, use a graphing utility to find one set of polar coordinates for the point given in rectangular coordinates.

27. $(3, -2)$ **28.** $(-5, 2)$

29. $\left(\sqrt{3}, 2\right)$ **30.** $\left(3\sqrt{2}, 3\sqrt{2}\right)$

31. $\left(\frac{5}{2}, \frac{4}{3}\right)$ **32.** $\left(\frac{7}{4}, \frac{3}{2}\right)$

In Exercises 33–48, convert the rectangular equation to polar form. Assume $a > 0$.

33. $x^2 + y^2 = 9$ **34.** $x^2 + y^2 = 16$

35. $y = 4$ **36.** $y = x$

37. $x = 10$ **38.** $x = 4a$

39. $3x - y + 2 = 0$ **40.** $3x + 5y - 2 = 0$

41. $xy = 16$ **42.** $2xy = 1$

43. $y^2 - 8x - 16 = 0$ **44.** $(x^2 + y^2)^2 = 9(x^2 - y^2)$

45. $x^2 + y^2 = a^2$ **46.** $x^2 + y^2 = 9a^2$

47. $x^2 + y^2 - 2ax = 0$ **48.** $x^2 + y^2 - 2ay = 0$

In Exercises 49–64, convert the polar equation to rectangular form.

49. $r = 4 \sin \theta$

50. $r = 2 \cos \theta$

51. $\theta = \dfrac{2\pi}{3}$

52. $\theta = \dfrac{5\pi}{3}$

53. $r = 4$

54. $r = 10$

55. $r = 4 \csc \theta$

56. $r = -3 \sec \theta$

57. $r^2 = \cos \theta$

58. $r^2 = \sin 2\theta$

59. $r = 2 \sin 3\theta$

60. $r = 3 \cos 2\theta$

61. $r = \dfrac{2}{1 + \sin \theta}$

62. $r = \dfrac{1}{1 - \cos \theta}$

63. $r = \dfrac{6}{2 - 3 \sin \theta}$

64. $r = \dfrac{6}{2 \cos \theta - 3 \sin \theta}$

In Exercises 65–70, describe the graph of the polar equation and find the corresponding rectangular equation. Sketch its graph.

65. $r = 6$

66. $r = 8$

67. $\theta = \dfrac{\pi}{6}$

68. $\theta = \dfrac{3\pi}{4}$

69. $r = 3 \sec \theta$

70. $r = 2 \csc \theta$

Synthesis

True or False? **In Exercises 71 and 72, determine whether the statement is true or false. Justify your answer.**

71. If $\theta_1 = \theta_2 + 2\pi n$ for some integer n, then (r, θ_1) and (r, θ_2) represent the same point on the polar coordinate system.

72. If $|r_1| = |r_2|$, then (r_1, θ) and (r_2, θ) represent the same point on the polar coordinate system.

73. Convert the polar equation $r = 2(h \cos \theta + k \sin \theta)$ to rectangular form and verify that it is the equation of a circle. Find the radius of the circle and the rectangular coordinates of the center of the circle.

74. Convert the polar equation $r = \cos \theta + 3 \sin \theta$ to rectangular form and identify the graph.

75. ***Think About It***

(a) Show that the distance between the points (r_1, θ_1) and (r_2, θ_2) is $\sqrt{r_1^2 + r_2^2 - 2r_1 r_2 \cos(\theta_1 - \theta_2)}$.

(b) Describe the positions of the points relative to each other for $\theta_1 = \theta_2$. Simplify the Distance Formula for this case. Is the simplification what you expected? Explain.

(c) Simplify the Distance Formula for $\theta_1 - \theta_2 = 90°$. Is the simplification what you expected? Explain.

(d) Choose two points on the polar coordinate system and find the distance between them. Then choose different polar representations of the same two points and apply the Distance Formula again. Discuss the result.

76. ***Exploration***

(a) Set the window format of your graphing utility on rectangular coordinates and locate the cursor at any position off the coordinate axes. Move the cursor horizontally and observe any changes in the displayed coordinates of the points. Explain the changes in the coordinates. Now repeat the process moving the cursor vertically.

(b) Set the window format of your graphing utility on polar coordinates and locate the cursor at any position off the coordinate axes. Move the cursor horizontally and observe any changes in the displayed coordinates of the points. Explain the changes in the coordinates. Now repeat the process moving the cursor vertically.

(c) Explain why the results of parts (a) and (b) are not the same.

Skills Review

In Exercises 77–80, use the properties of logarithms to expand the expression as a sum, difference, and/or constant multiple of logarithms. (Assume all variables are positive.)

77. $\log_6 \dfrac{x^2 z}{3y}$

78. $\log_4 \dfrac{\sqrt{2x}}{y}$

79. $\ln x(x + 4)^2$

80. $\ln 5x^2(x^2 + 1)$

In Exercises 81–84, condense the expression to the logarithm of a single quantity.

81. $\log_7 x - \log_7 3y$

82. $\log_5 a + 8 \log_5(x + 1)$

83. $\dfrac{1}{2} \ln x + \ln(x - 2)$

84. $\ln 6 + \ln y - \ln(x - 3)$

6.7 Graphs of Polar Equations

What you should learn

- Graph polar equations by point plotting.
- Use symmetry to sketch graphs of polar equations.
- Use zeros and maximum r-values to sketch graphs of polar equations.
- Recognize special polar graphs.

Why you should learn it

Equations of several common figures are simpler in polar form than in rectangular form. For instance, Exercise 6 on page 505 shows the graph of a circle and its polar equation.

Emphasize setting up the table of θ values. Your students will benefit from labeling the points as they plot them.

Introduction

In previous chapters, you spent a lot of time learning how to sketch graphs on rectangular coordinate systems. You began with the basic point-plotting method, which was then enhanced by sketching aids such as symmetry, intercepts, asymptotes, periods, and shifts. This section approaches curve sketching on the polar coordinate system similarly, beginning with a demonstration of point plotting.

Example 1 Graphing a Polar Equation by Point Plotting

Sketch the graph of the polar equation $r = 4 \sin \theta$.

Solution

The sine function is periodic, so you can get a full range of r-values by considering values of θ in the interval $0 \le \theta \le 2\pi$, as shown in the following table.

θ	0	$\frac{\pi}{6}$	$\frac{\pi}{3}$	$\frac{\pi}{2}$	$\frac{2\pi}{3}$	$\frac{5\pi}{6}$	π	$\frac{7\pi}{6}$	$\frac{3\pi}{2}$	$\frac{11\pi}{6}$	2π
r	0	2	$2\sqrt{3}$	4	$2\sqrt{3}$	2	0	-2	-4	-2	0

If you plot these points as shown in Figure 6.61, it appears that the graph is a circle of radius 2 whose center is at the point $(x, y) = (0, 2)$.

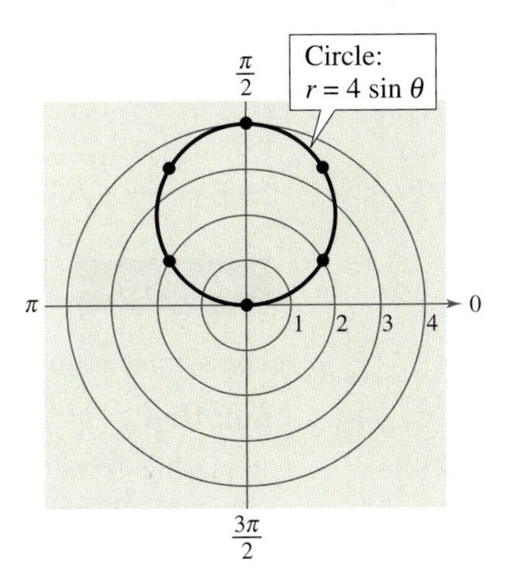

FIGURE 6.61

CHECKPOINT Now try Exercise 21.

You can confirm the graph in Figure 6.61 by converting the polar equation to rectangular form and then sketching the graph of the rectangular equation. You can also use a graphing utility set to *polar* mode and graph the polar equation or set the graphing utility to *parametric* mode and graph a parametric representation.

Symmetry

Point out to your students that these tests are sufficient for showing symmetry; however, they are not necessary. A polar graph can exhibit symmetry even when the tests fail to indicate symmetry.

In Figure 6.61, note that as θ increases from 0 to 2π the graph is traced out twice. Moreover, note that the graph is *symmetric with respect to the line* $\theta = \pi/2$. Had you known about this symmetry and retracing ahead of time, you could have used fewer points.

Symmetry with respect to the line $\theta = \pi/2$ is one of three important types of symmetry to consider in polar curve sketching. (See Figure 6.62.)

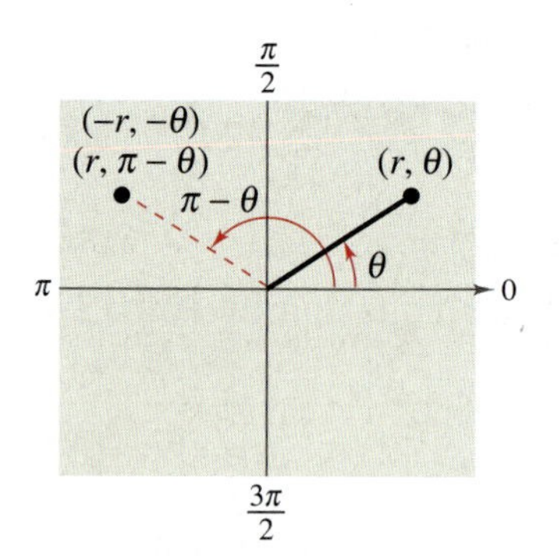

Symmetry with Respect to the Line $\theta = \dfrac{\pi}{2}$

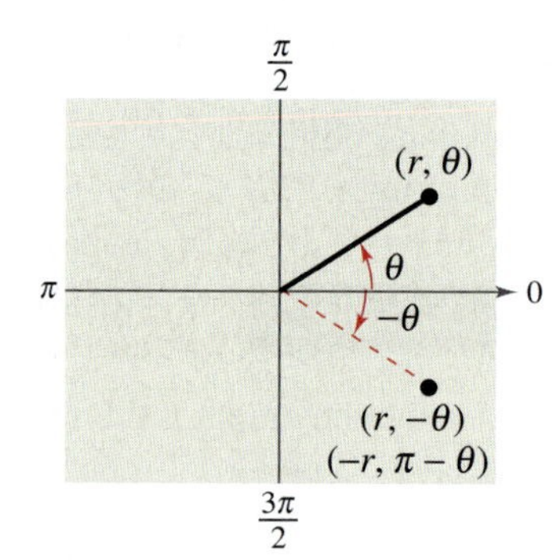

Symmetry with Respect to the Polar Axis

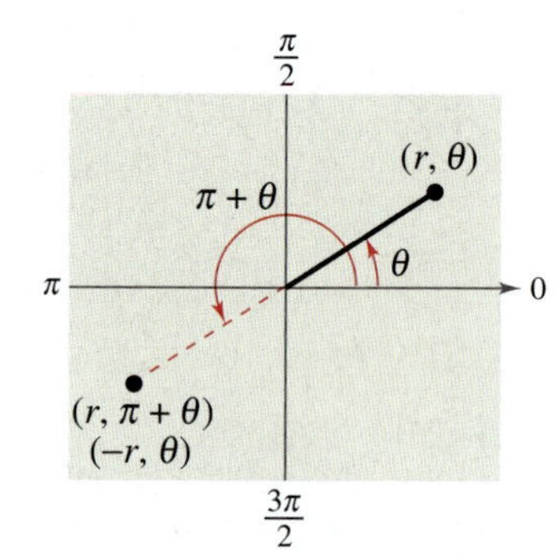

Symmetry with Respect to the Pole

FIGURE 6.62

STUDY TIP

Note in Example 2 that $\cos(-\theta) = \cos\theta$. This is because the cosine function is *even*. Recall from Section 4.2 that the cosine function is even and the sine function is odd. That is, $\sin(-\theta) = -\sin\theta$.

Tests for Symmetry in Polar Coordinates

The graph of a polar equation is symmetric with respect to the following if the given substitution yields an equivalent equation.

1. *The line* $\theta = \pi/2$: Replace (r, θ) by $(r, \pi - \theta)$ or $(-r, -\theta)$.
2. *The polar axis:* Replace (r, θ) by $(r, -\theta)$ or $(-r, \pi - \theta)$.
3. *The pole:* Replace (r, θ) by $(r, \pi + \theta)$ or $(-r, \theta)$.

Example 2 Using Symmetry to Sketch a Polar Graph

Use symmetry to sketch the graph of $r = 3 + 2\cos\theta$.

Solution

Replacing (r, θ) by $(r, -\theta)$ produces $r = 3 + 2\cos(-\theta) = 3 + 2\cos\theta$. So, you can conclude that the curve is symmetric with respect to the polar axis. Plotting the points in the table and using polar axis symmetry, you obtain the graph shown in Figure 6.63. This graph is called a **limaçon.**

θ	0	$\frac{\pi}{3}$	$\frac{\pi}{2}$	$\frac{2\pi}{3}$	π
r	5	4	3	2	1

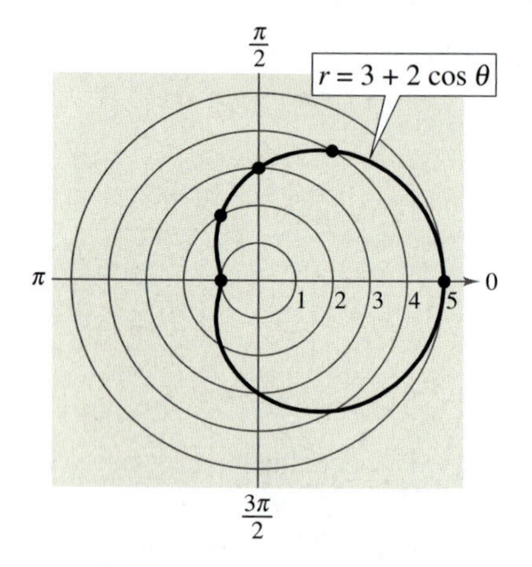

FIGURE 6.63

CHECKPOINT Now try Exercise 27.

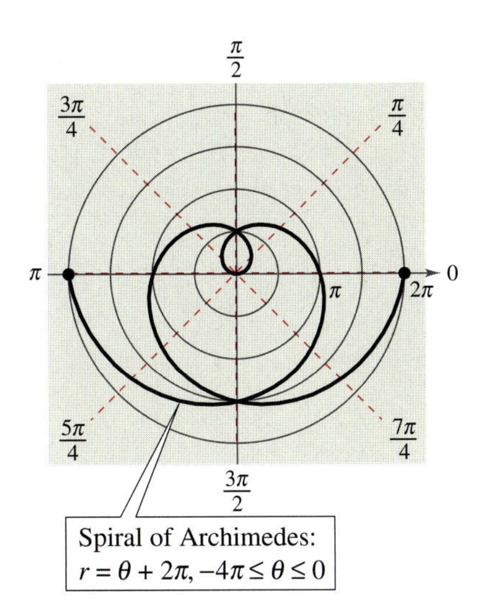

FIGURE 6.64

The three tests for symmetry in polar coordinates listed on page 500 are sufficient to guarantee symmetry, but they are not necessary. For instance, Figure 6.64 shows the graph of $r = \theta + 2\pi$ to be symmetric with respect to the line $\theta = \pi/2$, and yet the tests on page 500 fail to indicate symmetry because neither of the following replacements yields an equivalent equation.

Original Equation	*Replacement*	*New Equation*
$r = \theta + 2\pi$	(r, θ) by $(-r, -\theta)$	$-r = -\theta + 2\pi$
$r = \theta + 2\pi$	(r, θ) by $(r, \pi - \theta)$	$r = -\theta + 3\pi$

The equations discussed in Examples 1 and 2 are of the form

$r = 4 \sin \theta = f(\sin \theta)$ and $r = 3 + 2 \cos \theta = g(\cos \theta)$.

The graph of the first equation is symmetric with respect to the line $\theta = \pi/2$, and the graph of the second equation is symmetric with respect to the polar axis. This observation can be generalized to yield the following tests.

Quick Tests for Symmetry in Polar Coordinates

1. The graph of $r = f(\sin \theta)$ is symmetric with respect to the line $\theta = \dfrac{\pi}{2}$.
2. The graph of $r = g(\cos \theta)$ is symmetric with respect to the polar axis.

Zeros and Maximum *r*-Values

Two additional aids to graphing of polar equations involve knowing the θ-values for which $|r|$ is maximum and knowing the θ-values for which $r = 0$. For instance, in Example 1, the maximum value of $|r|$ for $r = 4 \sin \theta$ is $|r| = 4$, and this occurs when $\theta = \pi/2$, as shown in Figure 6.61. Moreover, $r = 0$ when $\theta = 0$.

Example 3 Sketching a Polar Graph

Sketch the graph of $r = 1 - 2 \cos \theta$.

Solution

From the equation $r = 1 - 2 \cos \theta$, you can obtain the following.

Symmetry: With respect to the polar axis

Maximum value of $|r|$*:* $r = 3$ when $\theta = \pi$

Zero of r: $r = 0$ when $\theta = \pi/3$

The table shows several θ-values in the interval $[0, \pi]$. By plotting the corresponding points, you can sketch the graph shown in Figure 6.65.

θ	0	$\frac{\pi}{6}$	$\frac{\pi}{3}$	$\frac{\pi}{2}$	$\frac{2\pi}{3}$	$\frac{5\pi}{6}$	π
r	-1	-0.73	0	1	2	2.73	3

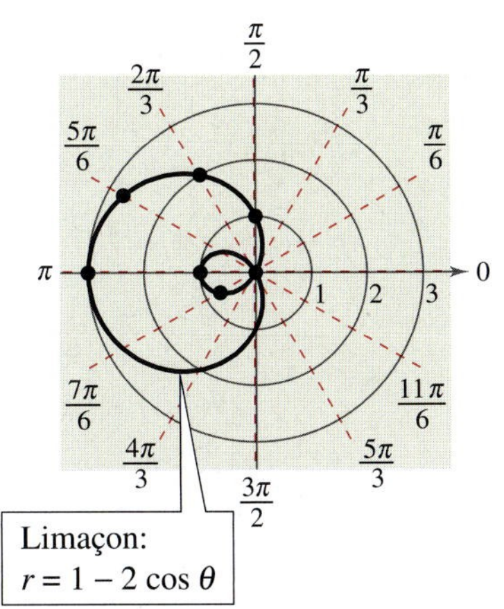

FIGURE 6.65

Note how the negative r-values determine the *inner loop* of the graph in Figure 6.65. This graph, like the one in Figure 6.63, is a limaçon.

CHECKPOINT Now try Exercise 29.

Some curves reach their zeros and maximum r-values at more than one point, as shown in Example 4.

Example 4 Sketching a Polar Graph

Sketch the graph of $r = 2\cos 3\theta$.

Solution

Symmetry: With respect to the polar axis

Maximum value of $|r|$*:* $|r| = 2$ when $3\theta = 0, \pi, 2\pi, 3\pi$ or $\theta = 0, \pi/3, 2\pi/3, \pi$

Zeros of r: $r = 0$ when $3\theta = \pi/2, 3\pi/2, 5\pi/2$ or $\theta = \pi/6, \pi/2, 5\pi/6$

θ	0	$\frac{\pi}{12}$	$\frac{\pi}{6}$	$\frac{\pi}{4}$	$\frac{\pi}{3}$	$\frac{5\pi}{12}$	$\frac{\pi}{2}$
r	2	$\sqrt{2}$	0	$-\sqrt{2}$	-2	$-\sqrt{2}$	0

By plotting these points and using the specified symmetry, zeros, and maximum values, you can obtain the graph shown in Figure 6.66. This graph is called a **rose curve,** and each of the loops on the graph is called a *petal* of the rose curve. Note how the entire curve is generated as θ increases from 0 to π.

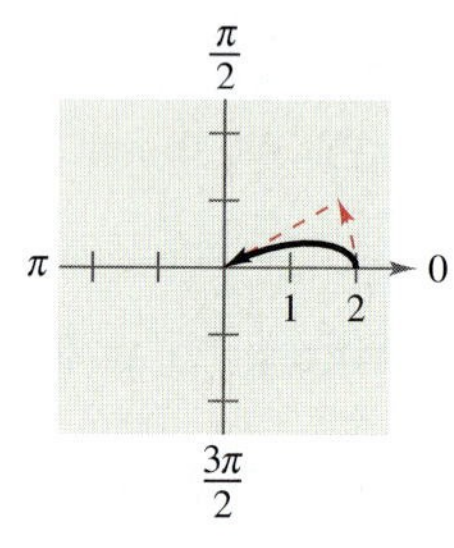

$0 \le \theta \le \frac{\pi}{6}$

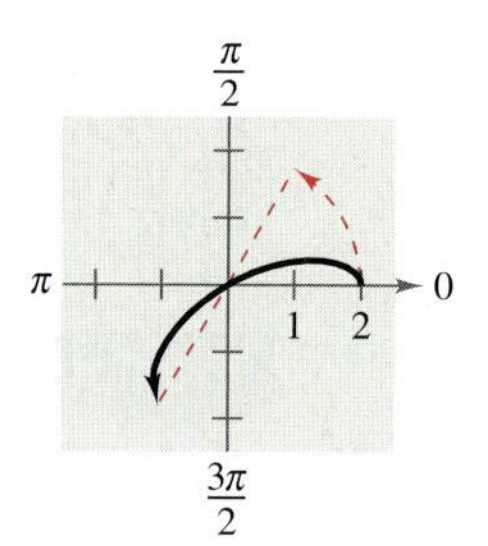

$0 \le \theta \le \frac{\pi}{3}$

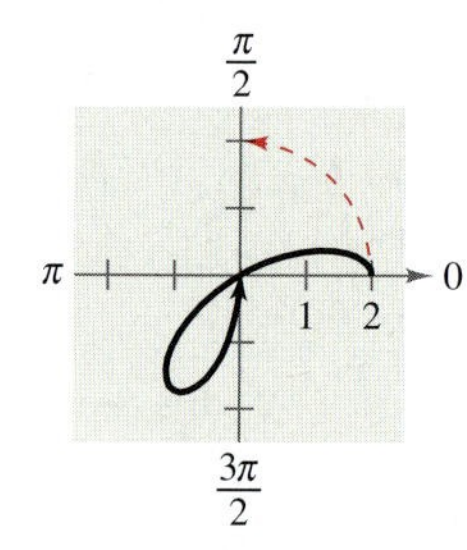

$0 \le \theta \le \frac{\pi}{2}$

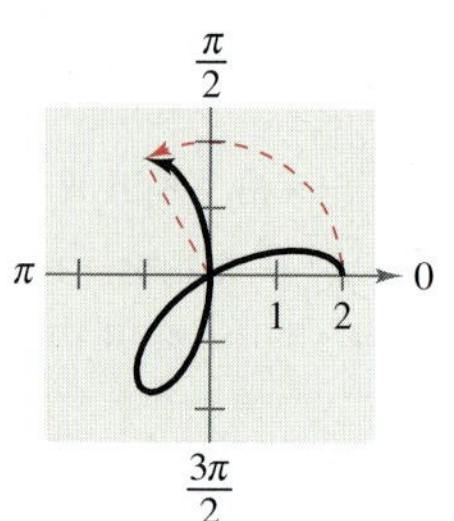

$0 \le \theta \le \frac{2\pi}{3}$

$0 \le \theta \le \frac{5\pi}{6}$

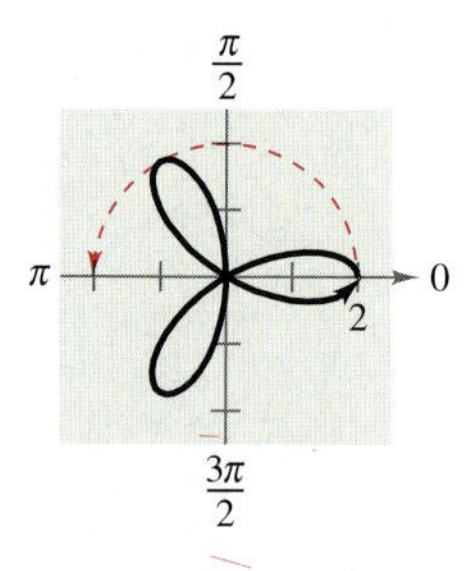

$0 \le \theta \le \pi$

FIGURE 6.66

Exploration

Notice that the rose curve in Example 4 has three petals. How many petals do the rose curves given by $r = 2\cos 4\theta$ and $r = 2\sin 3\theta$ have? Determine the numbers of petals for the curves given by $r = 2\cos n\theta$ and $r = 2\sin n\theta$, where n is a positive integer.

Technology

Use a graphing utility in *polar* mode to verify the graph of $r = 2\cos 3\theta$ shown in Figure 6.66.

CHECKPOINT Now try Exercise 33.

Special Polar Graphs

Several important types of graphs have equations that are simpler in polar form than in rectangular form. For example, the circle

$$r = 4 \sin \theta$$

in Example 1 has the more complicated rectangular equation

$$x^2 + (y - 2)^2 = 4.$$

Several other types of graphs that have simple polar equations are shown below.

Limaçons

$r = a \pm b \cos \theta$

$r = a \pm b \sin \theta$

$(a > 0, b > 0)$

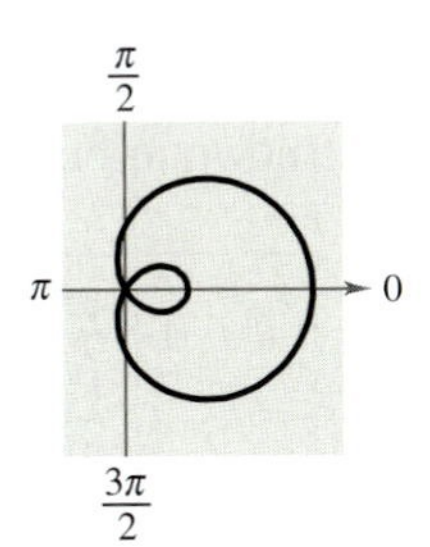

$\frac{a}{b} < 1$
Limaçon with inner loop

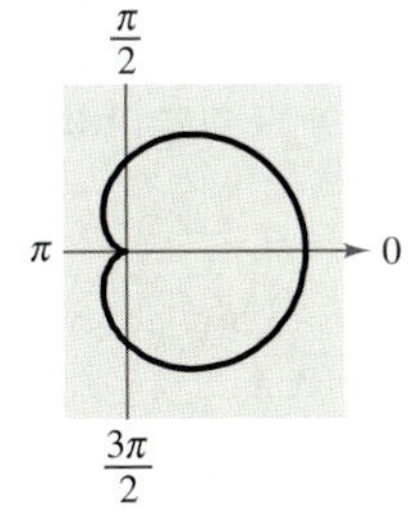

$\frac{a}{b} = 1$
Cardioid (heart-shaped)

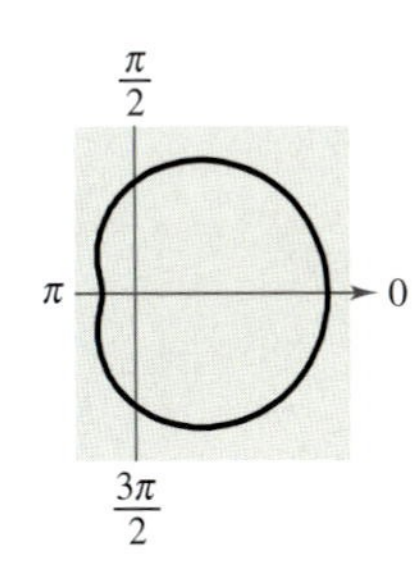

$1 < \frac{a}{b} < 2$
Dimpled limaçon

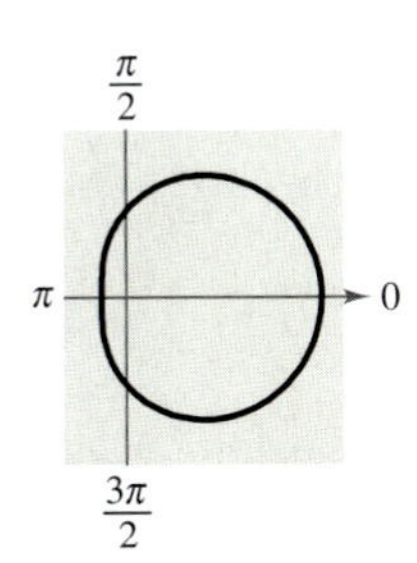

$\frac{a}{b} \geq 2$
Convex limaçon

Rose Curves

n petals if n is odd,
$2n$ petals if n is even
$(n \geq 2)$

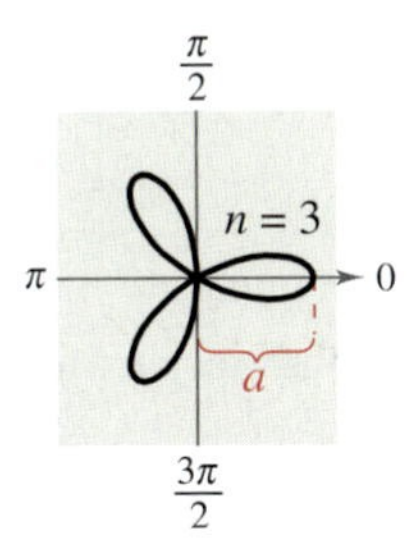

$r = a \cos n\theta$
Rose curve

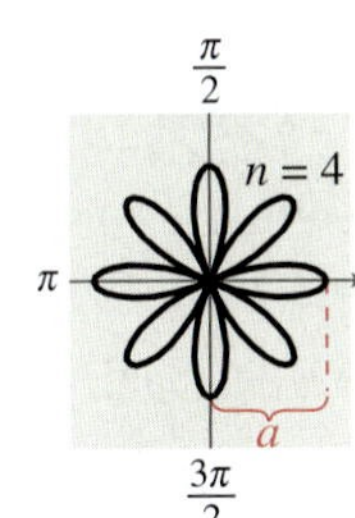

$r = a \cos n\theta$
Rose curve

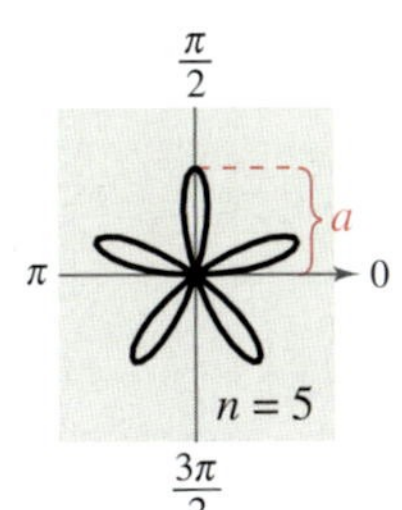

$r = a \sin n\theta$
Rose curve

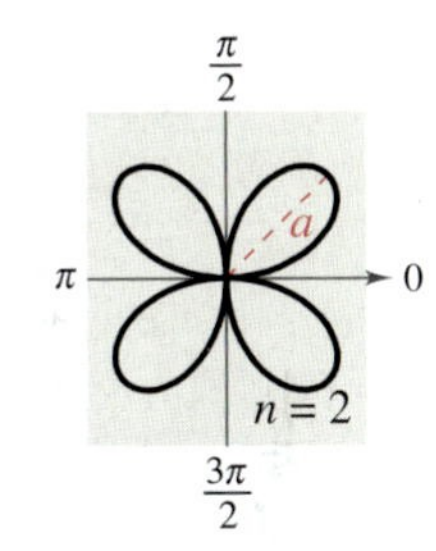

$r = a \sin n\theta$
Rose curve

Circles and Lemniscates

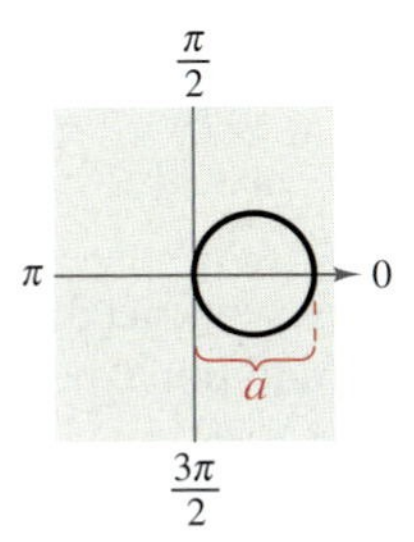

$r = a \cos \theta$
Circle

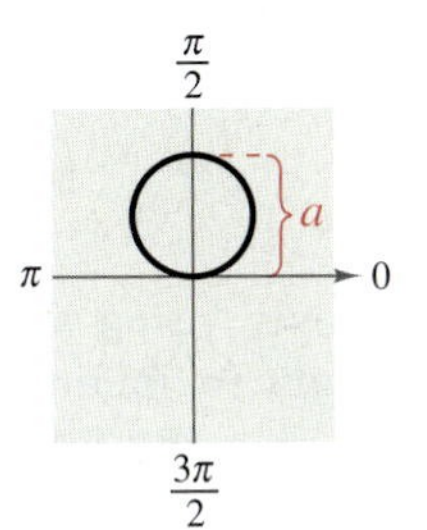

$r = a \sin \theta$
Circle

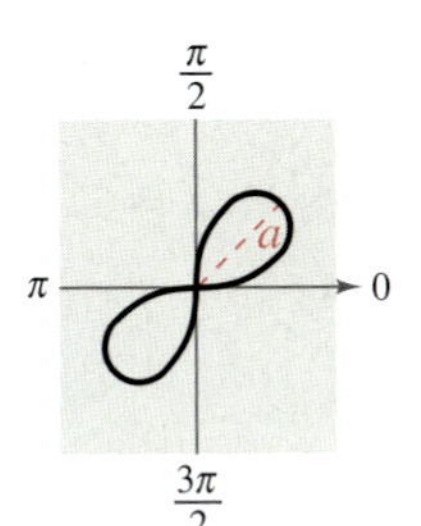

$r^2 = a^2 \sin 2\theta$
Lemniscate

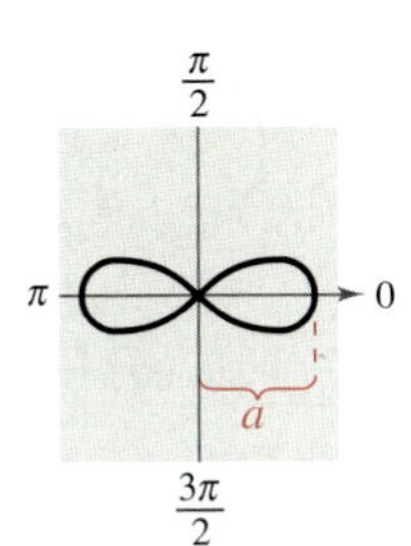

$r^2 = a^2 \cos 2\theta$
Lemniscate

The quick tests for symmetry presented in this section can be especially useful in graphing these special polar graphs.

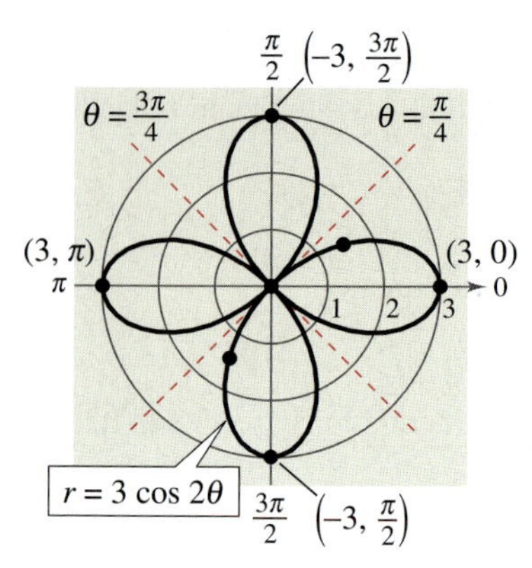

FIGURE 6.67

Example 5 Sketching a Rose Curve

Sketch the graph of $r = 3\cos 2\theta$.

Solution

Type of curve: Rose curve with $2n = 4$ petals

Symmetry: With respect to polar axis, the line $\theta = \pi/2$, and the pole

Maximum value of $|r|$: $|r| = 3$ when $\theta = 0, \pi/2, \pi, 3\pi/2$

Zeros of r: $r = 0$ when $\theta = \pi/4, 3\pi/4$

Using this information together with the additional points shown in the following table, you obtain the graph shown in Figure 6.67.

θ	0	$\frac{\pi}{6}$	$\frac{\pi}{4}$	$\frac{\pi}{3}$
r	3	$\frac{3}{2}$	0	$-\frac{3}{2}$

CHECKPOINT Now try Exercise 35.

Activities

1. Test $r^2 = 3\sin\theta$ for symmetry.
   Answer: Symmetric with respect to the pole
2. Find the maximum value of $|r|$ and any zeros of r for the polar equation $r = 2 + 2\cos\theta$.
   Answer: Maximum value of $|r| = 4$ when $\theta = 0$, and $r = 0$ when $\theta = \pi$
3. Identify the shape of the graph of the polar equation $r = 2\sin 3\theta$.
   Answer: Rose curve with 3 petals

Example 6 Sketching a Lemniscate

Sketch the graph of $r^2 = 9\sin 2\theta$.

Solution

Type of curve: Lemniscate

Symmetry: With respect to the pole

Maximum value of $|r|$: $|r| = 3$ when $\theta = \frac{\pi}{4}$

Zeros of r: $r = 0$ when $\theta = 0, \frac{\pi}{2}$

If $\sin 2\theta < 0$, this equation has no solution points. So, you restrict the values of θ to those for which $\sin 2\theta \ge 0$.

$$0 \le \theta \le \frac{\pi}{2} \quad \text{or} \quad \pi \le \theta \le \frac{3\pi}{2}$$

Moreover, using symmetry, you need to consider only the first of these two intervals. By finding a few additional points (see table below), you can obtain the graph shown in Figure 6.68.

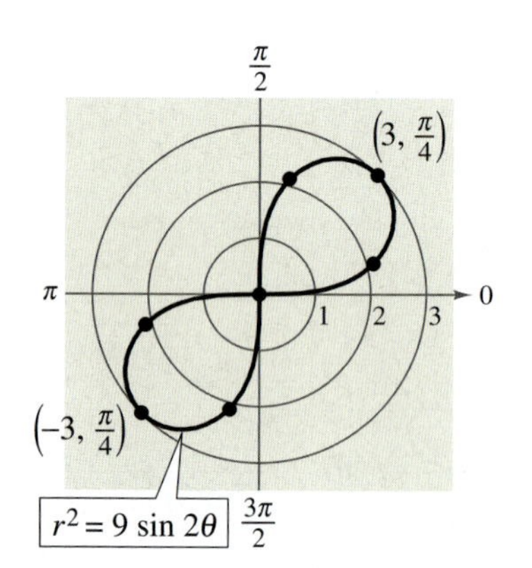

FIGURE 6.68

θ	0	$\frac{\pi}{12}$	$\frac{\pi}{4}$	$\frac{5\pi}{12}$	$\frac{\pi}{2}$
$r = \pm 3\sqrt{\sin 2\theta}$	0	$\frac{\pm 3}{\sqrt{2}}$	± 3	$\frac{\pm 3}{\sqrt{2}}$	0

CHECKPOINT Now try Exercise 39.

6.7 Exercises

VOCABULARY CHECK: Fill in the blanks.

1. The graph of $r = f(\sin \theta)$ is symmetric with respect to the line ________.
2. The graph of $r = g(\cos \theta)$ is symmetric with respect to the ________ ________.
3. The equation $r = 2 + \cos \theta$ represents a ________ ________.
4. The equation $r = 2 \cos \theta$ represents a ________.
5. The equation $r^2 = 4 \sin 2\theta$ represents a ________.
6. The equation $r = 1 + \sin \theta$ represents a ________.

PREREQUISITE SKILLS REVIEW: Practice and review algebra skills needed for this section at **www.Eduspace.com.**

In Exercises 1–6, identify the type of polar graph.

1.

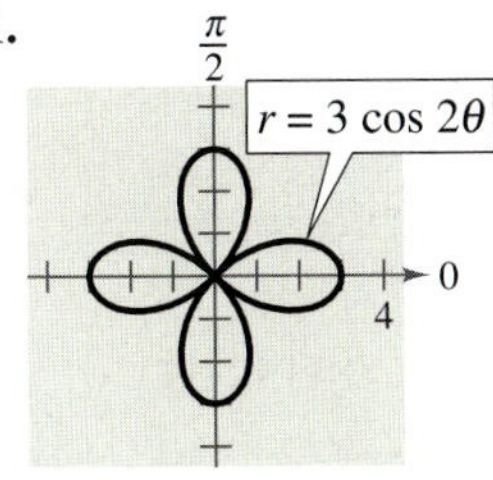

2.

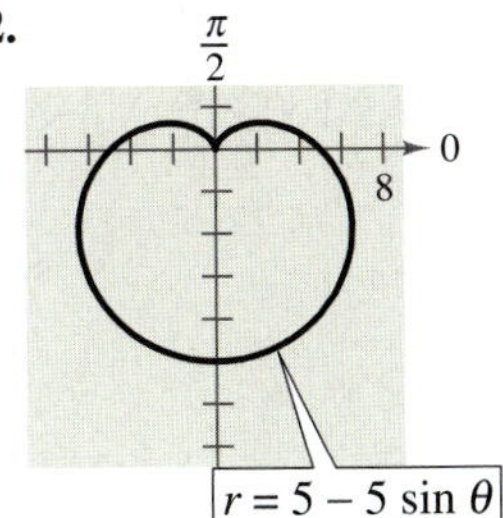

3.

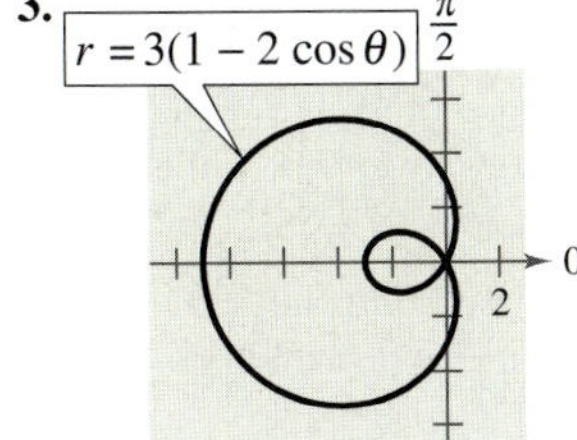

4.

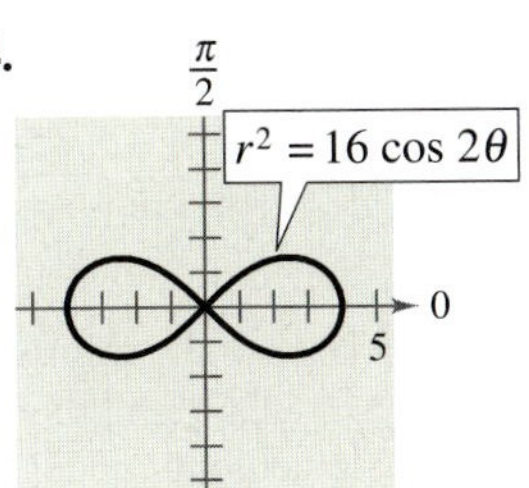

5.

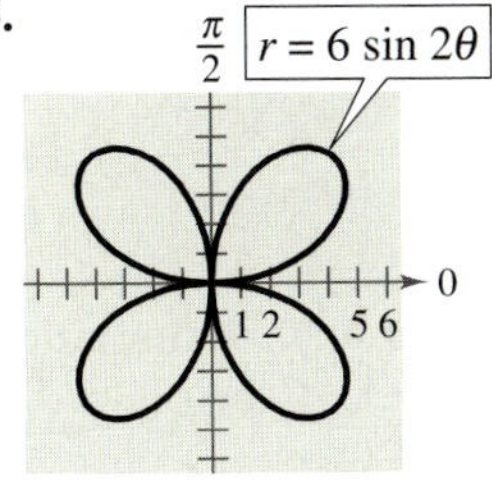

6.

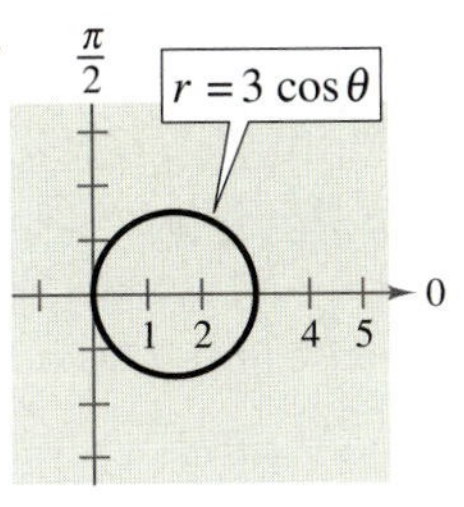

In Exercises 7–12, test for symmetry with respect to $\theta = \pi/2$, the polar axis, and the pole.

7. $r = 5 + 4 \cos \theta$

8. $r = 16 \cos 3\theta$

9. $r = \dfrac{2}{1 + \sin \theta}$

10. $r = \dfrac{3}{2 + \cos \theta}$

11. $r^2 = 16 \cos 2\theta$

12. $r^2 = 36 \sin 2\theta$

In Exercises 13–16, find the maximum value of $|r|$ and any zeros of r.

13. $r = 10(1 - \sin \theta)$

14. $r = 6 + 12 \cos \theta$

15. $r = 4 \cos 3\theta$

16. $r = 3 \sin 2\theta$

In Exercises 17–40, sketch the graph of the polar equation using symmetry, zeros, maximum r-values, and any other additional points.

17. $r = 5$

18. $r = 2$

19. $r = \dfrac{\pi}{6}$

20. $r = -\dfrac{3\pi}{4}$

21. $r = 3 \sin \theta$

22. $r = 4 \cos \theta$

23. $r = 3(1 - \cos \theta)$

24. $r = 4(1 - \sin \theta)$

25. $r = 4(1 + \sin \theta)$

26. $r = 2(1 + \cos \theta)$

27. $r = 3 + 6 \sin \theta$

28. $r = 4 - 3 \sin \theta$

29. $r = 1 - 2 \sin \theta$

30. $r = 1 - 2 \cos \theta$

31. $r = 3 - 4 \cos \theta$

32. $r = 4 + 3 \cos \theta$

33. $r = 5 \sin 2\theta$

34. $r = 3 \cos 2\theta$

35. $r = 2 \sec \theta$

36. $r = 5 \csc \theta$

37. $r = \dfrac{3}{\sin \theta - 2 \cos \theta}$

38. $r = \dfrac{6}{2 \sin \theta - 3 \cos \theta}$

39. $r^2 = 9 \cos 2\theta$

40. $r^2 = 4 \sin \theta$

In Exercises 41–46, use a graphing utility to graph the polar equation. Describe your viewing window.

41. $r = 8 \cos \theta$

42. $r = \cos 2\theta$

43. $r = 3(2 - \sin \theta)$

44. $r = 2 \cos(3\theta - 2)$

45. $r = 8 \sin \theta \cos^2 \theta$

46. $r = 2 \csc \theta + 5$

In Exercises 47–52, use a graphing utility to graph the polar equation. Find an interval for θ for which the graph is traced *only once*.

47. $r = 3 - 4 \cos \theta$

48. $r = 5 + 4 \cos \theta$

49. $r = 2\cos\left(\frac{3\theta}{2}\right)$

50. $r = 3\sin\left(\frac{5\theta}{2}\right)$

51. $r^2 = 9\sin 2\theta$

52. $r^2 = \frac{1}{\theta}$

In Exercises 53–56, use a graphing utility to graph the polar equation and show that the indicated line is an asymptote of the graph.

Name of Graph	*Polar Equation*	*Asymptote*
53. Conchoid	$r = 2 - \sec\theta$	$x = -1$
54. Conchoid	$r = 2 + \csc\theta$	$y = 1$
55. Hyperbolic spiral	$r = \frac{3}{\theta}$	$y = 3$
56. Strophoid	$r = 2\cos 2\theta \sec\theta$	$x = -2$

Synthesis

True or False? **In Exercises 57 and 58, determine whether the statement is true or false. Justify your answer.**

57. In the polar coordinate system, if a graph that has symmetry with respect to the polar axis were folded on the line $\theta = 0$, the portion of the graph above the polar axis would coincide with the portion of the graph below the polar axis.

58. In the polar coordinate system, if a graph that has symmetry with respect to the pole were folded on the line $\theta = 3\pi/4$, the portion of the graph on one side of the fold would coincide with the portion of the graph on the other side of the fold.

59. ***Exploration*** Sketch the graph of $r = 6\cos\theta$ over each interval. Describe the part of the graph obtained in each case.

(a) $0 \le \theta \le \frac{\pi}{2}$ (b) $\frac{\pi}{2} \le \theta \le \pi$

(c) $-\frac{\pi}{2} \le \theta \le \frac{\pi}{2}$ (d) $\frac{\pi}{4} \le \theta \le \frac{3\pi}{4}$

60. ***Graphical Reasoning*** Use a graphing utility to graph the polar equation $r = 6[1 + \cos(\theta - \phi)]$ for (a) $\phi = 0$, (b) $\phi = \pi/4$, and (c) $\phi = \pi/2$. Use the graphs to describe the effect of the angle ϕ. Write the equation as a function of $\sin\theta$ for part (c).

61. The graph of $r = f(\theta)$ is rotated about the pole through an angle ϕ. Show that the equation of the rotated graph is $r = f(\theta - \phi)$.

62. Consider the graph of $r = f(\sin\theta)$.

(a) Show that if the graph is rotated counterclockwise $\pi/2$ radians about the pole, the equation of the rotated graph is $r = f(-\cos\theta)$.

(b) Show that if the graph is rotated counterclockwise π radians about the pole, the equation of the rotated graph is $r = f(-\sin\theta)$.

(c) Show that if the graph is rotated counterclockwise $3\pi/2$ radians about the pole, the equation of the rotated graph is $r = f(\cos\theta)$.

In Exercises 63–66, use the results of Exercises 61 and 62.

63. Write an equation for the limaçon $r = 2 - \sin\theta$ after it has been rotated through the given angle.

(a) $\frac{\pi}{4}$ (b) $\frac{\pi}{2}$ (c) π (d) $\frac{3\pi}{2}$

64. Write an equation for the rose curve $r = 2\sin 2\theta$ after it has been rotated through the given angle.

(a) $\frac{\pi}{6}$ (b) $\frac{\pi}{2}$ (c) $\frac{2\pi}{3}$ (d) π

65. Sketch the graph of each equation.

(a) $r = 1 - \sin\theta$ (b) $r = 1 - \sin\left(\theta - \frac{\pi}{4}\right)$

66. Sketch the graph of each equation.

(a) $r = 3\sec\theta$ (b) $r = 3\sec\left(\theta - \frac{\pi}{4}\right)$

(c) $r = 3\sec\left(\theta + \frac{\pi}{3}\right)$ (d) $r = 3\sec\left(\theta - \frac{\pi}{2}\right)$

67. ***Exploration*** Use a graphing utility to graph and identify $r = 2 + k\sin\theta$ for $k = 0, 1, 2$, and 3.

68. ***Exploration*** Consider the equation $r = 3\sin k\theta$.

(a) Use a graphing utility to graph the equation for $k = 1.5$. Find the interval for θ over which the graph is traced only once.

(b) Use a graphing utility to graph the equation for $k = 2.5$. Find the interval for θ over which the graph is traced only once.

(c) Is it possible to find an interval for θ over which the graph is traced only once for any rational number k? Explain.

Skills Review

In Exercises 69–72, find the zeros (if any) of the rational function.

69. $f(x) = \frac{x^2 - 9}{x + 1}$

70. $f(x) = 6 + \frac{4}{x^2 + 4}$

71. $f(x) = 5 - \frac{3}{x - 2}$

72. $f(x) = \frac{x^3 - 27}{x^2 + 4}$

In Exercises 73 and 74, find the standard form of the equation of the ellipse with the given characteristics. Then sketch the ellipse.

73. Vertices: $(-4, 2), (2, 2)$; minor axis of length 4

74. Foci: $(3, 2), (3, -4)$; major axis of length 8

6.8 Polar Equations of Conics

What you should learn

- Define conics in terms of eccentricity.
- Write and graph equations of conics in polar form.
- Use equations of conics in polar form to model real-life problems.

Why you should learn it

The orbits of planets and satellites can be modeled with polar equations. For instance, in Exercise 58 on page 512, a polar equation is used to model the orbit of a satellite.

Alternative Definition of Conic

In Sections 6.3 and 6.4, you learned that the rectangular equations of ellipses and hyperbolas take simple forms when the origin lies at their *centers*. As it happens, there are many important applications of conics in which it is more convenient to use one of the *foci* as the origin. In this section, you will learn that polar equations of conics take simple forms if one of the foci lies at the pole.

To begin, consider the following alternative definition of conic that uses the concept of eccentricity.

> **Alternative Definition of Conic**
>
> The locus of a point in the plane that moves so that its distance from a fixed point (focus) is in a constant ratio to its distance from a fixed line (directrix) is a **conic.** The constant ratio is the **eccentricity** of the conic and is denoted by e. Moreover, the conic is an **ellipse** if $e < 1$, a **parabola** if $e = 1$, and a **hyperbola** if $e > 1$. (See Figure 6.69.)

In Figure 6.69, note that for each type of conic, the focus is at the pole.

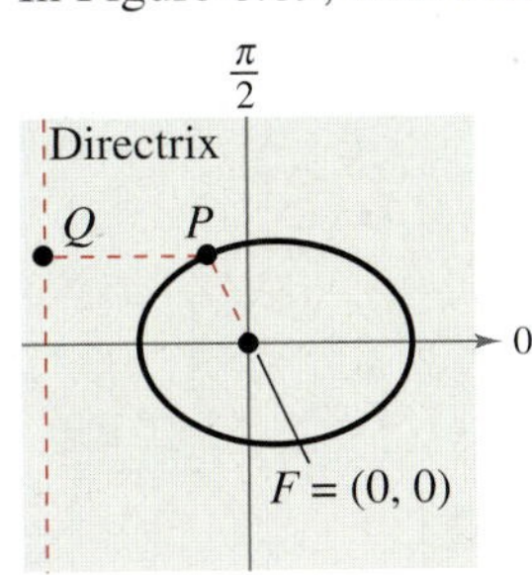

Ellipse: $0 < e < 1$

$\frac{PF}{PQ} < 1$

Parabola: $e = 1$

$\frac{PF}{PQ} = 1$

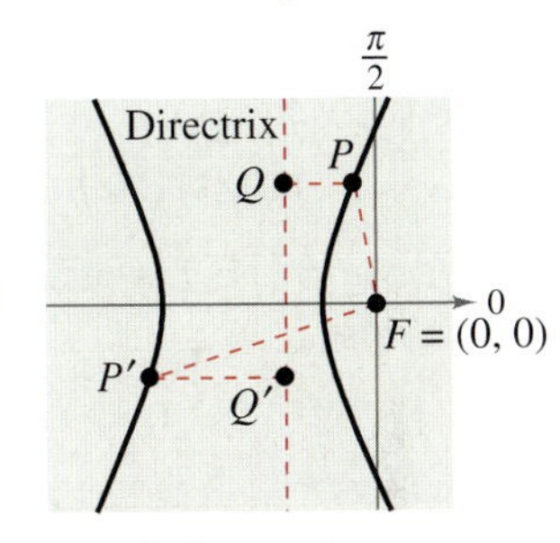

Hyperbola $e > 1$

$\frac{PF}{PQ} = \frac{P'F}{P'Q'} > 1$

FIGURE 6.69

Polar Equations of Conics

The benefit of locating a focus of a conic at the pole is that the equation of the conic takes on a simpler form. For a proof of the polar equations of conics, see Proofs in Mathematics on page 524.

> **Polar Equations of Conics**
>
> The graph of a polar equation of the form
>
> **1.** $r = \dfrac{ep}{1 \pm e \cos \theta}$ or **2.** $r = \dfrac{ep}{1 \pm e \sin \theta}$
>
> is a conic, where $e > 0$ is the eccentricity and $|p|$ is the distance between the focus (pole) and the directrix.

Equations of the form

$$r = \frac{ep}{1 \pm e \cos \theta} = g(\cos \theta)$$ Vertical directrix

correspond to conics with a vertical directrix and symmetry with respect to the polar axis. Equations of the form

$$r = \frac{ep}{1 \pm e \sin \theta} = g(\sin \theta)$$ Horizontal directrix

correspond to conics with a horizontal directrix and symmetry with respect to the line $\theta = \pi/2$. Moreover, the converse is also true—that is, any conic with a focus at the pole and having a horizontal or vertical directrix can be represented by one of the given equations.

Example 1 Identifying a Conic from Its Equation

Identify the type of conic represented by the equation $r = \dfrac{15}{3 - 2 \cos \theta}$.

Algebraic Solution

To identify the type of conic, rewrite the equation in the form $r = (ep)/(1 \pm e \cos \theta)$.

$$r = \frac{15}{3 - 2 \cos \theta}$$ Write original equation.

$$= \frac{5}{1 - (2/3) \cos \theta}$$ Divide numerator and denominator by 3.

Because $e = \frac{2}{3} < 1$, you can conclude that the graph is an ellipse.

CHECKPOINT Now try Exercise 11.

Graphical Solution

You can start sketching the graph by plotting points from $\theta = 0$ to $\theta = \pi$. Because the equation is of the form $r = g(\cos \theta)$, the graph of r is symmetric with respect to the polar axis. So, you can complete the sketch, as shown in Figure 6.70. From this, you can conclude that the graph is an ellipse.

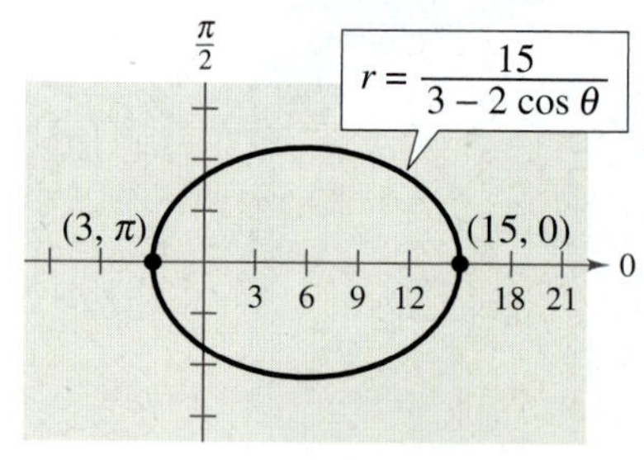

FIGURE 6.70

Additional Example
Identify the conic and sketch its graph.

$r = \dfrac{4}{2 + 2 \sin \theta}$

Solution

Parabola

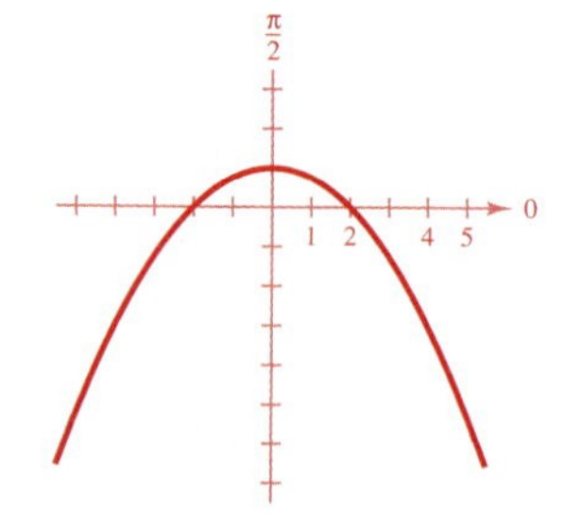

For the ellipse in Figure 6.70, the major axis is horizontal and the vertices lie at (15, 0) and $(3, \pi)$. So, the length of the *major* axis is $2a = 18$. To find the length of the *minor* axis, you can use the equations $e = c/a$ and $b^2 = a^2 - c^2$ to conclude that

$$\begin{aligned} b^2 &= a^2 - c^2 \\ &= a^2 - (ea)^2 \\ &= a^2(1 - e^2). \end{aligned}$$ Ellipse

Because $e = \frac{2}{3}$, you have $b^2 = 9^2\left[1 - \left(\frac{2}{3}\right)^2\right] = 45$, which implies that $b = \sqrt{45} = 3\sqrt{5}$. So, the length of the minor axis is $2b = 6\sqrt{5}$. A similar analysis for hyperbolas yields

$$\begin{aligned} b^2 &= c^2 - a^2 \\ &= (ea)^2 - a^2 \\ &= a^2(e^2 - 1). \end{aligned}$$ Hyperbola

Example 2 Sketching a Conic from Its Polar Equation

Identify the conic $r = \dfrac{32}{3 + 5\sin\theta}$ and sketch its graph.

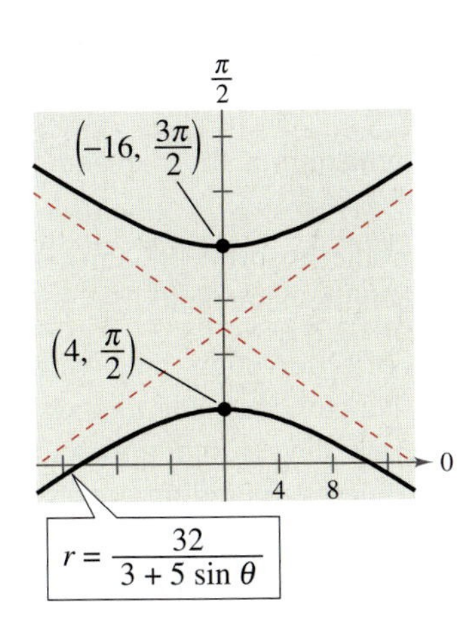

FIGURE 6.71

Solution

Dividing the numerator and denominator by 3, you have

$$r = \frac{32/3}{1 + (5/3)\sin\theta}.$$

Because $e = \frac{5}{3} > 1$, the graph is a hyperbola. The transverse axis of the hyperbola lies on the line $\theta = \pi/2$, and the vertices occur at $(4, \pi/2)$ and $(-16, 3\pi/2)$. Because the length of the transverse axis is 12, you can see that $a = 6$. To find b, write

$$b^2 = a^2(e^2 - 1) = 6^2\left[\left(\frac{5}{3}\right)^2 - 1\right] = 64.$$

So, $b = 8$. Finally, you can use a and b to determine that the asymptotes of the hyperbola are $y = 10 \pm \frac{3}{4}x$. The graph is shown in Figure 6.71.

CHECKPOINT Now try Exercise 19.

Technology

Use a graphing utility set in *polar* mode to verify the four orientations shown at the right. Remember that e must be positive, but p can be positive or negative.

In the next example, you are asked to find a polar equation of a specified conic. To do this, let p be the distance between the pole and the directrix.

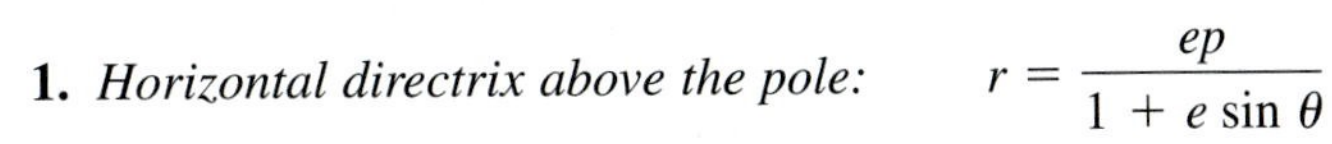

1. *Horizontal directrix above the pole:* $r = \dfrac{ep}{1 + e\sin\theta}$
2. *Horizontal directrix below the pole:* $r = \dfrac{ep}{1 - e\sin\theta}$
3. *Vertical directrix to the right of the pole:* $r = \dfrac{ep}{1 + e\cos\theta}$
4. *Vertical directrix to the left of the pole:* $r = \dfrac{ep}{1 - e\cos\theta}$

Example 3 Finding the Polar Equation of a Conic

Find the polar equation of the parabola whose focus is the pole and whose directrix is the line $y = 3$.

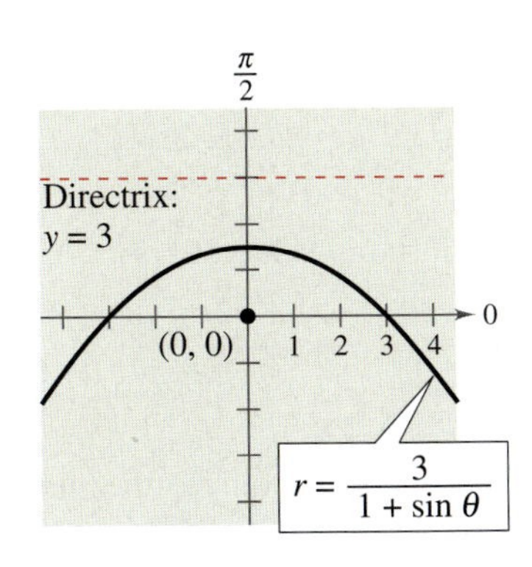

FIGURE 6.72

Solution

From Figure 6.72, you can see that the directrix is horizontal and above the pole, so you can choose an equation of the form

$$r = \frac{ep}{1 + e\sin\theta}.$$

Moreover, because the eccentricity of a parabola is $e = 1$ and the distance between the pole and the directrix is $p = 3$, you have the equation

$$r = \frac{3}{1 + \sin\theta}.$$

CHECKPOINT Now try Exercise 33.

Activities

1. Identify the conic and sketch its graph.

$$r = \frac{4}{3 - 2\cos\theta}$$

Answer: Ellipse

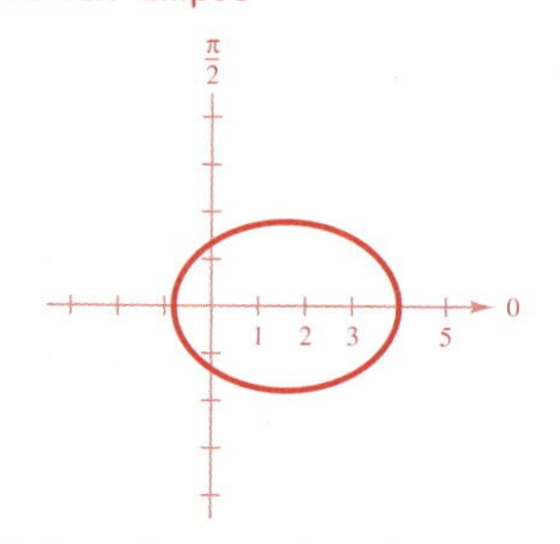

2. Find a polar equation of the parabola with focus at the pole and directrix $y = 2$.

Answer: $r = \dfrac{2}{1 + \sin\theta}$

Applications

Kepler's Laws (listed below), named after the German astronomer Johannes Kepler (1571–1630), can be used to describe the orbits of the planets about the sun.

1. Each planet moves in an elliptical orbit with the sun at one focus.
2. A ray from the sun to the planet sweeps out equal areas of the ellipse in equal times.
3. The square of the period (the time it takes for a planet to orbit the sun) is proportional to the cube of the mean distance between the planet and the sun.

Although Kepler simply stated these laws on the basis of observation, they were later validated by Isaac Newton (1642–1727). In fact, Newton was able to show that each law can be deduced from a set of universal laws of motion and gravitation that govern the movement of all heavenly bodies, including comets and satellites. This is illustrated in the next example, which involves the comet named after the English mathematician and physicist Edmund Halley (1656–1742).

If you use Earth as a reference with a period of 1 year and a distance of 1 astronomical unit (an *astronomical unit* is defined as the mean distance between Earth and the sun, or about 93 million miles), the proportionality constant in Kepler's third law is 1. For example, because Mars has a mean distance to the sun of $d = 1.524$ astronomical units, its period P is given by $d^3 = P^2$. So, the period of Mars is $P \approx 1.88$ years.

Example 4 Halley's Comet

Halley's comet has an elliptical orbit with an eccentricity of $e \approx 0.967$. The length of the major axis of the orbit is approximately 35.88 astronomical units. Find a polar equation for the orbit. How close does Halley's comet come to the sun?

Solution

Using a vertical axis, as shown in Figure 6.73, choose an equation of the form $r = ep/(1 + e\sin\theta)$. Because the vertices of the ellipse occur when $\theta = \pi/2$ and $\theta = 3\pi/2$, you can determine the length of the major axis to be the sum of the r-values of the vertices. That is,

$$2a = \frac{0.967p}{1 + 0.967} + \frac{0.967p}{1 - 0.967} \approx 29.79p \approx 35.88.$$

So, $p \approx 1.204$ and $ep \approx (0.967)(1.204) \approx 1.164$. Using this value of ep in the equation, you have

$$r = \frac{1.164}{1 + 0.967\sin\theta}$$

where r is measured in astronomical units. To find the closest point to the sun (the focus), substitute $\theta = \pi/2$ in this equation to obtain

$$r = \frac{1.164}{1 + 0.967\sin(\pi/2)} \approx 0.59 \text{ astronomical unit} \approx 55{,}000{,}000 \text{ miles.}$$

FIGURE 6.73

CHECKPOINT Now try Exercise 57.

6.8 Exercises

VOCABULARY CHECK:

In Exercises 1–3, fill in the blanks.

1. The locus of a point in the plane that moves so that its distance from a fixed point (focus) is in a constant ratio to its distance from a fixed line (directrix) is a ________.
2. The constant ratio is the ________ of the conic and is denoted by ________.
3. An equation of the form $r = \dfrac{ep}{1 + e\cos\theta}$ has a ________ directrix to the ________ of the pole.
4. Match the conic with its eccentricity.

   (a) $e < 1$ (b) $e = 1$ (c) $e > 1$

   (i) parabola (ii) hyperbola (iii) ellipse

PREREQUISITE SKILLS REVIEW: Practice and review algebra skills needed for this section at **www.Eduspace.com.**

In Exercises 1–4, write the polar equation of the conic for $e = 1$, $e = 0.5$, and $e = 1.5$. Identify the conic for each equation. Verify your answers with a graphing utility.

1. $r = \dfrac{4e}{1 + e\cos\theta}$
2. $r = \dfrac{4e}{1 - e\cos\theta}$
3. $r = \dfrac{4e}{1 - e\sin\theta}$
4. $r = \dfrac{4e}{1 + e\sin\theta}$

In Exercises 5–10, match the polar equation with its graph. [The graphs are labeled (a), (b), (c), (d), (e), and (f).]

(a)

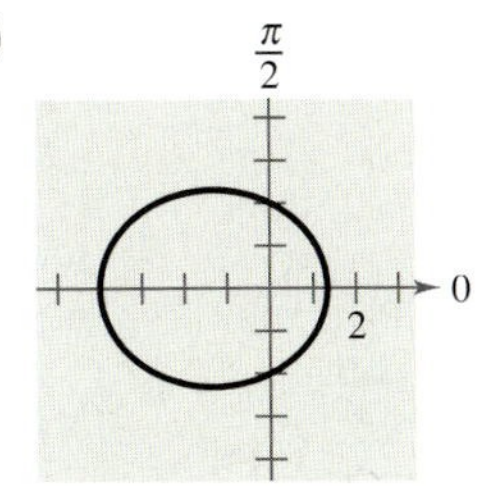

(b)

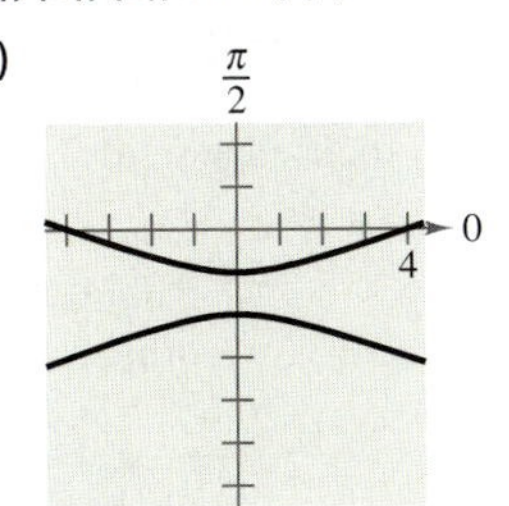

(c)

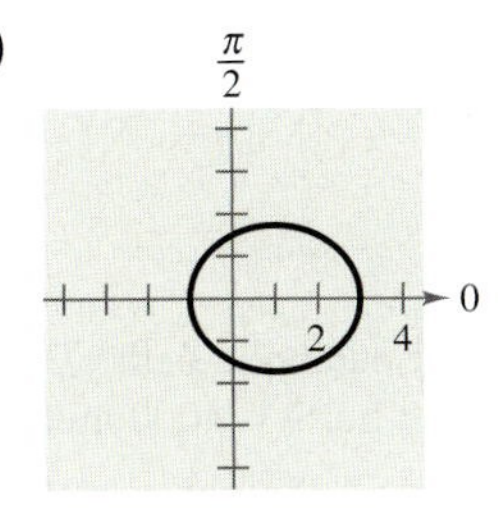

(d)

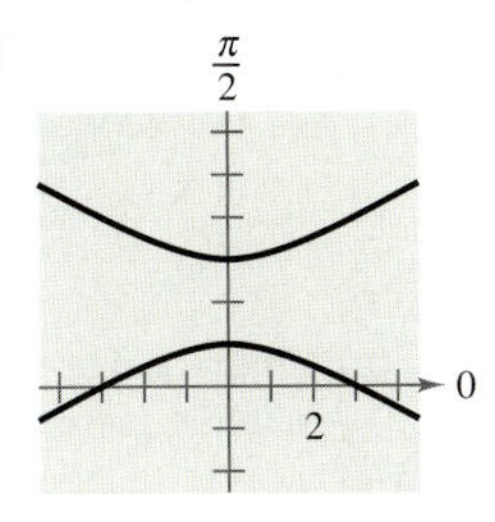

(e)

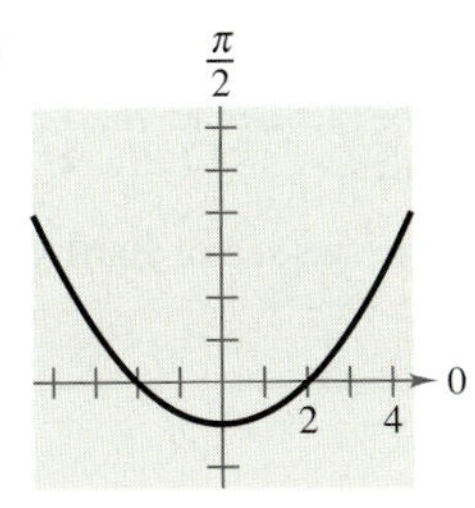

(f)

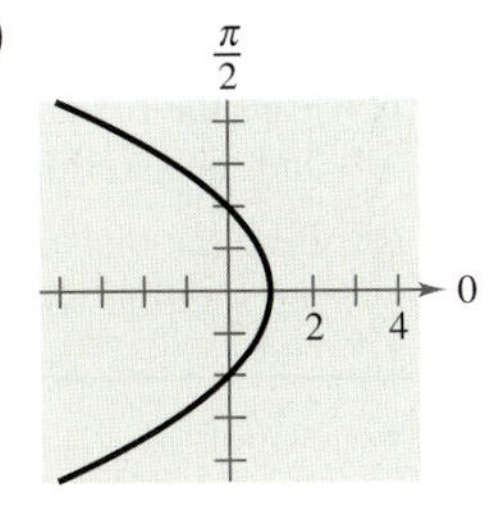

5. $r = \dfrac{2}{1 + \cos\theta}$
6. $r = \dfrac{3}{2 - \cos\theta}$
7. $r = \dfrac{3}{1 + 2\sin\theta}$
8. $r = \dfrac{2}{1 - \sin\theta}$
9. $r = \dfrac{4}{2 + \cos\theta}$
10. $r = \dfrac{4}{1 - 3\sin\theta}$

In Exercises 11–24, identify the conic and sketch its graph.

11. $r = \dfrac{2}{1 - \cos\theta}$
12. $r = \dfrac{3}{1 + \sin\theta}$
13. $r = \dfrac{5}{1 + \sin\theta}$
14. $r = \dfrac{6}{1 + \cos\theta}$
15. $r = \dfrac{2}{2 - \cos\theta}$
16. $r = \dfrac{3}{3 + \sin\theta}$
17. $r = \dfrac{6}{2 + \sin\theta}$
18. $r = \dfrac{9}{3 - 2\cos\theta}$
19. $r = \dfrac{3}{2 + 4\sin\theta}$
20. $r = \dfrac{5}{-1 + 2\cos\theta}$
21. $r = \dfrac{3}{2 - 6\cos\theta}$
22. $r = \dfrac{3}{2 + 6\sin\theta}$
23. $r = \dfrac{4}{2 - \cos\theta}$
24. $r = \dfrac{2}{2 + 3\sin\theta}$

In Exercises 25–28, use a graphing utility to graph the polar equation. Identify the graph.

25. $r = \dfrac{-1}{1 - \sin\theta}$
26. $r = \dfrac{-5}{2 + 4\sin\theta}$
27. $r = \dfrac{3}{-4 + 2\cos\theta}$
28. $r = \dfrac{4}{1 - 2\cos\theta}$

In Exercises 29–32, use a graphing utility to graph the rotated conic.

29. $r = \dfrac{2}{1 - \cos(\theta - \pi/4)}$ (See Exercise 11.)

30. $r = \dfrac{3}{3 + \sin(\theta - \pi/3)}$ (See Exercise 16.)

31. $r = \dfrac{6}{2 + \sin(\theta + \pi/6)}$ (See Exercise 17.)

32. $r = \dfrac{5}{-1 + 2\cos(\theta + 2\pi/3)}$ (See Exercise 20.)

In Exercises 33–48, find a polar equation of the conic with its focus at the pole.

	Conic	*Eccentricity*	*Directrix*
33.	Parabola	$e = 1$	$x = -1$
34.	Parabola	$e = 1$	$y = -2$
35.	Ellipse	$e = \frac{1}{2}$	$y = 1$
36.	Ellipse	$e = \frac{3}{4}$	$y = -3$
37.	Hyperbola	$e = 2$	$x = 1$
38.	Hyperbola	$e = \frac{3}{2}$	$x = -1$

	Conic	*Vertex or Vertices*
39.	Parabola	$(1, -\pi/2)$
40.	Parabola	$(6, 0)$
41.	Parabola	$(5, \pi)$
42.	Parabola	$(10, \pi/2)$
43.	Ellipse	$(2, 0), (10, \pi)$
44.	Ellipse	$(2, \pi/2), (4, 3\pi/2)$
45.	Ellipse	$(20, 0), (4, \pi)$
46.	Hyperbola	$(2, 0), (8, 0)$
47.	Hyperbola	$(1, 3\pi/2), (9, 3\pi/2)$
48.	Hyperbola	$(4, \pi/2), (1, \pi/2)$

49. ***Planetary Motion*** The planets travel in elliptical orbits with the sun at one focus. Assume that the focus is at the pole, the major axis lies on the polar axis, and the length of the major axis is $2a$ (see figure). Show that the polar equation of the orbit is $r = a(1 - e^2)/(1 - e\cos\theta)$ where e is the eccentricity.

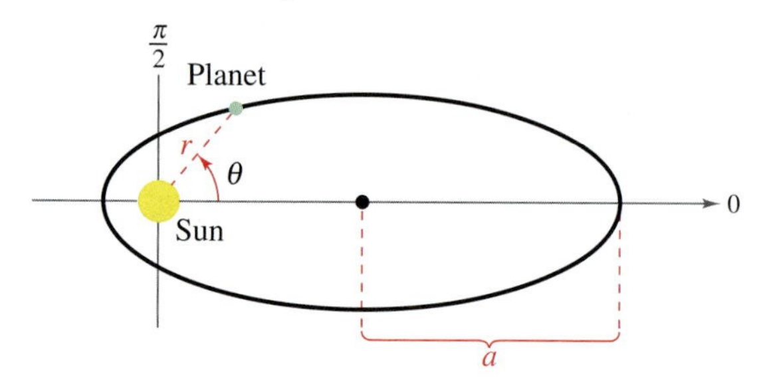

50. ***Planetary Motion*** Use the result of Exercise 49 to show that the minimum distance (*perihelion distance*) from the sun to the planet is $r = a(1 - e)$ and the maximum distance (*aphelion distance*) is $r = a(1 + e)$.

Planetary Motion **In Exercises 51–56, use the results of Exercises 49 and 50 to find the polar equation of the planet's orbit and the perihelion and aphelion distances.**

51. Earth $a = 95.956 \times 10^6$ miles, $e = 0.0167$

52. Saturn $a = 1.427 \times 10^9$ kilometers, $e = 0.0542$

53. Venus $a = 108.209 \times 10^6$ kilometers, $e = 0.0068$

54. Mercury $a = 35.98 \times 10^6$ miles, $e = 0.2056$

55. Mars $a = 141.63 \times 10^6$ miles, $e = 0.0934$

56. Jupiter $a = 778.41 \times 10^6$ kilometers, $e = 0.0484$

57. ***Astronomy*** The comet Encke has an elliptical orbit with an eccentricity of $e \approx 0.847$. The length of the major axis of the orbit is approximately 4.42 astronomical units. Find a polar equation for the orbit. How close does the comet come to the sun?

Model It

58. ***Satellite Tracking*** A satellite in a 100-mile-high circular orbit around Earth has a velocity of approximately 17,500 miles per hour. If this velocity is multiplied by $\sqrt{2}$, the satellite will have the minimum velocity necessary to escape Earth's gravity and it will follow a parabolic path with the center of Earth as the focus (see figure).

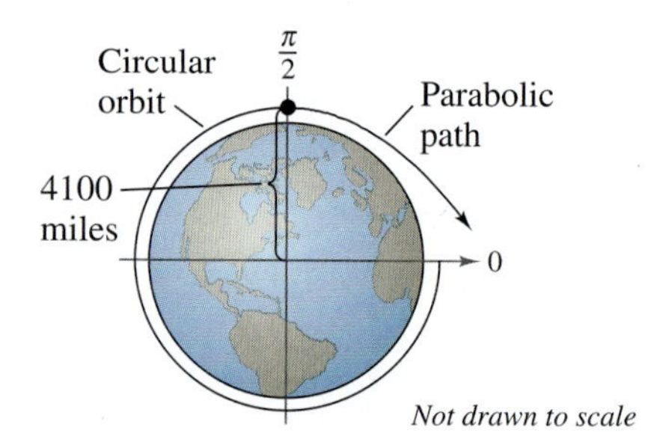

(a) Find a polar equation of the parabolic path of the satellite (assume the radius of Earth is 4000 miles).

(b) Use a graphing utility to graph the equation you found in part (a).

(c) Find the distance between the surface of the Earth and the satellite when $\theta = 30°$.

(d) Find the distance between the surface of Earth and the satellite when $\theta = 60°$.

Synthesis

True or False? **In Exercises 59–61, determine whether the statement is true or false. Justify your answer.**

59. For a given value of $e > 1$ over the interval $\theta = 0$ to $\theta = 2\pi$, the graph of

$$r = \frac{ex}{1 - e\cos\theta}$$

is the same as the graph of

$$r = \frac{e(-x)}{1 + e\cos\theta}.$$

60. The graph of

$$r = \frac{4}{-3 - 3\sin\theta}$$

has a horizontal directrix above the pole.

61. The conic represented by the following equation is an ellipse.

$$r^2 = \frac{16}{9 - 4\cos\left(\theta + \frac{\pi}{4}\right)}$$

62. ***Writing*** In your own words, define the term *eccentricity* and explain how it can be used to classify conics.

63. Show that the polar equation of the ellipse

$$\frac{x^2}{a^2} + \frac{y^2}{b^2} = 1 \quad \text{is} \quad r^2 = \frac{b^2}{1 - e^2\cos^2\theta}.$$

64. Show that the polar equation of the hyperbola

$$\frac{x^2}{a^2} - \frac{y^2}{b^2} = 1 \quad \text{is} \quad r^2 = \frac{-b^2}{1 - e^2\cos^2\theta}.$$

In Exercises 65–70, use the results of Exercises 63 and 64 to write the polar form of the equation of the conic.

65. $\dfrac{x^2}{169} + \dfrac{y^2}{144} = 1$

66. $\dfrac{x^2}{25} + \dfrac{y^2}{16} = 1$

67. $\dfrac{x^2}{9} - \dfrac{y^2}{16} = 1$

68. $\dfrac{x^2}{36} - \dfrac{y^2}{4} = 1$

69. Hyperbola — One focus: $(5, \pi/2)$; Vertices: $(4, \pi/2), (4, -\pi/2)$

70. Ellipse — One focus: $(4, 0)$; Vertices: $(5, 0), (5, \pi)$

71. ***Exploration*** Consider the polar equation

$$r = \frac{4}{1 - 0.4\cos\theta}.$$

(a) Identify the conic without graphing the equation.

(b) Without graphing the following polar equations, describe how each differs from the given polar equation.

$$r = \frac{4}{1 + 0.4\cos\theta}, \quad r_2 = \frac{4}{1 - 0.4\sin\theta}$$

(c) Use a graphing utility to verify your results in part (b).

72. ***Exploration*** The equation

$$r = \frac{ep}{1 \pm e\sin\theta}$$

is the equation of an ellipse with $e < 1$. What happens to the lengths of both the major axis and the minor axis when the value of e remains fixed and the value of p changes? Use an example to explain your reasoning.

Skills Review

In Exercises 73–78, solve the trigonometric equation.

73. $4\sqrt{3}\tan\theta - 3 = 1$

74. $6\cos x - 2 = 1$

75. $12\sin^2\theta = 9$

76. $9\csc^2 x - 10 = 2$

77. $2\cot x = 5\cos\dfrac{\pi}{2}$

78. $\sqrt{2}\sec\theta = 2\csc\dfrac{\pi}{4}$

In Exercises 79–82, find the exact value of the trigonometric function given that u and v are in Quadrant IV and $\sin u = -\frac{3}{5}$ and $\cos v = 1/\sqrt{2}$.

79. $\cos(u + v)$

80. $\sin(u + v)$

81. $\cos(u - v)$

82. $\sin(u - v)$

In Exercises 83 and 84, find the exact values of $\sin 2u$, $\cos 2u$, and $\tan 2u$ using the double-angle formulas.

83. $\sin u = \dfrac{4}{5}, \quad \dfrac{\pi}{2} < u < \pi$

84. $\tan u = -\sqrt{3}, \quad \dfrac{3\pi}{2} < u < 2\pi$

6 Chapter Summary

What did you learn?

Section 6.1	Review Exercises
☐ Find the inclination of a line *(p. 450)*.	1–4
☐ Find the angle between two lines *(p. 451)*.	5–8
☐ Find the distance between a point and a line *(p. 452)*.	9, 10
Section 6.2	
☐ Recognize a conic as the intersection of a plane and a double-napped cone *(p. 457)*.	11, 12
☐ Write equations of parabolas in standard form and graph parabolas *(p. 458)*.	13–16
☐ Use the reflective property of parabolas to solve real-life problems *(p. 460)*.	17–20
Section 6.3	
☐ Write equations of ellipses in standard form and graph ellipses *(p. 466)*.	21–24
☐ Use properties of ellipses to model and solve real-life problems *(p. 470)*.	25, 26
☐ Find the eccentricities of ellipses *(p. 470)*.	27–30
Section 6.4	
☐ Write equations of hyperbolas in standard form *(p. 475)*.	31–34
☐ Find asymptotes of and graph hyperbolas *(p. 477)*.	35–38
☐ Use properties of hyperbolas to solve real-life problems *(p. 480)*.	39, 40
☐ Classify conics from their general equations *(p. 481)*.	41–44
Section 6.5	
☐ Evaluate sets of parametric equations for given values of the parameter *(p. 485)*.	45, 46
☐ Sketch curves that are represented by sets of parametric equations *(p. 486)* and rewrite the equations as single rectangular equations *(p. 487)*.	47–52
☐ Find sets of parametric equations for graphs *(p. 488)*.	53–56
Section 6.6	
☐ Plot points on the polar coordinate system *(p. 493)*.	57–60
☐ Convert points from rectangular to polar form and vice versa *(p. 494)*.	61–68
☐ Convert equations from rectangular to polar form and vice versa *(p. 496)*.	69–80
Section 6.7	
☐ Graph polar equations by point plotting *(p. 499)*.	81–90
☐ Use symmetry *(p. 500)*, zeros, and maximum r-values *(p. 501)* to sketch graphs of polar equations.	81–90
☐ Recognize special polar graphs *(p. 503)*.	91–94
Section 6.8	
☐ Define conics in terms of eccentricity and write and graph equations of conics in polar form *(p. 507)*.	95–102
☐ Use equations of conics in polar form to model real-life problems *(p. 510)*.	103, 104

6 Review Exercises

6.1 In Exercises 1–4, find the inclination θ (in radians and degrees) of the line with the given characteristics.

1. Passes through the points $(-1, 2)$ and $(2, 5)$
2. Passes through the points $(3, 4)$ and $(-2, 7)$
3. Equation: $y = 2x + 4$
4. Equation: $6x - 7y - 5 = 0$

In Exercises 5–8, find the angle θ (in radians and degrees) between the lines.

5. $4x + y = 2$
   $-5x + y = -1$

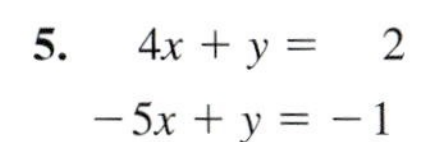

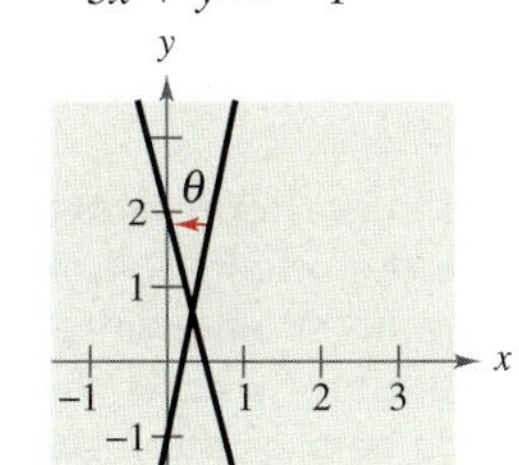

6. $-5x + 3y = 3$
   $-2x + 3y = 1$

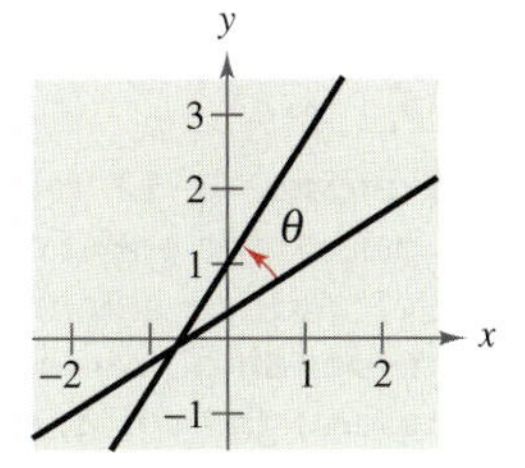

7. $2x - 7y = 8$
   $0.4x + y = 0$

8. $0.02x + 0.07y = 0.18$
   $0.09x - 0.04y = 0.17$

In Exercises 9 and 10, find the distance between the point and the line.

	Point	*Line*
9.	$(1, 2)$	$x - y - 3 = 0$
10.	$(0, 4)$	$x + 2y - 2 = 0$

6.2 In Exercises 11 and 12, state what type of conic is formed by the intersection of the plane and the double-napped cone.

11.

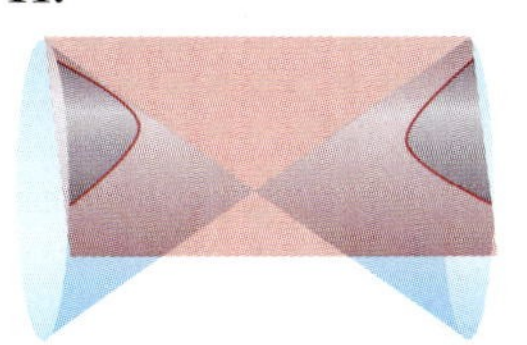

12.

In Exercises 13–16, find the standard form of the equation of the parabola with the given characteristics. Then graph the parabola.

13. Vertex: $(0, 0)$
    Focus: $(4, 0)$
14. Vertex: $(2, 0)$
    Focus: $(0, 0)$
15. Vertex: $(0, 2)$
    Directrix: $x = -3$
16. Vertex: $(2, 2)$
    Directrix: $y = 0$

In Exercises 17 and 18, find an equation of the tangent line to the parabola at the given point, and find the x-intercept of the line.

17. $x^2 = -2y, \ (2, -2)$
18. $x^2 = -2y, \ (-4, -8)$

19. ***Architecture*** A parabolic archway is 12 meters high at the vertex. At a height of 10 meters, the width of the archway is 8 meters (see figure). How wide is the archway at ground level?

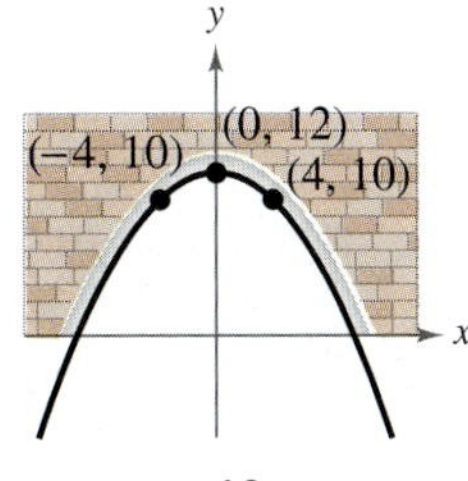

FIGURE FOR 19

FIGURE FOR 20

20. ***Flashlight*** The light bulb in a flashlight is at the focus of its parabolic reflector, 1.5 centimeters from the vertex of the reflector (see figure). Write an equation of a cross section of the flashlight's reflector with its focus on the positive x-axis and its vertex at the origin.

6.3 In Exercises 21–24, find the standard form of the equation of the ellipse with the given characteristics. Then graph the ellipse.

21. Vertices: $(-3, 0), (7, 0)$; foci: $(0, 0), (4, 0)$
22. Vertices: $(2, 0), (2, 4)$; foci: $(2, 1), (2, 3)$
23. Vertices: $(0, 1), (4, 1)$; endpoints of the minor axis: $(2, 0), (2, 2)$
24. Vertices: $(-4, -1), (-4, 11)$; endpoints of the minor axis: $(-6, 5), (-2, 5)$

25. ***Architecture*** A semielliptical archway is to be formed over the entrance to an estate. The arch is to be set on pillars that are 10 feet apart and is to have a height (atop the pillars) of 4 feet. Where should the foci be placed in order to sketch the arch?

26. ***Wading Pool*** You are building a wading pool that is in the shape of an ellipse. Your plans give an equation for the elliptical shape of the pool measured in feet as

$$\frac{x^2}{324} + \frac{y^2}{196} = 1.$$

Find the longest distance across the pool, the shortest distance, and the distance between the foci.

In Exercises 27–30, find the center, vertices, foci, and eccentricity of the ellipse.

27. $\dfrac{(x+2)^2}{81}+\dfrac{(y-1)^2}{100}=1$

28. $\dfrac{(x-5)^2}{1}+\dfrac{(y+3)^2}{36}=1$

29. $16x^2+9y^2-32x+72y+16=0$

30. $4x^2+25y^2+16x-150y+141=0$

6.4 In Exercises 31–34, find the standard form of the equation of the hyperbola with the given characteristics.

31. Vertices: $(0, \pm 1)$; foci: $(0, \pm 3)$

32. Vertices: $(2, 2), (-2, 2)$; foci: $(4, 2), (-4, 2)$

33. Foci: $(0, 0), (8, 0)$; asymptotes: $y = \pm 2(x-4)$

34. Foci: $(3, \pm 2)$; asymptotes: $y = \pm 2(x-3)$

In Exercises 35–38, find the center, vertices, foci, and the equations of the asymptotes of the hyperbola, and sketch its graph using the asymptotes as an aid.

35. $\dfrac{(x-3)^2}{16}-\dfrac{(y+5)^2}{4}=1$

36. $\dfrac{(y-1)^2}{4}-x^2=1$

37. $9x^2-16y^2-18x-32y-151=0$

38. $-4x^2+25y^2-8x+150y+121=0$

39. ***LORAN*** Radio transmitting station A is located 200 miles east of transmitting station B. A ship is in an area to the north and 40 miles west of station A. Synchronized radio pulses transmitted at 186,000 miles per second by the two stations are received 0.0005 second sooner from station A than from station B. How far north is the ship?

40. ***Locating an Explosion*** Two of your friends live 4 miles apart and on the same "east-west" street, and you live halfway between them. You are having a three-way phone conversation when you hear an explosion. Six seconds later, your friend to the east hears the explosion, and your friend to the west hears it 8 seconds after you do. Find equations of two hyperbolas that would locate the explosion. (Assume that the coordinate system is measured in feet and that sound travels at 1100 feet per second.)

In Exercises 41–44, classify the graph of the equation as a circle, a parabola, an ellipse, or a hyperbola.

41. $5x^2-2y^2+10x-4y+17=0$

42. $-4y^2+5x+3y+7=0$

43. $3x^2+2y^2-12x+12y+29=0$

44. $4x^2+4y^2-4x+8y-11=0$

6.5 In Exercises 45 and 46, complete the table for each set of parametric equations. Plot the points (x, y) and sketch a graph of the parametric equations.

45. $x = 3t - 2$ and $y = 7 - 4t$

t	-3	-2	-1	0	1	2	3
x							
y							

46. $x = \dfrac{1}{5}t$ and $y = \dfrac{4}{t-1}$

t	-1	0	2	3	4	5
x						
y						

In Exercises 47–52, (a) sketch the curve represented by the parametric equations (indicate the orientation of the curve) and (b) eliminate the parameter and write the corresponding rectangular equation whose graph represents the curve. Adjust the domain of the resulting rectangular equation, if necessary. (c) Verify your result with a graphing utility.

47. $x = 2t$
$y = 4t$

48. $x = 1 + 4t$
$y = 2 - 3t$

49. $x = t^2$
$y = \sqrt{t}$

50. $x = t + 4$
$y = t^2$

51. $x = 6\cos\theta$
$y = 6\sin\theta$

52. $x = 3 + 3\cos\theta$
$y = 2 + 5\sin\theta$

53. Find a parametric representation of the circle with center $(5, 4)$ and radius 6.

54. Find a parametric representation of the ellipse with center $(-3, 4)$, major axis horizontal and eight units in length, and minor axis six units in length.

55. Find a parametric representation of the hyperbola with vertices $(0, \pm 4)$ and foci $(0, \pm 5)$.

56. ***Involute of a Circle*** The *involute* of a circle is described by the endpoint P of a string that is held taut as it is unwound from a spool (see figure). The spool does not rotate. Show that a parametric representation of the involute of a circle is

$$x = r(\cos\theta + \theta\sin\theta)$$

$$y = r(\sin\theta - \theta\cos\theta).$$

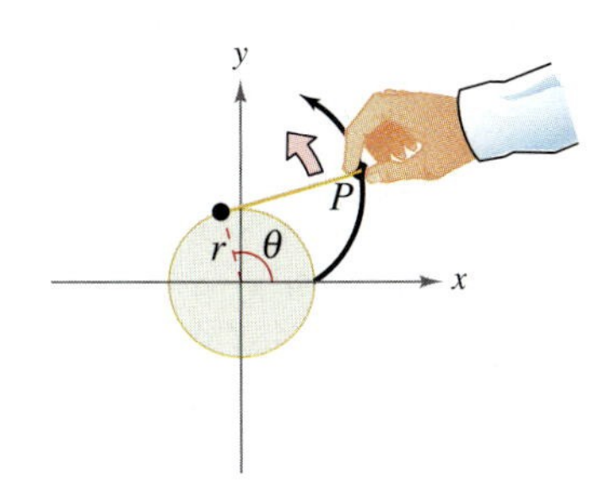

6.6 **In Exercises 57–60, plot the point given in polar coordinates and find two additional polar representations of the point, using $-2\pi < \theta < 2\pi$.**

57. $\left(2, \frac{\pi}{4}\right)$

58. $\left(-5, -\frac{\pi}{3}\right)$

59. $(-7, 4.19)$

60. $(\sqrt{3}, 2.62)$

In Exercises 61–64, a point in polar coordinates is given. Convert the point to rectangular coordinates.

61. $\left(-1, \frac{\pi}{3}\right)$ **62.** $\left(2, \frac{5\pi}{4}\right)$

63. $\left(3, \frac{3\pi}{4}\right)$ **64.** $\left(0, \frac{\pi}{2}\right)$

In Exercises 65–68, a point in rectangular coordinates is given. Convert the point to polar coordinates.

65. $(0, 2)$

66. $(-\sqrt{5}, \sqrt{5})$

67. $(4, 6)$

68. $(3, -4)$

In Exercises 69–74, convert the rectangular equation to polar form.

69. $x^2 + y^2 = 49$ **70.** $x^2 + y^2 = 20$

71. $x^2 + y^2 - 6y = 0$ **72.** $x^2 + y^2 - 4x = 0$

73. $xy = 5$ **74.** $xy = -2$

In Exercises 75–80, convert the polar equation to rectangular form.

75. $r = 5$ **76.** $r = 12$

77. $r = 3\cos\theta$ **78.** $r = 8\sin\theta$

79. $r^2 = \sin\theta$ **80.** $r^2 = \cos 2\theta$

6.7 **In Exercises 81–90, determine the symmetry of r, the maximum value of $|r|$, and any zeros of r. Then sketch the graph of the polar equation (plot additional points if necessary).**

81. $r = 4$

82. $r = 11$

83. $r = 4\sin 2\theta$

84. $r = \cos 5\theta$

85. $r = -2(1 + \cos\theta)$

86. $r = 3 - 4\cos\theta$

87. $r = 2 + 6\sin\theta$

88. $r = 5 - 5\cos\theta$

89. $r = -3\cos 2\theta$

90. $r = \cos 2\theta$

In Exercises 91–94, identify the type of polar graph and use a graphing utility to graph the equation.

91. $r = 3(2 - \cos\theta)$

92. $r = 3(1 - 2\cos\theta)$

93. $r = 4\cos 3\theta$

94. $r^2 = 9\cos 2\theta$

6.8 **In Exercises 95–98, identify the conic and sketch its graph.**

95. $r = \dfrac{1}{1 + 2\sin\theta}$

96. $r = \dfrac{2}{1 + \sin\theta}$

97. $r = \dfrac{4}{5 - 3\cos\theta}$

98. $r = \dfrac{16}{4 + 5\cos\theta}$

In Exercises 99–102, find a polar equation of the conic with its focus at the pole.

99. Parabola Vertex: $(2, \pi)$

100. Parabola Vertex: $(2, \pi/2)$

101. Ellipse Vertices: $(5, 0), (1, \pi)$

102. Hyperbola Vertices: $(1, 0), (7, 0)$

103. ***Explorer 18*** On November 26, 1963, the United States launched Explorer 18. Its low and high points above the surface of Earth were 119 miles and 122,800 miles, respectively (see figure). The center of Earth was at one focus of the orbit. Find the polar equation of the orbit and find the distance between the surface of Earth (assume Earth has a radius of 4000 miles) and the satellite when $\theta = \pi/3$.

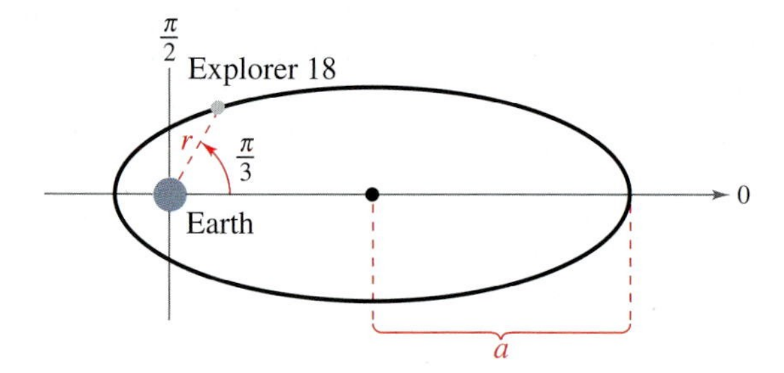

104. ***Asteroid*** An asteroid takes a parabolic path with Earth as its focus. It is about 6,000,000 miles from Earth at its closest approach. Write the polar equation of the path of the asteroid with its vertex at $\theta = \pi/2$. Find the distance between the asteroid and Earth when $\theta = -\pi/3$.

Synthesis

True or False? **In Exercises 105–107, determine whether the statement is true or false. Justify your answer.**

105. The graph of $\frac{1}{4}x^2 - y^4 = 1$ is a hyperbola.

106. Only one set of parametric equations can represent the line $y = 3 - 2x$.

107. There is a unique polar coordinate representation of each point in the plane.

108. Consider an ellipse with the major axis horizontal and 10 units in length. The number b in the standard form of the equation of the ellipse must be less than what real number? Explain the change in the shape of the ellipse as b approaches this number.

109. The graph of the parametric equations $x = 2 \sec t$ and $y = 3 \tan t$ is shown in the figure. How would the graph change for the equations $x = 2 \sec(-t)$ and $y = 3 \tan(-t)$?

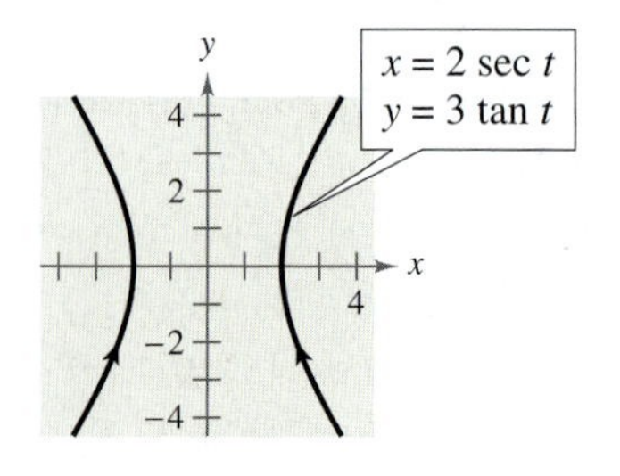

110. A moving object is modeled by the parametric equations $x = 4 \cos t$ and $y = 3 \sin t$, where t is time (see figure). How would the path change for the following?

(a) $x = 4 \cos 2t, \quad y = 3 \sin 2t$

(b) $x = 5 \cos t, \quad y = 3 \sin t$

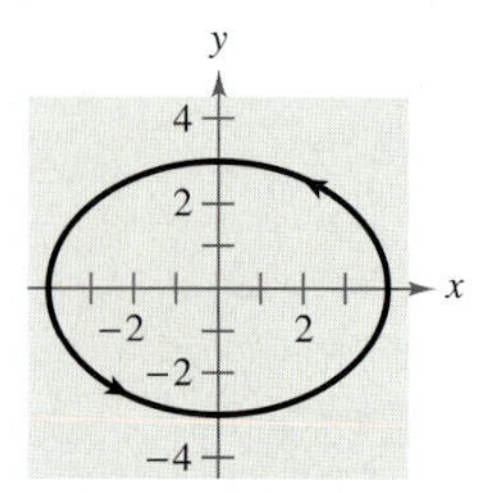

111. Identify the type of symmetry each of the following polar points has with the point in the figure.

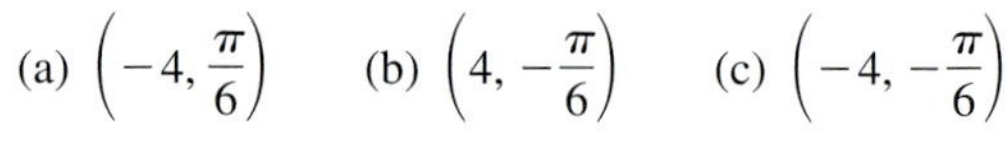
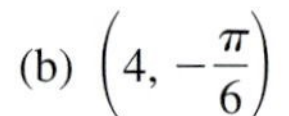
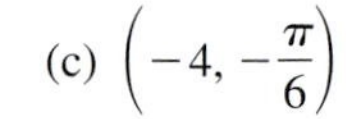

(a) $\left(-4, \frac{\pi}{6}\right)$ (b) $\left(4, -\frac{\pi}{6}\right)$ (c) $\left(-4, -\frac{\pi}{6}\right)$

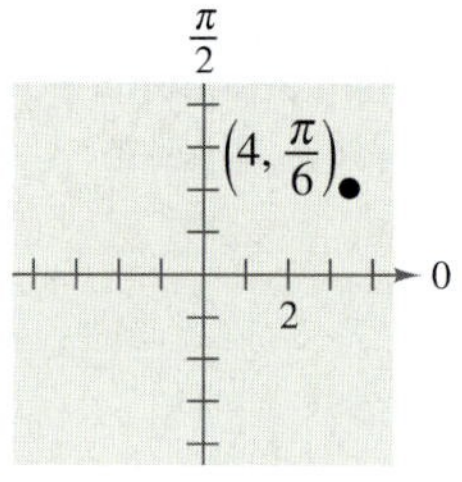

112. What is the relationship between the graphs of the rectangular and polar equations?

(a) $x^2 + y^2 = 25, \quad r = 5$ (b) $x - y = 0, \quad \theta = \frac{\pi}{4}$

113. ***Geometry*** The area of the ellipse in the figure is twice the area of the circle. What is the length of the major axis? (*Hint:* The area of an ellipse is $A = \pi ab$.)

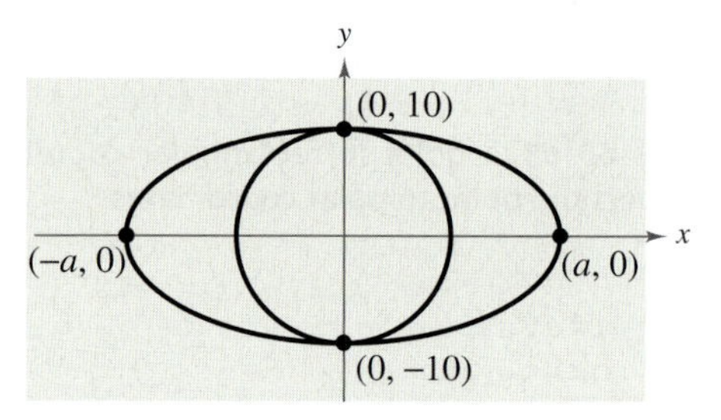

6 Chapter Test

Take this test as you would take a test in class. When you are finished, check your work against the answers given in the back of the book.

1. Find the inclination of the line $2x - 7y + 3 = 0$.
2. Find the angle between the lines $3x + 2y - 4 = 0$ and $4x - y + 6 = 0$.
3. Find the distance between the point $(7, 5)$ and the line $y = 5 - x$.

In Exercises 4–7, classify the conic and write the equation in standard form. Identify the center, vertices, foci, and asymptotes (if applicable). Then sketch the graph of the conic.

4. $y^2 - 4x + 4 = 0$
5. $x^2 - 4y^2 - 4x = 0$
6. $9x^2 + 16y^2 + 54x - 32y - 47 = 0$
7. $2x^2 + 2y^2 - 8x - 4y + 9 = 0$

8. Find the standard form of the equation of the parabola with vertex $(3, -2)$, with a vertical axis, and passing through the point $(0, 4)$.
9. Find the standard form of the equation of the hyperbola with foci $(0, 0)$ and $(0, 4)$ and asymptotes $y = \pm\frac{1}{2}x + 2$.
10. Sketch the curve represented by the parametric equations $x = 2 + 3\cos\theta$ and $y = 2\sin\theta$. Eliminate the parameter and write the corresponding rectangular equation.
11. Find a set of parametric equations of the line passing through the points $(2, -3)$ and $(6, 4)$. (There are many correct answers.)
12. Convert the polar coordinate $\left(-2, \dfrac{5\pi}{6}\right)$ to rectangular form.
13. Convert the rectangular coordinate $(2, -2)$ to polar form and find two additional polar representations of this point.
14. Convert the rectangular equation $x^2 + y^2 - 4y = 0$ to polar form.

In Exercises 15–18, sketch the graph of the polar equation. Identify the type of graph.

15. $r = \dfrac{4}{1 + \cos\theta}$
16. $r = \dfrac{4}{2 + \cos\theta}$
17. $r = 2 + 3\sin\theta$
18. $r = 3\sin 2\theta$

19. Find a polar equation of the ellipse with focus at the pole, eccentricity $e = \frac{1}{4}$, and directrix $y = 4$.
20. A straight road rises with an inclination of 0.15 radian from the horizontal. Find the slope of the road and the change in elevation over a one-mile stretch of the road.
21. A baseball is hit at a point 3 feet above the ground toward the left field fence. The fence is 10 feet high and 375 feet from home plate. The path of the baseball can be modeled by the parametric equations $x = (115\cos\theta)t$ and $y = 3 + (115\sin\theta)t - 16t^2$. Will the baseball go over the fence if it is hit at an angle of $\theta = 30°$? Will the baseball go over the fence if $\theta = 35°$?

6 Cumulative Test for Chapters 4–6

Take this test to review the material from earlier chapters. When you are finished, check your work against the answers given in the back of the book.

1. Consider the angle $\theta = -120°$.
   (a) Sketch the angle in standard position.
   (b) Determine a coterminal angle in the interval $[0°, 360°)$.
   (c) Convert the angle to radian measure.
   (d) Find the reference angle θ'.
   (e) Find the exact values of the six trigonometric functions of θ.
2. Convert the angle $\theta = 2.35$ radians to degrees. Round the answer to one decimal place.
3. Find $\cos\theta$ if $\tan\theta = -\frac{4}{3}$ and $\sin\theta < 0$.

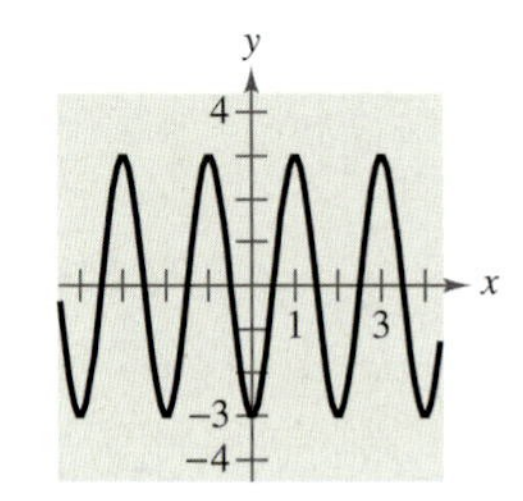

FIGURE FOR 7

In Exercises 4–6, sketch the graph of the function. (Include two full periods.)

4. $f(x) = 3 - 2\sin\pi x$
5. $g(x) = \frac{1}{2}\tan\left(x - \frac{\pi}{2}\right)$
6. $h(x) = -\sec(x + \pi)$

7. Find a, b, and c such that the graph of the function $h(x) = a\cos(bx + c)$ matches the graph in the figure.
8. Sketch the graph of the function $f(x) = \frac{1}{2}x\sin x$ over the interval $-3\pi \le x \le 3\pi$.

In Exercises 9 and 10, find the exact value of the expression without using a calculator.

9. $\tan(\arctan 6.7)$
10. $\tan\left(\arcsin\frac{3}{5}\right)$

11. Write an algebraic expression equivalent to $\sin(\arccos 2x)$.
12. Use the fundamental identities to simplify: $\cos\left(\frac{\pi}{2} - x\right)\csc x$.
13. Subtract and simplify: $\dfrac{\sin\theta - 1}{\cos\theta} - \dfrac{\cos\theta}{\sin\theta - 1}$.

In Exercises 14–16, verify the identity.

14. $\cot^2\alpha(\sec^2\alpha - 1) = 1$
15. $\sin(x + y)\sin(x - y) = \sin^2 x - \sin^2 y$
16. $\sin^2 x\cos^2 x = \frac{1}{8}(1 - \cos 4x)$

In Exercises 17 and 18, find all solutions of the equation in the interval $[0, 2\pi)$.

17. $2\cos^2\beta - \cos\beta = 0$
18. $3\tan\theta - \cot\theta = 0$

19. Use the Quadratic Formula to solve the equation in the interval $[0, 2\pi)$: $\sin^2 x + 2\sin x + 1 = 0$.
20. Given that $\sin u = \frac{12}{13}$, $\cos v = \frac{3}{5}$, and angles u and v are both in Quadrant I, find $\tan(u - v)$.
21. If $\tan\theta = \frac{1}{2}$, find the exact value of $\tan(2\theta)$.

22. If $\tan \theta = \frac{4}{3}$, find the exact value of $\sin \frac{\theta}{2}$.

23. Write the product $5 \sin \frac{3\pi}{4} \cdot \cos \frac{7\pi}{4}$ as a sum or difference.

24. Write $\cos 8x + \cos 4x$ as a product.

FIGURE FOR 25–28

In Exercises 25–28, use the information to solve the triangle shown in the figure. Round your answers to two decimal places.

25. $A = 30°,\ a = 9,\ b = 8$

26. $A = 30°,\ b = 8,\ c = 10$

27. $A = 30°,\ C = 90°,\ b = 10$

28. $a = 4,\ b = 8,\ c = 9$

29. Two sides of a triangle have lengths 7 inches and 12 inches. Their included angle measures 60°. Find the area of the triangle.

30. Find the area of a triangle with sides of lengths 11 inches, 16 inches, and 17 inches.

In Exercises 31 and 32, identify the conic and sketch its graph.

31. $\frac{(x-2)^2}{4} + \frac{(y+1)^2}{9} = 1$

32. $x^2 + y^2 - 2x - 4y + 1 = 0$

33. Find the standard form of the equation of the ellipse with vertices $(0, 0)$ and $(0, 4)$ and endpoints of the minor axis $(1, 2)$ and $(-1, 2)$.

34. Sketch the curve represented by parametric equations $x = 4 \ln t$ and $y = \frac{1}{2}t^2$. Then eliminate the parameter and write the corresponding rectangular equation whose graph represents the curve.

35. Plot the point $(-2, -3\pi/4)$ and find three additional polar representations for $-2\pi < \theta < 2\pi$.

36. Convert the rectangular equation $-8x - 3y + 5 = 0$ to polar form.

37. Convert the polar equation $r = \frac{2}{4 - 5\cos\theta}$ to rectangular form.

In Exercises 38–40, sketch the graph of the polar equation. Identify the type of graph.

38. $r = -\frac{\pi}{6}$

39. $r = 3 - 2\sin\theta$

40. $r = 2 + 5\cos\theta$

41. A ceiling fan with 21-inch blades makes 63 revolutions per minute. Find the angular speed of the fan in radians per minute. Find the linear speed of the tips of the blades in inches per minute.

42. Find the area of the sector of a circle with a radius of 8 yards and a central angle of 114°.

43. From a point 200 feet from a flagpole, the angles of elevation to the bottom and top of the flag are 16° 45′ and 18°, respectively. Approximate the height of the flag to the nearest foot.

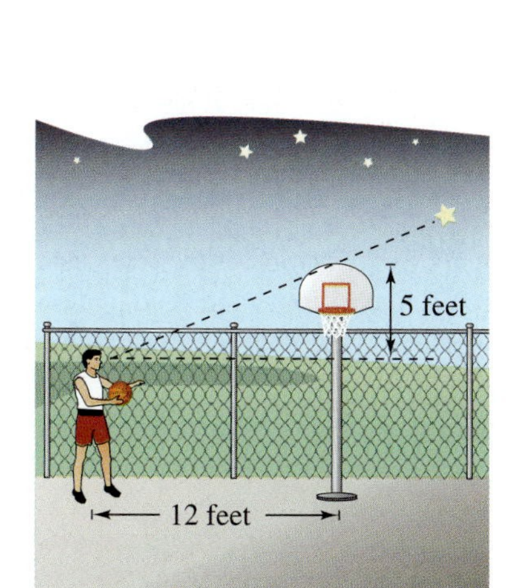

FIGURE FOR 44

44. To determine the angle of elevation of a star in the sky, you get the star in your line of vision with the backboard of a basketball hoop that is 5 feet higher than your eyes (see figure). Your horizontal distance from the backboard is 12 feet. What is the angle of elevation of the star?

45. Write a model for a particle in simple harmonic motion with a displacement of 4 inches and a period of 8 seconds.

Proofs in Mathematics

Inclination and Slope (p. 450)

If a nonvertical line has inclination θ and slope m, then $m = \tan \theta$.

Proof

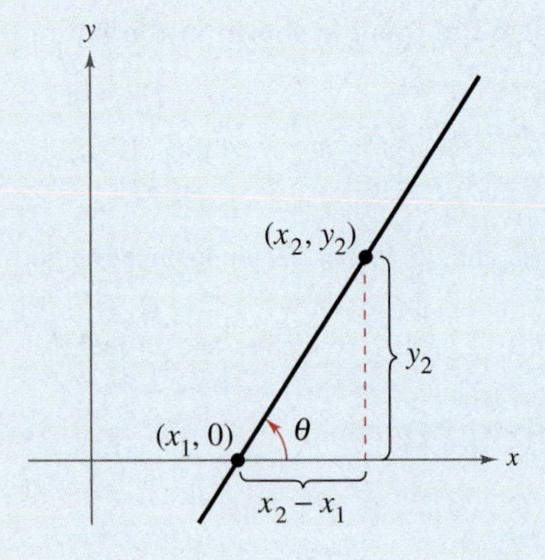

If $m = 0$, the line is horizontal and $\theta = 0$. So, the result is true for horizontal lines because $m = 0 = \tan 0$.

If the line has a positive slope, it will intersect the x-axis. Label this point $(x_1, 0)$, as shown in the figure. If (x_2, y_2) is a second point on the line, the slope is

$$m = \frac{y_2 - 0}{x_2 - x_1} = \frac{y_2}{x_2 - x_1} = \tan \theta.$$

The case in which the line has a negative slope can be proved in a similar manner.

Distance Between a Point and a Line (p. 452)

The distance between the point (x_1, y_1) and the line $Ax + By + C = 0$ is

$$d = \frac{|Ax_1 + By_1 + C|}{\sqrt{A^2 + B^2}}.$$

Proof

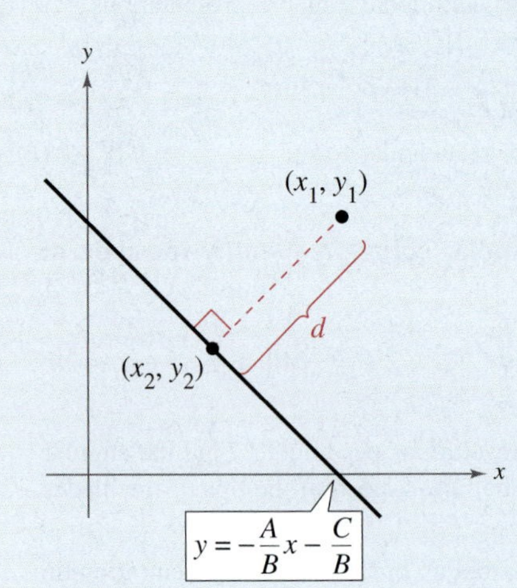

For simplicity's sake, assume that the given line is neither horizontal nor vertical (see figure). By writing the equation $Ax + By + C = 0$ in slope-intercept form

$$y = -\frac{A}{B}x - \frac{C}{B}$$

you can see that the line has a slope of $m = -A/B$. So, the slope of the line passing through (x_1, y_1) and perpendicular to the given line is B/A, and its equation is $y - y_1 = (B/A)(x - x_1)$. These two lines intersect at the point (x_2, y_2), where

$$x_2 = \frac{B(Bx_1 - Ay_1) - AC}{A^2 + B^2} \quad \text{and} \quad y_2 = \frac{A(-Bx_1 + Ay_1) - BC}{A^2 + B^2}.$$

Finally, the distance between (x_1, y_1) and (x_2, y_2) is

$$\begin{aligned} d &= \sqrt{(x_2 - x_1)^2 + (y_2 - y_1)^2} \\ &= \sqrt{\left(\frac{B^2x_1 - ABy_1 - AC}{A^2 + B^2} - x_1\right)^2 + \left(\frac{-ABx_1 + A^2y_1 - BC}{A^2 + B^2} - y_1\right)^2} \\ &= \sqrt{\frac{A^2(Ax_1 + By_1 + C)^2 + B^2(Ax_1 + By_1 + C)^2}{(A^2 + B^2)^2}} \\ &= \frac{|Ax_1 + By_1 + C|}{\sqrt{A^2 + B^2}}. \end{aligned}$$

Parabolic Paths

There are many natural occurrences of parabolas in real life. For instance, the famous astronomer Galileo discovered in the 17th century that an object that is projected upward and obliquely to the pull of gravity travels in a parabolic path. Examples of this are the center of gravity of a jumping dolphin and the path of water molecules in a drinking fountain.

Standard Equation of a Parabola *(p. 458)*

The standard form of the equation of a parabola with vertex at (h, k) is as follows.

$(x - h)^2 = 4p(y - k), \quad p \neq 0$ — Vertical axis, directrix: $y = k - p$

$(y - k)^2 = 4p(x - h), \quad p \neq 0$ — Horizontal axis, directrix: $x = h - p$

The focus lies on the axis p units (*directed distance*) from the vertex. If the vertex is at the origin $(0, 0)$, the equation takes one of the following forms.

$x^2 = 4py$ — Vertical axis

$y^2 = 4px$ — Horizontal axis

Proof

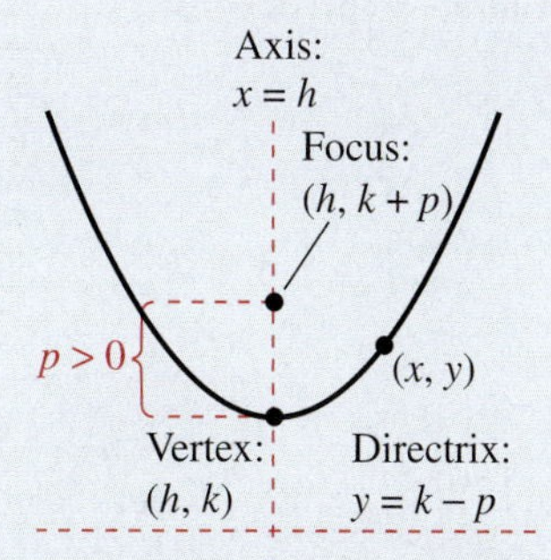

Parabola with vertical axis

For the case in which the directrix is parallel to the x-axis and the focus lies above the vertex, as shown in the top figure, if (x, y) is any point on the parabola, then, by definition, it is equidistant from the focus $(h, k + p)$ and the directrix $y = k - p$. So, you have

$$\sqrt{(x - h)^2 + [y - (k + p)]^2} = y - (k - p)$$
$$(x - h)^2 + [y - (k + p)]^2 = [y - (k - p)]^2$$
$$(x - h)^2 + y^2 - 2y(k + p) + (k + p)^2 = y^2 - 2y(k - p) + (k - p)^2$$
$$(x - h)^2 + y^2 - 2ky - 2py + k^2 + 2pk + p^2 = y^2 - 2ky + 2py + k^2 - 2pk + p^2$$
$$(x - h)^2 - 2py + 2pk = 2py - 2pk$$
$$(x - h)^2 = 4p(y - k).$$

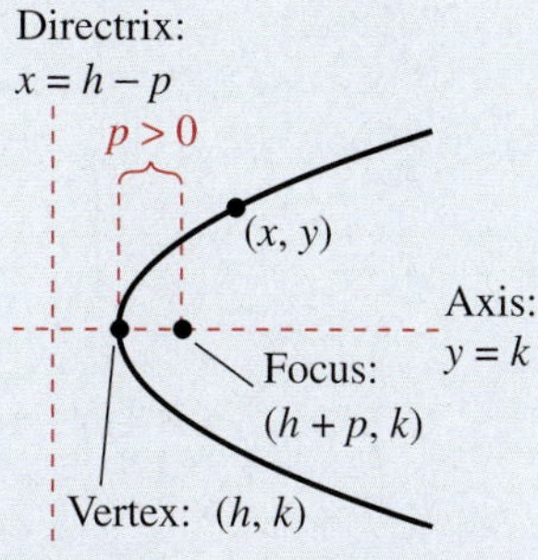

Parabola with horizontal axis

For the case in which the directrix is parallel to the y-axis and the focus lies to the right of the vertex, as shown in the bottom figure, if (x, y) is any point on the parabola, then, by definition, it is equidistant from the focus $(h + p, k)$ and the directrix $x = h - p$. So, you have

$$\sqrt{[x - (h + p)]^2 + (y - k)^2} = x - (h - p)$$
$$[x - (h + p)]^2 + (y - k)^2 = [x - (h - p)]^2$$
$$x^2 - 2x(h + p) + (h + p)^2 + (y - k)^2 = x^2 - 2x(h - p) + (h - p)^2$$
$$x^2 - 2hx - 2px + h^2 + 2ph + p^2 + (y - k)^2 = x^2 - 2hx + 2px + h^2 - 2ph + p^2$$
$$-2px + 2ph + (y - k)^2 = 2px - 2ph$$
$$(y - k)^2 = 4p(x - h).$$

Note that if a parabola is centered at the origin, then the two equations above would simplify to $x^2 = 4py$ and $y^2 = 4px$, respectively.

Polar Equations of Conics *(p. 507)*

The graph of a polar equation of the form

1. $r = \dfrac{ep}{1 \pm e \cos \theta}$

or

2. $r = \dfrac{ep}{1 \pm e \sin \theta}$

is a conic, where $e > 0$ is the eccentricity and $|p|$ is the distance between the focus (pole) and the directrix.

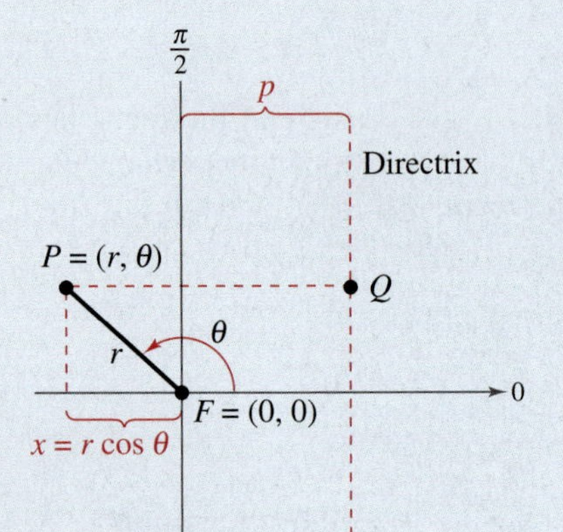

Proof

A proof for $r = ep/(1 + e \cos \theta)$ with $p > 0$ is shown here. The proofs of the other cases are similar. In the figure, consider a vertical directrix, p units to the right of the focus $F = (0, 0)$. If $P = (r, \theta)$ is a point on the graph of

$$r = \frac{ep}{1 + e \cos \theta}$$

the distance between P and the directrix is

$$\begin{aligned} PQ &= |p - x| \\ &= |p - r \cos \theta| \\ &= \left| p - \left(\frac{ep}{1 + e \cos \theta} \right) \cos \theta \right| \\ &= \left| p \left(1 - \frac{e \cos \theta}{1 + e \cos \theta} \right) \right| \\ &= \left| \frac{p}{1 + e \cos \theta} \right| \\ &= \left| \frac{r}{e} \right|. \end{aligned}$$

Moreover, because the distance between P and the pole is simply $PF = |r|$, the ratio of PF to PQ is

$$\begin{aligned} \frac{PF}{PQ} &= \frac{|r|}{\left| \frac{r}{e} \right|} \\ &= |e| \\ &= e \end{aligned}$$

and, by definition, the graph of the equation must be a conic.

P.S. Problem Solving

This collection of thought-provoking and challenging exercises further explores and expands upon concepts learned in this chapter.

1. Several mountain climbers are located in a mountain pass between two peaks. The angles of elevation to the two peaks are 0.84 radian and 1.10 radians. A range finder shows that the distances to the peaks are 3250 feet and 6700 feet, respectively (see figure).

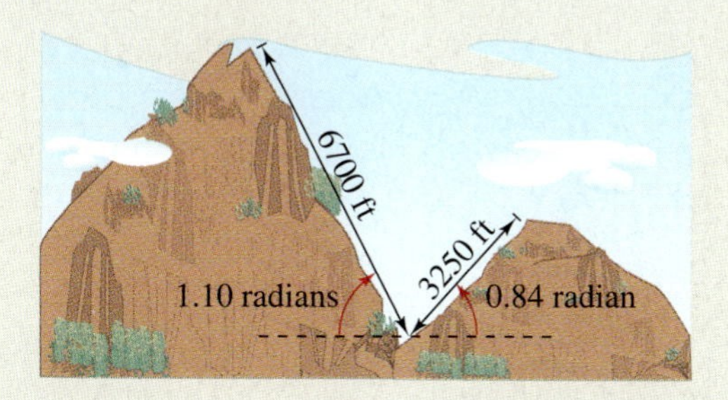

(a) Find the angle between the two lines of sight to the peaks.

(b) Approximate the amount of vertical climb that is necessary to reach the summit of each peak.

2. Statuary Hall is an elliptical room in the United States Capitol in Washington D.C. The room is also called the Whispering Gallery because a person standing at one focus of the room can hear even a whisper spoken by a person standing at the other focus. This occurs because any sound that is emitted from one focus of an ellipse will reflect off the side of the ellipse to the other focus. Statuary Hall is 46 feet wide and 97 feet long.

(a) Find an equation that models the shape of the room.

(b) How far apart are the two foci?

(c) What is the area of the floor of the room? (The area of an ellipse is $A = \pi ab$.)

3. Find the equation(s) of all parabolas that have the x-axis as the axis of symmetry and focus at the origin.

4. Find the area of the square inscribed in the ellipse below.

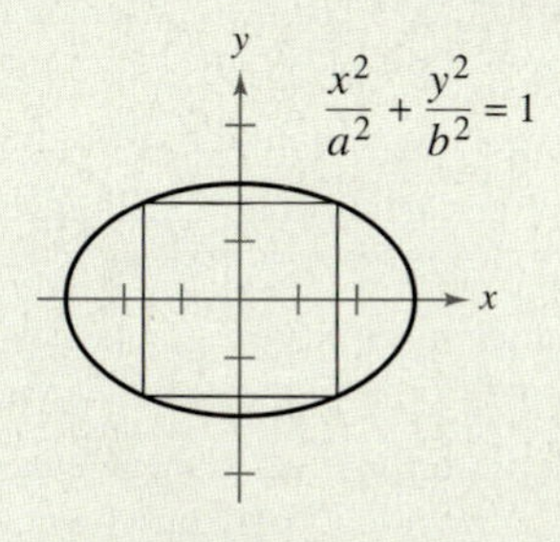

5. A tour boat travels between two islands that are 12 miles apart (see figure). For a trip between the islands, there is enough fuel for a 20-mile trip.

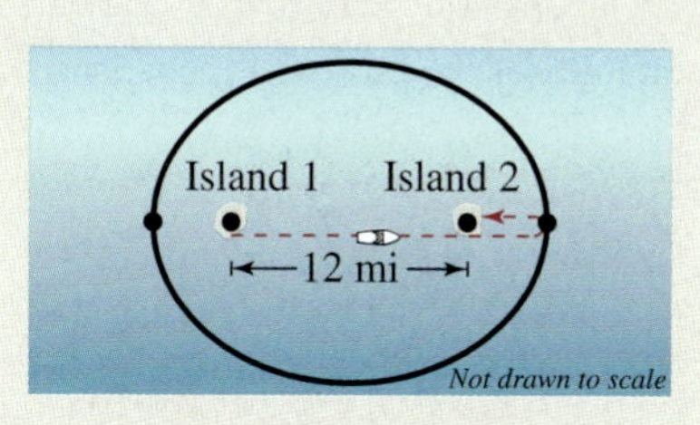

(a) Explain why the region in which the boat can travel is bounded by an ellipse.

(b) Let (0, 0) represent the center of the ellipse. Find the coordinates of each island.

(c) The boat travels from one island, straight past the other island to the vertex of the ellipse, and back to the second island. How many miles does the boat travel? Use your answer to find the coordinates of the vertex.

(d) Use the results from parts (b) and (c) to write an equation for the ellipse that bounds the region in which the boat can travel.

6. Find an equation of the hyperbola such that for any point on the hyperbola, the difference between its distances from the points (2, 2) and (10, 2) is 6.

7. Prove that the graph of the equation

$$Ax^2 + Cy^2 + Dx + Ey + F = 0$$

is one of the following (except in degenerate cases).

Conic	*Condition*
(a) Circle	$A = C$
(b) Parabola	$A = 0$ or $C = 0$ (but not both)
(c) Ellipse	$AC > 0$
(d) Hyperbola	$AC < 0$

8. The following sets of parametric equations model projectile motion.

$x = (v_0 \cos \theta)t \qquad x = (v_0 \cos \theta)t$

$y = (v_0 \sin \theta)t \qquad y = h + (v_0 \sin \theta)t - 16t^2$

(a) Under what circumstances would you use each model?

(b) Eliminate the parameter for each set of equations.

(c) In which case is the path of the moving object not affected by a change in the velocity v? Explain.

9. As t increases, the ellipse given by the parametric equations

$$x = \cos t \text{ and } y = 2 \sin t$$

is traced out *counterclockwise*. Find a parametric representation for which the same ellipse is traced out *clockwise*.

10. A **hypocycloid** has the parametric equations

$$x = (a - b) \cos t + b \cos\left(\frac{a - b}{b}t\right)$$

and

$$y = (a - b) \sin t - b \sin\left(\frac{a - b}{b}t\right).$$

Use a graphing utility to graph the hypocycloid for each value of a and b. Describe each graph.

(a) $a = 2, b = 1$ (b) $a = 3, b = 1$

(c) $a = 4, b = 1$ (d) $a = 10, b = 1$

(e) $a = 3, b = 2$ (f) $a = 4, b = 3$

11. The curve given by the parametric equations

$$x = \frac{1 - t^2}{1 + t^2}$$

and

$$y = \frac{t(1 - t^2)}{1 + t^2}$$

is called a **strophoid.**

(a) Find a rectangular equation of the strophoid.

(b) Find a polar equation of the strophoid.

(c) Use a graphing utility to graph the strophoid.

12. The rose curves described in this chapter are of the form

$$r = a \cos n\theta \qquad \text{or} \qquad r = a \sin n\theta$$

where n is a positive integer that is greater than or equal to 2. Use a graphing utility to graph $r = a \cos n\theta$ and $r = a \sin n\theta$ for some noninteger values of n. Describe the graphs.

13. What conic section is represented by the polar equation

$$r = a \sin \theta + b \cos \theta?$$

14. The graph of the polar equation

$$r = e^{\cos\theta} - 2 \cos 4\theta + \sin^5\left(\frac{\theta}{12}\right)$$

is called the *butterfly curve*, as shown in the figure.

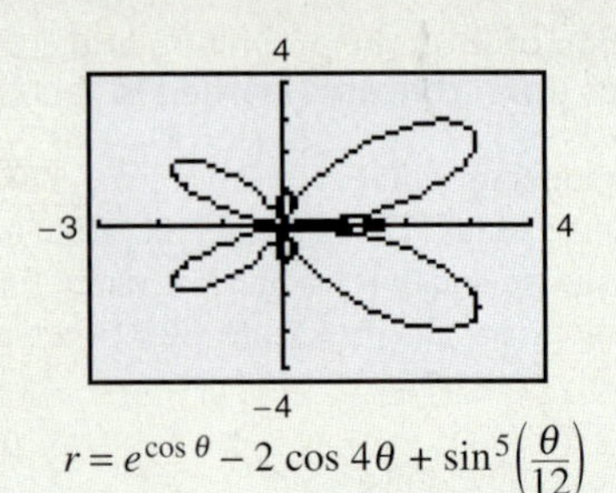

$r = e^{\cos\theta} - 2 \cos 4\theta + \sin^5\left(\frac{\theta}{12}\right)$

FIGURE FOR 14

(a) The graph above was produced using $0 \le \theta \le 2\pi$. Does this show the entire graph? Explain your reasoning.

(b) Approximate the maximum r-value of the graph. Does this value change if you use $0 \le \theta \le 4\pi$ instead of $0 \le \theta \le 2\pi$? Explain.

15. Use a graphing utility to graph the polar equation

$$r = \cos 5\theta + n \cos \theta$$

for $0 \le \theta \le \pi$ for the integers $n = -5$ to $n = 5$. As you graph these equations, you should see the graph change shape from a heart to a bell. Write a short paragraph explaining what values of n produce the heart portion of the curve and what values of n produce the bell portion.

16. The planets travel in elliptical orbits with the sun at one focus. The polar equation of the orbit of a planet with one focus at the pole and major axis of length $2a$ is

$$r = \frac{(1 - e^2)a}{1 - e \cos \theta}$$

where e is the eccentricity. The minimum distance (perihelion) from the sun to a planet is $r = a(1 - e)$ and the maximum distance (aphelion) is $r = a(1 + e)$. The length of the major axis for the planet Neptune is $a = 9.000 \times 10^9$ kilometers and the eccentricity is $e = 0.0086$. The length of the major axis for the planet Pluto is $a = 10.813 \times 10^9$ kilometers and the eccentricity is $e = 0.2488$.

(a) Find the polar equation of the orbit of each planet.

(b) Find the perihelion and aphelion distances for each planet.

(c) Use a graphing utility to graph the polar equation of each planet's orbit in the same viewing window.

(d) Do the orbits of the two planets intersect? Will the two planets ever collide? Why or why not?

(e) Is Pluto ever closer to the sun than Neptune? Why is Pluto called the ninth planet and Neptune the eighth planet?

Answers to All Exercises and Tests

Chapter 1

Section 1.1 *(page 9)*

Vocabulary Check *(page 9)*

1. (a) v (b) vi (c) i (d) iv (e) iii (f) ii
2. Cartesian **3.** Distance Formula
4. Midpoint Formula

1. A: $(2, 6)$, B: $(-6, -2)$, C: $(4, -4)$, D: $(-3, 2)$
2. A: $\left(\frac{3}{2}, -4\right)$, B: $(0, -2)$, C: $\left(-3, \frac{5}{2}\right)$, D: $(-6, 0)$

3.

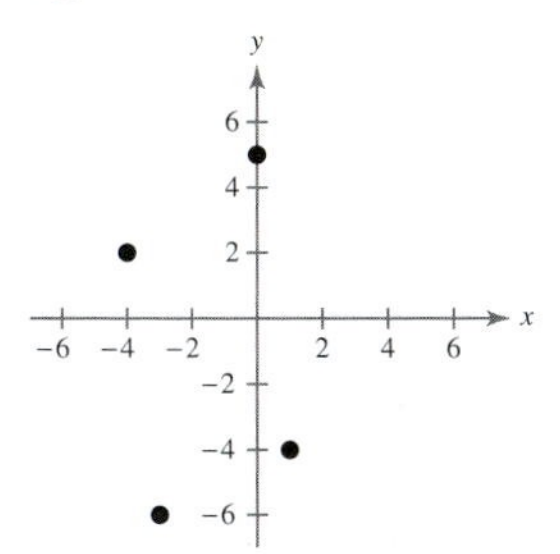

4.

5.

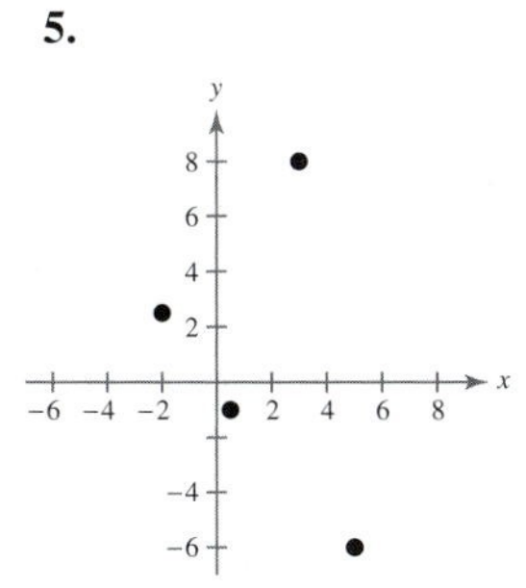

6.

7. $(-3, 4)$ **8.** $(4, -8)$ **9.** $(-5, -5)$
10. $(-12, 0)$ **11.** Quadrant IV **12.** Quadrant III
13. Quadrant II **14.** Quadrant I
15. Quadrant III or IV **16.** Quadrant I or IV
17. Quadrant III **18.** Quadrant III
19. Quadrant I or III **20.** Quadrant II or IV

21.

22.

23. 8 **24.** 7 **25.** 5 **26.** 10
27. (a) 4, 3, 5 (b) $4^2 + 3^2 = 5^2$
28. (a) 5, 12, 13 (b) $5^2 + 12^2 = 13^2$
29. (a) 10, 3, $\sqrt{109}$ (b) $10^2 + 3^2 = \left(\sqrt{109}\right)^2$
30. (a) 4, 7, $\sqrt{65}$ (b) $4^2 + 7^2 = \left(\sqrt{65}\right)^2$

31. (a)

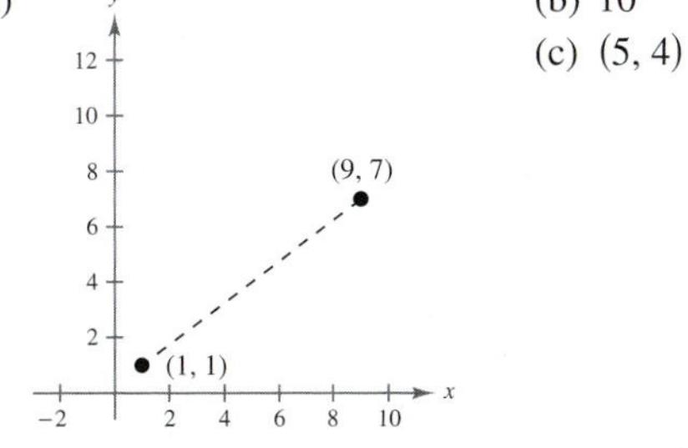

(b) 10
(c) $(5, 4)$

32. (a)

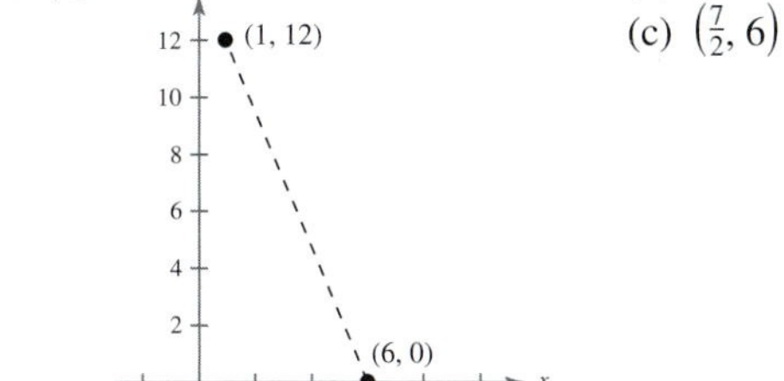

(b) 13
(c) $\left(\frac{7}{2}, 6\right)$

33. (a)

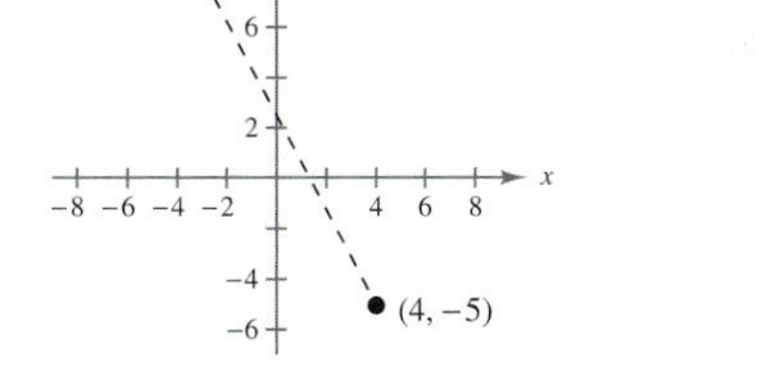

(b) 17
(c) $\left(0, \frac{5}{2}\right)$

34. (a)

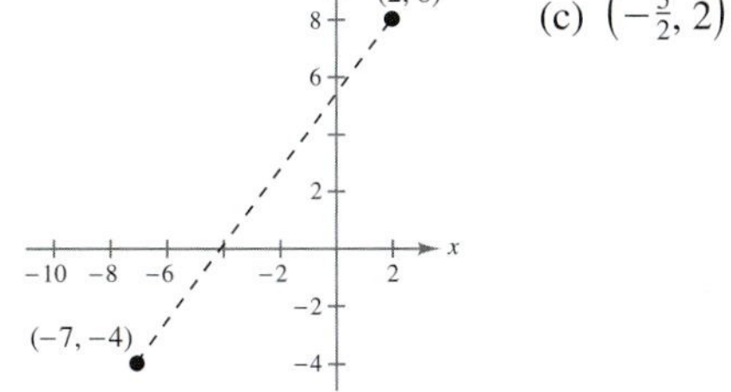

(b) 15
(c) $\left(-\frac{5}{2}, 2\right)$

35. (a)

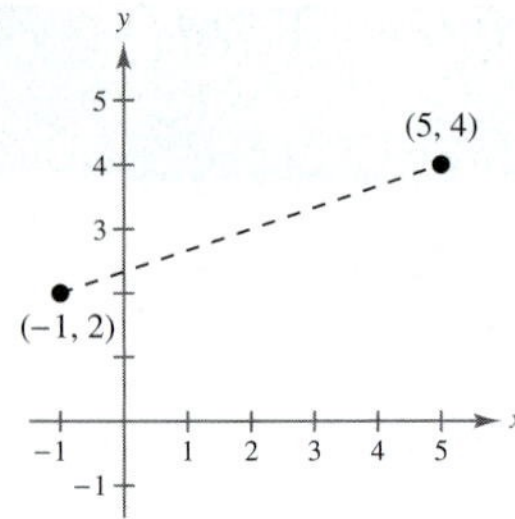

(b) $2\sqrt{10}$
(c) $(2, 3)$

36. (a)

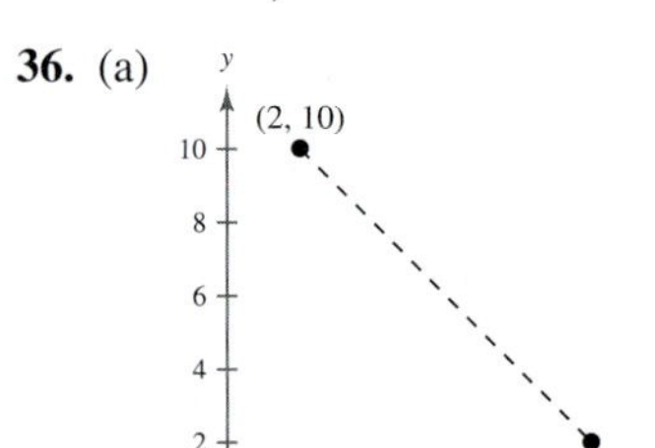

(b) $8\sqrt{2}$
(c) $(6, 6)$

37. (a)

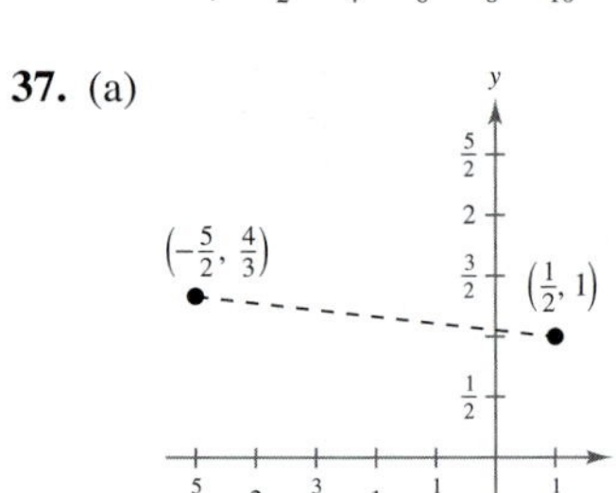

(b) $\dfrac{\sqrt{82}}{3}$
(c) $\left(-1, \frac{7}{6}\right)$

38. (a)

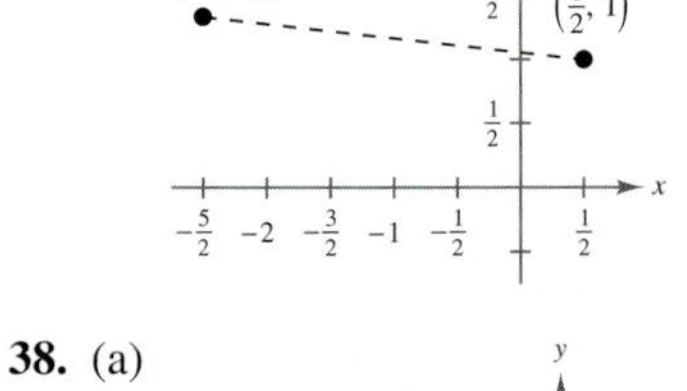

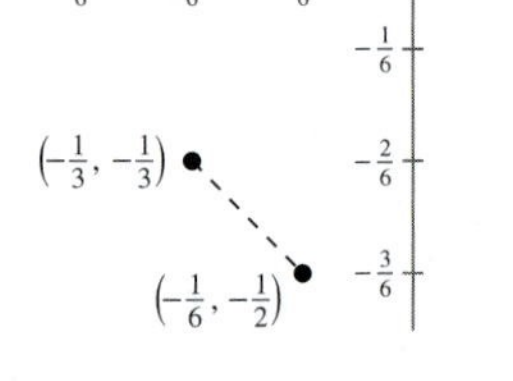

(b) $\dfrac{\sqrt{2}}{6}$
(c) $\left(-\frac{1}{4}, -\frac{5}{12}\right)$

39. (a)

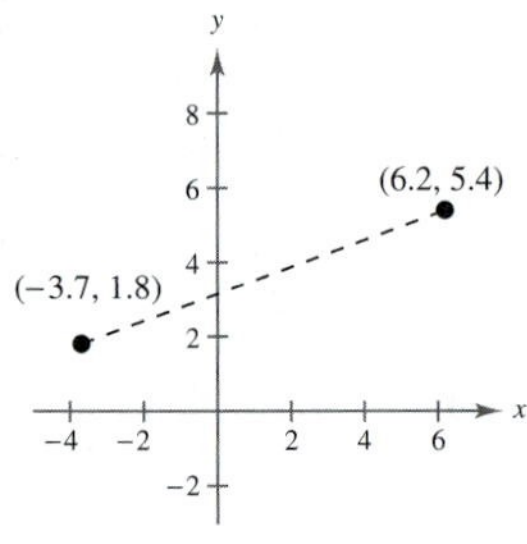

(b) $\sqrt{110.97}$
(c) $(1.25, 3.6)$

40. (a)

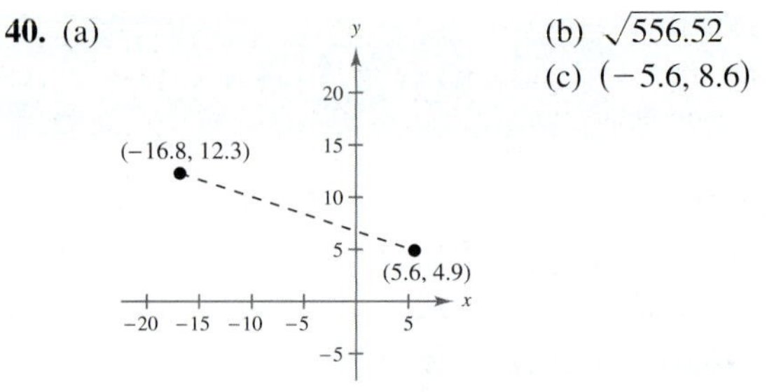

(b) $\sqrt{556.52}$
(c) $(-5.6, 8.6)$

41. $(\sqrt{5})^2 + (\sqrt{45})^2 = (\sqrt{50})^2$
42. Distances between the points: $\sqrt{29}$, $\sqrt{58}$, $\sqrt{29}$
43. $(2x_m - x_1, 2y_m - y_1)$
44. (a) $(7, 0)$ (b) $(9, -3)$
45. $\left(\dfrac{3x_1 + x_2}{4}, \dfrac{3y_1 + y_2}{4}\right), \left(\dfrac{x_1 + x_2}{2}, \dfrac{y_1 + y_2}{2}\right),$
$\left(\dfrac{x_1 + 3x_2}{4}, \dfrac{y_1 + 3y_2}{4}\right)$
46. (a) $\left(\frac{7}{4}, -\frac{7}{4}\right), \left(\frac{5}{2}, -\frac{3}{2}\right), \left(\frac{13}{4}, -\frac{5}{4}\right)$
(b) $\left(-\frac{3}{2}, -\frac{9}{4}\right), \left(-1, -\frac{3}{2}\right), \left(-\frac{1}{2}, -\frac{3}{4}\right)$
47. $2\sqrt{505} \approx 45$ yards **48.** $30\sqrt{41} \approx 192$ kilometers
49. \$3803.5 million **50.** \$2393.5 million
51. $(0, 1), (4, 2), (1, 4)$
52. $(3, 3), (1, 0), (3, -3), (5, 0)$
53. $(-3, 6), (2, 10), (2, 4), (-3, 4)$
54. $(-5, 2), (-7, 0), (-3, 0), (-5, -4)$
55. \$3.31 per pound; 2001 **56.** $\approx 91\%$ **57.** $\approx 250\%$
58. (a) $\approx 25\%$ (b) $\approx 147\%$
59. (a) The number of artists elected each year seems to be nearly steady except for the first few years. From six to eight artists will be elected in 2008.
(b) The Rock and Roll Hall of Fame was opened in 1986.
60. (a) 1990s (b) 26.5%; 21.2% (c) \$6.24
(d) Answers will vary.
61. 1998: \$19,384.5 million; 2000: \$20,223.0 million; 2002: \$21,061.5 million
62. (a)

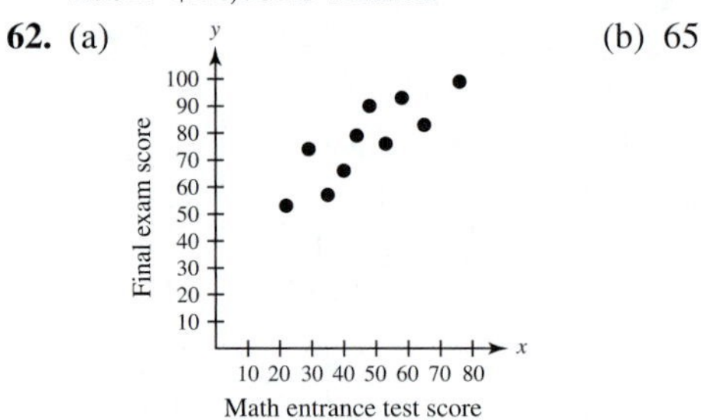

(b) 65
(c) No. There are many variables that will affect the final exam score.
63. $\sqrt[3]{\dfrac{4.47}{\pi}} \approx 1.12$ inches **64.** $\dfrac{603.2}{4\pi} \approx 48$ feet

65. Length of side = 43 centimeters;
area = 800.64 square centimeters

66. 34 centimeters

67. (a)
(b) $l = 1.5w$; $p = 5w$
(c) 7.5 meters × 5 meters

68. (a)
(b) $w = 1.25h$; $V = (20 \text{ inches})h^2$
(c) $h = 10$ inches, $w = 12.5$ inches, $l = 16$ inches

69. (a)
(b) 2002
(c) Answers will vary. Sample answer: Technology now enables us to transport information in many ways other than by mail. The Internet is one example.

70. (a)
(b) 1994
(c) 2003; 42 teams

71.

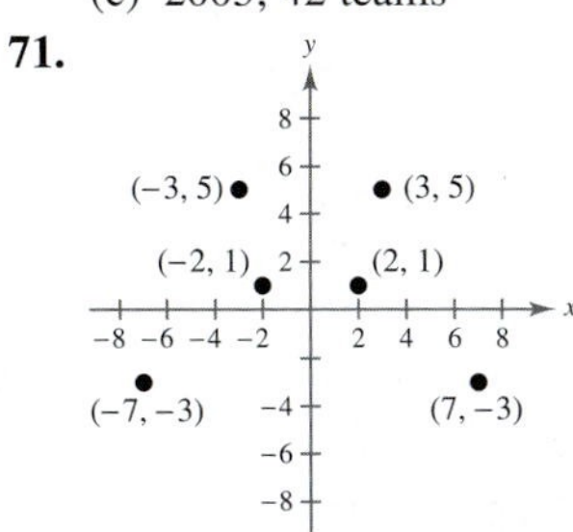

(a) The point is reflected through the y-axis.
(b) The point is reflected through the x-axis.
(c) The point is reflected through the origin.

72. (a)

	First Set	Second Set
Distance A to B	3	$\sqrt{10}$
Distance B to C	5	$\sqrt{10}$
Distance A to C	4	$\sqrt{40}$
	Right triangle	Isosceles triangle

(b)

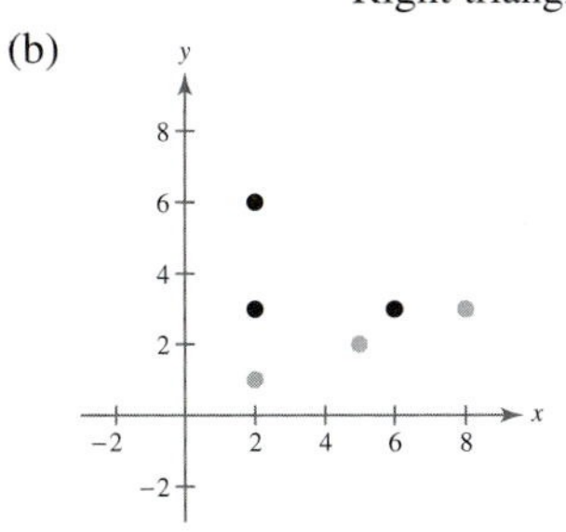

The first set of points is not collinear. The second set of points is collinear.
(c) A set of three points is collinear if the sum of two distances among the points is exactly equal to the third distance.

73. False. The Midpoint Formula would be used 15 times.

74. True. Two sides of the triangle have lengths of $\sqrt{149}$, and the third side has a length of $\sqrt{18}$.

75. No. It depends on the magnitudes of the quantities measured.

76. Use the Midpoint Formula to prove that the diagonals of the parallelogram bisect each other.

$$\left(\frac{b+a}{2}, \frac{c+0}{2}\right) = \left(\frac{a+b}{2}, \frac{c}{2}\right)$$

$$\left(\frac{a+b+0}{2}, \frac{c+0}{2}\right) = \left(\frac{a+b}{2}, \frac{c}{2}\right)$$

77. b **78.** c **79.** d

80. a **81.** $x = 1$

82. $x = 6$ **83.** $x = 2 \pm \sqrt{11}$

84. $x = \dfrac{-3 \pm \sqrt{73}}{4}$

85. $x < \frac{3}{5}$ **86.** $x \le -\frac{23}{4}$

87. $14 < x < 22$

88. $x \le -13$ or $x \ge -2$

Section 1.2 *(page 22)*

Vocabulary Check *(page 22)*

1. solution or solution point **2.** graph
3. intercepts **4.** y-axis **5.** circle; (h, k); r
6. numerical

1. (a) Yes (b) Yes **2.** (a) Yes (b) No
3. (a) No (b) Yes **4.** (a) Yes (b) No

CHAPTER 1

5.

x	-1	0	1	2	$\frac{5}{2}$
y	7	5	3	1	0
(x, y)	$(-1, 7)$	$(0, 5)$	$(1, 3)$	$(2, 1)$	$\left(\frac{5}{2}, 0\right)$

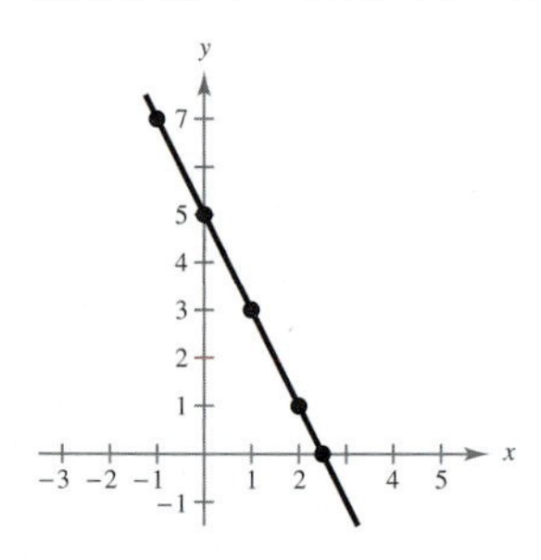

6.

x	-2	0	1	$\frac{4}{3}$	2
y	$-\frac{5}{2}$	-1	$-\frac{1}{4}$	0	$\frac{1}{2}$
(x, y)	$\left(-2, -\frac{5}{2}\right)$	$(0, -1)$	$\left(1, -\frac{1}{4}\right)$	$\left(\frac{4}{3}, 0\right)$	$\left(2, \frac{1}{2}\right)$

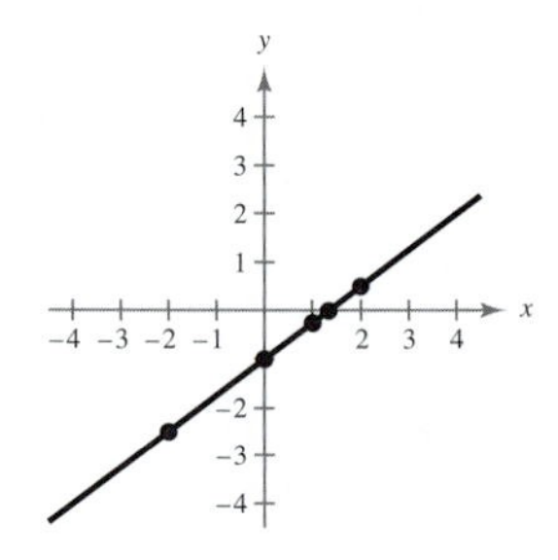

7.

x	-1	0	1	2	3
y	4	0	-2	-2	0
(x, y)	$(-1, 4)$	$(0, 0)$	$(1, -2)$	$(2, -2)$	$(3, 0)$

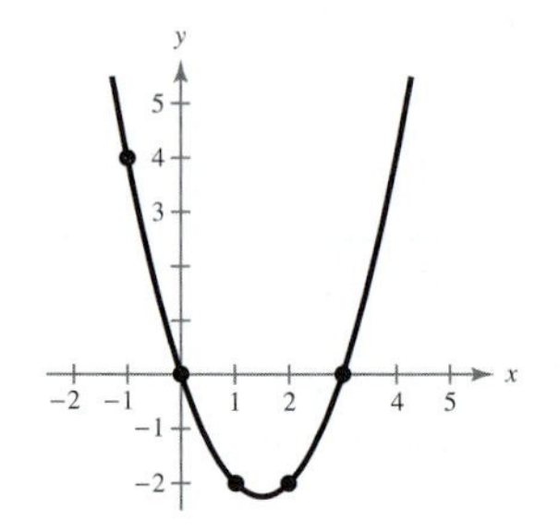

8.

x	-2	-1	0	1	2
y	1	4	5	4	1
(x, y)	$(-2, 1)$	$(-1, 4)$	$(0, 5)$	$(1, 4)$	$(2, 1)$

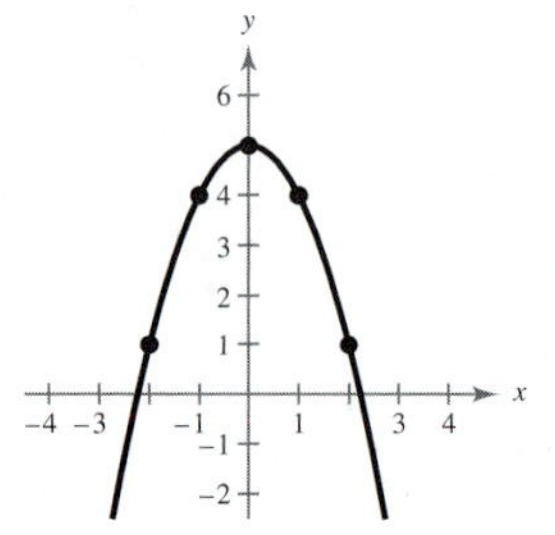

9. x-intercepts: $(\pm 2, 0)$
y-intercept: $(0, 16)$

10. x-intercept: $(-3, 0)$
y-intercept: $(0, 9)$

11. x-intercept: $\left(\frac{6}{5}, 0\right)$
y-intercept: $(0, -6)$

12. x-intercept: $\left(\frac{8}{3}, 0\right)$
y-intercept: $(0, 8)$

13. x-intercept: $(-4, 0)$
y-intercept: $(0, 2)$

14. x-intercept: $\left(\frac{1}{2}, 0\right)$

15. x-intercept: $\left(\frac{7}{3}, 0\right)$
y-intercept: $(0, 7)$

16. x-intercept: $(-10, 0)$
y-intercept: $(0, -10)$

17. x-intercepts: $(0, 0)$, $(2, 0)$
y-intercept: $(0, 0)$

18. x-intercepts: $\left(\pm\sqrt{5}, 0\right)$
y-intercept: $(0, -25)$

19. x-intercept: $(6, 0)$
y-intercepts: $\left(0, \pm\sqrt{6}\right)$

20. x-intercept: $(-1, 0)$
y-intercepts: $(0, \pm 1)$

21.

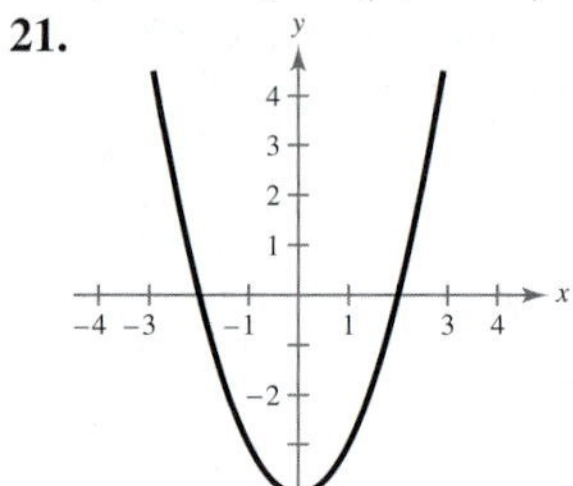

22.

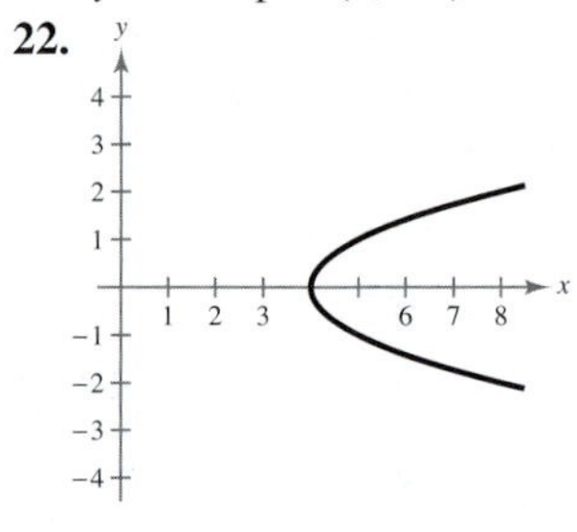

23.

24.

25. y-axis symmetry
26. x-axis symmetry
27. Origin symmetry
28. y-axis symmetry
29. Origin symmetry
30. y-axis symmetry
31. x-axis symmetry
32. Origin symmetry

33.

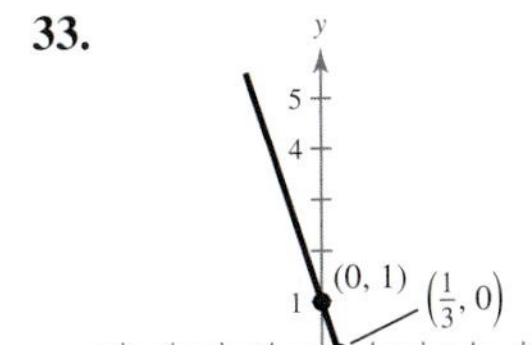

34.

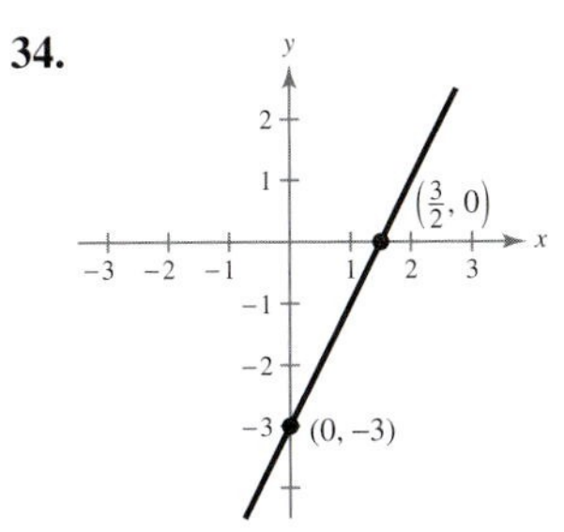

35.

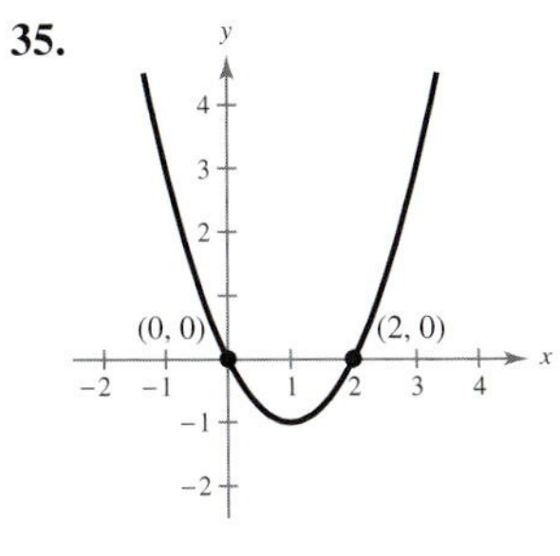

36.

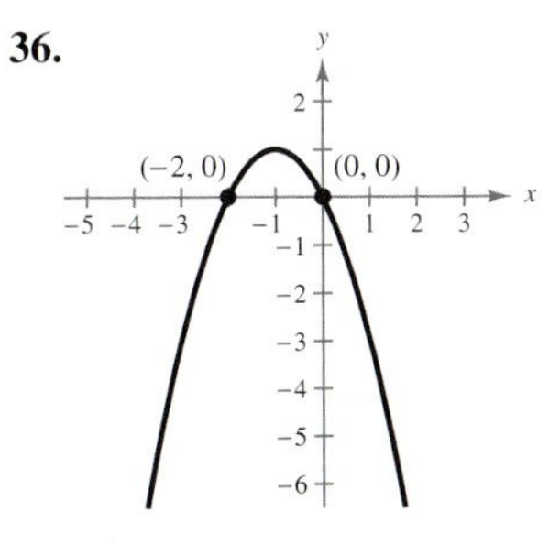

37.

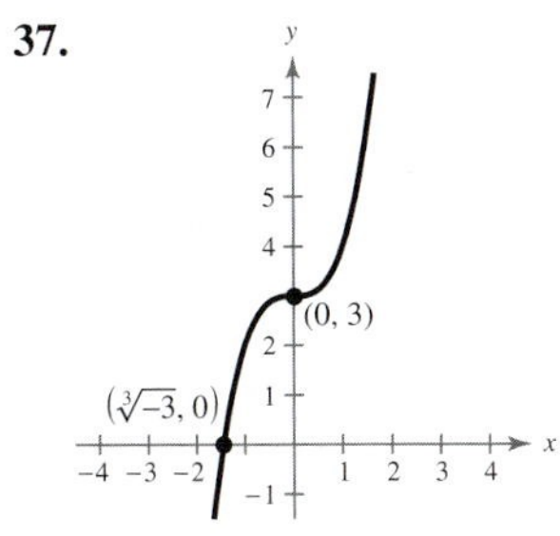

38.

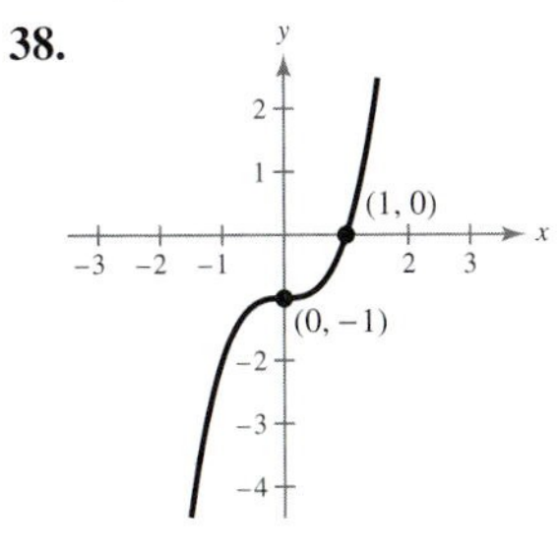

39.

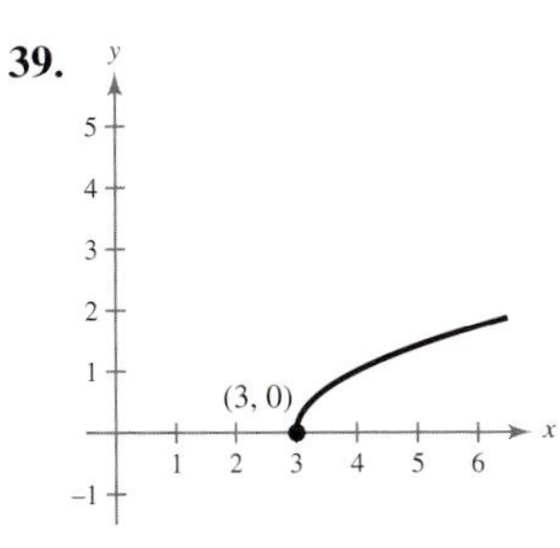

40.

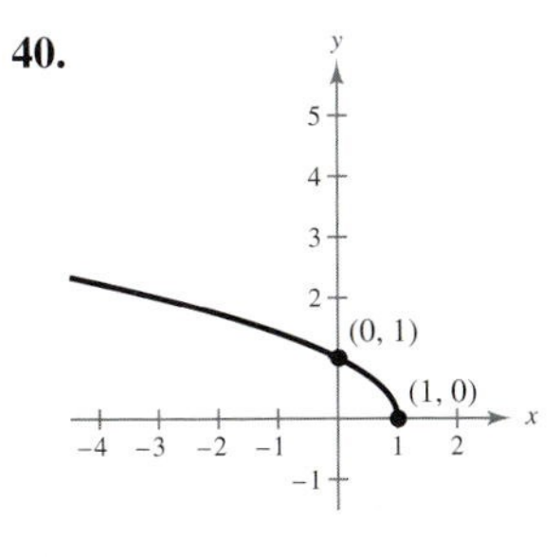

41.

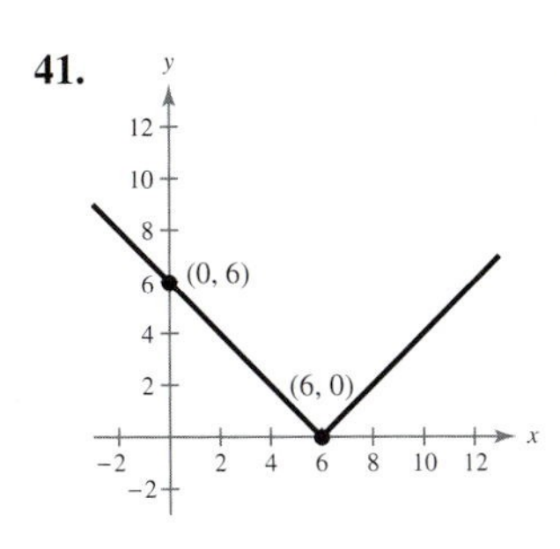

42.

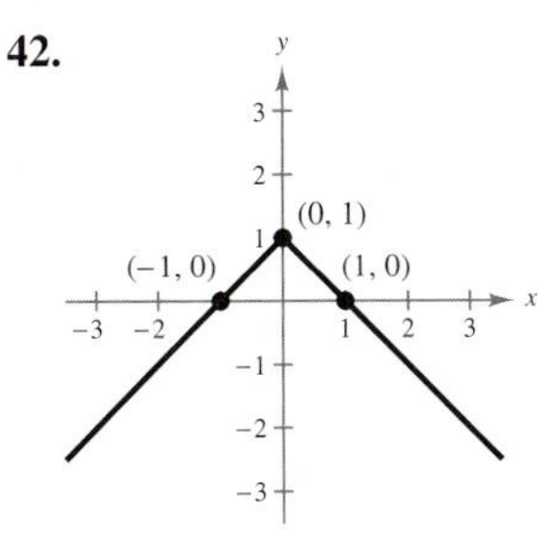

43.

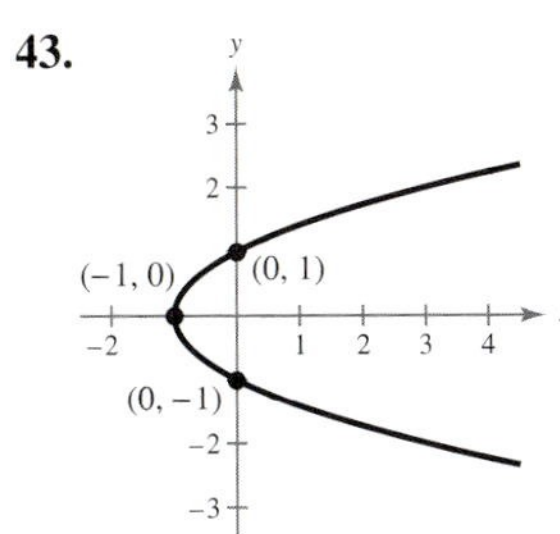

44.

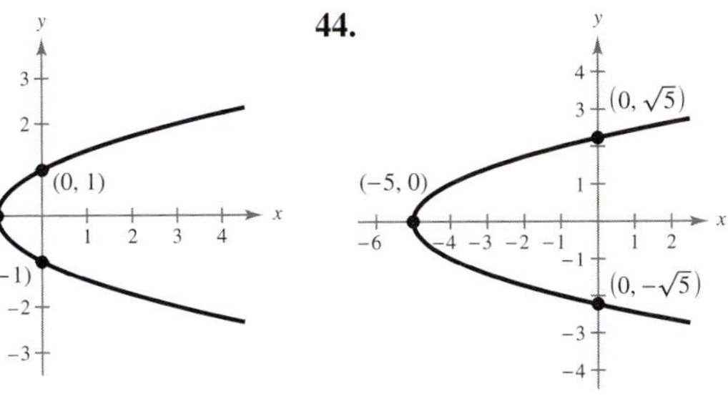

45.

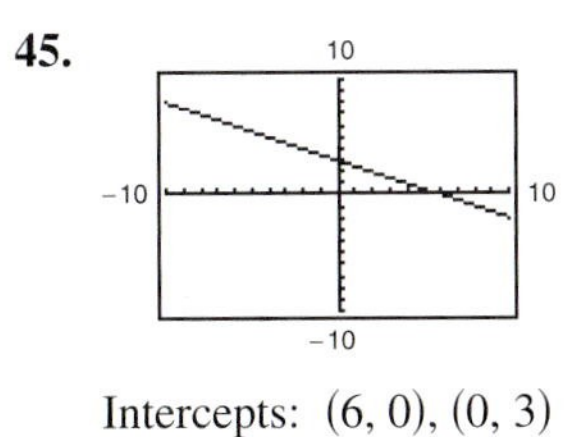

Intercepts: (6, 0), (0, 3)

46.

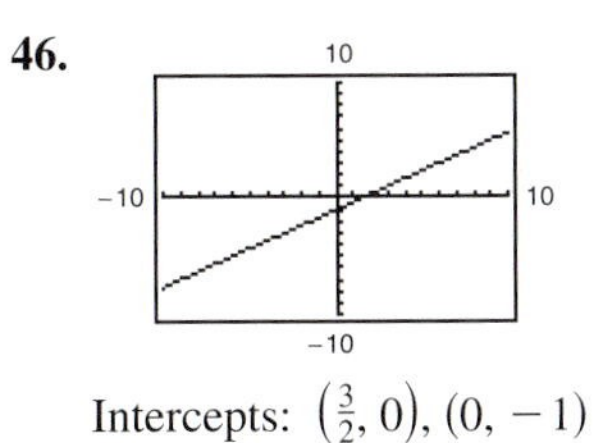

Intercepts: $\left(\frac{3}{2}, 0\right)$, (0, −1)

47.

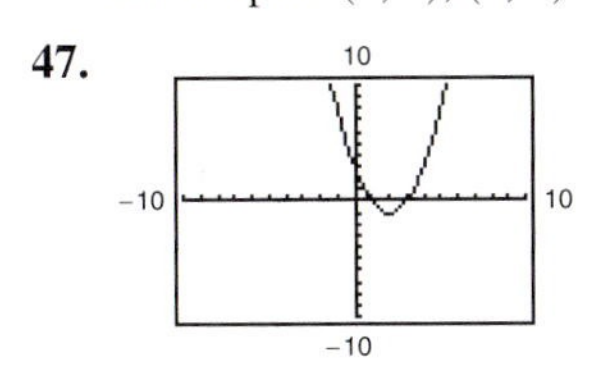

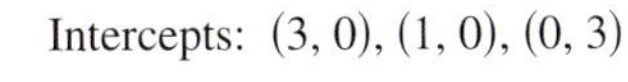

Intercepts: (3, 0), (1, 0), (0, 3)

48.

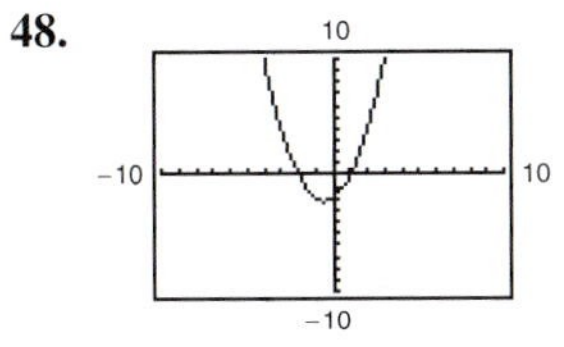

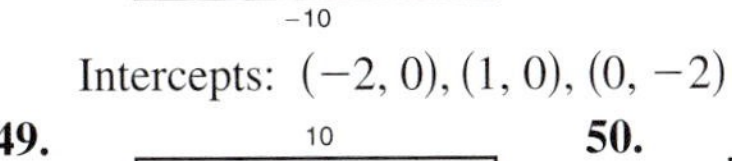

Intercepts: (−2, 0), (1, 0), (0, −2)

49.

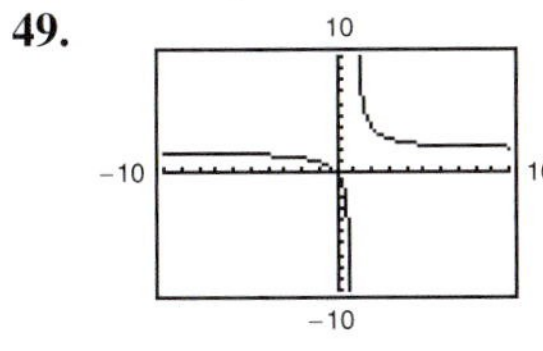

Intercept: (0, 0)

50.

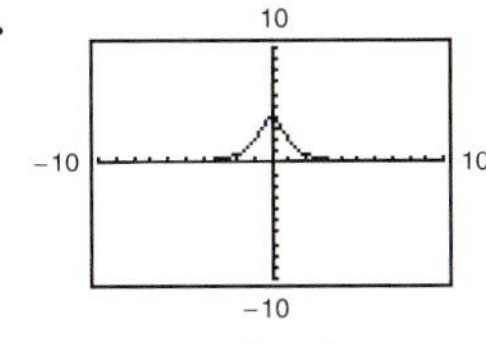

Intercept: (0, 4)

51.

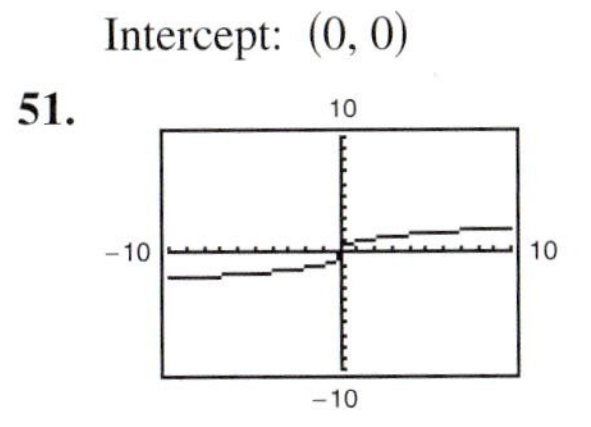

Intercept: (0, 0)

52.

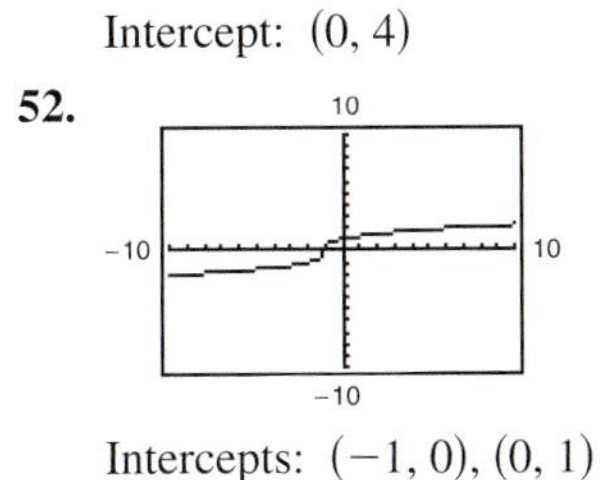

Intercepts: (−1, 0), (0, 1)

53.

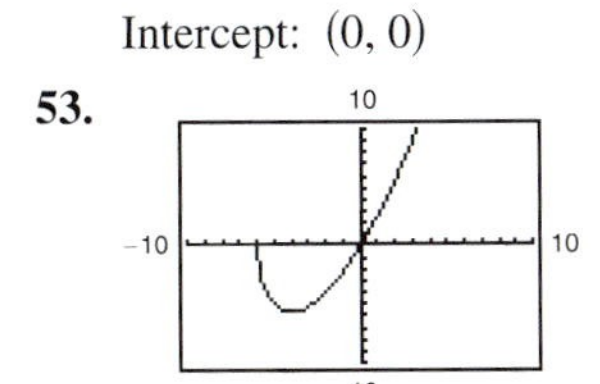

Intercepts: (0, 0), (−6, 0)

54.

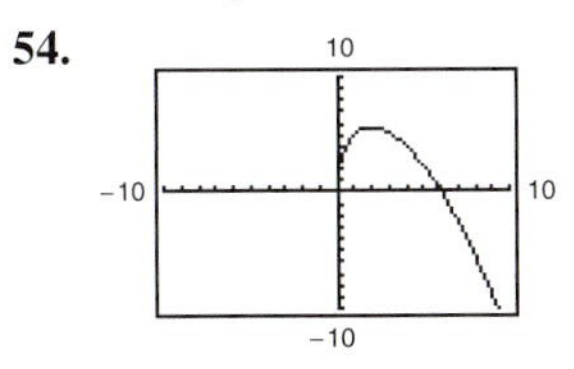

Intercepts: (0, 0), (6, 0)

19. m is undefined.
There is no y-intercept.

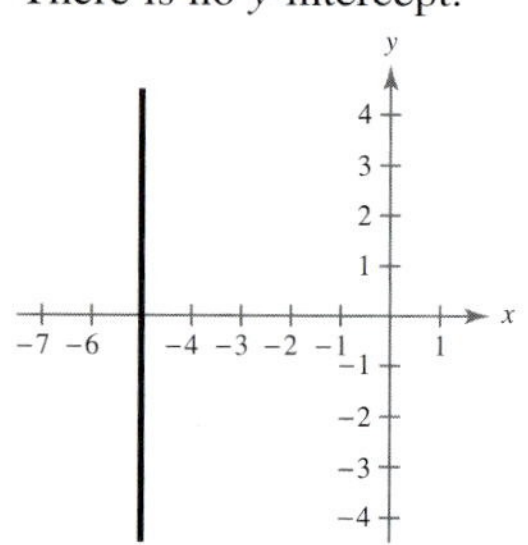

20. m is undefined.
There is no y-intercept.

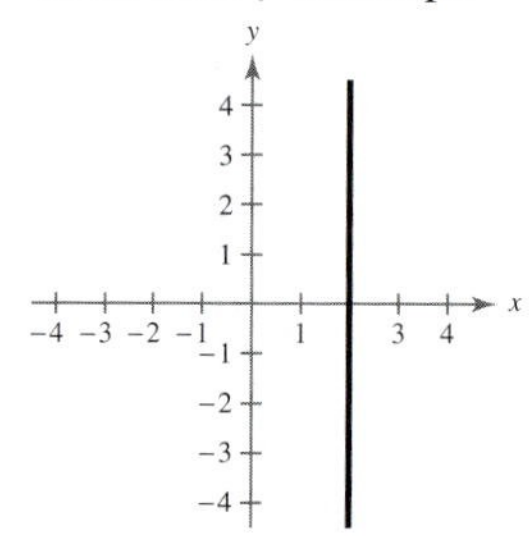

21.

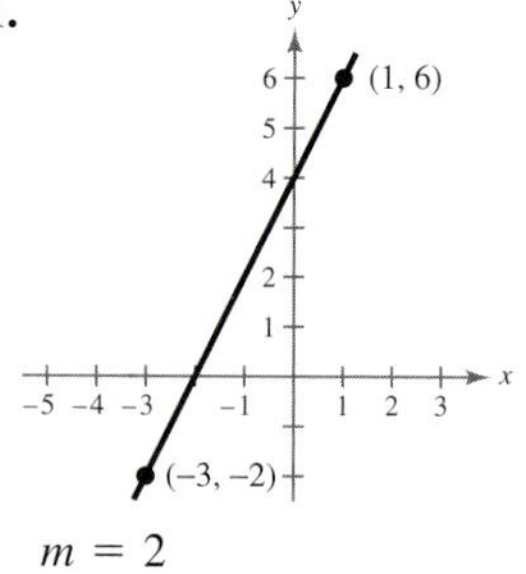

$m = 2$

22.

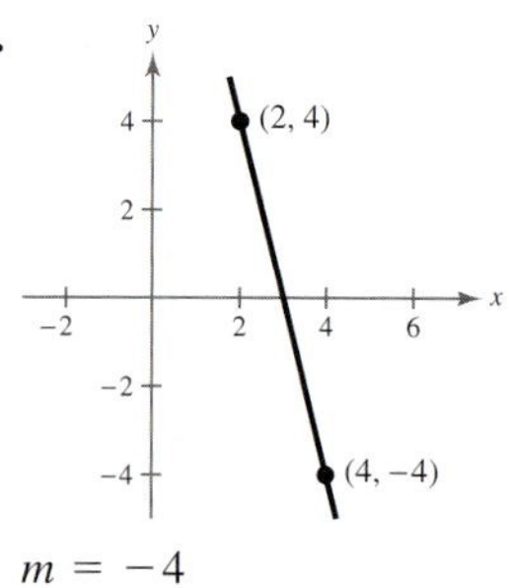

$m = -4$

23.

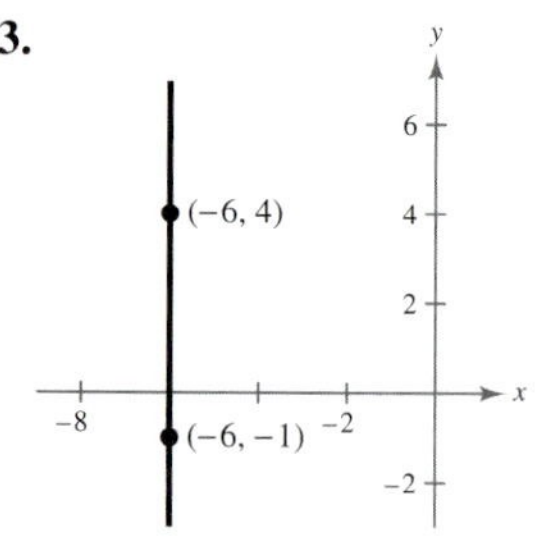

m is undefined.

24.

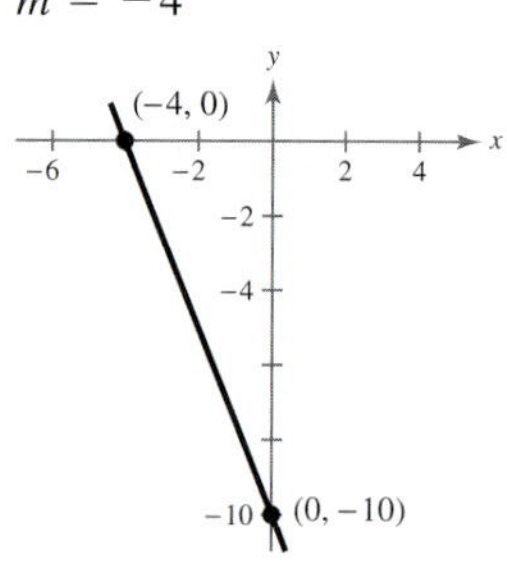

$m = -\frac{5}{2}$

25.

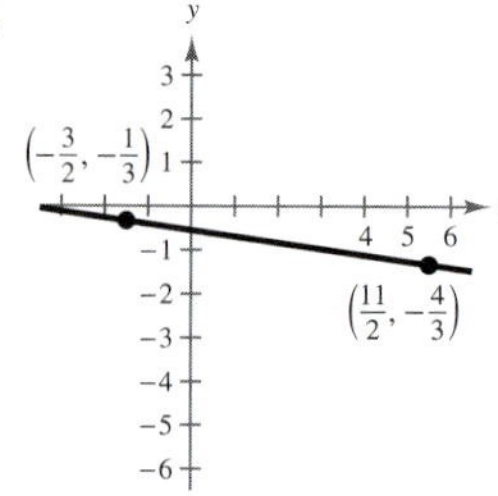

$m = -\frac{1}{7}$

26.

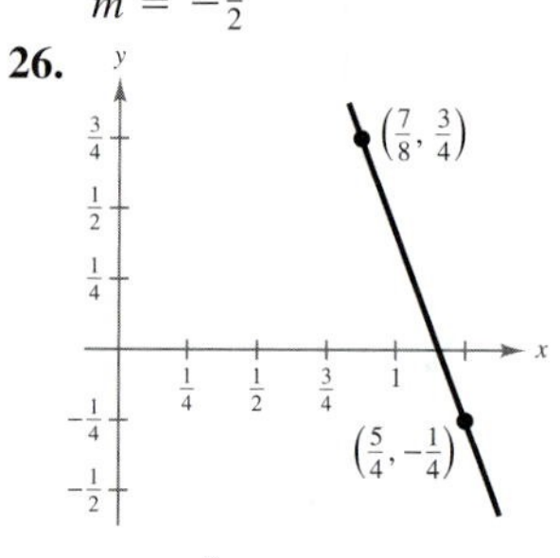

$m = -\frac{8}{3}$

27.

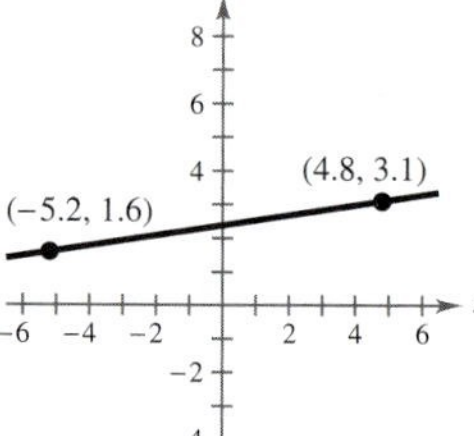

$m = 0.15$

28.

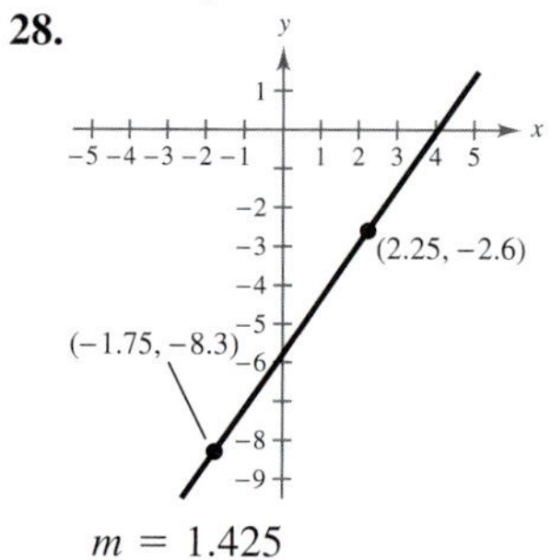

$m = 1.425$

29. $(0, 1), (3, 1), (-1, 1)$

30. $(-4, 0), (-4, 3), (-4, 5)$

31. $(6, -5), (7, -4), (8, -3)$

32. $(0, 4), (9, -5), (11, -7)$

33. $(-8, 0), (-8, 2), (-8, 3)$

34. $(-4, -1), (-2, -1), (0, -1)$

35. $(-4, 6), (-3, 8), (-2, 10)$

36. $(-2, -5), (1, -11), (3, -15)$

37. $(9, -1), (11, 0), (13, 1)$

38. $(-3, -5), (1, -7), (5, -9)$

39. $y = 3x - 2$

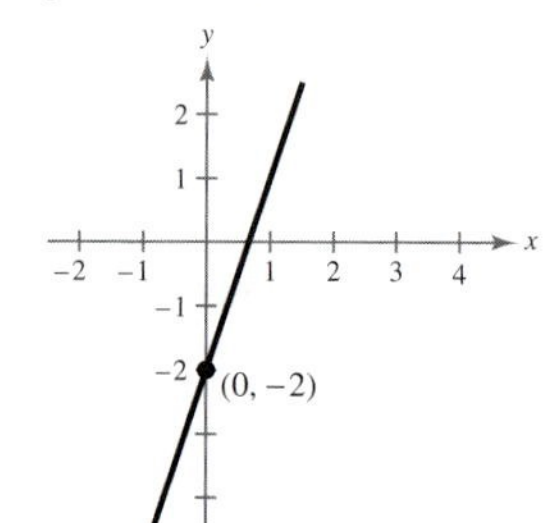

40. $y = -x + 10$

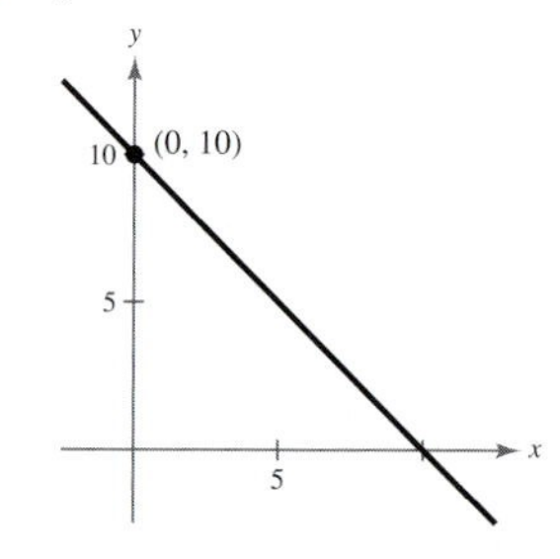

41. $y = -2x$

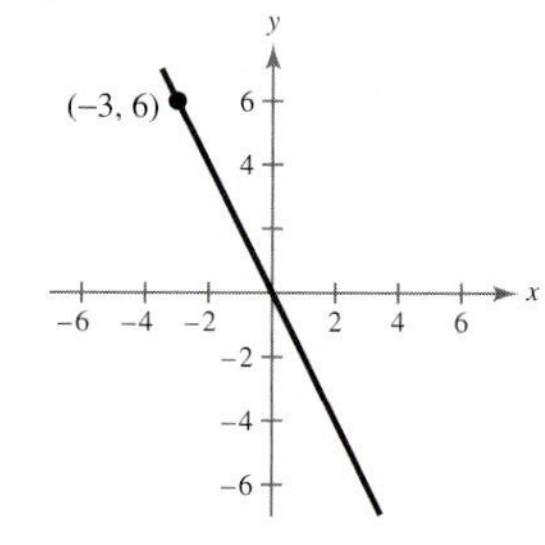

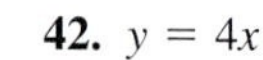

42. $y = 4x$

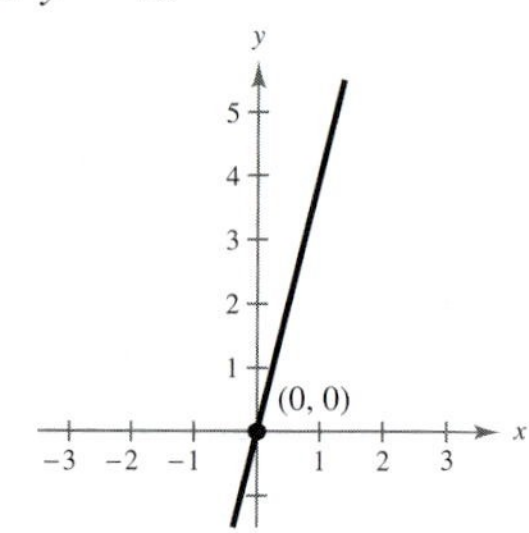

43. $y = -\frac{1}{3}x + \frac{4}{3}$

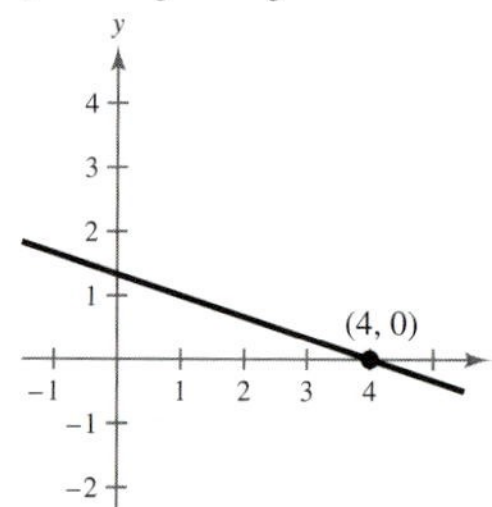

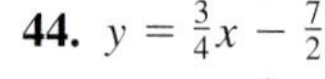

44. $y = \frac{3}{4}x - \frac{7}{2}$

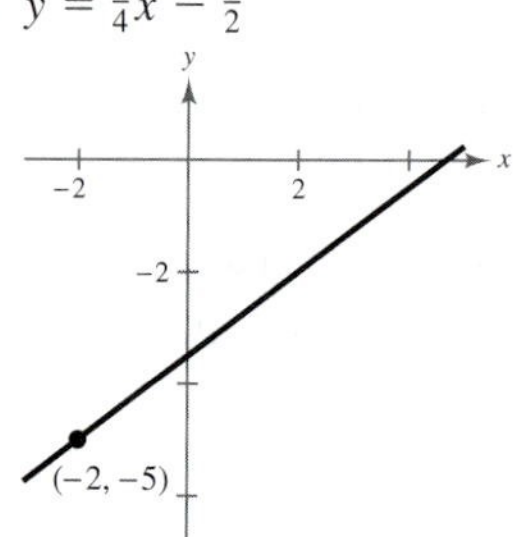

45. $x = 6$

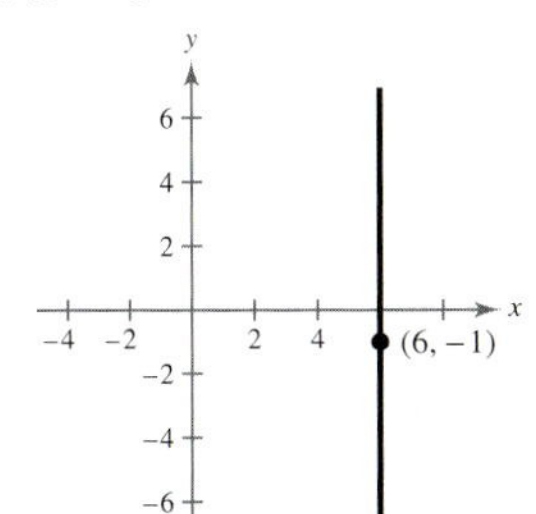

46. $x = -10$

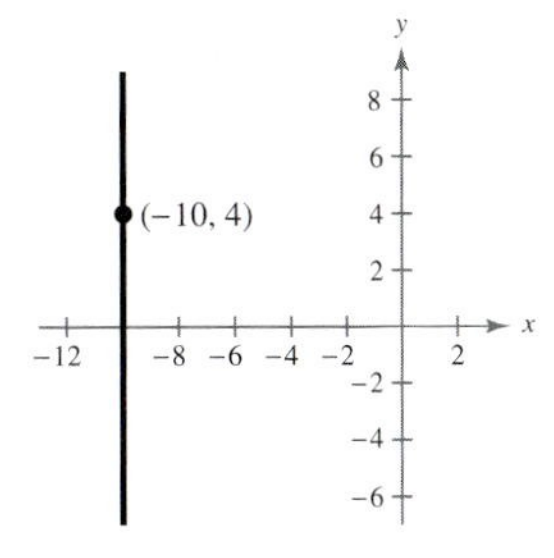

47. $y = \frac{5}{2}$

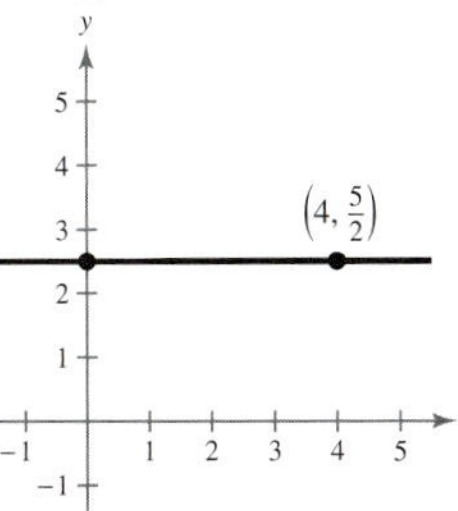

48. $y = \frac{3}{2}$

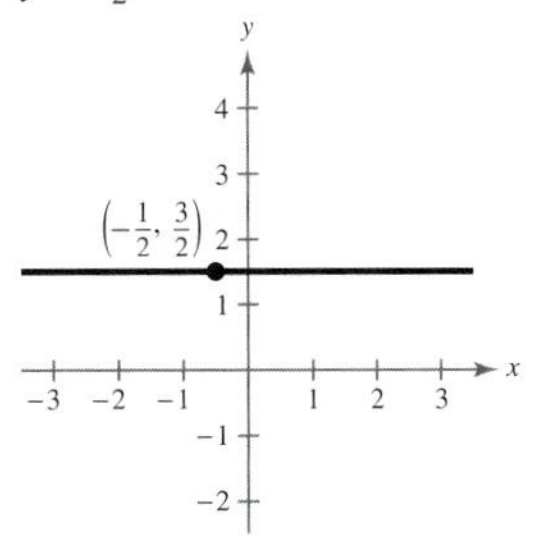

49. $y = 5x + 27.3$

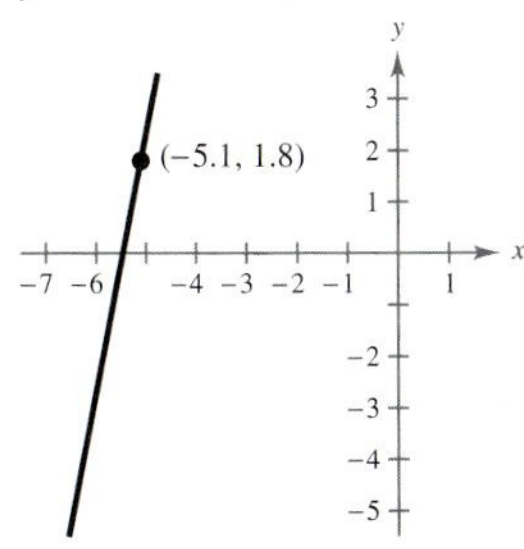

50. $y = -2.5x - 2.75$

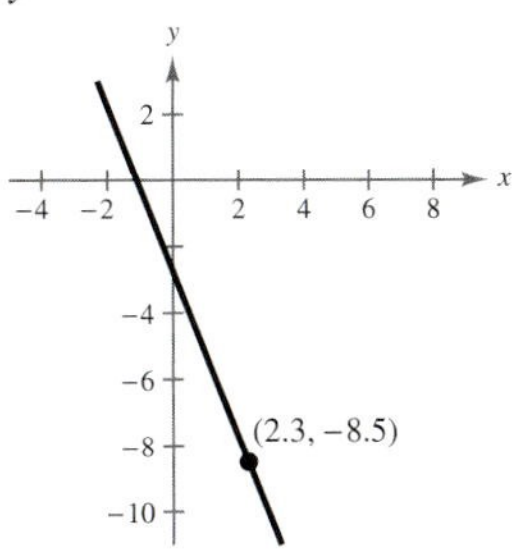

51. $y = -\frac{3}{5}x + 2$

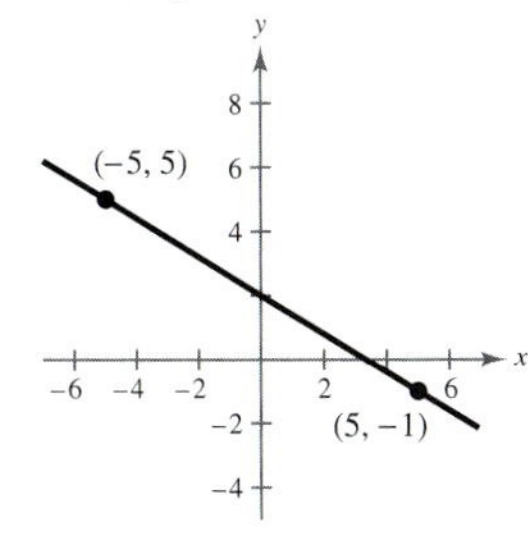

52. $y = \frac{7}{8}x - \frac{1}{2}$

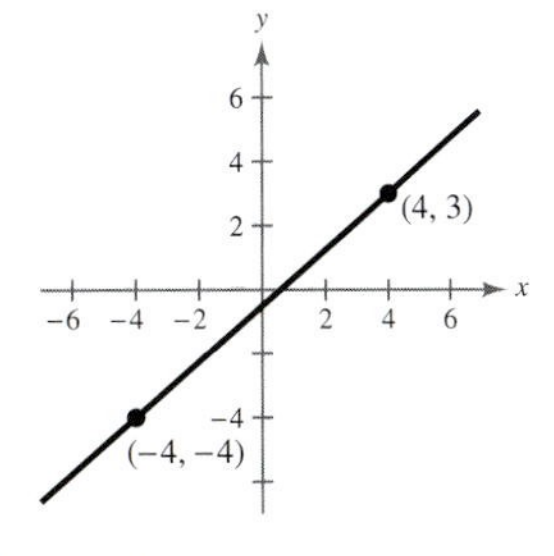

53. $x = -8$

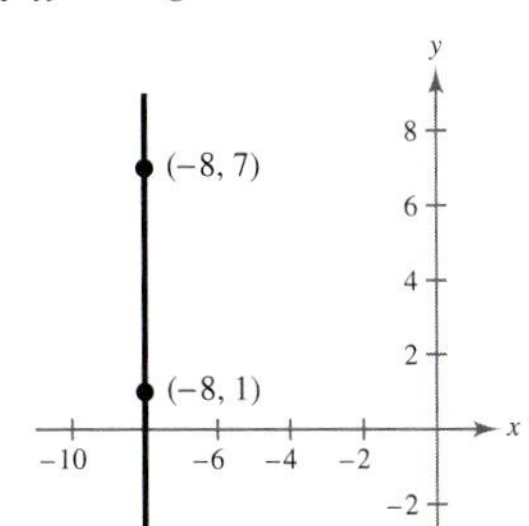

54. $y = 4$

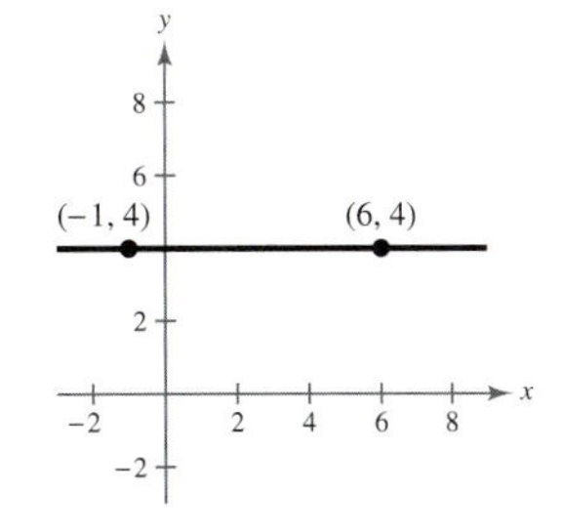

55. $y = -\frac{1}{2}x + \frac{3}{2}$

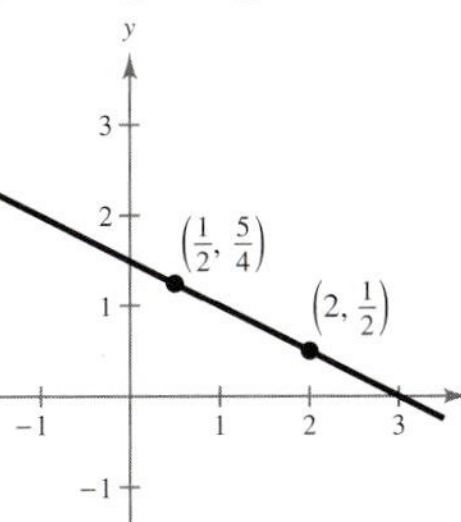

56. $y = -\frac{1}{3}x + \frac{4}{3}$

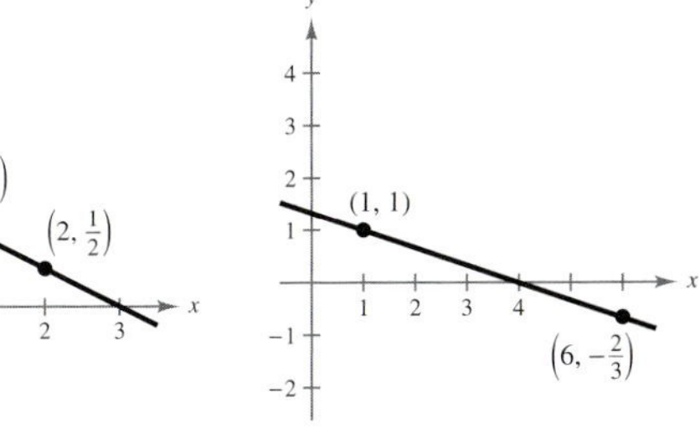

57. $y = -\frac{6}{5}x - \frac{18}{25}$

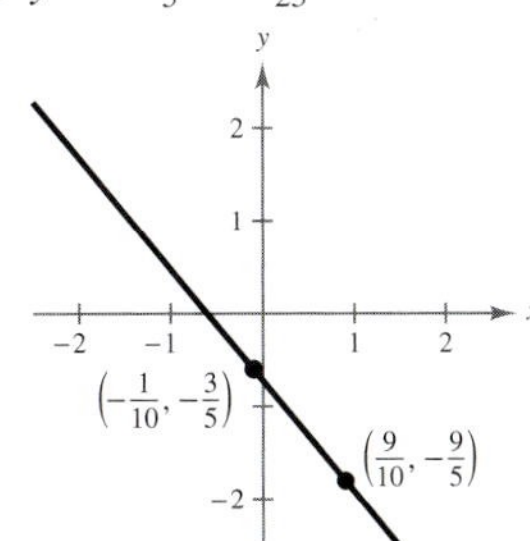

58. $y = -\frac{3}{25}x + \frac{159}{100}$

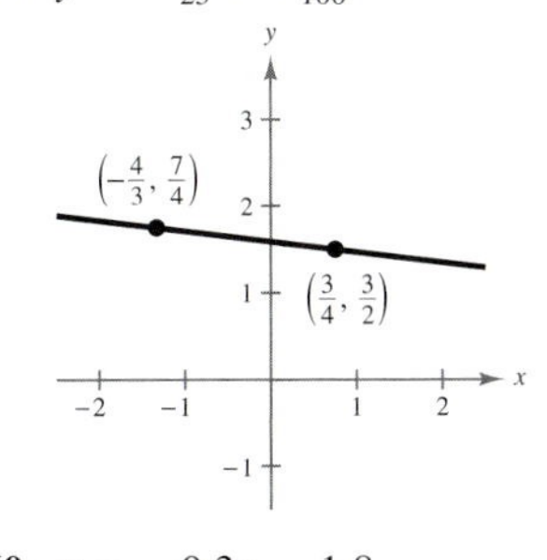

59. $y = 0.4x + 0.2$

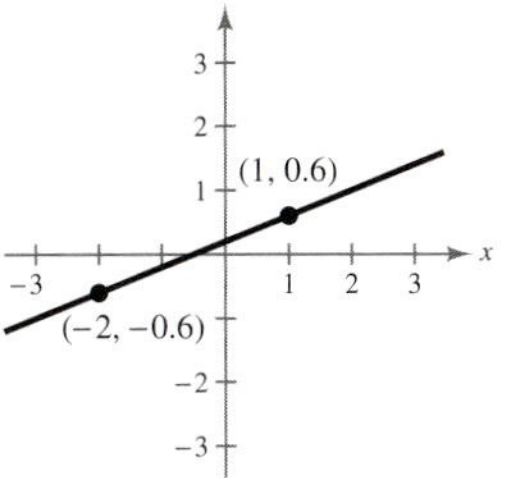

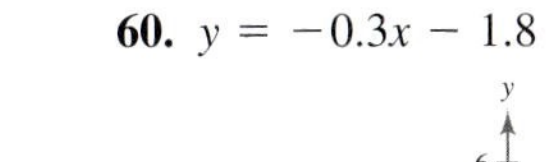

60. $y = -0.3x - 1.8$

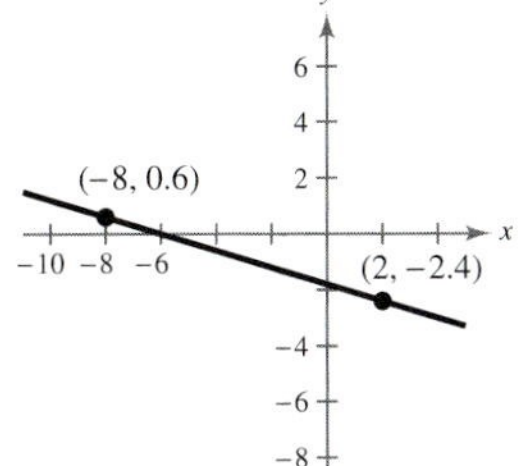

61. $y = -1$

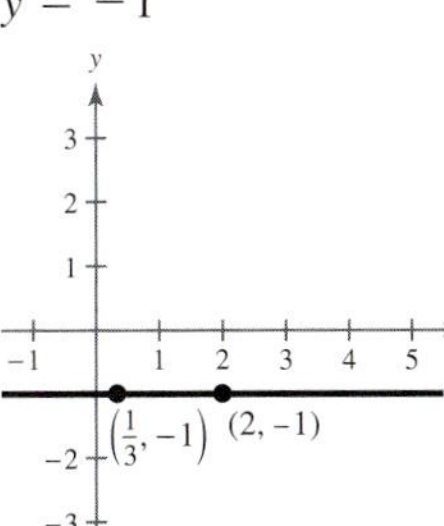

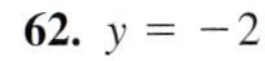

62. $y = -2$

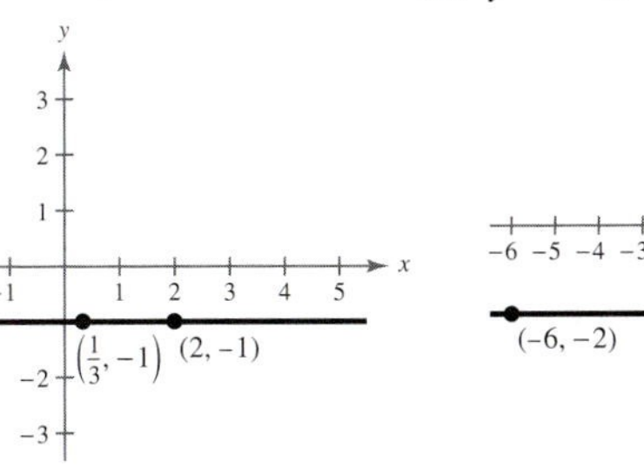

63. $x = \frac{7}{3}$

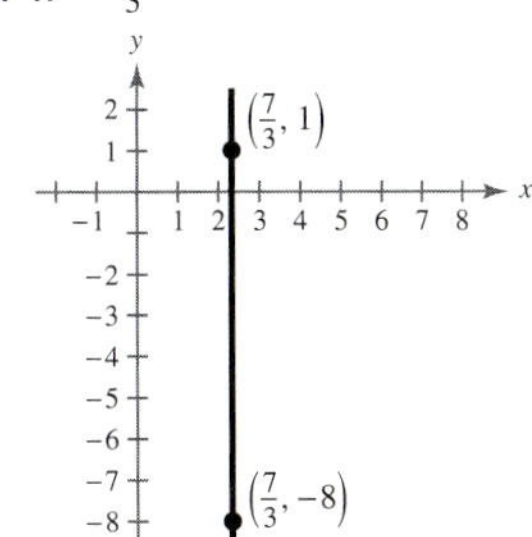

64. $x = 1.5$

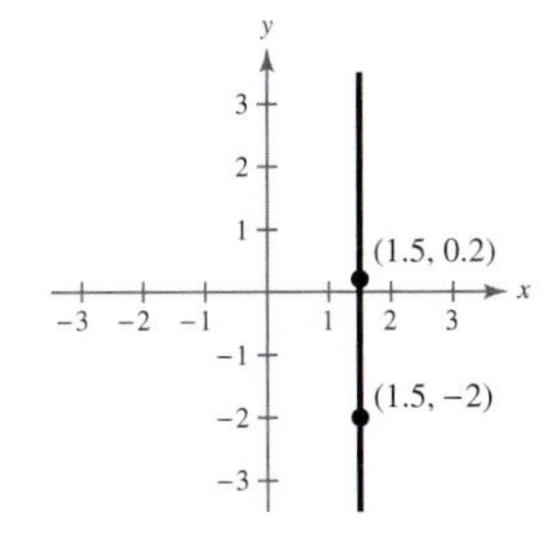

19. $0, \pm\sqrt{2}$ **20.** $\pm 3, 4$ **21.** $\pm\frac{1}{2}, 6$ **22.** $0, \pm\frac{5}{3}$

23. $\frac{1}{2}$ **24.** $-\frac{2}{3}$

25.

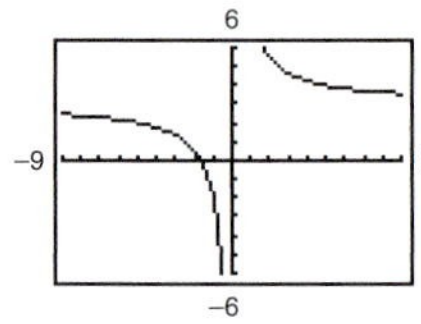

$-\frac{5}{3}$

26.

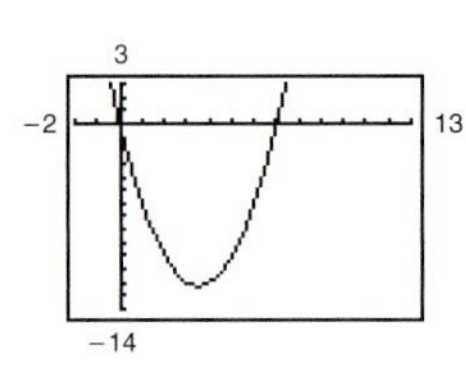

0, 7

27.

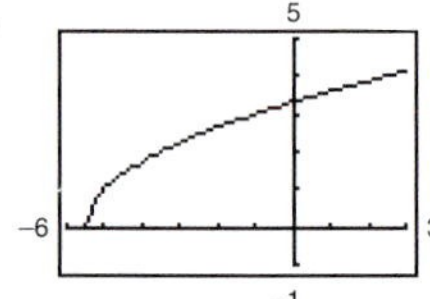

$-\frac{11}{2}$

28.

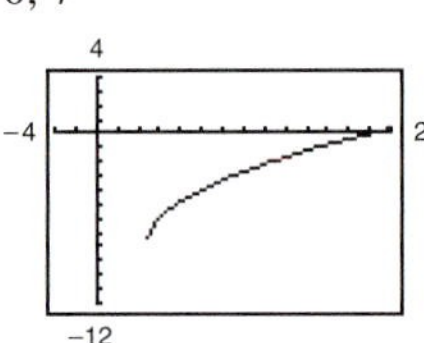

26

29.

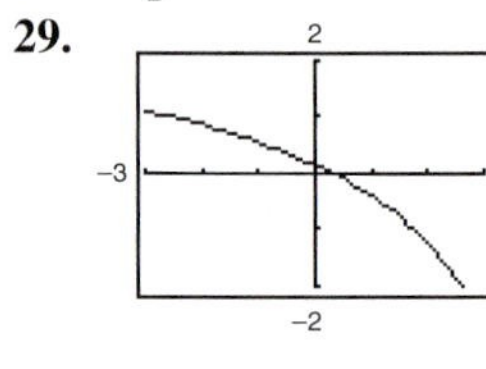

$\frac{1}{3}$

30.

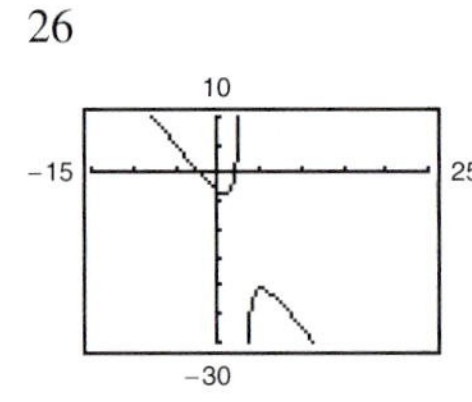

$\pm\frac{3\sqrt{2}}{2}$

31. Increasing on $(-\infty, \infty)$

32. Increasing on $(2, \infty)$; Decreasing on $(-\infty, 2)$

33. Increasing on $(-\infty, 0)$ and $(2, \infty)$
Decreasing on $(0, 2)$

34. Increasing on $(1, \infty)$; Decreasing on $(-\infty, -1)$

35. Increasing on $(-\infty, 0)$ and $(2, \infty)$; Constant on $(0, 2)$

36. Increasing on $(-\infty, -1)$, $(0, \infty)$; Decreasing on $(-1, 0)$

37. Increasing on $(1, \infty)$; Decreasing on $(-\infty, -1)$
Constant on $(-1, 1)$

38. Increasing on $(-\infty, -2)$, $(0, \infty)$
Decreasing on $(-2, -1)$, $(-1, 0)$

39.

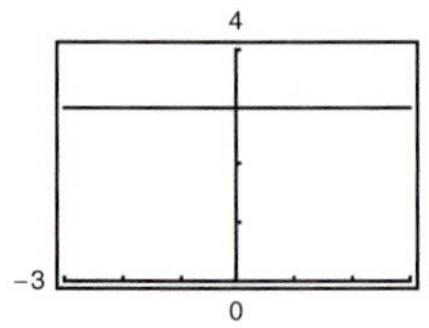

Constant on $(-\infty, \infty)$

40.

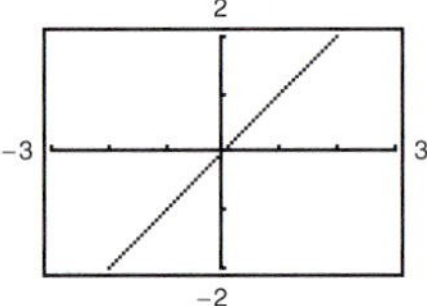

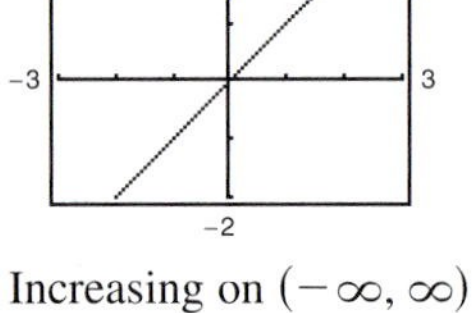

Increasing on $(-\infty, \infty)$

41.

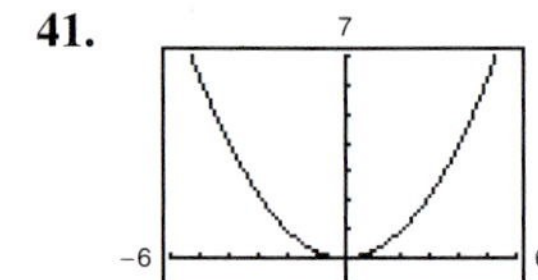

Decreasing on $(-\infty, 0)$
Increasing on $(0, \infty)$

42.

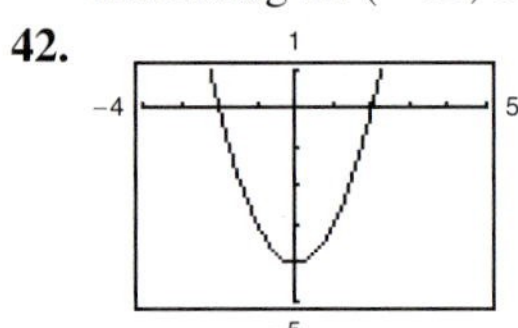

Increasing on $(0, \infty)$
Decreasing on $(-\infty, 0)$

43.

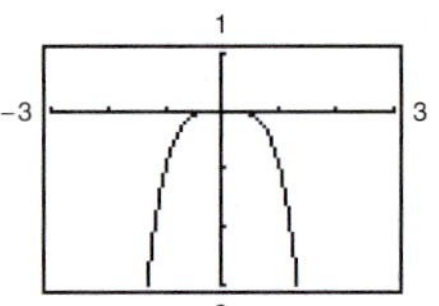

Increasing on $(-\infty, 0)$
Decreasing on $(0, \infty)$

44.

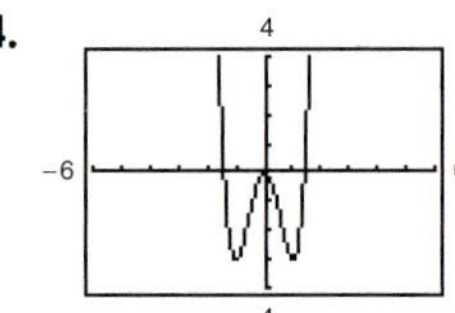

Increasing on $(-1, 0)$, $(1, \infty)$
Decreasing on $(-\infty, -1)$, $(0, 1)$

45.

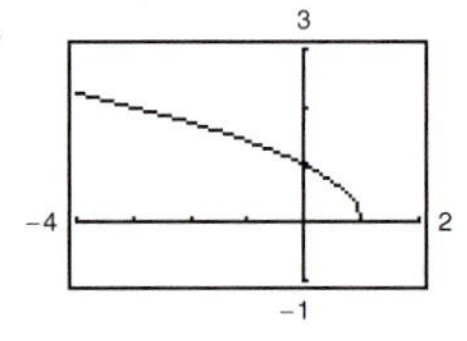

Decreasing on $(-\infty, 1)$

46.

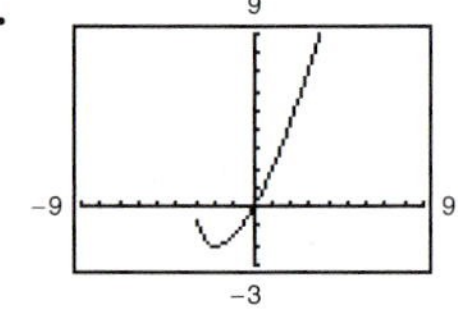

Increasing on $(-2, \infty)$
Decreasing on $(-3, -2)$

47.

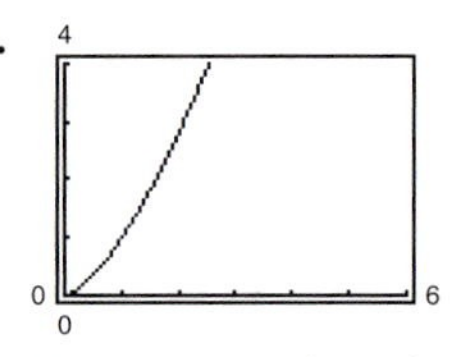

Increasing on $(0, \infty)$

48.

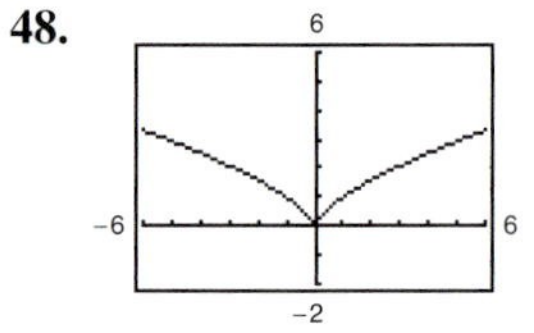

Decreasing on $(-\infty, 0)$
Increasing on $(0, \infty)$

49.

Relative minimum: $(1, -9)$

50.

Relative minimum: $\left(\frac{1}{3}, -\frac{16}{3}\right)$

51.

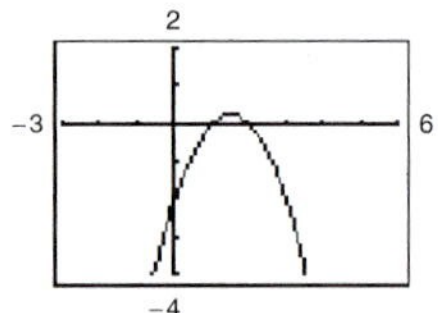

Relative maximum: $(1.5, 0.25)$

52.

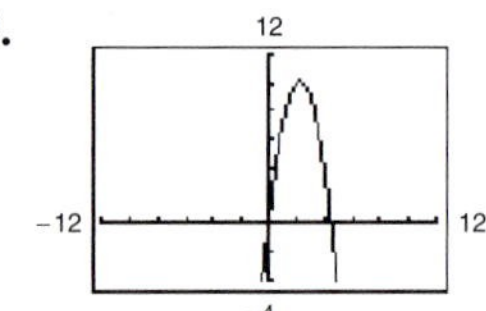

Relative maximum: $(2.25, 10.125)$

53.

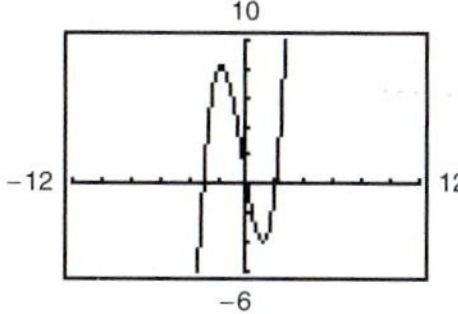

Relative maximum: $(-1.79, 8.21)$
Relative minimum: $(1.12, -4.06)$

54.

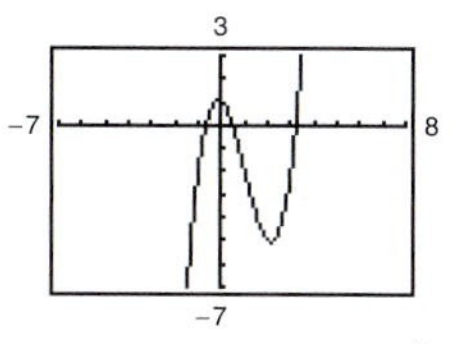

Relative maximum: $(-0.15, 1.08)$
Relative minimum: $(2.15, -5.08)$

55.

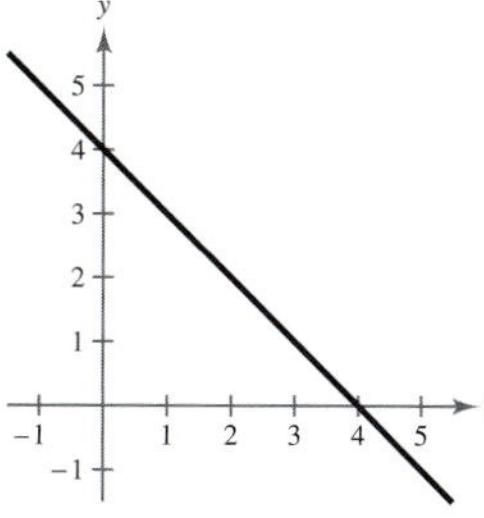

$(-\infty, 4]$

56.

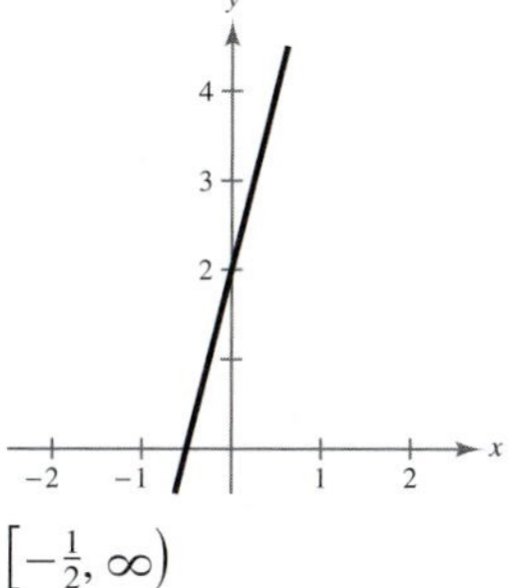

$\left[-\frac{1}{2}, \infty\right)$

57.

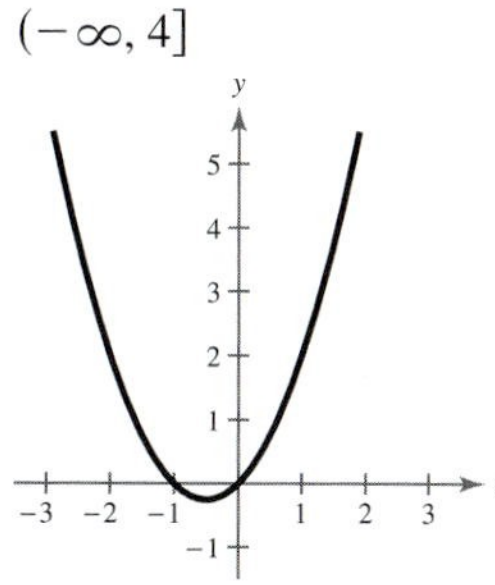

$(-\infty, -1], [0, \infty)$

58.

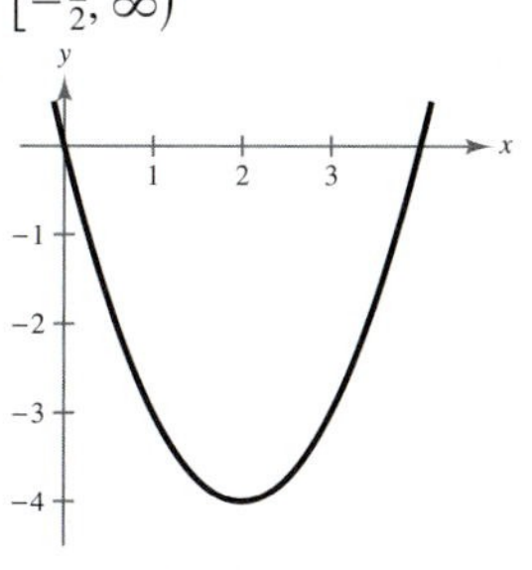

$(-\infty, 0], [4, \infty)$

59.

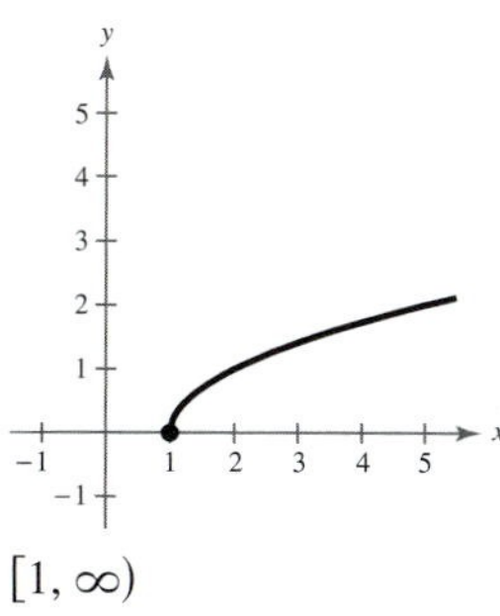

$[1, \infty)$

60.

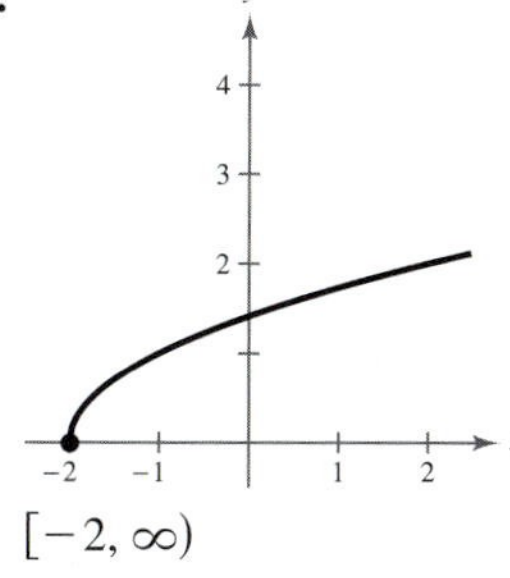

$[-2, \infty)$

61.

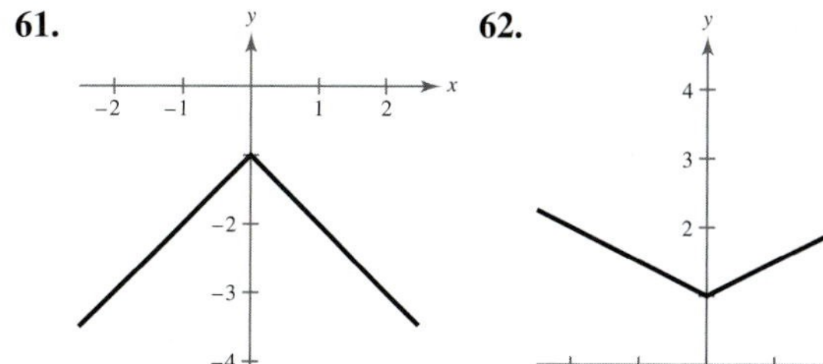

$f(x) < 0$ for all x

62.

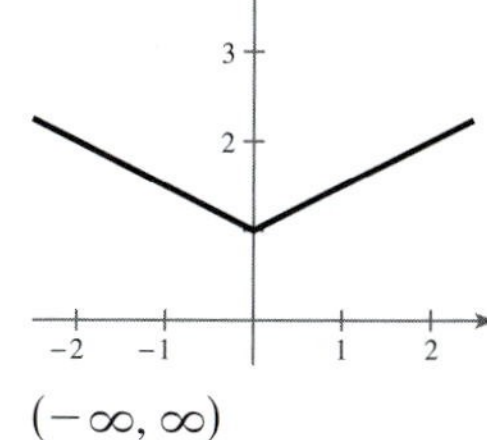

$(-\infty, \infty)$

63. The average rate of change from $x_1 = 0$ to $x_2 = 3$ is -2.
64. The average rate of change from $x_1 = 0$ to $x_2 = 3$ is 3.
65. The average rate of change from $x_1 = 1$ to $x_2 = 5$ is 18.
66. The average rate of change from $x_1 = 1$ to $x_2 = 5$ is 4.
67. The average rate of change from $x_1 = 1$ to $x_2 = 3$ is 0.
68. The average rate of change from $x_1 = 1$ to $x_2 = 6$ is 0.
69. The average rate of change from $x_1 = 3$ to $x_2 = 11$ is $-\frac{1}{4}$.
70. The average rate of change from $x_1 = 3$ to $x_2 = 8$ is $-\frac{1}{5}$.
71. Even; y-axis symmetry
72. Neither even nor odd; no symmetry
73. Odd; origin symmetry **74.** Odd; origin symmetry
75. Neither even nor odd; no symmetry
76. Even; y-axis symmetry
77. $h = -x^2 + 4x - 3$ **78.** $h = 3 - 4x + x^2$
79. $h = 2x - x^2$ **80.** $h = 2 - \sqrt[3]{x}$
81. $L = \frac{1}{2}y^2$ **82.** $L = 2 - \sqrt[3]{2y}$ **83.** $L = 4 - y^2$
84. $L = \dfrac{2}{y}$
85. (a)

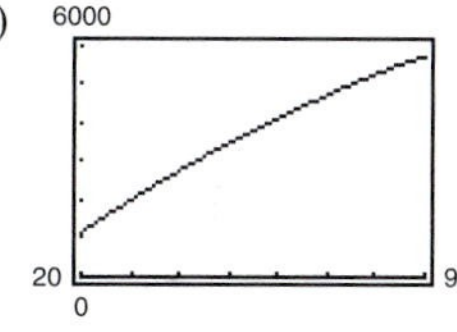

(b) 30 watts

86. (a)

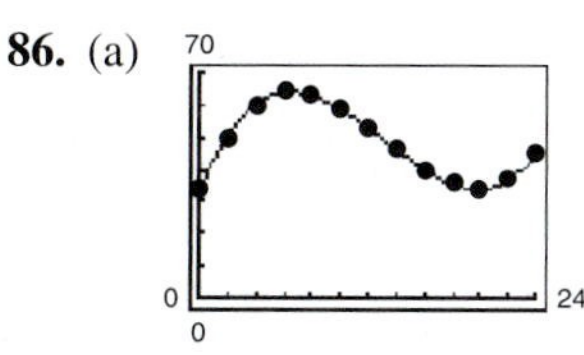

(b) The model fits the data very well.
(c) The temperature was increasing from 6 A.M. ($x = 0$) to noon ($x = 6$), and again from 2 A.M. ($x = 20$) to 6 A.M. ($x = 24$). The temperature was decreasing from noon to 2 A.M.
(d) The maximum temperature was 63.93°F and the minimum temperature was 33.98°F.
(e) Answers will vary.

87. (a) Ten thousands (b) Ten millions (c) Percents

88. (a) $A = 64 - 2x^2,\ 0 \le x \le 4$

(b) $32 \le A \le 64$

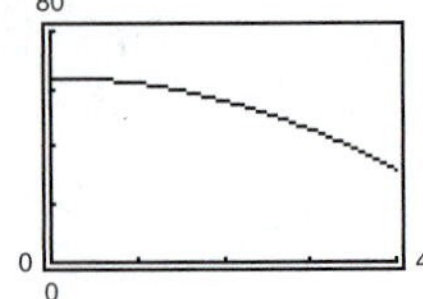

(c) Square with sides of $4\sqrt{2}$ meters

89. (a)

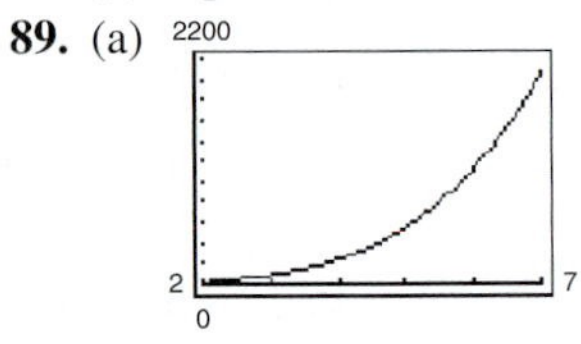

(b) The average rate of change from 2002 to 2007 is 408.56. The estimated revenue is increasing each year at a fast pace.

90. (a)

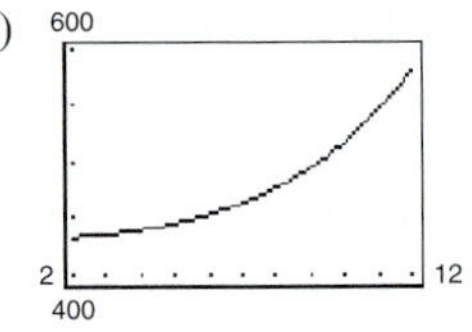

(b) The average rate of change of the model from 1992 to 2002 is 14.73. This shows that the number of students enrolling in college increased at a fairly steady rate.

(c) The five-year time period when the rate of change was the least was from 1992 to 1997 with a 6.05 average rate, and the greatest from 1997 to 2002 with a 23.41 average rate.

91. (a) $s = -16t^2 + 64t + 6$

(b)

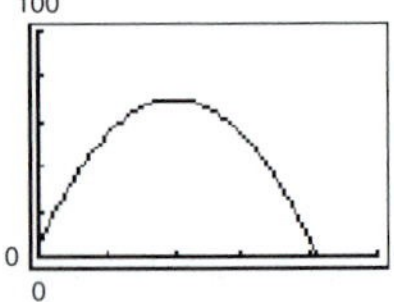

(c) Average rate of change $= 16$

(d) The slope of the secant line is positive.

(e) Secant line: $16t + 6$

(f)

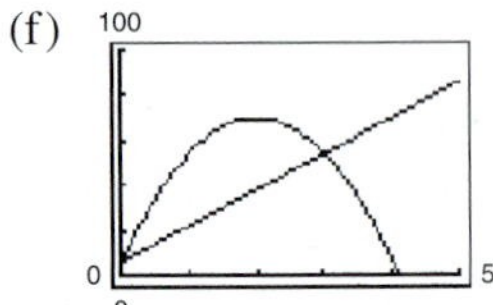

92. (a) $s = -16t^2 + 72t + 6.5$

(b)

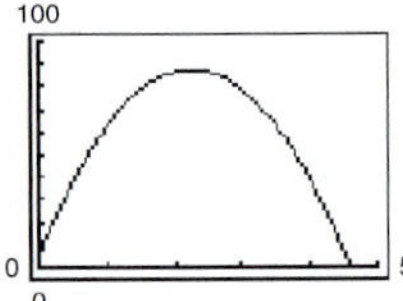

(c) Average rate of change $= 8$

(d) The slope of the secant line is positive.

(e) Secant line $= 8t + 6.5$

(f)

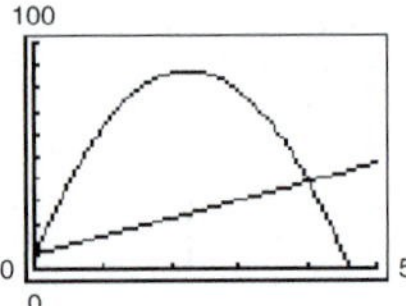

93. (a) $s = -16t^2 + 120t$

(b)

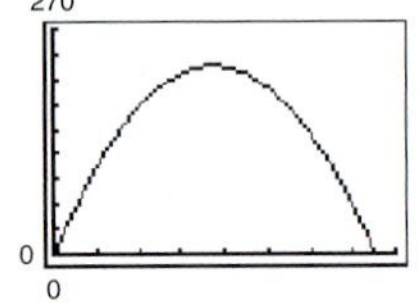

(c) Average rate of change $= -8$

(d) The slope of the secant line is negative.

(e) Secant line: $-8t + 240$

(f)

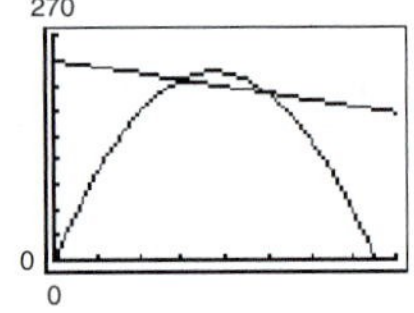

94. (a) $s = -16t^2 + 96t$

(b)

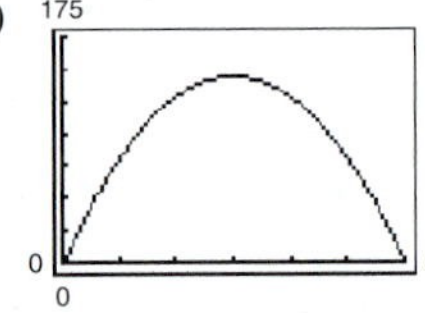

(c) Average rate of change $= -16$

(d) The slope of the secant line is negative.

(e) Secant line $= -16t + 160$

(f)

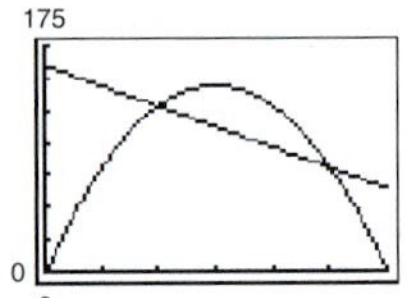

95. (a) $s = -16t^2 + 120$

(b)

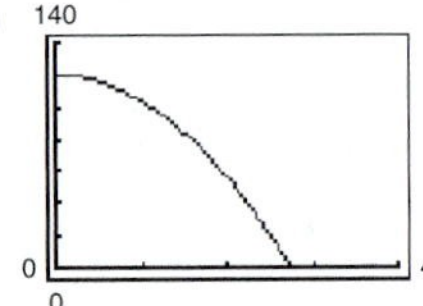

(c) Average rate of change $= -32$

(d) The slope of the secant line is negative.

(e) Secant line: $-32t + 120$

(f)

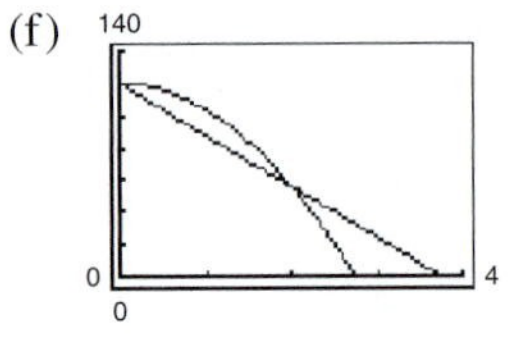

96. (a) $s = -16t^2 + 80$

(b) 120; 0; 3; 0

(c) Average rate of change $= -48$

(d) The slope of the secant line is negative.

(e) Secant line $= -48t + 112$

(f) 120; 0; 3; 0

97. False. The function $f(x) = \sqrt{x^2 + 1}$ has a domain of all real numbers.

98. False. An odd function is symmetric with respect to the origin, so its domain must include negative values.

99. (a) Even. The graph is a reflection in the x-axis.

(b) Even. The graph is a reflection in the y-axis.

(c) Even. The graph is a vertical translation of f.

(d) Neither. The graph is a horizontal translation of f.

100. Yes. For each value of y there corresponds one and only one value of x.

101. (a) $\left(\frac{3}{2}, 4\right)$ (b) $\left(\frac{3}{2}, -4\right)$

102. (a) $\left(\frac{5}{3}, -7\right)$ (b) $\left(\frac{5}{3}, 7\right)$

103. (a) $(-4, 9)$ (b) $(-4, -9)$

104. (a) $(-5, -1)$ (b) $(-5, 1)$

105. (a)

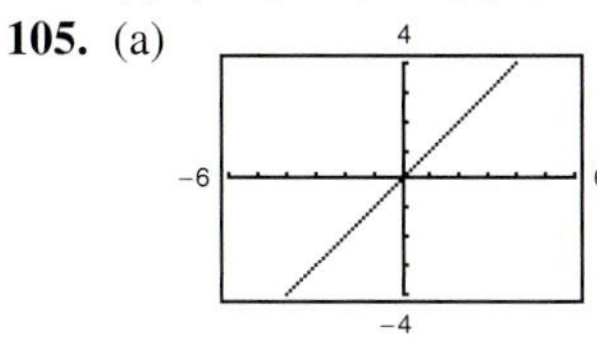

(b)

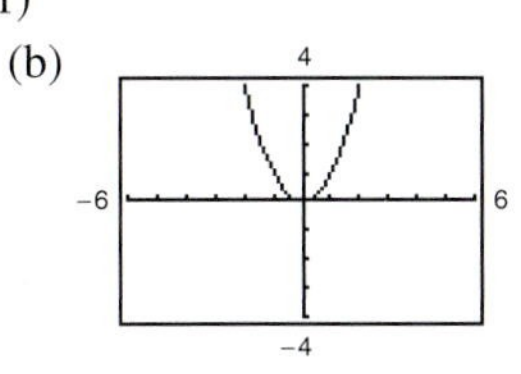

(c)

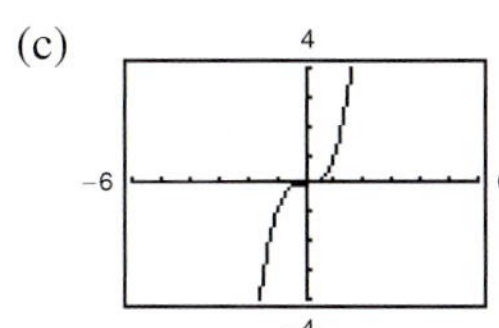

(d)

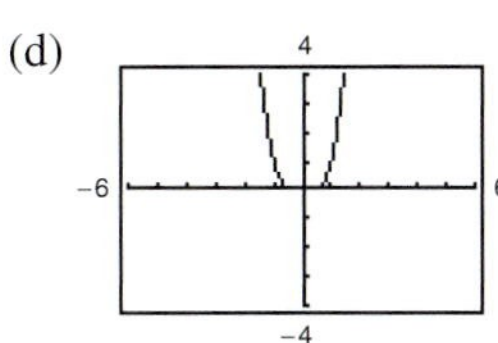

(e)

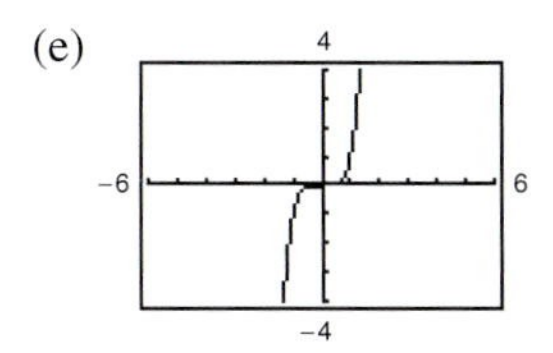

(f)

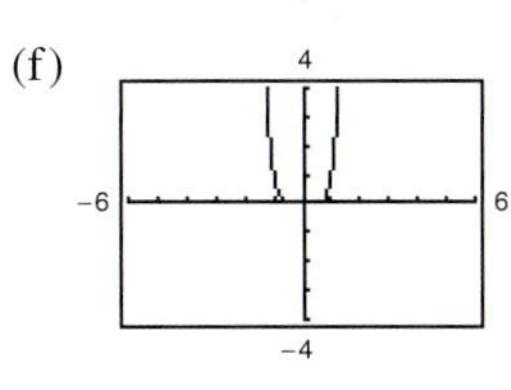

All the graphs pass through the origin. The graphs of the odd powers of x are symmetric with respect to the origin, and the graphs of the even powers are symmetric with respect to the y-axis. As the powers increase, the graphs become flatter in the interval $-1 < x < 1$.

106. Both graphs will pass through the origin. $y = x^7$ will be symmetric with respect to the origin, and $y = x^8$ will be symmetric with respect to the y-axis.

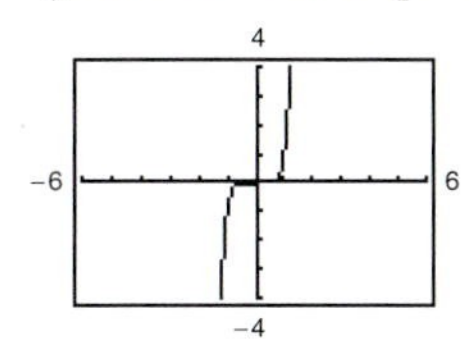

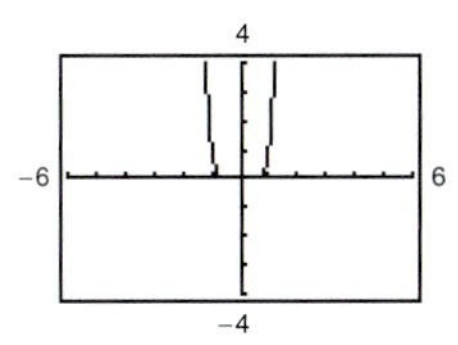

107. 0, 10 **108.** -5, 15 **109.** $0, \pm 1$ **110.** $\frac{5}{4}$

111. (a) 37 (b) -28 (c) $5x - 43$

112. (a) -24 (b) 144 (c) $x^2 - 18x + 56$

113. (a) -9 (b) $2\sqrt{7} - 9$

(c) The given value is not in the domain of the function.

114. (a) -3 (b) $-\frac{87}{16}$ (c) $139 - 2\sqrt{3}$

115. $h + 4,\ h \neq 0$ **116.** $-6 - h,\ h \neq 0$

Section 1.6 *(page 71)*

Vocabulary Check *(page 71)*

1. g **2.** i **3.** h **4.** a **5.** b
6. e **7.** f **8.** c **9.** d

1.

(a) $f(x) = -2x + 6$

(b)

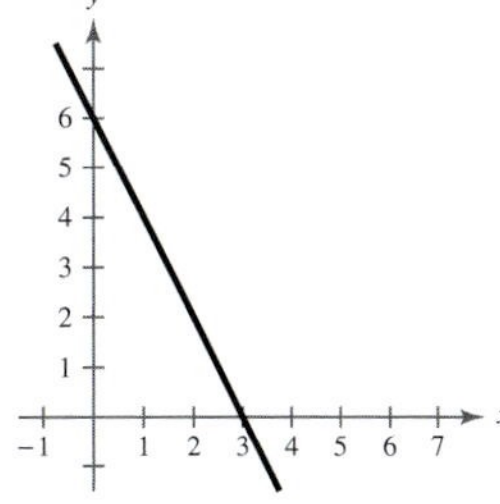

2.

(a) $f(x) = \frac{5}{2}x - \frac{1}{2}$

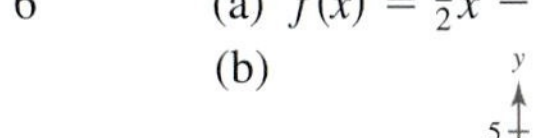

(b)

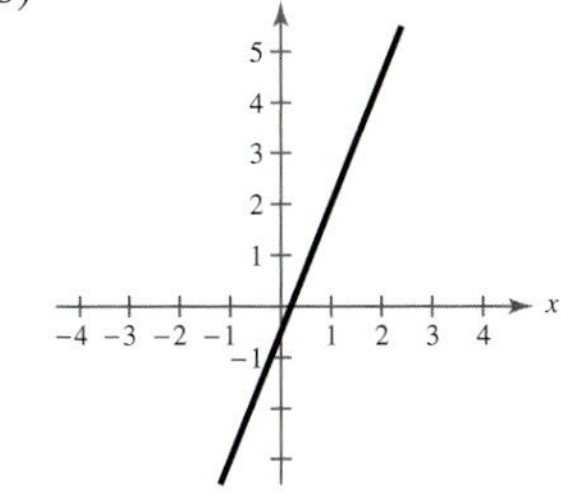

3.

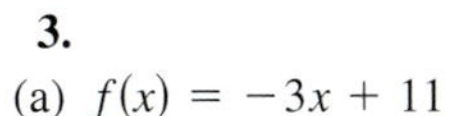

(a) $f(x) = -3x + 11$

(b)

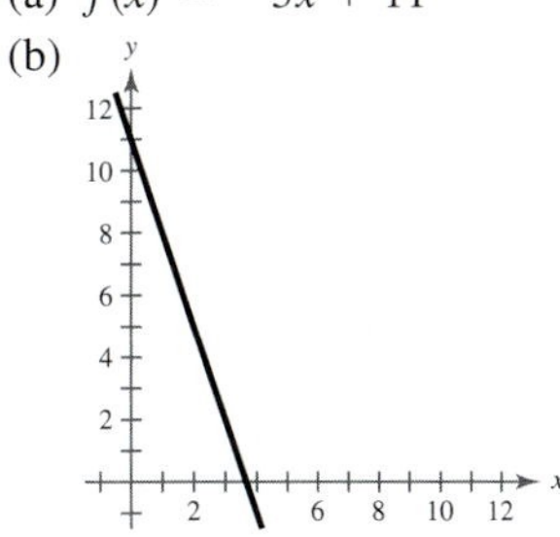

4.

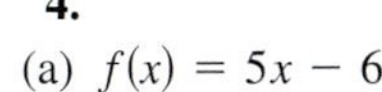

(a) $f(x) = 5x - 6$

(b)

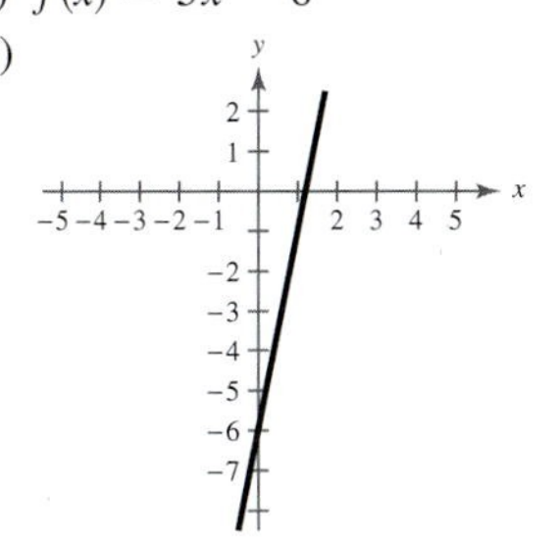

5.
(a) $f(x) = -1$
(b)
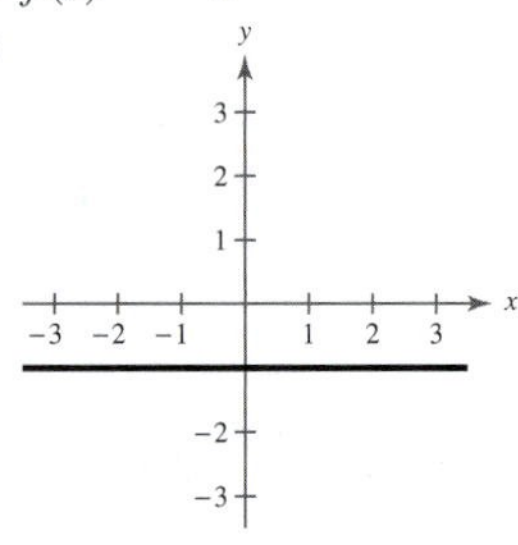

6.
(a) $f(x) = -\frac{1}{2}x + 7$
(b)
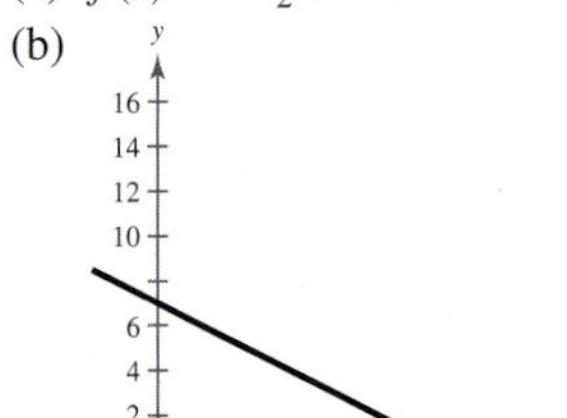

7.
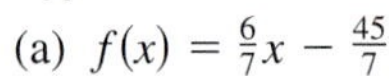
(a) $f(x) = \frac{6}{7}x - \frac{45}{7}$
(b)
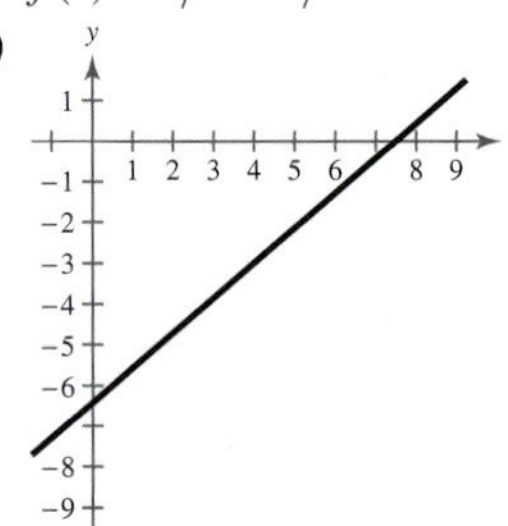

8.
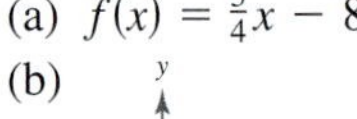
(a) $f(x) = \frac{3}{4}x - 8$
(b)
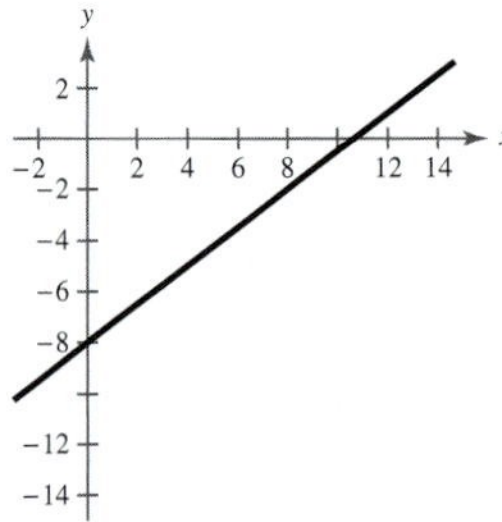

9.
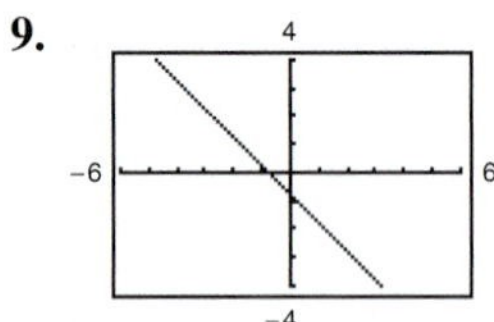

10.
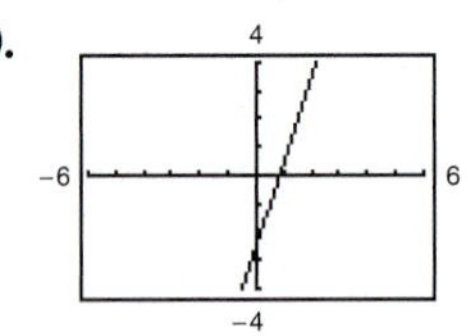

11.
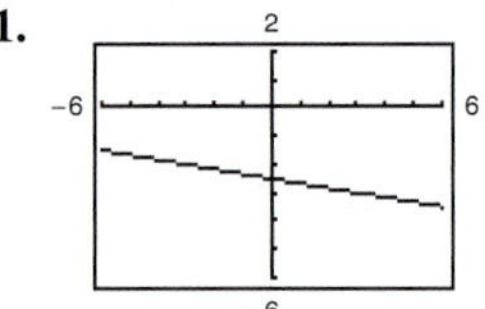

12.
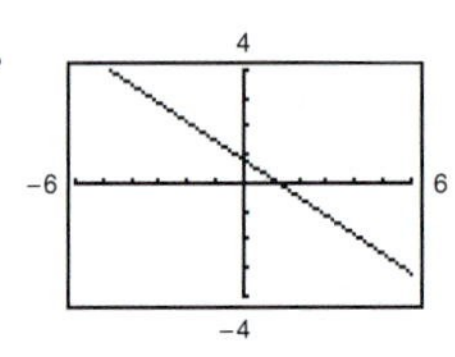

13.
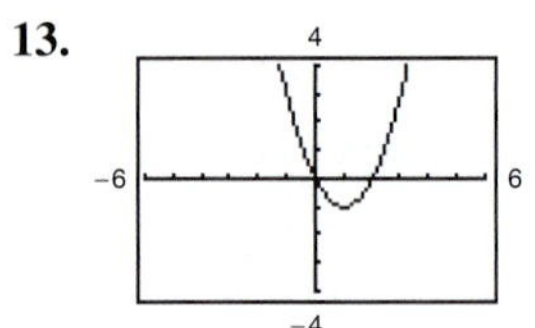

14.
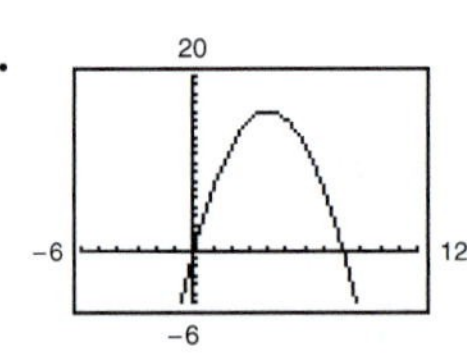

15.
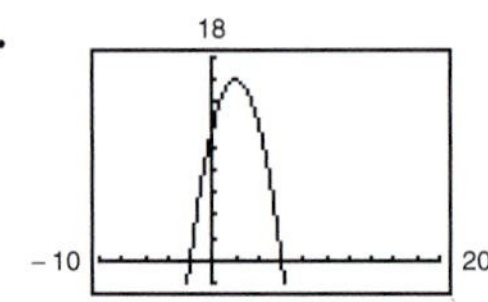

16.
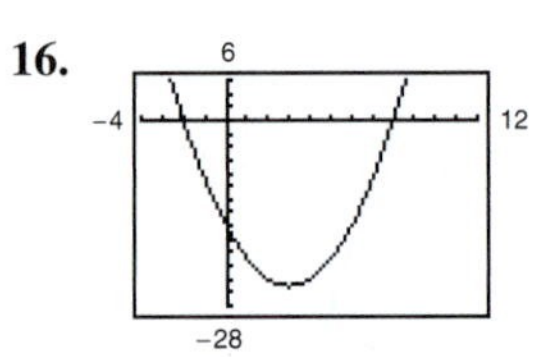

17.
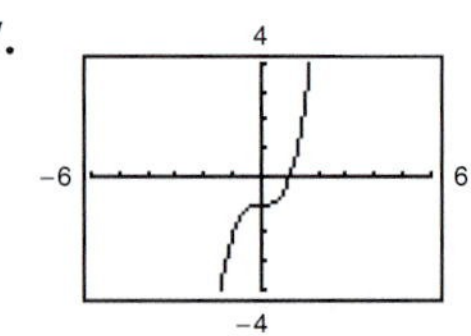

18.
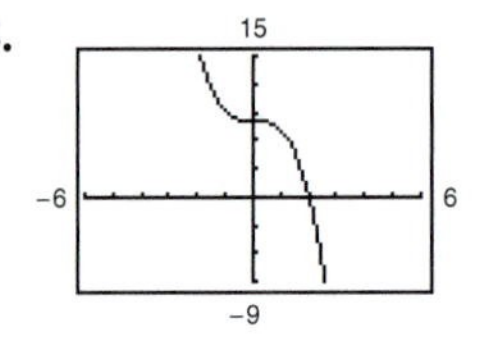

19.
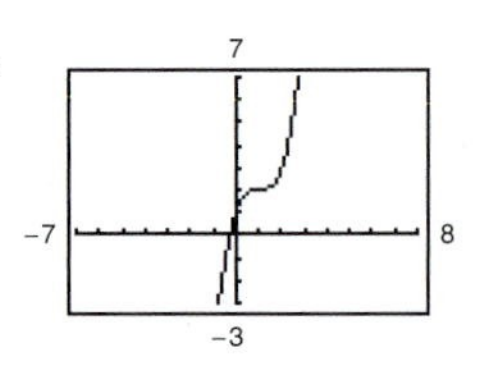

20.
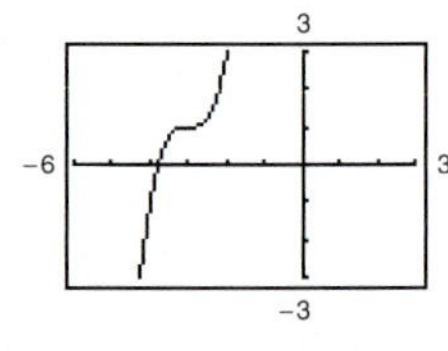

21.
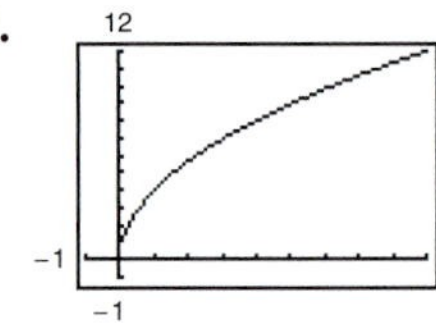

22.
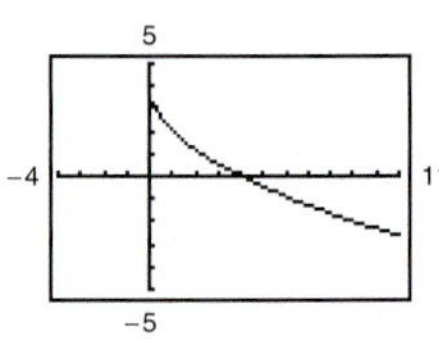

23.
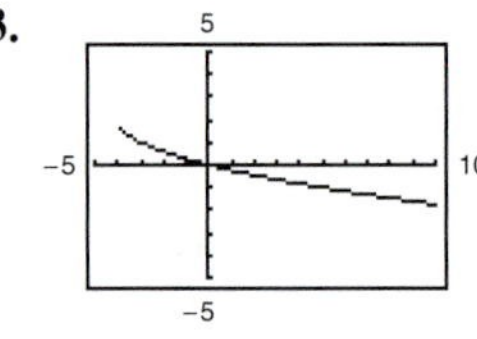

24.
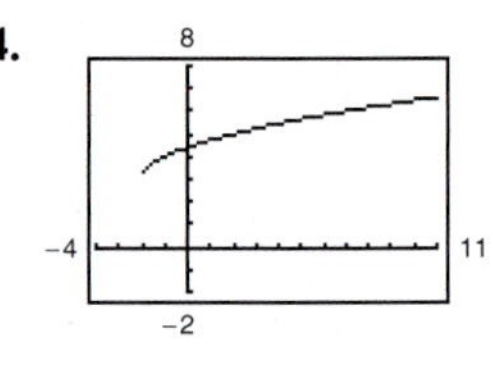

25.
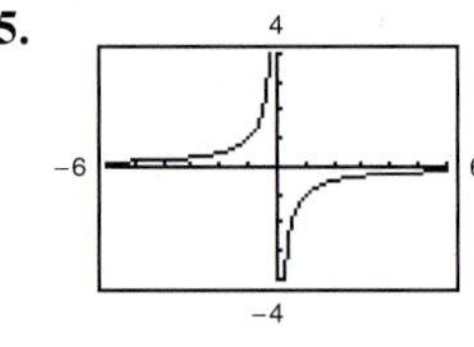

26.
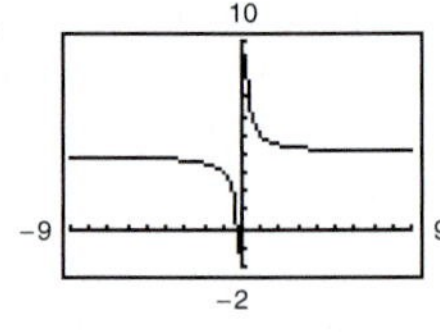

27.
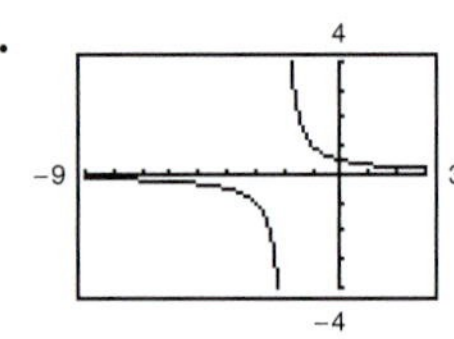

28.
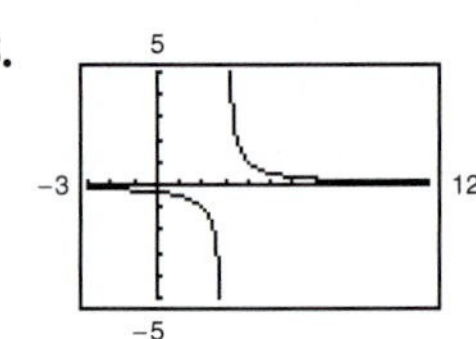

29. (a) 2 (b) 2 (c) -4 (d) 3
30. (a) -6 (b) 0 (c) 18 (d) 6
31. (a) 1 (b) 3 (c) 7 (d) -19
32. (a) 7 (b) -1 (c) 31 (d) 11
33. (a) 6 (b) -11 (c) 6 (d) -22
34. (a) 8 (b) 2 (c) 6 (d) 13
35. (a) -10 (b) -4 (c) -1 (d) 41
36. (a) -22 (b) -85 (c) 6 (d) -29

37.
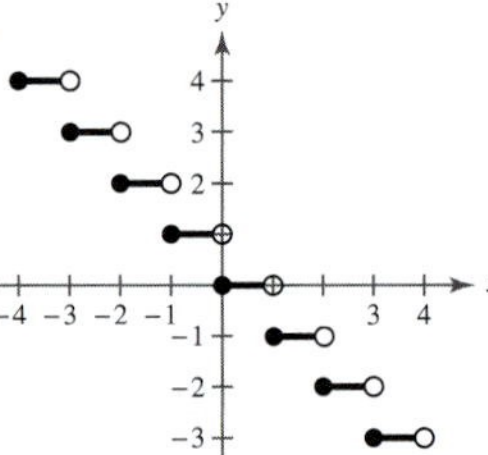

38.
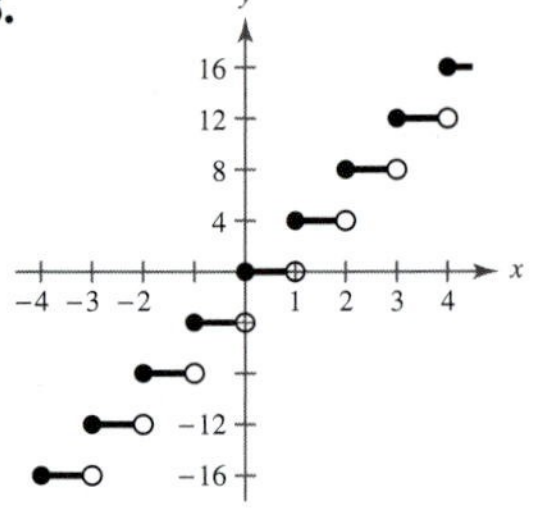

39.

40.

41.

42.

43.

44.

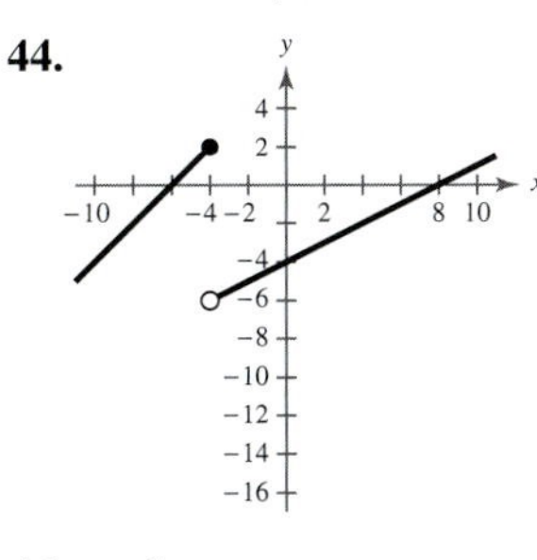

45.

46.

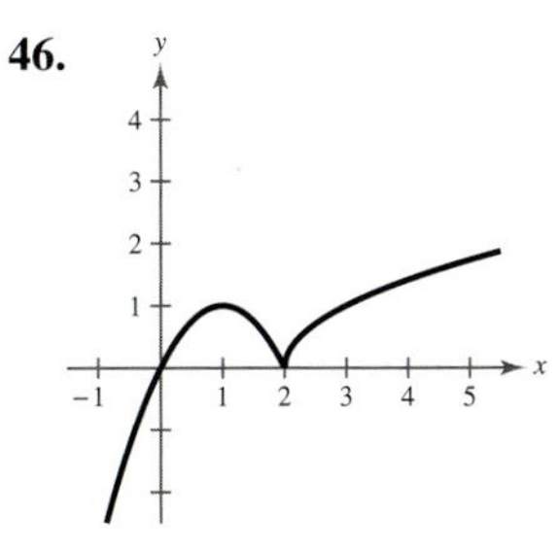

47.

48.

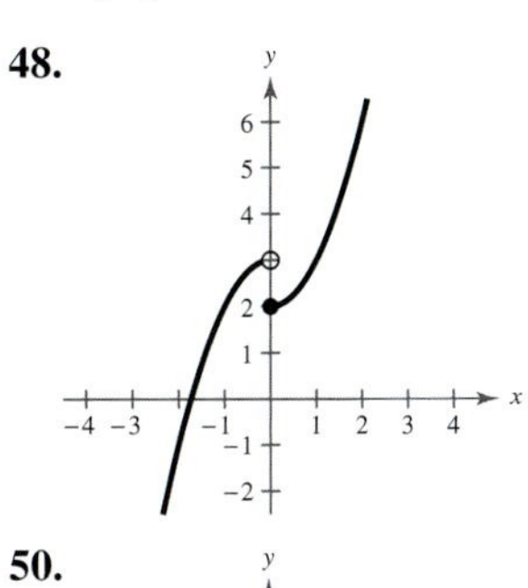

49.

50.

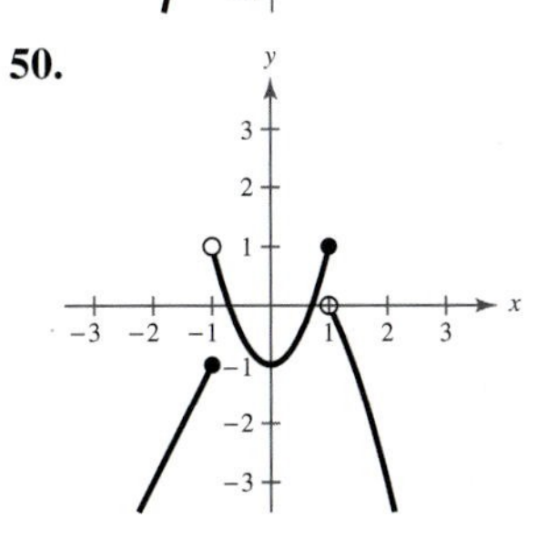

51. (a)

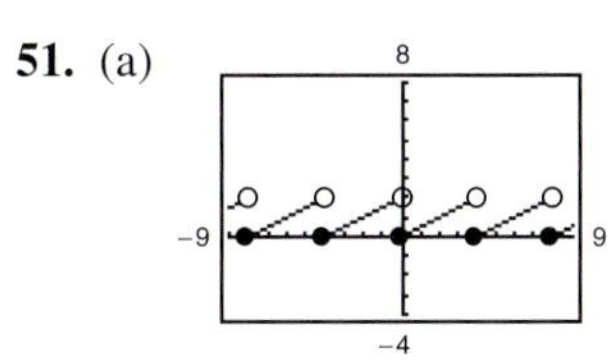

(b) Domain: $(-\infty, \infty)$; Range: $[0, 2)$
(c) Sawtooth pattern

52. (a)

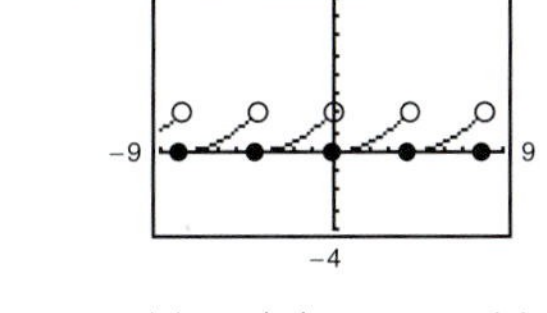

(b) Domain: $(-\infty, \infty)$; Range: $[0, 2)$
(c) Sawtooth pattern

53. (a) $f(x) = |x|$ (b) $g(x) = |x + 2| - 1$
54. (a) $f(x) = \sqrt{x}$ (b) $g(x) = 1 + \sqrt{x + 2}$
55. (a) $f(x) = x^3$ (b) $g(x) = (x - 1)^3 - 2$
56. (a) $f(x) = \dfrac{1}{x}$ (b) $g(x) = \dfrac{1}{x} - 2$
57. (a) $f(x) = 2$ (b) $g(x) = 2$
58. (a) $f(x) = x^2$ (b) $g(x) = 1 - (x + 2)^2$
59. (a) $f(x) = x$ (b) $g(x) = x - 2$
60. (a) $f(x) = [\![x]\!]$ (b) $g(x) = [\![x - 1]\!]$
61. (a) (b) $5.64

62. (a) C_2 is the appropriate model, because the cost does not increase until after the next minute of conversation has started.

(b)

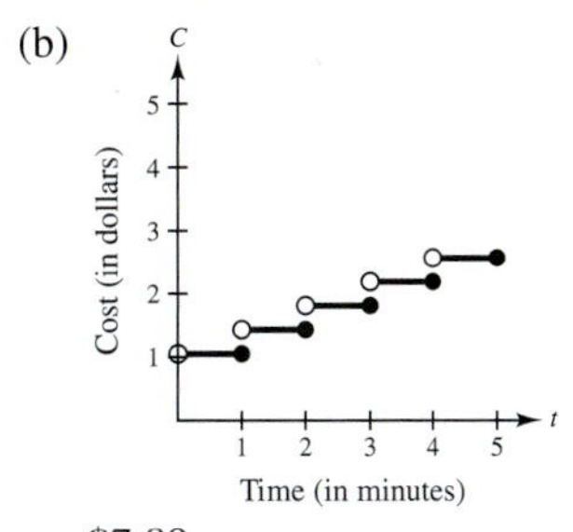

$7.89

63. (a)

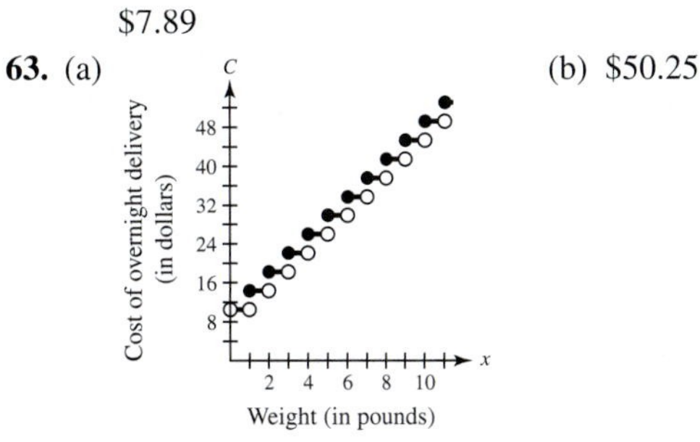

(b) $50.25

CHAPTER 1

64. (a) $C = 9.8 + 2.5[\![x]\!]$

(b)

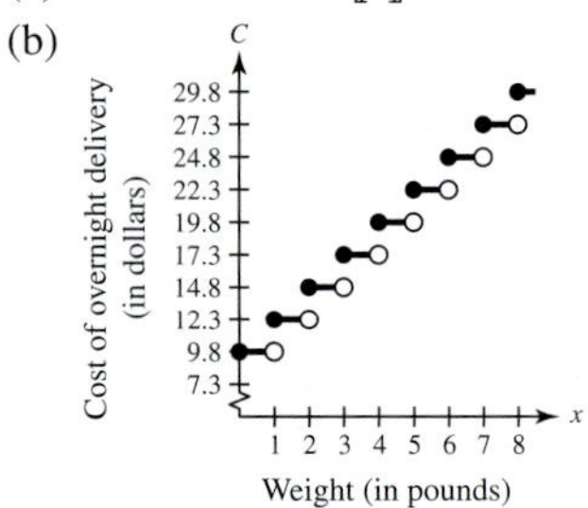

65. (a) $W(30) = 360;\ W(40) = 480;$
$W(45) = 570;\ W(50) = 660$

(b) $W(h) = \begin{cases} 12h, & 0 < h \le 45 \\ 18(h - 45) + 540, & h > 45 \end{cases}$

66. $f(t) = \begin{cases} t, & 0 \le t \le 2 \\ 2t - 2, & 2 < t \le 8 \\ \frac{1}{2}t + 10, & 8 < t \le 9 \end{cases}$

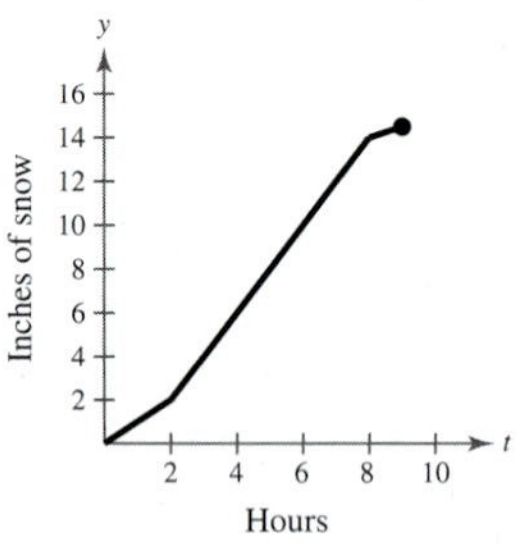

Total accumulation = 14.5 inches

67. (a) $f(x) = \begin{cases} 0.505x^2 - 1.47x + 6.3, & 1 \le x \le 6 \\ -1.97x + 26.3, & 6 < x \le 12 \end{cases}$

Answers will vary. Sample answer: The domain is determined by inspection of a graph of the data with the two models.

(b)

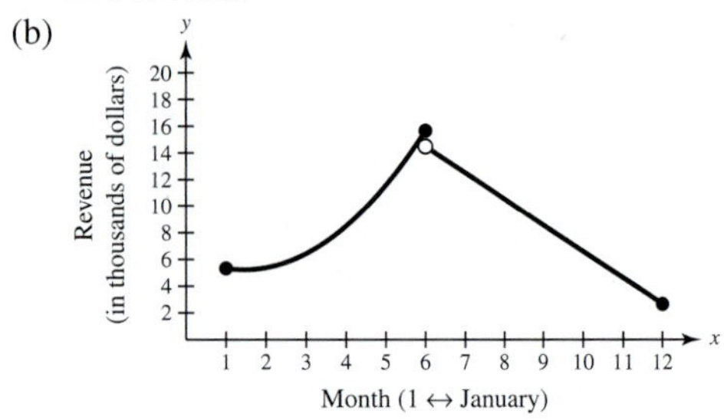

(c) $f(5) = 11.575, f(11) = 4.63$; These values represent the revenue for the months of May and November, respectively.

(d) These values are quite close to the actual data values.

68.

Interval	*Input Pipe*	*Drainpipe 1*	*Drainpipe 2*
[0, 5]	Open	Closed	Closed
[5, 10]	Open	Open	Closed
[10, 20]	Closed	Closed	Closed
[20, 30]	Closed	Closed	Open
[30, 40]	Open	Open	Open
[40, 45]	Open	Closed	Open
[45, 50]	Open	Open	Open
[50, 60]	Open	Open	Closed

69. False. A piecewise-defined function is a function that is defined by two or more equations over a specified domain. That domain may or may not include x- and y-intercepts.

70. True. The solution sets are the same.

71. $f(x) = \begin{cases} -\frac{4}{3}x + 6, & 0 \le x \le 3 \\ -\frac{2}{5}x + \frac{16}{5}, & 3 < x \le 8 \end{cases}$

72. $f(x) = \begin{cases} x^2, & x \le 2 \\ 7 - x, & x > 2 \end{cases}$

73. $x \le 1$

−3 −2 −1 0 1 2 3 x

74. $x < \frac{5}{2}$

$\frac{5}{2}$

−1 0 1 2 3 4 5 x

75. Neither **76.** Neither

Section 1.7 *(page 79)*

Vocabulary Check *(page 79)*

1. rigid **2.** $-f(x); f(-x)$ **3.** nonrigid
4. horizontal shrink; horizontal stretch
5. vertical stretch; vertical shrink
6. (a) iv (b) ii (c) iii (d) i

1. (a) (b)

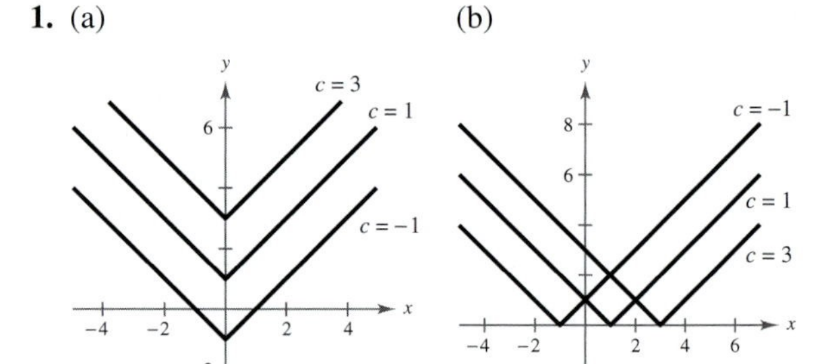

(c)

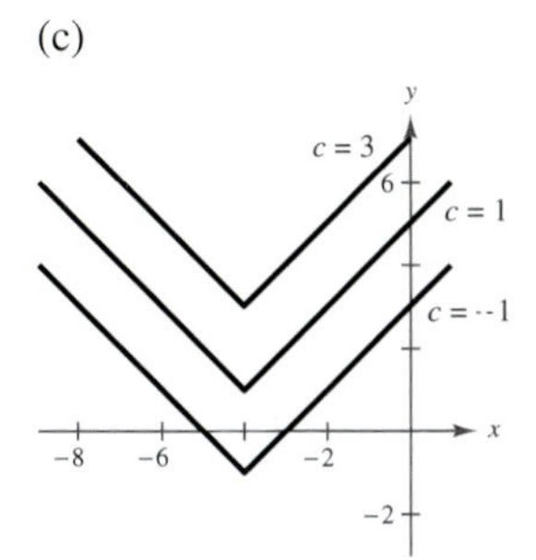

2. (a)

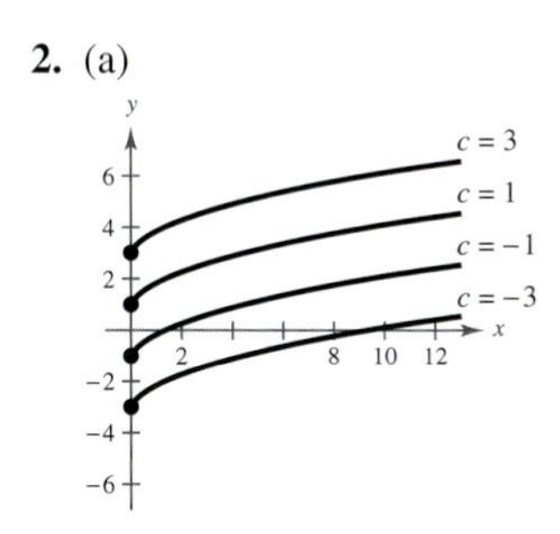

(b)

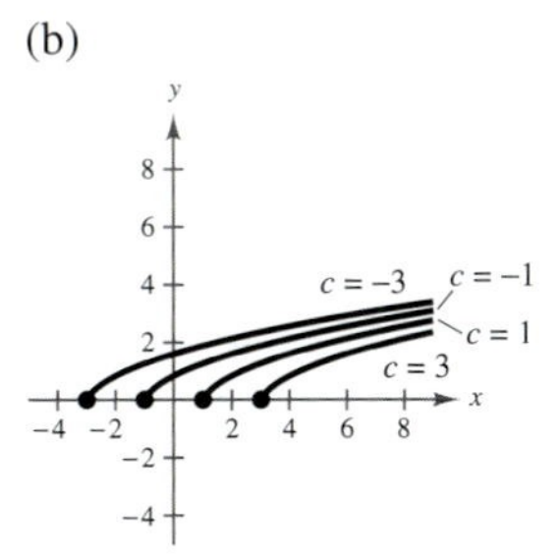

(c)

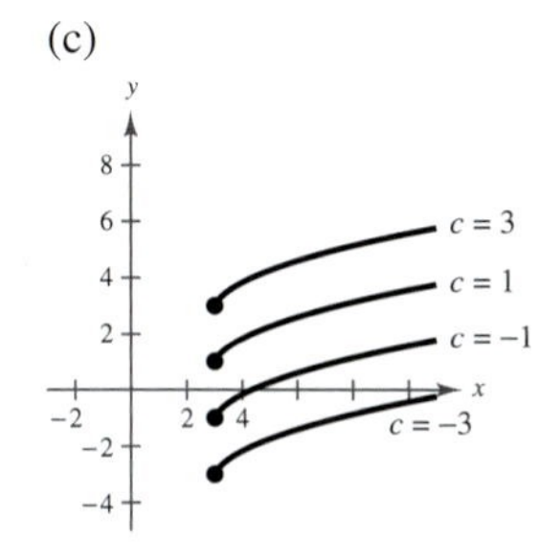

3. (a) (b)

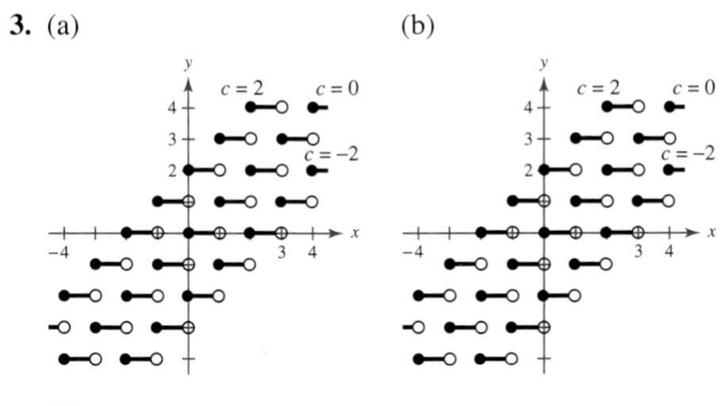

(c)

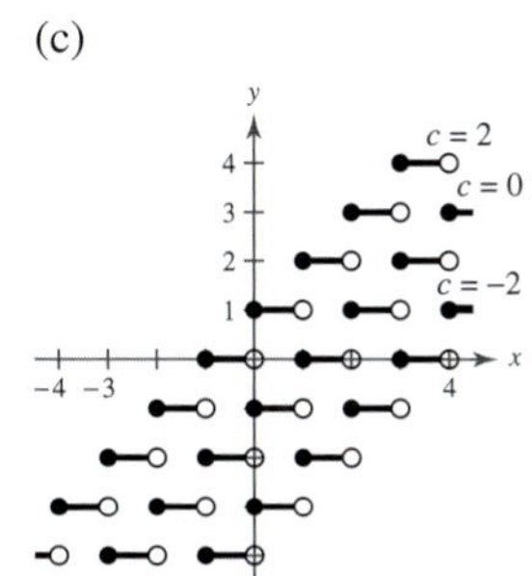

4. (a) (b)

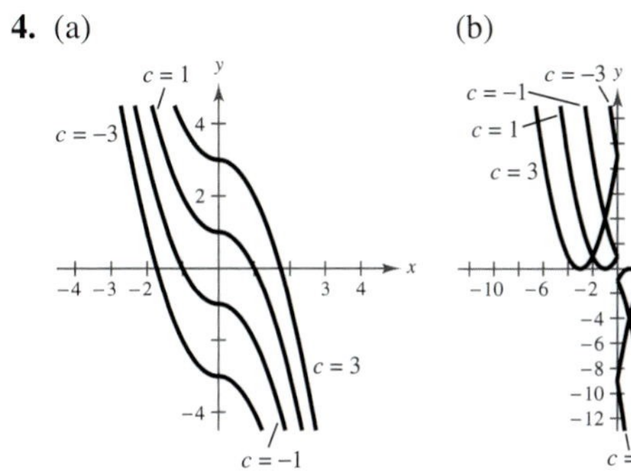

5. (a)

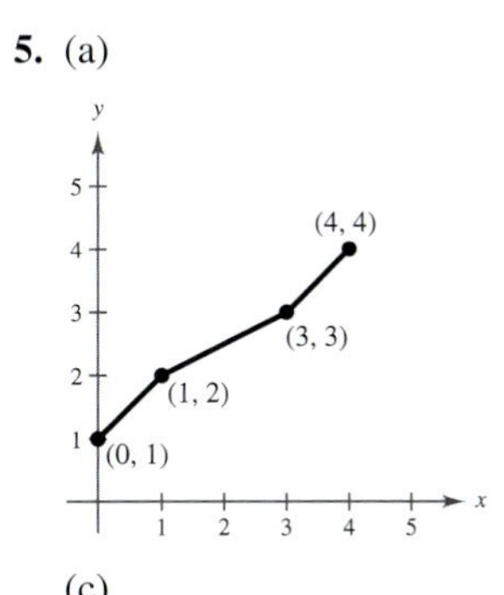

(b)

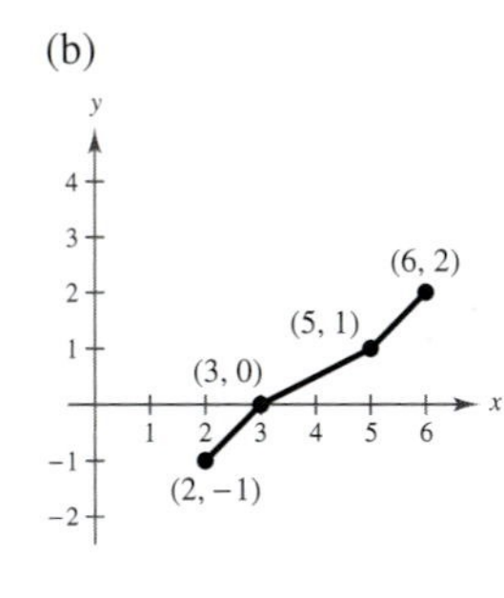

(c)

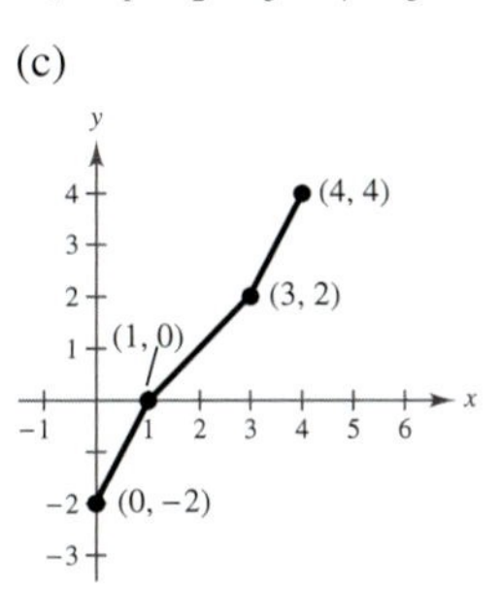

(d)

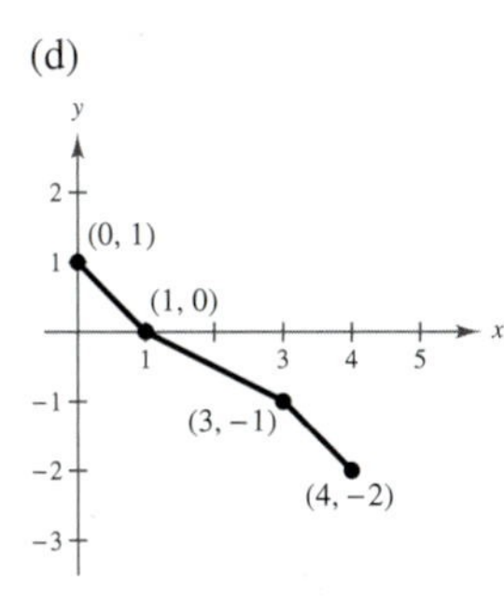

(e)

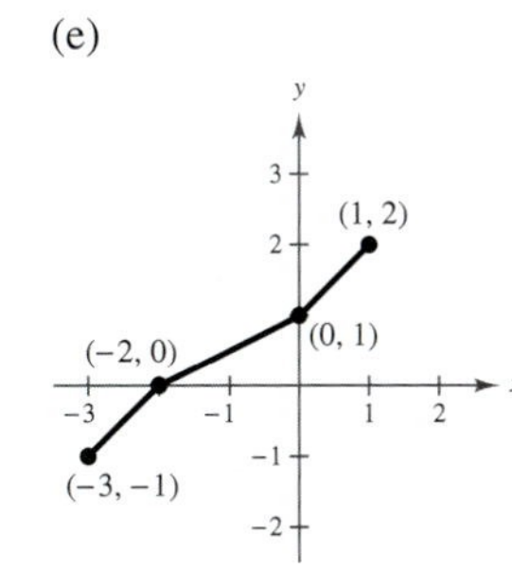

(f)

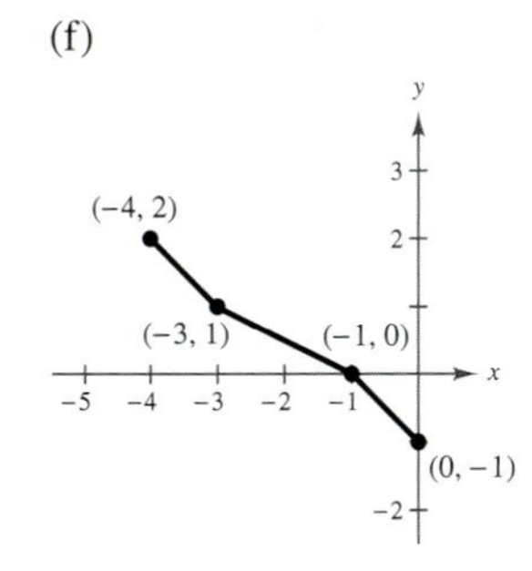

(g)

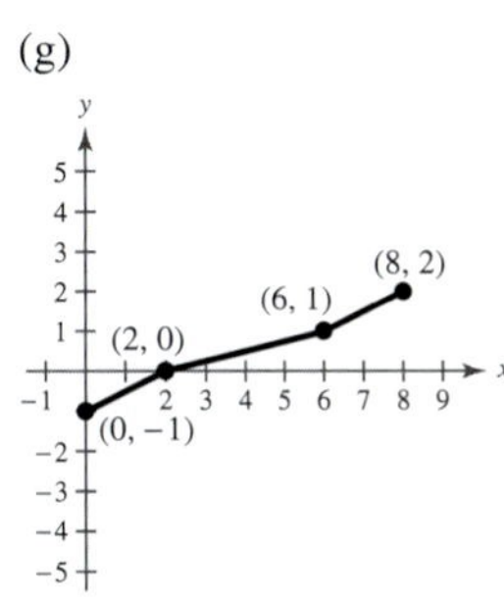

6. (a)

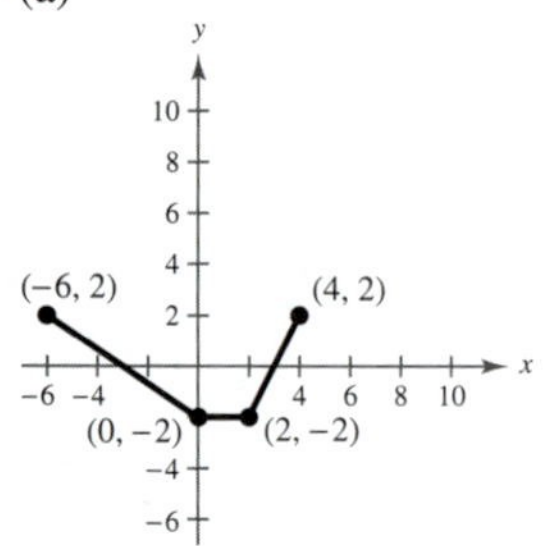

(b)

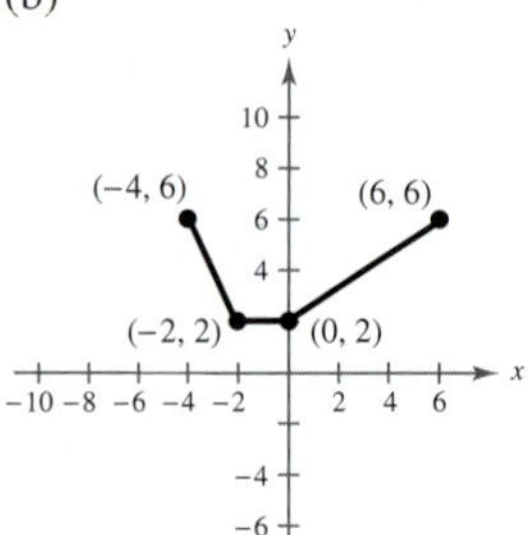

(c)

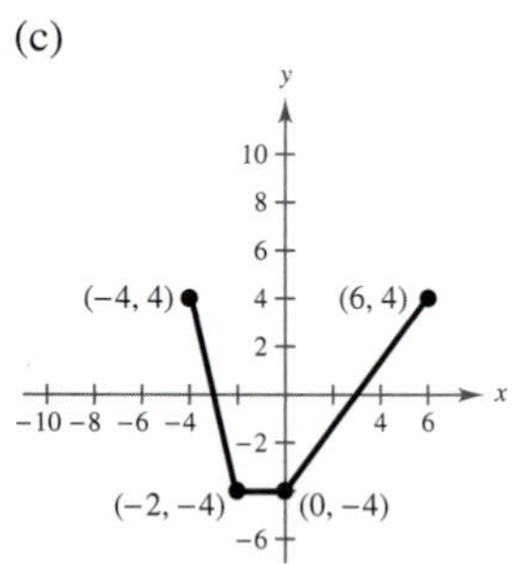

(d)

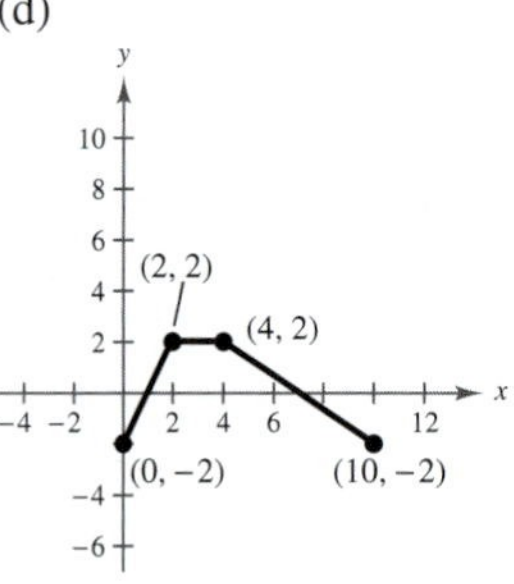

(e)

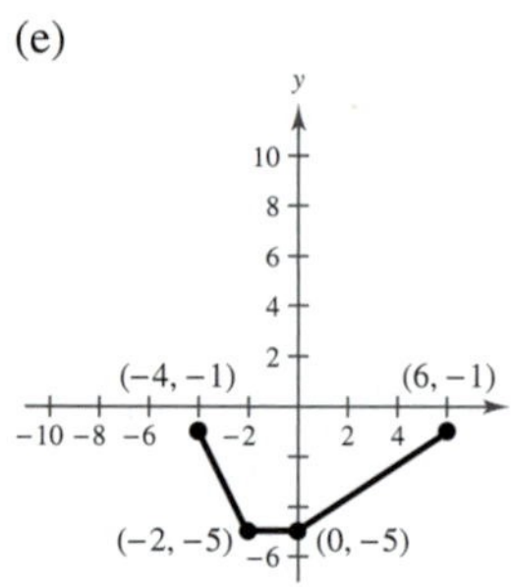

(f)

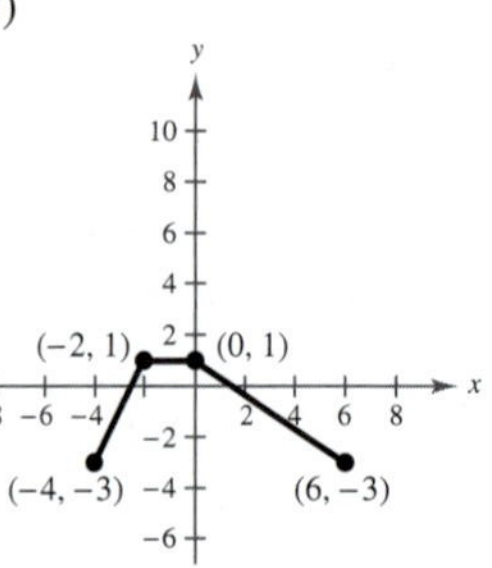

(g)

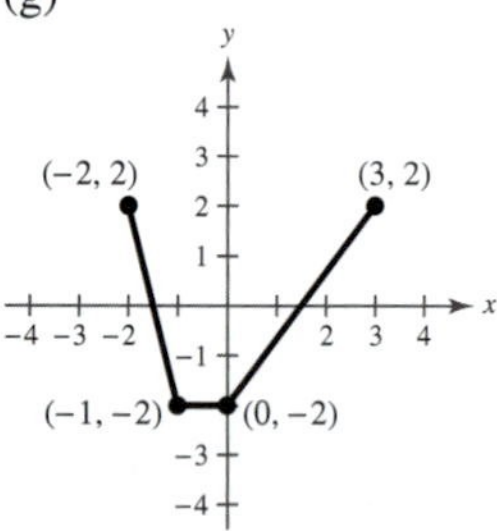

7. (a)

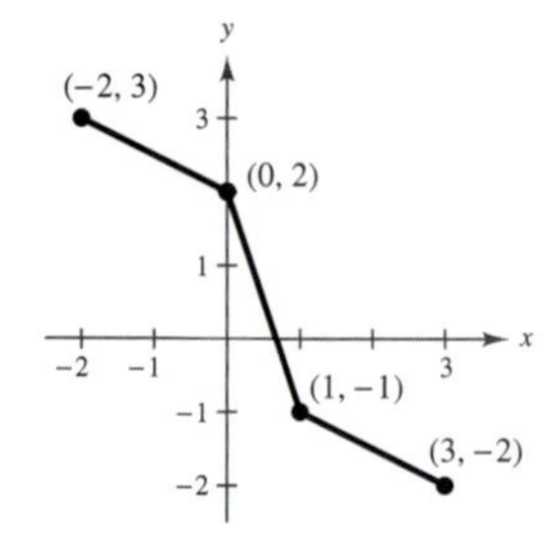

(b)

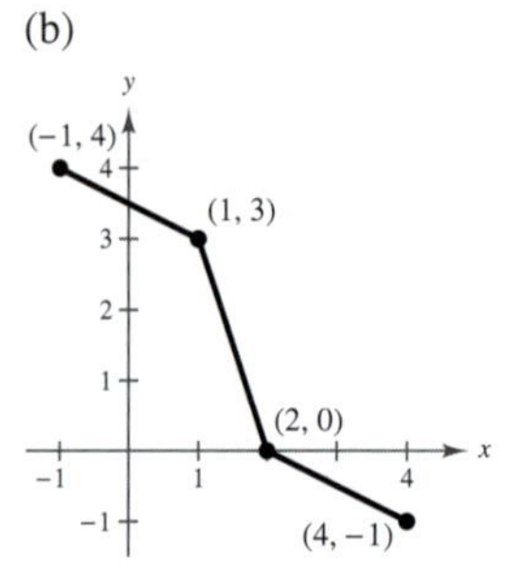

(c)

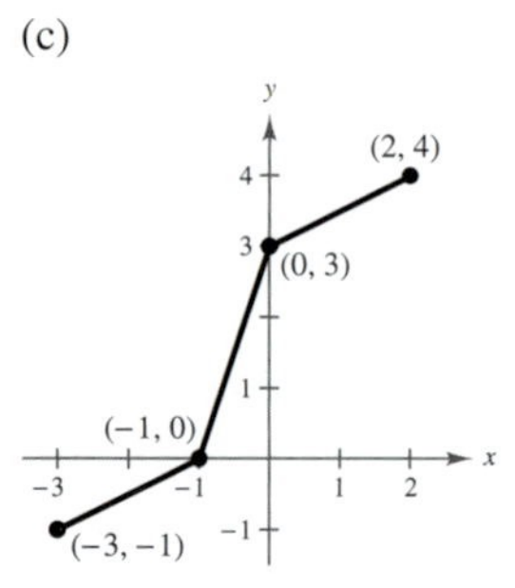

(d)

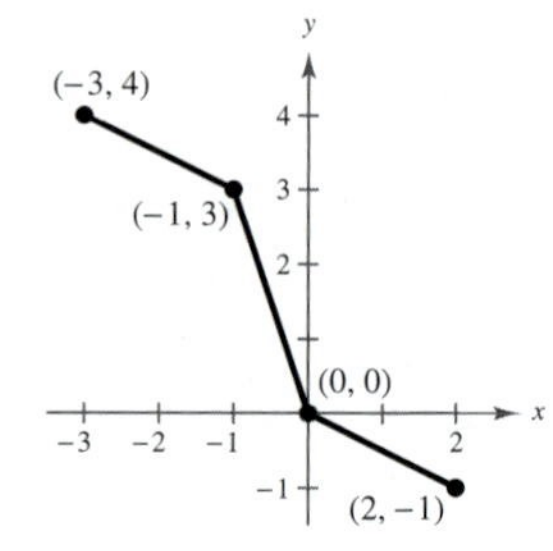

(e)

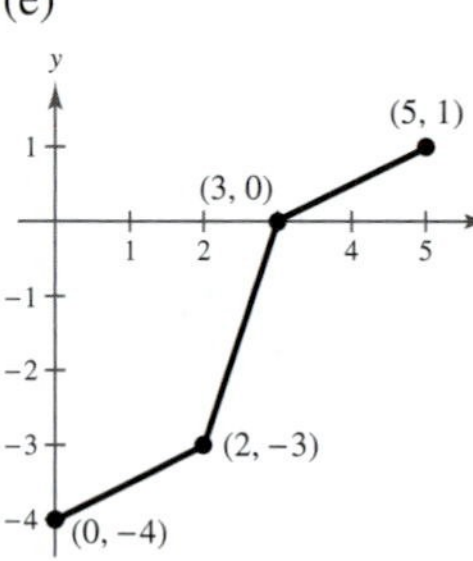

(f)

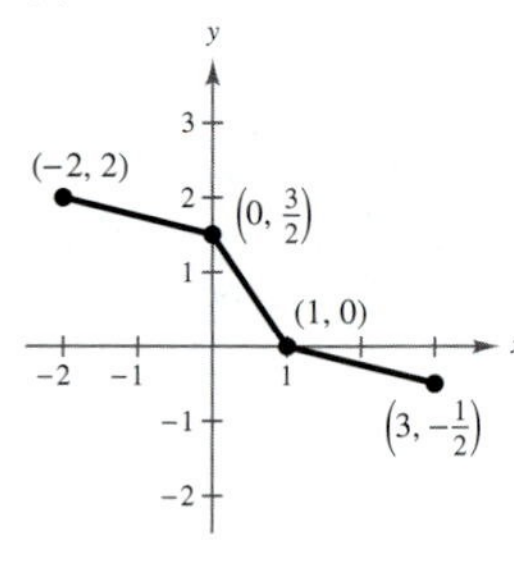

(g)

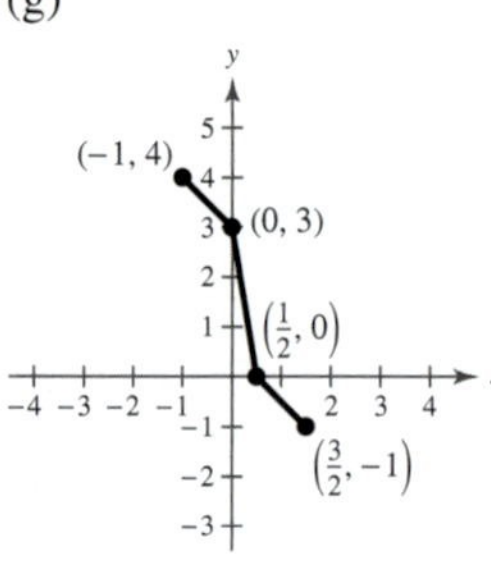

8. (a)

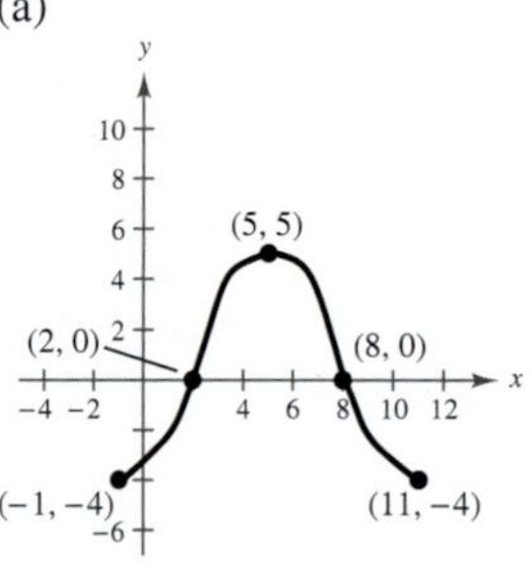

(b)

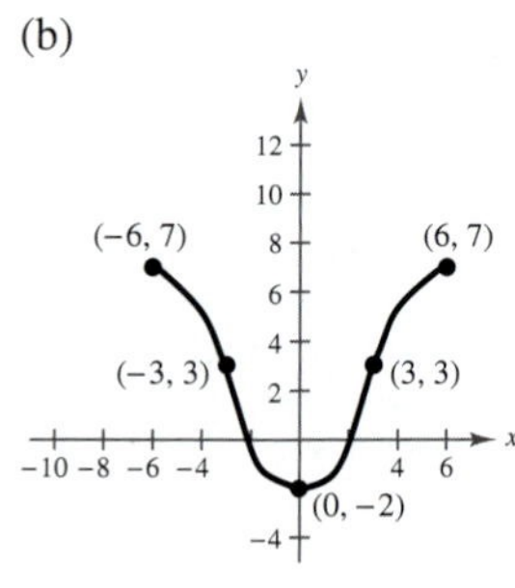

(c)

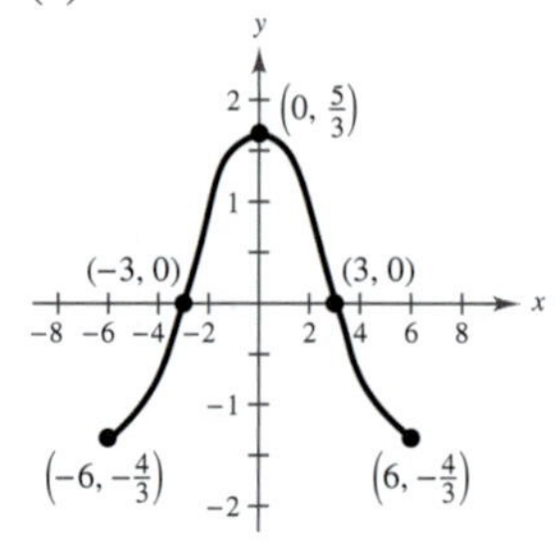

(d)

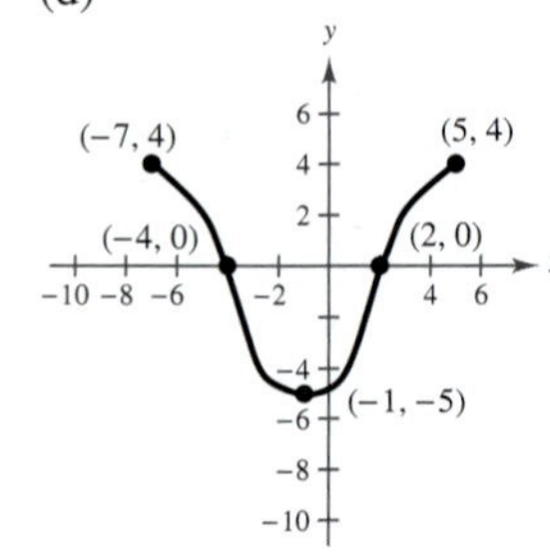

(e)

(f)

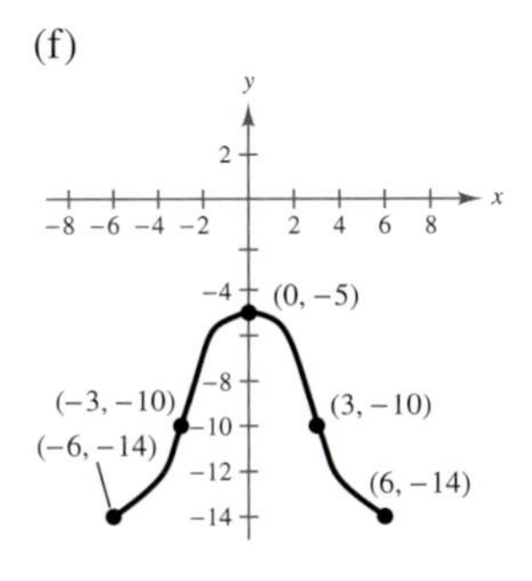

(g)

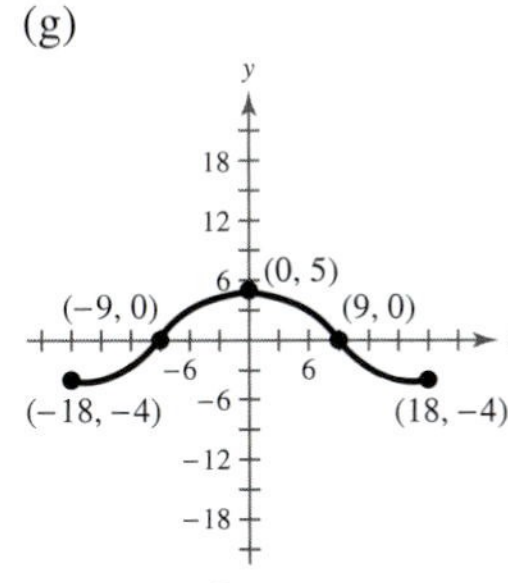

9. (a) $y = x^2 - 1$ (b) $y = 1 - (x + 1)^2$
(c) $y = -(x - 2)^2 + 6$ (d) $y = (x - 5)^2 - 3$

10. (a) $y = 1 - x^3$ (b) $y = (x - 1)^3 + 1$
(c) $y = -(x + 3)^3 - 1$ (d) $y = (x - 10)^3 - 4$

11. (a) $y = |x| + 5$ (b) $y = -|x + 3|$
(c) $y = |x - 2| - 4$ (d) $y = -|x - 6| - 1$

12. (a) $y = \sqrt{x} - 3$ (b) $y = \sqrt{x + 1} - 7$
(c) $y = -\sqrt{x - 5} + 5$ (d) $y = -\sqrt{-x + 3} - 4$

13. Horizontal shift of $y = x^3$; $y = (x - 2)^3$

14. Vertical shrink of $y = x$; $y = \frac{1}{2}x$

15. Reflection in the x-axis of $y = x^2$; $y = -x^2$

16. Vertical shift of $y = [\![x]\!]$; $y = [\![x]\!] + 4$

17. Reflection in the x-axis and vertical shift of $y = \sqrt{x}$; $y = 1 - \sqrt{x}$

18. Horizontal shift of $y = |x|$; $y = |x + 2|$

19. (a) $f(x) = x^2$
(b) Reflection in the x-axis, and vertical shift 12 units upward, of $f(x) = x^2$
(c)

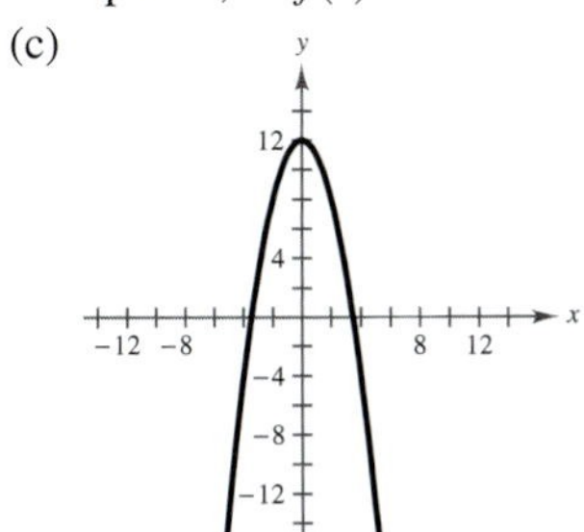

(d) $g(x) = 12 - f(x)$

20. (a) $f(x) = x^2$
(b) Horizontal shift eight units to the right, of $f(x) = x^2$
(c)

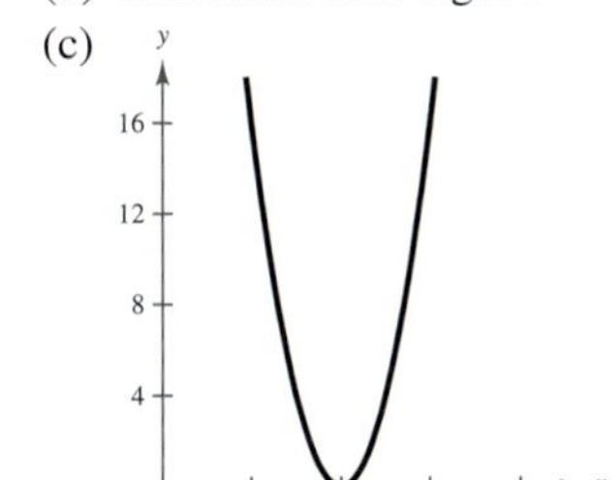

(d) $g(x) = f(x - 8)$

21. (a) $f(x) = x^3$
(b) Vertical shift seven units upward, of $f(x) = x^3$
(c)

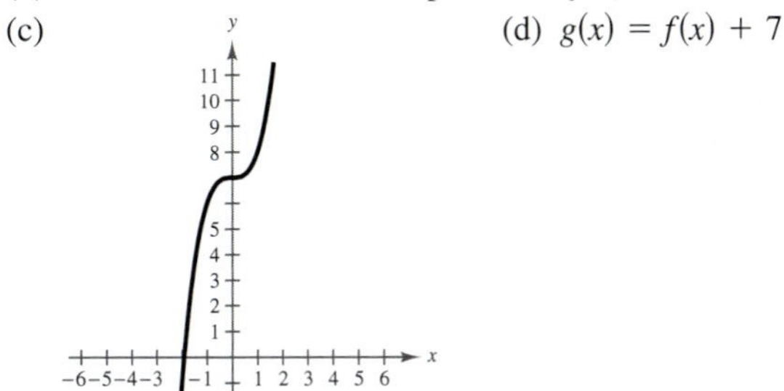

(d) $g(x) = f(x) + 7$

22. (a) $f(x) = x^3$
(b) Reflection in the x-axis, and a vertical shift of one unit downward, of $f(x) = x^3$
(c)

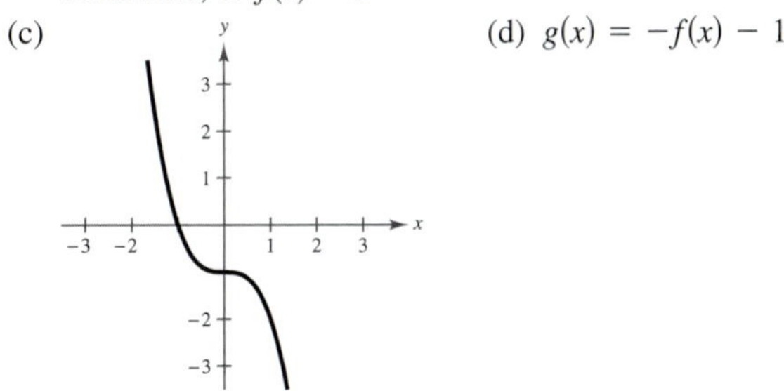

(d) $g(x) = -f(x) - 1$

23. (a) $f(x) = x^2$
(b) Vertical shrink of two-thirds, and vertical shift four units upward, of $f(x) = x^2$
(c)

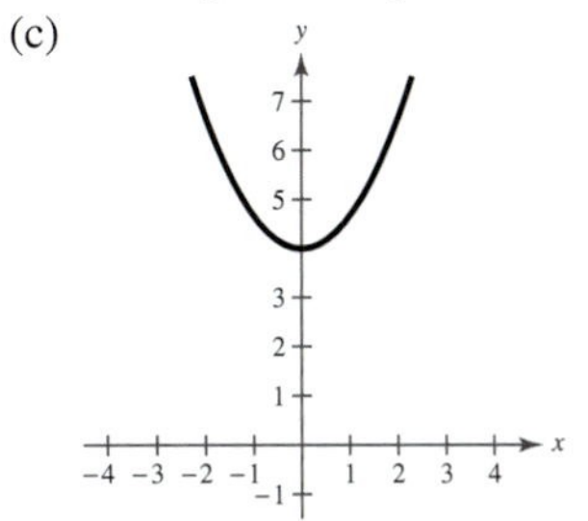

(d) $g(x) = \frac{2}{3}f(x) + 4$

24. (a) $f(x) = x^2$

(b) Vertical stretch of two, and horizontal shift seven units to the right, of $f(x) = x^2$

(c)

(d) $g(x) = 2f(x - 7)$

25. (a) $f(x) = x^2$

(b) Reflection in the x-axis, horizontal shift five units to the left, and vertical shift two units upward, of $f(x) = x^2$

(c)

(d) $g(x) = 2 - f(x + 5)$

26. (a) $f(x) = x^2$

(b) Reflection in the x-axis, horizontal shift 10 units to the left, and vertical shift five units upward, of $f(x) = x^2$

(c)

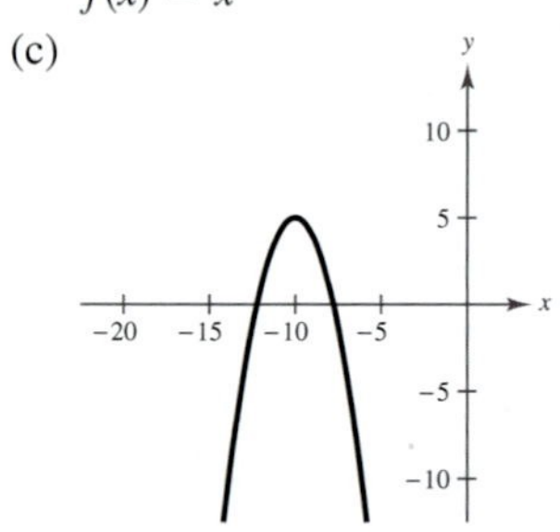

(d) $g(x) = -f(x + 10) + 5$

27. (a) $f(x) = \sqrt{x}$

(b) Horizontal shrink of $\frac{1}{3}$, of $f(x) = \sqrt{x}$

(c)

(d) $g(x) = f(3x)$

28. (a) $f(x) = \sqrt{x}$

(b) Horizontal stretch of four, of $f(x) = \sqrt{x}$

(c)

(d) $g(x) = f(\frac{1}{4}x)$

29. (a) $f(x) = x^3$

(b) Vertical shift two units upward, and horizontal shift one unit to the right, of $f(x) = x^3$

(c)

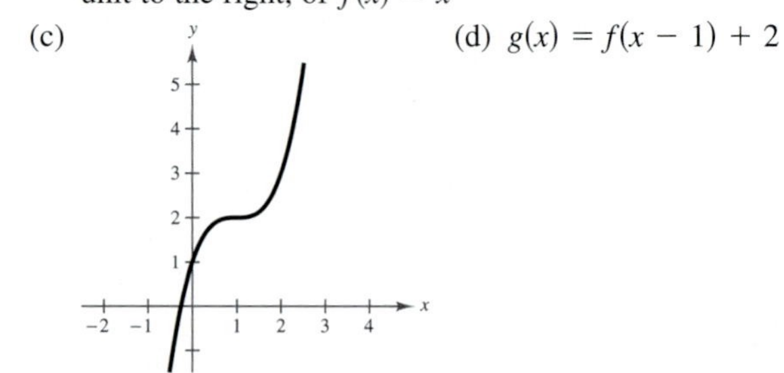

(d) $g(x) = f(x - 1) + 2$

30. (a) $f(x) = x^3$

(b) Vertical shift 10 units downward, and horizontal shift three units to the left, of $f(x) = x^3$

(c)

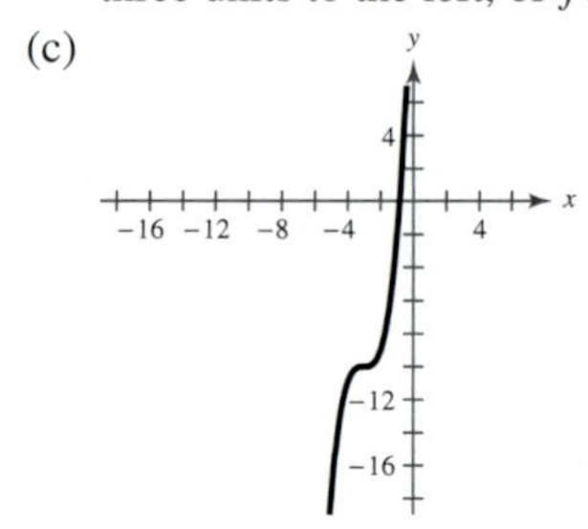

(d) $g(x) = f(x + 3) - 10$

31. (a) $f(x) = |x|$

(b) Reflection in the x-axis, and vertical shift two units downward, of $f(x) = |x|$

(c)

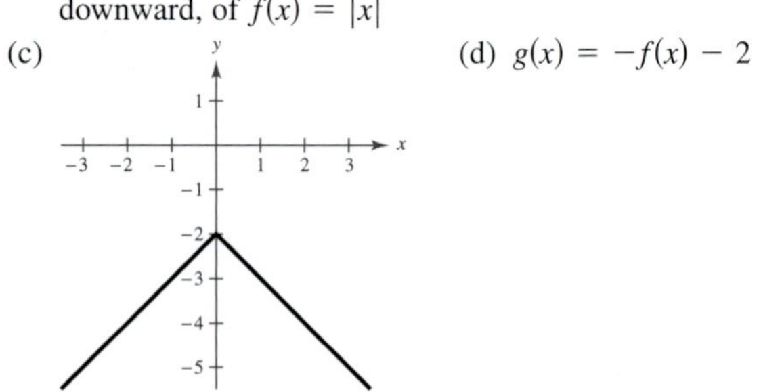

(d) $g(x) = -f(x) - 2$

32. (a) $f(x) = |x|$
(b) Reflection in the x-axis, horizontal shift five units to the left, and vertical shift six units upward, of $f(x) = |x|$
(c)
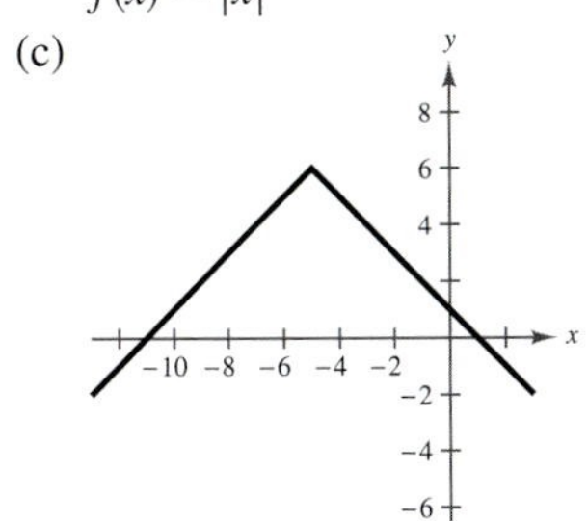
(d) $g(x) = 6 - f(x + 5)$

33. (a) $f(x) = |x|$
(b) Reflection in the x-axis, horizontal shift four units to the left, and vertical shift eight units upward, of $f(x) = |x|$
(c)
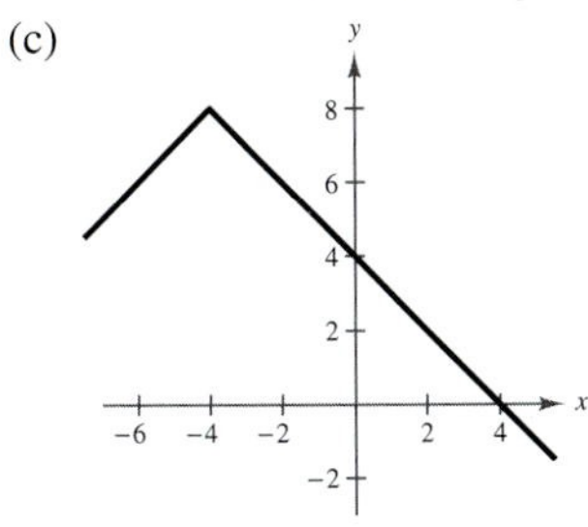
(d) $g(x) = -f(x + 4) + 8$

34. (a) $f(x) = |x|$
(b) Reflection in the y-axis, horizontal shift three units to the right, and vertical shift nine units upward, of $f(x) = |x|$
(c)
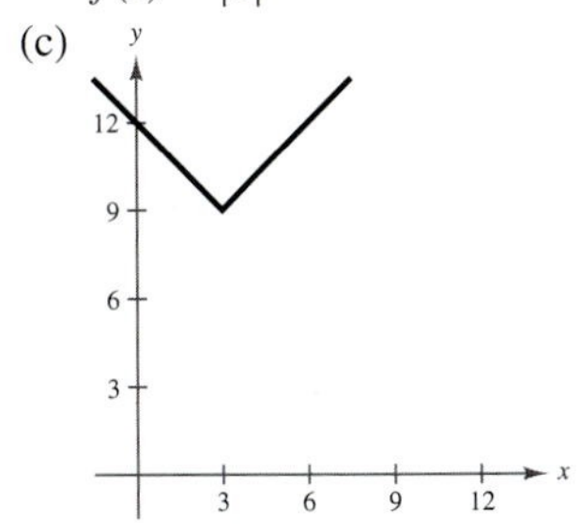
(d) $g(x) = f(-(x - 3)) + 9$

35. (a) $f(x) = [\![x]\!]$
(b) Reflection in the x-axis, and vertical shift three units upward, of $f(x) = [\![x]\!]$
(c)
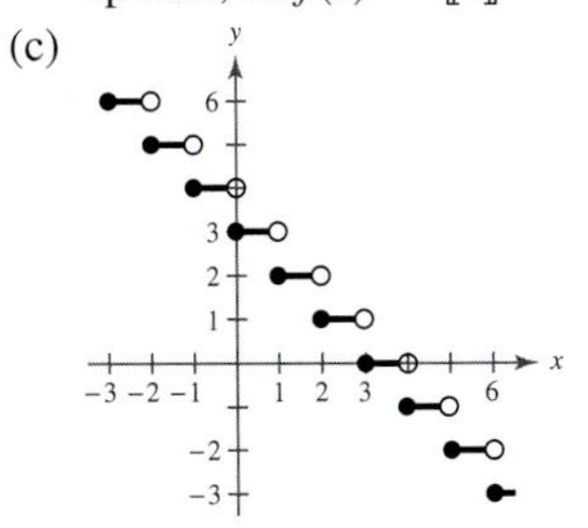
(d) $g(x) = 3 - f(x)$

36. (a) $f(x) = [\![x]\!]$
(b) Horizontal shift five units to the left, and vertical stretch of two, of $f(x) = [\![x]\!]$
(c)
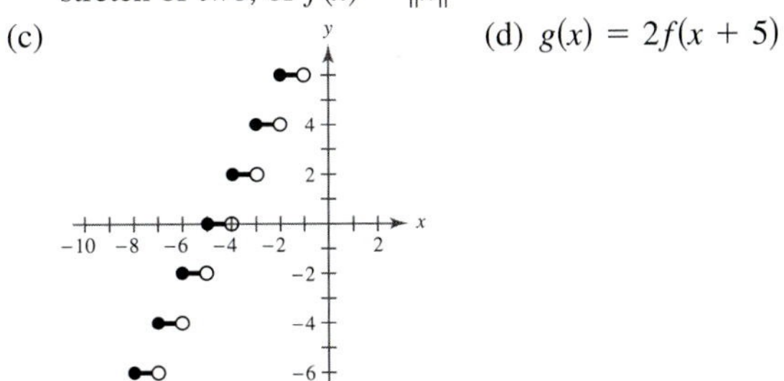
(d) $g(x) = 2f(x + 5)$

37. (a) $f(x) = \sqrt{x}$
(b) Horizontal shift of nine units to the right, of $f(x) = \sqrt{x}$
(c)
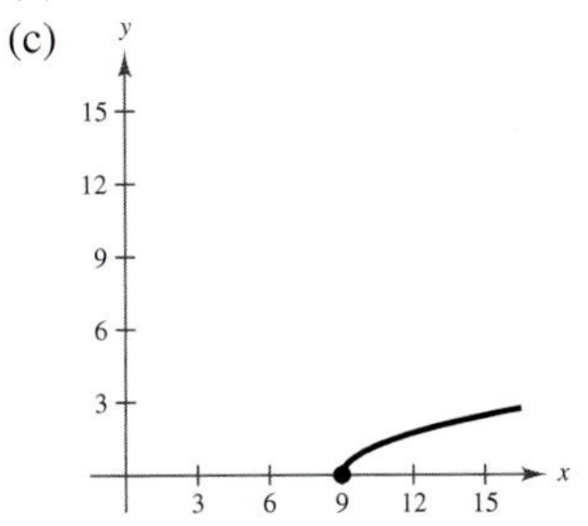
(d) $g(x) = f(x - 9)$

38. (a) $f(x) = \sqrt{x}$
(b) Horizontal shift of four units to the left, and vertical shift eight units upward, of $f(x) = \sqrt{x}$
(c)
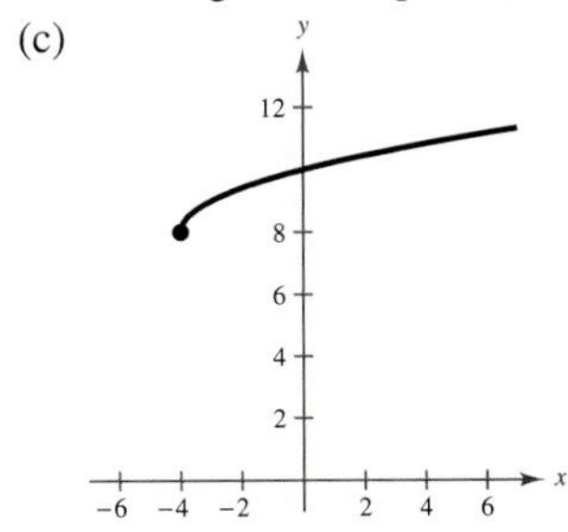
(d) $g(x) = f(x + 4) + 8$

39. (a) $f(x) = \sqrt{x}$
(b) Reflection in the y-axis, horizontal shift of seven units to the right, and vertical shift two units downward, of $f(x) = \sqrt{x}$
(c)
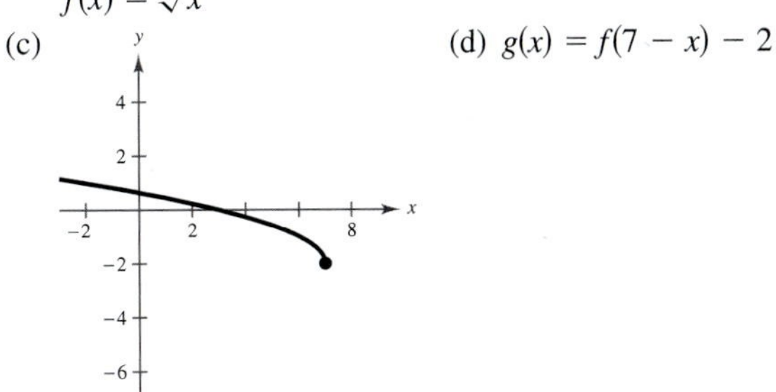
(d) $g(x) = f(7 - x) - 2$

40. (a) $f(x) = \sqrt{x}$
(b) Reflection in the x-axis, horizontal shift of one unit to the left, and vertical shift six units downward, of $f(x) = \sqrt{x}$
(c)

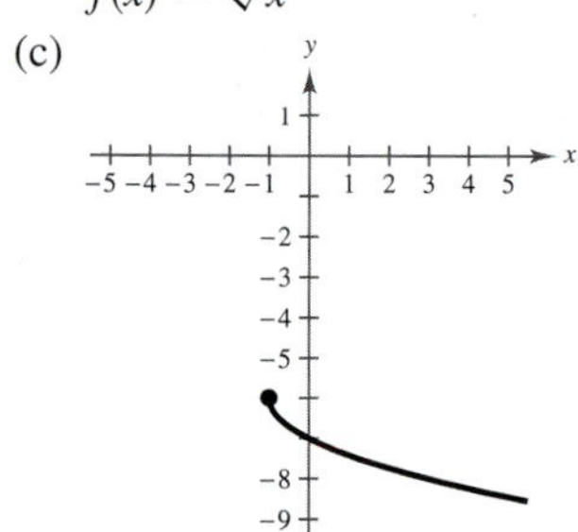

(d) $g(x) = -f(x + 1) - 6$

41. (a) $f(x) = \sqrt{x}$
(b) Horizontal stretch, and vertical shift four units downward, of $f(x) = \sqrt{x}$
(c)

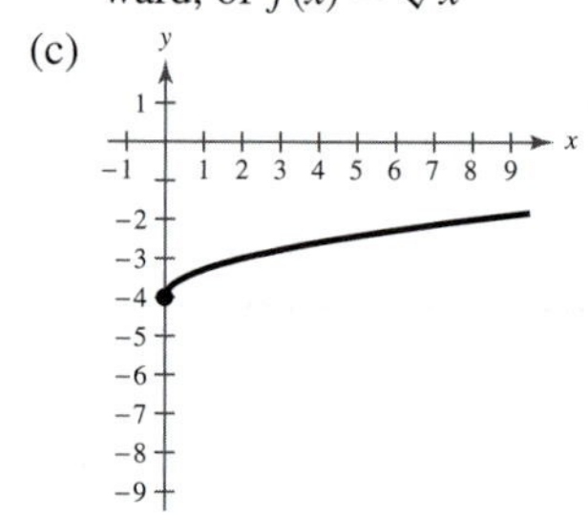

(d) $g(x) = f\left(\frac{1}{2}x\right) - 4$

42. (a) $f(x) = \sqrt{x}$
(b) Horizontal shrink of $\frac{1}{3}$, and vertical shift one unit upward, of $f(x) = \sqrt{x}$
(c)

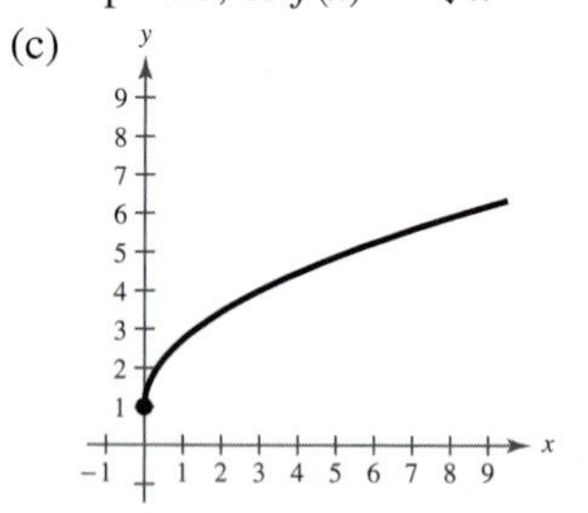

(d) $g(x) = f(3x) + 1$

43. $f(x) = (x - 2)^2 - 8$ **44.** $f(x) = -(x + 3)^2 - 7$
45. $f(x) = (x - 13)^3$ **46.** $f(x) = (-x + 6)^3 - 6$
47. $f(x) = -|x| - 10$ **48.** $f(x) = |x + 1| - 7$
49. $f(x) = -\sqrt{-x + 6}$ **50.** $f(x) = -\sqrt{-x + 9}$
51. (a) $y = -3x^2$ (b) $y = 4x^2 + 3$
52. (a) $y = \frac{1}{4}x^3$ (b) $y = -2x^3$
53. (a) $y = -\frac{1}{2}|x|$ (b) $y = 3|x| - 3$
54. (a) $y = 8\sqrt{x}$ (b) $y = -\frac{1}{4}\sqrt{x}$
55. Vertical stretch of $y = x^3$; $y = 2x^3$
56. Vertical stretch of $y = |x|$; $y = 6|x|$
57. Reflection in the x-axis and vertical shrink of $y = x^2$; $y = -\frac{1}{2}x^2$
58. Horizontal stretch of $y = [\![x]\!]$; $y = [\![\frac{1}{2}x]\!]$
59. Reflection in the y-axis and vertical shrink of $y = \sqrt{x}$; $y = \frac{1}{2}\sqrt{-x}$
60. Reflection in the x-axis, vertical stretch, and vertical shift two units downward, of $y = |x|$; $y = -2|x| - 2$
61. $y = -(x - 2)^3 + 2$ **62.** $y = |x + 4| - 2$
63. $y = -\sqrt{x} - 3$ **64.** $y = (x - 2)^2 + 4$
65. (a)

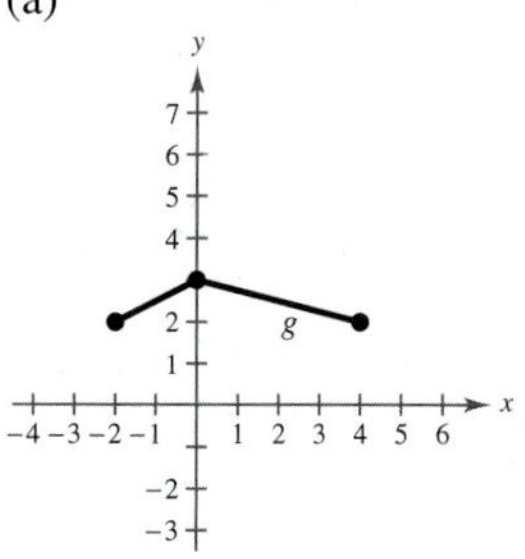

(b)

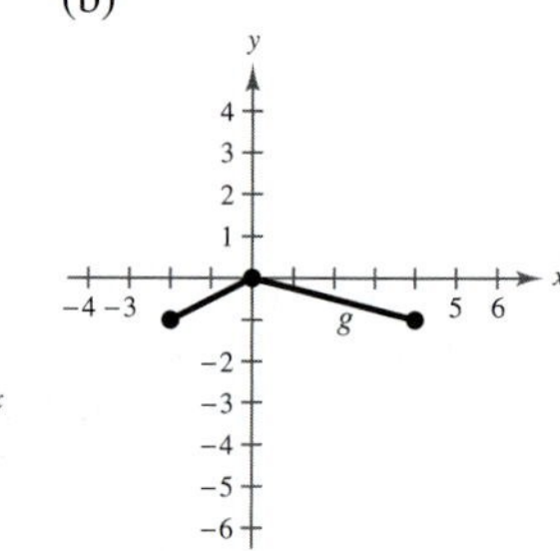

(c)

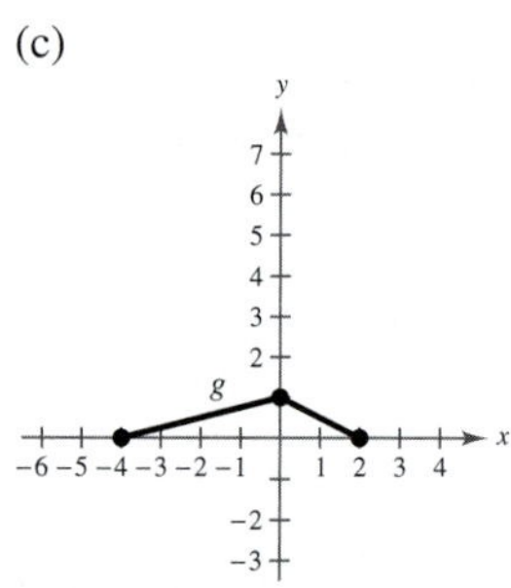

(d)

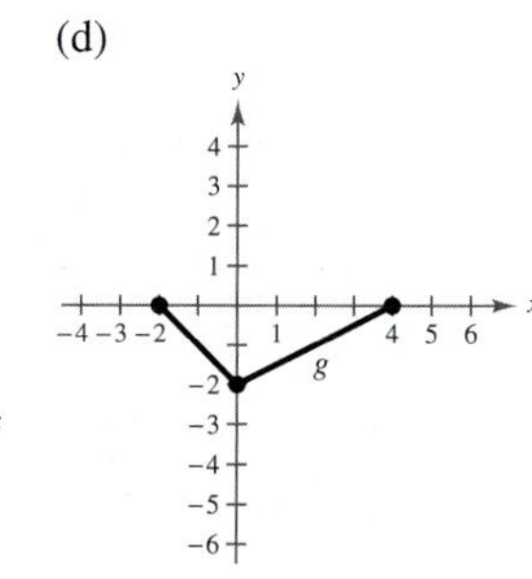

(e)

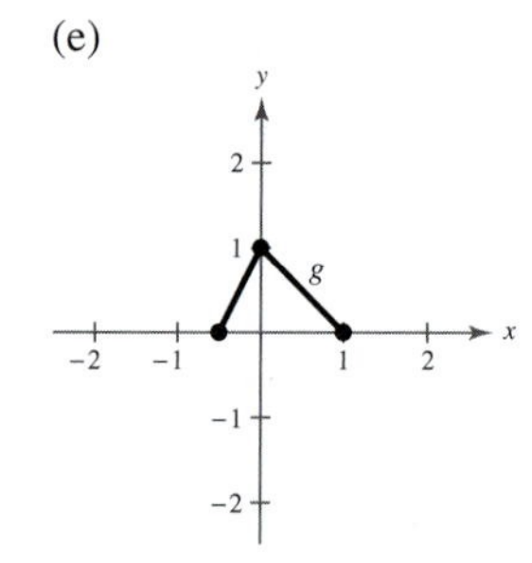

(f)

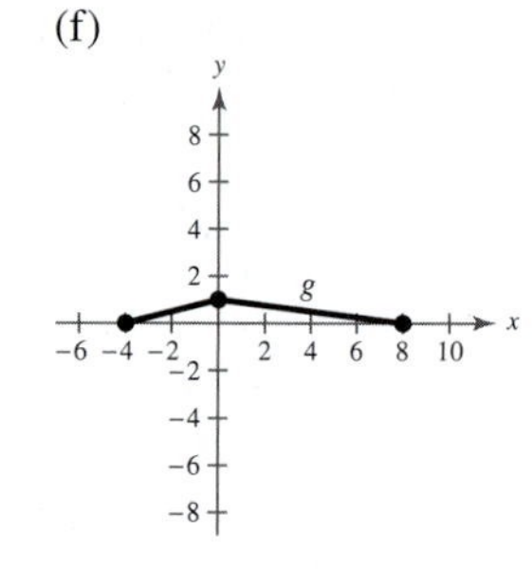

66. (a)

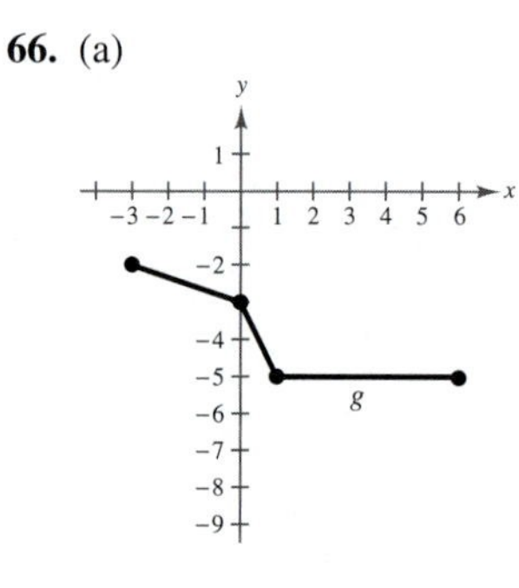

(b)

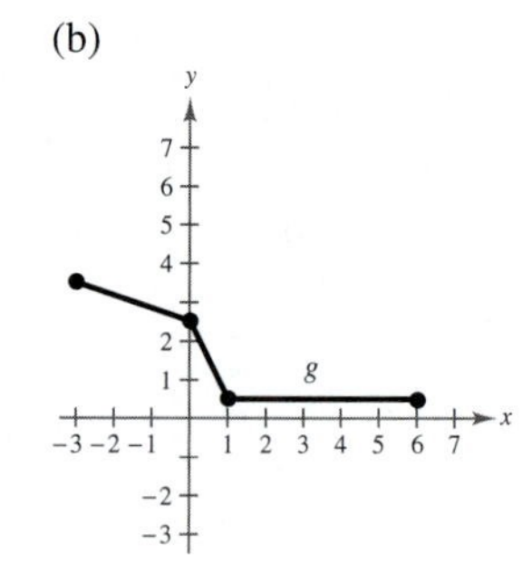

(c)

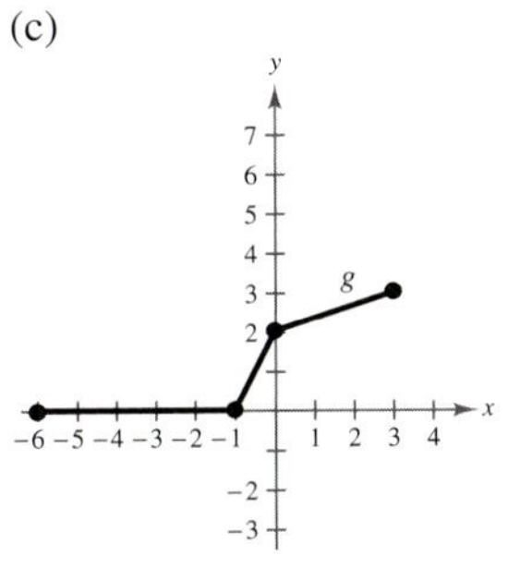

(d)

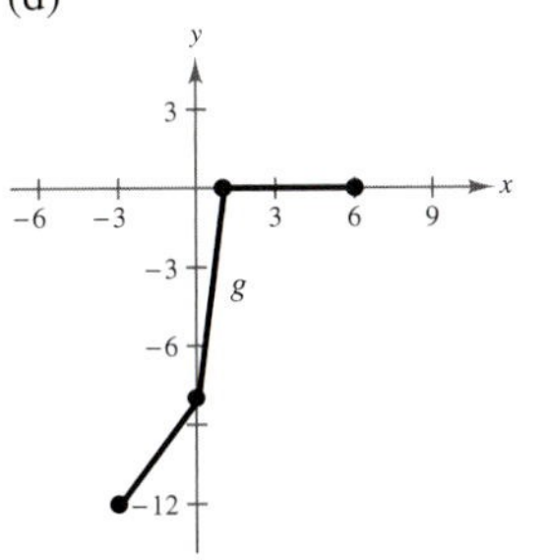

(e) (f)

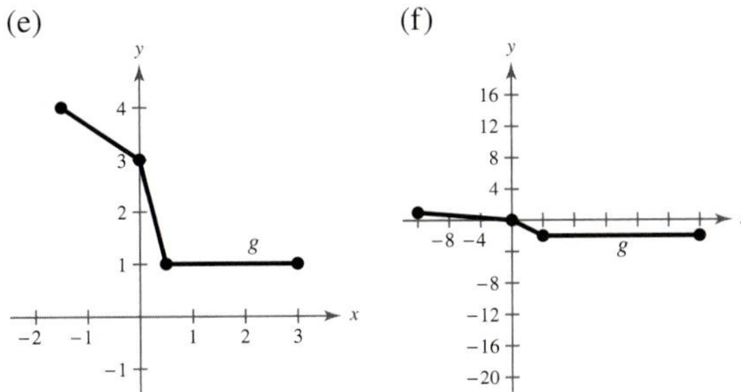

67. (a) Horizontal stretch of 0.035 and a vertical shift of 20.6 units upward.

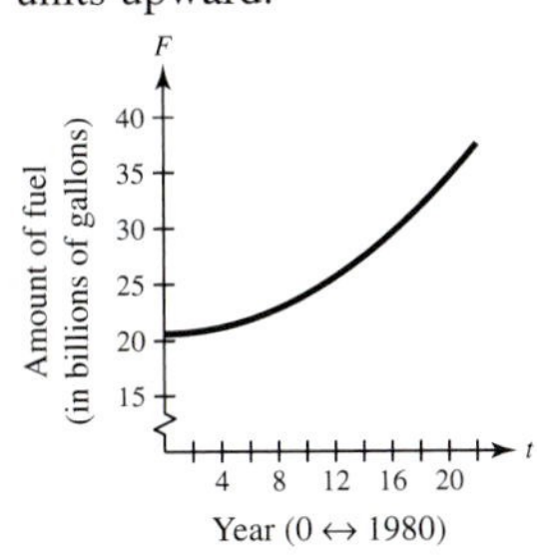

(b) 0.77-billion-gallon increase in fuel usage by trucks each year

(c) $f(t) = 20.6 + 0.035(t + 10)^2$. The graph is shifted 10 units to the left.

(d) 52.1 billion gallons. Yes.

68. (a) Horizontal shift of 20.396 units to the left and vertical shrink of 0.0054

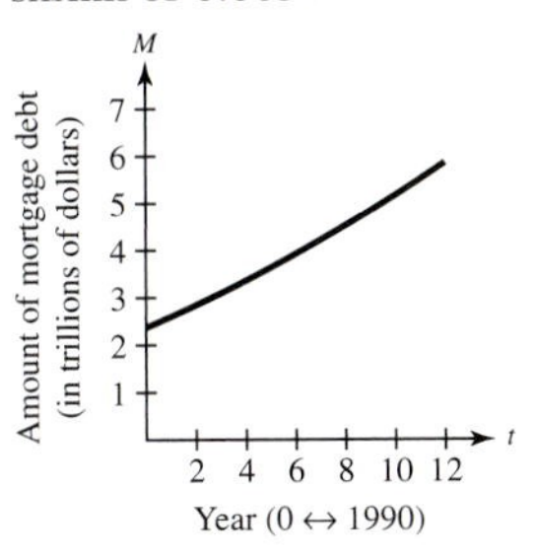

(b) $f(t) = 0.0054(t + 30.396)^2$. The graph is shifted 10 units to the left.

69. True. $|-x| = |x|$

70. False. The point $(-2, -67)$ will lie on the transformation.

71. (a) $g(t) = \frac{3}{4}f(t)$ (b) $g(t) = f(t) + 10{,}000$ (c) $g(t) = f(t - 2)$

72. If you consider the x-axis to be a mirror, the graph of $y = -f(x)$ is the mirror image of the graph of $y = f(x)$.

73. $(-2, 0), (-1, 1), (0, 2)$ **74.** Answers will vary.

75. $\dfrac{4}{x(1 - x)}$ **76.** $\dfrac{-20}{(x + 5)(x - 5)}$ **77.** $\dfrac{3x - 2}{x(x - 1)}$

78. $\dfrac{3x - 5}{2(x - 5)}$ **79.** $\dfrac{(x - 4)\sqrt{x^2 - 4}}{x^2 - 4}$

80. $\dfrac{x + 1}{x(x + 2)}, x \neq 2$ **81.** $5(x - 3), x \neq -3$

82. $\dfrac{x + 1}{(x - 7)(x + 3)}, x \neq 0, -1, -4$

83. (a) 38 (b) $\frac{57}{4}$ (c) $x^2 - 12x + 38$

84. (a) -3 (b) 3 (c) $\sqrt{x} - 3$

85. All real numbers x except $x = 1$

86. All real numbers x such that $x \geq 3$, except $x = 8$

87. All real numbers x such that $-9 \leq x \leq 9$

88. All real numbers x

Section 1.8 *(page 89)*

Vocabulary Check *(page 89)*

1. addition; subtraction; multiplication; division

2. composition **3.** $g(x)$ **4.** inner; outer

1.

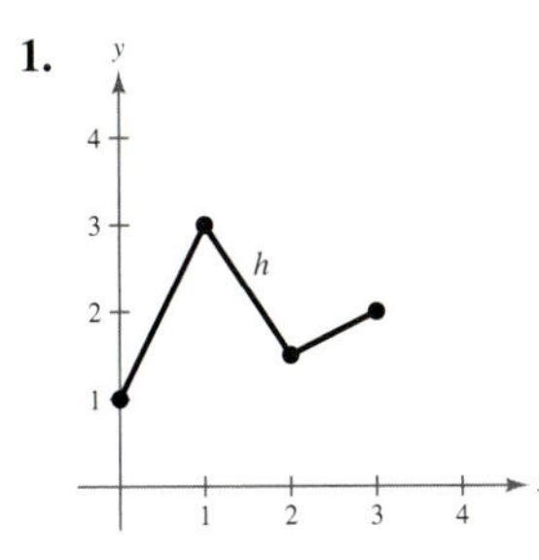

2.

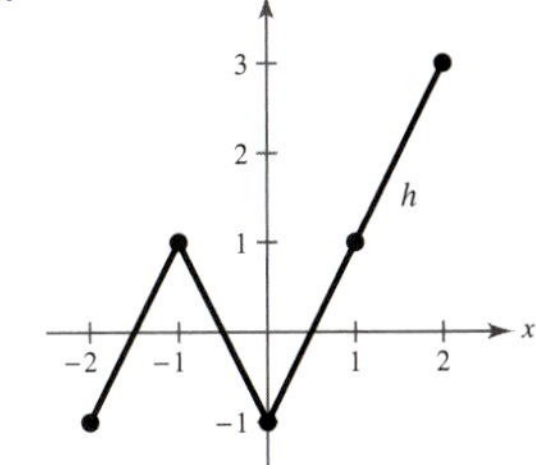

3.

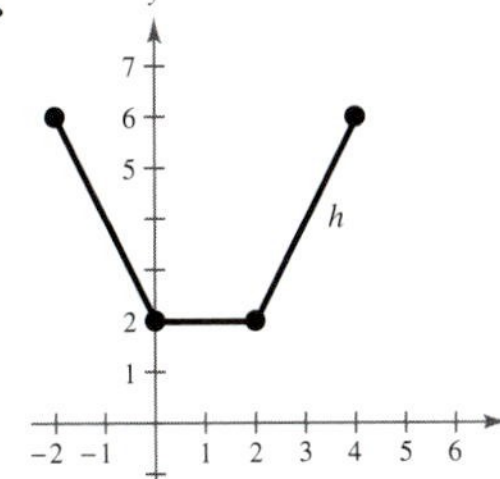

4.

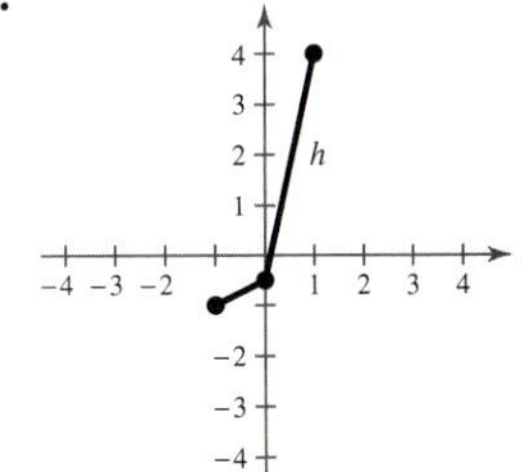

5. (a) $2x$ (b) 4 (c) $x^2 - 4$

(d) $\dfrac{x + 2}{x - 2}$; all real numbers x except $x = 2$

6. (a) $x - 3$ (b) $3x - 7$ (c) $-2x^2 + 9x - 10$

(d) $\dfrac{2x - 5}{2 - x}$; all real numbers x except $x = 2$

64. $\dfrac{k(2v)^2}{kv^2} = 4$ **65.** 506 feet

66. Diameter $= 2r = 0.054$ inch **67.** 1470 joules

68. No. The 15-inch pizza is the best buy.

69. The velocity is increased by one-third.

70. (a) The safe load is unchanged.
(b) The safe load is eight times as great.
(c) The safe load is four times as great.
(d) The safe load is one-fourth as great.

71. (a)

(b) Yes. $k_1 = 4200, k_2 = 3800, k_3 = 4200,$
$k_4 = 4800, k_5 = 4500$

(c) $C = \dfrac{4300}{d}$

(d)

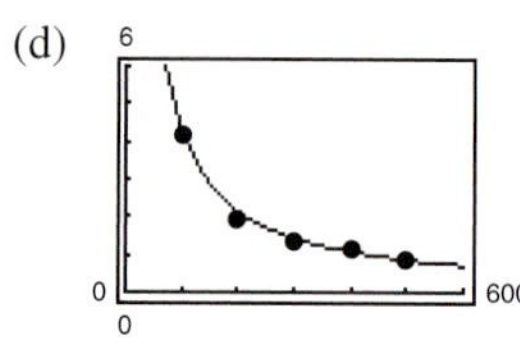

(e) ≈ 1433 meters

72. (a)

(b) Yes. $k = 0.575$
(c) 15.7 pounds

73. (a)

(b) 0.2857 microwatt per square centimeter

74. The illumination is reduced to one-fourth of its original value. Explanations will vary.

75. False. y will increase if k is positive and y will decrease if k is negative.

76. False. E is jointly proportional to the mass of an object and the square of its velocity.

77. True. The closer the value of $|r|$ is to 1, the better the fit.

78. (a) Good approximation (b) Poor approximation
(c) Poor approximation (d) Good approximation

79. The accuracy is questionable when based on such limited data.

80. Answers will vary.

81. $x > 5$ **82.** $x \geq \frac{11}{8}$

83. $-4 < x < 5$ **84.** $x \geq 3,\ x \leq -\frac{1}{3}$

85. (a) $-\frac{5}{3}$ (b) $-\frac{7}{3}$ (c) 21

86. (a) 6 (b) 9 (c) 383 **87.** Answers will vary.

Review Exercises *(page 117)*

1.

2.

3. Quadrant IV **4.** Quadrant I or II

5. (a)

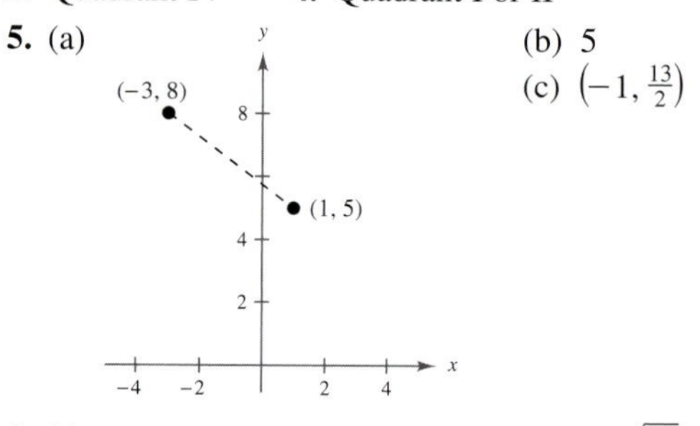

(b) 5
(c) $\left(-1, \frac{13}{2}\right)$

6. (a)

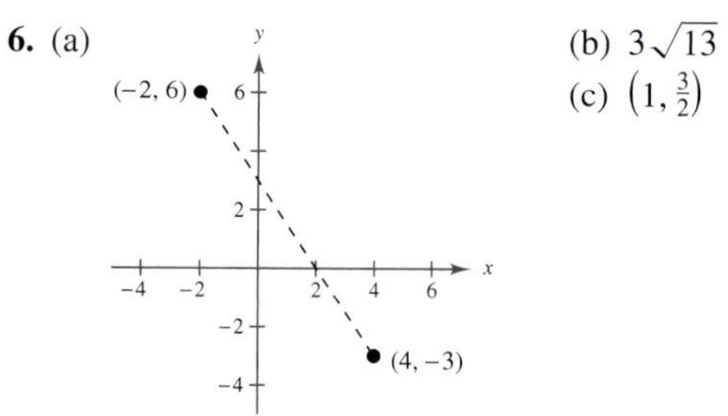

(b) $3\sqrt{13}$
(c) $\left(1, \frac{3}{2}\right)$

7. (a)

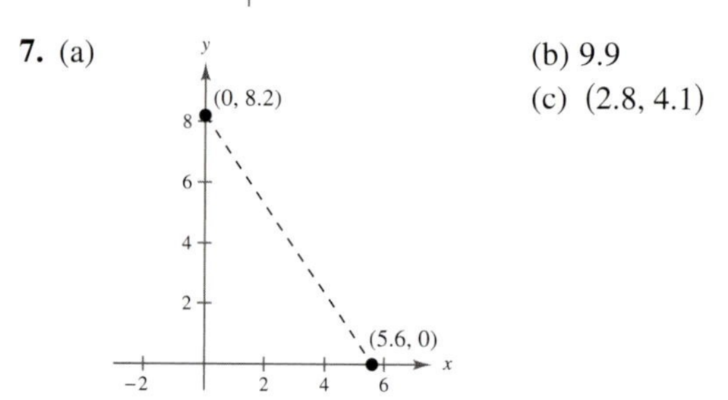

(b) 9.9
(c) (2.8, 4.1)

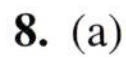

8. (a)

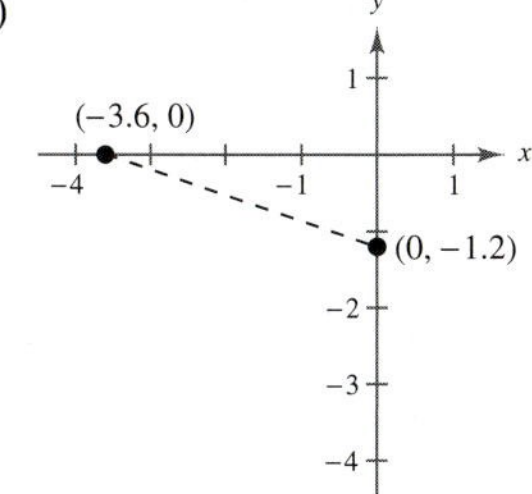

(b) $\sqrt{14.4}$

(c) $(-1.8, -0.6)$

9. (2, 5), (4, 5), (2, 0), (4, 0)

10. $(-4, 6), (-1, 8), (-4, 10), (-7, 8)$

11. \$656.45 million

12. (a)

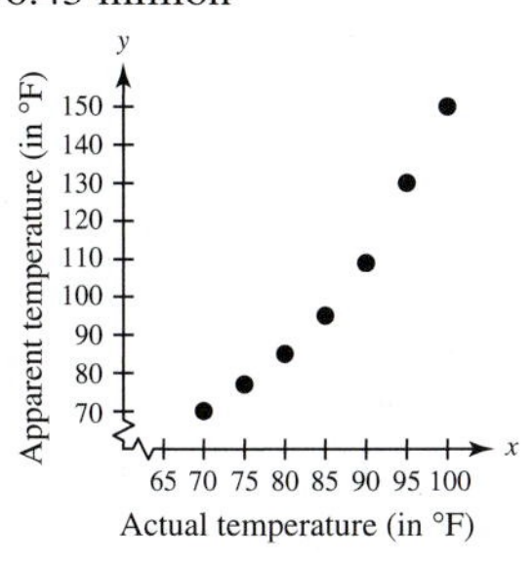

(b) 80°F

13. Radius $\approx$ 22.5 centimeters

14. (a)

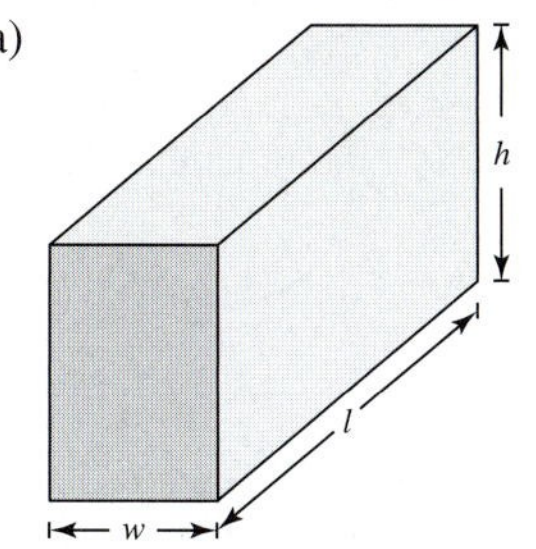

(b) $l = 24$ inches, $w = 8$ inches, $h = 12$ inches

15.

x	-2	-1	0	1	2
y	-11	-8	-5	-2	1

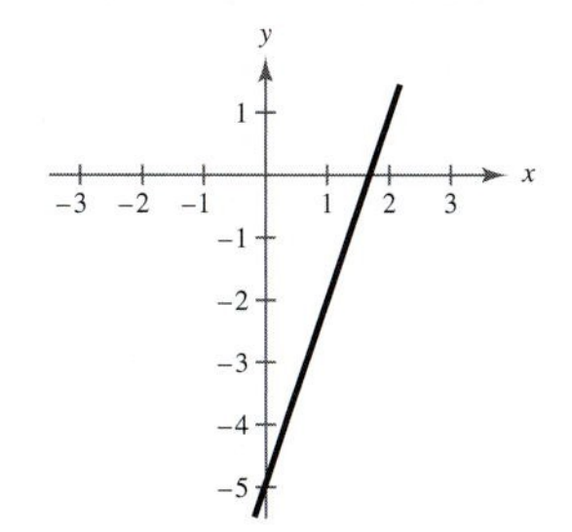

16.

x	-4	-2	0	2	4
y	4	3	2	1	0

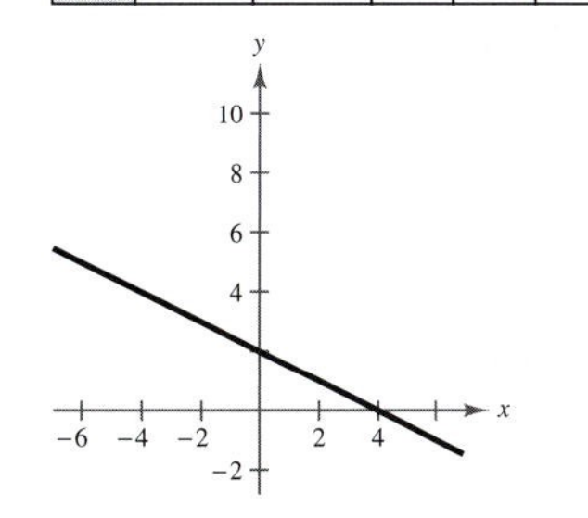

17.

x	-1	0	1	2	3	4
y	4	0	-2	-2	0	4

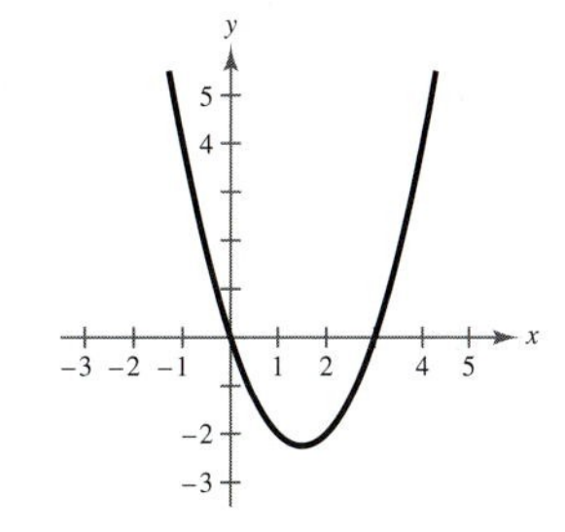

18.

x	-2	-1	0	1	2	3
y	1	-6	-9	-8	-3	6

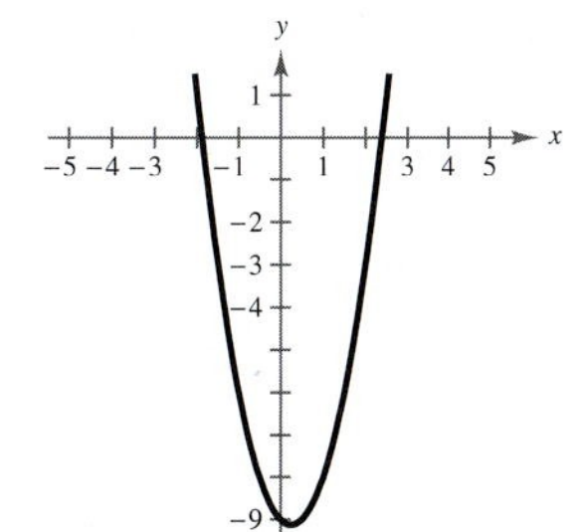

19. **20.**

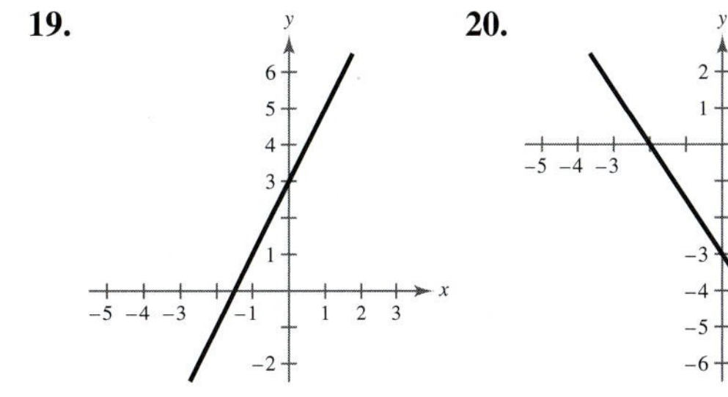

21.

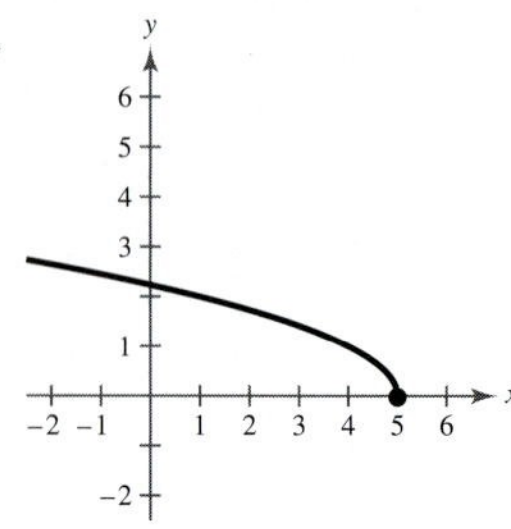

22.

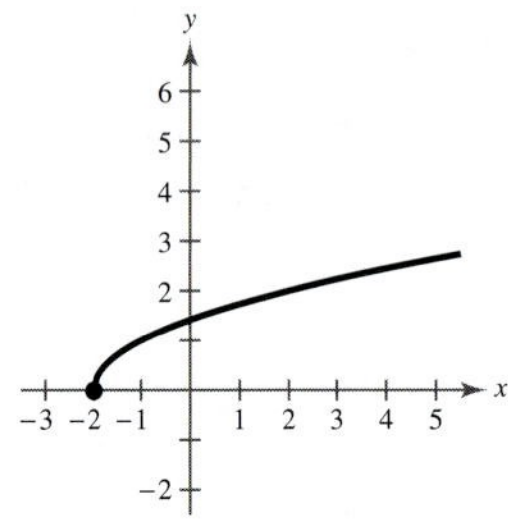

23.

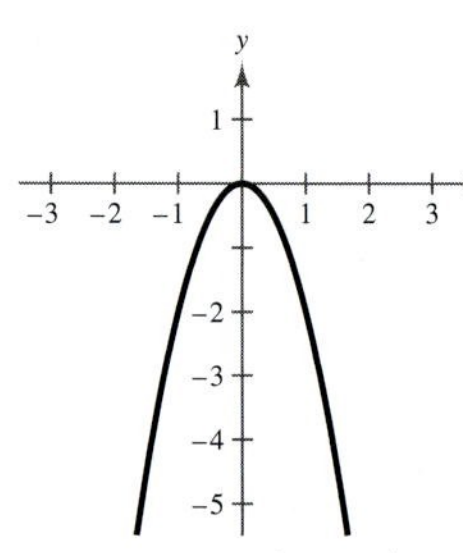

24.

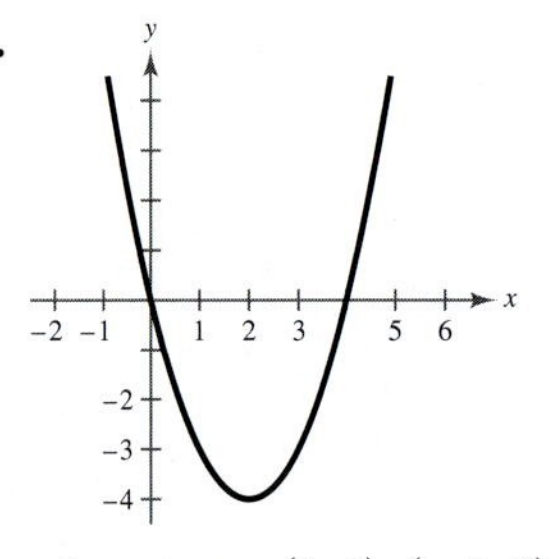

25. x-intercept: $\left(-\frac{7}{2}, 0\right)$
y-intercept: $(0, 7)$

26. x-intercepts: $(2, 0), (-4, 0)$
y-intercept: $(0, -2)$

27. x-intercepts: $(1, 0), (5, 0)$
y-intercept: $(0, 5)$

28. x-intercepts: $(2, 0), (-2, 0), (0, 0)$
y-intercept: $(0, 0)$

29. No symmetry

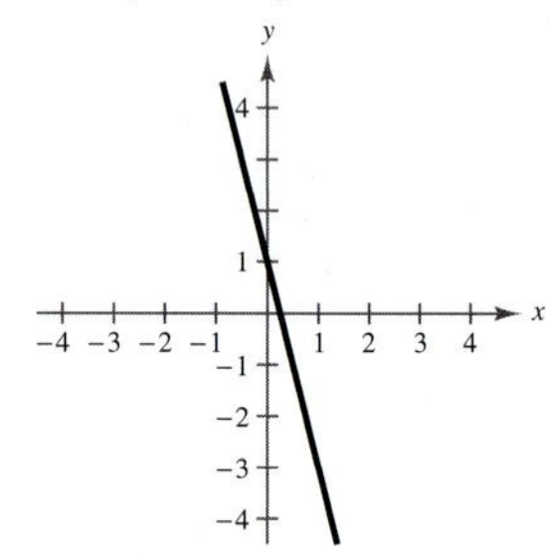

30. No symmetry

31. y-axis symmetry

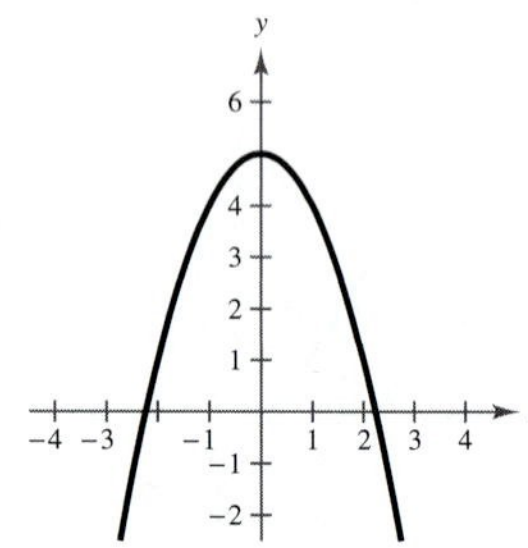

32. y-axis symmetry

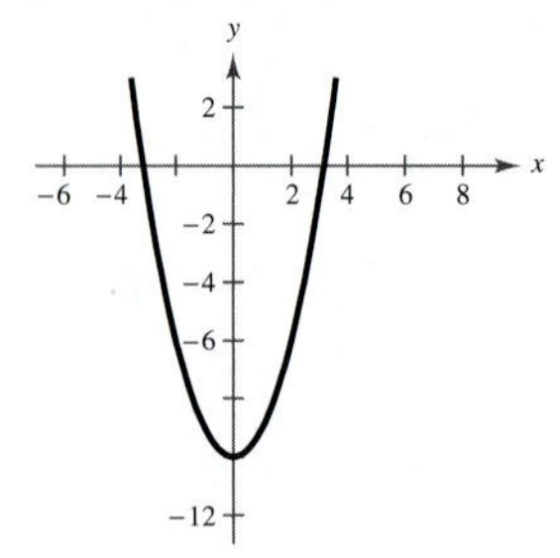

33. No symmetry

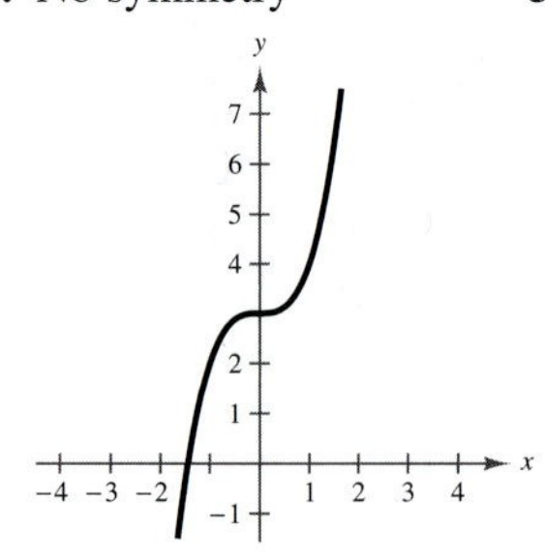

34. No symmetry

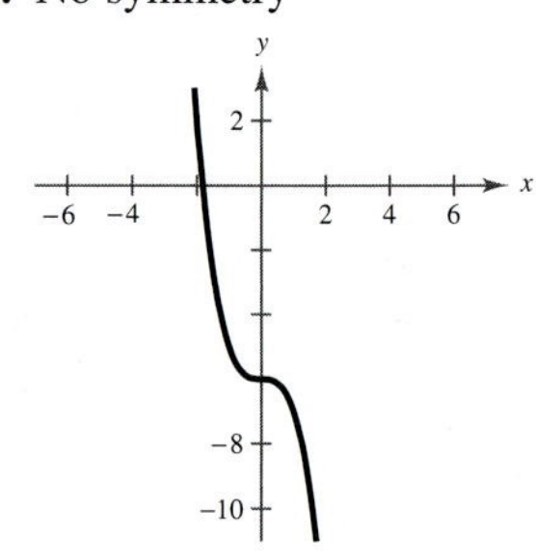

35. No symmetry

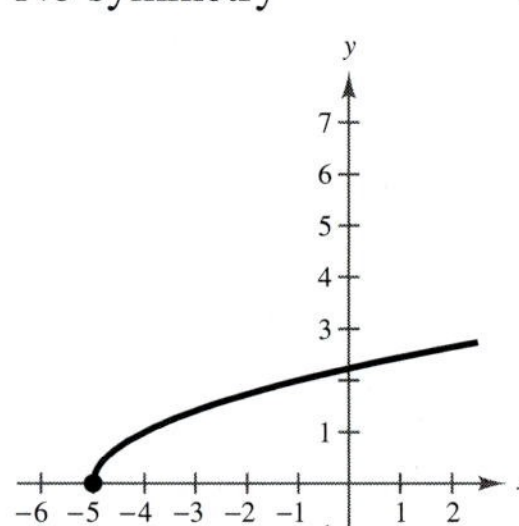

36. y-axis symmetry

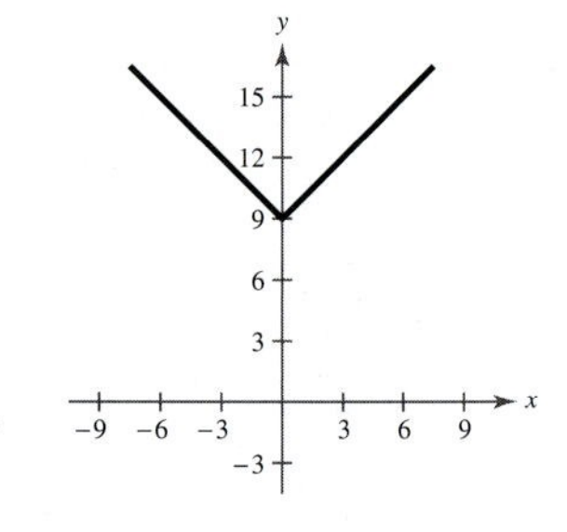

37. Center: $(0, 0)$;
Radius: 3

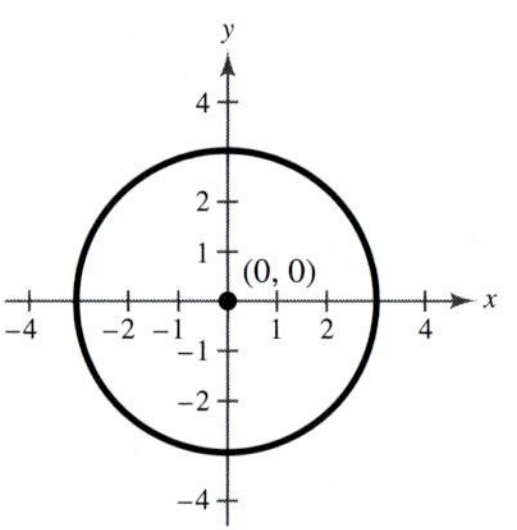

38. Center: $(0, 0)$;
Radius: 2

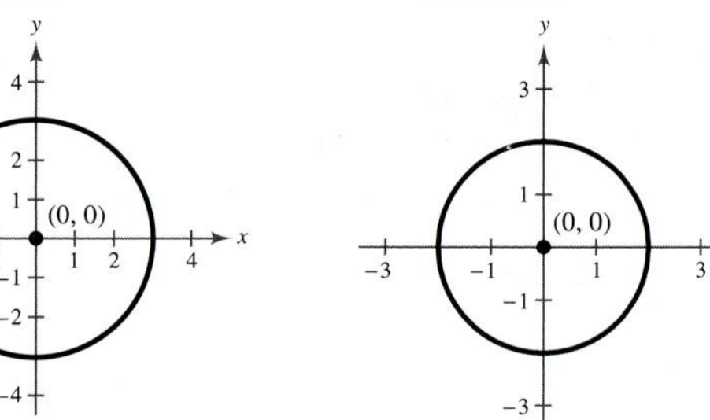

39. Center: $(-2, 0)$;
Radius: 4

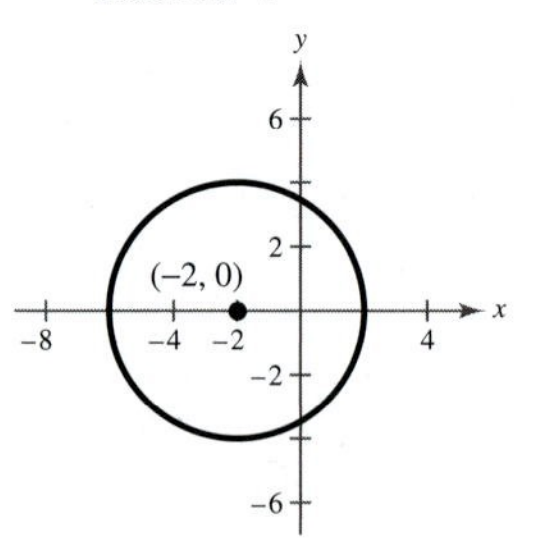

40. Center: $(0, 8)$;
Radius: 9

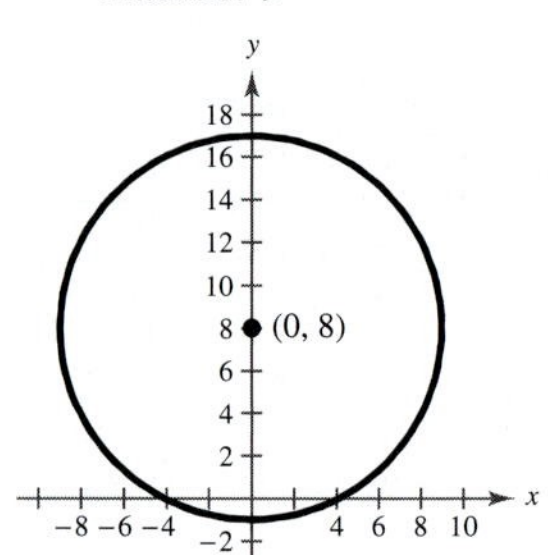

41. Center: $\left(\frac{1}{2}, -1\right)$; Radius: 6

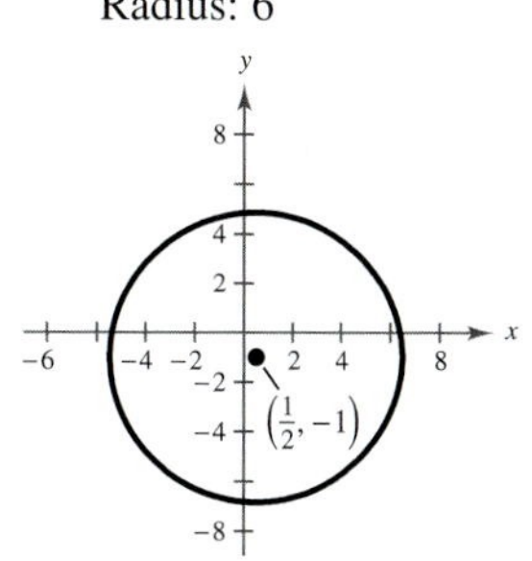

42. Center: $\left(-4, \frac{3}{2}\right)$; Radius: 10

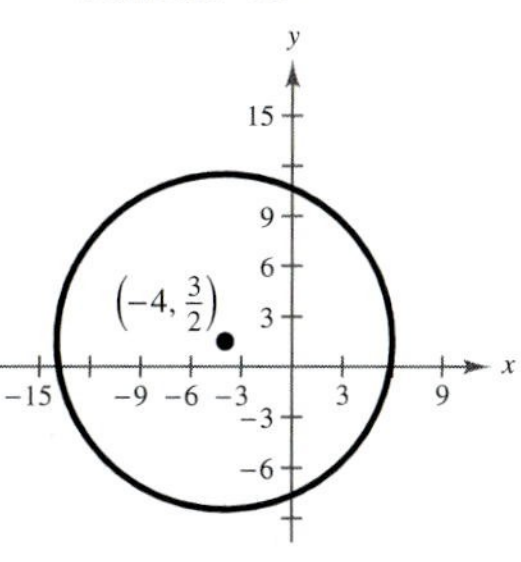

43. $(x - 2)^2 + (y + 3)^2 = 13$

44. $(x - 1)^2 + \left(y + \frac{13}{2}\right)^2 = \frac{85}{4}$

45. (a)

x	0	4	8	12	16	20
F	0	5	10	15	20	25

(b)

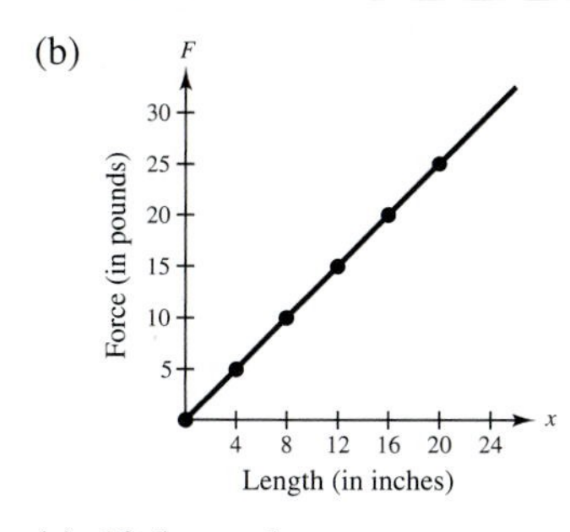

(c) 12.5 pounds

46. (a)

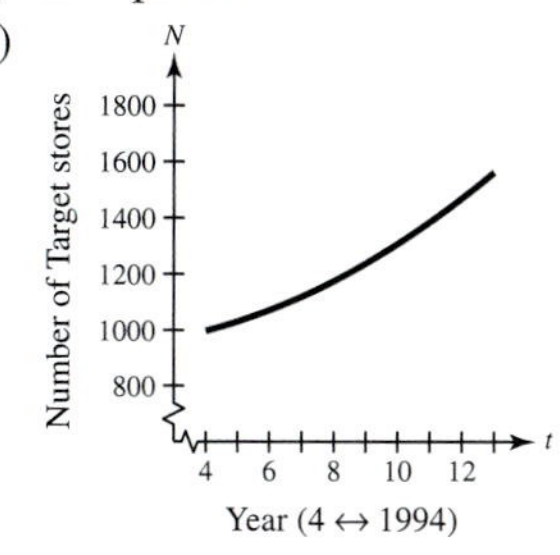

(b) 1999

47. slope: 0
y-intercept: 6

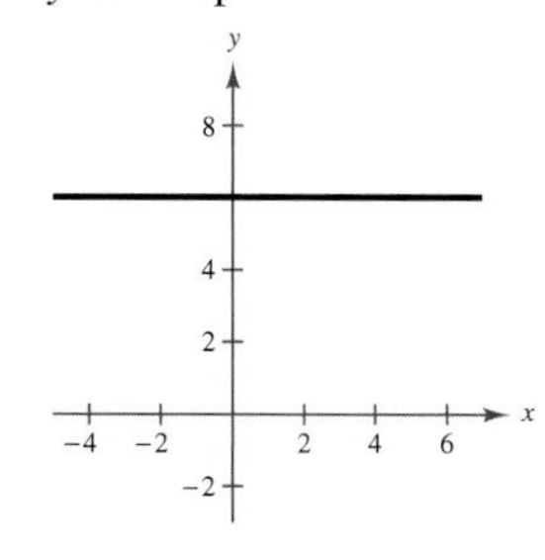

48. slope: undefined
y-intercept: none

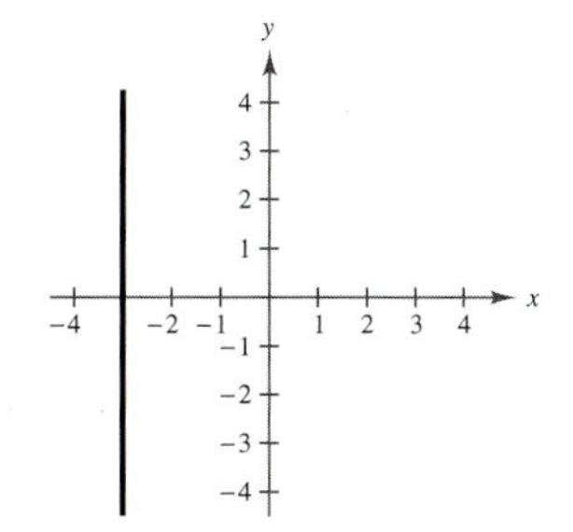

49. slope: 3
y-intercept: 13

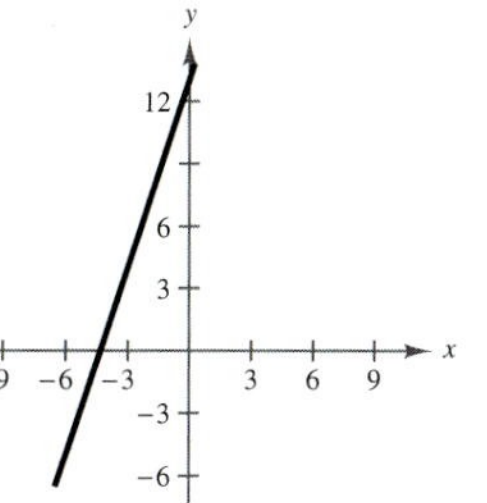

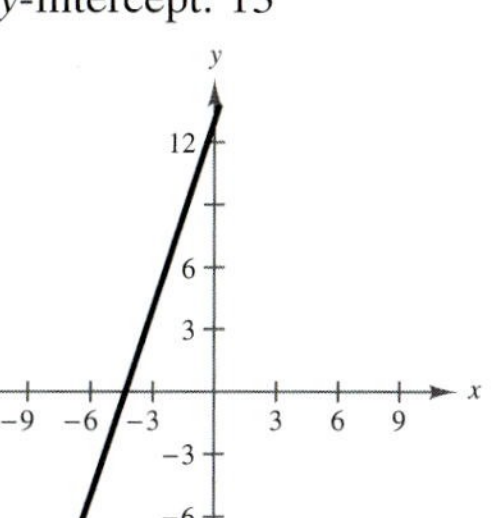

50. slope: -10
y-intercept: 9

51.

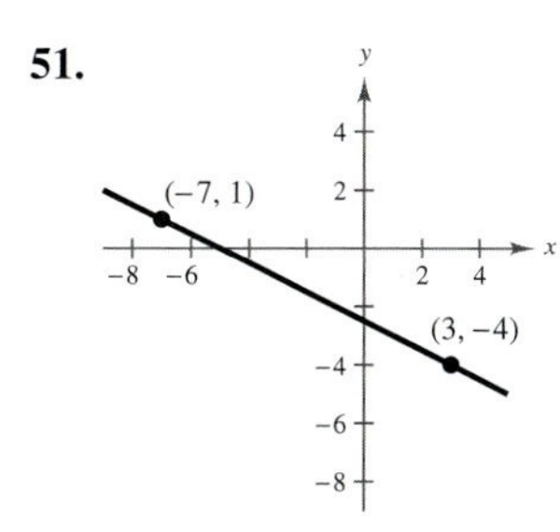

$m = -\frac{1}{2}$

52.

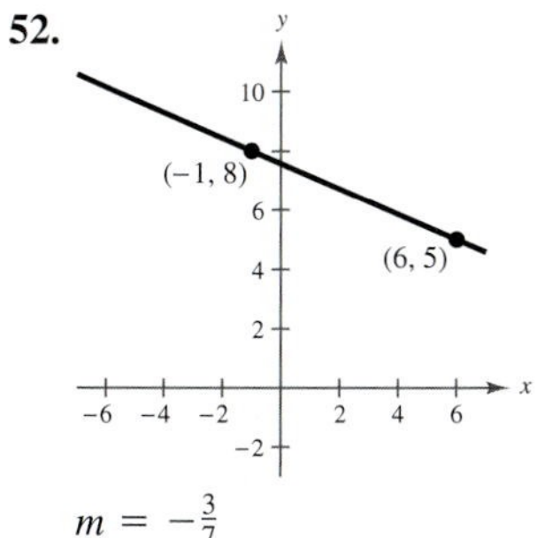

$m = -\frac{3}{7}$

53.

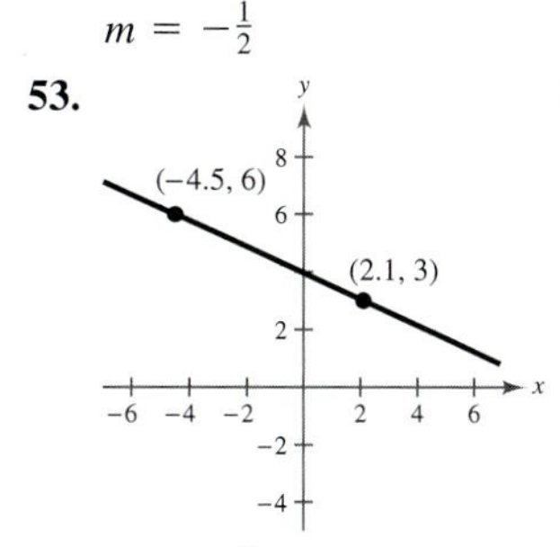

$m = -\frac{5}{11}$

54.

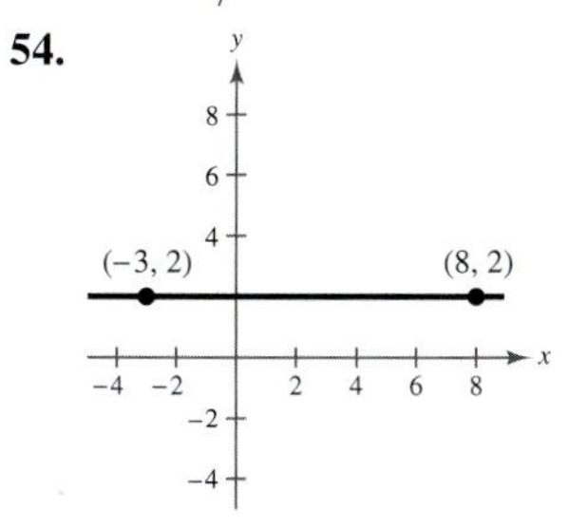

$m = 0$

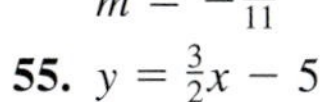

55. $y = \frac{3}{2}x - 5$

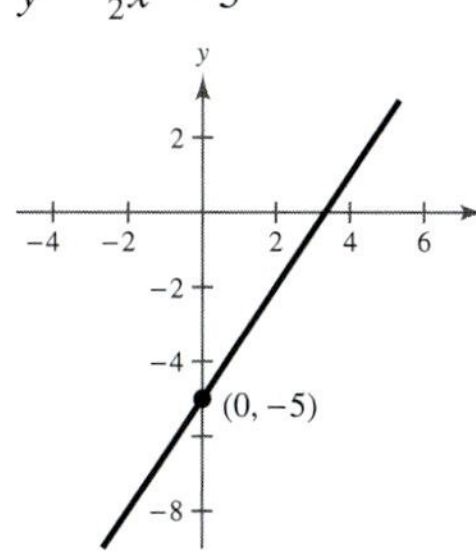

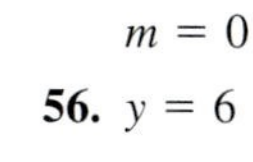

56. $y = 6$

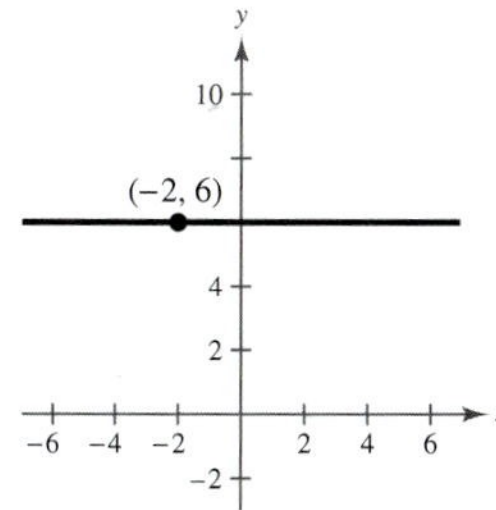

57. $y = -\frac{1}{2}x + 2$

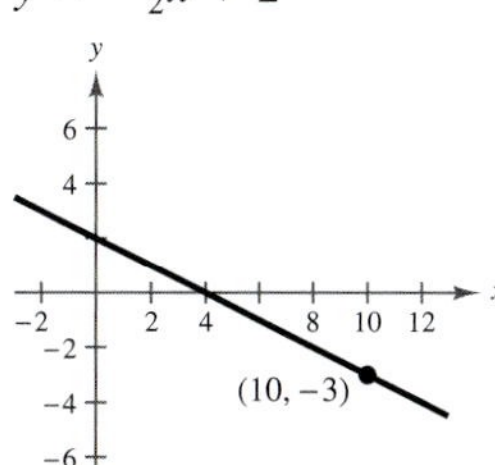

58. $x = -8$

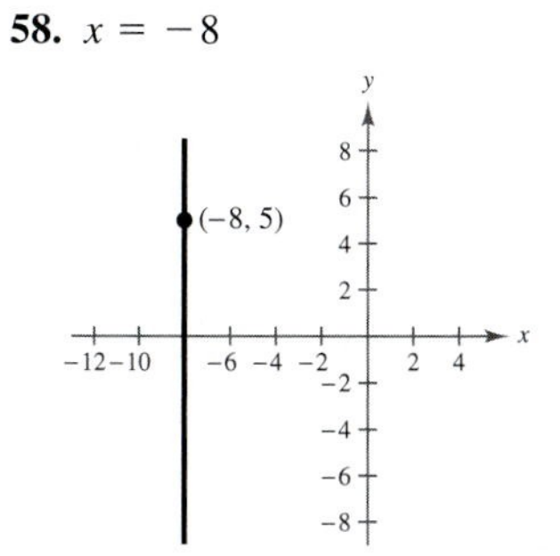

CHAPTER 1

59. $x = 0$ **60.** $y = \frac{3}{2}x + 2$

61. $y = -\frac{4}{3}x + \frac{8}{3}$ **62.** $y = -\frac{1}{5}x + \frac{1}{5}$

63. (a) $y = \frac{5}{4}x - \frac{23}{4}$ (b) $y = -\frac{4}{5}x + \frac{2}{5}$

64. (a) $y = -\frac{2}{3}x - \frac{7}{3}$ (b) $y = \frac{3}{2}x + 15$

65. $V = 850t + 7400,\quad 6 \le t \le 11$

66. $V = 5.15t + 42.05,\quad 6 \le t \le 11$

67. No **68.** Yes **69.** Yes **70.** No

71. (a) 5 (b) 17 (c) $t^4 + 1$ (d) $t^2 + 2t + 2$

72. (a) -3 (b) -1 (c) 2 (d) 6

73. All real numbers x such that $-5 \le x \le 5$

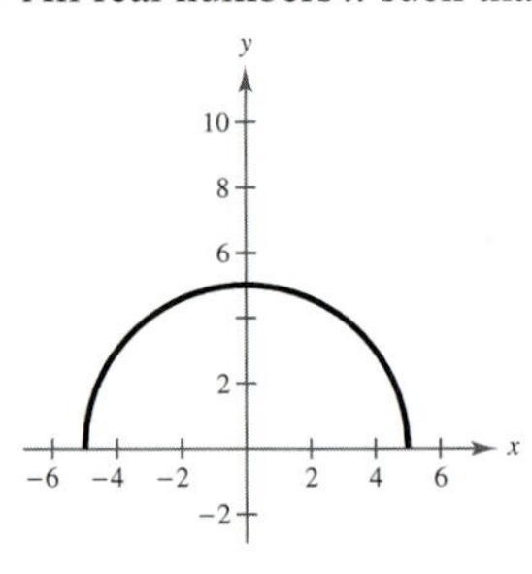

74. All real numbers

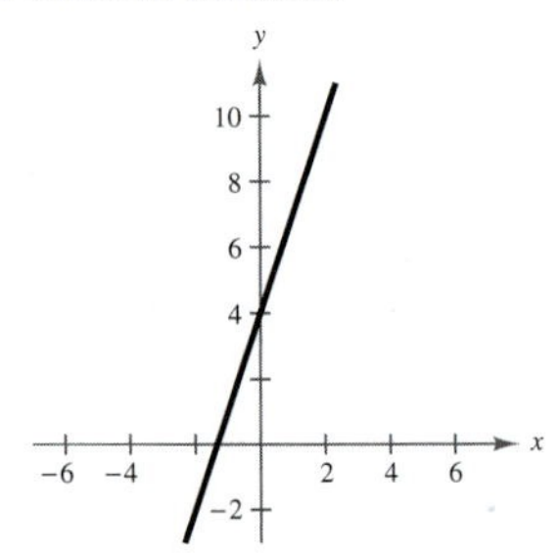

75. All real numbers x except $x = 3, -2$

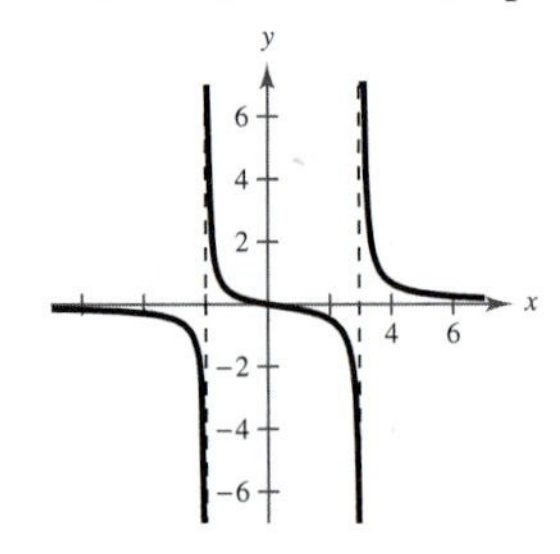

76. All real numbers

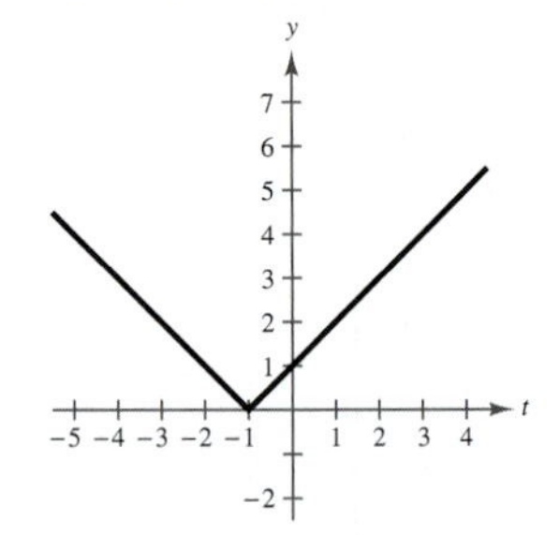

77. (a) 16 feet per second (b) 1.5 seconds
(c) -16 feet per second

78. (a) $f(x) = 0.4(50 - x) + x = 20 + 0.6x$
(b) Domain: $0 \le x \le 50$; Range: $20 \le y \le 50$
(c) $8\frac{1}{3}$ liters

79. $4x + 2h + 3,\quad h \ne 0$

80. $3x^2 + 3xh + h^2 - 10x - 5h + 1,\ h \ne 0$

81. Function **82.** Function **83.** Not a function

84. Not a function **85.** $\frac{7}{3}, 3$ **86.** $-1, \frac{1}{5}$

87. $-\frac{3}{8}$ **88.** $1, \pm 5$

89. Increasing on $(0, \infty)$
Decreasing on $(-\infty, -1)$
Constant on $(-1, 0)$

90. Increasing on $(-2, 0)$ and $(2, \infty)$
Decreasing on $(-\infty, -2)$ and $(0, 2)$

91.

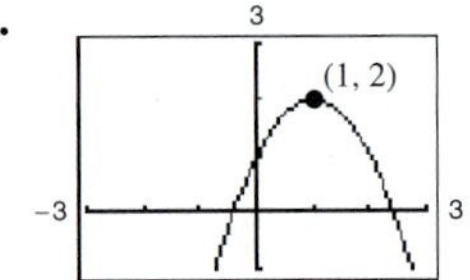

92.

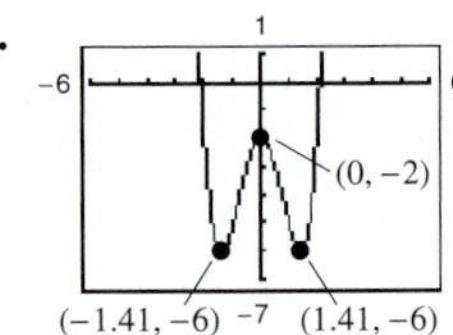

93.

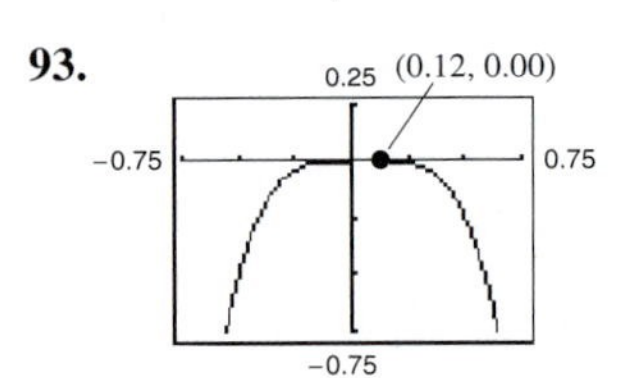

94.

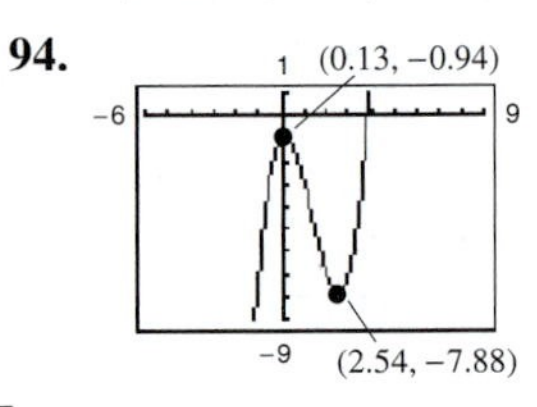

95. 4 **96.** 28 **97.** $\dfrac{1 - \sqrt{2}}{2}$

98. -0.2 **99.** Neither even nor odd

100. Even **101.** Odd **102.** Even

103. $f(x) = -3x$

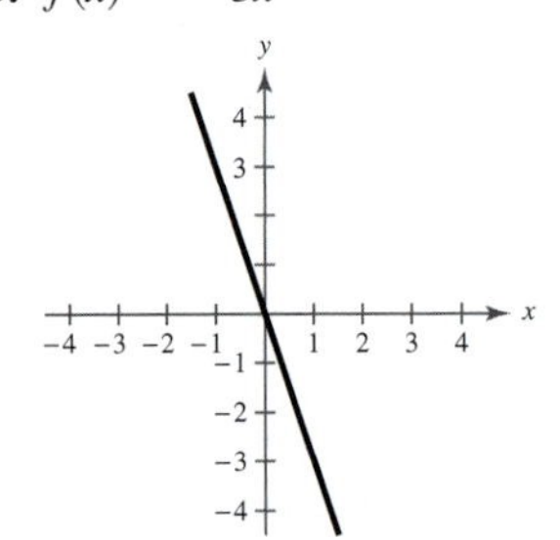

104. $f(x) = -\frac{3}{4}x - 5$

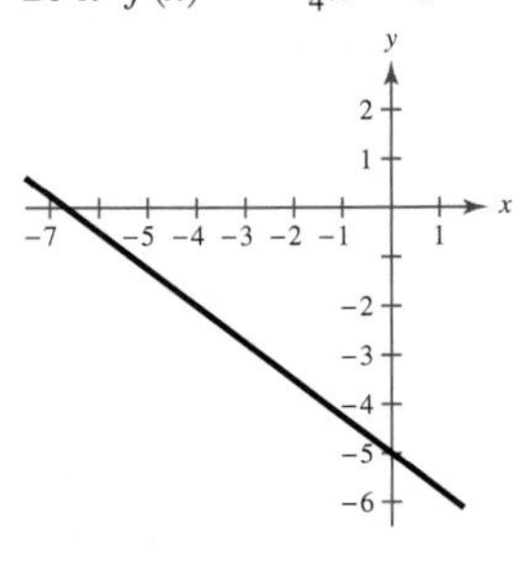

105. **106.**

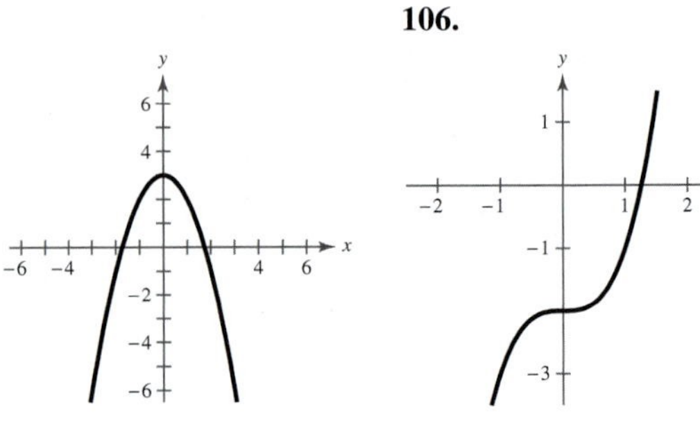

107.

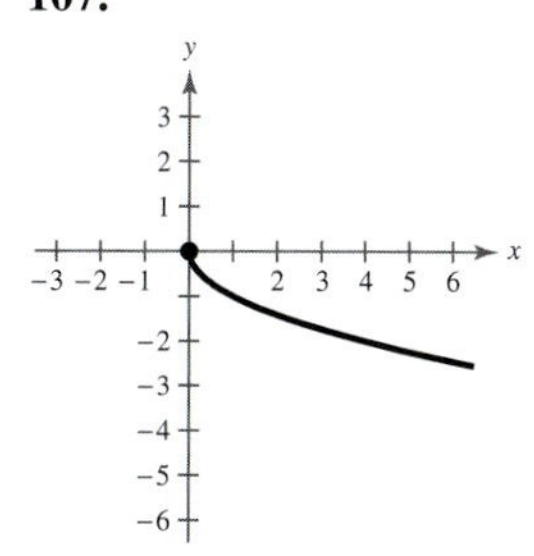

108.

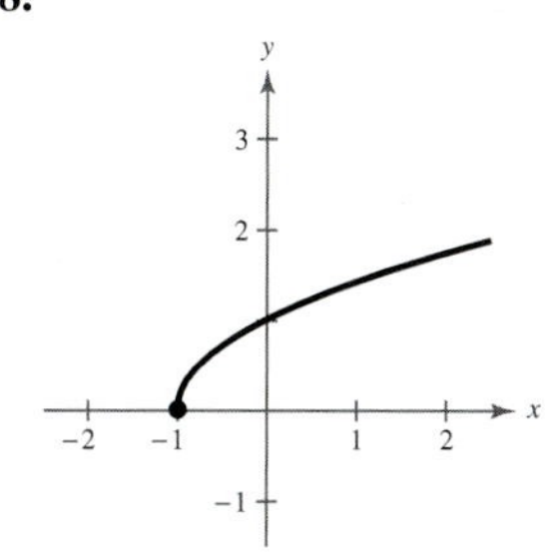

109.

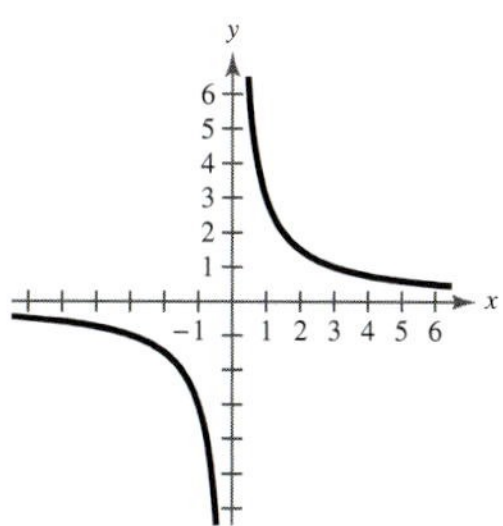

110.

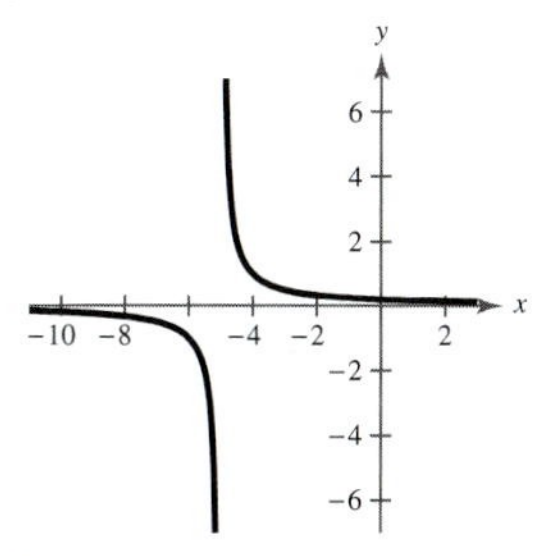

111.

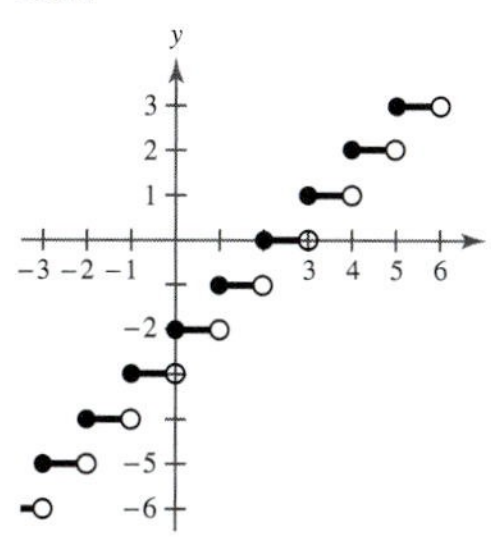

112.

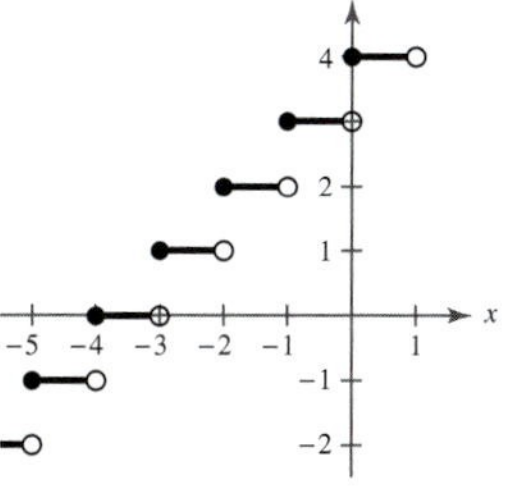

113.

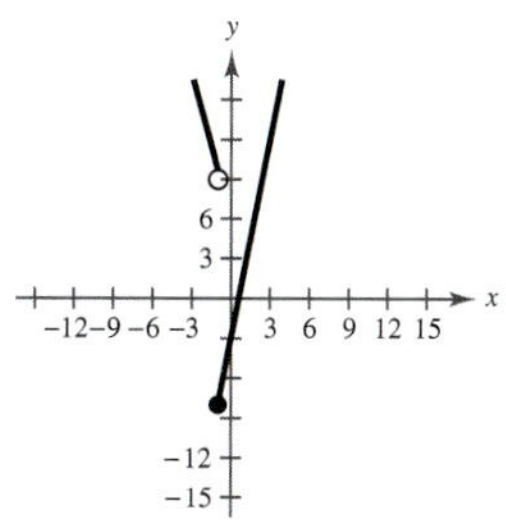

114.

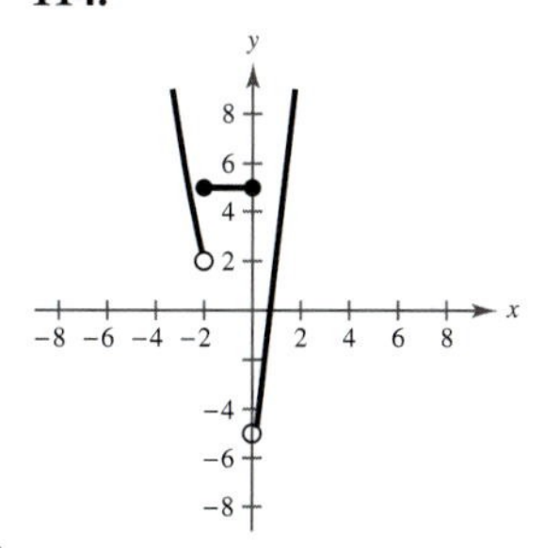

115. $y = x^3$ **116.** $y = \sqrt{x}$

117. (a) $f(x) = x^2$

(b) Vertical shift of nine units downward

(c)

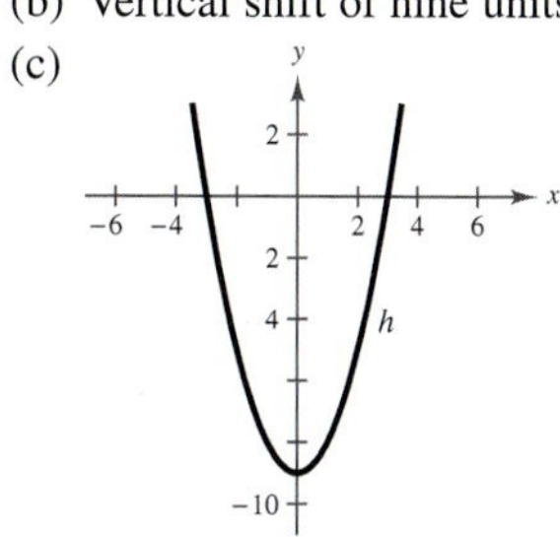

(d) $h(x) = f(x) - 9$

118. (a) $f(x) = x^3$

(b) Horizontal shift of two units to the right and vertical shift of two units upward

(c)

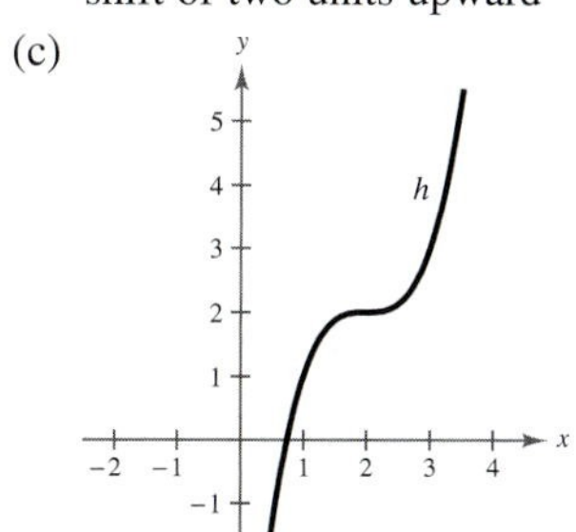

(d) $h(x) = f(x - 2) + 2$

119. (a) $f(x) = \sqrt{x}$

(b) Horizontal shift of seven units to the right

(c)

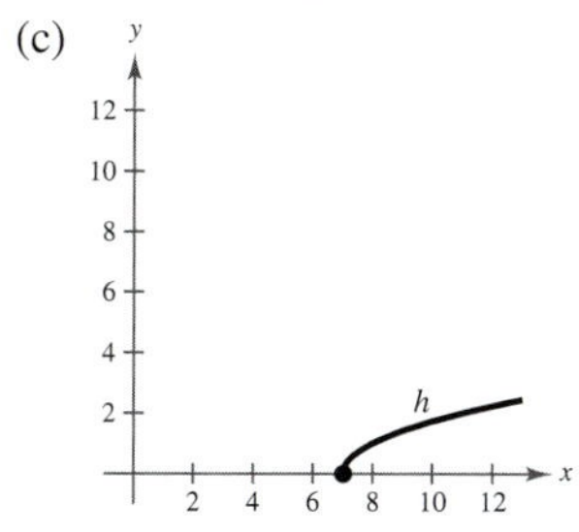

(d) $h(x) = f(x - 7)$

120. (a) $f(x) = |x|$

(b) Horizontal shift of three units to the left and vertical shift of five units downward

(c)

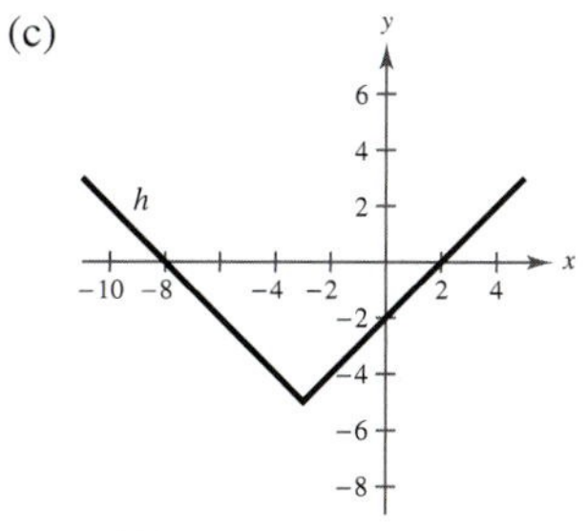

(d) $h(x) = f(x + 3) - 5$

121. (a) $f(x) = x^2$

(b) Reflection in the x-axis, horizontal shift of three units to the left, and vertical shift of one unit upward

(c)

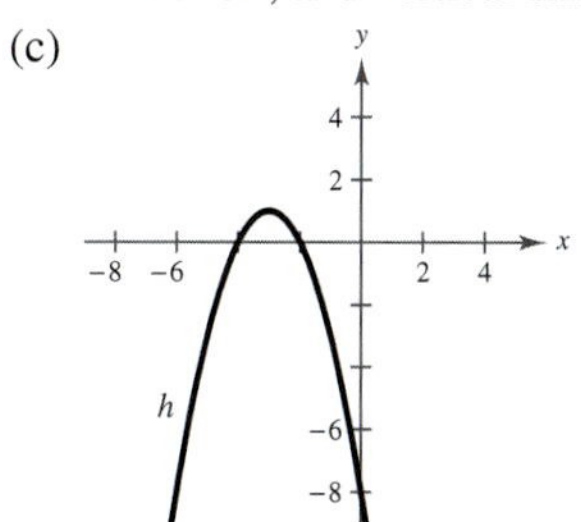

(d) $h(x) = -f(x + 3) + 1$

122. (a) $f(x) = x^3$

(b) Reflection in the x-axis, horizontal shift of five units to the right, and vertical shift of five units downward

(c)

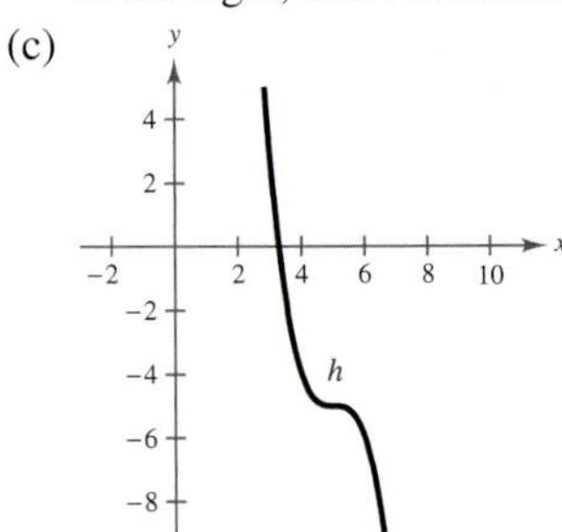

(d) $h(x) = -f(x - 5) - 5$

123. (a) $f(x) = [\![x]\!]$

(b) Reflection in the x-axis and vertical shift of six units upward

(c)

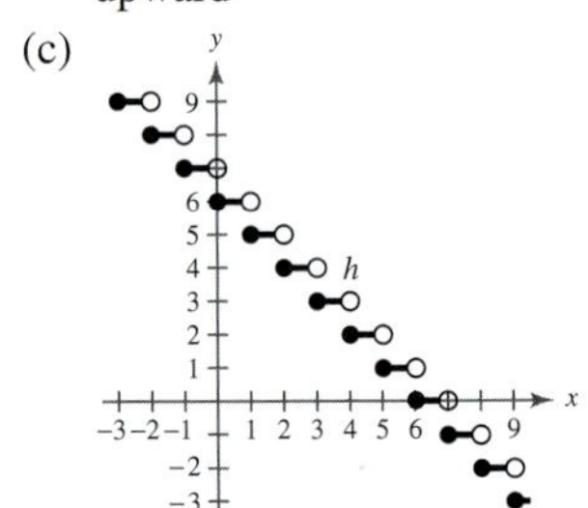

(d) $h(x) = -f(x) + 6$

124. (a) $f(x) = \sqrt{x}$

(b) Reflection in the x-axis, horizontal shift of one unit to the left, and vertical shift of nine units upward

(c)

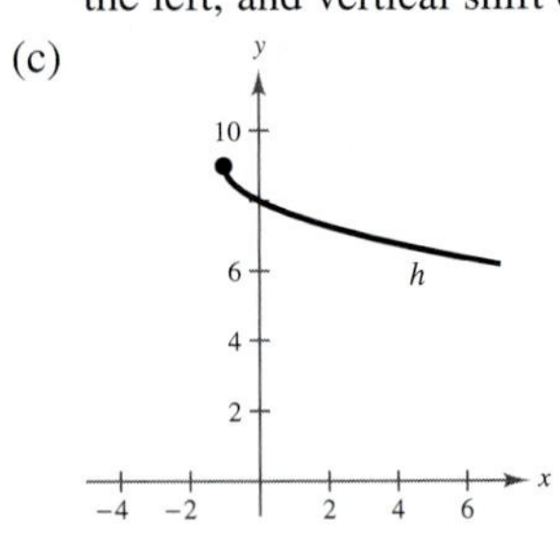

(d) $h(x) = -f(x + 1) + 9$

125. (a) $f(x) = |x|$

(b) Reflections in the x-axis and the y-axis, horizontal shift of four units to the right, and vertical shift of six units upward

(c)

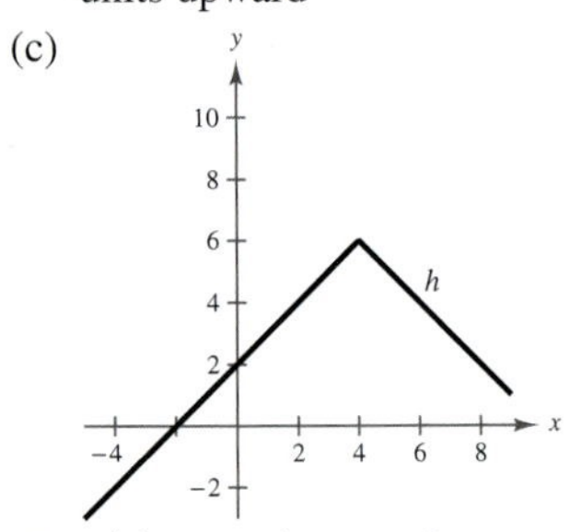

(d) $h(x) = -f(-x + 4) + 6$

126. (a) $f(x) = x^2$

(b) Reflection in the x-axis, horizontal shift of one unit to the left, and vertical shift of three units downward

(c)

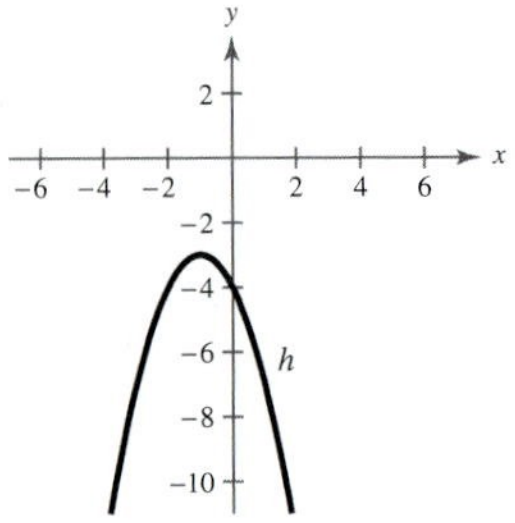

(d) $h(x) = -f(x + 1) - 3$

127. (a) $f(x) = [\![x]\!]$

(b) Horizontal shift of nine units to the right and vertical stretch

(c)

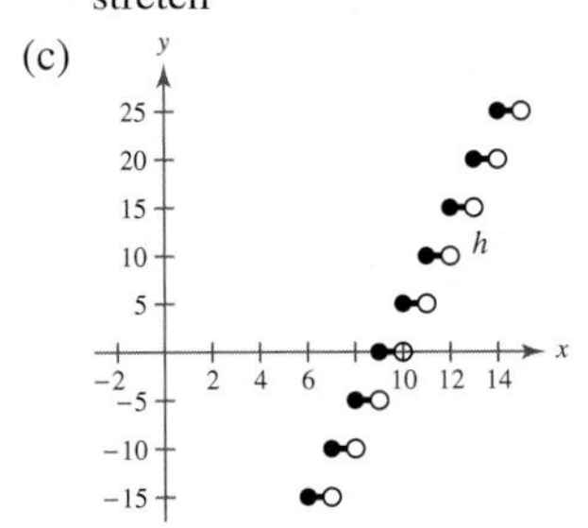

(d) $h(x) = 5f(x - 9)$

128. (a) $f(x) = x^3$

(b) Reflection in the x-axis and vertical shrink

(c)

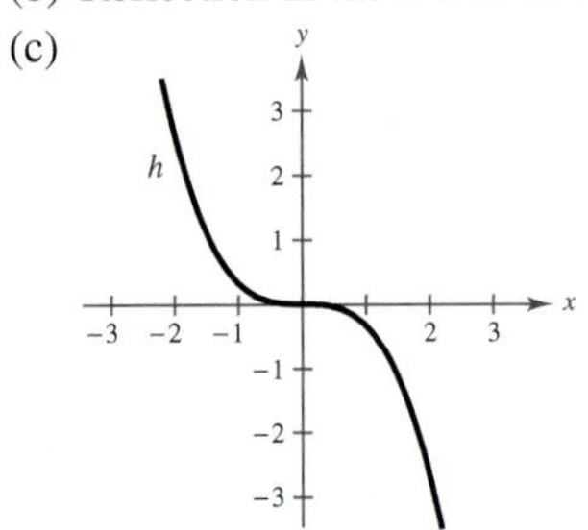

(d) $h(x) = -\frac{1}{3}f(x)$

129. (a) $f(x) = \sqrt{x}$

(b) Reflection in the x-axis, vertical stretch, and horizontal shift of four units to the right

(c)

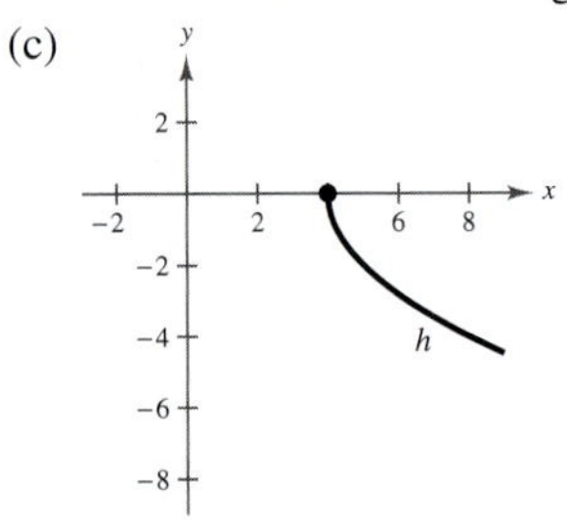

(d) $h(x) = -2f(x - 4)$

130. (a) $f(x) = |x|$
(b) Vertical shrink and vertical shift of one unit downward
(c)

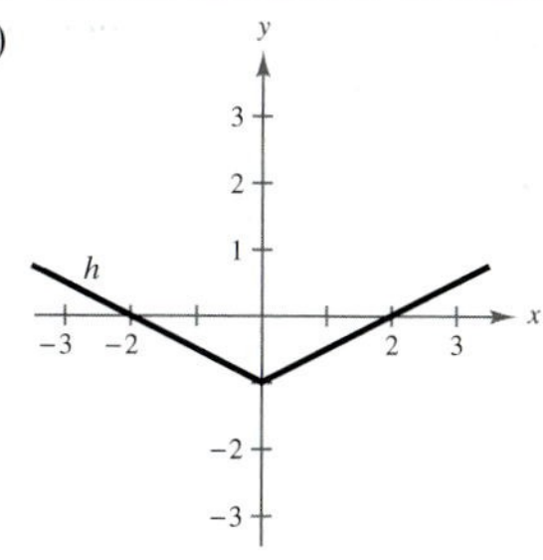

(d) $h(x) = \frac{1}{2}f(x) - 1$

131. (a) $x^2 + 2x + 2$ (b) $x^2 - 2x + 4$
(c) $2x^3 - x^2 + 6x - 3$
(d) $\dfrac{x^2 + 3}{2x - 1}$; all real numbers x except $x = \dfrac{1}{2}$

132. (a) $x^2 - 4 + \sqrt{3 - x}$ (b) $x^2 - 4 - \sqrt{3 - x}$
(c) $(x^2 - 4)\sqrt{3 - x}$
(d) $\dfrac{x^2 - 4}{\sqrt{3 - x}}$; all real numbers x such that $x < 3$

133. (a) $x - \frac{8}{3}$ (b) $x - 8$
Domains of $f, g, f \circ g$, and $g \circ f$: all real numbers

134. (a) $x + 3$ (b) $\sqrt[3]{x^3 + 3}$
Domains of $f, g, f \circ g$, and $g \circ f$: all real numbers

135. $f(x) = x^3, g(x) = 6x - 5$

136. $f(x) = \sqrt[3]{x}, g(x) = x + 2$

137. (a) $(v + d)(t) = -36.04t^2 + 804.6t - 1112$
(b)

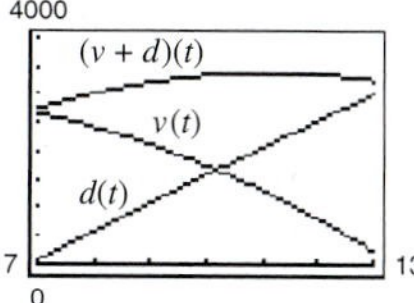

(c) $(v + d)(10) = 3330$

138. (a) $N(T(t)) = 100t^2 + 275$
The composition function $N(T(t))$ represents the number of bacteria in the food as a function of time.
(b) $t = 2.18$

139. $f^{-1}(x) = x + 7$ **140.** $f^{-1}(x) = x - 5$

141. The function has an inverse.

142. The function does not have an inverse.

143.

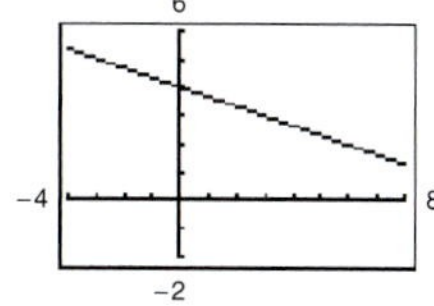

The function has an inverse.

144.

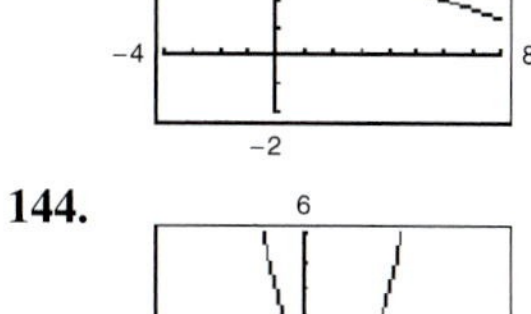

The function does not have an inverse.

145.

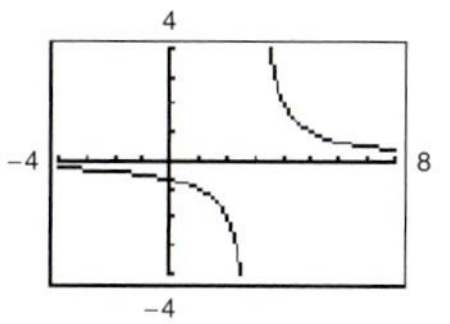

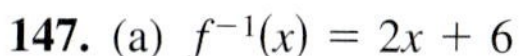

The function has an inverse.

146.

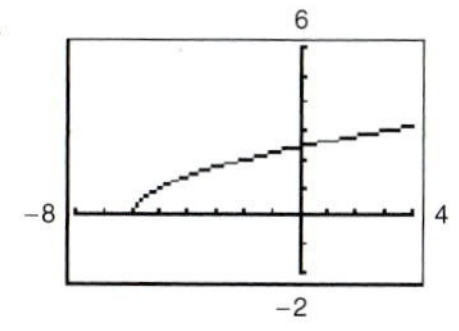

The function has an inverse.

147. (a) $f^{-1}(x) = 2x + 6$
(b)

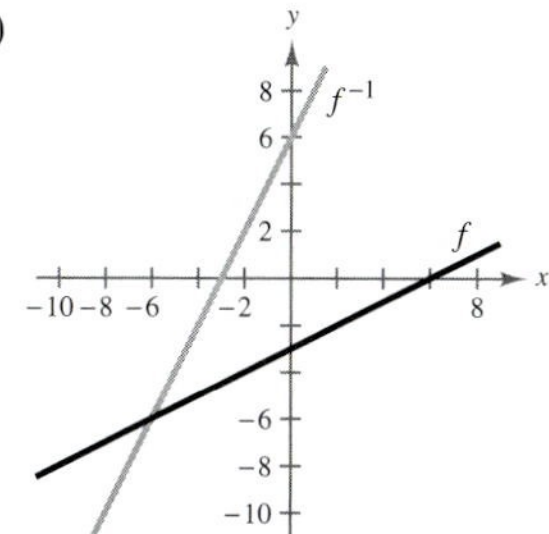

(c) The graph of f^{-1} is the reflection of the graph of f in the line $y = x$.
(d) Both f and f^{-1} have domains and ranges that are all real numbers.

148. (a) $f^{-1}(x) = \dfrac{x + 7}{5}$
(b)

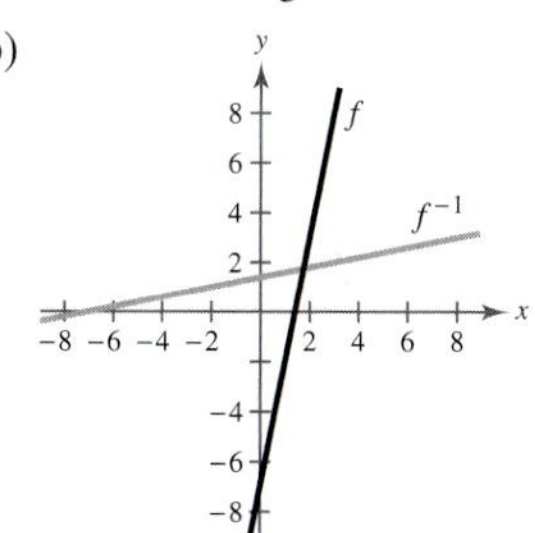

(c) The graphs are reflections of each other in the line $y = x$.
(d) Both f and f^{-1} have domains and ranges that are all real numbers.

149. (a) $f^{-1}(x) = x^2 - 1,\ x \geq 0$
(b)

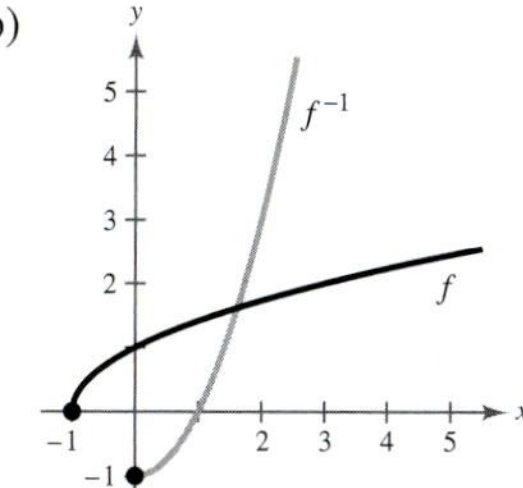

(c) The graph of f^{-1} is the reflection of the graph of f in the line $y = x$.
(d) f has a domain of $[-1, \infty)$ and a range of $[0, \infty)$; f^{-1} has a domain of $[0, \infty)$ and a range of $[-1, \infty)$.

150. (a) $f^{-1}(x) = \sqrt[3]{x - 2}$

(b)

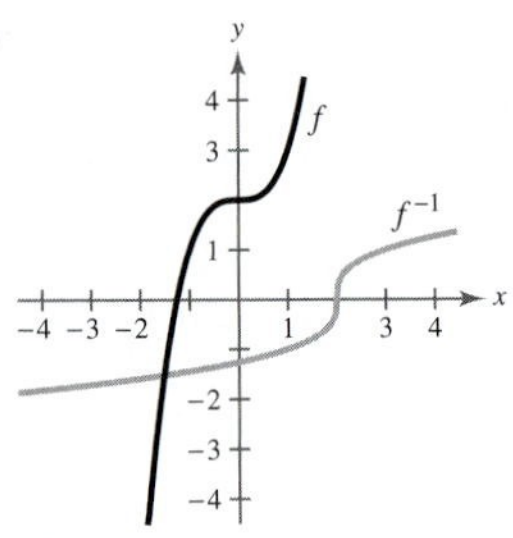

(c) The graphs are reflections of each other in the line $y = x$.

(d) Both f and f^{-1} have domains and ranges that are all real numbers.

151. $x \geq 4;\ f^{-1}(x) = \sqrt{\dfrac{x}{2}} + 4$

152. $x \geq 2;\ f^{-1}(x) = x + 2,\ x \geq 0$

153. (a)

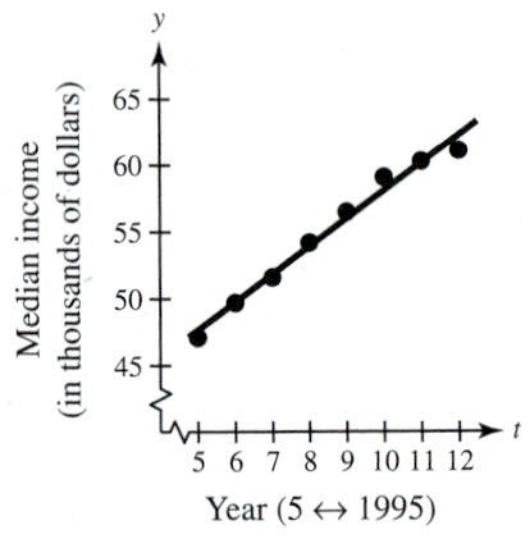

(b) The model is a good fit for the actual data.

154. (a) and (b)

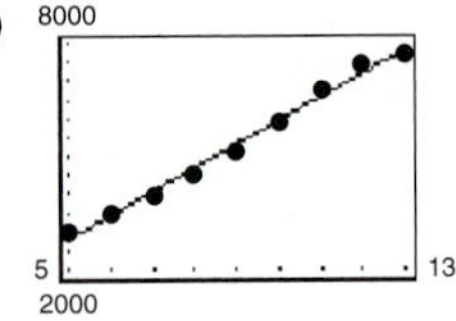

$S = 627.02t - 346$

This model is a good fit for the actual data.

(c) 2008: \$10,940.36

(d) The factory sales of electronic gaming software in the U.S. increases by \$627.02 million dollars each year.

155. Model: $m = \frac{8}{5}k$; 3.2 kilometers, 16 kilometers

156. 2438.7 kilowatts **157.** A factor of 4

158. 666 boxes **159.** $\approx$ 2 hours, 26 minutes

160. \$44.80

161. False. The graph is reflected in the x-axis, shifted nine units to the left, and then shifted 13 units downward.

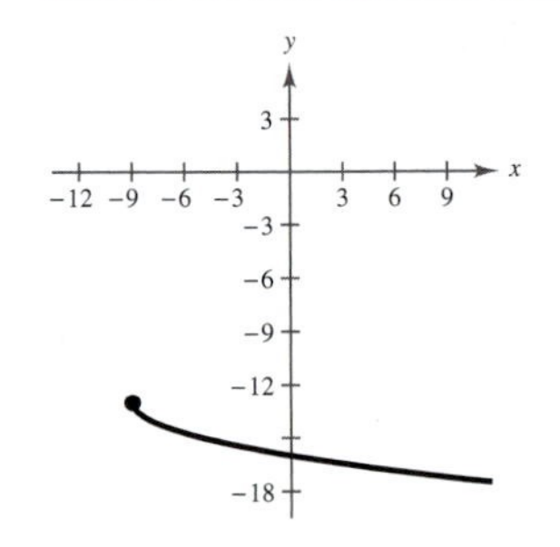

162. True. If $f(x) = x^3$ and $g(x) = \sqrt[3]{x}$, then the domain of g is all real numbers, which is equal to the range of f and vice versa.

163. True. If y is directly proportional to x, then $y = kx$, so $x = (1/k)y$. Therefore, x is directly proportional to y.

164. The Vertical Line Test is used to determine if the graph of y is a function of x. The Horizontal Line Test is used to determine if a function has an inverse function.

165. A function from a set A to a set B is a relation that assigns to each element x in the set A exactly one element y in the set B.

Chapter Test *(page 123)*

1.

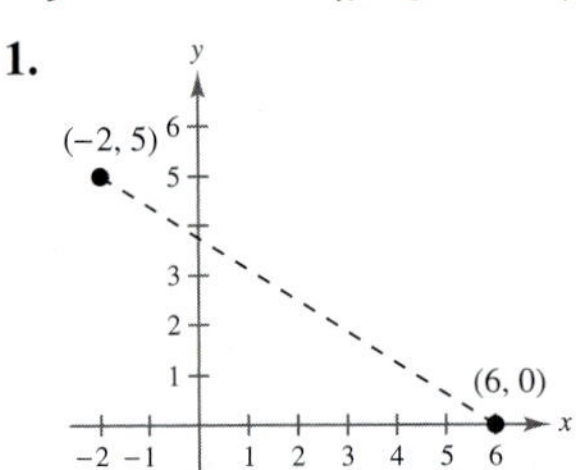

Midpoint: $\left(2, \frac{5}{2}\right)$; Distance: $\sqrt{89}$

2. $\approx$ 11.937 centimeters

3. No symmetry

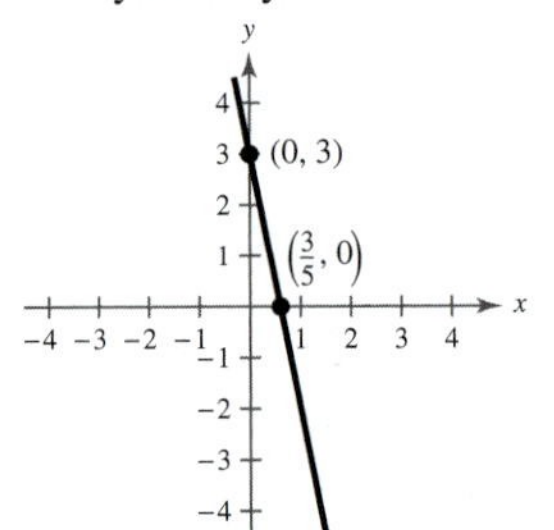

4. y-axis symmetry

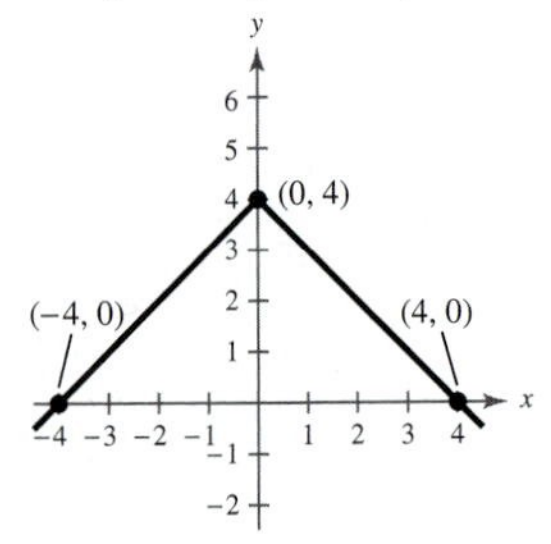

5. y-axis symmetry

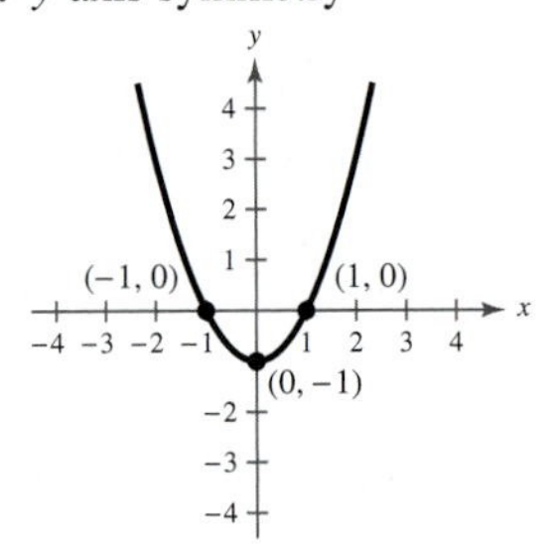

6. $(x - 1)^2 + (y - 3)^2 = 16$ **7.** $2x + y - 1 = 0$

8. $17x + 10y - 59 = 0$

9. (a) $4x - 7y + 44 = 0$ (b) $7x + 4y - 53 = 0$

10. (a) $-\dfrac{1}{8}$ (b) $-\dfrac{1}{28}$ (c) $\dfrac{\sqrt{x}}{x^2 - 18x}$

11. $-10 \leq x \leq 10$

12. (a) $0, \pm 0.4314$

(b)

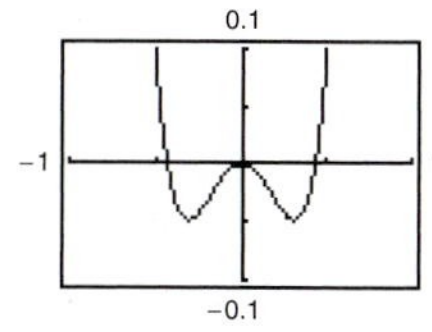

(c) Increasing on $(-0.31, 0), (0.31, \infty)$
Decreasing on $(-\infty, -0.31), (0, 0.31)$

(d) Even

13. (a) $0, 3$

(b)

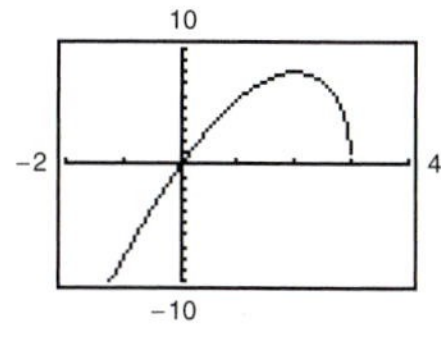

(c) Increasing on $(-\infty, 2)$
Decreasing on $(2, 3)$

(d) Neither even nor odd

14. (a) -5

(b)

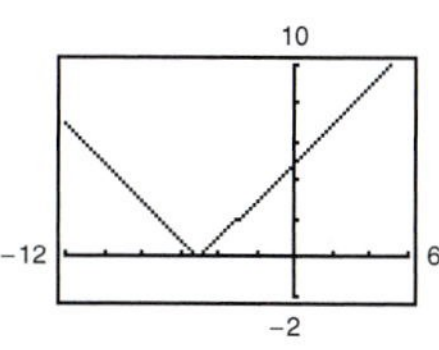

(c) Increasing on $(-5, \infty)$
Decreasing on $(-\infty, -5)$

(d) Neither even nor odd

15.

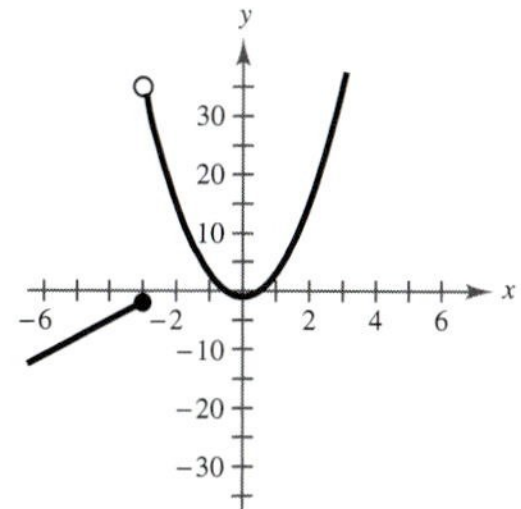

16. Reflection in the x-axis of $y = [\![x]\!]$

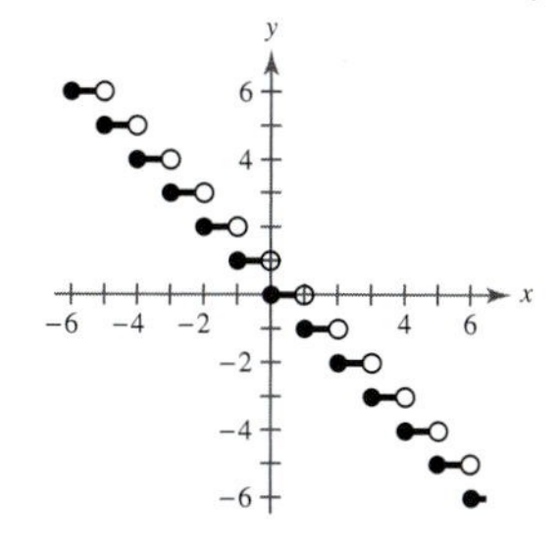

17. Reflection in the x-axis, horizontal shift, and vertical shift of $y = \sqrt{x}$

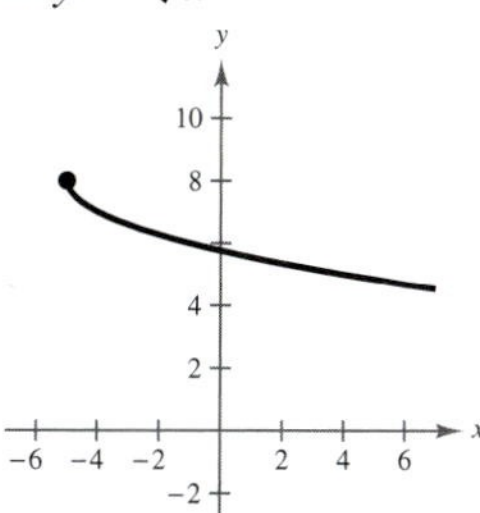

18. (a) $2x^2 - 4x - 2$ (b) $4x^2 + 4x - 12$

(c) $-3x^4 - 12x^3 + 22x^2 + 28x - 35$

(d) $\dfrac{3x^2 - 7}{-x^2 - 4x + 5}, \quad x \neq -5, 1$

(e) $3x^4 + 24x^3 + 18x^2 - 120x + 68$

(f) $-9x^4 + 30x^2 - 16$

19. (a) $\dfrac{1 + 2x^{3/2}}{x}, \quad x > 0$ (b) $\dfrac{1 - 2x^{3/2}}{x}, \quad x > 0$

(c) $\dfrac{2\sqrt{x}}{x}, \quad x > 0$ (d) $\dfrac{1}{2x^{3/2}}, \quad x > 0$

(e) $\dfrac{\sqrt{x}}{2x}, \quad x > 0$ (f) $\dfrac{2\sqrt{x}}{x}, \quad x > 0$

20. $f^{-1}(x) = \sqrt[3]{x - 8}$ **21.** No inverse

22. $f^{-1}(x) = \left(\frac{1}{3}x\right)^{2/3}, \ x \geq 0$ **23.** $v = 6\sqrt{s}$

24. $A = \dfrac{25}{6}xy$ **25.** $b = \dfrac{48}{a}$

Problem Solving *(page 125)*

1. (a) $W_1 = 2000 + 0.07S$ (b) $W_2 = 2300 + 0.05S$

(c)

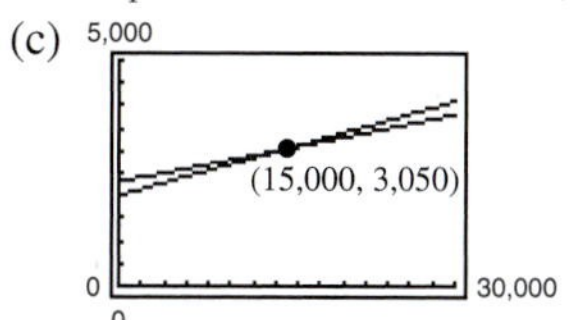

Both jobs pay the same monthly salary if sales equal \$15,000.

(d) No. Job 1 would pay \$3400 and job 2 would pay \$3300.

2. Mapping numbers onto letters is not a function since each number corresponds to three letters.
Mapping letters onto numbers is a function since every letter is assigned exactly one number.

3. (a) The function will be even.

(b) The function will be odd.

(c) The function will be neither even nor odd.

4. $f(x) = x$ $\quad$ $g(x) = -x$

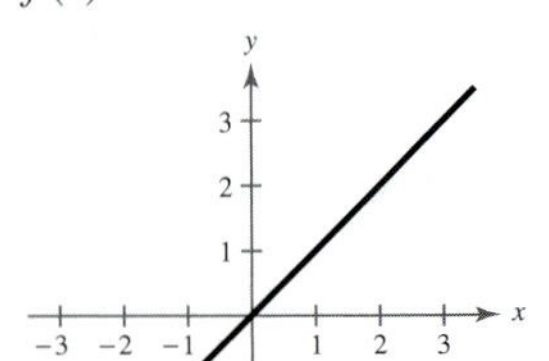

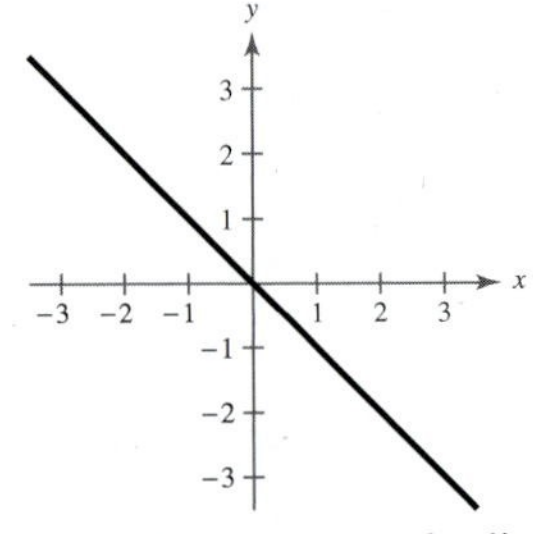

Both graphs are already symmetric with respect to the line $y = x$.

General formula: $y = -x + c,\ c \geq 0$

5. $f(x) = a_{2n}x^{2n} + a_{2n-2}x^{2n-2} + \cdots + a_2x^2 + a_0$

$f(-x) = a_{2n}(-x)^{2n} + a_{2n-2}(-x)^{2n-2} + \cdots + a_2(-x)^2 + a_0$

$= f(x)$

6. (6, 8)

$$f(x) = \begin{cases} \frac{12}{7}x - \frac{16}{7}, & 2.5 \leq x \leq 6 \\ -\frac{12}{7}x + \frac{128}{7}, & 6 < x \leq 9.5 \end{cases}$$

7. (a) $81\frac{2}{3}$ hours (b) $25\frac{5}{7}$ miles per hour

(c) $y = \dfrac{-180}{7}x + 3400$

Domain: $0 \leq x \leq \dfrac{1190}{9}$

Range: $0 \leq y \leq 3400$

(d)

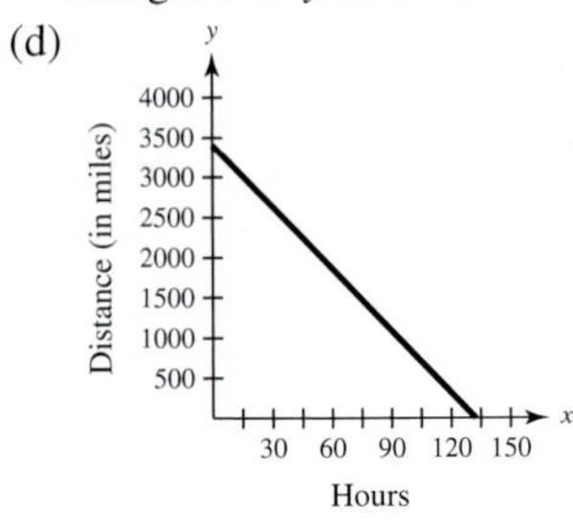

8. (a) 1 (b) 1.5 (c) 1.75 (d) 1.875 (e) 1.9375

(f) Yes, 2.

(g) a. $y = x - 1$

b. $y = 1.5x - 1.5$

c. $y = 1.75x - 1.75$

d. $y = 1.875x - 1.875$

e. $y = 1.9375x - 1.9375$

(h) $y = 2x - 2$

9. (a) $(f \circ g)(x) = 4x + 24$ (b) $(f \circ g)^{-1}(x) = \frac{1}{4}x - 6$

(c) $f^{-1}(x) = \frac{1}{4}x;\ g^{-1}(x) = x - 6$

(d) $(g^{-1} \circ f^{-1})(x) = \frac{1}{4}x - 6$

(e) $(f \circ g)(x) = 8x^3 + 1;\ (f \circ g)^{-1}(x) = \frac{1}{2}\sqrt[3]{x - 1};$

$f^{-1}(x) = \sqrt[3]{x - 1};\ g^{-1}(x) = \frac{1}{2}x;$

$(g^{-1} \circ f^{-1})(x) = \frac{1}{2}\sqrt[3]{x - 1}$

(f) Answers will vary.

(g) $(f \circ g)^{-1}(x) = (g^{-1} \circ f^{-1})(x)$

10. (a) $T = \frac{1}{2}\sqrt{4 + x^2} + \frac{1}{4}\sqrt{x^2 - 6x + 10}$

(b) $0 \leq x \leq 3$

(c)

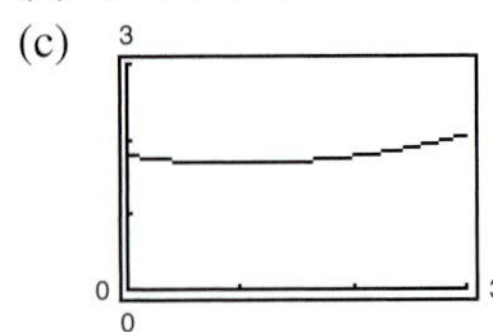

(d) $x = 1$

(e) The distance $x = 1$ yields a time of 1.68 hours.

11. (a) (b)

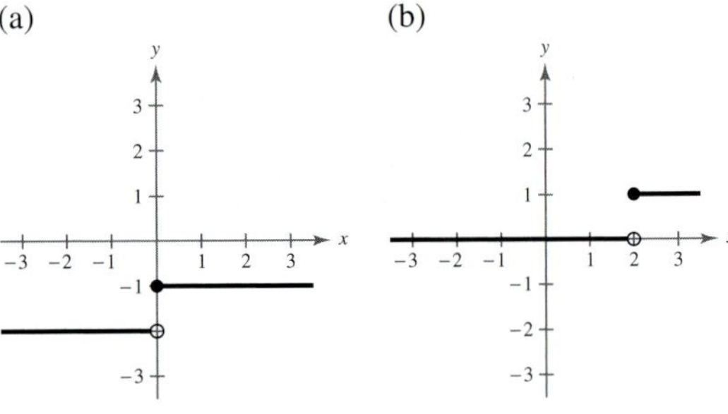

(c) (d)

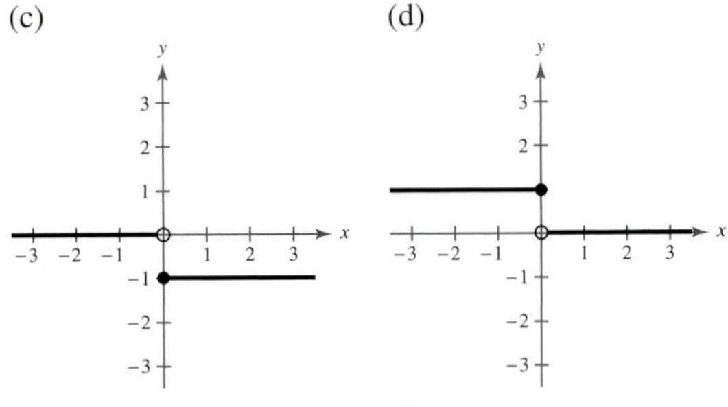

(e) (f)

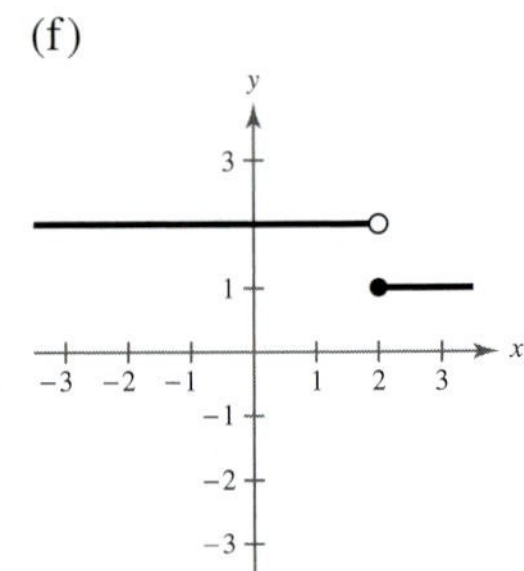

12. (a) Domain: all real numbers x except $x = 1$

Range: all real numbers y except $y = 0$

(b) $f(f(x)) = \dfrac{x - 1}{x}$

Domain: all real numbers x except $x = 0, 1$

(c) $f(f(f(x))) = x$

The graph is not a line because there are holes at $x = 0$ and $x = 1$.

13. Proof

14. (a)

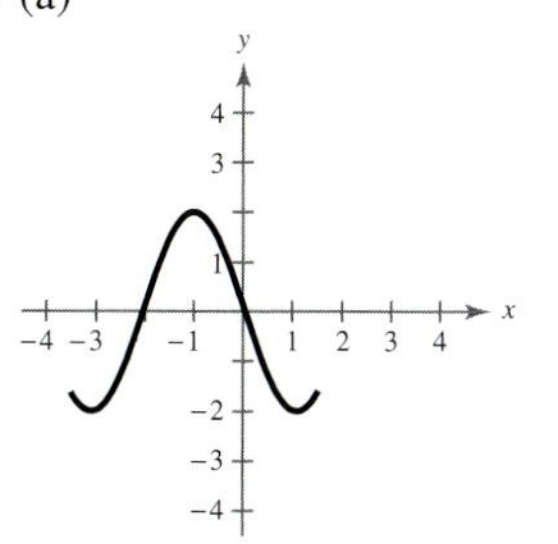

(b)

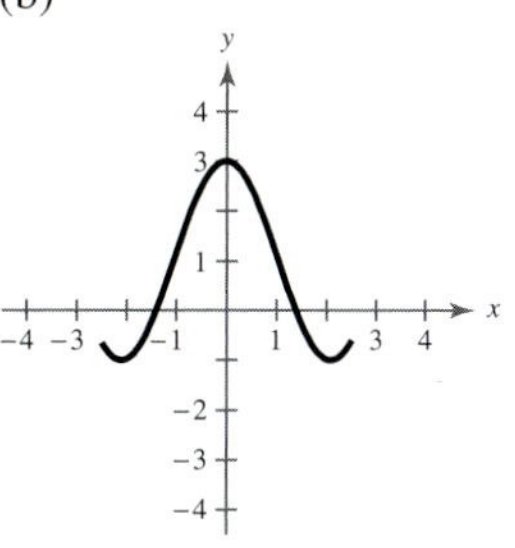

(c)

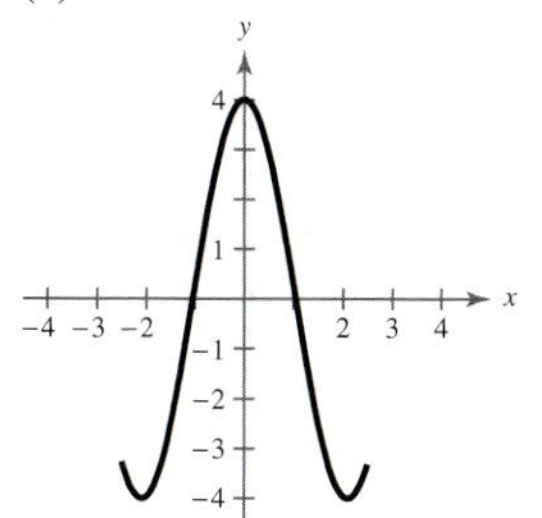

(d)

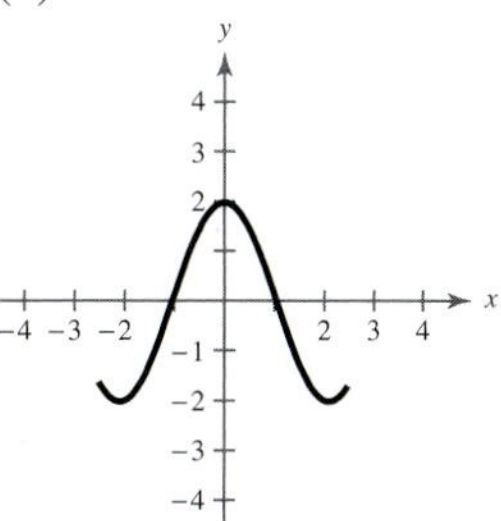

(e)

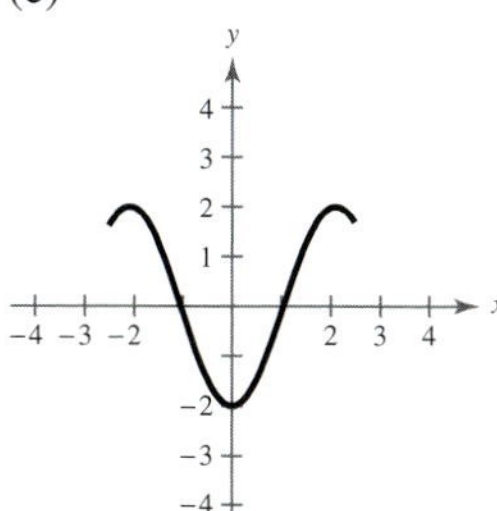

(f)

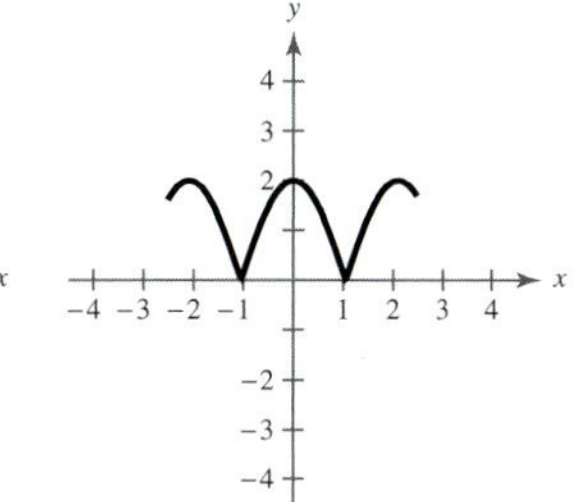

(g)

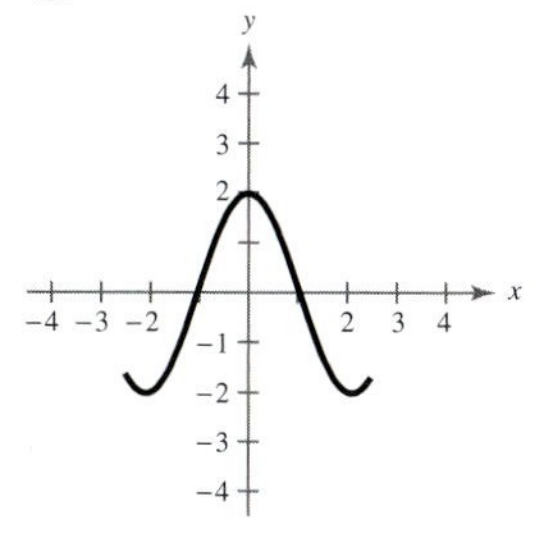

15. (a)

x	-4	-2	0	4
$f(f^{-1}(x))$	-4	-2	0	4

(b)

x	-3	-2	0	1
$(f + f^{-1})(x)$	5	1	-3	-5

(c)

x	-3	-2	0	1
$(f \circ f^{-1})(x)$	4	0	2	6

(d)

x	-4	-3	0	4
$\lvert f^{-1}(x)\rvert$	2	1	1	3

Chapter 2

Section 2.1 *(page 134)*

Vocabulary Check *(page 134)*

1. nonnegative integer; real **2.** quadratic; parabola
3. axis **4.** positive; minimum
5. negative; maximum

1. g **2.** c **3.** b **4.** h
5. f **6.** a **7.** e **8.** d

9. (a)

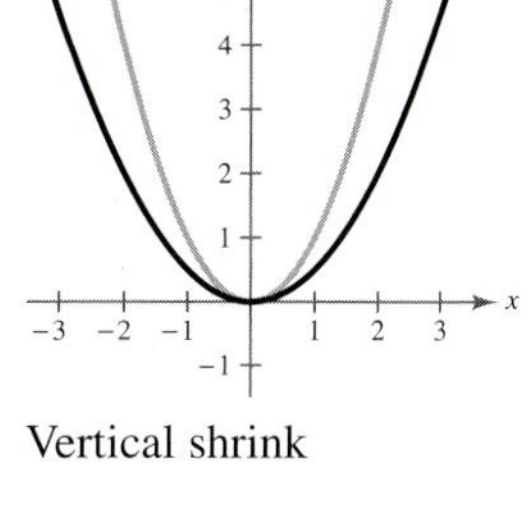

Vertical shrink

(b)

Vertical shrink and reflection in the x-axis

(c)

Vertical stretch

(d)

Vertical stretch and reflection in the x-axis

10. (a)

Vertical shift

(b)

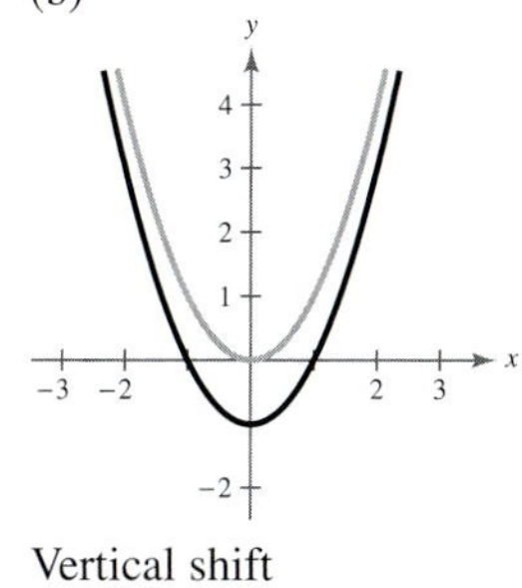

Vertical shift

(c)

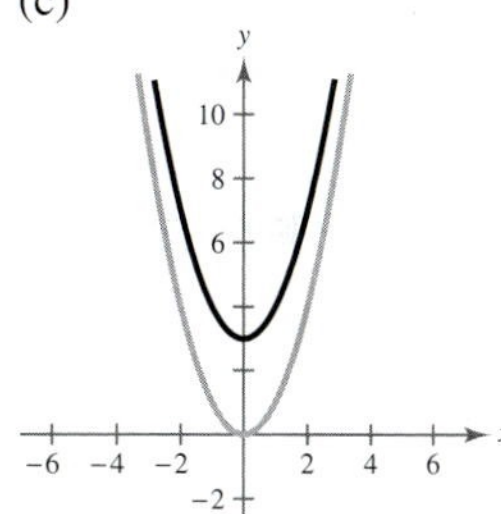

Vertical shift

(d)

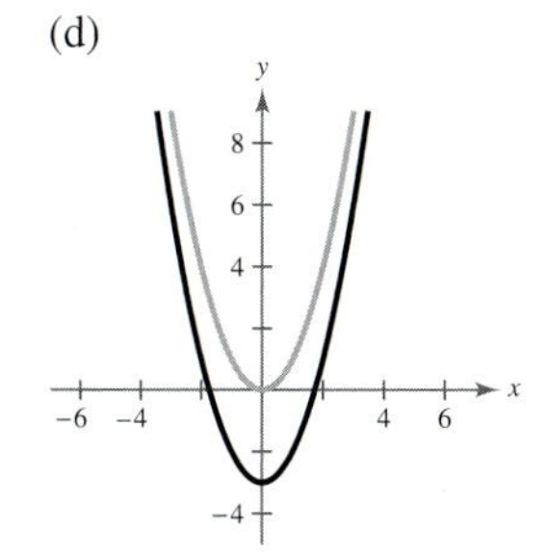

Vertical shift

11. (a)

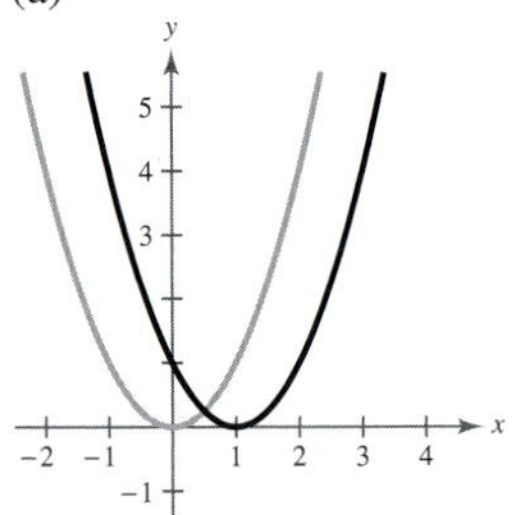

Horizontal shift

(b)

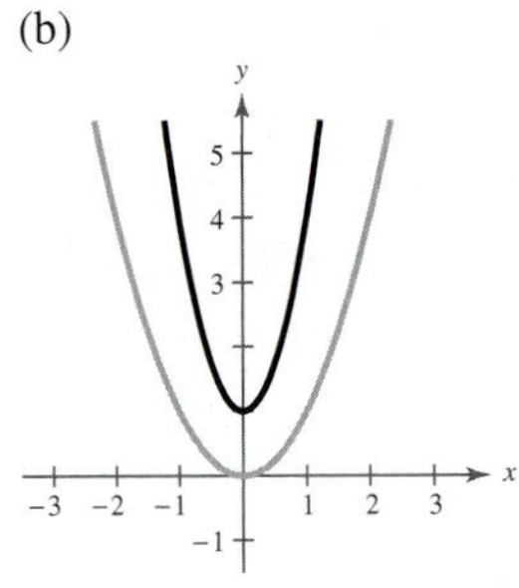

Horizontal shrink and vertical shift

(c)

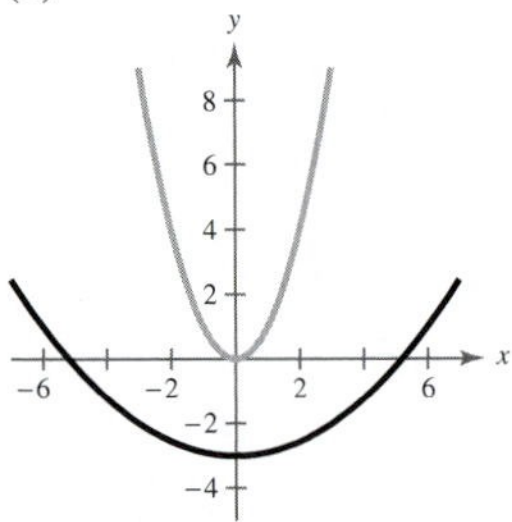

Horizontal stretch and vertical shift

(d)

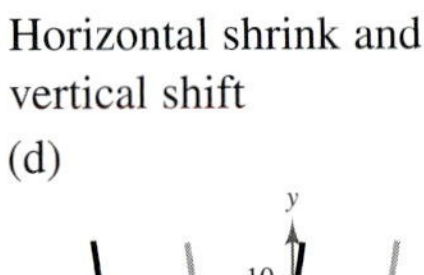

Horizontal shift

12. (a)

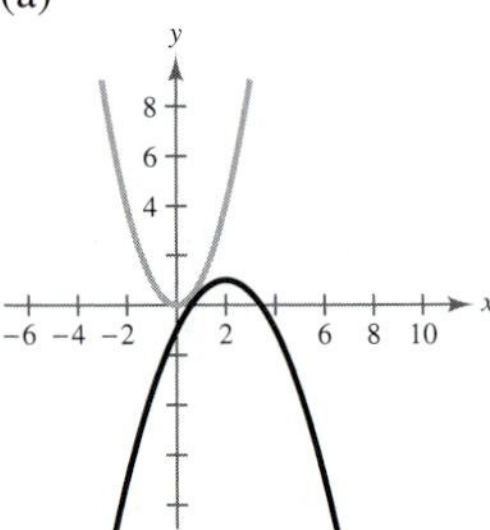

Horizontal shift, vertical shrink, reflection in the x-axis, and vertical shift

(b)

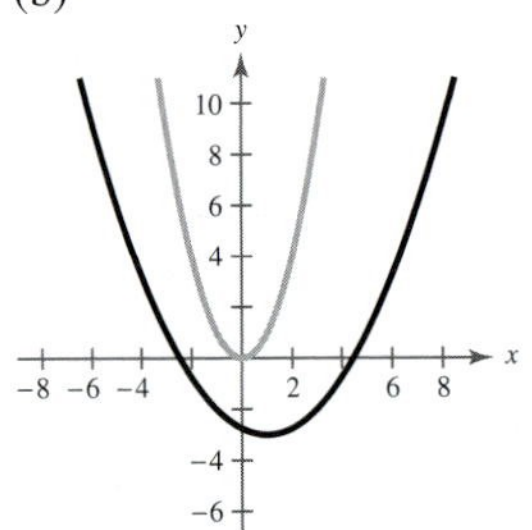

Horizontal shift, vertical shrink, and vertical shift

(c)

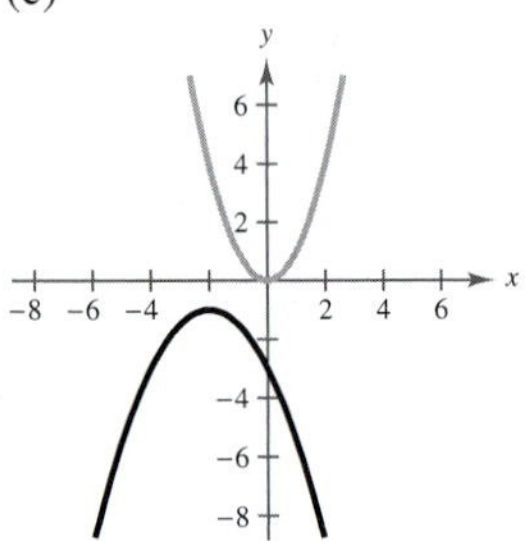

Reflection in the x-axis, vertical shrink, horizontal shift, and vertical shift

(d)

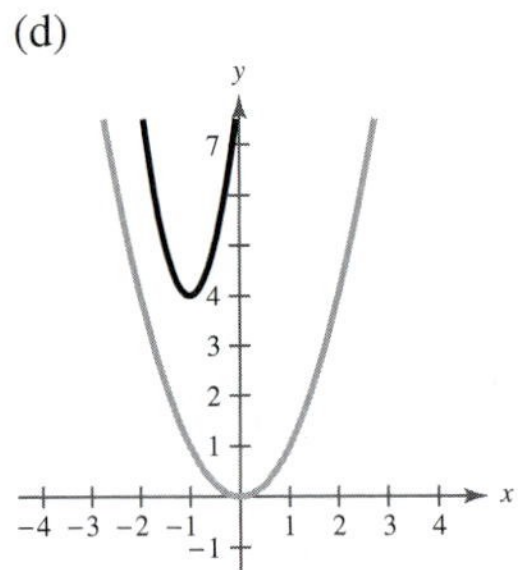

Horizontal shift, vertical stretch, and vertical shift

13.

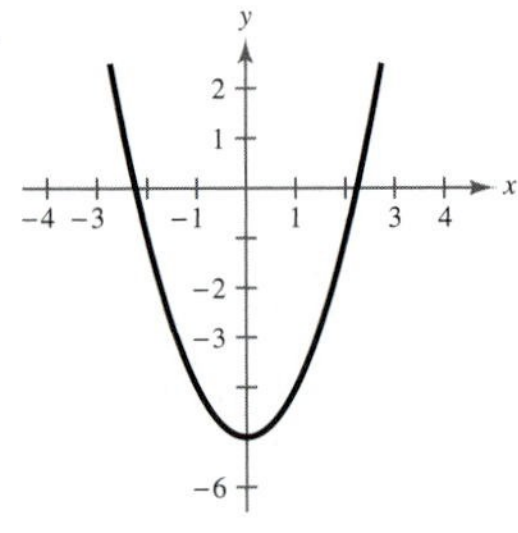

Vertex: $(0, -5)$
Axis of symmetry: y-axis
x-intercepts: $(\pm\sqrt{5}, 0)$

14.

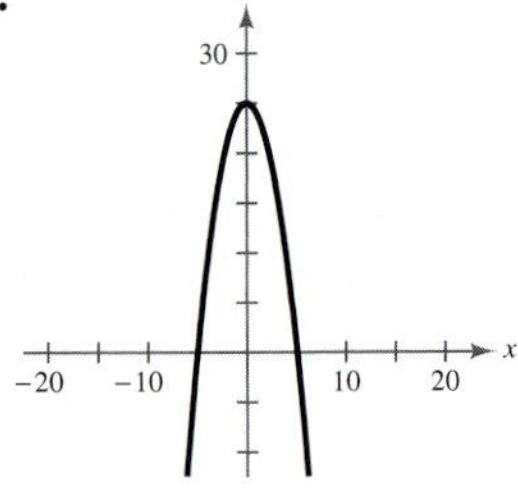

Vertex: $(0, 25)$
Axis of symmetry: $x = 0$
x-intercepts: $(\pm 5, 0)$

15.

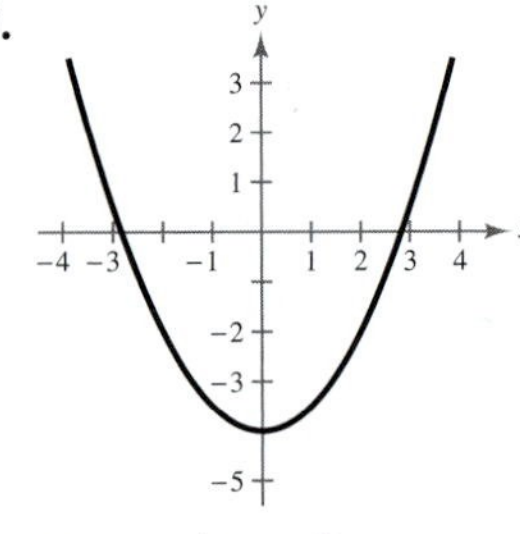

Vertex: $(0, -4)$
Axis of symmetry: y-axis
x-intercepts: $(\pm 2\sqrt{2}, 0)$

16.

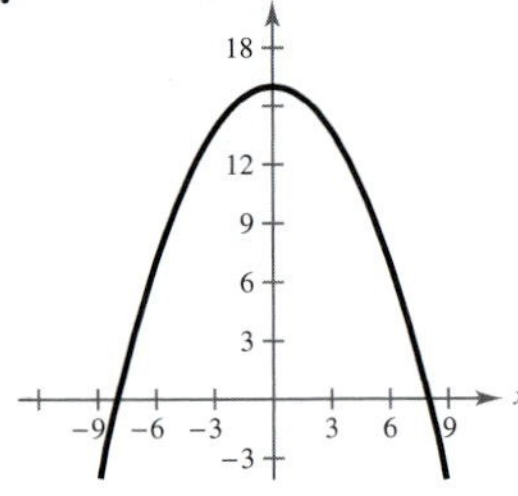

Vertex: $(0, 16)$
Axis of symmetry: $x = 0$
x-intercepts: $(\pm 8, 0)$

17.

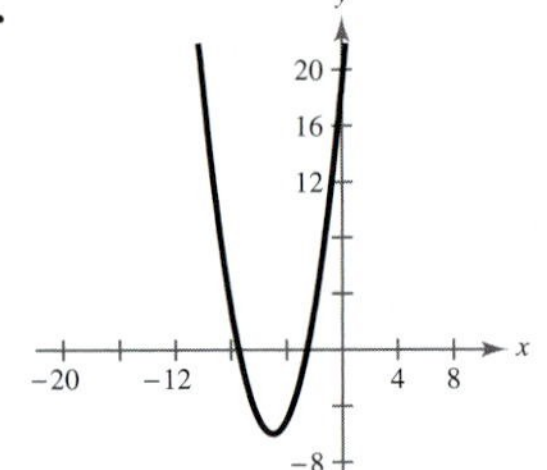

Vertex: $(-5, -6)$
Axis of symmetry: $x = -5$
x-intercepts: $(-5 \pm \sqrt{6}, 0)$

18.

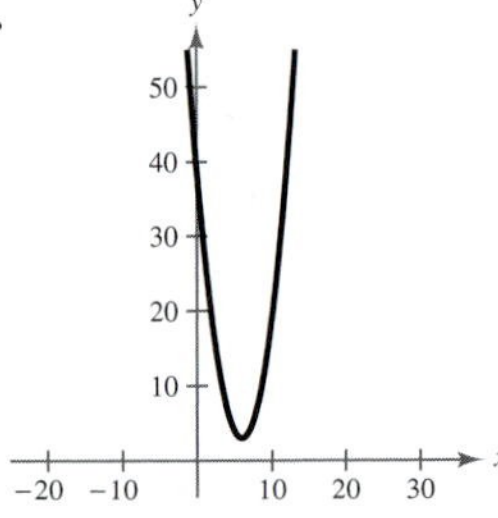

Vertex: $(6, 3)$
Axis of symmetry: $x = 6$
No x-intercept

19.

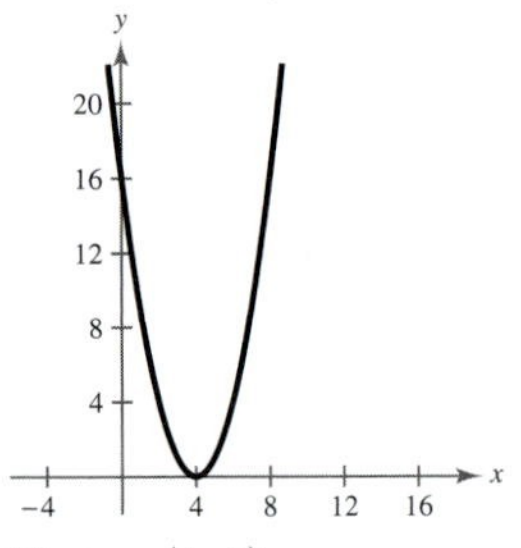

Vertex: $(4, 0)$
Axis of symmetry: $x = 4$
x-intercept: $(4, 0)$

20.

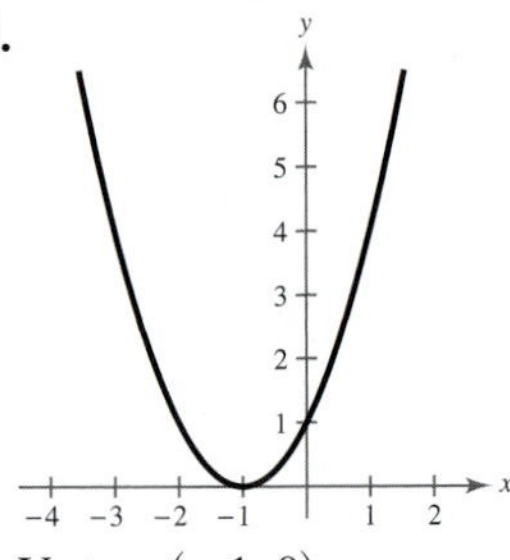

Vertex: $(-1, 0)$
Axis of symmetry: $x = -1$
x-intercept: $(-1, 0)$

21.
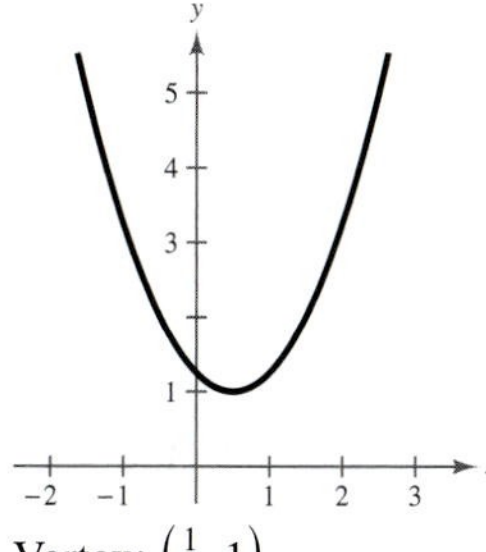

Vertex: $\left(\frac{1}{2}, 1\right)$
Axis of symmetry: $x = \frac{1}{2}$
No x-intercept

22.
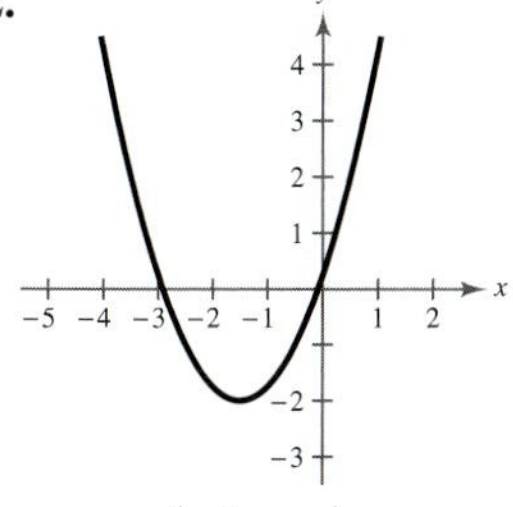

Vertex: $\left(-\frac{3}{2}, -2\right)$
Axis of symmetry: $x = -\frac{3}{2}$
x-intercepts: $\left(-\frac{3}{2} \pm \sqrt{2}, 0\right)$

23.
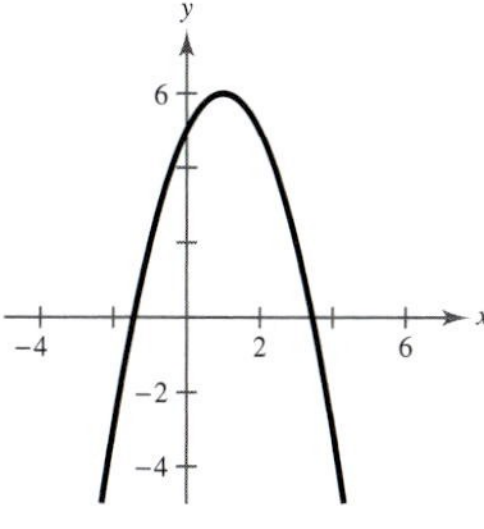

Vertex: $(1, 6)$
Axis of symmetry: $x = 1$
x-intercepts: $\left(1 \pm \sqrt{6}, 0\right)$

24.
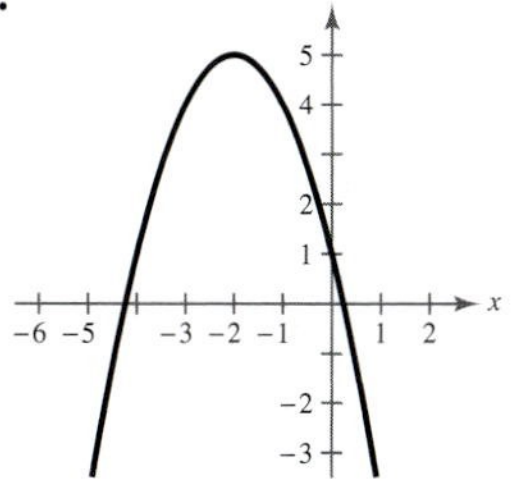

Vertex: $(-2, 5)$
Axis of symmetry: $x = -2$
x-intercepts: $\left(-2 \pm \sqrt{5}, 0\right)$

25.
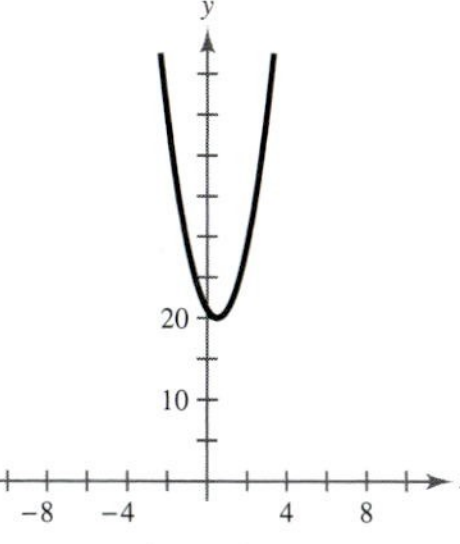

Vertex: $\left(\frac{1}{2}, 20\right)$
Axis of symmetry: $x = \frac{1}{2}$
No x-intercept

26.
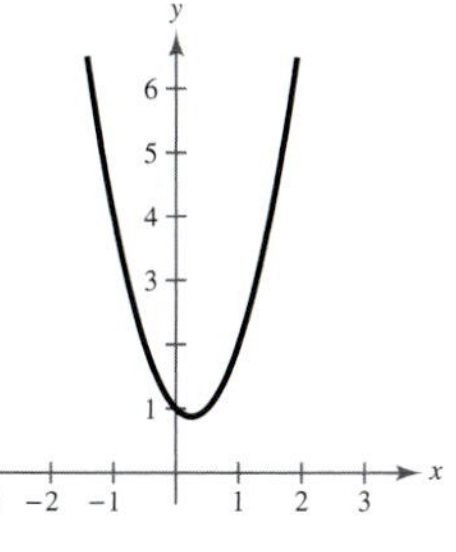

Vertex: $\left(\frac{1}{4}, \frac{7}{8}\right)$
Axis of symmetry: $x = \frac{1}{4}$
No x-intercept

27.
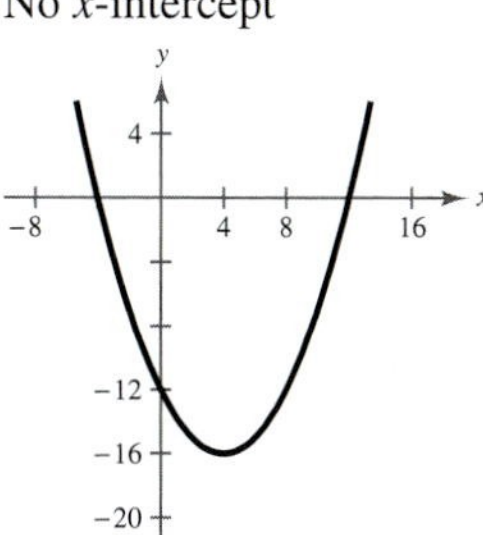

Vertex: $(4, -16)$
Axis of symmetry: $x = 4$
x-intercepts: $(-4, 0), (12, 0)$

28.
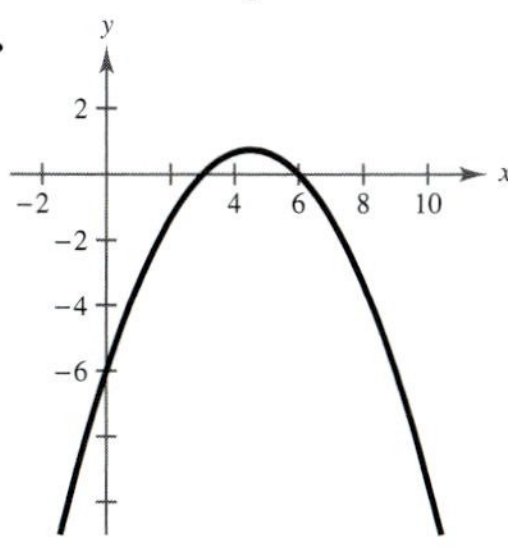

Vertex: $\left(\frac{9}{2}, \frac{3}{4}\right)$
Axis of symmetry: $x = \frac{9}{2}$
x-intercepts: $(3, 0), (6, 0)$

29.
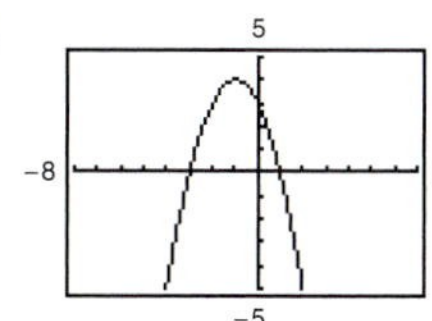

Vertex: $(-1, 4)$
Axis of symmetry: $x = -1$
x-intercepts: $(1, 0), (-3, 0)$

30.
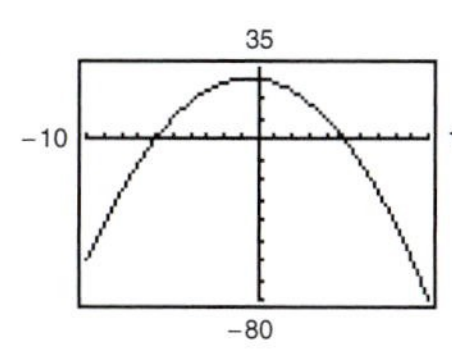

Vertex: $\left(-\frac{1}{2}, \frac{121}{4}\right)$
Axis of symmetry: $x = -\frac{1}{2}$
x-intercepts: $(-6, 0), (5, 0)$

31.
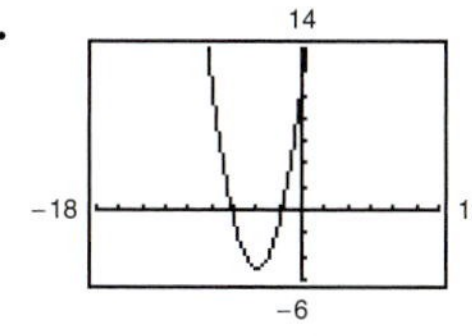

Vertex: $(-4, -5)$
Axis of symmetry: $x = -4$
x-intercepts: $\left(-4 \pm \sqrt{5}, 0\right)$

32.
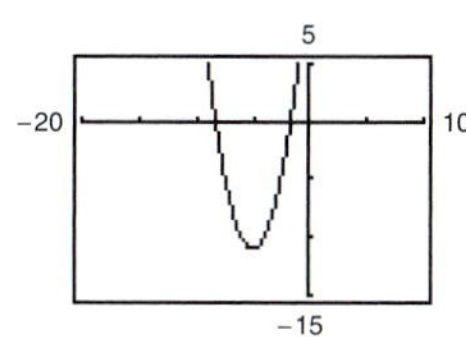

Vertex: $(-5, -11)$
Axis of symmetry: $x = -5$
x-intercepts: $\left(-5 \pm \sqrt{11}, 0\right)$

33.
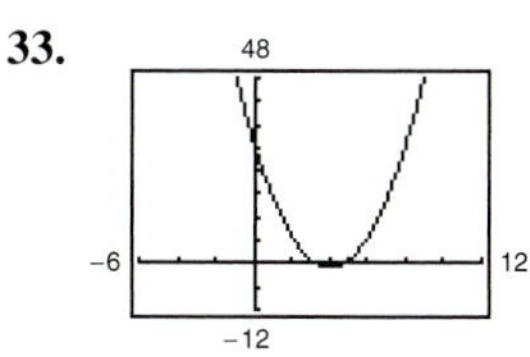

Vertex: $(4, -1)$
Axis of symmetry: $x = 4$
x-intercepts: $\left(4 \pm \frac{1}{2}\sqrt{2}, 0\right)$

34.
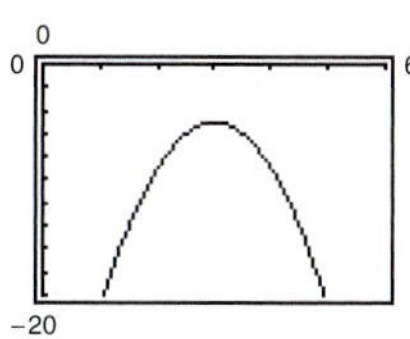

Vertex: $(3, -5)$
Axis of symmetry: $x = 3$
No x-intercept

35.
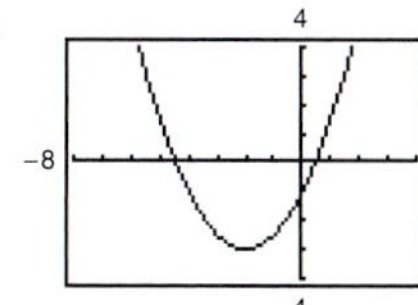

Vertex: $(-2, -3)$
Axis of symmetry: $x = -2$
x-intercepts: $\left(-2 \pm \sqrt{6}, 0\right)$

36.
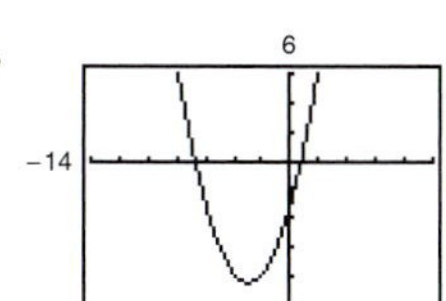

Vertex: $\left(-3, -\frac{42}{5}\right)$
Axis of symmetry: $x = -3$
x-intercepts: $\left(-3 \pm \sqrt{14}, 0\right)$

37. $y = (x - 1)^2$ **38.** $y = -x^2 + 1$
39. $y = -(x + 1)^2 + 4$ **40.** $y = (x + 2)^2 - 1$
41. $y = -2(x + 2)^2 + 2$ **42.** $y = 2(x - 2)^2$
43. $f(x) = (x + 2)^2 + 5$ **44.** $f(x) = (x - 4)^2 - 1$

45. $f(x) = -\frac{1}{2}(x - 3)^2 + 4$ **46.** $f(x) = -\frac{1}{4}(x - 2)^2 + 3$

47. $f(x) = \frac{3}{4}(x - 5)^2 + 12$ **48.** $f(x) = 2(x + 2)^2 - 2$

49. $f(x) = -\frac{24}{49}\left(x + \frac{1}{4}\right)^2 + \frac{3}{2}$ **50.** $f(x) = \frac{19}{81}\left(x - \frac{5}{2}\right)^2 - \frac{3}{4}$

51. $f(x) = -\frac{16}{3}\left(x + \frac{5}{2}\right)^2$ **52.** $f(x) = -450(x - 6)^2 + 6$

53. $(\pm 4, 0)$ **54.** $(3, 0)$ **55.** $(5, 0), (-1, 0)$

56. $(-3, 0), \left(\frac{1}{2}, 0\right)$

57.

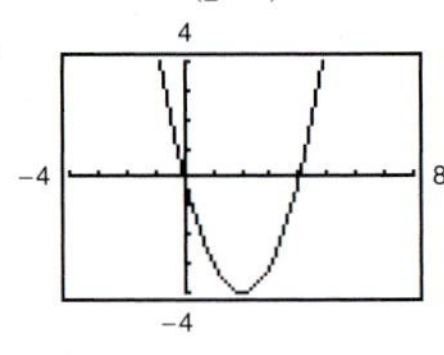

$(0, 0), (4, 0)$

58.

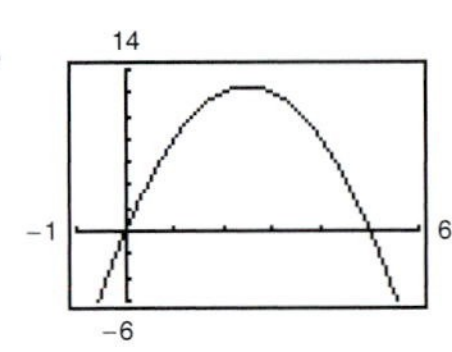

$(0, 0), (5, 0)$

59.

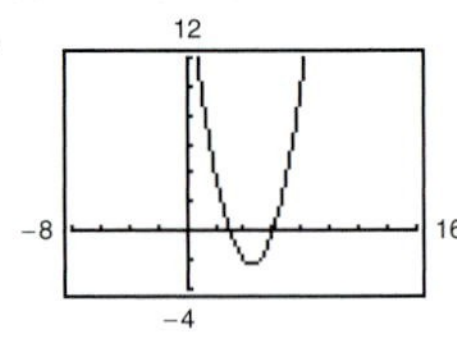

$(3, 0), (6, 0)$

60.

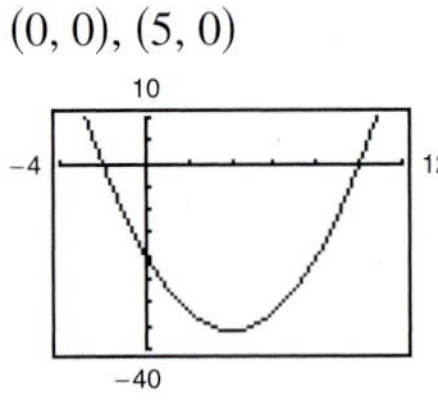

$(-2, 0), (10, 0)$

61.

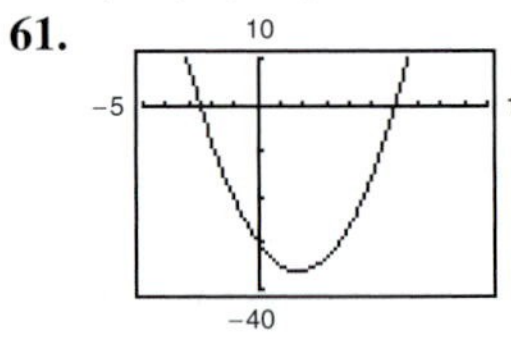

$\left(-\frac{5}{2}, 0\right), (6, 0)$

62.

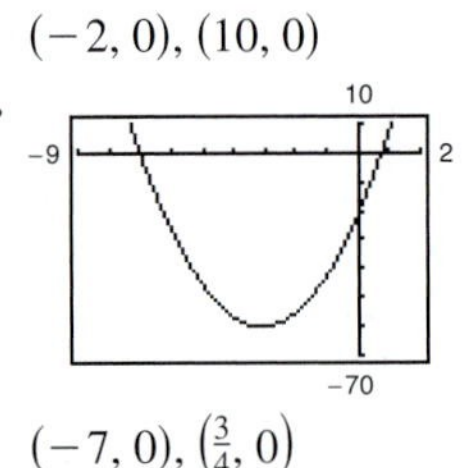

$(-7, 0), \left(\frac{3}{4}, 0\right)$

63.

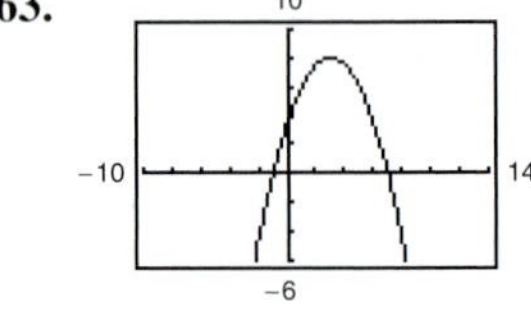

$(7, 0), (-1, 0)$

64.

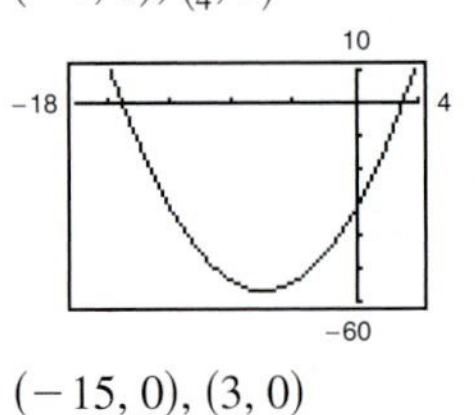

$(-15, 0), (3, 0)$

65. $f(x) = x^2 - 2x - 3$
$g(x) = -x^2 + 2x + 3$

66. $f(x) = x^2 - 25$
$g(x) = -x^2 + 25$

67. $f(x) = x^2 - 10x$
$g(x) = -x^2 + 10x$

68. $f(x) = x^2 - 12x + 32$
$g(x) = -x^2 + 12x - 32$

69. $f(x) = 2x^2 + 7x + 3$
$g(x) = -2x^2 - 7x - 3$

70. $f(x) = 2x^2 + x - 10$
$g(x) = -2x^2 - x + 10$

71. 55, 55 **72.** $\frac{S}{2}, \frac{S}{2}$ **73.** 12, 6 **74.** 21, 7

75. (a) $A = \dfrac{8x(50 - x)}{3}$

(b)

x	5	10	15	20	25	30
A	600	1067	1400	1600	1667	1600

$x = 25$ feet, $y = 33\frac{1}{3}$ feet

(c)

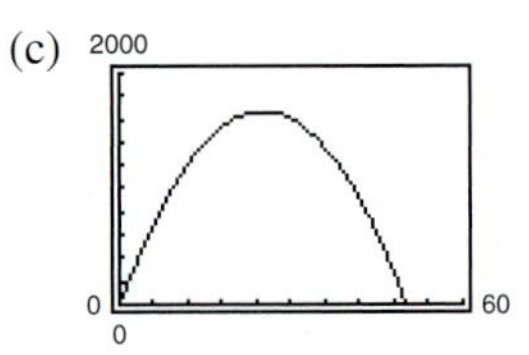

$x = 25$ feet, $y = 33\frac{1}{3}$ feet

(d) $A = -\frac{8}{3}(x - 25)^2 + \frac{5000}{3}$ (e) They are identical.

76. (a) $r = \dfrac{1}{2}y$; $d = y\pi$ (b) $y = \dfrac{200 - 2x}{\pi}$

(c) $A = x\left(\dfrac{200 - 2x}{\pi}\right)$; $x = 50$ meters, $y = \dfrac{100}{\pi}$ meters

77. 16 feet

78. (a) 1.5 feet (b) $\frac{6657}{64}$ feet $\approx$ 104.02 feet
(c) $\approx$228.64 feet

79. 20 fixtures **80.** 1222 units **81.** 350,000 units

82. \$2000

83. (a) \$14,000,000; \$14,375,000; \$13,500,000
(b) 24; \$14,400

84. (a) \$408; \$468; \$432
(b) \$6.25 per pet; \$468.75
Answers will vary.

85. (a)

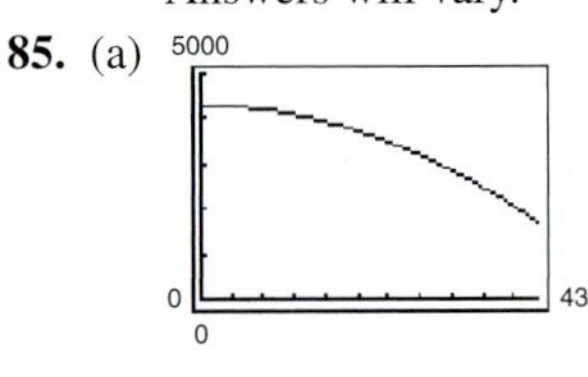

(b) 4299; answers will vary.
(c) 8879; 24

86. (a) and (c)

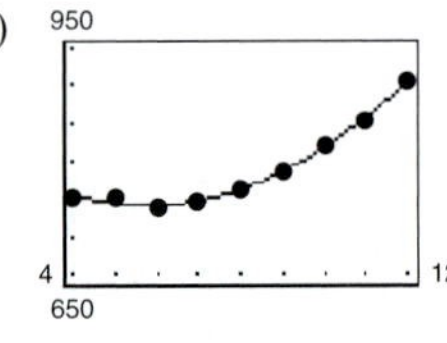

(b) $y = 4.30x^2 - 49.9x + 886$ (d) 1995
(e) Answers will vary. (f) $\approx$1,381,000

87. (a) 25 0 100 -5

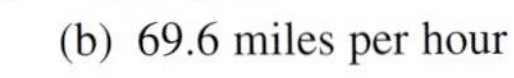
(b) 69.6 miles per hour

88. (a) and (c)

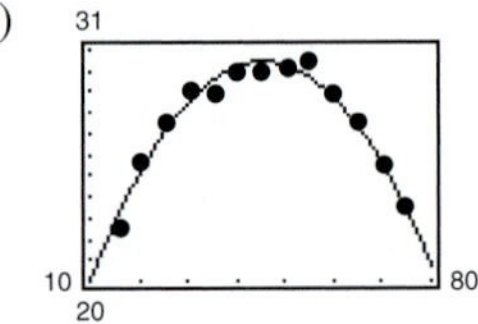

(b) $y = -0.0082x^2 + 0.746x + 13.47$
(d) 45.5 miles per hour

89. True. The equation has no real solutions, so the graph has no x-intercepts.

90. True. The vertex of $f(x)$ is $\left(-\frac{5}{4}, \frac{53}{4}\right)$ and the vertex of $g(x)$ is $\left(-\frac{5}{4}, -\frac{71}{4}\right)$.

91. $f(x) = a\left(x + \dfrac{b}{2a}\right)^2 + \dfrac{4ac - b^2}{4a}$

92. Conditions (a) and (d) are preferable because profits would be increasing.

93. Yes. A graph of a quadratic equation whose vertex is on the x-axis has only one x-intercept.

94. Answers will vary. **95.** $y = -\frac{1}{3}x + \frac{5}{3}$

96. $y = \frac{3}{2}x - \frac{13}{4}$ **97.** $y = \frac{5}{4}x + 3$

98. $y = -3x - 20$ **99.** 27 **100.** 7

101. $-\frac{1408}{49}$ **102.** $-\frac{4}{3}$ **103.** 109 **104.** 72

105. Answers will vary.

Section 2.2 *(page 148)*

Vocabulary Check *(page 148)*

1. continuous **2.** Leading Coefficient Test

3. n; $n - 1$ **4.** (a) solution; (b) $(x - a)$; (c) x-intercept

5. touches; crosses **6.** standard

7. Intermediate Value

1. c **2.** g **3.** h **4.** f

5. a **6.** e **7.** d **8.** b

9. (a) (b)

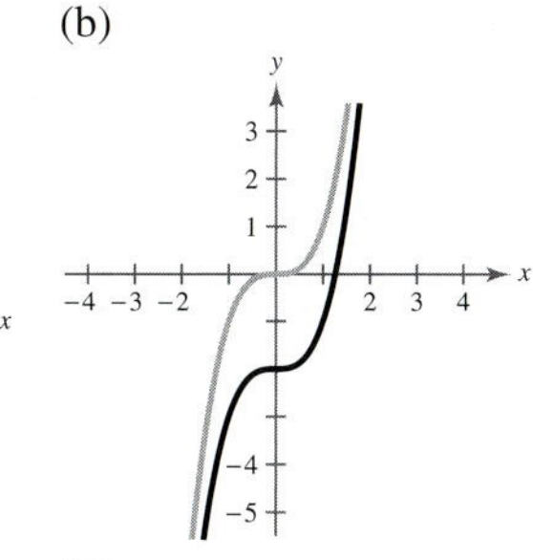

(c)

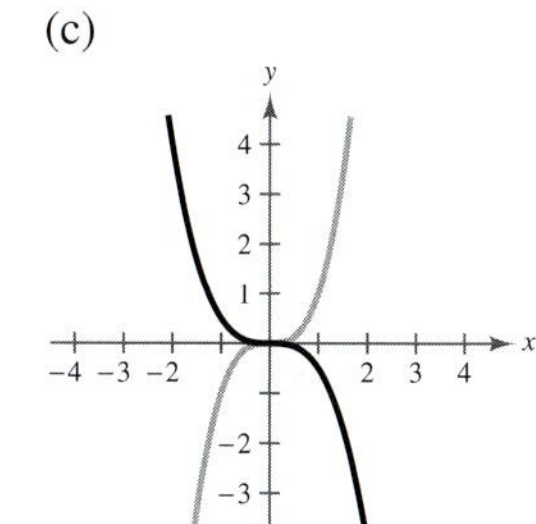

(d)

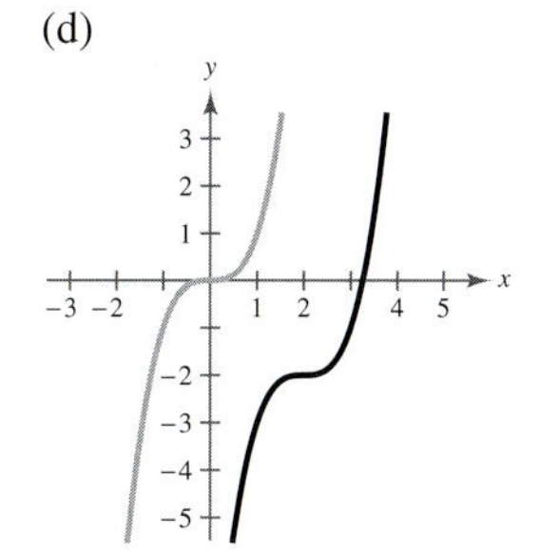

10. (a)

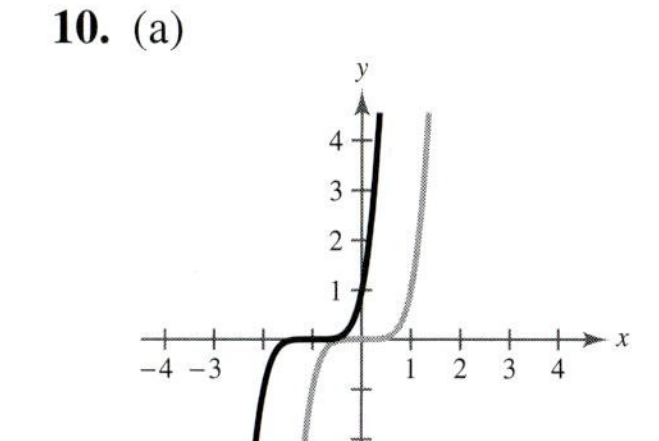

(b)

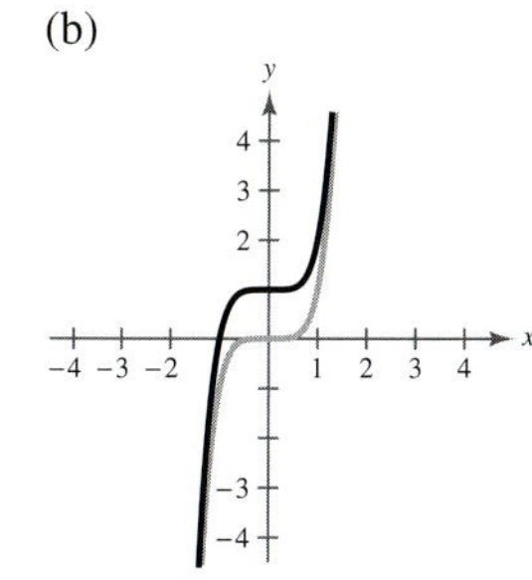

(c)

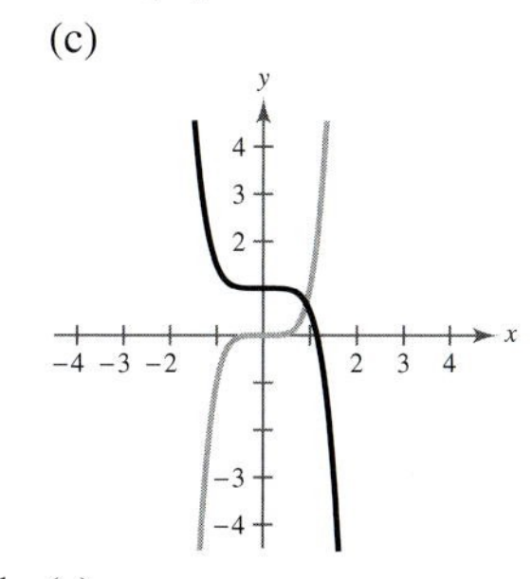

(d)

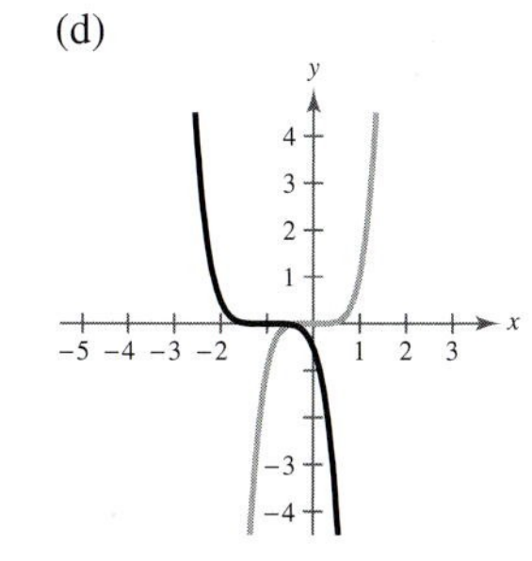

11. (a)

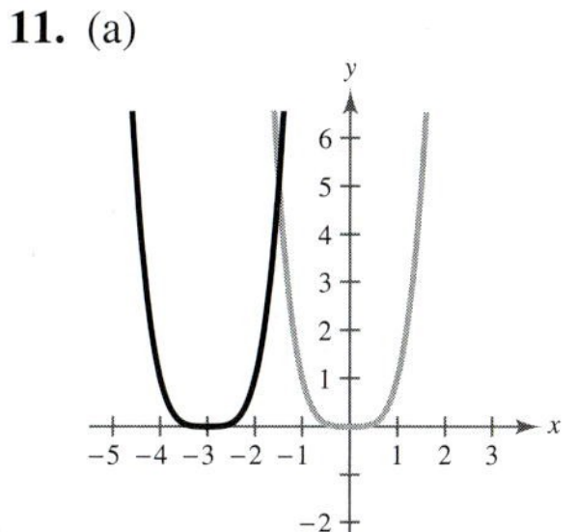

(b)

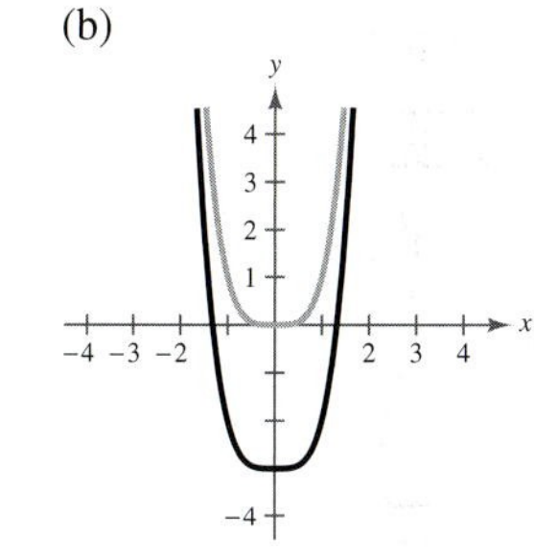

(c)

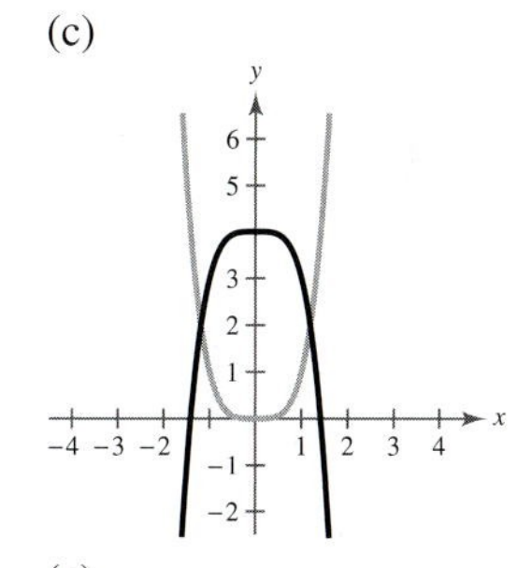

(d)

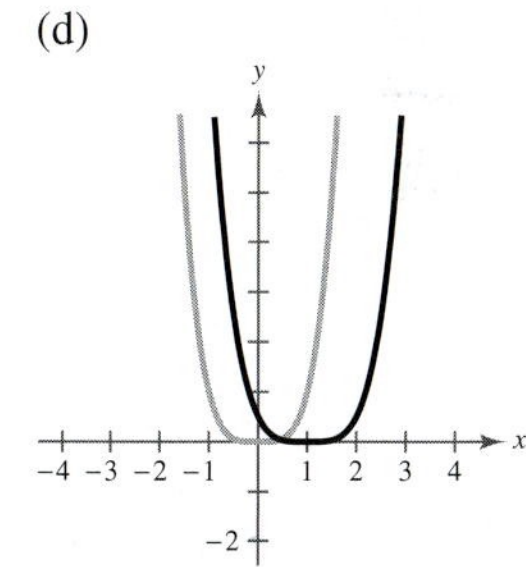

(e)

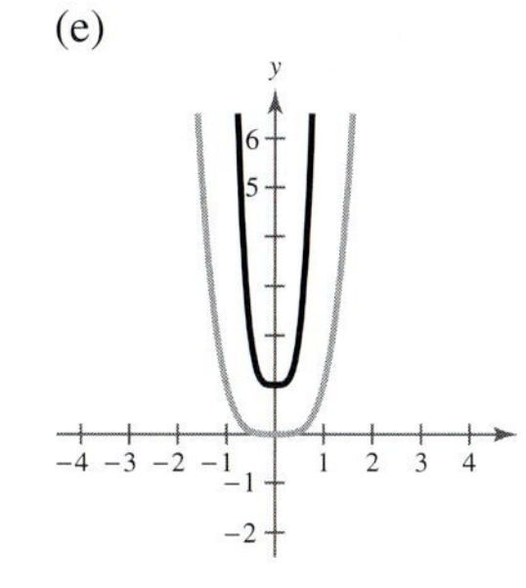

(f)

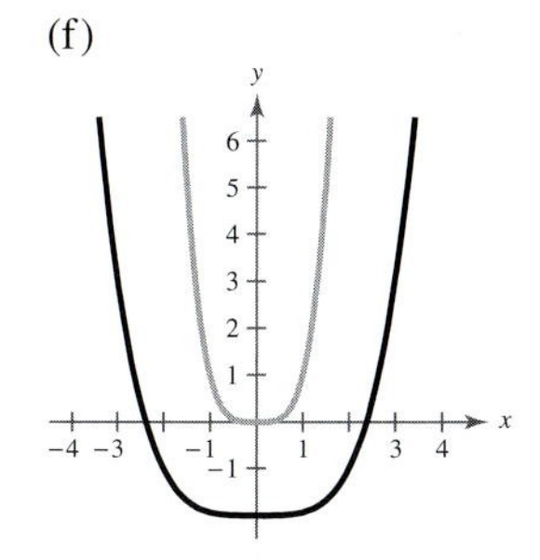

12. (a)

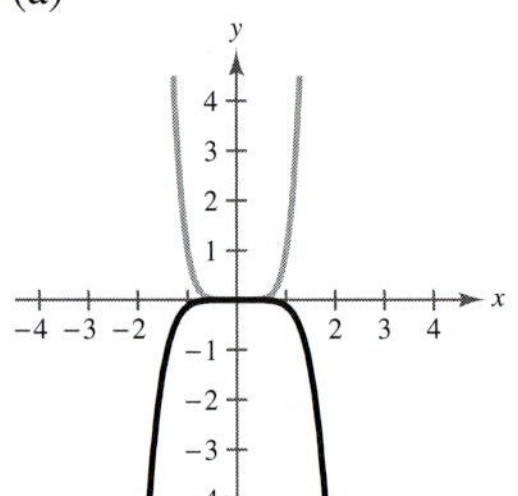

(b)

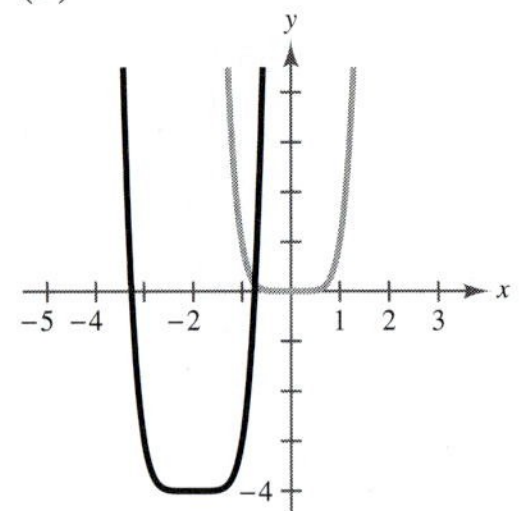

(c)

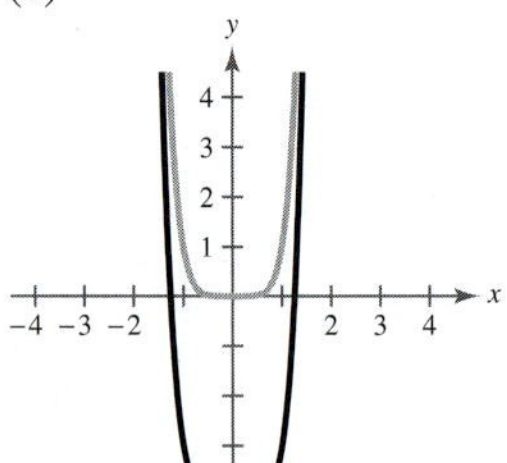

(d)

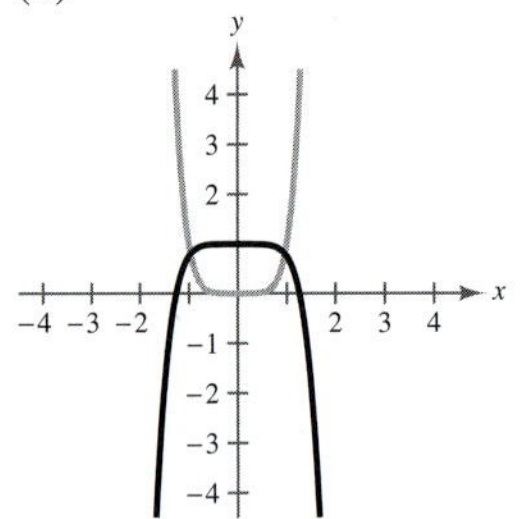

(e)

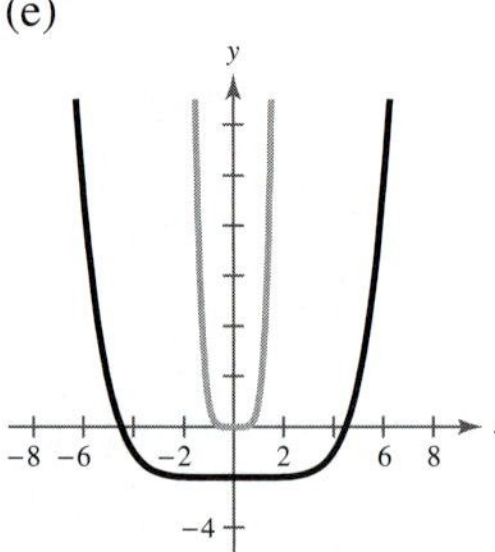

(f)

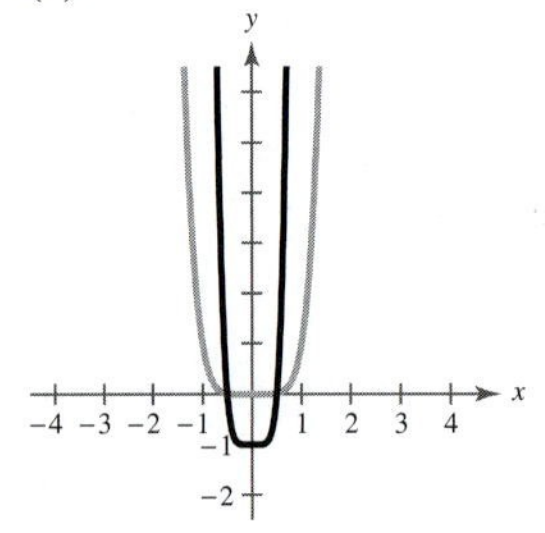

13. Falls to the left, rises to the right
14. Rises to the left, rises to the right
15. Falls to the left, falls to the right
16. Falls to the left, falls to the right
17. Rises to the left, falls to the right
18. Falls to the left, rises to the right
19. Rises to the left, falls to the right
20. Rises to the left, rises to the right
21. Falls to the left, falls to the right
22. Rises to the left, falls to the right

23.

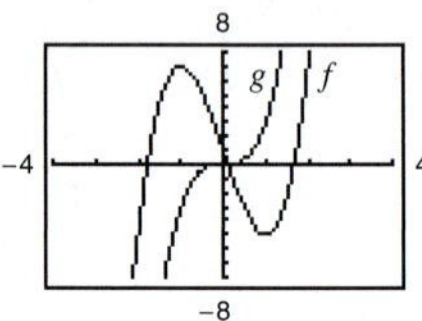

24.

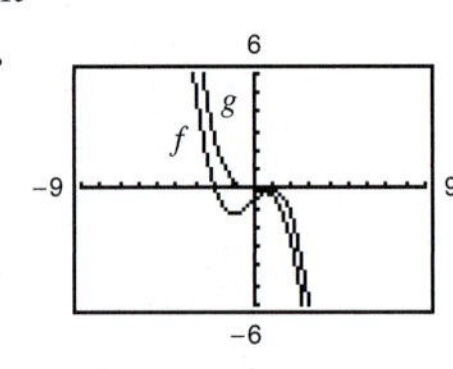

25.

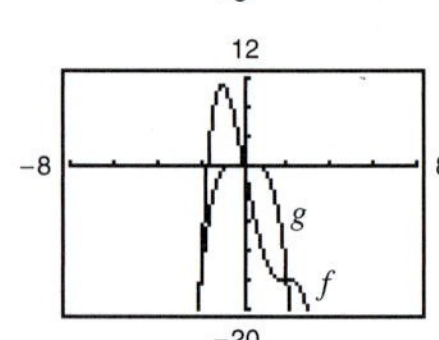

26.

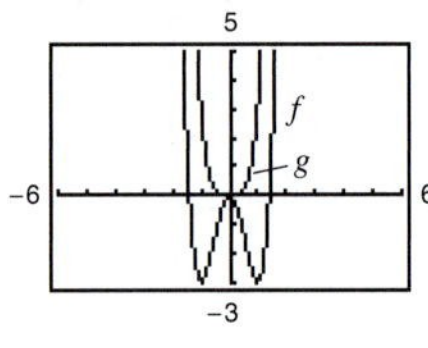

27. (a) ± 5
(b) odd multiplicity; number of turning points: 1
(c)

28. (a) ± 7
(b) odd multiplicity; number of turning points: 1
(c)

29. (a) 3
(b) even multiplicity; number of turning points: 1
(c)

30. (a) -5
(b) even multiplicity; number of turning points: 1
(c)

31. (a) $-2, 1$
(b) odd multiplicity; number of turning points: 1
(c)

32. (a) $\dfrac{-5 \pm \sqrt{37}}{2}$
(b) odd multiplicity; number of turning points: 1
(c)

33. (a) $0, 2 \pm \sqrt{3}$
(b) odd multiplicity; number of turning points: 2
(c)
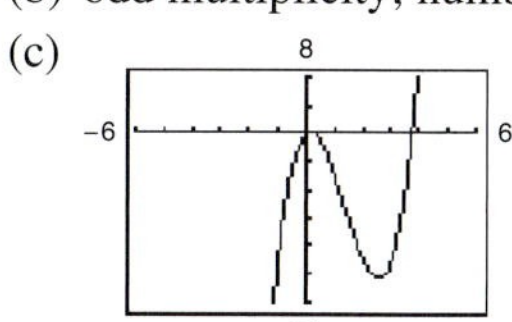

34. (a) $0, 1 \pm \sqrt{2}$
(b) odd multiplicity; number of turning points: 2
(c)
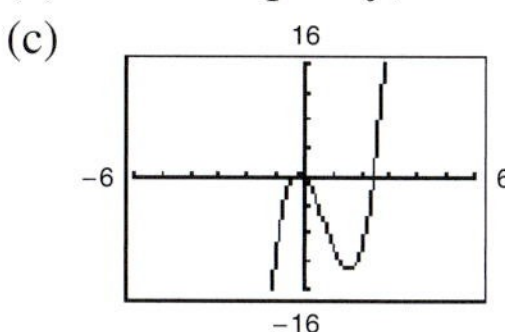

35. (a) 0, 2
(b) 0, odd multiplicity; 2, even multiplicity; number of turning points: 2
(c)
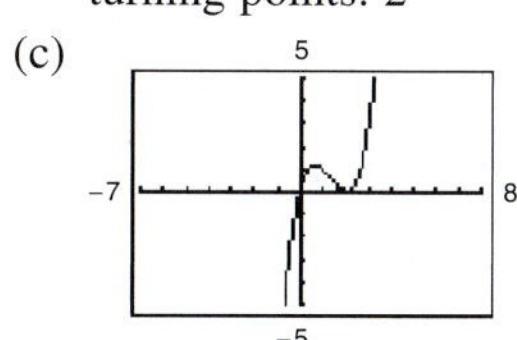

36. (a) $5, -4$
(b) $5, -4, 0$, odd multiplicity; 0, even multiplicity; number of turning points: 3
(c)
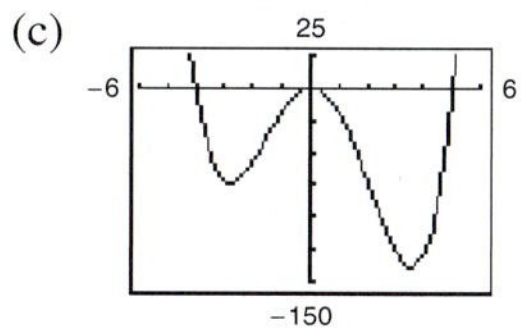

37. (a) $0, \pm\sqrt{3}$
(b) 0, odd multiplicity; $\pm\sqrt{3}$, even multiplicity; number of turning points: 4
(c)
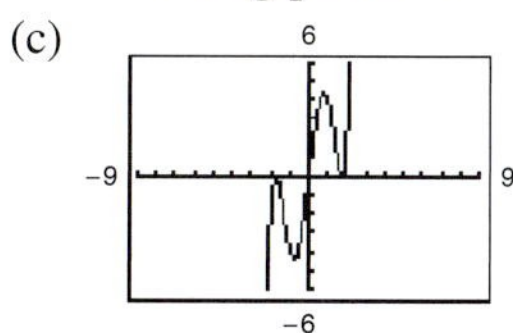

38. (a) $0, \pm\sqrt{2}$
(b) odd multiplicity; number of turning points: 2
(c)
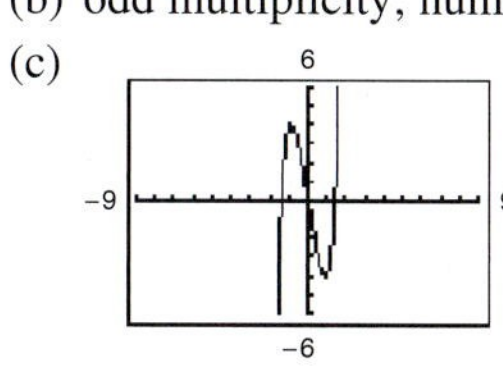

39. (a) No real zeros
(b) number of turning points: 1
(c)
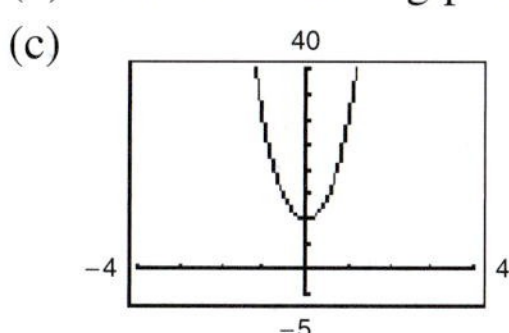

40. (a) $\pm\sqrt{5}$
(b) odd multiplicity; number of turning points: 3
(c)
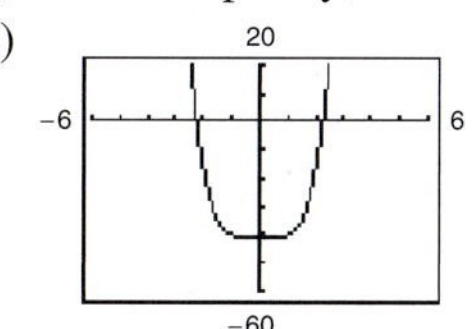

41. (a) $\pm 2, -3$
(b) odd multiplicity; number of turning points: 2
(c)
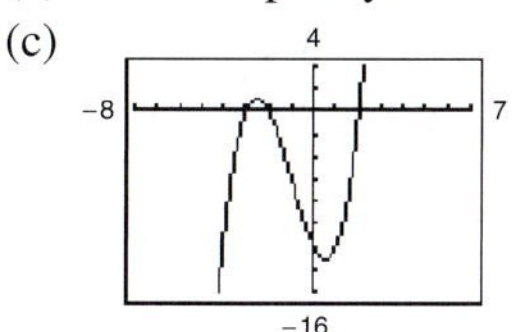

42. (a) $4, \pm 5$
(b) odd multiplicity; number of turning points: 2
(c)
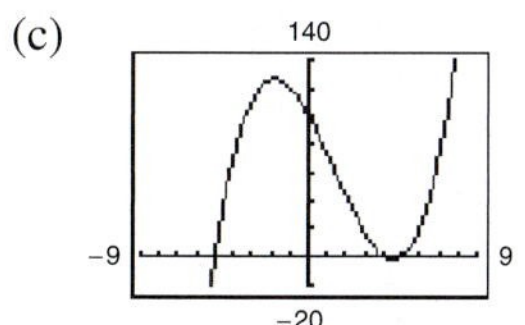

43. (a)
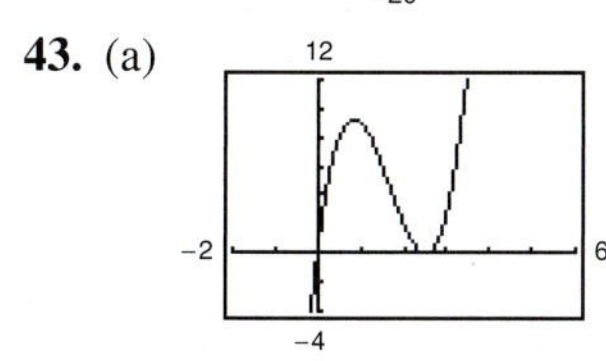

(b) x-intercepts: $(0, 0), \left(\frac{5}{2}, 0\right)$ (c) $x = 0, \frac{5}{2}$
(d) The answers in part (c) match the x-intercepts.

44. (a)
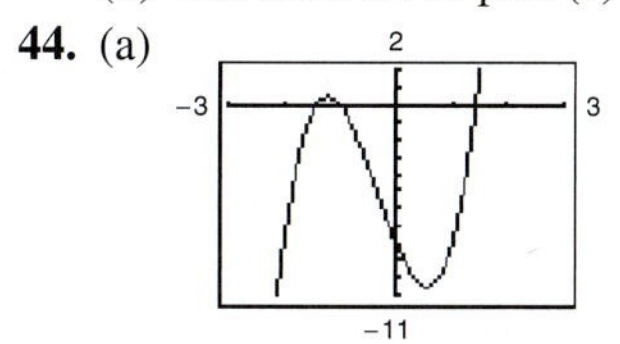

(b) x-intercepts: $\left(\sqrt{2}, 0\right), \left(-\sqrt{2}, 0\right), (-1, 0)$
(c) $x = \sqrt{2}, -\sqrt{2}, -1$
(d) The answers in part (c) match the x-intercepts.

45. (a)

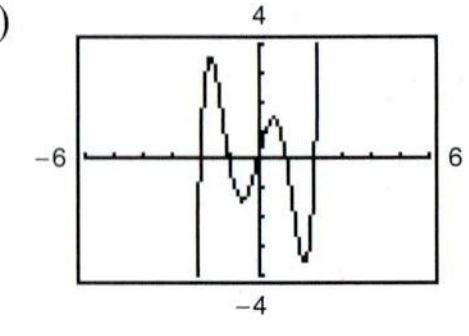

(b) x-intercepts: $(0, 0)$, $(\pm 1, 0)$, $(\pm 2, 0)$
(c) $x = 0, 1, -1, 2, -2$
(d) The answers in part (c) match the x-intercepts.

46. (a)

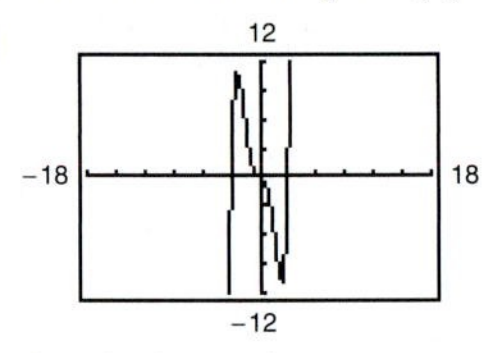

(b) $(0, 0)$, $(\pm 3, 0)$ (c) $x = 0, 3, -3$
(d) The answers in part (c) match the x-intercepts.

47. $f(x) = x^2 - 10x$ **48.** $f(x) = x^2 + 3x$
49. $f(x) = x^2 + 4x - 12$ **50.** $f(x) = x^2 - x - 20$
51. $f(x) = x^3 + 5x^2 + 6x$ **52.** $f(x) = x^3 - 7x^2 + 10x$
53. $f(x) = x^4 - 4x^3 - 9x^2 + 36x$
54. $f(x) = x^5 - 5x^3 + 4x$ **55.** $f(x) = x^2 - 2x - 2$
56. $f(x) = x^3 - 10x^2 + 27x - 22$
57. $f(x) = x^2 + 4x + 4$ **58.** $f(x) = x^2 + 12x + 32$
59. $f(x) = x^3 + 2x^2 - 3x$
60. $f(x) = x^3 - 9x^2 + 6x + 56$ **61.** $f(x) = x^3 - 3x$
62. $f(x) = x^3 - 27x^2 + 243x - 729$
63. $f(x) = x^4 + x^3 - 15x^2 + 23x - 10$
64. $f(x) = x^4 - 4x^3 - 23x^2 + 54x + 72$
65. $f(x) = x^5 + 16x^4 + 96x^3 + 256x^2 + 256x$
66. $f(x) = x^5 - 10x^4 + 14x^3 + 88x^2 - 183x + 90$

67. (a) Falls to the left, rises to the right
(b) $0, \pm 3$ (c) Answers will vary.
(d)

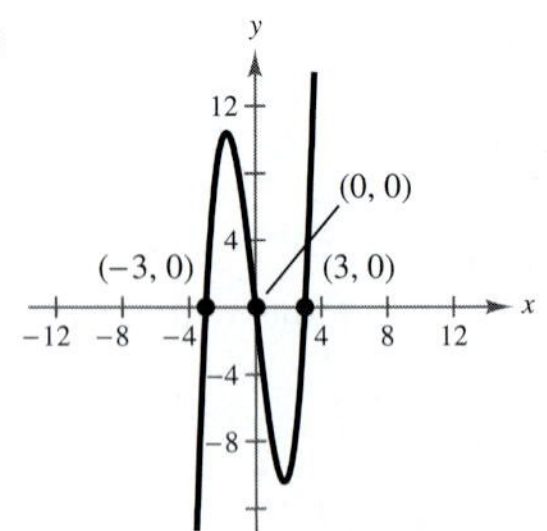

68. (a) Rises to the left, rises to the right
(b) $0, \pm 2$ (c) Answers will vary.
(d)

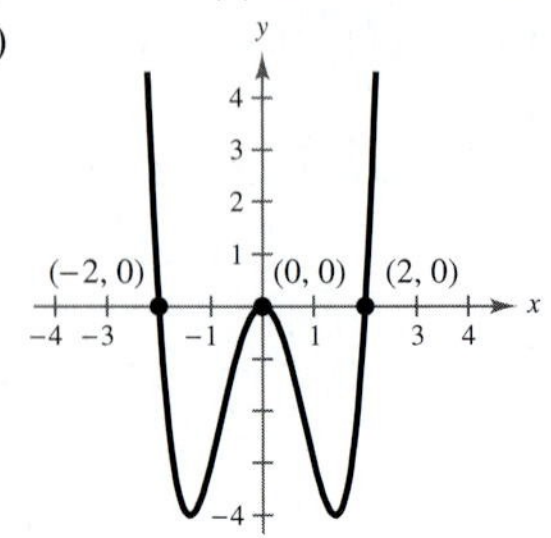

69. (a) Rises to the left, rises to the right
(b) No zeros (c) Answers will vary.
(d)

70. (a) Falls to the left, falls to the right
(b) 2, 8 (c) Answers will vary.
(d)

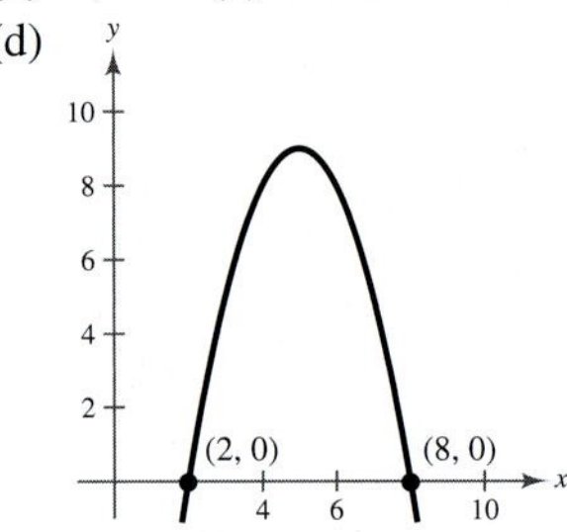

71. (a) Falls to the left, rises to the right
(b) 0, 3 (c) Answers will vary.
(d)

72. (a) Rises to the left, falls to the right
(b) 1 (c) Answers will vary.
(d)

73. (a) Falls to the left, rises to the right
(b) 0, 2, 3 (c) Answers will vary.
(d)

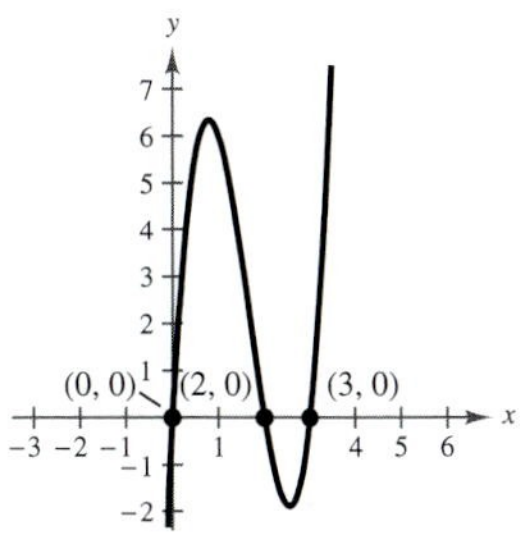

74. (a) Rises to the left, falls to the right
(b) $-\frac{3}{2}, 0, \frac{5}{2}$ (c) Answers will vary.
(d)

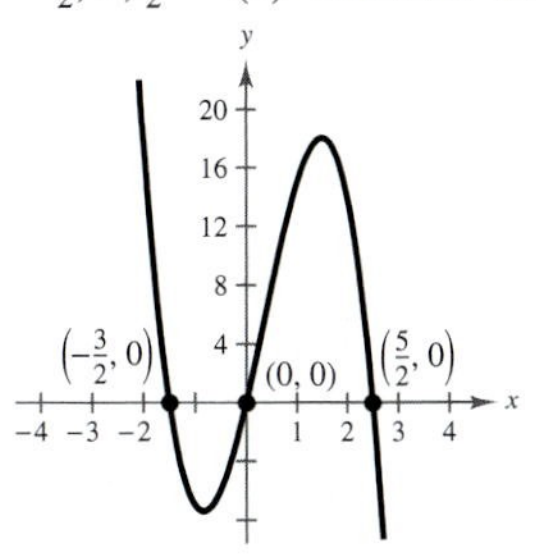

75. (a) Rises to the left, falls to the right
(b) $-5, 0$ (c) Answers will vary.
(d)

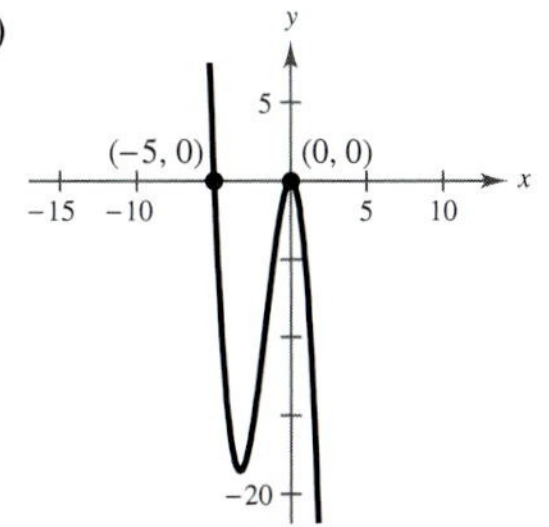

76. (a) Rises to the left, rises to the right
(b) $0, \pm 4$ (c) Answers will vary.
(d)

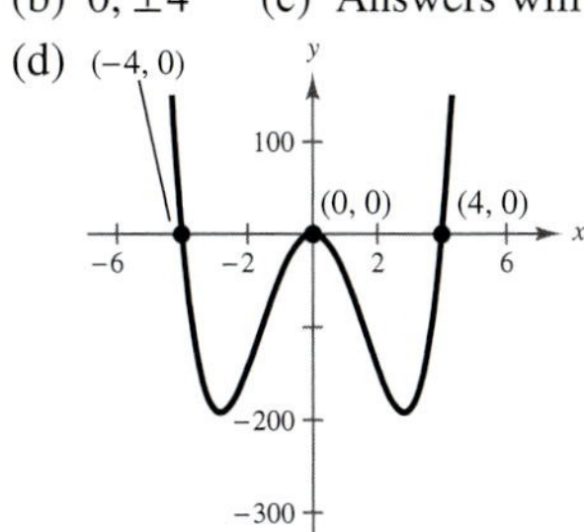

77. (a) Falls to the left, rises to the right
(b) 0, 4 (c) Answers will vary.
(d)

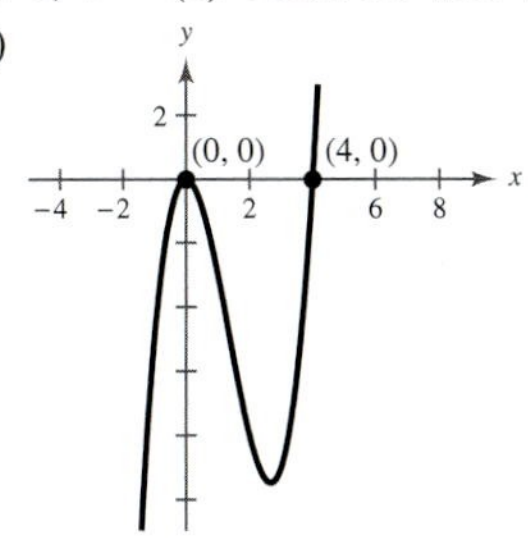

78. (a) Falls to the left, rises to the right
(b) 0, 4 (c) Answers will vary.
(d)

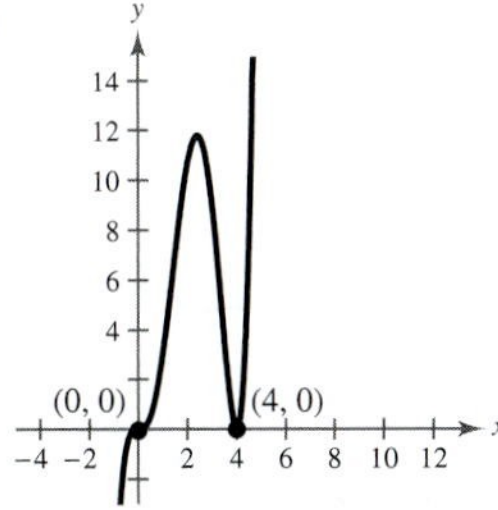

79. (a) Falls to the left, falls to the right
(b) ± 2 (c) Answers will vary.
(d)

80. (a) Falls to the left, rises to the right
(b) $-1, 3$ (c) Answers will vary.
(d)

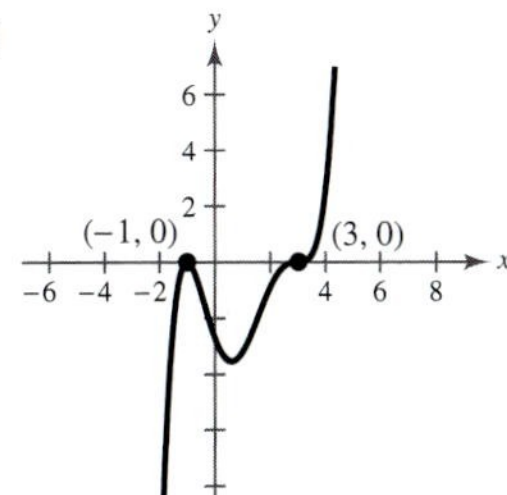

81.

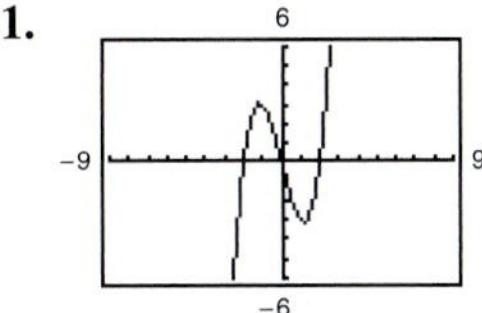

Zeros: $0, \pm 2$, odd multiplicity

82.

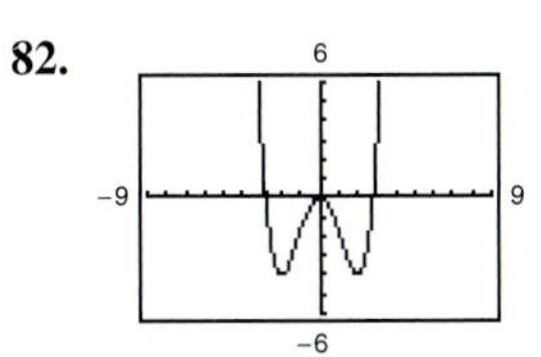

Zeros: $\pm 2\sqrt{2}$, odd multiplicity; 0, even multiplicity

CHAPTER 2

83. 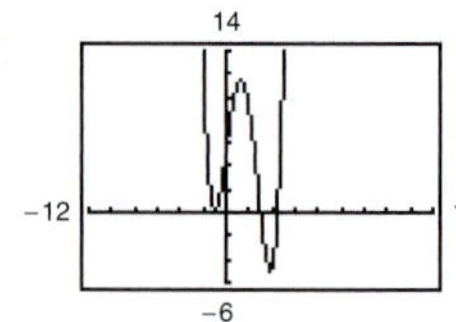

Zeros: -1, even multiplicity; $3, \frac{9}{2}$, odd multiplicity

84. 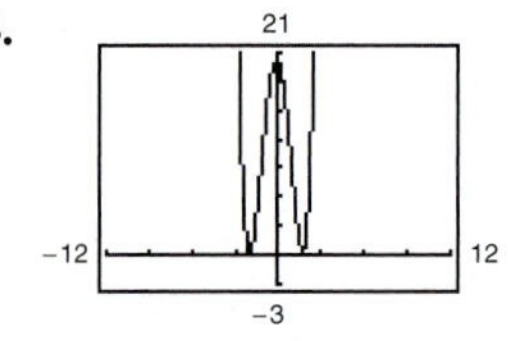

Zeros: $-2, \frac{5}{3}$, even multiplicity

85. $[-1, 0], [1, 2], [2, 3]$; $\approx -0.879, 1.347, 2.532$

86. $[0, 1], [6, 7], [11, 12]$; $\approx 0.845, 6.385, 11.588$

87. $[-2, -1], [0, 1]$; $\approx -1.585, 0.779$

88. $[-4, -3], [-1, 0], [0, 1], [3, 4]$; $\approx \pm 3.113, \pm 0.556$

89. (a) $V = l \times w \times h$
$= (36 - 2x)(36 - 2x)x$
$= x(36 - 2x)^2$

(b) Domain: $0 < x < 18$

(c)

x	1	2	3	4	5	6	7
V	1156	2048	2700	3136	3380	3456	3388

6 inches × 24 inches × 24 inches

(d) 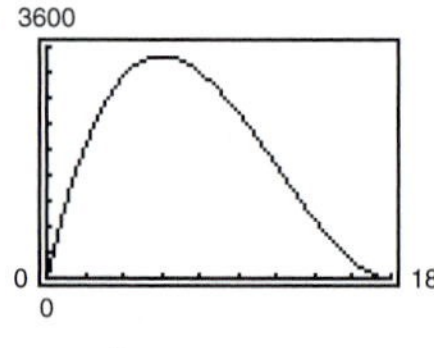

$x = 6$

90. (a) Answers will vary. (b) Domain: $0 < x < 6$

(c) 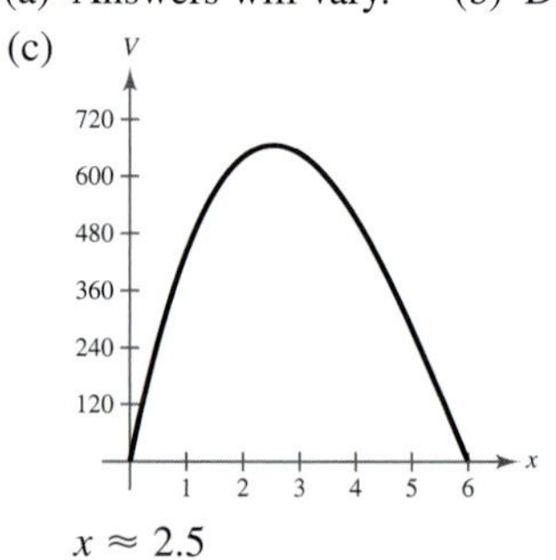

$x \approx 2.5$

91. (a) $A = -2x^2 + 12x$

(b) $V = -384x^2 + 2304x$

(c) 0 inches $< x <$ 6 inches

(d) 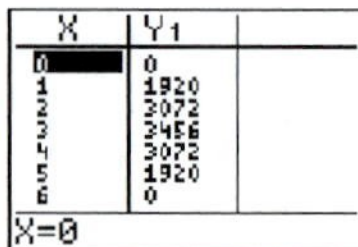

When $x = 3$, the volume is maximum at $V = 3456$; dimensions of gutter are 3 inches × 6 inches × 3 inches

(e) 4000, 0, 6, 0

The maximum value is the same.

(f) No. Answers will vary.

92. (a) $V = \frac{16}{3}\pi r^3$ (b) $r \geq 0$

(c) 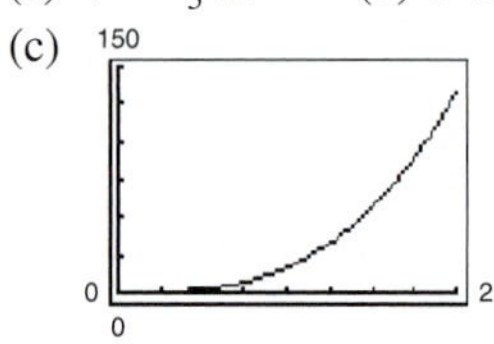

(d) Radius ≈ 1.93 feet; Length ≈ 7.72 feet

93. 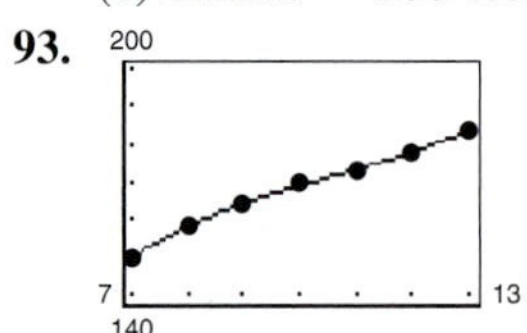

The model is a good fit.

94. 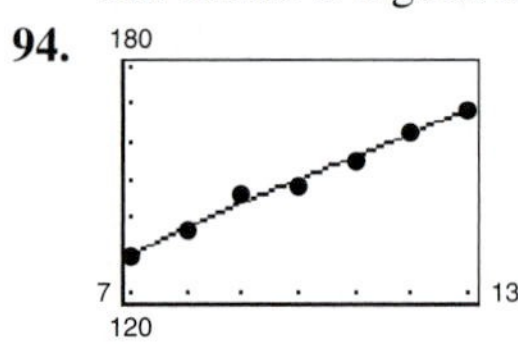

The model is a good fit.

95. Region 1: 259,370
Region 2: 223,470
Answers will vary.

96. Answers will vary.

97. (a) 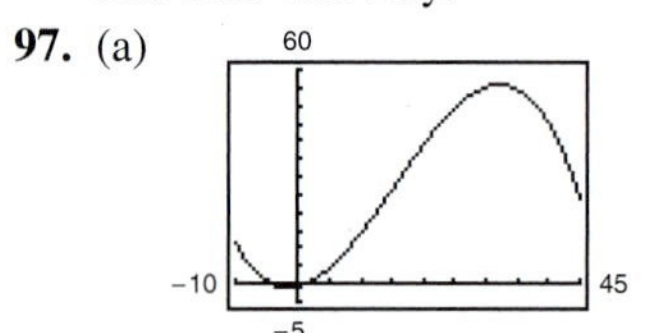

(b) $t \approx 15$

(c) Vertex: $(15.22, 2.54)$

(d) The results are approximately equal.

98. $x \approx 200$

99. False. A fifth-degree polynomial can have at most four turning points.

100. True. $f(x) = (x - 1)^6$ has one repeated solution.

101. True. The degree of the function is odd and its leading coefficient is negative, so the graph rises to the left and falls to the right.

102. (a) Degree: 3; Leading coefficient: positive

(b) Degree: 2; Leading coefficient: positive

(c) Degree: 4; Leading coefficient: positive

(d) Degree: 5; Leading coefficient: positive

103.

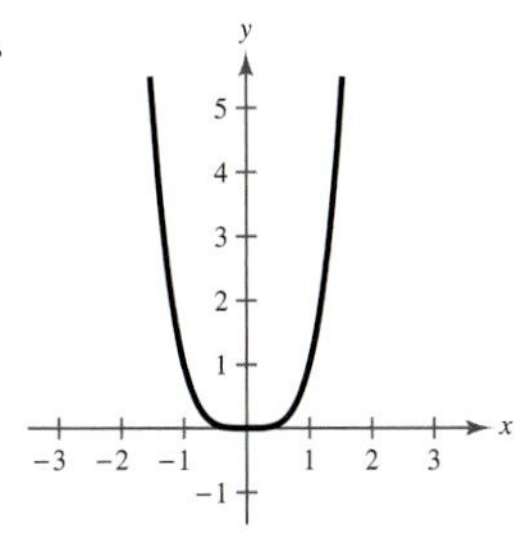

(a) Vertical shift of two units; Even
(b) Horizontal shift of two units; Neither even nor odd
(c) Reflection in the y-axis; Even
(d) Reflection in the x-axis; Even
(e) Horizontal stretch; Even
(f) Vertical shrink; Even
(g) $g(x) = x^3$; Neither odd nor even
(h) $g(x) = x^{16}$; Even

104. (a)

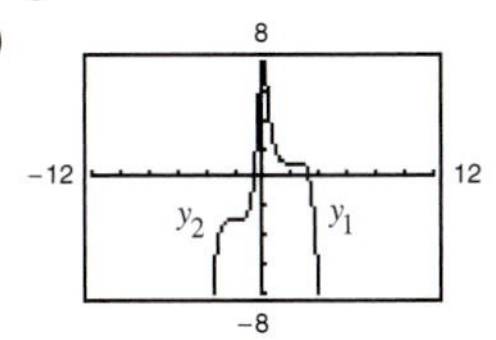

y_1 is decreasing. y_2 is increasing.

(b) Either always increasing or always decreasing. The behavior is determined by a.

(c)

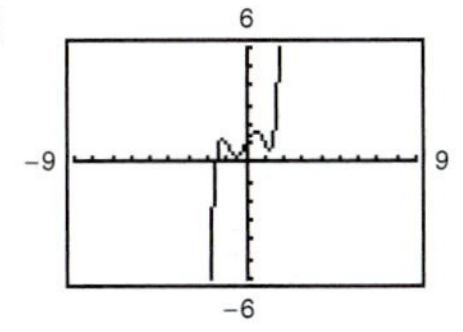

Because $H(x)$ is not always increasing or always decreasing, H cannot be written in the form $H(x) = a(x - h)^5 + k$.

105. $(5x - 8)(x + 3)$ **106.** $x(6x - 1)(x - 10)$

107. $x^2(4x + 5)(x - 3)$ **108.** $(y + 6)(y^2 - 6y + 36)$

109. $-\frac{7}{2}, 4$ **110.** $-\frac{2}{3}, 8$ **111.** $-\frac{5}{4}, \frac{1}{3}$

112. -12 **113.** $1 \pm \sqrt{22}$ **114.** $4 \pm \sqrt{14}$

115. $\dfrac{-5 \pm \sqrt{185}}{4}$ **116.** $\dfrac{-2 \pm \sqrt{31}}{3}$

117. Horizontal translation four units to the left of $y = x^2$

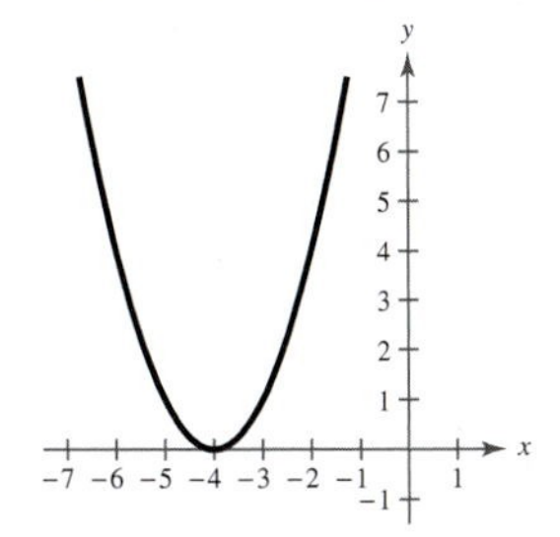

118. Reflection in the x-axis and vertical shift of three units up of $y = x^2$

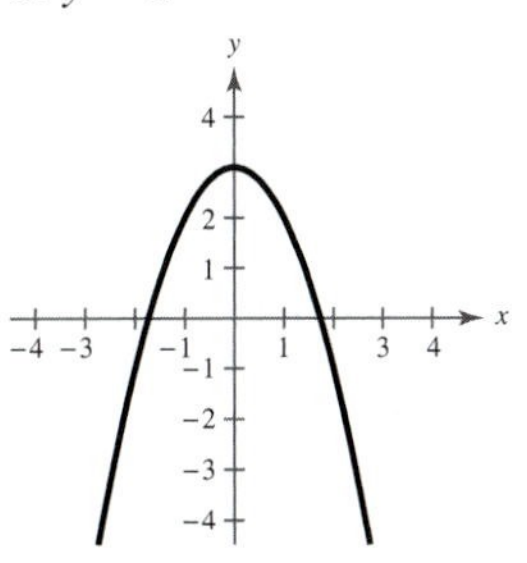

119. Horizontal translation one unit left and vertical translation five units down of $y = \sqrt{x}$

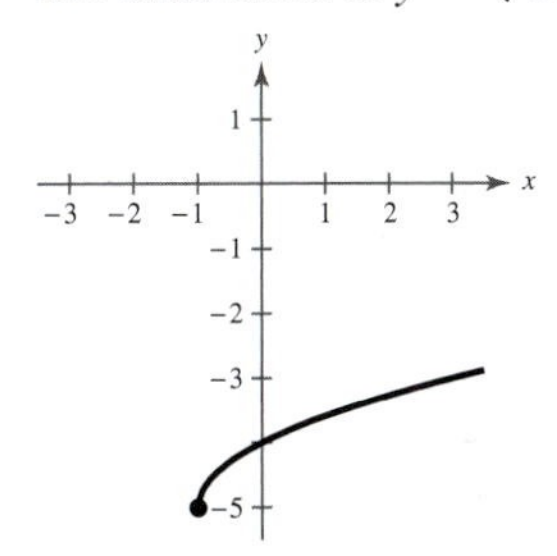

120. Reflection in the x-axis, horizontal translation, and vertical translation of $y = \sqrt{x}$

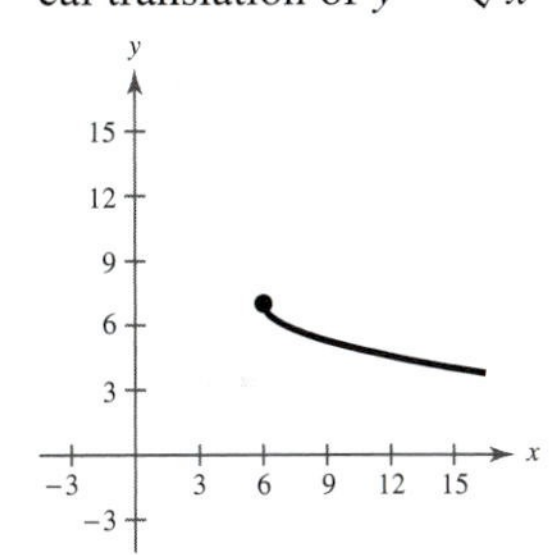

121. Vertical stretch by a factor of 2 and vertical translation nine units up of $y = [\![x]\!]$

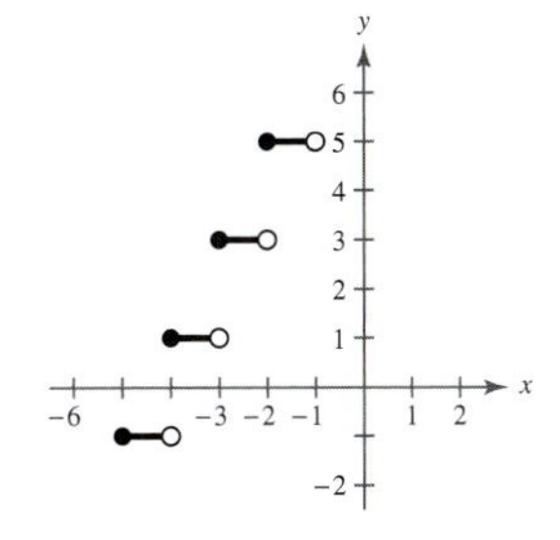

122. Vertical shrink, reflection in the x-axis, horizontal shift of 3 units to the left, and vertical shift of 10 units up of $y = [\![x]\!]$

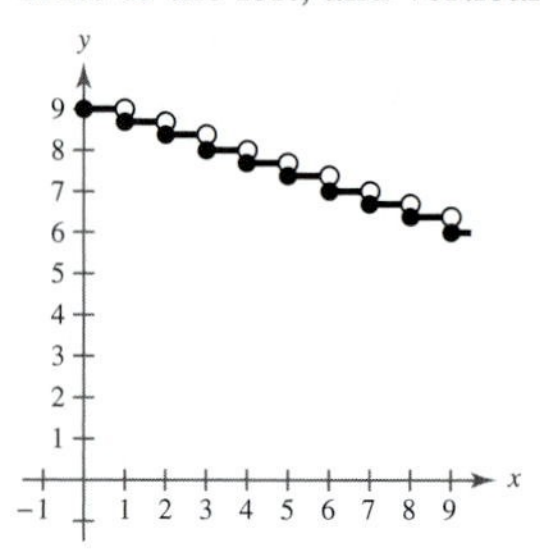

Section 2.3 *(page 159)*

Vocabulary Check *(page 159)*

1. dividend; divisor; quotient; remainder
2. improper; proper **3.** synthetic division
4. factor **5.** remainder

1. Answers will vary. **2.** Answers will vary.

3.

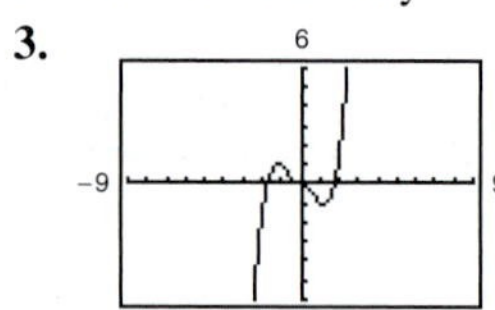

4.

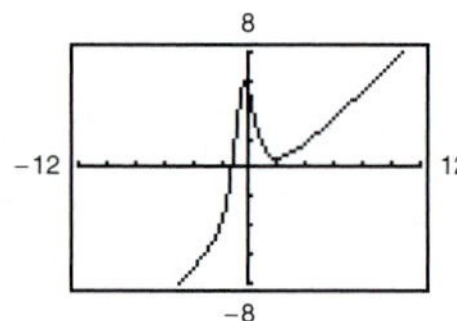

5. $2x + 4$ **6.** $5x + 3$ **7.** $x^2 - 3x + 1$

8. $2x^2 - 4x + 3$ **9.** $x^3 + 3x^2 - 1$

10. $x^2 + 7x + 18 + \dfrac{42}{x - 3}$ **11.** $7 - \dfrac{11}{x + 2}$

12. $4 - \dfrac{9}{2x + 1}$ **13.** $3x + 5 - \dfrac{2x - 3}{2x^2 + 1}$

14. $x - \dfrac{x + 9}{x^2 + 1}$ **15.** $x^2 + 2x + 4 + \dfrac{2x - 11}{x^2 - 2x + 3}$

16. $x^2 + \dfrac{x^2 + 7}{x^3 - 1}$ **17.** $x + 3 + \dfrac{6x^2 - 8x + 3}{(x - 1)^3}$

18. $2x - \dfrac{17x - 5}{x^2 - 2x + 1}$ **19.** $3x^2 - 2x + 5$

20. $5x^2 + 3x - 2$ **21.** $4x^2 - 9$ **22.** $9x^2 - 16$

23. $-x^2 + 10x - 25$ **24.** $3x^2 + 2x + 12$

25. $5x^2 + 14x + 56 + \dfrac{232}{x - 4}$

26. $5x^2 - 10x + 26 - \dfrac{44}{x + 2}$

27. $10x^3 + 10x^2 + 60x + 360 + \dfrac{1360}{x - 6}$

28. $x^4 - 16x^3 + 48x^2 - 144x + 312 - \dfrac{856}{x + 3}$

29. $x^2 - 8x + 64$ **30.** $x^2 + 9x + 81$

31. $-3x^3 - 6x^2 - 12x - 24 - \dfrac{48}{x - 2}$

32. $-3x^3 + 6x^2 - 12x + 24 - \dfrac{48}{x + 2}$

33. $-x^3 - 6x^2 - 36x - 36 - \dfrac{216}{x - 6}$

34. $-x^2 + 3x - 6 + \dfrac{11}{x + 1}$ **35.** $4x^2 + 14x - 30$

36. $3x^2 + \dfrac{1}{2}x + \dfrac{3}{4} + \dfrac{49}{8x - 12}$

37. $f(x) = (x - 4)(x^2 + 3x - 2) + 3,\ f(4) = 3$

38. $f(x) = (x + 2)(x^2 - 7x + 3) + 2, f(-2) = 2$

39. $f(x) = \left(x + \frac{2}{3}\right)(15x^3 - 6x + 4) + \frac{34}{3},\ f\left(-\frac{2}{3}\right) = \frac{34}{3}$

40. $f(x) = \left(x - \frac{1}{5}\right)(10x^2 - 20x - 7) + \frac{13}{5},\ f\left(\frac{1}{5}\right) = \frac{13}{5}$

41. $f(x) = \left(x - \sqrt{2}\right)\left[x^2 + \left(3 + \sqrt{2}\right)x + 3\sqrt{2}\right] - 8,$
$f\left(\sqrt{2}\right) = -8$

42. $f(x) = \left(x + \sqrt{5}\right)\left[x^2 + \left(2 - \sqrt{5}\right)x - 2\sqrt{5}\right] + 6,$
$f\left(-\sqrt{5}\right) = 6$

43. $f(x) = \left(x - 1 + \sqrt{3}\right)\left[-4x^2 + \left(2 + 4\sqrt{3}\right)x + \left(2 + 2\sqrt{3}\right)\right],$
$f\left(1 - \sqrt{3}\right) = 0$

44. $f(x) = \left(x - 2 - \sqrt{2}\right)\left[-3x^2 + \left(2 - 3\sqrt{2}\right)x + 8 - 4\sqrt{2}\right],$
$f\left(2 + \sqrt{2}\right) = 0$

45. (a) 1 (b) 4 (c) 4 (d) 1954

46. (a) 14 (b) 3122 (c) 434 (d) 2

47. (a) 97 (b) $-\frac{5}{3}$ (c) 17 (d) -199

48. (a) -2.5 (b) 20 (c) 65.5 (d) 5668

49. $(x - 2)(x + 3)(x - 1)$; Zeros: $2, -3, 1$

50. $(x + 4)(x + 2)(x - 6)$; Zeros: $-4, -2, 6$

51. $(2x - 1)(x - 5)(x - 2)$; Zeros: $\frac{1}{2}, 5, 2$

52. $(3x - 2)(4x - 3)(4x - 1)$; Zeros: $\frac{2}{3}, \frac{3}{4}, \frac{1}{4}$

53. $\left(x + \sqrt{3}\right)\left(x - \sqrt{3}\right)(x + 2)$; Zeros: $-\sqrt{3}, \sqrt{3}, -2$

54. $\left(x - \sqrt{2}\right)\left(x + \sqrt{2}\right)(x + 2)$; Zeros: $\sqrt{2}, -\sqrt{2}, -2$

55. $(x - 1)\left(x - 1 - \sqrt{3}\right)\left(x - 1 + \sqrt{3}\right)$;
Zeros: $1,\ 1 + \sqrt{3},\ 1 - \sqrt{3}$

56. $\left(x - 2 + \sqrt{5}\right)\left(x - 2 - \sqrt{5}\right)(x + 3)$;
Zeros: $2 - \sqrt{5},\ 2 + \sqrt{5},\ -3$

57. (a) Answers will vary. (b) $2x - 1$
(c) $f(x) = (2x - 1)(x + 2)(x - 1)$ (d) $\frac{1}{2}, -2, 1$
(e)

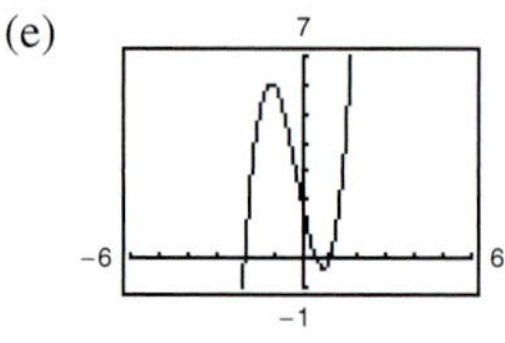

58. (a) Answers will vary. (b) $3x - 1$
(c) $f(x) = (3x - 1)(x + 3)(x - 2)$ (d) $\frac{1}{3}, -3, 2$
(e)

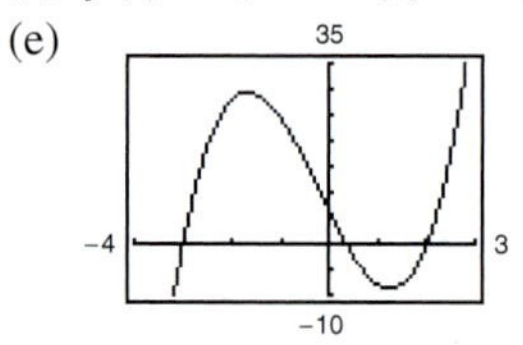

59. (a) Answers will vary. (b) $(x - 1), (x - 2)$
(c) $f(x) = (x - 1)(x - 2)(x - 5)(x + 4)$
(d) $1, 2, 5, -4$
(e)

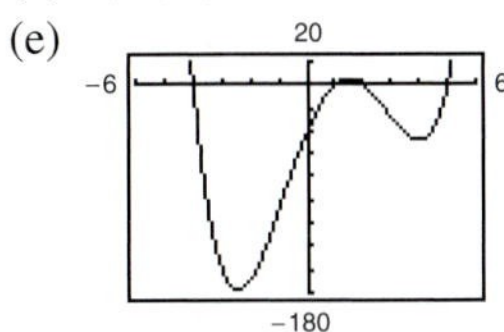

60. (a) Answers will vary. (b) $(4x + 3), (2x - 1)$
(c) $f(x) = (4x + 3)(2x - 1)(x + 2)(x - 4)$
(d) $-\frac{3}{4}, \frac{1}{2}, -2, 4$
(e)

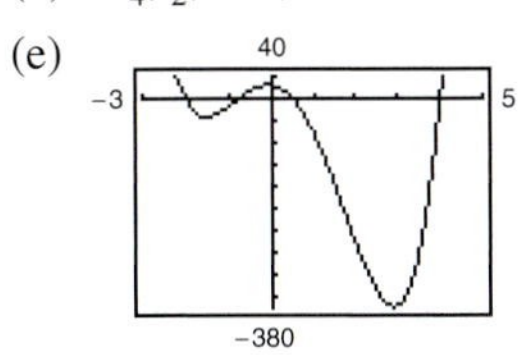

61. (a) Answers will vary. (b) $x + 7$
(c) $f(x) = (x + 7)(2x + 1)(3x - 2)$
(d) $-7, -\frac{1}{2}, \frac{2}{3}$
(e)

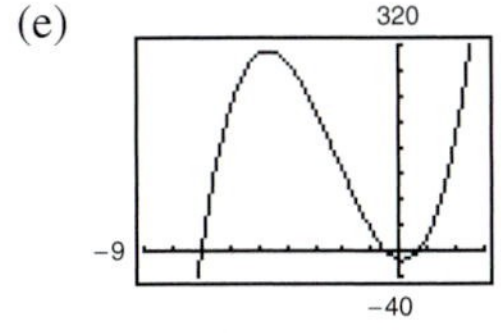

62. (a) Answers will vary. (b) $x - 3$
(c) $f(x) = (x - 3)(2x + 5)(5x - 3)$
(d) $3, -\frac{5}{2}, \frac{3}{5}$
(e)

63. (a) Answers will vary. (b) $(x - \sqrt{5})$
(c) $f(x) = (x - \sqrt{5})(x + \sqrt{5})(2x - 1)$ (d) $\pm\sqrt{5}, \frac{1}{2}$
(e)

64. (a) Answers will vary. (b) $x - 4\sqrt{3}$
(c) $f(x) = (x - 4\sqrt{3})(x + 4\sqrt{3})(x + 3)$
(d) $\pm 4\sqrt{3}, -3$
(e)

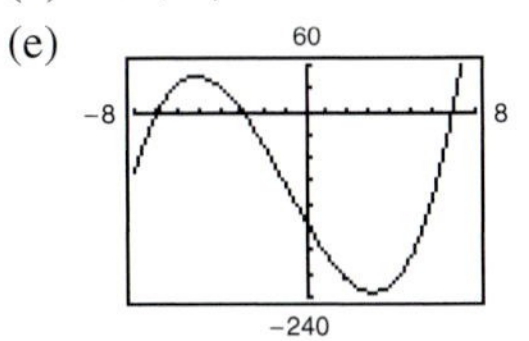

65. (a) Zeros are 2 and $\approx \pm 2.236$.
(b) $x = 2$ (c) $f(x) = (x - 2)(x - \sqrt{5})(x + \sqrt{5})$

66. (a) Zeros are 4 and $\approx \pm 1.414$.
(b) $x = 4$ (c) $g(x) = (x - 4)(x - \sqrt{2})(x + \sqrt{2})$

67. (a) Zeros are -2, ≈ 0.268, and ≈ 3.732.
(b) $x = -2$
(c) $h(t) = (t + 2)[t - (2 + \sqrt{3})][t - (2 - \sqrt{3})]$

68. (a) Zeros are 6, ≈ 0.764, and ≈ 5.236.
(b) $x = 6$
(c) $f(s) = (s - 6)[s - (3 + \sqrt{5})][s - (3 - \sqrt{5})]$

69. $2x^2 - x - 1, \ x \neq \frac{3}{2}$ **70.** $x^2 - 7x - 8, \ x \neq -8$

71. $x^2 + 3x, \ x \neq -2, -1$ **72.** $x^2 + 9x - 1, \ x \neq \pm 2$

73. (a) and (b)

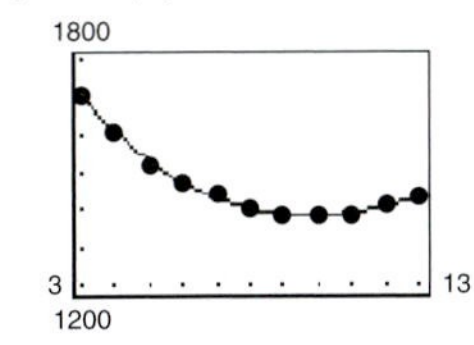

$M = -0.242x^3 + 12.43x^2 - 173.4x + 2118$

(c)

t	3	4	5	6	7	8
$M(t)$	1703	1608	1531	1473	1430	1402

t	9	10	11	12	13
$M(t)$	1388	1385	1392	1409	1433

Answers will vary.
(d) 1614 thousand. No, because the model will approach negative infinity quickly.

74. (a) and (b)

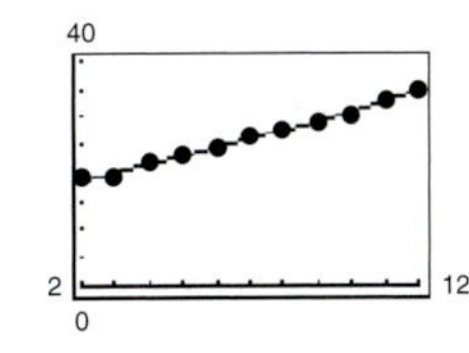

$R = 0.00260x^3 - 0.0292x^2 + 1.558x + 15.63$
(c) \$49.38

75. False. $-\frac{4}{7}$ is a zero of f.

76. True.
$f(x) = (2x - 1)(x + 1)(x - 2)(x - 3)(3x + 2)(x + 4)$

77. True. The degree of the numerator is greater than the degree of the denominator.

78. (a) $f(x) = (x - 2)x^2 + 5 = x^3 - 2x^2 + 5$
(b) $f(x) = -(x + 3)x^2 + 1 = -x^3 - 3x^2 + 1$

79. $x^{2n} + 6x^n + 9$ **80.** $x^{2n} - x^n + 3$

81. The remainder is 0.

82. Multiply the divisor and the quotient and add the remainder to obtain the dividend.

83. $c = -210$ **84.** $c = 42$

85. 0; $x + 3$ is a factor of f.

86. Because $f(x)$ is in factored form, it is easier to evaluate directly.

87. $\pm\frac{5}{3}$ **88.** $\pm\frac{\sqrt{21}}{4}$ **89.** $-\frac{7}{5}, 2$ **90.** $\frac{5}{4}, \frac{3}{2}$

91. $\frac{-3 \pm \sqrt{3}}{2}$ **92.** $\frac{-3 \pm \sqrt{21}}{2}$

93. $f(x) = x^3 - 7x^2 + 12x$

94. $f(x) = x^2 + 5x - 6$

95. $f(x) = x^3 + x^2 - 7x - 3$

96. $f(x) = x^4 - 3x^3 - 5x^2 + 9x - 2$

Section 2.4 *(page 167)*

Vocabulary Check *(page 167)*

1. (a) iii (b) i (c) ii **2.** $\sqrt{-1}; -1$

3. complex numbers; $a + bi$ **4.** principal square

5. complex conjugates

1. $a = -10,\ b = 6$ **2.** $a = 13,\ b = 4$

3. $a = 6,\ b = 5$ **4.** $a = 0,\ b = -\frac{5}{2}$ **5.** $4 + 3i$

6. $3 + 4i$ **7.** $2 - 3\sqrt{3}i$ **8.** $1 + 2\sqrt{2}i$

9. $5\sqrt{3}i$ **10.** $2i$ **11.** 8 **12.** 45

13. $-1 - 6i$ **14.** $4 + 2i$ **15.** $0.3i$ **16.** $0.02i$

17. $11 - i$ **18.** $8 + 4i$ **19.** 4 **20.** $-3 - 11i$

21. $3 - 3\sqrt{2}i$ **22.** 4 **23.** $-14 + 20i$

24. $17 + 18i$ **25.** $\frac{1}{6} + \frac{7}{6}i$ **26.** $-4.2 + 7.5i$

27. $5 + i$ **28.** $6 - 22i$ **29.** $12 + 30i$

30. $32 - 72i$ **31.** 24 **32.** 18 **33.** $-9 + 40i$

34. $-5 - 12i$ **35.** -10 **36.** $-8i$

37. $6 - 3i,\ 45$ **38.** $7 + 12i,\ 193$

39. $-1 + \sqrt{5}i,\ 6$ **40.** $-3 - \sqrt{2}i,\ 11$

41. $-2\sqrt{5}i,\ 20$ **42.** $-\sqrt{15}i,\ 15$

43. $\sqrt{8},\ 8$ **44.** $1 + \sqrt{8},\ 9 + 4\sqrt{2}$ **45.** $-5i$

46. $7i$ **47.** $\frac{8}{41} + \frac{10}{41}i$ **48.** $\frac{5}{2} + \frac{5}{2}i$ **49.** $\frac{4}{5} + \frac{3}{5}i$

50. $4 + i$ **51.** $-5 - 6i$ **52.** $8 - 4i$

53. $-\frac{120}{1681} - \frac{27}{1681}i$ **54.** $\frac{60}{169} - \frac{25}{169}i$ **55.** $-\frac{1}{2} - \frac{5}{2}i$

56. $\frac{12}{5} + \frac{9}{5}i$ **57.** $\frac{62}{949} + \frac{297}{949}i$ **58.** $\frac{5}{17} - \frac{20}{17}i$

59. $-2\sqrt{3}$ **60.** $-5\sqrt{2}$ **61.** -10 **62.** -75

63. $(21 + 5\sqrt{2}) + (7\sqrt{5} - 3\sqrt{10})i$ **64.** $-2 - 4\sqrt{6}i$

65. $1 \pm i$ **66.** $-3 \pm i$ **67.** $-2 \pm \frac{1}{2}i$

68. $\frac{1}{3} \pm 2i$ **69.** $-\frac{5}{2}, -\frac{3}{2}$ **70.** $\frac{1}{8} \pm \frac{\sqrt{11}}{8}i$

71. $2 \pm \sqrt{2}i$ **72.** $\frac{3}{7} \pm \frac{\sqrt{34}}{14}i$ **73.** $\frac{5}{7} \pm \frac{5\sqrt{15}}{7}$

74. $\frac{1}{3} \pm \frac{\sqrt{23}}{3}i$ **75.** $-1 + 6i$ **76.** $-4 + 2i$

77. $-5i$ **78.** $-i$ **79.** $-375\sqrt{3}i$ **80.** -8

81. i **82.** $\frac{1}{8}i$

83. (a) $z_1 = 9 + 16i,\ z_2 = 20 - 10i$

(b) $z = \frac{11{,}240}{877} + \frac{4630}{877}i$

84. (a) 8 (b) 8 (c) 8

85. (a) 16 (b) 16 (c) 16 (d) 16

86. (a) 1 (b) i (c) -1 (d) $-i$

87. False. If the complex number is real, the number equals its conjugate.

88. True. $x^4 - x^2 + 14 = 56$

$$(-i\sqrt{6})^4 - (-i\sqrt{6})^2 + 14 \stackrel{?}{=} 56$$
$$36 + 6 + 14 \stackrel{?}{=} 56$$
$$56 = 56$$

89. False.

$i^{44} + i^{150} - i^{74} - i^{109} + i^{61} = 1 - 1 + 1 - i + i = 1$

90. $\sqrt{-6}\sqrt{-6} = \sqrt{6}i\sqrt{6}i = 6i^2 = -6$

91–92. Proof **93.** $-x^2 - 3x + 12$

94. $x^3 + x^2 + 2x - 6$

95. $3x^2 + \frac{23}{2}x - 2$ **96.** $4x^2 - 20x + 25$ **97.** -31

98. 14 **99.** $\frac{27}{2}$ **100.** $-\frac{4}{3}$ **101.** $a = \frac{\sqrt{3V\pi b}}{2\pi b}$

102. $r = \frac{\sqrt{\alpha m_1 m_2 F}}{F}$ **103.** 1 liter

Section 2.5 *(page 179)*

Vocabulary Check *(page 179)*

1. Fundamental Theorem of Algebra

2. Linear Factorization Theorem **3.** Rational Zero

4. conjugate **5.** irreducible over the reals

6. Descartes' Rule of Signs **7.** lower; upper

1. 0, 6 **2.** $0, -3, \pm 1$ **3.** $2, -4$ **4.** $-5, 8$

5. $-6, \pm i$ **6.** $3, 2, \pm 3i$ **7.** $\pm 1, \pm 3$

8. $\pm 1, \pm 2, \pm 4, \pm 8, \pm 16$

9. $\pm 1, \pm 3, \pm 5, \pm 9, \pm 15, \pm 45, \pm\frac{1}{2}, \pm\frac{3}{2}, \pm\frac{5}{2}, \pm\frac{9}{2}, \pm\frac{15}{2}, \pm\frac{45}{2}$

10. $\pm 1, \pm 2, \pm\frac{1}{2}, \pm\frac{1}{4}$ **11.** 1, 2, 3 **12.** $-2, -1, 3$

13. $1, -1, 4$ **14.** 1, 2, 6 **15.** $-1, -10$

16. 3 **17.** $\frac{1}{2}, -1$ **18.** $\frac{1}{3}, 3$ **19.** $-2, 3, \pm\frac{2}{3}$

20. $\pm 1, 5, \frac{5}{2}$ **21.** $-1, 2$ **22.** $0, -1, -3, 4$

23. $-6, \frac{1}{2}, 1$ **24.** $-2, 0, 1$

25. (a) $\pm 1, \pm 2, \pm 4$

(b)

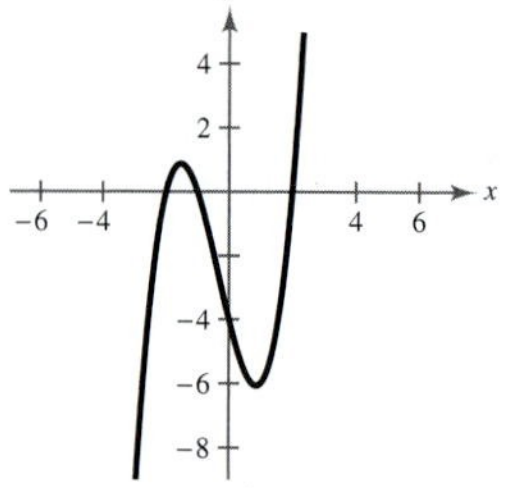

(c) $-2, -1, 2$

26. (a) $\pm 1, \pm 2, \pm 4, \pm 8, \pm 16, \pm\frac{1}{3}, \pm\frac{2}{3}, \pm\frac{4}{3}, \pm\frac{8}{3}, \pm\frac{16}{3}$
(b)
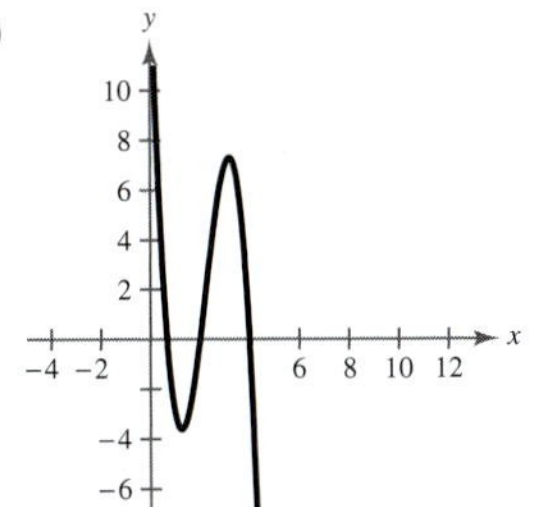

(c) $\frac{2}{3}, 2, 4$

27. (a) $\pm 1, \pm 3, \pm\frac{1}{2}, \pm\frac{3}{2}, \pm\frac{1}{4}, \pm\frac{3}{4}$
(b)
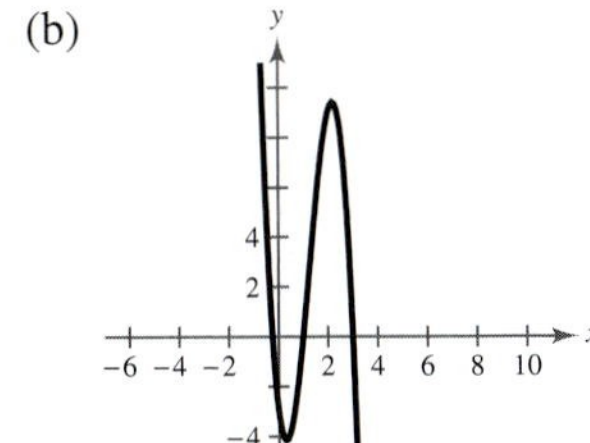

(c) $-\frac{1}{4}, 1, 3$

28. (a) $\pm 1, \pm 3, \pm 5, \pm 15, \pm\frac{1}{2}, \pm\frac{3}{2}, \pm\frac{5}{2}, \pm\frac{15}{2}, \pm\frac{1}{4}, \pm\frac{3}{4}, \pm\frac{5}{4}, \pm\frac{15}{4}$
(b)
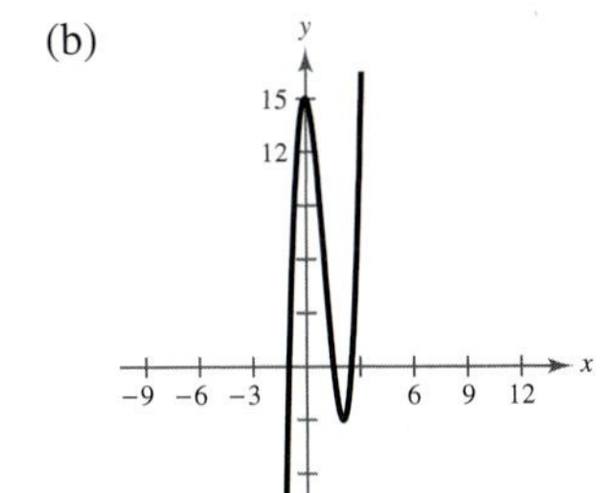

(c) $-1, \frac{3}{2}, \frac{5}{2}$

29. (a) $\pm 1, \pm 2, \pm 4, \pm 8, \pm\frac{1}{2}$
(b)
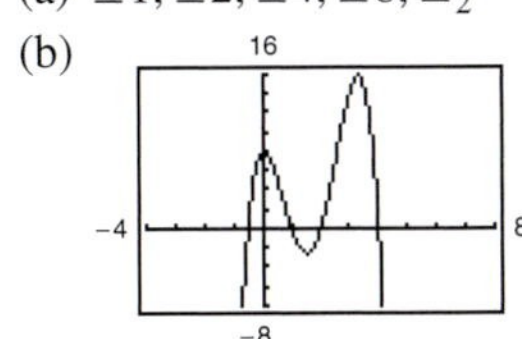

(c) $-\frac{1}{2}, 1, 2, 4$

30. (a) $\pm 1, \pm 2, \pm 4, \pm\frac{1}{2}, \pm\frac{1}{4}$
(b)
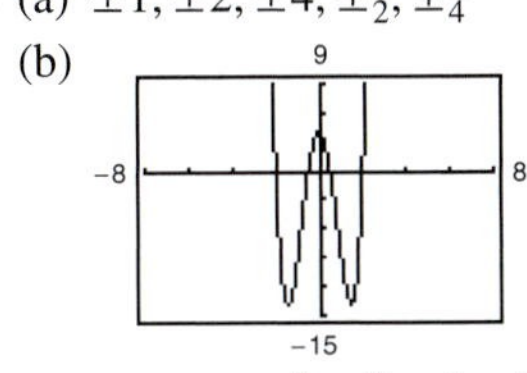

(c) $\pm 2, \pm\frac{1}{2}$

31. (a) $\pm 1, \pm 3, \pm\frac{1}{2}, \pm\frac{3}{2}, \pm\frac{1}{4}, \pm\frac{3}{4}, \pm\frac{1}{8}, \pm\frac{3}{8}, \pm\frac{1}{16}, \pm\frac{3}{16}, \pm\frac{1}{32}, \pm\frac{3}{32}$
(b)
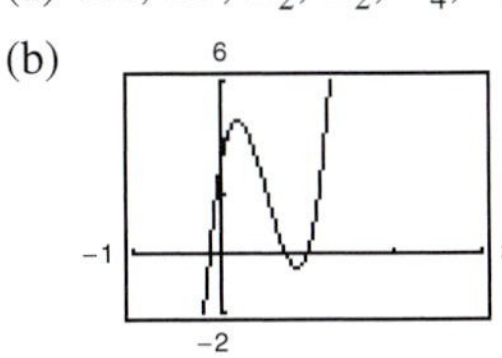

(c) $1, \frac{3}{4}, -\frac{1}{8}$

32. (a) $\pm 1, \pm 2, \pm 3, \pm 6, \pm 9, \pm 18, \pm\frac{1}{2}, \pm\frac{3}{2}, \pm\frac{9}{2}, \pm\frac{1}{4}, \pm\frac{3}{4}, \pm\frac{9}{4}$
(b)
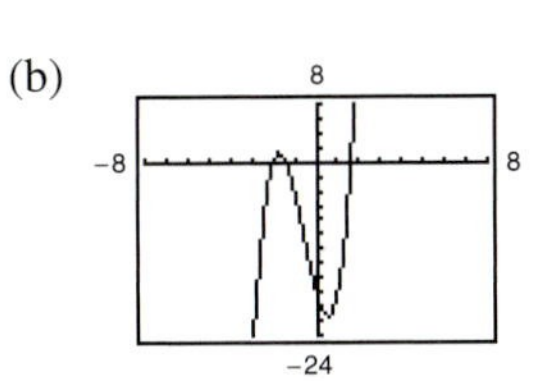

(c) $-2, \frac{1}{8} \pm \frac{\sqrt{145}}{8}$

33. (a) $\pm 1, \approx \pm 1.414$
(b) $f(x) = (x + 1)(x - 1)(x + \sqrt{2})(x - \sqrt{2})$

34. (a) $\pm 2, \approx \pm 1.732$
(b) $P(t) = (t + 2)(t - 2)(t + \sqrt{3})(t - \sqrt{3})$

35. (a) $0, 3, 4, \approx \pm 1.414$
(b) $h(x) = x(x - 3)(x - 4)(x + \sqrt{2})(x - \sqrt{2})$

36. (a) $\pm 3, 1.5, 0.333$
(b) $g(x) = (x + 3)(x - 3)(2x - 3)(3x - 1)$

37. $x^3 - x^2 + 25x - 25$ **38.** $x^3 - 4x^2 + 9x - 36$

39. $x^3 + 4x^2 - 31x - 174$ **40.** $x^3 - 10x^2 + 33x - 34$

41. $3x^4 - 17x^3 + 25x^2 + 23x - 22$

42. $x^4 + 8x^3 + 9x^2 - 10x + 100$

43. (a) $(x^2 + 9)(x^2 - 3)$ (b) $(x^2 + 9)(x + \sqrt{3})(x - \sqrt{3})$
(c) $(x + 3i)(x - 3i)(x + \sqrt{3})(x - \sqrt{3})$

44. (a) $(x^2 - 6)(x^2 - 2x + 3)$
(b) $(x + \sqrt{6})(x - \sqrt{6})(x^2 - 2x + 3)$
(c) $(x + \sqrt{6})(x - \sqrt{6})(x - 1 - \sqrt{2}i)(x - 1 + \sqrt{2}i)$

45. (a) $(x^2 - 2x - 2)(x^2 - 2x + 3)$
(b) $(x - 1 + \sqrt{3})(x - 1 - \sqrt{3})(x^2 - 2x + 3)$
(c) $(x - 1 + \sqrt{3})(x - 1 - \sqrt{3})(x - 1 + \sqrt{2}i)$
$(x - 1 - \sqrt{2}i)$

46. (a) $(x^2 + 4)(x^2 - 3x - 5)$
(b) $(x^2 + 4)\left(x - \frac{3 + \sqrt{29}}{2}\right)\left(x - \frac{3 - \sqrt{29}}{2}\right)$
(c) $(x + 2i)(x - 2i)\left(x - \frac{3 + \sqrt{29}}{2}\right)\left(x - \frac{3 - \sqrt{29}}{2}\right)$

47. $-\frac{3}{2}, \pm 5i$ **48.** $-1, \pm 3i$ **49.** $\pm 2i, 1, -\frac{1}{2}$

50. $-3, 5 \pm 2i$ **51.** $-3 \pm i, \frac{1}{4}$ **52.** $-\frac{2}{3}, 1 \pm \sqrt{3}i$

53. $2, -3 \pm \sqrt{2}i, 1$ **54.** $-2, -1 \pm 3i$

55. $\pm 5i$; $(x + 5i)(x - 5i)$

56. $\frac{1 \pm \sqrt{223}i}{2}$; $\left(x - \frac{1 - \sqrt{223}i}{2}\right)\left(x - \frac{1 + \sqrt{223}i}{2}\right)$

57. $2 \pm \sqrt{3}$; $(x - 2 - \sqrt{3})(x - 2 + \sqrt{3})$

58. $-5 \pm \sqrt{2}$; $(x + 5 + \sqrt{2})(x + 5 - \sqrt{2})$

59. $\pm 3, \pm 3i$; $(x + 3)(x - 3)(x + 3i)(x - 3i)$

60. $\pm 5, \pm 5i$; $(y + 5)(y - 5)(y + 5i)(y - 5i)$

61. $1 \pm i$; $(z - 1 + i)(z - 1 - i)$

62. $1 \pm i, 1$; $(x - 1)(x - 1 + i)(x - 1 - i)$

63. $2, 2 \pm i$; $(x - 2)(x - 2 + i)(x - 2 - i)$

64. $3 \pm 2i, -4$; $(x + 4)(x - 3 + 2i)(x - 3 - 2i)$

65. $-2, 1 \pm \sqrt{2}i$; $(x + 2)(x - 1 + \sqrt{2}i)(x - 1 - \sqrt{2}i)$

66. $-2 \pm \sqrt{3}i, -5;$
$(x + 5)(x + 2 + \sqrt{3}i)(x + 2 - \sqrt{3}i)$
67. $-\frac{1}{5}, 1 \pm \sqrt{5}i;$ $(5x + 1)(x - 1 + \sqrt{5}i)(x - 1 - \sqrt{5}i)$
68. $1 \pm \sqrt{3}i, -\frac{2}{3};$ $(3x + 2)(x - 1 + \sqrt{3}i)(x - 1 - \sqrt{3}i)$
69. $2, \pm 2i;$ $(x - 2)^2(x + 2i)(x - 2i)$
70. $-3, \pm i;$ $(x + 3)^2(x + i)(x - i)$
71. $\pm i, \pm 3i;$ $(x + i)(x - i)(x + 3i)(x - 3i)$
72. $\pm 2i, \pm 5i;$ $(x + 2i)(x - 2i)(x + 5i)(x - 5i)$
73. $-10, -7 \pm 5i$ **74.** $1 \pm 2i, \frac{1}{2}$ **75.** $-\frac{3}{4}, 1 \pm \frac{1}{2}i$
76. $\frac{1}{3} \pm \frac{2}{3}i, 1$ **77.** $-2, -\frac{1}{2}, \pm i$ **78.** $1 \pm \sqrt{3}i, 2$
79. No real zeros **80.** Two or no positive zeros
81. No real zeros **82.** Two or no positive zeros
83. One positive zero **84.** One or three positive zeros
85. One or three positive zeros
86. One or three negative zeros
87. Answers will vary. **88.** Answers will vary.
89. Answers will vary. **90.** Answers will vary.
91. $1, -\frac{1}{2}$ **92.** $-\frac{3}{2}, \frac{1}{3}, \frac{3}{2}$ **93.** $-\frac{3}{4}$ **94.** $\frac{2}{3}$
95. $\pm 2, \pm\frac{3}{2}$ **96.** $-3, \frac{1}{2}, 4$ **97.** $\pm 1, \frac{1}{4}$
98. $-2, -\frac{1}{3}, \frac{1}{2}$ **99.** d **100.** a **101.** b **102.** c
103. (a)

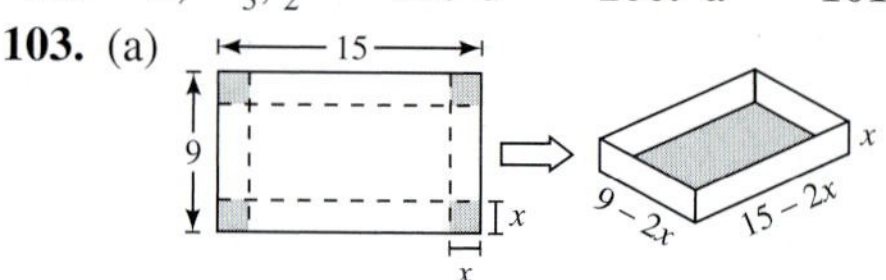

(b) $V = x(9 - 2x)(15 - 2x)$
Domain: $0 < x < \frac{9}{2}$
(c)

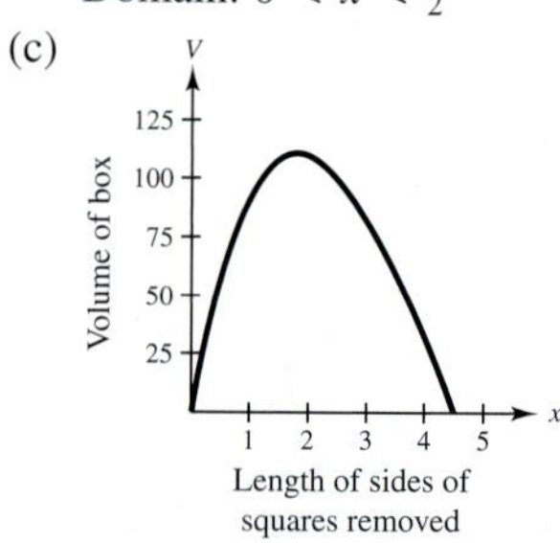

1.82 centimeters × 5.36 centimeters × 11.36 centimeters
(d) $\frac{1}{2}, \frac{7}{2}, 8;$ 8 is not in the domain of V.
104. (a) Answers will vary.
(b)

20 inches × 20 inches × 40 inches
(c) $15, \dfrac{15 \pm 15\sqrt{5}}{2};$ the value $x = \dfrac{15 - 15\sqrt{5}}{2}$ is physically impossible because x is negative.
105. $x \approx 38.4$, or \$384,000 **106.** $x \approx 31.5$, or \$315,000
107. (a) $V = x^3 + 9x^2 + 26x + 24 = 120$
(b) 4 feet by 5 feet by 6 feet

108. (a) $A = (160 + x)(250 + x) = 60{,}000$ feet squared
(b) $160 + \dfrac{-410 + \sqrt{248{,}100}}{2} \approx 204.05$ feet
$250 + \dfrac{-410 + \sqrt{248{,}100}}{2} \approx 294.05$ feet
(c) $A = (160 + x)(250 + 2x) = 60{,}000;$ the corral has dimensions 313.2 feet by 191.6 feet.
109. $x \approx 40$, or 4000 units
110. No. Setting $h = 64$ and solving the resulting equation yields imaginary roots.
111. No. Setting $p = 9{,}000{,}000$ and solving the resulting equation yields imaginary roots.
112. (a) $A = 0.0167x^3 - 0.508x^2 + 5.60x - 13.4$
(b)

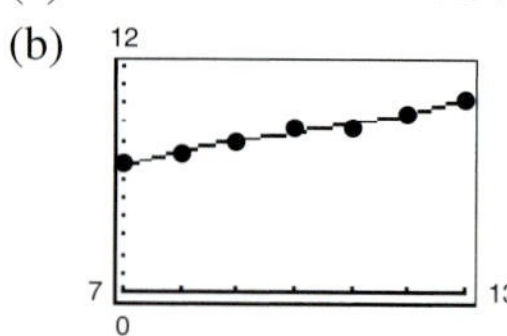

(c) 2000
(d) 2001
(e) Yes, the attendance will continue to grow.
113. False. The most complex zeros it can have is two, and the Linear Factorization Theorem guarantees that there are three linear factors, so one zero must be real.
114. False. f does not have real coefficients.
115. r_1, r_2, r_3 **116.** r_1, r_2, r_3
117. $5 + r_1, 5 + r_2, 5 + r_3$
118. $\dfrac{r_1}{2}, \dfrac{r_2}{2}, \dfrac{r_3}{2}$ **119.** The zeros cannot be determined.
120. $-r_1, -r_2, -r_3$
121. (a) $0 < k < 4$ (b) $k = 4$ (c) $k < 0$ (d) $k > 4$
122. (a) No (b) No
123. Answers will vary. There are infinitely many possible functions for f. Sample equation and graph:
$f(x) = -2x^3 + 3x^2 + 11x - 6$

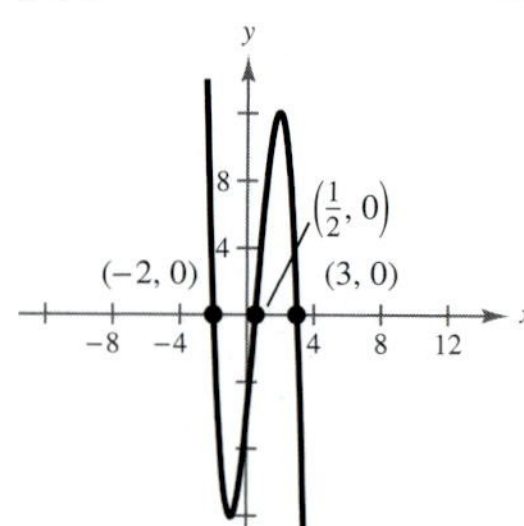

124. Answers will vary. Sample graph:

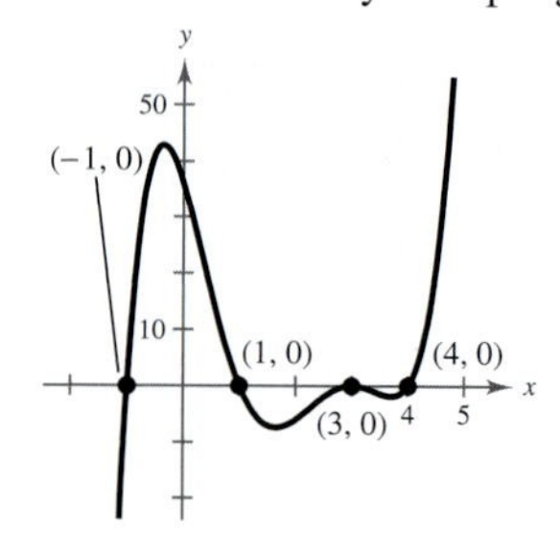

125. Answers will vary.

126. (a) $-2, 1, 4$
(b) The graph touches the x-axis at $x = 1$.
(c) The least possible degree of the function is 4, because there are at least four real zeros (1 is repeated) and a function can have at most the number of real zeros equal to the degree of the function. The degree cannot be odd by the definition of multiplicity.
(d) Positive. From the information in the table, it can be concluded that the graph will eventually rise to the left and rise to the right.
(e) $f(x) = x^4 - 4x^3 - 3x^2 + 14x - 8$
(f)

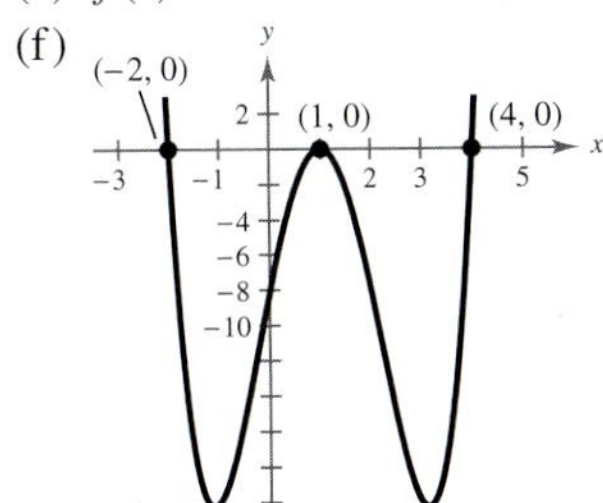

127. (a) $x^2 + b$ (b) $x^2 - 2ax + a^2 + b^2$

128. (a) Not correct because f has $(0, 0)$ as an intercept.
(b) Not correct because the function must be at least a fourth-degree polynomial.
(c) Correct function
(d) Not correct because k has $(-1, 0)$ as an intercept.

129. $-11 + 9i$ **130.** $12 + 11i$ **131.** $20 + 40i$

132. 106

133.

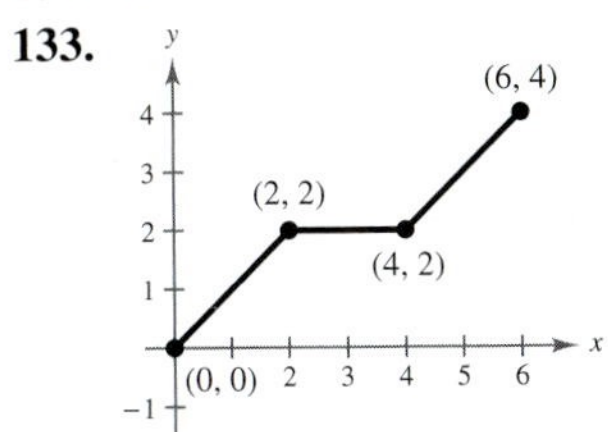

134.

135.

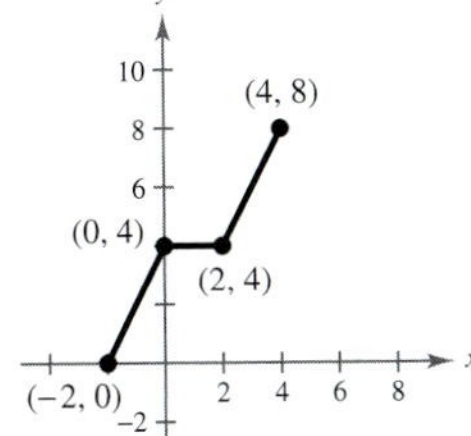

136.

137.

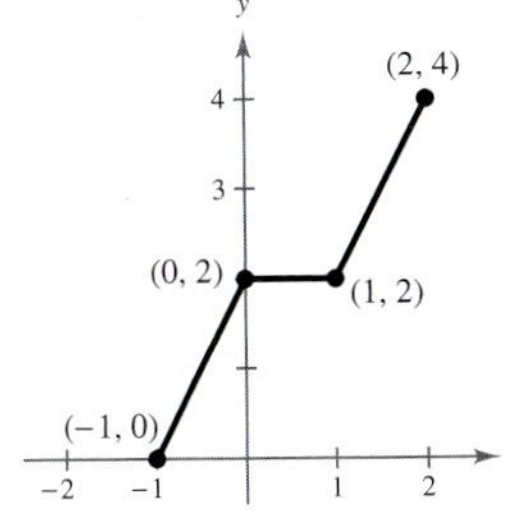

138.

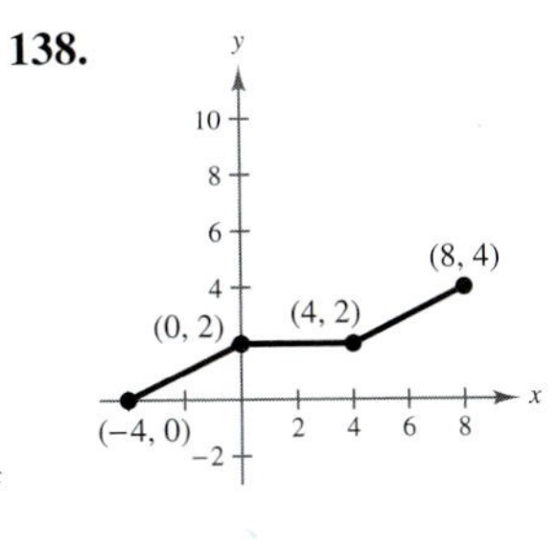

Section 2.6 *(page 193)*

Vocabulary Check *(page 193)*

1. rational functions **2.** vertical asymptote
3. horizontal asymptote **4.** slant asymptote

1. (a)

x	$f(x)$
0.5	-2
0.9	-10
0.99	-100
0.999	-1000

x	$f(x)$
1.5	2
1.1	10
1.01	100
1.001	1000

x	$f(x)$
5	0.25
10	$0.\overline{1}$
100	$0.\overline{01}$
1000	$0.\overline{001}$

(b) Vertical asymptote: $x = 1$
Horizontal asymptote: $y = 0$
(c) Domain: all real numbers x except $x = 1$

2. (a)

x	$f(x)$
0.5	-5
0.9	-45
0.99	-495
0.999	-4995

x	$f(x)$
1.5	15
1.1	55
1.01	505
1.001	5005

x	$f(x)$
5	6.25
10	$5.\overline{55}$
100	$5.\overline{05}$
1000	$5.\overline{005}$

(b) Vertical asymptote: $x = 1$
Horizontal asymptote: $y = 5$
(c) Domain: all real numbers x except $x = 1$

3. (a)

x	$f(x)$
0.5	-1
0.9	-12.79
0.99	-147.8
0.999	-1498

x	$f(x)$
1.5	5.4
1.1	17.29
1.01	152.3
1.001	1502

x	$f(x)$
5	3.125
10	$3.\overline{03}$
100	$3.\overline{0003}$
1000	3

(b) Vertical asymptotes: $x = \pm 1$
Horizontal asymptote: $y = 3$
(c) Domain: all real numbers x except $x = \pm 1$

4. (a)

x	$f(x)$
0.5	$-2.\overline{66}$
0.9	-18.95
0.99	-199
0.999	-1999

x	$f(x)$
1.5	4.8
1.1	20.95
1.01	201
1.001	2001

x	$f(x)$
5	$0.8\overline{33}$
10	$0.\overline{40}$
100	0.04
1000	0.004

(b) Vertical asymptotes: $x = \pm 1$
Horizontal asymptote: $y = 0$
(c) Domain: all real numbers x except $x = \pm 1$

5. Domain: all real numbers x except $x = 0$
Vertical asymptote: $x = 0$
Horizontal asymptote: $y = 0$

6. Domain: all real numbers x except $x = 2$
Vertical asymptote: $x = 2$
Horizontal asymptote: $y = 0$

7. Domain: all real numbers x except $x = 2$
Vertical asymptote: $x = 2$
Horizontal asymptote: $y = -1$

8. Domain: all real numbers x except $x = -\frac{1}{2}$
Vertical asymptote: $x = -\frac{1}{2}$
Horizontal asymptote: $y = -\frac{5}{2}$

9. Domain: all real numbers x except $x = \pm 1$
Vertical asymptotes: $x = \pm 1$

10. Domain: all real numbers x except $x = -1$
Vertical asymptote: $x = -1$
Horizontal asymptote: none

11. Domain: all real numbers x
Horizontal asymptote: $y = 3$

12. Domain: all real numbers x
Horizontal asymptote: $y = 3$
No vertical asymptote

13. d **14.** a **15.** c **16.** b

17. 1 **18.** None **19.** 6 **20.** 2

21. Domain: all real numbers x except $x = \pm 4$;
Vertical asymptote: $x = -4$; horizontal asymptote: $y = 0$

22. Domain: all real numbers x except $x = \pm 3$;
Vertical asymptote: $x = 3$; horizontal asymptote: $y = 0$

23. Domain: all real numbers x except $x = -1, 3$;
Vertical asymptote: $x = 3$; horizontal asymptote: $y = 1$

24. Domain: all real numbers x except $x = 1, 2$;
Vertical asymptote: $x = 1$; horizontal asymptote: $y = 1$

25. Domain: all real numbers x except $x = -1, \frac{1}{2}$;
Vertical asymptote: $x = \frac{1}{2}$; horizontal asymptote: $y = \frac{1}{2}$

26. Domain: all real numbers x except $x = -\frac{1}{3}, \frac{3}{2}$;
Vertical asymptote: $x = -\frac{1}{3}$; horizontal asymptote: $y = 1$

27. (a) Domain: all real numbers x except $x = -2$
(b) y-intercept: $\left(0, \frac{1}{2}\right)$
(c) Vertical asymptote: $x = -2$
Horizontal asymptote: $y = 0$
(d)

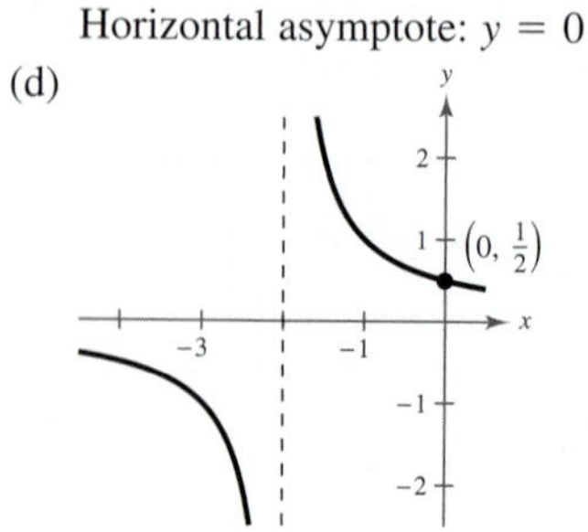

28. (a) Domain: all real numbers x except $x = 3$
(b) y-intercept: $\left(0, -\frac{1}{3}\right)$
(c) Vertical asymptote: $x = 3$
Horizontal asymptote: $y = 0$
(d)

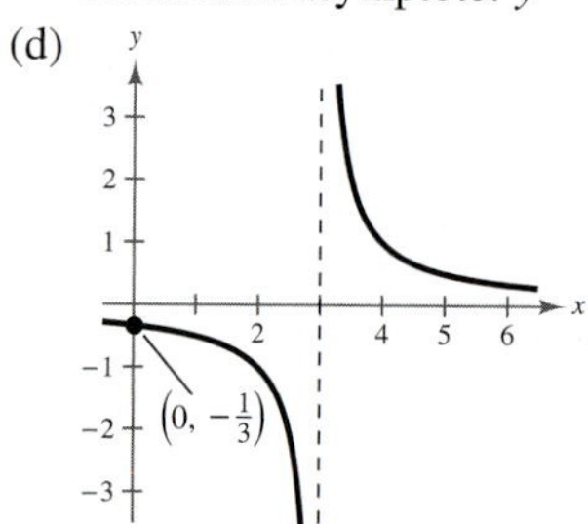

29. (a) Domain: all real numbers x except $x = -2$
(b) y-intercept: $\left(0, -\frac{1}{2}\right)$
(c) Vertical asymptote: $x = -2$
Horizontal asymptote: $y = 0$
(d)

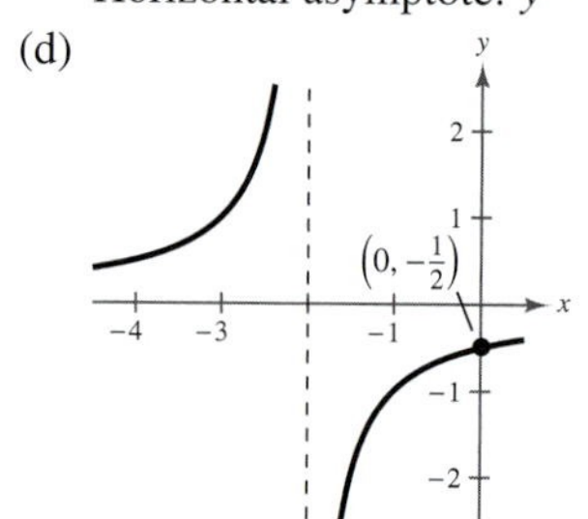

30. (a) Domain: all real numbers x except $x = 3$
(b) y-intercept: $\left(0, \frac{1}{3}\right)$
(c) Vertical asymptote: $x = 3$
Horizontal asymptote: $y = 0$
(d)

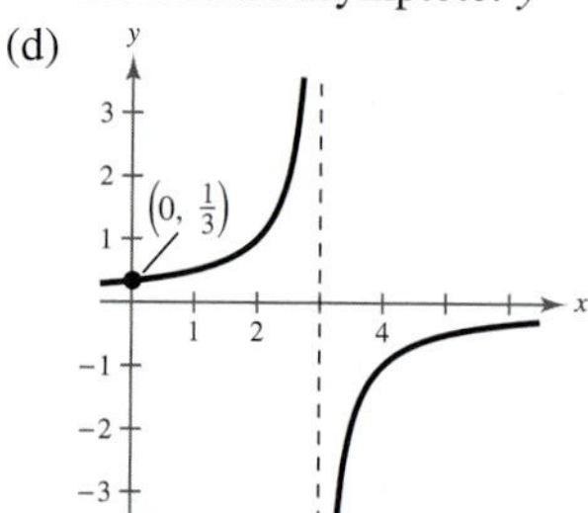

31. (a) Domain: all real numbers x except $x = -1$
(b) x-intercept: $\left(-\frac{5}{2}, 0\right)$
y-intercept: $(0, 5)$
(c) Vertical asymptote: $x = -1$
Horizontal asymptote: $y = 2$
(d)

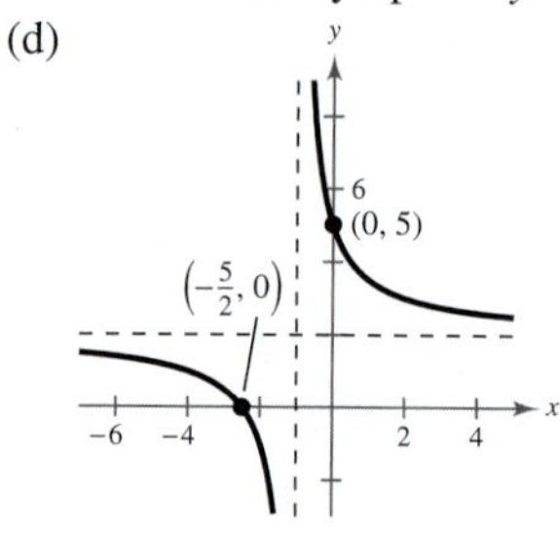

32. (a) Domain: all real numbers x except $x = 1$
(b) x-intercept: $\left(\frac{1}{3}, 0\right)$
y-intercept: $(0, 1)$
(c) Vertical asymptote: $x = 1$
Horizontal asymptote: $y = 3$
(d)

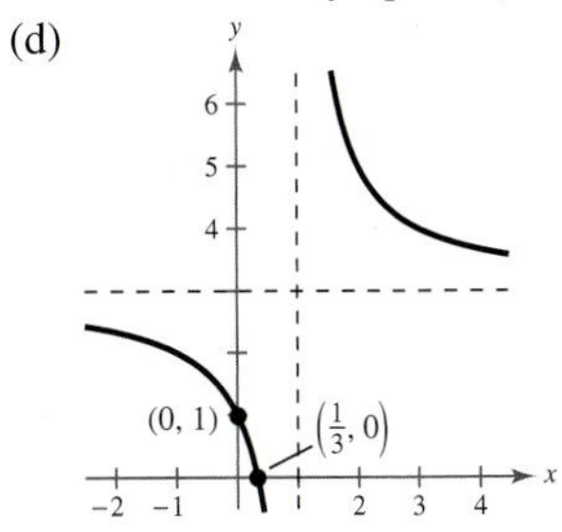

33. (a) Domain: all real numbers x
(b) Intercept: $(0, 0)$
(c) Horizontal asymptote: $y = 1$
(d)

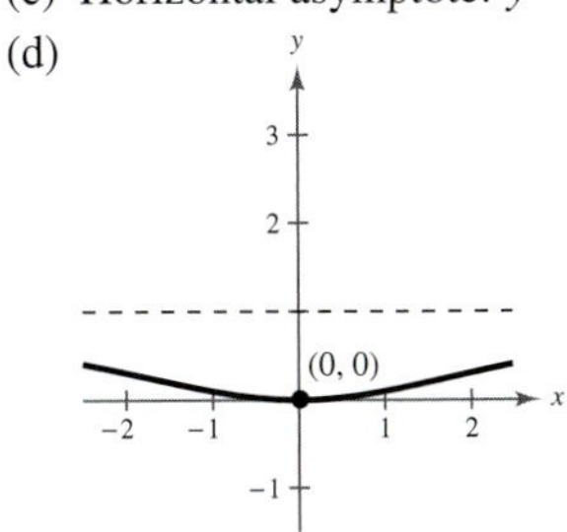

34. (a) Domain: all real numbers t except $t = 0$
(b) t-intercept: $\left(\frac{1}{2}, 0\right)$ (c) Vertical asymptote: $t = 0$
Horizontal asymptote: $y = -2$
(d)

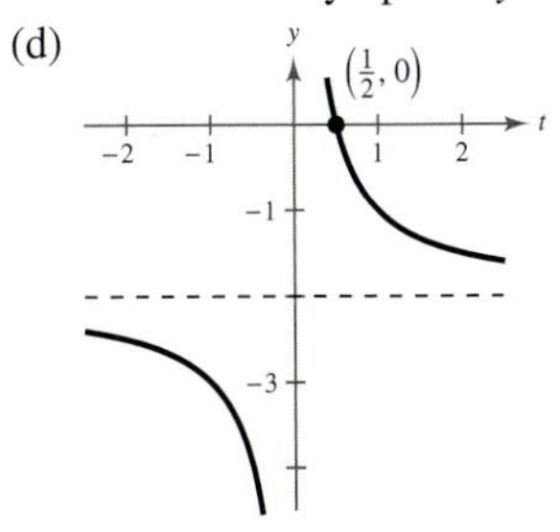

35. (a) Domain: all real numbers s
(b) Intercept: $(0, 0)$ (c) Horizontal asymptote: $y = 0$
(d)

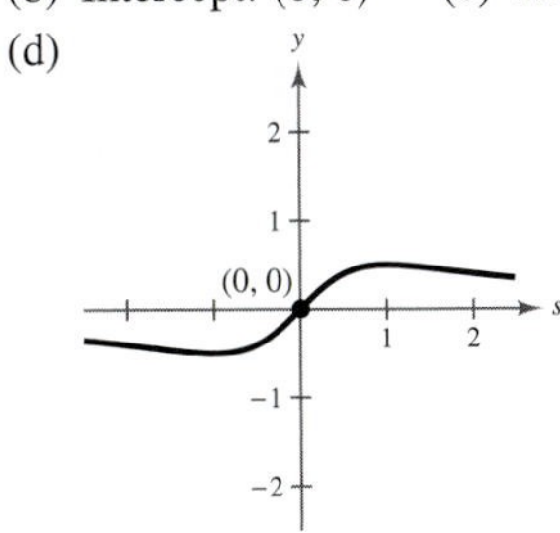

36. (a) Domain: all real numbers x except $x = 2$
(b) y-intercept: $\left(0, -\frac{1}{4}\right)$
(c) Vertical asymptote: $x = 2$
Horizontal asymptote: $y = 0$
(d)

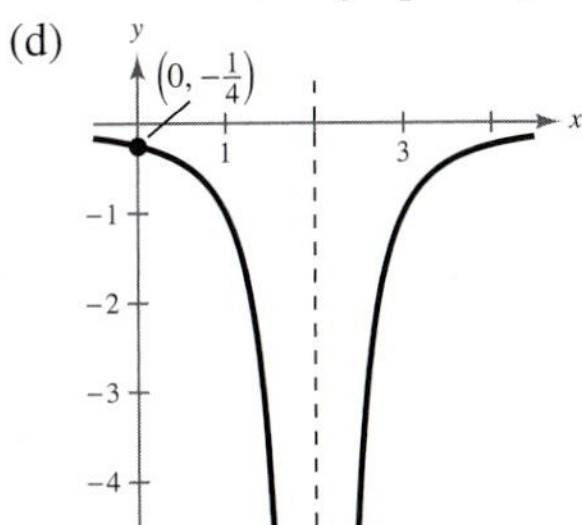

37. (a) Domain: all real numbers x except $x = \pm 2$
(b) x-intercepts: $(1, 0)$ and $(4, 0)$
y-intercept: $(0, -1)$
(c) Vertical asymptotes: $x = \pm 2$
Horizontal asymptote: $y = 1$
(d)

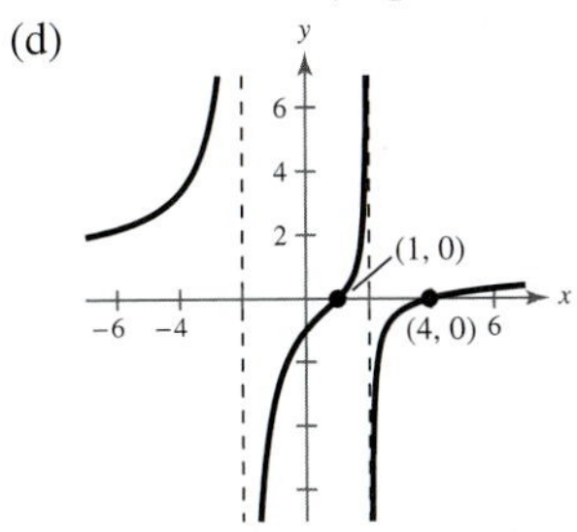

38. (a) Domain: all real numbers x except $x = \pm 3$
(b) x-intercepts: $(-2, 0)$, $(4, 0)$
y-intercept: $(0, 0.88)$
(c) Vertical asymptotes: $x = \pm 3$
Horizontal asymptote: $y = 1$
(d)

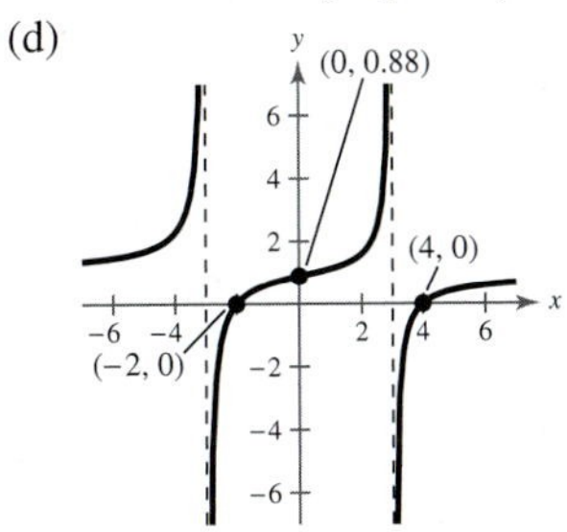

39. (a) Domain: all real numbers x except $x = \pm 1, 2$
(b) x-intercept: $(3, 0)$, $\left(-\frac{1}{2}, 0\right)$
y-intercept: $\left(0, -\frac{3}{2}\right)$
(c) Vertical asymptotes: $x = 2$, $x = \pm 1$
Horizontal asymptote: $y = 0$

(d)

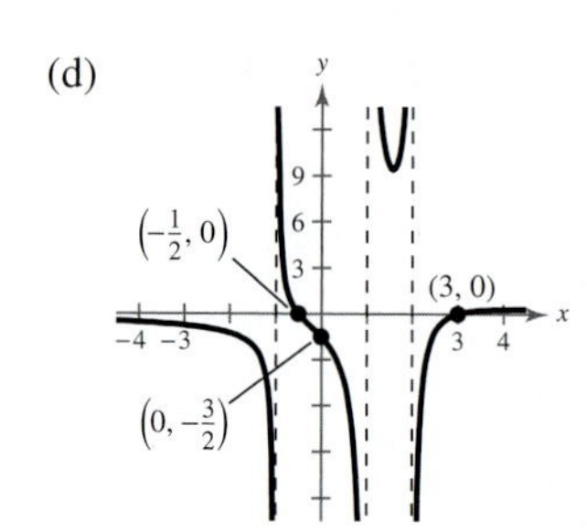

40. (a) Domain: all real numbers x except $x = -2, 1, 3$
(b) x-intercepts: $(-1, 0), (2, 0)$
y-intercept: $\left(0, -\frac{1}{3}\right)$
(c) Vertical asymptotes: $x = -2, x = 1, x = 3$
Horizontal asymptote: $y = 0$
(d)

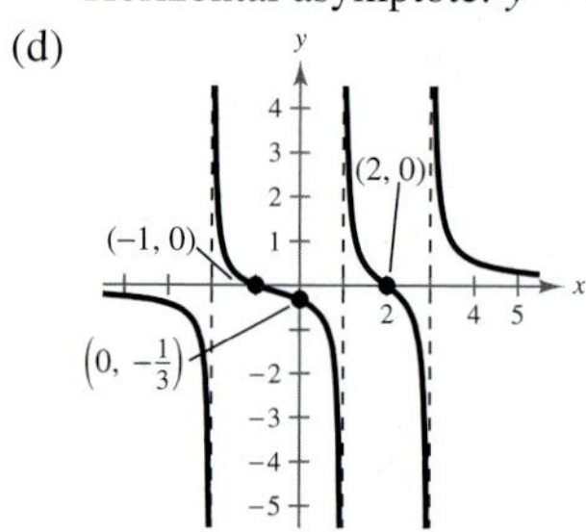

41. (a) Domain: all real numbers x except $x = 2, -3$
(b) Intercept: $(0, 0)$
(c) Vertical asymptote: $x = 2$
Horizontal asymptote: $y = 1$
(d)

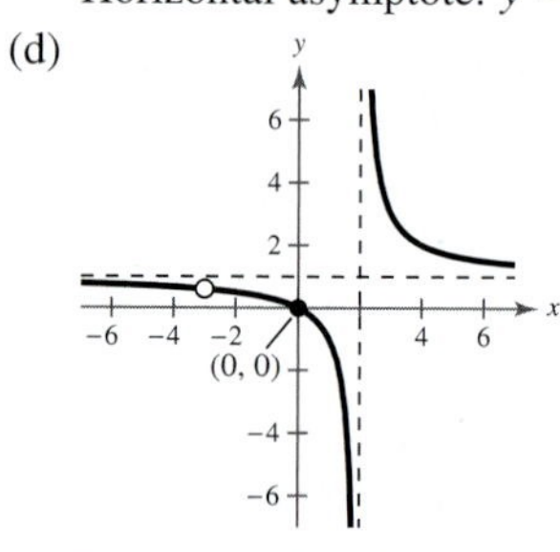

42. (a) Domain: all real numbers x except $x = 3, -4$
(b) y-intercept: $(0, -1.66)$
(c) Vertical asymptote: $x = 3$
Horizontal asymptote: $y = 0$
(d)

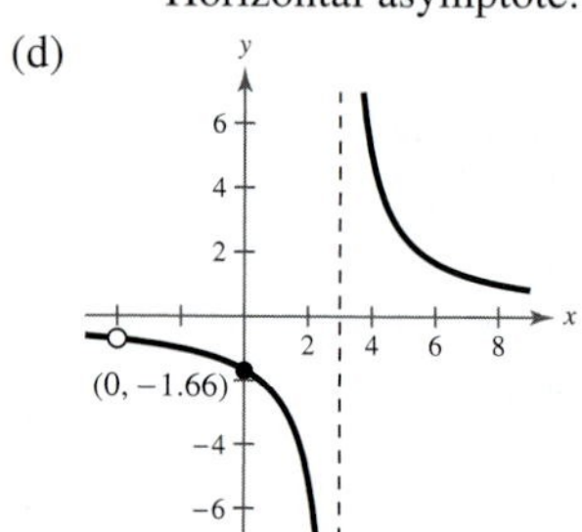

43. (a) Domain: all real numbers x except $x = -\frac{3}{2}, 2$
(b) x-intercept: $\left(\frac{1}{2}, 0\right)$
y-intercept: $\left(0, \frac{1}{3}\right)$
(c) Vertical asymptote: $x = -\frac{3}{2}$
Horizontal asymptote: $y = 1$
(d)

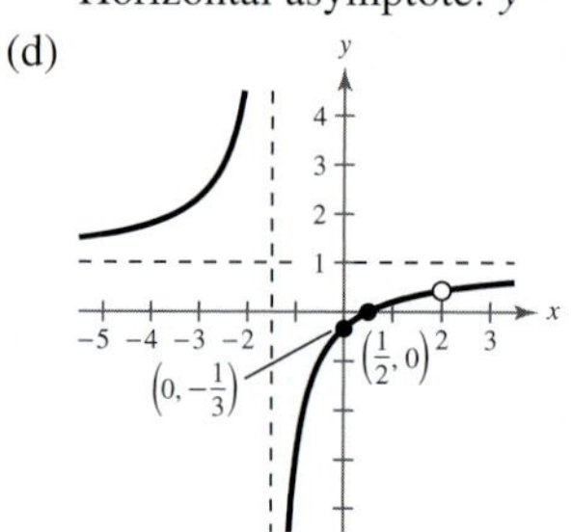

44. (a) Domain: all real numbers x except $x = -0.5, 2$
(b) x-intercept: $\left(\frac{2}{3}, 0\right)$
y-intercept: $(0, -2)$
(c) Vertical asymptote: $x = -0.5$
Horizontal asymptote: $y = \frac{3}{2}$
(d)

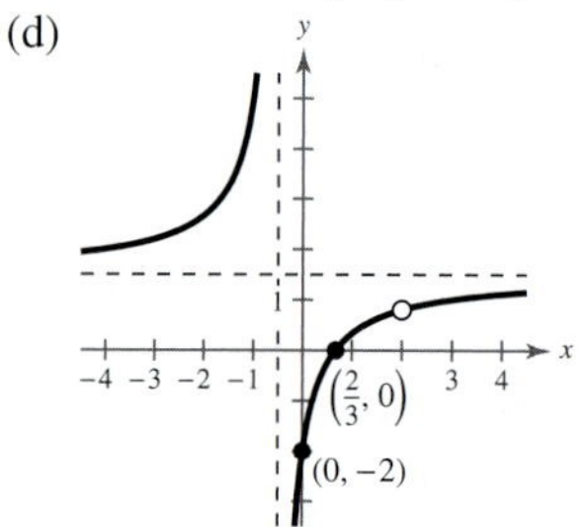

45. (a) Domain: all real numbers t except $t = -1$
(b) t-intercept: $(1, 0)$
y-intercept: $(0, -1)$
(c) Vertical asymptote: None
Horizontal asymptote: None
(d)

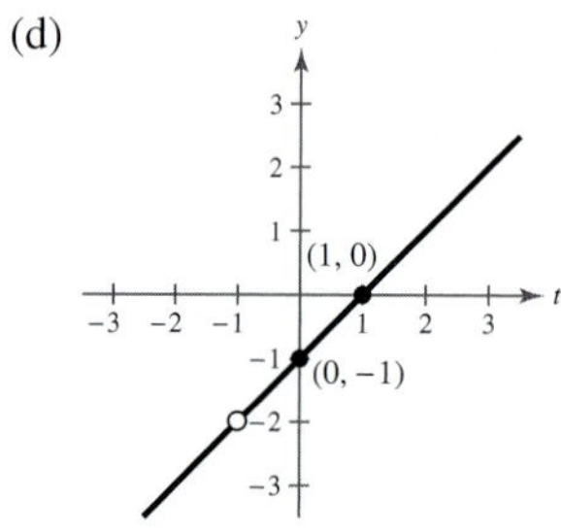

46. (a) Domain: all real numbers x except $x = 4$
(b) x-intercept: $(-4, 0)$
y-intercept: $(0, 4)$
(c) Vertical asymptote: None
Horizontal asymptote: None
(d)

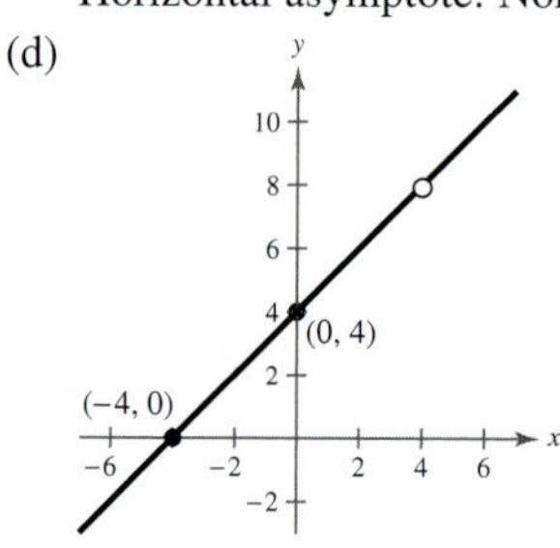

47. (a) Domain of f: all real numbers x except $x = -1$
Domain of g: all real numbers x
(b) $x - 1$; Vertical asymptotes: none
(c)

x	-3	-2	-1.5	-1	-0.5	0	1
$f(x)$	-4	-3	-2.5	Undef.	-1.5	-1	0
$g(x)$	-4	-3	-2.5	-2	-1.5	-1	0

(d)
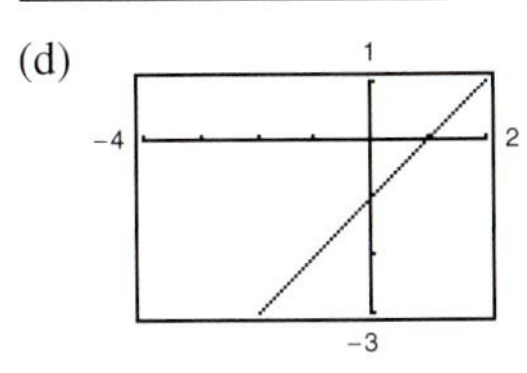

(e) Because there are only a finite number of pixels, the graphing utility may not attempt to evaluate the function where it does not exist.

48. (a) Domain of f: all real numbers x except $x = 0, 2$
Domain of g: all real numbers x
(b) x; Vertical asymptotes: none
(c)

x	-1	0	1	1.5	2	2.5	3
$f(x)$	-1	Undef.	1	1.5	Undef.	2.5	3
$g(x)$	-1	0	1	1.5	2	2.5	3

(d)
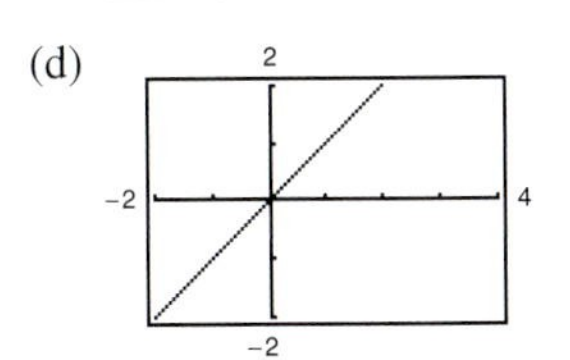

(e) Because there are only a finite number of pixels, the graphing utility may not attempt to evaluate the function where it does not exist.

49. (a) Domain of f: all real numbers x except $x = 0, 2$
Domain of g: all real numbers x except $x = 0$
(b) $\frac{1}{x}$; Vertical asymptote: $x = 0$
(c)

x	-0.5	0	0.5	1	1.5	2	3
$f(x)$	-2	Undef.	2	1	$\frac{2}{3}$	Undef.	$\frac{1}{3}$
$g(x)$	-2	Undef.	2	1	$\frac{2}{3}$	$\frac{1}{2}$	$\frac{1}{3}$

(d)
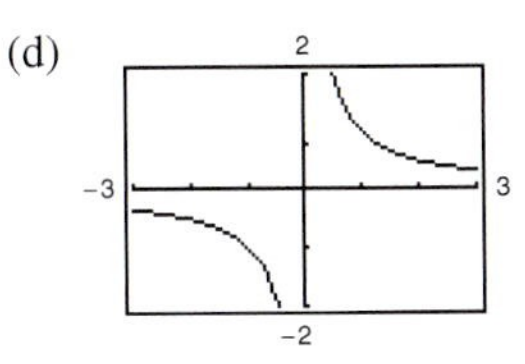

(e) Because there are only a finite number of pixels, the graphing utility may not attempt to evaluate the function where it does not exist.

50. (a) Domain of f: all real numbers x except $x = 3, 4$
Domain of g: all real numbers x except $x = 4$
(b) $\frac{2}{x - 4}$; Vertical asymptote: $x = 4$
(c)

x	0	1	2	3	4	5	6
$f(x)$	$-\frac{1}{2}$	$-\frac{2}{3}$	-1	Undef.	Undef.	2	1
$g(x)$	$-\frac{1}{2}$	$-\frac{2}{3}$	-1	-2	Undef.	2	1

(d)
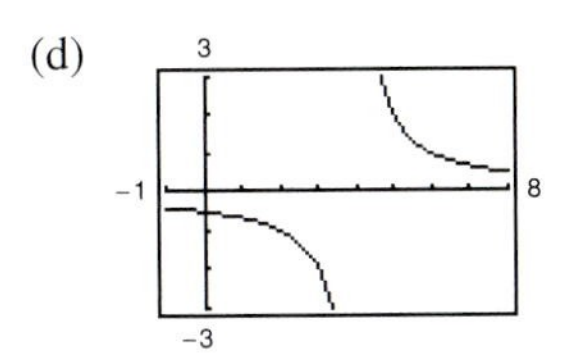

(e) Because there are only a finite number of pixels, the graphing utility may not attempt to evaluate the function where it does not exist.

51. (a) Domain: all real numbers x except $x = 0$
(b) x-intercepts: $(2, 0)$, $(-2, 0)$
(c) Vertical asymptote: $x = 0$
Slant asymptote: $y = x$
(d)
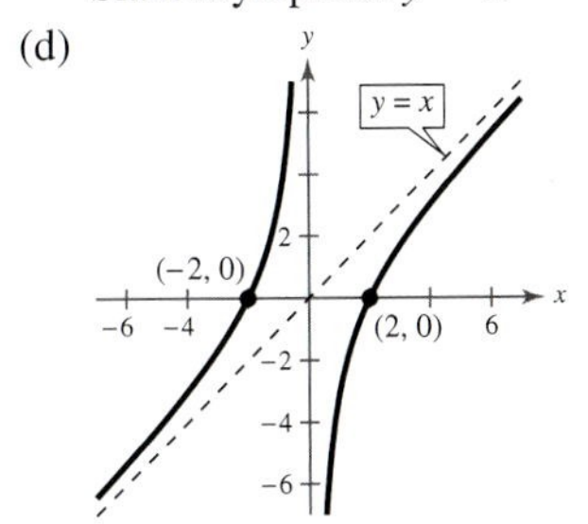

52. (a) Domain: all real numbers x except $x = 0$
(b) No intercepts
(c) Vertical asymptote: $x = 0$
Slant asymptote: $y = x$
(d)
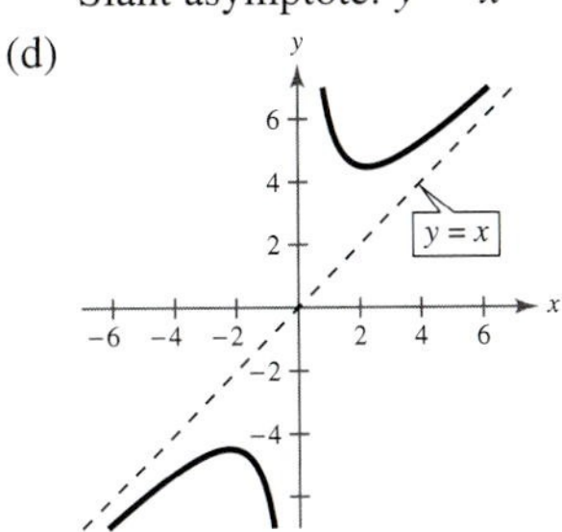

53. (a) Domain: all real numbers x except $x = 0$
(b) No intercepts
(c) Vertical asymptote: $x = 0$
Slant asymptote: $y = 2x$

74. (a)

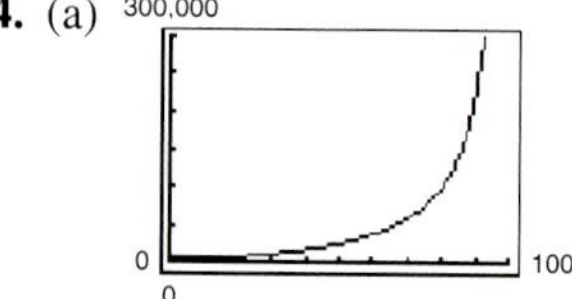

(b) \$4411.76; \$25,000; \$225,000

(c) No. The function is undefined at $p = 100$.

75. (a) 333 deer, 500 deer, 800 deer (b) 1500 deer

76. (a) Answers will vary. (b) $[0, 950]$

(c)

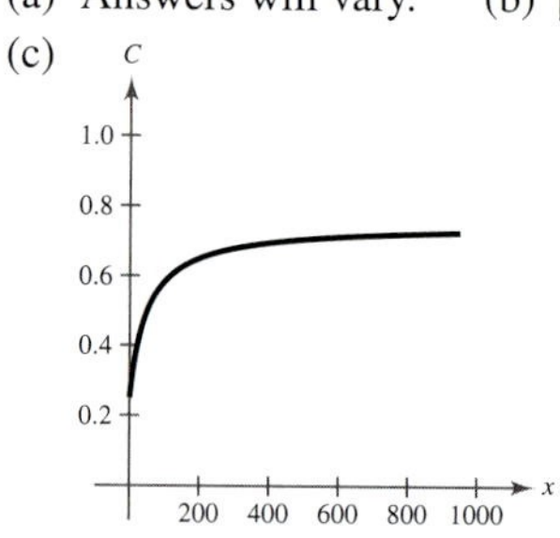

(d) Increases more slowly; 75%

77. (a) Answers will vary. (b) $(4, \infty)$

(c)

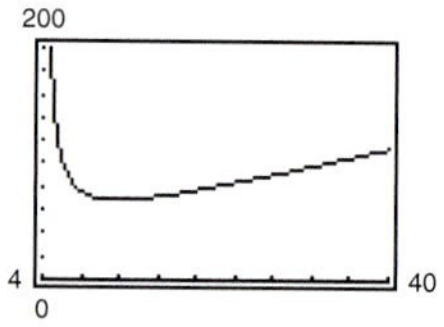

11.75 inches × 5.87 inches

78. 12.8 inches × 8.5 inches

79. (a) Answers will vary.

(b) Vertical asymptote: $x = 25$

Horizontal asymptote: $y = 25$

(c)

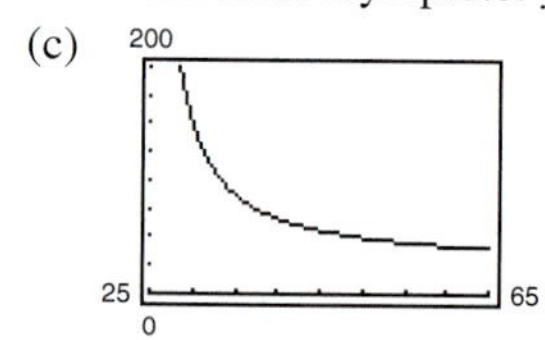

(d)

x	30	35	40	45	50	55	60
y	150	87.5	66.7	56.3	50	45.8	42.9

(e) Yes. You would expect the average speed for the round trip to be the average of the average speeds for the two parts of the trip.

(f) No. At 20 miles per hour you would use more time in one direction than is required for the round trip at an average speed of 50 miles per hour.

80. (a)

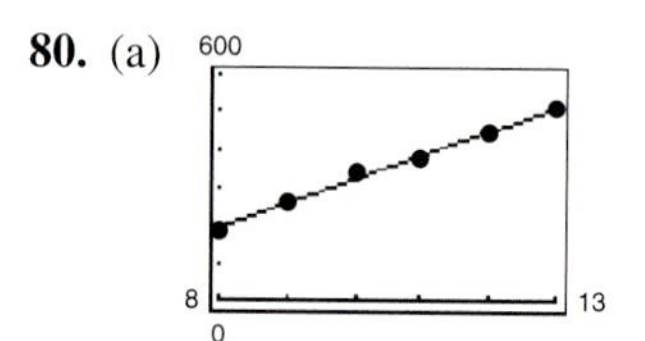

The model is a good fit for the data.

(b) \$763.8 million in sales

(c) No, horizontal asymptote at $y = 1454$.

81. False. Polynomials do not have vertical asymptotes.

82. False. The graph of $f(x) = \dfrac{x}{x^2 + 1}$ crosses $y = 0$, which is a horizontal asymptote.

83. $f(x) = \dfrac{2x^2}{x^2 + 1}$

84. $f(x) = \dfrac{x^3}{(x + 2)(x - 1)}$

85. $(x - 7)(x - 8)$ **86.** $(3x - 4)(x + 9)$

87. $(x - 5)(x + 2i)(x - 2i)$

88. $(x + 6)(x + \sqrt{2})(x - \sqrt{2})$

89. $x \geq \frac{10}{3}$

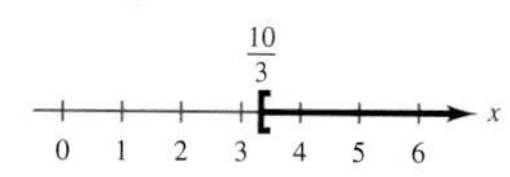

90. $x < 0$

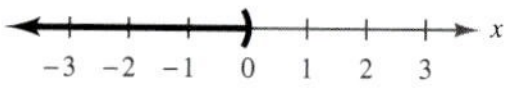

91. $-3 < x < 7$

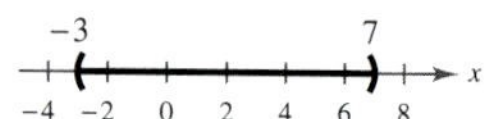

92. $x \leq -\frac{13}{2}, x \geq \frac{7}{2}$

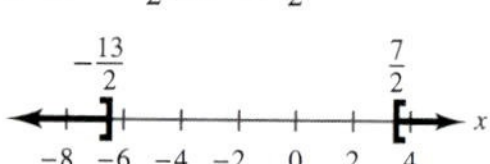

93. Answers will vary.

Section 2.7 *(page 204)*

Vocabulary Check *(page 204)*

1. critical; test intervals **2.** zeros; undefined values

3. $P = R - C$

1. (a) No (b) Yes (c) Yes (d) No

2. (a) Yes (b) No (c) Yes (d) Yes

3. (a) Yes (b) No (c) No (d) Yes

4. (a) No (b) Yes (c) Yes (d) No

5. $2, -\frac{3}{2}$ **6.** $0, \frac{25}{9}$ **7.** $\frac{7}{2}, 5$ **8.** $-2, -1, 1, 4$

9. $[-3, 3]$ **10.** $(-6, 6)$

11. $(-7, 3)$ **12.** $(-\infty, 2] \cup [4, \infty)$

13. $(-\infty, -5] \cup [1, \infty)$ **14.** $(-1, 7)$

15. $(-3, 2)$

16. $(-\infty, -3) \cup (1, \infty)$

17. $(-3, 1)$

18. $(-\infty, 2 - \sqrt{5}) \cup (2 + \sqrt{5}, \infty)$

19. $(-\infty, -4 - \sqrt{21}] \cup [-4 + \sqrt{21}, \infty)$

20. $\left(-\infty, \dfrac{3}{2} - \dfrac{\sqrt{39}}{2}\right] \cup \left[\dfrac{3}{2} + \dfrac{\sqrt{39}}{2}, \infty\right)$

21. $(-1, 1) \cup (3, \infty)$

22. $(-\infty, 2]$

23. $[-3, 2] \cup [3, \infty)$

24. $\left[-\frac{13}{2}, -2\right] \cup [2, \infty)$

25. $x = \frac{1}{2}$

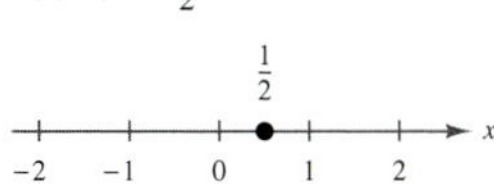

26. All real numbers

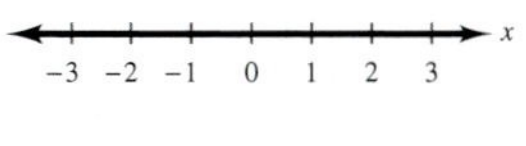

27. $(-\infty, 0) \cup \left(0, \frac{3}{2}\right)$ **28.** $(3, \infty)$

29. $[-2, 0] \cup [2, \infty)$ **30.** $(-\infty, 0] \cup [2, \infty)$

31. $[-2, \infty)$ **32.** $(-\infty, 3]$

33.

(a) $x \le -1,\ x \ge 3$

(b) $0 \le x \le 2$

34.

(a) $2 - \sqrt{2} \le x \le 2 + \sqrt{2}$

(b) $x \le -2,\ x \ge 6$

35.

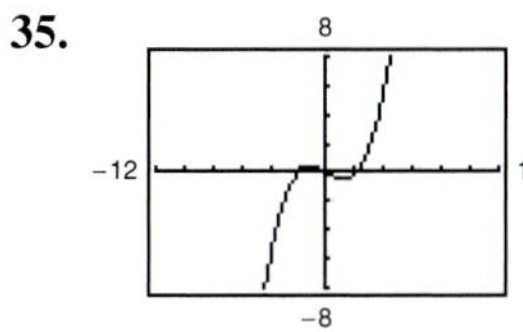

(a) $-2 \le x \le 0,$
$2 \le x < \infty$

(b) $x \le 4$

36.

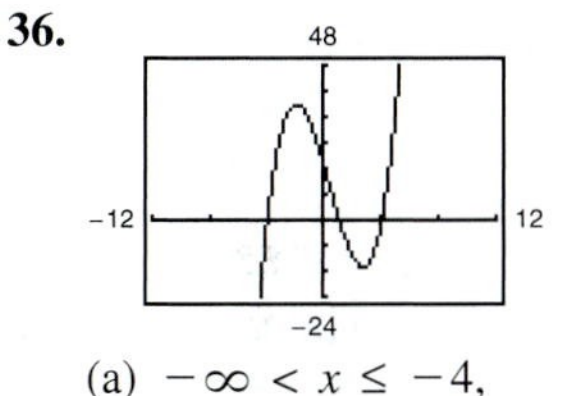

(a) $-\infty < x \le -4,$
$1 \le x \le 4$

(b) $x = -2,\ 5 \le x < \infty$

37. $(-\infty, -1) \cup (0, 1)$

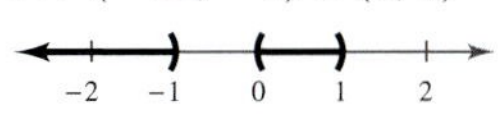

38. $(-\infty, 0) \cup \left(\frac{1}{4}, \infty\right)$

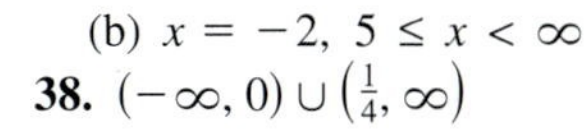

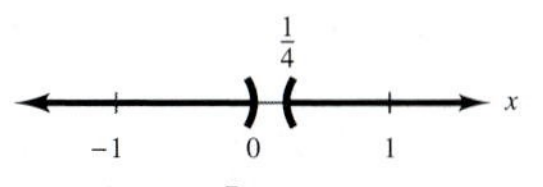

39. $(-\infty, -1) \cup (4, \infty)$ **40.** $(-2, 3]$

41. $(5, 15)$ **42.** $\left(-\infty, -\frac{1}{2}\right) \cup (1, \infty)$

43. $\left(-5, -\frac{3}{2}\right) \cup (-1, \infty)$ **44.** $(-14, -2) \cup (6, \infty)$

45. $\left(-\frac{3}{4}, 3\right) \cup [6, \infty)$ **46.** $(-\infty, -3) \cup (0, \infty)$

47. $(-3, -2] \cup [0, 3)$ **48.** $[-3, 0) \cup [2, \infty)$

49. $(-\infty, -1) \cup \left(-\frac{2}{3}, 1\right) \cup (3, \infty)$

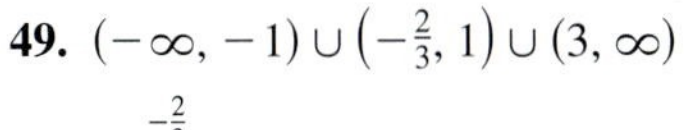

CHAPTER 2

50. $(-\infty, -4) \cup [-2, 1) \cup [6, \infty)$

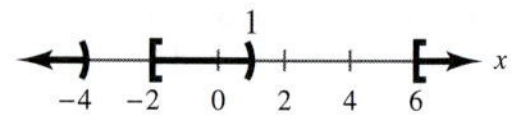

51.

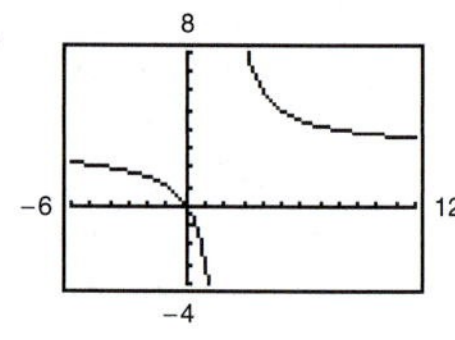

(a) $0 \le x < 2$

(b) $2 < x \le 4$

52.

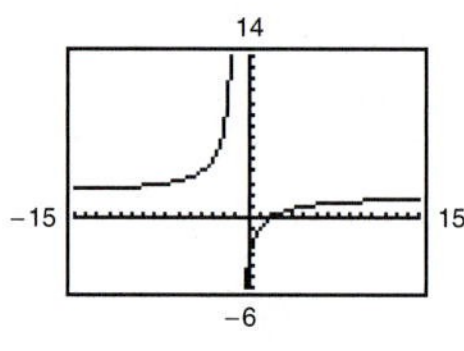

(a) $-1 < x \le 2$

(b) $-2 \le x < -1$

53.

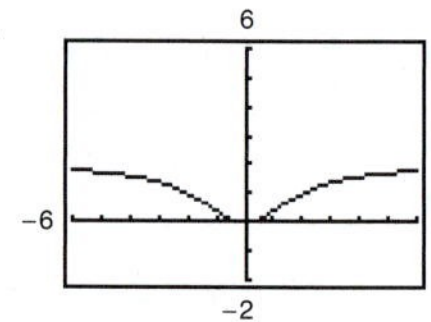

(a) $|x| \ge 2$

(b) $-\infty < x < \infty$

54.

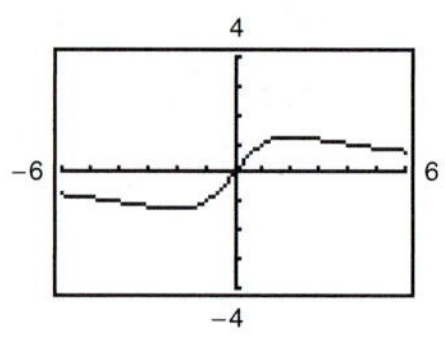

(a) $1 \le x \le 4$

(b) $-\infty < x \le 0$

55. $[-2, 2]$ **56.** $(-\infty, -2] \cup [2, \infty)$

57. $(-\infty, 3] \cup [4, \infty)$ **58.** $[-4, 4]$

59. $(-5, 0] \cup (7, \infty)$ **60.** $(-3, 0] \cup (3, \infty)$

61. $(-3.51, 3.51)$ **62.** $(-1.13, 1.13)$

63. $(-0.13, 25.13)$ **64.** $(-4.42, 0.42)$

65. $(2.26, 2.39)$ **66.** $(1.19, 1.30)$

67. (a) $t = 10$ seconds (b) 4 seconds $< t <$ 6 seconds

68. (a) $t = 8$ seconds

(b) 0 seconds $\le t < 4 - 2\sqrt{2}$ seconds and $4 + 2\sqrt{2}$ seconds $< t \le 8$ seconds

69. 13.8 meters $\le L \le$ 36.2 meters

70. 45.97 feet $\le L \le$ 174.03 feet

71. $40{,}000 \le x \le 50{,}000$; $50.00 \le p \le 55.00$

72. 90,000 units $\le x \le$ 100,000 units; $30.00 \le p \le 32.00$

73. (a)

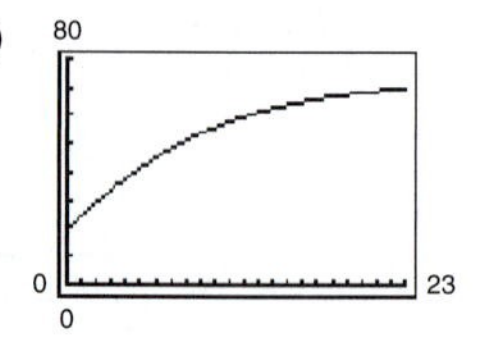

(b)

t	24	26	28	30	32	34
C	70.5	71.6	72.9	74.6	76.8	79.6

2011

(c) $t \approx 31$

(d)

t	36	37	38	39
C	83.2	85.4	87.8	90.5

t	40	41	42	43
C	93.5	96.8	100.4	104.4

2016 to 2021

(e) $37 \le t \le 41$

(f) Answers will vary.

74. (a)

d	4	6	8	10	12
Load	2223.9	5593.9	10,312	16,378	23,792

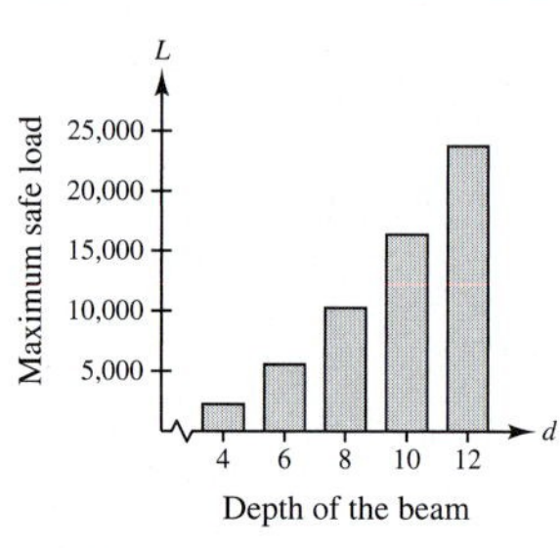

(b) 3.83 inches

75. $R_1 \ge 2$ ohms

76. (a) 1995 (b) $t \approx 5.1$ (c) 2006 (d) $t \approx 16.25$

77. True. The test intervals are $(-\infty, -3)$, $(-3, 1)$, $(1, 4)$, and $(4, \infty)$.

78. True. The y-values are greater than zero for all values of x.

79. $(-\infty, -4] \cup [4, \infty)$ **80.** $(-\infty, \infty)$

81. $\left(-\infty, -2\sqrt{30}\right] \cup \left[2\sqrt{30}, \infty\right)$

82. $\left(-\infty, -2\sqrt{10}\right] \cup \left[2\sqrt{10}, \infty\right)$

83. (a) If $a > 0$ and $c \le 0$, b can be any real number. If $a > 0$ and $c > 0$, $b < -2\sqrt{ac}$ or $b > 2\sqrt{ac}$.

(b) 0

84. (a) $x = a$, $x = b$

(b)

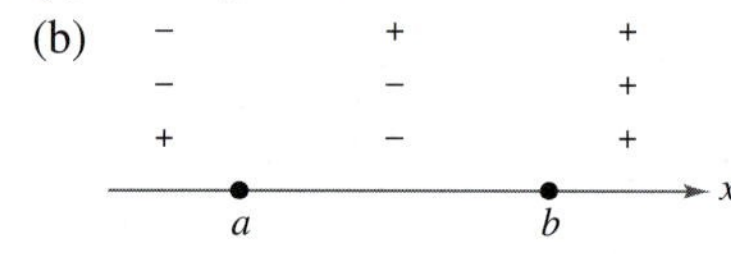

(c) The real zeros of the polynomial

85. $(2x + 5)^2$ **86.** $(x + 7)(x - 1)$

87. $(x + 3)(x + 2)(x - 2)$ **88.** $2x(x - 3)(x^2 + 3x + 9)$

89. $2x^2 + x$ **90.** $\frac{3}{2}b^2 + b$

Review Exercises *(page 208)*

1. (a)

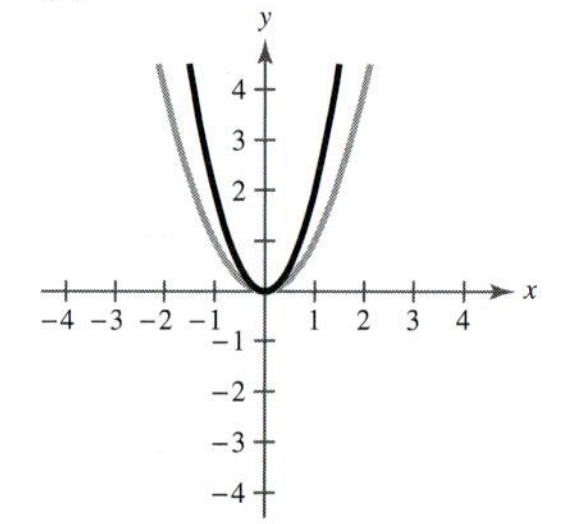

Vertical stretch

(b)

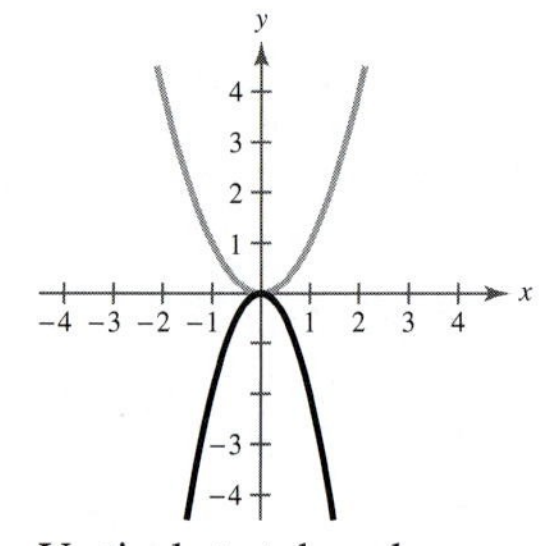

Vertical stretch and reflection in the x-axis

(c)

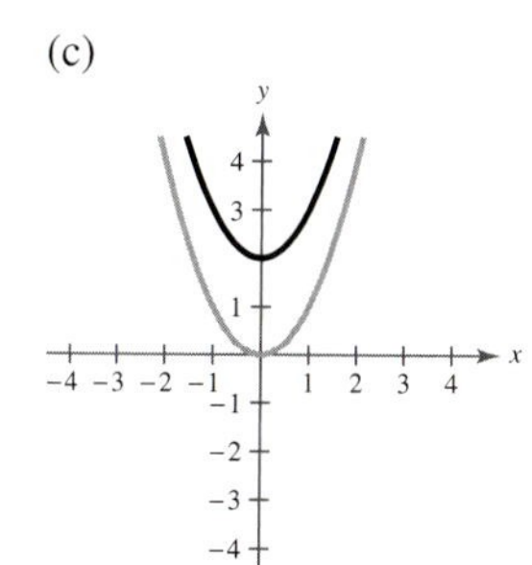

Vertical shift

(d)

Horizontal shift

2. (a)

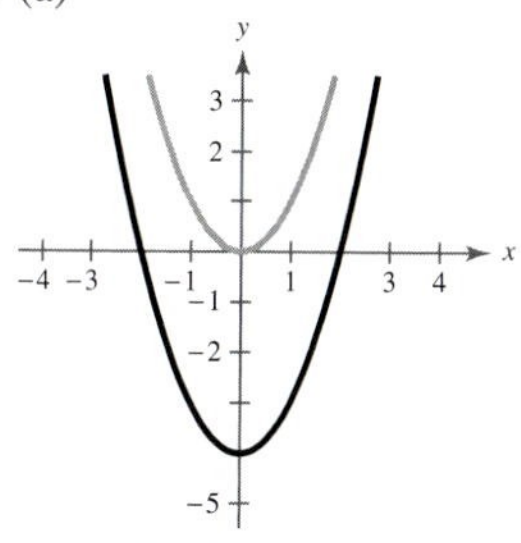

Vertical shift

(b)

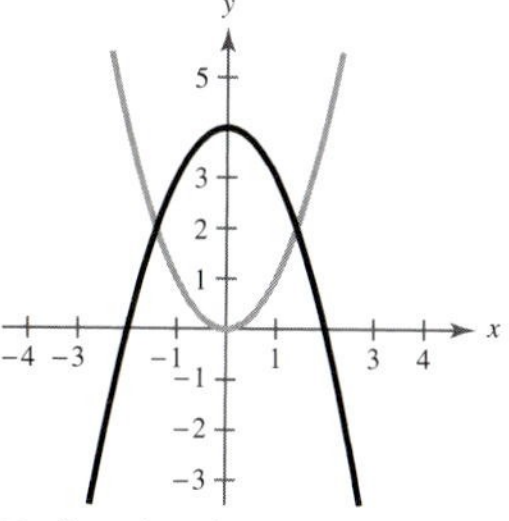

Reflection in the x-axis and vertical shift

(c)

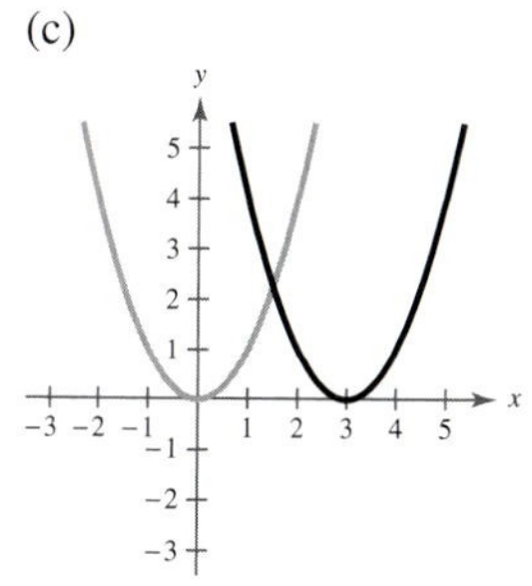

Horizontal shift

(d)

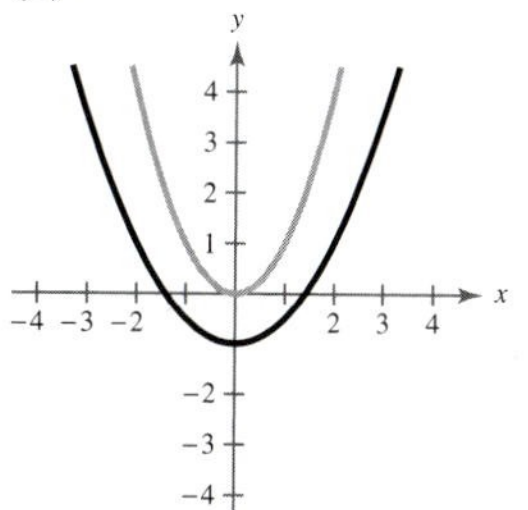

Vertical shrink and vertical shift

3. $g(x) = (x - 1)^2 - 1$

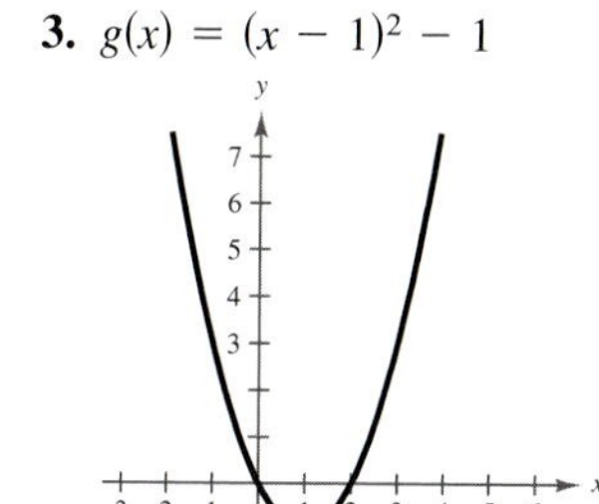

Vertex: $(1, -1)$
Axis of symmetry: $x = 1$
x-intercepts: $(0, 0)$, $(2, 0)$

4. $f(x) = -(x - 3)^2 + 9$

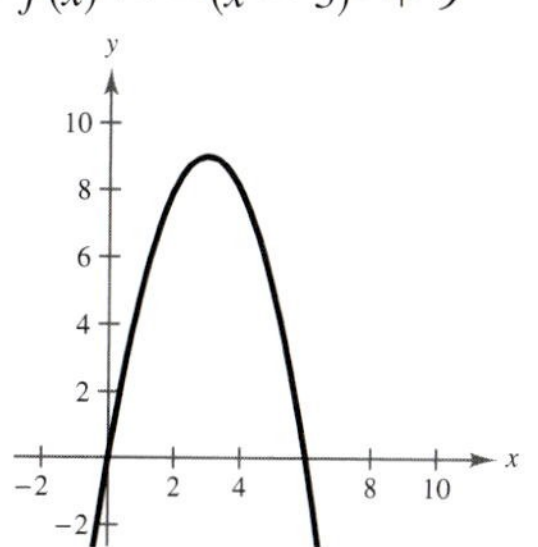

Vertex: $(3, 9)$
Axis of symmetry: $x = 3$
x-intercepts: $(0, 0)$, $(6, 0)$

5. $f(x) = (x + 4)^2 - 6$

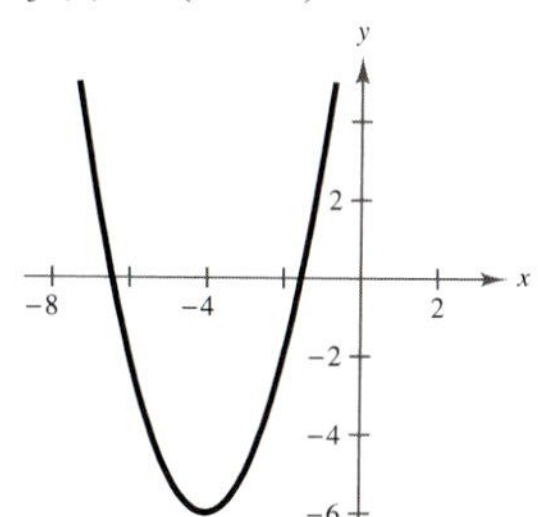

Vertex: $(-4, -6)$
Axis of symmetry: $x = -4$
x-intercepts: $\left(-4 \pm \sqrt{6}, 0\right)$

6. $h(x) = -(x - 2)^2 + 7$

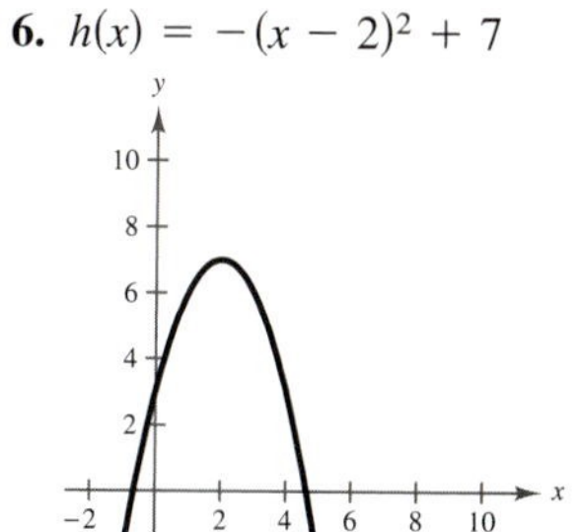

Vertex: $(2, 7)$
Axis of symmetry: $x = 2$
x-intercepts: $\left(2 \pm \sqrt{7}, 0\right)$

7. $f(t) = -2(t - 1)^2 + 3$

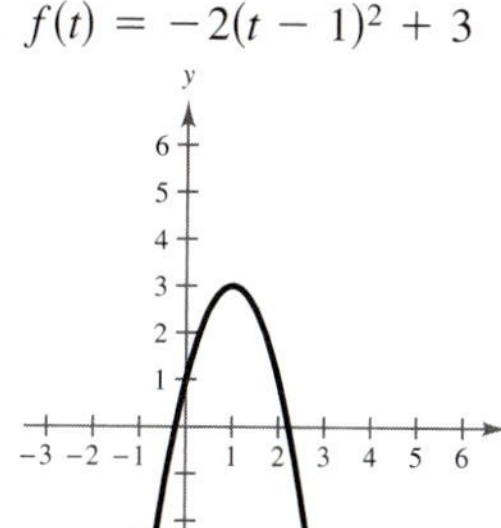

Vertex: $(1, 3)$
Axis of symmetry: $t = 1$
t-intercepts: $\left(1 \pm \dfrac{\sqrt{6}}{2}, 0\right)$

8. $f(x) = (x - 4)^2 - 4$

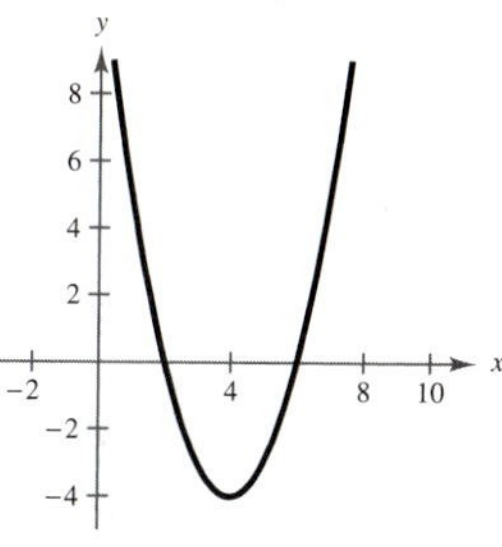

Vertex: $(4, -4)$
Axis of symmetry: $x = 4$
x-intercepts: $(2, 0)$, $(6, 0)$

9. $h(x) = 4\left(x + \frac{1}{2}\right)^2 + 12$

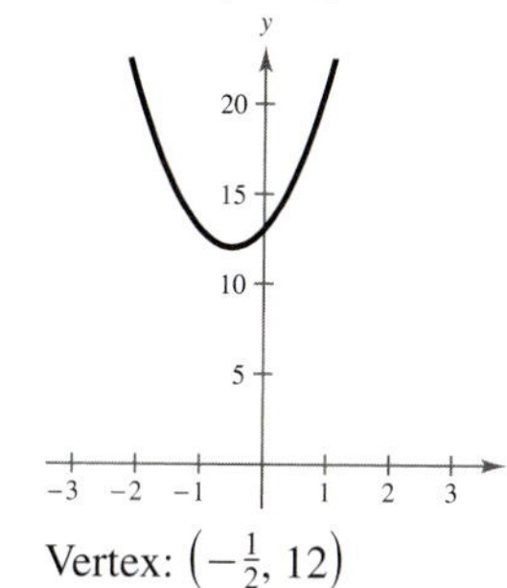

Vertex: $\left(-\frac{1}{2}, 12\right)$
Axis of symmetry: $x = -\frac{1}{2}$
No x-intercept

10. $f(x) = (x - 3)^2 - 8$

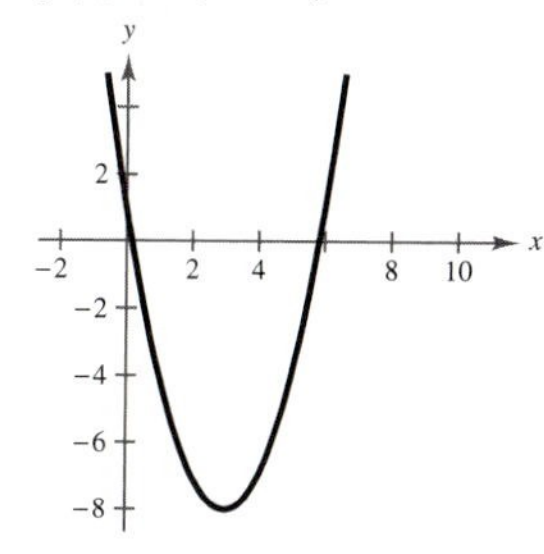

Vertex: $(3, -8)$
Axis of symmetry: $x = 3$
x-intercepts: $\left(3 \pm 2\sqrt{2}, 0\right)$

Section 3.2 *(page 236)*

Vocabulary Check *(page 236)*

1. logarithmic **2.** 10 **3.** natural; e
4. $a^{\log_a x} = x$ **5.** $x = y$

1. $4^3 = 64$ **2.** $3^4 = 81$ **3.** $7^{-2} = \frac{1}{49}$
4. $10^{-3} = \frac{1}{1000}$ **5.** $32^{2/5} = 4$ **6.** $16^{3/4} = 8$
7. $36^{1/2} = 6$ **8.** $8^{2/3} = 4$ **9.** $\log_5 125 = 3$
10. $\log_8 64 = 2$ **11.** $\log_{81} 3 = \frac{1}{4}$ **12.** $\log_9 27 = \frac{3}{2}$
13. $\log_6 \frac{1}{36} = -2$ **14.** $\log_4 \frac{1}{64} = -3$ **15.** $\log_7 1 = 0$
16. $\log 0.001 = -3$ **17.** 4 **18.** $\frac{1}{2}$ **19.** 0
20. 1 **21.** 2 **22.** -3 **23.** -0.097
24. -2.699 **25.** 1.097 **26.** 1.877 **27.** 4
28. 0 **29.** 1 **30.** 15

31.
Domain: $(0, \infty)$
x-intercept: $(1, 0)$
Vertical asymptote: $x = 0$

32.
Domain: $(0, \infty)$
x-intercept: $(1, 0)$
Vertical asymptote: $x = 0$

33.
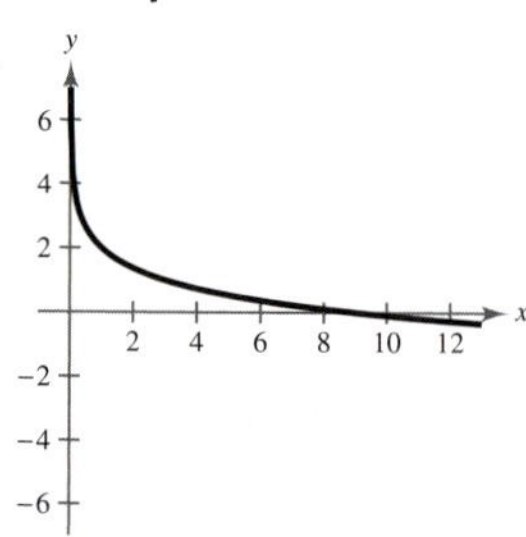
Domain: $(0, \infty)$
x-intercept: $(9, 0)$
Vertical asymptote: $x = 0$

34.
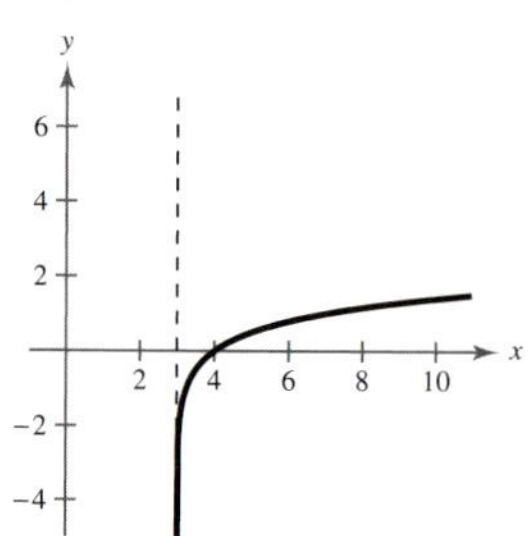
Domain: $(3, \infty)$
x-intercept: $(4, 0)$
Vertical asymptote: $x = 3$

35.
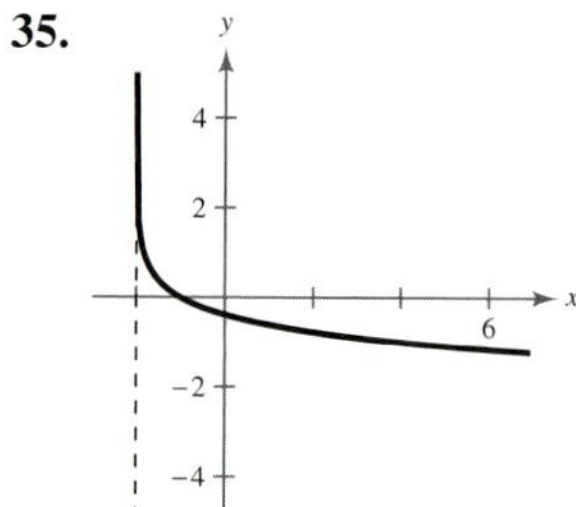
Domain: $(-2, \infty)$
x-intercept: $(-1, 0)$
Vertical asymptote: $x = -2$

36.
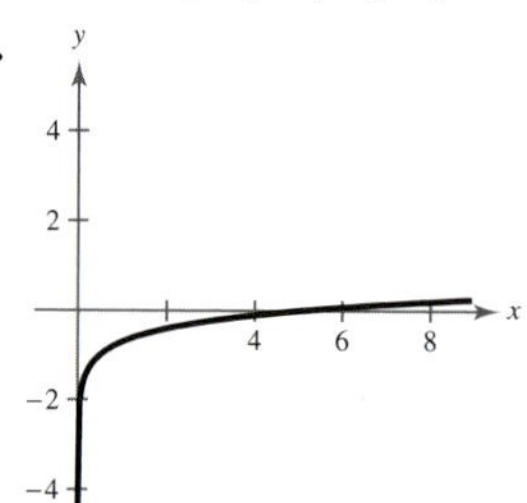
Domain: $(1, \infty)$
x-intercept: $\left(\frac{626}{625}, 0\right)$
Vertical asymptote: $x = 1$

37.
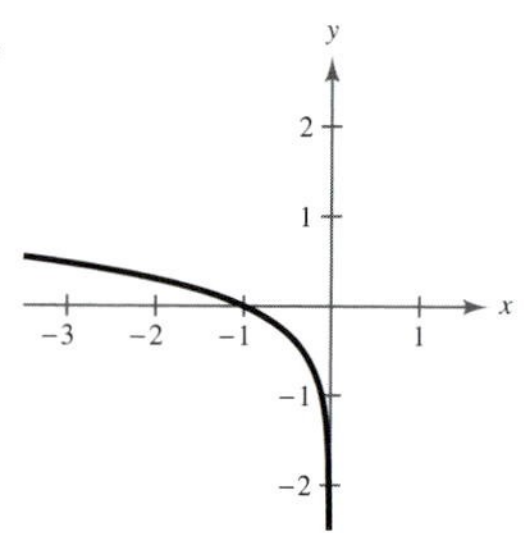
Domain: $(0, \infty)$
x-intercept: $(5, 0)$
Vertical asymptote: $x = 0$

38.
Domain: $(-\infty, 0)$
x-intercept: $(-1, 0)$
Vertical asymptote: $x = 0$

39. c **40.** f **41.** d
42. e **43.** b **44.** a
45. $e^{-0.693\ldots} = \frac{1}{2}$ **46.** $e^{-0.916\ldots} = \frac{2}{5}$
47. $e^{1.386\ldots} = 4$ **48.** $e^{2.302\ldots} = 10$
49. $e^{5.521\ldots} = 250$ **50.** $e^{6.520\ldots} = 679$
51. $e^0 = 1$ **52.** $e^1 = e$ **53.** $\ln 20.0855\ldots = 3$
54. $\ln 7.3890\ldots = 2$ **55.** $\ln 1.6487\ldots = \frac{1}{2}$
56. $\ln 1.3956\ldots = \frac{1}{3}$ **57.** $\ln 0.6065\ldots = -0.5$
58. $\ln 0.0165\ldots = -4.1$ **59.** $\ln 4 = x$
60. $\ln 3 = 2x$ **61.** 2.913
62. -3.418 **63.** -0.575
64. 0.693 **65.** 3 **66.** -2
67. $-\frac{2}{3}$ **68.** $-\frac{5}{2}$

69. 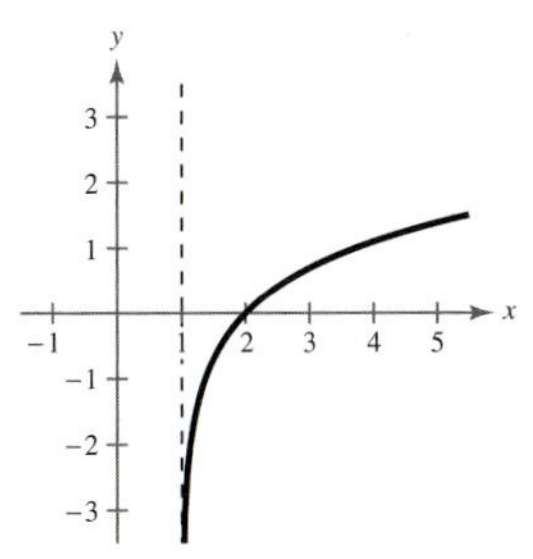

Domain: $(1, \infty)$
x-intercept: $(2, 0)$
Vertical asymptote: $x = 1$

70. 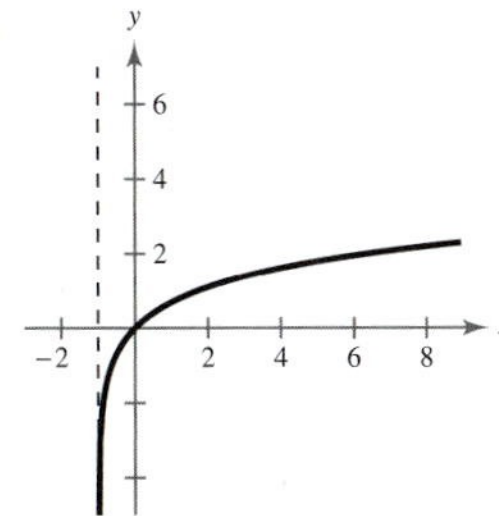

Domain: $(-1, \infty)$
x-intercept: $(0, 0)$
Vertical asymptote: $x = -1$

71. 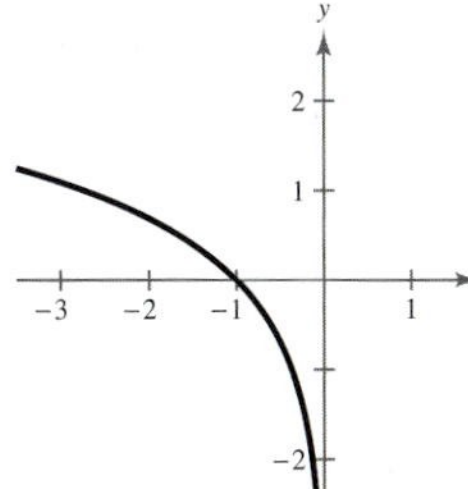

Domain: $(-\infty, 0)$
x-intercept: $(-1, 0)$
Vertical asymptote: $x = 0$

72. 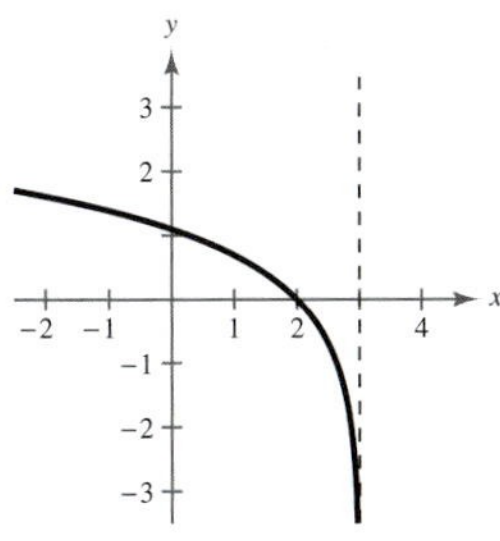

Domain: $(-\infty, 3)$
x-intercept: $(2, 0)$
Vertical asymptote: $x = 3$

73. 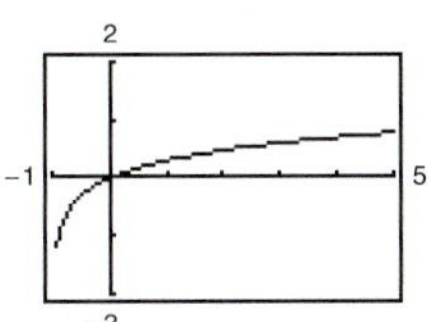

74. 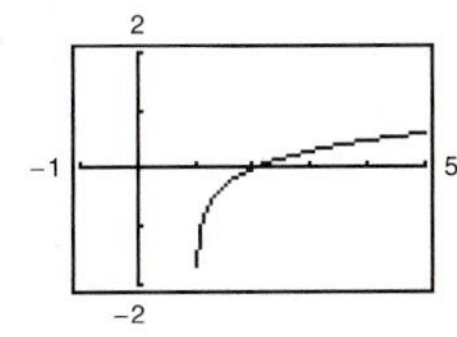

75. 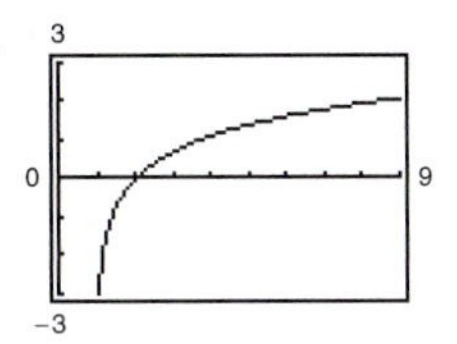

76. 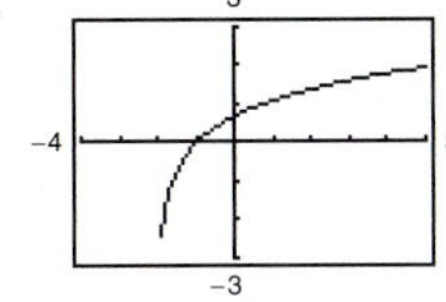

77. 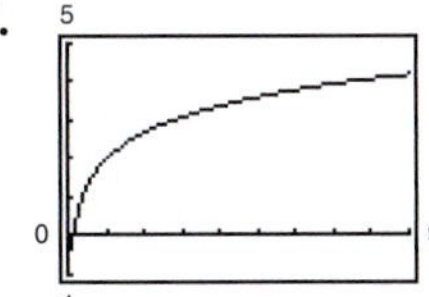

78. 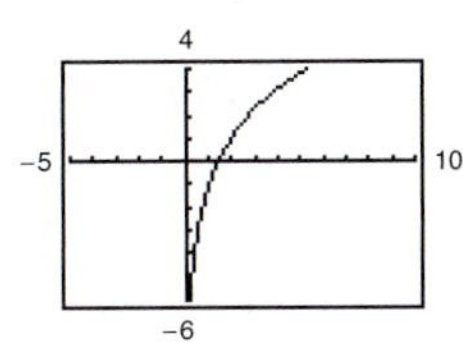

79. $x = 3$ **80.** $x = 12$ **81.** $x = 7$ **82.** $x = \frac{9}{5}$
83. $x = 4$ **84.** $x = 6$ **85.** $x = -5, 5$
86. $x = -2, 3$
87. (a) 30 years; 20 years (b) \$396,234; \$301,123.20
(c) \$246,234; \$151,123.20
(d) $x = 1000$; The monthly payment must be greater than \$1000.

88. (a)

K	1	2	4	6	8	10	12
t	0	7.3	14.6	18.9	21.9	24.2	26.2

The number of years required to multiply the original investment by K increases with K. However, the larger the value of K, the fewer the years required to increase the value of the investment by an additional multiple of the original investment.

(b) 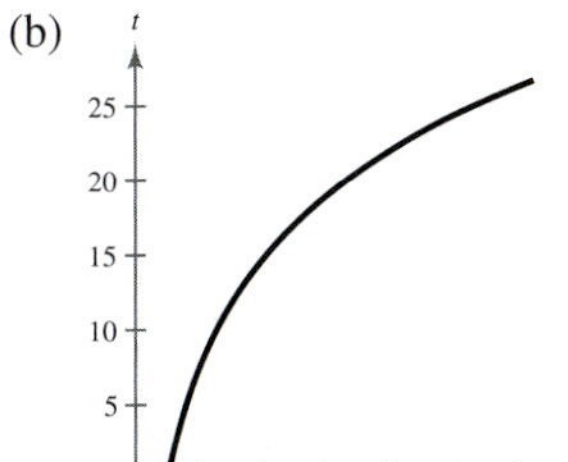

89. (a) 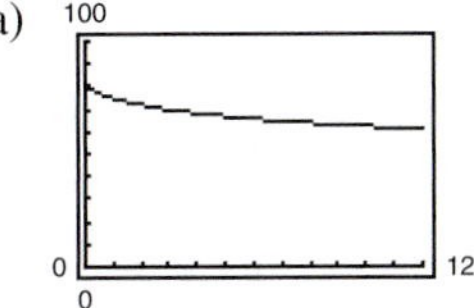

(b) 80 (c) 68.1 (d) 62.3
90. (a) 120 decibels
(b) 100 decibels
(c) No, the difference results from the logarithmic relationship between intensity and number of decibels.
91. False. Reflecting $g(x)$ about the line $y = x$ will determine the graph of $f(x)$.
92. True. $\log_3 27 = 3 \Rightarrow 3^3 = 27$

93. 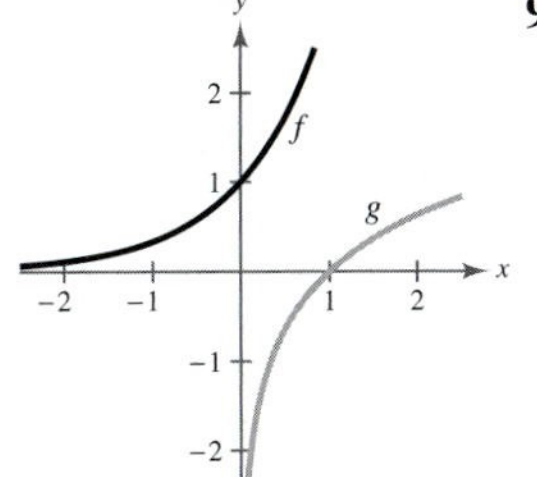

The functions f and g are inverses.

94. 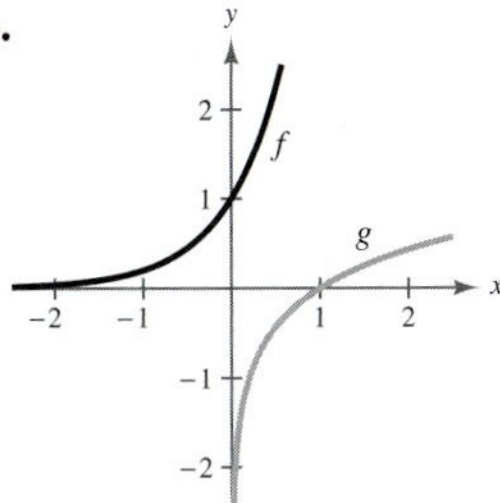

The functions f and g are inverses.

95.

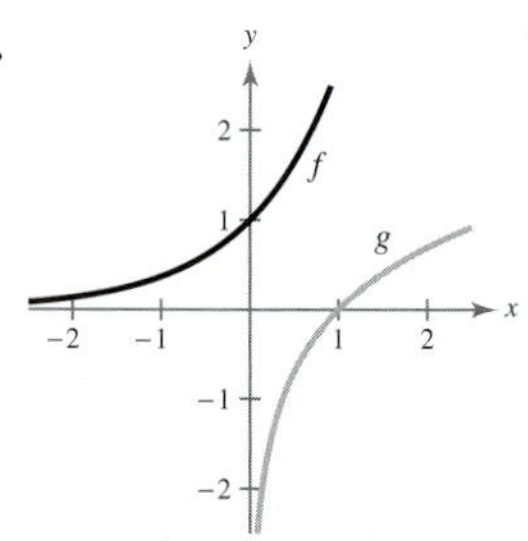

The functions f and g are inverses.

96.

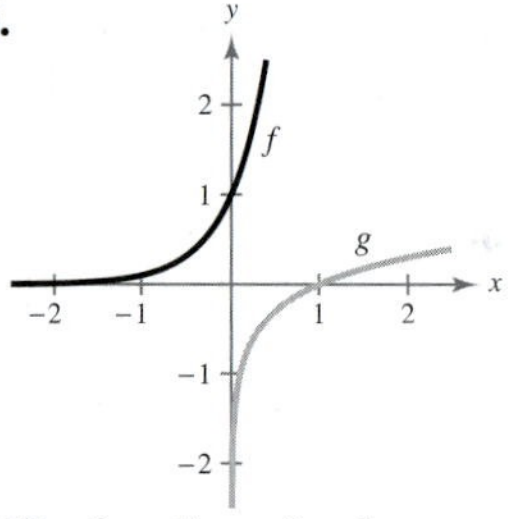

The functions f and g are inverses.

97. (a)

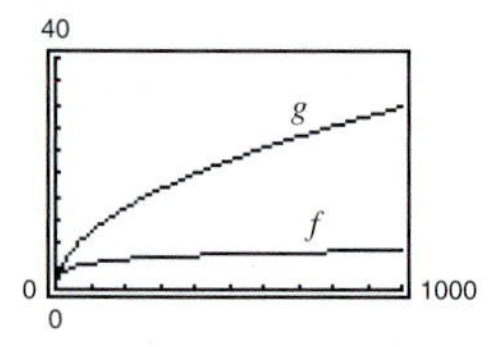

$g(x)$; The natural log function grows at a slower rate than the square root function.

(b)

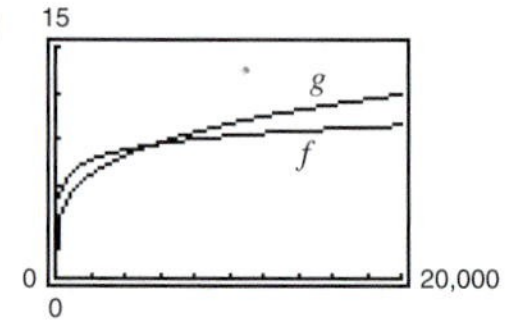

$g(x)$; The natural log function grows at a slower rate than the fourth root function.

98. (a)

x	1	5	10	10^2
$f(x)$	0	0.322	0.230	0.046

x	10^4	10^6
$f(x)$	0.00092	0.0000138

(b) 0

(c)

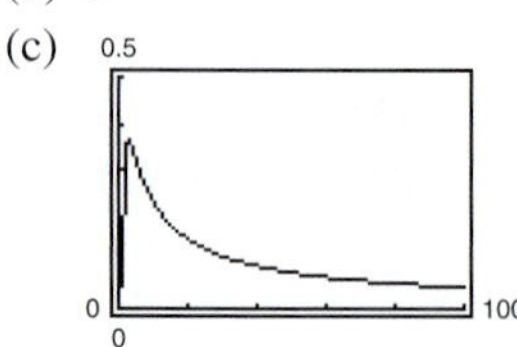

99. (a) False (b) True (c) True (d) False

100. Answers will vary.

101. (a)

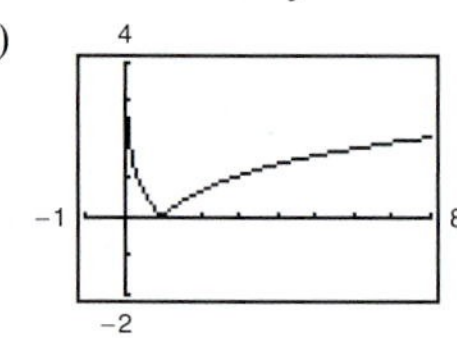

(b) Increasing: $(1, \infty)$
Decreasing: $(0, 1)$

(c) Relative minimum: $(1, 0)$

102. (a)

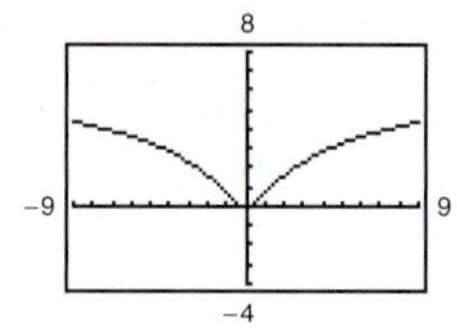

(b) Increasing: $(0, \infty)$
Decreasing: $(-\infty, 0)$

(c) Relative minimum: $(0, 0)$

103. 15 **104.** 1 **105.** 4300 **106.** -2

107. 1028 **108.** -344

Section 3.3 *(page 243)*

Vocabulary Check *(page 243)*

1. change-of-base **2.** $\dfrac{\log x}{\log a} = \dfrac{\ln x}{\ln a}$

3. c **4.** a **5.** b

1. (a) $\dfrac{\log x}{\log 5}$ (b) $\dfrac{\ln x}{\ln 5}$ **2.** (a) $\dfrac{\log x}{\log 3}$ (b) $\dfrac{\ln x}{\ln 3}$

3. (a) $\dfrac{\log x}{\log \frac{1}{5}}$ (b) $\dfrac{\ln x}{\ln \frac{1}{5}}$ **4.** (a) $\dfrac{\log x}{\log \frac{1}{3}}$ (b) $\dfrac{\ln x}{\ln \frac{1}{3}}$

5. (a) $\dfrac{\log \frac{3}{10}}{\log x}$ (b) $\dfrac{\ln \frac{3}{10}}{\ln x}$ **6.** (a) $\dfrac{\log \frac{3}{4}}{\log x}$ (b) $\dfrac{\ln \frac{3}{4}}{\ln x}$

7. (a) $\dfrac{\log x}{\log 2.6}$ (b) $\dfrac{\ln x}{\ln 2.6}$ **8.** (a) $\dfrac{\log x}{\log 7.1}$ (b) $\dfrac{\ln x}{\ln 7.1}$

9. 1.771 **10.** 0.712 **11.** -2.000 **12.** -1.161

13. -0.417 **14.** -0.694 **15.** 2.633 **16.** -3.823

17. $\frac{3}{2}$ **18.** $4 + 4\log_2 3$ **19.** $-3 - \log_5 2$

20. $\log 3 - 2$ **21.** $6 + \ln 5$ **22.** $\ln 6 - 2$ **23.** 2

24. -3 **25.** $\frac{3}{4}$ **26.** $\frac{1}{3}$ **27.** 2.4 **28.** -0.8

29. -9 is not in the domain of $\log_3 x$.

30. -16 is not in the domain of $\log_2 x$.

31. 4.5 **32.** 12 **33.** $-\frac{1}{2}$ **34.** $\frac{3}{4}$ **35.** 7

36. 7 **37.** 2 **38.** 3 **39.** $\log_4 5 + \log_4 x$

40. $\log_3 10 + \log_3 z$ **41.** $4\log_8 x$ **42.** $\log y - \log 2$

43. $1 - \log_5 x$ **44.** $-3\log_6 z$ **45.** $\frac{1}{2}\ln z$

46. $\frac{1}{3}\ln t$ **47.** $\ln x + \ln y + 2\ln z$

48. $\log 4 + 2\log x + \log y$ **49.** $\ln z + 2\ln(z - 1)$

50. $\ln(x + 1) + \ln(x - 1) - 3\ln x$

51. $\frac{1}{2}\log_2(a - 1) - 2\log_2 3$ **52.** $\ln 6 - \frac{1}{2}\ln(x^2 + 1)$

53. $\frac{1}{3}\ln x - \frac{1}{3}\ln y$ **54.** $\ln x - \frac{3}{2}\ln y$

55. $4\ln x + \frac{1}{2}\ln y - 5\ln z$

56. $\frac{1}{2}\log_2 x + 4\log_2 y - 4\log_2 z$

57. $2\log_5 x - 2\log_5 y - 3\log_5 z$

58. $\log x + 4\log y - 5\log z$

59. $\frac{3}{4}\ln x + \frac{1}{4}\ln(x^2 + 3)$ **60.** $\ln x + \frac{1}{2}\ln(x + 2)$

61. $\ln 3x$ **62.** $\ln yt$ **63.** $\log_4 \dfrac{z}{y}$ **64.** $\log_5 \dfrac{8}{t}$

65. $\log_2(x+4)^2$ **66.** $\log_7(z-2)^{2/3}$

67. $\log_3 \sqrt[4]{5x}$ **68.** $\log_6 \dfrac{1}{16x^4}$ **69.** $\ln \dfrac{x}{(x+1)^3}$

70. $\ln 64(z-4)^5$ **71.** $\log \dfrac{xz^3}{y^2}$ **72.** $\log_3 \dfrac{x^3y^4}{z^4}$

73. $\ln \dfrac{x}{(x^2-4)^4}$ **74.** $\ln \dfrac{z^4(z+5)^4}{(z-5)^2}$

75. $\ln \sqrt[3]{\dfrac{x(x+3)^2}{x^2-1}}$ **76.** $\ln\left(\dfrac{x^3}{x^2-1}\right)^2$

77. $\log_8 \dfrac{\sqrt[3]{y}(y+4)^2}{y-1}$ **78.** $\log_4\left[x^6(x-1)\sqrt{x+1}\right]$

79. $\log_2 \frac{32}{4} = \log_2 32 - \log_2 4$; Property 2

80. $\log_7\sqrt{70} = \frac{1}{2}(\log_7 7 + \log_7 10)$

$= \frac{1}{2} + \log_7\sqrt{10}$; Properties 1 and 3

81. $\beta = 10(\log I + 12)$; 60 dB **82.** 24 dB difference

83. ≈ 3

84. (a) $90 - \log(t+1)^{15}$ (b) 90

(c) 79.5 (d) 73.3

(e)

(f) 9 months (g) $90 - 15 \log 10 = 75$

85. $y = 256.24 - 20.8 \ln x$

86. (a) and (b)

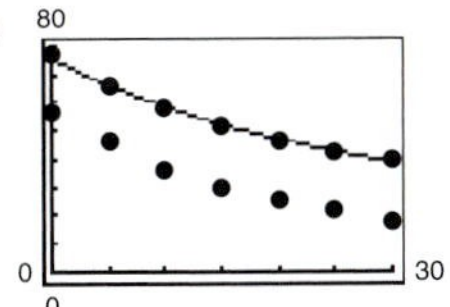

(c)

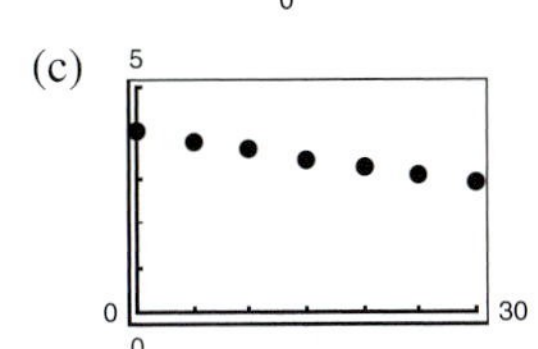

$T = 21 + e^{-0.037t + 3.997}$

The results are similar.

(d)

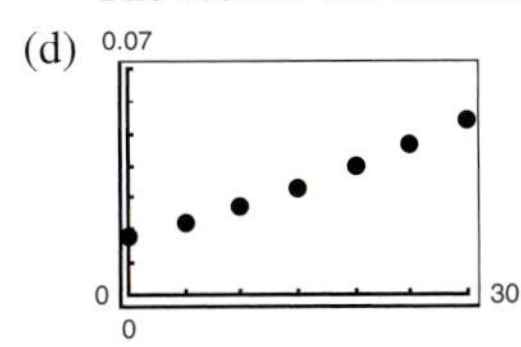

$T = 21 + \dfrac{1}{0.001t + 0.016}$

(e) Answers will vary.

87. False. $\ln 1 = 0$ **88.** True; Property 1

89. False. $\ln(x-2) \neq \ln x - \ln 2$

90. False. $f(\sqrt{x}) = \frac{1}{2}f(x)$

91. False. $u = v^2$ **92.** True

93–94. Answers will vary.

95. $f(x) = \dfrac{\log x}{\log 2} = \dfrac{\ln x}{\ln 2}$

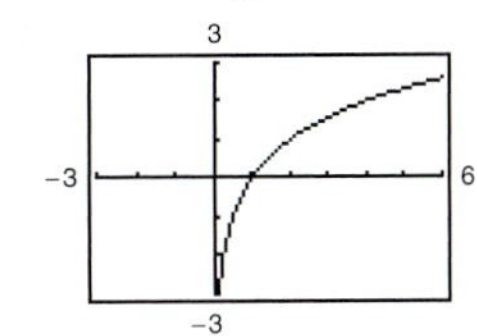

96. $f(x) = \dfrac{\log x}{\log 4} = \dfrac{\ln x}{\ln 4}$

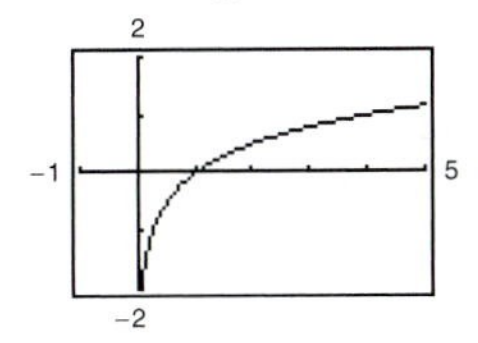

97. $f(x) = \dfrac{\log x}{\log \frac{1}{2}} = \dfrac{\ln x}{\ln \frac{1}{2}}$

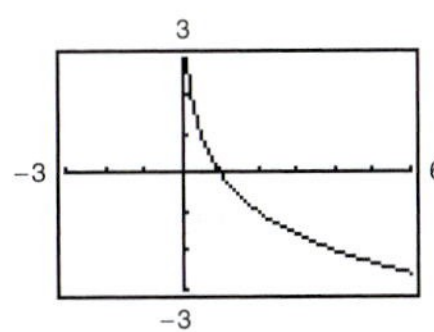

98. $f(x) = \dfrac{\log x}{\log \frac{1}{4}} = \dfrac{\ln x}{\ln \frac{1}{4}}$

99. $f(x) = \dfrac{\log x}{\log 11.8} = \dfrac{\ln x}{\ln 11.8}$

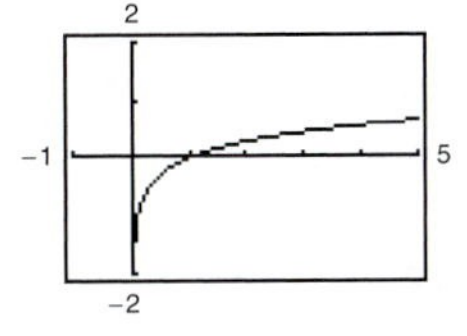

100. $f(x) = \dfrac{\log x}{\log 12.4} = \dfrac{\ln x}{\ln 12.4}$

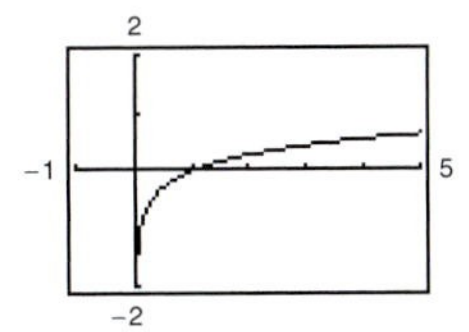

101. $f(x) = h(x)$; Property 2

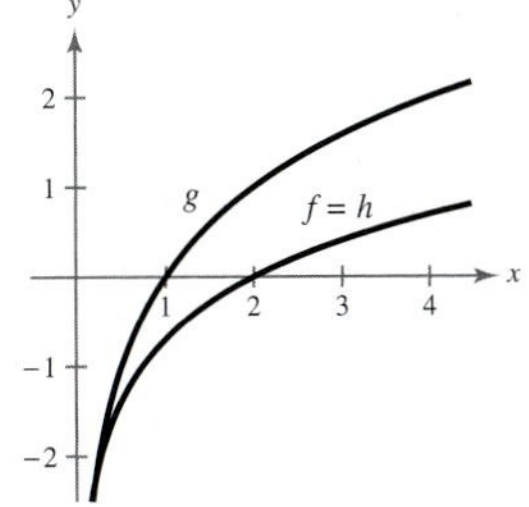

102. $\ln 2 \approx 0.6931$ $\ln 10 \approx 2.3025$

$\ln 3 \approx 1.0986$ $\ln 12 \approx 2.4848$

$\ln 4 \approx 1.3862$ $\ln 15 \approx 2.7080$

$\ln 5 \approx 1.6094$ $\ln 16 \approx 2.7724$

$\ln 6 \approx 1.7917$ $\ln 18 \approx 2.8903$

$\ln 8 \approx 2.0793$ $\ln 20 \approx 2.9956$

$\ln 9 \approx 2.1972$

103. $\frac{3x^4}{2y^3}, x \neq 0$ **104.** $\frac{27y^3}{8x^6}$ **105.** $1, x \neq 0, y \neq 0$

106. $\frac{(xy)^2}{x + y}$ **107.** $-1, \frac{1}{3}$ **108.** $1, \frac{1}{4}$

109. $\frac{-1 \pm \sqrt{97}}{6}$ **110.** $\frac{1 \pm \sqrt{31}}{2}$

Section 3.4 *(page 253)*

Vocabulary Check *(page 253)*

1. solve
2. (a) $x = y$ (b) $x = y$ (c) x (d) x
3. extraneous

1. (a) Yes (b) No **2.** (a) No (b) No
3. (a) No (b) Yes (c) Yes, approximate
4. (a) Yes (b) No (c) Yes, approximate
5. (a) Yes, approximate (b) No (c) Yes
6. (a) Yes (b) No (c) No
7. (a) No (b) Yes (c) Yes, approximate
8. (a) Yes (b) Yes, approximate (c) No
9. 2 **10.** 5 **11.** -5 **12.** -3 **13.** 2
14. 5 **15.** $\ln 2 \approx 0.693$ **16.** $\ln 4 \approx 1.386$
17. $e^{-1} \approx 0.368$ **18.** $e^{-7} \approx 0.000912$ **19.** 64
20. 0.008 **21.** $(3, 8)$ **22.** $(\frac{2}{3}, 9)$ **23.** $(9, 2)$
24. $(5, 0)$ **25.** $2, -1$ **26.** $-2, 4$
27. $\approx 1.618, \approx -0.618$ **28.** 0, 1 **29.** $\frac{\ln 5}{\ln 3} \approx 1.465$
30. $\frac{\ln 16}{\ln 5} \approx 1.723$ **31.** $\ln 5 \approx 1.609$
32. $\ln \frac{91}{4} \approx 3.125$ **33.** $\ln 28 \approx 3.332$
34. $\frac{\ln 37}{\ln 6} \approx 2.015$ **35.** $\frac{\ln 80}{2 \ln 3} \approx 1.994$
36. $\frac{\ln 3000}{5 \ln 6} \approx 0.894$ **37.** 2 **38.** $-\frac{\ln(0.10)}{3 \ln 4} \approx 0.554$
39. 4 **40.** 8 **41.** $3 - \frac{\ln 565}{\ln 2} \approx -6.142$
42. $\frac{-\ln 64 - \ln 431}{\ln 8} \approx -4.917$ **43.** $\frac{1}{3} \log\left(\frac{3}{2}\right) \approx 0.059$
44. $6 + \log \frac{7}{5} \approx 6.146$ **45.** $1 + \frac{\ln 7}{\ln 5} \approx 2.209$
46. $6 - \frac{\ln 5}{\ln 3} \approx 4.535$ **47.** $\frac{\ln 12}{3} \approx 0.828$
48. $\frac{\ln 50}{2} \approx 1.956$ **49.** $-\ln \frac{3}{5} \approx 0.511$
50. $-\frac{1}{4} \ln \frac{3}{40} \approx 0.648$ **51.** 0 **52.** $\ln \frac{25}{3} \approx 2.120$
53. $\frac{\ln \frac{8}{3}}{3 \ln 2} + \frac{1}{3} \approx 0.805$ **54.** $3 - \frac{\ln \frac{7}{2}}{2 \ln 4} \approx 2.548$

55. $\ln 5 \approx 1.609$ **56.** $\ln 2 \approx 0.693; \ln 3 \approx 1.099$
57. $\ln 4 \approx 1.386$ **58.** No solution
59. $2 \ln 75 \approx 8.635$ **60.** $\ln 7 \approx 1.946$
61. $\frac{1}{2} \ln 1498 \approx 3.656$ **62.** $\frac{\ln 31}{6} \approx 0.572$
63. $\frac{\ln 4}{365 \ln\left(1 + \frac{0.065}{365}\right)} \approx 21.330$
64. $\frac{\ln 21}{9 \ln 3.938225} \approx 0.247$
65. $\frac{\ln 2}{12 \ln\left(1 + \frac{0.10}{12}\right)} \approx 6.960$ **66.** $\frac{\ln 30}{3 \ln\left(16 - \frac{0.878}{26}\right)} \approx 0.409$

67. **68.**

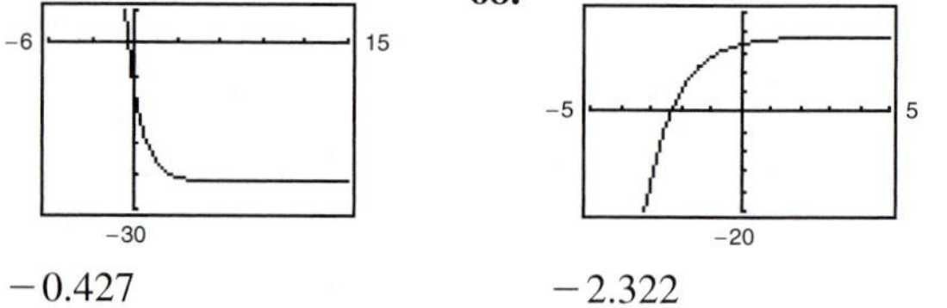

-0.427 -2.322

69. **70.**

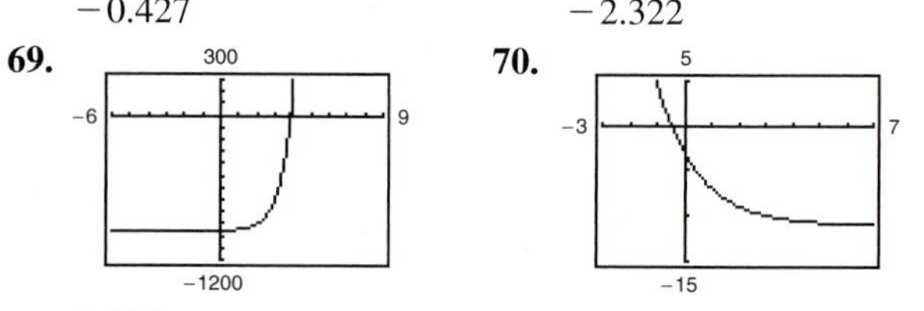

3.847 -0.478

71. **72.**

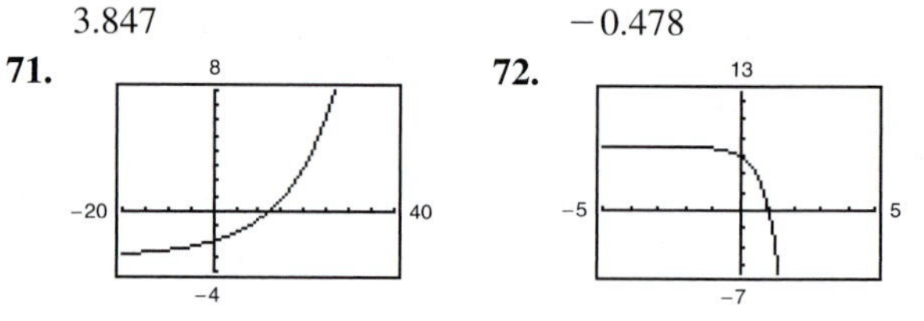

12.207 1.081

73. **74.**

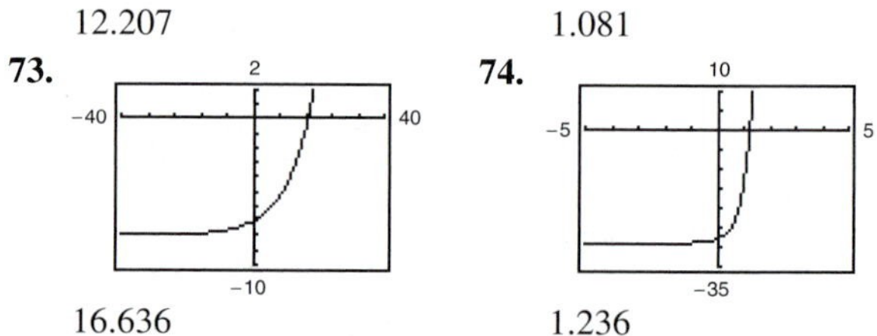

16.636 1.236

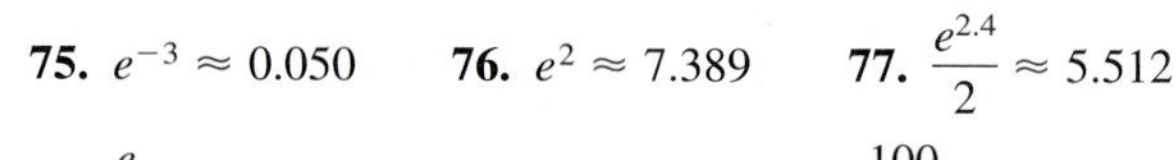

75. $e^{-3} \approx 0.050$ **76.** $e^2 \approx 7.389$ **77.** $\frac{e^{2.4}}{2} \approx 5.512$
78. $\frac{e}{4} \approx 0.680$ **79.** 1,000,000 **80.** $\frac{100}{3} \approx 33.333$
81. $\frac{e^{10/3}}{5} \approx 5.606$ **82.** $e^{7/2} \approx 33.115$
83. $e^2 - 2 \approx 5.389$ **84.** $e^{10} + 8 = 22{,}034.466$
85. $e^{-2/3} \approx 0.513$ **86.** $e^{-4/3} \approx 0.264$
87. $2(3^{11/6}) \approx 14.988$ **88.** $10^{11/5} + 2 \approx 160.489$
89. No solution **90.** $\frac{-1 + \sqrt{1 + 4e}}{2} \approx 1.223$
91. $1 + \sqrt{1 + e} \approx 2.928$ **92.** $\frac{-3 + \sqrt{9 + 4e}}{2} \approx 0.729$
93. No solution **94.** $\frac{3 + \sqrt{13}}{2} \approx 3.303$ **95.** 7

96. No solution **97.** $\frac{-1+\sqrt{17}}{2} \approx 1.562$ **98.** 2

99. 2 **100.** 9 **101.** $\frac{725+125\sqrt{33}}{8} \approx 180.384$

102. $\frac{1225+125\sqrt{73}}{2} \approx 1146.500$

103.

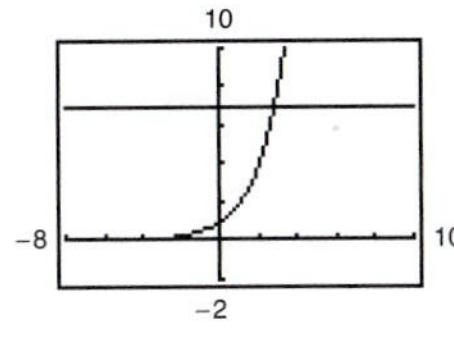

2.807

104.

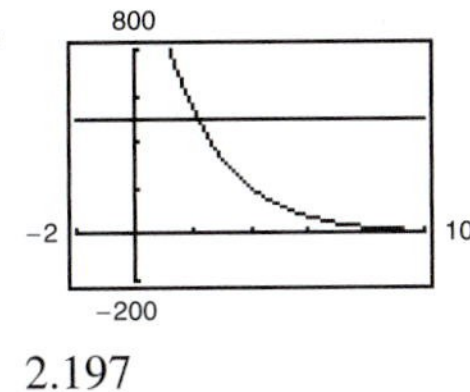

2.197

105.

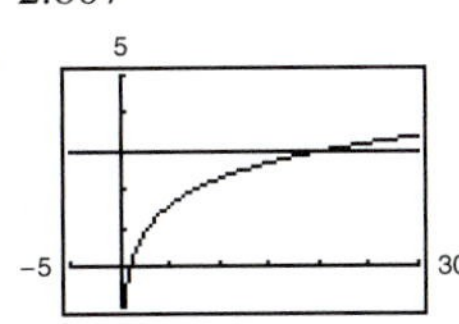

20.086

106.

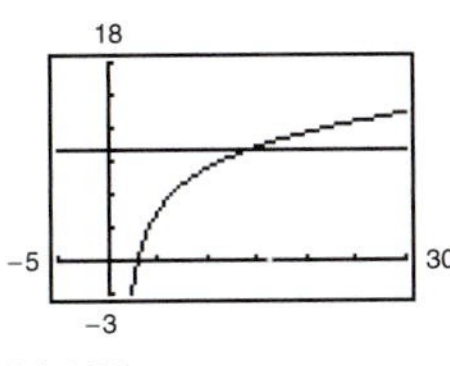

14.182

107. (a) 8.2 years (b) 12.9 years

108. (a) 5.8 years (b) 9.2 years

109. (a) 1426 units (b) 1498 units

110. (a) 303 units (b) 528 units

111. (a)

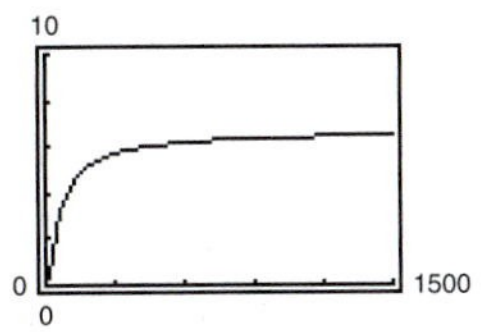

(b) $V = 6.7$; The yield will approach 6.7 million cubic feet per acre.

(c) 29.3 years

112. 12.76 inches **113.** 2001 **114.** 2001

115. (a) $y = 100$ and $y = 0$; The range falls between 0% and 100%.

(b) Males: 69.71 inches Females: 64.51 inches

116. (a)

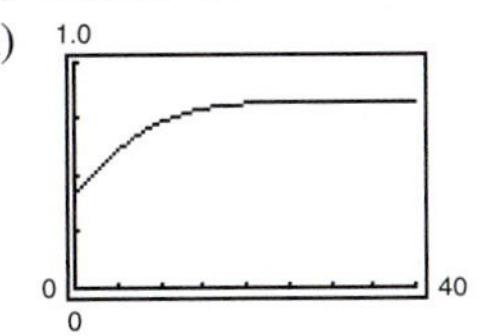

(b) Horizontal asymptotes: $P = 0$, $P = 0.83$

The proportion of correct responses will approach 0.83 as the number of trials increases.

(c) ≈ 5 trials

117. (a)

x	0.2	0.4	0.6	0.8	1.0
y	162.6	78.5	52.5	40.5	33.9

(b)

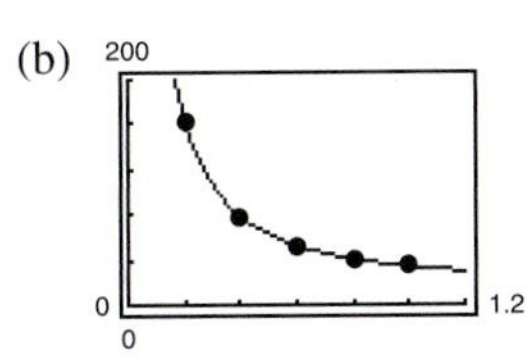

The model appears to fit the data well.

(c) 1.2 meters

(d) No. According to the model, when the number of g's is less than 23, x is between 2.276 meters and 4.404 meters, which isn't realistic in most vehicles.

118. (a) $T = 20$; Room temperature (b) ≈ 0.81 hour

119. $\log_b uv = \log_b u + \log_b v$

True by Property 1 in Section 5.3.

120. $\log_b(u+v) = (\log_b u)(\log_b v)$

False.

$2.04 \approx \log(10+100) \neq (\log 10)(\log 100) = 2$

121. $\log_b(u-v) = \log_b u - \log_b v$

False.

$1.95 \approx \log(100-10) \neq \log 100 - \log 10 = 1$

122. $\log_b \frac{u}{v} = \log_b u - \log_b v$

True by Property 2 in Section 5.3.

123. Yes. See Exercise 93.

124. For $rt < \ln 2$ years, double the amount you invest. For $rt > \ln 2$ years, double your interest rate or double the number of years, because either of these will double the exponent in the exponential function.

125. Yes. Time to double: $t = \frac{\ln 2}{r}$;

Time to quadruple: $t = \frac{\ln 4}{r} = 2\left(\frac{\ln 2}{r}\right)$

126. Answers will vary. **127.** $4|x|y^2\sqrt{3y}$

128. $4\sqrt{2} - 10$ **129.** $5\sqrt[3]{3}$ **130.** $\frac{1}{2}\sqrt{10} + 1$

131.

132.

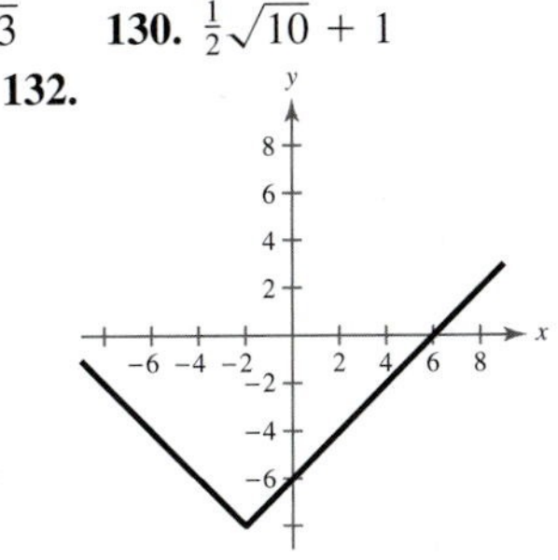

133.

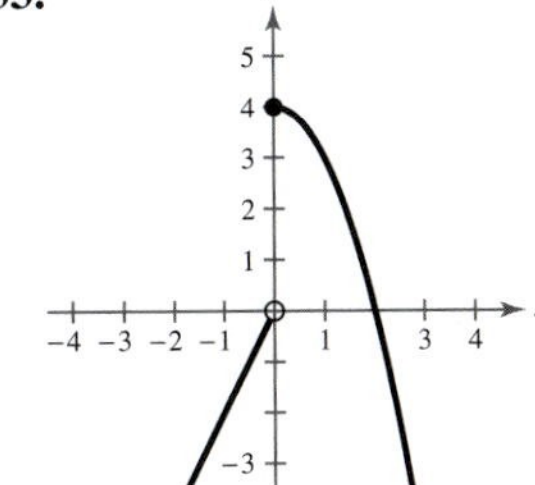

134.

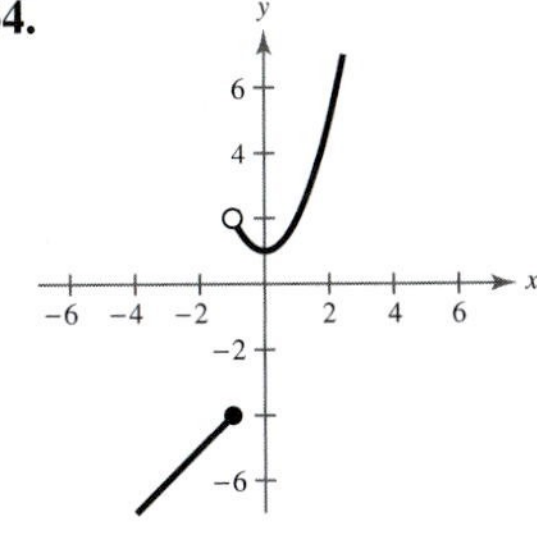

135. 1.226 **136.** 1.262 **137.** −5.595 **138.** 1.486

Section 3.5 *(page 264)*

Vocabulary Check *(page 264)*

1. $y = ae^{bx}$; $y = ae^{-bx}$
2. $y = a + b \ln x$; $y = a + b \log x$
3. normally distributed **4.** bell; average value
5. sigmoidal

1. c **2.** e **3.** b **4.** a **5.** d **6.** f

	Initial Investment	*Annual % Rate*	*Time to Double*	*Amount After 10 years*
7.	$1000	3.5%	19.8 yr	$1419.07
8.	$750	10.5%	6.60 yr	$2143.24
9.	$750	8.9438%	7.75 yr	$1834.33
10.	$10,000	5.7762%	12 yr	$17,817.93
11.	$500	11.0%	6.3 yr	$1505.00
12.	$600	34.66%	2 yr	$19,205.00
13.	$6376.28	4.5%	15.4 yr	$10,000.00
14.	$1637.46	2%	34.7 yr	$2000.00

15. $112,087.09 **16.** $4214.16

17. (a) 6.642 years (b) 6.330 years
(c) 6.302 years (d) 6.301 years

18. (a) 6.94 years (b) 6.63 years
(c) 6.602 years (d) 6.601 years

19.

r	2%	4%	6%	8%	10%	12%
t	54.93	27.47	18.31	13.73	10.99	9.16

20.

Use PwrReg: $t = 1.099r^{-1}$

21.

r	2%	4%	6%	8%	10%	12%
t	55.48	28.01	18.85	14.27	11.53	9.69

22.

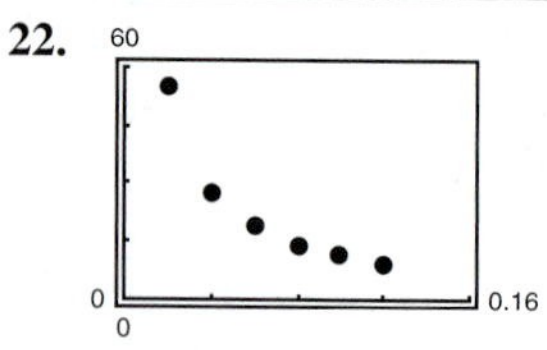

Use PwrReg: $t = 1.222r^{-1}$

23.

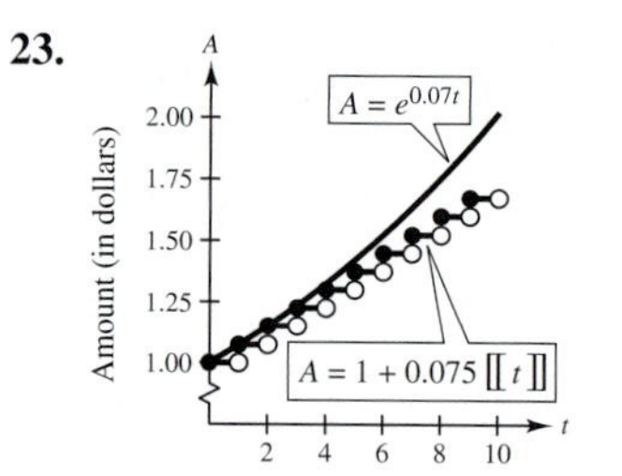

Continuous compounding

24.

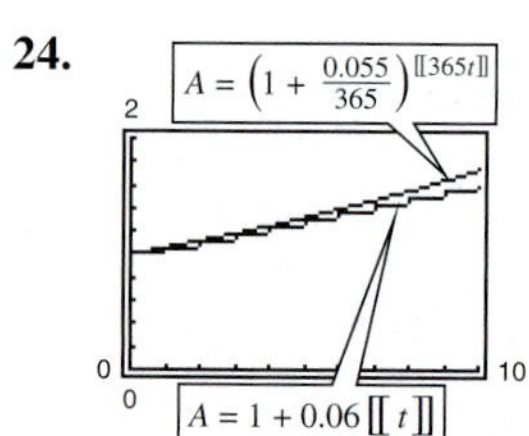

Daily compounding

	Half-life (years)	*Initial Quantity*	*Amount After 1000 Years*
25.	1599	10 g	6.48 g
26.	1599	2.31 g	1.5 g
27.	5715	2.26 g	2 g
28.	5715	3 g	2.66 g
29.	24,100	2.16 g	2.1 g
30.	24,100	0.41 g	0.4 g

31. $y = e^{0.7675x}$ **32.** $y = \frac{1}{2}e^{0.5756x}$

33. $y = 5e^{-0.4024x}$ **34.** $y = e^{-0.4621x}$

35. (a) Decreasing due to the negative exponent.
(b) 2000: population of 2430 thousand
2003: population of 2408.95 thousand
(c) 2018

36. (a) Bulgaria: $y = 7.8e^{-0.00940t}$; 5.9 million
Canada: $y = 31.3e^{0.00915t}$; 41.2 million
China: $y = 1268.9e^{0.00602t}$; 1520.1 million
United Kingdom: $y = 59.5e^{0.00282t}$; 64.8 million
United States: $y = 282.3e^{0.00910t}$; 370.9 million
(b) b; The greater the rate of growth, the greater the value of b.
(c) b determines whether the population is increasing ($b > 0$) or decreasing ($b < 0$).

37. $k = 0.2988$; ≈ 5,309,734 hits

38. $k = 0.1337$; $144.98 million **39.** 3.15 hours

40. 61.16 hours

41. (a) ≈ 12,180 years old (b) ≈ 4797 years old

42. 15,642 years

43. (a) $V = -6394t + 30,788$ (b) $V = 30,788e^{-0.268t}$
(c)

The exponential model depreciates faster.

(d)

t	1	3
$V = -6394t + 30,788$	24,394	11,606
$V = 30,788e^{-0.268t}$	23,550	13,779

(e) Answers will vary.

44. (a) $V = -300t + 1150$ (b) $V = 1150e^{-0.368799t}$

(c)

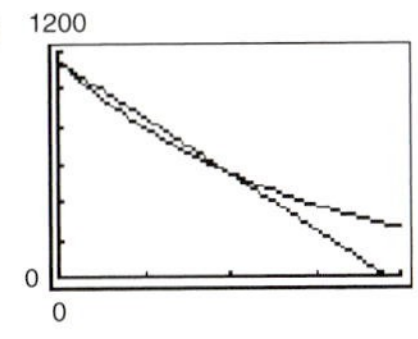

The exponential model depreciates faster.

(d)

t	1 year	3 years
$V = -300t + 1150$	850	250
$V = 1150e^{-0.368799t}$	795	380

(e) Answers will vary.

45. (a) $S(t) = 100(1 - e^{-0.1625t})$

(b)

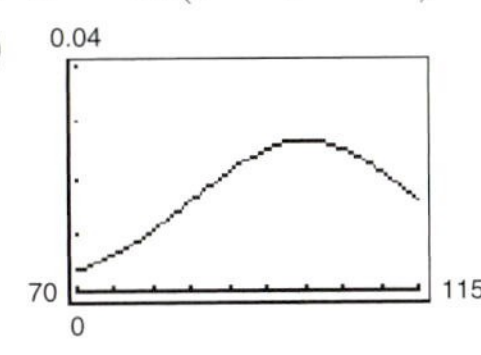

(c) 55,625

46. (a) $N = 30(1 - e^{-0.050t})$ (b) 36 days

47. (a) (b) 100

48. (a)

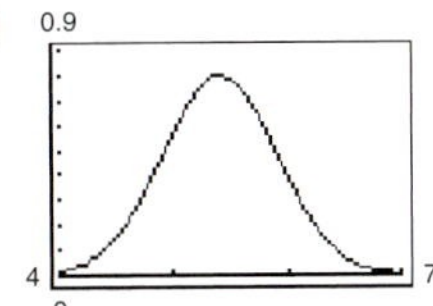

(b) 5.4 hours per week

49. (a) 203 animals (b) 13 years

(c)

Horizontal asymptotes: $y = 0$, $y = 1000$. The population size will approach 1000 as time increases.

50. (a) $S = \dfrac{500{,}000}{1 + 0.6e^{0.026t}}$ (b) 287,567 units sold

51. (a) $10^{7.9} \approx 79{,}432{,}823$ (b) $10^{8.3} \approx 199{,}526{,}231$

(c) $10^{4.2} \approx 15{,}849$

52. (a) 7.91 (b) 7.68 (c) 5.40

53. (a) 20 decibels (b) 70 decibels

(c) 40 decibels (d) 120 decibels

54. (a) 10 decibels (b) 140 decibels

(c) 80 decibels (d) 100 decibels

55. 95% **56.** 97% **57.** 4.64 **58.** 4.95

59. 1.58×10^{-6} moles per liter

60. $10^{-3.2} \approx 6.3 \times 10^{-4}$ moles per liter **61.** $10^{5.1}$

62. 10 **63.** 3:00 A.M.

64. (a) (b) Interest; $t \approx 26$ years

(c)

Interest; $t \approx 11$ years; The interest is still the majority of the monthly payment in the early years, but now the principal and interest are nearly equal when $t \approx 11$ years.

65. (a)

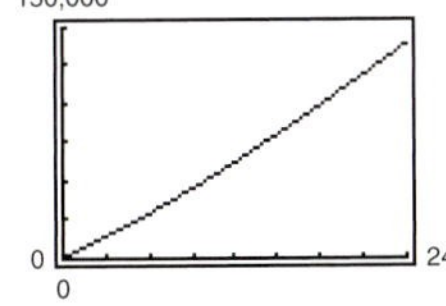

(b) ≈ 21 years; Yes

66. (a) $t_3 = 0.2729s - 6.0143$

$t_4 = 1.5385e^{0.02913s}$ or $t_4 = 1.5385(1.0296)^s$

(b)

(c)

s	30	40	50	60	70	80	90
t_1	3.6	4.6	6.7	9.4	12.5	15.9	19.6
t_2	3.3	4.9	7.0	9.5	12.5	15.9	19.9
t_3	2.2	4.9	7.6	10.4	13.1	15.8	18.5
t_4	3.7	4.9	6.6	8.8	11.8	15.8	21.2

(d) Model t_1: Sum = 2.0
Model t_2: Sum = 1.1
Model t_3: Sum = 5.6
Model t_4: Sum = 2.7
The quadratic (model t_2) fits best.

67. False. The domain can be the set of real numbers for a logistic growth function.

68. False. A logistic growth function never has an x-intercept.

69. False. The graph of $f(x)$ is the graph of $g(x)$ shifted upward five units.

70. True. The graph of a Gaussian model will never have an x-intercept.

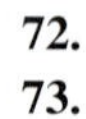

4.

5. $y = 2x + 2$

6. For some values of x there correspond two values of y.

7. (a) $\frac{3}{2}$ (b) Division by 0 is undefined. (c) $\frac{s+2}{s}$

8. (a) Vertical shrink by $\frac{1}{2}$
(b) Vertical shift of two units upward
(c) Horizontal shift of two units to the left

9. (a) $5x - 2$ (b) $-3x - 4$ (c) $4x^2 - 11x - 3$
(d) $\frac{x-3}{4x+1}$; Domain: all real numbers x except $x = -\frac{1}{4}$

10. (a) $\sqrt{x-1} + x^2 + 1$ (b) $\sqrt{x-1} - x^2 - 1$
(c) $x^2\sqrt{x-1} + \sqrt{x-1}$
(d) $\frac{\sqrt{x-1}}{x^2+1}$; Domain: all real numbers x such that $x \geq 1$

11. (a) $2x + 12$ (b) $\sqrt{2x^2 + 6}$
Domain of $f \circ g$: all real numbers x such that $x \geq -6$
Domain of $g \circ f$: all real numbers

12. (a) $|x| - 2$ (b) $|x - 2|$
Domain of $f \circ g$ and $g \circ f$: all real numbers

13. Yes; $h^{-1}(x) = \frac{1}{5}(x + 2)$ **14.** 2438.65 kilowatts

15. $y = -\frac{3}{4}(x + 8)^2 + 5$

16.

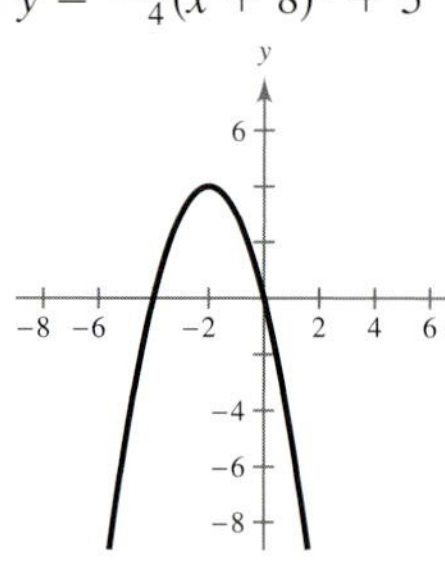

17.

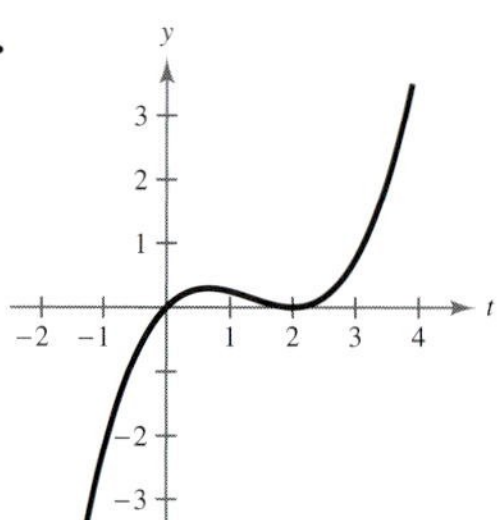

18.

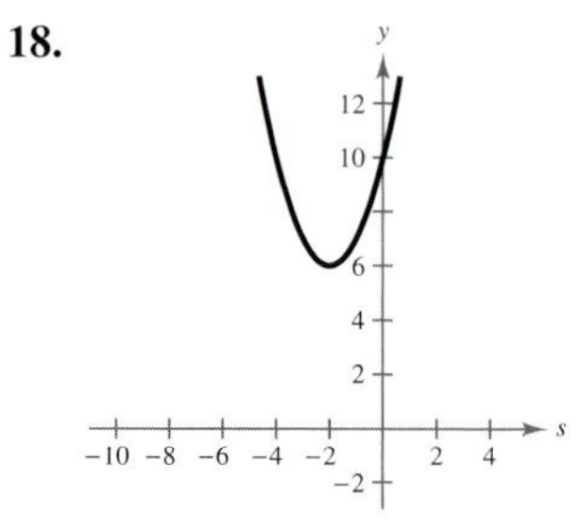

19. $-2, \pm 2i$; $(x + 2)(x + 2i)(x - 2i)$

20. $-7, 0, 3$; $x(x)(x - 3)(x + 7)$

21. $4, -\frac{1}{2}, 1 \pm 3i$; $(x - 4)(2x + 1)(x - 1 + 3i)(x - 1 - 3i)$

22. $3x - 2 - \frac{3x - 2}{2x^2 + 1}$

23. $2x^3 - x^2 + 2x - 10 + \frac{25}{x + 2}$

24.

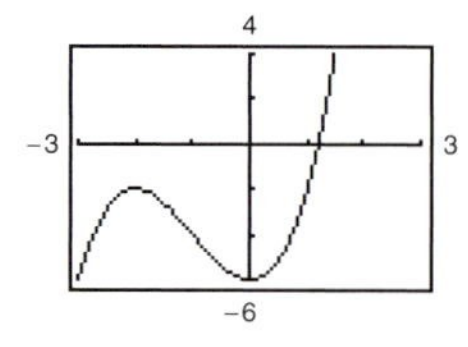

Interval: $[1, 2]$; 1.20

25. Intercept: $(0, 0)$
Vertical asymptotes: $x = \pm 3$
Horizontal asymptote: $y = 0$

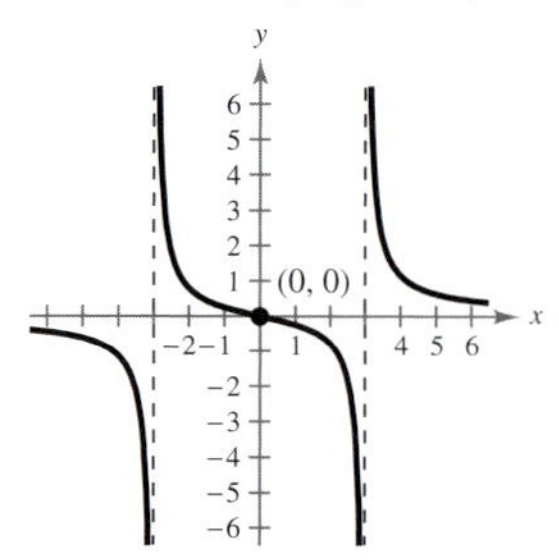

26. y-intercept: $(0, -1)$
x-intercept: $(1, 0)$
Horizontal asymptote: $y = 1$
Vertical asymptote: $x = -1$

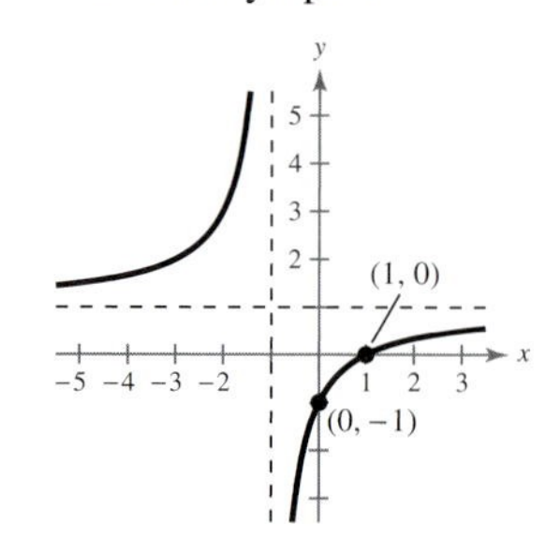

27. y-intercept: $(0, 6)$
x-intercepts: $(-2, 0), (-3, 0)$
Slant asymptote: $y = x + 4$
Vertical asymptote: $x = -1$

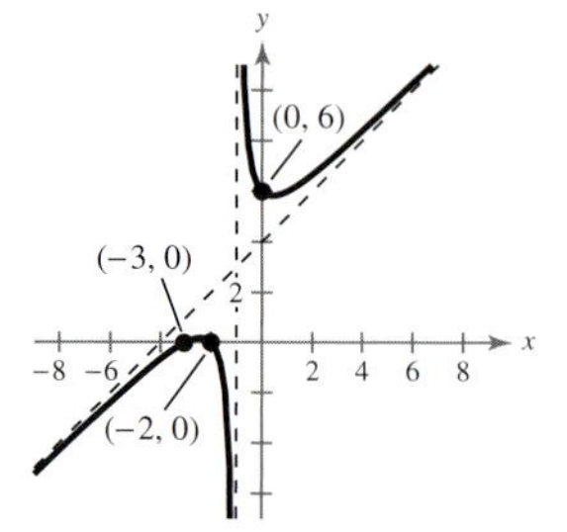

28. $x \leq -2$ or $0 \leq x \leq 2$

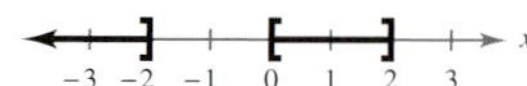

29. All real numbers x such that $x < -5$ or $x > -1$

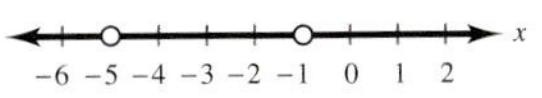

30. Reflect f in the x-axis and y-axis, and shift three units to the right.

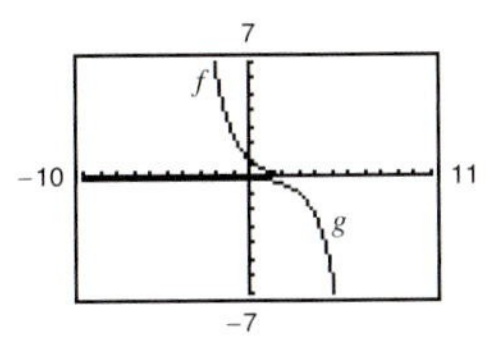

31. Reflect f in the x-axis, and shift four units upward.

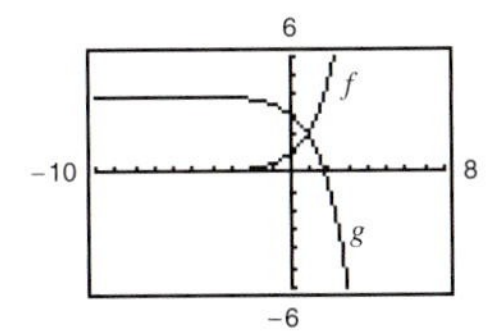

32. 1.991 **33.** -0.067 **34.** 1.717 **35.** 0.281

36. $\ln(x + 4) + \ln(x - 4) - 4 \ln x,\ x > 4$

37. $\ln \dfrac{x^2}{\sqrt{x + 5}},\ x > 0$ **38.** $x = \dfrac{\ln 12}{2} \approx 1.242$

39. $\ln 3 \approx 1.099$ or $3 \ln 2 \approx 2.079$

40. $e^6 - 2 \approx 401.429$

41. (a)

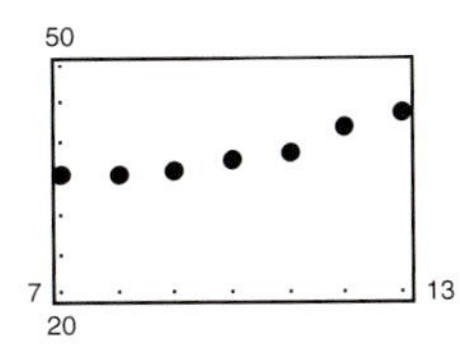

(b) $S = 0.274t^2 - 4.08t + 50.6$

(c)

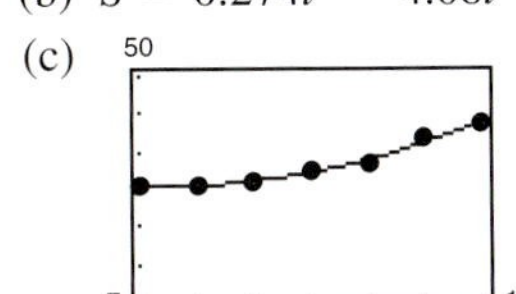

The model is a good fit for the data.

(d) 65.9 Yes, this is a reasonable answer.

42. 6.3 hours

Problem Solving (*page 279*)

1.

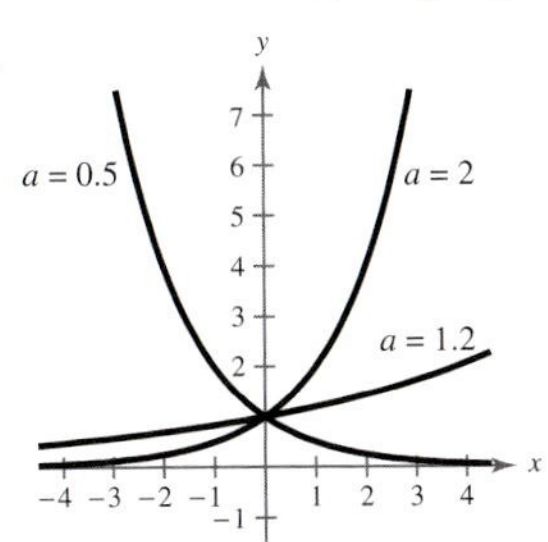

$y = 0.5^x$ and $y = 1.2^x$

$0 \le a \le 1.44$

2.

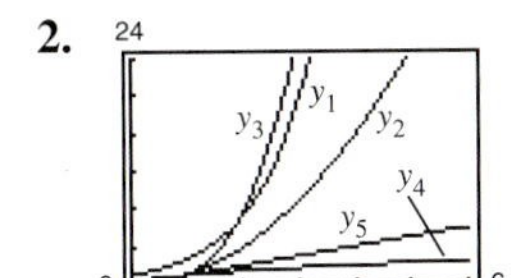

y_3 increases at the fastest rate.

3. As $x \to \infty$, the graph of e^x increases at a greater rate than the graph of x^n.

4–6. Answers will vary.

7. (a)

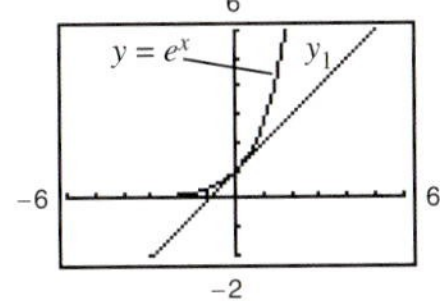

(b)

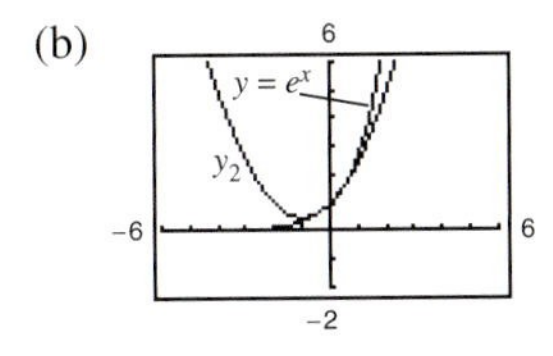

(c)

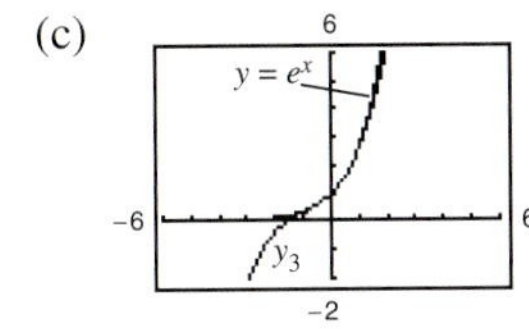

8. $y_4 = 1 + \dfrac{x}{1!} + \dfrac{x^2}{2!} + \dfrac{x^3}{3!} + \dfrac{x^4}{4!}$

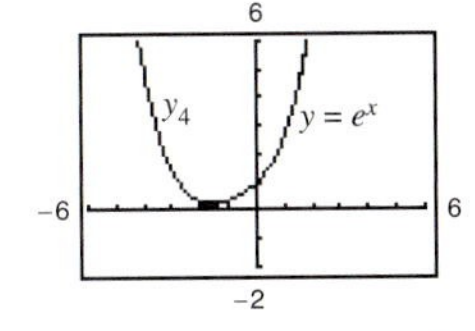

The pattern implies that

$$e^x = 1 + \frac{x}{1!} + \frac{x^2}{2!} + \frac{x^3}{3!} + \cdots$$

9.

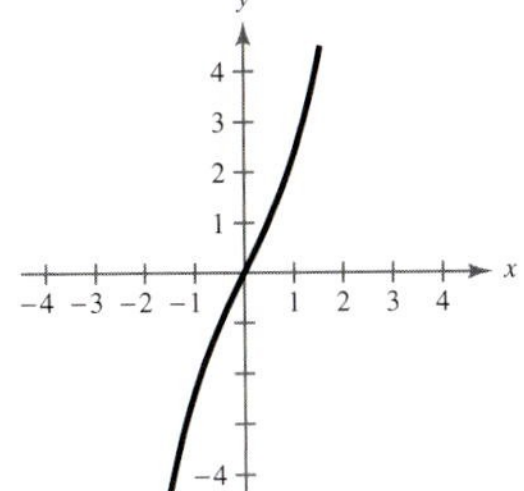

$$f^{-1}(x) = \ln\left(\frac{x + \sqrt{x^2 + 4}}{2}\right)$$

10. $f^{-1}(x) = \dfrac{\ln\left(\dfrac{x + 1}{x - 1}\right)}{\ln a}$ **11.** c

7. $\sin\theta = \frac{5\sqrt{29}}{29}$ $\csc\theta = \frac{\sqrt{29}}{5}$
$\cos\theta = -\frac{2\sqrt{29}}{29}$ $\sec\theta = -\frac{\sqrt{29}}{2}$
$\tan\theta = -\frac{5}{2}$ $\cot\theta = -\frac{2}{5}$

8. $\sin\theta = -\frac{2\sqrt{29}}{29}$ $\csc\theta = -\frac{\sqrt{29}}{2}$
$\cos\theta = -\frac{5\sqrt{29}}{29}$ $\sec\theta = -\frac{\sqrt{29}}{5}$
$\tan\theta = \frac{2}{5}$ $\cot\theta = \frac{5}{2}$

9. $\sin\theta = \frac{68\sqrt{5849}}{5849}$ $\csc\theta = \frac{\sqrt{5849}}{68}$
$\cos\theta = -\frac{35\sqrt{5849}}{5849}$ $\sec\theta = -\frac{\sqrt{5849}}{35}$
$\tan\theta = -\frac{68}{35}$ $\cot\theta = -\frac{35}{68}$

10. $\sin\theta = -\frac{31\sqrt{1157}}{1157}$ $\csc\theta = -\frac{\sqrt{1157}}{31}$
$\cos\theta = \frac{14\sqrt{1157}}{1157}$ $\sec\theta = \frac{\sqrt{1157}}{14}$
$\tan\theta = -\frac{31}{14}$ $\cot\theta = -\frac{14}{31}$

11. Quadrant III **12.** Quadrant I
13. Quadrant II **14.** Quadrant IV

15. $\sin\theta = \frac{3}{5}$ $\csc\theta = \frac{5}{3}$
$\cos\theta = -\frac{4}{5}$ $\sec\theta = -\frac{5}{4}$
$\tan\theta = -\frac{3}{4}$ $\cot\theta = -\frac{4}{3}$

16. $\sin\theta = -\frac{3}{5}$ $\csc\theta = -\frac{5}{3}$
$\cos\theta = -\frac{4}{5}$ $\sec\theta = -\frac{5}{4}$
$\tan\theta = \frac{3}{4}$ $\cot\theta = \frac{4}{3}$

17. $\sin\theta = -\frac{15}{17}$ $\csc\theta = -\frac{17}{15}$
$\cos\theta = \frac{8}{17}$ $\sec\theta = \frac{17}{8}$
$\tan\theta = -\frac{15}{8}$ $\cot\theta = -\frac{8}{15}$

18. $\sin\theta = -\frac{15}{17}$ $\csc\theta = -\frac{17}{15}$
$\cos\theta = \frac{8}{17}$ $\sec\theta = \frac{17}{8}$
$\tan\theta = -\frac{15}{8}$ $\cot\theta = -\frac{8}{15}$

19. $\sin\theta = -\frac{\sqrt{10}}{10}$ $\csc\theta = -\sqrt{10}$
$\cos\theta = \frac{3\sqrt{10}}{10}$ $\sec\theta = \frac{\sqrt{10}}{3}$
$\tan\theta = -\frac{1}{3}$ $\cot\theta = -3$

20. $\sin\theta = \frac{1}{4}$ $\csc\theta = 4$
$\cos\theta = -\frac{\sqrt{15}}{4}$ $\sec\theta = -\frac{4\sqrt{15}}{15}$
$\tan\theta = -\frac{\sqrt{15}}{15}$ $\cot\theta = -\sqrt{15}$

21. $\sin\theta = \frac{\sqrt{3}}{2}$ $\csc\theta = \frac{2\sqrt{3}}{3}$
$\cos\theta = -\frac{1}{2}$ $\sec\theta = -2$
$\tan\theta = -\sqrt{3}$ $\cot\theta = -\frac{\sqrt{3}}{3}$

22. $\sin\theta = 0$ $\csc\theta$ is undefined.
$\cos\theta = -1$ $\sec\theta = -1$
$\tan\theta = 0$ $\cot\theta$ is undefined.

23. $\sin\theta = 0$ $\csc\theta$ is undefined.
$\cos\theta = -1$ $\sec\theta = -1$
$\tan\theta = 0$ $\cot\theta$ is undefined.

24. $\sin\theta = -1$ $\csc\theta = -1$
$\cos\theta = 0$ $\sec\theta$ is undefined.
$\tan\theta$ is undefined. $\cot\theta = 0$

25. $\sin\theta = \frac{\sqrt{2}}{2}$ $\csc\theta = \sqrt{2}$
$\cos\theta = -\frac{\sqrt{2}}{2}$ $\sec\theta = -\sqrt{2}$
$\tan\theta = -1$ $\cot\theta = -1$

26. $\sin\theta = -\frac{\sqrt{10}}{10}$ $\csc\theta = -\sqrt{10}$
$\cos\theta = -\frac{3\sqrt{10}}{10}$ $\sec\theta = -\frac{\sqrt{10}}{3}$
$\tan\theta = \frac{1}{3}$ $\cot\theta = 3$

27. $\sin\theta = -\frac{2\sqrt{5}}{5}$ $\csc\theta = -\frac{\sqrt{5}}{2}$
$\cos\theta = -\frac{\sqrt{5}}{5}$ $\sec\theta = -\sqrt{5}$
$\tan\theta = 2$ $\cot\theta = \frac{1}{2}$

28. $\sin\theta = -\frac{4}{5}$ $\csc\theta = -\frac{5}{4}$
$\cos\theta = \frac{3}{5}$ $\sec\theta = \frac{5}{3}$
$\tan\theta = -\frac{4}{3}$ $\cot\theta = -\frac{3}{4}$

29. 0 **30.** -1 **31.** Undefined **32.** -1 **33.** 1
34. Undefined **35.** Undefined **36.** 0
37. $\theta' = 23°$ **38.** $\theta' = 51°$

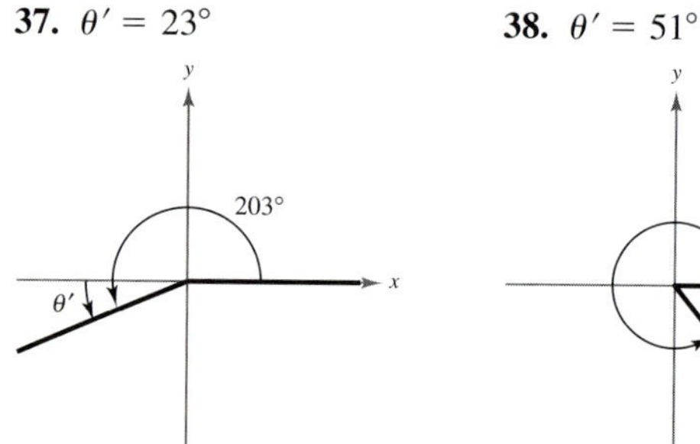

39. $\theta' = 65°$

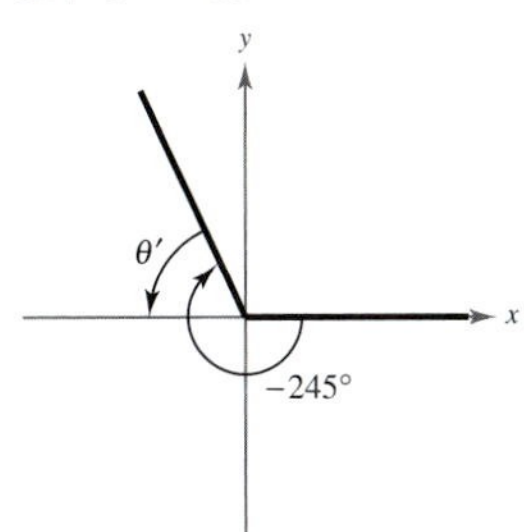

40. $\theta' = 35°$

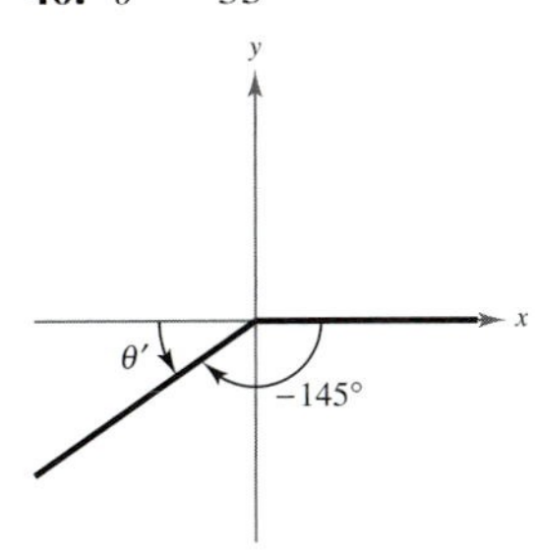

41. $\theta' = \dfrac{\pi}{3}$

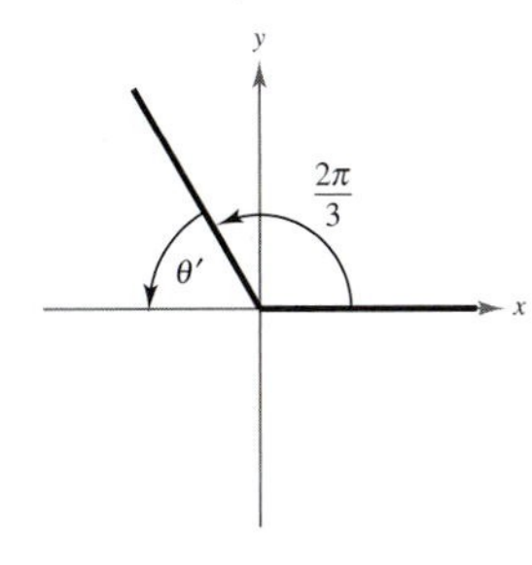

42. $\theta' = \dfrac{\pi}{4}$

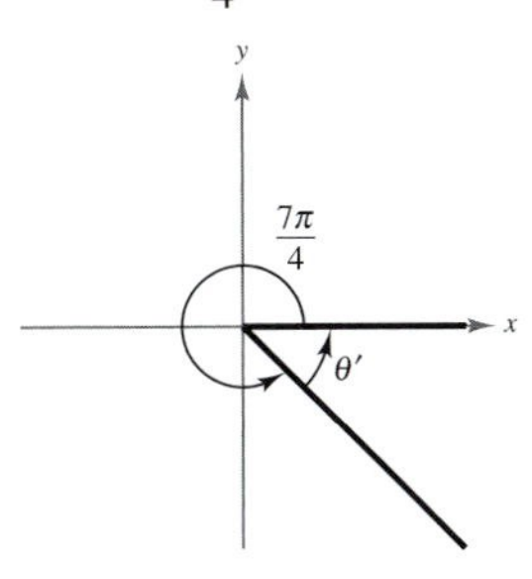

43. $\theta' = 3.5 - \pi$

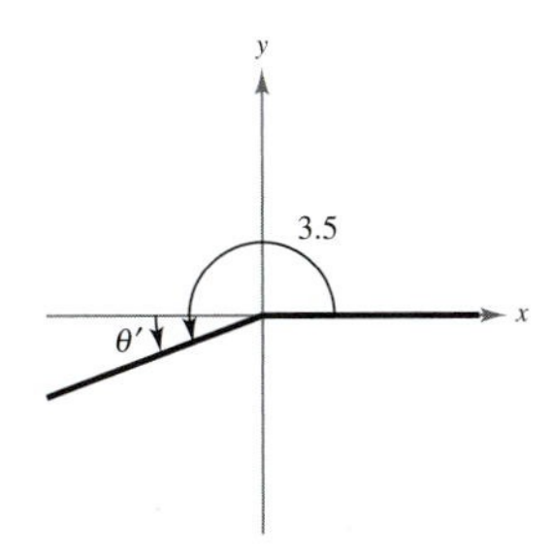

44. $\theta' = \dfrac{\pi}{3}$

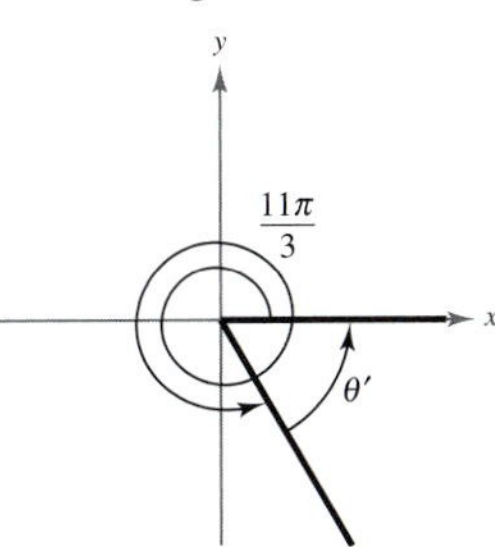

45. $\sin 225° = -\dfrac{\sqrt{2}}{2}$

$\cos 225° = -\dfrac{\sqrt{2}}{2}$

$\tan 225° = 1$

46. $\sin 300° = -\dfrac{\sqrt{3}}{2}$

$\cos 300° = \dfrac{1}{2}$

$\tan 300° = -\sqrt{3}$

47. $\sin 750° = \dfrac{1}{2}$

$\cos 750° = \dfrac{\sqrt{3}}{2}$

$\tan 750° = \dfrac{\sqrt{3}}{3}$

48. $\sin(-405°) = -\dfrac{\sqrt{2}}{2}$

$\cos(-405°) = \dfrac{\sqrt{2}}{2}$

$\tan(-405°) = -1$

49. $\sin(-150°) = -\dfrac{1}{2}$

$\cos(-150°) = -\dfrac{\sqrt{3}}{2}$

$\tan(-150°) = \dfrac{\sqrt{3}}{3}$

50. $\sin(-840°) = -\dfrac{\sqrt{3}}{2}$

$\cos(-840°) = -\dfrac{1}{2}$

$\tan(-840°) = \sqrt{3}$

51. $\sin \dfrac{4\pi}{3} = -\dfrac{\sqrt{3}}{2}$

$\cos \dfrac{4\pi}{3} = -\dfrac{1}{2}$

$\tan \dfrac{4\pi}{3} = \sqrt{3}$

52. $\sin \dfrac{\pi}{4} = \dfrac{\sqrt{2}}{2}$

$\cos \dfrac{\pi}{4} = \dfrac{\sqrt{2}}{2}$

$\tan \dfrac{\pi}{4} = 1$

53. $\sin\left(-\dfrac{\pi}{6}\right) = -\dfrac{1}{2}$

$\cos\left(-\dfrac{\pi}{6}\right) = \dfrac{\sqrt{3}}{2}$

$\tan\left(-\dfrac{\pi}{6}\right) = -\dfrac{\sqrt{3}}{3}$

54. $\sin\left(-\dfrac{\pi}{2}\right) = -1$

$\cos\left(-\dfrac{\pi}{2}\right) = 0$

$\tan\left(-\dfrac{\pi}{2}\right)$ is undefined.

55. $\sin \dfrac{11\pi}{4} = \dfrac{\sqrt{2}}{2}$

$\cos \dfrac{11\pi}{4} = -\dfrac{\sqrt{2}}{2}$

$\tan \dfrac{11\pi}{4} = -1$

56. $\sin \dfrac{10\pi}{3} = -\dfrac{\sqrt{3}}{2}$

$\cos \dfrac{10\pi}{3} = -\dfrac{1}{2}$

$\tan \dfrac{10\pi}{3} = \sqrt{3}$

57. $\sin\left(-\dfrac{3\pi}{2}\right) = 1$

$\cos\left(-\dfrac{3\pi}{2}\right) = 0$

$\tan\left(-\dfrac{3\pi}{2}\right)$ is undefined.

58. $\sin\left(-\dfrac{25\pi}{4}\right) = -\dfrac{\sqrt{2}}{2}$

$\cos\left(-\dfrac{25\pi}{4}\right) = \dfrac{\sqrt{2}}{2}$

$\tan\left(-\dfrac{25\pi}{4}\right) = -1$

59. $\dfrac{4}{5}$ **60.** $\dfrac{\sqrt{10}}{10}$ **61.** $-\dfrac{\sqrt{13}}{2}$ **62.** $-\sqrt{3}$ **63.** $\dfrac{8}{5}$

64. $\dfrac{\sqrt{65}}{4}$ **65.** 0.1736 **66.** -1.4142 **67.** -0.3420

68. 2.0000 **69.** -1.4826 **70.** -28.6363

71. 3.2361 **72.** -0.1405 **73.** 4.6373 **74.** 0.2245

75. 0.3640 **76.** -0.3640 **77.** -0.6052

78. 1.0436 **79.** -0.4142 **80.** 4.4940

81. (a) $30° = \dfrac{\pi}{6}$, $150° = \dfrac{5\pi}{6}$ (b) $210° = \dfrac{7\pi}{6}$, $330° = \dfrac{11\pi}{6}$

82. (a) $45° = \dfrac{\pi}{4}$, $315° = \dfrac{7\pi}{4}$ (b) $135° = \dfrac{3\pi}{4}$, $225° = \dfrac{5\pi}{4}$

83. (a) $60° = \dfrac{\pi}{3}$, $120° = \dfrac{2\pi}{3}$ (b) $135° = \dfrac{3\pi}{4}$, $315° = \dfrac{7\pi}{4}$

84. (a) $60° = \dfrac{\pi}{3}$, $300° = \dfrac{5\pi}{3}$ (b) $120° = \dfrac{2\pi}{3}$, $240° = \dfrac{4\pi}{3}$

85. (a) $45° = \dfrac{\pi}{4}$, $225° = \dfrac{5\pi}{4}$ (b) $150° = \dfrac{5\pi}{6}$, $330° = \dfrac{11\pi}{6}$

86. (a) $60° = \dfrac{\pi}{3}$, $120° = \dfrac{2\pi}{3}$ (b) $240° = \dfrac{4\pi}{3}$, $300° = \dfrac{5\pi}{3}$

87. (a) $N = 22.099 \sin(0.522t - 2.219) + 55.008$
$F = 36.641 \sin(0.502t - 1.831) + 25.610$
(b) February: $N = 34.6°$, $F = -1.4°$
March: $N = 41.6°$, $F = 13.9°$
May: $N = 63.4°$, $F = 48.6°$
June: $N = 72.5°$, $F = 59.5°$
August: $N = 75.5°$, $F = 55.6°$
September: $N = 68.6°$, $F = 41.7°$
November: $N = 46.8°$, $F = 6.5°$
(c) Answers will vary.

88. (a) 26,134 units (b) 31,438 units (c) 21,452 units
(d) 26,756 units

89. (a) 2 centimeters (b) 0.14 centimeter
(c) -1.98 centimeters

90. (a) 2 centimeters (b) 0.11 centimeter
(c) -1.2 centimeters

91. 0.79 ampere

92. (a) 12 miles (b) 6 miles (c) 6.9 miles

93. False. In each of the four quadrants, the signs of the secant function and cosine function will be the same, because these functions are reciprocals of each other.

94. False. For θ in Quadrant II, $\theta' = 180° - \theta$. For θ in Quadrant III, $\theta' = \theta - 180°$. For θ in Quadrant IV, $\theta' = 360° - \theta$.

95. As θ increases from 0° to 90°, x decreases from 12 cm to 0 cm and y increases from 0 cm to 12 cm. Therefore, $\sin\theta = y/12$ increases from 0 to 1 and $\cos\theta = x/12$ decreases from 1 to 0. Thus, $\tan\theta = y/x$ and increases without bound. When $\theta = 90°$, the tangent is undefined.

96. Determine the trigonometric function of the reference angle and prefix the appropriate sign.

97.

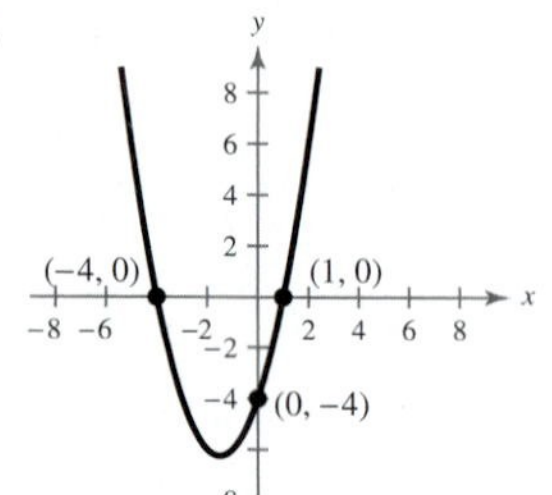

x-intercepts:
$(1, 0), (-4, 0)$
y-intercept: $(0, -4)$
Domain: all real numbers x

98.

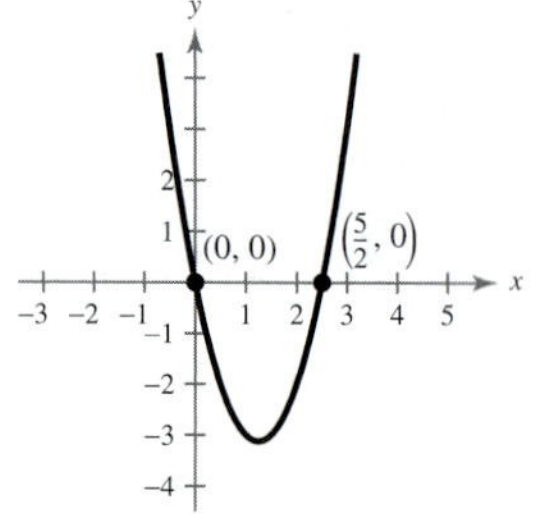

Intercept: $(0, 0)$
x-intercept: $\left(\frac{5}{2}, 0\right)$
Domain: all real numbers x

99.

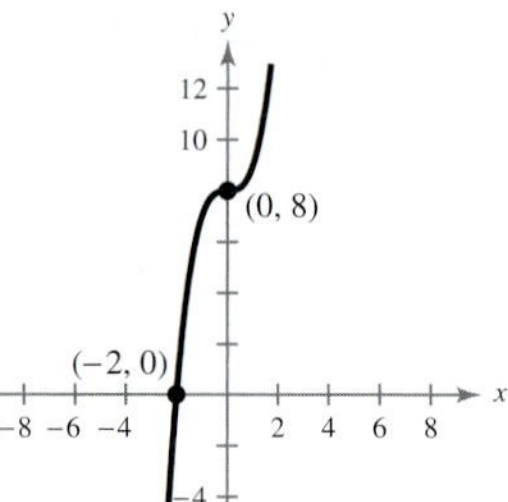

x-intercept: $(-2, 0)$
y-intercept: $(0, 8)$
Domain: all real numbers x

100.

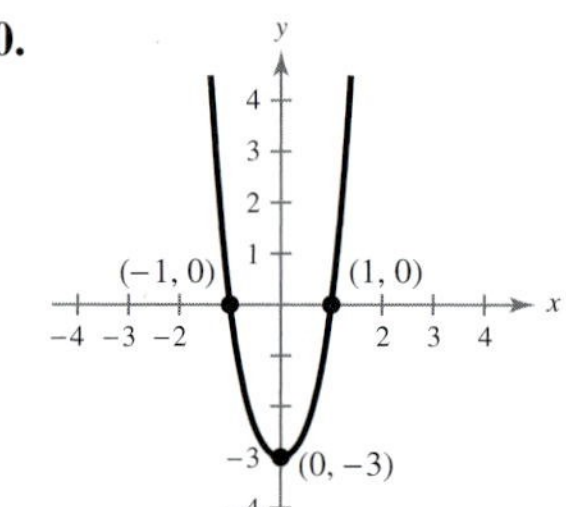

x-intercepts:
$(1, 0), (-1, 0)$
y-intercept: $(0, -3)$
Domain: all real numbers x

101.

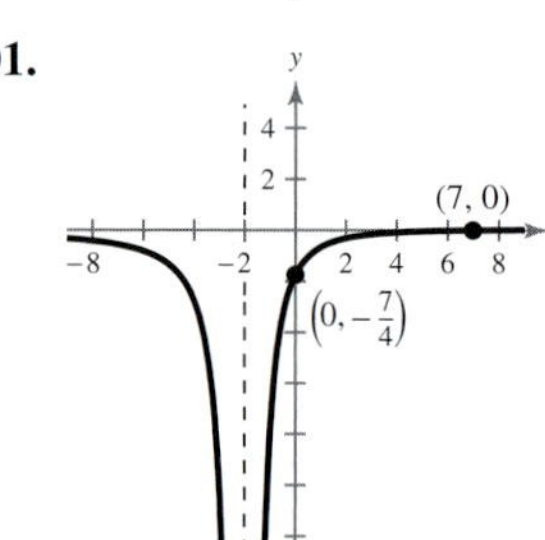

x-intercept: $(7, 0)$
y-intercept: $\left(0, -\frac{7}{4}\right)$
Vertical asymptote:
$x = -2$
Horizontal asymptote:
$y = 0$
Domain: all real numbers
x except $x = -2$

102.

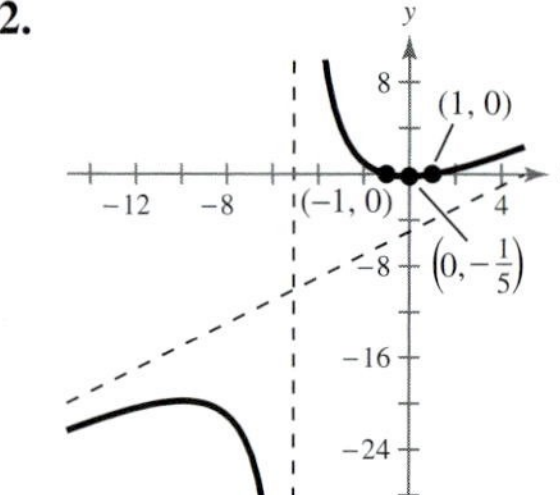

x-intercepts:
$(1, 0), (-1, 0)$
y-intercept: $\left(0, -\frac{1}{5}\right)$
Vertical asymptote:
$x = -5$
Slant asymptote:
$y = x - 5$
Domain: all real numbers
x except $x = -5$

103.

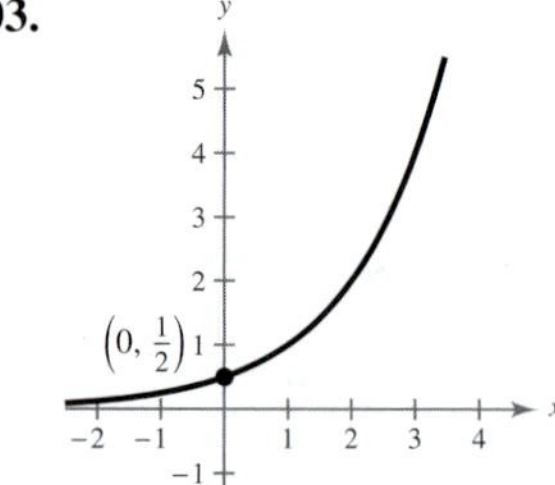

y-intercept: $\left(0, \frac{1}{2}\right)$
Horizontal asymptote:
$y = 0$
Domain: all real numbers x

104. y-intercept: $(0, 5)$
Horizontal asymptote: $y = 2$
Domain: all real numbers x

105. x-intercepts: $(\pm 1, 0)$
Vertical asymptote: $x = 0$
Domain: all real numbers x except $x = 0$

106.

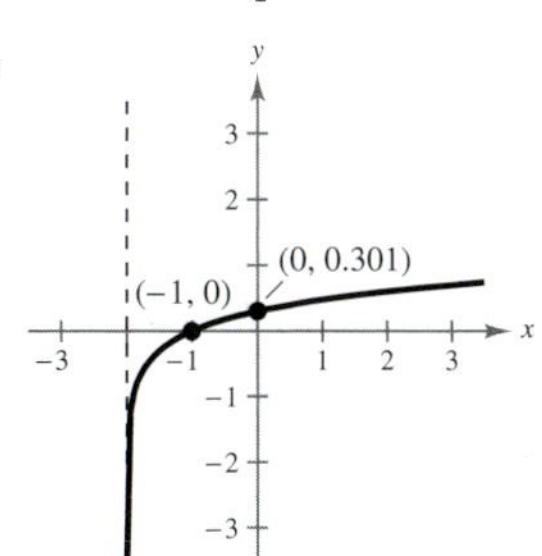

x-intercept: $(-1, 0)$
y-intercept: $(0, 0.301)$
Vertical asymptote: $x = -2$
Domain: all real numbers x such that $x > -2$

Section 4.5 *(page 328)*

Vocabulary Check *(page 328)*

1. cycle **2.** amplitude **3.** $\dfrac{2\pi}{b}$
4. phase shift **5.** vertical shift

1. Period: π
Amplitude: 3

2. Period: $\dfrac{2\pi}{3}$
Amplitude: 2

3. Period: 4π
Amplitude: $\frac{5}{2}$

4. Period: 6π
Amplitude: 3

5. Period: 6
Amplitude: $\frac{1}{2}$

6. Period: 4
Amplitude: $\frac{3}{2}$

7. Period: 2π
Amplitude: 3

8. Period: 3π
Amplitude: 1

9. Period: $\dfrac{\pi}{5}$
Amplitude: 3

10. Period: $\dfrac{\pi}{4}$
Amplitude: $\dfrac{1}{3}$

11. Period: 3π
Amplitude: $\dfrac{1}{2}$

12. Period: 8π
Amplitude: $\dfrac{5}{2}$

13. Period: 1
Amplitude: $\frac{1}{4}$

14. Period: 20
Amplitude: $\frac{2}{3}$

15. g is a shift of f π units to the right.

16. g is a shift of f π units to the left.

17. g is a reflection of f in the x-axis.

18. g is a reflection of f in the x-axis.

19. The period of f is twice the period of g.

20. The period of g is one-third the period of f.

21. g is a shift of f three units upward.

22. g is a shift of f two units downward.

23. The graph of g has twice the amplitude of the graph of f.

24. The period of g is $\frac{1}{3}$ the period of f.

25. The graph of g is a horizontal shift of the graph of f π units to the right.

26. g is a shift of f two units upward.

27.

28.

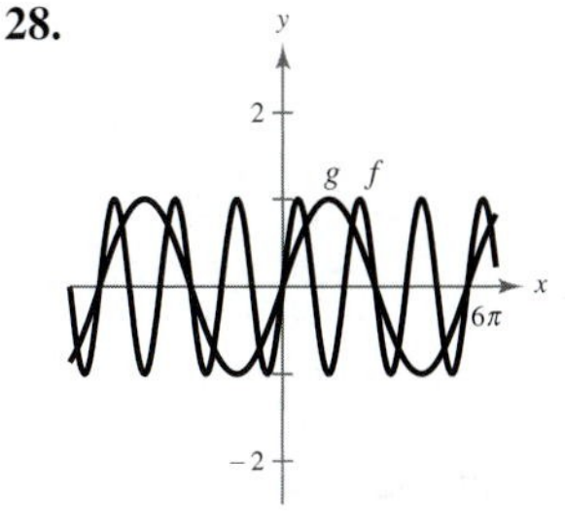

29.

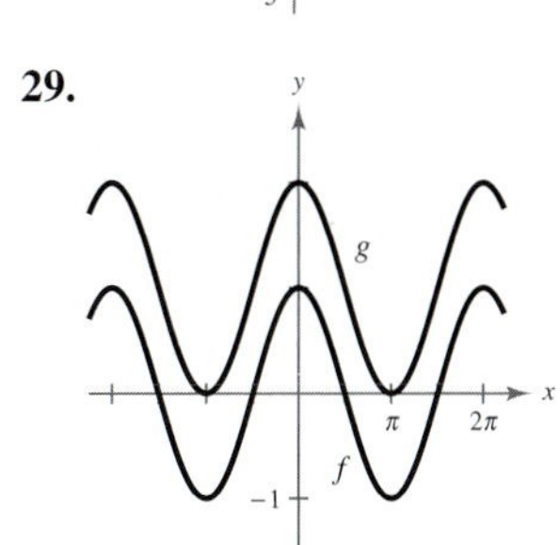

30.

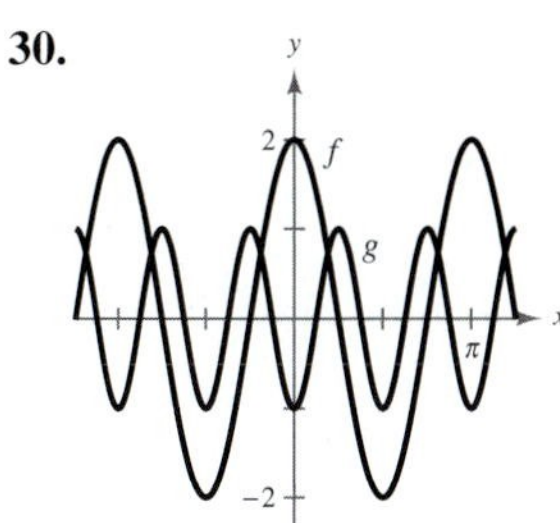

31.

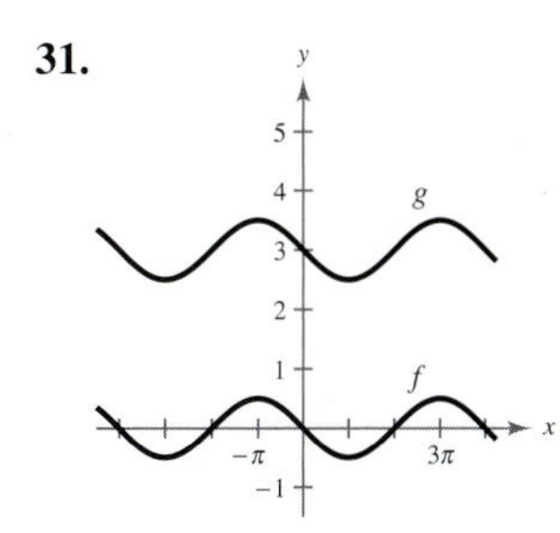

32.

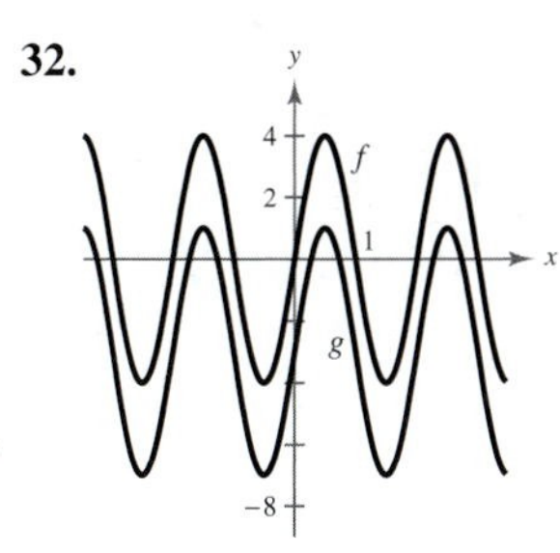

33.

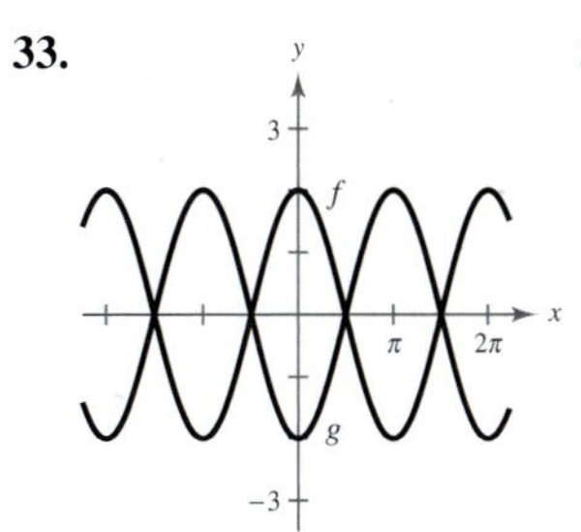

34.

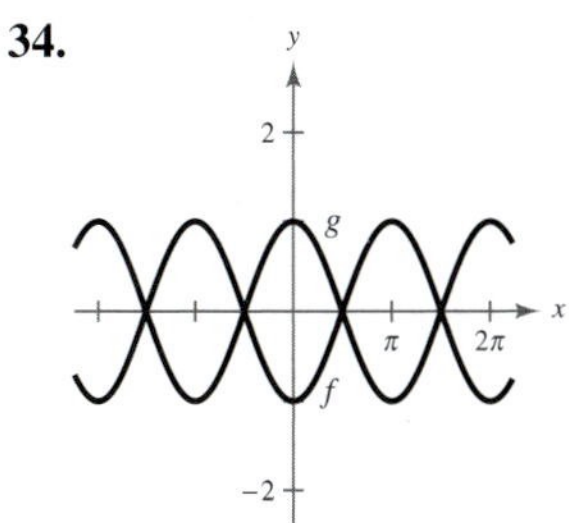

CHAPTER 4

35.

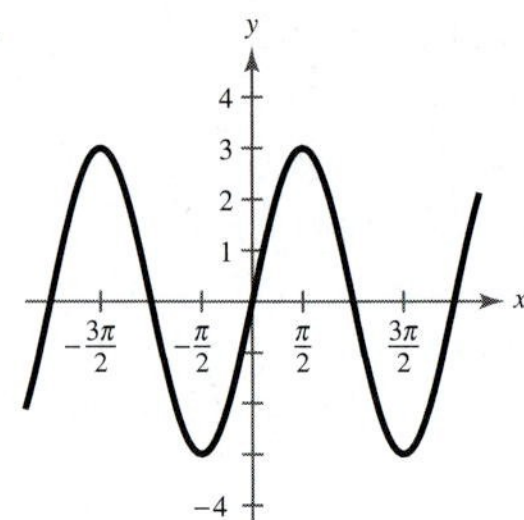

36.

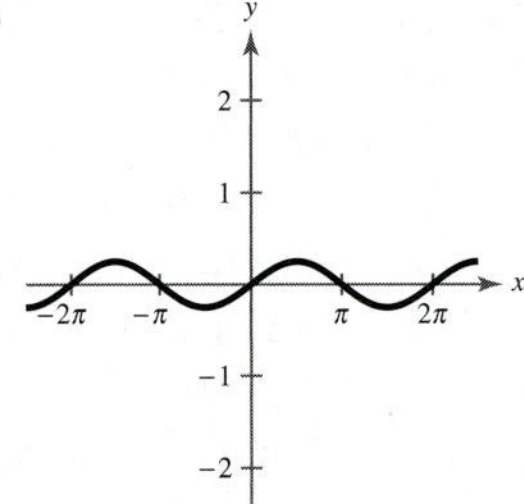

37.

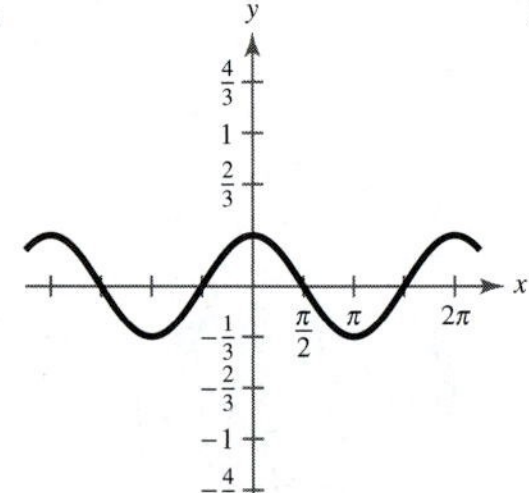

38.

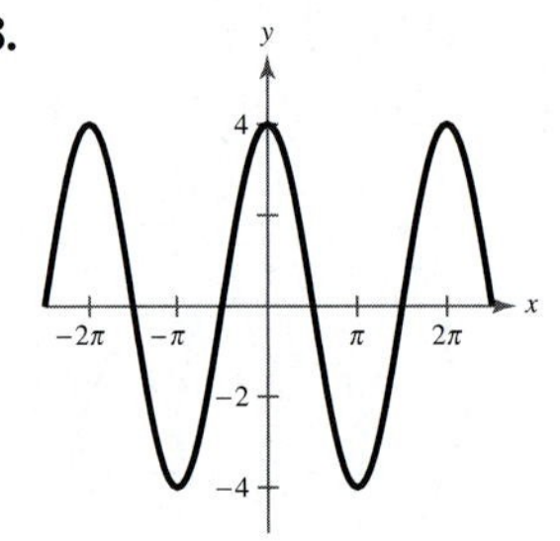

39.

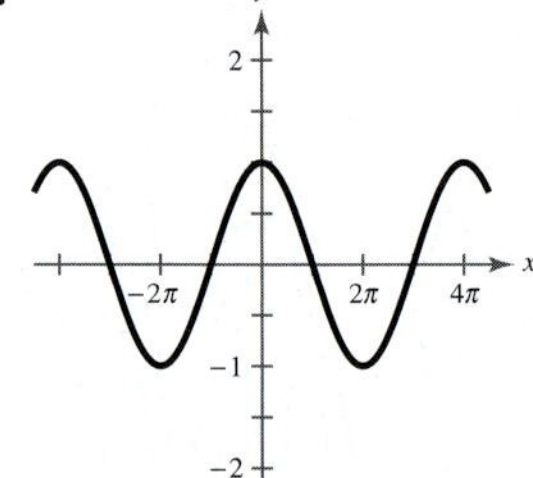

40.

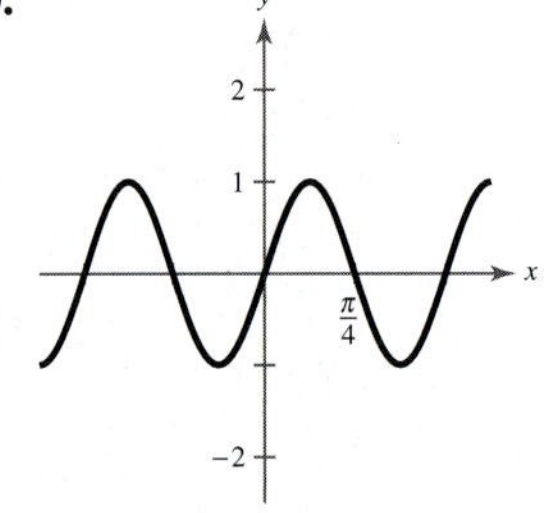

41.

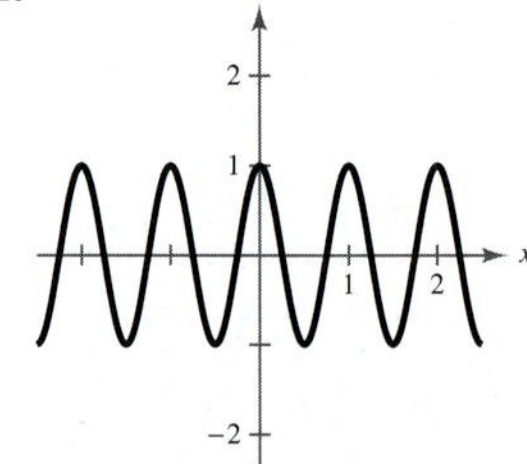

42.

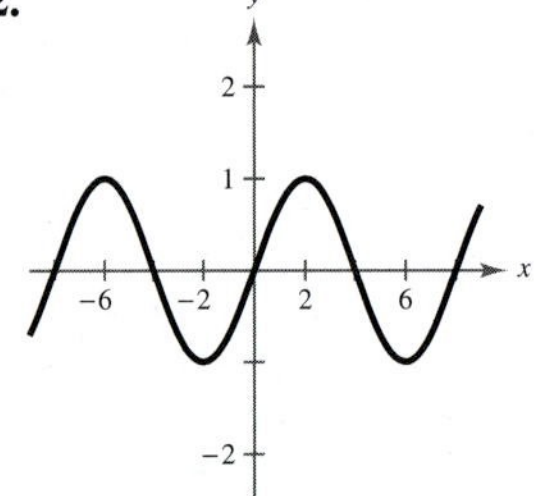

43.

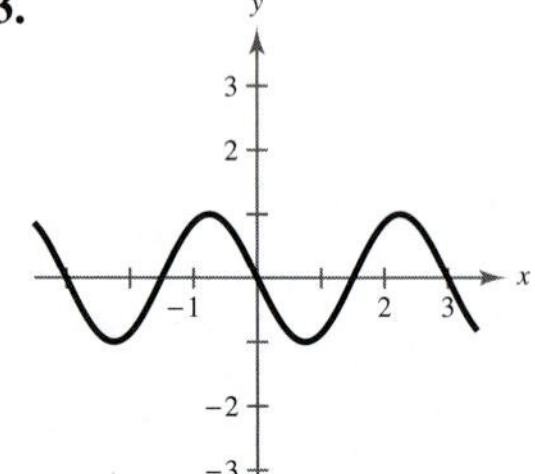

44.

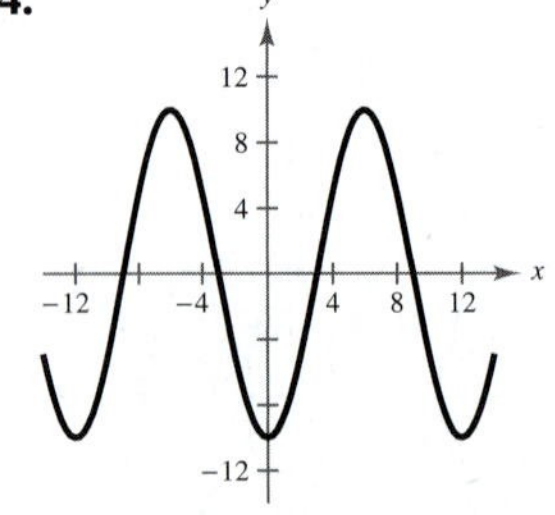

45.

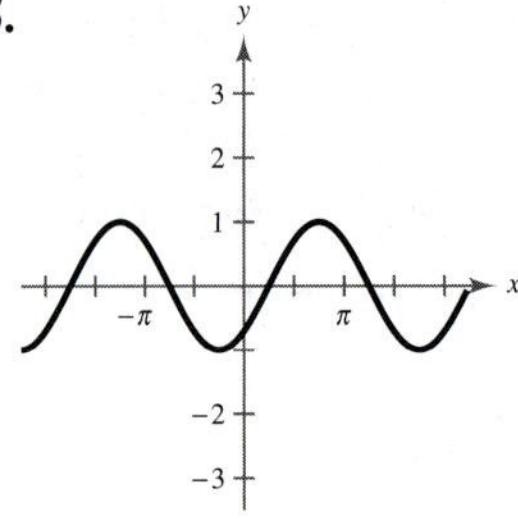

46.

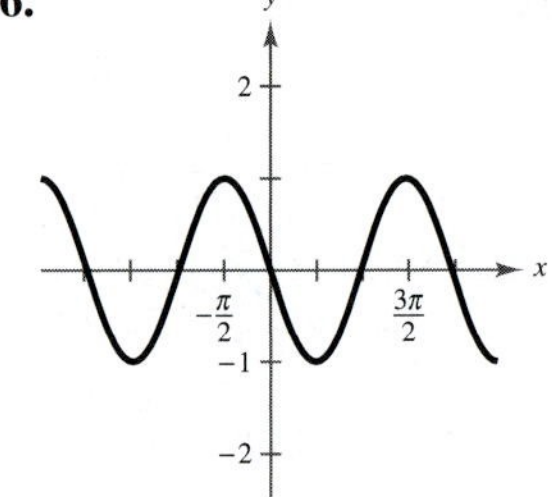

47.

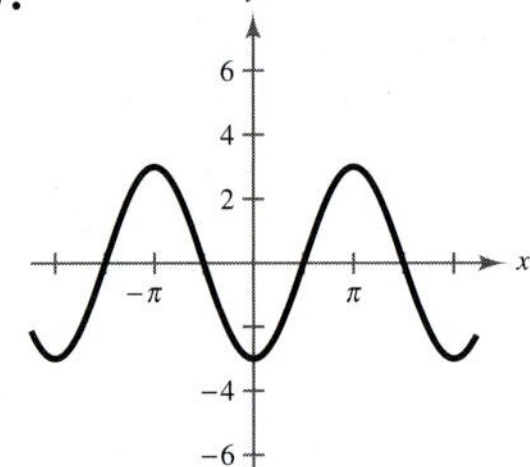

48.

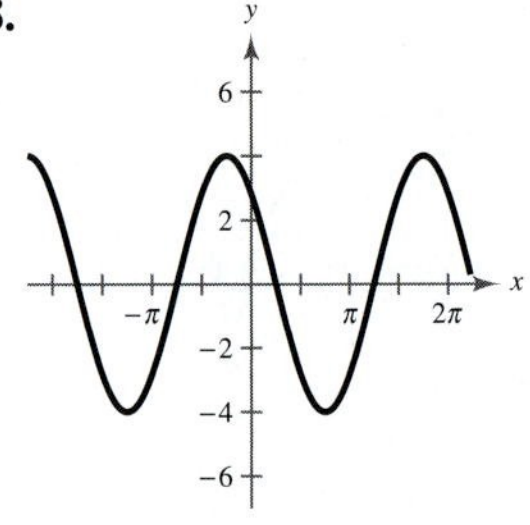

49.

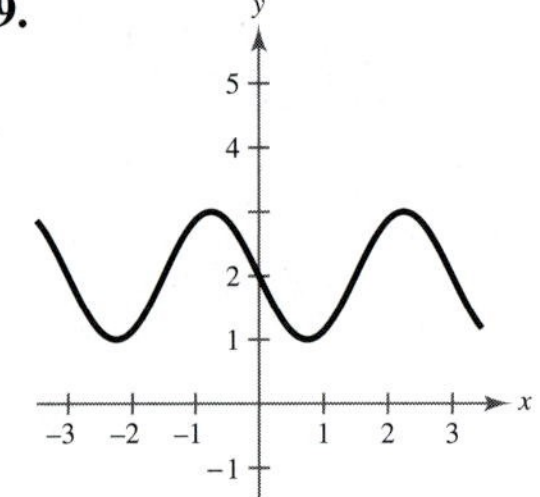

50.

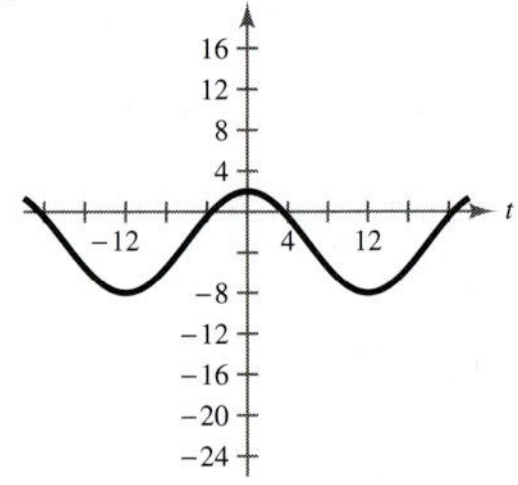

51.

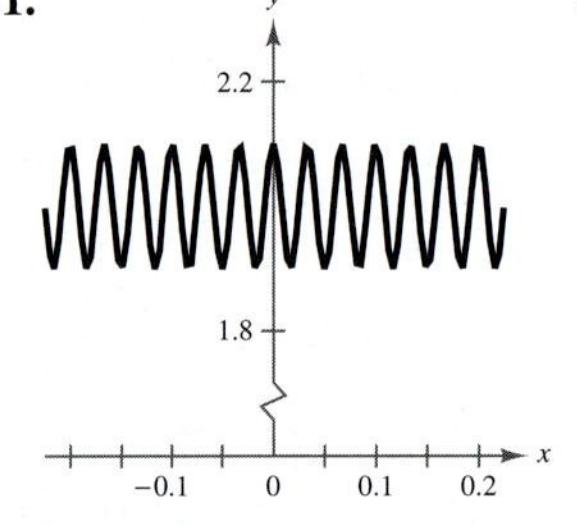

52.

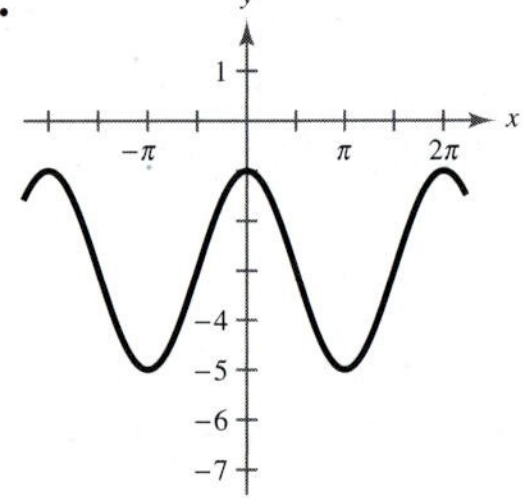

53.

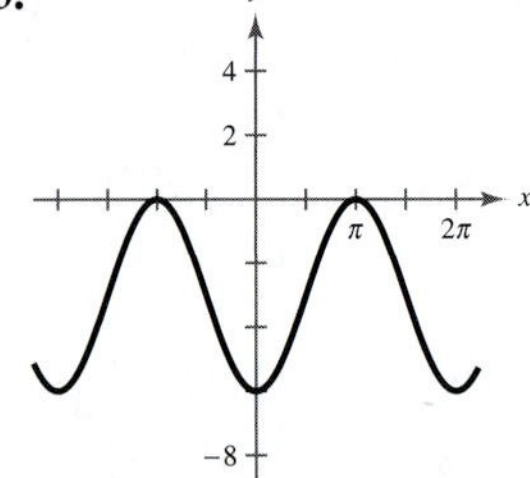

54.

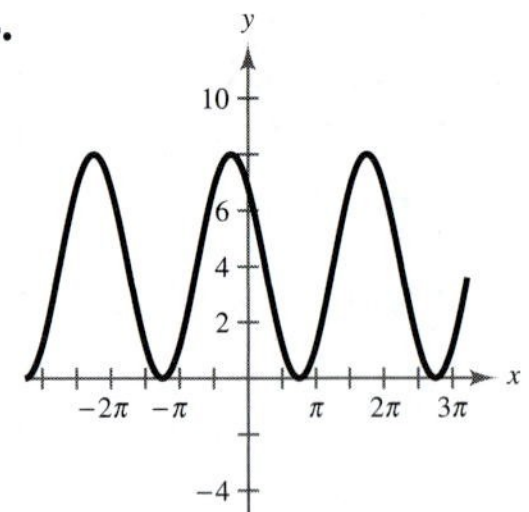

55.

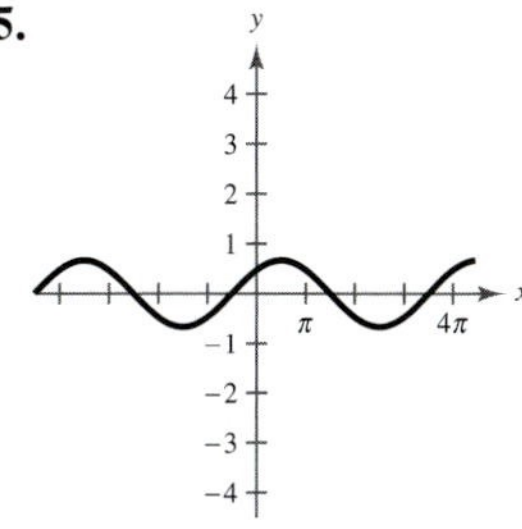

56.

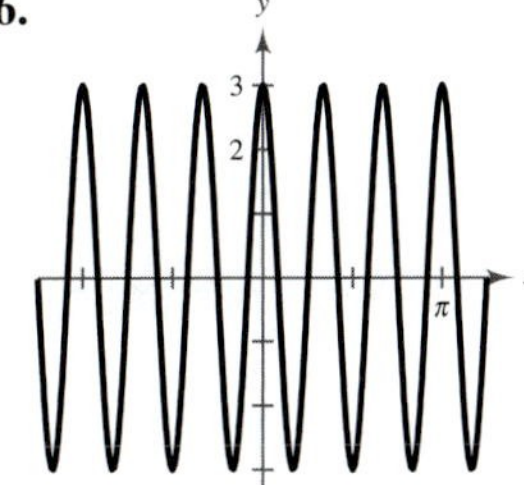

57.

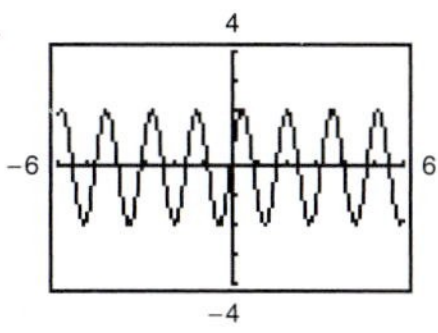

58.

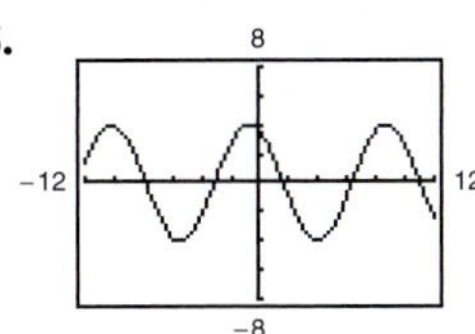

59.

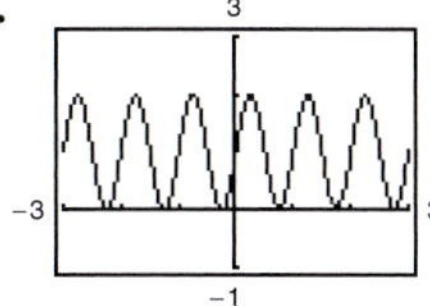

60.

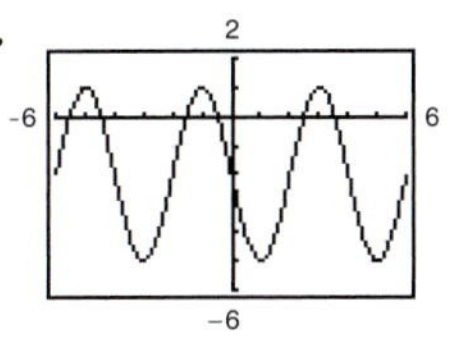

61.

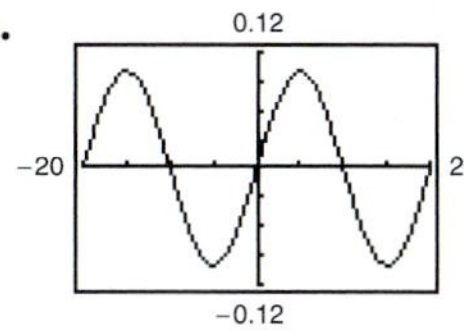

62.

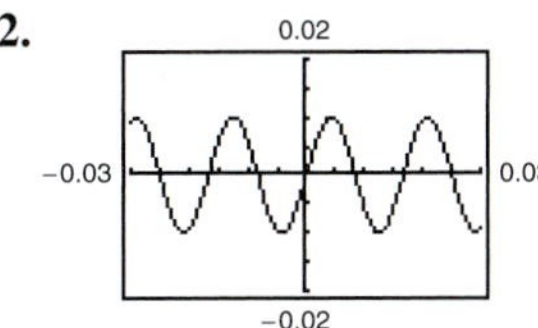

63. $a = 2, d = 1$ **64.** $a = 2, d = -1$

65. $a = -4, d = 4$ **66.** $a = -1, d = -3$

67. $a = -3, b = 2, c = 0$ **68.** $a = 2, b = \frac{1}{2}, c = 0$

69. $a = 2, b = 1, c = -\frac{\pi}{4}$ **70.** $a = 2, b = \frac{\pi}{2}, c = -\frac{\pi}{2}$

71.

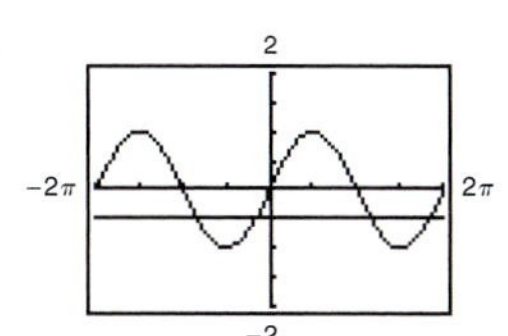

$x = -\frac{\pi}{6}, -\frac{5\pi}{6}, \frac{7\pi}{6}, \frac{11\pi}{6}$

72.

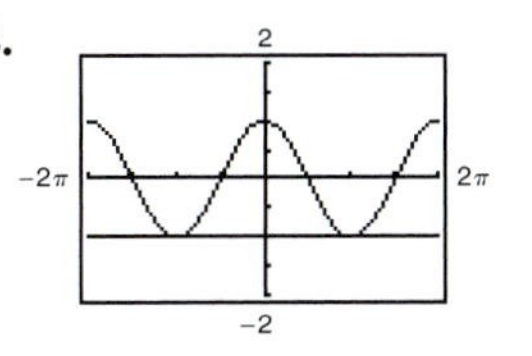

$x = \pi, -\pi$

73. (a) 6 seconds (b) 10 cycles per minute

(c)

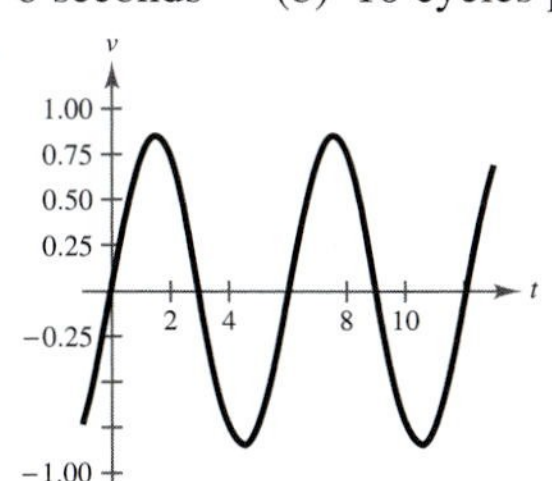

74. (a) 4 seconds (b) 15 cycles per minute

(c)

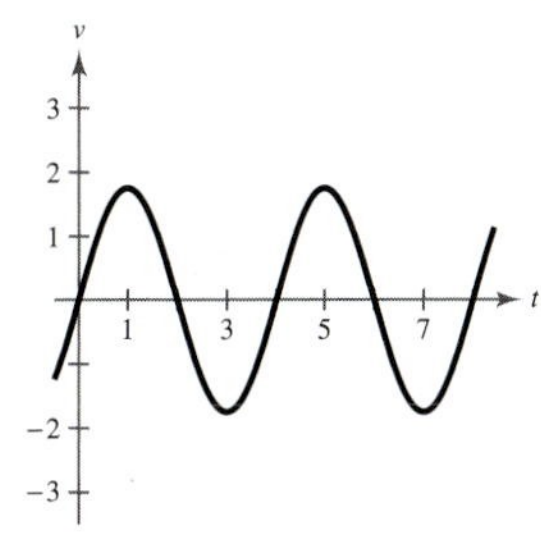

75. (a) $C(t) = 56.55 + 26.95 \cos\left(\frac{\pi}{6}t - 3.67\right)$

(b)

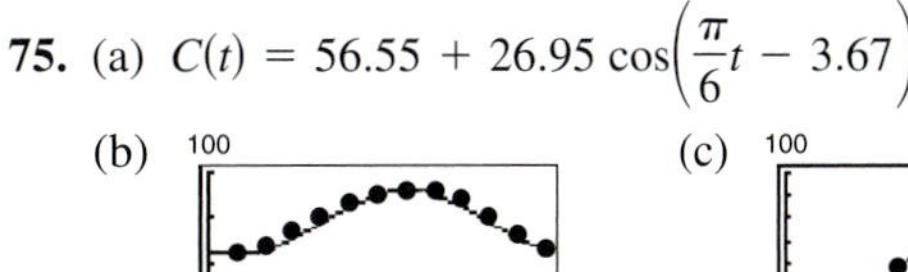

The model is a good fit.

(c)

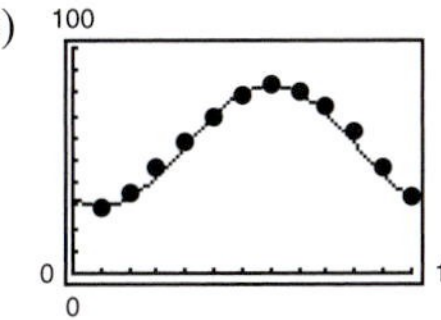

The model is a good fit.

(d) Tallahassee: 77.90°; Chicago: 56.55°

The constant term gives the annual average temperature.

(e) 12; yes; one full period is one year.

(f) Chicago; amplitude; the greater the amplitude, the greater the variability in temperature.

76. (a) $\frac{6}{5}$ seconds (b) 50 heartbeats per minute

77. (a) $\frac{1}{440}$ second (b) 440 cycles per second

78. (a)–(c)

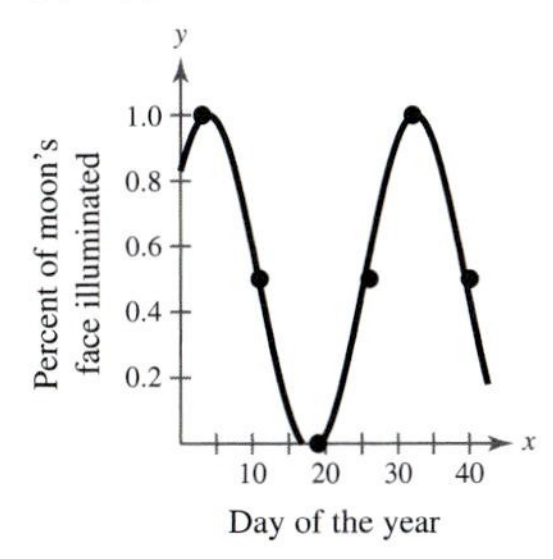

(b) $y = \frac{1}{2} + \frac{1}{2}\sin(0.21x + 0.92)$

The model is a good fit.

(d) 29 days

(e) 0.44

79. (a) 365; answers will vary.

(b) 30.3 gallons; the constant term

(c)

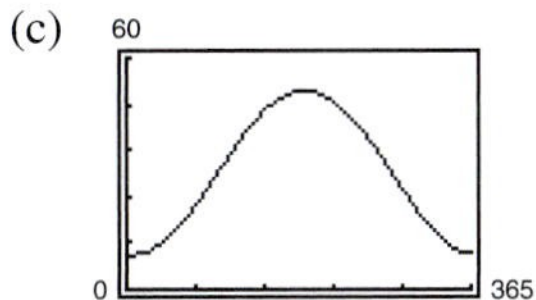

$124 < t < 252$

80. (a) 20 seconds; it takes 20 seconds to complete one revolution on the Ferris wheel.
(b) 50 feet; the diameter of the Ferris wheel is 100 feet.
(c)

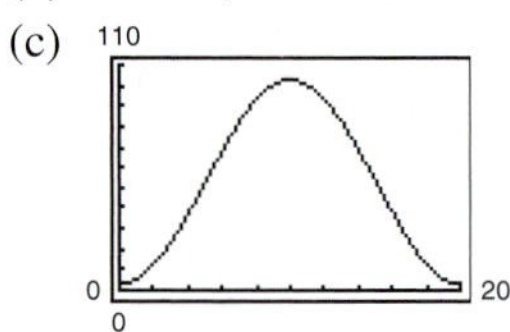

81. False. The graph of $f(x) = \sin(x + 2\pi)$ translates the graph of $f(x) = \sin x$ exactly one period to the left so that the two graphs look identical.

82. False. The function $y = \frac{1}{2}\cos 2x$ has an amplitude that is one-half that of $y = \cos x$. For $y = a\cos bx$, the amplitude is $|a|$.

83. True. Because $\cos x = \sin\left(x + \frac{\pi}{2}\right)$, $y = -\cos x$ is a reflection in the x-axis of $y = \sin\left(x + \frac{\pi}{2}\right)$.

84. Answers will vary.

85.

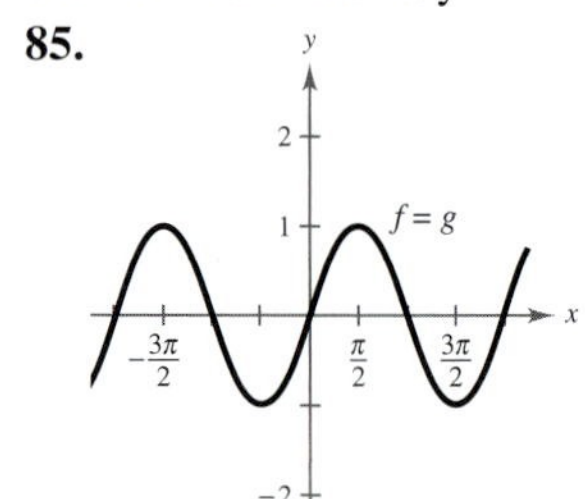

Conjecture:
$\sin x = \cos\left(x - \frac{\pi}{2}\right)$

86.

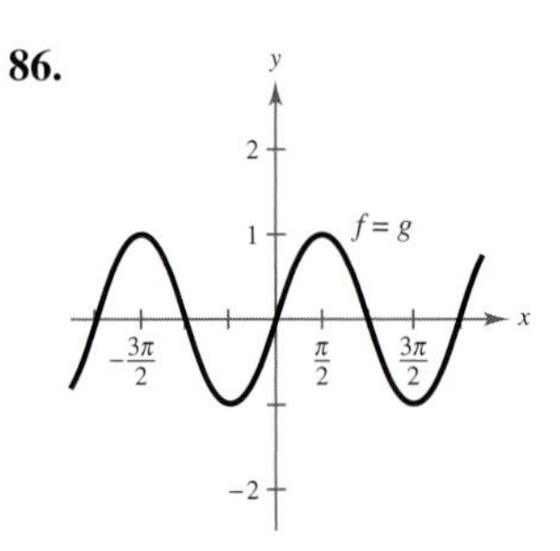

Conjecture:
$\sin x = -\cos\left(x + \frac{\pi}{2}\right)$

87. (a)

The graphs appear to coincide from $-\frac{\pi}{2}$ to $\frac{\pi}{2}$.

(b)

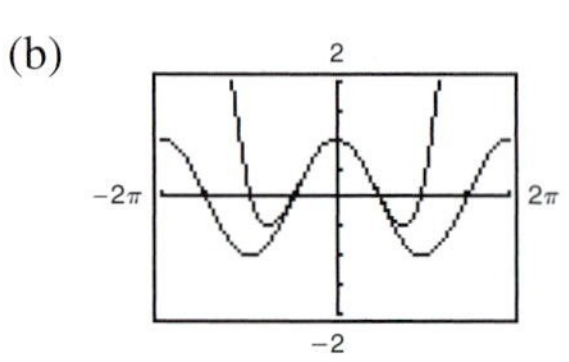

The graphs appear to coincide from $-\frac{\pi}{2}$ to $\frac{\pi}{2}$.

(c) $-\frac{x^7}{7!}, -\frac{x^6}{6!}$

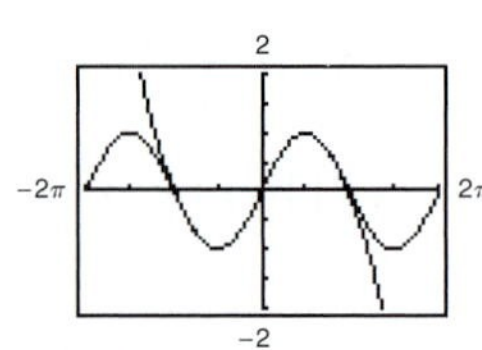

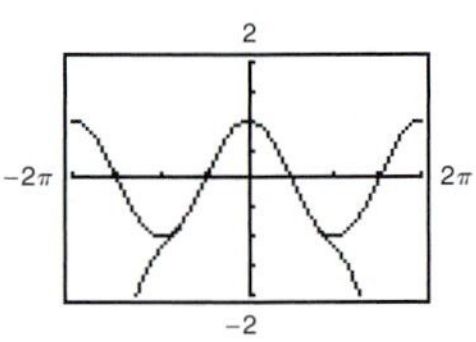

The interval of accuracy increased.

88. (a) 0.4794, 0.4794 (b) 0.8417, 0.8415 (c) 0.5, 0.5
(d) 0.8776, 0.8776 (e) 0.5417, 0.5403
(f) 0.7074, 0.7071
The error increases as x moves farther away from 0.

89. $\frac{1}{2}\log_{10}(x - 2)$ **90.** $2\log_2 x + \log_2(x - 3)$

91. $3\ln t - \ln(t - 1)$ **92.** $\frac{1}{2}\ln z - \frac{1}{2}\ln(z^2 + 1)$

93. $\log_{10}\sqrt{xy}$ **94.** $\log_2(x^3 y)$

95. $\ln\frac{3x}{y^4}$ **96.** $\ln\left(x^2\sqrt{2x}\right)$ **97.** Answers will vary.

Section 4.6 *(page 339)*

Vocabulary Check *(page 339)*

1. vertical **2.** reciprocal **3.** damping
4. π **5.** $x \neq n\pi$ **6.** $(-\infty, -1] \cup [1, \infty)$
7. 2π

1. e, π **2.** c, 2π **3.** a, 1 **4.** d, 2π
5. f, 4 **6.** b, 4

7.

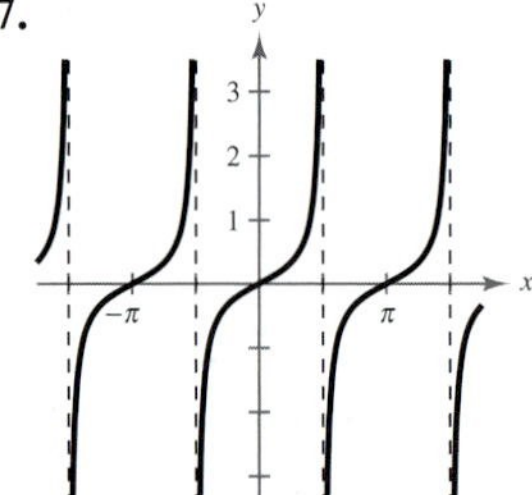

8.

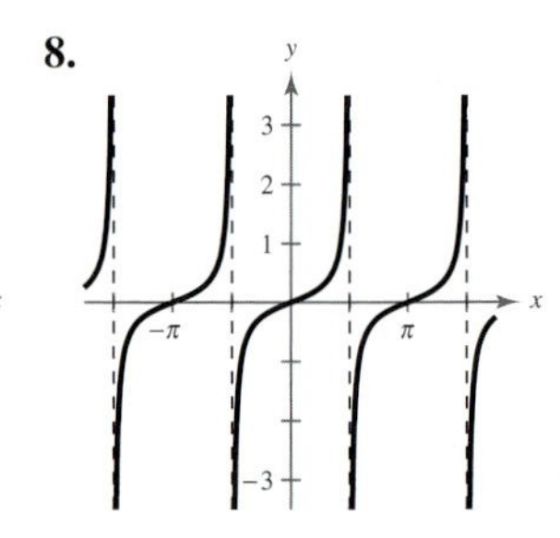

9. **10.**

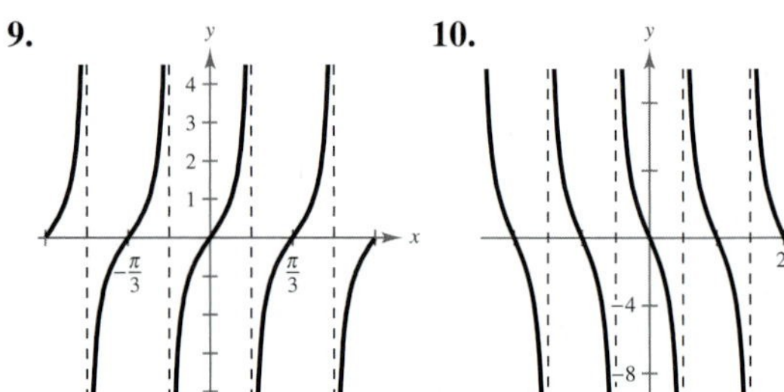

11. **12.**

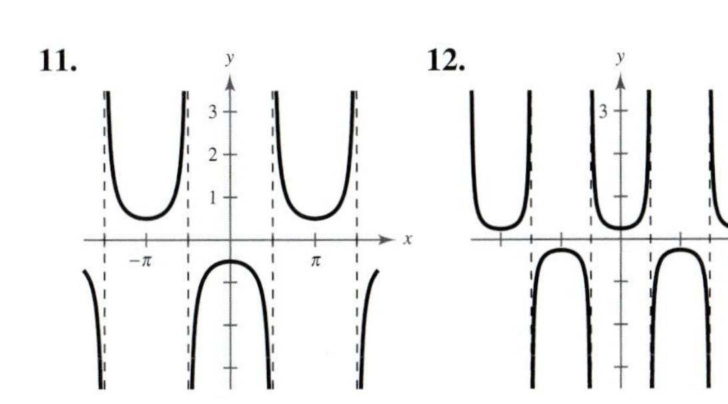

13. **14.**

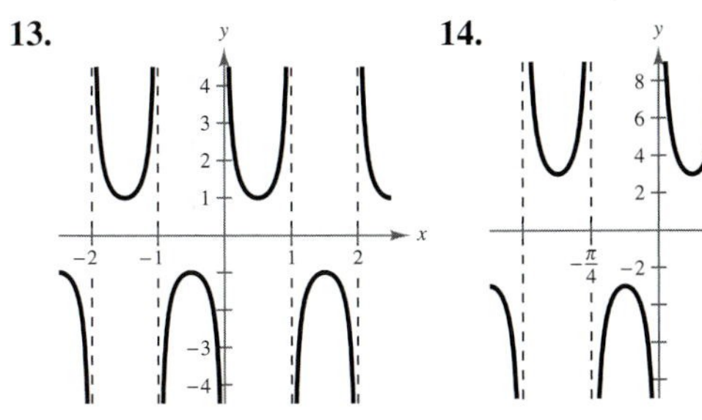

15. **16.**

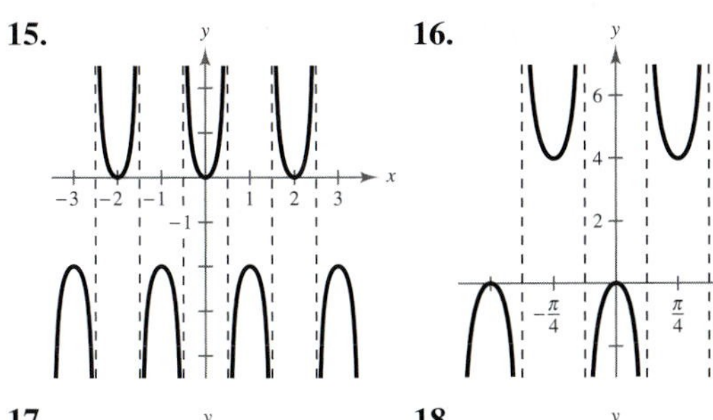

17. **18.**

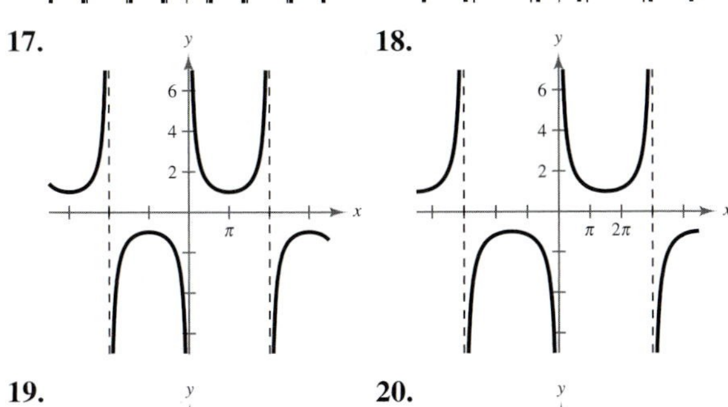

19. **20.**

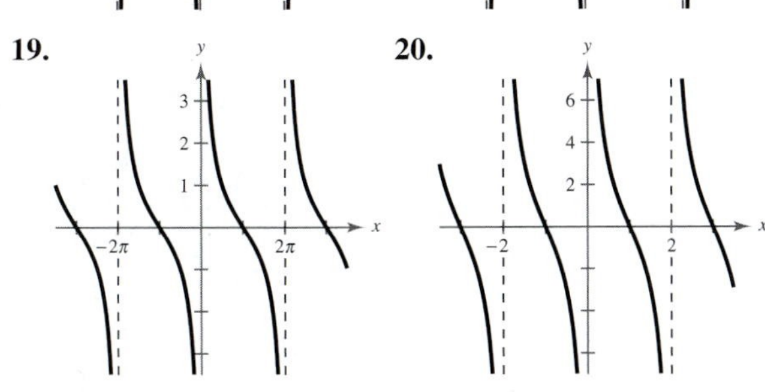

21. **22.** **23.** **24.** **25.** **26.** **27.** **28.** **29.** **30.** **31.** **32.**

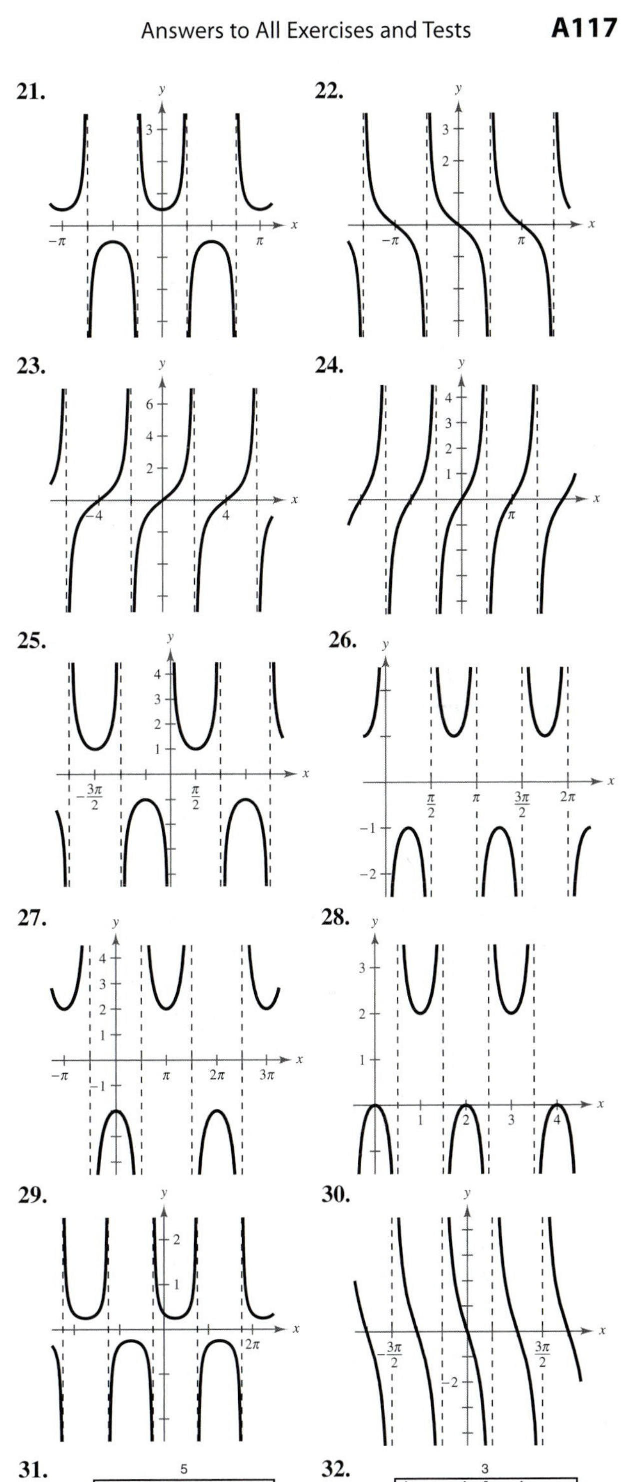

33.
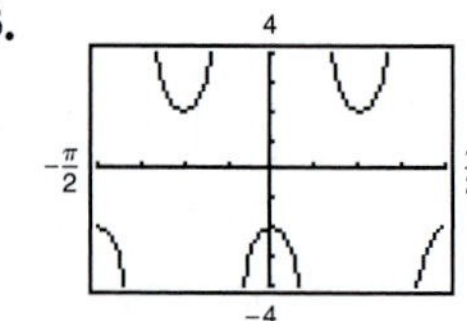

34.
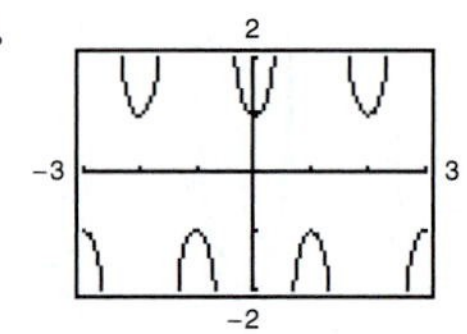

35.
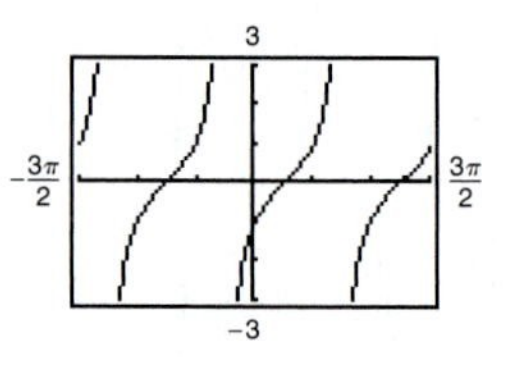

36.
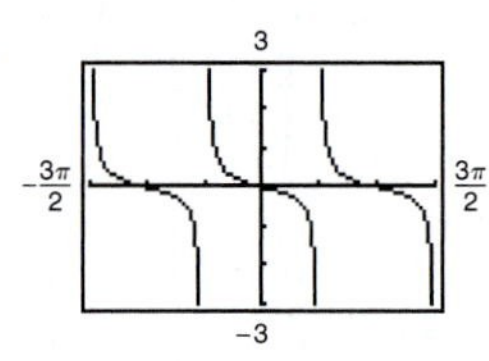

37.
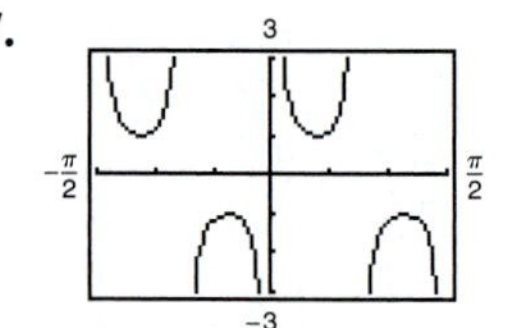

38.
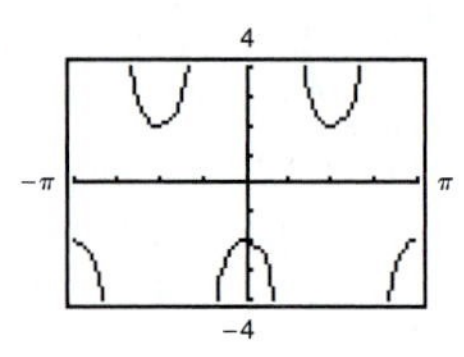

39.
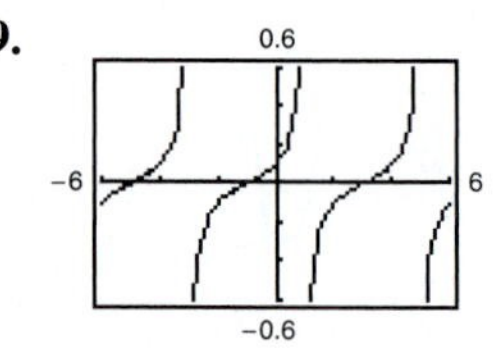

40.
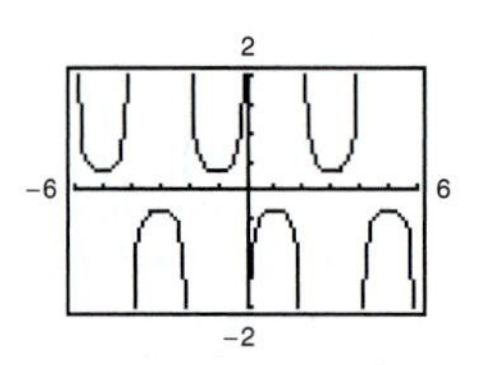

41. $-\frac{7\pi}{4}, -\frac{3\pi}{4}, \frac{\pi}{4}, \frac{5\pi}{4}$ **42.** $-\frac{5\pi}{3}, -\frac{2\pi}{3}, \frac{\pi}{3}, \frac{4\pi}{3}$

43. $-\frac{4\pi}{3}, -\frac{\pi}{3}, \frac{2\pi}{3}, \frac{5\pi}{3}$ **44.** $-\frac{7\pi}{4}, -\frac{3\pi}{4}, \frac{\pi}{4}, \frac{5\pi}{4}$

45. $-\frac{4\pi}{3}, -\frac{2\pi}{3}, \frac{2\pi}{3}, \frac{4\pi}{3}$ **46.** $-\frac{5\pi}{3}, -\frac{\pi}{3}, \frac{\pi}{3}, \frac{5\pi}{3}$

47. $-\frac{7\pi}{4}, -\frac{5\pi}{4}, \frac{\pi}{4}, \frac{3\pi}{4}$ **48.** $-\frac{2\pi}{3}, -\frac{\pi}{3}, \frac{4\pi}{3}, \frac{5\pi}{3}$

49. Even **50.** Odd

51. (a)
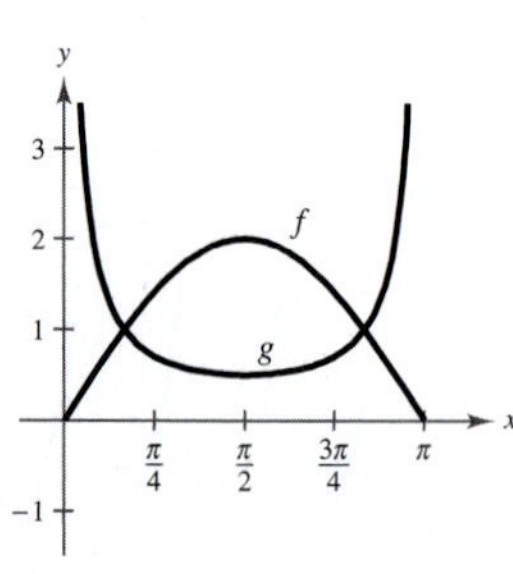

(b) $\frac{\pi}{6} < x < \frac{5\pi}{6}$

(c) f approaches 0 and g approaches $+\infty$ because the cosecant is the reciprocal of the sine.

52. (a)
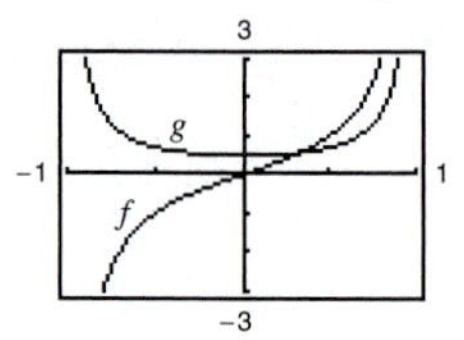

(b) $\left(-1, \frac{1}{3}\right)$ (c) $\left(-1, \frac{1}{3}\right)$; The intervals are the same.

53.
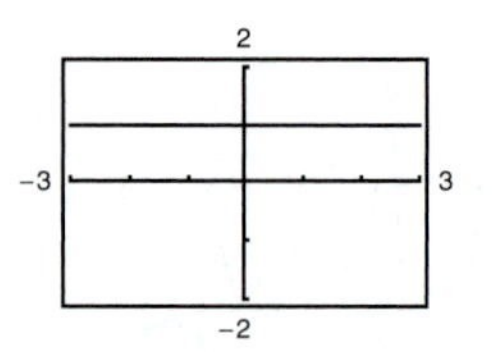

The expressions are equivalent except that when $\sin x = 0$, y_1 is undefined.

54.
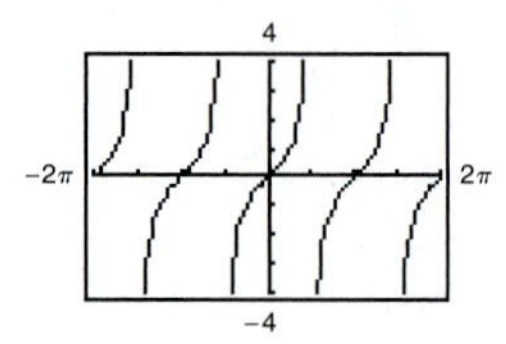

The expressions are equivalent.

55.
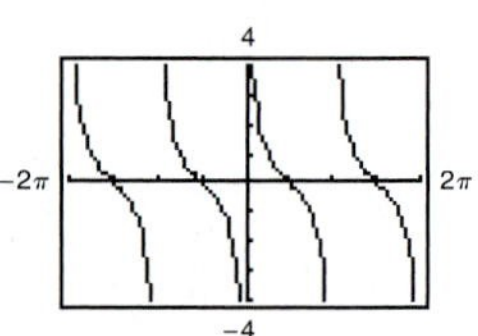

The expressions are equivalent.

56.
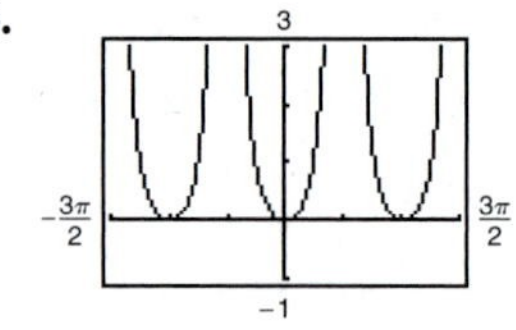

The expressions are equivalent.

57. d, $f \to 0$ as $x \to 0$. **58.** a, $f \to 0$ as $x \to 0$.

59. b, $g \to 0$ as $x \to 0$. **60.** c, $g \to 0$ as $x \to 0$.

61.
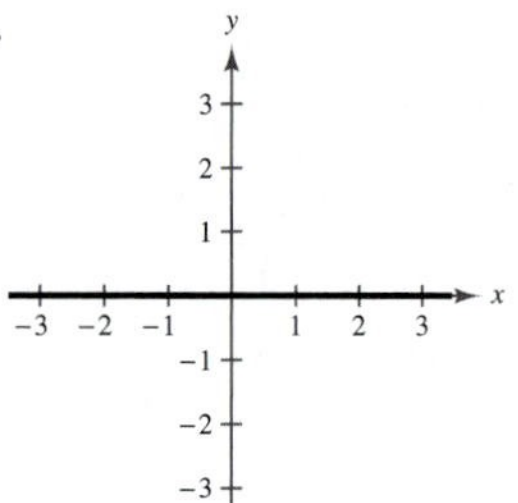

The functions are equal.

62.
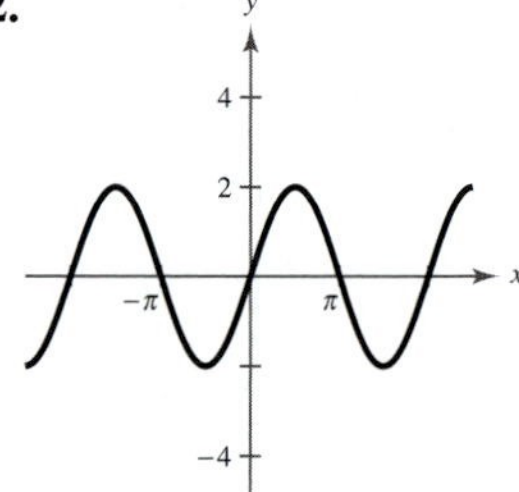

The functions are equal.

63.
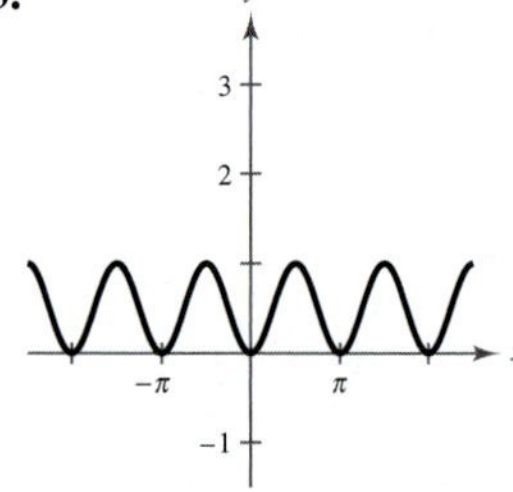

The functions are equal.

64.
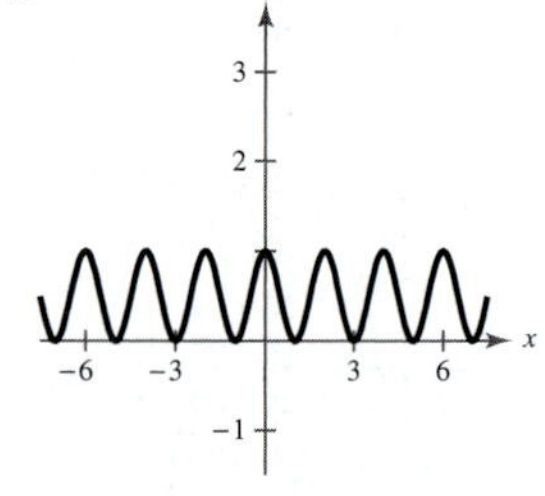

The functions are equal.

65.
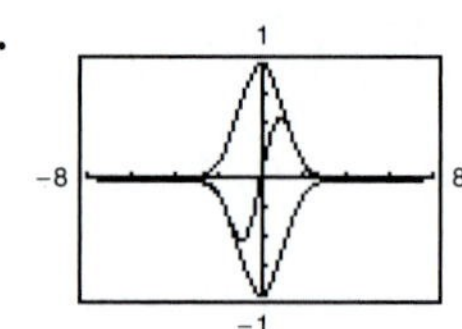

As $x \to \infty$, $g(x) \to 0$.

66.

As $x \to \infty$, $f(x) \to 0$.

67.

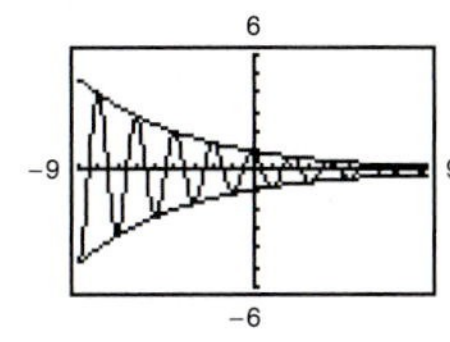

As $x \to \infty$, $f(x) \to 0$.

68.

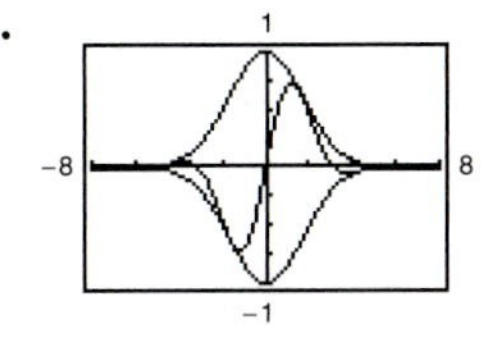

As $x \to \infty$, $h(x) \to 0$.

69.

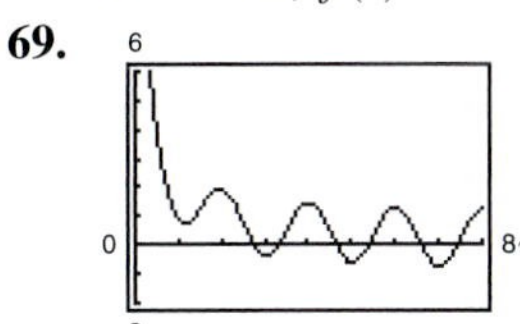

As $x \to 0$, $y \to \infty$.

70.

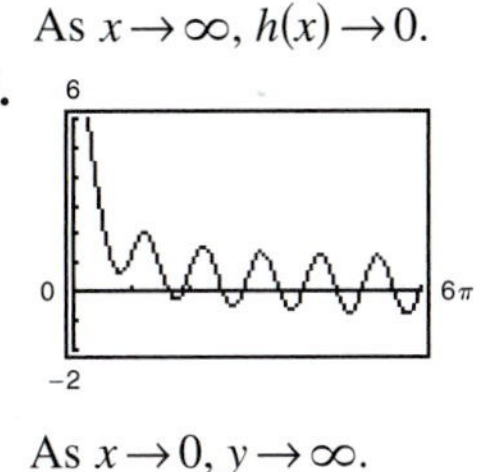

As $x \to 0$, $y \to \infty$.

71.

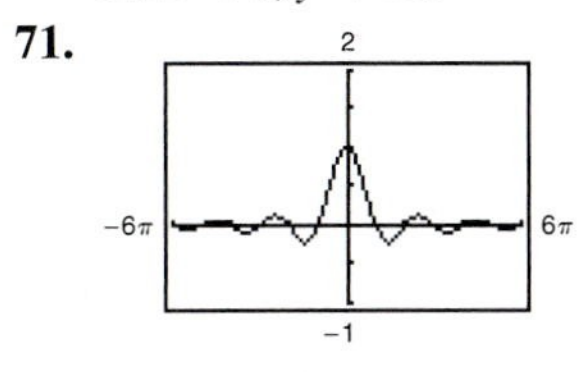

As $x \to 0$, $g(x) \to 1$.

72.

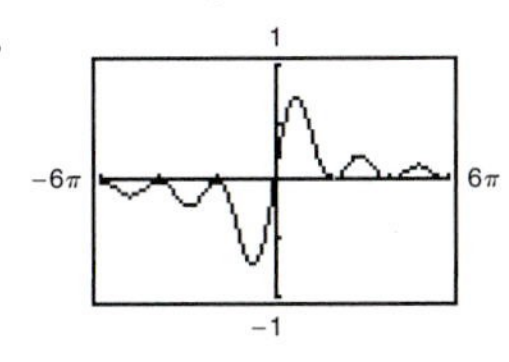

As $x \to 0$, $y \to \infty$.

73.

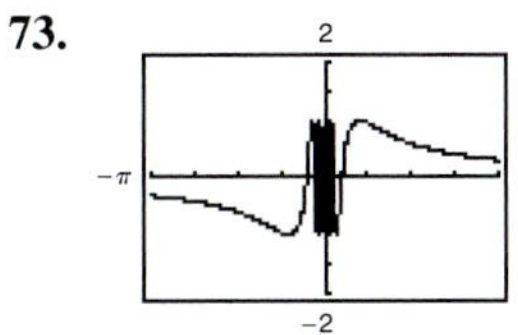

As $x \to 0$, $f(x)$ oscillates between 1 and -1.

74.

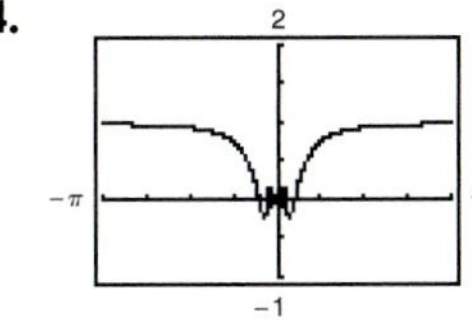

As $x \to 0$, $h(x)$ oscillates.

75. $d = 7 \cot x$

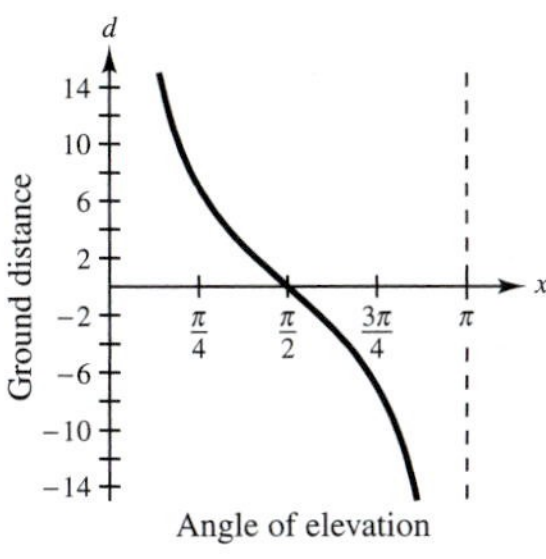

76. $d = 27 \sec x$

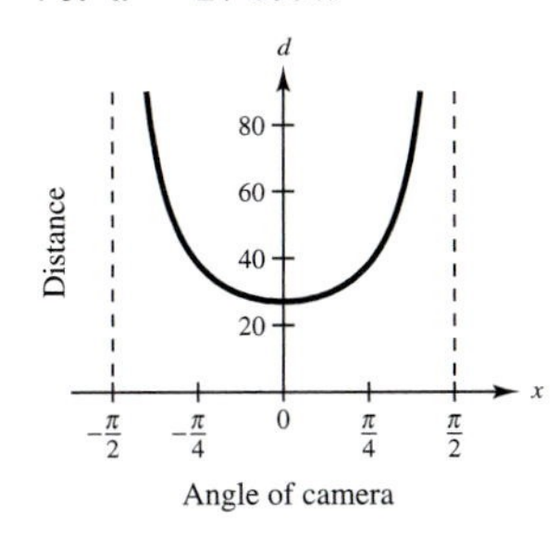

77. (a)

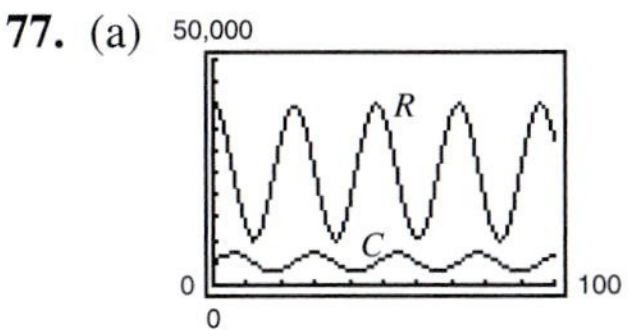

(b) As the predator population increases, the number of prey decreases. When the number of prey is small, the number of predators decreases.

(c) C: 24 months; R: 24 months

78.

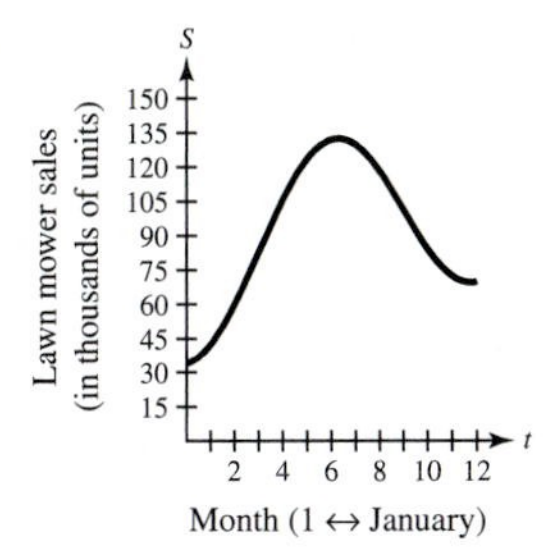

79. (a) H: 12 months; L: 12 months
(b) Summer; winter (c) 1 month

80. (a)

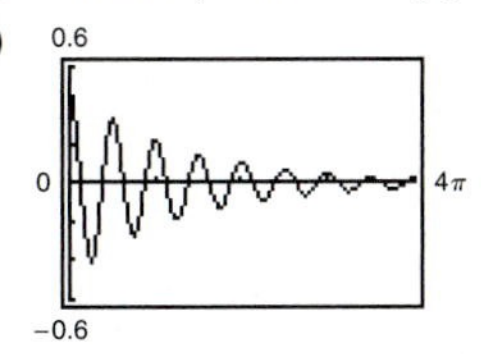

(b) y approaches 0 as t increases.

81. True. For a given value of x, the y-coordinate of $\csc x$ is the reciprocal of the y-coordinate of $\sin x$.

82. True. $y = \sec x$ is equal to $y = 1/\cos x$, and if the reciprocal of $y = \sin x$ is translated $\pi/2$ units to the left, then

$$\frac{1}{\sin\left(x + \dfrac{\pi}{2}\right)} = \frac{1}{\cos x} = \sec x.$$

83. As x approaches $\pi/2$ from the left, f approaches ∞. As x approaches $\pi/2$ from the right, f approaches $-\infty$.

84. As x approaches π from the left, f approaches ∞. As x approaches π from the right, f approaches $-\infty$.

85. (a)

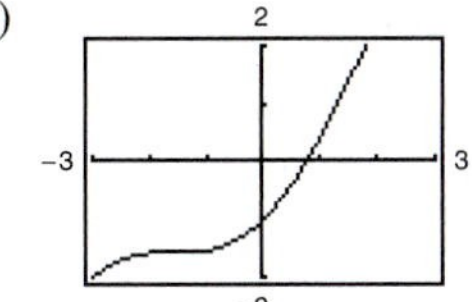

0.7391

(b) 1, 0.5403, 0.8576, 0.6543, 0.7935, 0.7014, 0.7640, 0.7221, 0.7504, 0.7314, . . . ; 0.7391

86.

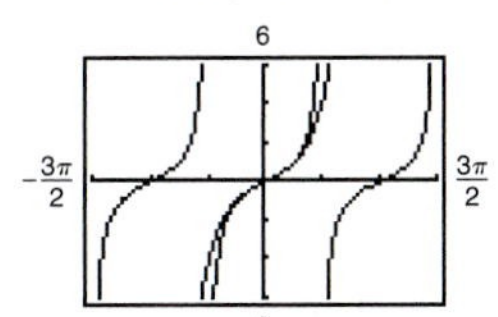

The graphs appear to coincide on the interval $-1.1 \le x \le 1.1$.

87.

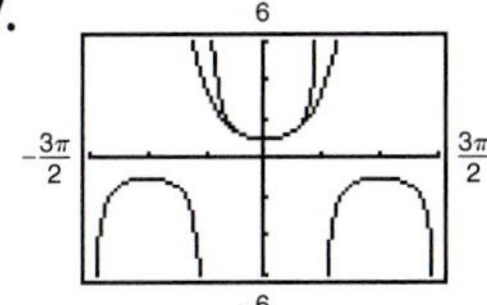

The graphs appear to coincide on the interval $-1.1 \le x \le 1.1$.

88. (a)

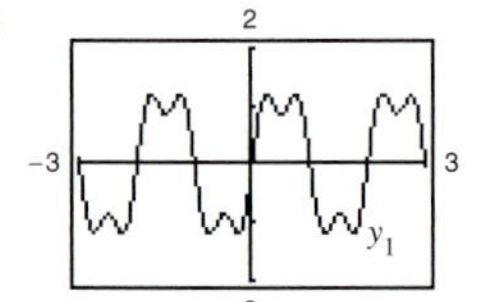

(b) $y_3 = \frac{4}{\pi}\left[\sin(\pi x) + \frac{1}{3}\sin(3\pi x) + \frac{1}{5}\sin(5\pi x) + \frac{1}{7}\sin(7\pi x)\right]$

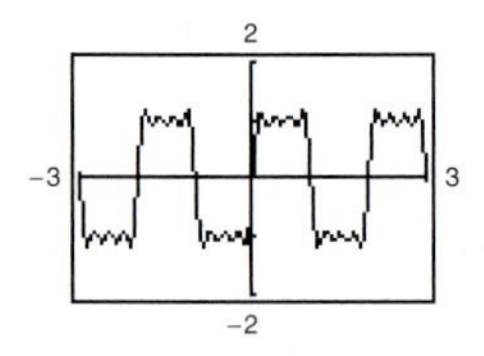

(c) $y_4 = \frac{4}{\pi}\left[\sin(\pi x) + \frac{1}{3}\sin(3\pi x) + \frac{1}{5}\sin(5\pi x) + \frac{1}{7}\sin(7\pi x) + \frac{1}{9}\sin(9\pi x)\right]$

89. $\frac{\ln 54}{2} \approx 1.994$ **90.** $\frac{1}{3}\left(\frac{\log_{10} 98}{\log_{10} 8}\right) \approx 0.735$

91. $-\ln 2 \approx -0.693$

92. $\frac{1}{365}\left(\frac{\log_{10} 5}{\log_{10} 1.00041096}\right) \approx 10.732$

93. $\frac{2 + e^{73}}{3} \approx 1.684 \times 10^{31}$

94. $\frac{14 - e^{68}}{2} \approx -1.702 \times 10^{29}$

95. $\pm\sqrt{e^{3.2} - 1} \approx \pm 4.851$ **96.** $e^{10} - 4 \approx 22{,}022.466$

97. 2 **98.** $\sqrt{65} \approx 8.062$

Section 4.7 *(page 349)*

Vocabulary Check *(page 349)*

1. $y = \sin^{-1} x;\ -1 \le x \le 1$
2. $y = \arccos x;\ 0 \le y \le \pi$
3. $y = \tan^{-1} x;\ -\infty < x < \infty;\ -\frac{\pi}{2} < y < \frac{\pi}{2}$

1. $\frac{\pi}{6}$ **2.** 0 **3.** $\frac{\pi}{3}$ **4.** $\frac{\pi}{2}$ **5.** $\frac{\pi}{6}$ **6.** $-\frac{\pi}{4}$

7. $\frac{5\pi}{6}$ **8.** $-\frac{\pi}{4}$ **9.** $-\frac{\pi}{3}$ **10.** $\frac{\pi}{3}$ **11.** $\frac{2\pi}{3}$

12. $\frac{\pi}{4}$ **13.** $\frac{\pi}{3}$ **14.** $-\frac{\pi}{6}$ **15.** 0 **16.** 0

17. **18.**

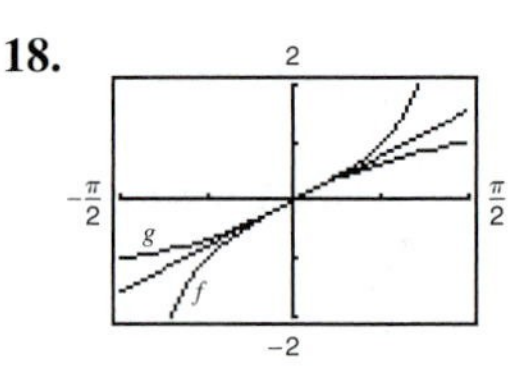

19. 1.29 **20.** 0.47 **21.** -0.85 **22.** 2.35

23. -1.25 **24.** 1.50 **25.** 0.32 **26.** 1.31

27. 1.99 **28.** -0.13 **29.** 0.74 **30.** 1.23

31. 0.85 **32.** 1.91 **33.** 1.29 **34.** -1.50

35. $-\frac{\pi}{3}, -\frac{\sqrt{3}}{3}, 1$ **36.** $\pi, \frac{2\pi}{3}, \frac{\sqrt{3}}{2}$

37. $\theta = \arctan \frac{x}{4}$ **38.** $\theta = \arccos \frac{4}{x}$

39. $\theta = \arcsin \frac{x + 2}{5}$ **40.** $\theta = \arctan \frac{x + 1}{10}$

41. $\theta = \arccos \frac{x + 3}{2x}$ **42.** $\theta = \arctan \frac{1}{x + 1},\ x \ne 1$

43. 0.3 **44.** 25 **45.** -0.1 **46.** -0.2 **47.** 0

48. $\frac{\pi}{2}$ **49.** $\frac{3}{5}$ **50.** $\frac{5}{3}$ **51.** $\frac{\sqrt{5}}{5}$ **52.** $\frac{2\sqrt{5}}{5}$

53. $\frac{12}{13}$ **54.** $-\frac{13}{5}$ **55.** $\frac{\sqrt{34}}{5}$ **56.** $-\frac{3\sqrt{7}}{7}$

57. $\frac{\sqrt{5}}{3}$ **58.** $\frac{8}{5}$ **59.** $\frac{1}{x}$ **60.** $\frac{x}{\sqrt{x^2 + 1}}$

61. $\sqrt{1 - 4x^2}$ **62.** $\sqrt{9x^2 + 1}$ **63.** $\sqrt{1 - x^2}$

64. $\frac{1}{\sqrt{2x - x^2}}$ **65.** $\frac{\sqrt{9 - x^2}}{x}$ **66.** x

67. $\frac{\sqrt{x^2 + 2}}{x}$ **68.** $\frac{\sqrt{r^2 - (x - h)^2}}{r}$

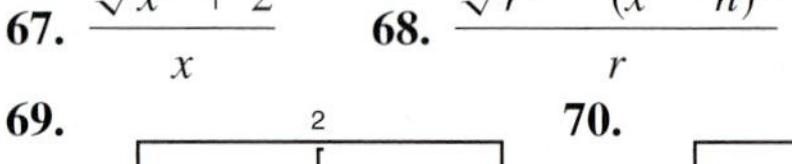

69. **70.**

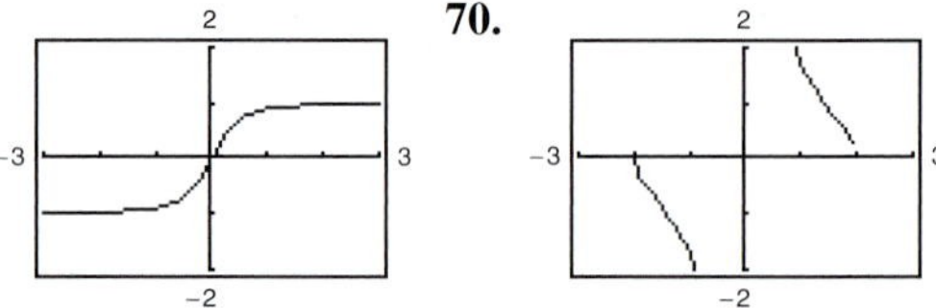

Asymptotes: $y = \pm 1$ Asymptote: $x = 0$

71. $\frac{9}{\sqrt{x^2 + 81}},\ x > 0;\ \frac{-9}{\sqrt{x^2 + 81}},\ x < 0$

72. $\frac{x}{6}$ **73.** $\frac{|x - 1|}{\sqrt{x^2 - 2x + 10}}$ **74.** $\frac{\sqrt{4x - x^2}}{x - 2}$

75. **76.**

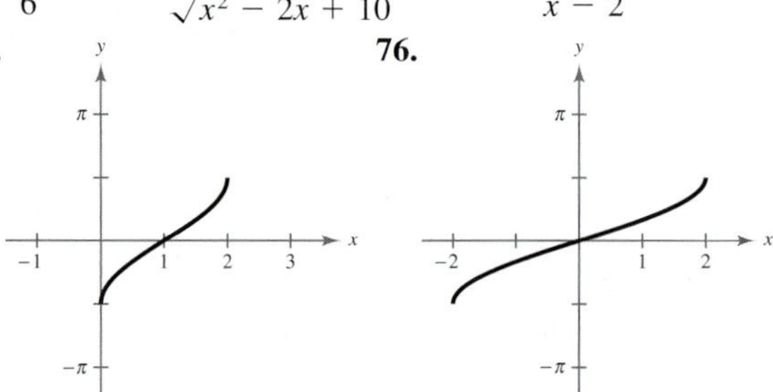

The graph of g is a horizontal shift one unit to the right of f. The graph of g is a horizontal stretch of the graph of f.

77.

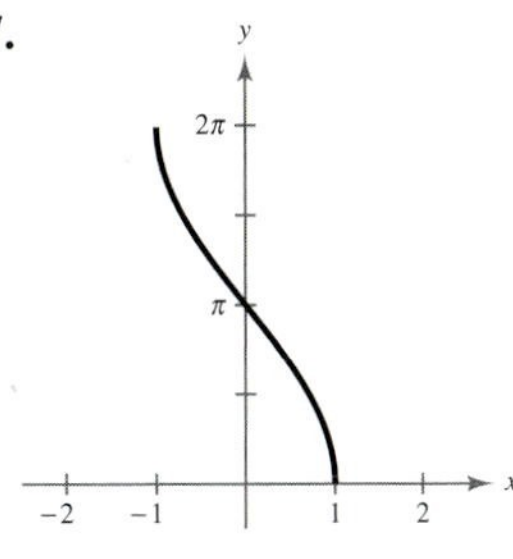

78.

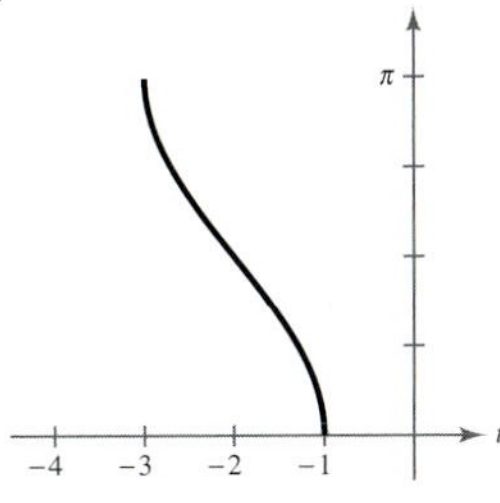

79.

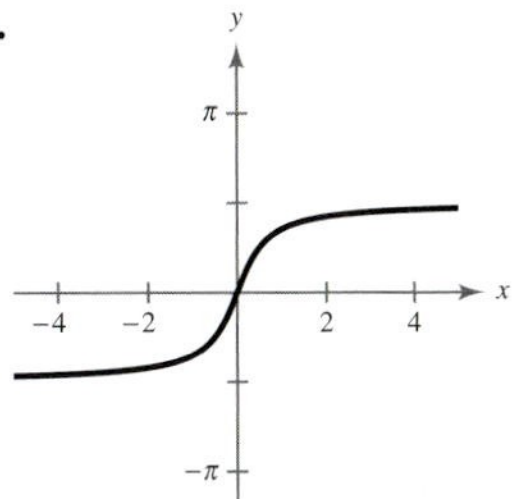

80.

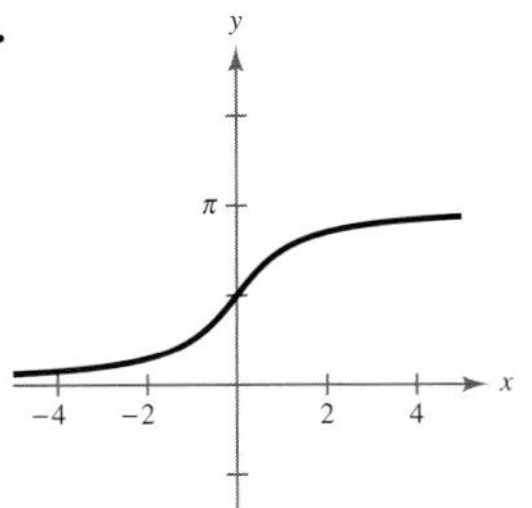

81.

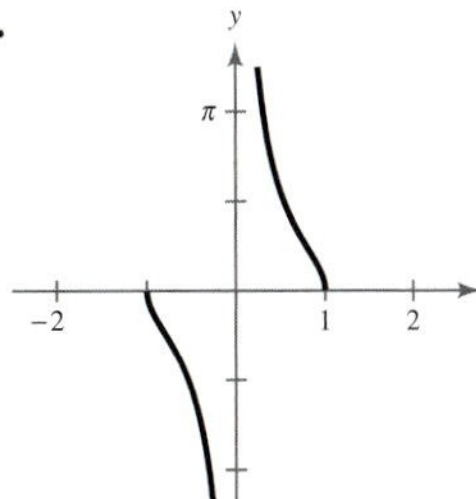

82.

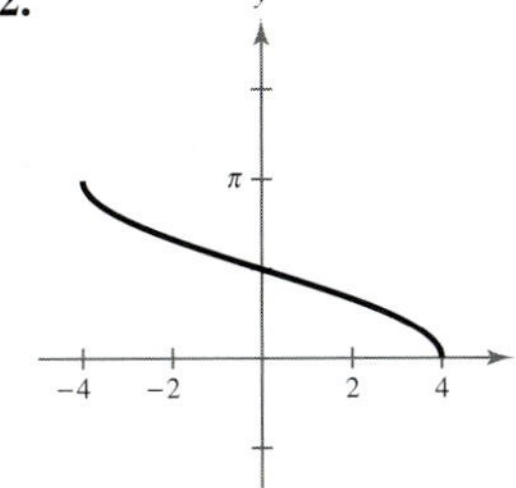

83.

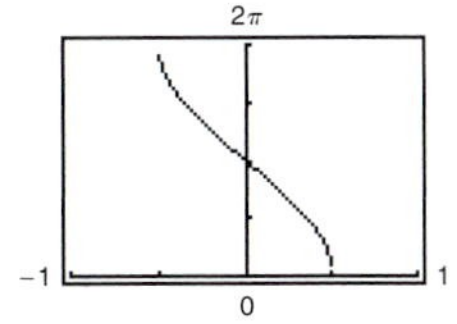

84.

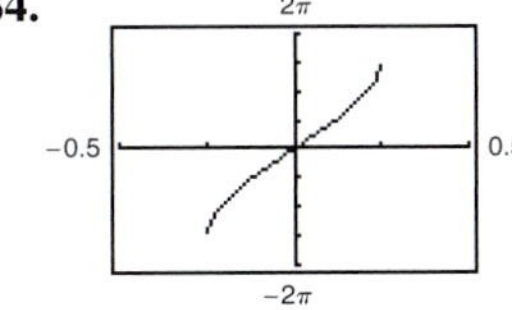

85.

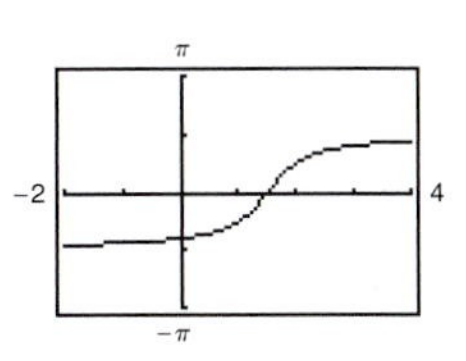

86.

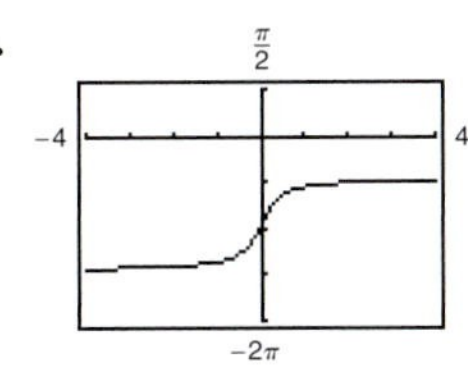

87.

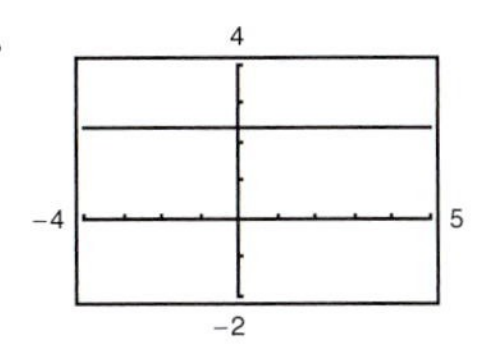

88.

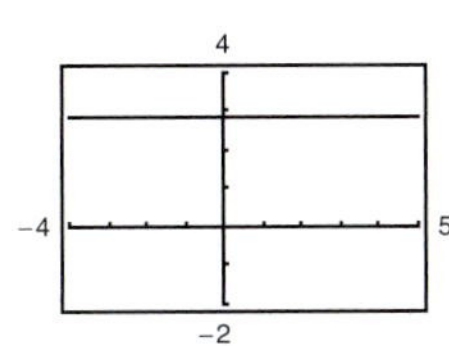

89. $3\sqrt{2}\sin\left(2t + \frac{\pi}{4}\right)$

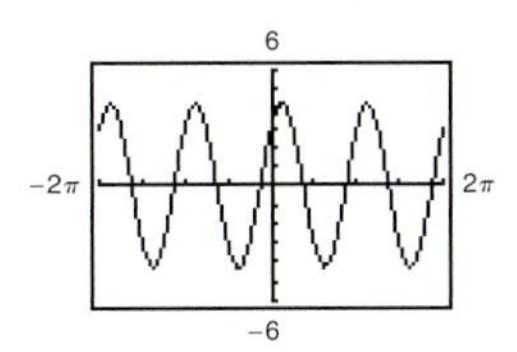

The graph implies that the identity is true.

90. $5\sin\left(\pi t + \arctan\frac{4}{3}\right)$

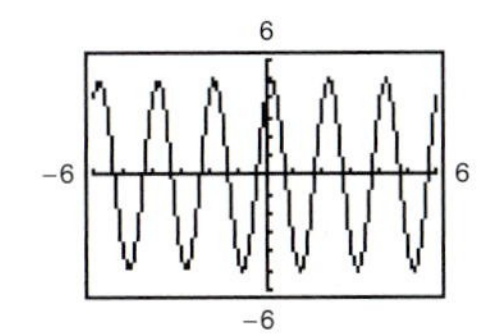

The graph implies that the identity is true.

91. (a) $\theta = \arcsin\dfrac{5}{s}$ (b) 0.13, 0.25

92. (a) $\theta = \arctan\dfrac{s}{750}$ (b) 21.8°, 58.0°

93. (a)

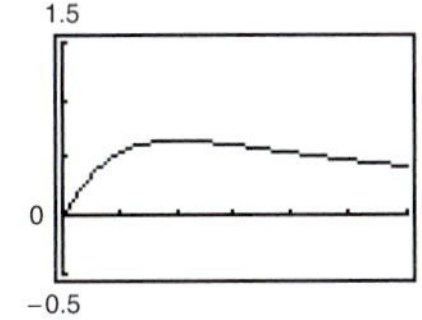

(b) 2 feet (c) $\beta = 0$; As x increases, β approaches 0.

94. (a) $\theta = 32.9°$ (b) 12.94 feet

95. (a) $\theta \approx 26.0°$ (b) 24.4 feet

96. (a) $\theta = \arctan\dfrac{6}{x}$ (b) 40.6°, 80.5°

97. (a) $\theta = \arctan\dfrac{x}{20}$ (b) 14.0°, 31.0°

98. False. $\dfrac{5\pi}{6}$ is not in the range of the arcsine.

99. False. $\dfrac{5\pi}{4}$ is not in the range of the arctangent.

100. False. The graphs are not the same.

101. Domain: $(-\infty, \infty)$

Range: $(0, \pi)$

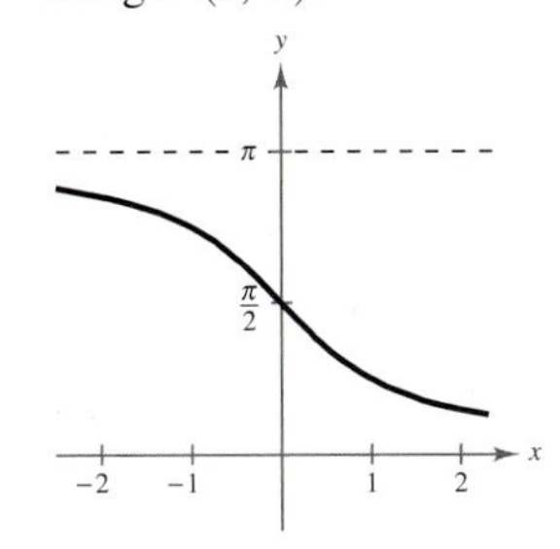

102. Domain: $(-\infty, -1] \cup [1, \infty)$
Range: $[0, \pi/2) \cup (\pi/2, \pi]$

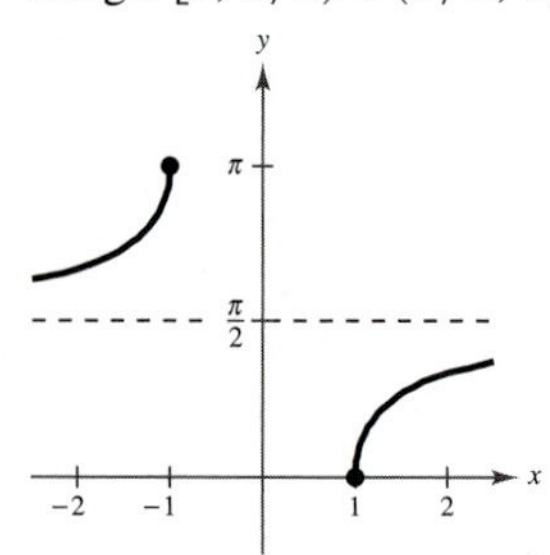

103. Domain: $(-\infty, -1] \cup [1, \infty)$
Range: $[-\pi/2, 0) \cup (0, \pi/2]$

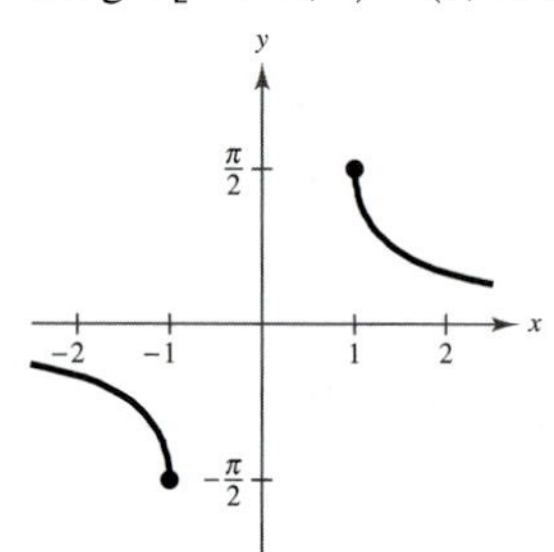

104. (a) $\frac{\pi}{4}$ (b) 0 (c) $\frac{5\pi}{6}$ (d) $\frac{\pi}{6}$

105. (a) $\frac{\pi}{4}$ (b) $\frac{\pi}{2}$ (c) 1.25 (d) 2.03

106.

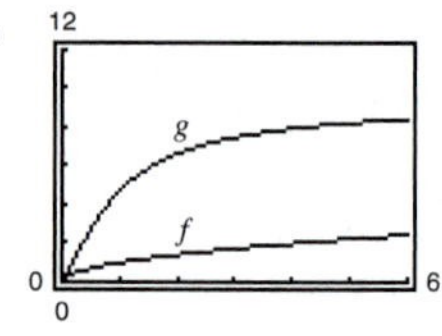

As x increases to infinity, g approaches 3π, but f has no maximum.
$a \approx 87.54$

107. (a) $f \circ f^{-1}$ $\quad$ $f^{-1} \circ f$

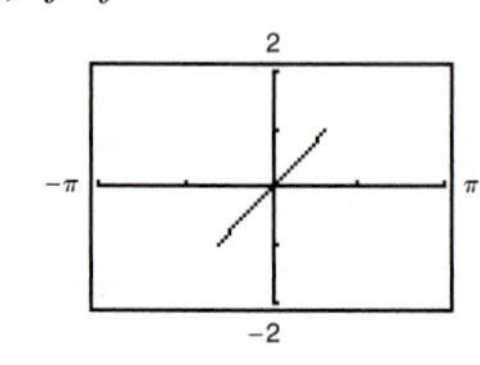

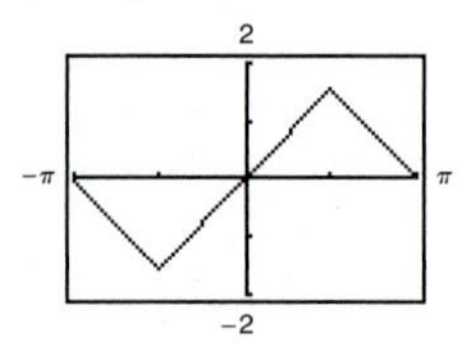

(b) The domains and ranges of the functions are restricted. The graphs of $f \circ f^{-1}$ and $f^{-1} \circ f$ differ because of the domains and ranges of f and f^{-1}.

108. (a)–(e) Answers will vary. **109.** 1279.284

110. 0.051 **111.** 117.391 **112.** 2.718×10^{-8}

113.

$\sqrt{7}$

$\cos\theta = \frac{\sqrt{7}}{4}$ $\quad$ $\sec\theta = \frac{4\sqrt{7}}{7}$

$\tan\theta = \frac{3\sqrt{7}}{7}$ $\quad$ $\cot\theta = \frac{\sqrt{7}}{3}$

$\csc\theta = \frac{4}{3}$

114.

$\sqrt{5}$

$\sin\theta = \frac{2\sqrt{5}}{5}$ $\quad$ $\sec\theta = \sqrt{5}$

$\cos\theta = \frac{\sqrt{5}}{5}$ $\quad$ $\cot\theta = \frac{1}{2}$

$\csc\theta = \frac{\sqrt{5}}{2}$

115.

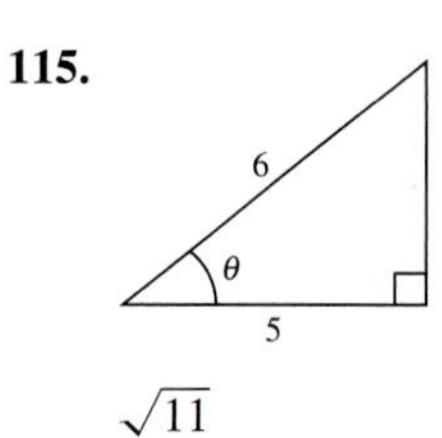

$\sqrt{11}$

$\sin\theta = \frac{\sqrt{11}}{6}$ $\quad$ $\sec\theta = \frac{6}{5}$

$\tan\theta = \frac{\sqrt{11}}{5}$ $\quad$ $\cot\theta = \frac{5\sqrt{11}}{11}$

$\csc\theta = \frac{6\sqrt{11}}{11}$

116.

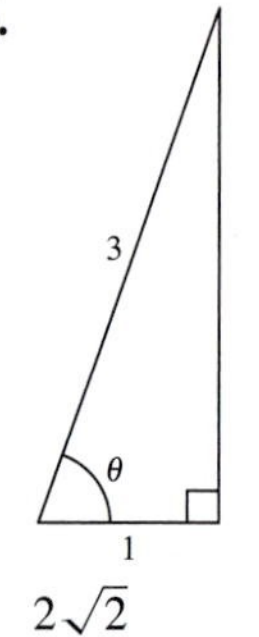

$2\sqrt{2}$

$\sin\theta = \frac{2\sqrt{2}}{3}$ $\quad$ $\csc\theta = \frac{3\sqrt{2}}{4}$

$\cos\theta = \frac{1}{3}$ $\quad$ $\cot\theta = \frac{\sqrt{2}}{4}$

$\tan\theta = \frac{2\sqrt{2}}{1}$

117. Eight people **118.** 3 miles per hour

119. (a) \$21,253.63 (b) \$21,275.17
(c) \$21,285.66 (d) \$21,286.01

120. 2008: \$458,504

Section 4.8 *(page 359)*

Vocabulary Check *(page 359)*

1. elevation; depression **2.** bearing
3. harmonic motion

1. $a \approx 3.64$, $c \approx 10.64$, $B = 70°$

2. $a \approx 8.82$, $b \approx 12.14$, $A = 36°$

3. $a \approx 8.26$, $c \approx 25.38$, $A = 19°$

4. $b \approx 274.27$
$c \approx 277.24$
$B = 81.6°$

5. $c \approx 11.66$
$A \approx 30.96°$
$B \approx 59.04°$

6. $b \approx 24.49$
$A \approx 45.58°$
$B \approx 44.42°$

7. $a \approx 49.48$
$A \approx 72.08°$
$B \approx 17.92°$

8. $a \approx 9.36$
$A \approx 81.97°$
$B \approx 8.03°$

9. $a \approx 91.34$
$b \approx 420.70$
$B = 77°45'$

10. $b \approx 30.73$
$c \approx 33.85$
$A = 24°48'$

11. 2.56 inches **12.** 1.62 meters

13. 19.99 inches **14.** 2.80 feet **15.** 107.2 feet

16. 1648.5 feet **17.** 19.7 feet **18.** 81.2 feet

19. (a)

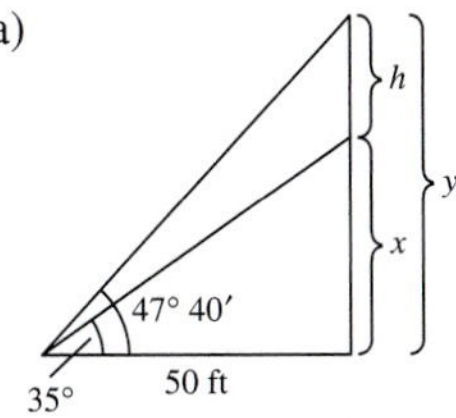

(b) $h = 50(\tan 47°40' - \tan 35°)$ (c) 19.9 feet

20. 123.5 feet **21.** 2236.8 feet **22.** 56.3°

23. (a)

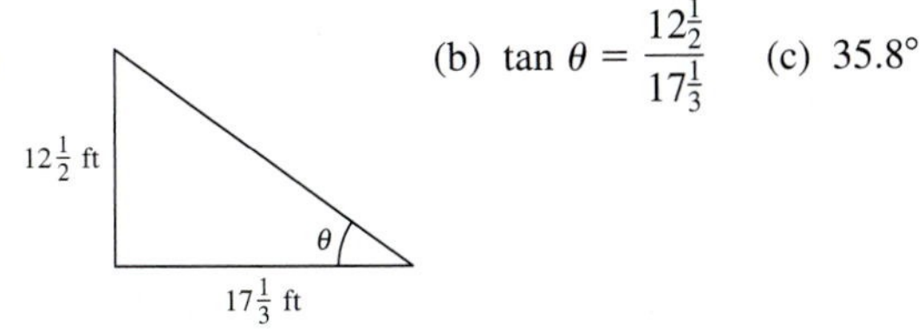

(b) $\tan\theta = \dfrac{12\frac{1}{2}}{17\frac{1}{3}}$ (c) 35.8°

24. 71.34° **25.** 2.06°

26. (a) 5099 feet (b) 117.7 seconds **27.** 0.73 mile

28. 6.8°; 2516.3 feet **29.** 554 miles north; 709 miles east

30. (a) 429.26 miles north; 2434.44 miles east
(b) 280°

31. (a) 58.18 nautical miles west; 104.95 nautical miles south
(b) S 36.7° W; distance = 130.9 nautical miles

32. (a) 21.4 hours
(b) 5.86 nautical miles east; 239.93 nautical miles south
(c) 178.6°

33. (a) N 58° E (b) 68.82 meters

34. $d \approx 5.46$ kilometers **35.** N 56.31° W

36. 208° **37.** 1933.3 feet **38.** 11.8 kilometers

39. ≈ 3.23 miles or $\approx 17{,}054$ feet

40. ≈ 1.025 miles or ≈ 5412 feet

41. 78.7° **42.** 52.1° **43.** 35.3° **44.** 54.7°

45. 29.4 inches **46.** 25 inches **47.** $y = \sqrt{3}r$

48. 9.06 centimeters **49.** $a \approx 12.2, b \approx 7$

50. $a \approx 21.6$ feet, $b \approx 7.2$ feet, $c \approx 13$ feet

51. $d = 4\sin(\pi t)$ **52.** $d = 3\sin\left(\dfrac{\pi t}{3}\right)$

53. $d = 3\cos\left(\dfrac{4\pi t}{3}\right)$ **54.** $d = 2\cos\left(\dfrac{\pi t}{5}\right)$

55. (a) 4 (b) 4 (c) 4 (d) $\frac{1}{16}$

56. (a) $\frac{1}{2}$ (b) 10 (c) $\frac{1}{2}$ (d) $\frac{1}{40}$

57. (a) $\frac{1}{16}$ (b) 60 (c) 0 (d) $\frac{1}{120}$

58. (a) $\frac{1}{64}$ (b) 396 (c) 0 (d) $\frac{1}{792}$

59. $\omega = 528\pi$ **60.** $d = \dfrac{7}{4}\cos\dfrac{\pi t}{5}$

61. (a)

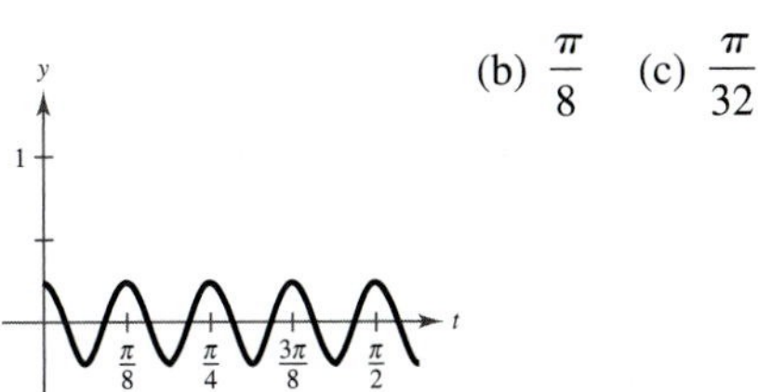

(b) $\dfrac{\pi}{8}$ (c) $\dfrac{\pi}{32}$

62. (a)

θ	L_1	L_2	$L_1 + L_2$
0.1	$\dfrac{2}{\sin 0.1}$	$\dfrac{3}{\cos 0.1}$	23.0
0.2	$\dfrac{2}{\sin 0.2}$	$\dfrac{3}{\cos 0.2}$	13.1
0.3	$\dfrac{2}{\sin 0.3}$	$\dfrac{3}{\cos 0.3}$	9.9
0.4	$\dfrac{2}{\sin 0.4}$	$\dfrac{3}{\cos 0.4}$	8.4

(b)

θ	L_1	L_2	$L_1 + L_2$
0.5	$\dfrac{2}{\sin 0.5}$	$\dfrac{3}{\cos 0.5}$	7.6
0.6	$\dfrac{2}{\sin 0.6}$	$\dfrac{3}{\cos 0.6}$	7.2
0.7	$\dfrac{2}{\sin 0.7}$	$\dfrac{3}{\cos 0.7}$	7.0
0.8	$\dfrac{2}{\sin 0.8}$	$\dfrac{3}{\cos 0.8}$	7.1

7.0 (minimum length)

(c) $L = L_1 + L_2 = \dfrac{2}{\sin\theta} + \dfrac{3}{\cos\theta}$

(d)

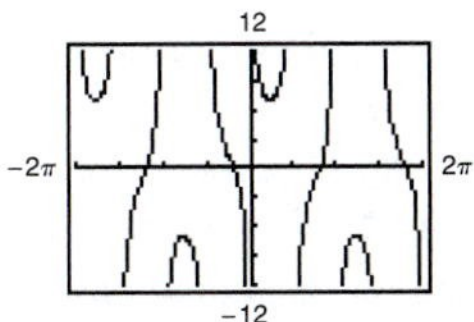

7.0 (minimum length)

63. (a)

Base 1	Base 2	Altitude	Area
8	$8 + 16 \cos 30°$	$8 \sin 30°$	59.7
8	$8 + 16 \cos 40°$	$8 \sin 40°$	72.7
8	$8 + 16 \cos 50°$	$8 \sin 50°$	80.5
8	$8 + 16 \cos 60°$	$8 \sin 60°$	83.1
8	$8 + 16 \cos 70°$	$8 \sin 70°$	80.7
8	$8 + 16 \cos 80°$	$8 \sin 80°$	74.0
8	$8 + 16 \cos 90°$	$8 \sin 90°$	64.0

(b)

Base 1	Base 2	Altitude	Area
8	$8 + 16 \cos 56°$	$8 \sin 56°$	82.73
8	$8 + 16 \cos 58°$	$8 \sin 58°$	83.04
8	$8 + 16 \cos 59°$	$8 \sin 59°$	83.11
8	$8 + 16 \cos 60°$	$8 \sin 60°$	83.14
8	$8 + 16 \cos 61°$	$8 \sin 61°$	83.11
8	$8 + 16 \cos 62°$	$8 \sin 62°$	83.04

83.14 square feet

(c) $A = 64(1 + \cos \theta)(\sin \theta)$

(d)

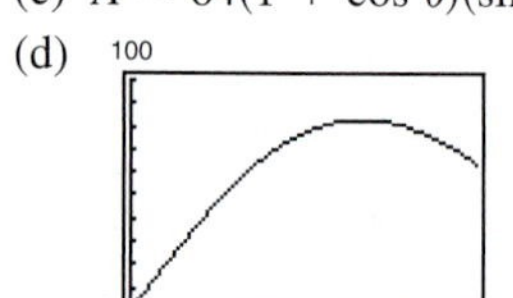

$\approx$ 83.1 square feet when $\theta = 60°$

The answers are the same.

64. (a)

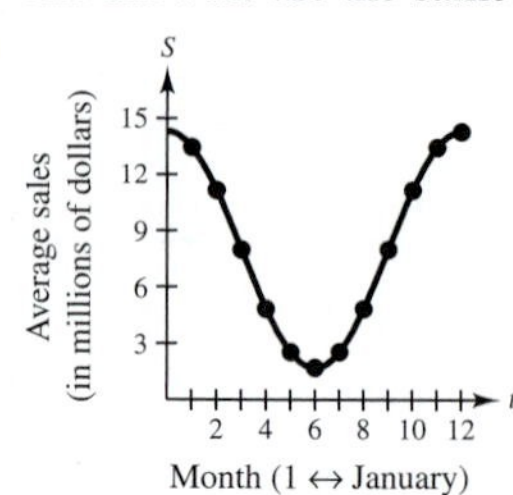

(b) $S = 8 + 6.3 \cos\left(\frac{\pi}{6}t\right)$ or $S = 8 + 6.3 \sin\left(\frac{\pi}{6}t + \frac{\pi}{2}\right)$

The model is a good fit.

(c) 12. Yes, sales of outerwear are seasonal.

(d) Maximum displacement from average sales of \$8 million

65. False. The tower is leaning, so it is not perfectly vertical and does not form a right angle with the ground.

66. False. One period is the time for one complete cycle of the motion.

67. No. N 24° E means 24 degrees east of north.

68. Air navigation is always measured clockwise from the north. Nautical navigation measures the acute angle a path makes with a fixed north-south line.

69. $y = 4x + 6$

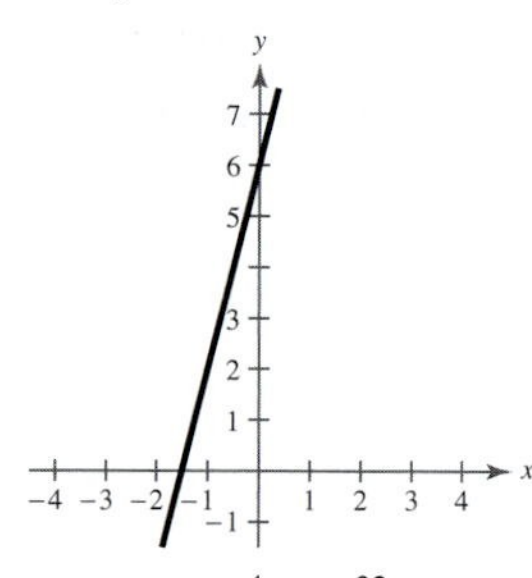

70. $y = -\frac{1}{2}x + \frac{1}{6}$

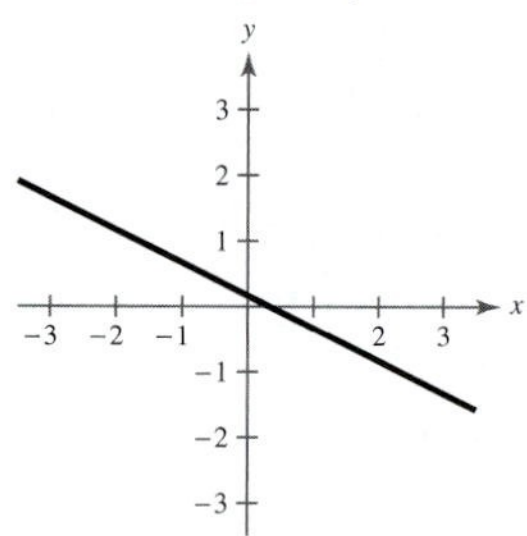

71. $y = -\frac{4}{5}x + \frac{22}{5}$

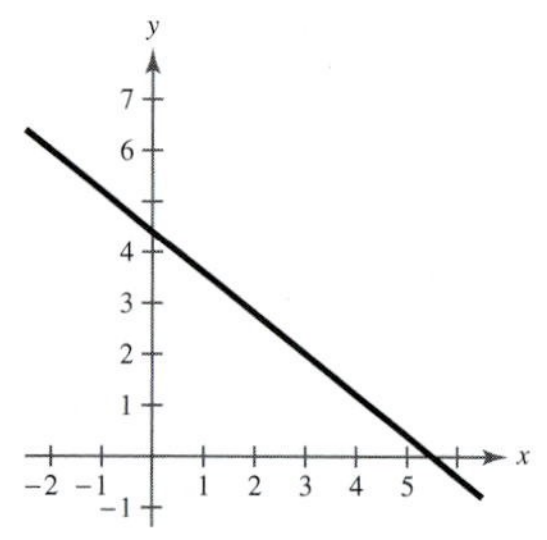

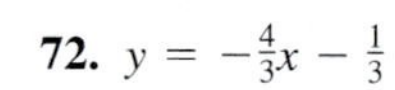

72. $y = -\frac{4}{3}x - \frac{1}{3}$

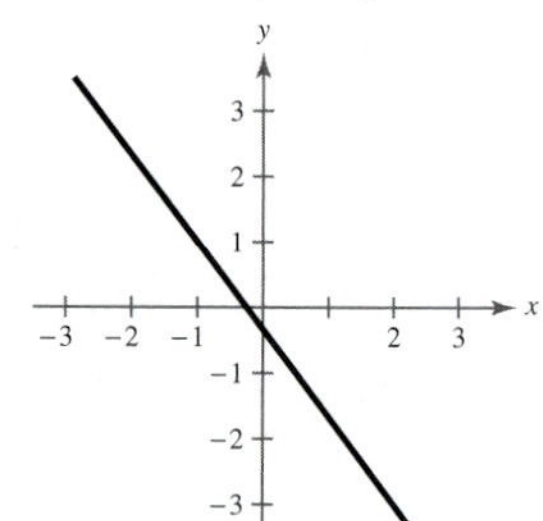

Review Exercises *(page 365)*

1. 0.5 radian

2. 4.5 radians

3.

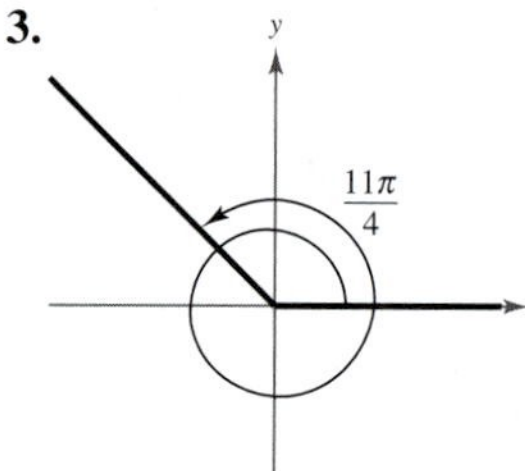

(b) Quadrant II

(c) $\frac{3\pi}{4}, -\frac{5\pi}{4}$

4.

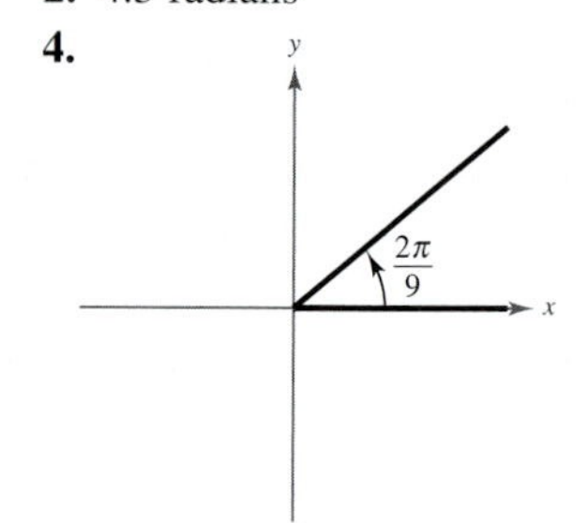

(b) Quadrant I

(c) $\frac{20\pi}{9}, -\frac{16\pi}{9}$

5.

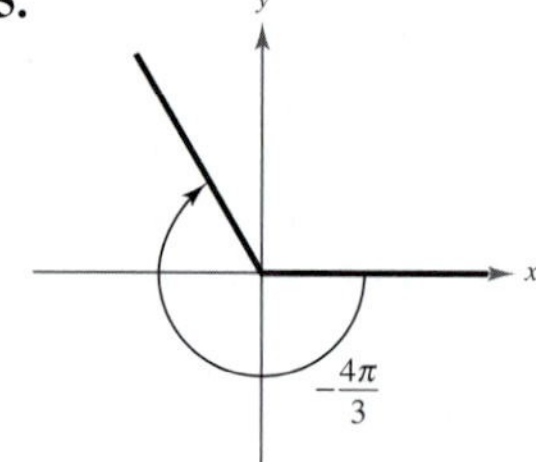

(b) Quadrant II

(c) $\frac{2\pi}{3}, -\frac{10\pi}{3}$

6.

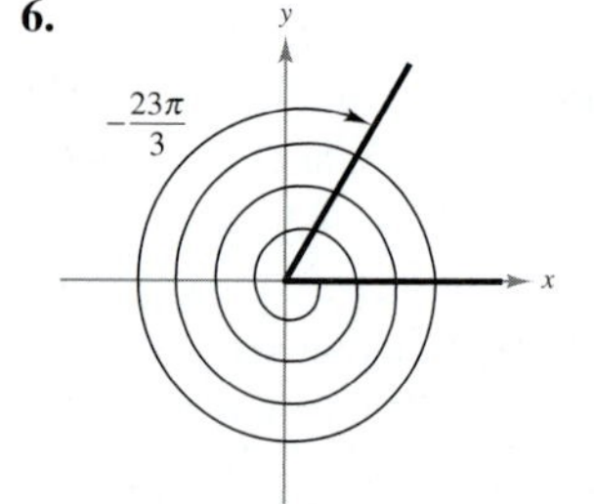

(b) Quadrant I

(c) $\frac{\pi}{3}, -\frac{17\pi}{3}$

7.

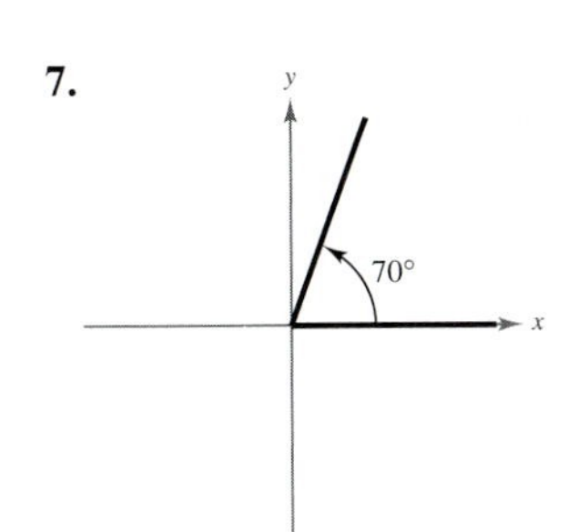

(b) Quadrant I
(c) 430°, −290°

8.

(b) Quadrant IV
(c) 640°, −80°

9.

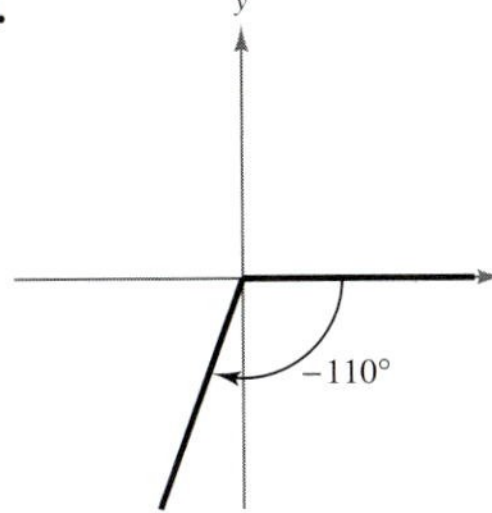

(b) Quadrant III
(c) 250°, −470°

10.

(b) Quadrant IV
(c) 315°, −45°

11. 8.378 **12.** −2.225 **13.** −0.589 **14.** 3.443
15. 128.571° **16.** −330.000° **17.** −200.535°
18. 326.586° **19.** 48.17 inches **20.** 11.52 meters
21. (a) $66\frac{2}{3}\pi$ radians per minute
(b) 400π inches per minute
22. 212.1 inches per second
23. Area = 339.29 square inches
24. Area = 55.31 square millimeters **25.** $\left(-\frac{1}{2}, \frac{\sqrt{3}}{2}\right)$
26. $\left(-\frac{\sqrt{2}}{2}, \frac{\sqrt{2}}{2}\right)$ **27.** $\left(-\frac{\sqrt{3}}{2}, \frac{1}{2}\right)$ **28.** $\left(-\frac{1}{2}, \frac{\sqrt{3}}{2}\right)$

29. $\sin\frac{7\pi}{6} = -\frac{1}{2}$ $\csc\frac{7\pi}{6} = -2$
$\cos\frac{7\pi}{6} = -\frac{\sqrt{3}}{2}$ $\sec\frac{7\pi}{6} = -\frac{2\sqrt{3}}{3}$
$\tan\frac{7\pi}{6} = \frac{\sqrt{3}}{3}$ $\cot\frac{7\pi}{6} = \sqrt{3}$

30. $\sin\frac{\pi}{4} = \frac{\sqrt{2}}{2}$ $\csc\frac{\pi}{4} = \sqrt{2}$
$\cos\frac{\pi}{4} = \frac{\sqrt{2}}{2}$ $\sec\frac{\pi}{4} = \sqrt{2}$
$\tan\frac{\pi}{4} = 1$ $\cot\frac{\pi}{4} = 1$

31. $\sin\left(-\frac{2\pi}{3}\right) = -\frac{\sqrt{3}}{2}$ $\csc\left(-\frac{2\pi}{3}\right) = -\frac{2\sqrt{3}}{3}$
$\cos\left(-\frac{2\pi}{3}\right) = -\frac{1}{2}$ $\sec\left(-\frac{2\pi}{3}\right) = -2$
$\tan\left(-\frac{2\pi}{3}\right) = \sqrt{3}$ $\cot\left(-\frac{2\pi}{3}\right) = \frac{\sqrt{3}}{3}$

32. $\sin 2\pi = 0$ $\csc 2\pi$ is undefined.
$\cos 2\pi = 1$ $\sec 2\pi = 1$
$\tan 2\pi = 0$ $\cot 2\pi$ is undefined.

33. $\sin\frac{11\pi}{4} = \sin\frac{3\pi}{4} = \frac{\sqrt{2}}{2}$

34. $\cos 4\pi = \cos 0 = 1$ **35.** $\sin\left(-\frac{17\pi}{6}\right) = \sin\frac{7\pi}{6} = -\frac{1}{2}$

36. $\cos\left(-\frac{13\pi}{3}\right) = \cos\frac{5\pi}{3} = \frac{1}{2}$ **37.** −75.3130

38. −1.1368 **39.** 3.2361 **40.** −0.3420

41. $\sin\theta = \frac{4\sqrt{41}}{41}$
$\cos\theta = \frac{5\sqrt{41}}{41}$
$\tan\theta = \frac{4}{5}$
$\csc\theta = \frac{\sqrt{41}}{4}$
$\sec\theta = \frac{\sqrt{41}}{5}$
$\cot\theta = \frac{5}{4}$

42. $\sin\theta = \frac{\sqrt{2}}{2}$
$\cos\theta = \frac{\sqrt{2}}{2}$
$\tan\theta = 1$
$\csc\theta = \sqrt{2}$
$\sec\theta = \sqrt{2}$
$\cot\theta = 1$

43. $\sin\theta = \frac{\sqrt{3}}{2}$
$\cos\theta = \frac{1}{2}$
$\tan\theta = \sqrt{3}$
$\csc\theta = \frac{2\sqrt{3}}{3}$
$\sec\theta = 2$
$\cot\theta = \frac{\sqrt{3}}{3}$

44. $\sin\theta = \frac{5}{9}$
$\cos\theta = \frac{2\sqrt{14}}{9}$
$\tan\theta = \frac{5\sqrt{14}}{28}$
$\csc\theta = \frac{9}{5}$
$\sec\theta = \frac{9\sqrt{14}}{28}$
$\cot\theta = \frac{2\sqrt{14}}{5}$

45. (a) 3 (b) $\frac{2\sqrt{2}}{3}$ (c) $\frac{3\sqrt{2}}{4}$ (d) $\frac{\sqrt{2}}{4}$

46. (a) $\frac{1}{4}$ (b) $\sqrt{17}$ (c) $\frac{\sqrt{17}}{17}$ (d) $\frac{\sqrt{17}}{4}$

47. (a) $\frac{1}{4}$ (b) $\frac{\sqrt{15}}{4}$ (c) $\frac{4\sqrt{15}}{15}$ (d) $\frac{\sqrt{15}}{15}$

48. (a) $\frac{1}{5}$ (b) $2\sqrt{6}$ (c) $\frac{\sqrt{6}}{12}$ (d) 5 **49.** 0.6494

50. 5.2408 **51.** 0.5621 **52.** 5.3860 **53.** 3.6722
54. 0.2045 **55.** 71.3 meters **56.** 19.5 feet

57. $\sin\theta = \frac{4}{5}$ $\csc\theta = \frac{5}{4}$
$\cos\theta = \frac{3}{5}$ $\sec\theta = \frac{5}{3}$
$\tan\theta = \frac{4}{3}$ $\cot\theta = \frac{3}{4}$

58. $\sin\theta = -\frac{4}{5}$ $\csc\theta = -\frac{5}{4}$
$\cos\theta = \frac{3}{5}$ $\sec\theta = \frac{5}{3}$
$\tan\theta = -\frac{4}{3}$ $\cot\theta = -\frac{3}{4}$

59. $\sin\theta = \frac{15\sqrt{241}}{241}$ $\csc\theta = \frac{\sqrt{241}}{15}$
$\cos\theta = \frac{4\sqrt{241}}{241}$ $\sec\theta = \frac{\sqrt{241}}{4}$
$\tan\theta = \frac{15}{4}$ $\cot\theta = \frac{4}{15}$

60. $\sin\theta = -\frac{\sqrt{26}}{26}$ $\csc\theta = -\sqrt{26}$
$\cos\theta = -\frac{5\sqrt{26}}{26}$ $\sec\theta = -\frac{\sqrt{26}}{5}$
$\tan\theta = \frac{1}{5}$ $\cot\theta = 5$

61. $\sin\theta = \frac{9\sqrt{82}}{82}$ $\csc\theta = \frac{\sqrt{82}}{9}$
$\cos\theta = \frac{-\sqrt{82}}{82}$ $\sec\theta = -\sqrt{82}$
$\tan\theta = -9$ $\cot\theta = -\frac{1}{9}$

62. $\sin\theta = \frac{4}{5}$ $\csc\theta = \frac{5}{4}$
$\cos\theta = \frac{3}{5}$ $\sec\theta = \frac{5}{3}$
$\tan\theta = \frac{4}{3}$ $\cot\theta = \frac{3}{4}$

63. $\sin\theta = \frac{4\sqrt{17}}{17}$ $\csc\theta = \frac{\sqrt{17}}{4}$
$\cos\theta = \frac{\sqrt{17}}{17}$ $\sec\theta = \sqrt{17}$
$\tan\theta = 4$ $\cot\theta = \frac{1}{4}$

64. $\sin\theta = -\frac{3\sqrt{13}}{13}$ $\csc\theta = -\frac{\sqrt{13}}{3}$
$\cos\theta = -\frac{2\sqrt{13}}{13}$ $\sec\theta = -\frac{\sqrt{13}}{2}$
$\tan\theta = \frac{3}{2}$ $\cot\theta = \frac{2}{3}$

65. $\sin\theta = -\frac{\sqrt{11}}{6}$
$\cos\theta = \frac{5}{6}$
$\tan\theta = -\frac{\sqrt{11}}{5}$
$\csc\theta = -\frac{6\sqrt{11}}{11}$
$\cot\theta = -\frac{5\sqrt{11}}{11}$

66. $\sin\theta = \frac{2}{3}$
$\cos\theta = -\frac{\sqrt{5}}{3}$
$\tan\theta = -\frac{2\sqrt{5}}{5}$
$\sec\theta = -\frac{3\sqrt{5}}{5}$
$\cot\theta = -\frac{\sqrt{5}}{2}$

67. $\cos\theta = -\frac{\sqrt{55}}{8}$
$\tan\theta = -\frac{3\sqrt{55}}{55}$
$\csc\theta = \frac{8}{3}$
$\sec\theta = -\frac{8\sqrt{55}}{55}$
$\cot\theta = -\frac{\sqrt{55}}{3}$

68. $\sin\theta = -\frac{5\sqrt{41}}{41}$
$\cos\theta = -\frac{4\sqrt{41}}{41}$
$\csc\theta = -\frac{\sqrt{41}}{5}$
$\sec\theta = -\frac{\sqrt{41}}{4}$
$\cot\theta = \frac{4}{5}$

69. $\sin\theta = \frac{\sqrt{21}}{5}$
$\tan\theta = -\frac{\sqrt{21}}{2}$
$\csc\theta = \frac{5\sqrt{21}}{21}$
$\sec\theta = -\frac{5}{2}$
$\cot\theta = -\frac{2\sqrt{21}}{21}$

70. $\cos\theta = \frac{\sqrt{3}}{2}$
$\tan\theta = -\frac{\sqrt{3}}{3}$
$\csc\theta = -2$
$\sec\theta = \frac{2\sqrt{3}}{3}$
$\cot\theta = -\sqrt{3}$

71. $\theta' = 84°$

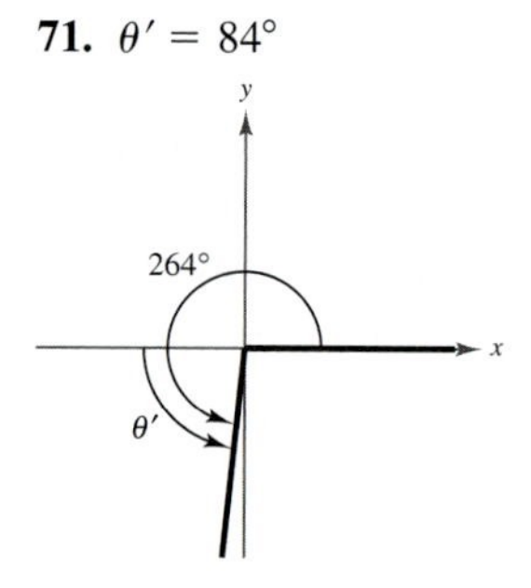

72. $\theta' = 85°$

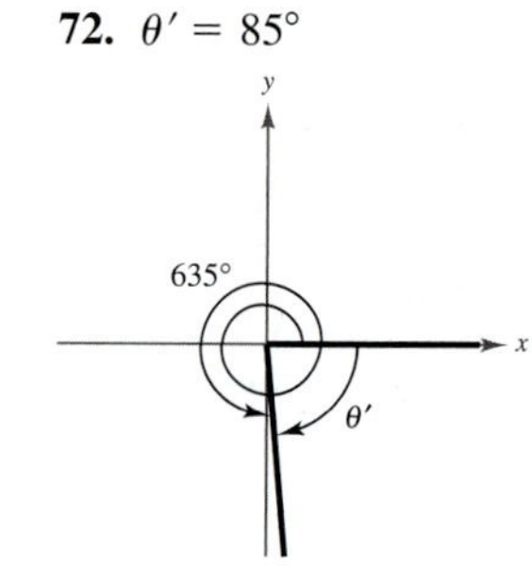

73. $\theta' = \frac{\pi}{5}$

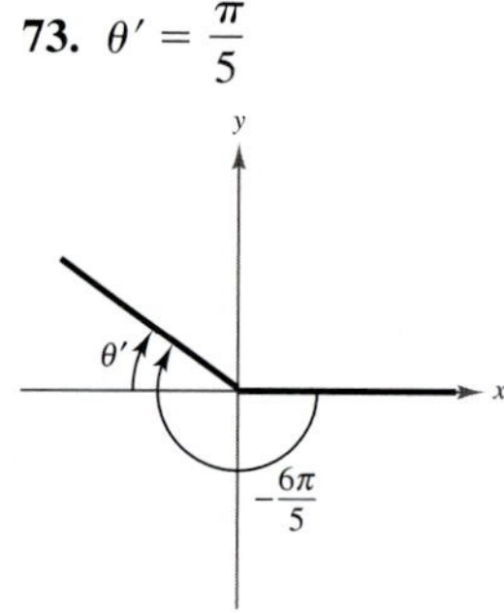

74. $\theta' = \frac{\pi}{3}$

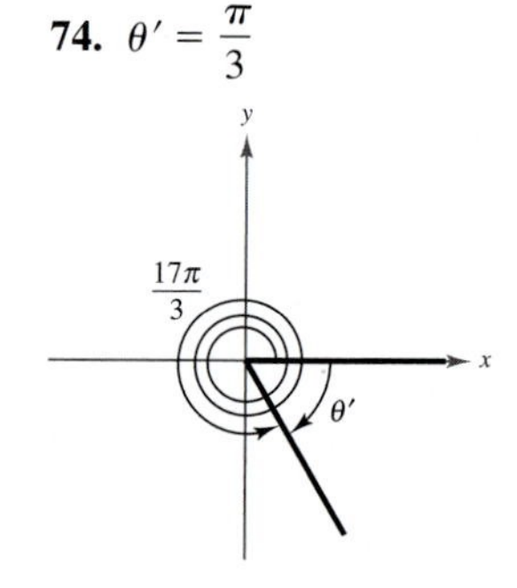

75. $\sin\frac{\pi}{3} = \frac{\sqrt{3}}{2}$; $\cos\frac{\pi}{3} = \frac{1}{2}$; $\tan\frac{\pi}{3} = \sqrt{3}$

76. $\sin\frac{\pi}{4} = \frac{\sqrt{2}}{2}$; $\cos\frac{\pi}{4} = \frac{\sqrt{2}}{2}$; $\tan\frac{\pi}{4} = 1$

77. $\sin\left(-\frac{7\pi}{3}\right) = -\frac{\sqrt{3}}{2}$; $\cos\left(-\frac{7\pi}{3}\right) = \frac{1}{2}$;
$\tan\left(-\frac{7\pi}{3}\right) = -\sqrt{3}$

78. $\sin\left(-\frac{5\pi}{4}\right) = \frac{\sqrt{2}}{2}$; $\cos\left(-\frac{5\pi}{4}\right) = -\frac{\sqrt{2}}{2}$;

$\tan\left(-\frac{5\pi}{4}\right) = -1$

79. $\sin 495° = \frac{\sqrt{2}}{2}$; $\cos 495° = -\frac{\sqrt{2}}{2}$; $\tan 495° = -1$

80. $\sin(-150°) = -\frac{1}{2}$; $\cos(-150°) = -\frac{\sqrt{3}}{2}$;

$\tan(-150°) = \frac{\sqrt{3}}{3}$

81. $\sin(-240°) = \frac{\sqrt{3}}{2}$; $\cos(-240°) = -\frac{1}{2}$;

$\tan(-240°) = -\sqrt{3}$

82. $\sin 315° = -\frac{\sqrt{2}}{2}$; $\cos 315° = \frac{\sqrt{2}}{2}$; $\tan 315° = -1$

83. -0.7568 **84.** -0.1425 **85.** 0.0584

86. 0.0878 **87.** 3.2361 **88.** 4.3813

89.

90.

91.

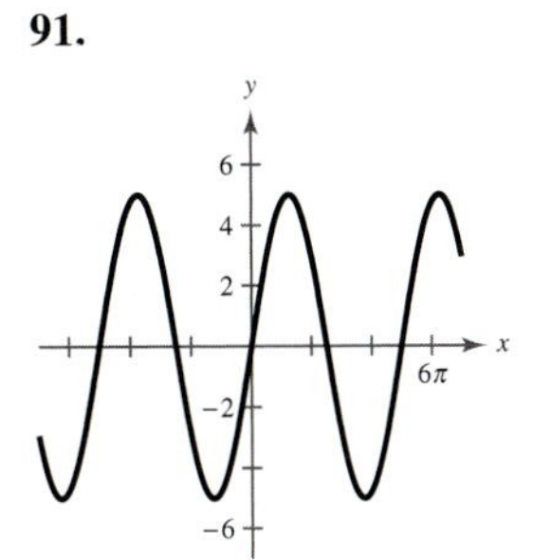

92.

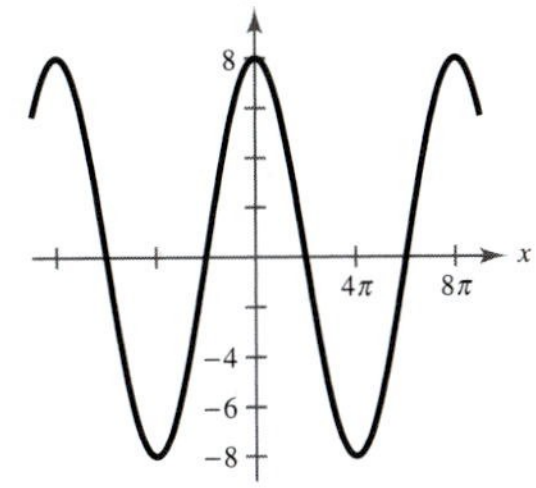

93.

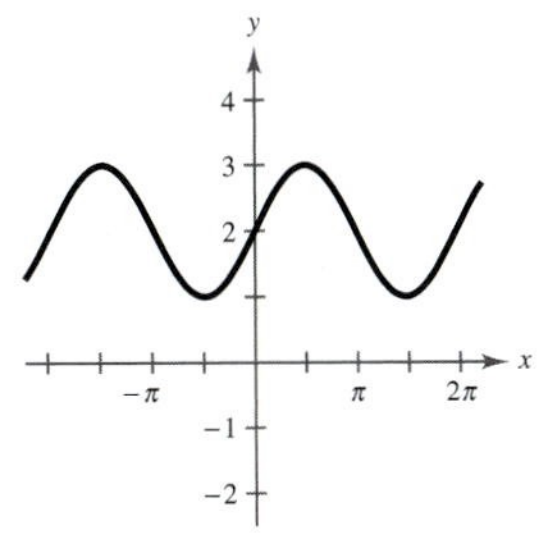

94.

95.

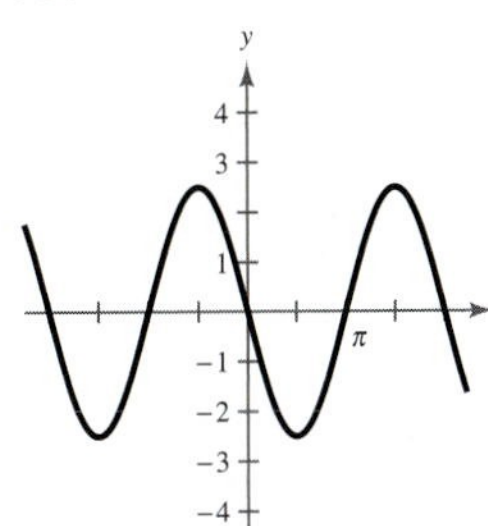

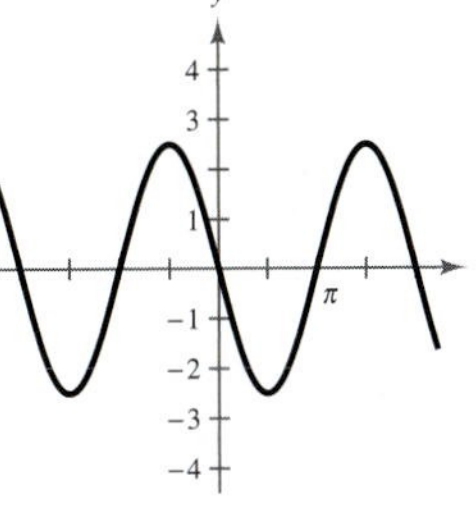

96.

97. (a) $y = 2 \sin 528\pi x$ (b) 264 cycles per second

98. (a)

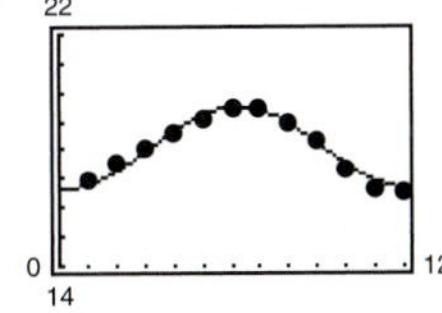

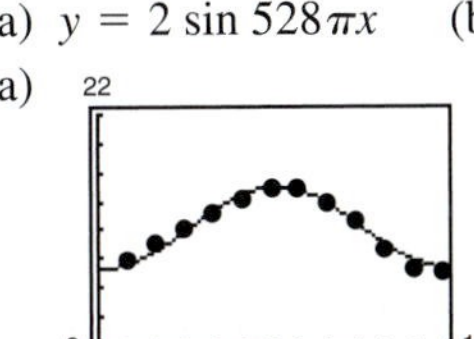

(b) 12. Yes. One period is one year.

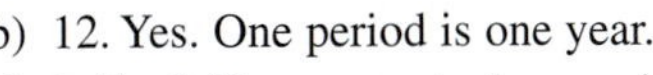

(c) 1.41. 1.41 represents the maximum change in time from the average time ($d = 18.09$) of sunset.

99.

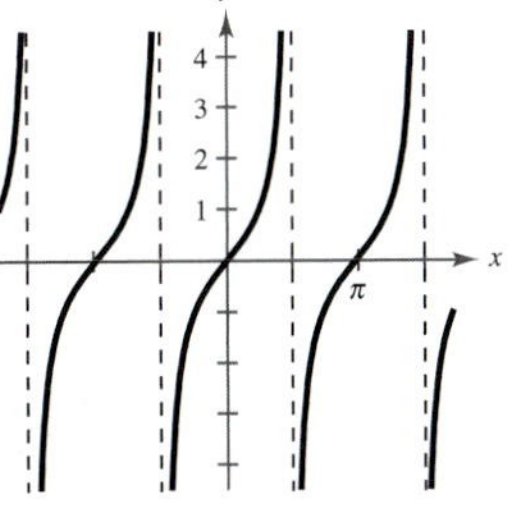

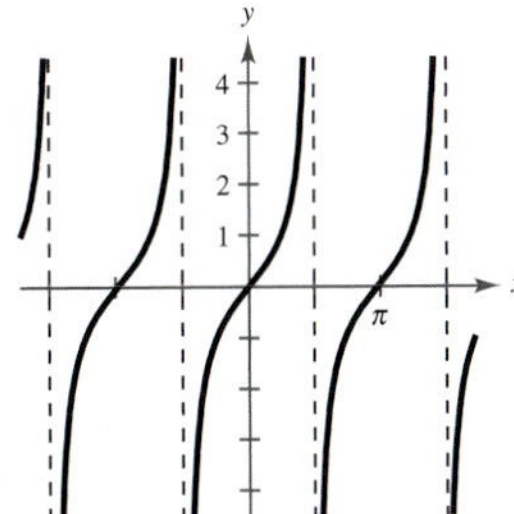

100.

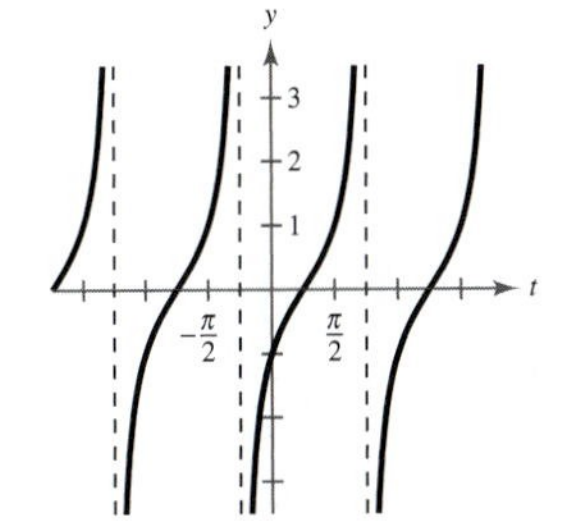

101.

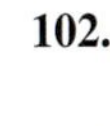

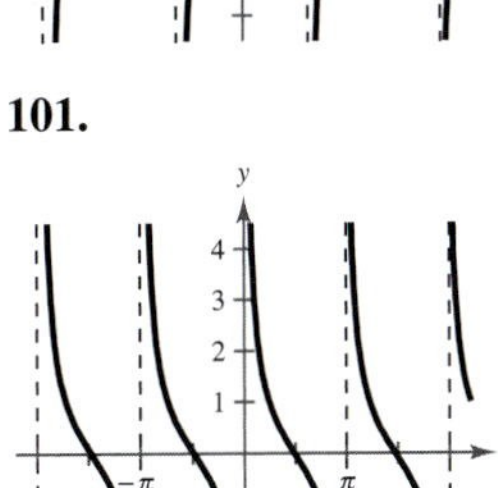

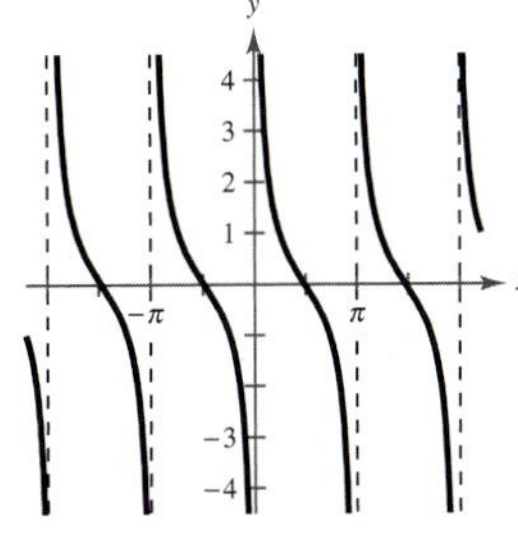

102.

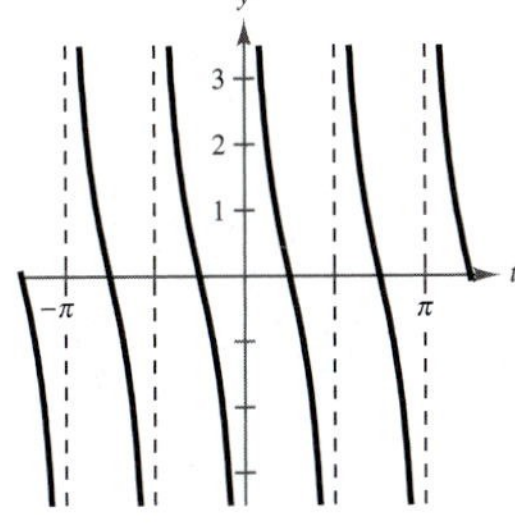

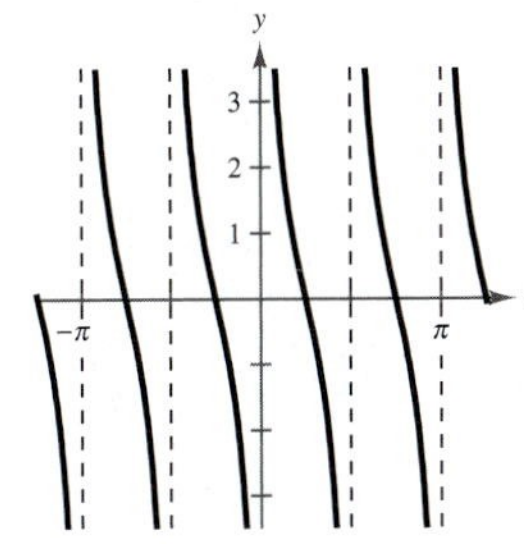

103.

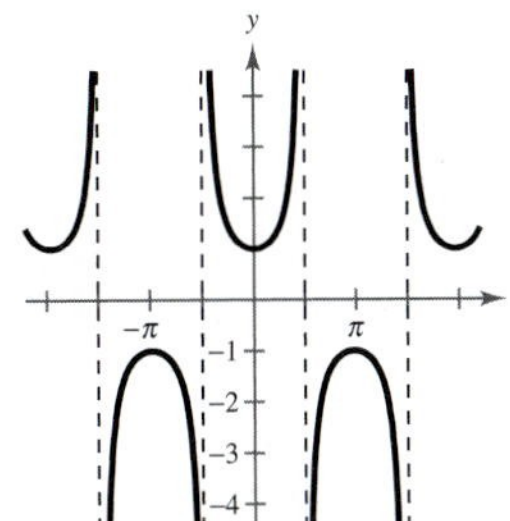

104.

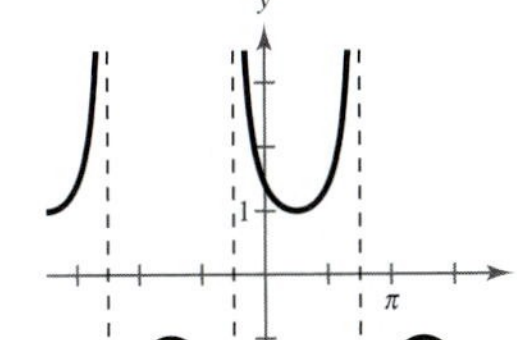

105.

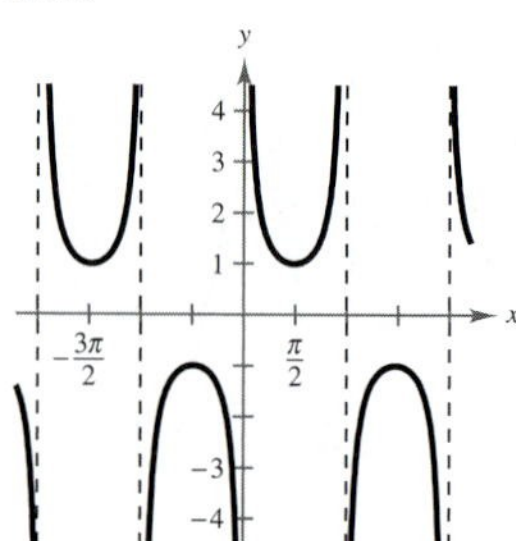

106.

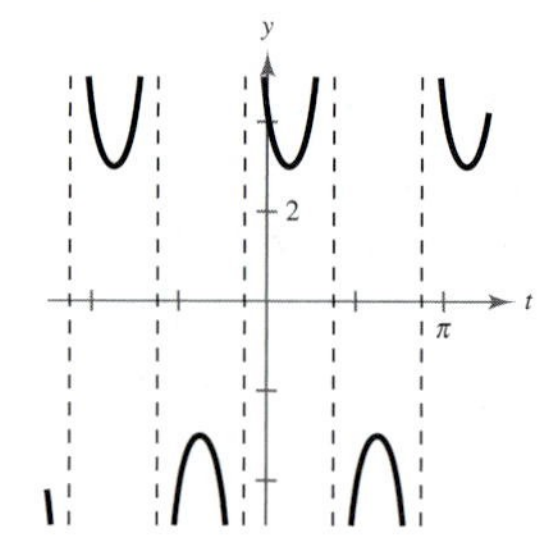

107.

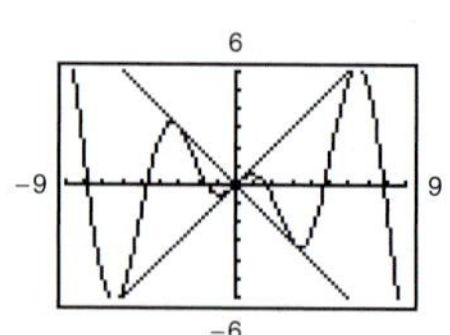

As $x \to +\infty, f(x) \to +\infty$

108.

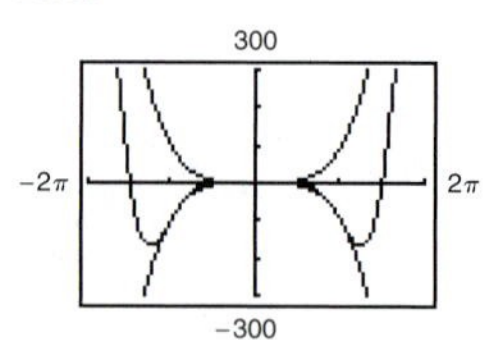

As $x \to +\infty, f(x) \to +\infty$

109. $-\frac{\pi}{6}$ **110.** $-\frac{\pi}{2}$ **111.** 0.41 **112.** 0.21

113. -0.46 **114.** 1.10 **115.** $\frac{\pi}{6}$ **116.** $\frac{\pi}{4}$

117. π **118.** $\frac{\pi}{6}$ **119.** 1.24 **120.** 2.66

121. -0.98 **122.** 1.45

123.

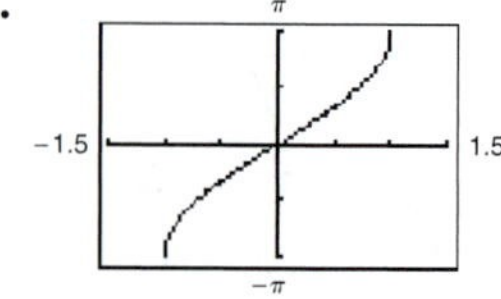

124.

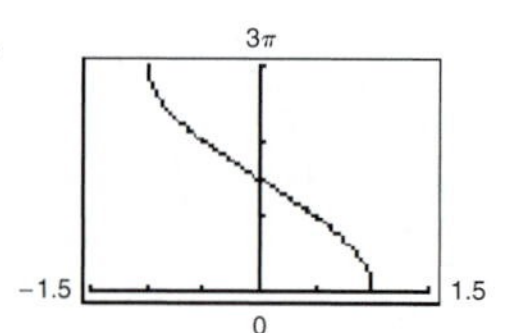

125.

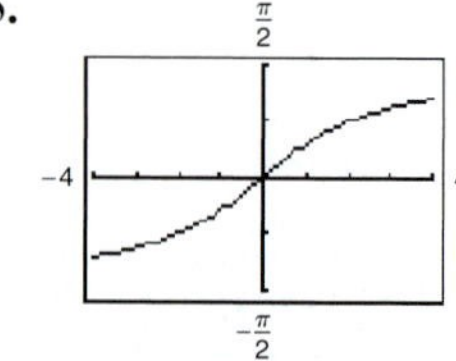

126.

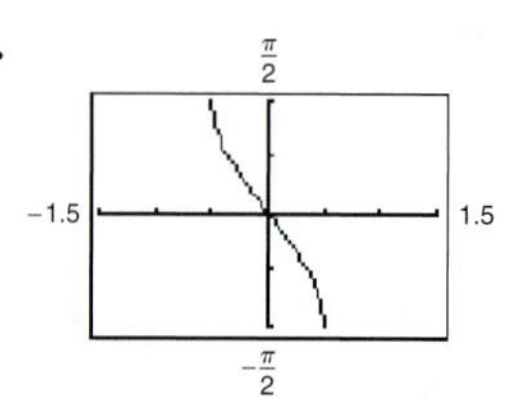

127. $\frac{4}{5}$ **128.** $\frac{4}{3}$ **129.** $\frac{13}{5}$ **130.** $-\frac{5}{12}$

131. $\frac{\sqrt{4-x^2}}{x}$ **132.** $\frac{1}{\sqrt{1-(x-1)^2}}$ **133.** 66.8°

134. 9.6 feet **135.** 1221 miles, 85.6°

136. $d = 0.75 \cos\left(\frac{2\pi t}{3}\right)$

137. False. The sine or cosine function is often useful for modeling simple harmonic motion.

138. True. The inverse sine, $y = \arcsin x$, is defined by $y = \arcsin x$ if and only if $\sin y = x$ where $-1 \le x \le 1$ and $-\pi/2 \le y \le \pi/2$.

139. False. For each θ there corresponds exactly one value of y.

140. False. $3\pi/4$ is not in the range of the arctangent function.

141. d; The period is 2π and the amplitude is 3.

142. a; The period is 2π and, because $a < 0$, the graph is reflected in the x-axis.

143. b; The period is 2 and the amplitude is 2.

144. c; The period is 4π and the amplitude is 2.

145. The function is undefined because $\sec \theta = 1/\cos \theta$.

146. (a)

θ	0.1	0.4	0.7
$\tan\left(\theta - \frac{\pi}{2}\right)$	-9.9666	-2.3652	-1.1872
$-\cot \theta$	-9.9666	-2.3652	-1.1872

θ	1.0	1.3
$\tan\left(\theta - \frac{\pi}{2}\right)$	-0.6421	-0.2776
$-\cot \theta$	-0.6421	-0.2776

(b) $\tan\left(\theta - \frac{\pi}{2}\right) = -\cot \theta$

147. The ranges of the other four trigonometric functions are $(-\infty, \infty)$ or $(-\infty, -1] \cup [1, \infty)$.

148. (a) The displacement is increased.
(b) The friction damps the oscillations more quickly.
(c) The frequency of the oscillations increases.

149. (a) $A = 0.4r^2, r > 0$; $s = 0.8r, r > 0$ (b) $A = 50\theta, \theta > 0$; $s = 10\theta, \theta > 0$

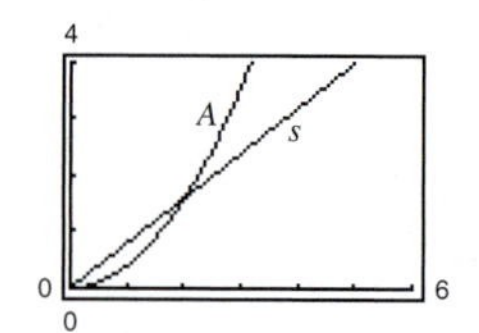

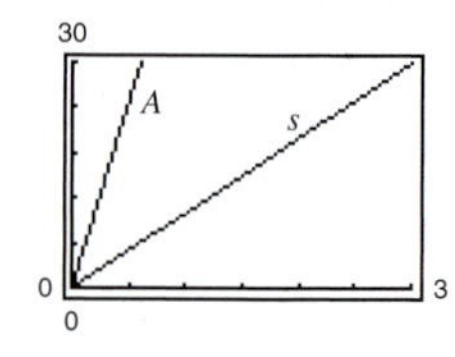

The area function increases more rapidly.

150. Answers will vary.

Chapter Test *(page 369)*

1. (a)

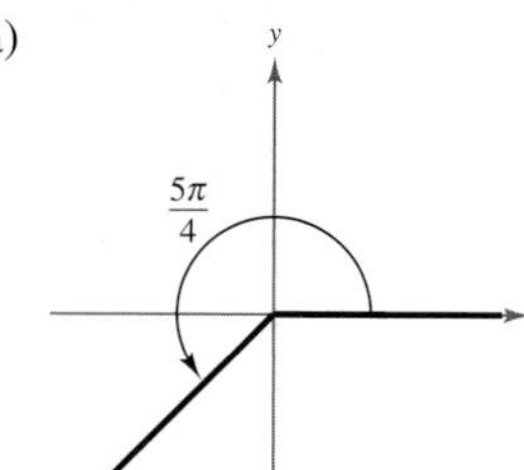

(b) $\frac{13\pi}{4}, -\frac{3\pi}{4}$
(c) 225°

2. 3000 radians per minute **3.** ≈ 709.04 square feet

4. $\sin \theta = \frac{3\sqrt{10}}{10}$ $\csc \theta = \frac{\sqrt{10}}{3}$

$\cos \theta = -\frac{\sqrt{10}}{10}$ $\sec \theta = -\sqrt{10}$

$\tan \theta = -3$ $\cot \theta = -\frac{1}{3}$

5. For $0 \le \theta < \frac{\pi}{2}$:

$\sin\theta = \frac{3\sqrt{13}}{13}$

$\cos\theta = \frac{2\sqrt{13}}{13}$

$\csc\theta = \frac{\sqrt{13}}{3}$

$\sec\theta = \frac{\sqrt{13}}{2}$

$\cot\theta = \frac{2}{3}$

For $\pi \le \theta < \frac{3\pi}{2}$:

$\sin\theta = -\frac{3\sqrt{13}}{13}$

$\cos\theta = -\frac{2\sqrt{13}}{13}$

$\csc\theta = -\frac{\sqrt{13}}{3}$

$\sec\theta = -\frac{\sqrt{13}}{2}$

$\cot\theta = \frac{2}{3}$

6. $\theta' = 70°$

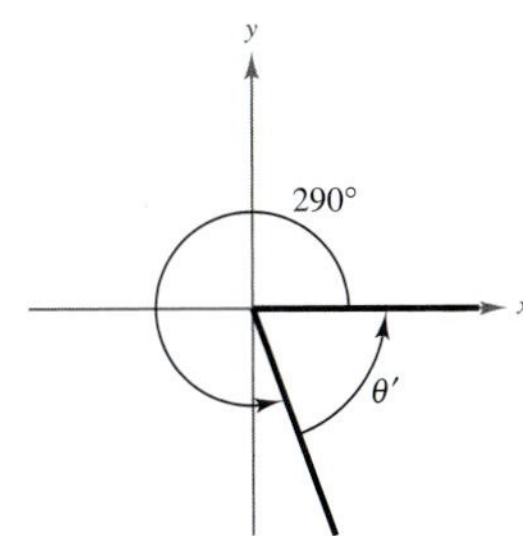

7. Quadrant III **8.** 150°, 210° **9.** 1.33, 1.81

10. $\sin\theta = -\frac{4}{5}$

$\tan\theta = -\frac{4}{3}$

$\csc\theta = -\frac{5}{4}$

$\sec\theta = \frac{5}{3}$

$\cot\theta = -\frac{3}{4}$

11. $\sin\theta = \frac{15}{17}$

$\cos\theta = -\frac{8}{17}$

$\tan\theta = -\frac{15}{8}$

$\csc\theta = \frac{17}{15}$

$\cot\theta = -\frac{8}{15}$

12.

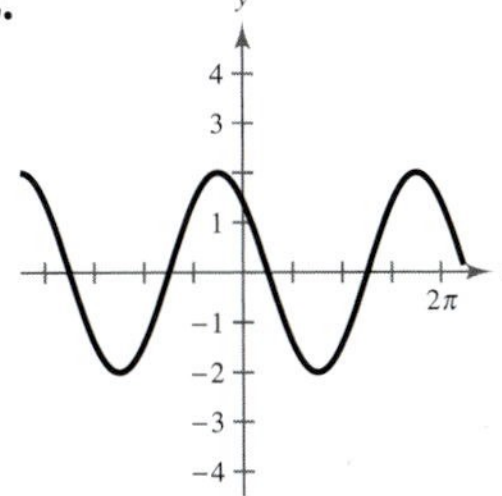

13.

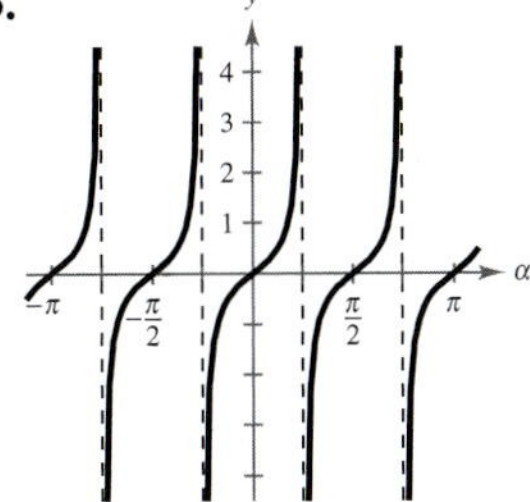

14.

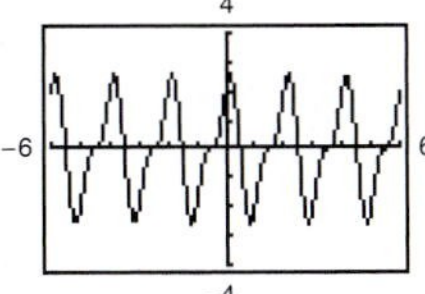

Period: 2

15.

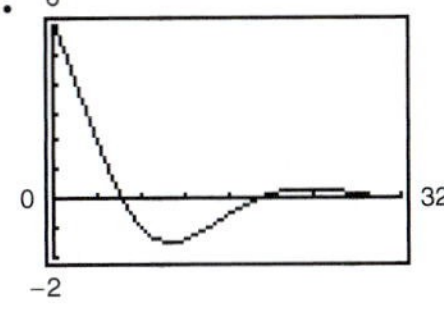

Not periodic

16. $a = -2, b = \frac{1}{2}, c = -\frac{\pi}{4}$ **17.** $\frac{\sqrt{5}}{2}$

18.

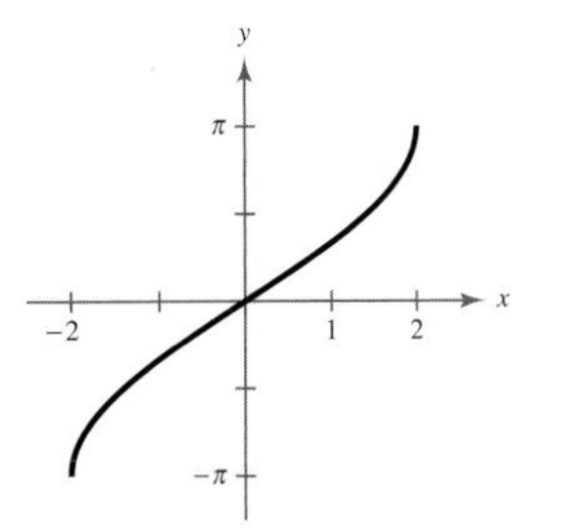

19. 310.1°

20. $d = -6\cos \pi t$

Problem Solving *(page 371)*

1. (a) $\frac{11\pi}{2}$ radians or 990° (b) $\approx$ 816.42 feet

2. Gear 1: 270°, $\frac{3\pi}{2}$ radians

Gear 2: $\approx$ 332.3°, $\frac{24\pi}{13}$ radians

Gear 3: $\approx$ 392.7°, $\frac{24\pi}{11}$ radians

Gear 4: 450°, $\frac{5\pi}{2}$ radians

Gear 5: $\approx$ 454.7°, $\frac{48\pi}{19}$ radians

3. (a) 4767 feet (b) 3705 feet

(c) $w = 2183$ feet,

$\tan 63° = \frac{w + 3705}{3000}$

4. (a) Answers will vary. (b) The ratios are equal.

(c) No; No

(d) Yes, because all six trigonometric functions are ratios of right triangles.

5. (a)

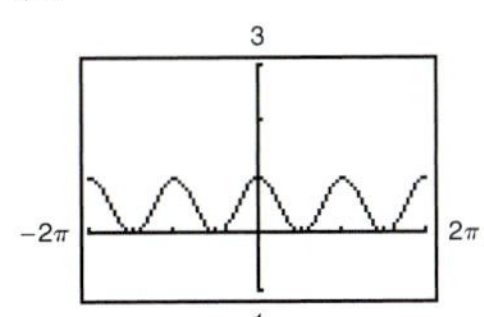

Even

(b)

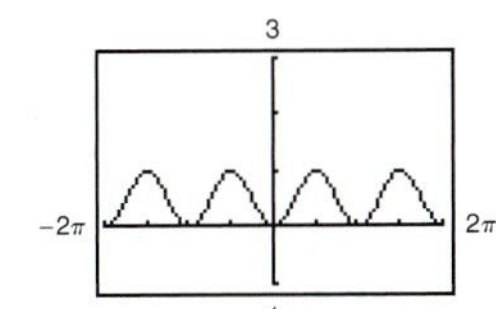

Even

6. (a) $h(x)$ is even. (b) $h(x)$ is even.

7. $h = 51 - 50\sin\left(8\pi t + \frac{\pi}{2}\right)$

8. (a)

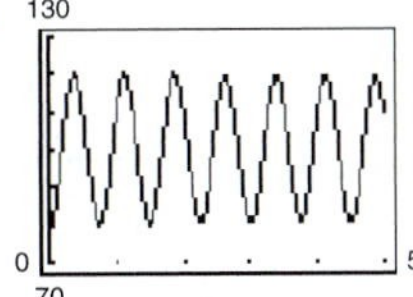

(b) Period = $\frac{3}{4}$ seconds; answers will vary.

(c) 20 millimeters; answers will vary.

(d) 80 beats per minute (e) Period = $\frac{15}{16}$ seconds; $\frac{32\pi}{15}$

CHAPTER 4

9. (a)

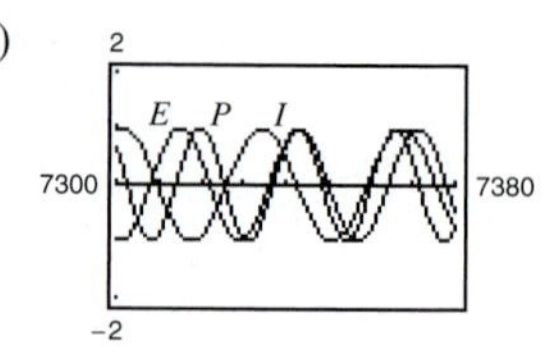

(b)

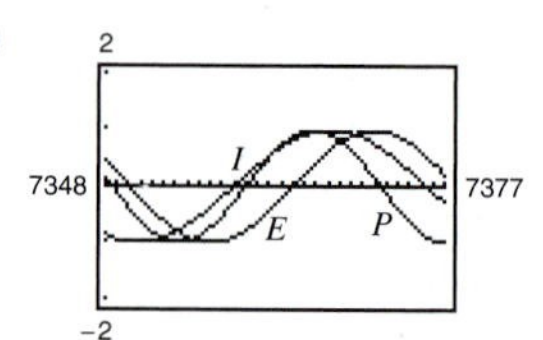

(c) $P(7369) = 0.631$
$E(7369) = 0.901$
$I(7369) = 0.945$

10. (a)

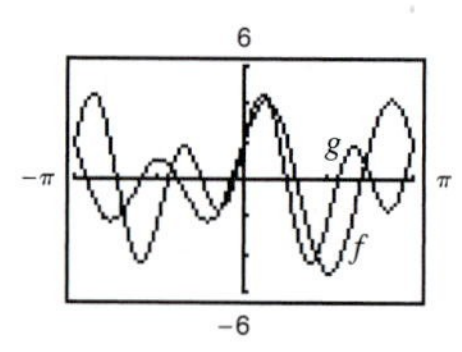

(b) Period of f: 2π; Period of g: π

(c) Yes, because the sine and cosine functions are periodic.

11. (a) 3.35, 7.35 (b) -0.65

(c) Yes. There is a difference of nine periods between the values.

12. (a) Equal; two-period shift

(b) Not equal; $f\left(t + \frac{1}{2}c\right)$ is a horizontal translation and $f\left(\frac{1}{2}t\right)$ is a period change.

(c) Not equal; For example, $\sin\left[\frac{1}{2}(\pi + 2\pi)\right] \neq \sin\left(\frac{1}{2}\pi\right)$.

13. (a) $40.5°$ (b) $x \approx 1.71$ feet; $y \approx 3.46$ feet

(c) ≈ 1.75 feet

(d) As you move closer to the rock, d must get smaller and smaller. The angles θ_1 and θ_2 will decrease along with the distance y, so d will decrease.

14. (a)

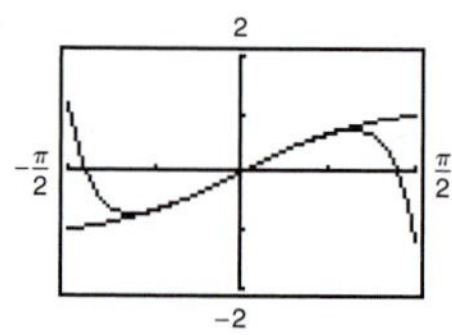

The approximation is accurate over the interval $-1 \leq x \leq 1$.

(b) $\dfrac{x^9}{9}$

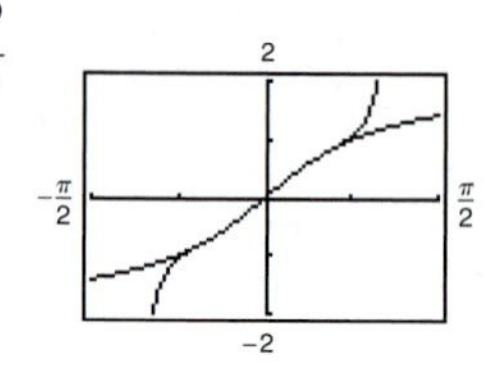

The accuracy improved.

Chapter 5

Section 5.1 *(page 379)*

Vocabulary Check *(page 379)*

1. $\tan u$ **2.** $\cos u$ **3.** $\cot u$ **4.** $\csc u$
5. $\cot^2 u$ **6.** $\sec^2 u$ **7.** $\cos u$ **8.** $\csc u$
9. $\cos u$ **10.** $-\tan u$

1. $\sin x = \dfrac{\sqrt{3}}{2}$, $\cos x = -\dfrac{1}{2}$, $\tan x = -\sqrt{3}$, $\csc x = \dfrac{2\sqrt{3}}{3}$, $\sec x = -2$, $\cot x = -\dfrac{\sqrt{3}}{3}$

2. $\sin x = -\dfrac{1}{2}$, $\cos x = -\dfrac{\sqrt{3}}{2}$, $\tan x = \dfrac{\sqrt{3}}{3}$, $\csc x = -2$, $\sec x = -\dfrac{2\sqrt{3}}{3}$, $\cot x = \sqrt{3}$

3. $\sin\theta = -\dfrac{\sqrt{2}}{2}$, $\cos\theta = \dfrac{\sqrt{2}}{2}$, $\tan\theta = -1$, $\sec\theta = \sqrt{2}$, $\csc\theta = -\sqrt{2}$, $\cot\theta = -1$

4. $\sin\theta = \dfrac{3}{5}$, $\cos\theta = \dfrac{4}{5}$, $\tan\theta = \dfrac{3}{4}$, $\sec\theta = \dfrac{5}{4}$, $\csc\theta = \dfrac{5}{3}$, $\cot\theta = \dfrac{4}{3}$

5. $\sin x = -\dfrac{5}{13}$, $\cos x = -\dfrac{12}{13}$, $\tan x = \dfrac{5}{12}$, $\sec x = -\dfrac{13}{12}$, $\csc x = -\dfrac{13}{5}$, $\cot x = \dfrac{12}{5}$

6. $\sin\phi = \dfrac{\sqrt{10}}{10}$, $\cos\phi = -\dfrac{3\sqrt{10}}{10}$, $\tan\phi = -\dfrac{1}{3}$, $\cot\phi = -3$, $\csc\phi = \sqrt{10}$, $\sec\phi = -\dfrac{\sqrt{10}}{3}$

7. $\sin\phi = -\dfrac{\sqrt{5}}{3}$, $\cos\phi = \dfrac{2}{3}$, $\tan\phi = -\dfrac{\sqrt{5}}{2}$, $\sec\phi = \dfrac{3}{2}$, $\csc\phi = -\dfrac{3\sqrt{5}}{5}$, $\cot\phi = -\dfrac{2\sqrt{5}}{5}$

8. $\sin x = \dfrac{3}{5}$, $\cos x = \dfrac{4}{5}$, $\tan x = \dfrac{3}{4}$, $\csc x = \dfrac{5}{3}$, $\sec x = \dfrac{5}{4}$, $\cot x = \dfrac{4}{3}$

9. $\sin x = \frac{1}{3}$
$\cos x = -\frac{2\sqrt{2}}{3}$
$\tan x = -\frac{\sqrt{2}}{4}$
$\csc x = 3$
$\sec x = -\frac{3\sqrt{2}}{4}$
$\cot x = -2\sqrt{2}$

10. $\sin x = \frac{\sqrt{15}}{4}$
$\cos x = \frac{1}{4}$
$\tan x = \sqrt{15}$
$\sec x = 4$
$\csc x = \frac{4\sqrt{15}}{15}$
$\cot x = \frac{\sqrt{15}}{15}$

11. $\sin \theta = -\frac{2\sqrt{5}}{5}$
$\cos \theta = -\frac{\sqrt{5}}{5}$
$\tan \theta = 2$
$\csc \theta = -\frac{\sqrt{5}}{2}$
$\sec \theta = -\sqrt{5}$
$\cot \theta = \frac{1}{2}$

12. $\sin \theta = -\frac{1}{5}$
$\cos \theta = -\frac{2\sqrt{6}}{5}$
$\tan \theta = \frac{\sqrt{6}}{12}$
$\sec \theta = -\frac{5\sqrt{6}}{12}$
$\csc \theta = -5$
$\cot \theta = 2\sqrt{6}$

13. $\sin \theta = -1$
$\cos \theta = 0$
$\tan \theta$ is undefined.
$\cot \theta = 0$
$\csc \theta = -1$
$\sec \theta$ is undefined.

14. $\sin \theta = 1$
$\cos \theta = 0$
$\tan \theta$ is undefined.
$\csc \theta = 1$
$\sec \theta$ is undefined.
$\cot \theta = 0$

15. d **16.** a **17.** b **18.** f **19.** e **20.** c
21. b **22.** c **23.** f **24.** a **25.** e **26.** d
27. $\csc \theta$ **28.** $\sin \beta$ **29.** $\cos^2 \phi$ **30.** 1
31. $\cos x$ **32.** $\cot \theta$ **33.** $\sin^2 x$ **34.** $\cos^2 x$
35. 1 **36.** $\sin^2 \theta$ **37.** $\tan x$ **38.** $\sin x$
39. $1 + \sin y$ **40.** $\sec t$ **41.** $\sec \beta$ **42.** $2 \sec \phi$
43. $\cos u + \sin u$ **44.** $\csc \theta \sec \theta$ **45.** $\sin^2 x$
46. $\cos^2 x$ **47.** $\sin^2 x \tan^2 x$ **48.** 1 **49.** $\sec x + 1$
50. $\cos x + 2$ **51.** $\sec^4 x$ **52.** $\sin^4 x$
53. $\sin^2 x - \cos^2 x$ **54.** $\sec^2 x + \tan^2 x$
55. $\cot^2 x(\csc x - 1)$ **56.** $\tan^2 x(\sec x - 1)$
57. $1 + 2 \sin x \cos x$ **58.** -1 **59.** $4 \cot^2 x$
60. $9 \cos^2 x$ **61.** $2 \csc^2 x$ **62.** $-2 \cot^2 x$ **63.** $2 \sec x$
64. $-\cot x$ **65.** $1 + \cos y$ **66.** $5(\sec x - \tan x)$
67. $3(\sec x + \tan x)$ **68.** $\tan^4 x(\csc x - 1)$

69.

x	0.2	0.4	0.6	0.8	1.0
y_1	0.1987	0.3894	0.5646	0.7174	0.8415
y_2	0.1987	0.3894	0.5646	0.7174	0.8415

x	1.2	1.4
y_1	0.9320	0.9854
y_2	0.9320	0.9854

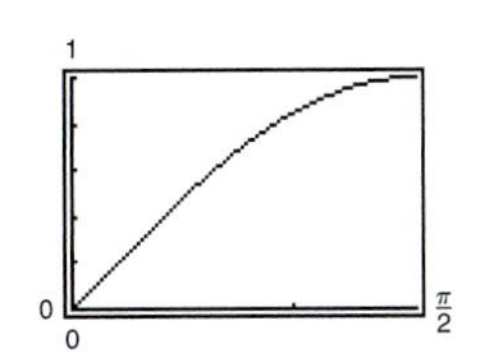

$y_1 = y_2$

70.

x	0.2	0.4	0.6	0.8	1.0
y_1	0.0403	0.1646	0.3863	0.7386	1.3105
y_2	0.0403	0.1646	0.3863	0.7386	1.3105

x	1.2	1.4
y_1	2.3973	5.7135
y_2	2.3973	5.7135

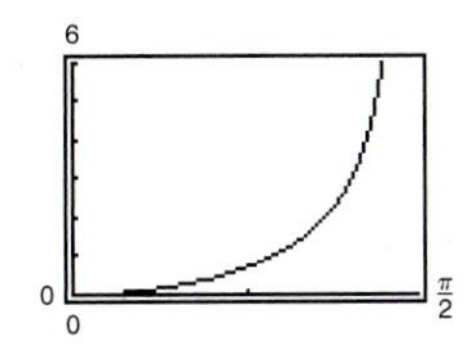

$y_1 = y_2$

71.

x	0.2	0.4	0.6	0.8	1.0
y_1	1.2230	1.5085	1.8958	2.4650	3.4082
y_2	1.2230	1.5085	1.8958	2.4650	3.4082

x	1.2	1.4
y_1	5.3319	11.6814
y_2	5.3319	11.6814

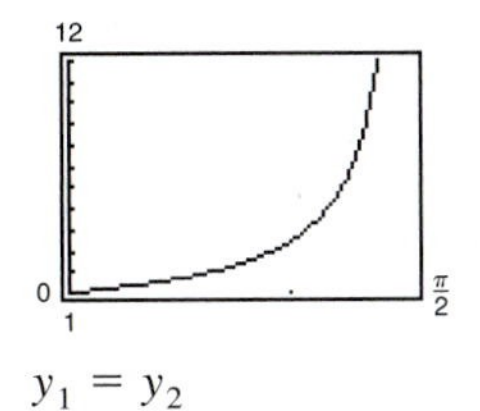

$y_1 = y_2$

72.

x	0.2	0.4	0.6	0.8	1.0
y_1	0.0428	0.2107	0.6871	2.1841	8.3087
y_2	0.0428	0.2107	0.6871	2.1841	8.3087

x	1.2	1.4
y_1	50.3869	1163.6143
y_2	50.3869	1163.6143

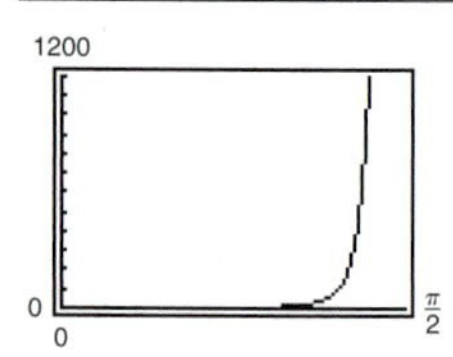

$y_1 = y_2$

73. $\csc x$ **74.** $\cot x$ **75.** $\tan x$ **76.** $\sec \theta$

77. $3 \sin \theta$ **78.** $8 \sin \theta$ **79.** $3 \tan \theta$ **80.** $2 \tan \theta$

81. $5 \sec \theta$ **82.** $10 \sec \theta$

83. $3 \cos \theta = 3$; $\sin \theta = 0$; $\cos \theta = 1$

84. $6 \cos \theta = 3$; $\sin \theta = \pm\dfrac{\sqrt{3}}{2}$; $\cos \theta = \dfrac{1}{2}$

85. $4 \sin \theta = 2\sqrt{2}$; $\sin \theta = \dfrac{\sqrt{2}}{2}$; $\cos \theta = \dfrac{\sqrt{2}}{2}$

86. $10 \sin \theta = -5\sqrt{3}$; $\sin \theta = -\dfrac{\sqrt{3}}{2}$; $\cos \theta = \dfrac{1}{2}$

87. $0 \le \theta \le \pi$ **88.** $\dfrac{\pi}{2} \le \theta \le \dfrac{3\pi}{2}$

89. $0 \le \theta < \dfrac{\pi}{2}, \dfrac{3\pi}{2} < \theta < 2\pi$ **90.** $0 < \theta < \pi$

91. $\ln|\cot x|$ **92.** $\ln|\tan x|$ **93.** $\ln|\csc t \sec t|$ **94.** 0

95. (a) $\csc^2 132° - \cot^2 132° \approx 1.8107 - 0.8107 = 1$

(b) $\csc^2 \dfrac{2\pi}{7} - \cot^2 \dfrac{2\pi}{7} \approx 1.6360 - 0.6360 = 1$

96. (a) $(\tan 346°)^2 + 1 = (\sec 346°)^2 \approx 1.0622$

(b) $(\tan 3.1)^2 + 1 = (\sec 3.1)^2 \approx 1.0017$

97. (a) $\cos(90° - 80°) = \sin 80° \approx 0.9848$

(b) $\cos\left(\dfrac{\pi}{2} - 0.8\right) = \sin 0.8 \approx 0.7174$

98. (a) $\sin(-250°) = -\sin 250° \approx 0.9397$

(b) $\sin\left(-\frac{1}{2}\right) = -\sin\left(\frac{1}{2}\right) \approx -0.4794$

99. $\mu = \tan \theta$ **100.** Answers will vary.

101. True. For example, $\sin(-x) = -\sin x$.

102. False. A cofunction identity can be used to transform a tangent function so that it can be represented by a cotangent function.

103. 1, 1 **104.** 1, 1 **105.** ∞, 0 **106.** 0, $-\infty$

107. Not an identity because $\cos \theta = \pm\sqrt{1 - \sin^2 \theta}$

108. Not an identity because $\cot \theta = \pm\sqrt{\csc^2 \theta - 1}$

109. Not an identity because $\dfrac{\sin k\theta}{\cos k\theta} = \tan k\theta$

110. Not an identity because $\dfrac{5}{\cos \theta} \neq \dfrac{1}{5 \cos \theta}$

111. An identity because $\sin \theta \cdot \dfrac{1}{\sin \theta} = 1$

112. Not an identity because $\dfrac{1}{\sin \theta} \neq 1$

113–114. Answers will vary.

115. $x - 25$ **116.** $4z + 12\sqrt{z} + 9$

117. $\dfrac{x^2 + 6x - 8}{(x + 5)(x - 8)}$ **118.** $\dfrac{3(2x + 1)}{x - 4}$

119. $\dfrac{-5x^2 + 8x + 28}{(x^2 - 4)(x + 4)}$ **120.** $\dfrac{x(x^2 + 5x + 1)}{x^2 - 25}$

121.

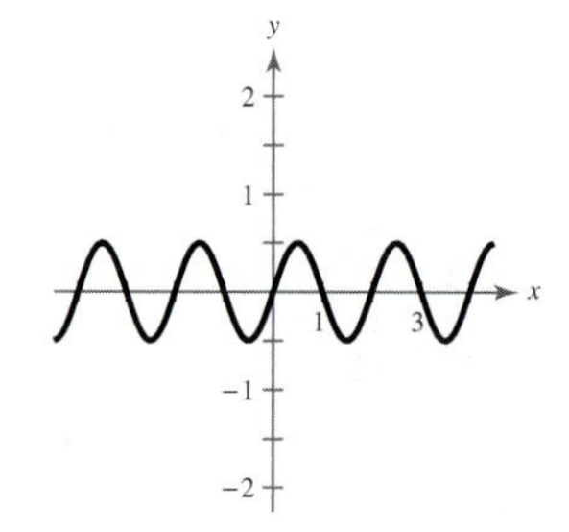

122.

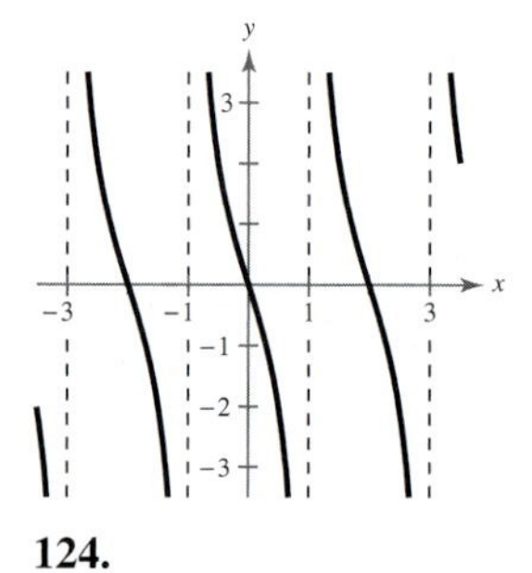

123.

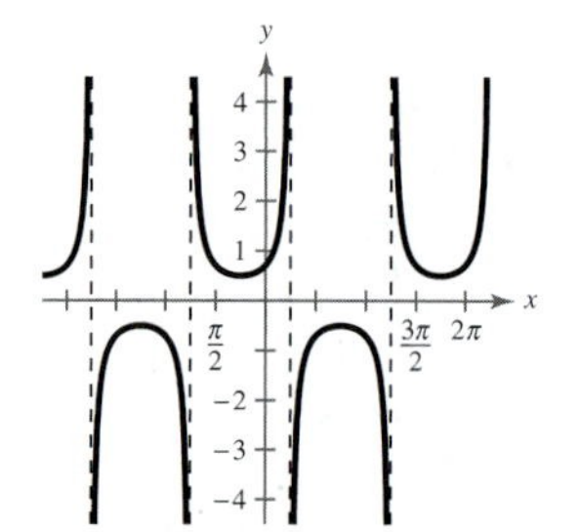

124.

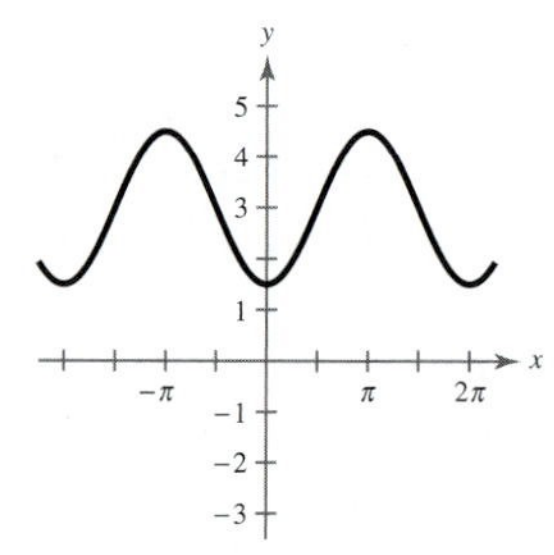

Section 5.2 *(page 387)*

Vocabulary Check *(page 387)*

1. identity **2.** conditional equation **3.** $\tan u$

4. $\cot u$ **5.** $\cos^2 u$ **6.** $\sin u$ **7.** $-\csc u$

8. $\sec u$

1–38. Answers will vary.

39. (a)

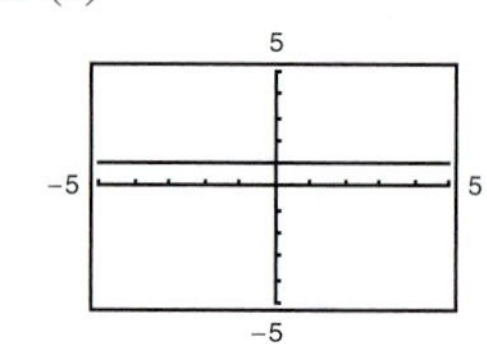

(b)

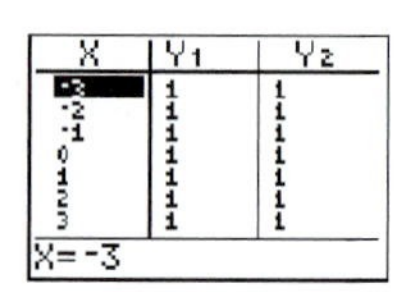

Identity

(c) Answers will vary.

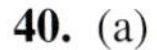

40. (a)

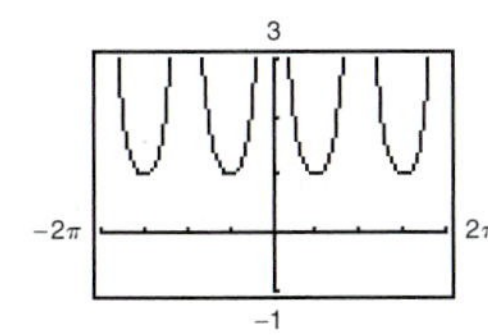

(b)

X	Y1	Y2
-3	50.214	50.214
-2	1.2095	1.2095
-1	1.4123	1.4123
0	ERROR	ERROR
1	1.4123	1.4123
2	1.2095	1.2095
3	50.214	50.214

X=-3

Identity

(c) Answers will vary.

41. (a)

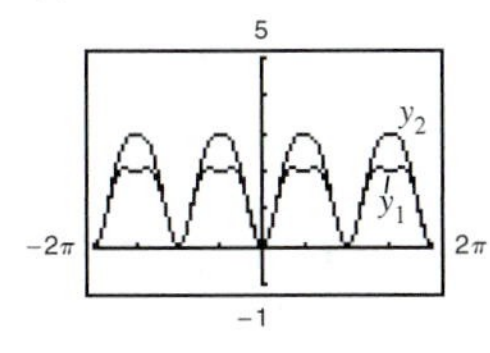

(b)

X	Y1	Y2
-4.712	2	3
-3.142	0	0
-1.571	2	3
0	0	0
1.5708	2	3
3.1416	0	0
4.7124	2	3

X=-4.71238898038

Not an identity

(c) Answers will vary.

42. (a)

5, −π, π, −5

(b)

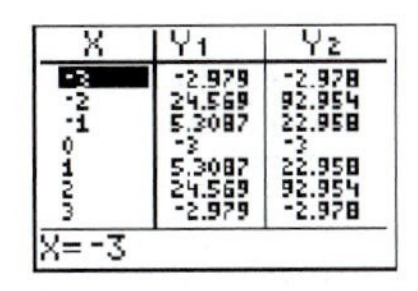

X	Y1	Y2
-3	-2.979	-2.978
-2	24.569	92.954
-1	5.3087	22.958
0	-3	-3
1	5.3087	22.958
2	24.569	92.954
3	-2.979	-2.978

X=-3

Not an identity

(c) Answers will vary.

43. (a)

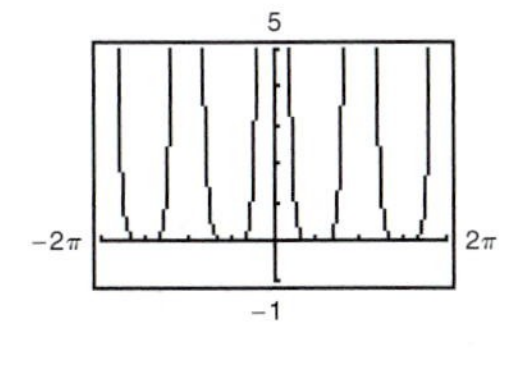

(b)

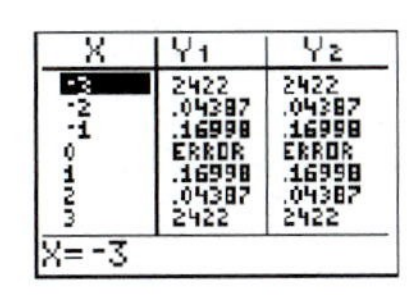

X	Y1	Y2
-3	2422	2422
-2	.04387	.04387
-1	.16998	.16998
0	ERROR	ERROR
1	.16998	.16998
2	.04387	.04387
3	2422	2422

X=-3

Identity

(c) Answers will vary.

44. (a)

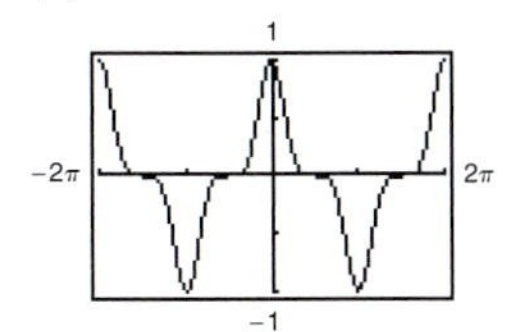

(b)

X	Y1	Y2
-4.712	0	0
-3.142	-1	-1
-1.571	0	0
0	1	1
1.5708	0	0
3.1416	-1	-1
4.7124	0	0

X=-4.71238898038

Identity

(c) Answers will vary.

45. (a)

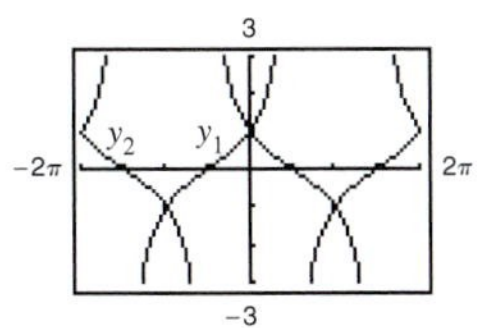

(b)

X	Y1	Y2
-3	-.8676	-1.153
-2	-.218	-4.588
-1	.29341	3.4082
0	1	1
1	3.4082	.29341
2	-4.588	-.218
3	-1.153	-.8676

X=-3

Not an identity

(c) Answers will vary.

46. (a)

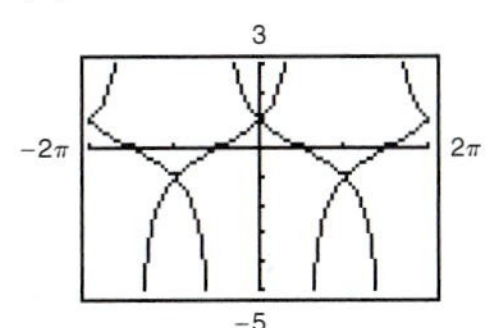

(b)

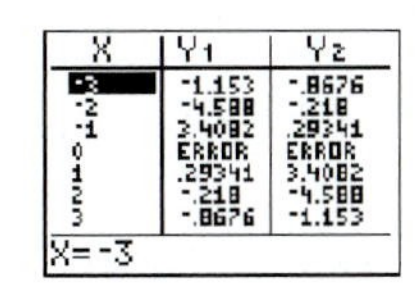

X	Y1	Y2
-3	-1.153	-.8676
-2	-4.588	-.218
-1	3.4082	.29341
0	ERROR	ERROR
1	.29341	3.4082
2	-.218	-4.588
3	-.8676	-1.153

X=-3

Not an identity

(c) Answers will vary.

47–50. Answers will vary. **51.** 1

52. 1 **53.** 2 **54.** 2 **55.** Answers will vary.

56. (a) Answers will vary.

(b)

θ	10°	20°	30°	40°	50°
s	28.36	13.74	8.66	5.96	4.20

θ	60°	70°	80°	90°
s	2.89	1.82	0.88	0

(c) Greatest: 10°; Least: 90° (d) Noon

57. False. An identity is an equation that is true for all real values of θ.

58. True. An identity is an equation that is true for all real values in the domain of the variable.

59. The equation is not an identity because $\sin\theta = \pm\sqrt{1-\cos^2\theta}$.

Possible answer: $\dfrac{7\pi}{4}$

60. The equation is not an identity because $\tan\theta = \pm\sqrt{\sec^2\theta - 1}$.

Possible answer: $\dfrac{3\pi}{4}$

61. $2 + (3 - \sqrt{26})i$ **62.** $-21 - 20i$ **63.** $-8 + 4i$

64. $-9 + 46i$ **65.** $-3 \pm \sqrt{21}$ **66.** $\dfrac{-5 \pm \sqrt{53}}{2}$

67. $1 \pm \sqrt{5}$ **68.** $\frac{1}{4}(1 \pm \sqrt{7})$

Section 5.3 *(page 396)*

Vocabulary Check *(page 396)*

1. general **2.** quadratic **3.** extraneous

1–6. Answers will vary. **7.** $\dfrac{2\pi}{3} + 2n\pi, \dfrac{4\pi}{3} + 2n\pi$

8. $\dfrac{7\pi}{6} + 2n\pi, \dfrac{11\pi}{6} + 2n\pi$ **9.** $\dfrac{\pi}{3} + 2n\pi, \dfrac{2\pi}{3} + 2n\pi$

10. $\dfrac{2\pi}{3} + n\pi$ **11.** $\dfrac{\pi}{6} + n\pi, \dfrac{5\pi}{6} + n\pi$

12. $\dfrac{\pi}{3} + n\pi, \dfrac{2\pi}{3} + n\pi$ **13.** $n\pi, \dfrac{3\pi}{2} + 2n\pi$

14. $\dfrac{\pi}{6} + n\pi, \dfrac{5\pi}{6} + n\pi, \dfrac{\pi}{3} + n\pi, \dfrac{2\pi}{3} + n\pi$

15. $\frac{\pi}{3}+n\pi, \frac{2\pi}{3}+n\pi$ **16.** $\frac{\pi}{3}+n\pi, \frac{2\pi}{3}+n\pi$

17. $\frac{\pi}{8}+\frac{n\pi}{2}, \frac{3\pi}{8}+\frac{n\pi}{2}$ **18.** $\frac{\pi}{9}+\frac{n\pi}{3}, \frac{2\pi}{9}+\frac{n\pi}{3}$

19. $\frac{n\pi}{3}, \frac{\pi}{4}+n\pi$ **20.** $\frac{\pi}{4}+\frac{n\pi}{2}, \frac{2\pi}{3}+2n\pi, \frac{4\pi}{3}+2n\pi$

21. $0, \frac{\pi}{2}, \pi, \frac{3\pi}{2}$ **22.** $0, \pi$ **23.** $0, \pi, \frac{\pi}{6}, \frac{5\pi}{6}, \frac{7\pi}{6}, \frac{11\pi}{6}$

24. $\frac{\pi}{2}, \frac{3\pi}{2}, \frac{2\pi}{3}, \frac{4\pi}{3}$ **25.** $\frac{\pi}{3}, \frac{5\pi}{3}, \pi$ **26.** $\frac{\pi}{3}, \frac{5\pi}{3}$

27. No solution **28.** 0 **29.** $\pi, \frac{\pi}{3}, \frac{5\pi}{3}$

30. $\frac{7\pi}{6}, \frac{3\pi}{2}, \frac{11\pi}{6}$ **31.** $\frac{\pi}{6}, \frac{5\pi}{6}, \frac{7\pi}{6}, \frac{11\pi}{6}$ **32.** $\frac{\pi}{3}, \frac{5\pi}{3}$

33. $\frac{\pi}{2}$ **34.** $\frac{\pi}{4}, \frac{5\pi}{4}$ **35.** $\frac{\pi}{6}+n\pi, \frac{5\pi}{6}+n\pi$

36. $\frac{2\pi}{3}+n\pi, \frac{5\pi}{6}+n\pi$ **37.** $\frac{\pi}{12}+\frac{n\pi}{3}$

38. $\frac{\pi}{12}+\frac{n\pi}{2}, \frac{5\pi}{12}+\frac{n\pi}{2}$ **39.** $\frac{\pi}{2}+4n\pi, \frac{7\pi}{2}+4n\pi$

40. $\frac{8\pi}{3}+4n\pi, \frac{10\pi}{3}+4n\pi$ **41.** $-1+4n$

42. $\frac{3}{4}+n$ **43.** $-2+6n, 2+6n$

44. $-2+8n, 2+8n$ **45.** 2.678, 5.820

46. 0.785, 2.356, 3.665, 3.927, 5.498, 5.760

47. 1.047, 5.236 **48.** 0.524, 2.618 **49.** 0.860, 3.426

50. 4.917 **51.** 0, 2.678, 3.142, 5.820

52. 0.515, 2.726, 3.657, 5.868

53. 0.983, 1.768, 4.124, 4.910

54. 0.524, 0.730, 2.412, 2.618

55. 0.3398, 0.8481, 2.2935, 2.8018

56. 0.5880, 2.0344, 3.7296, 5.1760

57. 1.9357, 2.7767, 5.0773, 5.9183 **58.** 1.7794, 4.5038

59. $\frac{\pi}{4}, \frac{5\pi}{4}$, arctan 5, arctan 5 $+\ \pi$

60. $\arctan(-2)+\pi, \arctan(-2)+2\pi, \frac{\pi}{4}, \frac{5\pi}{4}$

61. $\frac{\pi}{3}, \frac{5\pi}{3}$ **62.** $\frac{\pi}{6}, \frac{5\pi}{6}$

63. (a)

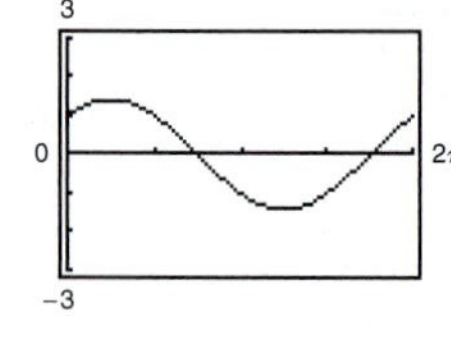

Maximum: (0.7854, 1.4142)
Minimum: (3.9270, −1.4142)

(b) $\frac{\pi}{4}\approx 0.7854$

$\frac{5\pi}{4}\approx 3.9270$

64. (a)

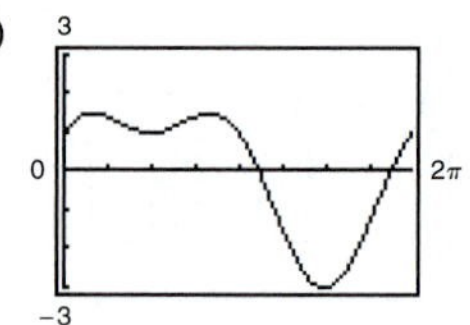

Maximum: (0.5236, 1.5)
Maximum: (2.6180, 1.5)
Minimum: (1.5708, 1)
Minimum: (4.7124, −3)

(b) $\frac{\pi}{6}\approx 0.5236$

$\frac{\pi}{2}\approx 1.5708$

$\frac{5\pi}{6}\approx 2.618$

$\frac{3\pi}{2}\approx 4.7124$

65. 1 **66.** 0.739

67. (a) All real numbers x except $x = 0$
(b) y-axis symmetry; Horizontal asymptote: $y = 1$
(c) Oscillates (d) Infinitely many solutions
(e) Yes, 0.6366

68. (a) All real numbers x except $x = 0$
(b) y-axis symmetry (c) y approaches 1.
(d) Four solutions: $\pm\pi, \pm 2\pi$

69. 0.04 second, 0.43 second, 0.83 second

70.

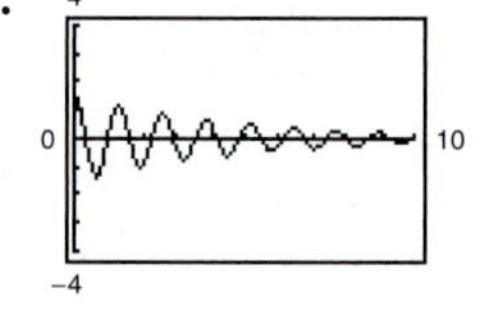

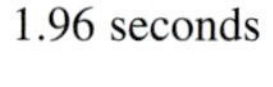

1.96 seconds

71. February, March, and April

72. January, October, November, December

73. 36.9°, 53.1° **74.** 1.9°

75. (a) Between $t = 8$ seconds and $t = 24$ seconds
(b) 5 times: $t = 16, 48, 80, 112, 144$ seconds

76. (a)

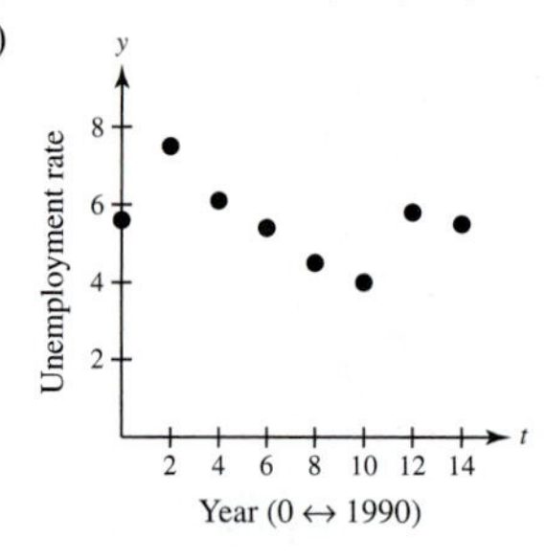

(b) 1
(c) The constant term, 5.45
(d) ≈ 13.37 years
(e) 2010

77. (a)

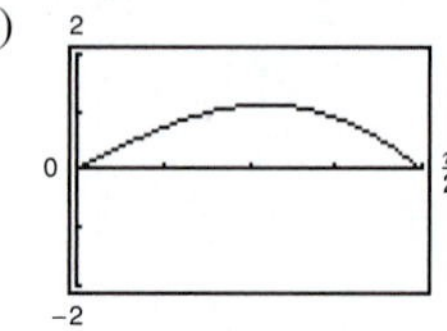

$A\approx 1.12$

(b) $0.6 < x < 1.1$

78. (a) $\frac{10}{3}$

(b)

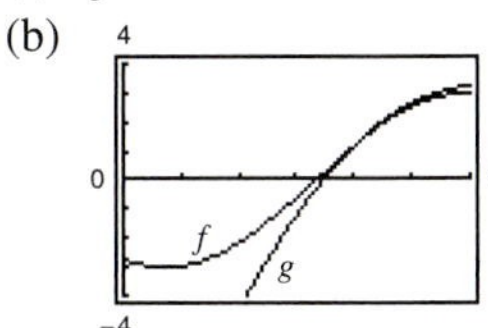

For $3.5 \le x \le 6$, the approximation appears to be good.

(c) 3.46, 8.81

3.46 is close to the zero of f in the interval $[0, 6]$.

79. True. The first equation has a smaller period than the second equation, so it will have more solutions in the interval $[0, 2\pi)$.

80. False. There is no value of x for which $\sin x = 3.4$.

81. 1 **82.** 3

83. $C = 24°$, $a \approx 54.8$, $b \approx 50.1$

84. $C = 19°$, $a \approx 15.4$, $c \approx 5.0$

85. $\sin 390° = \frac{1}{2}$, $\cos 390° = \frac{\sqrt{3}}{2}$, $\tan 390° = \frac{\sqrt{3}}{3}$

86. $\sin(600°) = -\frac{\sqrt{3}}{2}$, $\cos(600°) = -\frac{1}{2}$, $\tan(600°) = \sqrt{3}$

87. $\sin(-1845°) = -\frac{\sqrt{2}}{2}$, $\cos(-1845°) = \frac{\sqrt{2}}{2}$, $\tan(-1845°) = -1$

88. $\sin(-1410°) = \frac{1}{2}$, $\cos(-1410°) = \frac{\sqrt{3}}{2}$, $\tan(-1410°) = \frac{\sqrt{3}}{3}$

89. 1.36° **90.** 30 feet

91. Answers will vary.

Section 5.4 *(page 404)*

Vocabulary Check *(page 404)*

1. $\sin u \cos v - \cos u \sin v$
2. $\cos u \cos v - \sin u \sin v$
3. $\dfrac{\tan u + \tan v}{1 - \tan u \tan v}$
4. $\sin u \cos v + \cos u \sin v$
5. $\cos u \cos v + \sin u \sin v$
6. $\dfrac{\tan u - \tan v}{1 + \tan u \tan v}$

1. (a) $\dfrac{-\sqrt{2} - \sqrt{6}}{4}$ (b) $\dfrac{-1 + \sqrt{2}}{2}$

2. (a) $\dfrac{\sqrt{6} + \sqrt{2}}{4}$ (b) $\dfrac{\sqrt{2} - \sqrt{3}}{2}$

3. (a) $\dfrac{\sqrt{2} - \sqrt{6}}{4}$ (b) $\dfrac{\sqrt{2} + 1}{2}$

4. (a) $-\dfrac{\sqrt{6} + \sqrt{2}}{4}$ (b) $\dfrac{\sqrt{2} + 1}{2}$

5. (a) $\dfrac{1}{2}$ (b) $\dfrac{-\sqrt{3} - 1}{2}$

6. (a) $\dfrac{-2 - \sqrt{6}}{4}$ (b) $\dfrac{-\sqrt{2} - \sqrt{3}}{2}$

7. $\sin 105° = \frac{\sqrt{2}}{4}(\sqrt{3} + 1)$

$\cos 105° = \frac{\sqrt{2}}{4}(1 - \sqrt{3})$

$\tan 105° = -2 - \sqrt{3}$

8. $\sin 165° = \frac{\sqrt{2}}{4}(\sqrt{3} - 1)$

$\cos 165° = -\frac{\sqrt{2}}{4}(\sqrt{3} + 1)$

$\tan 165° = -2 + \sqrt{3}$

9. $\sin 195° = \frac{\sqrt{2}}{4}(1 - \sqrt{3})$

$\cos 195° = -\frac{\sqrt{2}}{4}(\sqrt{3} + 1)$

$\tan 195° = 2 - \sqrt{3}$

10. $\sin 255° = -\frac{\sqrt{2}}{4}(\sqrt{3} + 1)$

$\cos 255° = \frac{\sqrt{2}}{4}(1 - \sqrt{3})$

$\tan 255° = 2 + \sqrt{3}$

11. $\sin \frac{11\pi}{12} = \frac{\sqrt{2}}{4}(\sqrt{3} - 1)$

$\cos \frac{11\pi}{12} = -\frac{\sqrt{2}}{4}(\sqrt{3} + 1)$

$\tan \frac{11\pi}{12} = -2 + \sqrt{3}$

12. $\sin \frac{7\pi}{12} = \frac{\sqrt{2}}{4}(\sqrt{3} + 1)$

$\cos \frac{7\pi}{12} = \frac{\sqrt{2}}{4}(1 - \sqrt{3})$

$\tan \frac{7\pi}{12} = -2 - \sqrt{3}$

13. $\sin \frac{17\pi}{12} = -\frac{\sqrt{2}}{4}(\sqrt{3} + 1)$

$\cos \frac{17\pi}{12} = \frac{\sqrt{2}}{4}(1 - \sqrt{3})$

$\tan \frac{17\pi}{12} = 2 + \sqrt{3}$

14. $\sin\left(-\frac{\pi}{12}\right) = \frac{\sqrt{2}}{4}(1 - \sqrt{3})$

$\cos\left(-\frac{\pi}{12}\right) = \frac{\sqrt{2}}{4}(\sqrt{3} + 1)$

$\tan\left(-\frac{\pi}{12}\right) = -2 + \sqrt{3}$

50. $\sin\frac{u}{2} = \frac{\sqrt{5}}{5}$

$\cos\frac{u}{2} = \frac{2\sqrt{5}}{5}$

$\tan\frac{u}{2} = \frac{1}{2}$

51. $\sin\frac{u}{2} = \sqrt{\frac{89 - 8\sqrt{89}}{178}}$

$\cos\frac{u}{2} = -\sqrt{\frac{89 + 8\sqrt{89}}{178}}$

$\tan\frac{u}{2} = \frac{8 - \sqrt{89}}{5}$

52. $\sin\frac{u}{2} = \frac{1}{2}\sqrt{\frac{10 + 3\sqrt{10}}{5}}$

$\cos\frac{u}{2} = -\frac{1}{2}\sqrt{\frac{10 - 3\sqrt{10}}{5}}$

$\tan\frac{u}{2} = -3 - \sqrt{10}$

53. $\sin\frac{u}{2} = \frac{3\sqrt{10}}{10}$

$\cos\frac{u}{2} = -\frac{\sqrt{10}}{10}$

$\tan\frac{u}{2} = -3$

54. $\sin\frac{u}{2} = \frac{3\sqrt{14}}{14}$

$\cos\frac{u}{2} = \frac{\sqrt{70}}{14}$

$\tan\frac{u}{2} = \frac{3\sqrt{5}}{5}$

55. $|\sin 3x|$

56. $|\cos 2x|$ **57.** $-|\tan 4x|$ **58.** $-\left|\sin\left(\frac{x-1}{2}\right)\right|$

59. π

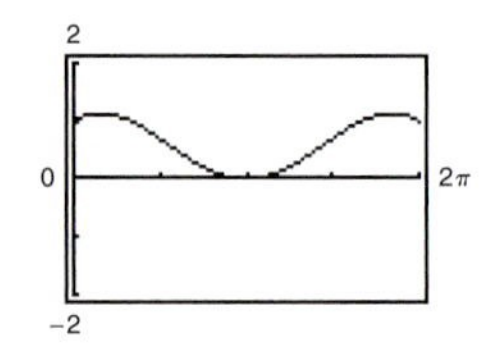

60. $0, \frac{\pi}{3}, \frac{5\pi}{3}$

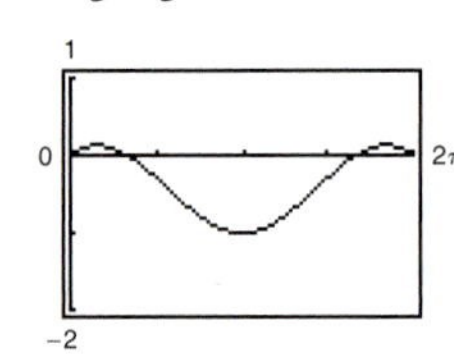

61. $\frac{\pi}{3}, \pi, \frac{5\pi}{3}$

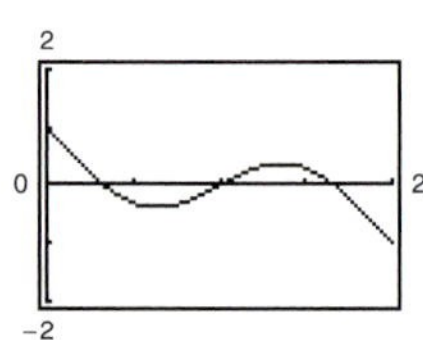

62. $0, \frac{\pi}{2}, \frac{3\pi}{2}$

63. $3\left(\sin\frac{\pi}{2} + \sin 0\right)$ **64.** $2\left(\sin\frac{7\pi}{6} + \sin\frac{\pi}{2}\right)$

65. $5(\cos 60° + \cos 90°)$ **66.** $3(\sin 60° + \sin 30°)$

67. $\frac{1}{2}(\sin 10\theta + \sin 2\theta)$ **68.** $\frac{3}{2}(\cos\alpha - \cos 5\alpha)$

69. $\frac{5}{2}(\cos 8\beta + \cos 2\beta)$ **70.** $\frac{1}{2}(\cos 2\theta + \cos 6\theta)$

71. $\frac{1}{2}(\cos 2y - \cos 2x)$ **72.** $\frac{1}{2}(\sin 2x + \sin 2y)$

73. $\frac{1}{2}(\sin 2\theta + \sin 2\pi)$

74. $\frac{1}{2}(\cos 2\pi - \cos 2\theta) = \frac{1}{2}(1 - \cos 2\theta)$ **75.** $2\cos 4\theta \sin\theta$

76. $2\sin 2\theta\cos\theta$ **77.** $2\cos 4x\cos 2x$

78. $2\sin 3x\cos 2x$ **79.** $2\cos\alpha\sin\beta$

80. $2\cos(\phi + \pi)\cos\pi = -2\cos(\phi + \pi)$

81. $-2\sin\theta\sin\frac{\pi}{2}$ **82.** $2\sin x\cos\frac{\pi}{2} = 0$

83. $\frac{\sqrt{3} + 1}{2}$ **84.** $\frac{\sqrt{3} - 1}{2}$

85. $-\sqrt{2}$ **86.** $-\sqrt{2}$

87. $0, \frac{\pi}{4}, \frac{\pi}{2}, \frac{3\pi}{4}, \pi, \frac{5\pi}{4}, \frac{3\pi}{2}, \frac{7\pi}{4}$

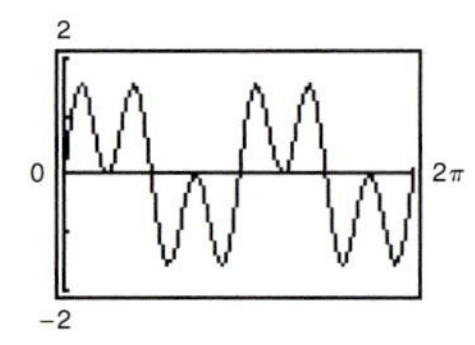

88. $0, \frac{\pi}{4}, \frac{\pi}{2}, \frac{3\pi}{4}, \pi, \frac{5\pi}{4}, \frac{3\pi}{2}, \frac{7\pi}{4}$

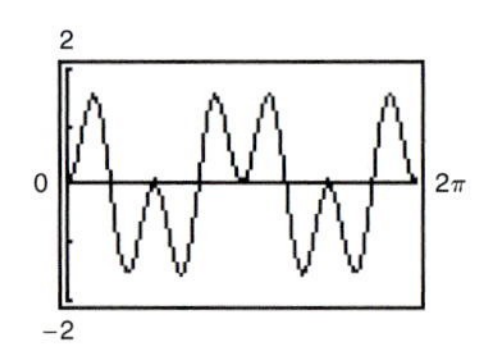

89. $\frac{\pi}{6}, \frac{5\pi}{6}$

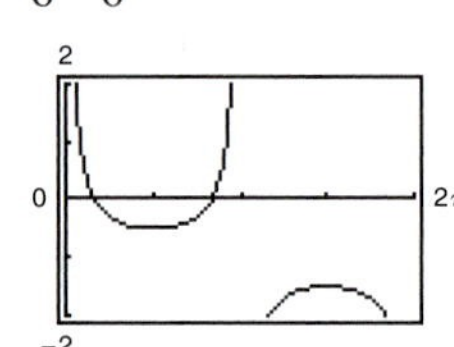

90. 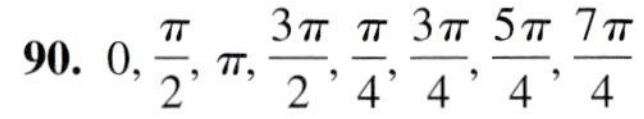
$0, \frac{\pi}{2}, \pi, \frac{3\pi}{2}, \frac{\pi}{4}, \frac{3\pi}{4}, \frac{5\pi}{4}, \frac{7\pi}{4}$

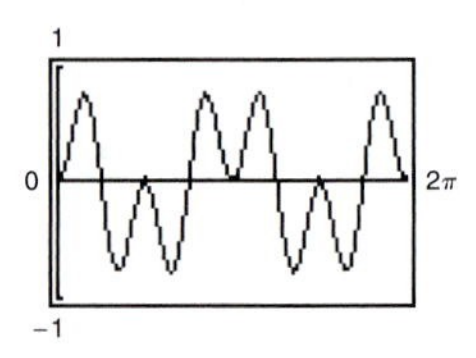

91. $\frac{25}{169}$ **92.** $\frac{144}{169}$ **93.** $\frac{4}{13}$ **94.** $\frac{36}{65}$

95–110. Answers will vary.

111. 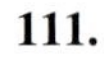

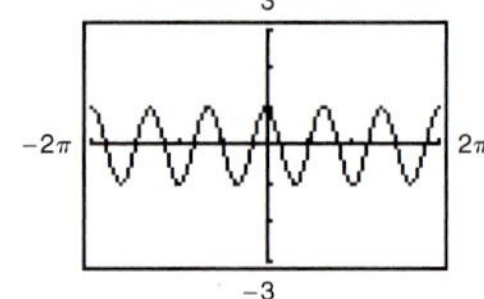

112.

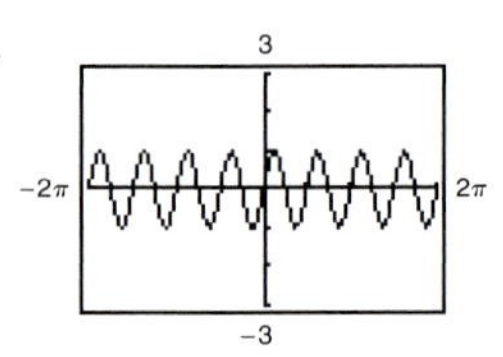

113.

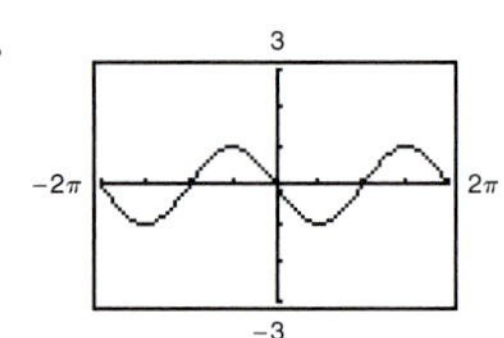

114.

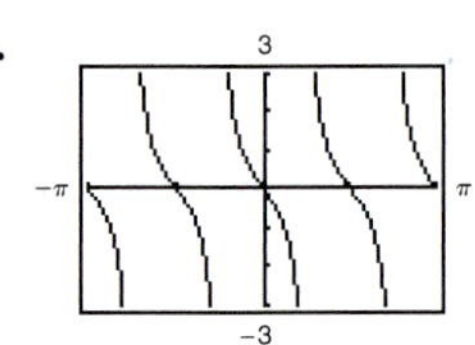

115.

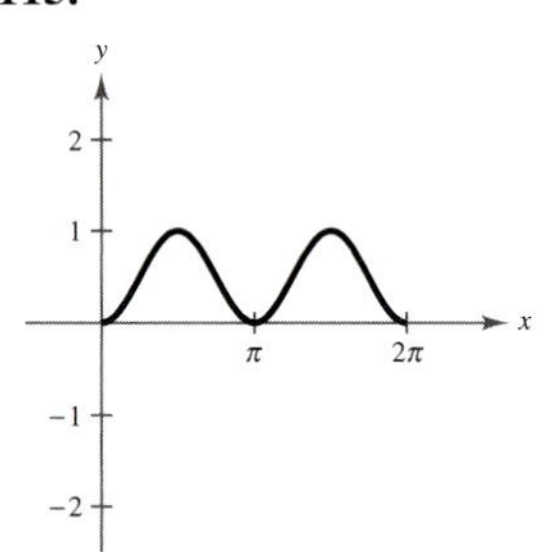

116.

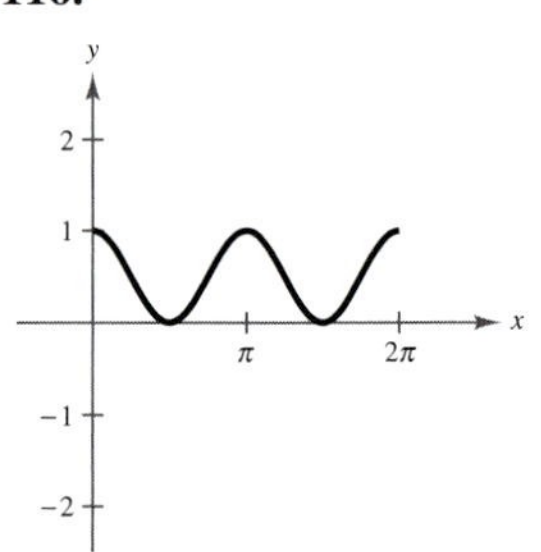

117. $2x\sqrt{1 - x^2}$ **118.** $2x^2 - 1$ **119.** 23.85°

120. (a) $A = 100 \sin \dfrac{\theta}{2} \cos \dfrac{\theta}{2}$

(b) $A = 50 \sin \theta$

The area is maximum when $\theta = \pi/2$.

121. (a) π (b) 0.4482

(c) 760 miles per hour; 3420 miles per hour

(d) $\theta = 2 \sin^{-1}\left(\dfrac{1}{M}\right)$

122. $x = 2r(1 - \cos \theta)$

123. False. For $u < 0$,

$$\begin{aligned} \sin 2u &= -\sin(-2u) \\ &= -2 \sin(-u) \cos(-u) \\ &= -2(-\sin u) \cos u \\ &= 2 \sin u \cos u. \end{aligned}$$

124. False. $\sin \dfrac{u}{2} = \sqrt{\dfrac{1 - \cos u}{2}}$ when $\dfrac{u}{2}$ is in the second quadrant.

125. (a) (b) π

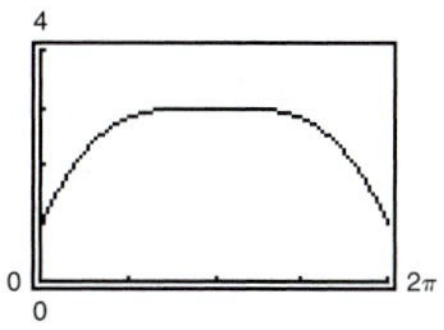

Maximum: $(\pi, 3)$

126. (a) (b) $\dfrac{\pi}{2}, \dfrac{7\pi}{6}, \dfrac{3\pi}{2}, \dfrac{11\pi}{6}$

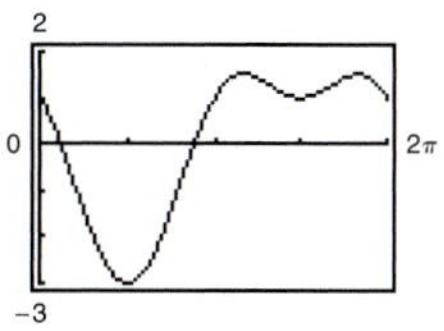

Minima: $\left(\dfrac{\pi}{2}, -3\right), \left(\dfrac{3\pi}{2}, 1\right)$

Maxima: $\left(\dfrac{7\pi}{6}, \dfrac{3}{2}\right), \left(\dfrac{11\pi}{6}, \dfrac{3}{2}\right)$

127. (a) $\frac{1}{4}(3 + \cos 4x)$ (b) $2 \cos^4 x - 2 \cos^2 x + 1$

(c) $1 - 2 \sin^2 x \cos^2 x$ (d) $1 - \frac{1}{2} \sin^2 2x$

(e) No. There is often more than one way to rewrite a trigonometric expression.

128. (a)

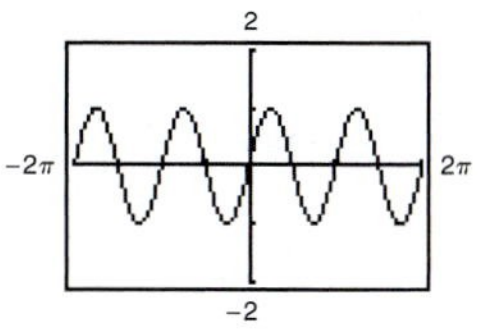

(b) $g(x) = \sin 2x$ (c) Answers will vary.

129. (a)

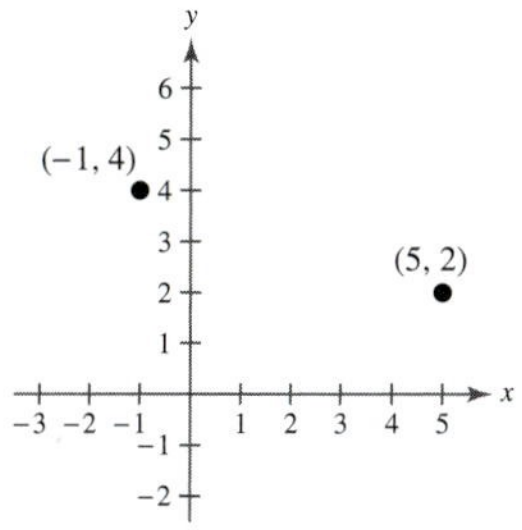

(b) Distance $= 2\sqrt{10}$ (c) Midpoint: $(2, 3)$

130. (a)

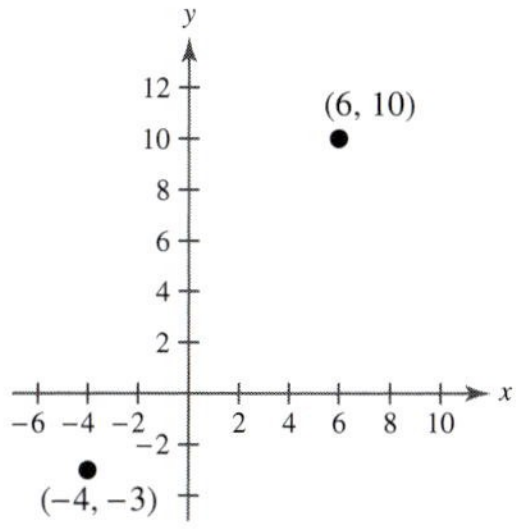

(b) Distance $= \sqrt{269}$ (c) Midpoint: $\left(1, \frac{7}{2}\right)$

131. (a)

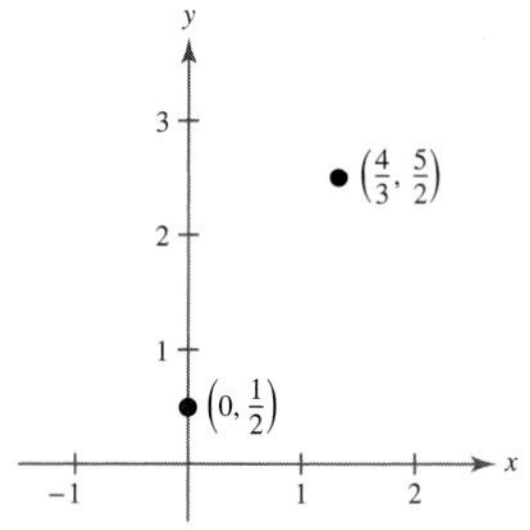

(b) Distance $= \frac{2}{3}\sqrt{13}$ (c) Midpoint: $\left(\frac{2}{3}, \frac{3}{2}\right)$

132. (a)

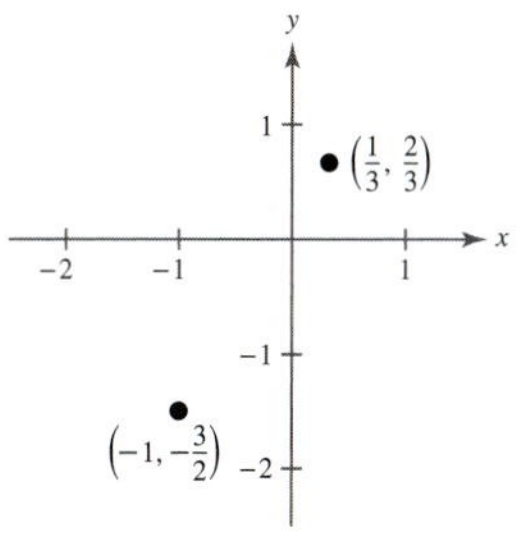

(b) Distance $= \frac{1}{6}\sqrt{233}$ (c) Midpoint: $\left(-\frac{1}{3}, -\frac{5}{12}\right)$

133. (a) Complement: 35°; supplement: 125°

(b) No complement; supplement: 18°

134. (a) No complement; supplement: 71°

(b) Complement: 12°; supplement: 102°

135. (a) Complement: $\dfrac{4\pi}{9}$; supplement: $\dfrac{17\pi}{18}$

(b) Complement: $\dfrac{\pi}{20}$; supplement: $\dfrac{11\pi}{20}$

136. (a) Complement: 0.62; supplement: 2.19

(b) No complement; supplement: 0.38

137. September: $235,000; October: $272,600

138. ≈15.7 gallons **139.** ≈127 feet

Section 5.6 *(page 425)*

Vocabulary Check *(page 425)*

1. oblique **2.** $\dfrac{b}{\sin B}$ **3.** $\dfrac{1}{2}ac \sin B$

1. $C = 105°, b \approx 28.28, c \approx 38.64$
2. $A = 35°, a \approx 11.88, b \approx 13.31$
3. $C = 120°, b \approx 4.75, c \approx 7.17$
4. $A = 35°, a \approx 36.50, b \approx 11.05$
5. $B \approx 21.55°, C \approx 122.45°, c \approx 11.49$
6. Two solutions:
$B \approx 45.79°, C \approx 74.21°, b \approx 7.45$
$B \approx 14.21°, C \approx 105.79°, b \approx 2.55$
7. $B = 60.9°, b \approx 19.32, c \approx 6.36$
8. $B = 101.1°, a \approx 1.35, b \approx 3.23$
9. $B = 42°4', a \approx 22.05, b \approx 14.88$
10. $C = 166°5', a \approx 3.30, c \approx 8.05$
11. $A \approx 10°11', C \approx 154°19', c \approx 11.03$
12. $A \approx 174°41', C \approx 2°34', a \approx 11.99$
13. $A \approx 25.57°, B \approx 9.43°, a \approx 10.53$
14. $B \approx 75.48°, C \approx 4.52°, b \approx 122.87$
15. $B \approx 18°13', C \approx 51°32', c \approx 40.06$
16. $A \approx 44°14', B \approx 50°26', b \approx 38.67$
17. $C = 83°, a \approx 0.62, b \approx 0.51$
18. $A = 48°, b \approx 2.29, c \approx 4.73$
19. $B \approx 48.74°, C \approx 21.26°, c \approx 48.23$
20. No solution **21.** No solution
22. $B \approx 36.82°, C \approx 67.18°, c \approx 32.30$
23. Two solutions:
$B \approx 72.21°, C \approx 49.79°, c \approx 10.27$
$B \approx 107.79°, C \approx 14.21°, c \approx 3.30$
24. No solution
25. (a) $b \le 5,\ b = \dfrac{5}{\sin 36°}$ (b) $5 < b < \dfrac{5}{\sin 36°}$
(c) $b > \dfrac{5}{\sin 36°}$
26. (a) $b \le 10,\ b = \dfrac{10}{\sin 60°}$ (b) $10 < b < \dfrac{10}{\sin 60°}$
(c) $b > \dfrac{10}{\sin 60°}$
27. (a) $b \le 10.8,\ b = \dfrac{10.8}{\sin 10°}$ (b) $10.8 < b < \dfrac{10.8}{\sin 10°}$
(c) $b > \dfrac{10.8}{\sin 10°}$
28. (a) $b \le 315.6,\ b = \dfrac{315.6}{\sin 88°}$ (b) $315.6 < b < \dfrac{315.6}{\sin 88°}$
(c) $b > \dfrac{315.6}{\sin 88°}$

29. 10.4 **30.** 474.9 **31.** 1675.2 **32.** 4.5
33. 3204.5 **34.** 159.3 **35.** 15.3 meters
36. (a)

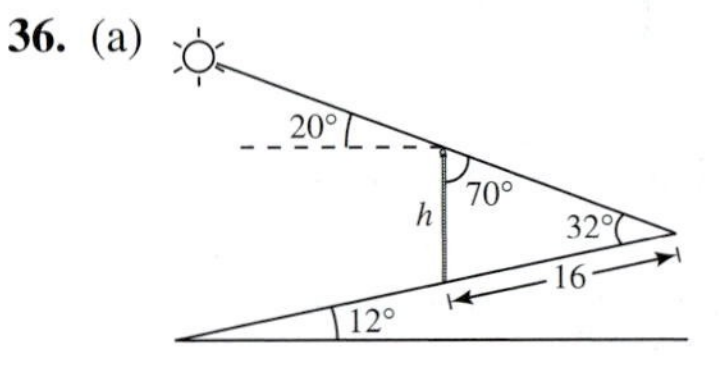

(b) $\dfrac{16}{\sin 70°} = \dfrac{h}{\sin 32°}$ (c) 9 meters
37. 16.1° **38.** 240° **39.** 77 meters
40. (a)

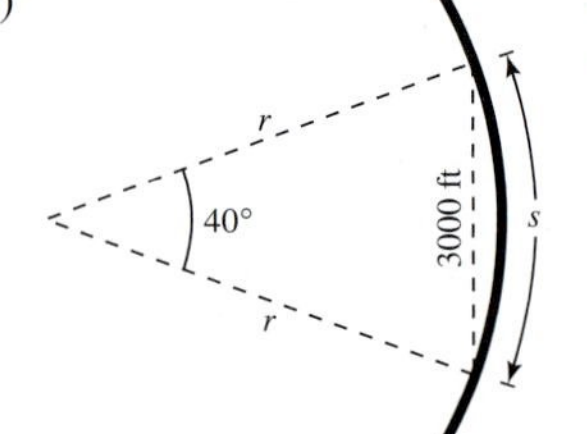

(b) 4385.71 feet
(c) 3061.80 feet
41. (a)

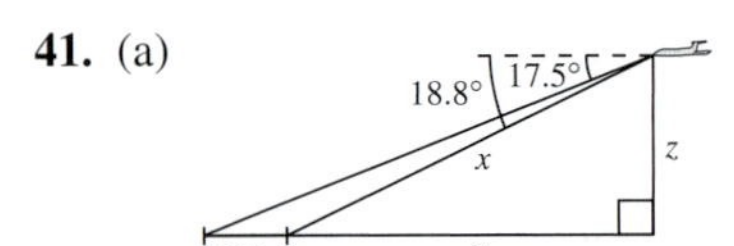

(b) 22.6 miles
(c) 21.4 miles
(d) 7.3 miles
42. From Pine Knob: 42.4 kilometers
From Colt Station: 15.5 kilometers
43. 3.2 miles
44. (a) $\alpha \approx 5.36°$ (b) $\beta = \sin^{-1}\left(\dfrac{d \sin \theta}{58.36}\right)$
(c) $d = \dfrac{58.36 \sin(84.64 - \theta)}{\sin \theta}$
(d)

θ	10°	20°	30°	40°	50°	60°
d	324.1	154.2	95.2	63.8	43.3	28.1

45. True. If an angle of a triangle is obtuse (greater than 90°), then the other two angles must be acute and therefore less than 90°. The triangle is oblique.
46. False. Two angles and one side determine a unique triangle, while two sides and one opposite angle do not necessarily determine a unique triangle.
47. (a) $\alpha = \arcsin(0.5 \sin \beta)$
(b)

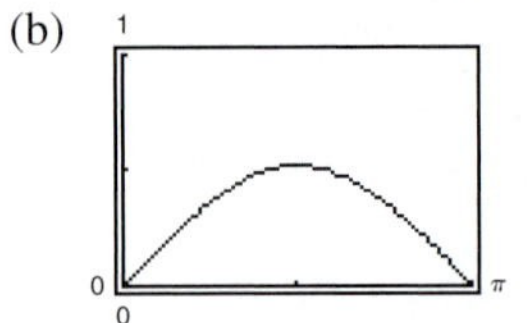

Domain: $0 < \beta < \pi$
Range: $0 < \alpha < \dfrac{\pi}{6}$
(c) $c = \dfrac{18 \sin[\pi - \beta - \arcsin(0.5 \sin \beta)]}{\sin \beta}$

(d) Domain: $0 < \beta < \pi$
Range: $9 < c < 27$

(e)

β	0.4	0.8	1.2	1.6
α	0.1960	0.3669	0.4848	0.5234
c	25.95	23.07	19.19	15.33

β	2.0	2.4	2.8
α	0.4720	0.3445	0.1683
c	12.29	10.31	9.27

As β increases from 0 to π, α increases and then decreases, and c decreases from 27 to 9.

48. (a) $A = 20\left(15 \sin \frac{3\theta}{2} - 4 \sin \frac{\theta}{2} - 6 \sin \theta\right)$

(b)

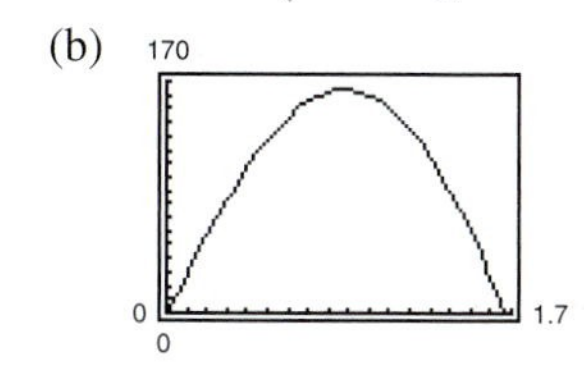

(c) Domain: $0 \le \theta \le 1.6690$
The area would increase and the domain would increase in length.

49. $\cos x$ **50.** $\tan x$ **51.** $\sin^2 x$ **52.** $\sec^2 x$

Section 5.7 *(page 432)*

Vocabulary Check *(page 432)*

1. Cosines 2. $b^2 = a^2 + c^2 - 2ac \cos B$
3. Heron's Area Formula

1. $A \approx 23.07°, B \approx 34.05°, C \approx 122.88°$
2. $A \approx 61.22°, B \approx 19.19°, C \approx 99.59°$
3. $B \approx 23.79°, C \approx 126.21°, a \approx 18.59$
4. $A \approx 53.73°, B \approx 21.27°, c \approx 11.98$
5. $A \approx 31.99°, B \approx 42.39°, C \approx 105.63°$
6. $A \approx 39.35°, B \approx 16.75°, C \approx 123.90°$
7. $A \approx 92.94°, B \approx 43.53°, C \approx 43.53°$
8. $A \approx 86.68°, B \approx 31.82°, C \approx 61.50°$
9. $B \approx 13.45°, C \approx 31.55°, a \approx 12.16$
10. $B \approx 16.53°, C \approx 108.47°, a \approx 8.64$
11. $A \approx 141°45', C \approx 27°40', b \approx 11.87$
12. $A \approx 37°6', C \approx 67°34', b \approx 9.94$
13. $A = 27°10', C = 27°10', b \approx 56.94$
14. $A \approx 157°2', B \approx 7°43', c \approx 4.21$
15. $A \approx 33.80°, B \approx 103.20°, c \approx 0.54$
16. $A \approx 23.65°, B \approx 53.35°, c \approx 0.91$

	a	b	c	d	θ	ϕ
17.	5	8	12.07	5.69	45°	135.1°
18.	25	35	52.20	31.22	60°	120°
19.	10	14	20	13.86	68.2°	111.8°
20.	40	60	63.30	80	104.5°	75.6°
21.	15	16.96	25	20	77.2°	102.8°
22.	35.18	25	50	35	68.8°	111.2°

23. 16.25 **24.** 54 **25.** 10.4 **26.** 1350.2
27. 52.11 **28.** 0.61

29. N 37.1° E, S 63.1° E

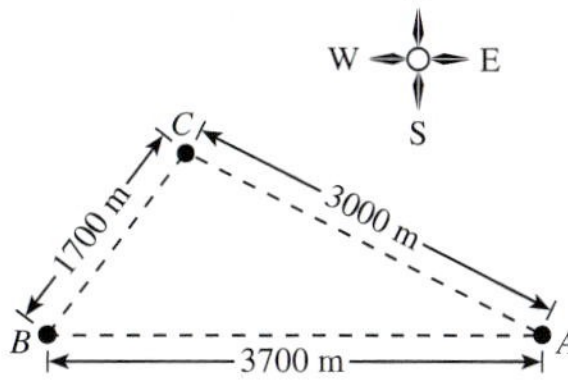

30. 1357.8 miles, 56°

31. 373.3 meters **32.** 41.2°, 52.9° **33.** 72.3°
34. 127.2° **35.** 43.3 miles
36. 131.1 feet, 118.6 feet
37. (a) N 58.4° W (b) S 81.5° W
38. (a) N 59.7° E (b) N 72.8° E **39.** 63.7 feet
40. 103.9 feet **41.** 24.2 miles **42.** 3.8 miles
43. $\overline{PQ} \approx 9.4, \overline{QS} = 5, \overline{RS} \approx 12.8$
44. (a) $x^2 - 3x \cos \theta - 46.75 = 0$
(b) $x = \frac{1}{2}\left(3 \cos \theta + \sqrt{9 \cos^2 \theta + 187}\right)$
(c)

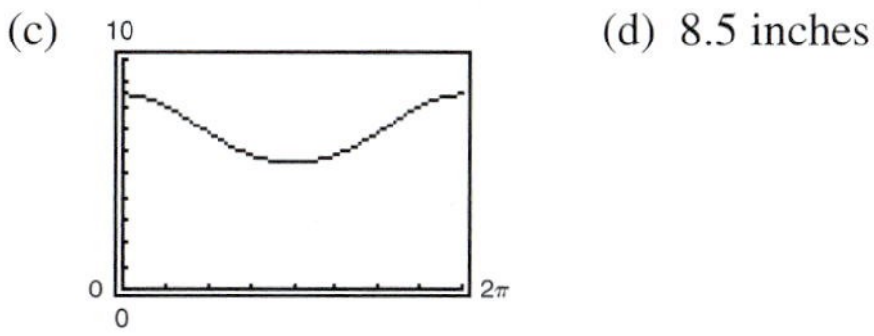

(d) 8.5 inches

45.

d (inches)	9	10	12	13	14
θ (degrees)	60.9°	69.5°	88.0°	98.2°	109.6°
s (inches)	20.88	20.28	18.99	18.28	17.48

d (inches)	15	16
θ (degrees)	122.9°	139.8°
s (inches)	16.55	15.37

46. 3.95 feet **47.** 46,837.5 square feet
48. 6577.8 square meters **49.** \$83,336.37
50. \$62,340.71
51. False. For s to be the average of the lengths of the three sides of the triangle, s would be equal to $(a + b + c)/3$.
52. False. To solve an SSA triangle, the Law of Sines is needed.
53. False. The three side lengths do not form a triangle.
54. (a) and (b) Answers will vary.
55. (a) 570.60 (b) 5910 (c) 177 **56.** 405.2 feet
57–58. Answers will vary. **59.** $-\frac{\pi}{2}$ **60.** $\frac{\pi}{2}$ **61.** $\frac{\pi}{3}$
62. $-\frac{\pi}{3}$ **63.** $-\frac{\pi}{3}$ **64.** $\frac{5\pi}{6}$ **65.** $\frac{1}{\sqrt{1-4x^2}}$
66. $\frac{\sqrt{1-9x^2}}{3x}$ **67.** $\frac{1}{x-2}$ **68.** $\frac{\sqrt{4-(x-1)^2}}{2}$

69. $\cos\theta = 1$
$\sec\theta = 1$
$\csc\theta$ is undefined.

70. $\sin\theta = -\frac{\sqrt{2}}{2}$
$\sec\theta = \sqrt{2}$
$\csc\theta = -\sqrt{2}$

71. $\tan\theta = -\frac{\sqrt{3}}{3}$
$\sec\theta = \frac{2\sqrt{3}}{3}$
$\csc\theta = -2$

72. $12 = 6\sec\theta$
$\sec\theta = 2$
$\csc\theta = \pm\frac{2\sqrt{3}}{3}$

73. $-2\sin\frac{7\pi}{12}\sin\frac{\pi}{4}$ **74.** $2\cos x\sin\left(-\frac{\pi}{2}\right)$

Review Exercises *(page 437)*

1. $\sec x$ **2.** $\csc x$ **3.** $\cos x$ **4.** $\cot x$
5. $\cot x$ **6.** $|\sec x|$

7. $\tan x = \frac{3}{4}$
$\csc x = \frac{5}{3}$
$\sec x = \frac{5}{4}$
$\cot x = \frac{4}{3}$

8. $\sin\theta = \frac{2\sqrt{13}}{13}$
$\cos\theta = \frac{3\sqrt{13}}{13}$
$\csc\theta = \frac{\sqrt{13}}{2}$
$\cot\theta = \frac{3}{2}$

9. $\cos x = \frac{\sqrt{2}}{2}$
$\tan x = -1$
$\csc x = -\sqrt{2}$
$\sec x = \sqrt{2}$
$\cot x = -1$

10. $\cos\theta = \frac{1}{9}$
$\tan\theta = 4\sqrt{5}$
$\csc\theta = \frac{9\sqrt{5}}{20}$
$\sec\theta = 9$
$\cot\theta = \frac{\sqrt{5}}{20}$

11. $\sin^2 x$ **12.** $\sec\theta\csc\theta$ **13.** 1 **14.** $\cos^2 x$
15. $\cot\theta$ **16.** $\tan u\sec u$ **17.** $\cot^2 x$ **18.** 1
19. $\sec x + 2\sin x$ **20.** $2\tan^2 x - 2\sec x\tan x + 1$
21. $-2\tan^2\theta$ **22.** $1 + \sin x$ **23–32.** Answers will vary.
33. $\frac{\pi}{3} + 2n\pi, \frac{2\pi}{3} + 2n\pi$ **34.** $\frac{\pi}{3} + 2n\pi, \frac{5\pi}{3} + 2n\pi$
35. $\frac{\pi}{6} + n\pi$ **36.** $\frac{\pi}{3} + 2n\pi, \frac{5\pi}{3} + 2n\pi$
37. $\frac{\pi}{3} + n\pi, \frac{2\pi}{3} + n\pi$ **38.** $\frac{\pi}{6} + n\pi, \frac{5\pi}{6} + n\pi$
39. $0, \frac{2\pi}{3}, \frac{4\pi}{3}$ **40.** $\frac{\pi}{6}, \frac{\pi}{2}, \frac{5\pi}{6}$ **41.** $0, \frac{\pi}{2}, \pi$ **42.** 0
43. $\frac{\pi}{8}, \frac{3\pi}{8}, \frac{9\pi}{8}, \frac{11\pi}{8}$ **44.** $0, \frac{\pi}{3}, \frac{2\pi}{3}, \pi, \frac{4\pi}{3}, \frac{5\pi}{3}$
45. $0, \frac{\pi}{8}, \frac{3\pi}{8}, \frac{5\pi}{8}, \frac{7\pi}{8}, \frac{9\pi}{8}, \frac{11\pi}{8}, \frac{13\pi}{8}, \frac{15\pi}{8}$
46. No solution **47.** $0, \pi$ **48.** $\frac{\pi}{2}, \frac{3\pi}{2}$
49. $\arctan(-4) + \pi, \arctan(-4) + 2\pi, \arctan 3, \pi + \arctan 3$
50. $\frac{3\pi}{4}, \frac{7\pi}{4}, \arctan(-5) + \pi, \arctan(-5) + 2\pi$

51. $\sin 285° = -\frac{\sqrt{2}}{4}(\sqrt{3} + 1)$
$\cos 285° = \frac{\sqrt{2}}{4}(\sqrt{3} - 1)$
$\tan 285° = -2 - \sqrt{3}$

52. $\sin 345° = \frac{\sqrt{2}}{4}(1 - \sqrt{3})$
$\cos 345° = \frac{\sqrt{2}}{4}(1 + \sqrt{3})$
$\tan 345° = -2 + \sqrt{3}$

53. $\sin\frac{25\pi}{12} = \frac{\sqrt{2}}{4}(\sqrt{3} - 1)$
$\cos\frac{25\pi}{12} = \frac{\sqrt{2}}{4}(\sqrt{3} + 1)$
$\tan\frac{25\pi}{12} = 2 - \sqrt{3}$

54. $\sin\frac{19\pi}{12} = -\frac{\sqrt{2}}{4}(\sqrt{3} + 1)$
$\cos\frac{19\pi}{12} = \frac{\sqrt{2}}{4}(\sqrt{3} - 1)$
$\tan\frac{19\pi}{12} = -2 - \sqrt{3}$

55. $\sin 15°$ **56.** $\cos 165°$ **57.** $\tan 35°$
58. $\tan(-47°)$ **59.** $-\frac{3}{52}(5 + 4\sqrt{7})$
60. $\frac{960 + 507\sqrt{7}}{1121}$ **61.** $\frac{1}{52}(5\sqrt{7} + 36)$
62. $\frac{12\sqrt{7} - 15}{52}$ **63.** $\frac{1}{52}(5\sqrt{7} - 36)$
64. $\frac{-960 + 507\sqrt{7}}{1121}$ **65–70.** Answers will vary.
71. $\frac{\pi}{4}, \frac{7\pi}{4}$ **72.** $\frac{3\pi}{2}$ **73.** $\frac{\pi}{6}, \frac{11\pi}{6}$ **74.** $0, \pi$

75. $\sin 2u = \frac{24}{25}$
$\cos 2u = -\frac{7}{25}$
$\tan 2u = -\frac{24}{7}$

76. $\sin 2u = -\frac{4}{5}$
$\cos 2u = \frac{3}{5}$
$\tan 2u = -\frac{4}{3}$

77.

78.

79. $\dfrac{1 - \cos 4x}{1 + \cos 4x}$ **80.** $\dfrac{1 + \cos 6x}{2}$

81. $\dfrac{3 - 4\cos 2x + \cos 4x}{4(1 + \cos 2x)}$ **82.** $\dfrac{1 - \cos 2x}{2}$

83. $\sin(-75°) = -\frac{1}{2}\sqrt{2 + \sqrt{3}}$
$\cos(-75°) = \frac{1}{2}\sqrt{2 - \sqrt{3}}$
$\tan(-75°) = -2 - \sqrt{3}$

84. $\sin 15° = \frac{1}{2}\sqrt{2 - \sqrt{3}}$
$\cos 15° = \frac{1}{2}\sqrt{2 + \sqrt{3}}$
$\tan 15° = 2 - \sqrt{3}$

85. $\sin \dfrac{19\pi}{12} = -\dfrac{1}{2}\sqrt{2 + \sqrt{3}}$
$\cos \dfrac{19\pi}{12} = \dfrac{1}{2}\sqrt{2 - \sqrt{3}}$
$\tan \dfrac{19\pi}{12} = -2 - \sqrt{3}$

86. $\sin\left(-\dfrac{17\pi}{12}\right) = \dfrac{1}{2}\sqrt{2 + \sqrt{3}}$
$\cos\left(-\dfrac{17\pi}{12}\right) = -\dfrac{1}{2}\sqrt{2 - \sqrt{3}}$
$\tan\left(-\dfrac{17\pi}{12}\right) = -2 - \sqrt{3}$

87. $\sin \dfrac{u}{2} = \dfrac{\sqrt{10}}{10}$
$\cos \dfrac{u}{2} = \dfrac{3\sqrt{10}}{10}$
$\tan \dfrac{u}{2} = \dfrac{1}{3}$

88. $\sin \dfrac{u}{2} = \sqrt{\dfrac{\sqrt{89} + 8}{2\sqrt{89}}} = \dfrac{\sqrt{178(8\sqrt{89} + 89)}}{178}$
$\cos \dfrac{u}{2} = -\sqrt{\dfrac{\sqrt{89} - 8}{2\sqrt{89}}} = \dfrac{\sqrt{-178(8\sqrt{89} - 89)}}{178}$
$\tan \dfrac{u}{2} = \dfrac{\sqrt{89} - 8}{5}$

89. $\sin \dfrac{u}{2} = \dfrac{3\sqrt{14}}{14}$
$\cos \dfrac{u}{2} = \dfrac{\sqrt{70}}{14}$
$\tan \dfrac{u}{2} = \dfrac{3\sqrt{5}}{5}$

90. $\sin \dfrac{u}{2} = \sqrt{\dfrac{7}{12}} = \dfrac{\sqrt{21}}{6}$
$\cos \dfrac{u}{2} = \sqrt{\dfrac{5}{12}} = \dfrac{\sqrt{15}}{6}$
$\tan \dfrac{u}{2} = \dfrac{7}{\sqrt{35}} = \dfrac{\sqrt{35}}{5}$

91. $-|\cos 5x|$ **92.** $\tan 3x$ **93.** $\dfrac{1}{2}\sin \dfrac{\pi}{3}$

94. $3[\cos(-30°) - \cos 60°]$ **95.** $\frac{1}{2}(\cos 2\theta + \cos 8\theta)$

96. $2(\sin 5\alpha + \sin \alpha)$ **97.** $2\cos 3\theta \sin \theta$

98. $2\cos \dfrac{5\theta}{2} \cos \dfrac{\theta}{2}$ **99.** $-2\sin x \sin \dfrac{\pi}{6}$

100. $2\cos x \sin \dfrac{\pi}{4}$ **101.** $\theta = 15°$ or $\dfrac{\pi}{12}$

102. (a) $V = \sin \dfrac{\theta}{2} \cos \dfrac{\theta}{2}$ cubic meters
(b) $V = \frac{1}{2}\sin \theta$ cubic meters
Volume is maximum when $\theta = \pi/2$.

103.

104. $y = \frac{1}{2}\sqrt{10}\sin\left(8t - \arctan \frac{1}{3}\right)$ **105.** $\frac{1}{2}\sqrt{10}$ feet

106. $\dfrac{4}{\pi}$ cycles per second

107. $A = 26°, a \approx 24.89, c \approx 56.23$

108. $A = 150°, a \approx 48.24, b \approx 16.75$

109. $C = 66°, a \approx 2.53, b \approx 9.11$

110. $C = 40°, a \approx 161.96\ b \approx 114.96$

111. $B = 108°, a \approx 11.76, c \approx 21.49$

112. $A = 80°, b \approx 334.95, c \approx 219.04$

113. $A \approx 20.41°, C \approx 9.59°, a \approx 20.92$

114. No solution

115. $B \approx 39.48°, C \approx 65.52°, c \approx 48.24$

116. Two solutions:
$A \approx 40.92°, C \approx 114.08°, c \approx 8.64$
$A \approx 139.08°, C \approx 15.92°, c \approx 2.60$

117. 7.9 **118.** 15.8 **119.** 33.5 **120.** 44.1

121. 31.1 meters **122.** 4.8 **123.** 31.01 feet

124. 586.4 feet

125. $A \approx 29.69°, B \approx 52.41°, C \approx 97.90°$

126. $A \approx 53.13°, B \approx 36.87°, C = 90°$

127. $A \approx 29.92°, B \approx 86.18°, C \approx 63.90°$

128. $A \approx 101.47°, B \approx 31.73°, C \approx 46.80°$

129. $A = 35°, C = 35°, b \approx 6.55$

130. $A \approx 9.90°, C \approx 20.10°, b \approx 29.09$

131. $A \approx 48.24°, B \approx 88.76°, c \approx 21.42$

132. $B \approx 35.20°, C \approx 82.80°, a \approx 17.37$

133. 615.1 meters

134.

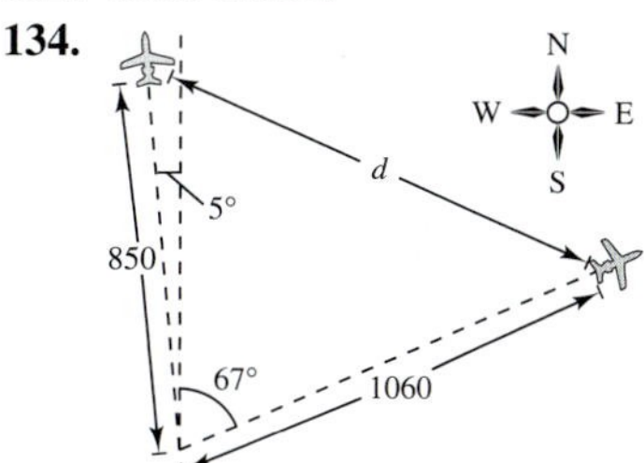

1135.5 miles

135. 9.80 **136.** 36.98 **137.** 8.36 **138.** 242.63

139. False. If $(\pi/2) < \theta < \pi$, then $\cos(\theta/2) > 0$. The sign of $\cos(\theta/2)$ depends on the quadrant in which $\theta/2$ lies.

140. False. Using the sum and difference formula, $\sin(x + y) = \sin x \cos y + \cos x \sin y$.

141. True. $4\sin(-x)\cos(-x) = 4(-\sin x)\cos x$
$= -4\sin x \cos x$
$= -2(2\sin x \cos x)$
$= -2\sin 2x$

CHAPTER 5

142. True by the product-to-sum formula

143. True. $\sin 90°$ is defined in the Law of Sines.

144. False. There may be no solution, one solution, or two solutions.

145. Reciprocal identities:

$$\sin\theta = \frac{1}{\csc\theta},\ \cos\theta = \frac{1}{\sec\theta},\ \tan\theta = \frac{1}{\cot\theta},$$

$$\csc\theta = \frac{1}{\sin\theta},\ \sec\theta = \frac{1}{\cos\theta},\ \cot\theta = \frac{1}{\tan\theta}$$

Quotient identities: $\tan\theta = \frac{\sin\theta}{\cos\theta},\ \cot\theta = \frac{\cos\theta}{\sin\theta}$

Pythagorean identities: $\sin^2\theta + \cos^2\theta = 1$,
$1 + \tan^2\theta = \sec^2\theta,\ 1 + \cot^2\theta = \csc^2\theta$

146. No. For an equation to be an identity, the equation must be true for all real numbers x. $\sin\theta = \frac{1}{2}$ has an infinite number of solutions but is not an identity.

147. $-1 \le \sin x \le 1$ for all x

148. (a) 49.91479°, 59.86118° (b) 54.735616°

149. $y_1 = y_2 + 1$ **150.** $y_1 = 1 - y_2$

151. $-1.8431, 2.1758, 3.9903, 8.8935, 9.8820$

152. $-3.1395, -2.0000, -0.4378, 2.0000$

Chapter Test *(page 441)*

1. $\sin\theta = -\frac{3\sqrt{13}}{13}$

$\cos\theta = -\frac{2\sqrt{13}}{13}$

$\csc\theta = -\frac{\sqrt{13}}{3}$

$\sec\theta = -\frac{\sqrt{13}}{2}$

$\cot\theta = \frac{2}{3}$

2. 1 **3.** 1 **4.** $\csc\theta\sec\theta$

5–10. Answers will vary.

11. $\frac{1}{16}\left(\frac{10 - 15\cos 2x + 6\cos 4x - \cos 6x}{1 + \cos 2x}\right)$ **12.** $\tan 2\theta$

13. $2(\sin 6\theta + \sin 2\theta)$ **14.** $-2\cos\frac{7\theta}{2}\sin\frac{\theta}{2}$

15. $0, \frac{3\pi}{4}, \pi, \frac{7\pi}{4}$ **16.** $\frac{\pi}{6}, \frac{\pi}{2}, \frac{5\pi}{6}, \frac{3\pi}{2}$

17. $\frac{\pi}{6}, \frac{5\pi}{6}, \frac{7\pi}{6}, \frac{11\pi}{6}$ **18.** $\frac{\pi}{6}, \frac{5\pi}{6}, \frac{3\pi}{2}$ **19.** $\frac{\sqrt{2} - \sqrt{6}}{4}$

20. $\sin 2u = \frac{4}{5}, \tan 2u = -\frac{4}{3}, \cos 2u = -\frac{3}{5}$

21. $C = 88°, b \approx 27.81, c \approx 29.98$

22. $A = 43°, b \approx 25.75, c \approx 14.45$

23. Two solutions:
$B \approx 29.12°, C \approx 126.88°, c \approx 22.03$
$B \approx 150.88°, C \approx 5.12°, c \approx 2.46$

24. No solution

25. $A \approx 39.96°, C \approx 40.04°, c \approx 15.02$

26. $A \approx 23.43°, B \approx 33.57°, c \approx 86.46$

27. Day 123 to day 223

28. 2052.5 square meters **29.** 606.3 miles; 29.1°

Problem Solving *(page 447)*

1.
(a) $\cos\theta = \pm\sqrt{1 - \sin^2\theta}$

$\tan\theta = \pm\frac{\sin\theta}{\sqrt{1 - \sin^2\theta}}$

$\cot\theta = \pm\frac{\sqrt{1 - \sin^2\theta}}{\sin\theta}$

$\sec\theta = \pm\frac{1}{\sqrt{1 - \sin^2\theta}}$

$\csc\theta = \frac{1}{\sin\theta}$

(b) $\sin\theta = \pm\sqrt{1 - \cos^2\theta}$

$\tan\theta = \pm\frac{\sqrt{1 - \cos^2\theta}}{\cos\theta}$

$\csc\theta = \pm\frac{1}{\sqrt{1 - \cos^2\theta}}$

$\sec\theta = \frac{1}{\cos\theta}$

$\cot\theta = \pm\frac{\cos\theta}{\sqrt{1 - \cos^2\theta}}$

2–3. Answers will vary.

4. (a) $p_1(t) = \sin(524\ \pi t)$
$p_2(t) = \frac{1}{2}\sin(1048\ \pi t)$
$p_3(t) = \frac{1}{3}\sin(1572\ \pi t)$
$p_5(t) = \frac{1}{5}\sin(2620\ \pi t)$
$p_6(t) = \frac{1}{6}\sin(3144\ \pi t)$

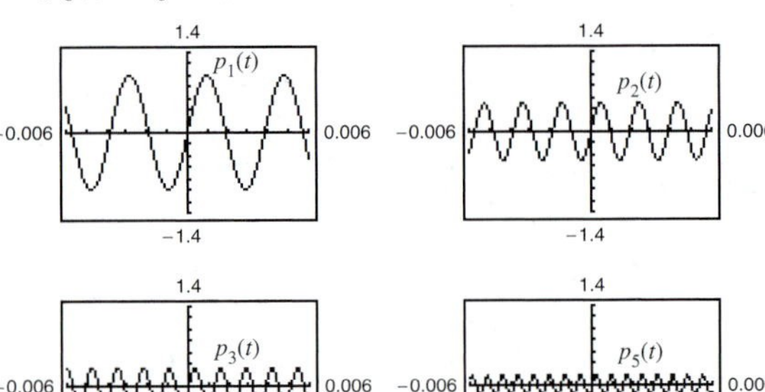

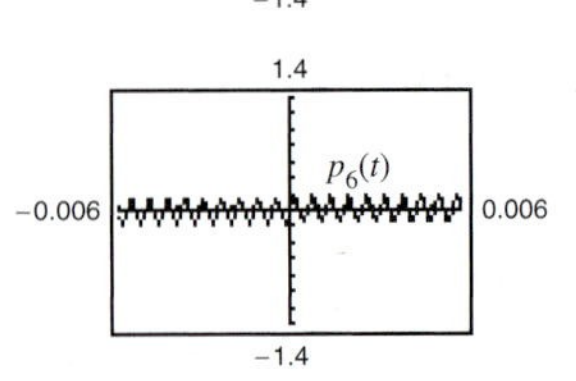

(b) p_1: $\frac{1}{262}$ p_5: $\frac{1}{1310}$
p_2: $\frac{1}{524}$ p_6: $\frac{1}{1572}$
p_3: $\frac{1}{786}$
p is periodic with period $\frac{1}{262}$.

(c) (0, 0), (0.00096, 0), (0.00191, 0)
(0.00285, 0), (0.00382, 0)

(d) Maximum: (0.00024, 1.19524)
Minimum: (0.00357, −1.19525)

5. $u + v = w$ **6.** $y = \frac{1}{64}v_0^2\sin^2\theta$

7. $\sin\frac{\theta}{2} = \sqrt{\frac{1 - \cos\theta}{2}}$

$\cos\frac{\theta}{2} = \sqrt{\frac{1 + \cos\theta}{2}}$

$\tan\frac{\theta}{2} = \frac{\sin\theta}{1 + \cos\theta}$

8. (a) $F = \dfrac{0.6\,W\cos\theta}{\sin 12^\circ}$

(b)

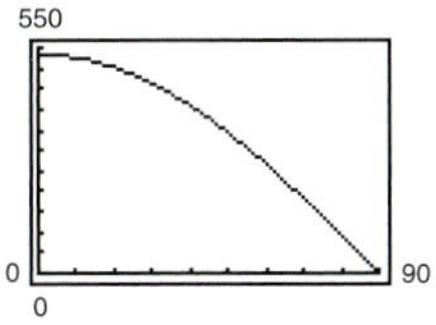

(c) Maximum: $\theta = 0^\circ$
Minimum: $\theta = 90^\circ$

9. (a)

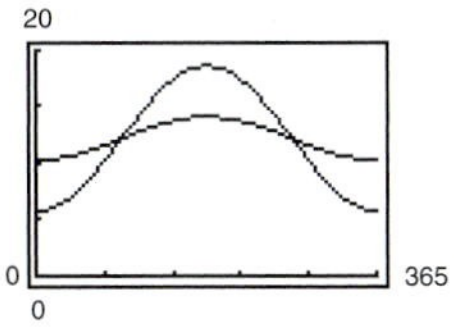

(b) $t = 91$, $t = 274$; Spring Equinox and Fall Equinox
(c) Seward; The amplitudes: 6.4 and 1.9
(d) 365.2 days

10. (a) High tides: 6:12 A.M., 6:36 P.M.
Low tides: 12:00 A.M., 12:24 P.M.
(b) The water depth never falls below 7 feet.
(c)

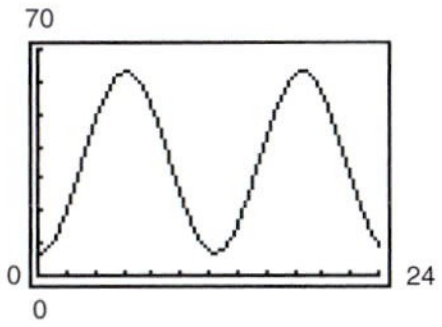

11. (a) $\dfrac{\pi}{6} \le x \le \dfrac{5\pi}{6}$ (b) $\dfrac{2\pi}{3} \le x \le \dfrac{4\pi}{3}$

(c) $\dfrac{\pi}{2} < x < \pi, \dfrac{3\pi}{2} < x < 2\pi$

(d) $0 \le x \le \dfrac{\pi}{4}, \dfrac{5\pi}{4} \le x \le 2\pi$

12. (a) $\sin(u + v + w)$
$= \sin u \cos v \cos w - \sin u \sin v \sin w + \cos u \sin v \cos w + \cos u \cos v \sin w$

(b) $\tan(u + v + w)$
$= \dfrac{\tan u + \tan v + \tan w - \tan u \tan v \tan w}{1 - \tan u \tan v - \tan u \tan w - \tan v \tan w}$

13. (a) $\cos 3\theta = \cos\theta - 4\sin^2\theta\cos\theta$
(b) $\cos 4\theta = \cos^4\theta - 6\sin^2\theta\cos^2\theta + \sin^4\theta$

14. (a)

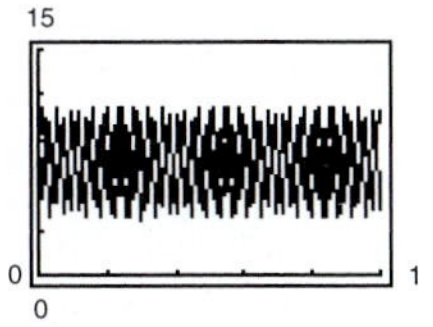

(b) 233.3 times per second

15. 2.01 feet **16.** S 22.09° E; 1025.88 yards

Chapter 6

Section 6.1 *(page 454)*

Vocabulary Check *(page 454)*

1. inclination **2.** $\tan\theta$

3. $\left|\dfrac{m_2 - m_1}{1 + m_1 m_2}\right|$ **4.** $\dfrac{|Ax_1 + By_1 + C|}{\sqrt{A^2 + B^2}}$

1. $\dfrac{\sqrt{3}}{3}$ **2.** 1 **3.** -1 **4.** $-\sqrt{3}$ **5.** $\sqrt{3}$

6. $-\dfrac{\sqrt{3}}{3}$ **7.** 3.2236 **8.** -0.2677

9. $\dfrac{3\pi}{4}$ radians, 135° **10.** 2.0344 radians, 116.6°

11. $\dfrac{\pi}{4}$ radian, 45° **12.** 1.1071 radians, 63.4°

13. 0.6435 radian, 36.9° **14.** 1.9513 radians, 111.8°
15. 1.0517 radians, 60.3° **16.** 0.6023 radian, 34.5°
17. 2.1112 radians, 121.0° **18.** 2.0344 radians, 116.6°
19. 1.2490 radians, 71.6° **20.** 2.4669 radians, 141.3°

21. 2.1112 radians, 121.0° **22.** $\dfrac{\pi}{4}$ radian, 45°

23. 1.1071 radians, 63.4° **24.** $\dfrac{\pi}{4}$ radian, 45°

25. 0.1974 radian, 11.3° **26.** 1.1071 radians, 63.4°
27. 1.4289 radians, 81.9° **28.** 1.4109 radians, 80.8°
29. 0.9273 radian, 53.1° **30.** 1.0808 radians, 61.9°
31. 0.8187 radian, 46.9° **32.** 1.0240 radians, 58.7°

33. $(2, 1) \leftrightarrow (4, 4)$: slope $= \frac{3}{2}$
$(4, 4) \leftrightarrow (6, 2)$: slope $= -1$
$(6, 2) \leftrightarrow (2, 1)$: slope $= \frac{1}{4}$
(2, 1): 42.3°; (4, 4): 78.7°; (6, 2): 59.0°

34. $(-3, 2) \leftrightarrow (1, 3)$: slope $= \frac{1}{4}$
$(1, 3) \leftrightarrow (2, 0)$: slope $= -3$
$(2, 0) \leftrightarrow (-3, 2)$: slope $= -\frac{2}{5}$
$(-3, 2)$: 35.8°; (1, 3): 94.4°; (2, 0): 49.8°

35. $(-4, -1) \leftrightarrow (3, 2)$: slope $= \frac{3}{7}$
$(3, 2) \leftrightarrow (1, 0)$: slope $= 1$
$(1, 0) \leftrightarrow (-4, -1)$: slope $= \frac{1}{5}$
$(-4, -1)$: 11.9°; (3, 2): 21.8°; (1, 0): 146.3°

36. $(-2, 2) \leftrightarrow (-3, 4)$: slope $= -2$
$(-3, 4) \leftrightarrow (2, 1)$: slope $= -\frac{3}{5}$
$(2, 1) \leftrightarrow (-2, 2)$: slope $= -\frac{1}{4}$
$(-3, 4)$: 32.5°; (2, 1): 16.9°; $(-2, 2)$: 130.6°

37. 0 **38.** $\dfrac{4\sqrt{5}}{5} \approx 1.7889$ **39.** $\dfrac{7}{5}$

40. $\dfrac{5\sqrt{2}}{2} \approx 3.5355$ **41.** 7 **42.** 4

43. $\frac{8\sqrt{37}}{37} \approx 1.3152$ **44.** $9\sqrt{2} \approx 12.7279$

45. (a)

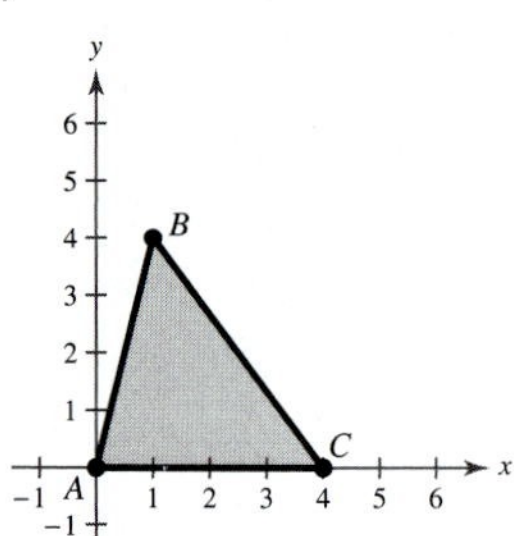

(b) 4 (c) 8

46. (a)

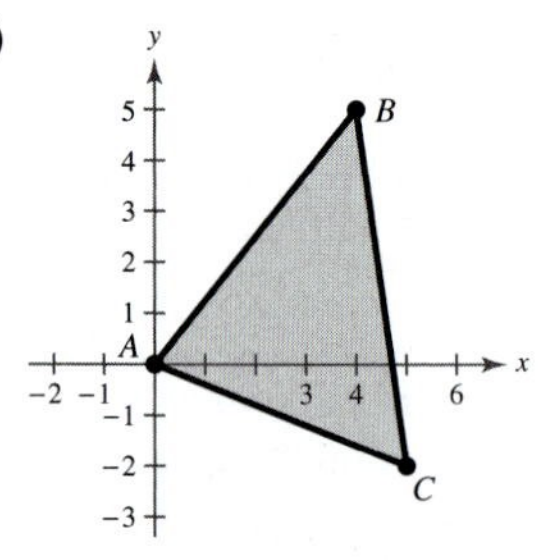

(b) $\frac{33\sqrt{29}}{29}$ (c) $\frac{33}{2}$

47. (a)

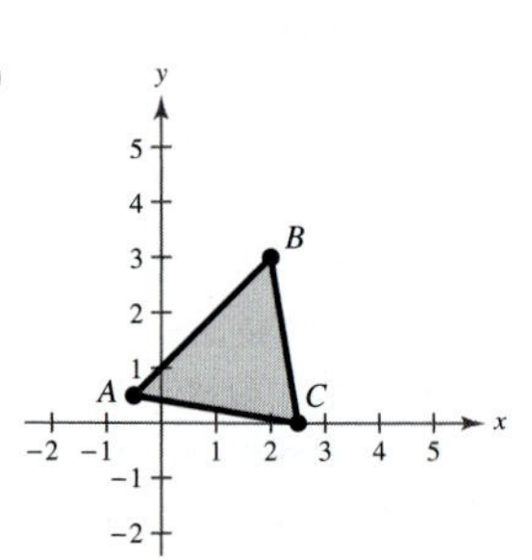

(b) $\frac{35\sqrt{37}}{74}$ (c) $\frac{35}{8}$

48. (a)

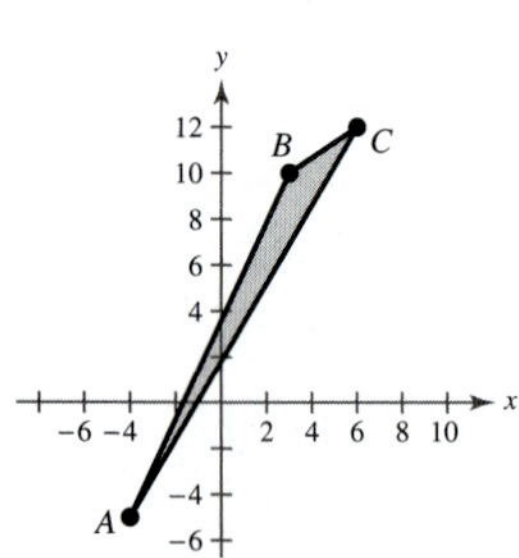

(b) $\frac{31\sqrt{389}}{389}$ (c) $\frac{31}{2}$

49. $2\sqrt{2}$ **50.** $\frac{9}{5}$ **51.** 0.1003, 1054 feet

52. 0.2027, 1049 feet **53.** 31.0°

54. (a)

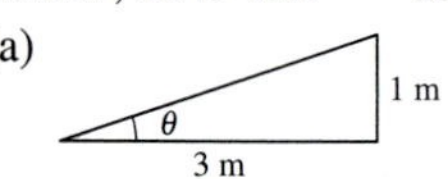

(b) 18.4°

(c) 15.8 meters

55. $\alpha \approx 33.69°$; $\beta \approx 56.31°$

56. (a) 0.6167 radian, 35.3° (b) 518.5 feet

(c) $y = 0.709x$

(d)

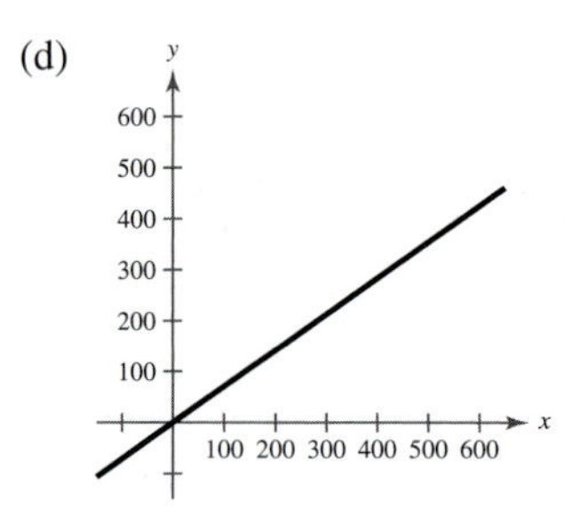

57. True. The inclination of a line is related to its slope by $m = \tan\theta$. If the angle is greater than $\pi/2$ but less than π, then the angle is in the second quadrant, where the tangent function is negative.

58. False. Substitute $\tan\theta_1$ and $\tan\theta_2$ for m_1 and m_2 in the formula for the angle between two lines.

59. (a) $d = \frac{4}{\sqrt{m^2 + 1}}$

(b)

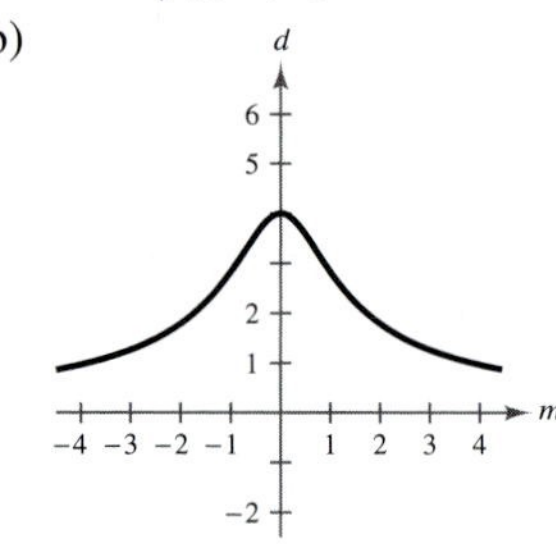

(c) $m = 0$

(d) The graph has a horizontal asymptote of $d = 0$. As the slope becomes larger, the distance between the origin and the line, $y = mx + 4$, becomes smaller and approaches 0.

60. (a) $d = \frac{3|m + 1|}{\sqrt{m^2 + 1}}$

(b)

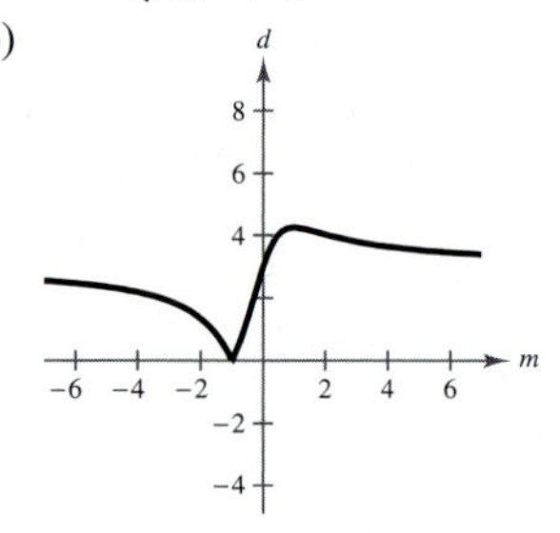

(c) $m = 1$ (d) Yes. $m = -1$

(e) $d = 3$. As the line approaches the vertical, the distance approaches 3.

61. x-intercept: $(7, 0)$
y-intercept: $(0, 49)$

62. x-intercept: $(-9, 0)$
y-intercept: $(0, 81)$

63. x-intercepts: $(5 \pm \sqrt{5}, 0)$
y-intercept: $(0, 20)$

64. No x-intercepts
y-intercept: $(0, 133)$

65. x-intercepts: $\left(\frac{7 \pm \sqrt{53}}{2}, 0\right)$
y-intercept: $(0, -1)$

66. x-intercepts: $(-11, 0), (2, 0)$
y-intercept: $(0, -22)$

67. $f(x) = 3\left(x + \frac{1}{3}\right)^2 - \frac{49}{3}$
Vertex: $\left(-\frac{1}{3}, -\frac{49}{3}\right)$

68. $f(x) = 2\left(x - \frac{1}{4}\right)^2 - \frac{169}{8}$
Vertex: $\left(\frac{1}{4}, -\frac{169}{8}\right)$

69. $f(x) = 5\left(x + \frac{17}{5}\right)^2 - \frac{324}{5}$
Vertex: $\left(-\frac{17}{5}, -\frac{324}{5}\right)$

70. $f(x) = -(x + 4)^2 + 1$
Vertex: $(-4, 1)$

71. $f(x) = 6\left(x - \frac{1}{12}\right)^2 - \frac{289}{24}$
Vertex: $\left(\frac{1}{12}, -\frac{289}{24}\right)$

72. $f(x) = -8\left(x + \frac{17}{8}\right)^2 + \frac{121}{8}$
Vertex: $\left(-\frac{17}{8}, \frac{121}{8}\right)$

73.

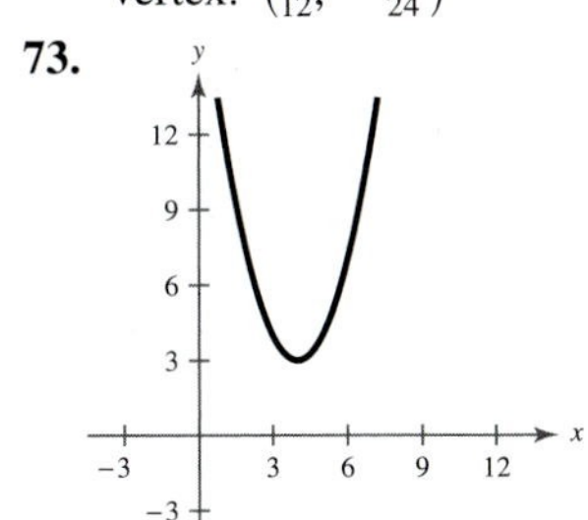

74.

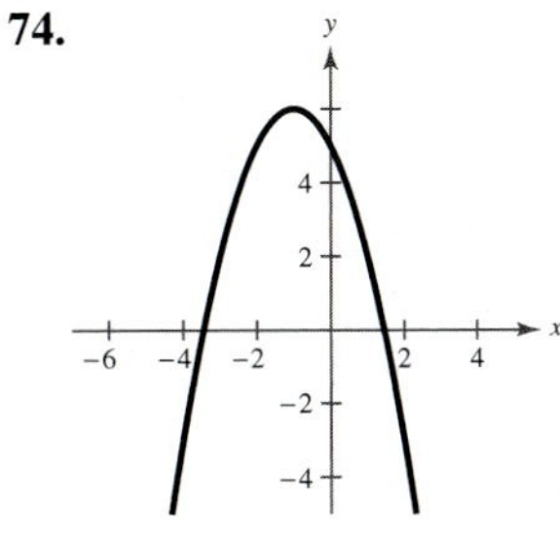

75.

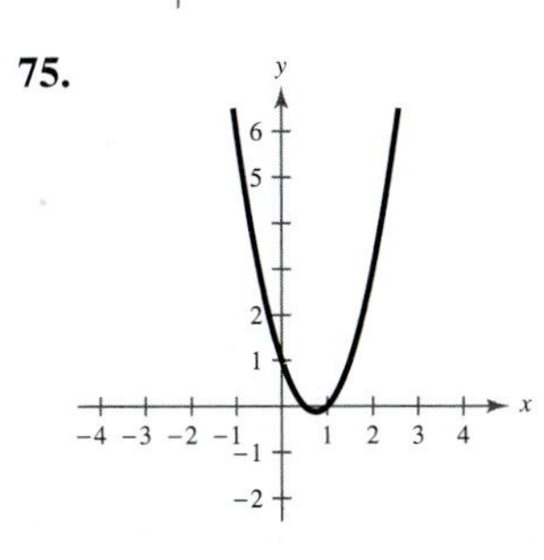

76.

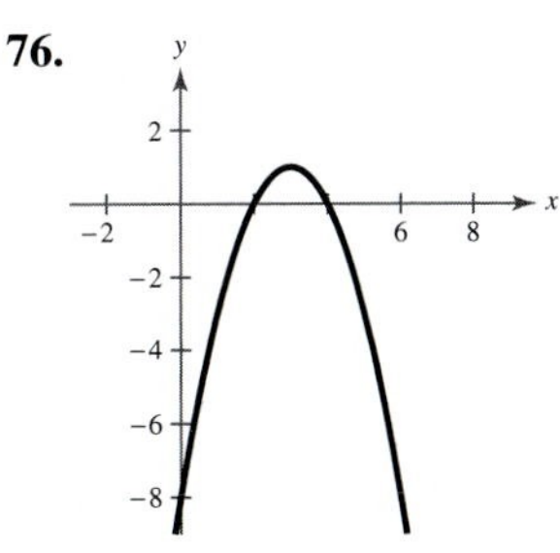

Section 6.2 *(page 462)*

Vocabulary Check *(page 462)*

1. conic **2.** locus **3.** parabola; directrix; focus
4. axis **5.** vertex **6.** focal chord **7.** tangent

1. A circle is formed when a plane intersects the top or bottom half of a double-napped cone and is perpendicular to the axis of the cone.

2. An ellipse is formed when a plane intersects only the top or bottom half of a double-napped cone but is not parallel or perpendicular to the axis of the cone, is not parallel to the side of the cone, and does not intersect the vertex.

3. A parabola is formed when a plane intersects the top or bottom half of a double-napped cone, is parallel to the side of the cone, and does not intersect the vertex.

4. A hyperbola is formed when a plane intersects both halves of a double-napped cone, is parallel to the axis of the cone, and does not intersect the vertex.

5. e **6.** b **7.** d **8.** f **9.** a **10.** c

11. Vertex: $(0, 0)$
Focus: $\left(0, \frac{1}{2}\right)$
Directrix: $y = -\frac{1}{2}$

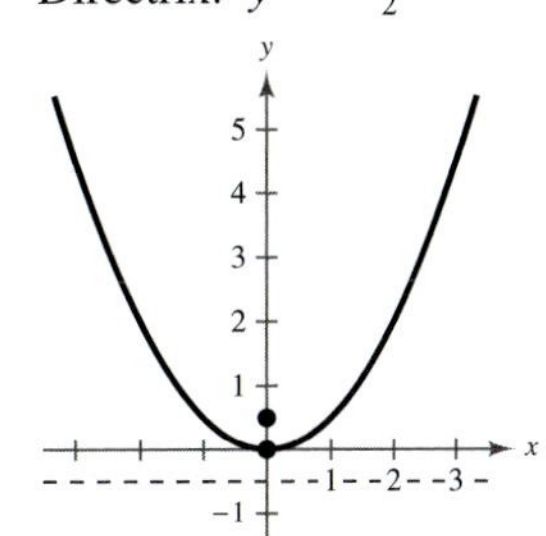

12. Vertex: $(0, 0)$
Focus: $\left(0, -\frac{1}{8}\right)$
Directrix: $y = \frac{1}{8}$

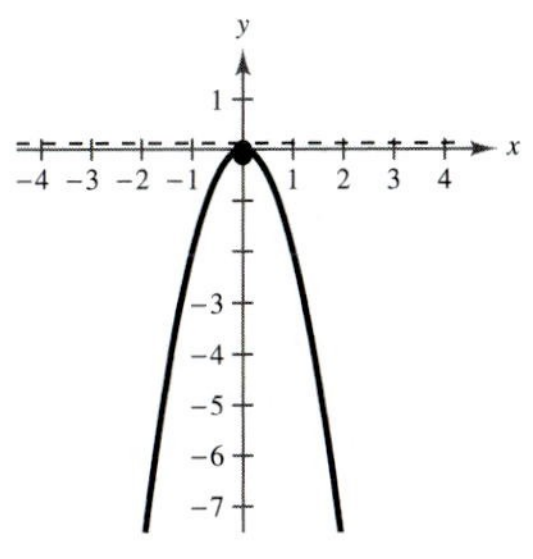

13. Vertex: $(0, 0)$
Focus: $\left(-\frac{3}{2}, 0\right)$
Directrix: $x = \frac{3}{2}$

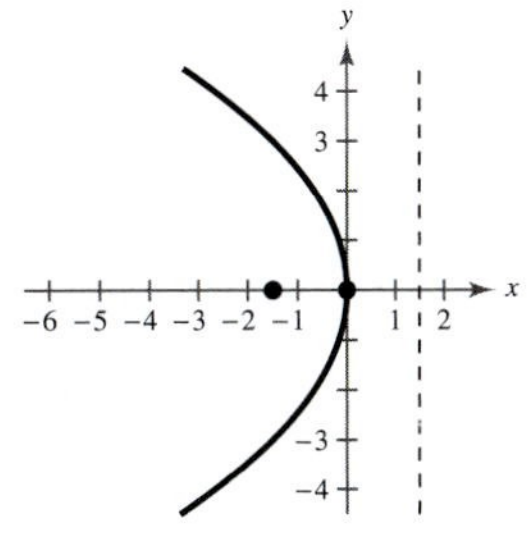

14. Vertex: $(0, 0)$
Focus: $\left(\frac{3}{4}, 0\right)$
Directrix: $x = -\frac{3}{4}$

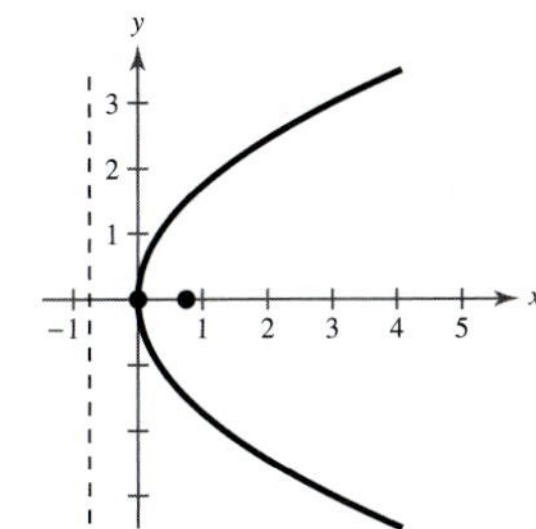

15. Vertex: $(0, 0)$
Focus: $\left(0, -\frac{3}{2}\right)$
Directrix: $y = \frac{3}{2}$

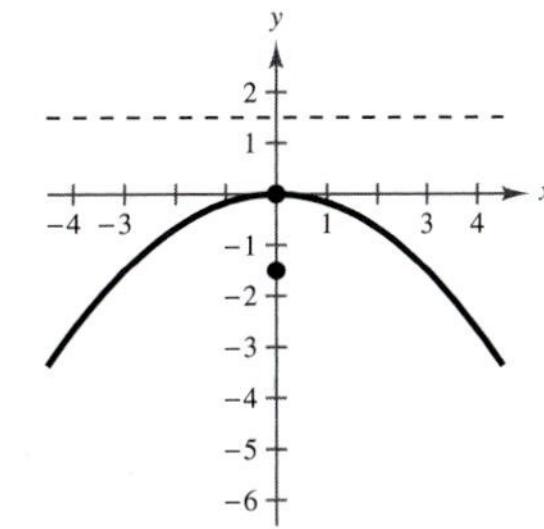

16. Vertex: $(0, 0)$
Focus: $\left(-\frac{1}{4}, 0\right)$
Directrix: $x = \frac{1}{4}$

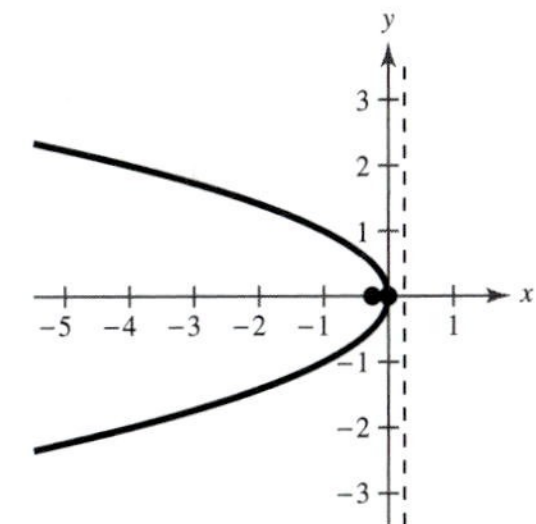

17. Vertex: $(1, -2)$
Focus: $(1, -4)$
Directrix: $y = 0$

18. Vertex: $(-5, 1)$
Focus: $\left(-\frac{21}{4}, 1\right)$
Directrix: $x = -\frac{19}{4}$

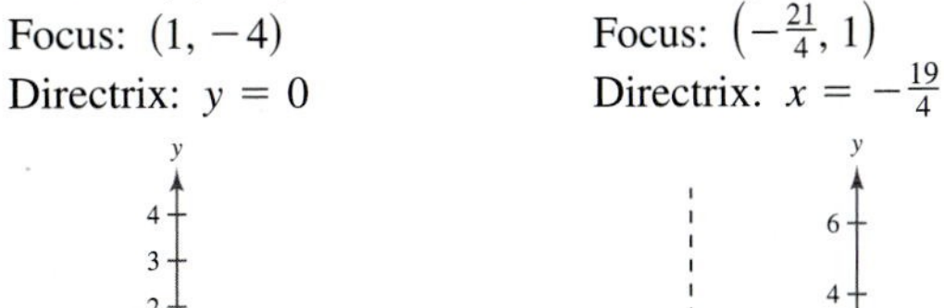

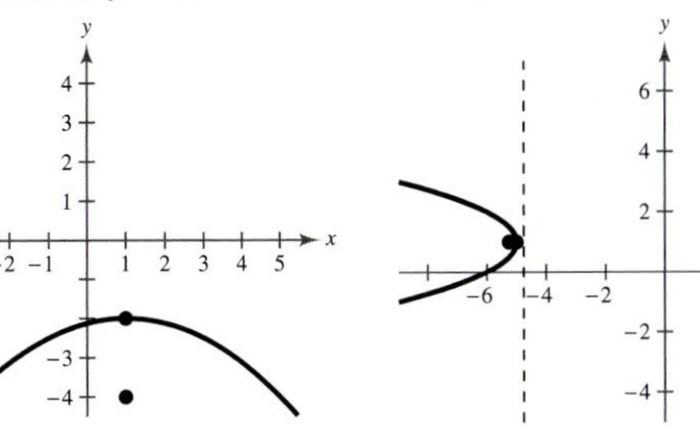

19. Vertex: $\left(-\frac{3}{2}, 2\right)$
Focus: $\left(-\frac{3}{2}, 3\right)$
Directrix: $y = 1$

20. Vertex: $\left(-\frac{1}{2}, 1\right)$
Focus: $\left(-\frac{1}{2}, 2\right)$
Directrix: $y = 0$

21. Vertex: $(1, 1)$
Focus: $(1, 2)$
Directrix: $y = 0$

22. Vertex: $(8, -1)$
Focus: $(9, -1)$
Directrix: $x = 7$

23. Vertex: $(-2, -3)$
Focus: $(-4, -3)$
Directrix: $x = 0$

24. Vertex: $(-1, 2)$
Focus: $(0, 2)$
Directrix: $x = -2$

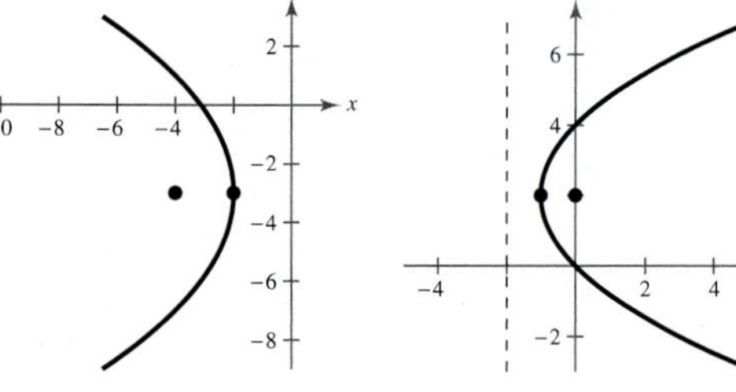

25. Vertex: $(-2, 1)$
Focus: $\left(-2, -\frac{1}{2}\right)$
Directrix: $y = \frac{5}{2}$

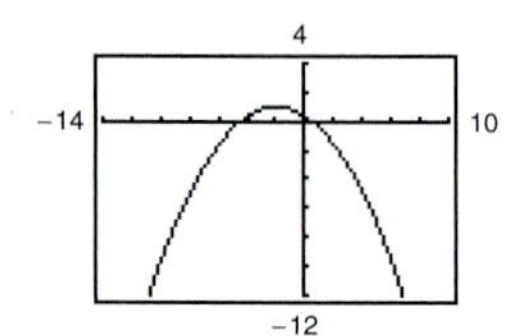

26. Vertex: $(1, -1)$
Focus: $(1, -3)$
Directrix: $y = 1$

27. Vertex: $\left(\frac{1}{4}, -\frac{1}{2}\right)$
Focus: $\left(0, -\frac{1}{2}\right)$
Directrix: $x = \frac{1}{2}$

28. Vertex: $(-1, 0)$
Focus: $(0, 0)$
Directrix: $x = -2$

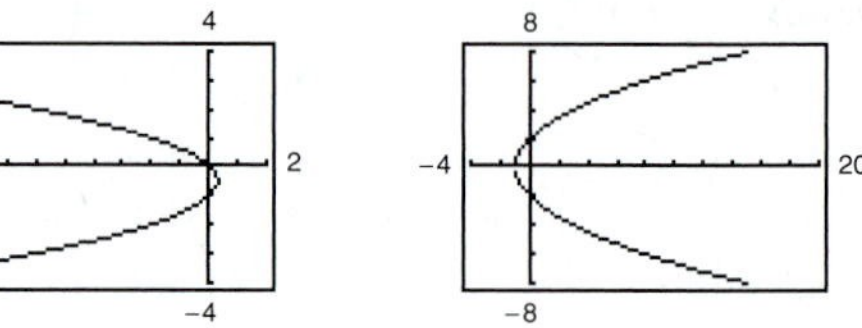

29. $x^2 = \frac{3}{2}y$ **30.** $y^2 = -18x$ **31.** $x^2 = -6y$
32. $y^2 = 10x$ **33.** $y^2 = -8x$ **34.** $x^2 = -8y$
35. $x^2 = 4y$ **36.** $x^2 = -12y$ **37.** $y^2 = -8x$
38. $y^2 = 12x$ **39.** $y^2 = 9x$ **40.** $x^2 = -3y$
41. $(x - 3)^2 = -(y - 1)$ **42.** $(y - 3)^2 = -2(x - 5)$
43. $y^2 = 4(x + 4)$ **44.** $(x - 3)^2 = 3(y + 3)$
45. $(y - 2)^2 = -8(x - 5)$ **46.** $(x + 1)^2 = -8(y - 2)$
47. $x^2 = 8(y - 4)$ **48.** $(y - 1)^2 = -12(x + 2)$
49. $(y - 2)^2 = 8x$ **50.** $x^2 = -16(y - 4)$
51. $y = \sqrt{6(x + 1)} + 3$ **52.** $y = -\sqrt{2(x - 4)} - 1$

53. $(2, 4)$ **54.** $(6, -3)$

55. $4x - y - 8 = 0; (2, 0)$ **56.** $6x + 2y + 9 = 0; \left(-\frac{3}{2}, 0\right)$
57. $4x - y + 2 = 0; \left(-\frac{1}{2}, 0\right)$ **58.** $8x + y - 8 = 0; (1, 0)$

59. $x = 106$ units

60.

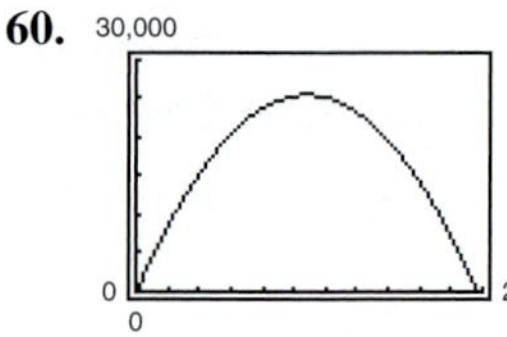

$x = 135$ units

61. $y = \frac{1}{18}x^2$

62. (a)

(b)

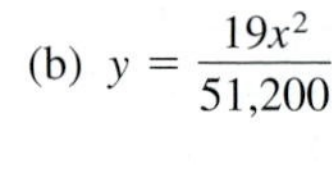

$y = \frac{19x^2}{51,200}$

(c)

Distance, x	0	250	400	500	1000
Height, y	0	23.19	59.38	92.77	371.09

63. (a) $y = -\frac{1}{640}x^2$ (b) 8 feet **64.** $y^2 = 640x$

65. (a) $17{,}500\sqrt{2}$ miles per hour $\approx 24{,}750$ miles per hour
(b) $x^2 = -16{,}400(y - 4100)$

66. (a)

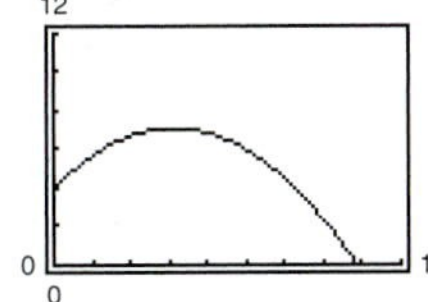

(b) Highest point: $(6.25, 7.125)$
Range: 15.69 feet

67. (a) $x^2 = -64(y - 75)$ (b) 69.3 feet

68. 34,294.6 feet

69. False. If the graph crossed the directrix, there would exist points closer to the directrix than the focus.

70. True. If the axis (line connecting the vertex and focus) is horizontal, then the direction must be vertical.

71. (a)

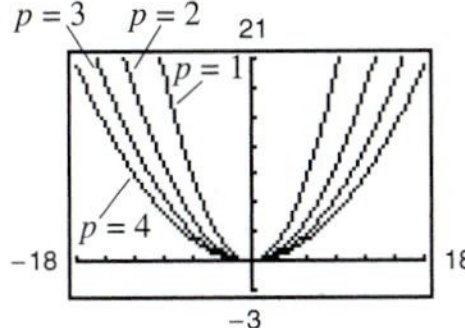

As p increases, the graph becomes wider.
(b) $(0, 1)$, $(0, 2)$, $(0, 3)$, $(0, 4)$
(c) 4, 8, 12, 16; $4|p|$
(d) Easy way to determine two additional points on the graph

72. (a) $\dfrac{64\sqrt{2}}{3} \approx 30.17$
(b) As p approaches zero, the parabola becomes narrower and narrower, thus the area becomes smaller and smaller.

73. $m = \dfrac{x_1}{2p}$ **74.** Answers will vary. **75.** $\pm 1, \pm 2, \pm 4$

76. $\pm\frac{1}{2}, \pm 1, \pm 2, \pm\frac{5}{2}, \pm 5, \pm 10$ **77.** $\pm\frac{1}{2}, \pm 1, \pm 2, \pm 4, \pm 8, \pm 16$

78. $\pm\frac{1}{3}, \pm\frac{2}{3}, \pm 1, \pm 2, \pm\frac{11}{3}, \pm\frac{22}{3}, \pm 11, \pm 22$

79. $f(x) = x^3 - 7x^2 + 17x - 15$ **80.** $\frac{3}{2}, \pm 5i$

81. $\frac{1}{2}, -\frac{5}{3}, \pm 2$

82.

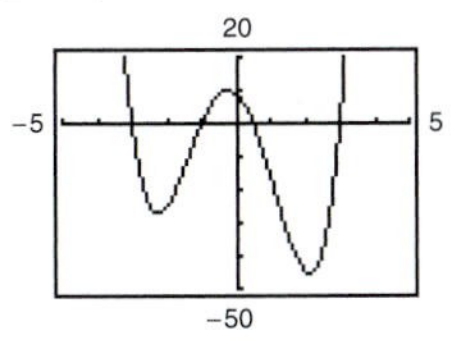

$\frac{1}{2}, -1, \pm 3$

83. $B \approx 23.67°$, $C \approx 121.33°$, $c \approx 14.89$

84. $A \approx 96.37°$, $C \approx 29.63°$, $a \approx 22.11$

85. $C = 89°$, $a \approx 1.93$, $b \approx 2.33$

86. $A = 50°$, $b \approx 10.87$, $c \approx 24.07$

87. $A \approx 16.39°$, $B \approx 23.77°$, $C \approx 139.84°$

88. $A \approx 43.53°$, $B \approx 19.42°$, $C \approx 117.05°$

89. $B \approx 24.62°$, $C \approx 90.38°$, $a \approx 10.88$

90. $A \approx 41.85°$, $C \approx 67.15°$, $b \approx 29.76$

Section 6.3 *(page 472)*

Vocabulary Check *(page 472)*

1. ellipse; foci **2.** major axis; center
3. minor axis **4.** eccentricity

1. b **2.** c **3.** d **4.** f **5.** a **6.** e

7. Ellipse
Center: $(0, 0)$
Vertices: $(\pm 5, 0)$
Foci: $(\pm 3, 0)$
Eccentricity: $\dfrac{3}{5}$

8. Ellipse
Center: $(0, 0)$
Vertices: $(0, \pm 12)$
Foci: $(0, \pm 3\sqrt{7})$
Eccentricity: $\dfrac{\sqrt{7}}{4}$

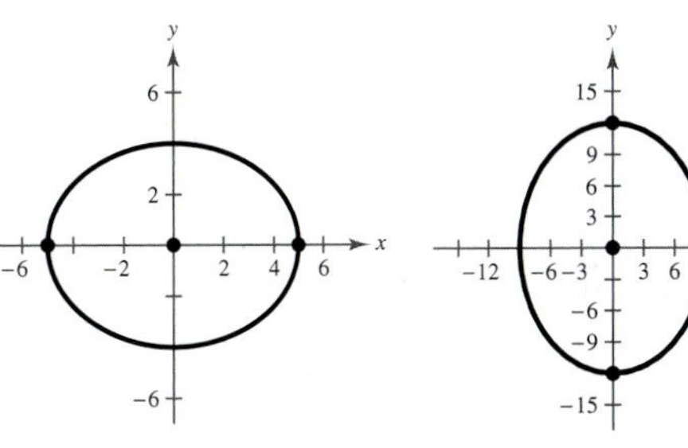

9. Circle
Center: $(0, 0)$
Radius: 5

10. Circle
Center: $(0, 0)$
Radius: 3

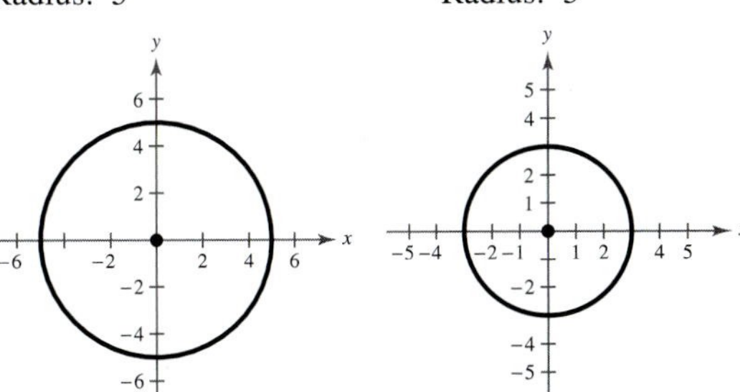

11. Ellipse
Center: $(0, 0)$
Vertices: $(0, \pm 3)$
Foci: $(0, \pm 2)$
Eccentricity: $\frac{2}{3}$

12. Ellipse
Center: $(0, 0)$
Vertices: $(\pm 8, 0)$
Foci: $(\pm 6, 0)$
Eccentricity: $\frac{3}{4}$

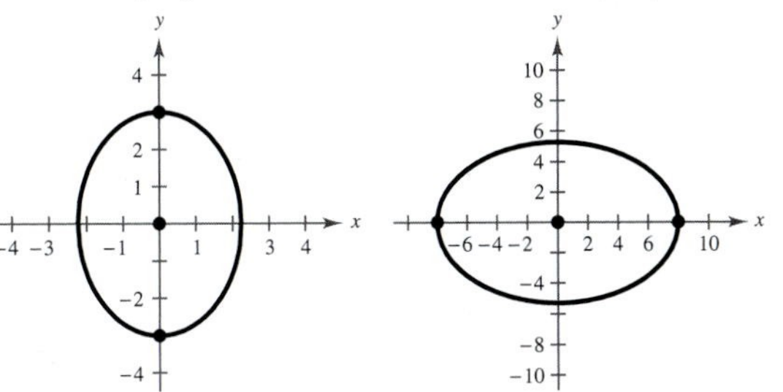

13. Ellipse
Center: $(-3, 5)$
Vertices: $(-3, 10), (-3, 0)$
Foci: $(-3, 8), (-3, 2)$
Eccentricity: $\frac{3}{5}$

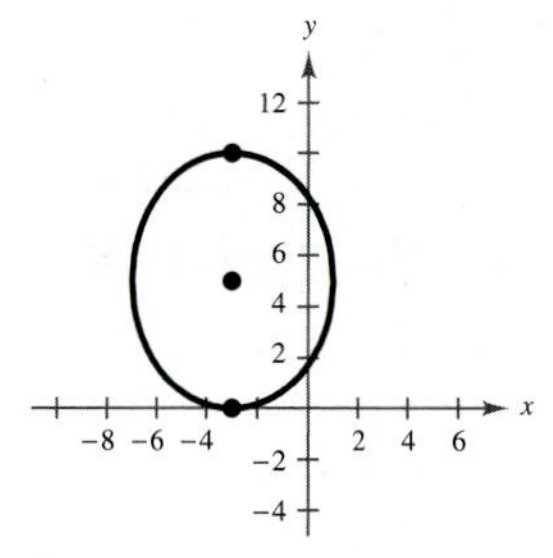

14. Ellipse
Center: $(4, -3)$
Vertices: $(4, 1), (4, -7)$
Foci: $(4, -1), (4, -5)$
Eccentricity: $\frac{1}{2}$

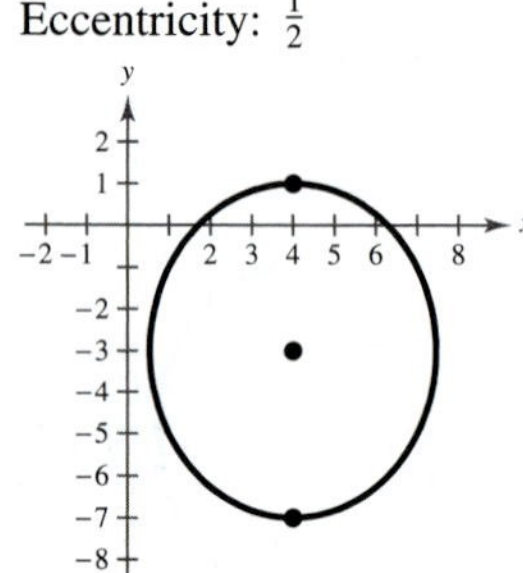

15. Circle
Center: $(0, -1)$
Radius: $\frac{2}{3}$

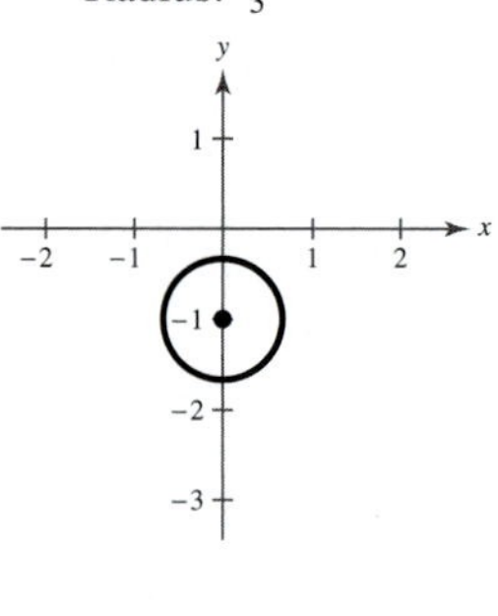

16. Ellipse
Center: $(-5, 1)$
Vertices: $\left(-\dfrac{7}{2}, 1\right), \left(-\dfrac{13}{2}, 1\right)$
Foci: $\left(-5 \pm \dfrac{\sqrt{5}}{2}, 1\right)$
Eccentricity: $\dfrac{\sqrt{5}}{3}$

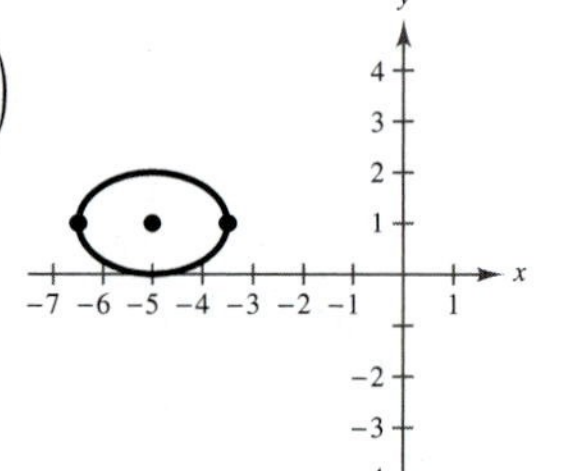

17. Ellipse
Center: $(-2, -4)$
Vertices: $(-3, -4), (-1, -4)$
Foci: $\left(\dfrac{-4 \pm \sqrt{3}}{2}, -4\right)$
Eccentricity: $\dfrac{\sqrt{3}}{2}$

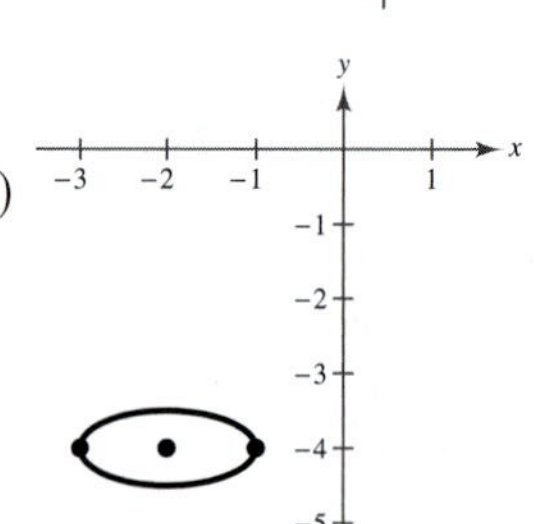

18. Circle
Center: $(3, 1)$
Radius: $\frac{5}{2}$

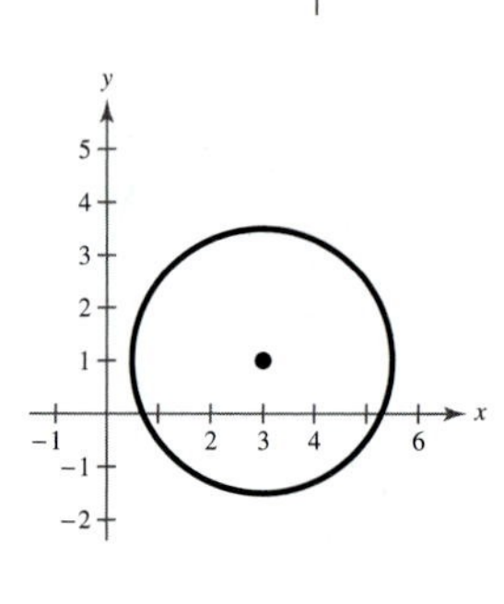

19. Ellipse
Center: $(-2, 3)$
Vertices: $(-2, 6), (-2, 0)$
Foci: $\left(-2, 3 \pm \sqrt{5}\right)$
Eccentricity: $\dfrac{\sqrt{5}}{3}$

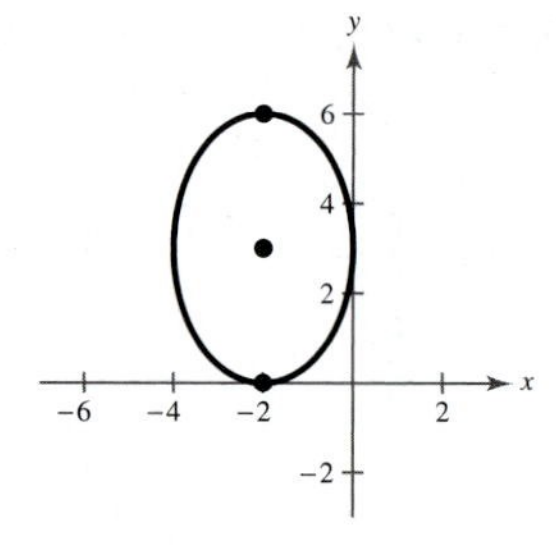

20. Ellipse
Center: $(3, -5)$
Vertices: $(3, 1), (3, -11)$
Foci: $\left(3, -5 \pm 2\sqrt{5}\right)$
Eccentricity: $\dfrac{\sqrt{5}}{3}$

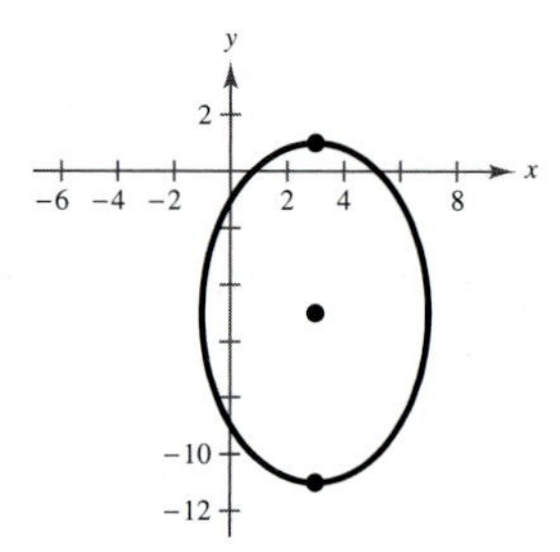

21. Circle
Center: $(1, -2)$
Radius: 6
Foci: $\left(3, -5 \pm 2\sqrt{5}\right)$

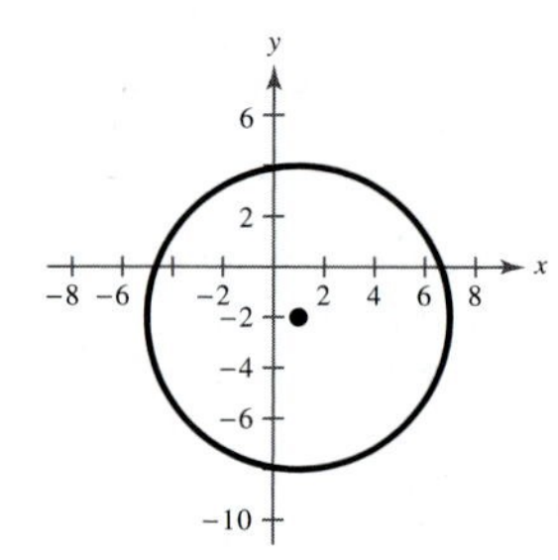

22. Ellipse
Center: $(4, 3)$
Vertices: $(14, 3), (-6, 3)$
Foci: $\left(4 \pm 4\sqrt{5}, 3\right)$
Eccentricity: $\dfrac{2\sqrt{5}}{5}$

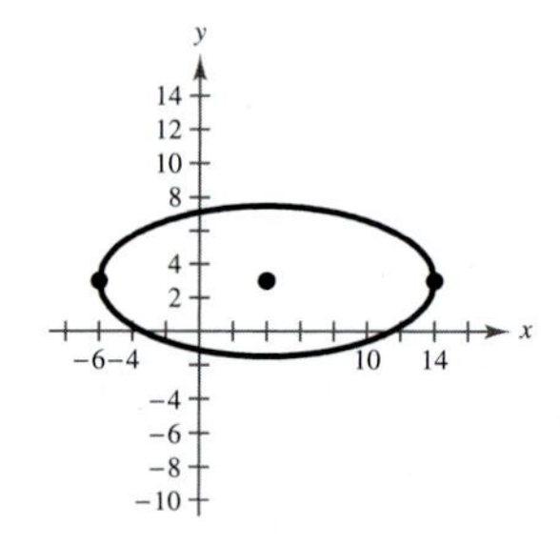

23. Ellipse
Center: $(-3, 1)$
Vertices: $(-3, 7), (-3, -5)$
Foci: $\left(-3, 1 \pm 2\sqrt{6}\right)$
Eccentricity: $\dfrac{\sqrt{6}}{3}$

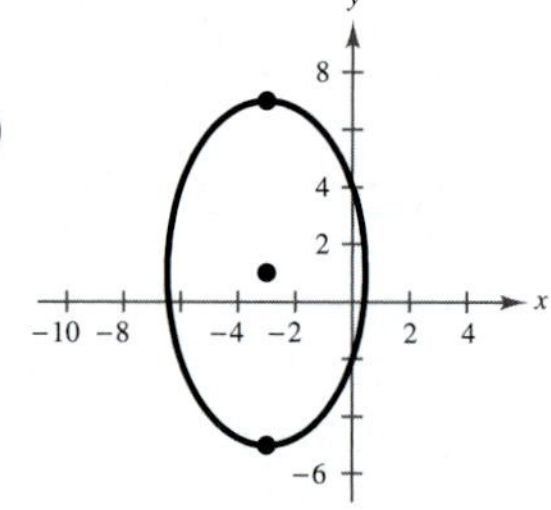

24. Ellipse

Center: $\left(-\frac{3}{2}, \frac{5}{2}\right)$

Vertices: $\left(-\frac{3}{2}, \frac{5}{2} \pm 2\sqrt{3}\right)$

Foci: $\left(-\frac{3}{2}, \frac{5}{2} \pm 2\sqrt{2}\right)$

Eccentricity: $\frac{\sqrt{6}}{3}$

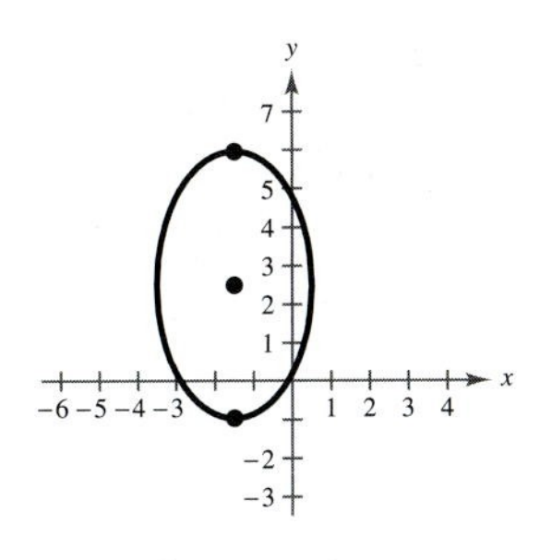

25. Ellipse

Center: $\left(3, -\frac{5}{2}\right)$

Vertices: $\left(9, -\frac{5}{2}\right), \left(-3, -\frac{5}{2}\right)$

Foci: $\left(3 \pm 3\sqrt{3}, -\frac{5}{2}\right)$

Eccentricity: $\frac{\sqrt{3}}{2}$

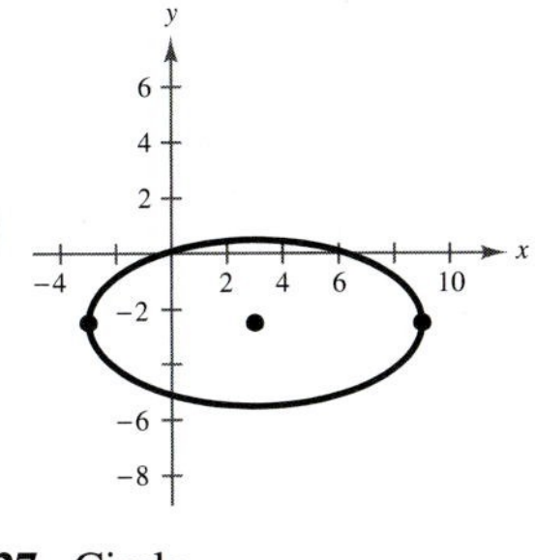

26. Circle
Center: $(2, -3)$
Radius: 4

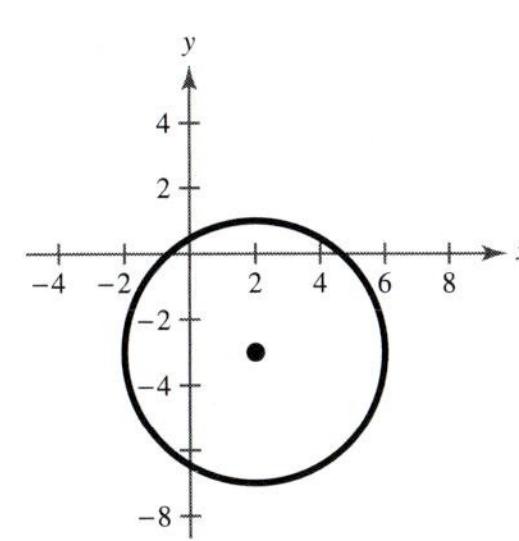

27. Circle
Center: $(-1, 1)$
Radius: $\frac{2}{3}$

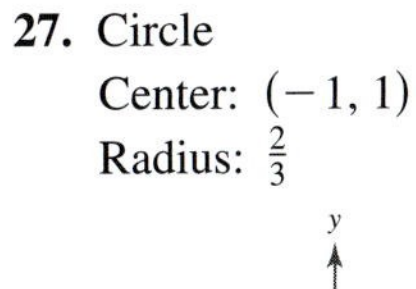

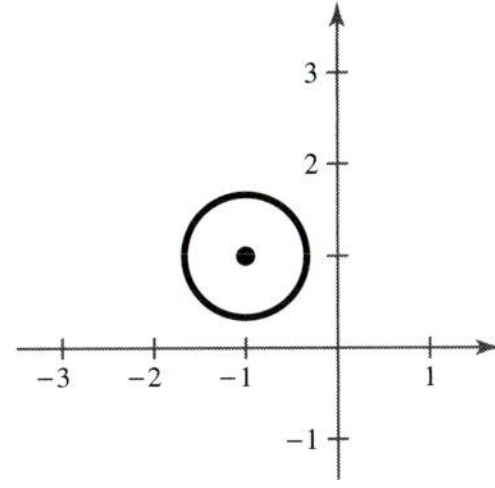

28. Ellipse
Center: $(1, -1)$
Vertices: $\left(\frac{9}{4}, -1\right), \left(-\frac{1}{4}, -1\right)$
Foci: $\left(\frac{7}{4}, -1\right), \left(\frac{1}{4}, -1\right)$
Eccentricity: $\frac{3}{5}$

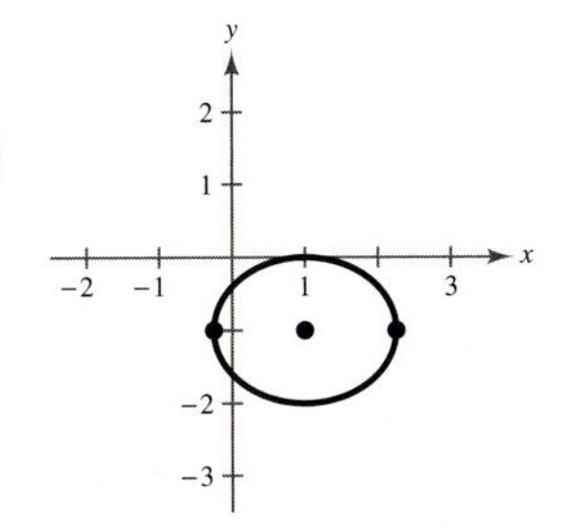

29. Ellipse
Center: $(2, 1)$
Vertices: $\left(\frac{7}{3}, 1\right), \left(\frac{5}{3}, 1\right)$
Foci: $\left(\frac{34}{15}, 1\right), \left(\frac{26}{15}, 1\right)$
Eccentricity: $\frac{4}{5}$

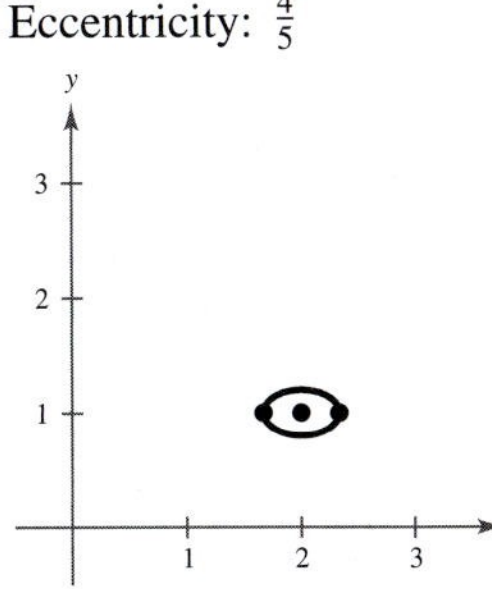

30. Circle
Center: $(2, -1)$
Radius: $\frac{5}{4}$

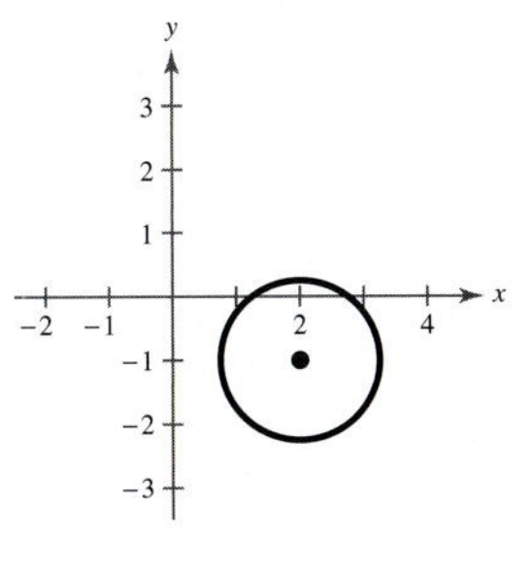

31.

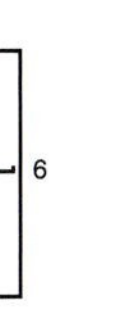

Center: $(0, 0)$
Vertices: $(0, \pm\sqrt{5})$
Foci: $(0, \pm\sqrt{2})$

32.

Center: $(0, 0)$
Vertices: $(\pm 2, 0)$
Foci: $(\pm 1, 0)$

33.

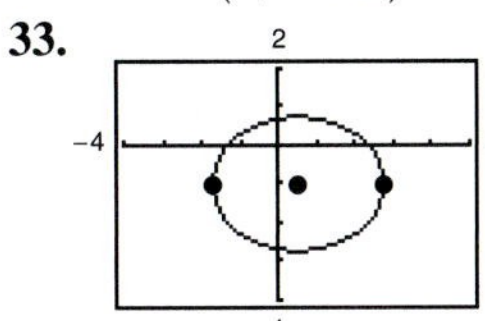

Center: $\left(\frac{1}{2}, -1\right)$
Vertices: $\left(\frac{1}{2} \pm \sqrt{5}, -1\right)$
Foci: $\left(\frac{1}{2} \pm \sqrt{2}, -1\right)$

34.

Center: $\left(-\frac{2}{3}, 2\right)$

Vertices: $\left(-\frac{2}{3}, 2 \pm \frac{2\sqrt{31}}{3}\right)$

Foci: $\left(-\frac{2}{3}, 2 \pm \frac{\sqrt{93}}{3}\right)$

35. $\frac{x^2}{4} + \frac{y^2}{16} = 1$ **36.** $\frac{x^2}{4} + \frac{4y^2}{9} = 1$

37. $\frac{x^2}{36} + \frac{y^2}{32} = 1$ **38.** $\frac{x^2}{48} + \frac{y^2}{64} = 1$ **39.** $\frac{x^2}{36} + \frac{y^2}{11} = 1$

40. $\frac{x^2}{16} + \frac{y^2}{12} = 1$ **41.** $\frac{21x^2}{400} + \frac{y^2}{25} = 1$

42. $\frac{x^2}{4} + \frac{y^2}{16} = 1$ **43.** $\frac{(x-2)^2}{1} + \frac{(y-3)^2}{9} = 1$

44. $\frac{(x-4)^2}{9} + \frac{y^2}{16} = 1$ **45.** $\frac{(x+2)^2}{16} + \frac{(y-3)^2}{9} = 1$

46. $\frac{(x-2)^2}{4} + \frac{(y+1)^2}{1} = 1$

47. $\frac{(x-2)^2}{4} + \frac{(y-4)^2}{1} = 1$ **48.** $\frac{(x-2)^2}{16} + \frac{y^2}{12} = 1$

49. $\frac{x^2}{48} + \frac{(y-4)^2}{64} = 1$ **50.** $\frac{(x-2)^2}{1} + \frac{4(y+1)^2}{9} = 1$

51. $\frac{x^2}{16} + \frac{(y-4)^2}{12} = 1$ **52.** $\frac{(x-3)^2}{36} + \frac{(y-2)^2}{32} = 1$

53. $\frac{(x-2)^2}{4} + \frac{(y-2)^2}{1} = 1$

54. $\frac{(x-5)^2}{16} + \frac{(y-6)^2}{36} = 1$ **55.** $\frac{x^2}{25} + \frac{y^2}{16} = 1$

56. $\frac{x^2}{48} + \frac{y^2}{64} = 1$

57. (a)

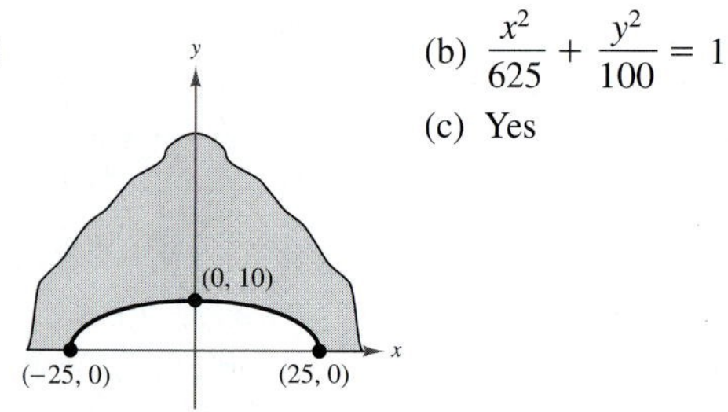

(b) $\dfrac{x^2}{625} + \dfrac{y^2}{100} = 1$

(c) Yes

58. Positions: $(\pm\sqrt{5}, 0)$;
Length of string: 6 feet

59. (a) $\dfrac{x^2}{321.84} + \dfrac{y^2}{20.89} = 1$

(b)

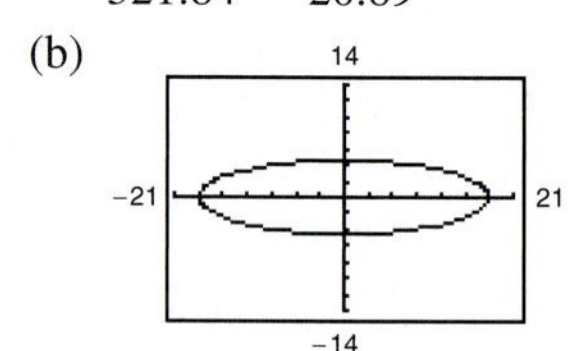

(c) Aphelion: 35.29 astronomical units
Perihelion: 0.59 astronomical unit

60. $e \approx 0.052$

61. (a) $\dfrac{x^2}{0.04} + \dfrac{y^2}{2.56} = 1$

(b)

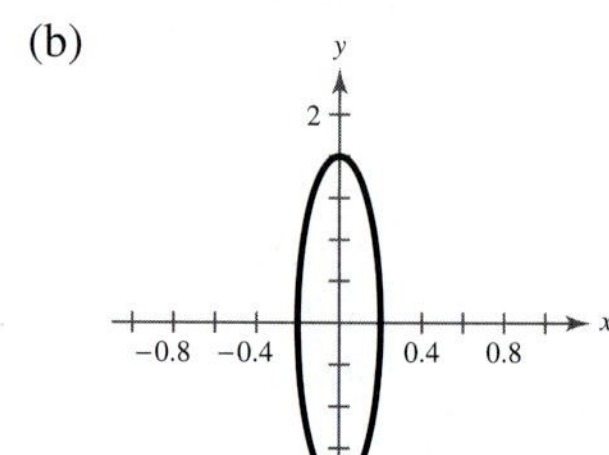

(c) The bottom half

62. Answers will vary.

63.

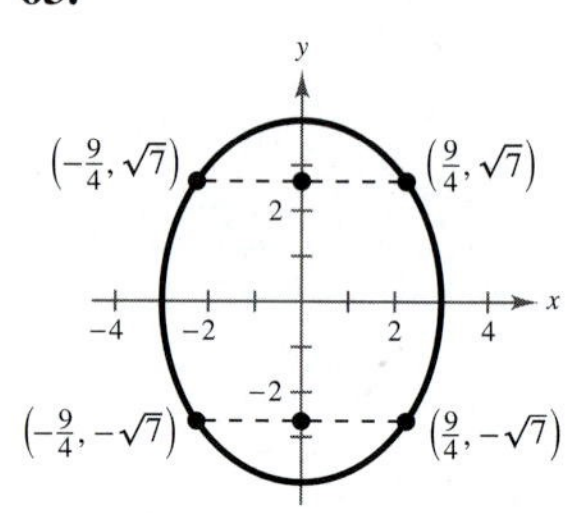

64.

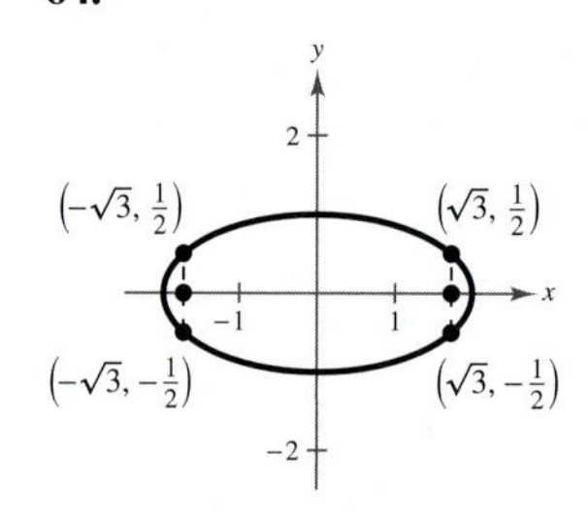

65.

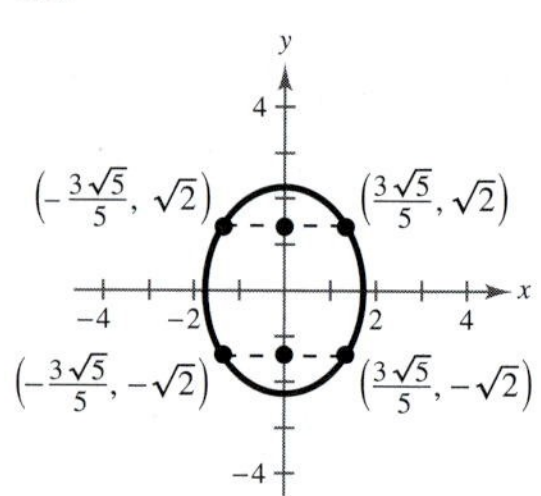

66.

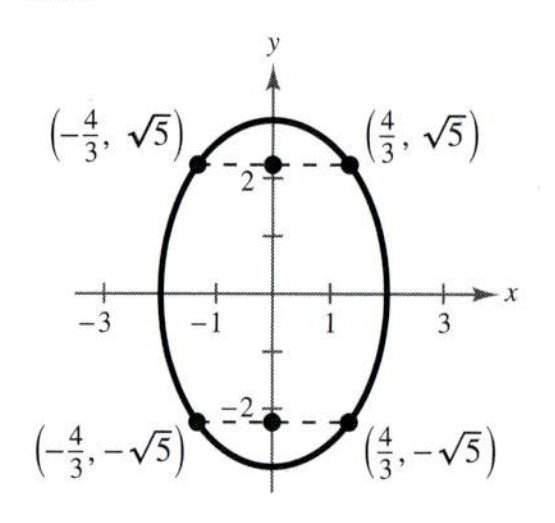

67. False. The graph of $x^2/4 + y^4 = 1$ is not an ellipse. The degree of y is 4, not 2.

68. True. If e is close to 1, the ellipse is elongated and the foci are close to the vertices.

69. (a) $A = \pi a(20 - a)$ (b) $\dfrac{x^2}{196} + \dfrac{y^2}{36} = 1$

(c)

a	8	9	10	11	12	13
A	301.6	311.0	314.2	311.0	301.6	285.9

$a = 10$, circle

(d)

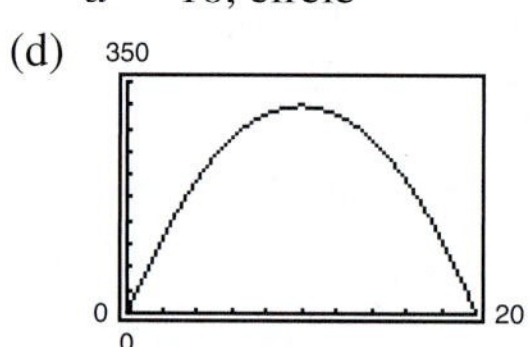

The shape of an ellipse with a maximum area is a circle. The maximum area is found when $a = 10$, (verified in part (c), and therefore $b = 10$, so the equation produces a circle.

70. (a) $2a$

(b) The sum of the distances from the two fixed points is constant.

71. $\frac{1}{32}(10 - 15\cos 2x + 6\cos 4x - \cos 6x)$

72. $\frac{1}{32}(10 + 15\cos 2x + 6\cos 4x + \cos 6x)$

73. $\frac{1}{4}\left(\frac{3}{2} + 2\cos 4x + \frac{1}{2}\cos 8x\right)$

74. $\frac{1}{8}(3 - 4\cos 4x + \cos 8x)$

75. Area = 357.9 **76.** Area = 15.23

77. Area = 47.9 **78.** Area = 45.65

Section 6.4 *(page 482)*

Vocabulary Check *(page 482)*

1. hyperbola; foci **2.** branches
3. transverse axis; center **4.** asymptotes
5. $Ax^2 + Cy^2 + Dx + Ey + F = 0$

1. b **2.** c **3.** a **4.** d

5. Center: $(0, 0)$
Vertices: $(\pm 1, 0)$
Foci: $(\pm\sqrt{2}, 0)$
Asymptotes: $y = \pm x$

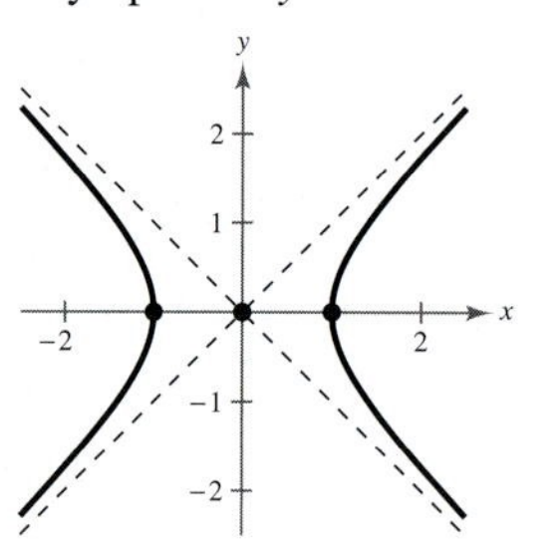

6. Center: $(0, 0)$
Vertices: $(\pm 3, 0)$
Foci: $(\pm\sqrt{34}, 0)$
Asymptotes: $y = \pm\frac{5}{3}x$

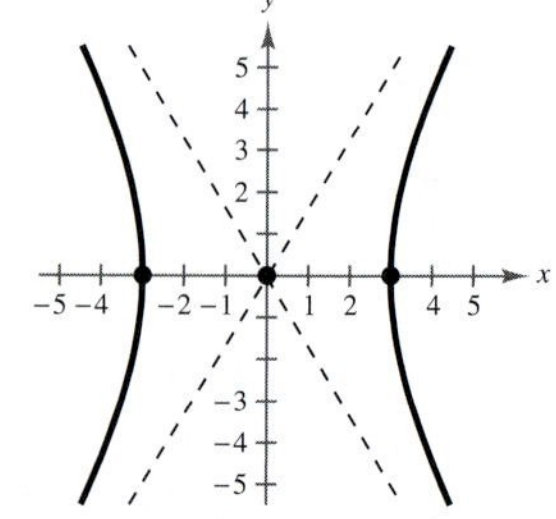

7. Center: $(0, 0)$
Vertices: $(0, \pm 5)$
Foci: $\left(0, \pm\sqrt{106}\right)$
Asymptotes: $y = \pm\frac{5}{9}x$

8. Center: $(0, 0)$
Vertices: $(\pm 6, 0)$
Foci: $\left(\pm 2\sqrt{10}, 0\right)$
Asymptotes: $y = \pm\frac{1}{3}x$

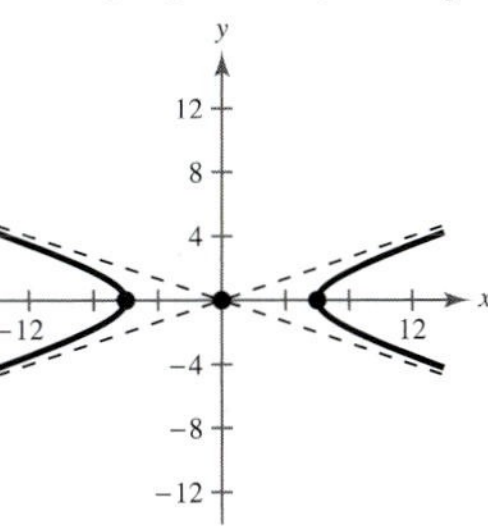

9. Center: $(1, -2)$
Vertices: $(3, -2), (-1, -2)$
Foci: $\left(1 \pm \sqrt{5}, -2\right)$
Asymptotes:
$y = -2 \pm \frac{1}{2}(x - 1)$

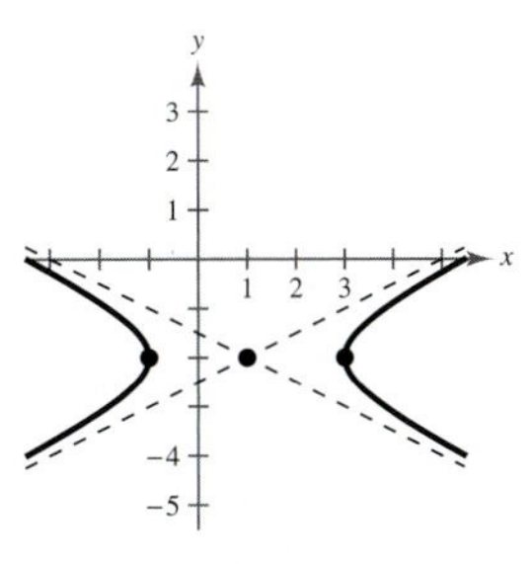

10. Center: $(-3, 2)$
Vertices: $(9, 2), (-15, 2)$
Foci: $(10, 2), (-16, 2)$
Asymptotes:
$y = 2 \pm \frac{5}{12}(x + 3)$

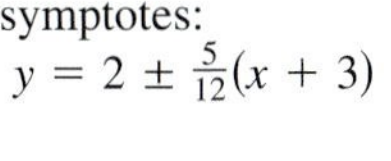

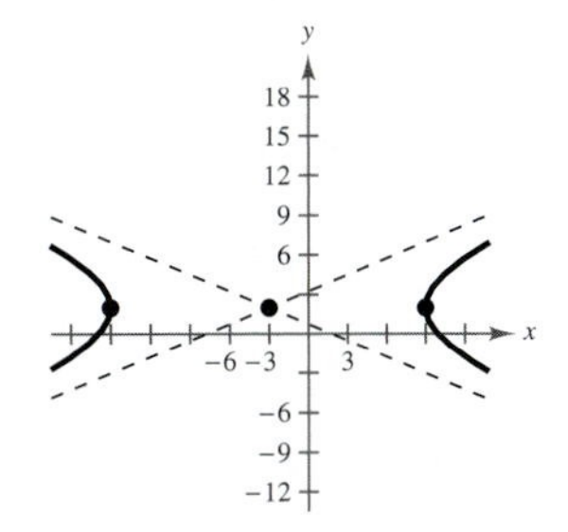

11. Center: $(2, -6)$
Vertices:
$\left(2, -\frac{17}{3}\right), \left(2, -\frac{19}{3}\right)$
Foci: $\left(2, -6 \pm \frac{\sqrt{13}}{6}\right)$

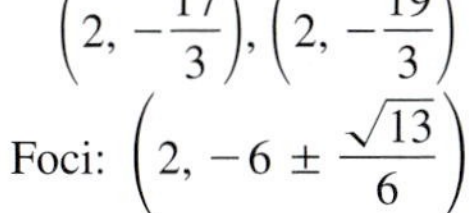

Asymptotes:
$y = -6 \pm \frac{2}{3}(x - 2)$

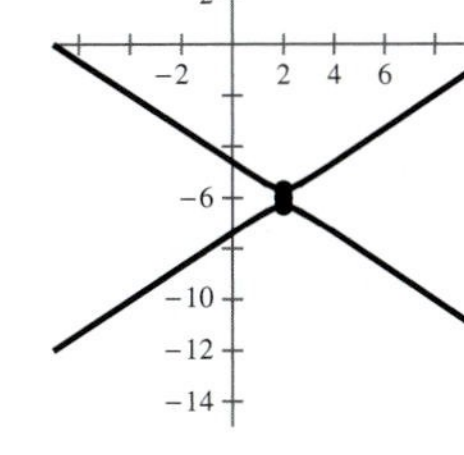

12. Center: $(-3, 1)$
Vertices: $\left(-3, \frac{3}{2}\right), \left(-3, \frac{1}{2}\right)$
Foci: $\left(-3, 1 \pm \frac{\sqrt{5}}{4}\right)$
Asymptotes:
$y = 1 \pm 2(x + 3)$

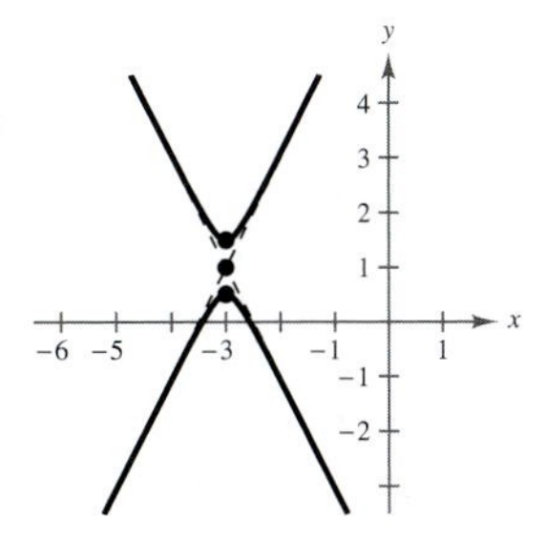

13. Center: $(2, -3)$
Vertices: $(3, -3), (1, -3)$
Foci: $\left(2 \pm \sqrt{10}, -3\right)$
Asymptotes:
$y = -3 \pm 3(x - 2)$

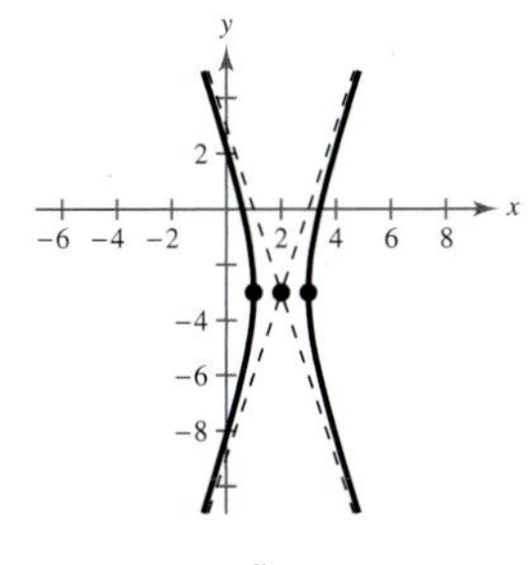

14. Center: $(0, 2)$
Vertices: $(\pm 6, 2)$
Foci: $\left(\pm 2\sqrt{10}, 2\right)$
Asymptotes: $y = 2 \pm \frac{1}{3}x$

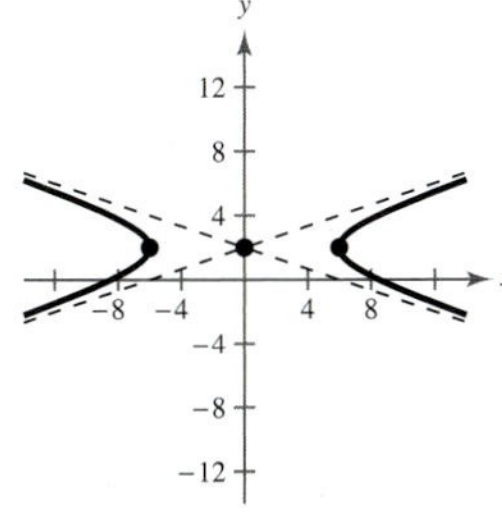

15. The graph of this equation is two lines intersecting at $(-1, -3)$.

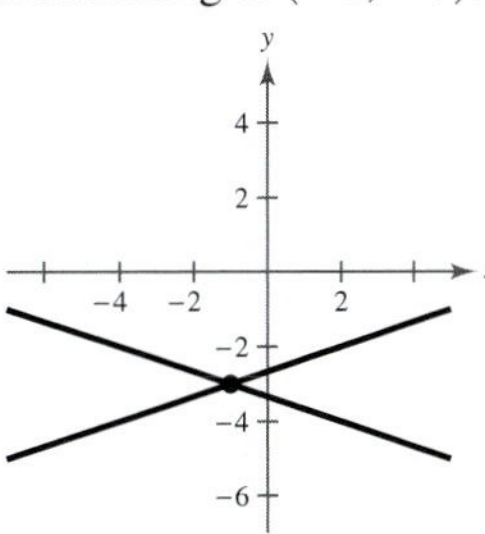

16. The graph of this equation is two lines intersecting at $(1, -2)$.

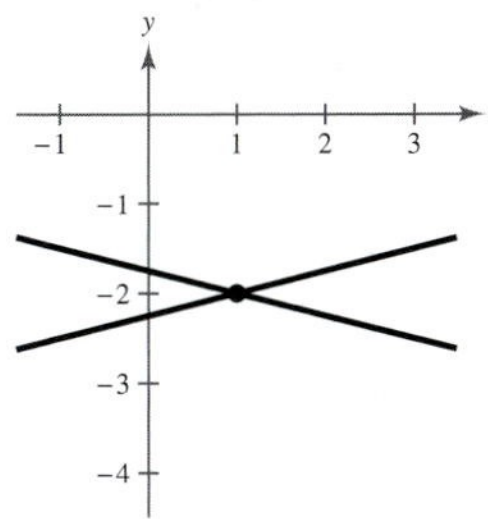

17. Center: $(0, 0)$
Vertices: $\left(\pm\sqrt{3}, 0\right)$
Foci: $\left(\pm\sqrt{5}, 0\right)$
Asymptotes: $y = \pm\frac{\sqrt{6}}{3}x$

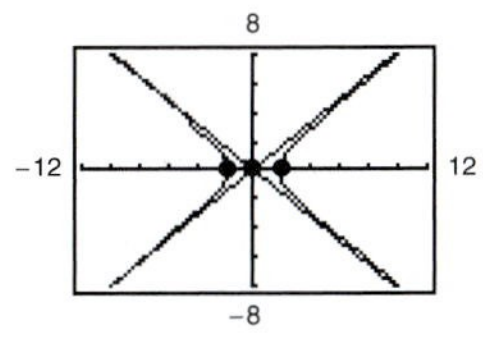

18. Center: $(0, 0)$
Vertices: $\left(0, \pm\sqrt{3}\right)$
Foci: $(0, \pm 3)$
Asymptotes: $y = \pm\frac{\sqrt{2}}{2}x$

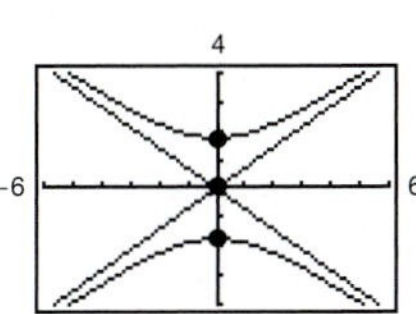

19. Center: $(1, -3)$
Vertices: $\left(1, -3 \pm \sqrt{2}\right)$
Foci: $\left(1, -3 \pm 2\sqrt{5}\right)$
Asymptotes:
$y = -3 \pm \frac{1}{3}(x - 1)$

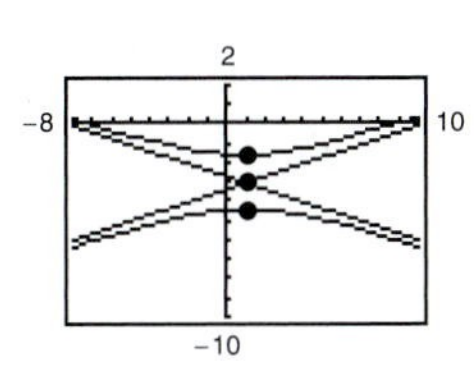

20. Center: $(-3, 5)$

Vertices: $\left(-\frac{10}{3}, 5\right), \left(-\frac{8}{3}, 5\right)$

Foci: $\left(-3 \pm \frac{\sqrt{10}}{3}, 5\right)$

Asymptotes: $y = 5 \pm 3(x + 3)$

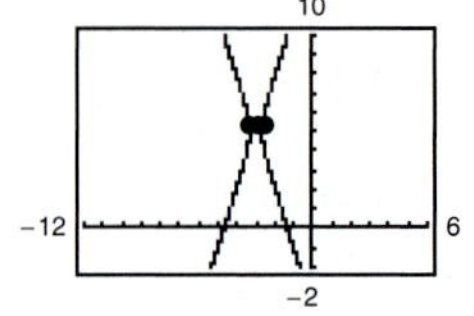

21. $\frac{y^2}{4} - \frac{x^2}{12} = 1$ **22.** $\frac{x^2}{16} - \frac{y^2}{20} = 1$

23. $\frac{x^2}{1} - \frac{y^2}{25} = 1$ **24.** $\frac{y^2}{9} - \frac{x^2}{1} = 1$

25. $\frac{17y^2}{1024} - \frac{17x^2}{64} = 1$ **26.** $\frac{x^2}{64} - \frac{y^2}{36} = 1$

27. $\frac{(x-4)^2}{4} - \frac{y^2}{12} = 1$ **28.** $\frac{y^2}{9} - \frac{(x-2)^2}{27} = 1$

29. $\frac{(y-5)^2}{16} - \frac{(x-4)^2}{9} = 1$ **30.** $\frac{x^2}{4} - \frac{(y-1)^2}{5} = 1$

31. $\frac{y^2}{9} - \frac{4(x-2)^2}{9} = 1$ **32.** $\frac{x^2}{4} - \frac{7(y-1)^2}{12} = 1$

33. $\frac{(y-2)^2}{4} - \frac{x^2}{4} = 1$ **34.** $\frac{y^2}{4} - \frac{(x-1)^2}{4} = 1$

35. $\frac{(x-2)^2}{1} - \frac{(y-2)^2}{1} = 1$

36. $\frac{(y-3)^2}{9} - \frac{(x-3)^2}{9} = 1$

37. $\frac{(x-3)^2}{9} - \frac{(y-2)^2}{4} = 1$

38. $\frac{(y-2)^2}{4} - \frac{(x-3)^2}{9} = 1$

39. (a) $\frac{x^2}{1} - \frac{y^2}{169/3} = 1$ (b) ≈ 2.403 feet

40. $\frac{x^2}{98{,}010{,}000} - \frac{y^2}{13{,}503{,}600} = 1$ **41.** $(3300, -2750)$

42. (a) $x \approx 110.3$ miles (b) 57.0 miles
(c) 0.00129 second
(d) The ship is at the position (144.2, 60).

43. $\left(12(\sqrt{5} - 1), 0\right) \approx (14.83, 0)$

44. (a) Circle
(b) $\frac{(x-100)^2}{62{,}500} + \frac{y^2}{62{,}500} = 1$

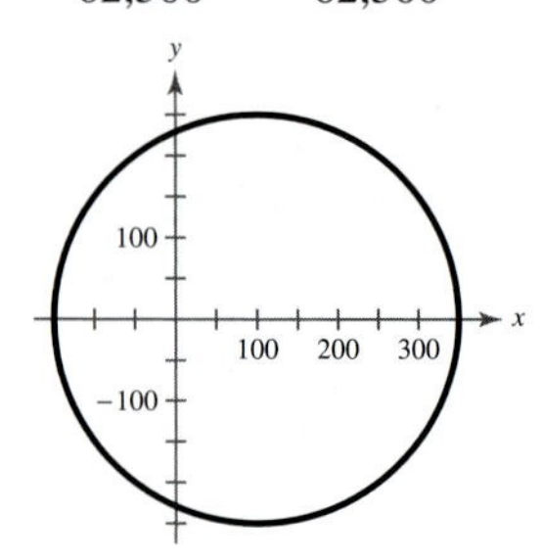

(c) Approximately 180.28 meters

45. Circle **46.** Ellipse **47.** Hyperbola
48. Parabola **49.** Hyperbola **50.** Circle

51. Parabola **52.** Ellipse **53.** Ellipse
54. Hyperbola **55.** Parabola **56.** Circle
57. Ellipse **58.** Parabola **59.** Circle **60.** Hyperbola

61. True. For a hyperbola, $c^2 = a^2 + b^2$. The larger the ratio of b to a, the larger the eccentricity of the hyperbola, $e = c/a$.

62. False. For the trivial solution of two intersecting lines to occur, the standard form of the equation of the hyperbola would be equal to zero,
$$\frac{(x-h)^2}{a^2} - \frac{(y-k)^2}{b^2} = 0 \text{ or } \frac{(y-k)^2}{a^2} - \frac{(x-h)^2}{b^2} = 0.$$

63. Answers will vary.

64. The extended diagonals of the central rectangle are asymptotes of the hyperbola.

65. $y = 1 - 3\sqrt{\frac{(x-3)^2}{4} - 1}$

66.

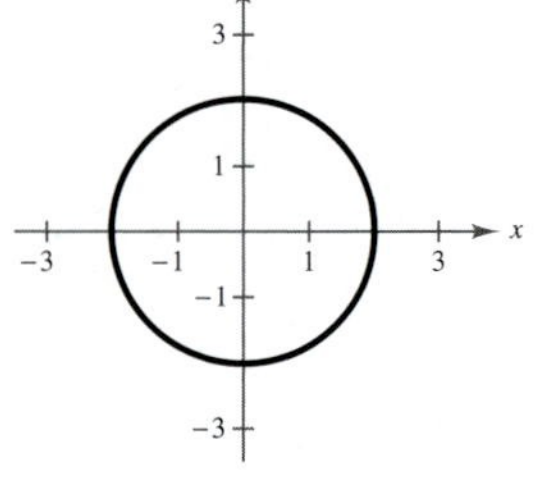

The equation $y = x^2 + C$ is a parabola that could intersect the circle in zero, one, two, three, or four places depending on its location on the y-axis.
(a) $C > 2$ and $C < -\frac{17}{4}$ (b) $C = 2$
(c) $-2 < C < 2$, $C = -\frac{17}{4}$ (d) $C = -2$
(d) $-\frac{17}{4} < C < -2$

67. $x(x + 4)(x - 4)$ **68.** $(x + 7)^2$ **69.** $2x(x - 6)^2$
70. $x(3x + 2)(2x - 5)$ **71.** $2(2x + 3)(4x^2 - 6x + 9)$
72. $-(x^2 + 1)(x - 4)$

73.

74.

75. **76.**

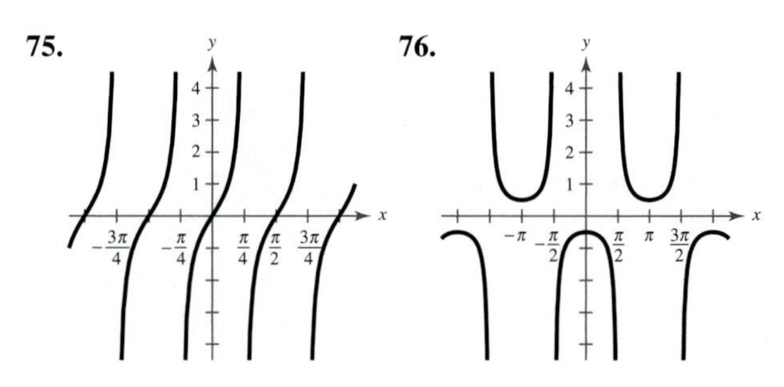

Section 6.5 *(page 490)*

Vocabulary Check *(page 490)*

1. plane curve; parametric; parameter
2. orientation **3.** eliminating the parameter

1. (a)

t	0	1	2	3	4
x	0	1	$\sqrt{2}$	$\sqrt{3}$	2
y	3	2	1	0	-1

(b)

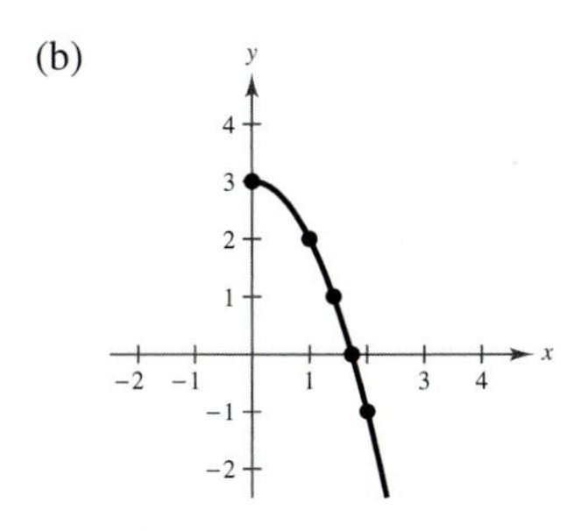

(c) $y = 3 - x^2$

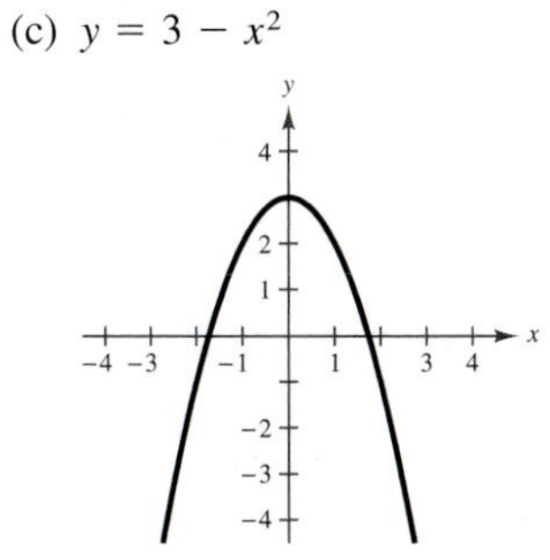

The graph of the rectangular equation shows the entire parabola rather than just the right half.

2. (a)

θ	$-\frac{\pi}{2}$	$-\frac{\pi}{4}$	0	$\frac{\pi}{4}$	$\frac{\pi}{2}$
x	0	2	4	2	0
y	-2	$-\sqrt{2}$	0	$\sqrt{2}$	2

(b)

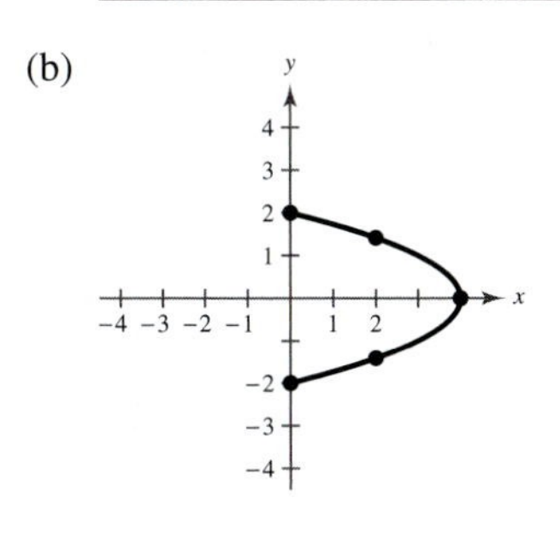

(c) $x = -y^2 + 4$

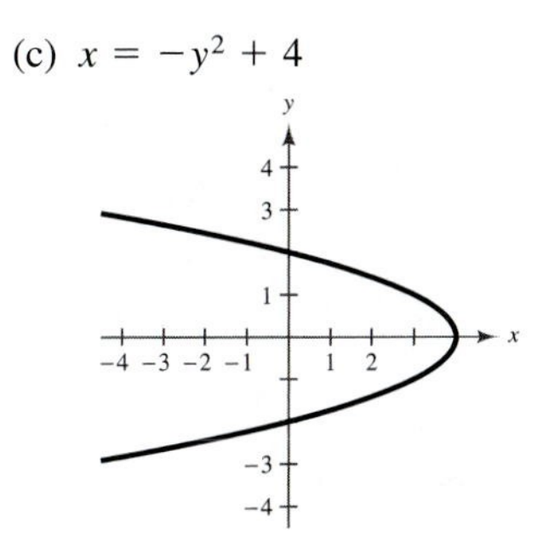

The graph of the rectangular equation continues the graph into the second and third quadrants.

3. (a)

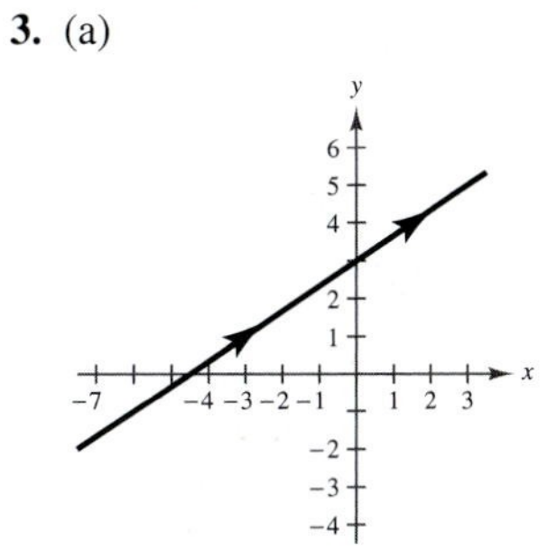

(b) $y = \frac{2}{3}x + 3$

4. (a)

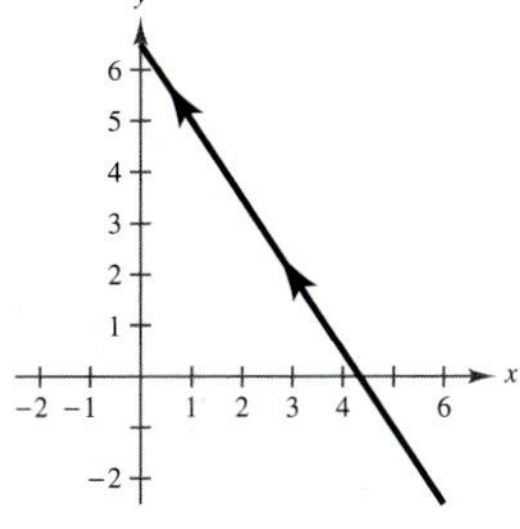

(b) $y = -\frac{3}{2}x + \frac{13}{2}$

5. (a)

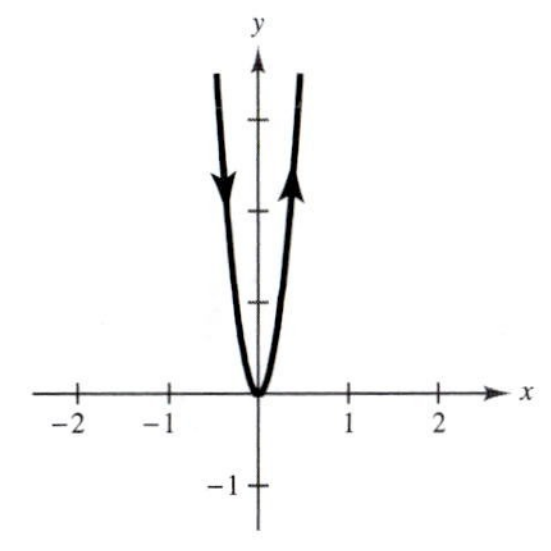

(b) $y = 16x^2$

6. (a)

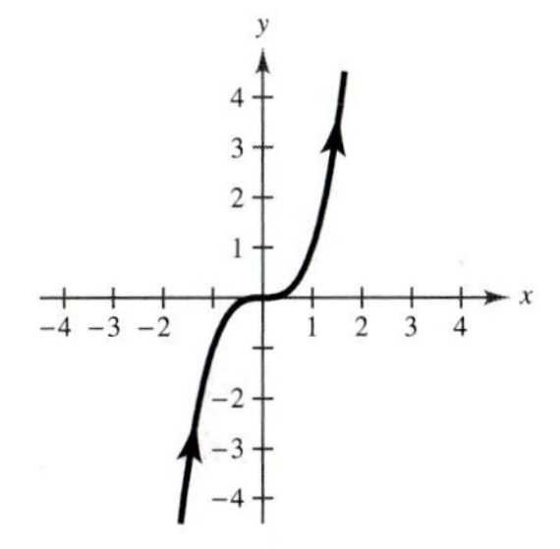

(b) $y = x^3$

7. (a)

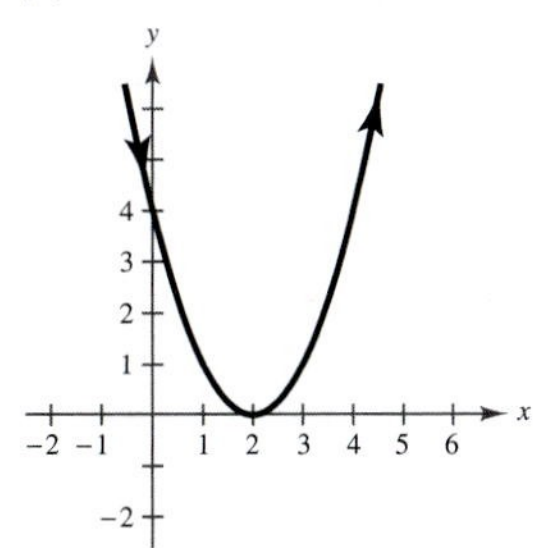

(b) $y = x^2 - 4x + 4$

8. (a)

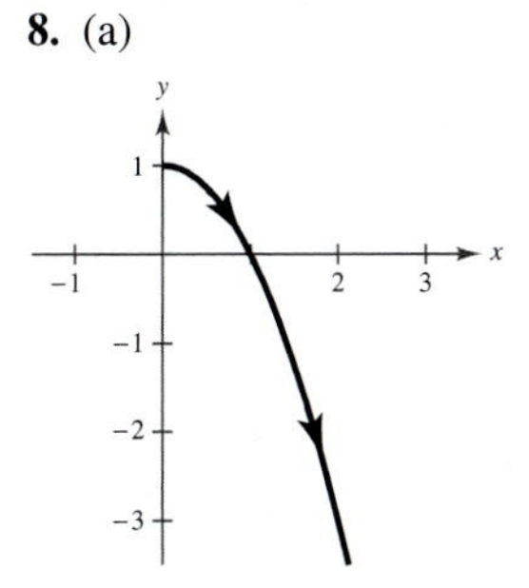

(b) $y = 1 - x^2$

9. (a)

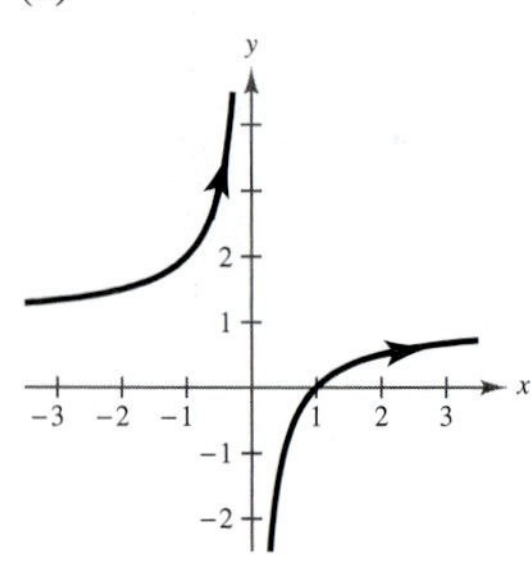

(b) $y = \dfrac{(x - 1)}{x}$

10. (a)

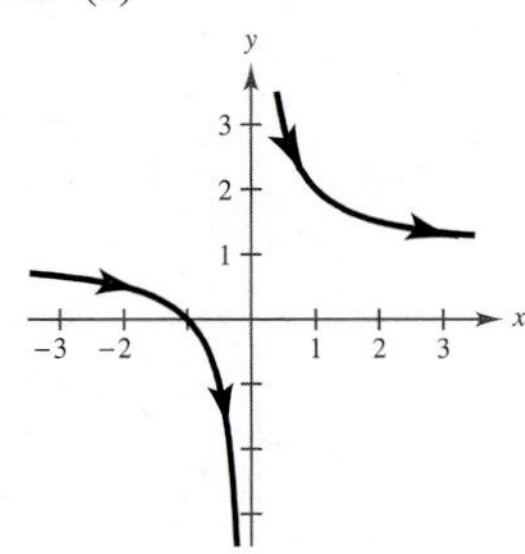

(b) $y = \dfrac{x + 1}{x}$

11. (a)

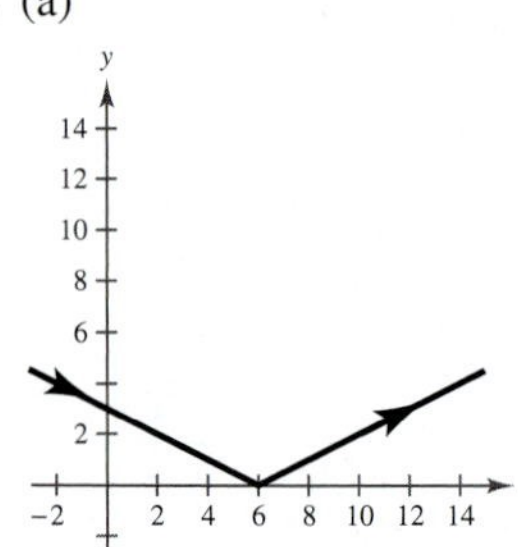

(b) $y = \left|\dfrac{x}{2} - 3\right|$

12. (a)

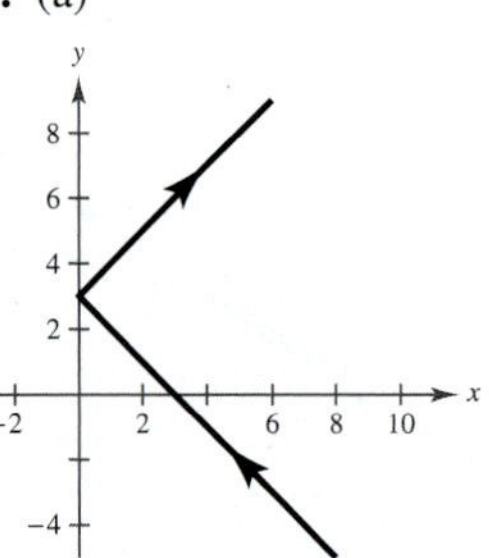

(b) $y = -x + 3,\ x > 0$ and $y = x + 3,\ x \geq 0$

13. (a)

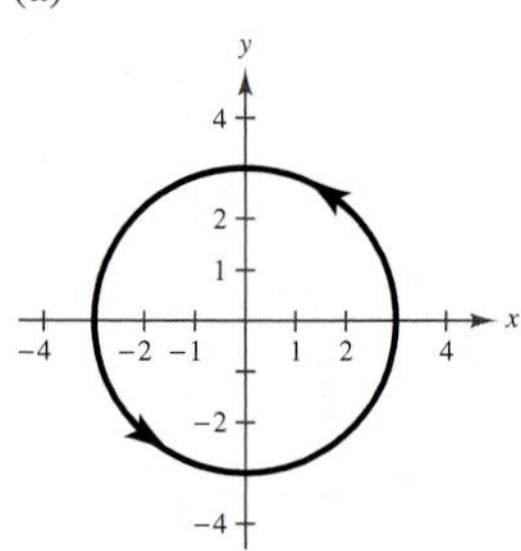

(b) $\dfrac{y^2}{9} + \dfrac{x^2}{9} = 1$

14. (a)

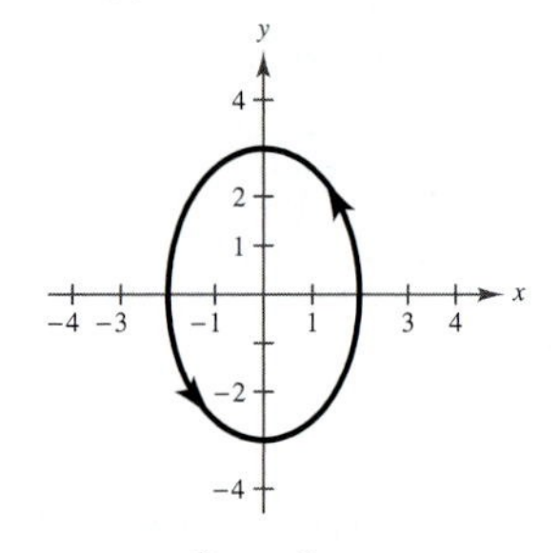

(b) $\dfrac{x^2}{4} + \dfrac{y^2}{9} = 1$

15. (a)

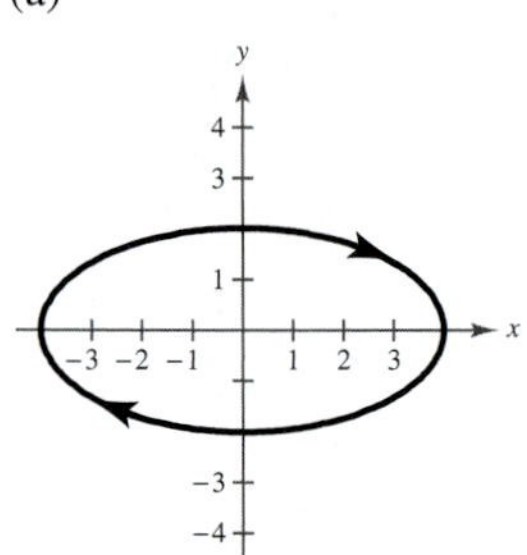

(b) $\dfrac{x^2}{16} + \dfrac{y^2}{4} = 1$

16. (a)

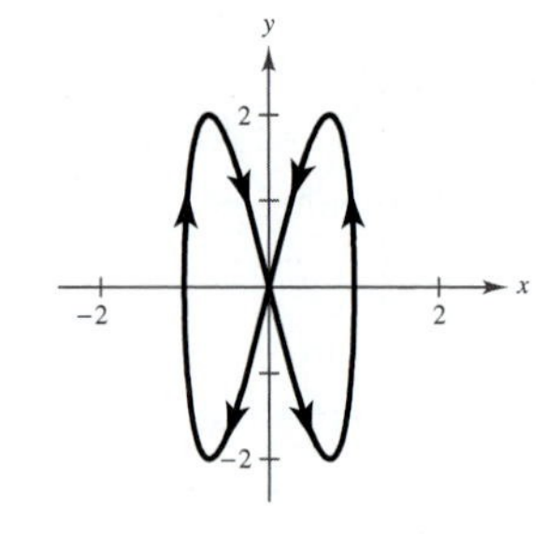

(b) $y = \pm 4x\sqrt{1 - x^2}$

17. (a)

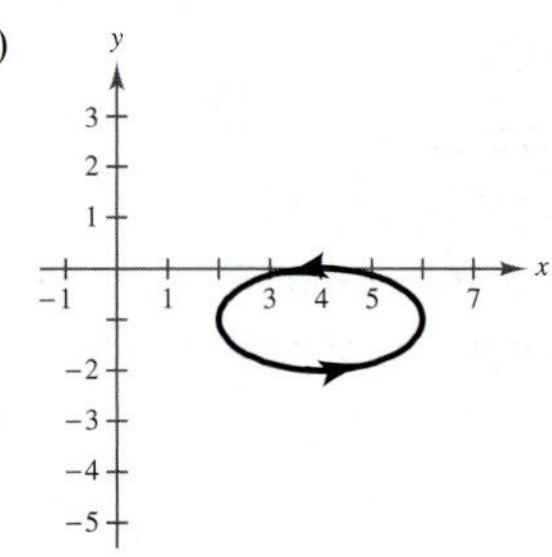

(b) $\dfrac{(x - 4)^2}{4} + (y + 1)^2 = 1$

18. (a)

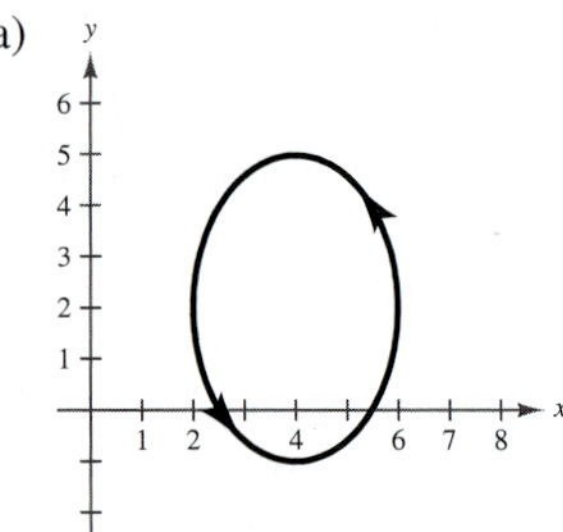

(b) $\dfrac{(x - 4)^2}{4} + \dfrac{(y - 2)^2}{9} = 1$

19. (a)

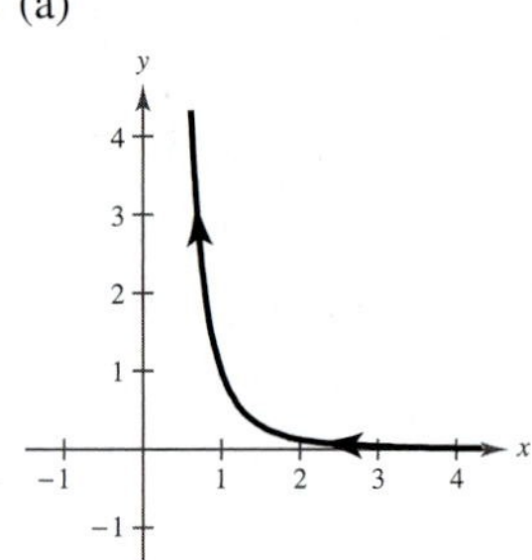

(b) $y = \dfrac{1}{x^3}$

20. (a)

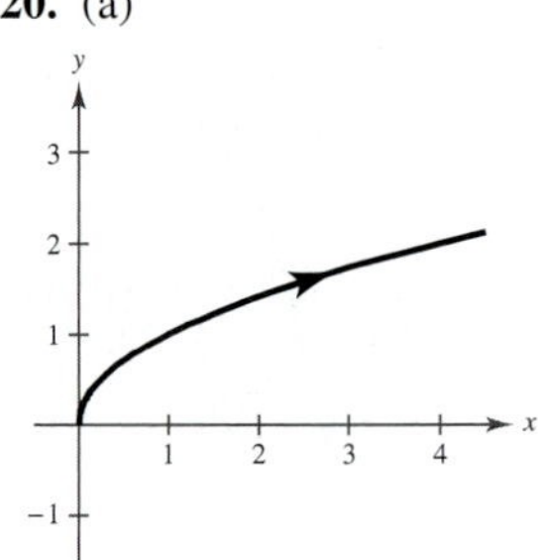

(b) $y = \sqrt{x}$

21. (a)

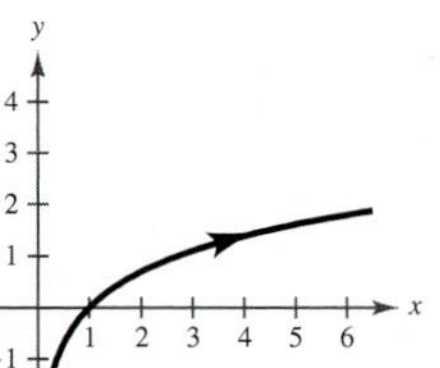

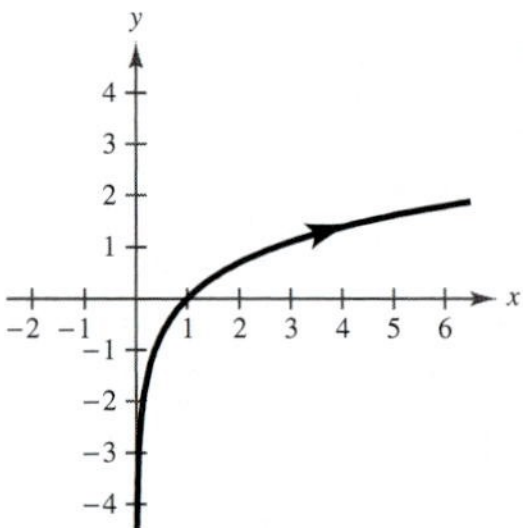

(b) $y = \ln x$

22. (a)

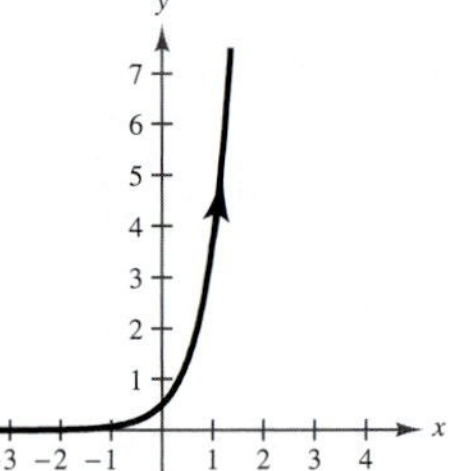

(b) $y = \dfrac{e^{2x}}{2}$

23. Each curve represents a portion of the line $y = 2x + 1$.

	Domain	*Orientation*
(a)	$(-\infty, \infty)$	Left to right
(b)	$[-1, 1]$	Depends on θ
(c)	$(0, \infty)$	Right to left
(d)	$(0, \infty)$	Left to right

24. Each curve represents a portion of the parabola $y = x^2 - 1$.

	Domain	*Orientation*
(a)	$(-\infty, \infty)$	Left to right
(b)	$[0, \infty)$	Depends on t
(c)	$[-1, 1]$	Depends on t
(d)	$(0, \infty)$	Left to right

25. $y - y_1 = m(x - x_1)$ **26.** $(x - h)^2 + (y - k)^2 = r^2$

27. $\dfrac{(x - h)^2}{a^2} + \dfrac{(y - k)^2}{b^2} = 1$

28. $\dfrac{(x - h)^2}{a^2} - \dfrac{(y - k)^2}{b^2} = 1$

29. $x = 6t$, $y = -3t$ **30.** $x = 2 + 4t$, $y = 3 - 6t$

31. $x = 3 + 4 \cos \theta$, $y = 2 + 4 \sin \theta$ **32.** $x = -3 + 5 \cos \theta$, $y = 2 + 5 \sin \theta$

33. $x = 4 \cos \theta$, $y = \sqrt{7} \sin \theta$ **34.** $x = 4 + 5 \cos \theta$, $y = 2 + 4 \sin \theta$

35. $x = 4 \sec \theta$, $y = 3 \tan \theta$ **36.** $x = 2 \sec \theta$, $y = 2\sqrt{3} \tan \theta$

37. (a) $x = t,\ y = 3t - 2$ (b) $x = -t + 2,\ y = -3t + 4$

38. (a) $x = t, y = \frac{1}{3}(t + 2)$ (b) $x = -t + 2, y = -\frac{1}{3}(t - 4)$

39. (a) $x = t, y = t^2$ (b) $x = -t + 2, y = t^2 - 4t + 4$

40. (a) $x = t, y = t^3$ (b) $x = -t + 2, y = (-t + 2)^3$

41. (a) $x = t, y = t^2 + 1$ (b) $x = -t + 2, y = t^2 - 4t + 5$

42. (a) $x = t, y = 2 - t$ (b) $x = -t + 2, y = t$

43. (a) $x = t, y = \dfrac{1}{t}$ (b) $x = -t + 2, y = -\dfrac{1}{t - 2}$

44. (a) $x = t, y = \dfrac{1}{2t}$ (b) $x = -t + 2, y = \dfrac{1}{-2t + 4}$

45.

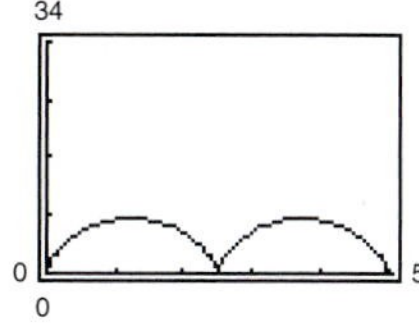

46.

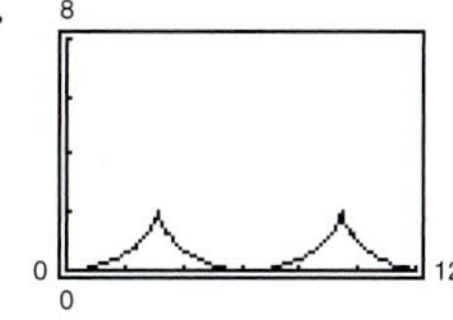

47.

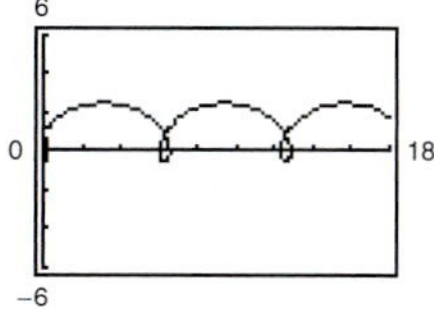

48.

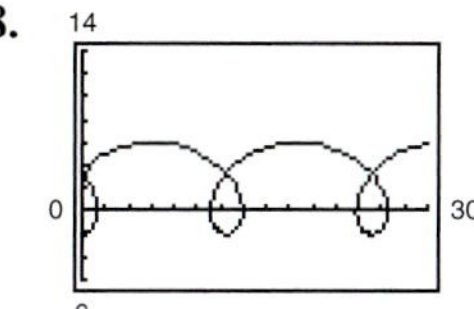

49.

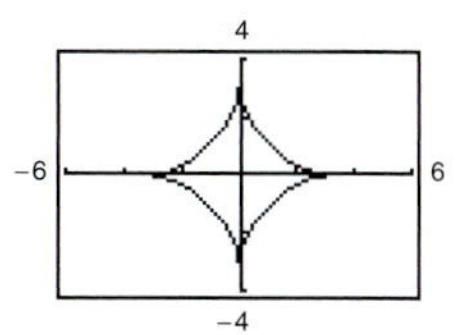

50.

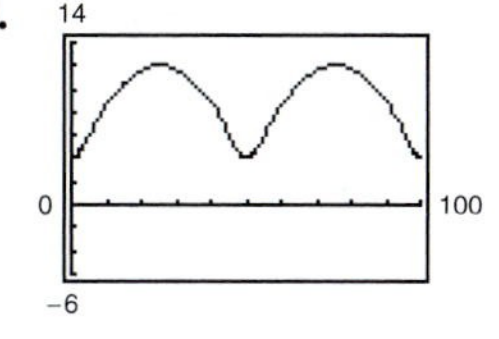

51.

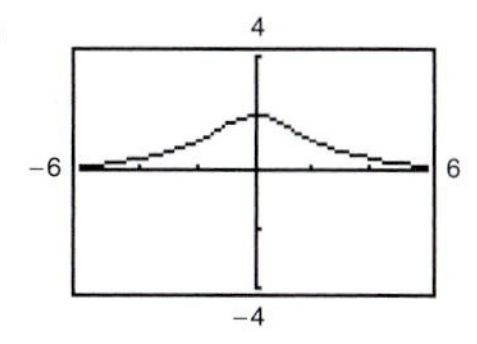

52.

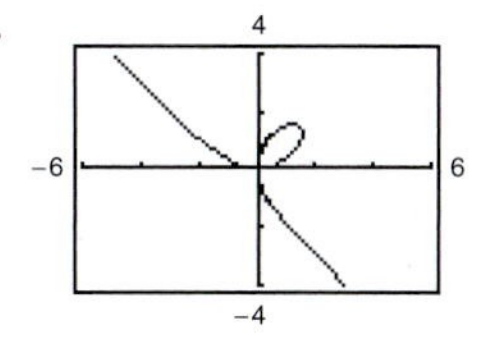

53. b
Domain: $[-2, 2]$
Range: $[-1, 1]$

54. c
Domain: $[-4, 4]$
Range: $[-6, 6]$

55. d

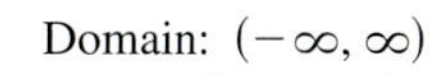

Domain: $(-\infty, \infty)$
Range: $(-\infty, \infty)$

56. a
Domain: $(-\infty, \infty)$
Range: $[-2, 2]$

57. (a)

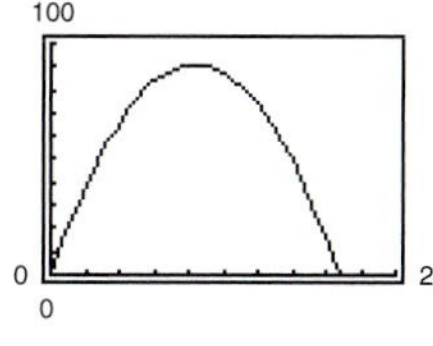

Maximum height: 90.7 feet
Range: 209.6 feet

(b)

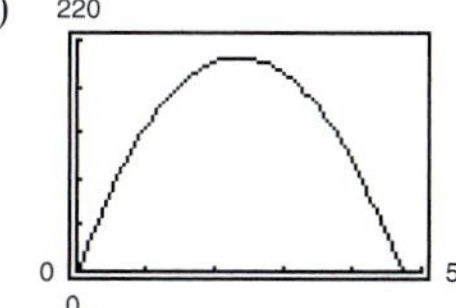

Maximum height: 204.2 feet
Range: 471.6 feet

(c)

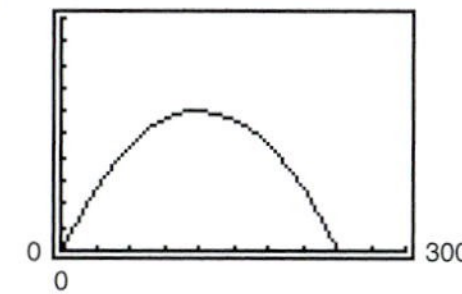

Maximum height: 60.5 feet
Range: 242.0 feet

(d)

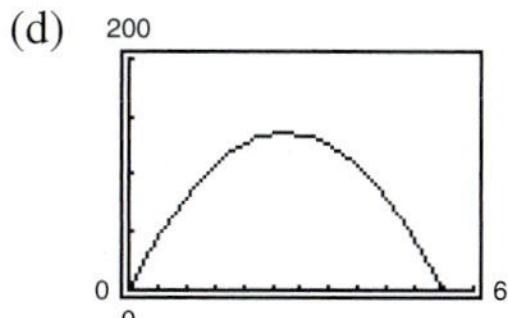

Maximum height: 136.1 feet
Range: 544.5 feet

58. (a)

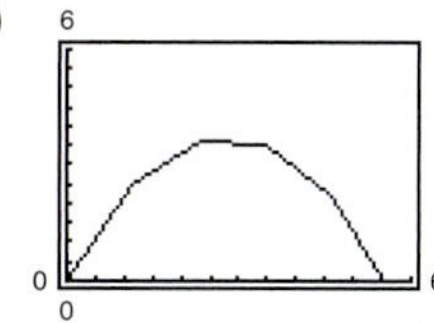

Maximum height: 3.8 feet
Range: 56.3 feet

(b)

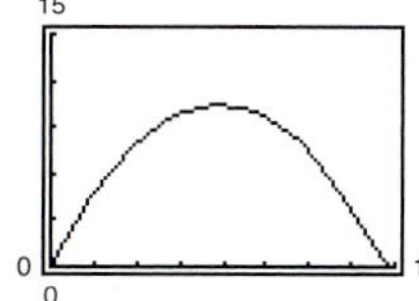

Maximum height: 10.5 feet
Range: 156.3 feet

(c)

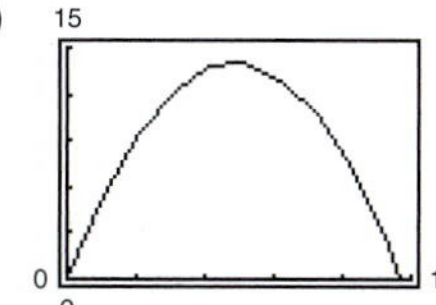

Maximum height: 14.1 feet
Range: 97.4 feet

(d)

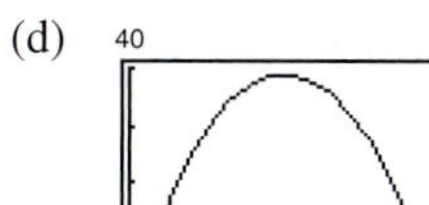

Maximum height: 39.1 feet
Range: 270.6 feet

59. (a) $x = (146.67 \cos \theta)t$
$y = 3 + (146.67 \sin \theta)t - 16t^2$

(b) No

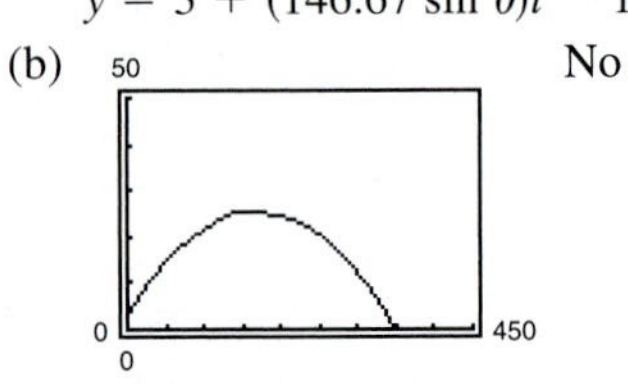

(c) Yes

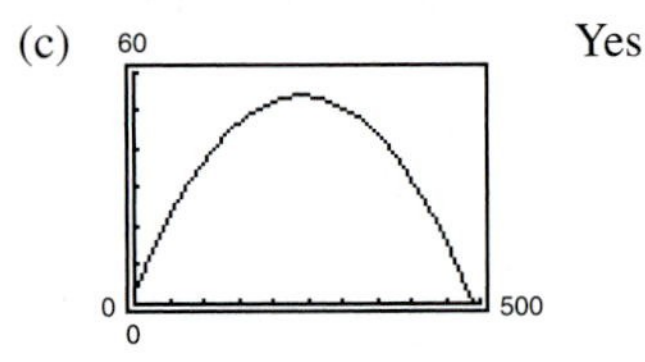

(d) 19.3°

60. (a) $x = (240 \cos 10°)t$
$y = 5 + (240 \sin 10°)t - 16t^2$

(b) 643 feet

(c) 32.1 feet

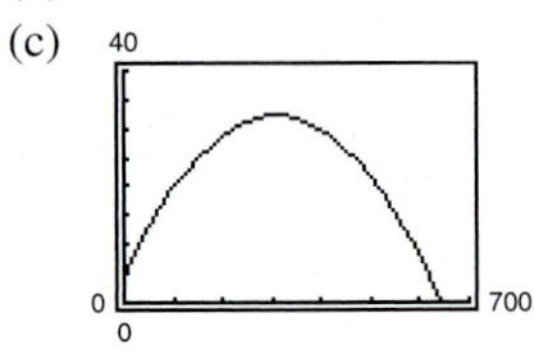

(d) 2.72 seconds

61. Answers will vary.

62. (a) $h = 7, v_0 = 40, \theta = 45°$
$x = (40 \cos 45°)t$
$y = 7 + (40 \sin 45°)t - 16t^2$

(b)

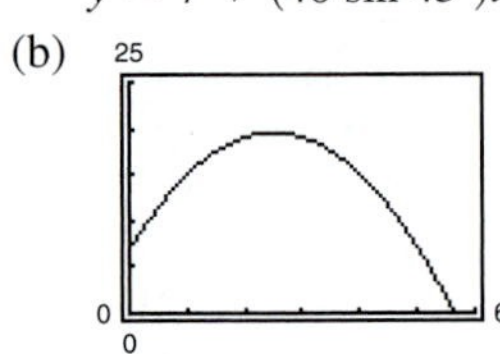

(c) Maximum height: 19.5 feet
Range: 56.2 feet

63. $x = a\theta - b \sin \theta$
$y = a - b \cos \theta$

64. $x = 3 \cos \theta - \cos 3\theta$
$y = 3 \sin \theta - \sin 3\theta$

65. True
$x = t$
$y = t^2 + 1 \Rightarrow y = x^2 + 1$
$x = 3t$
$y = 9t^2 + 1 \Rightarrow y = x^2 + 1$

66. False. $y = x$ for $x \geq 0$

67. Parametric equations are useful when graphing two functions simultaneously on the same coordinate system. For example, they are useful when tracking the path of an object so that the position and the time associated with that position can be determined.

68. Sketching a plane curve starts by choosing a numeric value for the parameter. Then, the coordinates can be determined from the value chosen for the parameter. Finally, plotting the resulting points in the order of increasing parameter values shows the direction, or orientation, of the curve.

69. $\theta' = 75°$

70. $\theta' = 50°$

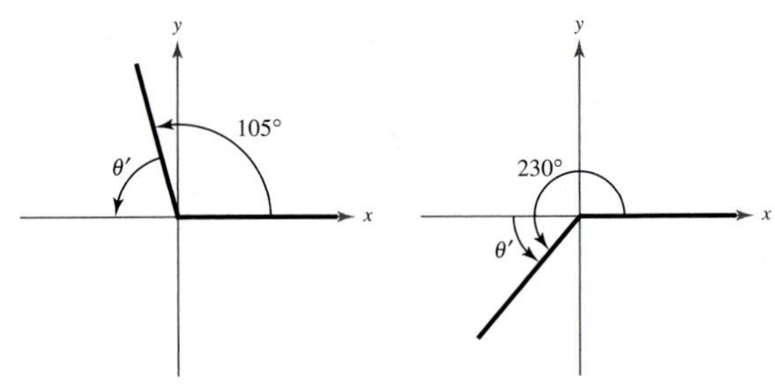

71. $\theta' = \frac{\pi}{3}$

72. $\theta' = \frac{\pi}{6}$

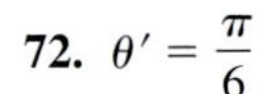

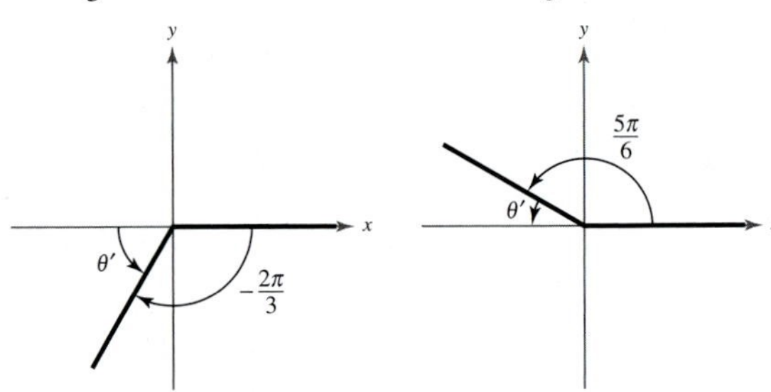

Section 6.6 *(page 497)*

Vocabulary Check *(page 497)*

1. pole **2.** directed distance; directed angle

3. polar **4.** $x = r \cos \theta$ $\tan \theta = \frac{y}{x}$
$y = r \sin \theta$ $r^2 = x^2 + y^2$

1.

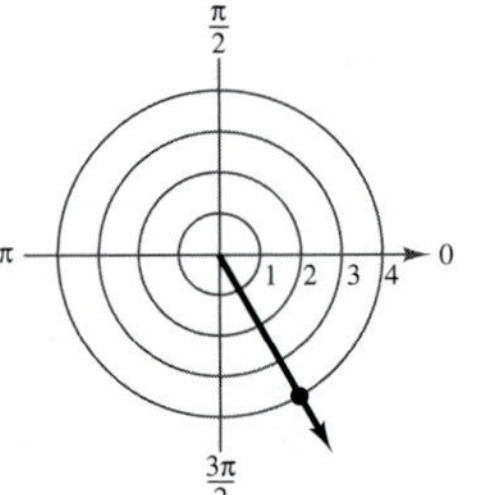

$\left(4, \frac{5\pi}{3}\right), \left(-4, -\frac{4\pi}{3}\right)$

2.

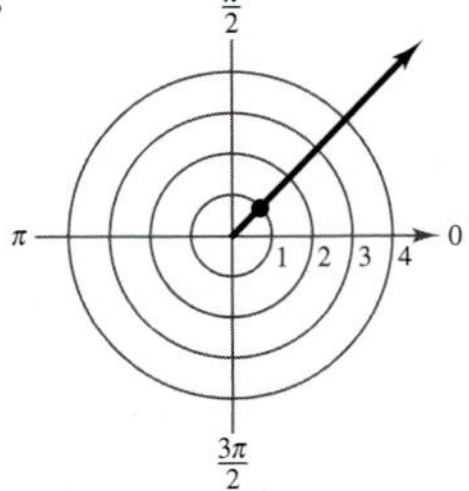

$\left(-1, \frac{5\pi}{4}\right), \left(1, \frac{\pi}{4}\right)$

3.

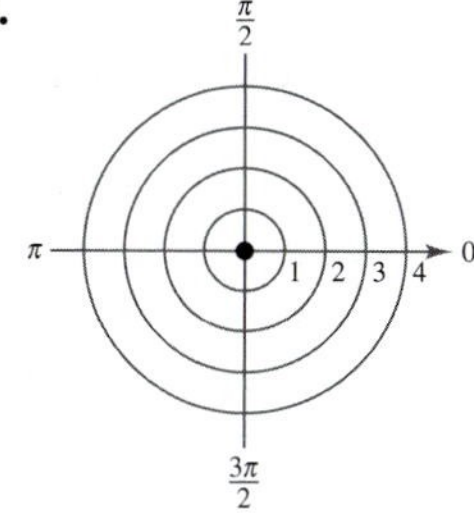

$\left(0, \frac{5\pi}{6}\right), \left(0, -\frac{13\pi}{6}\right)$

4.

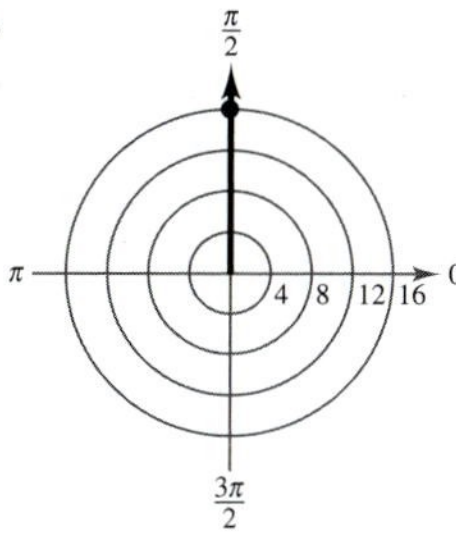

$\left(16, \frac{\pi}{2}\right), \left(-16, \frac{3\pi}{2}\right)$

5.

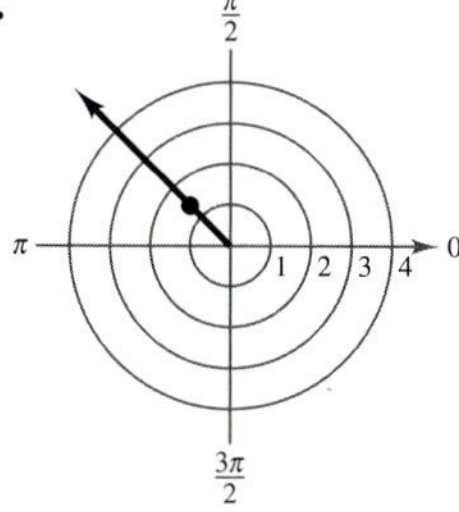

$(\sqrt{2}, 8.64), (-\sqrt{2}, -0.78)$

6.

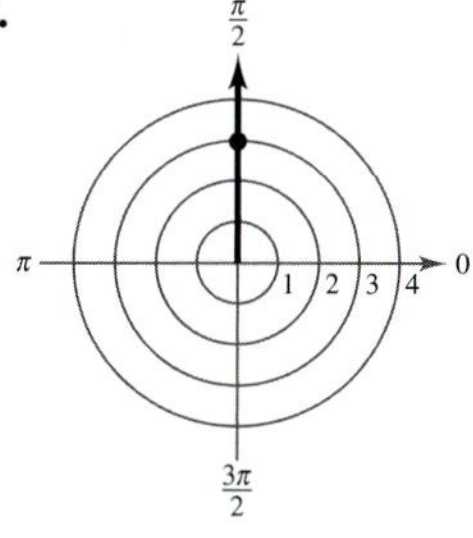

$(-3, 4.7132), (3, 1.5716)$

7.

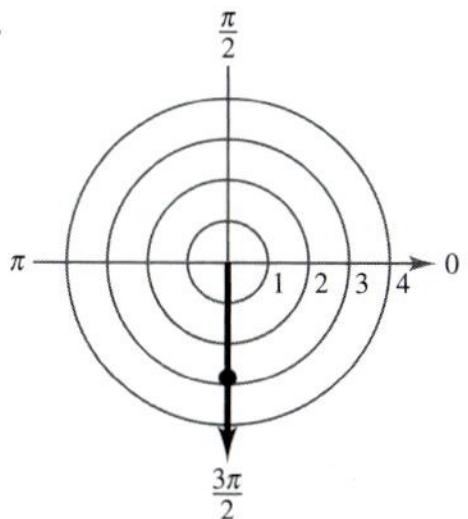

$(2\sqrt{2}, 10.99), (-2\sqrt{2}, 7.85)$

8.

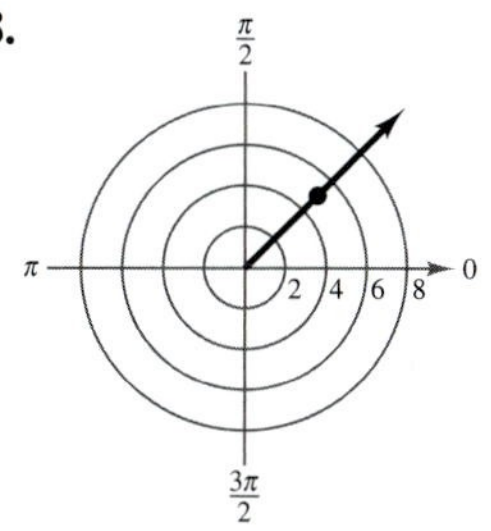

$(-5, 3.9232), (5, 0.7816)$

9. $(0, 3)$ **10.** $(0, -3)$ **11.** $\left(\frac{\sqrt{2}}{2}, \frac{\sqrt{2}}{2}\right)$ **12.** $(0, 0)$

13. $(-\sqrt{2}, \sqrt{2})$ **14.** $(\sqrt{3}, 1)$

15. $(-1.1340, -2.2280)$ **16.** $(-7.7258, -2.8940)$

17. $\left(\sqrt{2}, \frac{\pi}{4}\right)$ **18.** $\left(3\sqrt{2}, \frac{5\pi}{4}\right)$ **19.** $(6, \pi)$

20. $\left(5, \frac{3\pi}{2}\right)$ **21.** $(5, 2.2143)$ **22.** $(\sqrt{10}, 5.9614)$

23. $\left(\sqrt{6}, \frac{5\pi}{4}\right)$ **24.** $\left(2, \frac{11\pi}{6}\right)$ **25.** $(3\sqrt{13}, 0.9828)$

26. $(13, 1.1760)$ **27.** $(\sqrt{13}, 5.6952)$

28. $(\sqrt{29}, 2.7611)$ **29.** $(\sqrt{7}, 0.8571)$ **30.** $\left(6, \frac{\pi}{4}\right)$

31. $\left(\frac{17}{6}, 0.4900\right)$ **32.** $(2.3049, 0.7086)$ **33.** $r = 3$

34. $r = 4$ **35.** $r = 4 \csc \theta$ **36.** $\theta = \frac{\pi}{4}$

37. $r = 10 \sec \theta$ **38.** $r = 4a \sec \theta$

39. $r = \frac{-2}{3 \cos \theta - \sin \theta}$ **40.** $r = \frac{2}{3 \cos \theta + 5 \sin \theta}$

41. $r^2 = 16 \sec \theta \csc \theta = 32 \csc 2\theta$

42. $r^2 = \frac{1}{2} \sec \theta \csc \theta = \csc 2\theta$

43. $r = \frac{4}{1 - \cos \theta}$ or $-\frac{4}{1 + \cos \theta}$ **44.** $r^2 = 9 \cos 2\theta$

45. $r = a$ **46.** $r = 3a$ **47.** $r = 2a \cos \theta$

48. $r = 2a \sin \theta$ **49.** $x^2 + y^2 - 4y = 0$

50. $x^2 + y^2 - 2x = 0$ **51.** $\sqrt{3}x + y = 0$

52. $\sqrt{3}x + y = 0$ **53.** $x^2 + y^2 = 16$

54. $x^2 + y^2 = 100$ **55.** $y = 4$ **56.** $x = -3$

57. $x^2 + y^2 - x^{2/3} = 0$ **58.** $(x^2 + y^2)^2 = 2xy$

59. $(x^2 + y^2)^2 = 6x^2y - 2y^3$ **60.** $(x^2 + y^2)^3 = 9(x^2 - y^2)^2$

61. $x^2 + 4y - 4 = 0$ **62.** $y^2 = 2x + 1$

63. $4x^2 - 5y^2 - 36y - 36 = 0$ **64.** $2x - 3y = 6$

65. The graph of the polar equation consists of all points that are six units from the pole.
$x^2 + y^2 = 36$

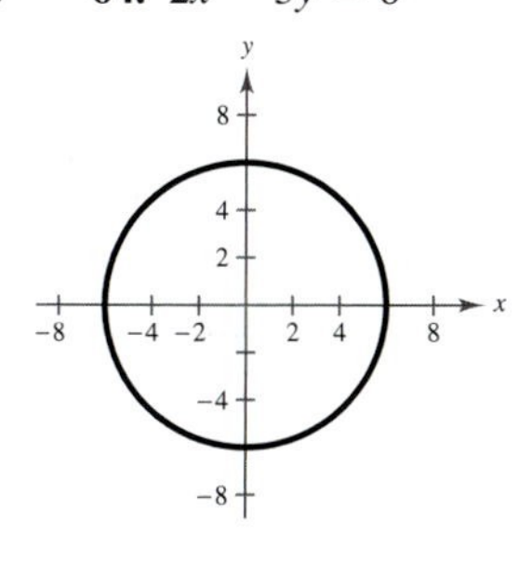

66. The graph of the polar equation consists of all points that are eight units from the pole.
$x^2 + y^2 = 64$

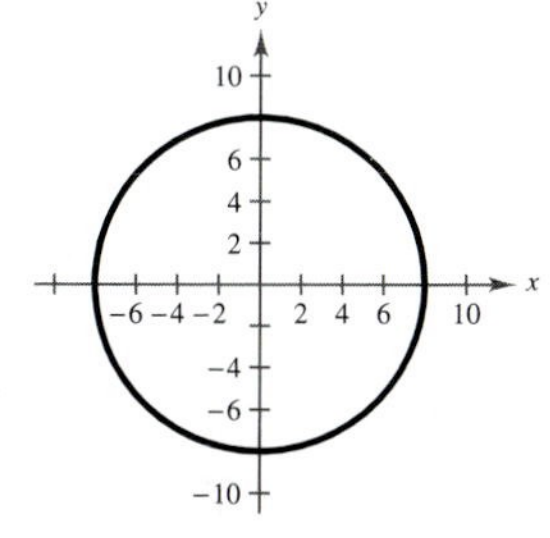

67. The graph of the polar equation consists of all points on the line that make an angle of $\pi/6$ with the positive polar axis.
$-\sqrt{3}x + 3y = 0$

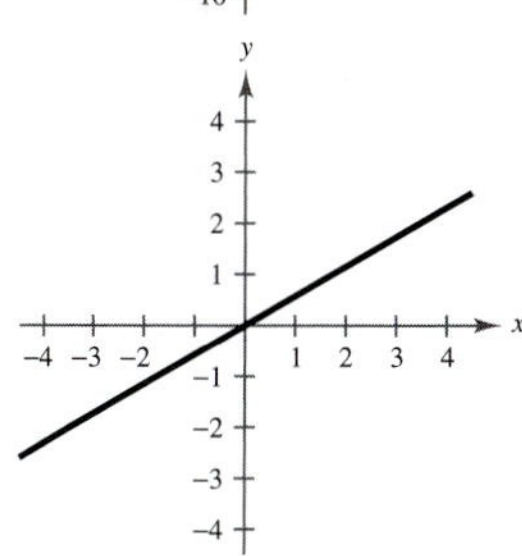

68. The graph of the polar equation consists of all points on the line that make an angle of $3\pi/4$ with the positive polar axis.
$x + y = 0$

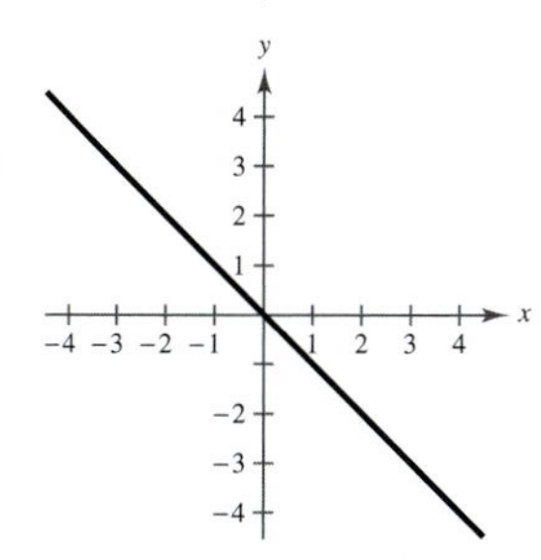

69. The graph of the polar equation is not evident by simple inspection, so convert to rectangular form.
$x - 3 = 0$

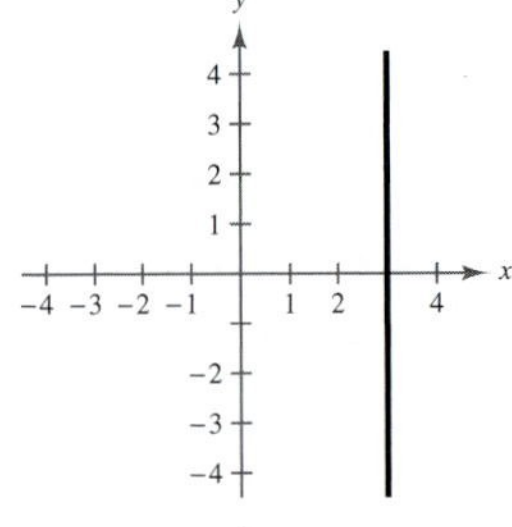

70. The graph of the polar equation is not evident by simple inspection, so convert to rectangular form.
$y - 2 = 0$

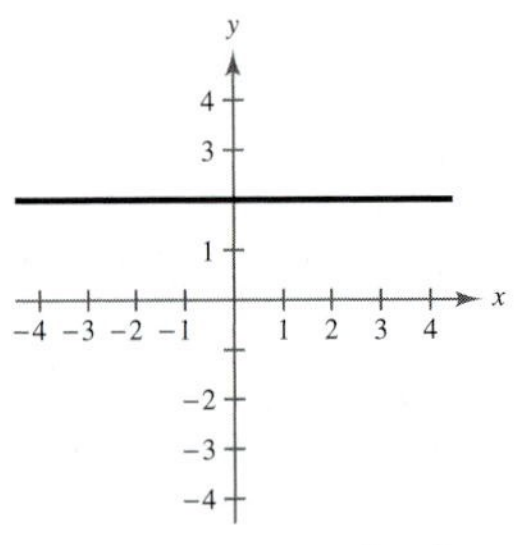

71. True. Because r is a directed distance, the point (r, θ) can be represented as $(r, \theta \pm 2\pi n)$.

72. False. If $r_1 = -r_2$, then (r_1, θ) and (r_2, θ) are different points.

73. $(x - h)^2 + (y - k)^2 = h^2 + k^2$
Radius: $\sqrt{h^2 + k^2}$
Center: (h, k)

74. $\left(x - \frac{1}{2}\right)^2 + \left(y - \frac{3}{2}\right)^2 = \frac{5}{2}$; circle

75. (a) Answers will vary.
(b) (r_1, θ_1), (r_2, θ_2) and the pole are collinear.
$d = \sqrt{r_1^2 + r_2^2 - 2r_1r_2} = |r_1 - r_2|$
This represents the distance between two points on the line $\theta = \theta_1 = \theta_2$.
(c) $d = \sqrt{r_1^2 + r_2^2}$
This is the result of the Pythagorean Theorem.
(d) Answers will vary. For example:
Points: $(3, \pi/6)$, $(4, \pi/3)$
Distance: 2.053
Points: $(-3, 7\pi/6)$, $(-4, 4\pi/3)$
Distance: 2.053

76. (a) Horizontal: x-coordinate changes
Vertical: y-coordinate changes
(b) Horizontal: r and θ both change
Vertical: r and θ both change
(c) Unlike r and θ, x and y measure horizontal and vertical changes, respectively.

77. $2 \log_6 x + \log_6 z - \log_6 3 - \log_6 y$

78. $\frac{1}{4} + \frac{1}{2} \log_4 x - \log_4 y$ **79.** $\ln x + 2 \ln(x + 4)$

80. $\ln 5 + 2 \ln x + \ln(x^2 + 1)$ **81.** $\log_7 \dfrac{x}{3y}$

82. $\log_5 a(x + 1)^8$ **83.** $\ln \sqrt{x}(x - 2)$ **84.** $\ln \dfrac{6y}{x - 3}$

Section 6.7 *(page 505)*

Vocabulary Check *(page 505)*

1. $\theta = \dfrac{\pi}{2}$ **2.** polar axis **3.** convex limaçon
4. circle **5.** lemniscate **6.** cardioid

1. Rose curve with 4 petals **2.** Cardioid
3. Limaçon with inner loop **4.** Lemniscate
5. Rose curve with 4 petals **6.** Circle

7. Polar axis **8.** Polar axis **9.** $\theta = \dfrac{\pi}{2}$

10. Polar axis **11.** $\theta = \dfrac{\pi}{2}$, polar axis, pole **12.** Pole

13. Maximum: $|r| = 20$ when $\theta = \dfrac{3\pi}{2}$
Zero: $r = 0$ when $\theta = \dfrac{\pi}{2}$

14. Maximum: $|r| = 18$ when $\theta = 0$
Zeros: $r = 0$ when $\theta = \dfrac{2\pi}{3}, \dfrac{4\pi}{3}$

15. Maximum: $|r| = 4$ when $\theta = 0, \dfrac{\pi}{3}, \dfrac{2\pi}{3}$
Zero: $r = 0$ when $\theta = \dfrac{\pi}{6}, \dfrac{\pi}{2}, \dfrac{5\pi}{6}$

16. Maximum: $|r| = 3$ when $\theta = \dfrac{\pi}{4}, \dfrac{3\pi}{4}, \dfrac{5\pi}{4}, \dfrac{7\pi}{4}$
Zeros: $r = 0$ when $\theta = 0, \dfrac{\pi}{2}, \pi, \dfrac{3\pi}{2}$

17.

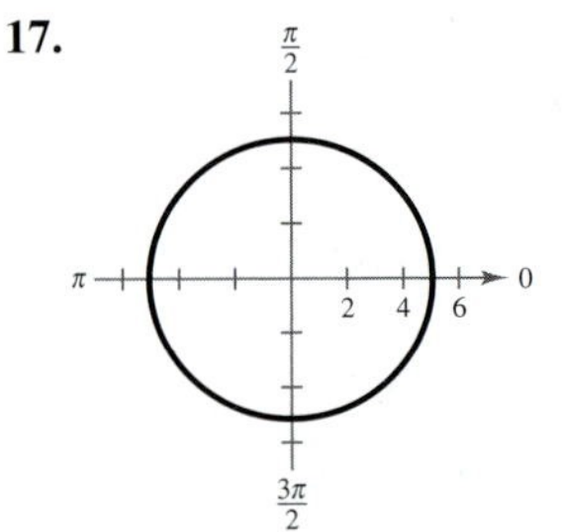

18.

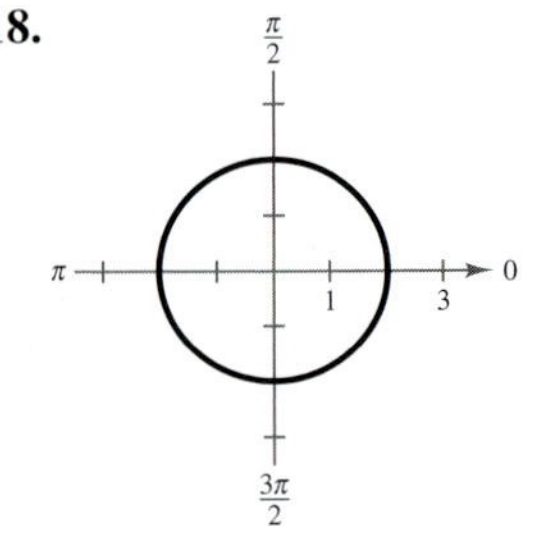

19.

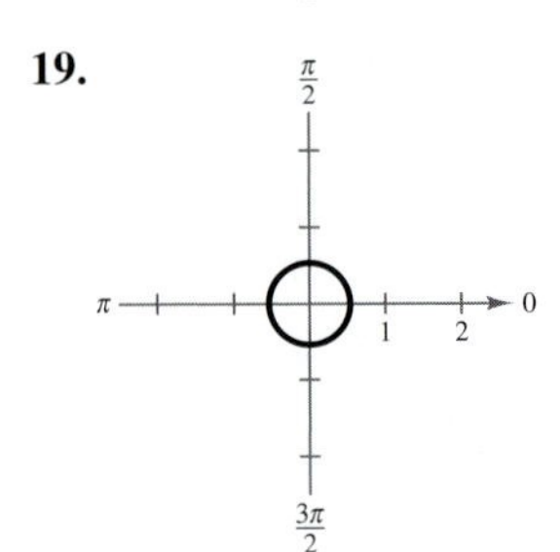

20.

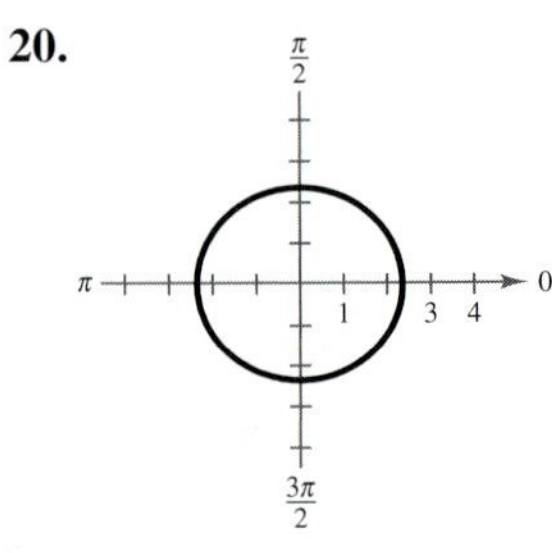

21.

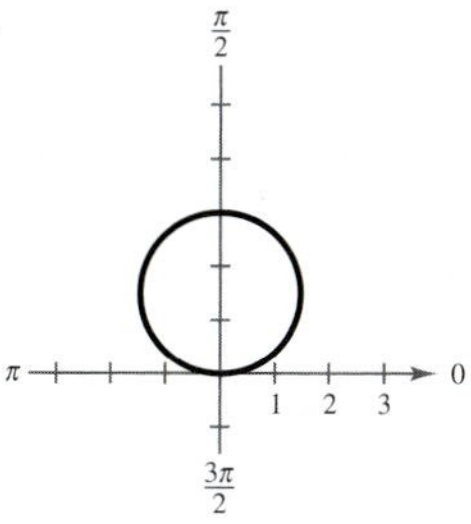

22.

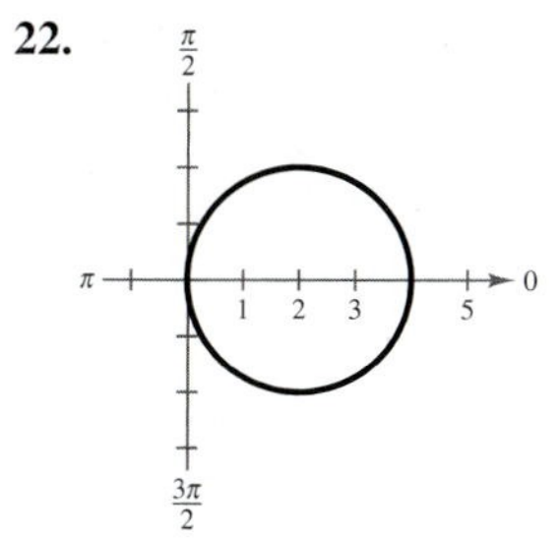

23.

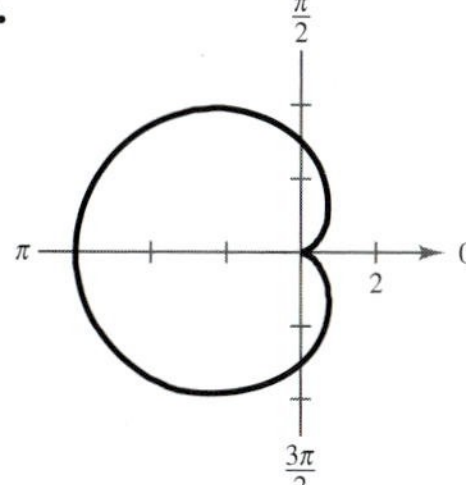

24.

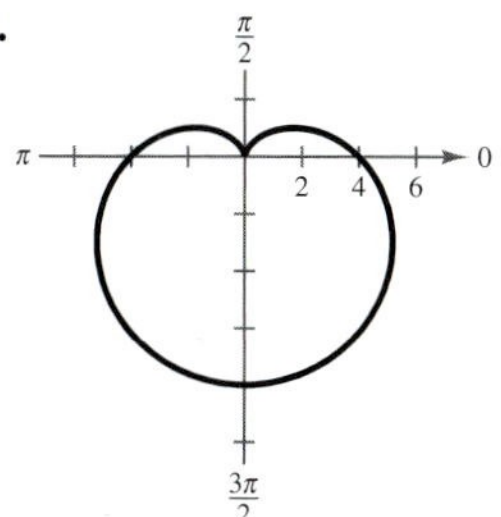

25.

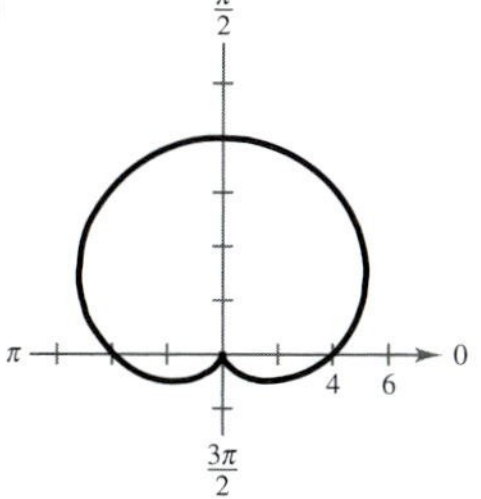

26.

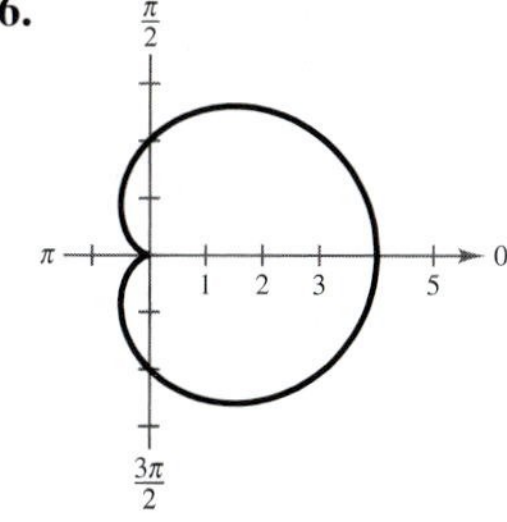

27.

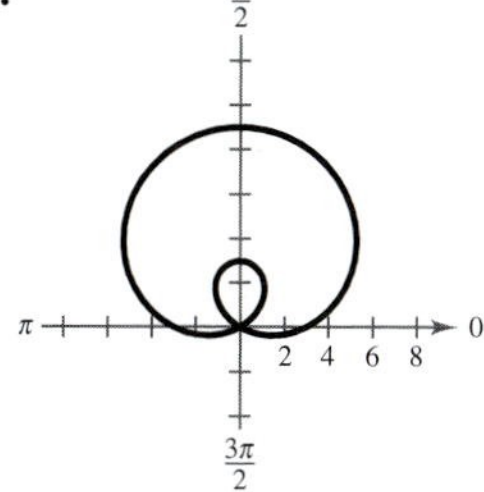

28.

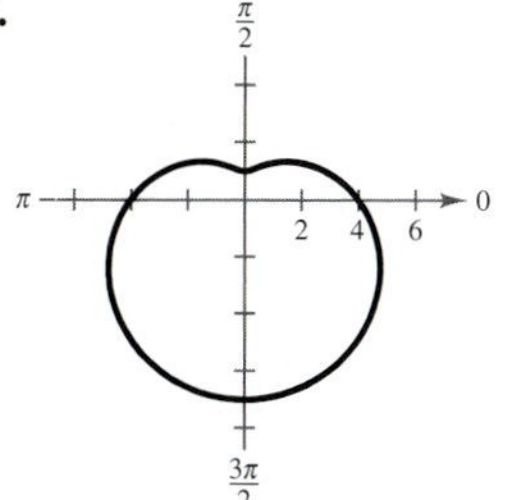

29.

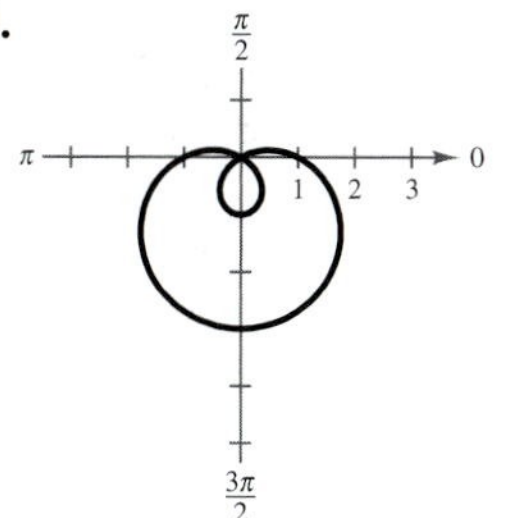

30.

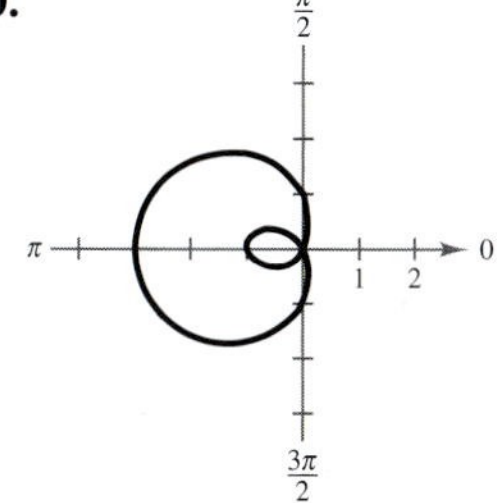

31.

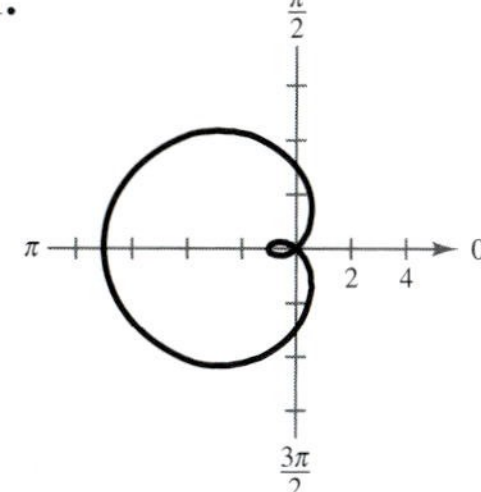

32.

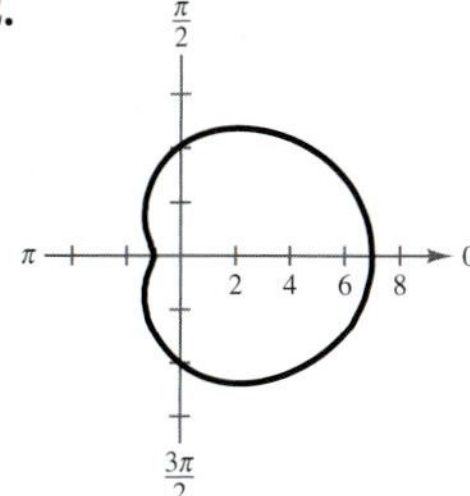

33.

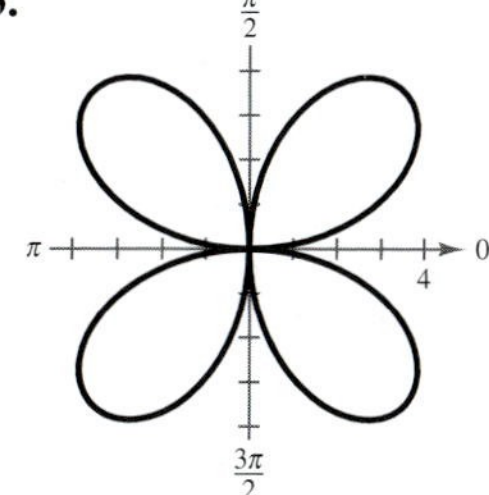

34.

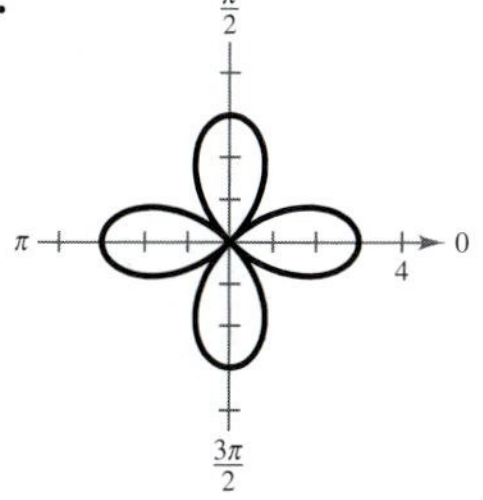

35.

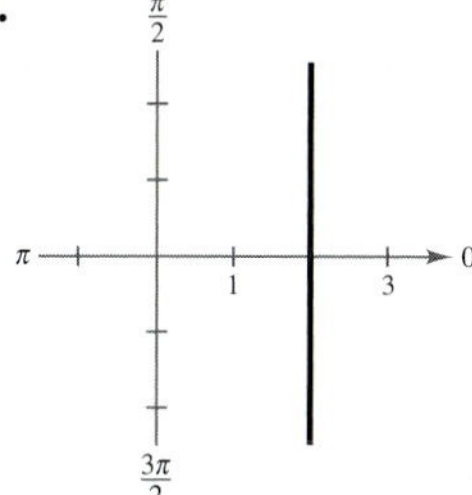

36.

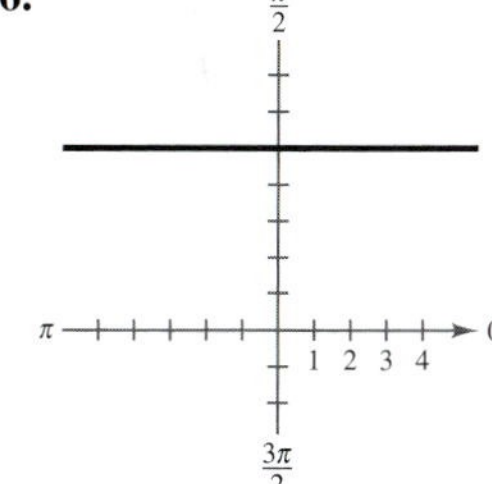

37.

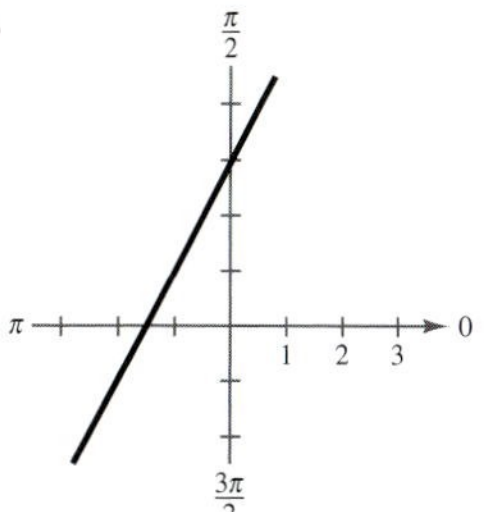

38.

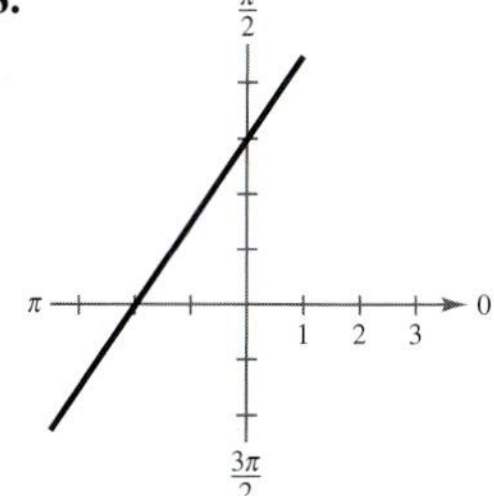

39.

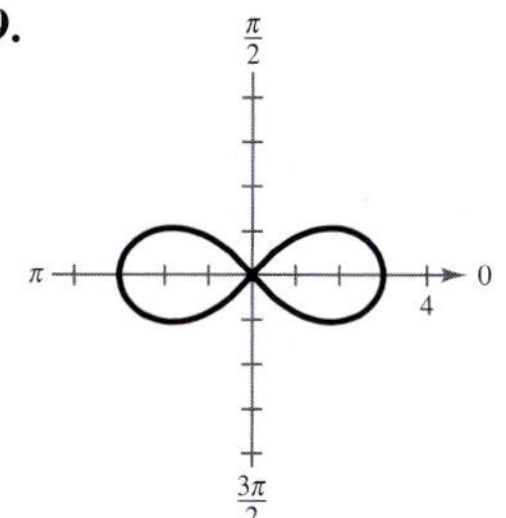

40.

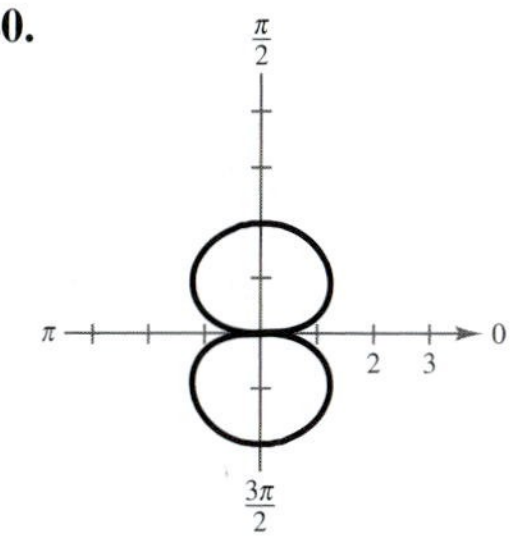

CHAPTER 6

41.
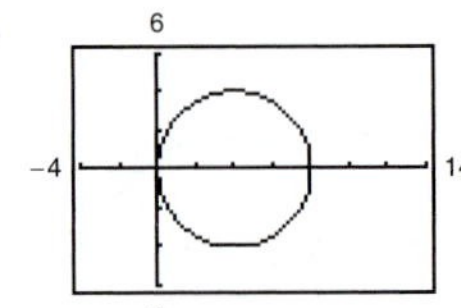

42.
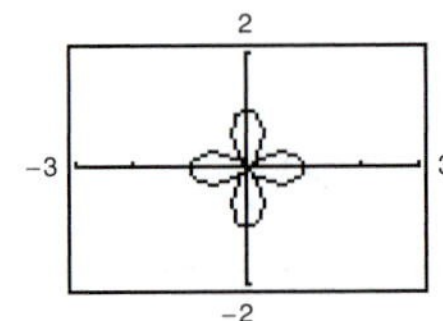

43.
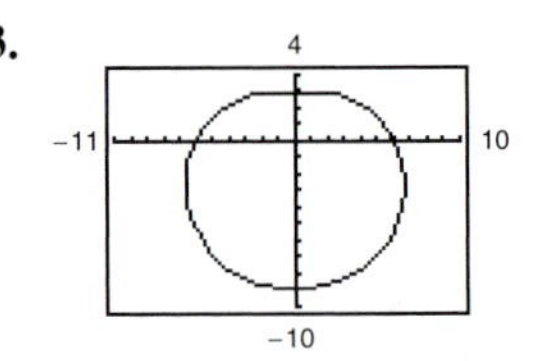

44.
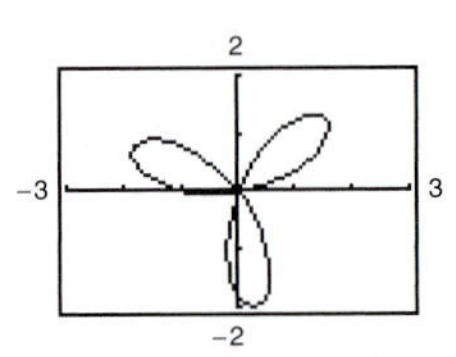

45.
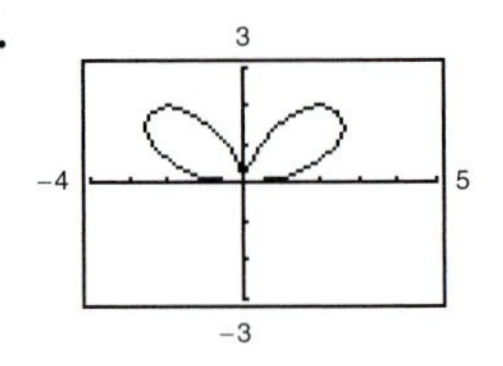

46.
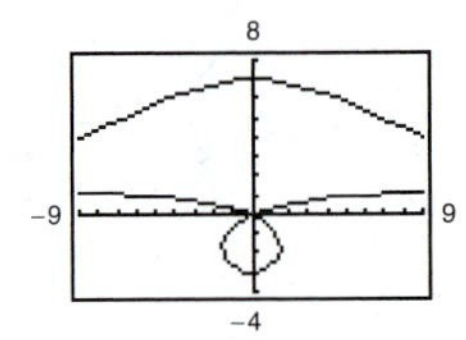

47.
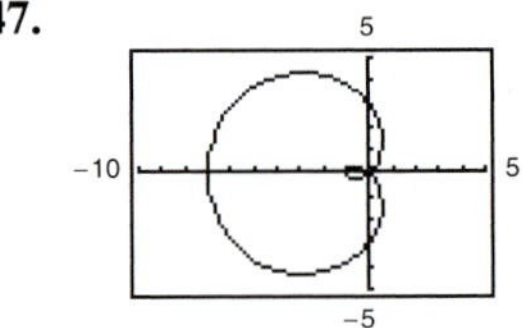

$0 \le \theta < 2\pi$

48.
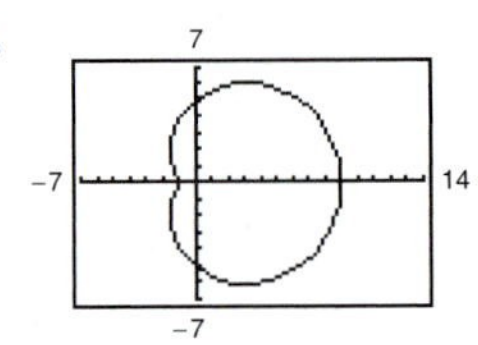

$0 \le \theta < 2\pi$

49.
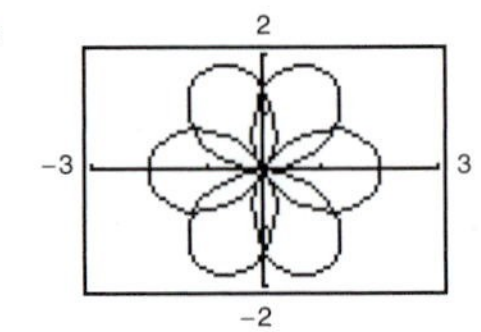

$0 \le \theta < 4\pi$

50.
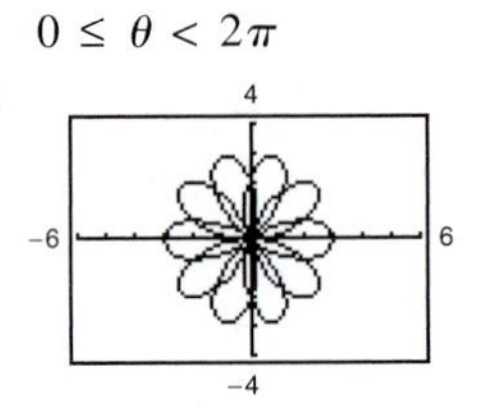

$0 \le \theta < 4\pi$

51.
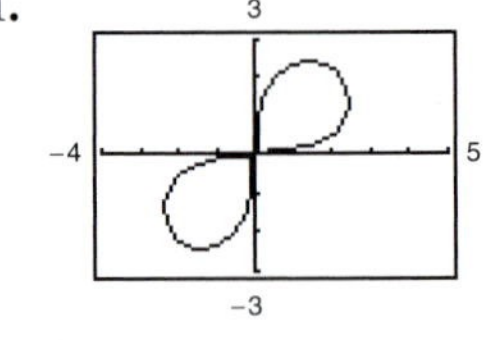

$0 \le \theta < \pi$

52.
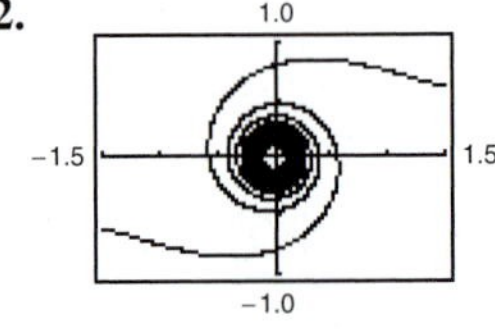

$0 \le \theta < \infty$

53.
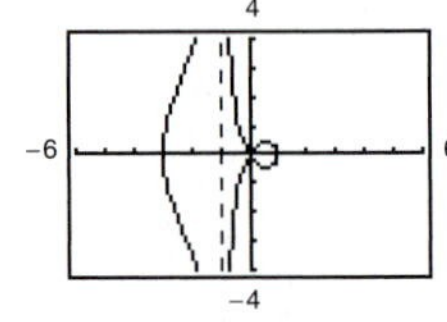

54.
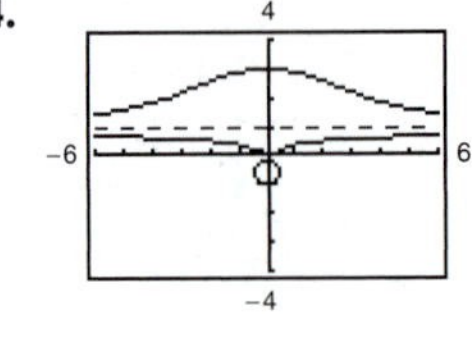

55.
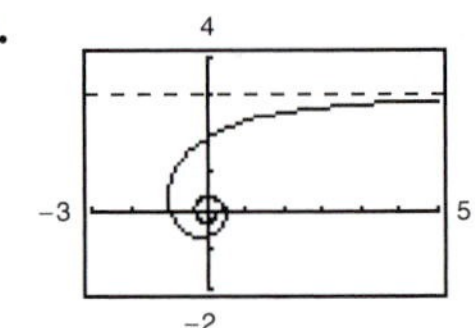

56.
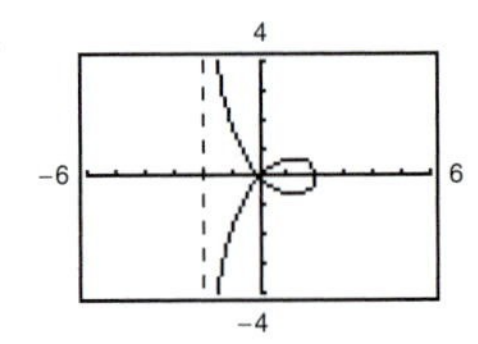

57. True. For a graph to have polar axis symmetry, replace (r, θ) by $(r, -\theta)$ or $(-r, \pi - \theta)$.

58. False. For a graph symmetric with respect to the pole, one portion of the graph coincides with the other portion when rotated π radians about the pole.

59. (a)
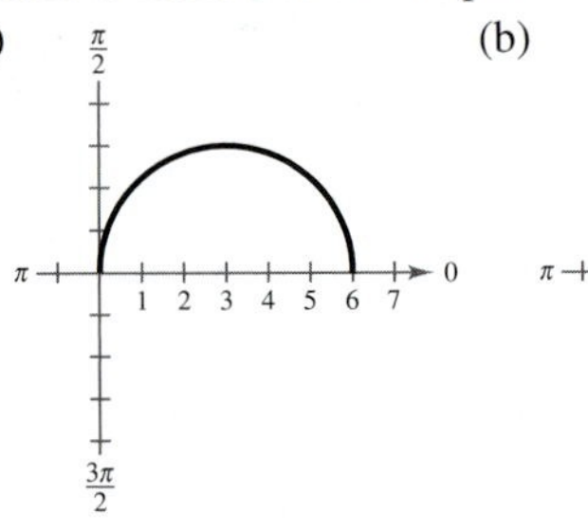

Upper half of circle

(b)
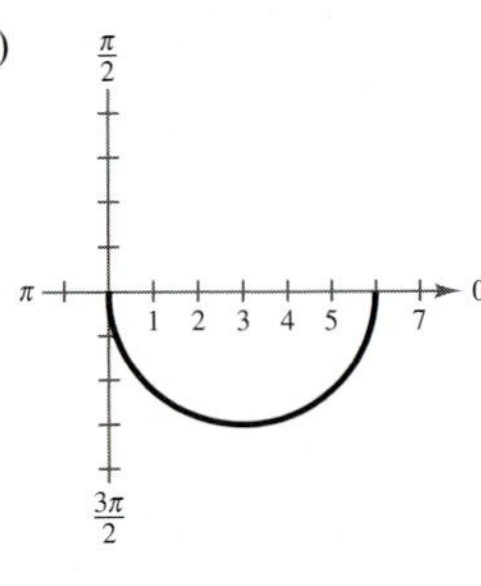

Lower half of circle

(c)
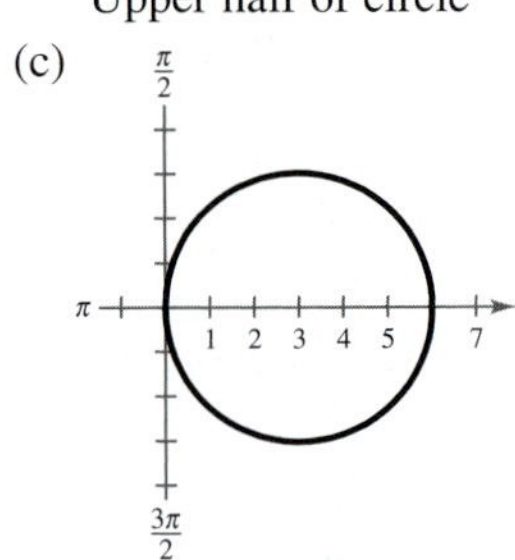

Full circle

(d)
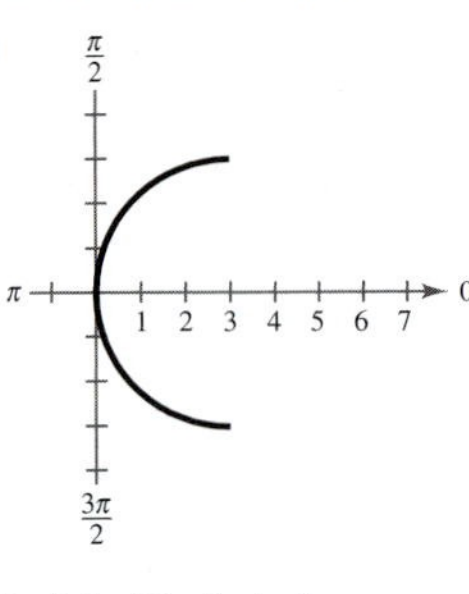

Left half of circle

60. (a)
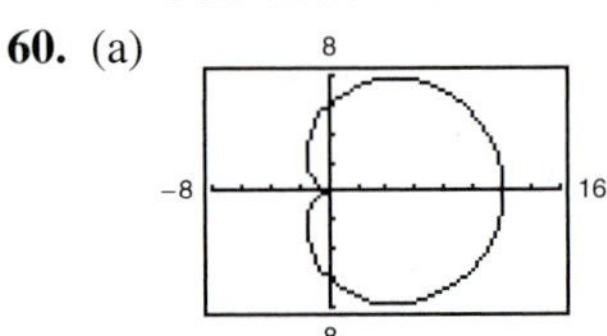

(b)
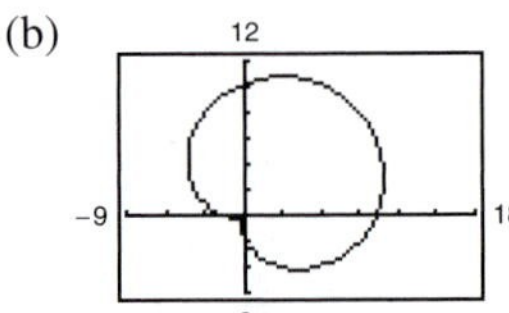

(c)
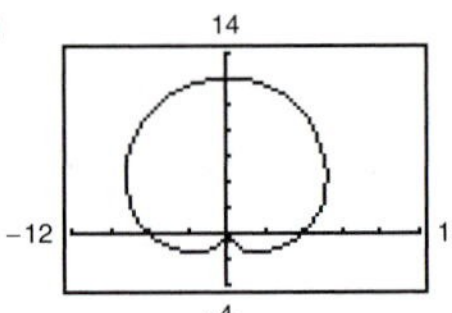

The angle ϕ controls rotation of the axis of symmetry.
$r = 6(1 + \sin\theta)$

61–62. Answers will vary.

63. (a) $r = 2 - \dfrac{\sqrt{2}}{2}(\sin\theta - \cos\theta)$ (b) $r = 2 + \cos\theta$

(c) $r = 2 + \sin\theta$ (d) $r = 2 - \cos\theta$

64. (a) $r = 4\sin\left(\theta - \dfrac{\pi}{6}\right)\cos\left(\theta - \dfrac{\pi}{6}\right)$

(b) $r = -4\sin\theta\cos\theta$

(c) $r = 4\sin\left(\theta - \dfrac{2\pi}{3}\right)\cos\left(\theta - \dfrac{2\pi}{3}\right)$

(d) $r = 4\sin\theta\cos\theta$

65. (a) (b)

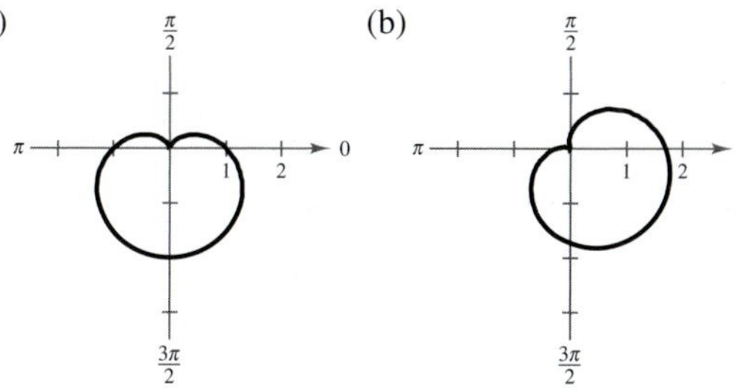

66. (a) (b)

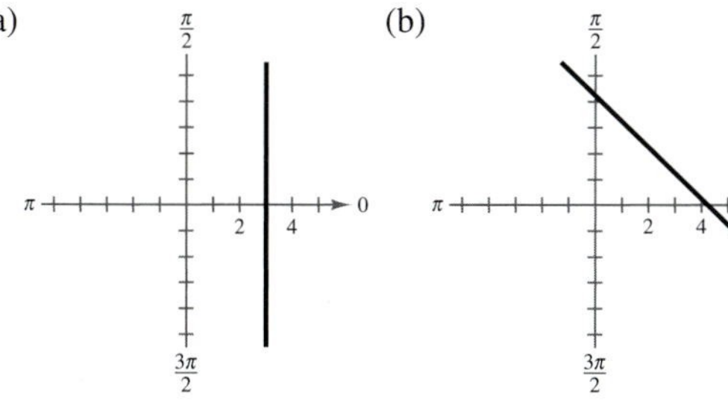

(c) (d)

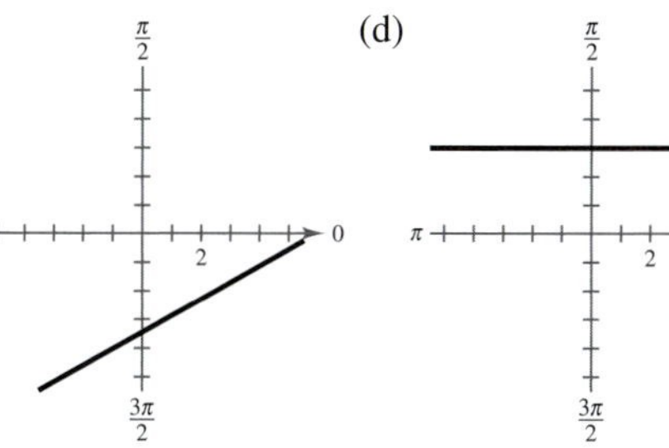

67.

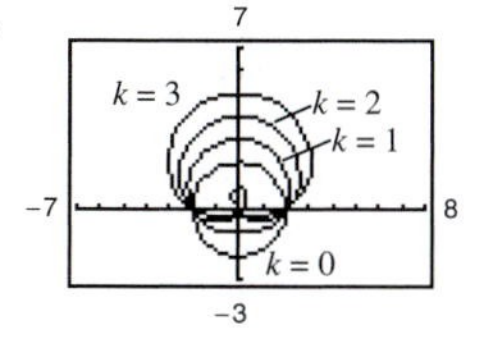

$k = 0$, circle
$k = 1$, convex limaçon
$k = 2$, cardioid
$k = 3$, limaçon with inner loop

68. (a)

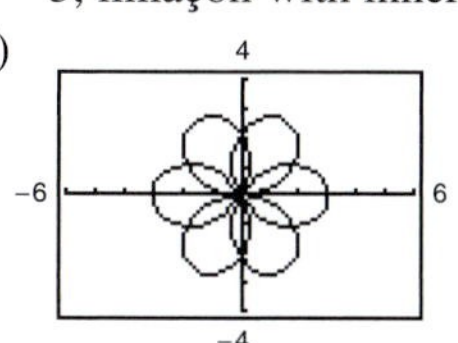

$0 \le \theta < 4\pi$

(b)

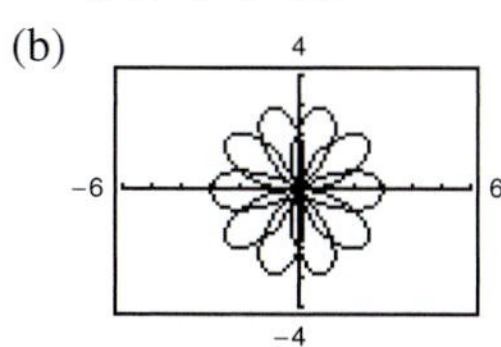

$0 \le \theta < 4\pi$

(c) Yes. Explanations will vary.

69. ± 3 **70.** No zeros **71.** $\frac{13}{5}$ **72.** 3

73. $\dfrac{(x+1)^2}{9} + \dfrac{(y-2)^2}{4} = 1$

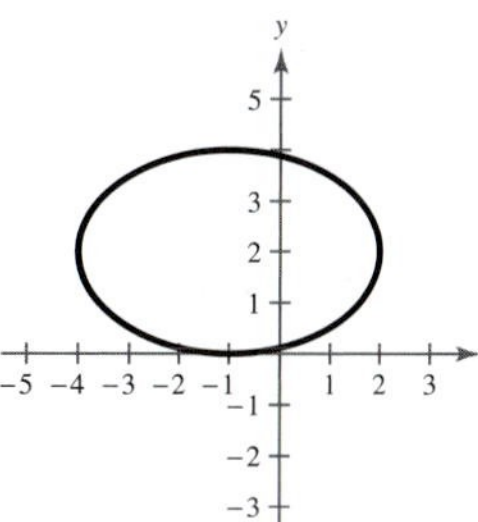

74. $\dfrac{(x-3)^2}{7} + \dfrac{(y+1)^2}{16} = 1$

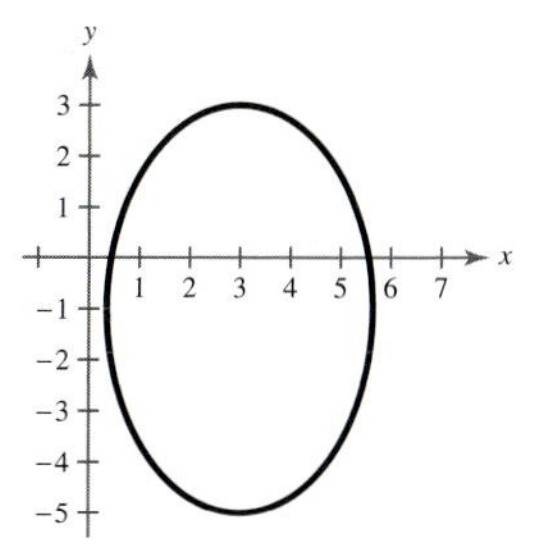

Section 6.8 *(page 511)*

Vocabulary Check *(page 511)*

1. conic **2.** eccentricity; e **3.** vertical; right
4. (a) iii (b) i (c) ii

1. $e = 1$: $r = \dfrac{4}{1 + \cos\theta}$, parabola

$e = 0.5$: $r = \dfrac{2}{1 + 0.5\cos\theta}$, ellipse

$e = 1.5$: $r = \dfrac{6}{1 + 1.5\cos\theta}$, hyperbola

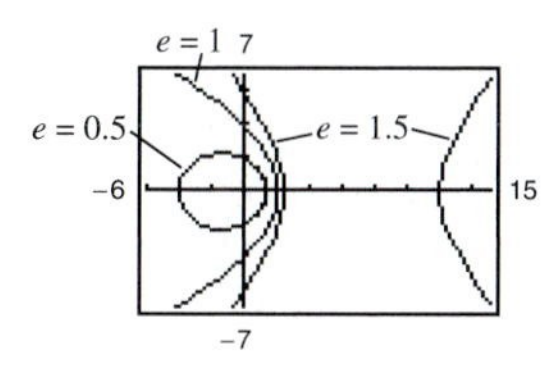

2. $e = 1$: $r = \dfrac{4}{1 - \cos\theta}$, parabola

$e = 0.5$: $r = \dfrac{2}{1 - 0.5\cos\theta}$, ellipse

$e = 1.5$: $r = \dfrac{6}{1 - 1.5\cos\theta}$, hyperbola

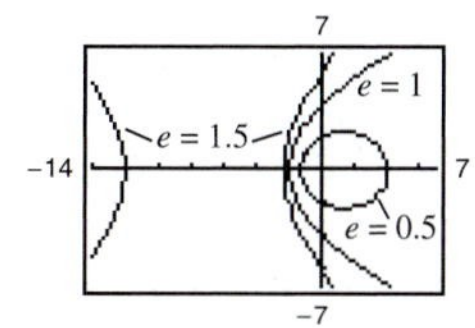

3. $e = 1$: $r = \dfrac{4}{1 - \sin\theta}$, parabola

$e = 0.5$: $r = \dfrac{2}{1 - 0.5\sin\theta}$, ellipse

$e = 1.5$: $r = \dfrac{6}{1 - 1.5\sin\theta}$, hyperbola

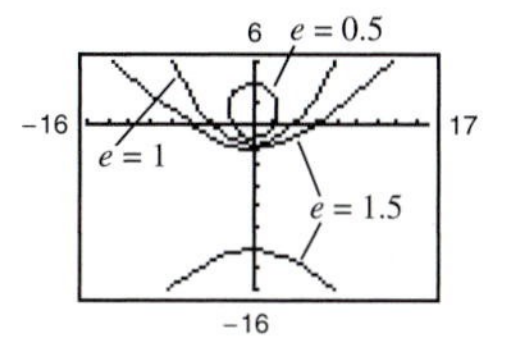

4. $e = 1$: $r = \dfrac{4}{1 + \sin\theta}$, parabola

$e = 0.5$: $r = \dfrac{2}{1 + 0.5\sin\theta}$, ellipse

$e = 1.5$: $r = \dfrac{6}{1 + 1.5\sin\theta}$, hyperbola

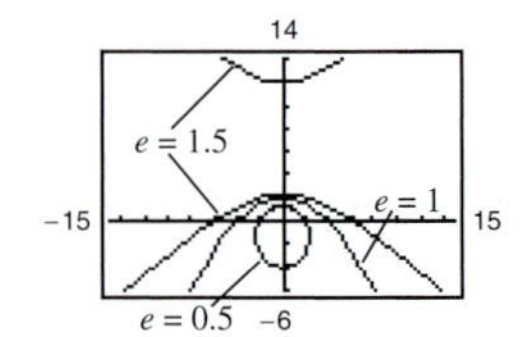

5. f **6.** c **7.** d **8.** e **9.** a **10.** b

11. Parabola

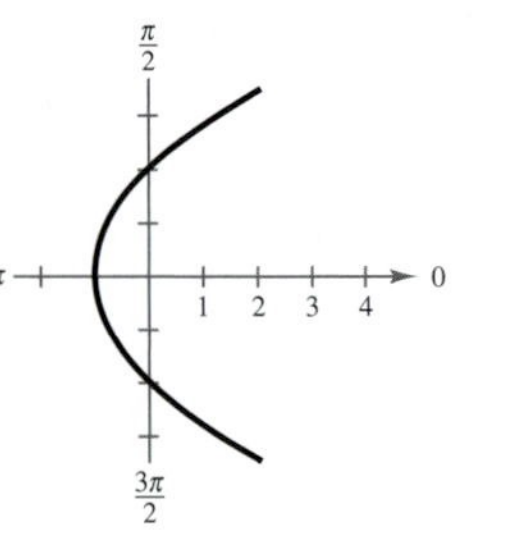

12. Parabola

13. Parabola

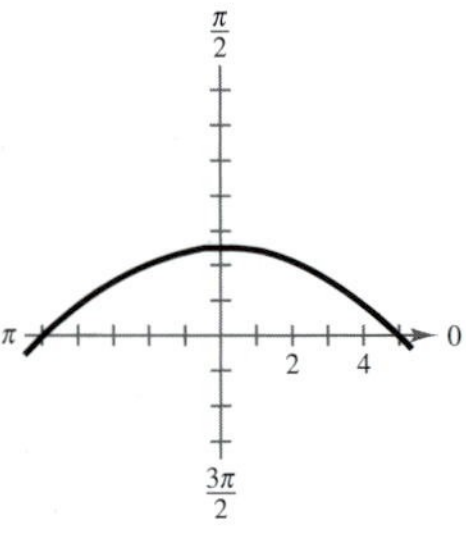

14. Parabola

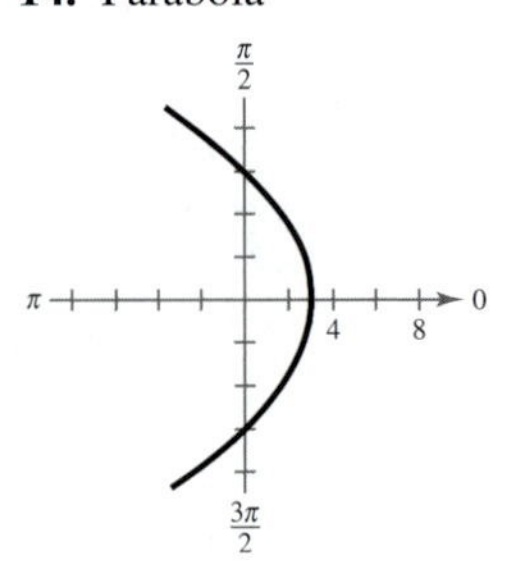

15. Ellipse

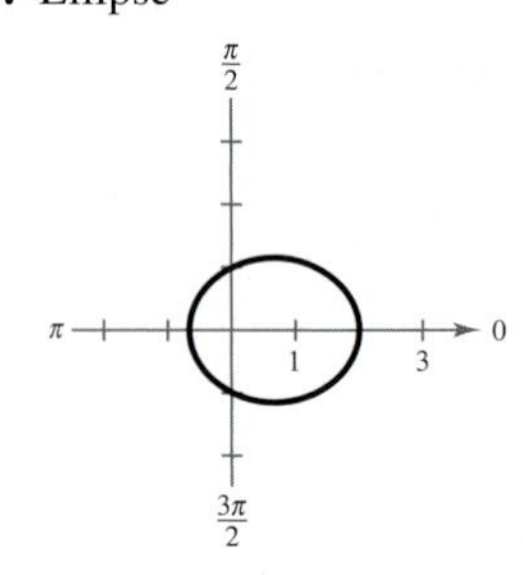

16. Ellipse

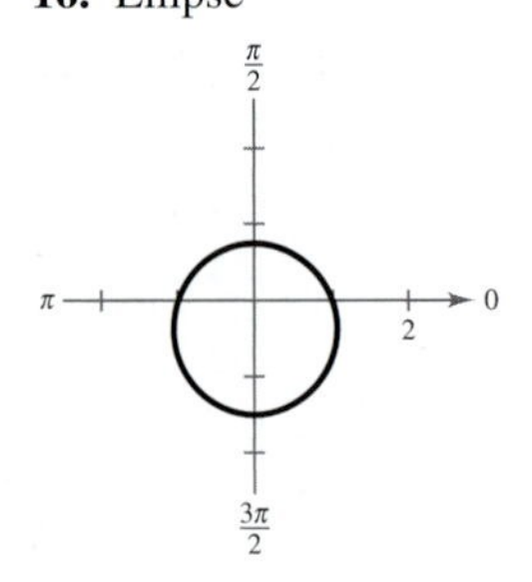

17. Ellipse

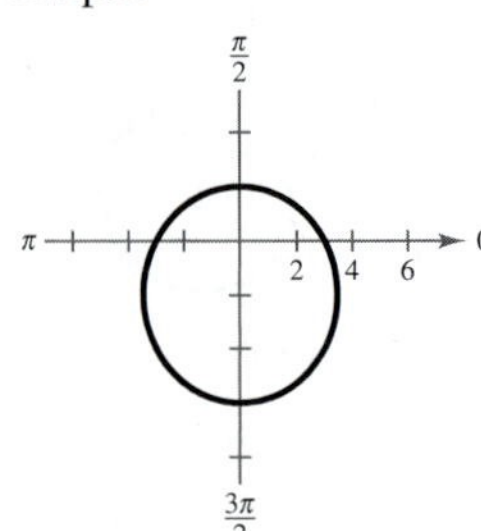

18. Ellipse

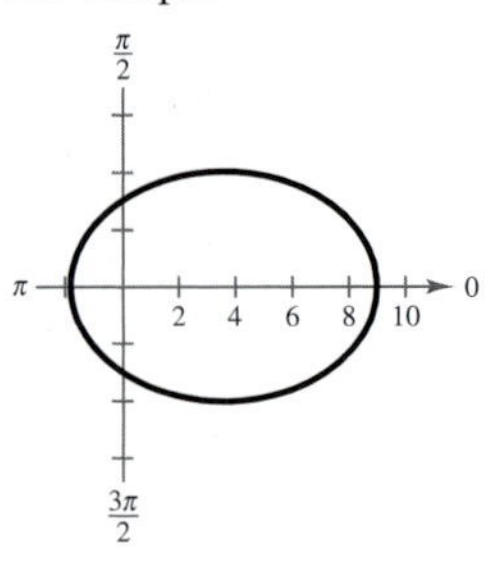

19. Hyperbola

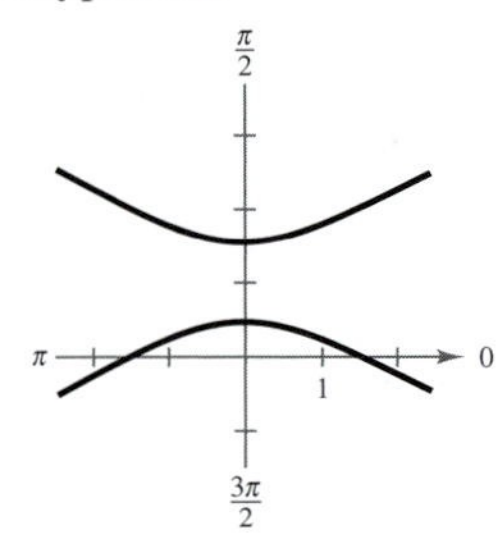

20. Hyperbola

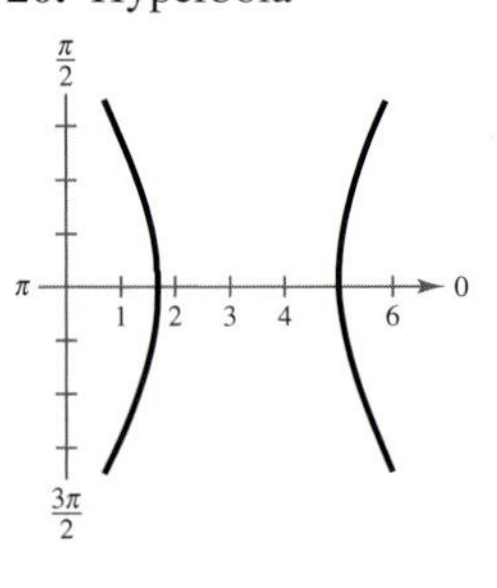

21. Hyperbola

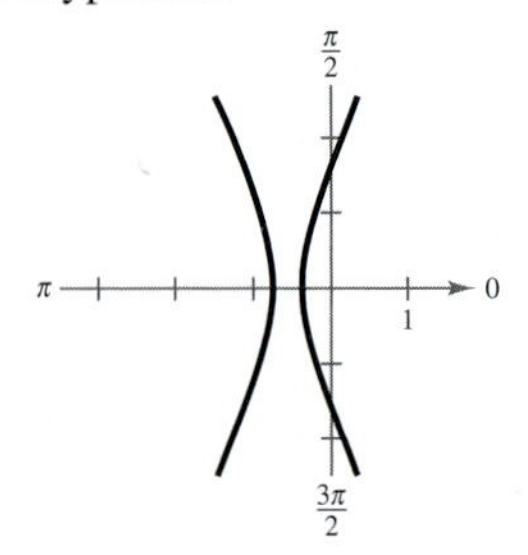

22. Hyperbola

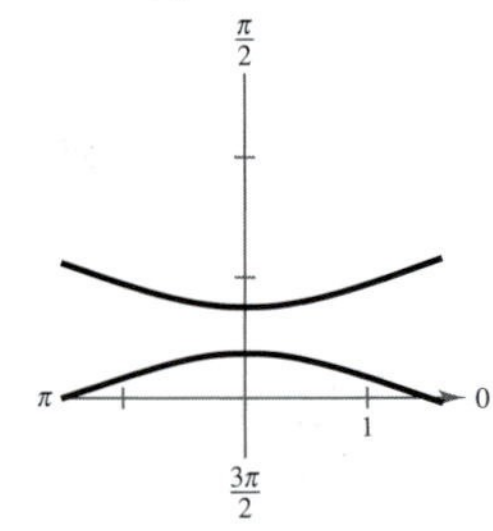

23. Ellipse

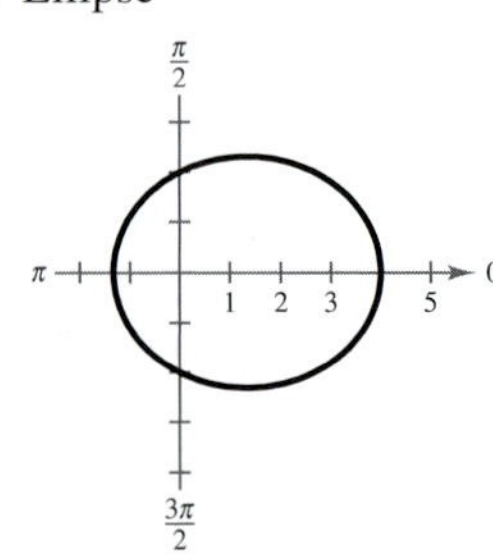

24. Hyperbola

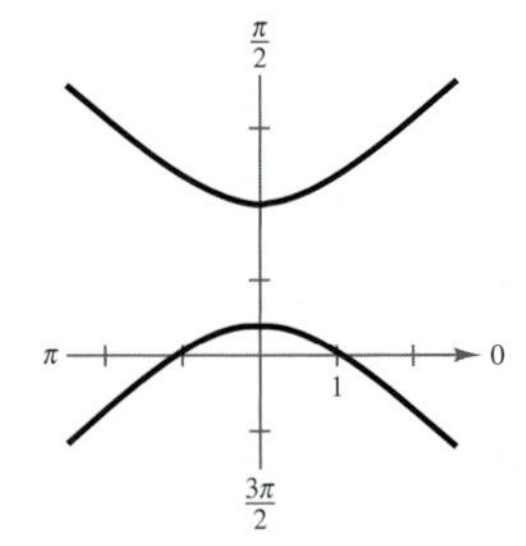

25.

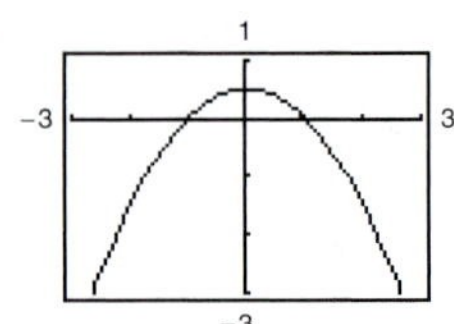

Parabola

26.

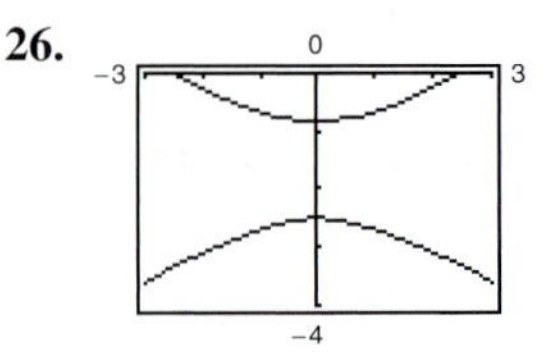

Hyperbola

27.

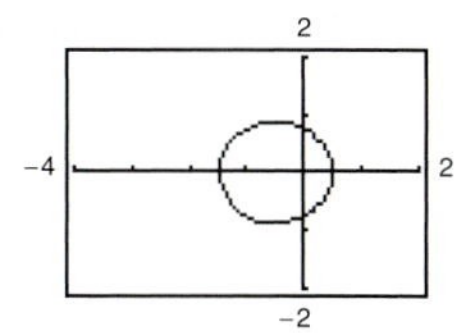

Ellipse

28.

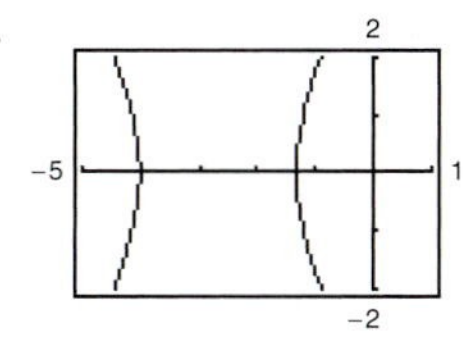

Hyperbola

29.

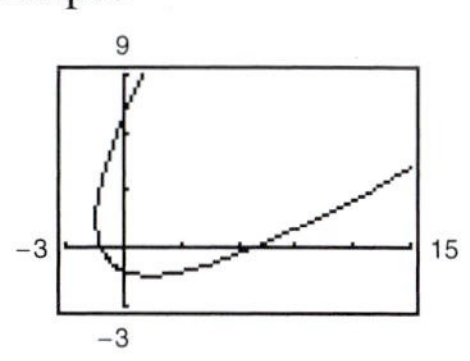

30.

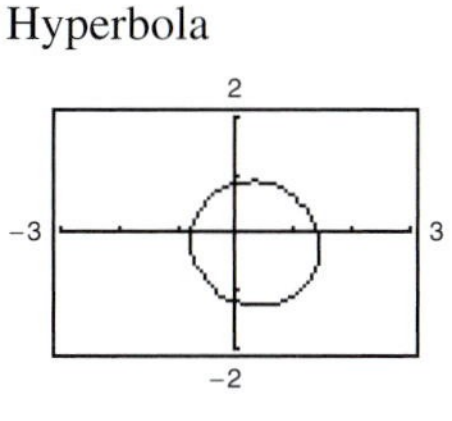

31.

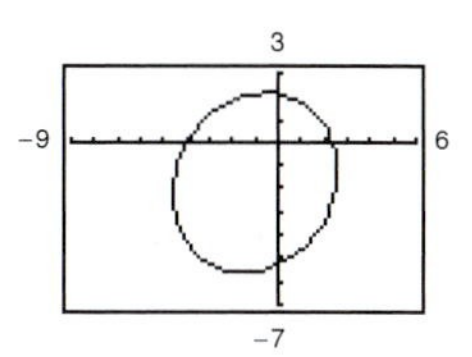

32.

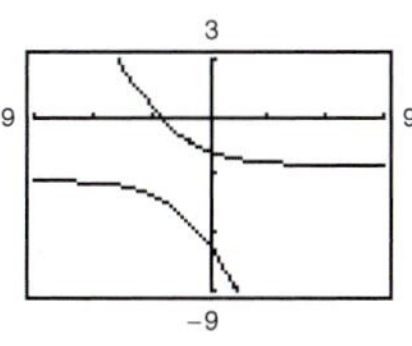

33. $r = \dfrac{1}{1 - \cos\theta}$ **34.** $r = \dfrac{2}{1 - \sin\theta}$

35. $r = \dfrac{1}{2 + \sin\theta}$ **36.** $r = \dfrac{9}{4 - 3\sin\theta}$

37. $r = \dfrac{2}{1 + 2\cos\theta}$ **38.** $r = \dfrac{3}{2 - 3\cos\theta}$

39. $r = \dfrac{2}{1 - \sin\theta}$ **40.** $r = \dfrac{12}{1 + \cos\theta}$

41. $r = \dfrac{10}{1 - \cos\theta}$ **42.** $r = \dfrac{20}{1 + \sin\theta}$

43. $r = \dfrac{10}{3 + 2\cos\theta}$ **44.** $r = \dfrac{8}{3 + \sin\theta}$

45. $r = \dfrac{20}{3 - 2\cos\theta}$ **46.** $r = \dfrac{16}{3 + 5\cos\theta}$

47. $r = \dfrac{9}{4 - 5\sin\theta}$ **48.** $r = \dfrac{8}{3 + 5\sin\theta}$

49–50. Answers will vary.

51. $r = \dfrac{9.5929 \times 10^7}{1 - 0.0167\cos\theta}$

Perihelion: 9.4354×10^7 miles

Aphelion: 9.7558×10^7 miles

52. $r = \dfrac{1.4228 \times 10^9}{1 - 0.0542\cos\theta}$

Perihelion: 1.3497×10^9 kilometers

Aphelion: 1.5043×10^9 kilometers

53. $r = \dfrac{1.0820 \times 10^8}{1 - 0.0068\cos\theta}$

Perihelion: 1.0747×10^8 kilometers

Aphelion: 1.0894×10^8 kilometers

54. $r = \dfrac{3.4459 \times 10^7}{1 - 0.2056\cos\theta}$

Perihelion: 2.8583×10^7 miles

Aphelion: 4.3377×10^7 miles

55. $r = \dfrac{1.4039 \times 10^8}{1 - 0.0934\cos\theta}$

Perihelion: 1.2840×10^8 miles

Aphelion: 1.5486×10^8 miles

56. $r = \dfrac{7.7659 \times 10^8}{1 - 0.0484\cos\theta}$

Perihelion: 7.4073×10^8 kilometers

Aphelion: 8.1609×10^8 kilometers

57. $r = \dfrac{0.624}{1 + 0.847\sin\pi/2}$; $r = 0.338$ astronomical unit

58. (a) $r = \dfrac{8200}{1 + \sin\theta}$

(b)

(c) 1467 miles (d) 394 miles

59. True. The graphs represent the same hyperbola.

60. False. The graph has a horizontal directrix below the pole.

61. True. The conic is an ellipse because the eccentricity is less than 1.

62–64. Answers will vary.

65. $r^2 = \dfrac{24{,}336}{169 - 25\cos^2\theta}$ **66.** $r^2 = \dfrac{400}{25 - 9\cos^2\theta}$

67. $r^2 = \dfrac{144}{25\cos^2\theta - 9}$ **68.** $r^2 = \dfrac{36}{10\cos^2\theta - 9}$

69. $r^2 = \dfrac{144}{25\sin^2\theta - 16}$ **70.** $r^2 = \dfrac{225}{25 - 16\cos^2\theta}$

71. (a) Ellipse

(b) The given polar equation, r, has a vertical directrix to the left of the pole. The equation, r_1, has a vertical directrix to the right of the pole, and the equation, r_2, has a horizontal directrix below the pole.

(c)

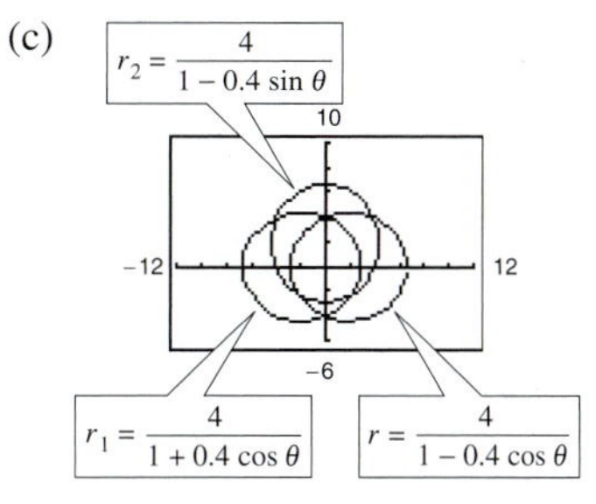

72. If e remains fixed and p changes, then the lengths of both the major axis and the minor axis change. For example, graph

$r = \dfrac{5}{1 - \frac{2}{3}\sin\theta}$, with $e = \frac{2}{3}$ and $p = \frac{15}{2}$, and graph

$r = \dfrac{6}{1 - \frac{2}{3}\sin\theta}$, with $e = \frac{2}{3}$ and $p = 9$, on the same set of coordinate axes.

CHAPTER 6

73. $\frac{\pi}{6} + n\pi$ **74.** $\frac{\pi}{3} + 2n\pi, \frac{5\pi}{3} + 2n\pi$

75. $\frac{\pi}{3} + n\pi, \frac{2\pi}{3} + n\pi$ **76.** $\frac{\pi}{3} + n\pi, \frac{2\pi}{3} + n\pi$

77. $\frac{\pi}{2} + n\pi$ **78.** $\frac{\pi}{3} + 2n\pi, \frac{5\pi}{3} + 2n\pi$ **79.** $\frac{\sqrt{2}}{10}$

80. $-\frac{7\sqrt{2}}{10}$ **81.** $\frac{7\sqrt{2}}{10}$ **82.** $\frac{\sqrt{2}}{10}$

83. $\sin 2u = -\frac{24}{25}$

$\cos 2u = -\frac{7}{25}$

$\tan 2u = \frac{24}{7}$

84. $\sin 2u = -\frac{\sqrt{3}}{2}$

$\cos 2u = -\frac{1}{2}$

$\tan 2u = \sqrt{3}$

Review Exercises *(page 515)*

1. $\frac{\pi}{4}$ radian, 45° **2.** 2.6012 radians, 149.04°

3. 1.1071 radians, 63.43° **4.** 0.7086 radian, 40.60°

5. 0.4424 radian, 25.35° **6.** 0.4424 radian, 25.35°

7. 0.6588 radian, 37.75° **8.** 1.4309 radians, 81.98°

9. $2\sqrt{2}$ **10.** $\frac{6\sqrt{5}}{5}$ **11.** Hyperbola **12.** Parabola

13. $y^2 = 16x$ **14.** $y^2 = -8(x - 2)$

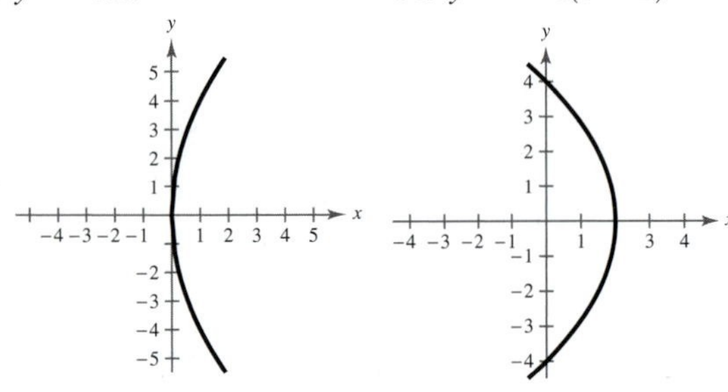

15. $(y - 2)^2 = 12x$ **16.** $(x - 2)^2 = 8(y - 2)$

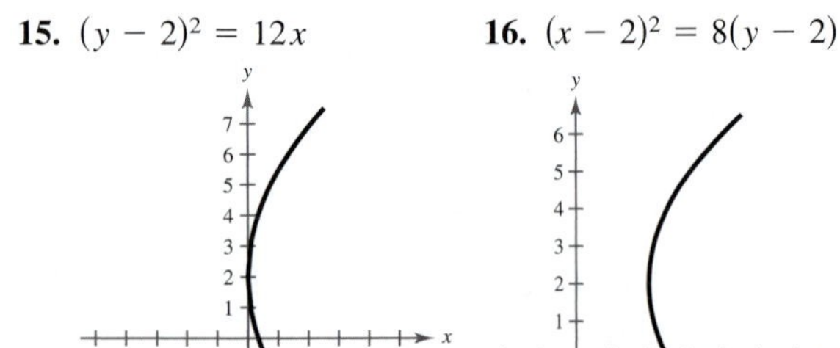

17. $y = -2x + 2; (1, 0)$ **18.** $y = 4x + 8; (-2, 0)$

19. $8\sqrt{6}$ meters **20.** $y^2 = 6x$

21. $\frac{(x - 2)^2}{25} + \frac{y^2}{21} = 1$ **22.** $\frac{(x - 2)^2}{3} + \frac{(y - 2)^2}{4} = 1$

23. $\frac{(x - 2)^2}{4} + (y - 1)^2 = 1$ **24.** $\frac{(x + 4)^2}{4} + \frac{(y - 5)^2}{36} = 1$

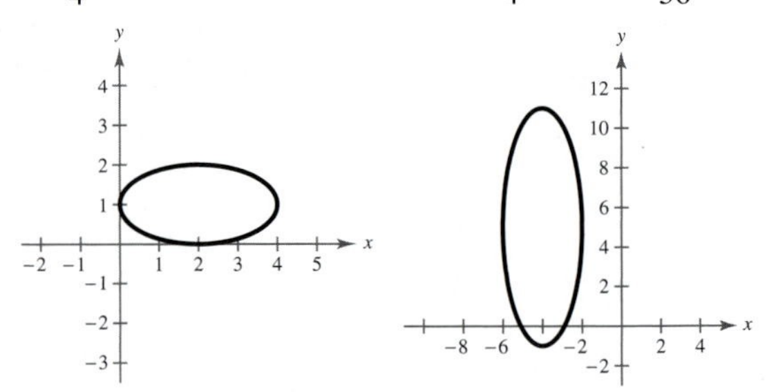

25. The foci occur 3 feet from the center of the arch on a line connecting the tops of the pillars.

26. Longest distance: 36 feet
Shortest distance: 28 feet
Distance between foci: $16\sqrt{2}$ feet

27. Center: $(-2, 1)$
Vertices: $(-2, 11), (-2, -9)$
Foci: $\left(-2, 1 \pm \sqrt{19}\right)$
Eccentricity: $\frac{\sqrt{19}}{10}$

28. Center: $(5, -3)$
Vertices: $(5, 3), (5, -9)$
Foci: $\left(5, -3 \pm \sqrt{35}\right)$
Eccentricity: $\frac{\sqrt{35}}{6}$

29. Center: $(1, -4)$
Vertices: $(1, 0), (1, -8)$
Foci: $\left(1, -4 \pm \sqrt{7}\right)$
Eccentricity: $\frac{\sqrt{7}}{4}$

30. Center: $(-2, 3)$
Vertices: $(3, 3), (-7, 3)$
Foci: $\left(-2 \pm \sqrt{21}, 3\right)$
Eccentricity: $\frac{\sqrt{21}}{5}$

31. $y^2 - \frac{x^2}{8} = 1$ **32.** $\frac{x^2}{4} - \frac{(y - 2)^2}{12} = 1$

33. $\frac{5(x - 4)^2}{16} - \frac{5y^2}{64} = 1$ **34.** $\frac{5y^2}{16} - \frac{5(x - 3)^2}{4} = 1$

35. Center: $(3, -5)$
Vertices: $(7, -5), (-1, -5)$
Foci: $\left(3 \pm 2\sqrt{5}, -5\right)$
Asymptotes:
$y = -5 \pm \frac{1}{2}(x - 3)$

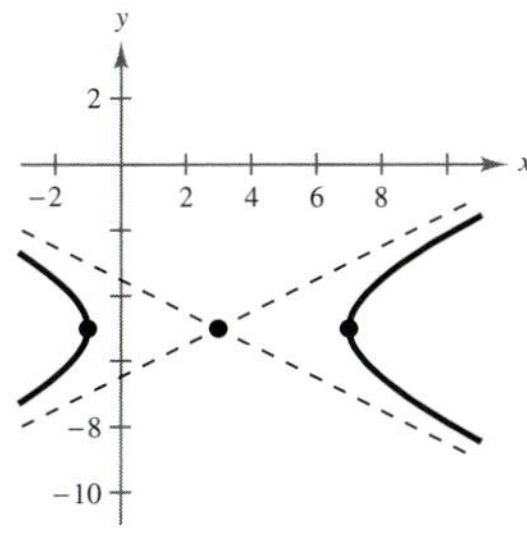

36. Center: $(0, 1)$
Vertices: $(0, 3), (0, -1)$
Foci: $\left(0, 1 \pm \sqrt{5}\right)$
Asymptotes: $y = 1 \pm 2x$

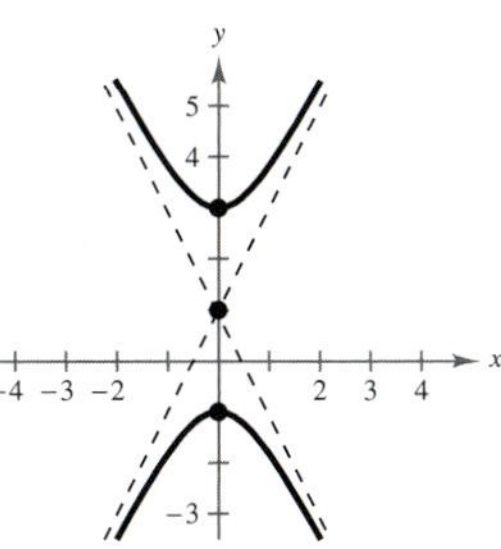

37. Center: $(1, -1)$
Vertices: $(5, -1), (-3, -1)$
Foci: $(6, -1), (-4, -1)$
Asymptotes:
$y = -1 \pm \frac{3}{4}(x - 1)$

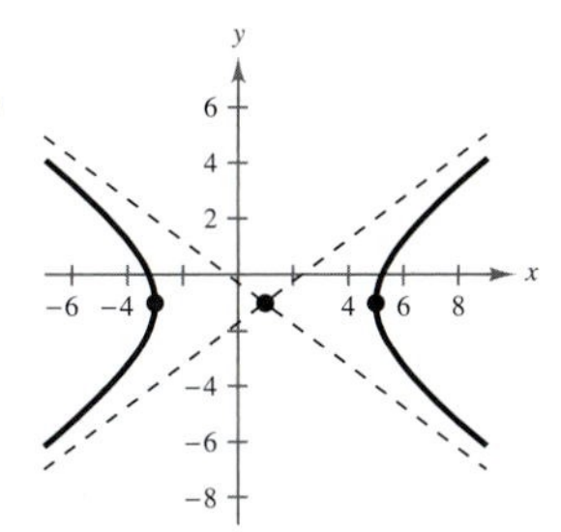

38. Center: $(-1, -3)$
Vertices: $(-1, -1), (-1, -5)$
Foci: $\left(-1, -3 \pm \sqrt{29}\right)$
Asymptotes:
$y = -3 \pm \frac{2}{5}(x + 1)$

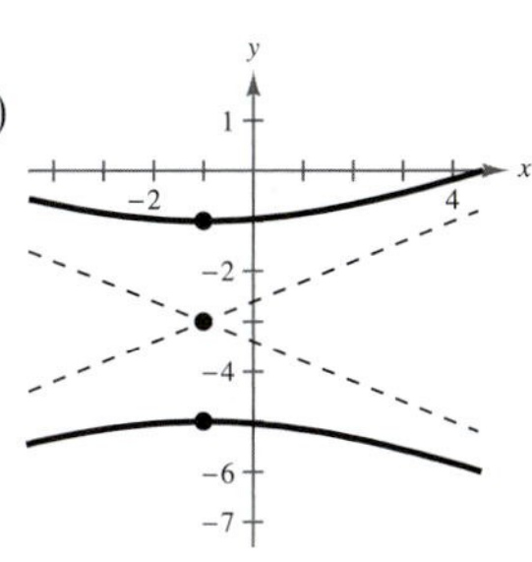

39. 72 miles

40. $\dfrac{576x^2}{25} - \dfrac{576y^2}{227} = 1$

$\dfrac{64(x - 1)^2}{25} - \dfrac{64y^2}{39} = 1$

41. Hyperbola **42.** Parabola
43. Ellipse **44.** Circle

45.

t	-3	-2	-1	0	1	2	3
x	-11	-8	-5	-2	1	4	7
y	19	15	11	7	3	-1	-5

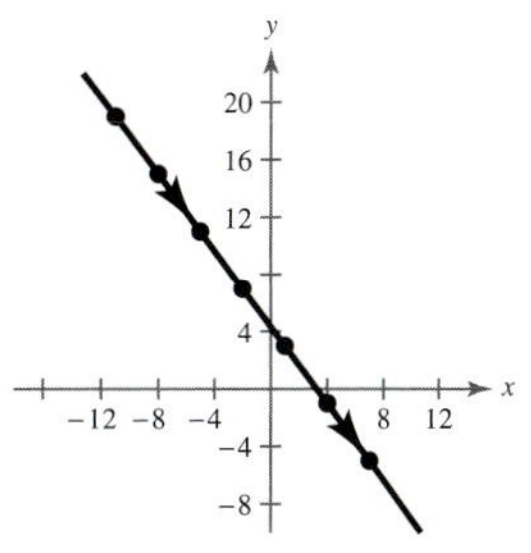

46.

t	-1	0	2	3	4	5
x	$-\frac{1}{5}$	0	$\frac{2}{5}$	$\frac{3}{5}$	$\frac{4}{5}$	1
y	-2	-4	4	2	$\frac{4}{3}$	1

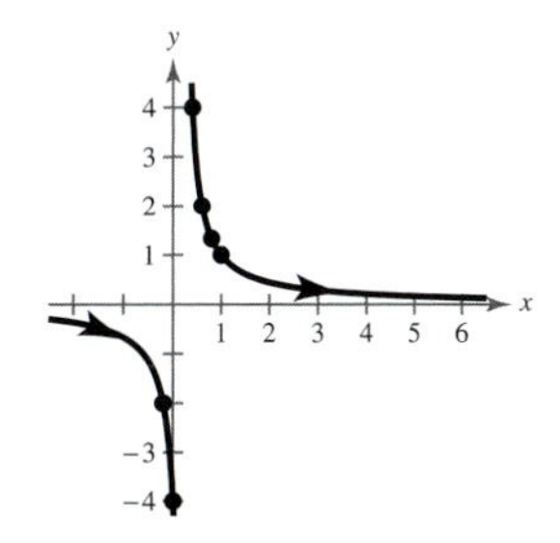

47. (a)

(b) $y = 2x$

48. (a)

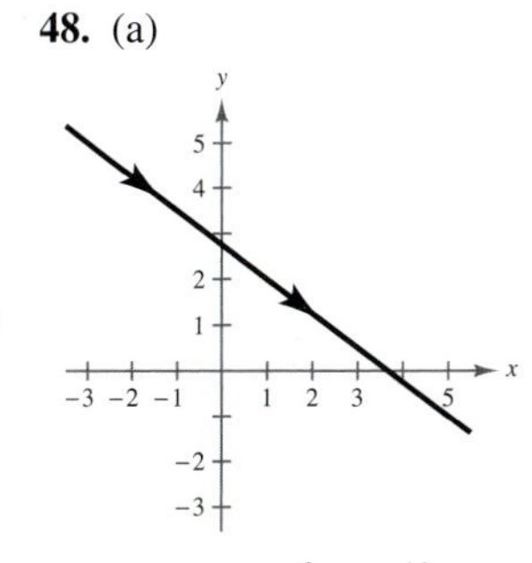

(b) $y = -\frac{3}{4}x + \frac{11}{4}$

49. (a)

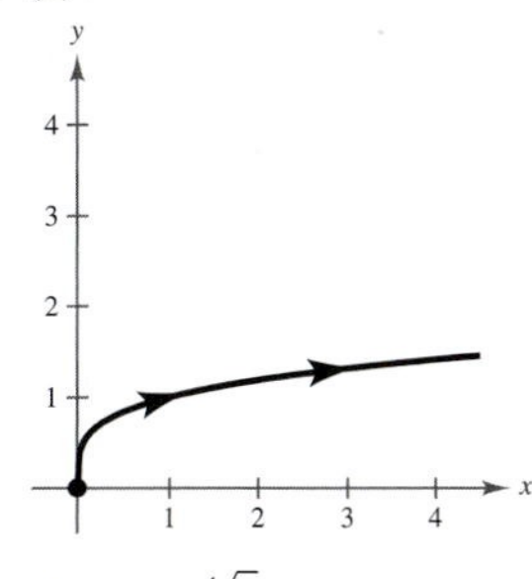

(b) $y = \sqrt[4]{x}$

50. (a)

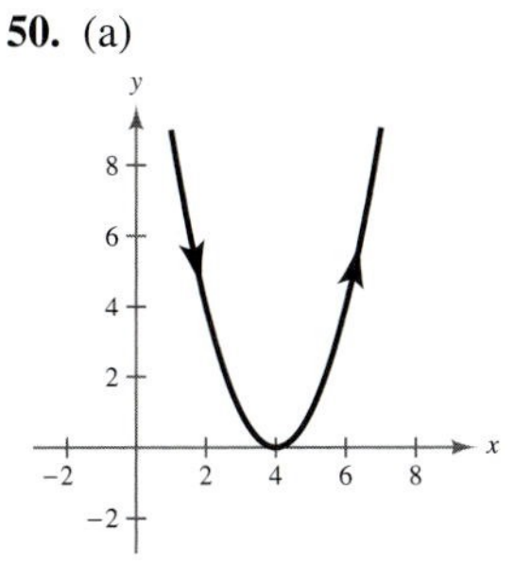

(b) $y = (x - 4)^2$

51. (a)

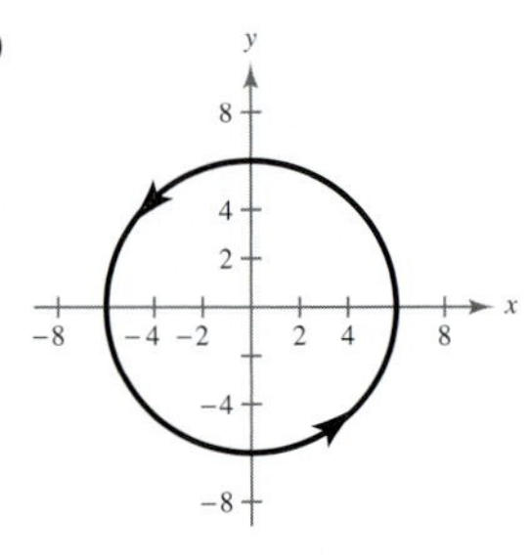

(b) $x^2 + y^2 = 36$

52. (a)

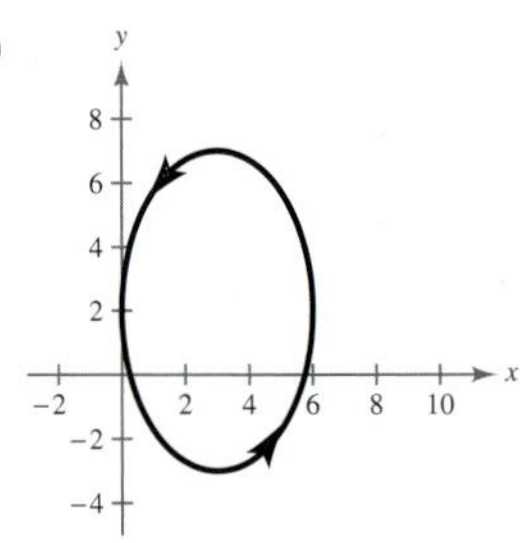

(b) $\dfrac{(x-3)^2}{9} + \dfrac{(y-2)^2}{25} = 1$

53. $x = 5 + 6\cos\theta$
$y = 4 + 6\sin\theta$

54. $x = -3 + 4\cos\theta$
$y = 4 + 3\sin\theta$

55. $x = 3\tan\theta$
$y = 4\sec\theta$

56. Answers will vary.

57.

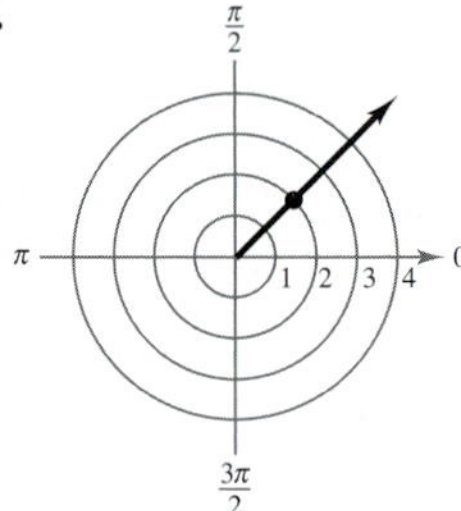

$\left(2, \frac{9\pi}{4}\right), \left(-2, \frac{5\pi}{4}\right)$

58.

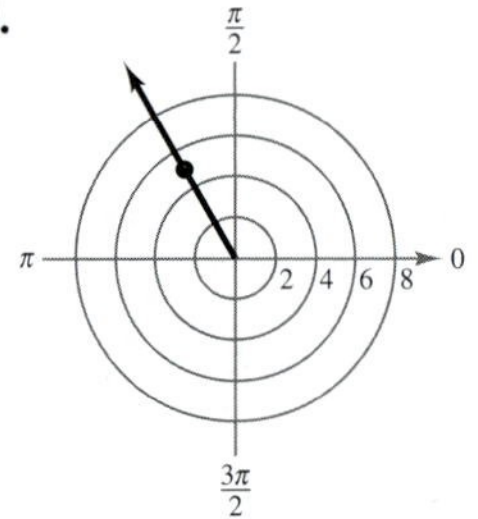

$\left(-5, \frac{5\pi}{3}\right), \left(5, \frac{2\pi}{3}\right)$

59.

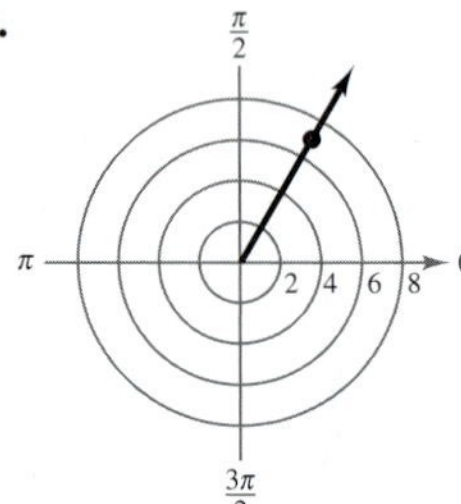

$(7, 1.05), (-7, 10.47)$

60.

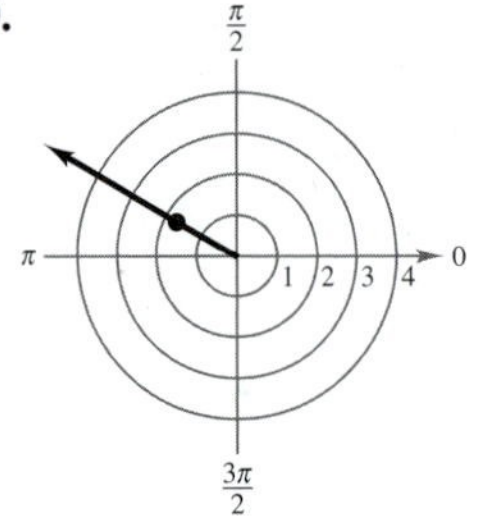

$(\sqrt{3}, 8.90), (-\sqrt{3}, 5.76)$

61. $\left(-\frac{1}{2}, -\frac{\sqrt{3}}{2}\right)$ **62.** $(-\sqrt{2}, -\sqrt{2})$

63. $\left(-\frac{3\sqrt{2}}{2}, \frac{3\sqrt{2}}{2}\right)$ **64.** $(0, 0)$ **65.** $\left(2, \frac{\pi}{2}\right)$

66. $\left(\sqrt{10}, \frac{3\pi}{4}\right)$ **67.** $(2\sqrt{13}, 0.9828)$ **68.** $(5, 5.3559)$

69. $r = 7$ **70.** $r = 2\sqrt{5}$ **71.** $r = 6\sin\theta$

72. $r = 4\cos\theta$ **73.** $r^2 = 10\csc 2\theta$

74. $r^2 = -4\csc 2\theta$ **75.** $x^2 + y^2 = 25$

76. $x^2 + y^2 = 144$ **77.** $x^2 + y^2 = 3x$

78. $x^2 + y^2 = 8y$ **79.** $x^2 + y^2 = y^{2/3}$

80. $(x^2 + y^2)^2 = x^2 - y^2$

81. Symmetry: $\theta = \frac{\pi}{2}$, polar axis, pole

Maximum value of $|r|$: $|r| = 4$ when $\theta = 0, \frac{\pi}{2}, \pi, \frac{3\pi}{2}$

No zeros of r

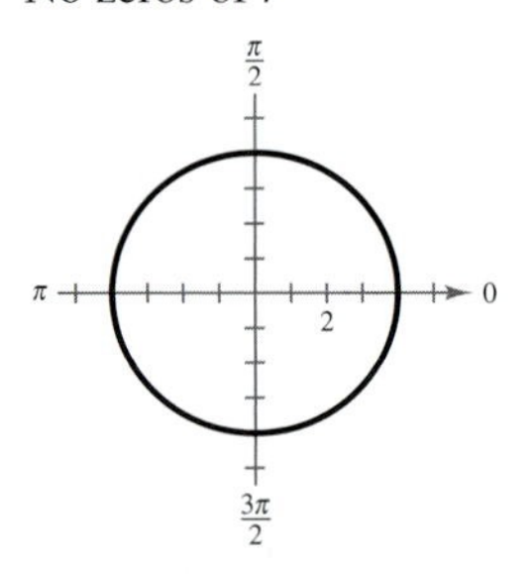

82. Symmetry: $\theta = \frac{\pi}{2}$, polar axis, pole

Maximum value of $|r|$: $|r| = 11$ when $\theta = 0, \frac{\pi}{2}, \pi, \frac{3\pi}{2}$

No zeros of r

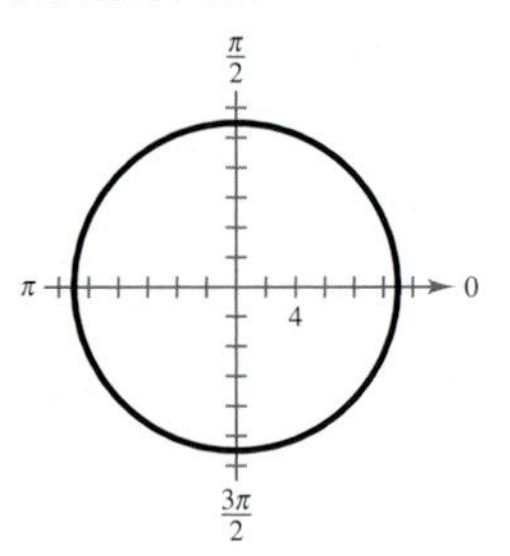

83. Symmetry: $\theta = \frac{\pi}{2}$, polar axis, pole

Maximum value of $|r|$: $|r| = 4$ when $\theta = \frac{\pi}{4}, \frac{3\pi}{4}, \frac{5\pi}{4}, \frac{7\pi}{4}$

Zeros of r: $r = 0$ when $\theta = 0, \frac{\pi}{2}, \pi, \frac{3\pi}{2}$

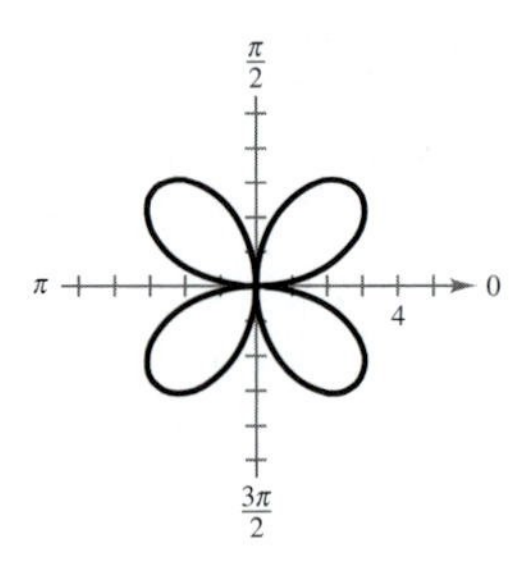

84. Symmetry: polar axis
Maximum value of $|r|$:

$$|r| = 1 \text{ when } \theta = 0, \frac{2\pi}{5}, \frac{4\pi}{5}, \frac{6\pi}{5}, \frac{8\pi}{5}$$

Zeros of r: $r = 0$ when $\theta = \frac{\pi}{10}, \frac{3\pi}{10}, \frac{5\pi}{10}, \frac{7\pi}{10}, \frac{9\pi}{10}$

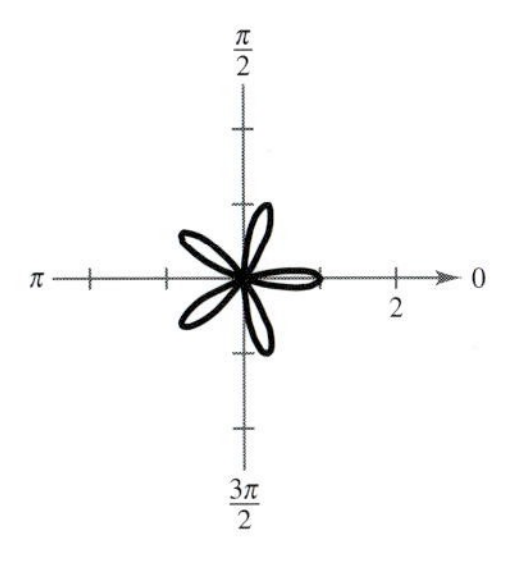

85. Symmetry: polar axis
Maximum value of $|r|$: $|r| = 4$ when $\theta = 0$
Zeros of r: $r = 0$ when $\theta = \pi$

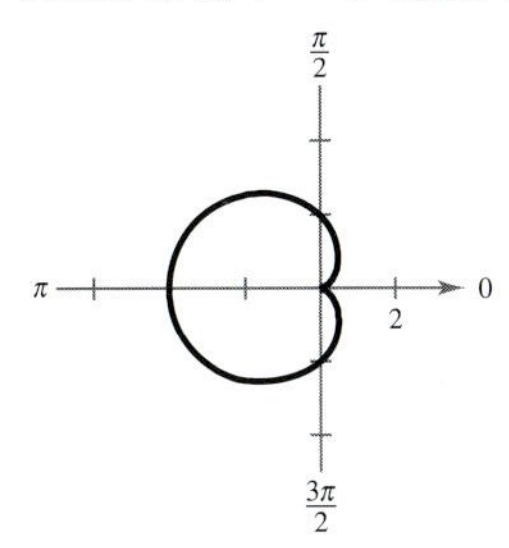

86. Symmetry: polar axis
Maximum value of $|r|$: $|r| = 7$ when $\theta = \pi$
Zeros of r: $r = 0$ when $\theta = \arccos \frac{3}{4}, 2\pi - \arccos \frac{3}{4}$

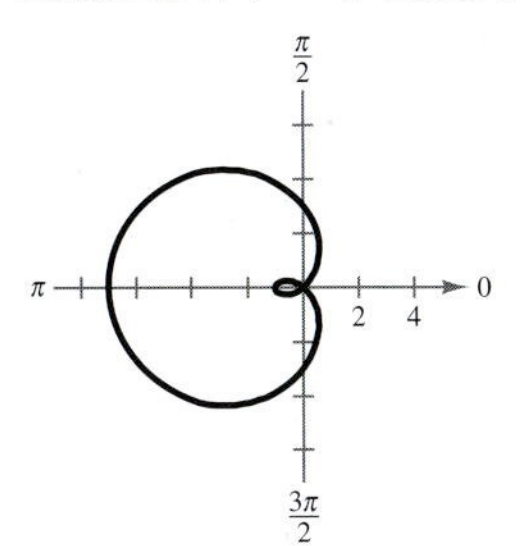

87. Symmetry: $\theta = \frac{\pi}{2}$

Maximum value of $|r|$: $|r| = 8$ when $\theta = \frac{\pi}{2}$

Zeros of r: $r = 0$ when $\theta = 3.4814, 5.9433$

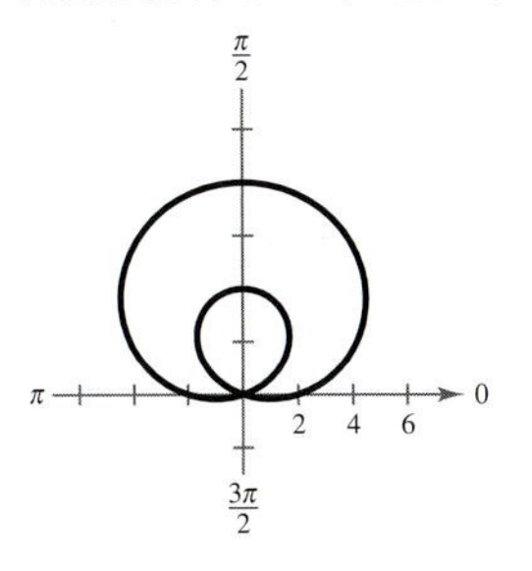

88. Symmetry: polar axis
Maximum value of $|r|$: $|r| = 10$ when $\theta = \pi$
Zero of r: $r = 0$ when $\theta = 0$

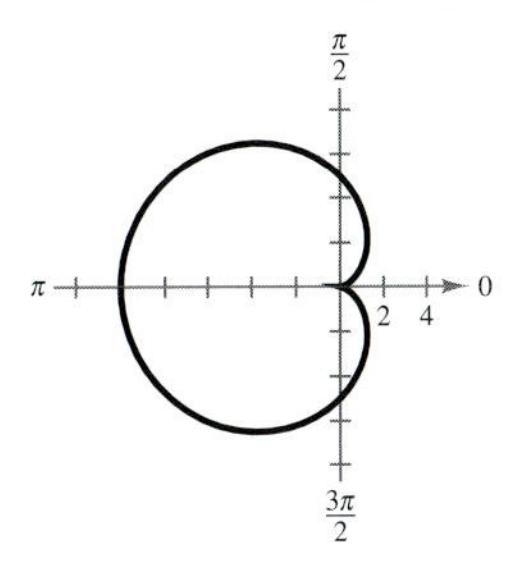

89. Symmetry: $\theta = \frac{\pi}{2}$, polar axis, pole

Maximum value of $|r|$: $|r| = 3$ when $\theta = 0, \frac{\pi}{2}, \pi, \frac{3\pi}{2}$

Zeros of r: $r = 0$ when $\theta = \frac{\pi}{4}, \frac{3\pi}{4}, \frac{5\pi}{4}, \frac{7\pi}{4}$

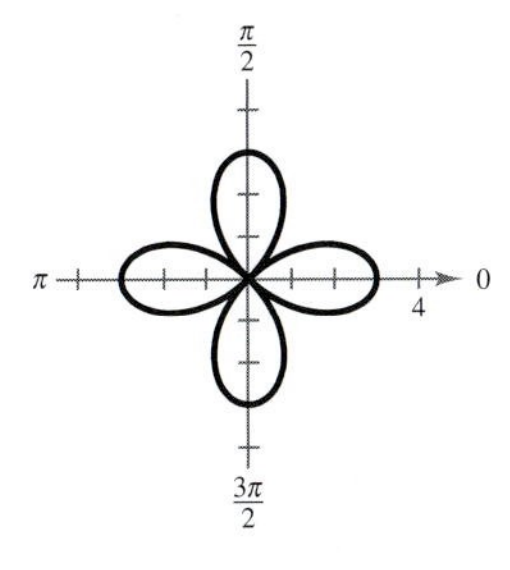

90. Symmetry: $\theta = \frac{\pi}{2}$, polar axis

Maximum value of $|r|$: $|r| = 1$ when $\theta = 0, \frac{\pi}{2}, \pi, \frac{3\pi}{2}$

Zeros of r: $r = 0$ when $\theta = \frac{\pi}{4}, \frac{3\pi}{4}, \frac{5\pi}{4}, \frac{7\pi}{4}$

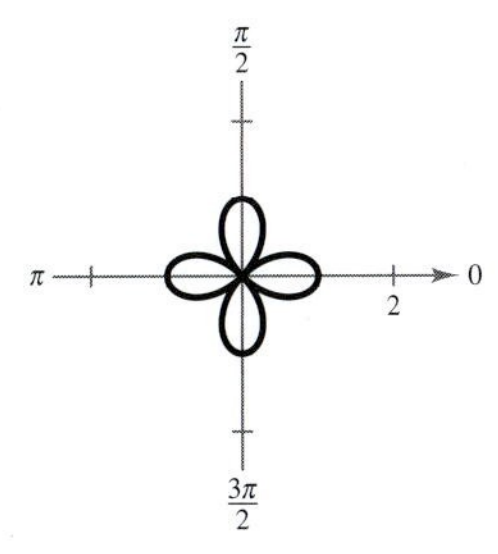

91. Limaçon

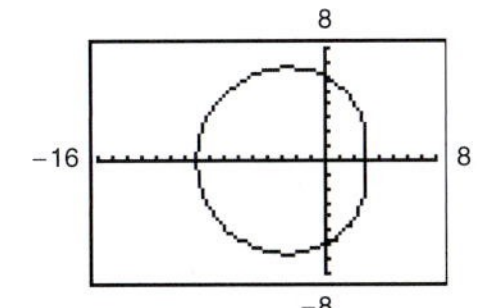

92. Limaçon

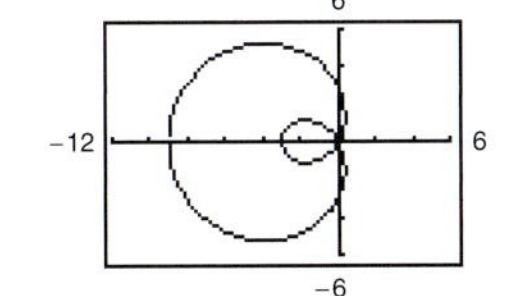

93. Rose curve

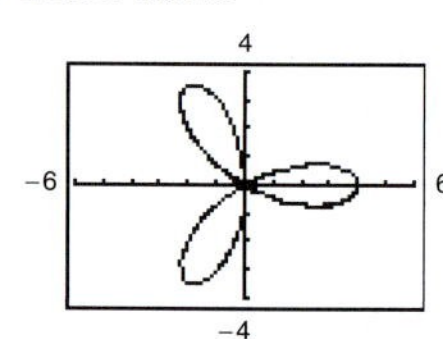

94. Lemniscate

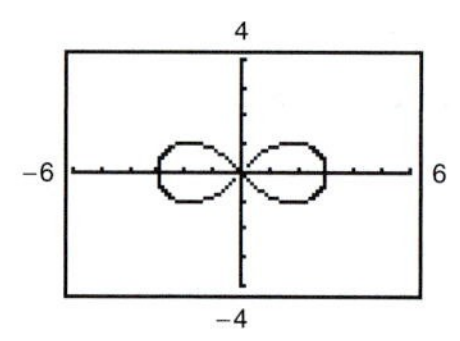

95. Hyperbola

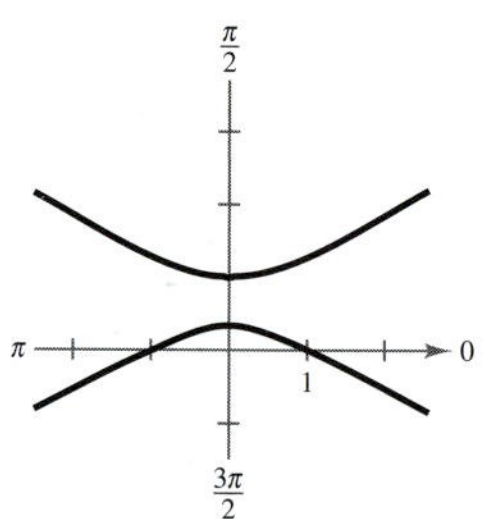

96. Parabola

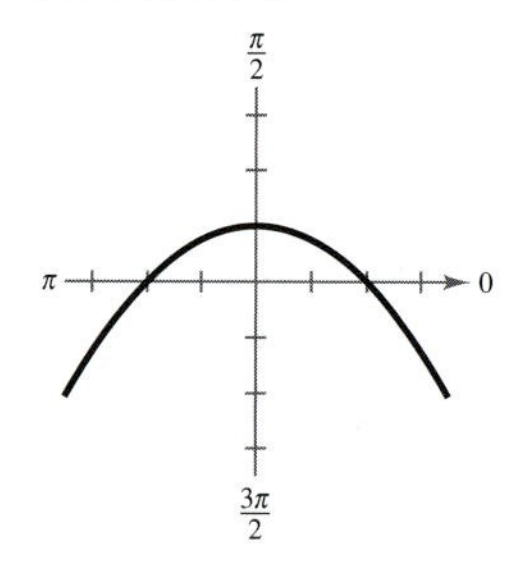

97. Ellipse

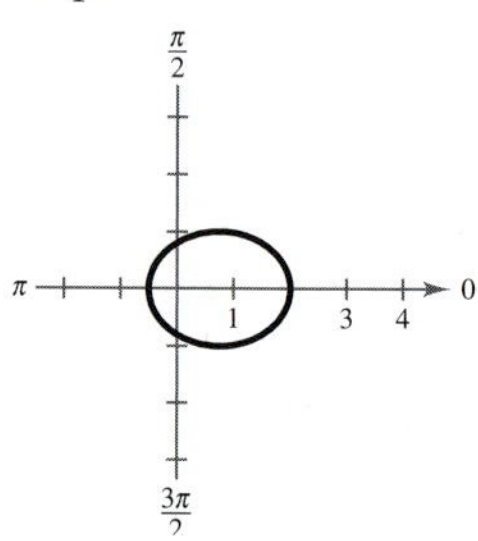

98. Hyperbola

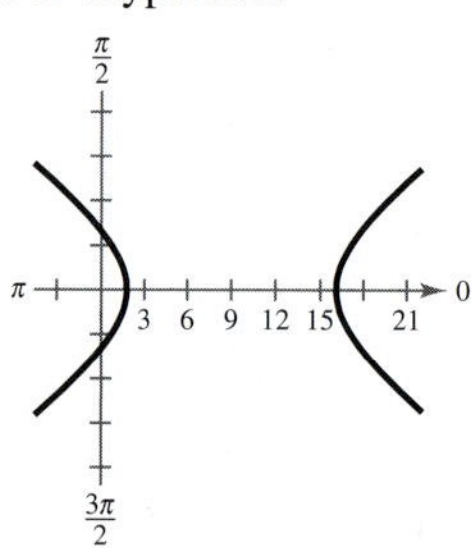

99. $r = \dfrac{4}{1 - \cos\theta}$ **100.** $r = \dfrac{4}{1 + \sin\theta}$

101. $r = \dfrac{5}{3 - 2\cos\theta}$ **102.** $r = \dfrac{7}{3 + 4\cos\theta}$

103. $r = \dfrac{7978.81}{1 - 0.937\cos\theta}$; 11,011.87 miles

104. $r = \dfrac{12{,}000{,}000}{1 + \sin\theta}$; 89,600,000 miles

105. False. The equation of a hyperbola is a second-degree equation.

106. False. The following are two sets of parametric equations for the line.

$x = t,\ y = 3 - 2t$

$x = 3t,\ y = 3 - 6t$

107. False. $(2, \pi/4)$, $(-2, 5\pi/4)$, and $(2, 9\pi/4)$ all represent the same point.

108. 5. The ellipse becomes more circular and approaches a circle of radius 5.

109. The orientation would be reversed.

110. (a) The speed would double.

(b) The elliptical orbit would be flatter; the length of the major axis would be greater.

111. (a) Symmetric to the pole

(b) Symmetric to the polar axis

(c) Symmetric to $\theta = \pi/2$

112. (a) The graphs are the same.

(b) The graphs are the same.

113. 40

Chapter Test *(page 519)*

1. 0.2783 radian, 15.9° **2.** 0.8330 radian, 47.7°

3. $\dfrac{7\sqrt{2}}{2}$

4. Parabola: $y^2 = 4(x - 1)$

Vertex: $(1, 0)$

Focus: $(2, 0)$

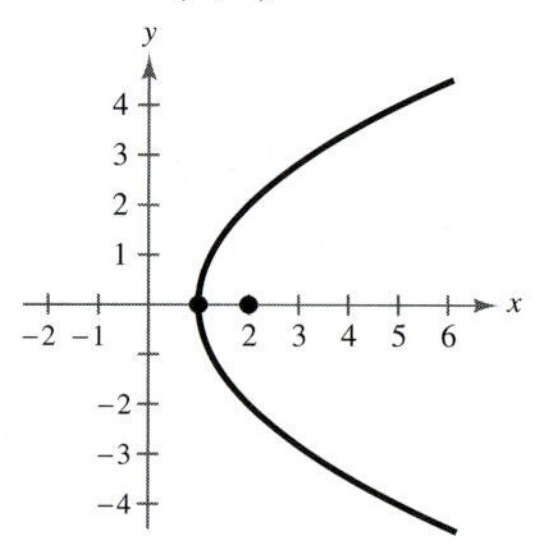

5. Hyperbola: $\dfrac{(x - 2)^2}{4} - y^2 = 1$

Center: $(2, 0)$

Vertices: $(0, 0), (4, 0)$

Foci: $\left(2 \pm \sqrt{5}, 0\right)$

Asymptotes: $y = \pm\frac{1}{2}(x - 2)$

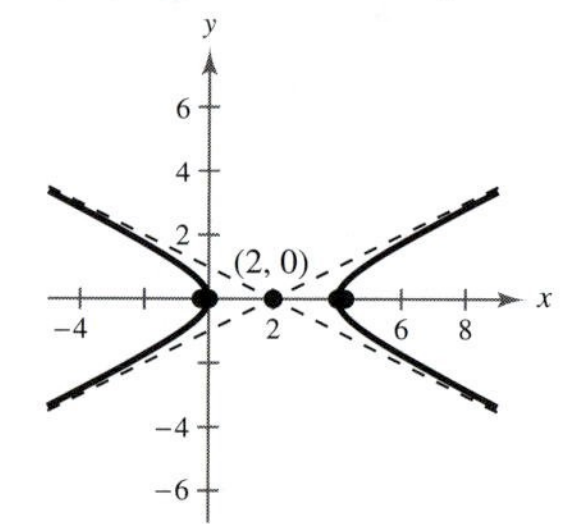

6. Ellipse: $\dfrac{(x + 3)^2}{16} + \dfrac{(y - 1)^2}{9} = 1$

Center: $(-3, 1)$

Vertices: $(1, 1), (-7, 1)$

Foci: $\left(-3 \pm \sqrt{7}, 1\right)$

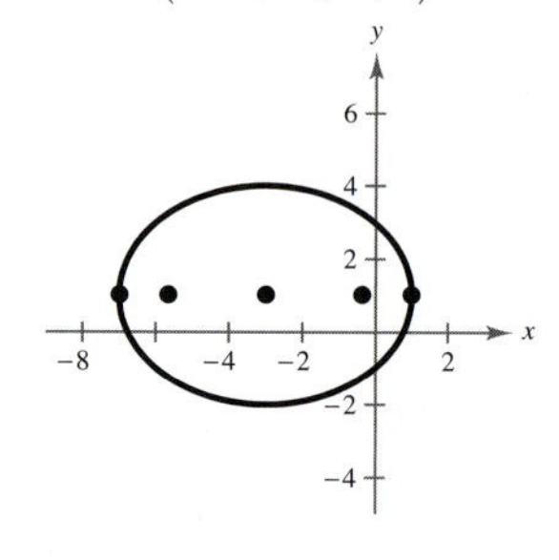

7. Circle: $(x - 2)^2 + (y - 1)^2 = \frac{1}{2}$
Center: $(2, 1)$

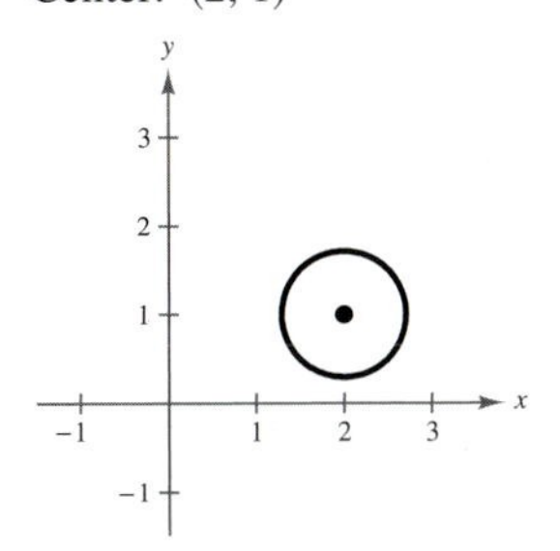

8. $(x - 3)^2 = \dfrac{3}{2}(y + 2)$ **9.** $\dfrac{5(y - 2)^2}{4} - \dfrac{5x^2}{16} = 1$

10.

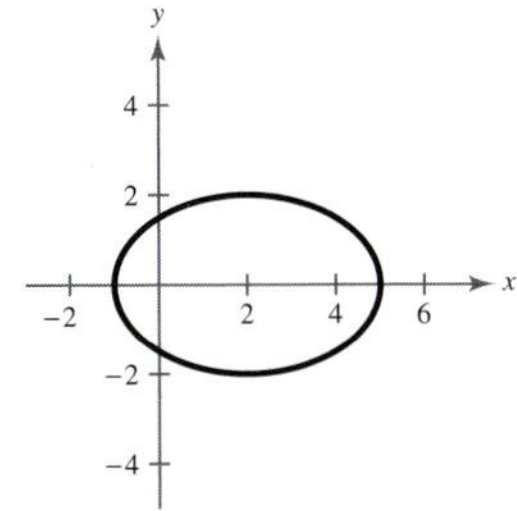

$\dfrac{(x - 2)^2}{9} + \dfrac{y^2}{4} = 1$

11. $x = 6 + 4t$
$y = 4 + 7t$

12. $(\sqrt{3}, -1)$

13. $\left(2\sqrt{2}, \dfrac{7\pi}{4}\right), \left(-2\sqrt{2}, \dfrac{3\pi}{4}\right), \left(2\sqrt{2}, -\dfrac{\pi}{4}\right)$

14. $r = 4 \sin \theta$

15.

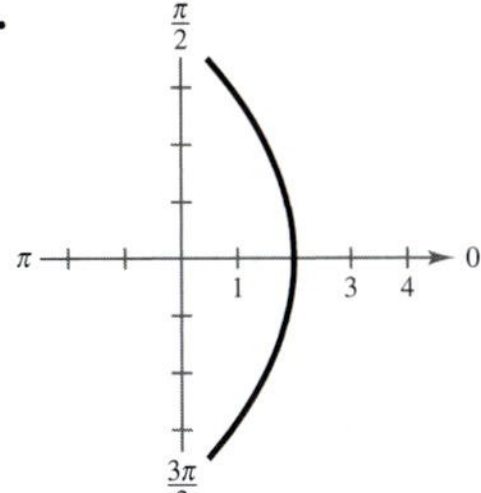

Parabola

16.

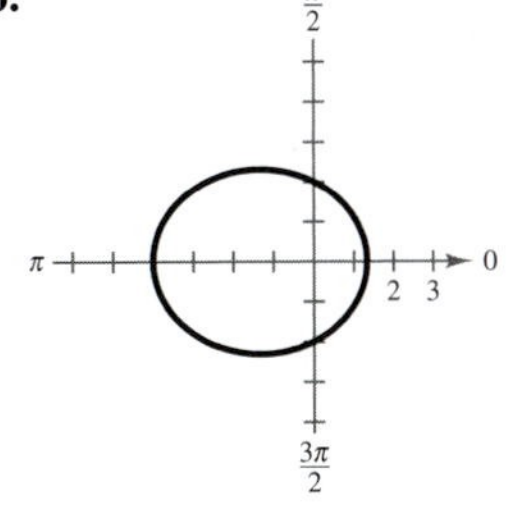

Ellipse

17.

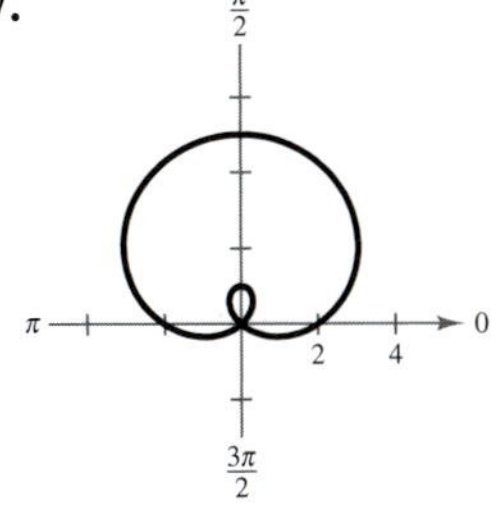

Limaçon with inner loop

18.

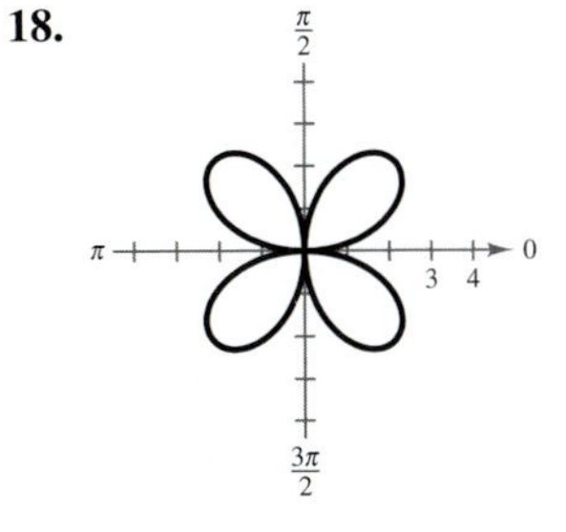

Rose curve

19. Answers will vary. For example: $r = \dfrac{1}{1 + 0.25 \sin \theta}$

20. Slope: 0.1511; Change in elevation: 789 feet

21. No; Yes

Cumulative Test for Chapters 4–6 *(page 520)*

1. (a)

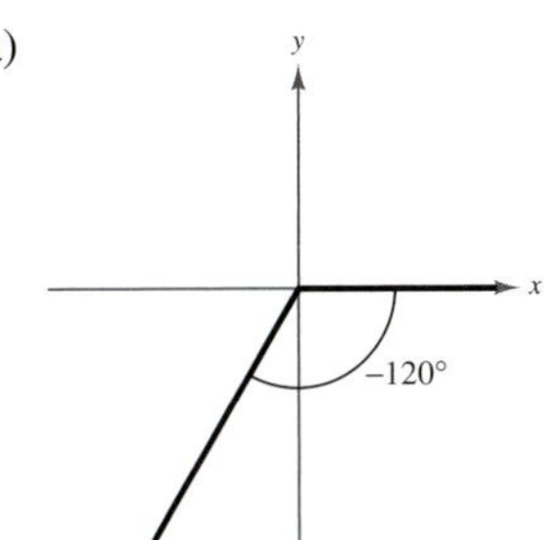

(b) 240°

(c) $-\dfrac{2\pi}{3}$

(d) 60°

(e) $\sin(-120^\circ) = -\dfrac{\sqrt{3}}{2}$ $\csc(-120^\circ) = -\dfrac{2\sqrt{3}}{3}$

$\cos(-120^\circ) = -\dfrac{1}{2}$ $\sec(-120^\circ) = -2$

$\tan(-120^\circ) = \sqrt{3}$ $\cot(-120^\circ) = \dfrac{\sqrt{3}}{3}$

2. 134.6° **3.** $\frac{3}{5}$

4. Ellipse

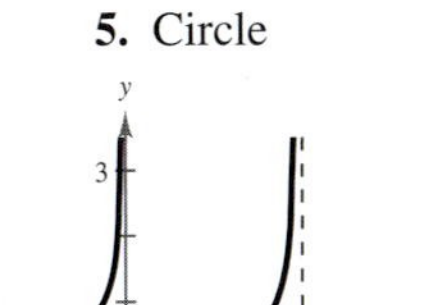

5. Circle

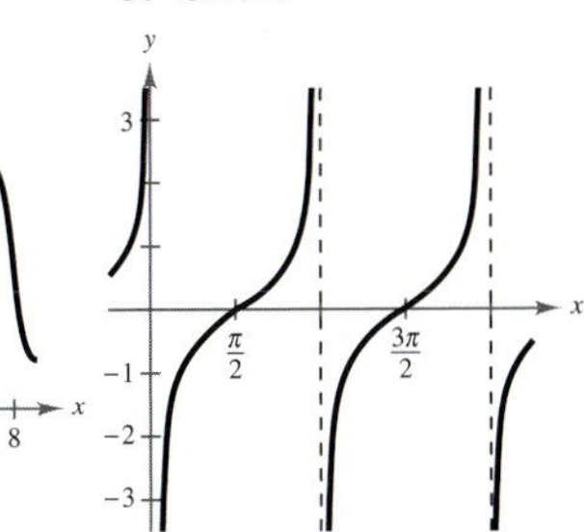

6.

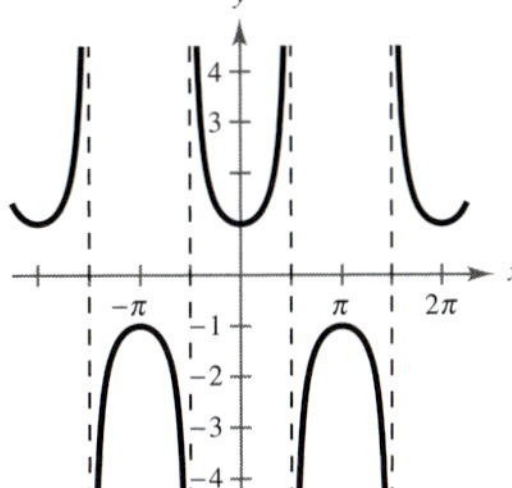

7. $a = -3, b = \pi, c = 0$

8.

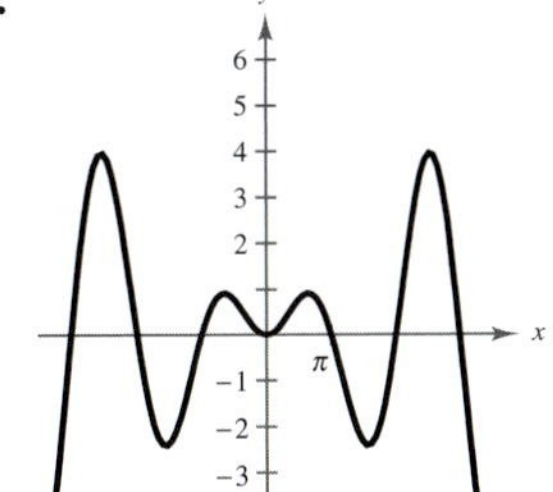

9. 6.7 **10.** $\frac{3}{4}$

11. $\sqrt{1 - 4x^2}$ **12.** 1 **13.** $2 \tan \theta$

14–16. Answers will vary. **17.** $\dfrac{\pi}{3}, \dfrac{\pi}{2}, \dfrac{3\pi}{2}, \dfrac{5\pi}{3}$

18. $\frac{\pi}{6}, \frac{5\pi}{6}, \frac{7\pi}{6}, \frac{11\pi}{6}$ **19.** $\frac{3\pi}{2}$ **20.** $\frac{16}{63}$ **21.** $\frac{4}{3}$

22. $\frac{\sqrt{5}}{5}, \frac{2\sqrt{5}}{5}$ **23.** $\frac{5}{2}\left(\sin\frac{5\pi}{2} - \sin\pi\right)$

24. $2\cos(6x)\cos(2x)$

25. $B \approx 26.39°, C \approx 123.61°, c \approx 15.0$

26. $B \approx 52.48°, C \approx 97.52°, a \approx 5.04$

27. $B = 60°, a \approx 5.77, c \approx 11.55$

28. $A = 26.38°, B \approx 62.72°, C \approx 90.90°$

29. 36.4 square inches **30.** 85.2 square inches

31. Ellipse

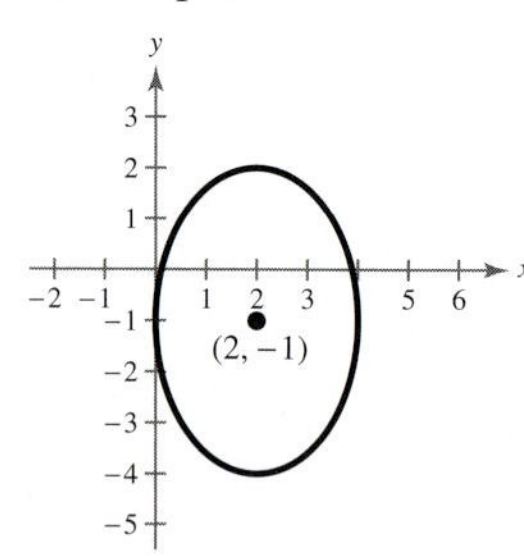

32. Circle

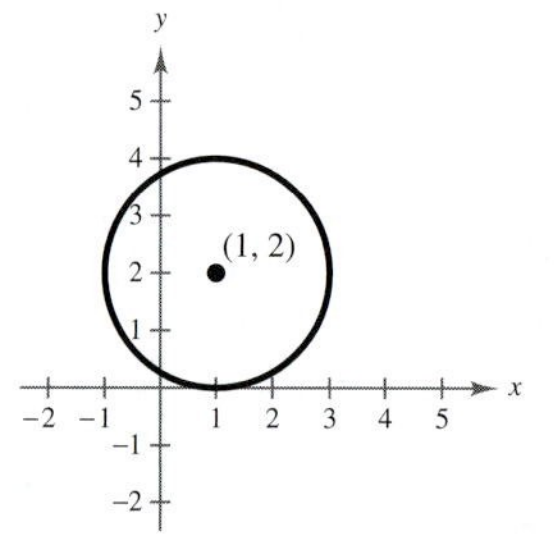

33. $\frac{x^2}{1} + \frac{(y-2)^2}{4} = 1$

34.

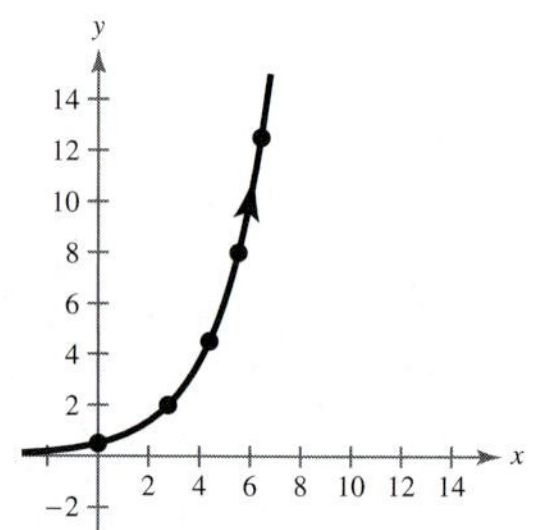

The corresponding rectangular equation is $y = \frac{\sqrt{e^x}}{2}$.

35. $\left(-2, \frac{5\pi}{4}\right), \left(-2, -\frac{7\pi}{4}\right), \left(-2, \frac{\pi}{4}\right)$

36. $-8r\cos\theta - 3r\sin\theta + 5 = 0$

37. $9x^2 + 20x - 16y^2 + 4 = 0$

38.

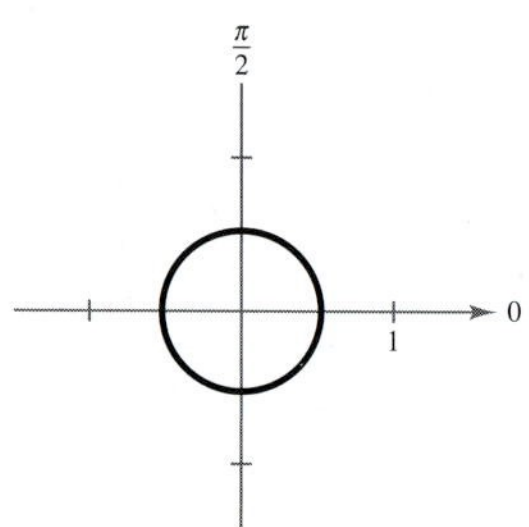

Circle

39.

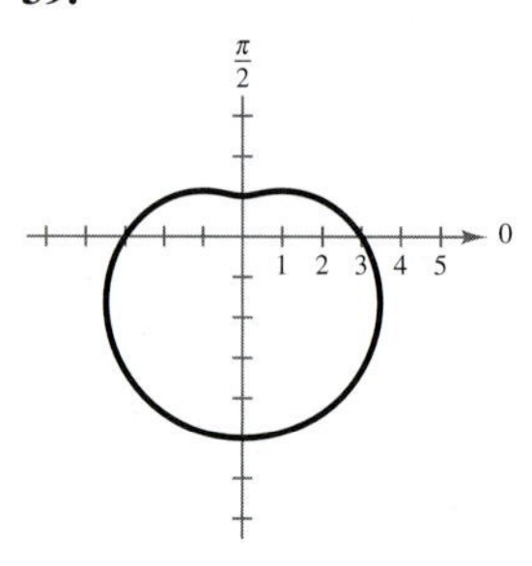

Dimpled limaçon

40.

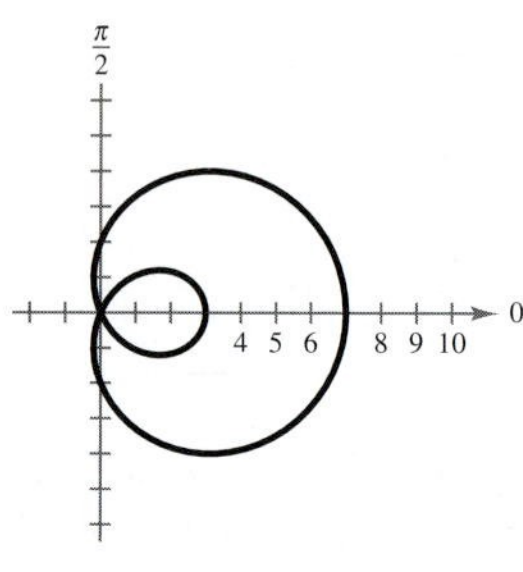

Limaçon with an open loop

41. $\approx$ 395.8 radians per minute; $\approx$ 8312.6 inches per minute

42. Area $\approx$ 63.67 square yards **43.** 5 feet **44.** 22.6°

45. $d = 4\cos\frac{\pi}{4}t$

Problem Solving *(page 525)*

1. (a) 1.2016 radians (b) 2420 feet, 5971 feet

2. (a) $\frac{x^2}{2352.25} + \frac{y^2}{529} = 1$ (b) $\approx$ 85.4 feet

(c) 1115.5π square feet $\approx$ 3504.45 square feet

3. $y^2 = 4p(x + p)$ **4.** $A = \frac{4a^2 b^2}{a^2 + b^2}$

5. (a) Since $d_1 + d_z \leq 20$, by definition, the outer bound that the boat can travel is an ellipse. The islands are the foci.

(b) Island 1: $(-6, 0)$;
Island 2: $(6, 0)$

(c) 20 miles; Vertex: $(10, 0)$

(d) $\frac{x^2}{100} + \frac{y^2}{64} = 1$

6. $\frac{(x-6)^2}{9} - \frac{(y-2)^2}{7} = 1$

7. Answers will vary.

8. (a) The first set of parametric equations models projectile motion along a straight line. The second set of parametric equations models projectile motion of an object launched at a height of h units above the ground that will eventually fall back to the ground.

(b) $y = (\tan\theta)x$; $y = h + x\tan\theta - \frac{16x^2\sec^2\theta}{v_0^2}$

(c) In the first case, the path of the moving object is not affected by changing the velocity because eliminating the parameter removes v_0.

9. Answers will vary. For example:
$x = \cos(-t)$
$y = 2\sin(-t)$

10. (a)

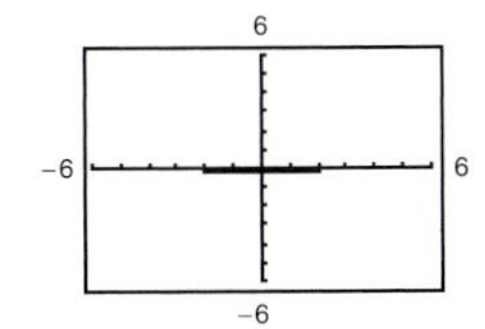

The graph is a line between −2 and 2 on the x-axis.

(b)

The graph is a three-sided figure with counterclockwise orientation.

(c)

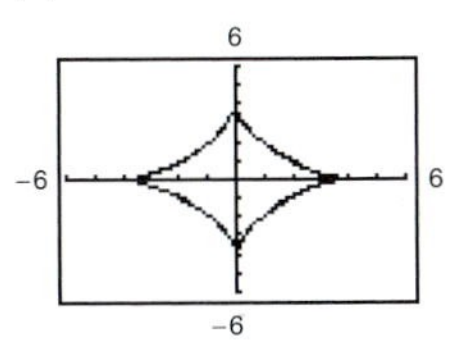

The graph is a four-sided figure with counterclockwise orientation.

(d)

The graph is a 10-sided figure with counterclockwise orientation.

(e)

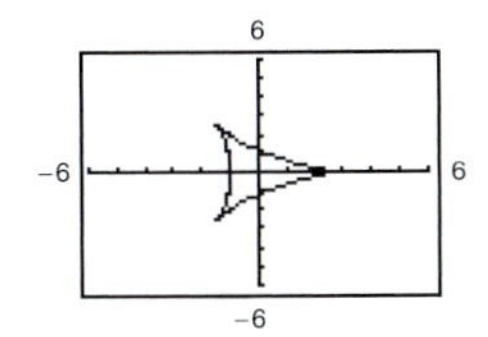

The graph is a three-sided figure with clockwise orientation.

(f)

The graph is a four-sided figure with clockwise orientation.

11. (a) $y^2 = x^2\left(\frac{1-x}{1+x}\right)$ (b) $r = \cos 2\theta \sec \theta$

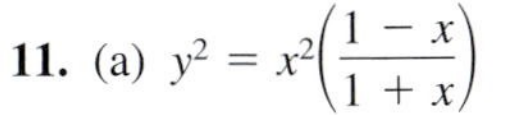

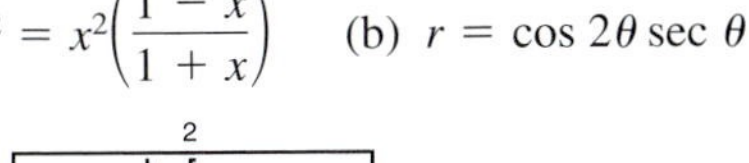

(c)

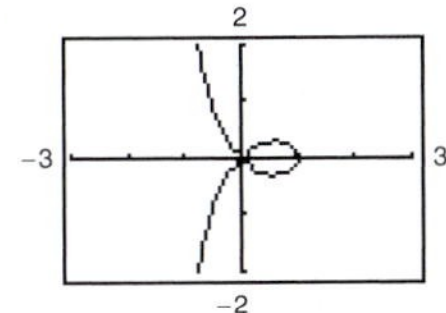

12.

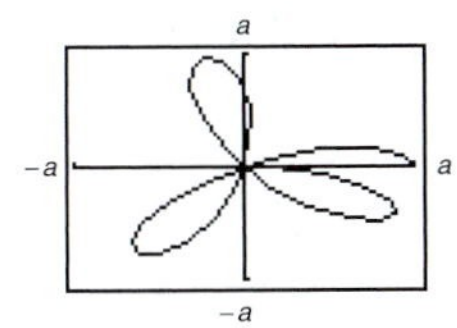

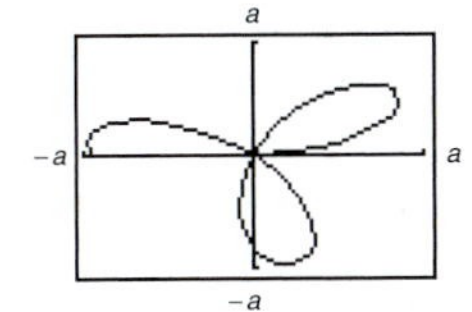

$r = a\cos(3.5\theta)$ $r = a\sin(2.63\theta)$

The graphs are rose curves where the petals never retrace, so there are infinitely many petals.

13. Circle

14. (a) No. Because of the exponential, the graph will continue to trace the butterfly curve at larger values of r.

(b) $r \approx 4.1$. This value will increase if θ is increased.

15.

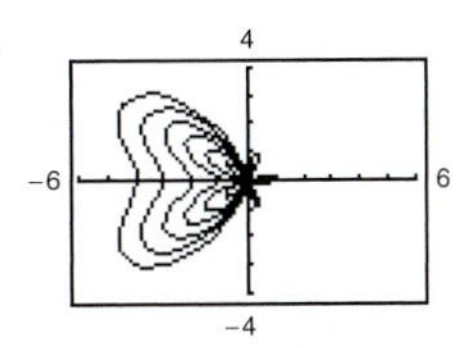

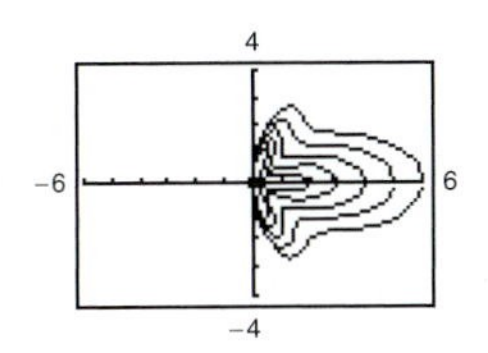

For $n \geq 1$, a bell is produced.

For $n \leq -1$, a heart is produced.

For $n = 0$, a rose curve is produced.

16. (a) $r_{\text{Neptune}} = \dfrac{4.4997 \times 10^9}{1 - 0.0086\cos\theta}$

$r_{\text{Pluto}} = \dfrac{5.07 \times 10^{10}}{1 - 0.2488\cos\theta}$

(b) Neptune: Aphelion $= 4.539 \times 10^9$ km
Perihelion $= 4.461 \times 10^9$ km

Pluto: Aphelion $= 6.752 \times 10^9$ km
Perihelion $= 4.061 \times 10^9$ km

(c)

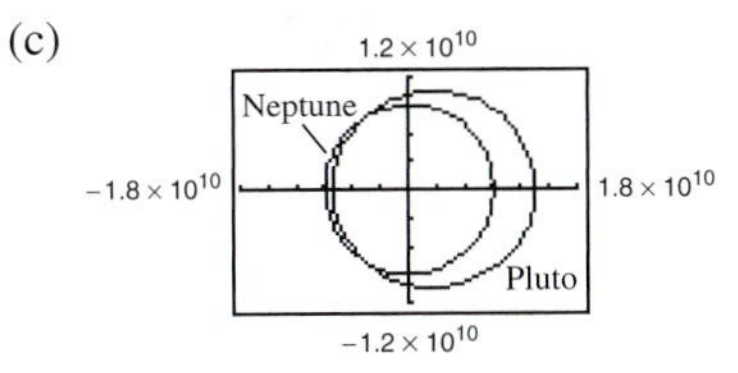

(d) If the orbits were in the same plane, then they would intersect. Furthermore, since the orbital periods differ (Neptune = 164.79 years, Pluto = 247.68 years), then the two planets would ultimately collide if the orbits intersect.

The orbital inclination of Pluto is significantly larger than that of Neptune (17.16° vs. 1.769°), so further analysis is required to determine if the orbits intersect.

(e) Yes, at times Pluto can be closer to the sun than Neptune. Pluto is called the ninth planet because it has the longest orbit around the sun and therefore also reaches the furthest distance away from the sun.

Explorations

Chapter 1

(page 25)

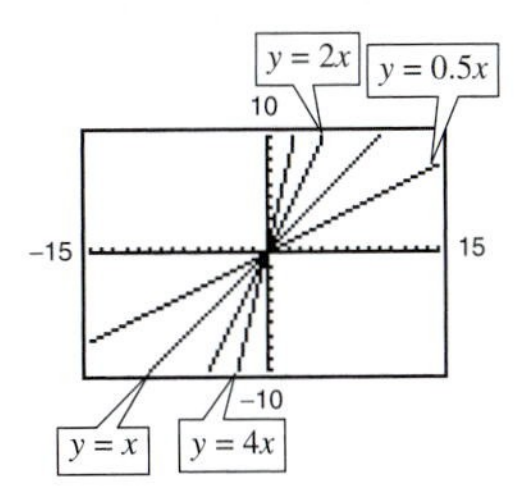

The line $y = 4x$.

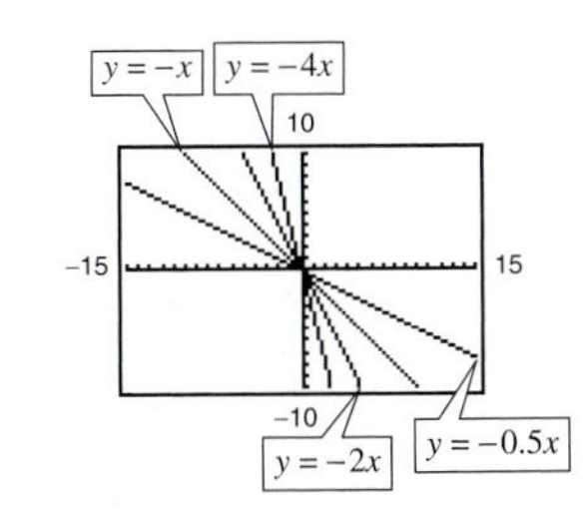

The line $y = -4x$.

As $|m|$ increases, the line rises or falls faster.

(page 30)

$d_1 = \sqrt{1 + m_1^2}, d_2 = \sqrt{1 + m_2^2}; m_1 = -\dfrac{1}{m_2}$

(page 59)

$\dfrac{s(9) - s(0)}{9 - 0} = \dfrac{540}{9} = 60$ feet per second

As the time traveled increases, the distance increases rapidly, causing the average speed to increase with each time increment. From $t = 0$ to $t = 4$ the average speed is less than from $t = 4$ to $t = 9$, therefore the overall average from $t = 0$ to $t = 9$ falls below the average found in part (b).

(page 60)

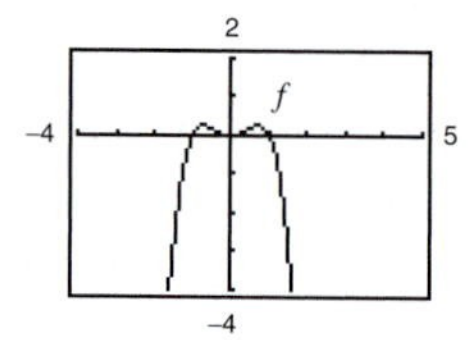

Even

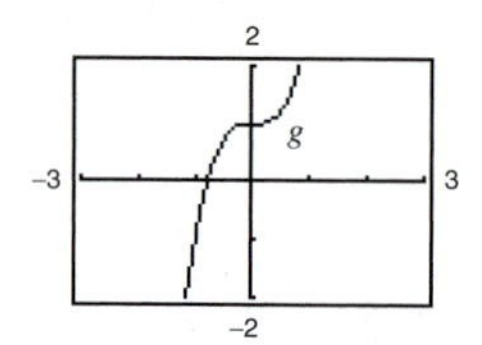

Neither

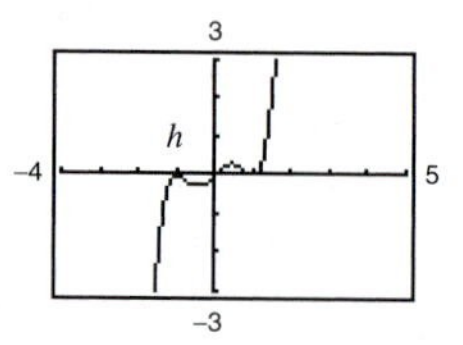

Odd

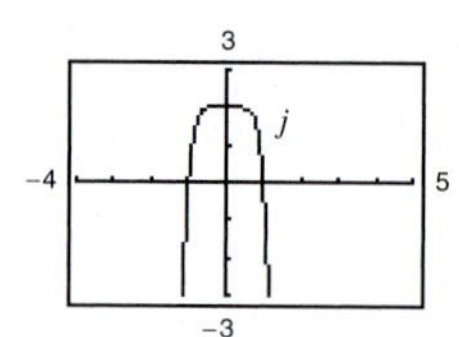

Even

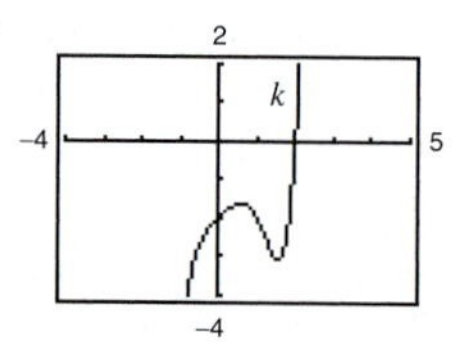

Neither

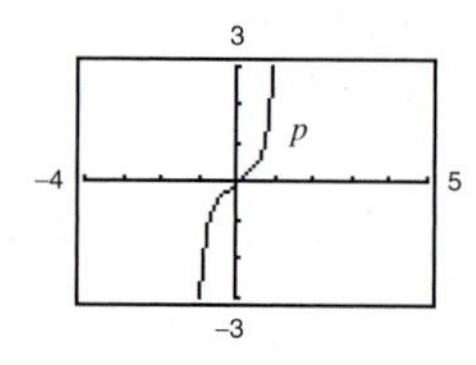

Odd

Equations of odd functions contain only odd powers of x. Equations of even functions contain only even powers of x. Odd functions have all variables raised to odd powers and even functions have all variables raised to even powers. A function that has variables raised to even and odd powers is neither odd nor even.

(page 75)

a.

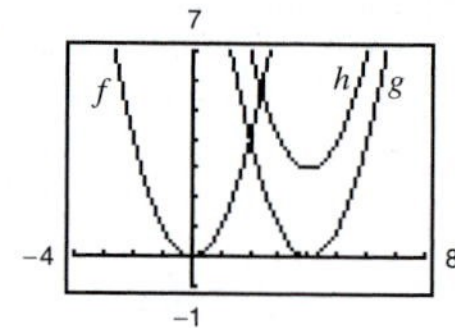

g is a right shift of four units. h is a right shift of four units and an upward shift of three units.

b.

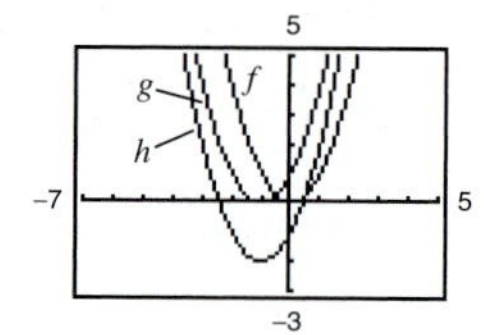

g is a left shift of one unit. h is a left shift of one unit and a downward shift of two units.

c.

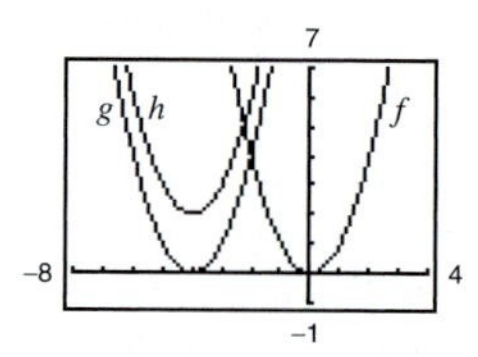

g is a left shift of four units. h is a left shift of four units and an upward shift of two units.

(page 76)

No. $g(x) = -x^4 - 2$. Yes. $h(x) = -(x - 3)^4$.

(page 94)

x	-10	0	7	45
$f(f^{-1}(x))$	-10	0	7	45
$f^{-1}(f(x))$	-10	0	7	45

The functions are inverses of each other.

(page 97)

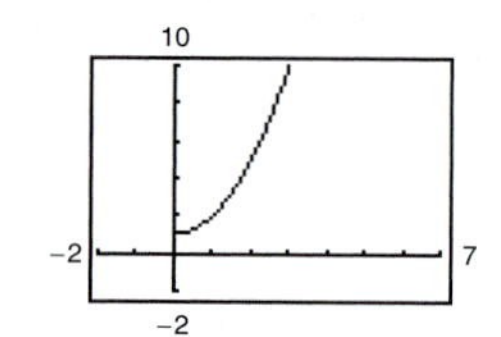

There is an inverse function $f^{-1}(x) = \sqrt{x - 1}$ since the domain of f is equal to the range of f^{-1} and the range of f is equal to the domain of f^{-1}.

Chapter 2

(page 129)

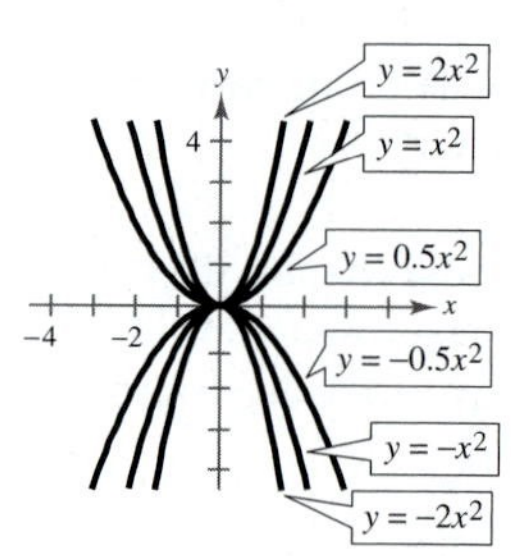

As $|a|$ gets larger, the parabola becomes narrower.

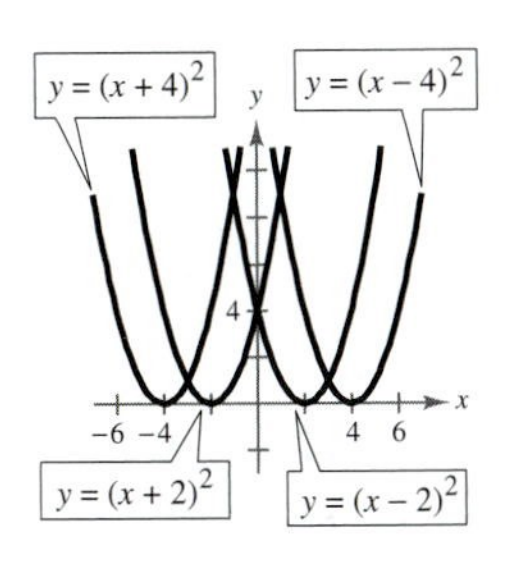

As h gets larger, the parabola shifts to the right.

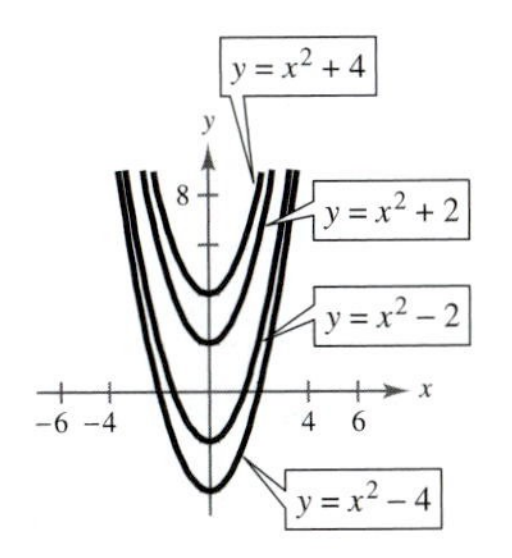

As k gets larger, the parabola shifts upward.

(page 141)

a. Third degree, odd; 1, greater than 0

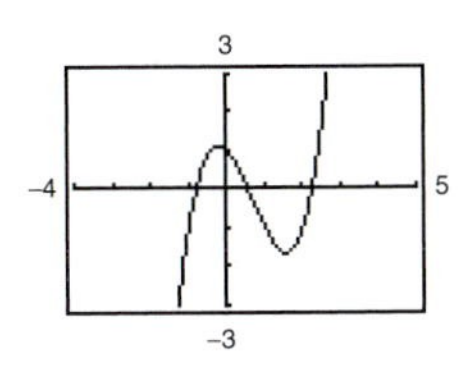

b. Fifth degree, odd; 2, greater than 0

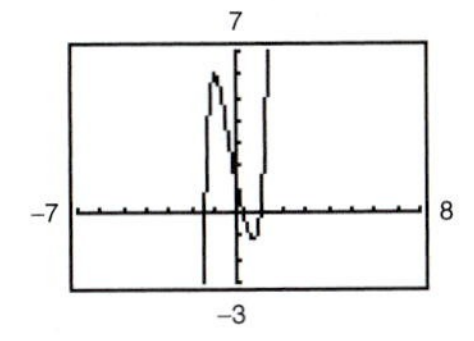

c. Fifth degree, odd; -2, less than 0

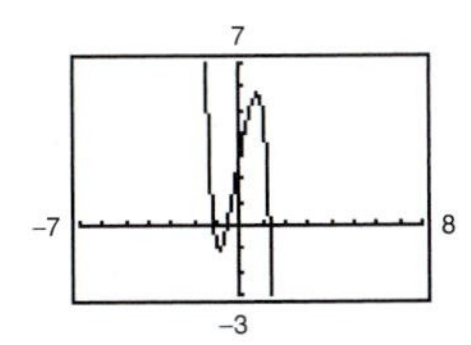

d. Third degree, odd; -1, less than 0

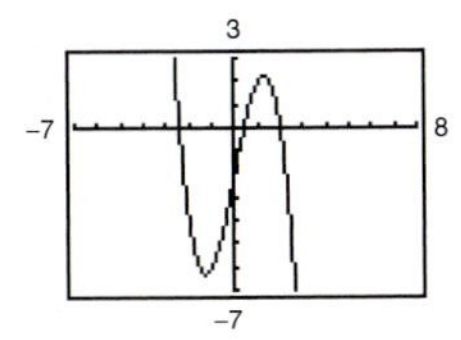

e. Second degree, even; 2, greater than 0

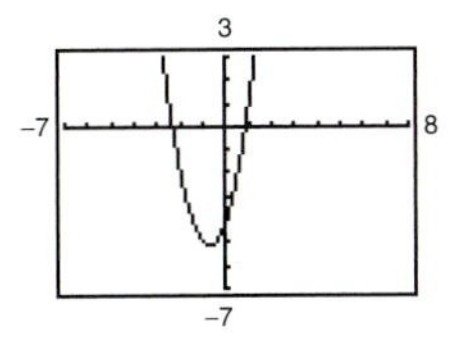

f. Fourth degree, even; 1, greater than 0

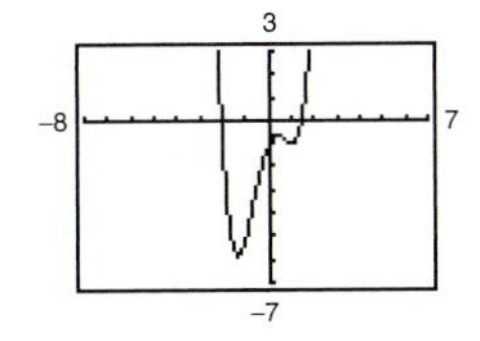

g. Second degree, even; 1, greater than 0

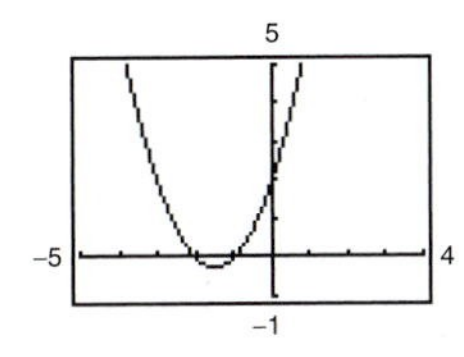

When the degree of the function is odd and the leading coefficient is positive, the graph falls to the left and rises to the right, but if the leading coefficient is negative, the graph falls to the right and rises to the left. When the degree of the function is even and the leading coefficient is positive, the graph rises to the left and right, but if the leading coefficient is negative, the graph falls to the left and right.

(page 142)

(a) 3 zeros; 1 relative minima and 1 relative maxima
The degree of function is 3, so the number of zeros matches the degree of the function.

(b) 4 zeros; 2 relative minima and 1 relative maxima
The degree of function is 4, so the number of zeros matches the degree of the function.

(c) 3 zeros; 1 relative minima and 1 relative maxima
The degree of function is 5, so the number of zeros does not exceed the degree of the function.

(page 164)

$i, -1, -i, 1, i, -1, -i, 1;$

The pattern repeats the first four results. Divide the exponent by 4.

If the remainder is 1, the result is i.

If the remainder is 2, the result is -1.

If the remainder is 3, the result is $-i$.

If the remainder is 0, the result is 1.

(page 200)

Graph of $y = x^2 + 2x + 1$

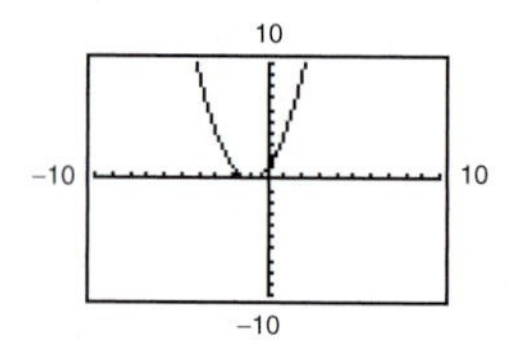

For part (b), the y-values that are less than or equal to 0 occur only at $x = -1$.

Graph of $y = x^2 + 3x + 5$

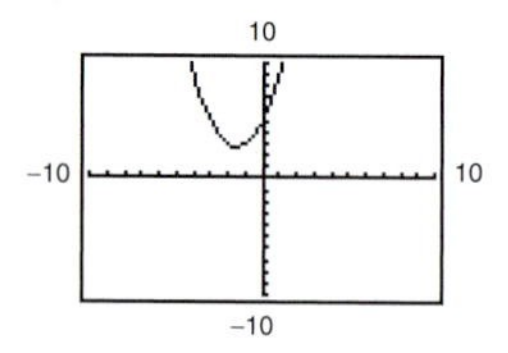

For part (c), there are no y-values that are less than 0.

Graph of $y = x^2 - 4x + 4$

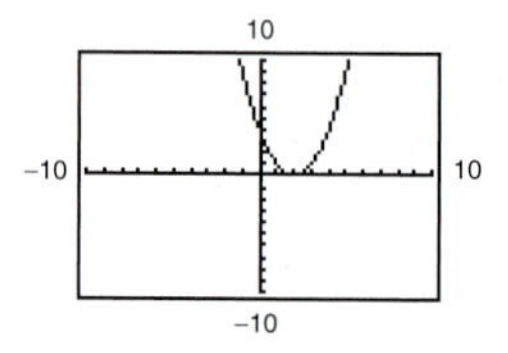

For part (d), the y-values that are greater than 0 occur for all values of x except 2.

Chapter 3

(page 219)

a.

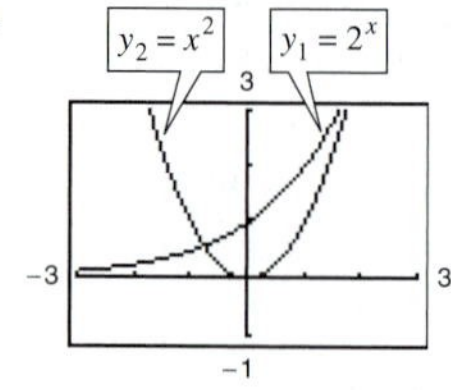

b.

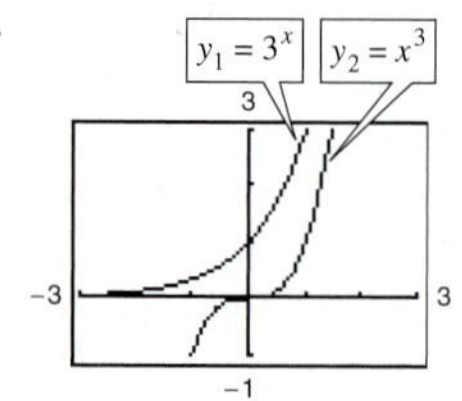

In both viewing windows, the constant raised to a variable power increases more rapidly than the function with a variable raised to a constant power.

(page 222)

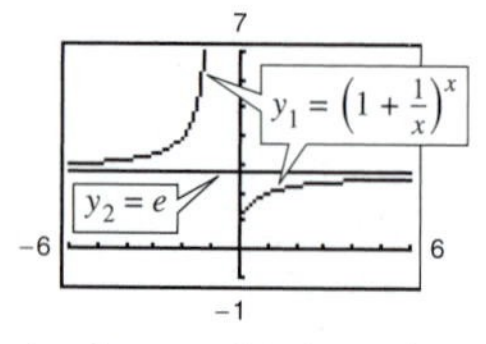

As the x-value in y_1 increases, y_1 approaches the value of e.

(page 223)

$A = \$5466.09$, $A = \$5466.35$, $A = \$5466.36$, $A = \$5466.38$.

No. Answers will vary.

(page 230)

x	-2	-1	0	1	2
$f(x) = 10^x$	$\frac{1}{100}$	$\frac{1}{10}$	1	10	100

x	$\frac{1}{100}$	$\frac{1}{10}$	1	10	100
$f(x) = \log x$	-2	-1	0	1	2

The domain of $f(x) = 10^x$ is equal to the range of $f(x) = \log x$ and vice versa. $f(x) = 10^x$ and $f(x) = \log x$ are inverses of each other.

(page 241)

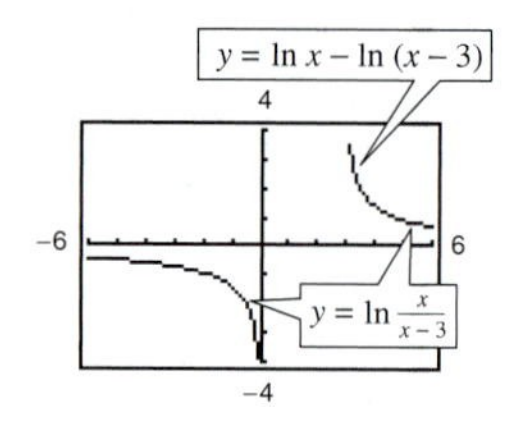

No; the domain of the first graph is $(3, \infty)$ and the domain of the second graph is $(-\infty, 0) \cup (3, \infty)$.

(page 251)

a. 7% **b.** 7.251% **c.** 7.186% **d.** 7.45%

Savings plan (d) will have the greatest effective yield. Savings plan (d) will have the highest balance after 5 years.

Chapter 4

(page 297)

a.

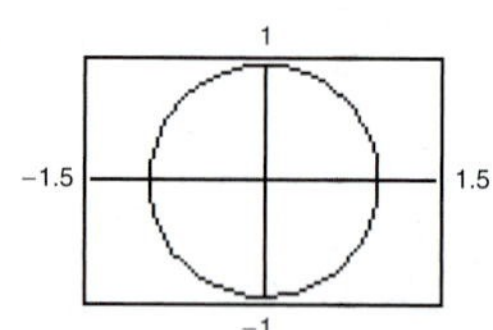

The graph is a circle.

b. The t-values represent the central angle in radians. The x- and y-values represent the location in the coordinate plane.

c. $-1 \le x \le 1, -1 \le y \le 1$

(page 323)

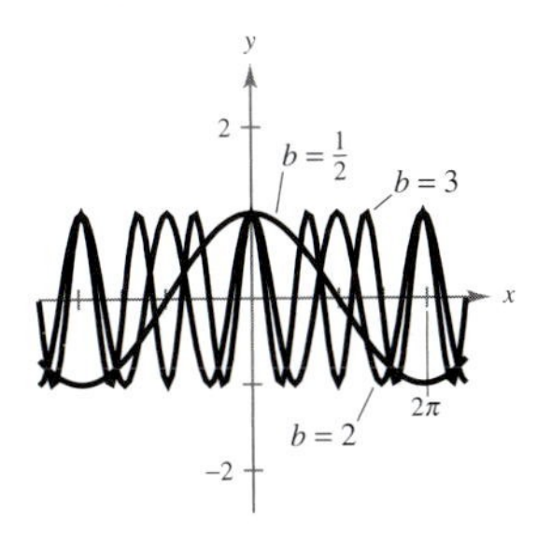

b affects the period of the graph.

$b = \frac{1}{2}$: $\frac{1}{2}$ cycle;

$b = 2$: 2 cycles;

$b = 3$: 3 cycles

(page 324)

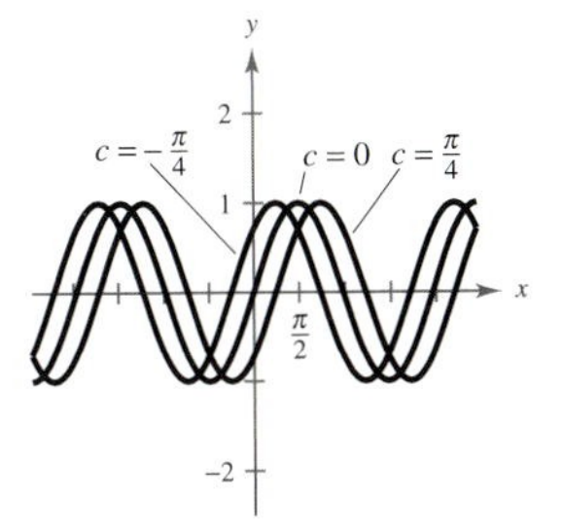

c shifts the graph horizontally.

Chapter 5

(page 391)

$\cos x = \pm\sqrt{2}$

No solution is obtained because $\pm\sqrt{2}$ are outside the range of the cosine function. No, first try to collect all the terms on one side and then try to separate the functions by factoring.

(page 393)

Yes. Preferences will vary.

(page 400)

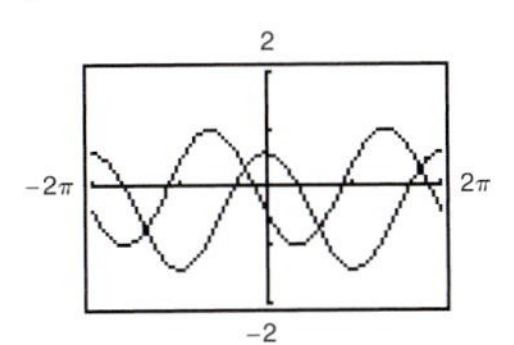

The graphs are different.
No, it is not true.

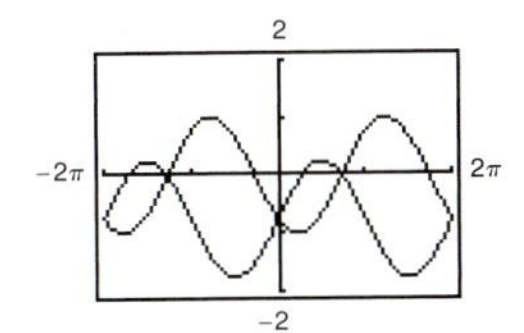

The graphs are different.
No, it is not true.

(page 429)

Pythagorean Theorem.

The Pythagorean Theorem is just a special case of the more general Law of Cosines.

Chapter 6

(page 487)

t should be greater than -1. The upper bound of t varies, but one possibility is $t = 20$.

(page 494)

a. Yes. $\theta \approx 3.927$, $x \approx -2.121$, $y \approx -2.121$

b. Yes. Answers and explanations will vary.

(page 502)

8 petals; 3 petals; For $r = 2\cos n\theta$ and $r = 2\sin n\theta$, there are n petals if n is odd, $2n$ petals if n is even.

Technology

Chapter 1

(page 3)

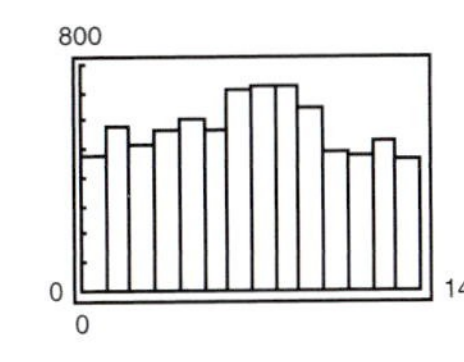

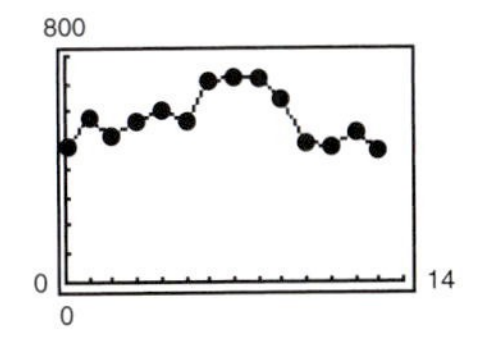

(page 30)

The lines appear perpendicular with the square setting.

(page 44)

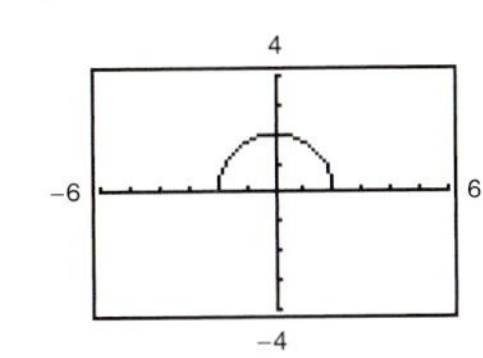

Domain: $[-2, 2]$

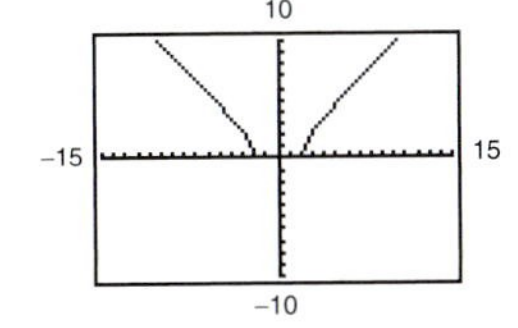

Domain: $(-\infty, -2] \cup [2, \infty)$
Yes, for -2 and 2.

Chapter 3

(page 258)

$D = 30.92(1.124)^t$. The models are very similar.

Chapter 4

(page 322)

No graph is visible. Try $-\pi \le x \le \pi$ and $-0.5 \le y \le 0.5$ as a viewing window.

(page 414)

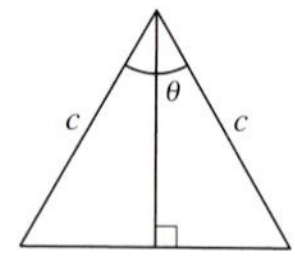

Let c be the length of the two equal sides of an isosceles triangle, and let θ be the angle between the two equal sides.

$$\begin{aligned}\text{Area} &= \left(\frac{1}{2}\text{ base}\right)(\text{height})\\ &= \left(c \sin\frac{\theta}{2}\right)\left(c \cos\frac{\theta}{2}\right)\\ &= c^2\sqrt{\frac{1-\cos\theta}{2}}\sqrt{\frac{1+\cos\theta}{2}}\\ &= c^2\sqrt{\frac{1-\cos^2\theta}{4}}\\ &= \frac{1}{2}c^2\sqrt{\sin^2\theta}\\ &= \frac{1}{2}c^2\sin\theta\end{aligned}$$

Examples will vary.

(page 424)

Yes

a. $A = 40°,\ a \approx 12.86,\ b \approx 15.32$

b. $A = 40°,\ b \approx 11.92,\ c \approx 15.56$

It is probably easier to use the right triangle definitions of sine, cosine, and tangent to solve the triangle.

(page 431)

a. Area $= \frac{1}{2}(2)(4)\sin 50° \approx 3.064$ square feet

b. Area $= \sqrt{\frac{9}{2}\left(\frac{9}{2}-2\right)\left(\frac{9}{2}-3\right)\left(\frac{9}{2}-4\right)} \approx 2.905$ square feet

c. Area $= \frac{1}{2}(2)(4) = 4$ square feet

d. Area $= \sqrt{6(6-3)(6-4)(6-5)} = 6$ square feet

Chapter 6

(page 453)

The angle of inclination is the positive angle made with the x-axis measured counterclockwise. Therefore, there are two cases: (1) a line with positive slope, implying $\theta < 90°$, and (2) a line with negative slope, implying $90° < \theta < 180°$. The angle between two lines is, by definition, the smaller of the two angles formed by the intersecting lines. The two angles formed are supplementary; therefore, one angle must be acute—i.e., less than 90°—or, if the lines are perpendicular, both angles equal 90°.

False. The inclination is the *positive angle measured counterclockwise* from the x-axis, not necessarily the angle between the line and the x-axis.

(page 461)

When a television signal hits the antenna dish, the parabolic shape of the dish reflects the signal inward to the focus of the parabola. The signal is then transferred to the receiving equipment.

(page 471)

a. Answers will vary.

b.

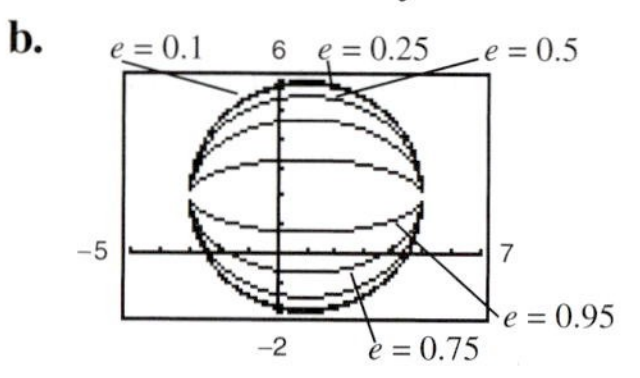

As e approaches 0, the shape of the ellipse approaches the shape of a circle.

c. When $e = 0$, the graph of the equation is circular. The equation of the ellipse is

$$\frac{(x-h)^2}{a^2} + \frac{(y-k)^2}{a^2} = 1,$$

otherwise known as the equation of a circle.

(page 481)

a.

b.

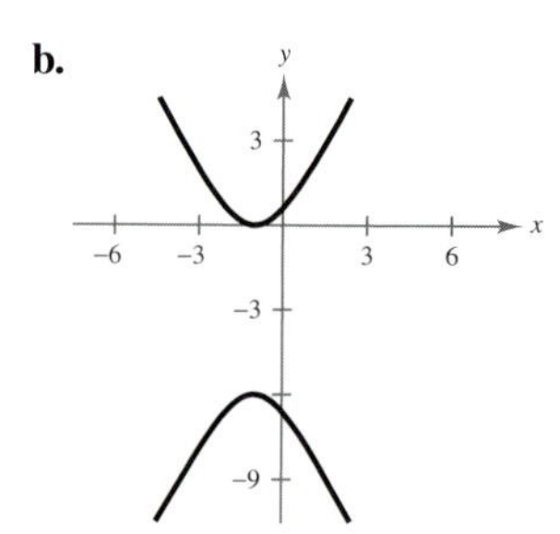

c.

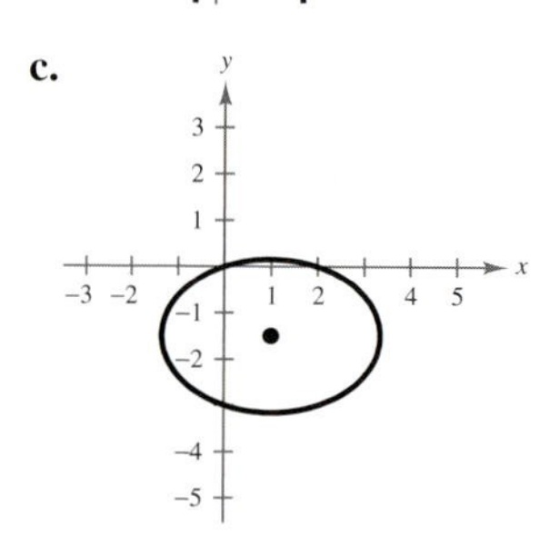

d.

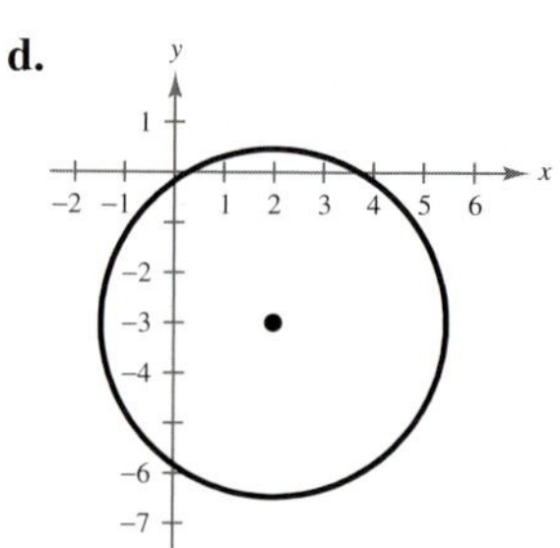

Discussions will vary.

Index

G

H

I

Q

R

S

Definition of the Six Trigonometric Functions

Right triangle definitions, where $0 < \theta < \pi/2$

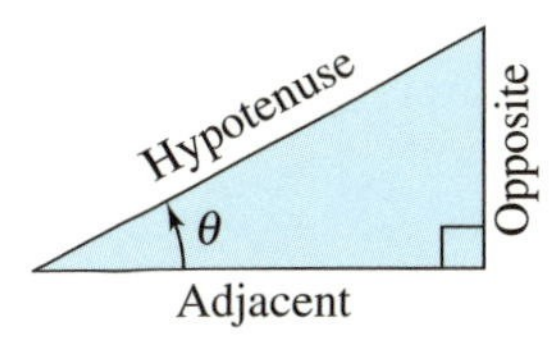

$\sin\theta = \dfrac{\text{opp.}}{\text{hyp.}}$ $\csc\theta = \dfrac{\text{hyp.}}{\text{opp.}}$

$\cos\theta = \dfrac{\text{adj.}}{\text{hyp.}}$ $\sec\theta = \dfrac{\text{hyp.}}{\text{adj.}}$

$\tan\theta = \dfrac{\text{opp.}}{\text{adj.}}$ $\cot\theta = \dfrac{\text{adj.}}{\text{opp.}}$

Circular function definitions, where θ *is any angle*

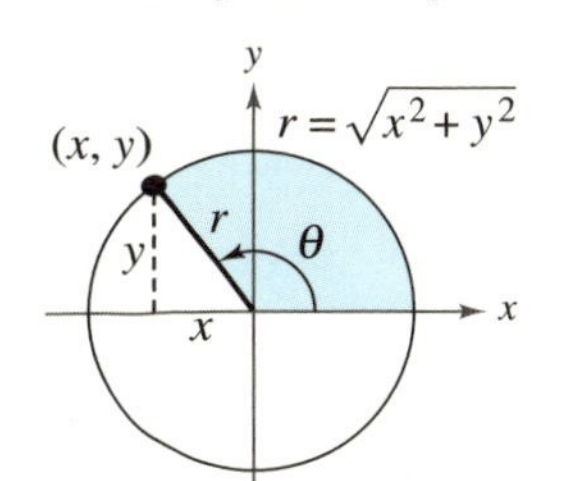

$\sin\theta = \dfrac{y}{r}$ $\csc\theta = \dfrac{r}{y}$

$\cos\theta = \dfrac{x}{r}$ $\sec\theta = \dfrac{r}{x}$

$\tan\theta = \dfrac{y}{x}$ $\cot\theta = \dfrac{x}{y}$

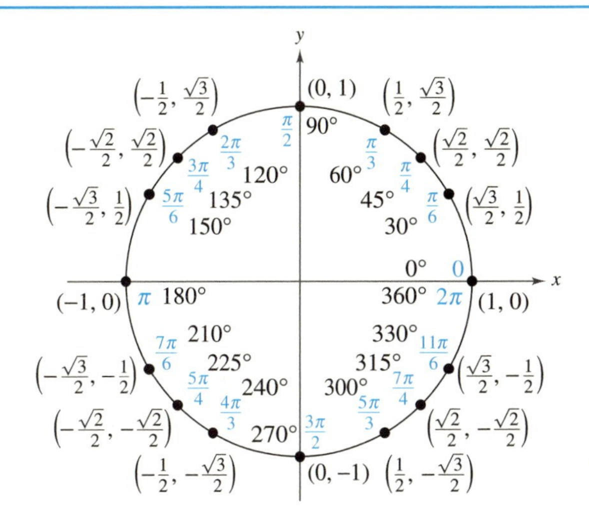

Reciprocal Identities

$\sin u = \dfrac{1}{\csc u}$ $\cos u = \dfrac{1}{\sec u}$ $\tan u = \dfrac{1}{\cot u}$

$\csc u = \dfrac{1}{\sin u}$ $\sec u = \dfrac{1}{\cos u}$ $\cot u = \dfrac{1}{\tan u}$

Quotient Identities

$\tan u = \dfrac{\sin u}{\cos u}$ $\cot u = \dfrac{\cos u}{\sin u}$

Pythagorean Identities

$\sin^2 u + \cos^2 u = 1$

$1 + \tan^2 u = \sec^2 u$ $1 + \cot^2 u = \csc^2 u$

Cofunction Identities

$\sin\left(\dfrac{\pi}{2} - u\right) = \cos u$ $\cot\left(\dfrac{\pi}{2} - u\right) = \tan u$

$\cos\left(\dfrac{\pi}{2} - u\right) = \sin u$ $\sec\left(\dfrac{\pi}{2} - u\right) = \csc u$

$\tan\left(\dfrac{\pi}{2} - u\right) = \cot u$ $\csc\left(\dfrac{\pi}{2} - u\right) = \sec u$

Even/Odd Identities

$\sin(-u) = -\sin u$ $\cot(-u) = -\cot u$

$\cos(-u) = \cos u$ $\sec(-u) = \sec u$

$\tan(-u) = -\tan u$ $\csc(-u) = -\csc u$

Sum and Difference Formulas

$\sin(u \pm v) = \sin u \cos v \pm \cos u \sin v$

$\cos(u \pm v) = \cos u \cos v \mp \sin u \sin v$

$\tan(u \pm v) = \dfrac{\tan u \pm \tan v}{1 \mp \tan u \tan v}$

Double-Angle Formulas

$\sin 2u = 2 \sin u \cos u$

$\cos 2u = \cos^2 u - \sin^2 u = 2\cos^2 u - 1 = 1 - 2\sin^2 u$

$\tan 2u = \dfrac{2 \tan u}{1 - \tan^2 u}$

Power-Reducing Formulas

$\sin^2 u = \dfrac{1 - \cos 2u}{2}$

$\cos^2 u = \dfrac{1 + \cos 2u}{2}$

$\tan^2 u = \dfrac{1 - \cos 2u}{1 + \cos 2u}$

Sum-to-Product Formulas

$\sin u + \sin v = 2 \sin\left(\dfrac{u + v}{2}\right) \cos\left(\dfrac{u - v}{2}\right)$

$\sin u - \sin v = 2 \cos\left(\dfrac{u + v}{2}\right) \sin\left(\dfrac{u - v}{2}\right)$

$\cos u + \cos v = 2 \cos\left(\dfrac{u + v}{2}\right) \cos\left(\dfrac{u - v}{2}\right)$

$\cos u - \cos v = -2 \sin\left(\dfrac{u + v}{2}\right) \sin\left(\dfrac{u - v}{2}\right)$

Product-to-Sum Formulas

$\sin u \sin v = \dfrac{1}{2}[\cos(u - v) - \cos(u + v)]$

$\cos u \cos v = \dfrac{1}{2}[\cos(u - v) + \cos(u + v)]$

$\sin u \cos v = \dfrac{1}{2}[\sin(u + v) + \sin(u - v)]$

$\cos u \sin v = \dfrac{1}{2}[\sin(u + v) - \sin(u - v)]$

FORMULAS FROM GEOMETRY

Triangle:

$h = a \sin \theta$

Area $= \frac{1}{2}bh$

$c^2 = a^2 + b^2 - 2ab \cos \theta$ (Law of Cosines)

Right Triangle:

Pythagorean Theorem
$c^2 = a^2 + b^2$

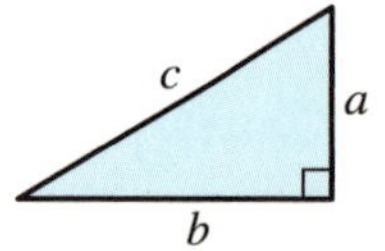

Equilateral Triangle:

$h = \frac{\sqrt{3}s}{2}$

Area $= \frac{\sqrt{3}s^2}{4}$

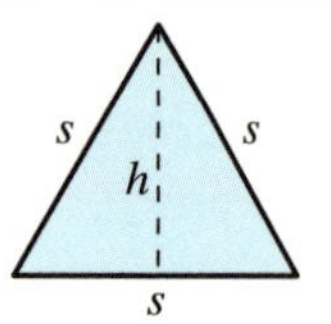

Parallelogram:

Area $= bh$

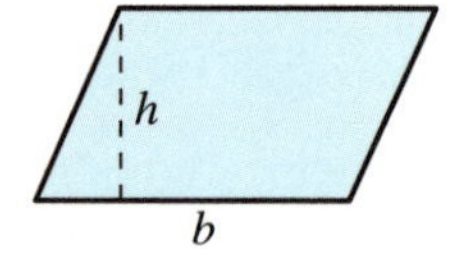

Trapezoid:

Area $= \frac{h}{2}(a + b)$

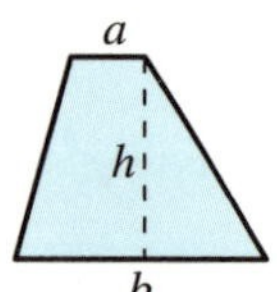

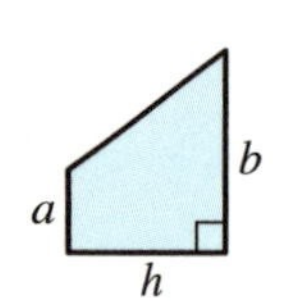

Circle:

Area $= \pi r^2$

Circumference $= 2\pi r$

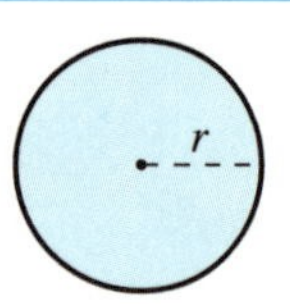

Sector of Circle:

Area $= \frac{\theta r^2}{2}$

$s = r\theta$

θ in radians

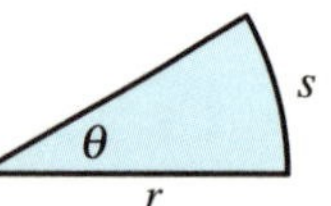

Circular Ring:

Area $= \pi(R^2 - r^2)$

$= 2\pi pw$

$p =$ average radius,

$w =$ width of ring

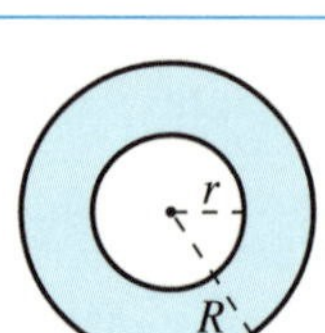

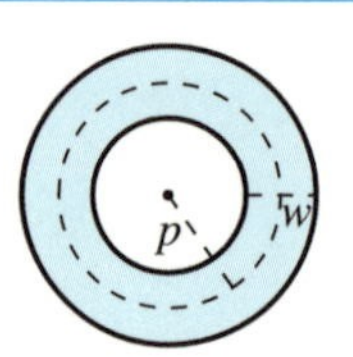

Sector of Circular Ring:

Area $= \theta pw$

$p =$ average radius,

$w =$ width of ring,

θ in radians

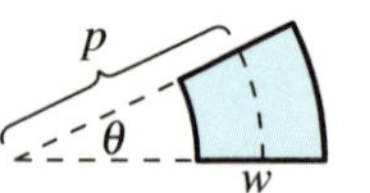

Ellipse:

Area $= \pi ab$

Circumference $\approx 2\pi\sqrt{\frac{a^2 + b^2}{2}}$

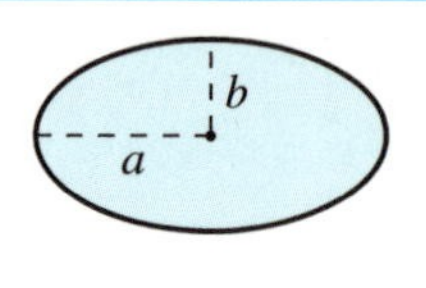

Cone:

Volume $= \frac{Ah}{3}$

$A =$ area of base

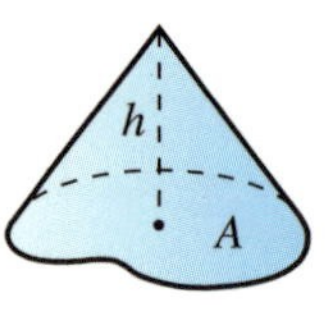

Right Circular Cone:

Volume $= \frac{\pi r^2 h}{3}$

Lateral Surface Area $= \pi r\sqrt{r^2 + h^2}$

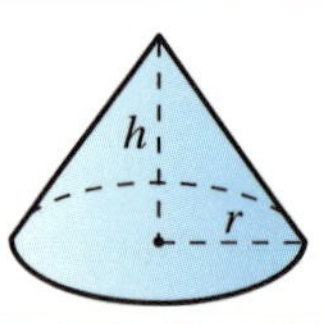

Frustum of Right Circular Cone:

Volume $= \frac{\pi(r^2 + rR + R^2)h}{3}$

Lateral Surface Area $= \pi s(R + r)$

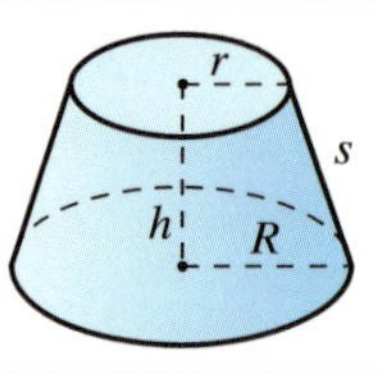

Right Circular Cylinder:

Volume $= \pi r^2 h$

Lateral Surface Area $= 2\pi rh$

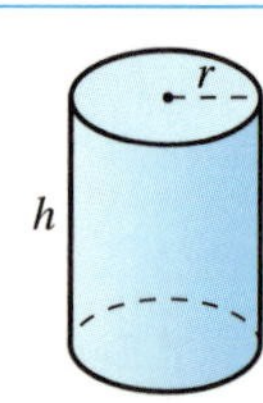

Sphere:

Volume $= \frac{4}{3}\pi r^3$

Surface Area $= 4\pi r^2$

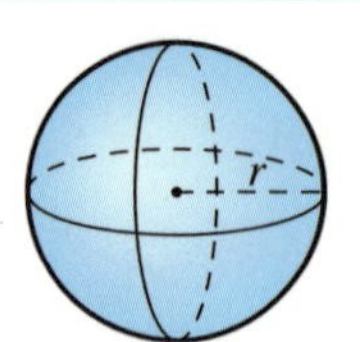

Wedge:

$A = B \sec \theta$

$A =$ area of upper face,

$B =$ area of base

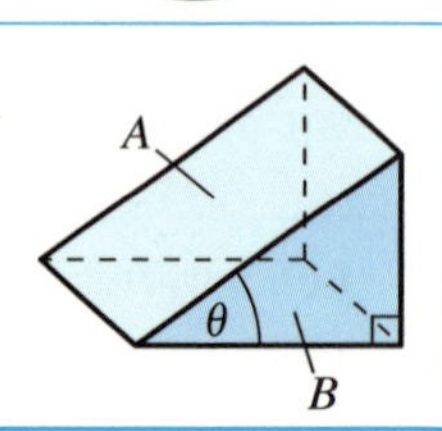